FORMULA INDEX

ORGANOMETALLIC COMPOUNDS

METHODS OF SYNTHESIS
PHYSICAL CONSTANTS AND CHEMICAL REACTIONS

Edited by
MICHAEL DUB

FORMULA INDEX
to the Second Edition of
VOLUMES I TO III
Covering the Literature from 1937 to 1964

Prepared by
MICHAEL DUB and RICHARD W. WEISS
Monsanto Company

SPRINGER SCIENCE+BUSINESS MEDIA, LLC
1969

ORGANOMETALLIC COMPOUNDS
Second Edition

Vol. I — Compounds of Transition Metals, published in 1966

Vol. II — Compounds of Germanium, Tin And Lead Including Biological Activity And Commercial Application, published in 1967

Vol. III — Compounds of Arsenic, Antimony, And Bismuth, Second Edition, Covering the Literature from 1937 to 1964, published in 1968

ISBN 978-3-642-50286-6 ISBN 978-3-642-50284-2 (eBook)
DOI 10.1007/978-3-642-50284-2

Originally published by Springer-Verlag New York Inc. in 1969
Library of Congress Catalog Card Number 66-28249
Title No. 104360

INTRODUCTION

This formula index contains the compounds of all three volumes.

The molecular formulae show metal atoms first, followed by carbon, hydrogen, and other nonmetal atoms arranged alphabetically. The monometallic and homopolymetallic compounds are followed by hetero-bimetallic, -trimetallic, and -polymetallic compounds. Heterometallic compounds are listed under each metal.

To make the use of the index easier, for the most part, simplified linear structural formulae are given after the molecular formulae. In the case of compounds containing complicated structures, such as fused carbocyclic, heterocyclic, or spiro rings, index names or a combination of group symbols and the parent compound names are used. Polymeric compounds are listed under their monomer formulae. Arseno, antimono, and bismutho compounds appear under their monomeric formulae. Arsenomethane and arsenobenzene, which were isolated as a pentamer and hexamer, respectively, and also as oligomers, are listed under their monomeric formulae and under As_5 and As_6, respectively.

The reference fiven at the end of each entry includes the volume number underscored, followed by the page number.

An asterisk following a reference page number signifies that the name or formula of this compound in the text was corrected. The lists of corrections and additions to the three volumes are appended to this volume.

Several compounds missed in the main body of this index are compiled in the "Additions" section.

We wish to express our appreciation to Mrs. Mary Alice Doiron for her perseverance shown in the preparation of this typescript.

Michael Dub

Richard W. Weiss

St. Louis, August 1969

GROUP AND LIGAND SYMBOLS

Ac acetyl
BiPy bipyridyl
Bu n-butyl
Bz benzoyl
C_2HN_2S thiadiazolyl
C_3H_2NS thiazolyl
C_3H_4NS thiazolinyl
C_3H_6NS tetrahydrothiazolyl
C_4H_3O furyl
C_4H_3S thienyl
C_4H_4N pyrrolyl
C_4H_7O tetrahydrofuryl
C_5H_4N pyridyl
C_5H_5 cyclopentadienyl
$C_5H_{5-n}Y_n$ Y_n-substituted cyclopentadienyl
C_5H_6 cyclopentadiene
C_5H_7 cyclopentenyl
$C_6H_{5-n}Y_n$ Y_n-substituted phenyl
C_6H_6 benzene
C_6H_{11} cyclohexyl
C_7H_4NS benzothiazolyl
$C_7H_5N_2$ benzimidazolyl
C_8H_8 cyclooctatetraene
C_9H_7 indenyl
$C_{10}H_7$ naphthyl
$C_{10}H_{7-n}Y_n$ Y_n-substituted naphthyl
$C_{10}H_8$ naphthalene
$C_{10}H_{15}OSO_3H$ camphorsulfonic acid
$C_{13}H_9$ fluorenyl
$C_{14}H_9$ phenanthryl
$C_{18}H_{11}$ benzanthracenyl
Et ethyl
Fc = $C_5H_5FeC_5H_4$ ferrocenyl
Fcd = $C_5H_4FeC_5H_4$ 1,1′-ferrocenylene
i- iso
i-Bu = Me_2CHCH_2 isobutyl
i-Pr = Me_2CH isopropyl
Me methyl
MeC_6H_4 tolyl
$(O_2N)_3C_6H_2O$ picrate
Ph phenyl
Pip piperidine
Pr n-Propyl
Py pyridine
s- secondary (also symmetric in triazine)
s-Bu = MeEtCH secondary butyl
t- tertiary
t-Bu = Me_3C tertiary butyl
THF tetrahydrofuran

$AgAsC_{14}H_{20}$ $[PhC{\equiv}CAgAsEt_3]$, 3, 11

$AgGeC_{72}H_{60}P_3$ $Ph_3GeAg \cdot 3PPh_3$, 2, 139

$Ag_4Ge_4C_{72}H_{100}I_4P_4$ $(Et_3GePPh_2 \cdot AgI)_4$, 2, 127

$AlCrC_{12}H_{12}Cl_4$ $[(C_6H_6)_2Cr]AlCl_4$, 1, 78

$AlCrC_{12}H_{15}O_3$ $Et_2Al(C_5H_5)Cr(CO)_3$, 1, 93

$AlCrC_{18}H_{16}Cl_4$ $[(C_6H_6)(Ph_2)Cr]AlCl_4$, 1, 87

$AlCrC_{24}H_{20}Br_4$ $[(Ph_2)_2Cr]AlBr_4$, 1, 83

$AlBrC_{24}H_{20}Cl_4$ $[(Ph_2)_2Cr]AlCl_4$, 1, 83

$AlCr_3C_{24}H_{15}O_9$ $Al[(C_5H_5)Cr(CO)_3]_3$, 1, 93

$AlFe_3C_{31}H_{27}Cl_4$ $[Fc_3C]AlCl_4$, 1, 431

$AlMnC_7H_4Cl_4O_5$ $[(C_2H_4)Mn(CO)_5]AlCl_4$, 1, 180

$AlReC_{12}H_{12}Cl_4$ $[(C_6H_6)_2Re]AlCl_4$, 1, 190

$AlReC_{18}H_{24}Cl_4$ $[(Me_3C_6H_3)_2Re]AlCl_4$, 1, 190

$AlSbC_9H_{24}$ $[Me_3Sb \cdot AlEt_3]$, 3, 654

$AlTiC_{10}H_{10}Cl_4$ $(C_5H_5)_2TiCl_2AlCl_2$, 1, 32

$AlTiC_{12}H_{15}Cl_3$ $(C_5H_5)_2TiCl_2AlEtCl$, 1, 33

$AlTiC_{12}H_{16}Cl$ $[(C_5H_5)_2Ti \cdot Me_2AlCl]$, 1, 33

$[(C_5H_5)_2Ti \cdot Me_2AlCl]$, 1, 33

$AlTiC_{12}H_{16}Cl_2$ $(C_5H_5)_2TiCl_2AlMe_2$, 1, 33

$AlTiC_{12}H_{27}Cl_3$ $i\text{-}Bu(Cl)TiCl_2Al(i\text{-}Bu)_2$, 1, 33

$AlTiC_{14}H_{20}Cl$ $[(C_5H_5)_2Ti \cdot Et_2AlCl]$, 1, 33

$AlTiC_{14}H_{20}Cl_2$ $(C_5H_5)_2TiCl_2AlEt_2$, 1, 33

$AlTiC_{18}H_{28}Cl_2$ $(C_5H_5)_2TiCl_2Al(i\text{-}Bu)_2$, 1, 34

$AlTiC_{22}H_{24}Br_2$ Diindenyl-$TiBr_2AlEt_2$, 1, 34

$AlTiC_{22}H_{24}Cl_2$ Diindenyl-$TiCl_2AlEt_2$, 1, 34

$AlTiC_{22}H_{24}I_2$ Diindenyl-TiI_2AlEt_2, 1, 34

$AlVC_{18}H_{24}Cl_4$ $[(1,3,5\text{-}Me_3C_6H_3)_2V]$-$AlCl_4$, 1, 51

$Al_2As_2C_{28}H_{32}$ $(Ph_2AsAlMe_2)_2$, 3, 298

$Al_2CoC_{15}H_{13}Cl_5$ $(Ph_2Al_2CoCl_5 \cdot 0.5C_6H_6)_n$, 1, 618

$Al_2SnC_{16}H_{40}$ $[Et_2Sn \cdot 2AlEt_3]$, 2, 253

$Al_2TiC_6H_6Cl_8$ $(C_6H_6)Al_2Ti^{II}Cl_8$, 1, 32

$Al_2TiC_7H_8Cl_8$ $(PhMe)Al_2Ti^{II}Cl_8$, 1, 32

$Al_2TiC_9H_{12}Cl_3$ $(1,3,5\text{-}Me_3C_6H_3)Al_2$-$Ti^{II}Cl_8$, 1, 32

$Al_2TiC_{26}H_{61}O_4$ $Et_2Al(OBu)_2TiEt(OBu)_2$-$AlEt_2$, 1, 34

$Al_2Ti_2C_{28}H_{40}$ $[(C_5H_5)_2TiAlEt_2]_2$, 1, 34

$Al_6MnC_{12}H_{12}Cl_{18}$ $[(C_6H_6)_2Mn(AlCl_3)_6]$, 1, 180

$Al_9Ti_2C_{18}H_{18}Cl_{27}$ $(C_6H_6)_3Ti_2(AlCl_3)_9$, 1, 32

$AsCBr_2F_3$ CF_3AsBr_2, 3, 188

$AsCCl_2F_3$ CF_3AsCl_2, 3, 177

$AsCH_2Cl_3$ $ClCH_2AsCl_2$, 3, 188

$AsCH_2F_3$ CF_3AsH_2, 3, 120

$AsCH_2F_3O_3$ $CF_3AsO_3H_2$, 3, 394

$AsCH_3$ Arsenomethane, 3, 130

$AsCH_3Br_2$ $MeAsBr_2$, 3, 188

$AsCH_3Cl_2$ $MeAsCl_2$, 3, 178

$AsCH_3F_2$ $MeAsF_2$, 3, 188

$AsCH_3I_2$ $MeAsI_2$, 3, 179

$AsCH_3O$ $MeAsO$, 3, 227

$AsCH_3O_2$ $[CH_2As(O)(OH)]_n$, 3, 373

$AsCH_3S$ $MeAsS$, 3, 264

$AsCH_3S_2$ $MeAsS_2$, 3, 550

$AsCH_4ClO_3$ $ClCH_2AsO_3H_2$, 3, 394

$AsCH_5$ $MeAsH_2$, 3, 120

$AsCH_5O_2$ $MeAs(OH)_2$, 3, 221

$AsCH_5O_3$ $MeAsO_3H_2$, 3, 390

$AsCH_5O_5S$ $MeAsO \cdot H_2SO_4$, 3, 228

$AsCH_5S_2$ $MeAs(SH)_2$, 3, 263

$AsCH_5S_3$ $MeAs(S)(SH)_2$, 3, 550

AsC_2BrF_6 $(CF_3)_2AsBr$, 3, 168

AsC_2ClF_6 $(CF_3)_2AsCl$, 3, 158

$AsC_2Cl_2F_3$ $CF_2{=}CFAsCl_2$, 3, 188

$AsC_2Cl_2F_5$ $CF_3CF_2AsCl_2$, 3, 179

$AsC_2Cl_3F_6$ $(CF_3)_2AsCl_3$, 3, 371

AsC_2F_6I $(CF_3)_2AsI$, 3, 159

AsC_2F_7 $(CF_3)_2AsF$, 3, 168

$CF_3CF_2AsF_2$, 3, 188

AsC_2HF_6 $(CF_3)_2AsH$, 3, 115

AsC_2HF_6O $(CF_3)_2AsOH$, 3, 207

$AsC_2HF_6O_2$ $(CF_3)_2AsO_2H$, 3, 373

$AsC_2H_2ClF_2$ $ClCH{=}CHAsF_2$, 3, 188

AsC_2H_2ClO $ClCH{=}CHAsO$, 3, 228
$AsC_2H_2Cl_3$ $ClCH{=}CHAsCl_2$, 3, 178
$AsC_2H_2F_5$ $CF_3CF_2AsH_2$, 3, 120
$AsC_2H_2F_5O_3$ $CF_3CF_2AsO_3H_2$, 3, 394
$AsC_2H_2F_6N$ $(CF_3)_2AsNH_2$, 3, 286
$AsC_2H_3Br_2$ $CH_2{=}CHAsBr_2$, 3, 188
$AsC_2H_3Br_2O_2$ $HO_2CCH_2AsBr_2$, 3, 188
$AsC_2H_3ClF_3$ $Me(CF_3)AsCl$, 3, 168
$AsC_2H_3Cl_2$ $CH_2{=}CHAsCl_2$, 3, 180
$AsC_2H_3Cl_2O_2$ $HO_2CCH_2AsCl_2$, 3, 188
$AsC_2H_3F_3I$ $Me(CF_3)AsI$, 3, 168
$AsC_2H_3O_2$ $(HO_2CCH_2As)_n$, 3, 136
$AsC_2H_3S_2$ $CH_2{=}CHAsS_2$, 3, 550
$AsC_2H_4ClO_3$ $ClCH{=}CHAsO_3H_2$, 3, 394
$AsC_2H_4Cl_3$ $(ClCH_2)_2AsCl$, 3, 168
  $ClCH_2CH_2AsCl_2$, 3, 189
$AsC_2H_5Br_2$ $EtAsBr_2$, 3, 189
$AsC_2H_5Cl_2$ $EtAsCl_2$, 3, 180
$AsC_2H_5Cl_2O_2$ $(ClCH_2)_2AsO_2H$, 3, 378
$AsC_2H_5F_2$ $EtAsF_2$, 3, 189
$AsC_2H_5I_2$ $EtAsI_2$, 3, 189
AsC_2H_5O $EtAsO$, 3, 228
$AsC_2H_5O_3$ $CH_2{=}CHAsO_3H_2$, 3, 394
$AsC_2H_5O_5$ $HO_2CCH_2AsO_3H_2$, 3, 394
AsC_2H_5S $EtAsS$, 3, 264
$AsC_2H_5S_2$ $EtAsS_2$, 3, 550
AsC_2H_6Br Me_2AsBr, 3, 159, 168
AsC_2H_6Cl Me_2AsCl, 3, 160
AsC_2H_6F Me_2AsF, 3, 168
$AsC_2H_6F_3$ Me_2AsF_3, 3, 371
AsC_2H_6I Me_2AsI, 3, 161
AsC_2H_7 Me_2AsH, 3, 115
$AsC_2H_7O_2$ $EtAs(OH)_2$, 3, 221
  Me_2AsO_2H 3, 373*
$AsC_2H_7O_3$ $EtAsO_3H_2$, 3, 392
$AsC_2H_7O_4$ $HOCH_2CH_2AsO_3H_2$, 3, 394
$AsC_2H_7S_2$ $EtAs(SH)_2$, 3, 263
$AsC_2H_{10}N_2O_3P$ $Me_2As(O)OP(O)(NH_2)_2$, 3, 388
$AsC_3Cl_2F_9$ $(CF_3)_3AsCl_2$, 3, 343
AsC_3F_6N $(CF_3)_2AsCN$, 3, 202
AsC_3F_6NS $(CF_3)_2AsSCN$, 3, 202
AsC_3F_9 $(CF_3)_3As$, 3, 2
AsC_3F_9S $(CF_3)_2AsSCF_3$, 3, 258
AsC_3F_9Se $(CF_3)_2AsSeCF_3$, 3, 285
AsC_3F_{11} $(CF_3)_3AsF_2$, 3, 343
$AsC_3H_3F_6$ $(CF_3)_2AsMe$, 3, 34
$AsC_3H_3N_2OS_2$ $MeAs(O)(SCN)_2$, 3, 390
$AsC_3H_3N_2S_2$ $MeAs(SCN)_2$, 3, 206
AsC_3H_4ClO $ClCH{=}CMeAsO$, 3, 234
$AsC_3H_4Cl_3$ $ClCH{=}CHCH_2AsCl_2$, 3, 189
$AsC_3H_4F_6N$ $(CF_3)_2AsNHMe$, 3, 286
$AsC_3H_4NS_2$ $NCAsSCH_2CH_2S$, 3, 583
$AsC_3H_5Cl_2$ $Me(ClCH{=}CH)AsCl$, 3, 161, 168
  $CH_2{=}CHCH_2AsCl_2$, 3, 189
$AsC_3H_5Cl_2O_2$ $HO_2CCH_2CH_2AsCl_2$, 3, 189
AsC_3H_5O $CH_2{=}CHCH_2AsO$, 3, 234
$AsC_3H_5O_5$ $CH_2{=}C(CO_2H)AsO_3H_2$, 3, 394
AsC_3H_5S $CH_2{=}CHCH_2AsS$, 3, 264
$AsC_3H_5S_2$ $CH_2{=}CHCH_2AsS_2$, 3, 550
AsC_3H_6Br $Me(CH_2{=}CH)AsBr$, 3, 168
AsC_3H_6Cl $ClCH{=}C(Me)AsH_2$, 3, 120
$AsC_3H_6ClO_2$ $Me(ClCH{=}CH)AsO_2H$, 3, 378
$AsC_3H_6Cl_3$ $Cl(CH_2)_3AsCl_2$, 3, 189
$AsC_3H_6F_3$ Me_2AsCF_3, 3, 35
$AsC_3H_6F_3S$ Me_2AsSCF_3, 3, 258
  $Me_2(CF_3)AsS$, 3, 364
AsC_3H_6N Me_2AsCN, 3, 202
$AsC_3H_7Cl_2$ $PrAsCl_2$, 3, 189
  $Me_2CHAsCl_2$, 3, 190
$AsC_3H_7Cl_2O$ $MeOCH_2CH_2AsCl_2$, 3, 190
$AsC_3H_7N_2S_4$ $MeAs[SC(S)NH_2]_2$, 3, 282
AsC_3H_7O $PrAsO$, 3, 234
  Me_2CHAsO, 3, 234
$AsC_3H_7O_3$ $CH_2{=}CHCH_2AsO_3H_2$, 3, 394
$AsC_3H_7O_5$ $HO_2CCH_2CH_2AsO_3H_2$, 3, 394
  $MeCH(CO_2H)AsO_3H_2$, 3, 394
AsC_3H_7S $PrAsS$, 3, 264
  Me_2CHAsS, 3, 264
$AsC_3H_7S_2$ Me_2CHAsS_2, 3, 550
  $MeAsSCH_2CH_2S$, 3, 583
AsC_3H_8Br $MeEtAsBr$, 3, 168
$AsC_3H_8ClO_3$ $Cl(CH_2)_3AsO_3H_2$, 3, 394
AsC_3H_9 Me_3As, 3, 3
$AsC_3H_9Br_2$ Me_3AsBr_2, 3, 343
$AsC_3H_9Cl_2$ Me_3AsCl_2, 3, 343
$AsC_3H_9Cl_3P$ Me_3AsPCl_3, 3, 5
$AsC_3H_9I_4$ Me_3AsI_4, 3, 343
AsC_3H_9O Me_2AsOMe, 3, 209
  Me_3AsO, 3, 356
$AsC_3H_9O_2$ $MeAs(OMe)_2$, 3, 251
$AsC_3H_9O_3$ $Me(HOCH_2CH_2)AsO_2H$, 3, 378
  $PrAsO_3H_2$, 3, 392
  $Me_2CHAsO_3H_2$, 3, 394

$MeAsO(OMe)_2$, 3, 544
$AsC_3H_9O_4$ $HOCH_2CH_2CH_2AsO_3H_2$, 3, 394
$MeOCH_2CH_2AsO_3H_2$, 3, 395
AsC_3H_9S Me_3AsS, 3, 364
$AsC_3H_9S_2$ $MeAs(SMe)_2$, 3, 266
AsC_3H_9Se Me_3AsSe, 3, 367
$AsC_3H_{10}NO_4$ $Me_3As(OH)NO_3$, 3, 348
AsC_3N_3 $As(CN)_3$, 3, 3
$AsC_4H_3Cl_2O$ $OCH{=}CHCH{=}CAsCl_2$, 3, 181
$AsC_4H_3Cl_2S$ $SCH{=}CHCH{=}CAsCl_2$, 3, 190
AsC_4H_3O $(OCH{=}CHCH{=}CAs)_n$, 3, 136
AsC_4H_3OS $OCH{=}CHCH{=}CAsS$, 3, 264
$AsC_4H_3O_2$ $OCH{=}CHCH{=}CAsO$, 3, 234
$AsC_4H_4ClO_3$ $HO_2CCCl{=}C(Me)AsO$, 3, 234
$AsC_4H_4Cl_3$ $(ClCH{=}CH)_2AsCl$, 3, 161
AsC_4H_5ClN $Me(ClCH{=}CH)AsCN$, 3, 202
$AsC_4H_5F_6$ $(CF_3)_2AsEt$, 3, 35
$AsC_4H_5F_6S$ $(CF_3)_2AsSEt$, 3, 258
$AsC_4H_5O_4$ $OCH{=}CHCH{=}CAsO_3H_2$, 3, 513
AsC_4H_6Br $(CH_2{=}CH)_2AsBr$, 3, 162
AsC_4H_6Cl $(CH_2{=}CH)_2AsCl$, 3, 162, 168
AsC_4H_6ClO $EtC({=}CHCl)AsO$, 3, 234
$AsC_4H_6ClS_2$ $ClCH{=}CHAsSCH_2CH_2S$, 3, 583
$AsC_4H_6Cl_3O$ $Me_2(CCl_3CO)As$, 3, 64
$AsC_4H_6F_3$ $Me_2AsCF{=}CF_2$, 3, 35
$AsC_4H_6F_3O_2$ $Me_2AsO_2CCF_3$, 3, 212
$AsC_4H_6F_5$ $Me_2AsCF_2CF_3$, 3, 35
$AsC_4H_6F_6N$ $(CF_3)_2AsNHEt$, 3, 286
$(CF_3)_2AsNMe_2$, 3, 286
AsC_4H_6I $(CH_2{=}CH)_2AsI$, 3, 168
$AsC_4H_7O_5$ $MeCH{=}C(CO_2H)AsO_3H_2$, 3, 395
$HO_2CCH{=}C(Me)AsO_3H_2$, 3, 395
AsC_4H_8Cl $Me_2(ClCH{=}CH)As$, 3, 64
1-Chloroarsolane, 3, 553
$AsC_4H_8ClO_2$ $ClCH{=}CHAs(OMe)_2$, 3, 251
$AsC_4H_8Cl_3$ $(MeCHCl)_2AsCl$, 3, 169
$AsC_4H_8Cl_3O$ $ClCH_2O(CH_2)_3AsCl_2$, 3, 190
AsC_4H_9 $Me_2(CH_2{=}CH)As$, 3, 64
Arsolane, 3, 553
$AsC_4H_9Cl_2$ $Et(ClCH_2CH_2)AsCl$, 3, 162
$BuAsCl_2$, 3, 190
$Me_2CHCH_2AsCl_2$, 3, 190
$AsC_4H_9Cl_2O$ $EtOCH_2CH_2AsCl_2$, 3, 190
$AsC_4H_9Cl_3I$ $[Me_3As\cdot CCl_3I]$, 3, 4
$AsC_4H_9F_3I$ $[Me_3(CF_3)As]I$, 3, 312
$AsC_4H_9I_2$ $BuAsI_2$, 3, 190
AsC_4H_9O Me_2AsAc, 3, 35
$BuAsO$, 3, 234
$AsC_4H_9OS_2$ $MeAsCH_2CH(CH_2OH)S$, 3, 583
$AsC_4H_9O_2$ $EtAsOCH_2CH_2O$, 3, 577
$AsC_4H_9O_3$ $CH_2{=}C(Me)CH_2AsO_3H_2$, 3, 395
$AsC_4H_9O_5$ $MeCH_2CH(CO_2H)AsO_3H_2$, 3, 395
AsC_4H_9S $BuAsS$, 3, 264
Me_2CHCH_2AsS, 3, 264
$AsC_4H_9S_2$ $BuAsS_2$, 3, 550
$EtAsSCH_2CH_2S$, 3, 583
$MeAsS(CH_2)_3S$, 3, 618
$AsC_4H_{10}Br$ Et_2AsBr, 3, 169
$AsC_4H_{10}Cl$ Et_2AsCl, 3, 162
$AsC_4H_{10}I$ Et_2AsI, 3, 162
$AsC_4H_{10}N$ $EtAs{=}NEt$, 3, 288
AsC_4H_{11} Me_2AsEt, 3, 64
Et_2AsH, 3, 117
$AsC_4H_{11}O_2$ $BuAs(OH)_2$, 3, 221
$EtAs(OMe)_2$, 3, 251
Et_2AsO_2H, 3, 374, 378
$MePrAsO_2H$, 3, 378
$AsC_4H_{11}O_3$ $BuAsO_3H_2$, 3, 393
$Me_2CHCH_2AsO_3H_2$, 3, 395
$EtAsO(OMe)_2$, 3, 545
$AsC_4H_{11}O_3S$ $EtSCH_2CH_2AsO_3H_2$, 3, 395
$AsC_4H_{11}O_4$ $EtOCH_2CH_2AsO_3H_2$, 3, 395
$AsC_4H_{11}S$ Me_2AsSEt, 3, 258
$AsC_4H_{11}S_2$ $Et_2As(S)SH$, 3, 389
$AsC_4H_{12}Br$ $[Me_4As]Br$, 3, 312
$AsC_4H_{12}Cl$ $[Me_4As]Cl$, 3, 312
$AsC_4H_{12}I$ $[Me_4As]I$, 3, 312
$AsC_4H_{12}I_3$ $[Me_4As]I_3$, 3, 312
$AsC_4H_{12}N$ Me_2AsNMe_2, 3, 286
$AsC_4H_{12}NO_3$ $[Me_4As]NO_3$, 3, 312
$AsC_4H_{12}O_2PS_2$ $Me_2AsSP(S)(OMe)_2$, 3, 261
$AsC_4H_{12}O_3P$ $Me_2AsPO(OMe)_2$, 3, 292
$AsC_4H_{12}PS$ $Me_2AsP(S)Me_2$, 3, 290
$AsC_4H_{13}O$ $[Me_4As]OH$, 3, 312
$AsC_4H_{13}O_4S$ $[Me_4As]OSO_3H$, 3, 312
$AsC_4H_{13}O_4S$ $[Me_4As]OSO_3H$, 3, 312
$AsC_4H_{14}N_2O_2PS$ $Me_2As(O)OP(S)(NHMe)_2$, 3, 388
$AsC_5F_{13}Se$ $(CF_3)_2AsSeCF_2CF_2CF_3$, 3, 285
$AsC_5H_4Cl_2N$ Pyridyl-$AsCl_2$ isomers, 3, 181, 190
$(ClCH{=}CH)_2AsCN$, 3, 202
$AsC_5H_4Cl_2NO$ $CH{=}CHCONHCH{=}CAsCl_2$, 3, 190

AsC_5H_4N $(\overline{CH=NCH=CHCH=C}As)_n$, 3, 136
AsC_5H_4NO 3-Pyridyl-AsO, 3, 234
$AsC_5H_4NO_2$ $\overline{CH=CHCONHCH=C}AsO$, 3, 234
$AsC_5H_5ClNO_2$ $\overline{CH=CHC(Cl)=NCHC}As(OH)_2$, 3, 221
$AsC_5H_5ClNO_3$ $H\overline{C=NCCl=CHCH=C}AsO_3H_2$, 3, 513
$AsC_5H_5Cl_2N_2$ $\overline{CH=NC(NH_2)=CHCH=C}AsCl_2$, 3, 190
$AsC_5H_6ClO_3$ $HO_2CCCl=C(Et)AsO$, 3, 234
$AsC_5H_6F_5$ $Me_2AsCF=CFCF_3$, 3, 64
$AsC_5H_6F_7$ $Me_2AsCF_2CF_2CF_3$, 3, 35, 64
$AsC_5H_6NO_3$ 3-Pyridinearsonic acid, 3, 513
$AsC_5H_6NO_4$ $H\overline{C=CHC(O)NHCH=C}AsO_3H_2$, 3, 513
$AsC_5H_7Cl_2$ $(ClCH=CH)_2AsMe$, 3, 64
$AsC_5H_7F_6$ $Me_2AsCF_2CHFCF_3$, 3, 64
$AsC_5H_7N_2$ $PrAs(CN)_2$, 3, 206
$AsC_5H_8ClOS_2$ $ClCH=CH\overline{AsSCH_2CH(CH_2OH)S}$, 3, 583
$AsC_5H_8ClO_3S_3$ ClCH=CH-$\overline{AsSCH_2CH(CH_2SO_3H)S}$, 3, 583
$AsC_5H_8ClS_2$ $ClCH=CH\overline{AsS(CH_2)_3S}$, 3, 618
$AsC_5H_9Cl_2$ $CH_2=CH(CH_2)_3AsCl_2$, 3, 190
$AsC_5H_9F_5I$ $[Me_3(CF_3CF_2)As]I$, 3, 312
$AsC_5H_9N_2S_4$ $Me\overline{AsSC(S)NHCH_2CH_2NHC(S)S}$, 3, 650
$AsC_5H_9O_2S_2$ $MeAs(SCOMe)_2$, 3, 279
$AsC_5H_9O_4S_2$ $MeAs(SCH_2CO_2H)_2$, 3, 266
$AsC_5H_{10}F_3$ Et_2AsCF_3, 3, 36, 64
$AsC_5H_{10}I$ 1-Iodoarsenane, 3, 589
$AsC_5H_{10}N$ Et_2AsCN, 3, 202
AsC_5H_{11} Arsenane, 3, 589
$AsC_5H_{11}Br_2$ $n-C_5H_{11}AsBr_2$, 3, 190
$AsC_5H_{11}Cl_2$ $n-C_5H_{11}AsCl_2$, 3, 190
$Me_2CHCH_2CH_2AsCl_2$, 3, 190
$EtMeCHCH_2AsCl_2$, 3, 191
$AsC_5H_{11}OS_2$ $Me_2AsSC(S)OEt$, 3, 261
$AsC_5H_{11}O_2$ $Et\overline{AsOCH_2CH(Me)O}$, 3, 577
$\overline{CH_2(CH_2)_4A}sO_2H$, 3, 589*
$AsC_5H_{11}O_3$ $CH_2=CH(CH_2)_3AsO_3H_2$, 3, 395
$AsC_5H_{11}O_4$ 5-Methyl-1,4,6,9-tetraoxa-5-arsaspiro[4.4]nonane, 3, 580
$AsC_5H_{11}O_5$ $PrCH(CO_2H)AsO_3H_2$, 3, 395
$AsC_5H_{11}S_2$ $Et\overline{AsS(CH_2)_3S}$, 3, 618
$AsC_5H_{12}Br$ MeBuAsBr, 3, 169
$AsC_5H_{12}Cl$ $Me_2As(CH_2)_3Cl$, 3, 64
EtPrAsCl, 3, 169
$AsC_5H_{12}I$ EtPrAsI, 3, 169
$Et(Me_2CH)AsI$, 3, 169
$AsC_5H_{12}NO_2S$ $Me_2AsSCH_2CH(NH_2)CO_2H$, 3, 259
$AsC_5H_{12}NS_2$ $Me_2AsSC(S)NMe_2$, 3, 262
AsC_5H_{13} Et_2AsMe, 3, 36
Me_2AsPr, 3, 64
$n-C_5H_{11}AsH_2$, 3, 120
$EtMe_2CAsH_2$, 3, 120
$AsC_5H_{13}O_2$ $PrAs(OMe)_2$, 3, 251
$MeAs(OEt)_2$, 3, 251
$MeBuAsO_2H$, 3, 375
$EtPrAsO_2H$, 3, 378
$AsC_5H_{13}O_3$ $Me(CH_2)_4AsO_3H_2$, 3, 395
$Me_2CHCH_2CH_2AsO_3H_2$, 3, 395
$MeEtCHCH_2AsO_3H_2$, 3, 395
$MeAsO(OEt)_2$, 3, 544
$AsC_5H_{13}S$ $Me_2As(CH_2)_3SH$, 3, 36
Et_2MeAsS, 3, 364
$AsC_5H_{13}S_2$ $MeAs(SEt)_2$, 3, 266
$AsC_5H_{13}Se$ $Et_2MeAsSe$, 3, 367
$AsC_5H_{14}NO_3$ $EtNH(CH_2)_3AsO_3H_2$, 3, 395
$AsC_5H_{15}IP$ $[Me_2AsPMe_3]I$, 3, 291
$AsC_5H_{15}N_2$ $(Me_2N)_2AsMe$, 3, 289
$AsC_5H_{15}O_2$ $[Me_3(HOCH_2CH_2)As]OH$, 3, 312
AsC_6Cl_7 Pentachlorophenyl-$AsCl_2$, 3, 191
AsC_6F_9 $(CF_2=CF)_3As$, 3, 6
AsC_6F_{15} $(CF_3CF_2)_3As$, 3, 30
AsC_6HCl_6 2,3,4,5-$Cl_4C_6HAsCl_2$, 3, 191
$AsC_6H_2Br_3O$ 2,4,6-$Br_3C_6H_2AsO$, 3, 234
$AsC_6H_2Cl_5$ 2,4,5-$Cl_3C_6H_2AsCl_2$, 3, 191
$AsC_6H_3ClNO_2S_2$ 4,3-$Cl(O_2N)C_6H_3AsS_2$, 3, 550
$AsC_6H_3ClNO_3$ 4,3-$Cl(O_2N)C_6H_3AsO$, 3, 235
$AsC_6H_3Cl_2N_2O_4$ 2,4-$(O_2N)_2C_6H_3AsCl_2$, 3, 191
$AsC_6H_3Cl_2O$ $Cl_2C_6H_3AsO$ isomers, 3, 235
$AsC_6H_3Cl_3NO_2$ 2,4-$Cl(O_2N)C_6H_3AsCl_2$, 3, 191
$AsC_6H_3Cl_4$ $Cl_2C_6H_3AsCl_2$ isomers, 3, 191
$AsC_6H_3N_2O$ $\overline{OCH=CHCH=C}As(CN)_2$, 3, 206
$AsC_6H_3N_2O_6$ 4,3,5-$HO(O_2N)_2C_6H_2AsO$,

3, 234
$AsC_6H_4BrCl_2$ $BrC_6H_4AsCl_2$ isomers, 3, 191
AsC_6H_4BrO 4-BrC_6H_4AsO, 3, 235
$AsC_6H_4Br_3$ 4-$BrC_6H_4AsBr_2$, 3, 191
$AsC_6H_4Br_3O_3$ $CH{=}C(CO_2Me)OC(Br){=}CAsBr_2$, 3, 192
AsC_6H_4ClO ClC_6H_4AsO isomers, 3, 235
$AsC_6H_4ClO_2$ 3,4-$Cl(HO)C_6H_3AsO$, 3, 235
$AsC_6H_4Cl_2NO_2$ $O_2NC_6H_4AsCl_2$ isomers, 3, 181, 182, 192
$AsC_6H_4Cl_2NO_3$ $HO(O_2N)C_6H_3AsCl_2$ isomers, 3, 192
$AsC_6H_4Cl_3$ $ClC_6H_4AsCl_2$ isomers, 3, 182, 192
$AsC_6H_4Cl_3N_2O$ 4,3-$HO(\overset{+}{N}{\equiv}N)C_6H_3\overset{-}{As}Cl_3$, 3, 389
$AsC_6H_4Cl_3O_2S$ 4-$ClO_2SC_6H_4AsCl_2$, 3, 192
$AsC_6H_4Cl_3O_3$ $Cl_3C_6H_2AsO_3H_2$ isomers, 3, 501
$AsC_6H_4Cl_3S$ 4-$ClSC_6H_4AsCl_2$, 3, 192
AsC_6H_4IO 2-IC_6H_4AsO, 3, 235
$AsC_6H_4NO_2S$ 3-$O_2NC_6H_4AsS$, 3, 264
4-$O_2NC_6H_4AsS$, 3, 264
$AsC_6H_4NO_3$ $O_2NC_6H_4AsO$ isomers, 3, 235
$AsC_6H_4NO_4$ $HO(O_2N)C_6H_3AsO$ isomers, 3, 236
4-$ONC_6H_4AsO_3H_2$, 3, 419
AsC_6H_5 Arsenobenzene, 3, 130
$AsC_6H_5BrNO_5$ 4,3-$Br(O_2N)C_6H_3AsO_3H_2$, 3, 479
$AsC_6H_5Br_2$ $PhAsBr_2$, 3, 192
$AsC_6H_5Br_2O_3S$ $HO_3SC_6H_4AsBr_2$ isomers, 3, 192
AsC_6H_5ClNO 3,4-$H_2N(Cl)C_6H_3AsO$, 3, 236
$AsC_6H_5ClNO_4$ 3,4-$O_2N(Cl)C_6H_3As(OH)_2$, 3, 221
$AsC_6H_5ClNO_5$ 4,3-$Cl(O_2N)C_6H_3AsO_3H_2$, 3, 474
$AsC_6H_5Cl_2$ $PhAsCl_2$, 3, 182
$AsC_6H_5Cl_2N_2O_2$ 4,3-$HO(\overset{+}{N}{\equiv}N)C_6H_3\overset{-}{As}Cl_2OH$, 3, 389 *
$AsC_6H_5Cl_2O$ $HOC_6H_4AsCl_2$ isomers, 3, 192
$AsC_6H_5Cl_2O_2$ 3,4-$(HO)_2C_6H_3AsCl_2$, 3, 192
$AsC_6H_5Cl_2O_3$ 2,4-$Cl_2C_6H_3AsO_3H_2$, 3, 479
$AsC_6H_5Cl_3N$ 3,4-$H_2N(Cl)C_6H_3AsCl_2$, 3, 192
AsC_6H_5FN $[3,4\text{-}H_2N(F)C_6H_3As]_n$, 3, 136
$AsC_6H_5FNO_5$ 4,3-$F(O_2N)C_6H_3AsO_3H_2$, 3, 479
$AsC_6H_5F_2$ $PhAsF_2$, 3, 192
$AsC_6H_5F_2O_3$ 3,4-$F_2C_6H_3AsO_3H_2$, 3, 479
$AsC_6H_5F_4$ $PhAsF_4$, 3, 389
$AsC_6H_5I_2$ $PhAsI_2$, 3, 184
$AsC_6H_5I_2O_3S$ 4-$HO_3SC_6H_4AsI_2$, 3, 192
$AsC_6H_5N_2O_4$ 2,4-$O_2N(H_2NOC)C_6H_3AsO$, 3, 237
$AsC_6H_5N_2O_8$ 4,3,5-$HO(O_2N)_2C_6H_2AsO_3H_2$, 3, 501
AsC_6H_5O $(4\text{-}HOC_6H_4As)_n$, 3, 136
$PhAsO$, 3, 228
AsC_6H_5OS $PhAsSO$, 3, 305
$AsC_6H_5O_2$ 4-HOC_6H_4AsO, 3, 229
$\{C_6H_4As(O)(OH)\}_n$, 3, 375
$AsC_6H_5O_4S$ $HO_3SC_6H_4AsO$ isomers, 3, 236
AsC_6H_5S $PhAsS$, 3, 264
$AsC_6H_6BrO_3$ $BrC_6H_4AsO_3H_2$ isomers, 3, 408
$AsC_6H_6Br_2N$ 4-$H_2NC_6H_4AsBr_2$, 3, 192
$AsC_6H_6Br_2NO_2S$ 4-$H_2NO_2SC_6H_4AsBr_2$, 3, 193
AsC_6H_6Cl 4-$ClC_6H_4AsH_2$, 3, 120
$AsC_6H_6ClF_4$ $Me_2AsC{=}CClCF_2CF_2$, 3, 37
$AsC_6H_6ClF_6$ $CF_3CCl{=}C(CF_3)AsMe_2$, 3, 64
$AsC_6H_6ClO_3$ $ClC_6H_4AsO_3H_2$ isomers, 3, 408, 409
$AsC_6H_6ClO_5S$ 4-$ClO_2SC_6H_4AsO_3H_2$, 3, 415
$AsC_6H_6Cl_2N$ 3-$H_2NC_6H_4AsCl_2$, 3, 193
$AsC_6H_6Cl_2NO$ $H_2N(HO)C_6H_3AsCl_2$ isomers, 3, 184, 193
$AsC_6H_6Cl_2NO_2S$ 4-$H_2NO_2SC_6H_4AsCl_2$, 3, 193
$AsC_6H_6Cl_3$ $(ClCH{=}CH)_3As$, 3, 6
$AsC_6H_6F_5$ $Me_2AsC{=}CFCF_2CF_2$, 3, 37
$AsC_6H_6IO_3$ $IC_6H_4AsO_3H_2$ isomers, 3, 409
$AsC_6H_6I_2NO_2S$ 4-$H_2NO_2SC_6H_4AsI_2$, 3, 193
AsC_6H_6N $(H_2NC_6H_4As)_n$ isomers, 3, 131, 136
AsC_6H_6NO $[3,4\text{-}H_2N(HO)C_6H_3As]_n$, 3, 131
$H_2NC_6H_4AsO$ isomers, 3, 229, 236
AsC_6H_6NOS 3,4-$H_2N(HO)C_6H_3AsS$, 3, 264

$AsC_6H_6NO_2$ $H_2N(HO)C_6H_3AsO$ isomers, 3, 229, 236
$AsC_6H_6NO_3S$ $H_2NO_2SC_6H_4AsO$ isomers, 3, 236
$AsC_6H_6NO_4$ $O_2NC_6H_4As(OH)_2$ isomers, 3, 221
$AsC_6H_6NO_5$ $O_2NC_6H_4AsO_3H_2$ isomers, 3, 419, 420
$AsC_6H_6NO_6$ $HO(O_2N)C_6H_3AsO_3H_2$ isomers, 3, 474, 479
$AsC_6H_6NO_8S$ 3,4-$O_2N(HO_3S)C_6H_3AsO_3H_2$, 3, 479
AsC_6H_6NS [3,4-$H_2N(HS)C_6H_3As]_n$, 3, 136
AsC_6H_7 $PhAsH_2$, 3, 122
$AsC_6H_7BrNO_3$ 3,4-$Br(H_2N)C_6H_3AsO_3H_2$, 3, 479
$AsC_6H_7ClNO_2$ 3,4-$H_2N(HO)C_6H_3AsO\cdot HCl$, 3, 230
4,3-$H_2N(HO)C_6H_3As(OH)Cl$, 3, 199
$AsC_6H_7ClNO_3$ 3,4-$H_2N(Cl)C_6H_3AsO_3H_2$, 3, 479
$AsC_6H_7Cl_2N_2$ 3,4-$(H_2N)_2C_6H_3AsCl_2$, 3, 193
$AsC_6H_7FNO_3$ 3,4-$H_2N(F)C_6H_3AsO_3H_2$, 3, 479
$AsC_6H_7INO_4$ 3,4,5-$H_2N(HO)(I)C_6H_2As$-O_3H_2, 3, 501
$AsC_6H_7N_2O_3S$ 2,4-$H_2N(H_2NSO_2)C_6H_3AsO$, 3, 236
$AsC_6H_7N_2O_5$ $H_2N(O_2N)C_6H_3AsO_3H_2$ isomers, 3, 475, 480
$AsC_6H_7N_2O_7S$ 2,4-$O_2N(H_2NO_2S)C_6H_3As$-O_3H_2, 3, 480
$AsC_6H_7O_2$ $PhAs(OH)_2$, 3, 221
$AsC_6H_7O_3$ $PhAsO_3H_2$, 3, 399
$AsC_6H_7O_3S$ 4-$HSC_6H_4AsO_3H_2$, 3, 415
$AsC_6H_7O_4$ $HOC_6H_4AsO_3H_2$ isomers, 3, 410
$AsC_6H_7O_5$ 2,4-$(HO)_2C_6H_3AsO_3H_2$, 3, 480
$AsC_6H_7O_6S$ $HO_3SC_6H_4AsO_3H_2$ isomers, 3, 415
$AsC_6H_8ClN_2$ $(NCCH_2CH_2)_2AsCl$, 3, 169
$AsC_6H_8ClN_2S_4$ ClCH=CH-$AsSC(S)NHCH_2CH_2NHC(S)S$, 3, 650
AsC_6H_8ClS $Me[SC(Me)=CHCH=C]AsCl$, 3, 169*

AsC_6H_8N 4-$H_2NC_6H_4AsH_2$, 3, 120
AsC_6H_8NO 3,4-$H_2N(HO)C_6H_3AsH_2$, 3, 120*
$AsC_6H_8NOS_2$ 3,4-$H_2N(HO)C_6H_3As(SH)_2$, 3, 263
$AsC_6H_8NO_2$ 4-$H_2NC_6H_4As(OH)_2$, 3, 221
$AsC_6H_8NO_2S$ 3,4-$H_2N(HO)C_6H_3As(SH)OH$, 3, 220
$AsC_6H_8NO_3$ 2-$H_2NC_6H_4AsO_3H_2$, 3, 421
3-$H_2NC_6H_4AsO_3H_2$, 3, 424
4-$H_2NC_6H_4AsO_3H_2$, 3, 425
$AsC_6H_8NO_3S$ 3,4-$H_2N(HS)C_6H_3AsO_3H_2$, 3, 475
$AsC_6H_8NO_4$ $HO(H_2N)C_6H_3AsO_3H_2$ isomers. 3, 475, 476, 480
$AsC_6H_8NO_4S$ 4-$H_2NO_2SC_6H_4As(OH)_2$, 3, 221
2-$H_2NO_2SC_6H_4As(OH)_2$, 3, 221
$AsC_6H_8NO_5S$ $H_2NO_2SC_6H_4AsO_3H_2$ isomers, 3, 415
AsC_6H_9 $(CH_2=CH)_3As$, 3, 6
$AsC_6H_9Cl_3N_3O_9$ $(O_2NOCH_2CHCl)_3As$, 3, 6
$AsC_6H_9Cl_6O_2$ $EtAs(OCH_2CCl_3)_2$, 3, 251
$AsC_6H_9F_6$ $(CF_3)_2AsBu$, 3, 37, 64
$AsC_6H_9I_2$ $(CH_2=CH)_3AsI_2$, 3, 343
$AsC_6H_9N_2$ $BuAs(CN)_2$, 3, 206
$AsC_6H_9N_2O_2$ 4-$H_2NNHC_6H_4As(OH)_2$, 3, 222
$(NCCH_2CH_2)_2AsO_2H$, 3, 378
$AsC_6H_9N_2O_3$ 4-$H_2NNHC_6H_4AsO_3H_2$, 3, 451
$AsC_6H_9N_2O_4$ 4,3-$HO(H_2NNH)C_6H_3AsO_3H_2$, 3, 480
$AsC_6H_9N_2O_5S$ 2,4-$H_2N(H_2NO_2S)C_6H_3As$-O_3H_2, 3, 480
$AsC_6H_{10}Cl$ $[(CH_2=CH)_3AsH]Cl$, 3, 342
$AsC_6H_{10}ClOS_2$ $ClCH=CHAsSCH_2CH(CH_2OMe)S$, 3, 583
AsC_6H_{11} 1-Vinylarsolane, 3, 553
$AsC_6H_{11}Cl_2$ cyclo-$C_6H_{11}AsCl_2$, 3, 193
$AsC_6H_{11}N_2S_4$ $EtAsSC(S)NHCH_2CH_2NHC(S)S$, 3, 650
$AsC_6H_{11}OS_2$ $Me_2AsSC(S)OCH_2CH=CH_2$, 3, 261
$AsC_6H_{11}O_4$ $EtAs(OAc)_2$, 3, 256
$AsC_6H_{12}ClN_4O_4P_2$ p-$ClC_6H_4As[OP(O)$-$(NH_2)_2]_2$, 3, 256
$AsC_6H_{12}ClO_2$ $ClCH=CHAs(OEt)_2$, 3, 251
$AsC_6H_{12}Cl_3$ $(MeCHCl)_3As$, 3, 7
$AsC_6H_{12}N$ $EtPrAsCN$, 3, 202

$AsC_6H_{12}NS$ EtPrAsSCN, 3, 202
$AsC_6H_{13}Cl_2$ $C_6H_{13}AsCl_2$, 3, 193
$AsC_6H_{13}N_2$ $(CH_2CH_2N)_2AsEt$, 3, 289
$AsC_6H_{13}OS_2$ $Me_2AsSC(S)OCHMe_2$, 3, 261
$AsC_6H_{13}O_3$ $EtAsOCH_2CH(CH_2OMe)O$, 3, 577
$AsC_6H_{13}O_5$ $Me(CH_2)_3CH(CO_2H)AsO_3H_2$, 3, 396
$AsC_6H_{14}Br$ Pr_2AsBr, 3, 170
$AsC_6H_{14}Cl$ Pr_2AsCl, 3, 163, 170
EtBuAsCl, 3, 169
$AsC_6H_{14}I$ EtBuAsI, 3, 169
$Et(Me_2CHCH_2)AsI$, 3, 169
Pr_2AsI, 3, 170
$AsC_6H_{14}N$ $Et_2AsNCH_2CH_2$, 3, 286
AsC_6H_{15} Et_3As, 3, 7
Me_2AsBu, 3, 64
MeEtPrAs, 3, 74
Pr_2AsH, 3, 117
$AsC_6H_{15}Br_2$ Et_3AsBr_2, 3, 343
$AsC_6H_{15}Cl_2$ Et_3AsCl_2, 3, 343
$AsC_6H_{15}I_4$ Et_3AsI_4, 3, 343
$AsC_6H_{15}O$ Et_2AsOEt, 3, 209
Et_3AsO, 3. 356
Me_2BuAsO, 3, 356
$AsC_6H_{15}O_2$ $EtAs(OEt)_2$, 3, 251
$BuAs(OMe)_2$, 3, 251
$\{Et_3AsO\text{-}O\}_n$, 3, 362
Pr_2AsO_2H, 3, 378
$AsC_6H_{15}O_3$ $[Me_2(HO_2CCH_2CH_2)As]OH$, 3, 312
$n\text{-}C_6H_{13}AsO_3H_2$, 3, 396
$EtAsO(OEt)_2$, 3, 545
$AsC_6H_{15}S$ $Me_2AsCH_2CH_2CH_2SMe$, 3, 37
Et_3AsS, 3, 364
$AsC_6H_{15}S_2$ $EtAs(SEt)_2$, 3, 266
$AsC_6H_{15}Se$ Et_3AsSe, 3, 367
$Me_2BuAsSe$, 3, 367
$AsC_6H_{16}I$ $[Et_2Me_2As]I$, 3, 313
$AsC_6H_{16}IO$ $[Et_2Me(MeO)As]I$, 3, 313
$AsC_6H_{16}NO_3$ $PrNH(CH_2)_3AsO_3H_2$, 3, 396
$AsC_6H_{16}O_2PS_2$ $Me_2AsSP(S)(OEt)_2$, 3, 261
$AsC_6H_{16}O_3P$ $Me_2AsPO(OEt)_2$, 3, 292
$Et_2AsPO(OMe)_2$, 3, 292
$AsC_6H_{16}O_4PS$ $Me_2As(O)OP(S)(OEt)_2$, 3, 388
$AsC_6H_{16}O_5P$ $Me_2As(O)OP(O)(OEt)_2$, 3, 388
$AsC_6H_{17}O$ $[Et_2Me_2As]OH$, 3, 313
$AsC_6H_{18}N_2O_2PS$ $Me_2As(O)OP(S)(NMe_2)_2$, 3, 388
$AsC_7H_4Cl_3O$ $4\text{-}ClOCC_6H_4AsCl_2$, 3, 184
AsC_7H_4N $(4\text{-}NCC_6H_4As)_n$, 3, 136
AsC_7H_4NO $4\text{-}NCC_6H_4AsO$, 3, 236
$AsC_7H_4NO_5$ $3,4\text{-}O_2N(HO_2C)C_6H_3AsO$, 3, 236
$AsC_7H_5BrF_3$ $(CF_3)PhAsBr$, 3, 170
$AsC_7H_5Br_2N_2$ $4\text{-}NCHNC_6H_4AsBr_2$, 3, 193
$AsC_7H_5ClF_3$ $Ph(CF_3)AsCl$, 3, 163
$AsC_7H_5Cl_2N_2O_4$ $4,3,5\text{-}Me(O_2N)_2C_6H_2AsCl_2$, 3, 193
$AsC_7H_5Cl_2O_2$ $HO_2CC_6H_4AsCl_2$ isomers, 3, 185, 193
$AsC_7H_5F_3I$ $Ph(CF_3)AsI$, 3, 163
$AsC_7H_5N_2O$ $4\text{-}NCNHC_6H_4AsO$, 3, 231
$AsC_7H_5N_2S$ Arsenoso-2-benzimidazolylthiol isomers, 3, 237
$AsC_7H_6Br_3O_3$ $CH{=}C(CO_2Et)OC(Br){=}CAsBr_2$, 3, 193
$AsC_7H_6ClF_6$ $Me_2AsC{=}CClCF_2CF_2CF_2$, 3, 38
$AsC_7H_6ClO_4$ $4\text{-}(ClCO)C_6H_4AsO_3H_2$, 3, 401
$AsC_7H_6Cl_2NO_2$ $H_2N(HO_2C)C_6H_3AsCl_2$ isomers, 3, 193
$AsC_7H_6Cl_3$ $4\text{-}ClCH_2C_6H_4AsCl_2$, 3, 194
$AsC_7H_6F_3O_3$ $3\text{-}CF_3C_6H_4AsO_3H_2$, 3, 401
$AsC_7H_6NO_2$ $4\text{-}NCC_6H_4As(OH)_2$, 3, 222
$AsC_7H_6NO_3$ $H_2N(HO_2C)C_6H_3AsO$ isomers, 3, 236, 237
$HO(H_2NCO)C_6H_3AsO$ isomers, 3, 237
$4\text{-}NCC_6H_4AsO_3H_2$, 3, 401
$AsC_7H_6NO_3S$ $4\text{-}NCSC_6H_4AsO_3H_2$, 3, 415
$AsC_7H_6NO_4$ $3,4\text{-}O_2N(MeO)C_6H_3AsO$, 3, 237
$AsC_7H_6NO_5$ 2,3-Dihydro-2-oxobenzoxazolearsonic acid isomers, 3, 513
$AsC_7H_6NO_7$ $HO_2C(O_2N)C_6H_3AsO_3H_2$ isomers, 3, 480
AsC_7H_7 $(4\text{-}MeC_6H_4As)_n$, 3, 136
$AsC_7H_7BrNO_2$ $Me(3\text{-}O_2NC_6H_4)AsBr$, 3, 170
$AsC_7H_7Br_2N_2$ $4\text{-}H_2NC({=}NH)C_6H_4AsBr_2$, 3, 194
$AsC_7H_7ClNO_2$ $Me(3\text{-}O_2NC_6H_4)AsCl$, 3, 170
$AsC_7H_7Cl_2$ $Me(2\text{-}ClC_6H_4)AsCl$, 3, 170
$MeC_6H_4AsCl_2$ isomers, 3, 185, 194
$AsC_7H_7Cl_2N_2$ $4\text{-}H_2NC({=}NH)C_6H_4AsCl_2$,

3, 194
$AsC_7H_7Cl_2N_2O_2$ 3,4,5-$H_2N(HO)(H_2NCO)$-$C_6H_2AsCl_2$, 3, 194
$AsC_7H_7Cl_2O$ 4-$HOCH_2C_6H_4AsCl_2$, 3, 194
$MeOC_6H_4AsCl_2$ isomers, 3, 194
$AsC_7H_7Cl_2S$ 2-$MeSC_6H_4AsCl_2$, 3, 194
$AsC_7H_7N_2$ [4-$H_2NC(=NH)C_6H_4As$]$_n$, 3, 137
$AsC_7H_7N_2O$ [4-$(H_2NCONH)C_6H_4As$]$_n$, 3, 137
$AsC_7H_7N_2O_2$ 4-$NCNHC_6H_4As(OH)_2$, 3, 222
4-$H_2NCONHC_6H_4AsO$, 3, 238
3,4-$H_2N(H_2NCO)C_6H_3AsO$, 3, 238
$AsC_7H_7N_2O_3$ 2,3-Dihydro-2-oxobenzimidazol-4-yl-$As(OH)_2$, 3, 222
4-$(NCNH)C_6H_4AsO_3H_2$, 3, 435
$AsC_7H_7N_2O_3S$ 2-Mercaptobenzimidazole-5-arsonic acid, 3, 513
$AsC_7H_7N_2O_6$ 3,4-$O_2N(H_2NCO)C_6H_3AsO_3H_2$, 3, 481
$AsC_7H_7N_2O_7$ 4,3,5-$Me(O_2N)_2C_6H_2AsO_3H_2$, 3, 501
AsC_7H_7O MeC_6H_4AsO isomers, 3, 231, 238
$AsC_7H_7O_2$ 4-$HOCH_2C_6H_4AsO$, 3, 238
$AsC_7H_7O_4$ $HO_2CC_6H_4As(OH)_2$ isomers, 3, 222
$AsC_7H_7O_5$ $HO_2CC_6H_4AsO_3H_2$ isomers, 3, 400, 408
AsC_7H_7S 4-MeC_6H_4AsS, 3, 265
AsC_7H_8Cl $MePhAsCl$, 3, 163
$AsC_7H_8ClN_2O_2$ $MeNH(3$-$O_2NC_6H_4)AsCl$, 3, 199
AsC_7H_8ClO $PhAs(OMe)Cl$, 3, 200
$AsC_7H_8ClO_3$ 4-$(ClCH_2)C_6H_4AsO_3H_2$, 3, 401
$AsC_7H_8Cl_2N$ $MeC{=}NC(Me){=}CHCH{=}CAsCl_2$, 3, 185
2-$MeNHC_6H_4AsCl_2$, 3, 194
AsC_7H_8I $MePhAsI$, 3, 163
$AsC_7H_8IN_2O_5$ 3,4,5-$H_2NCONH(HO)(I)$-$C_6H_2AsO_3H_2$, 3, 501
AsC_7H_8N [$CH{=}CHC(Me){=}NC(Me){=}CAs$]$_n$, 3, 136
[$MeAsNPh$]$_n$, 3, 288
AsC_7H_8NO 4-$MeNHC_6H_4AsO$, 3, 237
$MeC{=}NC(Me){=}CHCH{=}CAsO$, 3, 237
4-$H_2NCH_2C_6H_4AsO$, 3, 238
$AsC_7H_8NO_2$ 3,4-$H_2N(MeO)C_6H_3AsO$, 3, 238
$AsC_7H_8NO_3$ 2-$H_2NOCC_6H_4As(OH)_2$, 3, 222
$AsC_7H_8NO_3S$ 4-$MeHNO_2SC_6H_4AsO$, 3, 238
$AsC_7H_8NO_4$ 2,5-$Me(O_2N)C_6H_3As(OH)_2$, 3, 222
$Me(3$-$O_2NC_6H_4)AsO_2H$, 3, 378
4-$(H_2NCO)C_6H_4AsO_3H_2$, 3, 401
$AsC_7H_8NO_5$ $Me(O_2N)C_6H_3AsO_3H_2$ isomers, 3, 481
4,3-$HO(H_2NCO)C_6H_3AsO_3H_2$, 3, 481
$HO_2C(H_2N)C_6H_3AsO_3H_2$ isomers, 3, 481
$AsC_7H_8N_3O_2$ 3,4-$H_2N(H_2NCONH)C_6H_3AsO$, 3, 238
AsC_7H_9 $MePhAsH$, 3, 117
4-$MeC_6H_4AsH_2$, 3, 120
AsC_7H_9ClNO Me[3,4-$H_2N(HO)C_6H_3$]$AsCl$, 3, 170
AsC_7H_9INO Me[3,4-$H_2N(HO)C_6H_3$]AsI, 3, 170
$AsC_7H_9N_2O_2$ 4-[$H_2NC(=NH)$]$C_6H_4As(OH)_2$, 3, 222, 226
$AsC_7H_9N_2O_3$ 4-$H_2NCONHC_6H_4As(OH)_2$, 3, 222
4-[$H_2NC(=NH)$]$C_6H_4AsO_3H_2$, 3, 401
$AsC_7H_9N_2O_3S$ 4-$(H_2NCSNH)C_6H_4AsO_3H_2$, 3, 435
$AsC_7H_9N_2O_4$ 4-$(H_2NNHCO)C_6H_4AsO_3H_2$, 3, 401
3,4-$H_2N(H_2NCO)C_6H_3AsO_3H_2$, 3, 481
4-$(H_2NCONH)C_6H_4AsO_3H_2$, 3, 428
$AsC_7H_9N_2O_5$ $HO(H_2NCONH)C_6H_3AsO_3H_2$ isomers, 3, 481, 482
AsC_7H_9O 3-$MeOC_6H_4AsH_2$, 3, 121
$AsC_7H_9O_2$ $MePhAsO_2H$, 3, 375
$AsC_7H_9O_3$ $PhCH_2AsO_3H_2$, 3, 396
$MeC_6H_4AsO_3H_2$ isomers, 3, 401, 402
$AsC_7H_9O_3S$ $MeSC_6H_4AsO_3H_2$, 3, 415
$AsC_7H_9O_4$ 4-$(HOCH_2)C_6H_4AsO_3H_2$, 3, 402
$MeOC_6H_4AsO_3H_2$ isomers, 3, 411
$AsC_7H_9O_4S$ 4-$MeSO_2C_6H_4As(OH)_2$, 3, 222
$AsC_7H_9O_5S$ 4-$MeSO_2C_6H_4AsO_3H_2$, 3, 415
$AsC_7H_{10}N$ Pyridyl-$AsMe_2$ isomers, 3, 64
$AsC_7H_{10}NO$ 4,3-$HO(MeNH)C_6H_3AsH_2$, 3, 121
$AsC_7H_{10}NO_2$ 4-$H_2NCH_2C_6H_4As(OH)_2$, 3, 222
$AsC_7H_{10}NO_3$ Me[3,4-$H_2N(HO)C_6H_3$]AsO_2H,

3, 378
$4\text{-}(H_2NCH_2)C_6H_4AsO_3H_2$, 3, 402
$4\text{-}(MeNH)C_6H_4AsO_3H_2$, 3, 435
$2,4\text{-}H_2N(Me)C_6H_3AsO_3H_2$, 3, 482
$MeC{=}NC(Me){=}CHCH{=}CAsO_3H_2$, 3, 514
$AsC_7H_{10}NO_3S$ $HC{=}NC(SEt){=}CHCH{=}CAsO_3H_2$, 3, 514
$AsC_7H_{10}NO_4$ $4,3\text{-}HO(MeNH)C_6H_3AsO_3H_2$, 3, 482
$AsC_7H_{10}NO_6S$ $4,3\text{-}HO(HO_2SCH_2NH)C_6H_3\text{-}AsO_3H_2$, 3, 482
$AsC_7H_{10}N_3O_3$ $[H_2NC({=}NH)NH]C_6H_4AsO_3H_2$ isomers, 3, 422, 435
$AsC_7H_{10}N_3O_5$ $4,3\text{-}HO(H_2NNHCONH)C_6H_3\text{-}AsO_3H_2$, 3, 482
$AsC_7H_{11}ClNO_4$ $2\text{-}ClC_6H_4AsO_3H_2\cdot H_2NCH_2OH$, 3, 409
$4\text{-}ClC_6H_4AsO_3H_2\cdot H_2NCH_2OH$, 3, 409
$AsC_7H_{11}Cl_2$ $(ClCH{=}CH)_2AsPr$, 3, 65
$(ClCH{=}CH)_2AsCHMe_2$, 3, 65
$AsC_7H_{11}IN$ $[Me_2AsN{=}CHCH{=}CHCH{=}CH]I$, 3, 288
$AsC_7H_{11}N_2$ $MeAs(CH_2CH_2CN)_2$, 3, 38
$n\text{-}C_5H_{11}As(CN)_2$, 3, 206
$AsC_7H_{11}N_2O_3$ $HC{=}NC(NHEt){=}CHCH{=}CAsO_3H_2$, 3, 514
$AsC_7H_{12}I$ $[Me(CH_2{=}CH)_3As]I$, 3, 313
AsC_7H_{13} Hexahydro-1H-arsolo[1,2-a]-arsole, 3, 555
$AsC_7H_{13}N_2O_4$ $4\text{-}H_2NC_6H_4AsO_3H_2\cdot H_2NCH_2OH$, 3, 428
$AsC_7H_{13}O_2S_4$ $MeAs[SC(S)OEt]_2$, 3, 279
$AsC_7H_{13}O_4$ $PrAs(OAc)_2$, 3, 256
$AsC_7H_{14}N$ BuEtAsCN, 3, 202
$AsC_7H_{14}NOS_2$ $Me_2AsSC(S)N{=}(CH_2CH_2)_2{=}O$, 3, 262
AsC_7H_{15} $Me_2As(CH_2)_3CH{=}CH_2$, 3, 38
$AsC_7H_{15}Br_2$ $Me_2[CH_2{=}CH(CH_2)_3]AsBr_2$, 3, 343
$AsC_7H_{15}Cl_2$ $n\text{-}C_7H_{15}AsCl_2$, 3, 194
$AsC_7H_{15}N_2$ $(CH_2CH_2N)_2AsPr$, 3, 289
$AsC_7H_{15}N_2S_4$ $MeAs[SC(S)NMe_2]_2$, 3, 282
$AsC_7H_{15}O$ $Et_2AsOCH_2CH{=}CH_2$, 3, 209
$AsC_7H_{15}O_2$ $MeEtAsCH_2CO_2Et$, 3, 74
$AsC_7H_{15}O_3$ $EtAsOCH_2CH(CH_2OEt)O$, 3, 577
$AsC_7H_{15}O_4$ 2,5,7-Trimethyl-1,4,6,9-tetraoxa-5-arsaspiro[4.4]nonane, 3, 580
$AsC_7H_{15}O_5$ $Me(CH_2)_4CH(CO_2H)AsO_3H_2$, 3, 396
AsC_7H_{16} 1,2-Dimethylarsenane, 3, 589
$AsC_7H_{16}I$ $Et(Me_2CHCH_2CH_2)AsI$, 3, 170
$AsC_7H_{16}NS_2$ $Me_2AsSC(S)NEt_2$, 3, 262
$Me_2AsSC(S)NEt_2$, 3, 262
$AsC_7H_{17}O$ MeBuAsOEt, 3, 209
Et_2AsOPr, 3, 209
$AsC_7H_{17}O_2$ $PrAs(OEt)_2$, 3, 251
$MeAs(OCHMe_2)_2$, 3, 251
MeBuAs(O)OEt, 3, 388
$AsC_7H_{17}O_3$ $n\text{-}C_7H_{15}AsO_3H_2$, 3, 396
$MeAsO(OPr)_2$, 3, 544
$MeAsO(OCHMe_2)_2$, 3, 544
$PrAsO(OEt)_2$, 3, 546
$AsC_7H_{17}S_2$ $MeAs(SPr)_2$, 3, 266
$PrAs(SEt)_2$, 3, 266
$AsC_7H_{18}I$ $[BuMe_3As]I$, 3, 313
$AsC_7H_{18}NO_3$ $BuNH(CH_2)_3AsO_3H_2$, 3, 396
$AsC_8H_4Cl_4NS_2$ Chloromethylthioarsinous acid ester with 4,6,7-trichloro-2-mercaptobenzothiazole, 3, 305
$AsC_8H_5F_6$ $(CF_3)_2AsPh$, 3, 65
$AsC_8H_5N_2O_2$ 7-Arsenoso-4-hydroxyquinazoline, 3, 238
$AsC_8H_5O_5$ $3,5\text{-}(HO_2C)_2C_6H_3AsO$, 3, 238
AsC_8H_6Cl 1-Chloroarsindole, 3, 555
$AsC_8H_6ClO_2$ $(OCH{=}CHCH{=}C)_2AsCl$, 3, 163
$AsC_8H_6ClS_2$ $(SCH{=}CHCH{=}C)_2AsCl$, 3, 164
$AsC_8H_6Cl_2N$ $4\text{-}NCCH_2C_6H_4AsCl_2$, 3, 194
$AsC_8H_6Cl_2NO_2$ $ClCH{=}CH(2\text{-}O_2NC_6H_4)AsCl$, 3, 170
$AsC_8H_6Cl_3O$ $3\text{-}ClCH_2COC_6H_4AsCl_2$, 3, 194
AsC_8H_6NO $4\text{-}NCCH_2C_6H_4AsO$, 3, 239
$AsC_8H_6NO_4$ $HO_2C(H_2NOC)C_6H_3AsO$ isomers, 3, 239
$AsC_8H_6NO_5$ $HO_2C(NC)C_6H_3AsO_3H_2$ isomers, 3, 482
AsC_8H_7ClI 5-Chloro-2-iodoisoarsindoline, 3, 556
$AsC_8H_7ClNO_3$ $4,3\text{-}HO(ClCH_2CONH)C_6H_3AsO$, 3, 239
$AsC_8H_7Cl_2$ $Ph(ClCH{=}CH)AsCl$, 3, 164
$PhCH{=}CHAsCl_2$, 3, 194

2-CH_2=$CHC_6H_4AsCl_2$, 3, 194
$AsC_8H_7Cl_2N_2O_5$ 4,3,5-EtO(O_2N)$_2C_6H_2$-$AsCl_2$, 3, 194
$AsC_8H_7Cl_2O$ 4-$AcC_6H_4AsCl_2$, 3, 195
$AsC_8H_7Cl_2O_2$ 4-$HO_2CCH_2C_6H_4AsCl_2$, 3, 195
$AsC_8H_7Cl_2O_3$ 4-(HO_2CCHOH)$C_6H_4AsCl_2$, 3, 195
$AsC_8H_7Cl_3NO_4$ 4-(Cl_3CCONH)$C_6H_4AsO_3H_2$, 3, 435
$AsC_8H_7N_2O_2$ Me(3-$O_2NC_6H_4$)AsCN, 3, 202
$AsC_8H_7N_2O_3$ 3,5-(H_2NOC)$_2C_6H_3AsO$, 3, 239
4-$H_2NCONHCOC_6H_4AsO$, 3, 239
$AsC_8H_7N_2O_4$ 4-Hydroxyquinazolinearsonic acid isomers, 3, 514
$AsC_8H_7O_2$ [2,5-HO(Ac)C_6H_3As]$_n$, 3, 137
4-AcC_6H_4AsO, 3, 239
$AsC_8H_7O_2S_2$ ($\overline{SCH=CHCH=C}$)$_2AsO_2H$, 3, 378
$AsC_8H_7O_3$ 4-$AcOC_6H_4AsO$, 3, 239
4-$HO_2CCH_2C_6H_4AsO$, 3, 239
4-$HOCH_2COC_6H_4AsO$, 3, 239
2,5-HO(Ac)C_6H_3AsO, 3, 239
$AsC_8H_7O_4$ 4-$HO_2CCH_2OC_6H_4AsO$, 3, 239
($\overline{OCH=CHCH=C}$)$_2AsO_2H$, 3, 378
AsC_8H_8Cl 2-Chloroisoarsindoline, 3, 556
$AsC_8H_8ClO_2$ Ph(HO_2CCH_2)AsCl, 3, 170
(ClCH=CH)$PhAsO_2H$, 3, 378
$AsC_8H_8ClO_3$ 3,4-Cl($HOCH_2CH_2O$)C_6H_3AsO, 3, 239
$AsC_8H_8ClO_4$ $ClCH_2COC_6H_4AsO_3H_2$ isomers, 3, 402
$AsC_8H_8ClS_2$ 4-$ClC_6H_4\overline{AsSCH_2CH_2S}$, 3, 583
$AsC_8H_8Cl_2NO$ 4-(H_2NCOCH_2)$C_6H_4AsCl_2$, 3, 195
$AsC_8H_8Cl_2NO_4$ 4-($Cl_2CHCONH$)$C_6H_4AsO_3H_2$, 3, 435
$AsC_8H_8Cl_3S$ 4-($ClCH_2CH_2S$)$C_6H_4AsCl_2$, 3, 195
$AsC_8H_8F_3$ Me(CF_3)AsPh, 3, 74
AsC_8H_8I 2-Iodoisoarsindoline, 3, 557
AsC_8H_8N MePhAsCN, 3, 202
$AsC_8H_8NO_2$ 4-$H_2NOCCH_2C_6H_4AsO$, 3, 239
$AcNHC_6H_4AsO$ isomers, 3, 239
4-MeC(=NOH)C_6H_4AsO, 3, 239
$AsC_8H_8NO_3$ 3,4-H_2N(HO)$C_6H_3As(OH)_2$, 3, 220
HO(AcNH)C_6H_3AsO isomers, 3, 231, 239
4-($NCCH_2$)$C_6H_4AsO_3H_2$, 3, 402
$AsC_8H_8NO_4$ 4-$H_2NOCCH_2OC_6H_4As(OH)_2$, 3, 223
2-HO-isoarsindoline oxide nitrate, 3, 557
4-$O_2NC_6H_4\overline{AsOCH_2CH_2O}$, 3, 577
$AsC_8H_8NO_4S$ 3-Hydroxy-1,4,2-benzothiazine-6-arsonic acid, 3, 514
$AsC_8H_8NO_5$ O_2N($HOCH_2CH_2O$)C_6H_3AsO isomers, 3, 239
3-Hydroxy-1,4,2-benzoxazine-6-arsonic acid, 3, 514
$AsC_8H_8NO_6$ 4-(HO_2CCONH)$C_6H_4AsO_3H_2$, 3, 429
HO_2C(H_2NCO)$C_6H_3AsO_3H_2$ isomers, 3, 482
$AsC_8H_8NO_7$ 3,4-O_2N(AcO)$C_6H_3AsO_3H_2$, 3, 482
$AsC_8H_8NO_8$ 3,4-O_2N(HO_2CCH_2O)$C_6H_3AsO_3H_2$, 3, 477
AsC_8H_8NS MePhAsSCN, 3, 202
$AsC_8H_8N_3O_3$ 4-($\overline{N=CHCH=NN}$)$C_6H_4AsO_3H_2$, 3, 469
$AsC_8H_9ClNO_3$ 3,4-H_2N(HO)$C_6H_3As(OH)_2$·HCl, 3, 220
$AsC_8H_9ClNO_4$ 4-($ClCH_2CONH$)$C_6H_4AsO_3H_2$, 3, 435
$AsC_8H_9Cl_2$ 2,4-$Me_2C_6H_3AsCl_2$, 3, 195
$AsC_8H_9Cl_2N_2O$ 4-(H_2NCH_2CONH)$C_6H_4AsCl_2$, 3, 195
$AsC_8H_9Cl_2O_2$ 4-($HOCH_2CH_2O$)$C_6H_4AsCl_2$, 3, 195
$AsC_8H_9FNO_4$ 4,3-F(AcNH)$C_6H_3AsO_3H_2$, 3, 482
$AsC_8H_9INO_4$ 4-(ICH_2CONH)$C_6H_4AsO_3H_2$, 3, 435
$AsC_8H_9N_2O_2$ 4-$H_2NCONHCH_2C_6H_4AsO$, 3, 239
4-$H_2NCOCH_2NHC_6H_4AsO$, 3, 240
$AsC_8H_9N_2O_4S$ 4-($H_2NCOCH_2NHO_2S$)C_6H_4AsO, 3, 240
$AsC_8H_9N_2O_5$ 4-($H_2NCONHCO$)$C_6H_4AsO_3H_2$, 3, 402
4-($H_2NCOCONH$)$C_6H_4AsO_3H_2$, 3, 429
$AsC_8H_9N_2O_7$ 3,4-O_2N(H_2NCOCH_2O)C_6H_3-

AsO$_3$H$_2$, 3, 482
3,4,5-AcNH(HO)(O$_2$N)C$_6$H$_2$AsO$_3$H$_2$, 3, 501
AsC$_8$H$_9$N$_2$O$_8$ 4,3,5-EtO(O$_2$N)$_2$C$_6$H$_2$AsO$_3$H$_2$, 3, 501
AsC$_8$H$_9$N$_2$O$_9$ 3,4-O$_2$N(O$_2$NOCH$_2$CH$_2$O)C$_6$H$_3$-AsO$_3$H$_2$, 3, 482
4,2,6-HOCH$_2$CH$_2$O(O$_2$N)$_2$C$_6$H$_2$AsO$_3$H$_2$, 3, 501
AsC$_8$H$_9$O Me$_2$C$_6$H$_3$AsO isomers, 3, 240
AsC$_8$H$_9$O$_2$ EtAsOC$_6$H$_4$O-o, 3, 577
PhAsOCH$_2$CH$_2$O, 3, 577
AsC$_8$H$_9$O$_3$ HOCH$_2$CH$_2$OC$_6$H$_4$AsO isomers, 3, 240
PhCH=CHAsO$_3$H$_2$, 3, 396
2-(CH$_2$=CH)C$_6$H$_4$AsO$_3$H$_2$, 3, 402
Poly(p-arsonostyrene), 3, 402
AsC$_8$H$_9$O$_4$ 4-AcOC$_6$H$_4$As(OH)$_2$, 3, 223
AcC$_6$H$_4$AsO$_3$H$_2$ isomers, 3, 402, 411
Poly(4-hydroxystyrene-3-arsonic acid), 3, 542
AsC$_8$H$_9$O$_5$ OCH=CHCH=CAs(OAc)$_2$, 3, 256
PhCH(CO$_2$H)AsO$_3$H$_2$, 3, 396
4-(HO$_2$CCH$_2$)C$_6$H$_4$AsO$_3$H$_2$, 3, 402
4-(HOCH$_2$CO)C$_6$H$_4$AsO$_3$H$_2$, 3, 402
2,5-HO(Ac)C$_6$H$_3$AsO$_3$H$_2$, 3, 482
AsC$_8$H$_9$O$_5$S 4-HO$_2$CCH$_2$SC$_6$H$_4$AsO$_3$H$_2$, 3, 416
AsC$_8$H$_9$O$_6$ 4-(HO$_2$CCHOH)C$_6$H$_4$AsO$_3$H$_2$, 3, 402
4,3-HO(MeO$_2$C)C$_6$H$_3$AsO$_3$H$_2$, 3, 482
AsC$_8$H$_9$O$_7$S 4-HO$_2$CCH$_2$SO$_2$C$_6$H$_4$AsO$_3$H$_2$, 3, 416
AsC$_8$H$_9$S$_2$ PhAsSCH$_2$CH$_2$S, 3, 583
AsC$_8$H$_{10}$Br BrC$_6$H$_4$AsMe$_2$ isomers, 3, 39
AsC$_8$H$_{10}$Cl EtPhAsCl, 3, 170
AsC$_8$H$_{10}$ClO PhAs(OEt)Cl, 3, 200
AsC$_8$H$_{10}$ClO$_5$ 3,4-Cl(HOCH$_2$CH$_2$O)C$_6$H$_3$-AsO$_3$H$_2$, 3, 477
AsC$_8$H$_{10}$Cl$_2$N 4-Me$_2$NC$_6$H$_4$AsCl$_2$, 3, 195
AsC$_8$H$_{10}$Cl$_2$NO$_2$ 3,4-H$_2$N(HOCH$_2$CH$_2$O)C$_6$H$_3$-AsCl$_2$, 3, 195
AsC$_8$H$_{10}$I EtPhAsI, 3, 164
AsC$_8$H$_{10}$I$_2$NO$_2$S 4-Me$_2$NO$_2$SC$_6$H$_4$AsI$_2$, 3, 195
AsC$_8$H$_{10}$NO 4-Me$_2$NC$_6$H$_4$AsO, 3, 240
AsC$_8$H$_{10}$NO$_2$ O$_2$NC$_6$H$_4$AsMe$_2$ isomers, 3, 65
[3,4-H$_2$N(HOCH$_2$CH$_2$O)C$_6$H$_3$As]$_n$, 3, 137
3,4-H$_2$N(EtO)C$_6$H$_3$AsO, 3, 240
H$_2$N(HOCH$_2$CH$_2$)C$_6$H$_3$AsO isomers, 3, 240
AsC$_8$H$_{10}$NO$_2$S Me$_2$(4-O$_2$NC$_6$H$_4$)AsS, 3, 364
AsC$_8$H$_{10}$NO$_2$S$_2$ Me$_2$As(S)SC$_6$H$_4$NO$_2$-4, 3, 389
AsC$_8$H$_{10}$NO$_3$S 4-Me$_2$NO$_2$SC$_6$H$_4$AsO, 3, 240
4-EtNHO$_2$SC$_6$H$_4$AsO, 3, 240
AsC$_8$H$_{10}$NO$_4$ Me$_2$[2,4-HO(O$_2$N)C$_6$H$_3$]AsO, 3, 356
HO$_2$CCH$_2$(4-H$_2$NC$_6$H$_4$)AsO$_2$H, 3, 379
4-(H$_2$NCOCH$_2$)C$_6$H$_4$AsO$_3$H$_2$, 3, 402
AcNHC$_6$H$_4$AsO$_3$H$_2$ isomers, 3, 424, 429
AsC$_8$H$_{10}$NO$_4$S 4-(HOCH$_2$CH$_2$NHO$_2$S)C$_6$H$_3$AsO, 3, 240
4,3-HS(AcNH)C$_6$H$_3$AsO$_3$H$_2$, 3, 482
AsC$_8$H$_{10}$NO$_5$ Me[3,4-O$_2$N(MeO)C$_6$H$_3$]AsO$_2$H, 3, 379
AcNH(HO)C$_6$H$_3$AsO$_3$H$_2$ isomers, 3, 477
4-(HOCH$_2$CONH)C$_6$H$_4$AsO$_3$H$_2$, 3, 429
4-(HO$_2$CCH$_2$NH)C$_6$H$_4$AsO$_3$H$_2$, 3, 430
AsC$_8$H$_{10}$NO$_6$ 3,4-O$_2$N(HOCH$_2$CH$_2$)C$_6$H$_3$AsO$_3$H$_2$, 3, 482
3,4-O$_2$N(EtO)C$_6$H$_3$AsO$_3$H$_2$, 3, 483
AsC$_8$H$_{10}$NO$_7$ O$_2$N(HOCH$_2$CH$_2$O)C$_6$H$_3$AsO$_3$H$_2$ isomers, 3, 483
AsC$_8$H$_{10}$N$_5$ {4-[H$_2$NC(=NH)NHC(=NH)NH]-C$_6$H$_4$As}$_n$, 3, 137
AsC$_8$H$_{11}$ Me$_2$AsPh, 3, 39
EtPhAsH, 3, 117
AsC$_8$H$_{11}$N$_2$O$_4$ 4-(H$_2$NCOCH$_2$NH)C$_6$H$_4$AsO$_3$H$_2$, 3, 435
AsC$_8$H$_{11}$N$_2$O$_5$ 4-(HOCH$_2$NHCONH)C$_6$H$_4$AsO$_3$H$_2$, 3, 435
4,3-HO(MeNHCONH)C$_6$H$_3$AsO$_3$H$_2$, 3, 484
3,4,5-AcNH(HO)(H$_2$N)C$_6$H$_2$AsO$_3$H$_2$, 3, 501
AsC$_8$H$_{11}$N$_2$O$_6$ 3,4-O$_2$N(HOCH$_2$CH$_2$NH)C$_6$H$_3$-AsO$_3$H$_2$, 3, 477
AsC$_8$H$_{11}$N$_4$O$_4$ 3,4-(H$_2$NCONH)$_2$C$_6$H$_3$As(OH)$_2$, 3, 223
[H$_2$NC(=NH)NHCONH]C$_6$H$_4$AsO$_3$H$_2$ isomers, 3, 422, 430
AsC$_8$H$_{11}$O HOC$_6$H$_4$AsMe$_2$ isomers, 3, 65
Me(3-MeOC$_6$H$_4$)AsH, 3, 117
MePhAsOMe, 3, 209
Me$_2$AsOPh, 3, 209
Me$_2$PhAsO, 3, 356

$AsC_8H_{11}O_2$ $PhAs(OMe)_2$, 3, 252
$EtPhAsO_2H$, 3, 379
$Me(4\text{-}MeC_6H_4)AsO_2H$, 3, 379
$MePhAs(O)OMe$, 3, 388
$AsC_8H_{11}O_3$ $PhCH_2CH_2AsO_3H_2$, 3, 396
$Me_2C_6H_3AsO_3H_2$ isomers, 3, 483
$4\text{-}EtC_6H_4AsO_3H_2$, 3, 411
$AsC_8H_{11}O_4$ $PhOCH_2CH_2AsO_3H_2$, 3, 396
$HOCH_2CH_2C_6H_4AsO_3H_2$ isomers, 3, 411
$AsC_8H_{11}O_4S$ $4\text{-}HOCH_2CH_2SC_6H_4AsO_3H_2$, 3, 416
$AsC_8H_{11}O_6S$ $4\text{-}HOCH_2CH_2SO_2C_6H_4AsO_3H_2$, 3, 416
$AsC_8H_{11}S$ $Me(CH_2{=}CH)AsC{=}CHCH{=}C(Me)S$, 3, 74
$AsC_8H_{12}ClO$ $Me_2PhAs(OH)Cl$, 3, 346
$AsC_8H_{12}ClO_2$ $ClCH{=}CHAs(OCH_2CH{=}CH_2)_2$, 3, 252
$[Me(4\text{-}MeC_6H_4)As(OH)_2]Cl$, 3, 371
$AsC_8H_{12}N$ $Me_2AsCH_2C{=}CHCH{=}CHCH{=}N$, 3, 42
$AsC_8H_{12}NO$ $Me_2(N{=}CHCH{=}CHCH{=}CCH_2)AsO$, 3, 356
$AsC_8H_{12}NO_3$ $4\text{-}Me_2NC_6H_4AsO_3H_2$, 3, 430
$4\text{-}EtNHC_6H_4AsO_3H_2$, 3, 435
$AsC_8H_{12}NO_3S$ $4\text{-}Me_2NSO_2C_6H_4AsO_3H_2$, 3, 416
$AsC_8H_{12}NO_4$ $4\text{-}(HOCH_2CH_2NH)C_6H_4AsO_3H_2$, 3, 435
$3,4\text{-}H_2N(EtO)C_6H_3AsO_3H_2$, 3, 483
$AsC_8H_{12}NO_4S$ $4\text{-}Me_2NO_2SC_6H_4As(OH)_2$, 3, 223
$AsC_8H_{12}NO_5$ $4\text{-}[(HOCH_2)_2N]C_6H_4AsO_3H_2$, 3, 430
$H_2N(HOCH_2CH_2O)C_6H_3AsO_3H_2$ isomers, 3, 478, 483
$AsC_8H_{12}N_3O_5$ $3,4\text{-}O_2N(H_2NCH_2CH_2NH)\text{-}C_6H_3AsO_3H_2$, 3, 484
$AsC_8H_{12}N_5O_3$ $[H_2NC({=}NH)NHC({=}NH)NH]\text{-}C_6H_4AsO_3H_2$ isomers, 3, 422, 430
$AsC_8H_{12}N_5O_4$ $4,3\text{-}HO[H_2NC({=}NH)NHC\text{-}({=}NH)NH]C_6H_3AsO_3H_2$, 3, 484
$AsC_8H_{13}ClNO_4$ $3\text{-}ClC_6H_4AsO_3H_2\cdot H_2NCH_2\text{-}CH_2OH$, 3, 409
$AsC_8H_{13}Cl_2$ $(ClCH{=}CH)_2AsBu$, 3, 65
$AsC_8H_{13}N_2$ $n\text{-}C_6H_{13}As(CN)_2$, 3, 206*
$AsC_8H_{13}O_2$ $Me_2PhAs(OH)_2$, 3, 350
$AsC_8H_{14}Cl_3$ $Me(CH_2)_5CCl{=}CHAsCl_2$, 3, 195
$AsC_8H_{14}I$ $[Et(CH_2{=}CH)_3As]I$, 3, 313
$AsC_8H_{14}IN_2$ $[Me_2(NCCH_2CH_2)_2As]I$, 3, 313
AsC_8H_{15} $(CH_2{=}CH)_2AsBu$, 3, 65
$AsC_8H_{15}N_2S_4$ $BuAsSC(S)NHCH_2CH_2NHC(S)S$, 3, 650
$AsC_8H_{15}O_2$ $EtAs(OCH_2CH{=}CH_2)_2$, 3, 252
$AsC_8H_{15}S_2$ $EtAs(SCH_2CH{=}CH_2)_2$, 3, 266
$AsC_8H_{16}ClO$ $EtAs(OC_6H_{11})Cl$, 3, 200
$AsC_8H_{16}ClO_2$ $ClCH{=}CHAs(OCHMe_2)_2$, 3, 252
$AsC_8H_{16}ClO_3$ $Me(CH_2)_5CCl{=}CHAsO_3H_2$, 3, 396
$AsC_8H_{16}NS_2$ $Me_2AsSC(S)N(CH_2)_4CH_2$, 3, 262
$AsC_8H_{17}N_2S_4$ $EtAs[SC(S)NMe_2]_2$, 3, 282
$AsC_8H_{17}OS_2$ $Me_2AsSC(S)OC_5H_{11}$, 3, 261
$AsC_8H_{17}O_2$ Pr_2AsOAc, 3, 212
$AsC_8H_{17}O_3$ $EtAsOCH_2CH(CH_2OPr)O$, 3, 577
$AsC_8H_{17}O_4$ 5-Butyl-1,4,6,9-tetraoxa-5-arsaspiro[4.4]nonane, 3, 580
$AsC_8H_{17}O_5$ $Me(CH_2)_5CH(CO_2H)AsO_3H_2$, 3, 396
$AsC_8H_{18}Br$ Bu_2AsBr, 3, 170
$[Et_3(CH_2{=}CH)As]Br$, 3, 313
$AsC_8H_{18}Cl$ Bu_2AsCl, 3, 171
$(Me_2CHCH_2)_2AsCl$, 3, 171
$[MeCH(CH_2)_4AsMe_2]Cl$, 3, 589
$AsC_8H_{18}ClO$ $EtAs(OC_6H_{13})Cl$, 3, 200
$AsC_8H_{18}I$ Bu_2AsI, 3, 164, 171
$\{Me_3[CH_2{=}CH(CH_2)_3]As\}I$, 3, 313
$[MeCH(CH_2)_4AsMe_2]I$, 3, 589
$AsC_8H_{18}N$ $Pr_2AsNCH_2CH_2$, 3, 287
AsC_8H_{19} Et_2AsBu, 3, 65
Bu_2AsH, 3, 117
$AsC_8H_{19}Br_2$ $[Et_3(BrCH_2CH_2)As]Br$, 3, 313
$AsC_8H_{19}O$ Et_2AsOBu, 3, 209
Pr_2AsOEt, 3, 209
$MeBuAsOPr$, 3, 210
$AsC_8H_{19}O_2$ $EtAs(OPr)_2$, 3, 252
$BuAs(OEt)_2$, 3, 252
Bu_2AsO_2H, 3, 375, 379
$(Me_2CHCH_2)_2AsO_2H$, 3, 379
$Et_2As(O)OBu$, 3, 388
$Pr_2As(O)OEt$, 3, 388
$AsC_8H_{19}S_2$ $EtAs(SPr)_2$, 3, 266
$BuAs(SEt)_2$, 3, 266
$AsC_8H_{20}I$ $(Et_4As)I$, 3, 313

$AsC_8H_{20}I_3$ $[Et_4As]I_3$, 3, 314
$AsC_8H_{20}N$ Et_2AsNEt_2, 3, 287
$AsC_8H_{20}O_3P$ $Me_2AsPO(OPr)_2$, 3, 292
  $Me_2AsPO(OCHMe_2)_2$, 3, 292
  $Et_2AsPO(OEt)_2$, 3, 292
$AsC_8H_{21}BrP$ $[Me_2AsPEt_3]Br$, 3, 291
$AsC_8H_{21}ClP$ $[Me_2AsPEt_3]Cl$, 3, 291
$AsC_8H_{21}IP$ $[Me_2AsPEt_3]I$, 3, 291
  $[Me_2AsPEt_3]I$, 3, 291
$AsC_9H_5Cl_2N_4O$ 4,3-HO-[N=C(Cl)N=C(Cl)N=CNH]C_6H_3AsO, 3, 231
$AsC_9H_5Cl_2N_6O_7$ 4,3,5-N=CClN=CClN=CNH-$(O_2N)_2C_6H_2AsO_3H_2$, 3, 502
$AsC_9H_5Cl_3NO_3$ 2,6,8-Trichloroquinoline-5-arsonic acid, 3, 515
$AsC_9H_6BrClNO_3$ 6-Bromo-2-chloroquinoline-5-arsonic acid, 3, 515
$AsC_9H_6ClN_5O$ 4-[N=C(NH_2)N=C(Cl)N=CNH]-C_6H_3AsO, 3, 240
$AsC_9H_6Cl_2NO_3$ Dichloroquinolinearsonic acid isomers, 3, 515
$AsC_9H_6Cl_2N_3O_3S$ 4-(N=CClN=CClN=CS)-$C_6H_4AsO_3H_2$, 3, 416
$AsC_9H_6Cl_2N_5$ [3,4-H_2N(N=CClN=CClN=CNH)-$C_6H_3As]_n$, 3, 137
$AsC_9H_6Cl_3N_4O_3$ 3,4-Cl(N=CClN=CClN=CNH)-$C_6H_3AsO_3H_2$, 3, 484
$AsC_9H_6NO_2$ $(OCH{=}CHCH{=}C)_2AsCN$, 3, 203
  4-Hydroxyquinoline-6-AsO, 3, 240
$AsC_9H_6NO_3S$ 5,5'-Arsenobis(2-benzoxazolylmercaptoacetic acid), 3, 132
$AsC_9H_6N_3O_2S_2$ NC(Me)AsS_2H ester with 6,2-O_2N(HS)benzothiazole, 3, 305
$AsC_9H_7BrNO_4$ 6,2-Br(HO)-5-quinolinearsonic acid, 3, 515
AsC_9H_7ClN Ph(ClCH=CH)AsCN, 3, 203
$AsC_9H_7ClNO_3$ 2-Chloroquinolinearsonic acid isomers, 3, 515, 516
$AsC_9H_7ClNO_4$ 6,2-Cl(HO)-5-quinolinearsonic acid, 3, 516
$AsC_9H_7ClNO_6S$ (5-Arsono-7-chlorobenzoxazol-2-ylthio)acetic acid, 3, 516
$AsC_9H_7ClN_5$ {4-[N=C(NH_2)N=C(Cl)N=CNH]-$C_6H_4As\}_n$, 3, 137
$AsC_9H_7Cl_2N_4O_3$ N-(4,6-Dichloro-s-triazin-2-yl)arsanilic acid, 3, 430
$AsC_9H_7Cl_3NO_4$ 4-(Cl_2C=CClCONH)C_6H_4-AsO_3H_2, 3, 435
$AsC_9H_7Cl_4N_6$ 3,5,4-$(H_2N)_2$-(N=CClN=CClN=CNH)$C_6H_2AsCl_2$, 3, 195
$AsC_9H_7INO_4$ 8-Hydroxy-7-iodoquinoline-5-arsonic acid, 3, 516
$AsC_9H_7N_2O_2$ 4-($NCCH_2NHOC$)C_6H_4AsO, 3, 240
$AsC_9H_7N_2O_2S_2$ [4-(SCH=CHN=CNHSO_2)-$C_6H_4As]_n$, 3, 138
$AsC_9H_7N_2O_3S_2$ 4-(SCH=CHN=CNHO_2S)C_6H_4-AsO, 3, 241
$AsC_9H_7N_2O_6$ 8-Hydroxy-7-nitroquinoline-5-arsonic acid, 3, 516
$AsC_9H_7N_2S_2$ 2-Methyl-1,3,2'-dithiarsolo-[4,5-b]quinoxaline, 3, 582
$AsC_9H_7O_3$ 3-(HO_2CCH=CH)C_6H_4AsO, 3, 240
$AsC_9H_8ClN_4O_3S$ 3-[N=C(NH_2)N=CClN=CS]-$C_6H_4AsO_3H_2$, 3, 416
$AsC_9H_8ClN_4O_4$ 4-[N=C(OH)N=CClN=CNH]-$C_6H_4AsO_3H_2$, 3, 445
$AsC_9H_8Cl_2N_5O_3$ 4-(N=CClN=CClN=CNHNH)-$C_6H_4AsO_3H_2$, 3, 451
$AsC_9H_8Cl_2N_5O_4$ 4,3-HO(N=CClN=CClN=C-NHNH)$C_6H_3AsO_3H_2$, 3, 484
$AsC_9H_8Cl_2N_5S$ N=C(NH_2)N=C(NH_2)N=CS-$C_6H_4AsCl_2$, 3, 195
$AsC_9H_8NO_2$ 4-(H_2NOCCH=CH)C_6H_4AsO, 3, 240
$AsC_9H_8NO_3S$ 4-(N=CHSCH=C)$C_6H_4AsO_3H_2$, 3, 469
$AsC_9H_8NO_4$ 3,4-HO_2C(AcNH)C_6H_3AsO, 3, 241
  Hydroxyquinolinearsonic acid isomers, 3, 617
$AsC_9H_8NO_6S$ (5-Arsonobenzoxazol-2-ylthio)acetic acid, 3, 517
$AsC_9H_8N_3O_4$ 4-($H_2NCONHCONHCO$)C_6H_4AsO, 3, 241
$AsC_9H_8N_3O_4S_2$ 4-[OCNHC(S)SCHN=N]-$C_6H_4AsO_3H_2$, 3, 457
$AsC_9H_8N_3O_5$ 4-[N=C(CO_2H)CH=NN]C_6H_4-AsO_3H_2, 3, 469
$AsC_9H_8N_3O_9$ {3,4-O_2N[$O_2NOCH_2CH(ONO_2)$-CH_2O]$C_6H_3As\}_n$, 3, 138
$AsC_9H_8N_3S_2$ Cyanothioarsinous acid ester with 6-amino-2-mercapto-4-

methylbenzothiazole, 3, 306

$AsC_9H_8N_5OS$ 4-[N=C(NH$_2$)N=C(NH$_2$)N=CS]-C$_6$H$_4$AsO, 3, 241

$AsC_9H_8N_5S$ {4-[N=C(NH$_2$)N=C(NH$_2$)N=CS]-C$_6$H$_4$As}$_n$, 3, 138

AsC_9H_9 1-Methylarsindole, 3, 555

$AsC_9H_9ClNO_2$ 4-NCCH(Cl)CH$_2$C$_6$H$_4$As(OH)$_2$, 3, 223

$AsC_9H_9ClNO_3$ 4-(NCCHClCH$_2$)C$_6$H$_4$AsO$_3$H$_2$, 3, 403

$AsC_9H_9ClN_5O_3$ 4-[N=C(NH$_2$)N=CClN=CNH]-C$_6$H$_4$AsO$_3$H$_2$, 3, 445

$AsC_9H_9Cl_2N_6O$
4,3-HO[N=C(NH$_2$)N=C(NH$_2$)N=CNH]C$_6$H$_3$-AsCl$_2$, 3, 195

$AsC_9H_9Cl_2O_2$ 4-MeO$_2$CCH$_2$C$_6$H$_4$AsCl$_2$, 3, 195

$AsC_9H_9N_2O_3$ 3,4-AcNH(H$_2$NOC)C$_6$H$_3$AsO, 3, 241
4-(H$_2$NCONHCOCH$_2$)C$_6$H$_4$AsO, 3, 241
4-(H$_2$NCOCH$_2$NHCO)C$_6$H$_4$AsO, 3, 241
4-(HNCH=NCH=C)C$_6$H$_4$AsO$_3$H$_2$, 3, 469

$AsC_9H_9N_2O_3S$ (SCH=CHN=CNH)C$_6$H$_4$AsO$_3$H$_2$ isomers, 3, 424, 445

$AsC_9H_9N_2O_4$ 4-(NCCH$_2$NHCO)C$_6$H$_4$AsO$_3$H$_2$, 3, 403
4-(NCCH$_2$CONH)C$_6$H$_4$AsO$_3$H$_2$, 3, 436
7,8-H$_2$N(HO)-5-quinolinearsonic acid, 3, 517

$AsC_9H_9N_2O_5S_2$ 4-(SCH=CHN=CNHO$_2$S)C$_6$H$_4$-AsO$_3$H$_2$, 3, 416

$AsC_9H_9N_4O_5$ 4-[N=C(OH)N=C(OH)N=CNH]-C$_6$H$_4$AsO$_3$H$_2$, 3, 445

$AsC_9H_9N_6$ {4-[N=C(NH$_2$)N=C(NH$_2$)N=CNH]-C$_6$H$_4$As}$_n$, 3, 138

$AsC_9H_9N_6O$ {4,2-HO[N=C(NH$_2$)N=C(NH$_2$)N=C-NH]C$_6$H$_3$As}$_n$, 3, 138
4-[N=C(NH$_2$)N=C(NH$_2$)N=CNH]C$_6$H$_4$AsO, 3, 231

$AsC_9H_9N_6O_2$ 4,3-HO[N=C(NH$_2$)N=C(NH$_2$)N=C-NH]C$_6$H$_3$AsO, 3, 241

$AsC_9H_9N_6O_6$
3,4-O$_2$N[N=C(NH$_2$)N=C(NH$_2$)N=CO]C$_6$H$_3$-AsO$_3$H$_2$, 3, 484

AsC_9H_9O 2,2'-Arsenobis(4-vinylanisole) polymer, 3, 132

$AsC_9H_9O_2$ [2,5-MeO(Ac)C$_6$H$_3$As]$_n$, 3, 138
[4-(AcCH$_2$O)C$_6$H$_4$As]$_n$, 3, 138

$AsC_9H_9O_3$ [4,2,5-HO(MeO)(Ac)C$_6$H$_2$As]$_n$, 3, 138
4-(AcCH$_2$O)C$_6$H$_4$AsO, 3, 241
4-(HO$_2$CCH$_2$CH$_2$)C$_6$H$_4$AsO, 3, 241
3-(EtO$_2$C)C$_6$H$_4$AsO, 3, 241
5,2-Ac(MeO)C$_6$H$_3$AsO, 3, 241

$AsC_9H_9O_4$ 4-HO$_2$CCH=CHC$_6$H$_4$As(OH)$_2$, 3, 223
4-(MeO$_2$CCH$_2$O)C$_6$H$_4$AsO, 3, 240, 241
2,4,5-MeO(HO)(Ac)C$_6$H$_2$AsO, 3, 241

$AsC_9H_9O_5$ HO$_2$CCH=CHC$_6$H$_4$AsO$_3$H$_2$ isomers, 3, 403

$AsC_9H_{10}Br$ (CH$_2$=CHCH$_2$)PhAsBr, 3, 171
1-Br-1,2,3,4-tetrahydroarsinoline, 3, 592

$AsC_9H_{10}BrO_5$ 2-(HO$_2$CCHBrCH$_2$)C$_6$H$_4$AsO$_3$H$_2$, 3, 403

$AsC_9H_{10}Cl$ MePhAsCH=CHCl, 3, 74
5,2-Cl(Me)-isoarsindoline, 3, 557*
2-Cl-1,2,3,4-tetrahydroisoarsinoline, 3, 594

$AsC_9H_{10}ClN_6O_3$ 4-[N=C(NHNH$_2$)N=CClN=CNH]-C$_6$H$_4$AsO$_3$H$_2$, 3, 445
4-[N=C(NH$_2$)N=CClN=CNHNH]C$_6$H$_4$AsO$_3$H$_2$, 3, 451

$AsC_9H_{10}ClN_6O_4$
4,3-HO[N=C(NHNH$_2$)N=CClN=CNH]C$_6$H$_3$-AsO$_3$H$_2$, 3, 478

$AsC_9H_{10}Cl_2NO$ 4-EtOC(=NH)C$_6$H$_4$AsCl$_2$, 3, 196

$AsC_9H_{10}Cl_2N_2$ N-[H$_2$N(Cl$_2$As)C$_6$H$_3$]-melamine isomers, 3, 185, 186

$AsC_9H_{10}Cl_2N_7O$ 2-(4,6-H$_2$N(H$_2$NNH)-s-triazin-2-ylamino)-4-(dichloroarsino)phenol, 3, 186

$AsC_9H_{10}Cl_3O$ PhAs[OCH(CH$_2$Cl)$_2$]Cl, 3, 200

$AsC_9H_{10}F_3$ 4-CF$_3$C$_6$H$_4$AsMe$_2$, 3, 66

$AsC_9H_{10}I$ 1,2,3,4-Tetrahydro-2-iodoisoarsinoline, 3, 594

$AsC_9H_{10}NO_2$ Me(4-O$_2$NC$_6$H$_4$)AsCH=CH$_2$, 3, 74
{4-[HO$_2$CCH(NH$_2$)CH$_2$]C$_6$H$_4$As}$_n$, 3, 138
4-(AcNHCH$_2$)C$_6$H$_4$AsO, 3, 241
4-(Me$_2$NOC)C$_6$H$_4$AsO, 3, 241
4-[EtOC(=NH)]C$_6$H$_4$AsO, 3, 241

$AsC_9H_{10}NO_3$ 4-[$HO_2CCH(NH_2)CH_2$]C_6H_4AsO, 3, 242

$AsC_9H_{10}NO_3S_2$ p-$O_2NC_6H_4$-$\overline{AsSCH_2CH(CH_2OH)S}$, 3, 584

$AsC_9H_{10}NO_4$ 2,4-$HO(EtO_2CNH)C_6H_3AsO$, 3, 242

3-Methyl-1,4,2-benzoxazine-6-arsonic acid, 3, 518

$AsC_9H_{10}NO_5$ [3,4-$O_2N(HOCH_2CHOHCH_2O)$-C_6H_3As]$_n$, 3, 138

4-(HO_2CCH_2NHOC)$C_6H_4As(OH)_2$, 3, 223

2,5-$O_2N(MeCHOHCH_2O)C_6H_3AsO$, 3, 242

$AsC_9H_{10}NO_6$ 2,4-$O_2N(HOCH_2CHOHCH_2O)$-C_6H_3AsO, 3, 242

4-(MeO_2CCONH)$C_6H_4AsO_3H_2$, 3, 436

$AsC_9H_{10}NO_7$ 3,4-$O_2N(MeCOCH_2O)C_6H_3$-AsO_3H_2, 3, 484

$AsC_9H_{10}N_3O_3$ 4-($H_2NCONHCOCH_2NH$)C_6H_4-AsO, 3, 242

$AsC_9H_{10}N_3O_6$ 4-($H_2NCONHCONHCO$)C_6H_4-AsO_3H_2, 3, 403

$AsC_9H_{10}N_3O_{12}$ 3,4-$O_2N[O_2NOCH_2CH(ONO_2)$-$CH_2O]C_6H_3AsO_3H_2$, 3, 484

$AsC_9H_{10}N_5O_3$ 4-[$\overline{N{=}C(NH_2)N{=}C(NH_2)N{=}C}$]-$C_6H_4AsO_3H_2$, 3, 411

$AsC_9H_{10}N_5O_3S$ 4-[$\overline{N{=}C(NH_2)N{=}C(NH_2)N{=}CS}$]-$C_6H_4AsO_3H_2$, 3, 416

$AsC_9H_{10}N_5O_4$ 4-[$\overline{N{=}C(NH_2)N{=}C(OH)N{=}C}NH$]-$C_6H_4AsO_3H_2$, 3, 445

$AsC_9H_{10}N_7$ {4-[$\overline{N{=}C(NH_2)N{=}C(NH_2)N{=}C}NH$-$NH]C_6H_4As$}$_n$, 3, 138

AsC_9H_{11} 2-Methylisoarsindoline, 3, 557

$AsC_9H_{11}ClNO_2$ 4-[$EtOC(=NH\cdot HCl)$]C_6H_4AsO, 3, 241

$AsC_9H_{11}ClN_7O_3$ 4-[$\overline{N{=}C(NHNH_2)N{=}CClN{=}C}NH$-$NH]C_6H_4AsO_3H_2$, 3, 451

$AsC_9H_{11}Cl_2N_6$ N-[2,4-Diamino-4-(dichloroarsino)phenyl]melamine, 3, 186

$AsC_9H_{11}Cl_2O$ $PhAs(OCH_2CHClMe)Cl$, 3, 200

$AsC_9H_{11}N_2O_2$ 4-($H_2NCOCH_2NHCH_2$)C_6H_4AsO, 3, 242

$AsC_9H_{11}N_2O_3$ 4-($HOCH_2CH_2NHCONH$)C_6H_4AsO, 3, 242

$AsC_9H_{11}N_2O_4S$ 4-(H_2NCSCH_2CONH)C_6H_4-AsO_3H_2, 3, 436

$AsC_9H_{11}N_2O_5$ 4-(H_2NCOCH_2NHCO)$C_6H_4AsO_3H_2$, 3, 403

3,4-$O_2N(CH_2{=}CHCH_2NH)C_6H_3AsO_3H_2$, 3, 484

4-($MeNHOCCONH$)$C_6H_4AsO_3H_2$, 3, 429

$AsC_9H_{11}N_2O_5S$ 4-[$H_2NC(O)SCH_2CONH$]C_6H_4-AsO_3H_2, 3, 436

$AsC_9H_{11}N_2O_8$ 2,3,4-$HO(O_2N)(EtOCONH)$-$C_6H_2AsO_3H_2$, 3, 502

$AsC_9H_{11}N_2O_9$ 5,2-$O_2N[MeCH(ONO_2)CH_2O]$-$C_6H_3AsO_3H_2$, 3, 484

$AsC_9H_{11}N_6O_3$ N-(4,6-Diamino-s-triazin-2-yl)arsanilic acid, 3, 431

$AsC_9H_{11}N_6O_4$ $HO[\overline{N{=}C(NH_2)N{=}C(NH_2)N{=}C}NH]$-$C_6H_3AsO_3H_2$ isomers, 3, 484

$AsC_9H_{11}OS_2$ $Ph\overline{AsSCH_2CH(CH_2OH)S}$, 3, 584

4-$HOC_6H_4\overline{AsSCH_2CH(Me)S}$, 3, 584

$AsC_9H_{11}O_2$ 4-$HO_2CC_6H_4AsMe_2$, 3, 42

$MePhAsOAc$, 3, 212

[2-($MeCHOHCH_2O$)C_6H_4As]$_n$, 3, 138

[2,5-$MeO(MeCHOH)C_6H_3As$]$_n$, 3, 138

$Ph(CH_2{=}CHCH_2)AsO_2H$, 3, 379

$Ph\overline{AsOCH_2CH(Me)O}$, 3, 577

4-$MeC_6H_4\overline{AsOCH_2CH_2O}$, 3, 577

$AsC_9H_{11}O_2S_2$ 4-$HOC_6H_4\overline{AsSCH_2CH(CH_2OH)S}$, 3, 584

$AsC_9H_{11}O_3$ [4-($HOCH_2CHOHCH_2O$)C_6H_4As]$_n$, 3, 138

2-($MeCHOHCH_2O$)C_6H_4AsO, 3, 242

$MePh(HO_2CCH_2)AsO$, 3, 356

Poly(3-arsono-4-methoxystyrene), 3, 542

$AsC_9H_{11}O_3S_3$ $Ph\overline{AsSCH_2CH(CH_2SO_3H)S}$, 3, 584

$AsC_9H_{11}O_4$ 4-($HOCH_2CHOHCH_2O$)C_6H_4AsO, 3, 242

4-$AcCH_2C_6H_4AsO_3H_2$, 3, 411

$AsC_9H_{11}O_4S$ 4-$AcCH_2SC_6H_4AsO_3H_2$, 3, 416

$AsC_9H_{11}O_5$ 4-(MeO_2CCH_2)$C_6H_4AsO_3H_2$, 3, 403, 411

4-(EtO_2C)$C_6H_4AsO_3H_2$, 3, 403

4-($HO_2CCH_2CH_2$)$C_6H_4AsO_3H_2$, 3, 403, 412

4-[$HO_2CCH(Me)$]$C_6H_4AsO_3H_2$, 3, 412

5,2-$Ac(MeO)C_6H_3AsO_3H_2$, 3, 485

$AsC_9H_{11}O_6$ 2,4,5-$OCH(MeO)_2C_6H_2AsO_3H_2$, 3, 502

3,4,6-$Ac(HO)(MeO)C_6H_2AsO_3H_2$, 3, 502

$AsC_9H_{11}O_6S$ 4-$(AcCH_2O_2S)C_6H_4AsO_3H_2$, 3, 416

$AsC_9H_{11}O_7$ 2,4,5-$HO_2C(MeO)_2C_6H_2AsO_3H_2$, 3, 502

$AsC_9H_{11}S_2$ $Ph\overline{AsSCH_2CH(Me)S}$, 3, 584

4-$MeC_6H_4\overline{AsSCH_2CH_2S}$, 3, 584

$Ph\overline{AsS(CH_2)_3S}$, 3, 618

$AsC_9H_{12}BrN_2O_5$ $\{Me_3[4,3\text{-}Br(O_2N)C_6H_3]As\}NO_3$, 3, 314

$AsC_9H_{12}Br_2N$ $EtMeNC_6H_4AsBr_2$, 3, 196

$AsC_9H_{12}Br_2NO_2$ $\{Me_3[4,3\text{-}Br(O_2N)C_6H_3]As\}Br$, 3, 314

$AsC_9H_{12}Cl$ Et(4-MeC_6H_4)AsCl, 3, 171

$AsC_9H_{12}ClO$ PhAs(OPr)Cl, 3, 200

$AsC_9H_{12}Cl_2N$ $EtMeNC_6H_4AsCl_2$, 3, 196

$AsC_9H_{12}Cl_2N_9$

4-$[\overline{N{=}C(NHNH_2)N{=}C(NHNH_2)N{=}C}NHNH]C_6H_4AsCl_2$, 3, 196

$AsC_9H_{12}I$ Et(4-MeC_6H_4)AsI, 3, 171

$AsC_9H_{12}I_2N$ 4-$EtMeNC_6H_4AsI_2$, 3, 196

$AsC_9H_{12}NO$ 4-$MeEtNC_6H_4AsO$, 3, 242

$Me(Me_2N)C_6H_3AsO$ isomers, 3, 242

$AsC_9H_{12}NOS_2$ 3,4-$H_2N(HO)C_6H_3\overline{AsSCH_2CH(Me)S}$, 3, 584

4-$H_2NC_6H_4\overline{AsSCH_2CH(CH_2OH)S}$, 3, 584

$AsC_9H_{12}NO_2$ $[3,4\text{-}H_2N(MeCHOHCH_2O)C_6H_3As]_n$, 3, 138

$[3,4\text{-}H_2N(HOCH_2CH_2CH_2O)C_6H_3As]_n$, 3, 139

$AsC_9H_{12}NO_2S_2$ 3,4-$H_2N(HO)C_6H_3\overline{AsSCH_2CH(CH_2OH)S}$, 3, 584

$AsC_9H_{12}NO_3$ $[3,4\text{-}H_2N(HOCH_2CHOHCH_2O)C_6H_3As]_n$, 3, 139

$AsC_9H_{12}NO_3$ $H_2N(MeCHOHCH_2O)C_6H_3AsO$ isomers, 3, 242

4-$(CH_2{=}CHCH_2NH)C_6H_4AsO_3H_2$, 3, 436

$AsC_9H_{12}NO_4$ 3,4-$H_2N(HOCH_2CHOHCH_2O)C_6H_3AsO$, 3, 242

4-$(MeNHCOCH_2)C_6H_4AsO_3H_2$, 3, 403

4-$(H_2NCOCH_2CH_2)C_6H_4AsO_3H_2$, 3, 403

4-$(AcNHCH_2)C_6H_4AsO_3H_2$, 3, 403

$AsC_9H_{12}NO_4S_3$ 3,4-$H_2N(HO)C_6H_3\overline{AsSCH_2CH(CH_2SO_3H)S}$, 3, 584

$AsC_9H_{12}NO_5$ 4-$(HO_2CCH_2NHCH_2)C_6H_4AsO_3H_2$, 3, 404

4-$HOCH_2NHCOCH_2C_6H_4AsO_3H_2$, 3, 412

2,4-$HO(EtCONH)C_6H_3AsO_3H_2$, 3, 485

2,5-$MeO[MeC({=}NOH)]C_6H_3AsO_3H_2$, 3, 485

4-$(MeO_2CCH_2NH)C_6H_4AsO_3H_2$, 3, 436

4-$(HO_2CCH_2CH_2NH)C_6H_4AsO_3H_2$, 3, 436

4-$[MeCH(CO_2H)NH]C_6H_4AsO_3H_2$, 3, 436

4-$(EtO_2CNH)C_6H_4AsO_3H_2$, 3, 436

$AsC_9H_{12}NO_6$ $HO(EtOCONH)C_6H_3AsO_3H_2$ isomers, 3, 478, 485

$AsC_9H_{12}NO_7$ 5,2-$O_2N(MeCHOHCH_2O)C_6H_3AsO_3H_2$, 3, 485

$AsC_9H_{12}NO_8$ 3,4-$O_2N(HOCH_2CHOHCH_2O)C_6H_3AsO_3H_2$, 3, 485

$AsC_9H_{12}NS$ 4-$(MeEtN)C_6H_4AsS$, 3, 265

$AsC_9H_{12}NS_2$ $H_2NC_6H_4\overline{AsSCH_2CH(Me)S}$ isomers, 3, 584

$AsC_9H_{12}N_3O_4$ 4-$[H_2NC({=}NH)CH_2CONH]C_6H_4AsO_3H_2$, 3, 436

$AsC_9H_{12}N_7O_3$ 4-$[\overline{N{=}C(NHNH_2)N{=}C(NH_2)N{=}C}NH]C_6H_4AsO_3H_2$, 3, 431, 445

4-$[\overline{N{=}C(NH_2)N{=}C(NH_2)N{=}C}NHNH]C_6H_4AsO_3H_2$, 3, 451

$AsC_9H_{12}N_7O_4$

4,3-$HO[\overline{N{=}C(NH_2)N{=}C(NHNH_2)N{=}C}NH]C_6H_3AsO_3H_2$, 3, 485

$AsC_9H_{12}N_9O$ 4-$[\overline{N{=}C(NHNH_2)N{=}C(NHNH_2)N{=}C}NHNH]C_6H_4AsO$, 3, 242

AsC_9H_{13} $Me_2AsC_6H_4Me$ isomers, 3, 43, 66

Me_2AsCH_2Ph, 3, 66

MeEtPhAs, 3, 74

$AsC_9H_{13}BrI$ $[Me_3(4\text{-}BrC_6H_4)As]I$, 3, 314

$AsC_9H_{13}BrNO_3$ $[Me_3(4\text{-}BrC_6H_4)As]NO_3$, 3, 314

$\{Me_3[4,3\text{-}HO(O_2N)C_6H_3]As\}Br$, 3, 314

$AsC_9H_{13}ClN$ Me(4-$Me_2NC_6H_4$)AsCl, 3, 171

$AsC_9H_{13}ClNO_2$ $[Me_3(3\text{-}O_2NC_6H_4)As]Cl$, 3, 314

$AsC_9H_{13}ClNO_4$ 4-$(ClCH_2CHOHCH_2NH)C_6H_4AsO_3H_2$, 3, 436

$AsC_9H_{13}ClNO_5$ 4-$[HO_2CCH(NH_2\cdot HCl)CH_2]C_6H_4AsO_3H_2$, 3, 404

$AsC_9H_{13}Cl_2$ $MeEtPhAsCl_2$, 3, 344

$AsC_9H_{13}INO_2$ $[Me_3(3\text{-}O_2NC_6H_4)As]I$, 3, 314

$AsC_9H_{13}N_2O_3$ 4-$(Me_2NCH{=}N)C_6H_4AsO_3H_2$, 3, 442

4-$(Me_2C{=}NNH)C_6H_4AsO_3H_2$, 3, 452

3,4-$H_2N(CH_2{=}CHCH_2NH)C_6H_3AsO_3H_2$, 3,

485

$AsC_9H_{13}N_2O_4$ 4-$(H_2NCOCH_2NHCH_2)C_6H_4$-AsO_3H_2, 3, 404

$AsC_9H_{13}N_2O_5$ $[Me_3(3\text{-}O_2NC_6H_4)As]NO_3$, 3, 314

4-$(HOCH_2NHCOCH_2NH)C_6H_4AsO_3H_2$, 3, 436

$AsC_9H_{13}N_2O_6$ $\{Me_3[4,3\text{-}HO(O_2N)C_6H_3]As\}$-$NO_3$, 3, 314

$HO(HOCH_2CH_2NHCONH)C_6H_3AsO_3H_2$ isomers, 3, 485

3,4-$H_2NCONH(HOCH_2CH_2O)C_6H_3AsO_3H_2$, 3, 485

3,4-$O_2N(MeCHOHCH_2NH)C_6H_3AsO_3H_2$, 3, 485

2,3,4-$HO(H_2N)(EtOCONH)C_6H_2AsO_3H_2$, 3, 502

$AsC_9H_{13}N_8O_3$ 4-$[N{=}C(NH_2)N{=}C(NHNH_2)N{=}C$-$NHNH]C_6H_4AsO_3H_2$, 3, 452

4-$[N{=}C(NHNH_2)N{=}C(NHNH_2)N{=}CNH]C_6H_4$-$AsO_3H_2$, 3, 445

$AsC_9H_{13}N_8O_4$

4,3-$HO[N{=}C(NHNH_2)N{=}C(NHNH_2)N{=}CNH]$-$C_6H_3AsO_3H_2$, 3, 486

$AsC_9H_{13}O$ 3-$MeOC_6H_4AsMe_2$, 3, 66

MePhAsOEt, 3, 210

EtPhAsOMe, 3, 210

$AsC_9H_{13}O_3$ 4-$PrC_6H_4AsO_3H_2$, 3, 412

$AsC_9H_{13}O_4$ 4-$HOCH_2CH_2CH_2C_6H_4AsO_3H_2$, 3, 412

2-$MeCHOHCH_2C_6H_4AsO_3H_2$, 3, 412

$AsC_9H_{13}O_4S$ 4-$[HO(CH_2)_3S]C_6H_4AsO_3H_2$, 3, 416

$AsC_9H_{13}O_5$ 4-$(HOCH_2CHOHCH_2)C_6H_4AsO_3H_2$, 3, 412

2,5-$MeO(MeCHOH)C_6H_3AsO_3H_2$, 3, 486

$AsC_9H_{13}O_6S$ 4-$[HO(CH_2)_3O_2S]C_6H_4AsO_3H_2$, 3, 416

$AsC_9H_{13}S$ 2-$MeSC_6H_4AsMe_2$, 3, 44

MeEtPhAsS, 3, 364

$AsC_9H_{14}Br$ $[Me_3PhAs]Br$, 3, 314

$AsC_9H_{14}BrIN$ $\{Me_3[3,4\text{-}H_2N(Br)C_6H_3]As\}I$, 3, 315

$AsC_9H_{14}I$ $[Me_3PhAs]I$, 3, 314

$AsC_9H_{14}IO$ $[Me_3(HOC_6H_4)As]I$ isomers, 3, 315

$Me_2PhAs(OMe)I$, 3, 346

$AsC_9H_{14}NO_3$ 4-$(EtMeN)C_6H_4AsO_3H_2$, 3, 431

4,3-$HO[HOCH_2CH(OH)CH_2NH]C_6H_3AsH_2$, 3, 121

$[Me_3PhAs]NO_3$, 3, 314

$Me(Me_2N)C_6H_3AsO_3H_2$ isomers, 3, 486

$AsC_9H_{14}NO_4$ 4-$(MeCHOHCH_2NH)C_6H_4AsO_3H_2$, 3, 436

$[Me_2PhAs(OMe)]NO_3$, 3, 350

4-$(HOCH_2CH_2NHCH_2)C_6H_4AsO_3H_2$, 3, 404

2,5-$MeO[MeCH(NH_2)]C_6H_3AsO_3H_2$, 3, 486

$AsC_9H_{14}NO_5$ $H_2N(MeCHOHCH_2O)C_6H_3AsO_3H_2$ isomers, 3, 486

4,3-$HO(HOCH_2CHOHCH_2NH)C_6H_3AsO_3H_2$, 3, 486

$AsC_9H_{14}NO_6$ 3,4-$H_2N(HOCH_2CHOHCH_2O)$-$C_6H_3AsO_3H_2$, 3, 486

$AsC_9H_{14}N_2O_6$ 3,4-$O_2N(H_2NCH_2CHOHCH_2NH)$-$C_6H_3AsO_3H_2$, 3, 486

$AsC_9H_{14}N_3O_5$ 3,4-$O_2N[MeCH(NH_2)CH_2NH]$-$C_6H_3AsO_3H_2$, 3, 486

$AsC_9H_{14}N_9O_3$ 4-$[N{=}C(NHNH_2)N{=}C(NHNH_2)N{=}C$-$NHNH]C_6H_4AsO_3H_2$, 3, 452

AsC_9H_{15} $(CH_2{=}CHCH_2)_3As$ isomers, 3, 11, 12

$(CH_2{=}CMe)_3As$, 3, 12

$AsC_9H_{15}Br_2$ $(CH_2{=}CMe)_3AsBr_2$, 3, 344

$(MeCH{=}CH)_3AsBr_2$ isomers, 3, 344

$AsC_9H_{15}ClN$ $[Me_3(3\text{-}H_2NC_6H_4)As]Cl$, 3, 315

$AsC_9H_{15}ClNO$ $\{Me_3[3,4\text{-}H_2N(HO)C_6H_3]As\}$-$Cl$, 3, 315

$AsC_9H_{15}Cl_2$ $(ClCH{=}CH)_2AsCH_2CH_2CHMe_2$, 3, 66

$AsC_9H_{15}IN$ $[Me_3(3\text{-}H_2NC_6H_4)As]I$, 3, 315

$AsC_9H_{15}N_2O_3$ $HC{=}CHC(NEt_2){=}NCH{=}CAsO_3H_2$, 3, 518

$AsC_9H_{15}O$ $[Me_3PhAs]OH$, 3, 314

$AsC_9H_{16}N_3O_3$ 3,4-$H_2N[MeCH(NH_2)CH_2NH]$-$C_6H_3AsO_3H_2$, 3, 486

$AsC_9H_{17}OS_2$ $Me_2AsSC(S)OC_6H_{11}$, 3, 261

$AsC_9H_{17}O_2S_4$ $MeAs[SC(S)OPr]_2$, 3, 279

$PrAs[SC(S)OEt]_2$, 3, 279

$AsC_9H_{18}ClO$ $PrAs(OC_6H_{11})Cl$, 3, 200

$AsC_9H_{18}F_3$ Bu_2AsCF_3, 3, 66

$AsC_9H_{19}Br_2$ $[Br(CH_2)_3]_2CHAsMe_2$, 3, 66

$AsC_9H_{19}O_2$ $EtPrAsCH_2CO_2Et$, 3, 74

$AsC_9H_{19}O_3$ $EtAsOCH_2CH(CH_2OBu)O$, 3, 577

$AsC_9H_{19}O_4$ 2,3,5,7,8-Pentamethyl-1,4,-6,9-tetraoxa-5-arsaspiro[4.4]-nonane, 3, 580
$AsC_9H_{19}O_5$ $Me(CH_2)_6CH(CO_2H)AsO_3H_2$, 3, 396
$AsC_9H_{20}N$ $Et_2AsN(CH_2)_4CH_2$, 3, 287
$AsC_9H_{20}NO_3$ cyclo-$C_6H_{11}NH(CH_2)_3AsO_3H_2$, 3, 396
AsC_9H_{21} Pr_3As, 3, 12
$(Me_2CH)_3As$, 3, 30
Bu_2AsMe, 3, 66
$AsC_9H_{21}Cl_2$ Pr_3AsCl_2, 3, 344
$AsC_9H_{21}O$ Pr_2AsOPr, 3, 210
Bu_2AsOMe, 3, 210
MeBuAsOBu, 3, 210
Pr_3AsO, 3, 356
$MeBu_2AsO$, 3, 356
$AsC_9H_{21}O_2$ $MeAs(OBu)_2$, 3, 252
$PrAs(OPr)_2$, 3, 252
$\{Pr_3AsO\text{-}O\}_n$, 3, 362
MeBuAs(O)OBu, 3, 388
$AsC_9H_{21}O_2S_2$ $MeAs(OCH_2CH_2SEt)_2$, 3, 252
$AsC_9H_{21}O_3$ $MeAsO(OBu)_2$, 3, 544
$PrAsO(OPr)_2$, 3, 546
$AsC_9H_{21}S$ Pr_3AsS, 3, 364
Bu_2MeAsS, 3, 364
$AsC_9H_{21}S_2$ $MeAs(SBu)_2$, 3, 266
$PrAs(SPr)_2$, 3, 266
$AsC_9H_{21}S_4$ $MeAs(SCH_2CH_2SEt)_2$, 3, 266
$AsC_9H_{21}Se$ Pr_3AsSe, 3, 367
$Bu_2MeAsSe$, 3, 367
$AsC_9H_{22}O_3P$ $MeEtAsPO(OCHMe_2)_2$, 3, 292
$AsC_9H_{23}N_2$ $(Me_3CNH)_2AsMe$, 3, 289
$AsC_9H_{23}O_4P_2S_4$ $MeAs[SP(S)(OEt)_2]_2$, 3, 281
$AsC_9H_{23}O_7P_2S_2$ $MeAsO[OP(S)(OEt)_2]_2$, 3, 550*
$AsC_9H_{27}N_4O_3P_2S_2$ $MeAsO[OP(S)(NMe_2)_2]_2$, 3, 550
$AsC_{10}H_7ClNO_3S$ 7,7'-Arsenobis[3(4-chloro-2-benzoxazolylmercapto)propionic acid], 3, 132
$AsC_{10}H_7ClN_4O$ 4-$[N{=}C(NH_2)N{=}C(Cl)C{=}CNH]$-$C_6H_4AsO$, 3, 243
$AsC_{10}H_7Cl_2$ $C_{10}H_7AsCl_2$ isomers, 3, 196
$AsC_{10}H_7Cl_2N_2OS$ 2-(4-Dichloroarsinobenzamido)thiazole, 3, 186
$AsC_{10}H_7Cl_2N_4$ $\{3,4\text{-}Me[N{=}C(Cl)N{=}C(Cl)N{=}C\text{-}NH]C_6H_3As\}_n$, 3, 139
$AsC_{10}H_7Cl_2N_4O$
$\{3,4\text{-}MeO[N{=}C(Cl)N{=}C(Cl)N{=}CNH]C_6H_3\text{-}As\}_n$, 3, 139
$AsC_{10}H_7N_2O_2S$ 4-$(SCH{=}CHN{=}CNHCO)C_6H_4AsO$, 3, 242
$AsC_{10}H_7O$ 1-$C_{10}H_7AsO$, 3, 242
2-$C_{10}H_7AsO$, 3, 243
$AsC_{10}H_7S$ Poly(2,5-thienylenephenylarsine), 3, 113
$AsC_{10}H_8BrClNO_3$ 8,2,6-Br(Cl)(Me)-5-quinolinearsonic acid, 3, 518
$AsC_{10}H_8ClN_2$ Di-3-pyridyl-AsCl, 3, 164
$AsC_{10}H_8ClN_4$ $\{4\text{-}[N{=}C(NH_2)N{=}C(Cl)CH{=}CNH]\text{-}C_6H_4As\}_n$, 3, 139
$AsC_{10}H_8ClO_3$ $ClC_{10}H_6AsO_3H_2$ isomers, 3, 505
$AsC_{10}H_8ClS$ $Ph(SCH{=}CHCH{=}C)AsCl$, 3, 171
$AsC_{10}H_8Cl_2NO_3$ 2,8,6-Cl_2(Me)-5-quinolinearsonic acid, 3, 518
$AsC_{10}H_8F_7$ $MePhAsCF_2CF_2CF_3$, 3, 75
$AsC_{10}H_8NO$ 8-Methylquinolyl-5-AsO, 3, 243*
$AsC_{10}H_8NO_2$ 6,4,2-OAs(HO)(Me)-quinoline, 3, 232
$AsC_{10}H_8NO_3$ 3,3'-Arsenobis(6-hydroxysuccinanilic acid)dilactone, 3, 133
$AsC_{10}H_8NO_3S$ 6,6'-Arsenobis[3-(2-benzoxazolylmercapto)propionic acid], 3, 133
$AsC_{10}H_8NO_5$ $O_2NC_{10}H_6AsO_3H_2$ isomers, 3, 505
$AsC_{10}H_8NO_5$ $O_2NC_{10}H_6AsO_3H_2$ isomers, 3, 505
$AsC_{10}H_8N_3O_7$ 4-$[N{=}C(CO_2H)C(CO_2H){=}NN]$-$C_6H_4AsO_3H_2$, 3, 469
$AsC_{10}H_9BrNO_3$ Bromo(methyl)quinoline-5-arsonic acid isomers, 3, 518
$AsC_{10}H_9BrNO_4$ 8,2,6-Br(HO)(Me)-5-quinolinearsonic acid, 3, 519
$AsC_{10}H_9ClNO_3$ Chloro(methyl)quinolinearsonic acid isomers, 3, 519, 520
$AsC_{10}H_9ClNO_6S$ 3(7-Arsono-4-chlorobenzoxazol-2-ylthio)propionic acid, 3, 520
$AsC_{10}H_9ClN_3O_2S_4$ 2,4-$Cl(O_2N)C_6H_3$-

$AsSC(S)NHCH_2CH_2NHC(S)S$, 3, 650

$AsC_{10}H_{9}Cl_{2}$ $(ClCH{=}CH)_2AsPh$, 3, 46

$AsC_{10}H_{9}Cl_{2}N_{2}OS$ 2-(4-$Cl_2AsC_6H_4CONH$)-thiazoline, 3, 186

$AsC_{10}H_{9}Cl_{2}N_{4}$ 4-[$N{=}C(NH_2)N{=}CHCH{=}CNH$]-$C_6H_4AsCl_2$, 3, 196
4-[$N{=}CHCH{=}C(NH_2)N{=}CNH$]$C_6H_4AsCl_2$, 3, 196

$AsC_{10}H_{9}Cl_{2}N_{4}O$
4,3-HO[$N{=}C(NH_2)N{=}CHCH{=}CNH$]$C_6H_3AsCl_2$, 3, 196

$AsC_{10}H_{9}Cl_{2}N_{4}O_{3}$ 3,4-Me($N{=}CClN{=}CClN{=}C$-NH)$C_6H_3AsO_3H_2$, 3, 486

$AsC_{10}H_{9}Cl_{6}O_{2}$ $PhAs(OCH_2CCl_3)_2$, 3, 252

$AsC_{10}H_{9}INO_{4}$ 8,7,2-HO(I)(Me)-5-quinolinearsonic acid, 3, 520

$AsC_{10}H_{9}N_{2}O_{2}$ ($CH{=}NCH{=}CHCH{=}C$)$_2AsO_2H$, 3, 379

$AsC_{10}H_{9}N_{2}O_{6}$ 4-[$N{=}C(CO_2H)C(OH){=}CHN$]-$C_6H_4AsO_3H_2$, 3, 469

$AsC_{10}H_{9}N_{4}$ {4-[$N{=}C(NH_2)N{=}CHCH{=}CNH$]-$C_6H_4As$}$_n$, 3, 139

$AsC_{10}H_{9}N_{4}O$ 4-[$N{=}C(NH_2)N{=}CHCH{=}CNH$]-$C_6H_4AsO$, 3, 243

$AsC_{10}H_{9}N_{4}O_{2}$ HO[$N{=}C(NH_2)N{=}CHCH{=}CNH$]-$C_6H_3AsO$ isomers, 3, 243

$AsC_{10}H_{9}N_{4}O_{5}$ 4-[$N{=}CHC(NO_2)CHN{=}CNH$]-$C_6H_4AsO_3H_2$, 3, 445
3,4-O_2N($N{=}CHCH{=}CHN{=}CNH$)$C_6H_3AsO_3H_2$, 3, 486

$AsC_{10}H_{9}N_{4}O_{7}$
3,4-O_2N[$N{=}C(CO_2H)C(Me){=}NN$]C_6H_3-AsO_3H_2, 3, 487

$AsC_{10}H_{9}O_{2}$ 2-$C_{10}H_7As(OH)_2$, 3, 224
$OCCH_2CH_2COAsPh$, 3, 553

$AsC_{10}H_{9}O_{3}$ $C_{10}H_7AsO_3H_2$ isomers, 3, 505

$AsC_{10}H_{9}O_{3}S$ 4-($SCH{=}CHCH{=}C$)$C_6H_4AsO_3H_2$, 3, 469

$AsC_{10}H_{9}O_{4}$ ($OCH{=}CHCH{=}C$)$_2AsOAc$, 3, 212
4-($OCH{=}CHCH{=}C$)$C_6H_4AsO_3H_2$, 3, 469
4-$HOC_{10}H_6AsO_3H_2$, 3, 505

$AsC_{10}H_{9}O_{5}$ 3,5-$(MeO_2C)_2C_6H_3AsO$, 3, 243

$AsC_{10}H_{10}ClN_{2}S_{4}$
$ClC_6H_4AsSC(S)NHCH_2CH_2NHC(S)S$ isomers, 3, 650

$AsC_{10}H_{10}ClN_{4}O_{3}$ [$N{=}C(NH_2)N{=}CClCH{=}CNH$]-$C_6H_4AsO_3H_2$ isomers, 3, 424, 445

$AsC_{10}H_{10}ClN_{4}O_{4}$
2,5-HO[$N{=}C(NH_2)N{=}CClCH{=}CNH$]$C_6H_3$-$AsO_3H_2$, 3, 487
2,5-HO[$N{=}C(NH_2)N{=}CClCH{=}CNH$]$C_6H_3$-$AsO_3H_2$, 3, 487

$AsC_{10}H_{10}NO_{2}$ 4-($CH_2{=}CHCH_2NHOC$)C_6H_4AsO, 3, 243

$AsC_{10}H_{10}NO_{3}$ $H_2NC_{10}H_6AsO_3H_2$ isomers, 3, 505

Methylquinolinearsonic acid isomers, 3, 520, 521

$AsC_{10}H_{10}NO_{4}$ [3,4-AcNH(HO_2CCH_2O)C_6H_3-As]$_n$, 3, 139
4-($HO_2CCH_2CH_2CONH$)C_6H_4AsO, 3, 243
3,4-$H_2N(HO)C_{10}H_5$-1-AsO_3H_2, 3, 505
Hydroxymethylquinolinearsonic acid isomers, 3, 521, 522

$AsC_{10}H_{10}NO_{4}S_{2}$ 4-$H_2NC_6H_4$-$AsSCH(CO_2H)CH(CO_2H)S$, 3, 584

$AsC_{10}H_{10}NO_{5}$ 4,3-HO($HO_2CCH_2CH_2CONH$)-C_6H_3AsO, 3, 243

$AsC_{10}H_{10}NO_{5}S_{2}$ 4,2-HO(H_2N)C_6H_3-$AsSCH(CO_2H)CH(CO_2H)S$, 3, 584

$AsC_{10}H_{10}NO_{6}S$ 3-(Arsonobenzoxazol-2-ylthio)propionic acid isomers, 3, 522

$AsC_{10}H_{10}NO_{7}S$ 2-(5-Arsono-7-methoxybenzoxazol-2-ylthio)acetic acid, 3, 522

$AsC_{10}H_{10}NO_{7}S_{2}$ 3,4-$O_2N(HO)C_6H_3As$-$(SCH_2CO_2H)_2$, 3, 266

$AsC_{10}H_{10}N_{3}O_{2}S_{4}$ 4-$O_2NC_6H_4$-$AsSC(S)NHCH_2CH_2NHC(S)S$, 3, 650

$AsC_{10}H_{10}N_{3}O_{3}S$ 4-[$N{=}C(NH_2)N{=}CHCH{=}CS$]-$C_6H_4AsO_3H_2$, 3, 416

$AsC_{10}H_{10}N_{3}O_{5}$ 4-[$N{=}C(CO_2H)C(Me){=}NN$]-$C_6H_4AsO_3H_2$, 3, 469
[$OCON{=}C(Me)CHN{=}N$]$C_6H_4AsO_3H_2$ isomers, 3, 457

$AsC_{10}H_{10}N_{5}$ {4-[$N{=}C(NH_2)N{=}C(NH_2)CH{=}C$-NH]$C_6H_4As$}$_n$, 3, 139

$AsC_{10}H_{10}N_{5}O$ 4-[$N{=}C(NH_2)N{=}C(NH_2)CH{=}C$-NH]$C_6H_4AsO$, 3, 243*

$AsC_{10}H_{10}N_{5}O_{6}$ 3,4-O_2N[$CONHN{=}C(Me)CH$-$N{=}N$]$C_6H_3AsO_3H_2$, 3, 487

$AsC_{10}H_{10}N_{5}O_{8}S$ 3,4-O_2N-

$AsC_{10}H_{10}N_5O_6S$
$[N{=}C(NH_2)N{=}C(OMe)N{=}CS]C_6H_3AsO_3H_2$, 3, 487

$AsC_{10}H_{11}$ 1-Ethylarsindole, 3, 555

$AsC_{10}H_{11}N_2O_3$ $4\text{-}(H_2NOCCH_2CH_2CONH)\text{-}C_6H_4AsO$, 3, 243
$4\text{-}(H_2NOCCH_2NHOCCH_2)C_6H_4AsO$, 3, 243
$4\text{-}(H_2NOCCH_2CH_2NHOC)C_6H_4AsO$, 3, 243
$3,4\text{-}(AcNH)_2C_6H_3AsO$, 3, 243
$4\text{-}[HNC(Me){=}NCH{=}C]C_6H_4AsO_3H_2$, 3, 469

$AsC_{10}H_{11}N_2O_3S$ $[SCH{=}C(Me)N{=}CNH]C_6H_4\text{-}AsO_3H_2$ isomers, 3, 424, 446

$AsC_{10}H_{11}N_2O_4$ $[3,4\text{-}(H_2NOCCH_2NH)(HO_2\text{-}CCH_2O)C_6H_3As]_n$, 3, 139
$4\text{-}(CH_2CH_2CONHN{=}C)C_6H_4AsO_3H_2$, 3, 469

$AsC_{10}H_{11}N_2O_5$ $4\text{-}(OCOCH_2CH_2C{=}NNH)C_6H_4\text{-}AsO_3H_2$, 3, 452

$AsC_{10}H_{11}N_2O_6$ $4\text{-}[HO_2CC(Ac){=}NNH]C_6H_4\text{-}AsO_3H_2$, 3, 452
$4\text{-}[HO_2CCH(Ac)N{=}N]C_6H_4AsO_3H_2$, 3, 457

$AsC_{10}H_{11}N_2O_6S_3$ $2,4\text{-}H_2N(H_2NSO_2)C_6H_3\text{-}AsSCH(CO_2H)CH(CO_2H)S$, 3, 584

$AsC_{10}H_{11}N_2S_4$ $PhAsSC(S)NHCH_2CH_2NHC(S)S$, 3, 650

$AsC_{10}H_{11}N_4O$ $3,4\text{-}H_2N[N{=}C(Me)C(Me){=}NN]\text{-}C_6H_3AsO$, 3, 243

$AsC_{10}H_{11}N_4O_3$ N-(2-H_2N-4-pyrimidyl)-p-arsanilic acid, 3, 432, 446
N-(2-H_2N-4-pyrimidyl)-m-arsanilic acid, 3, 424
N-(4-H_2N-2-pyrimidyl)-p-arsanilic acid, 3, 446
$3,4\text{-}H_2N(N{=}CHCH{=}CHN{=}CNH)C_6H_3AsO_3H_2$, 3, 487

$AsC_{10}H_{11}N_4O_4$ $2\text{-}[OCNHN{=}C(Me)CHN{=}N]\text{-}C_6H_4AsO_3H_2$ isomers, 3, 457
$HO[N{=}C(NH_2)N{=}CHCH{=}CNH]C_6H_3AsO_3H_2$ isomers, 3, 487

$AsC_{10}H_{11}N_4O_5$
$3,4\text{-}H_2N[N{=}C(CO_2H)C(Me){=}NN]C_6H_3AsO_3H_2$, 3, 487
$3,4\text{-}O_2N[N{=}C(Me)C(Me){=}NN]C_6H_3AsO_3H_2$, 3, 487

$AsC_{10}H_{11}O_3$ $4\text{-}[MeCH_2CH(CO_2H)]C_6H_4AsO$, 3, 243
$4\text{-}[HO_2C(CH_2)_3]C_6H_4AsO$, 3, 243

$AsC_{10}H_{11}O_3S_2$ $4\text{-}HO_2CC_6H_4\text{-}AsSCH_2CH(CH_2OH)S$, 3, 585

$AsC_{10}H_{11}O_4$ $PhAs(CH_2CO_2H)_2$, 3, 47
$PhAs(OAc)_2$, 3, 256

$AsC_{10}H_{11}O_4S_2$ $PhAs(SCH_2CO_2H)_2$, 3, 266

$AsC_{10}H_{11}O_5$ $Ph(HO_2CCH_2)_2AsO$, 3, 356

$AsC_{10}H_{11}O_6$ $3,4\text{-}(MeO_2C)_2C_6H_3As(OH)_2$, 3, 224
$4\text{-}(HO_2CCH_2CH_2CO)C_6H_4AsO_3H_2$, 3, 404

$AsC_{10}H_{12}ClN_2O_5$ $4\text{-}[EtO_2CC(Cl){=}NNH]\text{-}C_6H_4AsO_3H_2$, 3, 452

$AsC_{10}H_{12}ClO$ $PhAs[OCH(Me)CH{=}CH_2]Cl$, 3, 200
$(ClCH{=}CH)PhAsOEt$, 3, 210

$AsC_{10}H_{12}ClO_2$ $Ph[HO_2C(CH_2)_3]AsCl$, 3, 171

$AsC_{10}H_{12}Cl_2NO_2$ $3,4\text{-}H_2N[HO_2C(CH_2)_3]\text{-}C_6H_3AsCl_2$, 3, 196

$AsC_{10}H_{12}Cl_2N_7O$
$2,5,4\text{-}H_2N(MeO)[N{=}C(NH_2)N{=}C(NH_2)N{=}CNH]\text{-}C_6H_2AsCl_2$, 3, 196

$AsC_{10}H_{12}I$ 1,1-Me_2-arsindolinium I^-, 3, 555

$AsC_{10}H_{12}I_2NO_3S$ $4\text{-}[O(CH_2CH_2)_2NO_2S]\text{-}C_6H_4AsI_2$, 3, 196

$AsC_{10}H_{12}N$ $MePhAsCH_2CH_2CN$, 3, 75

$AsC_{10}H_{12}NO_2$ $4\text{-}[H_2NOC(CH_2)_3]C_6H_4AsO$, 3, 243

$AsC_{10}H_{12}NO_2S$ 2,2'-(Arsenodi-p-phenylene)bis(tetrahydro-2H-1,2-triazine)1,1,1',1'-tetraoxide, 3, 133

$AsC_{10}H_{12}NO_3$ $4\text{-}(MeCHOHCH_2NHOC)C_6H_4AsO$, 3, 243
$2,4\text{-}HO(PrCONH)C_6H_3AsO$, 3, 244

$AsC_{10}H_{12}NO_4$ $[3,4\text{-}HO_2CCH_2NH(HOCH_2CH_2O)\text{-}C_6H_3As]_n$, 3, 139
$4\text{-}(HOCH_2CHOHCH_2NHOC)C_6H_4AsO$, 3, 244
$2,4\text{-}MeO(EtO_2CNH)C_6H_3AsO$, 3, 244

$AsC_{10}H_{12}NO_4S$ $4\text{-}[O(CH_2CH_2)_2NO_2S]C_6H_4\text{-}AsO$, 3, 244

$AsC_{10}H_{12}NO_5$
$4\text{-}O_2NC_6H_4AsOCH_2CH(CH_2OMe)O$, 3, 578
$4\text{-}O_2NC_6H_4AsOCH_2CH(CH_2OMe)O$, 3, 578

$AsC_{10}H_{12}NO_5S_2$ $3,4\text{-}H_2N(HO)C_6H_3As\text{-}(SCH_2CO_2H)_2$, 3, 267

$AsC_{10}H_{12}NO_6$ $4\text{-}(HO_2CCH_2CH_2CONH)C_6H_4\text{-}AsO_3H_2$, 3, 436
$4\text{-}(AcOCH_2CONH)C_6H_4AsO_3H_2$, 3, 436
$4\text{-}(EtO_2CCONH)C_6H_4AsO_3H_2$, 3, 436

$AsC_{10}H_{12}NO_7$ $2\text{-}[(HO_2CCH_2)_2N]C_6H_4AsO_3H_2$, 3, 422

$AsC_{10}H_{12}N_3O_3$ $[N{=}C(Me)C(Me){=}NN]C_6H_4\text{-}AsO_3H_2$ isomers, 3, 469

$AsC_{10}H_{12}N_3O_4S_4$ $o\text{-}O_2NC_6H_4As[SC(S)NH\text{-}CH_2OH]_2$, 3, 282

$AsC_{10}H_{12}N_3O_5$ $4\text{-}[N{=}NC(CH_2OH){=}C(CH_2OH)N]C_6H_4AsO_3H_2$, 3, 469

$AsC_{10}H_{12}N_5O_3$ $4\text{-}[N{=}C(NH_2)N{=}C(NH_2)CH{=}C\text{-}NH]C_6H_4AsO_3H_2$, 3, 446

$AsC_{10}H_{12}N_7O_3$ $4\text{-}[C(NH_2){=}NC(NH_2){=}NC(NH_2){=}CN{=}N]C_6H_4\text{-}AsO_3H_2$, 3, 457

$AsC_{10}H_{12}N_7O_6$ $2,4,5\text{-}O_2N[N{=}C(NH_2)N{=}C(NH_2)N{=}CNH]\text{-}(MeO)C_6H_2AsO_3H_2$, 3, 502

$AsC_{10}H_{13}$ $Me_2AsCH{=}CHPh$, 3, 47
$Me(CH_2{=}CHCH_2)AsPh$, 3, 75
$Me(CH_2{=}CHCH_2)AsPh$, 3, 75
1-Phenylarsolane, 3, 553
1,2,3,4-Tetrahydro-1-methylarsinoline, 3, 592
1,2,3,4-Tetrahydro-2-methylisoarsinoline, 3, 594*

$AsC_{10}H_{13}ClI$ 5-Chloro-2-methylisoarsindoline methiodide, 3, 557

$AsC_{10}H_{13}ClNO_2S$ $\{4\text{-}[Cl(CH_2)_4SO_2NH]\text{-}C_6H_4As\}_n$, 3, 139

$AsC_{10}H_{13}ClNO_6$ $3,4\text{-}ClCH_2CONH(HOCH_2\text{-}CH_2O)C_6H_3AsO_3H_2$, 3, 487

$AsC_{10}H_{13}Cl_2$ $Me_2(PhCH{=}CH)AsCl_2$, 3, 344
1,2,3,4-Tetrahydro-2-methylisoarsinoline 2,2-dichloride, 3, 594

$AsC_{10}H_{13}Cl_2N_8$ $4,5,2\text{-}MeNH(H_2N)\text{-}[N{=}C(NH_2)N{=}C(NH_2)N{=}CNH]C_6H_2AsCl_2$, 3, 196

$AsC_{10}H_{13}Cl_2O_2$ $PhAs(OCH_2CH_2Cl)_2$, 3, 252

$AsC_{10}H_{13}F_3I$ $[Me_3(4\text{-}CF_3C_6H_4)As]I$, 3, 315
$[Me_3(4\text{-}CF_3C_6H_4)As]I$, 3, 315

$AsC_{10}H_{13}N_2$ $(CH_2CH_2N)_2AsPh$, 3, 290

$AsC_{10}H_{13}N_2O_2S$ 2,2′-[Arsenobis(2-amino-p-phenylene)]bis(tetrahydro-2H-1,2-thiazine)1,1,1′,1′-tetraoxide, 3, 133

$AsC_{10}H_{13}N_2O_2S_2$ $4\text{-}(H_2NCONH)C_6H_4\text{-}AsSCH_2CH(CH_2OH)S$, 3, 585

$AsC_{10}H_{13}N_2O_3$ $4\text{-}(NCCH_2CH_2NHCH_2)C_6H_4\text{-}AsO_3H_2$, 3, 404

$AsC_{10}H_{13}N_2O_4$ $Me(NCCH_2CH_2)PhAs(OH)NO_3$, 3, 348

$AsC_{10}H_{13}N_2O_5$ $4\text{-}(Me_2NOCCONH)C_6H_4AsO_3H_2$, 3, 429
$4\text{-}(EtNHOCCONH)C_6H_4AsO_3H_2$, 3, 429
$4,3\text{-}HO(CH_2{=}CHCH_2NHCONH)C_6H_3AsO_3H_2$, 3, 487

$AsC_{10}H_{13}N_2O_5S$ $3,4\text{-}O_2N[S(CH_2CH_2)_2N]\text{-}C_6H_3AsO_3H_2$, 3, 487

$AsC_{10}H_{13}N_2O_6$ $3,4\text{-}O_2N[O(CH_2CH_2)_2N]\text{-}C_6H_3AsO_3H_2$, 3, 487, 488

$AsC_{10}H_{13}N_2O_7$ $3,4\text{-}(H_2NCOCH_2NH)(HO_2C\text{-}CH_2O)C_6H_3AsO_3H_2$, 3, 488

$AsC_{10}H_{13}N_2O_7S$ $3,4\text{-}O_2N[O_2S(CH_2)_4N]\text{-}C_6H_3AsO_3H_2$, 3, 488

$AsC_{10}H_{13}N_4O_3$ $3,4\text{-}H_2N[N{=}C(Me)C(Me){=}NN]\text{-}C_6H_3AsO_3H_2$, 3, 488

$AsC_{10}H_{13}N_6O_3$ $4\text{-}[N{=}C(NHMe)N{=}C(NH_2)N{=}C\text{-}NH]C_6H_4AsO_3H_2$, 3, 446

$AsC_{10}H_{13}N_8O_5$ $2,4,5\text{-}N{=}C(NH_2)N{=}C(NH_2)N{=}C\text{-}NH(MeNH)(O_2N)C_6H_2AsO_3H_2$, 3, 502

$AsC_{10}H_{13}O$ $CH_2CH_2CH_2CH_2As(O)Ph$, 3, 553
$PhAsCH_2CH_2OCH_2CH_2$, 3, 609

$AsC_{10}H_{13}OS$ $CH_2CH_2OCH_2CH_2As(S)Ph$, 3, 610

$AsC_{10}H_{13}O_2$ $MePhAsCH_2CH_2CO_2H$, 3, 75
$[4\text{-}(Me_2COHCH_2O)C_6H_4As]_n$, 3, 140

$AsC_{10}H_{13}O_3$ $PhAsOCH_2CH(CH_2OMe)O$, 3, 578

$AsC_{10}H_{13}O_4$ $Ph[HO_2C(CH_2)_3]AsO_2H$, 3, 379
$PhOC_{10}H_6AsO_3H_2$ isomers, 3, 510
5-Ph-1,4,6,9-tetraoxa-5-arsaspiro-[4.4]nonane, 3, 580

$AsC_{10}H_{13}O_5$ $4\text{-}(MeO_2CCH_2CH_2)C_6H_4AsO_3H_2$, 3, 404
$4\text{-}[HO_2C(CH_2)_3]C_6H_4AsO_3H_2$, 3, 404
$4\text{-}[EtCH(CO_2H)]C_6H_4AsO_3H_2$, 3, 404

$AsC_{10}H_{13}O_5S$ $Me(3\text{-}HO_3SC_6H_4)AsCH_2CH_2\text{-}CO_2H$, 3, 75
$4\text{-}EtO_2CCH_2SC_6H_4AsO_3H_2$, 3, 416

$AsC_{10}H_{13}O_6$ $2,4,5\text{-}(MeO)_2(Ac)C_6H_2AsO_3H_2$, 3, 502

$AsC_{10}H_{13}O_7S$ 4 $EtO_2CCH_2O_2SC_6H_4AsO_3H_2$, 3, 416

$AsC_{10}H_{14}Br$ 2-$BrC_6H_4AsEt_2$, 3, 66
BuPhAsBr, 3, 171

$AsC_{10}H_{14}BrO_4$ 2,3,4,5-Me(Br)(HO)(Me_2-CH)$C_6HAsO_3H_2$, 3, 504

$AsC_{10}H_{14}Cl$ BuPhAsCl, 3, 171

$AsC_{10}H_{14}ClO$ PhAs(OBu)Cl, 3, 200

$AsC_{10}H_{14}ClO_2$ $\overline{CH_2CH_2OCH_2CH_2As}$(Ph)(OH)-Cl, 3, 610

$AsC_{10}H_{14}I$ 2-Methylisoarsindoline methiodide, 3, 557

$AsC_{10}H_{14}I_2N$ [PhN(CH_2CH_2)$_2$AsI]·HI, 3, 619

$AsC_{10}H_{14}I_4N$ [PhN(CH_2CH_2)$_2$$AsI_3$]·HI, 3, 619

$AsC_{10}H_{14}NO$ [2,5-BuO(H_2N)C_6H_3As]$_n$, 3, 140
[3,4-H_2N(BuO)C_6H_3As]$_n$, 3, 140
[3,4-H_2N(BuO)C_6H_3As]$_n$, 3, 140

$AsC_{10}H_{14}NO_2$ [3,4-H_2N(Me_2COHCH_2O)-C_6H_3As]$_n$, 3, 140

$AsC_{10}H_{14}NO_3$ 3,4-H_2N[$Me_2C(OH)CH_2O$]-C_6H_3AsO, 3, 244

$AsC_{10}H_{14}NO_4$ 4-($MeNHCOCH_2CH_2$)C_6H_4-AsO_3H_2, 3, 404
4-(Me_2NOCCH_2)$C_6H_4AsO_3H_2$, 3, 404
4-($EtNHCOCH_2$)$C_6H_4AsO_3H_2$, 3, 404

$AsC_{10}H_{14}NO_5$ 4-($HOCH_2CH_2NHCOCH_2O$)-$C_6H_4As(OH)_2$, 3, 224
4-($HO_2CCH_2CH_2NHCH_2$)$C_6H_4AsO_3H_2$, 3, 404
4-($HOCH_2CH_2NCOCH_2$)$C_6H_4AsO_3H_2$, 3, 412
3,4-H_2N[$HO_2C(CH_2)_3$]$C_6H_3AsO_3H_2$, 3, 488
4,3-HO($Me_2CHCONH$)$C_6H_3AsO_3H_2$, 3, 488
3,4-AcNH(EtO)$C_6H_3AsO_3H_2$, 3, 488

$AsC_{10}H_{14}NO_5S$ 4-[$O_2\overline{S(CH_2)_4N}$]$C_6H_4AsO_3H_2$, 3, 470

$AsC_{10}H_{14}NO_5S_2$ 4-[$S(CH_2CH_2)_2NO_2S$]-$C_6H_4AsO_3H_2$, 3, 417

$AsC_{10}H_{14}NO_6$ 2,4-MeO(EtOCONH)C_6H_3-AsO_3H_2, 3, 488
2,4-HO(PrOCONH)$C_6H_3AsO_3H_2$, 3, 488
O_2N(BuO)$C_6H_3AsO_3H_2$ isomers, 3, 488

$AsC_{10}H_{14}NO_7$ 3,4-O_2N(Me_2COHCH_2O)C_6H_3-AsO_3H_2, 3, 488
3,4-HO_2CCH_2NH($HOCH_2CH_2O$)$C_6H_3AsO_3H_2$, 3, 488

$AsC_{10}H_{14}NO_9S$ 3,4-HO_3SCH_2CONH($HOCH_2$-CH_2O)$C_6H_3AsO_3H_2$, 3, 488

$AsC_{10}H_{14}N_3O_3S_2$ 3,4-H_2N(HO)C_6H_3As-(SCH_2CONH_2)$_2$, 3, 267

$AsC_{10}H_{14}N_3O_6$ 2,4,5-$H_2NCONHN{=}CH$(MeO)$_2$-$C_6H_2AsO_3H_2$, 3, 502

$AsC_{10}H_{14}N_3O_7$ [Me_4As]$OC_6H_2(NO_2)_3$, 3, 312

$AsC_{10}H_{14}O_3$ $\overline{CH_2CH_2OCH_2CH_2As}$(Ph)(O)OH, 3, 610

$AsC_{10}H_{15}$ Et_2AsPh, 3, 47, 66
MeEt(4-MeC_6H_4)As, 3, 75
MePrPhAs, 3, 75
BuPhAsH, 3, 117

$AsC_{10}H_{15}ClNO_2$ [$Me_3(O_2NC_6H_4CH_2)As$]Cl isomers, 3, 315

$AsC_{10}H_{15}ClNO_5S$ 4-[$Cl(CH_2)_4SO_2NH$]-$C_6H_4AsO_3H_2$, 3, 444

$AsC_{10}H_{15}Cl_2O$ β-(Dichloroarsino)camphor, 3, 186

$AsC_{10}H_{15}N_2O_3S$ 3,4-H_2N[$S(CH_2CH_2)_2N$]-$C_6H_3AsO_3H_2$, 3, 489

$AsC_{10}H_{15}N_2O_4$ 4-($HOCH_2CH_2NHOCCH_2NH$)-$C_6H_4As(OH)_2$, 3, 224
4-($H_2NCOCH_2CH_2NHCH_2$)$C_6H_4AsO_3H_2$, 3, 404
3,4-H_2N[$O(CH_2CH_2)_2N$]$C_6H_3AsO_3H_2$, 3, 489

$AsC_{10}H_{15}N_2O_5$ 4-[Me($HOCH_2CH_2NHCO$)N]-$C_6H_4AsO_3H_2$, 3, 432
4-($HOCH_2CH_2NHCOCH_2NH$)$C_6H_4AsO_3H_2$, 3, 436

$AsC_{10}H_{15}N_2O_5S$ 3,4-H_2N[$O_2\overline{S(CH_2)_4N}$]-$C_6H_3AsO_3H_2$, 3, 489

$AsC_{10}H_{15}N_2O_6$ 4-($HOCH_2CHOHCH_2NHCONH$)-$C_6H_4AsO_3H_2$, 3, 436
3,4-O_2N($HOCH_2CMe_2NH$)$C_6H_3AsO_3H_2$, 3, 489*
4,3-HO[$HO(CH_2)_3NHCONH$]$C_6H_3AsO_3H_2$, 3, 489
3,4-H_2NCOCH_2NH($HOCH_2CH_2O$)$C_6H_3AsO_3H_2$, 3, 489

$AsC_{10}H_{15}N_2O_7$ 3,4-O_2N[($HOCH_2CH_2$)$_2$N]-$C_6H_3AsO_3H_2$, 3, 489

$AsC_{10}H_{15}O$ 3-$HOC_6H_4AsEt_2$, 3, 66

MePhAsOPr, 3, 210
EtPhAsOEt, 3, 210
PrPhAsOMe, 3, 210
$AsC_{10}H_{15}O_2$ $PhAs(OEt)_2$, 3, 252
$AsC_{10}H_{15}O_3$ $4\text{-}BuC_6H_4AsO_3H_2$, 3, 412
$AsC_{10}H_{15}O_4$ $4\text{-}(Me_2COHCH_2)C_6H_4AsO_3H_2$, 3, 412
$2,4,5\text{-}Me_2CH(HO)(Me)C_6H_2AsO_3H_2$, 3, 502
$AsC_{10}H_{15}O_4S$ $4\text{-}(EtOCH_2CH_2S)C_6H_4AsO_3H_2$, 3, 416
$AsC_{10}H_{15}O_5$ $4\text{-}HOCH_2CH_2OCH_2CH_2C_6H_4\text{-}AsO_3H_2$, 3, 412
$AsC_{10}H_{15}O_5S$ $4\text{-}(HOCH_2CH_2OCH_2CH_2S)\text{-}C_6H_4AsO_3H_2$, 3, 416
$AsC_{10}H_{15}O_6S$ $4\text{-}(EtOCH_2CH_2O_2S)C_6H_4\text{-}AsO_3H_2$, 3, 416
$AsC_{10}H_{15}O_7S$ $4\text{-}(HOCH_2CH_2OCH_2CH_2O_2S)\text{-}C_6H_4AsO_3H_2$, 3, 416
$AsC_{10}H_{15}S_2$ $PhAs(SEt)_2$, 3, 267
$PhAs(SEt)_2$, 3, 267

$AsC_{10}H_{16}I$ $[Me_3(PhCH_2)As]I$, 3, 315
$[Me_2EtPhAs]I$, 3, 316
$AsC_{10}H_{16}IO$ $[Me_3(3\text{-}MeOC_6H_4)As]I$, 3, 316
$[MeEtPhAs(OMe)]I$, 3, 346
$AsC_{10}H_{16}N$ $Me_2NC_6H_4AsMe_2$, 3, 48*, 49
$Me(4\text{-}MeC_6H_4)AsNMe_2$, 3, 287
$AsC_{10}H_{16}NO_3$ $PhCH_2NH(CH_2)_3AsO_3H_2$, 3, 396
$AsC_{10}H_{16}NO_4S$ $4\text{-}Et_2NO_2SC_6H_4As(OH)_2$, 3, 224
$AsC_{10}H_{16}NO_5$ $4\text{-}(HOCH_2CH_2OCH_2CH_2NH)\text{-}C_6H_4AsO_3H_2$, 3, 436
$3,4\text{-}H_2N[Me_2C(OH)CH_2O]C_6H_3AsO_3H_2$, 3, 488*
$AsC_{10}H_{16}NO_6$ $3,4\text{-}HO[(HOCH_2CH_2)_2N]\text{-}C_6H_3AsO_3H_2$, 3, 489
$AsC_{10}H_{16}NO_8P_2S_2$ $4\text{-}O_2NC_6H_4As[SP(O)\text{-}(OMe)_2]_2$, 3, 281
$AsC_{10}H_{16}N_3O_3$ $4\text{-}[Me_2CHNHC(=NH)NH]\text{-}C_6H_4AsO_3H_2$, 3, 436
$AsC_{10}H_{16}N_3O_6$ $3,4\text{-}O_2N(HOCH_2CH_2NHCH_2\text{-}CH_2NH)C_6H_3AsO_3H_2$, 3, 489
$AsC_{10}H_{16}N_5O_5$ $4,3\text{-}HOCH_2CH_2O[H_2NC(=NH)\text{-}NHC(=NH)NH]C_6H_3AsO_3H_2$, 3, 489
$AsC_{10}H_{17}BrP$ $[Me_2AsPPhMe_2]Br$, 3, 291
$AsC_{10}H_{17}ClP$ $[Me_2AsPPhMe_2]Cl$, 3, 291
$AsC_{10}H_{17}IP$ $[Me_2AsPPhMe_2]I$, 3, 291
$AsC_{10}H_{17}N_2$ $(Me_2N)_2AsPh$, 3, 290
$AsC_{10}H_{17}N_2O_4$ $3,4\text{-}H_2N(HOCH_2CMe_2NH)\text{-}C_6H_3AsO_3H_2$, 3, 489
$AsC_{10}H_{17}N_4O_5$ $3,4\text{-}O_2N(H_2NCH_2CH_2NHCH_2\text{-}CH_2NH)C_6H_3AsO_3H_2$, 3, 489
$AsC_{10}H_{17}O_3$ β-Camphorarsonous acid, 3, 226
$AsC_{10}H_{17}O_4$ β-Camphor-AsO_3H_2, 3, 396
$AsC_{10}H_{18}I$ $[Me(CH_2{=}CMe)_3As]I$, 3, 316
$[Me(MeCH{=}CH)_3As]I$ isomers, 3, 316
$AsC_{10}H_{18}N_3O_5S_2$ $3,4\text{-}H_2N(HO)C_6H_3As\text{-}(SCH_2CO_2NH_4)_2$, 3, 267
$AsC_{10}H_{19}N_4O_3$ $3,4\text{-}H_2N(H_2NCH_2CH_2NHCH_2\text{-}CH_2NH)C_6H_3AsO_3H_2$, 3, 489
$AsC_{10}H_{19}O_2$ $4,5,2\text{-}HO(Me)(Me_2CH)C_6H_8AsO$, 3, 244
$AsC_{10}H_{19}O_2S_4$ $BuAs[SC(S)OEt]_2$, 3, 279
$AsC_{10}H_{20}ClO$ $BuAs(OC_6H_{11})Cl$, 3, 200
$AsC_{10}H_{20}ClO_4$ $ClCH{=}CHAs(OCH_2CH_2OEt)_2$, 3, 252
$AsC_{10}H_{21}$ $Bu_2AsCH{=}CH_2$, 3, 66
$AsC_{10}H_{21}O_4$ $5,2,7\text{-}Bu(Me)_2\text{-}1,4,6,9\text{-}$tetraoxa-5-arsaspiro[4.4]-nonane, 3, 580
$AsC_{10}H_{22}N$ $Bu_2As\overline{NCH_2CH_2}$, 3, 287
$AsC_{10}H_{23}$ Pr_2AsBu, 3, 66
Bu_2AsEt, 3, 67
$n\text{-}C_8H_{17}AsMe_2$, 3, 67
$AsC_{10}H_{23}O$ Pr_2AsOBu, 3, 210
$Pr_2AsOCH_2CHMe_2$, 3, 210
Bu_2AsOEt, 3, 210
$AsC_{10}H_{23}O_2$ $BuAs(OPr)_2$, 3, 252
$EtAs(OBu)_2$, 3, 252
$(Me_2CHCH_2CH_2)_2AsO_2H$, 3, 379
$AsC_{10}H_{23}O_3$ $EtAsO(OBu)_2$, 3, 545
$BuAsO(OPr)_2$, 3, 546
$AsC_{10}H_{23}S_2$ $EtAs(SBu)_2$, 3, 267
$BuAs(SPr)_2$, 3, 267
$AsC_{10}H_{24}IO$ $Et_2PrAs(OPr)I$, 3, 346
$AsC_{10}H_{24}N$ Pr_2AsNEt_2, 3, 287
$AsC_{10}H_{24}O_3P$ $Me_2AsPO(OBu)_2$, 3, 292
$Et_2AsPO(OPr)_2$, 3, 292
$Et_2AsPO(OCHMe_2)_2$, 3, 292
$EtBuAsPO(OEt)_2$, 3, 292

$AsC_{10}H_{26}N_2O_6PS$ $Me_2As(O)OP(S)[N(CH_2$-$CH_2OH)_2]_2$, 3, 388

$AsC_{11}H_7Cl_2OS$ $SC(Bz)=CHCH=CAsCl_2$, 3, 186

$AsC_{11}H_7O_2S$ $SC(Bz)=CHCH=CAsO$, 3, 244

$AsC_{11}H_8NS$ $Ph(SCH=CHCH=C)AsCN$, 3, 203

$AsC_{11}H_8N_3O_2$ 4-$(N=CHCH=CHN=CNHCO)C_6H_4$-AsO, 3, 244

$AsC_{11}H_8O_2$ 6-$H_2NCOC_{10}H_6AsO$, 3, 244

$AsC_{11}H_9N_2O_2S$ 4-$[SCH=C(Me)N=CNHCO]$-C_6H_4AsO, 3, 244

$AsC_{11}H_9N_2O_3S$ 4-$(N=CHCH=CHCH=CNHO_2S)$-C_6H_4AsO, 3, 244

$AsC_{11}H_9N_4O_7$ 4-$[O_2NC=CHC(NO_2)=CHN=CNH]$-$C_6H_4AsO_3H_2$, 3, 446

4-$[O_2NC=CHC(NO_2)=CHN=CNH]C_6H_4AsO_3H_2$, 3, 446

$AsC_{11}H_9O_4S$ 5-Benzoylthiophene-2-arsonic acid, 3, 522

$AsC_{11}H_9O_5$ 6-$HO_2CC_{10}H_6$-2-AsO_3H_2, 3, 505

$AsC_{11}H_{10}Br$ $Me(1$-$C_{10}H_7)AsBr$, 3, 171

$AsC_{11}H_{10}Cl$ $Me(1$-$C_{10}H_7)AsCl$, 3, 165

$AsC_{11}H_{10}F$ $Me(1$-$C_{10}H_7)AsF$, 3, 172

$AsC_{11}H_{10}I$ $Me(1$-$C_{10}H_7)AsI$, 3, 172

$AsC_{11}H_{10}NO_4$ 2-$(OCH=CHCH=CCH=N)$-$C_6H_4AsO_3H_2$, 3, 422

$AsC_{11}H_{10}NO_5S_2$ 4-$H_2NCOC_6H_4$-$AsSCH(CO_2H)CH(CO_2H)S$, 3, 585

$AsC_{11}H_{10}N_2O_5$ 4-$(HO_2CCH_2NHOCCH_2NOC)$-C_6H_4AsO, 3, 244

$AsC_{11}H_{10}N_3O$ 4-$[N=CHC(NH_2)=CHCH=CNH]$-C_6H_4AsO, 3, 244

$AsC_{11}H_{10}N_3O_5$ $CH=CHC(NO_2)=CHN=CNH$-$C_6H_4AsO_3H_2$ isomers, 3, 422, 425, 446

$AsC_{11}H_{10}N_3O_6$ $HO[N=CHC(NO_2)=CHCH=CNH]$-$C_6H_3AsO_3H_2$ isomers, 3, 490

$AsC_{11}H_{11}$ $MeC_{10}H_6AsH_2$, 3, 121

$AsC_{11}H_{11}Cl_2$ 4-$MeC_6H_4As(CH=CHCl)_2$, 3, 67

$AsC_{11}H_{11}Cl_2O_3$ $(ClCH=CH)_2As(O)OC_6H_4$-OMe-2, 3, 388

$AsC_{11}H_{11}N_2O_2S_4$ 2-$HO_2CC_6H_4$-$AsSC(S)NHCH_2CH_2NHC(S)S$, 3, 650

$AsC_{11}H_{11}N_2O_3$ N-2-Pyridyl-m-arsanilic acid, 3, 425

N-2-Pyridyl-p-arsanilic acid, 3, 446*

$AsC_{11}H_{11}N_2O_4S_2$ 4-$NCNHC_6H_4As(SCH_2CO_2H)_2$, 3, 267

$AsC_{11}H_{11}N_2O_5S$ 4-$(N=CHCH=CHCH=CNHO_2S)$-$C_6H_4AsO_3H_2$, 3, 417

$AsC_{11}H_{11}N_2O_5S_2$ 4-$(H_2NCONH)C_6H_4$-$AsSCH(CO_2H)CH(CO_2H)S$, 3, 585

$AsC_{11}H_{11}N_4$ $\{4$-$[N=C(NH_2)CH=C(Me)N=CNH]$-$C_6H_4As\}_n$, 3, 140

$AsC_{11}H_{11}N_4O$ 4-$[N=C(NH_2)N=C(Me)CH=CNH]$-C_6H_4AsO, 3, 244

4-$[N=CHN=C(NHMe)CH=CNH]C_6H_4AsO$, 3, 244

$AsC_{11}H_{11}O_2$ $Me(1$-$C_{10}H_7)AsO_2H$, 3, 379

$AsC_{11}H_{12}Cl_2I$ $[MePh(ClCH=CH)_2As]I$, 3, 316

$AsC_{11}H_{12}N$ 3-Quinolyl-$AsMe_2$, 3, 67

$AsC_{11}H_{12}NO_3$ $\{3$-$[HO_2CCH(NHAc)CH_2]$-$C_6H_4As\}_n$, 3, 140

$AsC_{11}H_{12}NO_4$ 4-$[HO_2CCH(NHAc)CH_2]$-C_6H_4AsO, 3, 244

$AsC_{11}H_{12}NO_5S_2$ 4-$H_2NOCC_6H_4As(SCH_2CO_2H)_2$, 3, 267

$AsC_{11}H_{12}NO_6S$ 2-(5-Arsonobenzoxazol-2-ylthio)butyric acid, 3, 522

$AsC_{11}H_{12}N_3O_3$ $[CH=CHC(NH_2)=CHN=CNH]$-$C_6H_4AsO_3H_2$ isomers, 3, 422, 446

$AsC_{11}H_{12}N_3O_4$ 4-$[H_2NOCCH_2NHOCCH_2NHCO]$-C_6H_4AsO, 3, 245

$AsC_{11}H_{12}N_3O_5S$ 2-$(4$-$H_2NO_2SC_6H_4NH)$-pyridine-5-arsonic acid, 3, 522

$AsC_{11}H_{12}N_5O$ 4-$[N=C(NH_2)N=C(NHMe)CH=C$-$NH]C_6H_4AsO$, 3, 245

4-$[N=C(NHMe)N=C(NH_2)CH=CNH]C_6H_4AsO$, 3, 245

$AsC_{11}H_{13}N_2O_3$ 4-$(AcNHCH_2CH_2NHOC)C_6H_4$-AsO, 3, 245

4-$[HNC(Et)=NCH=C]C_6H_4AsO_3H_2$, 3, 470

$AsC_{11}H_{13}N_2S_4$

$MeC_6H_4AsSC(S)NHCH_2CH_2NHC(S)S$ isomers, 3, 650

$AsC_{11}H_{13}N_4O_3$ 4-$[CH=C(Me)N=C(NH_2)N=CNH]$-$C_6H_4AsO_3H_2$, 3, 446

4-$[CH=C(NHMe)N=CHN=CNH]C_6H_4AsO_3H_2$, 3, 446

3-$[N=C(NH_2)N=C(Me)CH=CNH]C_6H_4AsO_3H_2$, 3, 425

$AsC_{11}H_{13}N_4O_4$ $HO[N=C(NH_2)N=C(Me)CH=CNH]$-

$C_6H_3AsO_3H_2$ isomers, 3, 490

$AsC_{11}H_{13}N_6$ {4-[N=C(NHMe)N=C(NHMe)N=C-NH]C_6H_4As}$_n$, 3, 140

$AsC_{11}H_{13}O$ 1-Phenyl-4-arsenanone, 3, 589

$AsC_{11}H_{13}O_2$ 1,2,3,4-Tetrahydro-7-methoxy-1-methyl-4-oxoarsinoline, 3, 592

$AsC_{11}H_{13}O_3$ 4-($HO_2CCH_2CMe_2$)C_6H_4AsO, 3, 245

4-[$HO_2C(CH_2)_4$]C_6H_4AsO, 3, 245

$(CH_2{=}CH)_2As(O)OC_6H_4OMe$-4, 3, 388

$AsC_{11}H_{14}BrD_2O$ [$Me_3(PhCOCD_2)As$]Br, 3, 316

$AsC_{11}H_{14}ClO$ PhAs[OCH(Et)CH=CH_2]Cl, 3, 201

$AsC_{11}H_{14}Cl_2NO_6$ 4-[$Cl_2CHCONHCH(CH_2OH)$-CH(OH)]$C_6H_4AsO_3H_2$, 3, 404

$AsC_{11}H_{14}Cl_2N_7$ N-(2-Dichloroarsino-5-dimethylaminophenyl)melamine, 3, 187

$AsC_{11}H_{14}I_2NO_2S$ 4-[$CH_2(CH_2)_4NO_2S$]-$C_6H_4AsI_2$, 3, 197

$AsC_{11}H_{14}NO$ Me(3-$MeOC_6H_4$)$AsCH_2CH_2CN$, 3, 75

$AsC_{11}H_{14}NO_2$ 4-$Et_2NOCC_6H_4AsO$, 3, 245

$AsC_{11}H_{14}NO_3S$ 4-[$CH_2(CH_2)_4NO_2S$]-C_6H_4AsO, 3, 245

$AsC_{11}H_{14}NO_3S_2$ HO(AcNH)C_6H_3-$AsSCH_2CH(CH_2OH)S$ isomers, 3, 585*

$AsC_{11}H_{14}NO_4$ [2,4-$HOCH_2CH_2O(EtO_2CNH)$-C_6H_3As]$_n$, 3, 140

$AsC_{11}H_{14}NO_5$ 2,4-($HOCH_2CH_2O$)(EtO_2CNH)-C_6H_3AsO, 3, 245

4-$O_2NC_6H_4AsOCH_2CH(CH_2OEt)O$, 3, 578

$AsC_{11}H_{14}NO_6$ 4-[$HO_2CCH(NHAc)CH_2$]-$C_6H_4AsO_3H_2$, 3, 405

2,4-Me($AcOCH_2CONH$)$C_6H_3AsO_3H_2$, 3, 490

$AsC_{11}H_{14}NO_7S_3$ 4,3-HO(HO_2SCH_2NH)C_6H_3-$As(SCH_2CO_2H)_2$, 3, 267

$AsC_{11}H_{14}NO_8S_3$ 4,3-HO(HO_3SCH_2NH)C_6H_3-$As(SCH_2CO_2H)_2$, 3, 267

$AsC_{11}H_{14}N_5O_3$ 4-N=C(NH_2)N=C(NHEt)N=C-$C_6H_4AsO_3H_2$, 3, 412

4-[CH=C(NHMe)N=C(NH_2)N=CNH]C_6H_4-AsO_3H_2, 3, 446

$AsC_{11}H_{15}ClNO_6$ 4-($ClCH_2CHOHCH_2O_2CCH_2$-NH)$C_6H_4AsO_3H_2$, 3, 437

$AsC_{11}H_{15}ClN_5O_3$ {4-[Me^+N=C(NH_2)N=C(NH_2)CH=CNH]C_6H_4-AsO_3H_2}Cl^-, 3, 446

{4-[CH=C(NH_2)N^+(Me)=C(NH_2)N=CNH]-$C_6H_4AsO_3H_2$}Cl^-, 3, 446

$AsC_{11}H_{15}IN$ [$Me_2(NCCH_2CH_2)PhAs$]I, 3, 316

$AsC_{11}H_{15}N_2O_2S_2$ 4-(H_2NCOCH_2NH)C_6H_4-$AsSCH_2CH(CH_2OH)S$, 3, 585

$AsC_{11}H_{15}N_2O_3$ 4-($HOCH_2CH_2NHCH_2CH_2NHOC$)-C_6H_4AsO, 3, 245

$AsC_{11}H_{15}N_2O_5$ 4-(PrNHOCCONH)$C_6H_4AsO_3H_2$, 3, 429

4,3-HO[$CH_2(CH_2)_3NCONH$]$C_6H_3AsO_3H_2$, 3, 490

2,4-O_2N[$CH_2(CH_2)_4N$]$C_6H_3AsO_3H_2$, 3, 490

$AsC_{11}H_{15}N_6O_3$ 4-[N=C(NHMe)N=C(NHMe)N=C-NH]$C_6H_4AsO_3H_2$, 3, 447

$AsC_{11}H_{15}O_2$ $MePhAsCH_2CO_2Et$, 3, 75

$AsC_{11}H_{15}O_3$ Me(3-$MeOC_6H_4$)$AsCH_2CH_2CO_2H$, 3, 75

$PhAsOCH_2CH(CH_2OEt)O$, 3, 578

4-$MeC_6H_4AsOCH_2CH(CH_2OMe)O$, 3, 578

$AsC_{11}H_{15}O_5$ 4-[$HO_2C(CH_2)_4$]$C_6H_4AsO_3H_2$, 3, 405

4-($HO_2CCH_2CMe_2$)$C_6H_4AsO_3H_2$, 3, 405

$AsC_{11}H_{15}O_6$ 2,5-MeO[AcOCH(Me)]C_6H_3-AsO_3H_2, 3, 490

2,4,5-$(MeO)_2$(EtCO)$C_6H_2AsO_3H_2$, 3, 503

$AsC_{11}H_{16}BrO$ [$Me_3(PhCOCH_2)As$]Br, 3, 316

$AsC_{11}H_{16}I$ [Me_3(PhCH=CH)As]I, 3, 316

1,2,3,4-Tetrahydro-2,2-dimethyliso-arsinolinium iodide, 3, 594

$AsC_{11}H_{16}IO$ [$CH_2CH_2OCH_2CH_2As$(Ph)(Me)]I, 3, 610

$AsC_{11}H_{16}IO_2$ [$Me_2(HO_2CCH_2CH_2)PhAs$]I, 3, 316

$AsC_{11}H_{16}NO$ [$H_2N(Me_2CHCH_2CH_2O)C_6H_3As$]$_n$ isomers, 3, 140

$AsC_{11}H_{16}NO_2S_2$ 4-($HOCH_2CH_2NH$)C_6H_4-$AsSCH_2CH(CH_2OH)S$, 3, 585

$AsC_{11}H_{16}NO_4$ {4,3-HO[$HOCH_2CH_2$($HOCH_2$-$CHOHCH_2$)N]C_6H_3As}$_n$, 3, 140

4-($PrNHCOCH_2$)$C_6H_4AsO_3H_2$, 3, 405

$AsC_{11}H_{16}NO_4$

4-$(Me_2NCOCH_2CH_2)C_6H_4AsO_3H_2$, 3, 405

4-$(EtNHCOCH_2CH_2)C_6H_4AsO_3H_2$, 3, 405

$AsC_{11}H_{16}NO_5$ HO[$HOCH_2CH_2(HOCH_2CHOH$-$CH_2)N$]C_6H_3AsO isomers, 3, 245

4-$(MeCHOHCH_2NHCOCH_2)C_6H_4AsO_3H_2$, 3, 412

2,5-MeO[AcNHCH(Me)]$C_6H_3AsO_3H_2$, 3, 490

$HO(BuCONH)C_6H_3AsO_3H_2$ isomers, 3, 490

$HO(Me_2CHCH_2CONH)C_6H_3AsO_3H_2$ isomers, 3, 490

$AsC_{11}H_{16}NO_5S$ 4-[$\overline{CH_2(CH_2)_4N}O_2S$]$C_6H_4$-$AsO_3H_2$, 3, 417

$AsC_{11}H_{16}NO_6$ 5,2-$O_2N(Me_2CHCH_2CH_2O)$-$C_6H_3AsO_3H_2$, 3, 490

$AsC_{11}H_{16}NO_7$ 4-$(HOCH_2CHOHCH_2O_2CCH_2NH)$-$C_6H_4AsO_3H_2$, 3, 437

N-Arabinosylarsanilic acid, 3, 437

2,4-$HOCH_2CH_2O(EtOCONH)C_6H_3AsO_3H_2$, 3, 490

$AsC_{11}H_{16}N_3O_2$ 4-[$(H_2NCH_2CH_2)_2NOC$]-C_6H_4AsO, 3, 245

$AsC_{11}H_{16}N_3O_8$ [$Me_3(HOCH_2CH_2)As$]OC_6H_2-$(NO_2)_3$, 3, 312

$AsC_{11}H_{16}N_7O_3$

4,2-Me_2N[$\overline{N=C(NH_2)N=C(NH_2)N=C}NH$]-$C_6H_3AsO_3H_2$, 3, 490

$AsC_{11}H_{17}$ 4-$MeC_6H_4AsEt_2$, 3, 67

$AsC_{11}H_{17}BrI$ [$MeEt_2(2$-$BrC_6H_4)As$]I, 3, 316

$AsC_{11}H_{17}ClNO$ [$Me_3(3$-$AcNHC_6H_4)As$]Cl, 3, 316

$AsC_{11}H_{17}INO$ [$Me_3(3$-$AcNHC_6H_4)As$]I, 3, 316

$AsC_{11}H_{17}INO_2$ {Me_3[3-$(MeNHCO_2)C_6H_4$]-As}I, 3, 317

$AsC_{11}H_{17}N_2O_3$ 4,2-H_2N[$\overline{CH_2(CH_2)_4N}$]-$C_6H_3AsO_3H_2$, 3, 490

$AsC_{11}H_{17}N_2O_5$ 4,3-$HO(BuNHCONH)C_6H_3$-AsO_3H_2, 3, 490

$AsC_{11}H_{17}N_2O_6$ 4-$(H_2NCH_2CHOHCH_2O_2CCH_2$-$NH)C_6H_4AsO_3H_2$, 3, 437

4-$(HOCH_2CHOHCH_2NHCOCH_2NH)C_6H_4AsO_3H_2$, 3, 437

$AsC_{11}H_{17}N_2O_7$ 3,4-HO[$(HOCH_2CH_2)_2NCO$-NH]$C_6H_3AsO_3H_2$, 3, 490

$AsC_{11}H_{17}O$ $PhCH(OH)(CH_2)_4AsH_2$, 3, 121

MePhAsOBu, 3, 210

EtPhAsOPr, 3, 210

PrPhAsOEt, 3, 210

BuPhAsOMe, 3, 210

$AsC_{11}H_{17}O_2S_4$ PrAs[$SC(S)OCH_2CH=CH_2$]$_2$, 3, 279

$AsC_{11}H_{17}S$ (+)-MeBuPhAsS, 3, 365

$AsC_{11}H_{18}IO$ [$MeEt_2(3$-$HOC_6H_4)As$]I, 3, 317

$AsC_{11}H_{18}NO_3$ $PhCH_2CH_2NH(CH_2)_3AsO_3H_2$, 3, 397

$AsC_{11}H_{18}NO_4$ 3,4-$H_2N(Me_2CHCH_2CH_2O)$-$C_6H_3AsO_3H_2$, 3, 491

$AsC_{11}H_{18}NO_7$ HO[$HOCH_2CHOHCH_2(HOCH_2$-$CH_2)N$]$C_6H_3AsO_3H_2$ isomers, 3, 491

$AsC_{11}H_{19}BrN$ [$Me_3(2$-$Me_2NC_6H_4)As$]Br, 3, 317

$AsC_{11}H_{19}IN$ [$Me_3(2$-$Me_2NC_6H_4)As$]I, 3, 317

$AsC_{11}H_{19}N_2O_2S_4$ MeAs[$SC(S)N(CH_2CH_2)_2O$]$_2$, 3, 282

$AsC_{11}H_{19}N_2S_4$ MeAs[$SC(S)\overline{N(CH_2)_3C}H_2$]$_2$, 3, 282

$AsC_{11}H_{19}O_4$ (2-Methyl-2-arsolanyl)-succinic acid di-Me ester, 3, 553

$AsC_{11}H_{19}O_4P_2S_4$ 4-MeC_6H_4As[SP(S)-$(OMe)_2$]$_2$, 3, 281

4-MeC_6H_4As[$SP(S)(OMe)_2$]$_2$, 3, 281

$AsC_{11}H_{23}N_2S_4$ MeAs[$SC(S)NEt_2$]$_2$, 3, 282

$AsC_{11}H_{24}BrO_2$ {Et_3[$MeCH(CO_2Et)$]As}Br, 3, 317

$AsC_{11}H_{24}N$ $Pr_2As\overline{N(CH_2)_4C}H_2$, 3, 287

$AsC_{11}H_{25}O_2$ $PrAs(OBu)_2$, 3, 253

$AsC_{11}H_{25}O_3$ $PrAsO(OBu)_2$, 3, 546

$AsC_{11}H_{25}S_2$ $MeAs(SC_5H_{11})_2$, 3, 267

$PrAs(SBu)_2$, 3, 267

$AsC_{11}H_{26}I$ [$Me_3(n$-$C_8H_{17})As$]I, 3, 317

$AsC_{11}H_{26}O_3P$ $MeEtAsPO(OBu)_2$, 3, 292

$Et(Me_2CHCH_2CH_2)AsPO(OEt)_2$, 3, 292

$AsC_{11}H_{27}IP$ [Me_2AsPPr_3]I, 3, 291

$AsC_{11}H_{27}O_4$ $Me_3As(OOCMe_3)_2$, 3, 361

$AsC_{11}H_{27}O_6P_2S_2$ PrAs[$OP(S)(OEt)_2$]$_2$, 3, 256

$AsC_{12}H_5F_{14}$ $(CF_3CF_2F_2)_2AsPh$, 3, 67

$AsC_{12}H_6Br_3O_3$ [$\overline{OC(Br)=CHCH=C}$]$_3$As, 3, 30

$AsC_{12}H_6Cl_2NO_2$ 3-(Dichloroarsino)-8-

nitrodibenzofuran, **3**, 187

$AsC_{12}H_6Cl_3O_3$ Tri(5-chloro-2-furyl)-arsine, **3**, 14

$AsC_{12}H_6NO_4$ 3-Arsenoso-8-nitrodibenzo-furan, **3**, 245

$AsC_{12}H_7BrCl$ Bromochlorodibenzarsole isomers, **3**, 561

$AsC_{12}H_7BrClO$ 2,10-Br(Cl)-phenoxar-sine, **3**, 612

$AsC_{12}H_7BrNO_2$ 5,3-Br(O_2N)-dibenzarsole, **3**, 561

$AsC_{12}H_7ClNO_2$ 5,3-Cl(O_2N)-dibenzarsole, **3**, 561

$AsC_{12}H_7Cl_2$ Dichlorodibenzarsole iso-mers, **3**, 561

$AsC_{12}H_7Cl_2O$ 3-(Dichloroarsino)benzo-furan, **3**, 187

2,10-Cl_2-phenoxarsine, **3**, 612

$AsC_{12}H_7N_2O_6$ 5,3,7-HO(O_2N)$_2$-dibenzar-sole 5-oxide, **3**, 561

$AsC_{12}H_7O_2$ Arsenosodibenzofuran iso-mers, **3**, 245

$AsC_{12}H_7O_5S$ 3-Arsenosodibenzofuran-8-SO_3H, **3**, 246

$AsC_{12}H_8Br$ 5-Br-dibenzarsole, **3**, 561

$AsC_{12}H_8BrCl_2O$ 2-(4-BrC_6H_4O)$C_6H_4AsCl_2$, **3**, 196

$AsC_{12}H_8BrN_2O_6$ [4,3-HO(O_2N)C_6H_3]$_2$AsBr, **3**, 172

$AsC_{12}H_8BrO_2$ 3,5-Br(HO)-dibenzarsole 5-oxide, **3**, 561

$AsC_{12}H_8Br_2Cl$ (4-BrC_6H_4)$_2$AsCl, **3**, 172

$AsC_{12}H_8Cl$ 5-Cl-dibenzarsole, **3**, 562

$AsC_{12}H_8ClF_2$ (4-FC_6H_4)$_2$AsCl, **3**, 172

$AsC_{12}H_8ClN_2O_4$ (3-$O_2NC_6H_4$)$_2$AsCl, **3**, 172

$AsC_{12}H_8ClN_2O_6$ [4,3-HO(O_2N)C_6H_3]$_2$AsCl, **3**, 172

$AsC_{12}H_8ClO$ 10-Cl-phenoxarsine, **3**, 610

$AsC_{12}H_8ClO_2$ Chlorohydroxydibenzarsole 5-oxide isomers, **3**, 562

$AsC_{12}H_8Cl_2N$ 3-Carbazolyl-$AsCl_2$, **3**, 196

$AsC_{12}H_8Cl_2NO_2$ (4-ClC_6H_4)(4-$O_2NC_6H_4$)-AsCl, **3**, 172

2-(4-$O_2NC_6H_4$)$C_6H_4AsCl_2$, **3**, 197

2,4-Ph(O_2N)$C_6H_4AsCl_2$, **3**, 197

$AsC_{12}H_8Cl_3$ (ClC_6H_4)$_2$AsCl isomers, **3**, 172

$AsC_{12}H_8IN_2O_6$ [4,3-HO(O_2N)C_6H_3]$_2$AsI, **3**, 172

$AsC_{12}H_8NO_4$ Hydroxynitrodibenzarsole 5-oxide isomers, **3**, 562

$AsC_{12}H_8NO_6$ 8-Nitrodibenzofuran-3-arsonic acid, **3**, 523

$AsC_{12}H_9$ Poly(p-phenylenephenylarsine), **3**, 113

$AsC_{12}H_9BrCl$ Ph(4-BrC_6H_4)AsCl, **3**, 172

$AsC_{12}H_9Br_2$ Ph(4-BrC_6H_4)AsBr, **3**, 172

5-Acenaphthenyl-$AsBr_2$, **3**, 197

$AsC_{12}H_9Br_2O_2$ (BrC_6H_4)$_2AsO_2H$ isomers, **3**, 380

$AsC_{12}H_9ClN$ 10-Chloro-5,10-dihydro-phenarsazine, **3**, 620

$AsC_{12}H_9ClNO_2$ Ph(2-$O_2NC_6H_4$)AsCl, **3**, 165

$AsC_{12}H_9ClNO_4$ (4-ClC_6H_4)[3,4-HO(O_2N)-C_6H_3]AsOH, **3**, 207

(4-ClC_6H_4)(4-$O_2NC_6H_4$)AsO_2H, **3**, 380

$AsC_{12}H_9ClNO_5$ (4-ClC_6H_4)[4,3-HO(O_2N)-C_6H_3]AsO_2H, **3**, 380

$AsC_{12}H_9ClNO_8S$ 3,4-O_2N(4-$ClC_6H_4SO_3$)-$C_6H_3AsO_3H_2$, **3**, 491

$AsC_{12}H_9Cl_2$ Ph(ClC_6H_4)AsCl isomers, **3**, 172

5-Acenaphthenyl-$AsCl_2$, **3**, 197

2-$PhC_6H_4AsCl_2$, **3**, 197

$AsC_{12}H_9Cl_2O$ $PhOC_6H_4AsCl_2$ isomers, **3**, 197

$AsC_{12}H_9Cl_2O_2$ (2-ClC_6H_4)$_2AsO_2H$, **3**, 380

(3-ClC_6H_4)$_2AsO_2H$, **3**, 380

(4-ClC_6H_4)$_2AsO_2H$, **3**, 380

$AsC_{12}H_9Cl_2O_3$ ($\overline{OCH=CHCH=C}$)$_3AsCl_2$, **3**, 344

$AsC_{12}H_9F_2O_2$ (4-FC_6H_4)$_2AsO_2H$, **3**, 380

$AsC_{12}H_9IN$ 10-Iodo-5,10-dihydrophen-arsazine, **3**, 622

$AsC_{12}H_9INO_2$ Ph(2-$O_2NC_6H_4$)AsI, **3**, 173

$AsC_{12}H_9I_2$ 5-Acenaphthenyl-AsI_2, **3**, 197

$AsC_{12}H_9I_2O_3S$ 4-(4-$HO_3SC_6H_4$)$C_6H_4AsI_2$, **3**, 197

$AsC_{12}H_9N_2O$ 4-$PhN_2C_6H_4AsO$, **3**, 246

$AsC_{12}H_9N_2O_2$ 4-($\overline{N=CHCH=CHCH=C}$NHOC)-C_6H_4AsO, **3**, 246

4-(4-$HOC_6H_4N_2$)C_6H_4AsO, 3, 246
4,3-HO(PhN_2)C_6H_3AsO, 3, 246
$AsC_{12}H_9N_2O_5$ (4-$O_2NC_6H_4$)$_2$AsOH, 3, 207
$AsC_{12}H_9N_2O_6$ ($O_2NC_6H_4$)$_2AsO_2H$ isomers, 3, 375, 380
$AsC_{12}H_9N_2O_7$ (4-$O_2NC_6H_4$)[4,3-HO(O_2N)-C_6H_3]AsO_2H, 3, 380
[4,2-O_2N(4-$O_2NC_6H_4$)C_6H_3]AsO_3H_2, 3, 506
$AsC_{12}H_9N_2O_8$ [4,3-HO(O_2N)C_6H_3]$_2AsO_2H$, 3, 380
$AsC_{12}H_9OS_3$ (SCH=CHCH=C)$_3$AsO, 3, 356
$AsC_{12}H_9O_2$ 5-Hydroxydibenzarsole 5-oxide, 3, 562
PhAsOC$_6H_4$O-o, 3, 578
$AsC_{12}H_9O_3$ Trifurylarsine isomers, 3, 14, 30
$AsC_{12}H_9O_4$ (OCH=CHCH=C)$_3$AsO, 3, 356
Dibenzofuranarsonic acid isomers, 3, 523*
$AsC_{12}H_9O_4S$ 4-(4-$HO_3SC_6H_4$)C_6H_4AsO, 3, 246
$AsC_{12}H_9O_7S$ 8-Sulfodibenzofuran-3-arsonic acid, 3, 523*
$AsC_{12}H_9S_3$ Tri-2-thienylarsine, 3, 14, 30
$AsC_{12}H_{10}Br$ Ph_2AsBr, 3, 173
$AsC_{12}H_{10}BrO_2$ Ph(4-BrC_6H_4)AsO_2H, 3, 380
$AsC_{12}H_{10}BrO_3$ (4-BrC_6H_4)$C_6H_4AsO_3H_2$ isomers, 3, 412
$AsC_{12}H_{10}Br_3$ Ph_2AsBr_3, 3, 371
$AsC_{12}H_{10}Cl$ Ph_2AsCl, 3, 165
$AsC_{12}H_{10}ClN_2O_6$ 2,3,5-HO(Cl)[2,4-(HO)$_2C_6H_3$N=N]$C_6H_2AsO_3H_2$, 3, 503
$AsC_{12}H_{10}ClO_2$ Ph(ClC_6H_4)AsO_2H isomers, 3, 380, 381
$AsC_{12}H_{10}ClO_3$ 4-(4-ClC_6H_4)$C_6H_4AsO_3H_2$, 3, 412
$AsC_{12}H_{10}ClO_5S$ 4-(4-$ClC_6H_4SO_2$)C_6H_4-AsO_3H_2, 3, 412
$AsC_{12}H_{10}Cl_3$ Ph_2AsCl_3, 3, 371
$AsC_{12}H_{10}F$ Ph_2AsF, 3, 173
$AsC_{12}H_{10}F_3$ Ph_2AsF_3, 3, 371
$AsC_{12}H_{10}I$ Ph_2AsI, 3, 173
2,3-Dihydro-2-iodo-1H-naphth[1,8-cd]arsenin, 3, 599
$AsC_{12}H_{10}I_2NO_2S$ 4-($H_2NO_2SC_6H_4$)C_6H_4-AsI_2, 3, 197
4-($PhNHO_2S$)$C_6H_4AsI_2$, 3, 197
$AsC_{12}H_{10}N$ Me(1-$C_{10}H_7$)AsCN, 3, 203
5,10-Dihydrophenarsazine, 3, 622
$AsC_{12}H_{10}NO$ 4-(4-$H_2NC_6H_4$)C_6H_4AsO, 3, 246
$AsC_{12}H_{10}NOS_2$ 3,4-H_2N(HO)C_6H_3AsSC$_6H_4$S-o, 3, 585
$AsC_{12}H_{10}NO_2$ AcNHC$_{10}H_6$AsO isomers, 3, 246
3-Carbazole-As(OH)$_2$, 3, 224
Aminohydroxydibenzarsole 5-oxide isomers, 3, 562, 563
10-HO-5,10-dihydrophenarsazine 10-oxide, 3, 622
$AsC_{12}H_{10}NO_2S$ [4-($PhNHO_2S$)C_6H_4As]$_n$, 3, 140
$AsC_{12}H_{10}NO_3$ Ph($O_2NC_6H_4$)AsOH isomers, 3, 207
$AsC_{12}H_{10}NO_3S$ 4-(4-$H_2NO_2SC_6H_4$)C_6H_4AsO, 3, 246
4-$PhNHO_2SC_6H_4AsO$, 3, 246
$AsC_{12}H_{10}NO_4$ Ph[3,4-O_2N(HO)C_6H_3]AsOH, 3, 207
Ph($O_2NC_6H_4$)AsO_2H isomers, 3, 381
8-H_2N-3-dibenzofuranarsonic acid, 3, 523
$AsC_{12}H_{10}NO_4S$ 2,3-Dihydro-3-oxo-2H-naphtho[1,2-b]-1,4-thiazine-6-arsonic acid, 3, 524
$AsC_{12}H_{10}NO_5$ Ph[3,4-O_2N(HO)C_6H_3]AsO_2H, 3, 381
[(4-$O_2NC_6H_4$)C_6H_4]AsO_3H_2 isomers, 3, 506
4,2-O_2N(Ph)$C_6H_3AsO_3H_2$, 3, 506
6-O_2N-acenaphthene-5-arsonic acid, 3, 506
3-Hydroxynaphth[1,2-b]-1,4-oxazin-6-arsonic acid, 3, 524
$AsC_{12}H_{10}NO_5S$ 4-(4-$O_2NC_6H_4S$)$C_6H_4AsO_3H_2$, 3, 417
$AsC_{12}H_{10}NO_6$ 4-[3,4-O_2N(HO)C_6H_3]C_6H_4-AsO_3H_2, 3, 506
$AsC_{12}H_{10}NO_7S$ 4-(4-$O_2NC_6H_4O_2S$)C_6H_4-AsO_3H_2, 3, 417

$AsC_{12}H_{10}NS$ Me(1-$C_{10}H_7$)AsSCN, 3, 203

$AsC_{12}H_{10}N_3$ Ph_2AsN_3, 3, 203

$AsC_{12}H_{10}N_3O_2$ 4-[N=C(Me)CH=CHN=CNHOC]-C_6H_4AsO, 3, 246

$AsC_{12}H_{10}N_3O_7$ 4-[2,4-$(O_2N)_2C_6H_3$NH]-$C_6H_4AsO_3H_2$, 3, 442*

$AsC_{12}H_{11}$ Ph_2AsH, 3, 116

$AsC_{12}H_{11}BrNO_2S$ Ph(4-$H_2NO_2SC_6H_4$)AsBr, 3, 173

$AsC_{12}H_{11}ClN$ Ph($H_2NC_6H_4$)AsCl isomers, 3, 173

$AsC_{12}H_{11}ClNO_2S$ Ph(4-$H_2NO_2SC_6H_4$)AsCl, 3, 173

$AsC_{12}H_{11}ClNO_4S$ Ph[4-(ClNHO_2S)C_6H_4]-AsO_2H, 3, 381

$AsC_{12}H_{11}ClNO_6S$ 3,4-H_2N(4-Cl$C_6H_4SO_3$)-$C_6H_3AsO_3H_2$, 3, 491

$AsC_{12}H_{11}Cl_2$ $(ClCH=CH)_2$AsCH=CHPh, 3, 67

$AsC_{12}H_{11}INO_2S$ Ph(4-$H_2NO_2SC_6H_4$)AsI, 3, 173

$AsC_{12}H_{11}I_2N_2O_4S_2$ 4-(4-$H_2NO_2SC_6H_4$-NHO_2S)$C_6H_4AsI_2$, 3, 197

$AsC_{12}H_{11}N_2O_2S$ [4-(4-$H_2NC_6H_4SO_2$NH)-C_6H_4As$]_n$, 3, 141

$AsC_{12}H_{11}N_2O_3$ 4-PhN=N$C_6H_4AsO_3H_2$, 3, 457

$AsC_{12}H_{11}N_2O_3S$ {4-[3,4-H_2N(HO)C_6H_3-SO_2NH]C_6H_4As$\}_n$, 3, 141
  [3,4-H_2N(4-$H_2NC_6H_4SO_3$)C_6H_3As$]_n$, 3, 141

$AsC_{12}H_{11}N_2O_3S$ 4-(4-$H_2NC_6H_4SO_2$NH)-C_6H_4AsO, 3, 232

$AsC_{12}H_{11}N_2O_4$ (4-$H_2NC_6H_4$)(4-$O_2NC_6H_4$)-AsO_2H, 3, 381
  4-(4-HOC_6H_4N=N)$C_6H_4AsO_3H_2$, 3, 458
  HO(PhN=N)$C_6H_3AsO_3H_2$ isomers, 3, 491

$AsC_{12}H_{11}N_2O_4S$ 4,3-HO(4-$H_2NC_6H_4SO_2$NH)-C_6H_3AsO, 3, 246
  3,4-H_2N(4-$H_2NC_6H_4SO_3$)C_6H_3AsO, 3, 246

$AsC_{12}H_{11}N_2O_5$ 4-[2,4-$(HO)_2C_6H_3$N=N]-$C_6H_4AsO_3H_2$, 3, 458
  3,4-O_2N(PhNH)$C_6H_3AsO_3H_2$, 3, 491
  4-[3,4-O_2N(H_2N)C_6H_3]$C_6H_4AsO_3H_2$, 3, 506
  4-(5-Arsono-2-pyridylamino)-C_6H_4-CO_2H, 3, 524

$AsC_{12}H_{11}N_2O_6$ 4-[N=C(CO_2H)COCH=C(Me)N]-$C_6H_4AsO_3H_2$, 3, 470
  3,4-O_2N(HOC_6H_4NH)$C_6H_3AsO_3H_2$ isomers, 3, 491

$AsC_{12}H_{11}N_2O_7S$ $O_2NC_6H_4NHO_2SC_6H_4AsO_3H_2$ isomers, 3, 417

$AsC_{12}H_{11}N_2O_8S$ 3,4-O_2N(4-$H_2NC_6H_4SO_3$)-$C_6H_3AsO_3H_2$, 3, 491

$AsC_{12}H_{11}N_4O_7$ N-[3-(5-Nitrofurfurylidene)carbazoyl]arsanilic acid, 3, 432

$AsC_{12}H_{11}O_2$ Ph_2AsO_2H, 3, 376*
  1-$C_{10}H_7$AsOCH$_2$CH$_2$O, 3, 578

$AsC_{12}H_{11}O_3$ (4-HO$C_6H_4)_2$AsOH, 3, 207
  Ph(HOC_6H_4)AsO_2H isomers, 3, 381
  Ph$C_6H_4AsO_3H_2$ isomers, 3, 412, 506
  Acenaphthene-5-arsonic acid, 3, 506

$AsC_{12}H_{11}O_3S$ 4-(SCH=CHCH=CCH=CH)C_6H_4-AsO_3H_2, 3, 405

$AsC_{12}H_{11}O_4$ (HO$C_6H_4)_2AsO_2H$ isomers, 3, 381
  4-(OCH=CHCH=CCH=CH)$C_6H_4AsO_3H_2$, 3, 405

$AsC_{12}H_{11}O_5S$ Ph(4-$HO_3SC_6H_4$)AsO_2H, 3, 382
  4-Ph$SO_2C_6H_4AsO_3H_2$, 3, 413

$AsC_{12}H_{11}O_6S$ 4-(4-$HO_3SC_6H_4$)$C_6H_4AsO_3H_2$, 3, 506

$AsC_{12}H_{12}BrN_2O_6$
  4-[N=C(CO_2Et)C(OH)=CBrN]$C_6H_4AsO_3H_2$, 3, 470

$AsC_{12}H_{12}Br_2N$ 4-$Me_2NC_{10}H_6AsBr_2$, 3, 197

$AsC_{12}H_{12}Cl$ Et(1-$C_{10}H_7$)AsCl, 3, 173

$AsC_{12}H_{12}ClN_2$ 4-ClC_6H_4As$(CH_2CH_2CN)_2$, 3, 67
  (4-$H_2NC_6H_4)_2$AsCl, 3, 173

$AsC_{12}H_{12}ClN_2O_2$ [3,4-H_2N(HO)$C_6H_3]_2$AsCl, 3, 173

$AsC_{12}H_{12}ClO_2$ [Ph_2As$(OH)_2$]Cl, 3, 371*

$AsC_{12}H_{12}ClO_4$ 5-(p-ClC_6H_4)-1,6-dioxa-5-arsaspiro[4,4]nonane-2,7-dione, 3, 579

$AsC_{12}H_{12}Cl_2N$ 4-$Me_2NC_{10}H_6AsCl_2$, 3, 197

$AsC_{12}H_{12}I_2N$ 4-$Me_2NC_{10}H_6AsI_2$, 3, 197

$AsC_{12}H_{12}N$ Ph(2-$H_2NC_6H_4$)AsH, 3, 117

$AsC_{12}H_{12}NO$ Ph($H_2NC_6H_4$)AsOH isomers, 3, 207
  4-$Me_2NC_{10}H_6$AsO, 3, 246

$AsC_{12}H_{12}NO_2$ Ph[3,4-$H_2N(HO)C_6H_3$]AsOH, 3, 207

Ph($H_2NC_6H_4$)AsO_2H isomers, 3, 382

$AsC_{12}H_{12}NO_3$ Ph[$H_2N(HO)C_6H_3$]AsO_2H isomers, 3, 382

4,2-$H_2N(Ph)C_6H_3AsO_3H_2$, 3, 506

[(4-$H_2NC_6H_4$)C_6H_4]AsO_3H_2 isomers, 3, 506

$AsC_{12}H_{12}NO_3S$ 4-(4-$H_2NC_6H_4S$)C_6H_4-AsO_3H_2, 3, 417

$AsC_{12}H_{12}NO_4$ 3-$AcNHC_{10}H_6$-1-AsO_3H_2, 3, 507

$AsC_{12}H_{12}NO_4S$ Ph(4-$H_2NO_2SC_6H_4$)AsO_2H, 3, 382

$AsC_{12}H_{12}NO_5S$ 4-($PhNHO_2S$)$C_6H_4AsO_3H_2$, 3, 417

4-(4-$H_2NC_6H_4O_2S$)$C_6H_4AsO_3H_2$, 3, 417

4-(4-$H_2NC_6H_4SO_2$)$C_6H_4AsO_3H_2$, 3, 413

4-(4-$H_2NO_2SC_6H_4$)$C_6H_4AsO_3H_2$, 3, 507

$AsC_{12}H_{12}NO_6S$ ($HOC_6H_4NHO_2S$)$C_6H_4AsO_3H_2$ isomers, 3, 417

3,4-$H_2N(PhSO_3)C_6H_3AsO_3H_2$, 3, 491

$AsC_{12}H_{12}N_3O_3$ 4-(4-$H_2NC_6H_4N$=N)C_6H_4-AsO_3H_2, 3, 458

$AsC_{12}H_{12}N_3O_4$ 4-[$H_2N(HO)C_6H_3N$=N]C_6H_4-AsO_3H_2 isomers, 3, 458

$AsC_{12}H_{12}N_3O_5$ 4,3-HO[2,4-HO(H_2N)C_6H_3-N=N]$C_6H_3AsO_3H_2$, 3, 491

$AsC_{12}H_{12}N_3O_6S$ 4-(4-$HO_3SC_6H_4NHN$=N)-$C_6H_4AsO_3H_2$, 3, 468

$AsC_{12}H_{12}N_3O_9S_2$ 4-[2,5-$(HO_3S)_2C_6H_3$-NHN=N]$C_6H_4AsO_3H_2$, 3, 468

$AsC_{12}H_{13}$ 1-$C_{10}H_7AsMe_2$, 3, 50

$AsC_{12}H_{13}BrNO_6$ 4-(MeCH=$CBrCO_2CH_2CONH$)-$C_6H_4AsO_3H_2$, 3, 437

$AsC_{12}H_{13}Br_2$ Me_2(1-$C_{10}H_7$)$AsBr_2$, 3, 344

$AsC_{12}H_{13}ClNO_6$ 4,3-Cl(MeCH=$CHCO_2CH_2$-CONH)$C_6H_3AsO_3H_2$, 3, 491

$AsC_{12}H_{13}Cl_2$ Me_2(1-$C_{10}H_7$)$AsCl_2$, 3, 344

$AsC_{12}H_{13}Cl_2O_2S_4$ 3,4-$Cl_2C_6H_3As$[SC(S)-OEt$]_2$, 3, 280

$AsC_{12}H_{13}NO_6S_2$

4,3-HO(AcNH)C_6H_4AsSCH(CO_2H)CH(CO_2H)S, 3, 586

$AsC_{12}H_{13}N_2$ PhAs$(CH_2CH_2CN)_2$, 3, 50

$AsC_{12}H_{13}N_2O$ (4-$H_2NC_6H_4)_2$AsOH, 3, 207

$AsC_{12}H_{13}N_2O_2$ ($H_2NC_6H_4)_2AsO_2H$ isomers, 3, 382

$AsC_{12}H_{13}N_2O_3$ 4-(4-$H_2NC_6H_4NH$)C_6H_4As-O_3H_2, 3, 442

3,4-$H_2N(PhNH)C_6H_3AsO_3H_2$, 3, 491

$AsC_{12}H_{13}N_2O_4$ [3,4-$H_2N(HO)C_6H_3]_2AsO_2H$, 3, 382

3,4-H_2N(4-HOC_6H_4NH)$C_6H_3AsO_3H_2$, 3, 491

$AsC_{12}H_{13}N_2O_5$ 4-($MeO_2CCH_2NHCOCH_2NHCO$)-C_6H_4AsO, 3, 232

$AsC_{12}H_{13}N_2O_5S$ ($H_2NC_6H_4NHO_2S$)C_6H_4As-O_3H_2 isomers, 3, 417

$AsC_{12}H_{13}N_2O_6$ 4-[N=C(CO_2Et)C(OH)=CHN]-$C_6H_4AsO_3H_2$, 3, 470

$AsC_{12}H_{13}N_2O_6S$ 3,4-H_2N(4-$H_2NC_6H_4SO_3$)-$C_6H_3AsO_3H_2$, 3, 491

4,3-HO(4-$H_2NC_6H_4SO_2NH$)$C_6H_3AsO_3H_2$, 3, 491

$AsC_{12}H_{13}N_2O_6S_2$ 4-(4-$H_2NO_2SC_6H_4NHO_2S$)-$C_6H_4As(OH)_2$, 3, 224

$AsC_{12}H_{13}N_2O_7S_2$ 4-(4-$H_2NO_2SC_6H_4NHO_2S$)-$C_6H_4AsO_3H_2$, 3, 417

$AsC_{12}H_{13}N_4O_5S$ 4,3-H_2N(4-$H_2NO_2SC_6H_4$-N=N)$C_6H_3AsO_3H_2$, 3, 492*

$AsC_{12}H_{13}N_4O_7$

3,4-O_2N[N=C(CO_2Et)C(Me)=NN]C_6H_3-AsO_3H_2, 3, 492

$AsC_{12}H_{13}O_4$ 5-Ph-1,6-dioxa-5-arsa-spiro[4,4]nonane-2,7-dione, 3, 579

$AsC_{12}H_{14}Br$ 4-$BrC_6H_4As(CH_2CH$=$CH_2)_2$, 3, 67

$AsC_{12}H_{14}BrN_2O_6$ 4-[$BrCH_2COC(CO_2Et)$=N-NH]$C_6H_4AsO_3H_2$, 3, 453

$AsC_{12}H_{14}ClO_2S_4$ 4-ClC_6H_4As[SC(S)OEt$]_2$, 3, 280

$AsC_{12}H_{14}ClO_4$ 4-$ClC_6H_4As(CH_2CH_2CO_2H)_2$, 3, 67

$AsC_{12}H_{14}NO_4S_4$ 4-$O_2NC_6H_4As$[SC(S)OEt$]_2$, 3, 280

$AsC_{12}H_{14}NO_5$ 3-(MeCH=$CHCOCH_2CONH$)C_6H_4-AsO_3H_2, 3, 425

$AsC_{12}H_{14}NO_6$ 4-(MeCH=$CHCO_2CH_2CONH$)-$C_6H_4AsO_3H_2$, 3, 437*

$AsC_{12}H_{14}N_2O_5$ [3,4-$H_2N(HO)C_6H_3$As-(OH)$]_2$O, 3, 257

$AsC_{12}H_{14}N_3$ 4-$H_2NC_6H_4As(CH_2CH_2CN)_2$, 3, 50
$AsC_{12}H_{14}N_3O_3$ 4-[2,4-$(H_2N)_2C_6H_3NH$]-$C_6H_4AsO_3H_2$, 3, 442
4-[N=C(Me)CH=C(Me)N=CNH]$C_6H_4AsO_3H_2$, 3, 447
$AsC_{12}H_{14}N_3O_5$ 4-[N=C(CO_2Et)C(Me)=NN]-$C_6H_4AsO_3H_2$, 3, 470
$AsC_{12}H_{14}N_3O_8$ 3,4-O_2N[AcC(CO_2Et)=NNH]-$C_6H_3AsO_3H_2$, 3, 492
$AsC_{12}H_{14}N_5O$
4-[N=C(NHMe)N=C(NHMe)CH=CNH]C_6H_4-AsO, 3, 246
$AsC_{12}H_{14}N_5O_5$
3,4-O_2N[C(Me)=NC(NH_2)=NC(Me)=CNH]-$C_6H_3AsO_3H_2$, 3, 492
$AsC_{12}H_{15}$ $(CH_2{=}CHCH_2)_2AsPh$, 3, 67
$AsC_{12}H_{15}NO_3$ 4-[MeCH=CHCH=C(Me)N]-$C_6H_4AsO_3H_2$, 3, 470
$AsC_{12}H_{15}N_2O_3$ 4-[HNC(Pr)=NCH=C]C_6H_4-AsO_3H_2, 3, 470
$AsC_{12}H_{15}N_2O_6$ 2-[AcC(CO_2Et)=NNH]C_6H_4-AsO_3H_2, 3, 453
3-[AcC(CO_2Et)=NNH]$C_6H_4AsO_3H_2$, 3, 453
4-[AcC(CO_2Et)=NNH]$C_6H_4AsO_3H_2$, 3, 453
$AsC_{12}H_{15}N_4O_4S_4$ 2,5-$(O_2N)_2C_6H_3As$-[SC(S)NMe_2]$_2$, 3, 283
$AsC_{12}H_{15}N_4O_5$
4,3-$HOCH_2CH_2O$[N=C(NH_2)N=CHCH=CNH]-$C_6H_3AsO_3H_2$, 3, 492
$AsC_{12}H_{15}N_6OS_2$ 4-(4,6-Diamino-s-triazin-2-ylamino)dithiobenzenearsonous acid 3-hydroxypropylene ester, 3, 586
$AsC_{12}H_{15}N_6O_2S_2$ 3-(4,6-Diamino-s-triazin-2-ylamino)4-hydroxydithiobenzenearsonous acid 3-hydroxypropylene ester, 3, 586
$AsC_{12}H_{15}O_2$ $PhAs(OCH_2CH{=}CH_2)_2$, 3, 253
$AsC_{12}H_{15}O_2S_4$ PhAs[SC(S)OEt]$_2$, 3, 280
$AsC_{12}H_{15}O_3$ 4-[$HO_2C(CH_2)_5$]C_6H_4AsO, 3, 246
$AsC_{12}H_{15}O_4$ $PhAs(CH_2CH_2CO_2H)_2$, 3, 67
EtAs($OCH_2C{=}CHCH{=}CHO$)$_2$, 3, 253
$AsC_{12}H_{16}ClN_2S_4$ 2-ClC_6H_4As[SC(S)NMe_2]$_2$, 3, 283
$AsC_{12}H_{16}ClN_4O_3$ {4-[Me-C=CHN$^+$(Me)=C(NH_2)N=CNH]$C_6H_4AsO_3H_2$}-Cl^-, 3, 447
$AsC_{12}H_{16}ClO$ PhAs[OCH(Pr)CH=CH_2]Cl, 3, 201
$AsC_{12}H_{16}Cl_2N_2O_5$ [4,3-HO(HCl·H_2N)C_6H_3-As(OH)]$_2$O, 3, 257
$AsC_{12}H_{16}IO$ 1-Phenyl-4-arsenanone methiodide, 3, 590
$AsC_{12}H_{16}IO_2$ 1,2,3,4-Tetrahydro-7-methoxy-1,1-dimethyl-4-oxoarsinolinium iodide, 3, 592
$AsC_{12}H_{16}NO_3S_2$ 3,4-$H_2N(HO)C_6H_3As(SCH_2$-Ac)$_2$, 3, 267
$AsC_{12}H_{16}NO_5$ 2,4-($MeCHOHCH_2O$(EtO_2CNH)-C_6H_3AsO, 3, 246
4-[$O(CH_2CH_2)_2NCH_2CO$]$C_6H_4AsO_3H_2$, 3, 405
4-($CH_2CH_2OCH_2CH_2NCOCH_2$)$C_6H_4AsO_3H_2$, 3, 413
4-[EtO_2CCH=C(Me)NH]$C_6H_4AsO_3H_2$, 3, 437*
$AsC_{12}H_{16}NO_6$ 4,3-HO[$(EtCO)_2N$]C_6H_3-AsO_3H_2, 3, 492
$AsC_{12}H_{16}NO_6S_2$ 3,4-$H_2N(HOCH_2CH_2O)C_6H_3$-As$(SCH_2CO_2H)_2$, 3, 267
3,4-$H_2N(HOCH_2CH_2O)C_6H_3As(SCH_2CO_2H)_2$, 3, 267
$AsC_{12}H_{16}N_3O_2$ 1,2,3,4-Tetrahydro-7-methoxy-1-methyl-4-semicarbazonoarsinoline, 3, 592
$AsC_{12}H_{16}N_3O_7S_2$ 3,4-$O_2N(HO)C_6H_3As$-[$SCH_2CH(NH_2)CO_2H$]$_2$, 3, 267
$AsC_{12}H_{16}N_3S_4$ 4-$O_2NC_6H_4As$[SC(S)NMe_2]$_2$, 3, 283
$AsC_{12}H_{16}N_4O_4P$ 4,4'-[Arsenobis(phenyleneimino)]bis(2-amino-1,6-dimethylpyrimidinium phosphate), 3, 133*
$AsC_{12}H_{16}N_5O_3$ 4-[CH=C(NHMe)N=C(NHMe)N=C-NH]$C_6H_4AsO_3H_2$, 3, 447
$AsC_{12}H_{17}$ Ph(cyclo-C_6H_{11})AsH, 3, 118
$AsC_{12}H_{17}BrN$ {Me_2[$NC(CH_2)_3$]PhAs}Br, 3, 317
$AsC_{12}H_{17}INO$ [$Me_2(NCCH_2CH_2)(3\text{-}MeOC_6H_4)$-As]I, 3, 317
$AsC_{12}H_{17}N_2O_4$ 4-[$CH_2(CH_2)_3NCH_2CONH$]-$C_6H_4AsO_3H_2$, 3, 437

AsC$_{12}$H$_{17}$N$_2$O$_5$ HO[CH$_2$(CH$_2$)$_3$NCH$_2$CONH]-C$_6$H$_3$AsO$_3$H$_2$ isomers, 3, 492
AsC$_{12}$H$_{17}$N$_2$S$_4$ PhAs[SC(S)NMe$_2$]$_2$, 3, 283
AsC$_{12}$H$_{17}$O 4-AcC$_6$H$_4$AsEt$_2$, 3, 67
AsC$_{12}$H$_{17}$O$_3$ 4-MeC$_6$H$_4$AsOCH$_2$CH(CH$_2$OEt)O, 3, 578
AsC$_{12}$H$_{17}$O$_4$ Ph[HO$_2$C(CH$_2$)$_5$]AsO$_2$H, 3, 382
AsC$_{12}$H$_{17}$O$_5$ 4-[HO$_2$C(CH$_2$)$_5$]C$_6$H$_4$AsO$_3$H$_2$, 3, 405
AsC$_{12}$H$_{17}$O$_6$ 4-AcOCH$_2$CH$_2$OCH$_2$CH$_2$C$_6$H$_4$-AsO$_3$H$_2$, 3, 413
AsC$_{12}$H$_{18}$Br 4-BrC$_6$H$_4$AsPr$_2$, 3, 67
[MeEt(CH$_2$=CHCH$_2$)PhAs]Br, 3, 317
AsC$_{12}$H$_{18}$ClNO$_6$PS Et$_2$As(O)OP(S)(OEt)-[OC$_6$H$_3$(NO$_2$)Cl-4,3], 3, 388
AsC$_{12}$H$_{18}$ClN$_2$O$_2$S$_4$ ClCH=CHAs[SC(S)N-(CH$_2$CH$_2$)$_2$O]$_2$, 3, 283
AsC$_{12}$H$_{18}$ClN$_4$ 4-ClC$_6$H$_4$As[CH$_2$CH$_2$C(=NH)-NH$_2$]$_2$, 3, 50
AsC$_{12}$H$_{18}$IO$_2$ [Me$_2$(EtO$_2$CCH$_2$)PhAs]I, 3, 317
[CH$_2$CH$_2$C(OH)$_2$CH$_2$CH$_2$AsMePh]I, 3, 590
AsC$_{12}$H$_{18}$IO$_3$ [Me$_2$(HO$_2$CCH$_2$CH$_2$)(3-MeO-C$_6$H$_4$)As]I, 3, 317
AsC$_{12}$H$_{18}$NO$_2$ 3-(MeNHCO$_2$)C$_6$H$_4$AsEt$_2$, 3, 67
AsC$_{12}$H$_{18}$NO$_2$S$_2$ 4-[HO(CH$_2$)$_3$NH]C$_6$H$_4$-AsCH$_2$CH(CH$_2$OH)S, 3, 586
AsC$_{12}$H$_{18}$NO$_4$ 2-O$_2$NC$_6$H$_4$As(OCHMe$_2$)$_2$, 3, 253
4-(Et$_2$NCH$_2$CO)C$_6$H$_4$AsO$_3$H$_2$, 3, 405
4-(PrNHCOCH$_2$CH$_2$)C$_6$H$_4$AsO$_3$H$_2$, 3, 405
AsC$_{12}$H$_{18}$NO$_5$ {4,3-HO[(HOCH$_2$CHOH-CH$_2$)$_2$N]C$_6$H$_3$As}$_n$, 3, 141
AsC$_{12}$H$_{18}$NO$_6$ 4-[(HOCH$_2$CH$_2$)$_2$NOCCH$_2$O]-C$_6$H$_4$As(OH)$_2$, 3, 224
HO[(HOCH$_2$CHOHCH$_2$)$_2$N]C$_6$H$_3$AsO isomers, 3, 247
AsC$_{12}$H$_{18}$NO$_7$ 4-[(HOCH$_2$)$_3$CNHOCCH$_2$O]-C$_6$H$_4$As(OH)$_2$, 3, 224
4-[(HOCH$_2$)$_3$CNHCOCH$_2$]C$_6$H$_4$AsO$_3$H$_2$, 3, 413
4,2-EtOCONH(MeCHOHCH$_2$O)C$_6$H$_3$AsO$_3$H$_2$, 3, 492
AsC$_{12}$H$_{18}$NO$_8$ 4,3-HO(glucosyl-NH)C$_6$H$_3$-As(OH)$_2$, 3, 224
N-Fructosylarsanilic acid, 3, 437
N-Galactosylarsanilic acid, 3, 437
N-Glucosylarsanilic acid, 3, 437
N-Mannosylarsanilic acid, 3, 437
AsC$_{12}$H$_{18}$N$_3$O$_4$S$_2$ 3,4-H$_2$N(HOCH$_2$CH$_2$O)-C$_6$H$_3$As(SCH$_2$CONH$_2$)$_2$, 3, 268
AsC$_{12}$H$_{18}$N$_3$O$_6$ 3,4-O$_2$N[O(CH$_2$CH$_2$)$_2$NCH$_2$-CH$_2$NH]C$_6$H$_3$AsO$_3$H$_2$, 3, 492
AsC$_{12}$H$_{18}$N$_3$O$_8$ MeEtPrAs(OH)OC$_6$H$_2$(NO$_2$)$_3$, 3, 348
AsC$_{12}$H$_{18}$N$_3$S$_4$ 4-H$_2$NC$_6$H$_4$As[SC(S)NMe$_2$]$_2$, 3, 283
AsC$_{12}$H$_{19}$ EtBuPhAs, 3, 76
EtPr(4-MeC$_6$H$_4$)As, 3, 76
AsC$_{12}$H$_{19}$Br$_2$ [Et$_2$(BrCH$_2$CH$_2$)PhAs]Br, 3, 317
AsC$_{12}$H$_{19}$INO$_2$ {Me$_3$[4-(Me$_2$NCO$_2$)C$_6$H$_4$]-As}I, 3, 317
AsC$_{12}$H$_{19}$N$_2$O$_5$ 4-[(HOCH$_2$CH$_2$)$_2$NOCCH$_2$NH]-C$_6$H$_4$As(OH)$_2$, 3, 224
AsC$_{12}$H$_{19}$N$_2$O$_6$ 4-[(HOCH$_2$CH$_2$)$_2$NCOCH$_2$NH]-C$_6$H$_4$AsO$_3$H$_2$, 3, 437
AsC$_{12}$H$_{19}$N$_4$ PhAs[CH$_2$CH$_2$C(=NH)NH$_2$]$_2$, 3, 50
AsC$_{12}$H$_{19}$O EtPhAsOBu, 3, 210
PrPhAsOPr, 3, 210
BuPhAsOEt, 3, 210
Pr$_2$AsOPh, 3, 211
AsC$_{12}$H$_{19}$O$_2$ PhAs(OPr)$_2$, 3, 253
PhAs(OCHMe)$_2$, 3, 253
AsC$_{12}$H$_{19}$S$_2$ PhAs(SPr)$_2$, 3, 268
AsC$_{12}$H$_{20}$Br [(CH$_2$=CHCH$_2$)$_4$As]Br, 3, 317
AsC$_{12}$H$_{20}$IO Et$_2$PhAs(OEt)I, 3, 346
AsC$_{12}$H$_{20}$NO$_8$ 3,4-HO[(HOCH$_2$CHOHCH$_2$)$_2$N]-C$_6$H$_3$AsO$_3$H$_2$, 3, 492
AsC$_{12}$H$_{20}$N$_5$ 4-H$_2$NC$_6$H$_4$As[CH$_2$CH$_2$C(=NH)-NH$_2$]$_2$, 3, 51
AsC$_{12}$H$_{20}$O$_3$P EtPhAsPO(OEt)$_2$, 3, 292
AsC$_{12}$H$_{20}$P 2-Me$_2$AsC$_6$H$_4$PEt$_2$, 3, 51
AsC$_{12}$H$_{21}$ (CH$_2$=CMeCH$_2$)$_3$As, 3, 14
(EtCH=CH)$_3$As, 3, 30
AsC$_{12}$H$_{21}$BrP [Me$_2$AsPPhEt$_2$]Br, 3, 291
AsC$_{12}$H$_{21}$ClN$_3$O$_5$ 4-(HOCH$_2$CONH)C$_6$H$_4$-AsO$_3$H$_2$·HN(CH$_2$CH$_2$)$_2$NH·HCl, 3, 429
AsC$_{12}$H$_{21}$IP [Et$_2$AsPPhMe$_2$]I, 3, 291
AsC$_{12}$H$_{21}$N$_2$O$_2$S$_4$ EtAs[SC(S)N(CH$_2$CH$_2$)$_2$O]$_2$, 3, 283

$AsC_{12}H_{21}N_2O_5$ 3,4-AcNH(HO)$C_6H_3AsO_3H_2\cdot HNEt_2$, 3, 477

$AsC_{12}H_{21}N_6O_8$ PhAs[CH_2CH_2C(=NH)$NH_3\cdot NO_3$]$_2$, 3, 50

$AsC_{12}H_{21}O$ [(CH_2=$CHCH_2$)$_4$As]OH, 3, 317

$AsC_{12}H_{21}O_6P_2S_2$ 2,4-$Me_2C_6H_3$As[SP(O)-(OMe)$_2$]$_2$, 3, 281

$AsC_{12}H_{22}Cl$ (cyclo-C_6H_{11})$_2$AsCl, 3, 173

$AsC_{12}H_{22}ClN_2S_4$ ClCH=CHAs[SC(S)NEt_2]$_2$, 3, 283

$AsC_{12}H_{22}N_5O_5$ 3,4-O_2N[H_2N(CH_2CH_2NH)$_3$]-$C_6H_3AsO_3H_2$, 3, 492

$AsC_{12}H_{23}$ (cyclo-C_6H_{11})$_2$AsH, 3, 118

$AsC_{12}H_{24}ClO_2$ ClCH=CHAs($OCH_2CH_2CHMe_2$)$_2$, 3, 253

$AsC_{12}H_{24}N_5O_3$ 3,4-H_2N[H_2N(CH_2CH_2NH)$_3$]-$C_6H_3AsO_3H_2$, 3, 492

$AsC_{12}H_{25}N_2S_4$ EtAs[SC(S)NEt_2]$_2$, 3, 283

$AsC_{12}H_{25}O_4$ 5,2,3,7,8-Bu(Me)$_4$-1,4,6,-9-tetraoxa-5-arsaspiro[4,4]nonane, 3, 580

$AsC_{12}H_{27}$ Bu_3As, 3, 15
(Me_2CHCH_2)$_3$As, 3, 16

$AsC_{12}H_{27}Cl_2$ Bu_3AsCl_2, 3, 344

$AsC_{12}H_{27}O$ $Pr_2AsOC_6H_{13}$, 3, 211
Bu_2AsOBu, 3, 211
Bu_3AsO, 3, 357

$AsC_{12}H_{27}O_2$ BuAs(OBu)$_2$, 3, 253

$AsC_{12}H_{27}O_3$ BuAsO(OBu)$_2$, 3, 546

$AsC_{12}H_{27}S$ Bu_3AsS, 3, 365

$AsC_{12}H_{27}S_2$ EtAs(SC_5H_{11})$_2$, 3, 268
BuAs(SBu)$_2$, 3, 268
EtAs($SCH_2CH_2CHMe_2$)$_2$, 3, 268

$AsC_{12}H_{27}Se$ Bu_3AsSe, 3, 367

$AsC_{12}H_{28}N$ Bu_2AsNEt_2, 3, 287

$AsC_{12}H_{28}O_3P$ Et_2AsPO(OBu)$_2$, 3, 292

$AsC_{12}H_{29}IO_3P$ [EtBuAsP(OEt)$_3$]I, 3, 291

$AsC_{13}H_6Cl_2NO_3$ x,9-(Cl_2As)(O_2N)-9-fluorenone isomers, 3, 187

$AsC_{13}H_7Cl_2O$ 3-(Cl_2As)-9-fluorenone, 3, 197

$AsC_{13}H_7O_2$ 4-Arsenoso-9-fluorenone, 3, 247
1-Arsenoso-9-fluorenone, 3, 247

$AsC_{13}H_8Cl_2N$ (ClC_6H_4)$_2$AsCN isomers, 3, 203

$AsC_{13}H_8NO$ 7,7′-Diamino-2,2′-arseno-9-fluorenone, 3, 134

$AsC_{13}H_8NO_2$ 3,5-NC(HO)-dibenzarsole 5-oxide, 3, 563

$AsC_{13}H_8NO_6$ Nitro-9-oxofluorenearsonic acid isomers, 3, 507

$AsC_{13}H_8N_3O_4$ (3-$O_2NC_6H_4$)$_2$AsCN, 3, 203

$AsC_{13}H_8N_3O_4S$ (3-$O_2NC_6H_4$)$_2$AsSCN, 3, 203

$AsC_{13}H_9Br_2N_2O_4$ [3,4-Br(O_2N)C_6H_3]$_2$-AsMe, 3, 68

$AsC_{13}H_9Br_2N_2O_5$ Me[4,3-Br(O_2N)C_6H_3]$_2$-AsO, 3, 357

$AsC_{13}H_9ClN$ Ph(ClC_6H_4)AsCN isomers, 3, 203

$AsC_{13}H_9Cl_2$ 2,5-Cl_2-5,10-dihydroacridarsine, 3, 596

$AsC_{13}H_9Cl_2O$ 8,10,2-Cl_2(Me)-phenoxarsine, 3, 612

$AsC_{13}H_9N_2$ 10-Cyano-5,10-dihydrophenarsazine, 3, 622

$AsC_{13}H_9N_2O_3S$ 4-(4-$NCC_6H_4SO_2$NH)C_6H_4-AsO, 3, 247

$AsC_{13}H_9N_2O_6S$ 3,4-O_2N(4-$NCC_6H_4SO_3$)-$C_6H_3AsO_3H_2$, 3, 492

$AsC_{13}H_9N_2S$ 10-Thiocyanato-5,10-dihydrophenarsazine, 3, 622

$AsC_{13}H_9O_4$ 4-Oxofluorenearsonic acid isomers, 3, 507, 508

$AsC_{13}H_{10}BrO$ Poly[(5-bromo-2-methoxy-m-phenylene)phenylarsine], 3, 113

$AsC_{13}H_{10}Br_2F_3$ $CF_3Ph_2AsBr_2$, 3, 344

$AsC_{13}H_{10}Cl$ 5-Chloro-5,10-dihydro-acridarsine, 3, 596

$AsC_{13}H_{10}ClFN$ Cl(F)(Me)-5,10-dihydro-phenarsazine isomers, 3, 622

$AsC_{13}H_{10}ClFNO$ Cl(F)(MeO)-5,10-dihydrophenarsazine isomers, 3, 622

$AsC_{13}H_{10}ClO$ 5,3-Cl(MeO)-dibenzarsole, 3, 563
Cl(Me)-phenoxarsine isomers, 3, 612

$AsC_{13}H_{10}ClO_2$ Ph($HO_2CC_6H_4$)AsCl isomers, 3, 173
2-Cl-5,10-dihydro-5-hydroxyacridarsine 5-oxide, 3, 596
10,2-Cl(MeO)-phenoxarsine, 3, 612

$AsC_{13}H_{10}ClO_3$ 2,8,10-Cl(Me)(HO)-phenoxarsine 10-oxide, 3, 612

$AsC_{13}H_{10}F_3$ Ph_2AsCF_3, 3, 51
$AsC_{13}H_{10}I_2NO_4S$ 4-(4-$HO_2CC_6H_4NHO_2S$)-$C_6H_4AsI_2$, 3, 197
$AsC_{13}H_{10}N$ Ph_2AsCN, 3, 203
$AsC_{13}H_{10}NO_2$ 4-$BzNHC_6H_4AsO$, 3, 247
4-(4-$H_2NOCC_6H_4$)C_6H_4AsO, 3, 247
$AsC_{13}H_{10}NO_3$ 4,3-HO(BzNH)C_6H_3AsO, 3, 247
4-(4-NCC_6H_4S)$C_6H_4AsO_3H_2$, 3, 418
$AsC_{13}H_{10}NO_4$ 3,4-O_2N($PhCH_2O$)C_6H_3AsO, 3, 247
Amino-9-oxofluorenearsonic acid isomers, 3, 508
9,10-$(HO)_2$-9-oxoacridine-3-arsonic acid, 3, 524
$AsC_{13}H_{10}NO_5$ (4-$O_2NC_6H_4$)(4-$HO_2CC_6H_4$)-AsOH, 3, 207
$AsC_{13}H_{10}NO_5S$ 4-(4-$NCC_6H_4SO_2$)C_6H_4-AsO_3H_2, 3, 413
$AsC_{13}H_{10}NO_6$ (4-$O_2NC_6H_4$)(4-$HO_2CC_6H_4$)-AsO_2H, 3, 382
$AsC_{13}H_{10}N_3O_2$ 4-(4-$H_2NOCC_6H_4N_2$)C_6H_4AsO, 3, 247
$AsC_{13}H_{10}N_3O_8$ 4-[3,5-$(O_2N)_2C_6H_3CONH$]-$C_6H_4AsO_3H_2$, 3, 437
$AsC_{13}H_{11}$ 5-Methyldibenzarsole, 3, 563
$AsC_{13}H_{11}Br_2$ (4-BrC_6H_4)$_2$AsMe, 3, 51
$AsC_{13}H_{11}Br_2O$ Me(4-BrC_6H_4)$_2$AsO, 3, 357
$AsC_{13}H_{11}ClNO_2$ ($O_2NC_6H_4$)(MeC_6H_4)AsCl isomers, 3, 173
$AsC_{13}H_{11}Cl_2O$ 2-(MeC_6H_4O)$C_6H_4AsCl_2$ isomers, 3, 198
$AsC_{13}H_{11}N_2$ Ph(2-$H_2NC_6H_4$)AsCN, 3, 203
$AsC_{13}H_{11}N_2O_2$ 4-(4-$H_2NC_6H_4CONH$)C_6H_4-AsO, 3, 247
$AsC_{13}H_{11}N_2O_3S$ N-2-Benzothiazolylarsanilic acid, 3, 447
$AsC_{13}H_{11}N_2O_4$ 4-[2-(2-Furyl)imidazol-4-yl]benzenearsonic acid, 3, 470
$AsC_{13}H_{11}N_2O_4S$ 4-[4-H_2NC(=NH)$C_6H_4SO_3$]-C_6H_4AsO, 3, 247
$AsC_{13}H_{11}N_2O_5$ 4-(4-$HO_2CC_6H_4N$=N)C_6H_4-AsO_3H_2, 3, 458
$AsC_{13}H_{11}N_2O_5S$ 4-(4-$NCC_6H_4SO_2NH$)C_6H_4-AsO_3H_2, 3, 444
$AsC_{13}H_{11}N_2O_6$ 4-[4,3-HO(HO_2C)C_6H_3N=N]-$C_6H_4AsO_3H_2$, 3, 458
$AsC_{13}H_{11}N_2O_6S$ MeAs(S)($OC_6H_4NO_2$-p)$_2$, 3, 550
3,4-H_2N(4-$NCC_6H_4SO_3$)$C_6H_3AsO_3H_2$, 3, 492
$AsC_{13}H_{11}N_2O_7$ Me[4,3-HO(O_2N)C_6H_3]$_2$AsO, 3, 357
MeAsO($OC_6H_4NO_2$-p)$_2$, 3, 545
$AsC_{13}H_{11}N_4O_3S$ 4-(2-Benzothiazolyl-NHN=N)$C_6H_4AsO_3H_2$, 3, 468
$AsC_{13}H_{11}O$ 10-Methylphenoxarsine, 3, 612
$AsC_{13}H_{11}O_2$ 4-$MeC_6H_4AsOC_6H_4O$-o, 3, 578
10-Me-phenoxarsine 10-oxide, 3, 612
5-HO-5,10-dihydroacridarsine 5-oxide, 3, 596*
$AsC_{13}H_{11}O_3$ Ph($HO_2CC_6H_4$)AsOH isomers, 3, 207
Ph(3-$OCHC_6H_4$)AsO_2H, 3, 382
$AsC_{13}H_{11}O_4$ (4-HOC_6H_4)(2-$HO_2CC_6H_4$)AsOH, 3, 207
Ph($HO_2CC_6H_4$)AsO_2H isomers, 3, 377, 382, 383
2-Me-2,2'-spirobi-1,3,2-benzodioxarsole, 3, 581
$AsC_{13}H_{11}O_5$ 4-(4-$HO_2CC_6H_4$)$C_6H_4AsO_3H_2$, 3, 508
$AsC_{13}H_{11}O_6$ 4-[OC(CH=$CHCO_2H$)=CHCH=C]-$C_6H_4AsO_3H_2$, 3, 470
$AsC_{13}H_{11}S_2$ 2,5-Ph(Me)-1,3,2-dithiaarsindane, 3, 587
$AsC_{13}H_{12}BrN_6O_4$ N-[(2-Amino-4-hydroxy-6-pteridyl)methyl]-3-bromoarsanilic acid, 3, 504
$AsC_{13}H_{12}Cl$ Ph(MeC_6H_4)AsCl isomers, 3, 174
$AsC_{13}H_{12}ClO_2$ (ClC_6H_4)(MeC_6H_4)AsO_2H isomers, 3, 383
$AsC_{13}H_{12}ClO_3$ 4,2-Cl($PhCH_2$)$C_6H_3AsO_3H_2$, 3, 492
$AsC_{13}H_{12}ClO_4$ 5,2-Cl(4-MeC_6H_4O)C_6H_3-AsO_3H_2, 3, 492
$AsC_{13}H_{12}IO$ Ph(2-$MeOC_6H_4$)AsI, 3, 174
$AsC_{13}H_{12}I_2NO_2S$ 4-($MePhNO_2S$)$C_6H_4AsI_2$, 3, 198
$AsC_{13}H_{12}N$ 10-Methyl-5,10-dihydrophenarsazine, 3, 622
$AsC_{13}H_{12}NO$ [3,4-H_2N($PhCH_2O$)C_6H_3As]$_n$,

3, 141
10-Methoxy-5,10-dihydrophenarsazine, 3, 622

AsC$_{13}$H$_{12}$NO$_3$ 4-PhNHOCC$_6$H$_4$As(OH)$_2$, 3, 224
4-(PhCH=N)C$_6$H$_4$AsO$_3$H$_2$, 3, 442

AsC$_{13}$H$_{12}$NO$_3$S 4-(MePhNO$_2$S)C$_6$H$_4$AsO, 3, 247

AsC$_{13}$H$_{12}$NO$_4$ (O$_2$NC$_6$H$_4$)(MeC$_6$H$_4$)AsO$_2$H isomers, 3, 383
Ph[4,3-Me(O$_2$N)C$_6$H$_3$]AsO$_2$H, 3, 383
(2-HOC$_6$H$_4$CH=N)C$_6$H$_4$AsO$_3$H$_2$ isomers, 3, 422, 442

AsC$_{13}$H$_{12}$NO$_5$ 4-(2-HO$_2$CC$_6$H$_4$NH)C$_6$H$_4$AsO$_3$H$_2$, 3, 442
2,4-HO(2-HOC$_6$H$_4$CH=N)C$_6$H$_3$AsO$_3$H$_2$, 3, 492

AsC$_{13}$H$_{12}$NO$_6$ 3,2-O$_2$N(PhCH$_2$O)C$_6$H$_3$AsO$_3$H$_2$, 3, 492

AsC$_{13}$H$_{12}$NO$_6$S 4-(4-H$_2$NOCC$_6$H$_4$O$_2$S)C$_6$H$_4$AsO$_3$H$_2$, 3, 418
4-(4-HO$_2$CC$_6$H$_4$NHO$_2$S)C$_6$H$_4$As(OH)$_2$, 3, 224

AsC$_{13}$H$_{12}$NO$_7$S 4-(4-HO$_2$CC$_6$H$_4$NHO$_2$S)C$_6$H$_4$AsO$_3$H$_2$, 3, 418

AsC$_{13}$H$_{12}$N$_2$O$_3$S CS[NHC$_6$H$_4$As(OH)]$_2$O, 3, 257

AsC$_{13}$H$_{12}$N$_2$O$_4$ CO[NHC$_6$H$_4$As(OH)]$_2$O, 3, 257

AsC$_{13}$H$_{12}$N$_3$ (3-H$_2$NC$_6$H$_4$)$_2$AsCN, 3, 203

AsC$_{13}$H$_{12}$N$_3$O$_2$ 4-[N=C(Me)CH=C(Me)N=C-NHCO]C$_6$H$_4$AsO, 3, 247

AsC$_{13}$H$_{12}$N$_3$O$_2$S {4-[4-H$_2$NC(=NH)C$_6$H$_4$SO$_2$NH]C$_6$H$_4$As}$_n$, 3, 141

AsC$_{13}$H$_{12}$N$_3$O$_3$S [4-H$_2$NC(=NH)C$_6$H$_4$SO$_2$NH]C$_6$H$_4$AsO, 3, 232

AsC$_{13}$H$_{12}$N$_3$O$_5$ 4-[4,3-H$_2$N(HO$_2$C)C$_6$H$_3$N=N]C$_6$H$_4$AsO$_3$H$_2$, 3, 458

AsC$_{13}$H$_{12}$N$_3$O$_6$ 4-[3,5-H$_2$N(O$_2$N)C$_6$H$_3$CONH]C$_6$H$_4$AsO$_3$H$_2$, 3, 437
5,2-O$_2$N[HOC=CHCH=C(Me)CH=CN=N]C$_6$H$_3$AsO$_3$H$_2$, 3, 493

AsC$_{13}$H$_{12}$N$_3$O$_8$S 4-[HO$_2$C(HO$_3$S)C$_6$H$_3$NHN=N]C$_6$H$_4$AsO$_3$H$_2$ isomers, 3, 468

AsC$_{13}$H$_{13}$ Ph$_2$AsMe, 3, 51

AsC$_{13}$H$_{13}$BrN$_3$O$_3$ 4-[4-BrC$_6$H$_4$NHC(=NH)NH]C$_6$H$_4$AsO$_3$H$_2$, 3, 437

AsC$_{13}$H$_{13}$Br$_2$ MePh$_2$AsBr$_2$, 3, 344

AsC$_{13}$H$_{13}$Br$_4$ MePh$_2$AsBr$_4$, 3, 344

AsC$_{13}$H$_{13}$ClN$_3$O$_3$ 4-[4-ClC$_6$H$_4$NHC(=NH)NH]C$_6$H$_4$AsO$_3$H$_2$, 3, 437

AsC$_{13}$H$_{13}$Cl$_2$ MePh$_2$AsCl$_2$, 3, 344

AsC$_{13}$H$_{13}$I$_2$ MePh$_2$AsI$_2$, 3, 344

AsC$_{13}$H$_{13}$N$_2$O$_2$S [4-(4-H$_2$NC$_6$H$_4$SO$_2$NHCH$_2$)C$_6$H$_4$As]$_n$, 3, 141

AsC$_{13}$H$_{13}$N$_2$O$_3$ 4-(4-MeC$_6$H$_4$N=N)C$_6$H$_4$AsO$_3$H$_2$, 3, 458

AsC$_{13}$H$_{13}$N$_2$O$_3$S {4-[3,4-H$_2$N(MeO)C$_6$H$_3$SO$_2$NH]C$_6$H$_4$As}$_n$, 3, 141
4-(4-H$_2$NC$_6$H$_4$SO$_2$NHCH$_2$)C$_6$H$_4$AsO, 3, 247

AsC$_{13}$H$_{13}$N$_2$O$_4$ 4-(4-H$_2$NC$_6$H$_4$CONH)C$_6$H$_4$AsO$_3$H$_2$, 3, 437
4-[HO(Me)C$_6$H$_3$N=N]C$_6$H$_4$AsO$_3$H$_2$ isomers, 3, 458

AsC$_{13}$H$_{13}$N$_2$O$_5$S 4-[4-H$_2$NC(=NH)C$_6$H$_4$SO$_2$]C$_6$H$_4$AsO$_3$H$_2$, 3, 413

AsC$_{13}$H$_{13}$N$_2$O$_6$ 4-[2,5-HO(O$_2$N)C$_6$H$_3$CH$_2$NH]C$_6$H$_4$AsO$_3$H$_2$, 3, 437
O$_2$N(MeOC$_6$H$_4$NH)C$_6$H$_3$AsO$_3$H$_2$ isomers, 3, 493

AsC$_{13}$H$_{13}$N$_2$O$_8$S 4-[3,4-O$_2$N(MeO)C$_6$H$_3$SO$_2$NH]C$_6$H$_4$AsO$_3$H$_2$, 3, 444

AsC$_{13}$H$_{13}$N$_4$O$_2$S$_4$ MeAs[SC(S)NHN=CH-C=CHCH=CHO]$_2$, 3, 283*

AsC$_{13}$H$_{13}$N$_4$O$_2$S$_6$ MeAs[SC(S)NHNHCO-C=CHCH=CHS]$_2$, 3, 283

AsC$_{13}$H$_{13}$N$_6$O$_4$ 2,4-H$_2$N(HO)-5,8-dihydropyrimido[4,5-b]-pyrazin-6-yl-o-arsanilic acid, 3, 422
N-[(2,4-H$_2$N(HO)-6-pteridyl)methyl]-arsanilic acid, 3, 432

AsC$_{13}$H$_{13}$N$_6$O$_4$S$_2$ 2,3-Dimercaptosuccinic acid cyclic p-[(4,6-diamino-s-triazin-2-yl)amino]benzenearsonite, 3, 586

AsC$_{13}$H$_{13}$O Ph$_2$AsOMe, 3, 211
MePh$_2$AsO, 3, 357

AsC$_{13}$H$_{13}$O$_2$ MeAs(OPh)$_2$, 3, 253
Ph(MeC$_6$H$_4$)AsO$_2$H isomers, 3, 383
Ph(PhCH$_2$)AsO$_2$H, 3, 383

AsC$_{13}$H$_{13}$O$_3$ Ph(MeOC$_6$H$_4$)AsO$_2$H isomers, 3, 383, 384
2-(PhCH$_2$)C$_6$H$_4$AsO$_3$H$_2$, 3, 406
MeC$_6$H$_4$C$_6$H$_4$AsO$_3$H$_2$ isomers, 3, 413

8-MeO-2,3,5,6-tetrahydro-1H,7H-benz[i,j]arsinolizine-1,7-dione, 3, 599

$AsC_{13}H_{13}O_3S$ 4-$PhCH_2SC_6H_4AsO_3H_2$, 3, 418

$AsC_{13}H_{13}O_4$ 2-(4-$MeOC_6H_4$)$C_6H_4AsO_3H_2$, 3, 413

$AsC_{13}H_{13}O_5S$ 4-(4-$MeC_6H_4SO_2$)C_6H_4-AsO_3H_2, 3, 413

$AsC_{13}H_{13}S_2$ MeAs(SPh)$_2$, 3, 268

$AsC_{13}H_{14}ClN_4OS_2$ 4-(2-Amino-4-chloro-6-pyrimidinylamino)dithiobenzenearsonous acid 3-hydroxypropylene ester, 3, 586

$AsC_{13}H_{14}ClO$ $MePh_2As(OH)Cl$, 3, 346

$AsC_{13}H_{14}NO_3$ 2,3,5,6-Tetrahydro-8-methoxy-1H,7H-benz[i,j]arsinolizine-1,7-dione oxime, 3, 598

$AsC_{13}H_{14}NO_4$ $MePh_2As(OH)NO_3$, 3, 348
3,4-$H_2N(PhCH_2O)C_6H_3AsO_3H_2$, 3, 493

$AsC_{13}H_{14}NO_4S$ 4-($MePhNO_2S$)$C_6H_4As(OH)_2$, 3, 224

$AsC_{13}H_{14}NO_5S$ ($MeC_6H_4NHO_2S$)$C_6H_4AsO_3H_2$ isomers, 3, 418
4-($MePhNO_2S$)$C_6H_4AsO_3H_2$, 3, 418

$AsC_{13}H_{14}NO_6S$ ($MeOC_6H_4NHO_2S$)$C_6H_4AsO_3H_2$ isomers, 3, 418
3,4-H_2N(4-$MeC_6H_4SO_3$)$C_6H_3AsO_3H_2$, 3, 493

$AsC_{13}H_{14}N_3O_3$ 4-[PhNHC(=NH)NH]C_6H_4-AsO_3H_2, 3, 438

$AsC_{13}H_{14}N_3O_4$
5,2-H_2N[HOC=CHCH=C(Me)CH=CN=N]-$C_6H_3AsO_3H_2$, 3, 493

$AsC_{13}H_{14}N_3O_5S$ 4-[4-H_2NC(=NH)$C_6H_4SO_2$-NH]$C_6H_4AsO_3H_2$, 3, 444

$AsC_{13}H_{14}N_5O_5S_2$
4-[N=C(NH_2)N=C(NH_2)N=CO]C_6H_4-$As(SCH_2CO_2H)_2$, 3, 268
4-(4-$H_2NO_2SC_6H_4NHCSNH$)$C_6H_4AsO_3H_2$, 3, 438

$AsC_{13}H_{14}N_7O_3$ N-[(2,4-Diamino-6-pteridyl)methyl]arsanilic acid, 3, 432

$AsC_{13}H_{14}N_7O_4$ N-[4-Amino-6-(4,5-dihydro-3-methyl-5-oxopyrazolyl)-2-s-triazinyl]arsanilic acid, 3, 447

$AsC_{13}H_{15}$ MeEt(1-$C_{10}H_7$)As, 3, 76

$AsC_{13}H_{15}ClN_7O_2S_2$
4-[N=C(NH_2)N=C(Cl)N=CNH]C_6H_4As-$(SCH_2CONH_2)_2$, 3, 268

$AsC_{13}H_{15}N_2O$ 3-$MeOC_6H_4As(CH_2CH_2CN)_2$, 3, 68

$AsC_{13}H_{15}N_2O_4$ 2-(4-phenetylamino)pyridine-5-arsonic acid, 3, 524*

$AsC_{13}H_{15}N_2O_5S$ 4-(4-$H_2NC_6H_4SO_2NHCH_2$)-$C_6H_4AsO_3H_2$, 3, 406

$AsC_{13}H_{15}N_4O$ 4-(H_2NCONH)$C_6H_4As(CH_2$-$CH_2CN)_2$, 3, 57

$AsC_{13}H_{15}O_2$ $MePh_2As(OH)_2$, 3, 350

$AsC_{13}H_{15}O_4S_4$ 4-$HO_2CC_6H_4As$[SC(S)OEt]$_2$, 3, 280

$AsC_{13}H_{15}O_6S_2$ 4-($MeCOCH_2O$)$C_6H_4As(SCH_2$-$CO_2H)_2$, 3, 268

$AsC_{13}H_{16}I$ [Me_3(1-$C_{10}H_7$)As]I, 3, 318

$AsC_{13}H_{16}NO_5$ 2-[2-(Ethoxy)ethoxy]quinoline-5-arsonic acid, 3, 524*

$AsC_{13}H_{16}NO_6$ 2,4-Me($MeCH=CHCO_2CH_2CO$-NH)$C_6H_3AsO_3H_2$, 3, 493

$AsC_{13}H_{16}N_3O_3S$ CH=C(Me)N=C(SEt)N=CNH-$C_6H_4AsO_3H_2$ isomers, 3, 425, 447

$AsC_{13}H_{16}N_3O_4S$ HO[N=C(SEt)N=C(Me)CH=C-NH]$C_6H_3AsO_3H_2$ isomers, 3, 493

$AsC_{13}H_{16}N_5O_3$ N-(4,6-Dimethyl-2-pyrimidylamidino)arsanilic acid, 3, 432

$AsC_{13}H_{17}N_2O$ {4-[$CH_2(CH_2)_3NCH_2CH_2CONH$]-$C_6H_4As$}$_n$, 3, 141

$AsC_{13}H_{17}N_2O_2$ 4-[$CH_2(CH_2)_3NCH_2CH_2CONH$]-$C_6H_4AsO$, 3, 247

$AsC_{13}H_{17}N_2O_3$ 4-[HNC(Bu)=NCH=C]C_6H_4-AsO_3H_2, 3, 470

$AsC_{13}H_{17}N_2O_5$ 4-[$CH_2(CH_2)_4NOCCONH$]-$C_6H_4AsO_3H_2$, 3, 429

$AsC_{13}H_{17}N_4O_4S_2$ 4-$NCNHC_6H_4As$[SCH_2CH-(NH_2)CO_2H]$_2$, 3, 268

$AsC_{13}H_{17}O_5$ 3-$MeOC_6H_4As(CH_2CH_2CO_2H)_2$, 3, 57

$AsC_{13}H_{18}ClO$ PhAs[OCH(Bu)CH=CH_2)Cl, 3, 201

$AsC_{13}H_{18}NO_4$ 4-(1-Piperidyl-$COCH_2$)-$C_6H_4AsO_3H_2$, 3, 406
4-(1-Piperidyl-CH_2CO)$C_6H_4AsO_3H_2$, 3, 406

$AsC_{13}H_{18}NO_5$ Ph[$Me_2CHCH(CO_2H)NHCOCH_2$]-

AsO_2H, 3, 384
$AsC_{13}H_{18}NO_6$ 4-$(Me_2CHCH_2CO_2CH_2CONH)$-$C_6H_4AsO_3H_2$, 3, 438
$AsC_{13}H_{18}N_3O_7S_2$ 3,4-$O_2N(MeO)C_6H_3As$-$[SCH_2CH(NH_2)CO_2H]_2$, 3, 268
$AsC_{13}H_{18}N_3O_9$ $[Me_3(AcOCH_2CH_2)As]$-$OC_6H_2(NO_2)_3$, 3, 313
$AsC_{13}H_{18}N_5O_3$
4-$[N{=}C(NHEt)N{=}C(NHEt)N{=}C]C_6H_4$-$AsO_3H_2$, 3, 413
$AsC_{13}H_{18}N_5O_3S$
4-$[N{=}C(NH_2)N{=}C(NEt_2)N{=}CS]C_6H_4$-$AsO_3H_2$, 3, 418
$AsC_{13}H_{19}Cl_2N_8$ 2-[4,6-Bis(dimethylamino)-s-triazin-2-ylamino]-5-dichloroarsino-1,3-phenylenediamine, 3, 187
$AsC_{13}H_{19}N_2O_4$ 4-$[CH_2(CH_2)_3NCH_2CH_2$-$CONH]C_6H_4AsO_3H_2$, 3, 438
4,2-$AcNH[CH_2(CH_2)_4N]C_6H_3AsO_3H_2$, 3, 493
$AsC_{13}H_{19}N_2O_6$ $MeAs(O_2CCH{=}CHCONMe_2)_2$, 3, 256
$MeAs(O_2CCH{=}CHCONMe_2)_2$, 3, 256
$AsC_{13}H_{19}N_6O_3$
4-$[N{=}C(NMe_2)N{=}C(NMe_2)N{=}CNH]C_6H_4$-$AsO_3H_2$, 3, 447
$AsC_{13}H_{19}N_6O_5$
4-$[N{=}C(NHCH_2CH_2OH)N{=}C(NHCH_2CH_2OH)N{=}C$-$NH]C_6H_4AsO_3H_2$, 3, 447
$AsC_{13}H_{19}O_2$ $EtPhAsCH_2CH_2CO_2Et$, 3, 76
$AsC_{13}H_{19}O_3$ $PhAsOCH_2CH(CH_2OBu)O$, 3, 578
$AsC_{13}H_{20}Br$ $[Et_2(CH_2{=}CHCH_2)PhAs]Br$, 3, 318
$AsC_{13}H_{20}O_3P$ $(CH_2{=}CHCH_2)PhAsPO(OEt)_2$, 3, 293
$AsC_{13}H_{21}$ Et(n-C_5H_{11})PhAs, 3, 76
EtBu(4-MeC_6H_4)As, 3, 76
$AsC_{13}H_{21}N_6O$ 4-$(H_2NCONH)C_6H_4As[CH_2$-$CH_2C({=}NH)NH_2]_2$, 3, 57
$AsC_{13}H_{21}O$ PrPhAsOBu, 3, 211
BuPhAsOPr, 3, 211
$AsC_{13}H_{23}ClNO_8$ 3-$ClC_6H_4AsO_3H_2$·methylglucamine salt, 3, 409
4-$ClC_6H_4AsO_3H_2$ methylglucamine salt, 3, 409
$AsC_{13}H_{23}N_2O_{11}$ 4,3-$HO(O_2N)C_6H_3AsO_3H_2$·-methylglucamine salt, 3, 475
$AsC_{13}H_{23}N_2S_4$ $MeAs[SC(S)N(CH_2)_4CH_2]_2$, 3, 283
$AsC_{13}H_{24}I$ $[Me(CH_2{=}CMeCH_2)_3As]I$, 3, 318
$AsC_{13}H_{25}$ (cyclo-$C_6H_{11})_2AsMe$, 3, 68
$AsC_{13}H_{25}O_2$ $MeAs(OC_6H_{11})_2$, 3, 253
$AsC_{13}H_{28}N$ $Bu_2AsN(CH_2)_4CH_2$, 3, 287
$AsC_{13}H_{29}$ $(C_6H_{13})_2AsMe$, 3, 68
$AsC_{13}H_{29}O_2$ $[EtO(CH_2)_3]_2CHAsMe_2$, 3, 57
$AsC_{13}H_{29}S_2$ $MeAs(SC_6H_{13})_2$, 3, 268
$PrAs(SC_5H_{11})_2$, 3, 268
$AsC_{14}H_8Cl_4NO_6$ 4-$[2,3,4,5,6\text{-}Cl_4(HO_2C)$-$C_6CONH]C_6H_4AsO_3H_2$, 3, 438
$AsC_{14}H_9ClN$ 2,5-Cl(NC)-5,10-dihydro-acridarsine, 3, 596
$AsC_{14}H_9Cl_3NO_2$ x-(Cl_3CCOO)-5,10-dihydrophenarsazine isomers, 3, 622
$AsC_{14}H_9F_3N$ 10-$(CF_2{=}CF)$-5,10-dihydro-phenarsazine, 3, 622
$AsC_{14}H_9O_5$ 1-Anthraquinone-AsO_3H_2, 3, 509
$AsC_{14}H_{10}Cl$ $Ph_2AsC{\equiv}CCl$, 3, 57
3,1-Cl(Ph)-arsindole, 3, 556
$AsC_{14}H_{10}Cl_2NO_2$ (Cl_2CHCOO)-5,10-dihydrophenarsazine isomers, 3, 623
$AsC_{14}H_{10}Cl_3O$ 2-(4-$ClCH_2COC_6H_4)C_6H_4$-$AsCl_2$, 3, 198
$AsC_{14}H_{10}N$ 5-Cyano-5,10-dihydroacridarsine, 3, 596
$AsC_{14}H_{10}NO$ Ph(3-$OCHC_6H_4$)AsCN, 3, 204
$AsC_{14}H_{10}NO_5$ 4-(o-$COC_6H_4CON)C_6H_4AsO_3H_2$, 3, 471
$AsC_{14}H_{11}$ $[4\text{-}(PhCH{=}CH)C_6H_4As]_n$, 3, 141
1-Phenylarsindole, 3, 556
$AsC_{14}H_{11}BrNO$ 5,10-Ac(Br)-5,10-dihydrophenarsazine, 3, 623
$AsC_{14}H_{11}ClN$ $(ClC_6H_4)(MeC_6H_4)AsCN$ isomers, 3, 204
$AsC_{14}H_{11}ClNO$ 5,10-Ac(Cl)-5,10-dihydrophenarsazine, 3, 623
$AsC_{14}H_{11}ClNO_2$ x-$(ClCH_2COO)$-5,10-dihydrophenarsazine isomers, 3, 623
$AsC_{14}H_{11}Cl_2$ $(4\text{-}ClC_6H_4)_2AsCH{=}CH_2$, 3, 68
4-$(PhCH{=}CH)C_6H_4AsCl_2$, 3, 187
$AsC_{14}H_{11}Cl_2O$ Ph(3-$ClCH_2COC_6H_4$)AsCl,

3, 174

$AsC_{14}H_{11}INO$ 5,10-Ac(I)-5,10-dihydrophenarsazine, 3, 623

$AsC_{14}H_{11}NO_4$ 2,4-HO(BzCONH)C_6H_4AsO, 3, 247

$AsC_{14}H_{11}N_2O_2$ ($O_2NC_6H_4$)(MeC_6H_4)AsCN isomers, 3, 204

$AsC_{14}H_{11}N_2O_3$ 4-(4-$H_2NCOC_6H_4$CONH)C_6H_4-AsO, 3, 232

$AsC_{14}H_{11}N_2O_4$ 4-(4-NCC_6H_4CONH)C_6H_4-AsO_3H_2, 3, 438

$AsC_{14}H_{11}N_2O_6$ 5,3,7,2,8-HO$(Me)_2(O_2N)_2$-dibenzarsole 5-oxide, 3, 563

$AsC_{14}H_{11}O$ 4-(PhCH=CH)C_6H_4AsO, 3, 248

$AsC_{14}H_{11}O_2S_2$ 4-(5-Methyl-1,3,2-dithiaarsindan-2-yl)benzoic acid, 3, 587*

o-[MeOC(S)S]-phenoxarsine, 3, 612

$AsC_{14}H_{11}O_3$ 4-(PhC≡C)$C_6H_4AsO_3H_2$, 3, 406

$AsC_{14}H_{11}O_3S$ (10-Phenoxarsinylthio)-acetic acid, 3, 613

$AsC_{14}H_{11}O_5$ $(HO_2CC_6H_4)_2$AsOH isomers, 3, 207

$AsC_{14}H_{11}O_6$ (4-$HO_2CC_6H_4$)AsO_2H, 3, 384

$AsC_{14}H_{12}Br_2IN_2O_4$ {Me_2[4,3-Br(O_2N)-C_6H_3]$_2$As}I, 3, 318

$AsC_{14}H_{12}Br_2N_3O_7$ {Me_2[4,3-Br(O_2N)-C_6H_3]$_2$As}NO_3, 3, 318

$AsC_{14}H_{12}Br_3N_2O_4$ {Me_2[4,3-Br(O_2N)-C_6H_3]$_2$As}Br, 3, 318

$AsC_{14}H_{12}Cl$ Ph_2AsCH=CHCl, 3, 58

5,2-Cl(Ph)-isoarsindoline, 3, 557

2-(p-ClC_6H_4)-isoarsindoline, 3, 557

Cl(Me)-dihydroacridarsine isomers, 3, 596, 597

$AsC_{14}H_{12}ClO$ Ph(3-AcC_6H_4)AsCl, 3, 174

Chlorodimethylphenoxarsine isomers, 3, 613

$AsC_{14}H_{12}ClO_2$ Ph($MeO_2CC_6H_4$)AsCl isomers, 3, 174

$AsC_{14}H_{12}ClO_4$ 2-(4-Cl$CH_2COC_6H_4$)C_6H_4-AsO_3H_2, 3, 509

$AsC_{14}H_{12}ClS_2$ ClCH=CHAs$(SPh)_2$, 3, 269

$AsC_{14}H_{12}Cl_3O$ $Ph_2AsOCH_2CCl_3$, 3, 211

$AsC_{14}H_{12}I$ 5,7-Dihydro-6-iodo-6H-dibenz[c,e]arsepin, 3, 642

$AsC_{14}H_{12}N$ Ph(MeC_6H_4)AsCN isomers, 3, 204

$AsC_{14}H_{12}NO_2$ 4-(4-AcNHC_6H_4)C_6H_4AsO, 3, 248

4-($PhCH_2NHOC$)C_6H_4AsO, 3, 248

$AsC_{14}H_{12}NO_3$ 4-(4-$NCC_6H_4CH_2$)$C_6H_4AsO_3H_2$, 3, 413

2,5-AcNH(HO)-dibenzarsole 5-oxide, 3, 563

5,10-Ac(HO)-5,10-dihydrophenarsazine 10 oxide, 3, 623

$AsC_{14}H_{12}NO_4$ 4,3-HO[3,4-HO(MeO)C_6H_3-CH=N]C_6H_3AsO, 3, 248

2,4-HO($PhCH_2O_2CNH$)C_6H_3AsO, 3, 248

$AsC_{14}H_{12}NO_6$ 4-(4-$HO_2CC_6H_4$CONH)C_6H_4-AsO_3H_2, 3, 438

4-[4,3-HO(O_2N)C_6H_3CH=CH]$C_6H_4AsO_3H_2$, 3, 406

2,4-O_2N(4-HOC_6H_4CH=CH)$C_6H_3AsO_3H_2$, 3, 493

$AsC_{14}H_{12}N_3O_3$ N-4-Quinazolinylarsanilic acid, 3, 448

$AsC_{14}H_{12}O_2$ x-(AcO)-5,10-dihydrophenarsazine isomers, 3, 623

$AsC_{14}H_{13}$ Ph_2AsCH=CH_2, 3, 68

1-Phenylarsindoline, 3, 556

2-Phenylisoarsindoline, 3, 558

5,6-Dihydro-5-methylarsanthridin, 3, 598

$AsC_{14}H_{13}ClNO$ Ph(2-AcNHC_6H_4)AsCl, 3, 174

5,2,2-Cl(HO)(Ph)-isoarsindolinium NO_3^-, 3, 557

2,2-(p-ClC_6H_4)(HO)-isoarsindolinium NO_3^-, 3, 558

$AsC_{14}H_{13}ClNO_7S$ 3,4-AcNH(4-Cl$C_6H_4SO_3$)-$C_6H_3AsO_3H_2$, 3, 493

$AsC_{14}H_{13}ClN_5$ {4-[4-ClC_6H_4NHC(=NH)NHC-(=NH)NH]C_6H_4As}$_n$, 3, 141

$AsC_{14}H_{13}Cl_2O$ 2-($Me_2C_6H_3O$)$C_6H_4AsCl_2$ isomers, 3, 198

$AsC_{14}H_{13}INO$ Ph(AcNHC_6H_4)AsI isomers, 3, 174

$AsC_{14}H_{13}N_2$ {4-[PhN=C(Me)NH]C_6H_4As}$_n$, 3, 141

$AsC_{14}H_{13}N_2O_3$ 4-[CH_2N(Ph)N=C]C_6H_4-AsO_3H_2, 3, 471

2,5-H_2NCOCH_2NH(HO)-dibenzarsole 5-oxide, 3, 564

$AsC_{14}H_{13}N_2O_4$ (4-$AcNHC_6H_4$)(4-$O_2NC_6H_4$)-AsOH, 3, 207
4-(4-$H_2NOCC_6H_4NHOC$)$C_6H_4As(OH)_2$, 3, 225

$AsC_{14}H_{13}N_2O_4S$ [3,4-H_2N(4-$AcNHC_6H_4$-SO_3)C_6H_3As]$_n$, 3, 141
[4,3-HO(4-$AcNHC_6H_4SO_2NH$)C_6H_3As]$_n$, 3, 141
4-(4-$AcNHC_6H_4SO_2NH$)C_6H_4AsO, 3, 248

$AsC_{14}H_{13}N_2O_5$ (4-$AcNHC_6H_4$)(4-$O_2NC_6H_4$)-AsO_2H, 3, 384
4-(PhNHOCCONH)$C_6H_4AsO_3H_2$, 3, 429
4-(4-$H_2NCOC_6H_4CONH$)$C_6H_4AsO_3H_2$, 3, 438

$AsC_{14}H_{13}N_2O_5S$ 4,3-HO($AcNHC_6H_4SO_2NH$)-C_6H_3AsO, 3, 248
3,4-H_2N(4-$AcNHC_6H_4SO_3$)C_6H_3AsO, 3, 248

$AsC_{14}H_{13}N_2O_7$ 3,4-O_2N($PhNHCOCH_2O$)-$C_6H_3AsO_3H_2$, 3, 493

$AsC_{14}H_{13}N_2O_8$ [3,4-O_2N(MeO)C_6H_3]$_2$$AsO_2H$, 3, 384

$AsC_{14}H_{13}N_2O_9S$ 3,4-O_2N(4-$AcNHC_6H_4SO_3$)-$C_6H_3AsO_3H_2$, 3, 493

$AsC_{14}H_{13}N_3O_4$ 3,3'-Bis[5,2,3-HO(Me)-(EtO_2C)-pyrryl-4-azo]-p-arsenophenol, 3, 134
3,3'-Dihydroxy-4,4'-bis[5,2,3-HO-(Me)(EtO_2C)-pyrryl-4-azo]arsenobenzene, 3, 134

$AsC_{14}H_{13}O$ 2-Phenylisoarsindoline 2-oxide, 3, 558

$AsC_{14}H_{13}O_2$ MePh($HO_2CC_6H_4$)As isomers, 3, 76
Poly[(4,6-dimethoxy-m-phenylene)-phenylarsine], 3, 113
5,3,7-HO(Me)$_2$-dibenzarsole 5-oxide, 3, 564
5,2-HO(Me)-5,10-dihydroacridarsine oxide, 3, 597
10-Ethoxyphenoxarsine, 3, 612*, 613

$AsC_{14}H_{13}O_2S$ $Ph_2AsSCH_2CO_2H$, 3, 259
MePh($HO_2CC_6H_4$)AsS isomers, 3, 364

$AsC_{14}H_{13}O_3$ MePh(2-$HO_2CC_6H_4$)AsO isomers, 3, 357
$PhCH{=}CHC_6H_4AsO_3H_2$ isomers, 3, 406

$AsC_{14}H_{13}O_3S$ Ph(4-HOC_6H_4)$AsSCH_2CO_2H$, 3, 259

$AsC_{14}H_{13}O_4$ Ph(2-$MeO_2CC_6H_4$)AsO_2H, 3, 384
4-(4-$HOC_6H_4CH{=}CH$)$C_6H_4AsO_3H_2$, 3, 406
4-($BzCH_2$)$C_6H_4AsO_3H_2$, 3, 413
2-(4-AcC_6H_4)$C_6H_4AsO_3H_2$, 3, 509

$AsC_{14}H_{13}S$ 2-Phenylisoarsindoline 2-sulfide, 3, 558

$AsC_{14}H_{14}BrO$ 2,2-HO(Ph)-isoarsindolinium Br^-, 3, 558

$AsC_{14}H_{14}BrO_2$ (4-$MeOC_6H_4$)$_2$AsBr, 3, 175

$AsC_{14}H_{14}Br_2I$ [Me_2(4-BrC_6H_4)$_2$As]I, 3, 318

$AsC_{14}H_{14}Br_2NO_3$ [Me_2(4-BrC_6H_4)$_2$As]NO_3, 3, 318

$AsC_{14}H_{14}Cl$ $Ph_2AsCH_2CH_2Cl$, 3, 68
Ph($PhCH_2CH_2$)AsCl, 3, 174
Et(2-PhC_6H_4)AsCl, 3, 174

$AsC_{14}H_{14}ClN_2O_2$ 4-[4-$ClC_6H_4N{=}C$(Me)NH]-$C_6H_4As(OH)_2$, 3, 225

$AsC_{14}H_{14}ClO$ Ph(4-$EtOC_6H_4$)AsCl, 3, 174

$AsC_{14}H_{14}ClO_2$ 2-(4-Chlorophenyl)isoarsindoline 2,2-dihydroxide, 3, 558

$AsC_{14}H_{14}ClO_3$ MePh($HO_2CC_6H_4$)As(OH)Cl isomers, 3, 246

$AsC_{14}H_{14}N$ 10-Et-5,10-dihydrophenarsazine, 3, 623

$AsC_{14}H_{14}NOS$ 10-($H_2NCH_2CH_2S$)-phenoxarsine, 3, 613
10-($HOCH_2CH_2S$)-5,10-dihydrophenarsazine, 3, 623

$AsC_{14}H_{14}NO_3$ Ph(2-$AcNHC_6H_4$)AsO_2H, 3, 384
4-(4-$H_2NC_6H_4CH{=}CH$)$C_6H_4AsO_3H_2$, 3, 406

$AsC_{14}H_{14}NO_3S$ Ph[3,4-H_2N(HO)C_6H_3]-$AsSCH_2CO_2H$, 3, 259

$AsC_{14}H_{14}NO_4$ 4-($PhNHCOCH_2$)$C_6H_4AsO_3H_2$, 3, 406, 413
($AcNHC_6H_4$)$C_6H_4AsO_3H_2$ isomers, 3, 509
2,2-HO(Ph)-isoarsindolinium NO_3^-, 3, 558
4-(4-$MeOC_6H_4CH{=}N$)$C_6H_4AsO_3H_2$, 3, 442
4-(4-MeC_6H_4CONH)$C_6H_4AsO_3H_2$, 3, 438

$AsC_{14}H_{14}NO_6$ 2,4-HO($PhCH_2OCONH$)C_6H_3-AsO_3H_2, 3, 493

AsC$_{14}$H$_{14}$NO$_6$S 4-(4-AcNHC$_6$H$_4$SO$_2$)C$_6$H$_4$-AsO$_3$H$_2$, 3, 413
4-(4-AcNHC$_6$H$_4$O$_2$S)C$_6$H$_4$AsO$_3$H$_2$, 3, 418

AsC$_{14}$H$_{14}$NO$_7$
4-[MeC=C(CO$_2$H)C(CO$_2$H)=C(Me)N]C$_6$H$_4$-AsO$_3$H$_2$, 3, 471

AsC$_{14}$H$_{14}$NO$_7$S 3,4-AcNH(PhSO$_3$)C$_6$H$_3$-AsO$_3$H$_2$, 3, 493

AsC$_{14}$H$_{14}$N$_3$O$_3$ 4,4′-Bis[5,2,3-HO(Me)-(EtO$_2$C)-pyrryl-4-azo]arsenobenzene, 3, 134

AsC$_{14}$H$_{14}$N$_3$O$_4$ 4-[4-O$_2$NC$_6$H$_4$N=C(Me)NH]-C$_6$H$_4$As(OH)$_2$, 3, 225

AsC$_{14}$H$_{14}$N$_3$O$_6$ 3,4-O$_2$N(AcNHC$_6$H$_4$NH)-C$_6$H$_3$AsO$_3$H$_2$ isomers, 3, 493, 494

AsC$_{14}$H$_{14}$N$_3$O$_8$S 4-(4-O$_2$NC$_6$H$_4$SO$_2$NHCOCH$_2$-NH)C$_6$H$_4$AsO$_3$H$_2$, 3, 438

AsC$_{14}$H$_{14}$O$_2$PS$_2$ 5-Dibenzarsolyl-SP(S)-(OMe)$_2$, 3, 564

AsC$_{14}$H$_{14}$O$_3$PS$_2$ 10-Phenoxarsinyl-SP(S)-(OMe)$_2$, 3, 613

AsC$_{14}$H$_{15}$ 2-PhC$_6$H$_4$AsMe$_2$, 3, 58
Ph$_2$AsEt, 3, 68
MePh(MeC$_6$H$_4$)As isomers, 3, 76
MePh(PhCH$_2$)As, 3, 76
(4-MeC$_6$H$_4$)$_2$AsH, 3, 118

AsC$_{14}$H$_{15}$ClN$_3$O 4-(ClCH$_2$CONH)C$_6$H$_4$-As(CH$_2$CH$_2$CN)$_2$, 3, 58

AsC$_{14}$H$_{15}$ClN$_5$O$_3$ 4-[4-ClC$_6$H$_4$NHC(=NH)-NHC(=NH)NH]C$_6$H$_4$AsO$_3$H$_2$, 3, 438

AsC$_{14}$H$_{15}$Cl$_2$ MePh(3-MeC$_6$H$_4$)AsCl$_2$, 3, 344

AsC$_{14}$H$_{15}$I$_2$O Me$_2$(4-PhOC$_6$H$_4$)AsI$_2$, 3, 344

AsC$_{14}$H$_{15}$N$_2$O$_2$ 4-[PhN=C(Me)NH]C$_6$H$_4$-As(OH)$_2$, 3, 225

AsC$_{14}$H$_{15}$N$_2$O$_3$ (4-H$_2$NC$_6$H$_4$)(4-AcNHC$_6$H$_4$)-AsO$_2$H, 3, 384
4-[4-H$_2$NC(=NH)C$_6$H$_4$CH$_2$]C$_6$H$_4$AsO$_3$H$_2$, 3, 413

AsC$_{14}$H$_{15}$N$_2$O$_3$S 4-(4-MeC$_6$H$_4$NHCSNH)-C$_6$H$_4$AsO$_3$H$_2$, 3, 438

AsC$_{14}$H$_{15}$N$_2$O$_4$S$_3$ 3-[SCH=C(Me)N=CNH]-C$_6$H$_4$As(SCH$_2$CO$_2$H)$_2$, 3, 269

AsC$_{14}$H$_{15}$N$_2$O$_5$ 3,4-H$_2$N(PhNHCOCH$_2$O)-C$_6$H$_3$AsO$_3$H$_2$, 3, 494

AsC$_{14}$H$_{15}$N$_2$O$_6$ 3,4-O$_2$N(4-EtOC$_6$H$_4$NH)-C$_6$H$_3$AsO$_3$H$_2$, 3, 494

AsC$_{14}$H$_{15}$N$_2$O$_6$S 4-(4-AcNHC$_6$H$_4$SO$_2$NH)-C$_6$H$_4$AsO$_3$H$_2$, 3, 444

AsC$_{14}$H$_{15}$N$_2$O$_7$S 4,3-HO(4-AcNHC$_6$H$_4$SO$_2$NH)-C$_6$H$_3$AsO$_3$H$_2$, 3, 494
3,4-H$_2$N(4-AcNHC$_6$H$_4$SO$_3$)C$_6$H$_3$AsO$_3$H$_2$, 3, 494

AsC$_{14}$H$_{15}$N$_3$O$_5$ 4-[4,3-HO(AcNH)C$_6$H$_4$N=N]-C$_6$H$_4$AsO$_3$H$_2$, 3, 458

AsC$_{14}$H$_{15}$N$_4$O$_4$ (3-H$_2$NCONHC$_6$H$_4$)$_2$AsO$_2$H, 3, 384

AsC$_{14}$H$_{15}$N$_4$O$_4$S$_2$ 4-[N=C(NH$_2$)N=CHCH=CNH]-C$_6$H$_4$As(SCH$_2$CO$_2$H)$_2$, 3, 269

AsC$_{14}$H$_{15}$O 2-PhOC$_6$H$_4$AsMe$_2$, 3, 68
4-PhOC$_6$H$_4$AsMe$_2$, 3, 68
Ph$_2$AsCH$_2$CH$_2$OH, 3, 68
Ph$_2$AsOEt, 3, 211

AsC$_{14}$H$_{15}$O$_2$ (MeC$_6$H$_4$)$_2$AsO$_2$H isomers, 3, 384
(PhCH$_2$)$_2$AsO$_2$H, 3, 384
Ph(PhCH$_2$CH$_2$)AsO$_2$H, 3, 384
Et(2-PhC$_6$H$_4$)AsO$_2$H, 3, 384
Ph(2,4-Me$_2$C$_6$H$_3$)AsO$_2$H, 3, 384
2,2-HO(Ph)-isoarsindolinium OH$^-$, 3, 558

AsC$_{14}$H$_{15}$O$_2$S$_4$ PhAs[SC(S)OCH$_2$CH=CH$_2$]$_2$, 3, 280

AsC$_{14}$H$_{15}$O$_3$ Ph(EtOC$_6$H$_4$)AsO$_2$H isomers, 3, 384, 385
[(Me$_2$C$_6$H$_3$)C$_6$H$_4$]AsO$_3$H$_2$ isomers, 3, 413, 414
4,2-Me(PhCH$_2$)C$_6$H$_3$AsO$_3$H$_2$, 3, 494
4,2-Me(4-MeC$_6$H$_4$)C$_6$H$_3$AsO$_3$H$_2$, 3, 509
1-C$_{10}$H$_7$AsOCH$_2$CH(CH$_2$OMe)O, 3, 578

AsC$_{14}$H$_{15}$O$_4$ (4-MeOC$_6$H$_4$)$_2$AsO$_2$H, 3, 385

AsC$_{14}$H$_{16}$I [Me$_2$Ph$_2$As]I, 3, 318

AsC$_{14}$H$_{16}$N Ph$_2$AsNMe$_2$, 3, 287

AsC$_{14}$H$_{16}$NO$_3$ [Me$_2$Ph$_2$As]NO$_3$, 3, 318

AsC$_{14}$H$_{16}$NO$_9$S$_2$ 3,4-H$_2$N(HO)C$_6$H$_3$As-[SCH(CO$_2$H)CH$_2$CO$_2$H]$_2$, 3, 269

AsC$_{14}$H$_{16}$N$_3$O 4-AcNHC$_6$H$_4$As(CH$_2$CH$_2$CN)$_2$, 3, 58

AsC$_{14}$H$_{16}$N$_3$O$_3$ 4-[4-MeC$_6$H$_4$NHC(=NH)NH]-C$_6$H$_4$AsO$_3$H$_2$, 3, 438
4-(4-Me$_2$NC$_6$H$_4$N=N)C$_6$H$_4$AsO$_3$H$_2$, 3, 458
4-(4-EtNHC$_6$H$_4$N=N)C$_6$H$_4$AsO$_3$H$_2$, 3, 459

AsC$_{14}$H$_{16}$N$_3$O$_4$ 3,4-H$_2$N(AcNHC$_6$H$_4$NH)-

$C_6H_3AsO_3H_2$ isomers, 3, 494
$AsC_{14}H_{16}N_3O_5$ 3,4-O_2N(4-$Me_2NC_6H_4NH$)-$C_6H_3AsO_3H_2$, 3, 494
$AsC_{14}H_{16}N_3O_6$
4-[EtO_2CC=C(Me)NHC(OH)=CN=N]C_6H_4-AsO_3H_2, 3, 459
$AsC_{14}H_{16}N_3O_6S$ 4-(4-$H_2NC_6H_4SO_2NHCO$-CH_2NH)$C_6H_4AsO_3H_2$, 3, 438
$AsC_{14}H_{16}N_3O_7$
HO[EtO_2CC=C(Me)NHC(OH)=CN=N]C_6H_3-AsO_3H_2 isomers, 3, 494
$AsC_{14}H_{16}N_5O_7$ [$Me_2(NCCH_2CH_2)_2As$]-$OC_6H_2(NO_2)_3$, 3, 313
$AsC_{14}H_{17}N_2O_3$ 4-(4-$Me_2NC_6H_4NH$)C_6H_4-AsO_3H_2, 3, 442
$AsC_{14}H_{17}N_2O_4$ [3,4-$H_2N(MeO)C_6H_3$]$_2$-AsO_2H, 3, 385
$AsC_{14}H_{17}N_4O_3S$ 2-[2-(Thiomorpholino)-5-pyridylamino]-5-pyridinearsonic acid, 3, 524
$AsC_{14}H_{17}N_4O_4$ 2-(4-Morpholino-5-pyridylamino)-5-pyridinearsonic acid, 3, 525
$AsC_{14}H_{17}N_6O_5S$ 4-[4-H_2NC(=NH)NHO_2S-C_6H_4NHC(=NH)NH]$C_6H_4AsO_3H_2$, 3, 438
$AsC_{14}H_{17}O_3$ O=$CCH_2CH_2As(Ph)CH_2CHCO_2Et$, 3, 590
$AsC_{14}H_{17}O_6$ 2,2,7,7-Tetramethyl-3,8-dioxo-5-phenyl-1,4,6,9-tetraoxa-5-arsaspiro[4,4]nonane, 3, 580
$AsC_{14}H_{18}NO_4$ 4-($CH_2CMe_2CH_2COCH$=CNH)-$C_6H_4AsO_3H_2$, 3, 443
$AsC_{14}H_{18}NO_5$ 2-[5,2-Me($EtOCH_2CH_2O$]-5-quinolinearsonic acid, 3, 525
$AsC_{14}H_{18}N_2O_2PS$ $Me_2As(O)OP(S)(NHPh)_2$, 3, 388
$AsC_{14}H_{19}$ (CH_2=$CMeCH_2)_2AsPh$, 3, 68
$AsC_{14}H_{19}Cl_2O_4$ [$HO_2C(CH_2)_3$]$_2PhAsCl_2$, 3, 345
$AsC_{14}H_{19}N_2O_4$
[$CH_2CH_2C(Me)_2CH_2C(OH)$=CN=N]C_6H_4-AsO_3H_2 isomers, 3, 459
$AsC_{14}H_{19}N_6OS_2$
4-[N=C(NH_2)N=C(NH_2)N=CNH]C_6H_4-$AsSCH_2CH(CH_2OEt)S$, 3, 587
$AsC_{14}H_{19}O_2$ PhAs($OCHMeCH$=$CH_2)_2$, 3, 253
$AsC_{14}H_{19}O_2S_4$ PhAs[SC(S)OPr]$_2$, 3, 280
PhAs[SC(S)$OCHMe_2$]$_2$, 3, 280
$AsC_{14}H_{19}O_4$ ($EtO_2CCH_2)_2AsPh$, 3, 68
PhAs[$(CH_2)_3CO_2H$]$_2$, 3, 68
$AsC_{14}H_{19}O_5$ Ph[$HO_2C(CH_2)_3$]$_2AsO$, 3, 357*
$AsC_{14}H_{20}ClO$ PhAs[OCH(C_5H_{11})CH=CH_2]Cl, 3, 201
$AsC_{14}H_{20}IO_5$ [Me($HO_2CCH_2CH_2)_2$(3-MeO-C_6H_4)As]I, 3, 319
$AsC_{14}H_{20}NO_4$ 4-(1-Piperidyl-$COCH_2CH_2$)-$C_6H_4AsO_3H_2$, 3, 407
$AsC_{14}H_{20}NO_6S_2$ 3,4-$H_2N(HOCH_2CH_2O)C_6H_3$-As($SCH_2CO_2Me)_2$, 3, 269
$AsC_{14}H_{20}NO_9$ N-(4-Arsonophenyl)glycine ester with 1,5-mannitan, 3, 433
$AsC_{14}H_{20}N_3O_7$ [$Et_3(CH_2$=CH)As]OC_6H_2-$(NO_2)_3$, 3, 313
$AsC_{14}H_{21}BrN$ {Me_2[NC$(CH_2)_5$]PhAs}Br, 3, 319
$AsC_{14}H_{21}N_2O_5$ 4,2-EtOCONH[$CH_2(CH_2)_4N$]-$C_6H_3AsO_3H_2$, 3, 494
$AsC_{14}H_{21}N_2O_6$ {3,4,5-AcNH(HO)[($HOCH_2$-$CHOHCH_2)_2N$]C_6H_2As}$_n$, 3, 142
$AsC_{14}H_{21}O_2$ PhAs(OEt)(OC_6H_{11}), 3, 253
$AsC_{14}H_{21}O_3$ 4-$MeC_6H_4AsOCH_2CH(CH_2OBu)O$, 3, 578
$AsC_{14}H_{21}O_4$ 5,2,3,7,8-Ph$(Me)_4$-1,4,6,9-tetraoxa-5-arsaspiro[4,4]nonane, 3, 580
$AsC_{14}H_{21}O_6$ 4-($PrCO_2CH_2CH_2OCH_2CH_2$)-$C_6H_4AsO_3H_2$, 3, 414
$AsC_{14}H_{22}Br$ [Et(CH_2=$CHCH_2$)PrPhAs]Br, 3, 319
$AsC_{14}H_{22}IO_2$ [$CH_2CH_2C(OMe)_2CH_2CH_2As$-MePh]I, 3, 590
$AsC_{14}H_{22}N_5O$ 4-$AcNHC_6H_4As$[CH_2CH_2C-(NH)NH_2]$_2$, 3, 58
$AsC_{14}H_{23}$ Bu_2AsPh, 3, 58
Et(n-C_5H_{11})(4-MeC_6H_4)As, 3, 77
Et(n-C_6H_{13})PhAs, 3, 77
$AsC_{14}H_{23}BrP$ [$CH_2CH_2As(Et)C_6H_4PEt_2$]Br, 3, 639
$AsC_{14}H_{23}ClNO_6P_2S_4$ 2,4-Cl(O_2N)C_6H_3As-[SP(S)$(OEt)_2$]$_2$, 3, 281
$AsC_{14}H_{23}N_2O_9$ 3,4,5-AcNH(HO)[($HOCH_2$-$CHOHCH_2)_2N$]$C_6H_2AsO_3H_2$, 3, 503
$AsC_{14}H_{23}N_2O_{10}$ 4,3-HO[$HOCH_2(CHOH)_4CH_2$-$NHCOCH_2NH$]$C_6H_3AsO_3H_2$, 3, 494

$AsC_{14}H_{23}O$ BuPhAsOBu, 3, 211
$AsC_{14}H_{23}O_2$ $PhAs(OBu)_2$, 3, 253
$PhAs(OCH_2CHMe_2)_2$, 3, 253
$AsC_{14}H_{23}O_6$ 2,2,7,7-Tetramethyl-3,8-dioxo-5-phenyl-1,4,6,9-tetraoxa-5-arsaspiro[4,4]nonane, 3, 580
$AsC_{14}H_{23}S_2$ $PhAs(SBu)_2$, 3, 269
$AsC_{14}H_{24}Br_2P$ $[CH_2CH_2As(Me)_2C_6H_4PEt_2]Br_2$, 3, 639
$AsC_{14}H_{24}ClN_6O_7$ $4\text{-}ClC_6H_4As[CH_2CH_2C(=NH)NH_3 \cdot NO_3]_2 \cdot EtOH$, 3, 50
$AsC_{14}H_{24}ClO_2$ $ClCH{=}CHAs(OC_6H_{11})_2$, 3, 253
$AsC_{14}H_{24}NO_9P_2S_2$ $4\text{-}O_2NC_6H_4AsO[OP(S)(OEt)_2]_2$, 3, 550
$AsC_{14}H_{24}N_3O_5$ $4\text{-}(Et_2NCH_2CHOHCH_2NHCONH)C_6H_4AsO_3H_2$, 3, 438
$AsC_{14}H_{24}N_7O_7$ $4\text{-}AcNHC_6H_4As[CH_2CH_2C(=NH)NH_3 \cdot NO_3]_2$, 3, 58
$AsC_{14}H_{24}O_3P$ $EtPhAsPO(OPr)_2$, 3, 293
$BuPhAsPO(OEt)_2$, 3, 293
$AsC_{14}H_{24}P$ $2\text{-}(Et_2As)C_6H_4PEt_2$, 3, 59
$AsC_{14}H_{25}N_2O_6$ $3,4\text{-}AcNH(HO)C_6H_3AsO_3H_2 \cdot Et_2NCH_2CH_2OH$, 3, 477
$AsC_{14}H_{25}O_4P_2S_4$ $PhAs[SP(S)(OEt)_2]_2$, 3, 281
$AsC_{14}H_{27}N_6O_5$ $3,4\text{-}O_2N[H_2N(CH_2CH_2NH)_4]C_6H_3AsO_3H_2$, 3, 494
$AsC_{14}H_{27}O_2$ $EtAs(OC_6H_{11})_2$, 3, 254
$AsC_{14}H_{28}I$ $[Me_2(cyclo\text{-}C_6H_{11})_2As]I$, 3, 319
$AsC_{14}H_{28}NO_3$ $[Me_2(cyclo\text{-}C_6H_{11})_2As]NO_3$, 3, 319
$AsC_{14}H_{31}$ $n\text{-}C_{12}H_{25}AsMe_2$, 3, 68
$AsC_{14}H_{31}O$ $Me_2(n\text{-}C_{12}H_{25})AsO$, 3, 357
$AsC_{14}H_{31}O_2$ $EtAs(OC_6H_{13})_2$, 3, 254
$AsC_{14}H_{31}S_2$ $EtAs(SC_6H_{13})_2$, 3, 269
$BuAs(SC_5H_{11})_2$, 3, 269
$AsC_{14}H_{32}I$ $[Et_3(n\text{-}C_8H_{17})As]I$, 3, 319
$AsC_{14}H_{32}O_3P$ $EtBuAsPO(OBu)_2$, 3, 293
$Et(Me_2CHCH_2)AsPO(OBu)_2$, 3, 293
$AsC_{14}H_{33}N_2S_2$ $EtAs(SCH_2CH_2NEt_2)_2$, 3, 269
$AsC_{14}H_{33}O_4$ $Et_3As(OOCMe_3)_2$, 3, 361
$AsC_{14}H_{35}Cl_2N_2S_2$ $[EtAs(SCH_2CH_2NHEt_2)_2]Cl_2$, 3, 269
$AsC_{15}H_2Cl_6N_3S_5$ SNCAs[S-2-(4,6,7-trichlorobenzothiazole)]$_2$, 3, 269
$AsC_{15}H_3Cl_8N_2S_4$ MeAs[S-2-(4,5,6,7-tetrachlorobenzothiazole)]$_2$, 3, 269
$AsC_{15}H_3Cl_{10}O_2S_2$ $MeAs(SCOC_6Cl_5)_2$, 3, 279
$AsC_{15}H_8N_3S_4$ NCAs(S-2-benzothiazole)$_2$, 3, 270, 306
$AsC_{15}H_8N_3S_5$ SNCAs(S-2-benzothiazole)$_2$, 3, 270
$AsC_{15}H_{10}ClN$ 12-Chloro-7,12-dihydropyrido[3,2-a]phenarsazine, 3, 639
$AsC_{15}H_{10}F_7$ $Ph_2AsCF_2CF_2CF_3$, 3, 68
$AsC_{15}H_{10}N$ 6,6'-Arsenobis(2-phenylquinoline), 3, 134
$AsC_{15}H_{11}Cl_2N_4O$ $Ph[4\text{-}(N{=}CClN{=}CClN{=}CNH)C_6H_4]AsOH$, 3, 208
$AsC_{15}H_{11}Cl_2N_4O_2$ $Ph[4\text{-}(N{=}CClN{=}CClN{=}CNH)C_6H_4]AsO_2H$, 3, 385
$AsC_{15}H_{11}Cl_2O_2S_2$ $MeAs(SCOC_6H_4Cl)_2$, 3, 279
$AsC_{15}H_{11}N_2S_4$ MeAs(S-2-benzothiazole)$_2$, 3, 270
$AsC_{15}H_{11}O_5$ 4-(2-Oxo-1,2-benzopyran-3-yl)-$C_6H_4AsO_3H_2$, 3, 471
$AsC_{15}H_{12}BrN_2O_3$ N-(4-Bromo-1-isoquinolyl)-m-arsanilic acid, 3, 425
N-(4-Bromo-1-isoquinolyl)-p-arsanilic acid, 3, 448
$AsC_{15}H_{12}BrN_4O_3$ (Aminobromoisoquinolylazo)benzenearsonic acid isomers, 3, 459
$AsC_{15}H_{12}ClN_2O_4S$
$4\text{-}[SC(=NC_6H_4Cl\text{-}2)NHCOCH]C_6H_4AsO_3H_2$, 3, 471
$4\text{-}[SC(=NC_6H_4Cl\text{-}4)NHCOCH]C_6H_4AsO_3H_2$, 3, 471
$AsC_{15}H_{12}ClN_4O_2$
$Ph\{4\text{-}[N{=}CClN{=}C(OH)N{=}CNH]C_6H_4\}AsOH$, 3, 208
$AsC_{15}H_{12}ClN_4O_3$
$Ph\{4\text{-}[N{=}CClN{=}C(OH)N{=}CNH]C_6H_4\}AsO_2H$, 3, 385
$4\text{-}[5,1\text{-}H_2N(Cl)\text{-}8\text{-}isoquinolylazo]C_6H_4AsO_3H_2$, 3, 459
$AsC_{15}H_{12}ClN_4O_6S$ $N\text{-}[4,6\text{-}Cl(HO_3SC_6H_4)\text{-}s\text{-}triazin\text{-}2\text{-}yl]NHC_6H_4AsO_3H_2$, 3, 448
$AsC_{15}H_{12}NO$ $Ph(3\text{-}AcC_6H_4)AsCN$, 3, 204

$AsC_{15}H_{12}NO_2$ Ph($MeO_2CC_6H_4$)AsCN isomers, 3, 204
$AsC_{15}H_{12}NO_3$ 4-[PhCH=C(CN)]$C_6H_4AsO_3H_2$, 3, 407
2-Ph-6-quinolinearsonic acid, 3, 525
$AsC_{15}H_{12}NO_4$ 8,2-HO(Ph)-5-quinolinearsonic acid, 3, 525
$AsC_{15}H_{12}NO_5$ 7-AcNH-9-oxo-4-fluorenearsonic acid, 3, 509
$AsC_{15}H_{12}N_3$ Tri-2-pyridylarsine, 3, 17
$AsC_{15}H_{12}N_3O$ (N=CHCH=CHCH=C)$_3$AsO, 3, 357
$AsC_{15}H_{12}N_3O_4$ 4-(8-Hydroxy-5-quinolylazo)-$C_6H_4AsO_3H_2$, 3, 459*
$AsC_{15}H_{12}N_3O_5$ N-(5-O_2N-4-isoquinolyl)-arsanilic acid, 3, 448
N-(8-O_2N-4-quinolyl)arsanilic acid, 3, 448
N-(5-O_2N-1-isoquinolyl)-o-arsanilic acid, 3, 423
N-(5-O_2N-1-isoquinolyl)-m-arsanilic acid, 3, 425
4-[HNC($C_6H_4NO_2$-6)=NCH=C]$C_6H_4AsO_3H_2$, 3, 471
$AsC_{15}H_{12}N_3O_6S$
4-[SC(=$NC_6H_4NO_2$-4)NHCOCH]$C_6H_4AsO_3H_2$, 3, 471
4-[SC(=$NC_6H_4NO_2$-3)NHCOCH]$C_6H_4AsO_3H_2$, 3, 471
$AsC_{15}H_{12}N_5O_5$ 2,4-O_2N(5-H_2N-8-isoquinolylazo)-$C_6H_3AsO_3H_2$, 3, 495
$AsC_{15}H_{13}ClI$ 3,1,1-Cl(Me)(Ph)arsindolinium I^-, 3, 556
$AsC_{15}H_{13}ClNO_6$ 4,3-Cl($BzOCH_2CONH$)C_6H_3-AsO_3H_2, 3, 495
$AsC_{15}H_{13}ClN_5O_2$
Ph{4-[N=C(NH_2)N=CClN=CNH]C_6H_4}-AsO_2H, 3, 385
$AsC_{15}H_{13}N_2O_3$ 4-(4-$AcNHC_6H_4$CONH)-C_6H_4AsO, 3, 248
4-(4-$AcNHC_6H_4$NHCO)C_6H_4AsO, 3, 248
N-(1-Isoquinolyl)-o-arsanilic acid, 3, 423
N-(1-Isoquinolyl)-p-arsanilic acid, 3, 448
N-(1-Isoquinolyl)-m-arsanilic acid, 3, 425
4-[HNC(Ph)=NCH=C]$C_6H_4AsO_3H_2$, 3, 471
$AsC_{15}H_{13}N_2O_4$ 4,3-HO(1-isoquinolylamino)-$C_6H_3AsO_3H_2$, 3, 495
$AsC_{15}H_{13}N_2O_4S$ 4-[SC(=NPh)NHCOCH]C_6H_4-AsO_3H_2, 3, 471
$AsC_{15}H_{13}N_2O_5$ 7-(H_2NCOCH_2NH)-9-oxofluorenearsonic acid isomers, 3, 509
7-(H_2NCOCH_2NH)-9-oxofluorenearsonic acid isomers, 3, 509
$AsC_{15}H_{13}N_2O_7S$ 3,4-AcNH(4-$NCC_6H_4SO_3$)-$C_6H_3AsO_3H_2$, 3, 495
$AsC_{15}H_{13}N_4O_3$ (5-Amino-8-isoquinolylazo)-$C_6H_4AsO_3H_2$ isomers, 3, 459
$AsC_{15}H_{13}O_2$ Ph_2AsO_2CCH=CH_2, 3, 212
$AsC_{15}H_{13}O_2S_2$ MeAs($SCOPh$)$_2$, 3, 279
10-[EtOC(S)S]-phenoxarsine, 3, 613
$AsC_{15}H_{13}O_3S$ 10-Phenoxarsinyl-SCH_2CH_2-CO_2H, 3, 613
$AsC_{15}H_{14}ClO$ $Ph_2AsCH_2CH_2COCl$, 3, 69
$AsC_{15}H_{14}N$ $Ph_2AsCH_2CH_2CN$, 3, 61
(MeC_6H_4)$_2$AsCN isomers, 3, 204
Ph(2,4-$Me_2C_6H_3$)AsCN, 3, 204
$AsC_{15}H_{14}NO$ 5,10-Ac(Me)-5,10-dihydrophenarsazine, 3, 623
$AsC_{15}H_{14}NOS_2$ 10-Phenoxarsinyl-SC(S)-NMe_2, 3, 613
10-[EtOC(S)S]-5,10-dihydrophenarsazine, 3, 624
$AsC_{15}H_{14}NO_3$ 4-(PhCH=CHCH=N)C_6H_4-AsO_3H_2, 3, 442
$AsC_{15}H_{14}NO_6$ 4-($PhCO_2CH_2CONH$)C_6H_4-AsO_3H_2, 3, 438
$AsC_{15}H_{14}NS_2$ S-(Dibenzarsol-5-yl)-N,N-dimethyldithiourethan, 3, 564
$AsC_{15}H_{14}N_3O_2$ 4-[4-NCC_6H_4N=C(Me)NH]-C_6H_4As(OH)$_2$, 3, 225
$AsC_{15}H_{14}N_3O_3$ 4-(4-$HOCH_2CH_2NHOCC_6H_4N_2$)-C_6H_4AsO, 3, 248
$AsC_{15}H_{14}N_3O_4$ 4-[PhNHCOCH(CN)NH]C_6H_4-AsO_3H_2, 3, 438
$AsC_{15}H_{14}N_7O_4S$ 2,2'-Bis[2-(4-amino-6-p-sulfoanilino-s-triazin-2-yl)-hydrazo]-p-arsenophenol, 3, 135
$AsC_{15}H_{15}$ (C_5H_5)$_3$As, 3, 17
2-p-Tolylisoarsindoline, 3, 558

1,2,3,4-Tetrahydro-2-phenylisoarsinoline, **3**, 595

$AsC_{15}H_{15}ClI$ 2,2-Me(4-ClC_6H_4)-isoarsindolinium I^-, **3**, 558

$AsC_{15}H_{15}ClN$ 10,5-Cl(Pr)-5,10-dihydrophenarsazine, **3**, 623

$AsC_{15}H_{15}Cl_2$ 1,2,3,4-Tetrahydro-2-phenylisoarsinoline 2,2-dichloride, **3**, 595

$AsC_{15}H_{15}N_2$ ($NCCH_2CH_2$)Ph(2-$H_2NC_6H_4$)As, **3**, 77

$AsC_{15}H_{15}N_2O$ {4-[4-$MeOC_6H_4N$=C(Me)NH]-C_6H_4As}$_n$, **3**, 142

$AsC_{15}H_{15}N_2O_2$ 4-[BzNHCH(NH_2)CH_2]C_6H_4-AsO, **3**, 248

$AsC_{15}H_{15}N_2O_3S$ [4-(4-$AcNHC_6H_4SO_2NH$-CH_2)C_6H_4As]$_n$, **3**, 142

$AsC_{15}H_{15}N_2O_4S$ 4-(4-$AcNHC_6H_4SO_2NHCH_2$)-C_6H_4AsO, **3**, 248*

$AsC_{15}H_{15}N_2O_5$ 4-(4-$AcNHC_6H_4CONH$)C_6H_4-AsO_3H_2, **3**, 438

$AsC_{15}H_{15}N_4$ {4-[H_2NC(=NH)C_6H_4N=C(Me)-NH]C_6H_4As}$_n$, **3**, 142

$AsC_{15}H_{15}N_6O_5$ 4-{N-[(2-Amino-4-hydroxy-6-pteridyl)methyl]amino}-phenylarsinicoacetic acid, **3**, 377

$AsC_{15}H_{15}N_6O_6$ N-[(2-Amino-6-hydroxy-6-pteridyl)methyl]-N-(carboxymethyl)arsanilic acid, **3**, 433

$AsC_{15}H_{15}O$ 2-p-Tolylisoarsindoline 2-oxide, **3**, 559

2-(4-$MeOC_6H_4$)-isoarsindoline, **3**, 559

$AsC_{15}H_{15}OS_2$ Ph_2AsSC(S)OEt, **3**, 261

$AsC_{15}H_{15}O_2$ $Ph_2AsCH_2CH_2CO_2H$, **3**, 61

EtPh(4-$HO_2CC_6H_4$)As, **3**, 77

2-(4-$MeOC_6H_4$)-isoarsindoline 2-oxide, **3**, 559

$AsC_{15}H_{15}O_2S$ Ph_2($HO_2CCH_2CH_2$)AsS, **3**, 365

$AsC_{15}H_{15}O_3$ Ph_2($HO_2CCH_2CH_2$)AsO, **3**, 357

4-(4-MeC_6H_4CH=CH)$C_6H_4AsO_3H_2$, **3**, 407

EtPh(4-$HO_2CC_6H_4$)AsO, **3**, 358

$AsC_{15}H_{15}S$ 1,2,3,4-Tetrahydro-2-phenylisoarsinoline 2-sulfide, **3**, 595*

$AsC_{15}H_{16}I$ 2,2-Me(Ph)-isoarsindolinium I^-, **3**, 558

5,5-Me_2-5,6-dihydroarsanthridinium I^-, **3**, 598

$AsC_{15}H_{16}NO_2S$ Ph_2AsSCH_2CH(NH_2)CO_2H, **3**, 259

$AsC_{15}H_{16}NO_3$ 4-(PhMeCHN=CH)$C_6H_4AsO_3H_2$, **3**, 407

$AsC_{15}H_{16}NO_3S$ Ph(4-HOC_6H_4)$AsSCH_2CH$-(NH_2)CO_2H, **3**, 259

$AsC_{15}H_{16}NO_4$ 4-($PhNHCOCH_2CH_2$)C_6H_4-AsO_3H_2, **3**, 407

2,2-HO(p-MeC_6H_4)-isoarsindolinium NO_3^-, **3**, 559

2,2-HO(Ph)-1,2,3,4-tetrahydroisoarsinolinium NO_3^-, **3**, 595

$AsC_{15}H_{16}NO_5$ 2-(4-$EtCO_2NHC_6H_4$)C_6H_4-AsO_3H_2, **3**, 510

2,2-HO(p-$MeOC_6H_4$)-isoarsindolinium NO_3^-, **3**, 559

$AsC_{15}H_{16}NO_7S$ 3,4-AcNH(4-$MeC_6H_4SO_3$)-$C_6H_3AsO_3H_2$, **3**, 495

$AsC_{15}H_{16}N_2O_4$ CO[$NHCH_2C_6H_4As$(OH)]$_2$O, **3**, 257

$AsC_{15}H_{16}N_3O_4S_2$ 4-[CH=CHC(NH_2)=CHN=CNH]-C_6H_4As(SCH_2CO_2H)$_2$, **3**, 270

$AsC_{15}H_{16}N_3O_7$ [Me_3PhAs]OC_6H_2(NO_2)$_3$, **3**, 314

$AsC_{15}H_{16}N_3O_7S$ 3,4-AcNH[4-H_2NC(=NH)-$C_6H_4SO_3$]$C_6H_3AsO_3H_2$, **3**, 495

$AsC_{15}H_{16}N_5$ {4-[4-MeC_6H_4NHC(=NH)NHC-(=NH)NH]C_6H_4As}$_n$, **3**, 142

$AsC_{15}H_{16}N_5O$ {4-[4-$MeOC_6H_4NHC$(=NH)NHC-(=NH)NH]C_6H_4As}$_n$, **3**, 142

$AsC_{15}H_{16}N_7O_5$ N-[(2-Amino-4-hydroxy-6-pteridyl)methyl]-N-(carbamoylmethyl)arsanilic acid, **3**, 433

$AsC_{15}H_{16}N_7O_7S$ 4-Hydroxy-3-{2-[4-amino-6-(4-sulfoanilino)-s-triazin-2-yl]hydrazo}-$C_6H_3AsO_3H_2$, **3**, 495

$AsC_{15}H_{17}$ 2-$PhC_6H_4CH_2AsMe_2$, **3**, 61

Ph_2AsPr, **3**, 69

(4-MeC_6H_4)$_2$AsMe, **3**, 69

EtPh(4-MeC_6H_4)As, **3**, 77

EtPh(4-MeC_6H_4)As, **3**, 77

$AsC_{15}H_{17}N_2$ $Ph_2AsCH_2CH_2C$(=NH)NH_2, **3**, 61

$AsC_{15}H_{17}N_2O_3$ 4-[4-$MeOC_6H_4N$=C(Me)NH]-C_6H_4As(OH)$_2$, **3**, 225

4-(4-$Me_2NC_6H_4CH$=N)$C_6H_4AsO_3H_2$, **3**, 442

$AsC_{15}H_{17}N_2O_3S$ Ph[3,4-$H_2N(HO)C_6H_3$]-$AsSCH_2CH(NH_2)CO_2H$, 3, 259

$AsC_{15}H_{17}N_2O_6S$ 4-(4-$AcNHC_6H_4SO_2NHCH_2$)-$C_6H_4AsO_3H_2$, 3, 407
3,4-Me($AcNHC_6H_4SO_2NH$)$C_6H_3AsO_3H_2$, 3, 495

$AsC_{15}H_{17}N_4O_2$ 4-[1-(4-Guanylphenylimino)ethylamino]benzenearsonous acid, 3, 226

$AsC_{15}H_{17}N_4O_4S_2$
4-[N=C(NH_2)N=C(Me)CH=CNH]C_6H_4-As$(SCH_2CO_2H)_2$, 3, 270

$AsC_{15}H_{17}O$ Ph_2AsOPr, 3, 211
Ph(4-MeC_6H_4)AsOEt, 3, 211

$AsC_{15}H_{17}O_2$ (MeC_6H_4)(2,4-$Me_2C_6H_3$)-AsO_2H isomers, 3, 385

$AsC_{15}H_{17}O_3$ Ph($PhCH_2OCH_2CH_2$)AsO_2H, 3, 385
2,2-HO(p-$MeOC_6H_4$)-isoarsindolinium hydroxide, 3, 559

$AsC_{15}H_{17}S_2$ MeAs$(SC_6H_4Me\text{-}p)_2$, 3, 270

$AsC_{15}H_{17}S_4$ MeAs$(SC_6H_4SMe\text{-}4)_2$, 3, 270

$AsC_{15}H_{18}I$ [Me_3(2-PhC_6H_4)As]I, 3, 319
[$MeEtPh_2As$]I, 3, 319

$AsC_{15}H_{18}IO$ [Me($HOCH_2CH_2$)Ph_2As]I, 3, 319
[Me_3(4-$PhOC_6H_4$)As]I, 3, 319
1,1-Me(Ph)-arsindolinium $I^-\cdot H_2O$, 3, 556

$AsC_{15}H_{18}N_3O_3$ $Ph_2AsCH_2CH_2C(=NH)NH_3\cdot NO_3$, 3, 61
4-(4-$PrNHC_6H_4N$=N)$C_6H_4AsO_3H_2$, 3, 459

$AsC_{15}H_{18}N_3O_5$
4-[MeC=C(CO_2Et)NHC(Me)=CN=N]C_6H_4-AsO_3H_2, 3, 459
4-[HNC(Me)=C(CO_2Et)C(Me)=CN=N]-$C_6H_4AsO_3H_2$, 3, 459*

$AsC_{15}H_{18}N_3O_6$
4,3-HO[HNC(Me)=C(CO_2Et)C(Me)=CN=N]-$C_6H_3AsO_3H_2$, 3, 495

$AsC_{15}H_{18}N_5O_3$ 4-[4-MeC_6H_4NHC(=NH)NHC-(=NH)NH]$C_6H_4AsO_3H_2$, 3, 438

$AsC_{15}H_{18}N_5O_4$ 4-(4-$MeOC_6H_4NHC$(=NH)-NHC(=NH)NH]$C_6H_4AsO_3H_2$, 3, 439

$AsC_{15}H_{18}N_5O_5$ N-[(1,2,3,6-Tetrahydro-1,3,7-trimethyl-2,4-dioxopurin-8-yl)methyl]arsanilic acid, 3, 433

$AsC_{15}H_{19}$ EtPr(1-$C_{10}H_7$)As, 3, 77

$AsC_{15}H_{19}IP$ [Me_2AsPPh_2Me]I, 3, 291

$AsC_{15}H_{19}N_6O_4S_2$
4-[N=C(NH_2)N=C(NMe_2)N=CNH]C_6H_4-As$(SCH_2CO_2H)_2$, 3, 271*

$AsC_{15}H_{19}O_6$ 5,2,2,7,7-$PhCH_2(Me)_4$-3,8-dioxo-1,4,6,9-tetraoxa-5-arsaspiro[4.4]nonane, 3, 581

$AsC_{15}H_{20}N_3O_4$ 4-[$EtO_2CN(CH_2CH_2)_2NCH_2$-CONH]C_6H_4AsO, 3, 249*

$AsC_{15}H_{21}N_2O_3$ 4-[HNC(C_6H_{13})=NCH=C]-$C_6H_4AsO_3H_2$, 3, 471

$AsC_{15}H_{21}N_6O_4S_2$
4-[N=C(NH_2)N=C(NH_2)N=CNH]C_6H_4-As[$SCH_2CH(NH_2)CO_2H$]$_2$, 3, 270

$AsC_{15}H_{22}I$ [Me$(CH_2{=}CMeCH_2)_2$PhAs]I, 3, 319

$AsC_{15}H_{22}NO_6$ 4-[$HO_2C(CH_2)_7CONH$]C_6H_4-AsO_3H_2, 3, 439

$AsC_{15}H_{22}NO_6S_2$
4-$H_2NC_6H_4AsSCH_2CH$[CH_2O(glucoside)S, 3, 587*

$AsC_{15}H_{22}N_3O_6$ 3,4,5-[$CH_2(CH_2)_3NCH_2$-CH_2CONH](HO)(AcNH)$C_6H_2AsO_3H_2$, 3, 503
4-[$EtO_2CN(CH_2CH_2)_2NCH_2CONH$]$C_6H_4AsO_3H_2$ 3, 439

$AsC_{15}H_{23}Cl_4O_6P_2S_2$ 2-MeC_6H_4As[OP(S)-$(OCH_2CH_2Cl)_2$]$_2$, 3, 257

$AsC_{15}H_{23}N_2O_5$ 4-[$H_2NOC(CH_2)_7CONH$]C_6H_4-AsO_3H_2, 3, 439

$AsC_{15}H_{23}N_2O_6$ {4-[$HOCH_2(CHOH)_4CH_2N$(Me)-$COCH_2NH$]C_6H_4As}$_n$, 3, 142

$AsC_{15}H_{23}N_6O_5$ N-[4,6-Bis(2-hydroxypropylamino)-s-triazin-2-yl]arsanilic acid, 3, 448

$AsC_{15}H_{23}O_2$ PhAs(OPr)(OC_6H_{11}), 3, 254

$AsC_{15}H_{23}O_6$ 8-Methyl-7,9,17,18-tetraoxa-8-arsatrispiro[5.1.1.5.2.2]-nonadecane-16,19-dione, 3, 581

$AsC_{15}H_{24}Br$ [Et($CH_2{=}CHCH_2$)BuPhAs]Br, 3, 319
[Et($CH_2{=}CHCH_2$)Pr(4-MeC_6H_4)As]Br, 3, 319

$AsC_{15}H_{24}Cl_2N_8O_5S_2$
4,3-HO[NH=C(NH_2)N=C(NH_2)N=CNH]C_6H_3-As[$SCH_2CH(NH_2\cdot HCl)CO_2H$]$_2$, 3, 270

$AsC_{15}H_{24}NO_9$ 4-[$HOCH_2(CHOH)_4CH_2N(Me)$-$COCH_2$]$C_6H_4AsO_3H_2$, 3, 414
$AsC_{15}H_{25}$ $Et(C_7H_{15})PhAs$, 3, 77
$Et(C_6H_{13})(4$-$MeC_6H_4)As$, 3, 77
$AsC_{15}H_{25}IP$ 1,1,5-Et_3-2,3,4,5-tetrahydro-1H-1,5-benzophosphoniaarsepin iodide, 3, 644
$AsC_{15}H_{25}N_2O_9$ 4-[$HOCH_2(CHOH)_4CH_2N(Me)$-$COCH_2NH$]$C_6H_4AsO_3H_2$, 3, 433
$AsC_{15}H_{25}N_2O_{10}$ 4,3-HO[$HOCH_2(CHOH)_4$-$CH_2N(Me)COCH_2NH$]$C_6H_3AsO_3H_2$, 3, 495
$AsC_{15}H_{26}Br_2P$ 1,1,5,5-$Et_2(Me)_2$-2,3,4,-5-tetrahydro-1H-1,5-benzophosphoniaarsepinium dibromide, 3, 644
$AsC_{15}H_{26}IO$ $Pr_2PhAs(OPr)I$, 3, 346
$MeBuPhAs(OBu)I$, 3, 346
$AsC_{15}H_{26}NO_4$ [$Pr_2PhAs(OPr)$]NO_3, 3, 350
$AsC_{15}H_{27}O_4P_2S_4$ 3-MeC_6H_4As[$SP(S)$-$(OEt)_2$]$_2$, 3, 281
$AsC_{15}H_{27}O_6P_2S_2$ 4-MeC_6H_4As[$SP(O)$-$(OEt)_2$]$_2$, 3, 281
$AsC_{15}H_{29}O_2$ $PrAs(OC_6H_{11})_2$, 3, 254
$AsC_{15}H_{31}OS_2$ $Me_2AsSC(S)OC_{12}H_{25}$, 3, 261
$AsC_{15}H_{33}$ $(C_5H_{11})_3As$, 3, 30
$AsC_{15}H_{33}O$ $Pr_2AsOC_9H_{19}$, 3, 211
$AsC_{15}H_{33}S$ $(n$-$C_5H_{11})_3AsS$, 3, 365
$AsC_{15}H_{33}S_2$ $MeAs(SC_7H_{15})_2$, 3, 270
$PrAs(SC_6H_{13})_2$, 3, 270
$AsC_{15}H_{34}I$ [$Me_3(n$-$C_{12}H_{25})As$]I, 3, 319
$AsC_{16}H_{10}ClFN$ 12,10-Cl(F)-7,12-dihydrobenzo[a]phenarsazine, 3, 627
7,x-Cl(F)-7,12-dihydrobenzo[c]phenarsazine isomers, 3, 632
$AsC_{16}H_{10}ClN_2S_4$ ClCH=CHAs(S-2-benzothiazole)$_2$, 3, 270
$AsC_{16}H_{10}ClO_3$ Chlorohydroxy-γ-benzophenoxarsine 12-oxide isomers, 3, 611
$AsC_{16}H_{10}Cl_2N$ Cl_2-7,12-dihydrobenzo-[a]phenarsazine isomers, 3, 627
7,9-Cl_2-7,12-dihydrobenzo[c]phenarsazine, 3, 632
$AsC_{16}H_{11}ClNO_2$ x,12-Cl(HO)-7,12-dihydrobenzo[a]phenarsazine 12-oxide isomers, 3, 627
$AsC_{16}H_{11}N_2O_8S_2$ 3,3′-Bis[1,4,8-HO-$(HO_3S)_2$-2-naphthylazo]arsenophenol, 3, 135
$AsC_{16}H_{11}ClNO_5$ N-(3-Chloro-1,4-dihydro-1,4-dioxonaphthyl)arsanilic acid, 3, 443
$AsC_{16}H_{11}Cl_2N_2S_4$ EtAs[S-2-(6-chlorobenzothiazole)]$_2$, 3, 270
$AsC_{16}H_{11}N_2O_9S_2$ 2-(5-Arsenoso-2-hydroxyphenylazo)-1-naphthol-4,8-disulfonic acid, 3, 232
$AsC_{16}H_{11}O_3$ Hydroxybenzophenoxarsine oxide isomers, 3, 611
$AsC_{16}H_{12}ClN_2$ 12-Chloro-7,12-dihydro-10-methylpyrido[3,2-a]phenarsazine, 3, 639
$AsC_{16}H_{12}ClO_4$ 2-$(ClC_6H_4O)C_{10}H_6$-1-AsO_3H_2 isomers, 3, 510
$AsC_{16}H_{12}I_2N_3O_{10}S_2$ 4-[4,5,2,7-$H_2N(HO)$-$(HO_3S)_2$-3-naphthylazo]-2,6-diiodobenzenearsonic acid, 3, 503
$AsC_{16}H_{12}NO_5$ N-(1,2-Dihydro-1,2-dioxonaphthyl)arsanilic acid, 3, 443*
2-(4-Arsonophenyl)quinoline-4-CO_2H, 3, 471
$AsC_{16}H_{12}NO_8S$ 3,4-$O_2N(2$-$C_{10}H_7SO_3)C_6H_3$-AsO_3H_2, 3, 495
N-(3,4-Dihydro-3,4-dioxo-x-sulfo-1-naphthyl)arsanilic acid isomers, 3, 423, 443
$AsC_{16}H_{12}NO_9S$ x-Hydroxy-y-(3,4-dihydro-3,4-dioxo-x-sulfo-1-naphthylamino)benzenearsonic acid isomers, 3, 496, 497
$AsC_{16}H_{12}NO_{11}S_2$ N-(3,4-Dihydro-3,4-dioxo-5,7-disulfo-4-naphthyl)arsanilic acid, 3, 443
$AsC_{16}H_{12}NO_{12}S_2$ 2-Hydroxy-4-(3,4-dihydro-3,4-dioxo-5,7-disulfo-1-naphthylamino)benzenearsonic acid, 3, 497
$AsC_{16}H_{12}N_3O_8S_2$ 8,2-$H_2N(4$-$OAsC_6H_4N$=$N)$-1-naphthol-3,6-disulfonic acid, 3, 233
$AsC_{16}H_{12}N_3O_{10}S$ 5,2-O_2N[2,3,6-$(HO)_2$-(HO_3S)-1-naphthylazo]$C_6H_3AsO_3H_2$, 3, 497
$AsC_{16}H_{12}N_3O_{12}S_2$ 5,2-O_2N[2,3,6-HO-$(HO_3S)_2$-1-naphthylazo]$C_6H_3AsO_3H_2$,

3, 497

$AsC_{16}H_{12}N_3O_{13}S_2$ 5,2-O_2N[4,5,2,7-$(HO)_2$-$(HO_3S)_2$-3-naphthylazo]$C_6H_3AsO_3H_2$, 3, 497

5,2-O_2N[4,5,2,7-$(HO)_2(HO_3S)_2$-3-naphthylazo]$C_6H_3AsO_3H_2$, 3, 497

$AsC_{16}H_{13}N_2$ Di-2-pyridyl-AsPh, 3, 69

$AsC_{16}H_{13}N_2O_4$ $(HOC_{10}H_6N{=}N)C_6H_4AsO_3H_2$ isomers, 3, 460

4,3-HO(PhN=N)naphthalenearsonic acid, 3, 510

$AsC_{16}H_{13}N_2O_5$

4-[$\overline{HNC}(C_6H_4CO_2H$-4)=NCH=$\overline{C}$]$C_6H_4AsO_3H_2$, 3, 472

$AsC_{16}H_{13}N_2O_6$ 5-[4-(4-Arsonophenyl)-imidazol-2-yl]salicylic acid, 3, 472

$AsC_{16}H_{13}N_2O_6S$

4-[$\overline{SC}(=NC_6H_4CO_2H)NHCO\overline{C}H$]$C_6H_4AsO_3H_2$ isomers, 3, 472

$AsC_{16}H_{13}N_2O_8S$ 2-[2,3,6-$(HO)_2(HO_3S)$-1-naphthylazo]benzenearsonic acid, 3, 460

2-[2,3,6-$(HO)_2(HO_3S)$-1-naphthylazo]-benzenearsonic acid, 3, 460

$AsC_{16}H_{13}N_2O_8S_2$ 2,2'-Bis[1,4,8-HO-$(HO_3S)_2$-2-naphthylhydrazo]-4-arsenophenol, 3, 135

$AsC_{16}H_{13}N_2O_9S_2$ 2-[2-(5-Arsenoso-2-hydroxyphenyl)hydrazo]-1-naphthol-4,8-disulfonic acid, 3, 233

$AsC_{16}H_{13}N_2O_{10}S_2$ (Hydroxydisulfonaphthylazo)benzenearsonic acid isomers, 3, 455, 460

$AsC_{16}H_{13}N_2O_{11}S_2$ (Dihydroxydisulfo-2-naphthylazo)benzenearsonic acid isomers, 3, 455, 460

4,3-HO[4,1,5-$HO(HO_3S)_2$-3-naphthylazo]benzenearsonic acid, 3, 498

$AsC_{16}H_{13}N_2O_{11}S_4$ 2-[2-(2-Hydroxy-5-thioarsenosophenyl)hydrazo]-1-naphthol-4,6,8-trisulfonic acid, 3, 265*

$AsC_{16}H_{13}N_2O_{12}S_3$ 2-[2-(5-Arsenoso-2-hydroxyphenyl)hydrazo]-1-naphthol-4,6,8-trisulfonic acid, 3, 233*

$AsC_{16}H_{13}O_2$ $Ph(C_{10}H_7)AsO_2H$ isomers, 3, 386

$AsC_{16}H_{13}O_2S_2$ 10-Phenoxarsinyl-SC(S)-$OCH_2CH{=}CH_2$, 3, 613

$AsC_{16}H_{13}O_5S$ 4-$(2\text{-}C_{10}H_7SO_2)C_6H_4AsO_3H_2$, 3, 414

10-Phenoxarsinyl-$SCH(CO_2H)CH_2CO_2H$, 3, 613

$AsC_{16}H_{13}O_6S$ $(4\text{-}HO_2CC_6H_4)_2AsSCH_2CO_2H$, 3, 259

$AsC_{16}H_{14}BrCl_2$ 2,2'-Spirobi(5-chloro-isoarsindolinium)Br^-, 3, 574

$AsC_{16}H_{14}ClO_2$ $(4\text{-}AcC_6H_4)_2AsCl$, 3, 174

$AsC_{16}H_{14}Cl_2I$ 2,2'-Spirobi(5-chloro-isoarsindolinium)I^-, 3, 574

$AsC_{16}H_{14}NO_3$ {4-[$HO_2CCH(NHBz)CH_2$]-C_6H_4As}$_n$, 3, 142*

$AsC_{16}H_{14}NO_6S$ 3,4-$H_2N(2\text{-}C_{10}H_7SO_3)C_6H_3$-$AsO_3H_2$, 3, 498

$AsC_{16}H_{14}N_3O_3$ 4-$(4,1\text{-}H_2NC_{10}H_6N{=}N)C_6H_4$-$AsO_3H_2$, 3, 461

4-$(2,1\text{-}H_2NC_{10}H_6N{=}N)C_6H_4AsO_3H_2$, 3, 461

$AsC_{16}H_{14}N_3O_5$ N-(2-Methyl-6-nitro-4-quinolyl)arsanilic acid, 3, 449

$AsC_{16}H_{14}N_3O_7$ 1-Ethylarsindole picrate, 3, 555

$AsC_{16}H_{14}N_3O_8S$ 6-[2,3,6-$(HO)_2(HO_3S)$-1-naphthylazo]-m-arsanilic acid, 3, 498

$AsC_{16}H_{14}N_3O_{10}S_2$ 4-[2,8,3,6-$H_2N(HO)$-$(HO_3S)_2$-naphthylazo]-$C_6H_4AsO_3H_2$, 3, 461

4-[8,1,3,6-$H_2N(HO)(HO_3S)_2$-2-naphthylazo]-$C_6H_4AsO_3H_2$, 3, 461

$AsC_{16}H_{14}N_3O_{11}S_2$ 6-[4,5,2,7-$(HO)_2$-$(HO_3S)_2$-2-naphthylazo]-m-arsanilic acid, 3, 498

$AsC_{16}H_{14}N_3O_{12}S_3$ 4-[1,3,6,8-$H_2N(HO_3S)_3$-2-naphthylazo]-$C_6H_4AsO_3H_2$, 3, 461

$AsC_{16}H_{15}ClI$ 3,1,1-Cl(Et)(Ph)-arsindolium I^-, 3, 556

$AsC_{16}H_{15}ClN_3O_7$ 5,2,2-$Cl(Me)_2$-isoarsindolium picrate, 3, 557

$AsC_{16}H_{15}N_2O_4$ [4,3-$HO_2CCH_2O(PhNHCOCH_2$-$NH)C_6H_3As$]$_n$, 3, 142

$AsC_{16}H_{15}N_2O_4$ 4-[$\overline{HNC}(C_6H_4OMe$-4)=NCH=$\overline{C}$]-

$C_6H_4AsO_3H_2$, 3, 472

$AsC_{16}H_{15}N_2O_4S$ 4-[$SC(=NC_6H_4Me)NHCOCH$]-$C_6H_4AsO_3H_2$ isomers, 3, 472

$AsC_{16}H_{15}N_2O_5$ [4,3-HO_2CCH_2O(4-HOC_6H_4-$NHCOCH_2NH$)C_6H_3As]$_n$, 3, 142

$AsC_{16}H_{15}N_2O_{10}S_2$ 4-[1,4,8-$HO(HO_3S)_2$-2-naphthylhydrazo]-$C_6H_4AsO_3H_2$, 3, 453

$AsC_{16}H_{15}O_2$ $Ph_2AsO_2CC(Me)=CH_2$, 3, 212

$AsC_{16}H_{15}O_2S_2$ 10-Phenoxarsinyl-SC(S)OPr, 3, 614

$AsC_{16}H_{15}O_3S$ 10-Phenoxarsinyl-S$(CH_2)_3$-CO_2H, 3, 614

10-phenoxarsinyl-SCH_2CO_2Et, 3, 614

$AsC_{16}H_{15}O_4$ $(4\text{-}AcC_6H_4)_2AsO_2H$, 3, 386

$AsC_{16}H_{15}O_4S$ $Ph_2AsSCH(CO_2H)CH_2CO_2H$, 3, 259

$AsC_{16}H_{15}O_5S$ Ph(4-HOC_6H_4)$AsSCH(CO_2H)$-CH_2CO_2H, 3, 259

$AsC_{16}H_{15}O_6$ 4-[4-$MeOC_6H_4CH=C(CO_2H)$]-$C_6H_4AsO_3H_2$, 3, 407

$AsC_{16}H_{16}Br$ 2,2′-Spirobi(isoarsindolium)Br^-, 3, 574

$AsC_{16}H_{16}BrD_2O$ [$Me_2Ph(PhCOCD_2)As$]Br, 3, 320

$AsC_{16}H_{16}I$ 2,2′-Spirobi(isoarsindolinium)I^-, 3, 574

$AsC_{16}H_{16}N$ $Ph_2AsCH_2CH(Me)CN$, 3, 61

$(MeC_6H_4)(2,4\text{-}Me_2C_6H_3)AsCN$ isomers, 3, 205

$AsC_{16}H_{16}NO$ 5,10-Ac(Et)-5,10-dihydrophenarsazine, 3, 624

$AsC_{16}H_{16}NO_3S$ 10-Phenoxarsinyl-$SCH_2CH_2CH(NH_2)CO_2H$, 3, 614

$AsC_{16}H_{16}NO_4$ 4-(4-$AcNHC_6H_4CH=CH$)C_6H_4-AsO_3H_2, 3, 407

$AsC_{16}H_{16}NO_5$ 2,4-$HOCH_2CH_2O$($PhCH_2O_2$-CNH)C_6H_3AsO, 3, 248

$AsC_{16}H_{16}NO_5S$ Ph[3,4-$H_2N(HO)C_6H_3$]-$AsSCH(CO_2H)CH_2CO_2H$, 3, 259

$AsC_{16}H_{16}NO_6$ 4-[$HO_2CCH(NHBz)CH_2$]C_6H_4-AsO_3H_2, 3, 407

$AsC_{16}H_{16}N_3O_4$ 4-[4-$MeC_6H_4NHCOCH(CN)$-NH]$C_6H_4AsO_3H_2$, 3, 439

$AsC_{16}H_{16}N_3O_7$ 2,2-Me_2-isoarsindolinium picrate, 3, 557

$AsC_{16}H_{16}N_3O_8$ 2,2-HO(Me)-1,2,3,4-tetrahydroisoarsinolinium picrate, 3, 594

$AsC_{16}H_{16}N_5O_5S_2$ 4-[$SCH=CHN=CNHO_2SC_6H_4$-NHC(=NH)NH]$C_6H_4AsO_3H_2$, 3, 439

$AsC_{16}H_{17}$ 4-($PhCH=CH$)$C_6H_4AsMe_2$, 3, 69

(9-Fluorenylidene)trimethylarsenic, 3, 308

$AsC_{16}H_{17}N_2O$ {4-[4-$EtOC_6H_4N=C(Me)NH$]-C_6H_4As}$_n$, 3, 142

$AsC_{16}H_{17}N_2O_2$ [4,3-EtO($PhNHCOCH_2NH$)-C_6H_3As]$_n$, 3, 142

$AsC_{16}H_{17}N_2O_5$ 4-[PhNHCOCH(Ac)NH]C_6H_4-AsO_3H_2, 3, 439

$AsC_{16}H_{17}N_2O_6$ 3,4-AcNH($PhNHCOCH_2O$)-$C_6H_3AsO_3H_2$, 3, 498

$AsC_{16}H_{17}N_2O_7$ 4,3-HO_2CCH_2O(PhNHCO-CH_2NH)$C_6H_3AsO_3H_2$, 3, 498

$AsC_{16}H_{17}N_4O_9$ [$Me_3(O_2NC_6H_4CH_2)As$]OC_6H_2-$(NO_2)_3$ isomers, 3, 315

$AsC_{16}H_{17}N_6O$ {4-[4-$AcNHC_6H_4NHC(=NH)NHC$-(=NH)NH]C_6H_4As}$_n$, 3, 142

$AsC_{16}H_{17}OS_2$ $Ph_2AsSC(S)OCHMe_2$, 3, 261

$AsC_{16}H_{17}O_2$ $Ph_2AsCH_2CH(Me)CO_2H$, 3, 69

$AsC_{16}H_{17}O_3$ Ph_2[$MeCH(CO_2H)CH_2$]AsO, 3, 358

$AsC_{16}H_{17}O_4$ 2-Butyl-2,2′-spirobi-1,3,2-benzodioxarsole, 3, 581*

$AsC_{16}H_{18}Br$ [Me_3(9-fluorenyl)As]Br, 3, 320

[$Me(CH_2=CHCH_2)Ph_2As$]Br, 3, 320

[$Me_2(2\text{-}PhC_6H_4)(CH_2=CH)As$]Br, 3, 320

$AsC_{16}H_{18}BrO$ [$Me_2Ph(PhCOCH_2)As$]Br, 3, 320

$AsC_{16}H_{18}Cl$ $Ph_2As(CH_2)_4Cl$, 3, 61

$AsC_{16}H_{18}I$ [Me_3(9-fluorenyl)As]I, 3, 320

$AsC_{16}H_{18}I$ 2,2-Me(p-MeC_6H_4)-isoarsindolinium I^-, 3, 559

2,2-Me(Ph)-1,2,3,4-tetrahydroisoarsinolinium I^-, 3, 595

$AsC_{16}H_{18}IO$ 2-(4-Methoxyphenyl)isoarsindoline methiodide, 3, 559

$AsC_{16}H_{18}N$ 2-(2-$Me_2NC_6H_4$)-isoarsindoline, 3, 559

$PhNCH_2CH_2As(Ph)CH_2CH_2$, 3, 619

$AsC_{16}H_{18}NO_2$ Hexahydro-1,4-diphenyl-1,4-azarsine 1,4-dioxide, 3, 620

$AsC_{16}H_{18}NO_6$ 4-$[BzNHCH(CH_2OH)CH(OH)]$-$C_6H_4AsO_3H_2$, 3, 407

$AsC_{16}H_{18}NO_7$ 2,4-$HOCH_2CH_2O(PhCH_2OCONH)$-$C_6H_3AsO_3H_2$, 3, 499

$AsC_{16}H_{18}N_3O_7$ $[Me_3(PhCH_2)As]OC_6H_2(NO_2)_3$, 3, 316

$AsC_{16}H_{18}N_3O_7S$ 4-[4-$AcNHC_6H_4SO_2NHCOCH_2NH]C_6H_4AsO_3H_2$, 3, 439

$AsC_{16}H_{18}O_2PS_2$ 5-Dibenzarsolyl-$SP(S)(OEt)_2$, 3, 564

$AsC_{16}H_{18}O_3S_2P$ 10-Phenoxarsinyl-$SP(S)(OEt)_2$, 3, 614

$AsC_{16}H_{19}$ (4-$MeC_6H_4)_2AsEt$, 3, 62

$Me_2AsCH(Ph)CH_2Ph$, 3, 62

$AsC_{16}H_{19}Br_2$ $[Me_2(BrCH_2CH_2)$(2-$PhC_6H_4)As]Br$, 3, 320

$AsC_{16}H_{19}ClNO_4$ 2-(2-$Me_2NC_6H_4$)-isoarsindoline perchlorate, 3, 559

$AsC_{16}H_{19}N_2O_3$ 4-[4-$EtOC_6H_4N{=}C(Me)NH]$-$C_6H_4As(OH)_2$, 3, 225

$AsC_{16}H_{19}N_6O_4$ 4-[4-$AcNHC_6H_4NHC({=}NH)$-$NHC({=}NH)NH]C_6H_4AsO_3H_2$, 3, 439

$AsC_{16}H_{19}O$ Ph(4-$MeC_6H_4)AsOPr$, 3, 211

$AsC_{16}H_{20}Br$ $[Me_2(PhCH_2)_2As]Br$, 3, 320

$[MeEt(CH_2{=}CHCH_2)$(1-$C_{10}H_7)As]Br$, 3, 321

$AsC_{16}H_{20}Cl$ $[MeEtPh(PhCH_2)As]Cl$, 3, 320

$AsC_{16}H_{20}ClO_4$ [(+)-$MeEtPh(PhCH_2)As]ClO_4$, 3, 321

[(-)-$MeEtPh(PhCH_2)As]ClO_4$, 3, 321

$AsC_{16}H_{20}Cl_2N$ $[PhAs(CH_2CH_2)_2NPh]\cdot 2HCl$, 3, 619

$AsC_{16}H_{20}I$ $[Me_3$(2-$PhC_6H_4CH_2)As]I$, 3, 319*

$[Et_2Ph_2As]I$, 3, 320

$[MeEtPh(PhCH_2)As]I$, 3, 320

$AsC_{16}H_{20}IO$ $[Me_2Et$(4-$PhOC_6H_4)As]I$, 3, 321

$AsC_{16}H_{20}NO_2$ $CH_2CH_2N(Ph)CH_2CH_2As(Ph)$-$O\cdot H_2O$, 3, 619, 620

$AsC_{16}H_{20}N_3O_3$ 4-(4-$Et_2NC_6H_4N{=}N)C_6H_4AsO_3H_2$, 3, 461

4-(4-$BuNHC_6H_4N{=}N)C_6H_4AsO_3H_2$, 3, 461

4-[4-$(Me_2CHCH_2NH)C_6H_4N{=}N]C_6H_4AsO_3H_2$, 3, 461

4-[4-$(MeEtCHNH)C_6H_4N{=}N]C_6H_4AsO_3H_2$, 3, 461

$AsC_{16}H_{20}N_3O_5$ 3,4-O_2N(4-$Et_2NC_6H_4NH)$-$C_6H_3AsO_3H_2$, 3, 499

$AsC_{16}H_{20}O_2PS_2$ $Ph_2AsSP(S)(OEt)_2$, 3, 261

$AsC_{16}H_{20}O_3P$ $Ph_2AsPO(OEt)_2$, 3, 293

$AsC_{16}H_{21}$ EtBu(1-$C_{10}H_7$)As, 3, 77

$AsC_{16}H_{21}N_2O_2S_4$ $PhAs[SC(S)N(CH_2CH_2)_2O]_2$, 3, 283

$AsC_{16}H_{21}N_2O_3$ 4-(4-$Et_2NC_6H_4NH)C_6H_4$-AsO_3H_2, 3, 442

$AsC_{16}H_{21}N_6O_4S_2$ 4-$[N{=}C(NH_2)N{=}CHCH{=}CNH]$-$C_6H_4As[SCH_2CH(NH_2)CO_2H]_2$, 3, 270

$AsC_{16}H_{21}O_6$ 2,7,2,7-$Et_2(Me)_2$-3,8-dioxo-5-phenyl-1,4,6,9-tetraoxa-5-arsaspiro[4,4]nonane, 3, 581

$AsC_{16}H_{22}Cl_3N_2O_4S_4$ 3,4,6-$Cl_3C_6H_2As$-$[SC(S)N(CH_2CH_2OH)_2]_2$, 3, 283

$AsC_{16}H_{23}Cl_2N_2S_4$ 2,4-$Cl_2C_6H_3As[SC(S)$-$NEt_2]_2$, 3, 283

$AsC_{16}H_{23}N_2O_7$ Arecoline·3,4-$AcNH(HO)$-$C_6H_3AsO_3H_2$ salt, 3, 477

$AsC_{16}H_{23}O_2$ $PhAs[OCH(Et)CH{=}CH_2]_2$, 3, 254

$AsC_{16}H_{23}O_2S_4$ $PhAs[SC(S)OBu]_2$, 3, 280

$AsC_{16}H_{23}O_4$ $PhAs(CH_2CH_2CO_2Et)_2$, 3, 62

$AsC_{16}H_{24}NO_8$ 4-$[MeO_2C(CH_2)_7CONH]C_6H_4$-$AsO_3H_2$, 3, 439

4-$[HO_2C(CH_2)_8CONH]C_6H_4AsO_3H_2$, 3, 439

$AsC_{16}H_{24}NO_8S_2$ 3,4-$H_2N(HOCH_2CH_2O)C_6H_3$-$As(SCH_2CO_2Et)_2$, 3, 270

$AsC_{16}H_{25}N_2O_5$ 4-$[MeNHOC(CH_2)_7CONH]$-$C_6H_4AsO_3H_2$, 3, 439

4-$[H_2NOC(CH_2)_8CONH]C_6H_4AsO_3H_2$, 3, 439

$AsC_{16}H_{25}N_2S_4$ $PhAs[SC(S)NEt_2]_2$, 3, 283

$AsC_{16}H_{25}O_2$ $PhAs(OBu)(OC_6H_{11})$, 3, 254

$AsC_{16}H_{27}N_2O_9$ 4-$[HOCH_2(CHOH)_4CH_2N(Me)$-$COCH(Me)NH]C_6H_4AsO_3H_2$, 3, 439

2,4-$Me[HOCH_2(CHOH)_4CH_2N(Me)COCH_2NH]$-$C_6H_3AsO_3H_2$, 3, 499

$AsC_{16}H_{27}N_2O_{10}$ 4-$[HOCH_2(CHOH)_4CH_2N$-$(CH_2CH_2OH)COCH_2NH]C_6H_4AsO_3H_2$, 3, 439

3,4-$MeO[HOCH_2(CHOH)_4CH_2N(Me)COCH_2NH]$-$C_6H_3AsO_3H_2$, 3, 499

$AsC_{16}H_{27}O_2$ $PhAs(OCH_2CH_2CHMe_2)_2$, 3, 254

$AsC_{16}H_{27}S_2$ $PhAs(SC_5H_{11})_2$, 3, 270

$AsC_{16}H_{28}O_3P$ $EtPhAsPO(OBu)_2$, 3, 293
$AsC_{16}H_{28}PBr_2$ $[\overline{CH_2CH_2AsEt_2C_6H_4PEt_2}]Br_2$, 3, 640
$AsC_{16}H_{30}I_2P$ $[2\text{-}(MeEt_2P)C_6H_4AsEt_2Me]I_2$, 3, 321
$AsC_{16}H_{31}N_2O_3$ 4-$H_2NC_6H_4AsO_3H_2 \cdot H_2NC_{10}H_{21}$, 3, 428
$AsC_{16}H_{31}O_2$ $BuAs(OC_6H_{11})_2$, 3, 254
$AsC_{16}H_{34}Br$ $(C_8H_{17})_2AsBr$, 3, 167
$AsC_{16}H_{34}Cl$ $(C_8H_{17})_2AsCl$, 3, 175
$AsC_{16}H_{35}O$ $Bu_2AsOC_8H_{17}$, 3, 211
$AsC_{16}H_{35}O_2$ $EtAs(OC_7H_{15})_2$, 3, 254
$AsC_{16}H_{35}O_2$ $EtAs(OC_7H_{15})_2$, 3, 254
$AsC_{16}H_{35}S_2$ $EtAs(SC_7H_{15})_2$, 3, 270
  $BuAs(SC_6H_{13})_2$, 3, 270
$AsC_{17}H_{10}ClF_3N$ 12,9-Cl(CF_3)-7,12-dihydrobenzo[a]phenarsazine, 3, 627
  7,10-Cl(CF_3)-7,12-dihydrobenzo[c]phenarsazine, 3, 632
$AsC_{17}H_{12}ClFN$ 12,10,x-Cl(F)(Me)-7,12-dihydrobenzo[a]phenarsazine isomers, 3, 627
$AsC_{17}H_{12}Cl_2N$ 9,12,8-Cl_2(Me)-7,12-dihydrobenzo[a]phenarsazine, 3, 627
  7,10,11-Cl_2(Me)-7,12-dihydrobenzo[c]phenarsazine, 3, 632
$AsC_{17}H_{12}N$ $Ph(C_{10}H_7)AsCN$ isomers, 3, 205
$AsC_{17}H_{13}ClN$ 12,x-Cl(Me)-7,12-dihydrobenzo[a]phenarsazine isomers, 3, 627
  12,6-Cl(Me)-7,12-dihydrobenzo[b]phenarsazine, 3, 630
  Cl(Me)-7,12-dihydrobenzo[c]phenarsazine isomers, 3, 632
$AsC_{17}H_{13}ClNO$ 12,10-Cl(MeO)-7,12-dihydrobenzo[a]phenarsazine, 3, 627
  7,x-Cl(MeO)-7,12-dihydrobenzo[c]phenarsazine isomers, 3, 632
$AsC_{17}H_{13}FN$ 10,12-F(Me)-7,12-dihydrobenzo[a]phenarsazine, 3, 627
  9,7-F(Me)-7,12-dihydrobenzo[c]phenarsazine, 3, 632
$AsC_{17}H_{13}N_2O_5$ 3-(4-$O_2NC_6H_4CH{=}N)C_{10}H_6$-1-$AsO_3H_2$, 3, 510

$AsC_{17}H_{13}O_3$ Hydroxymethylbenzophenoxarsine oxide isomers, 3, 611
$AsC_{17}H_{13}O_5$ 4-$PhCO_2C_{10}H_6$-1-AsO_3H_2, 3, 510
$AsC_{17}H_{14}Br_2N$ $Ph_2(\overline{N{=}CHCH{=}CHCH{=}C})AsBr_2$, 3, 345
$AsC_{17}H_{14}Cl_2NS$ 2,2'-Spirobi(5-chloroisoarsindolinium)$(SCN)^-$, 3, 574
$AsC_{17}H_{14}N$ 2-Pyridyl-$AsPh_2$, 3, 69
$AsC_{17}H_{14}NO$ $Ph_2(\overline{N{=}CHCH{=}CHCH{=}C})AsO$, 3, 358
$AsC_{17}H_{14}NO_2$ 12,x-HO(Me)-7,12-dihydrobenzo[a]phenarsazine 12-oxide isomers, 3, 628
  12,6-HO(Me)-7,12-dihydrobenzo[b]phenarsazine, 3, 628
  7,x-HO(Me)-7,12-dihydrobenzo[c]phenarsazine isomers, 3, 632
$AsC_{17}H_{14}N_5O_4$ 4-(4-Hydroxy-2-methylpyrimido[1.2-a]benzimidazol-3-ylazo)benzenearsonic acid, 3, 462
$AsC_{17}H_{15}ClIN_2O_6$ 5,8,7-Cl(HO)(I)-quinoline$\cdot H_2O_3AsC_6H_4(NHCOCH_2OH)$-p salt, 3, 429
$AsC_{17}H_{15}I_2N_2O_6$ 8,5,7-HO$(I)_2$-quinoline$\cdot H_2O_3AsC_6H_4(NHCOCH_2OH)$-p salt, 3, 429
$AsC_{17}H_{15}N_2O_4$ 4-(4,1-$MeOC_{10}H_6N{=}N$)-$C_6H_4AsO_3H_2$, 3, 462
$AsC_{17}H_{15}N_2O_4S_3$ 4-(2-Benzothiazolylamino)$C_6H_4As(SCH_2CO_2H)_2$, 3, 271
$AsC_{17}H_{15}O_2$ $(MeC_6H_4)(C_{10}H_7)AsO_2H$ isomers, 3, 386
$AsC_{17}H_{15}O_5$ $(MeOC_6H_4O)C_{10}H_6AsO_3H_2$ isomers, 3, 510
$AsC_{17}H_{16}NO$ 5,6-Dihydro-3-methoxy-5-methyl-11H-arsinolino[4,3-b]indole, 3, 593
$AsC_{17}H_{16}NOS_2$ 10-Phenoxarsinyl-SC(S)-$\overline{N(CH_2)_3C}H_2$, 3, 614
$AsC_{17}H_{16}NO_2S_2$ 10-Phenoxarsinyl-SC(S)-$N(CH_2CH_2)_2O$, 3, 614
$AsC_{17}H_{16}NO_7S$ 3-(8-Ethoxy-5-quinolylsulfonyl)-4-hydroxybenzenearsonic acid, 3, 499
$AsC_{17}H_{16}N_3O_3$ 4-[1,2-$H_2N(Me)C_{10}H_5$-3-$N{=}N]C_6H_4AsO_3H_2$, 3, 462

$AsC_{17}H_{16}N_3O_9$ 4-[3,5-$O_2N(HO_2CCH_2CH_2$-$CONH)C_6H_3CONH]C_6H_4AsO_3H_2$, 3, 439

$AsC_{17}H_{16}N_5O_5S$ 4,x-H_2N(N=CHCH=CHCH=C-$NHSO_2C_6H_4N$=N)$C_6H_3AsO_3H_2$, 3, 499

$AsC_{17}H_{17}Br_2Cl_2$ 5,2,2-Cl(Me)[4,2-Cl-$(BrCH_2)C_6H_4CH_2$]isoarsindolinium Br^-, 3, 557*

$AsC_{17}H_{17}N_2O_4$ [4,3-HO_2CCH_2O(4-MeC_6H_4-$NHCOCH_2NH)C_6H_3As]_n$, 3, 142*

$AsC_{17}H_{17}N_2O_6$ 8-Hydroxyquinoline·H_2O_3-$AsC_6H_4(NHCOCH_2OH)$-p, 3, 429

$AsC_{17}H_{17}N_2O_6S$ 4-(8-Ethoxy-5-quinolyl-$SO_2NH)C_6H_4AsO_3H_2$, 3, 444

$AsC_{17}H_{17}N_4O_5$ 1,2,3,4-Tetrahydro-7-methoxy-1-methyl-4-(2,4-dinitrophenylhydrazono)arsinoline, 3, 592

$AsC_{17}H_{17}N_4O_7$ [$Me_2(NCCH_2CH_2)PhAs$]-$OC_6H_2(NO_2)_3$, 3, 316

$AsC_{17}H_{17}N_6O_5S$
4-{4-[H_2NC=NC(NH_2)=CHCH=CN=N]C_6H_4-SO_2NH}$C_6H_4AsO_3H_2$, 3, 444

$AsC_{17}H_{17}OS_2$ 4-(PhCH=CH)C_6H_4-$AsSCH_2CH(CH_2OH)S$, 3, 587

$AsC_{17}H_{18}NOS_2$ 10-Phenoxarsinyl-SC(S)-NEt_2, 3, 614

$AsC_{17}H_{18}N_3O_4$ 4-[$Me_2C_6H_3NHCOCH(CN)NH$]-$C_6H_4AsO_3H_2$, 3, 440

$AsC_{17}H_{18}N_3O_7$ 4-[3,5-$H_2N(HO_2CCH_2CH_2$-$CONH)C_6H_3CONH]C_6H_4AsO_3H_2$, 3, 440
1,2,3,4-Tetrahydro-2,2-dimethylisoarsinolinium picrate, 3, 594

$AsC_{17}H_{19}Br_2$ 2,2-Me(2-$BrCH_2C_6H_4CH_2$)-isoarsindolinium Br^-, 3, 557

$AsC_{17}H_{19}N_2O$ 1,2,3,4-Tetrahydro-7-methoxy-1-methyl-4-(phenylhydrazono)arsinoline, 3, 592

$AsC_{17}H_{19}N_2O_5$ 4-[4-$MeC_6H_4NHCOCH(Ac)$-$NH]C_6H_4AsO_3H_2$, 3, 440

$AsC_{17}H_{19}N_2O_5S_2$
CH=NC(NHC_6H_4OEt)=CHCH=CAs($SCH_2CO_2H)_2$, 3, 271

$AsC_{17}H_{19}N_2S_2$ HN=$(C_6H_4)_2AsSC(S)NEt_2$, 3, 262

$AsC_{17}H_{19}N_2S_4$ MeAs[SC(S)NMePh]$_2$, 3, 283

$AsC_{17}H_{20}Br$ 1,1-Ph_2-arsenanium Br^-, 3, 591

$AsC_{17}H_{20}Cl$ $Ph_2As(CH_2)_5Cl$, 3, 69
$Ph_2As(CH_2)_5Cl$, 3, 69

$AsC_{17}H_{20}ClO_4$ [(+)-Me(CH_2=$CHCH_2$)Ph-$(PhCH_2)As]ClO_4$, 3, 321

$AsC_{17}H_{20}I$ [Me_3(4-PhCH=CHC_6H_4)As]I, 3, 231

$AsC_{17}H_{20}NO_7$ 2,4-$MeCHOHCH_2O(PhCH_2O$-$CONH)C_6H_3AsO_3H_2$, 3, 499

$AsC_{17}H_{20}NS_2$ $Ph_2AsSC(S)NEt_2$, 3, 262

$AsC_{17}H_{20}N_3O_7$ [$MeEt_2PhAs]OC_6H_2(NO_2)_3$, 3, 316

$AsC_{17}H_{21}$ $Ph_2AsC_5H_{11}$, 3, 69

$AsC_{17}H_{21}BrN$ 2,2-Me(2-$Me_2NC_6H_4$)-isoarsindolinium Br^-, 3, 559

$AsC_{17}H_{21}IN$ [$CH_2CH_2As(Ph)CH_2CH_2N$(Ph)-Me]I, 3, 619

$AsC_{17}H_{21}N_4O_7$ [Me_3(2-$Me_2NC_6H_4$)As]O-$C_6H_2(NO_2)_3$, 3, 317

$AsC_{17}H_{22}Br$ [MeEt($PhCH_2$)(4-MeC_6H_4)As]Br, 3, 321

$AsC_{17}H_{22}ClO_4$ [(+)-MePrPh($PhCH_2$)As]ClO_4, 3, 321

$AsC_{17}H_{22}I$ {Me_3[$PhCH_2CH$(Ph)]As}I, 3, 321

$AsC_{17}H_{22}N_3O_3$ 4-[4-(n-$C_5H_{11}NH)C_6H_4N$=N]-$C_6H_4AsO_3H_2$, 3, 462
4-[4-($Me_2CHCH_2CH_2NH)C_6H_4N$=N]C_6H_4-AsO_3H_2, 3, 462
4-[4-(sec-$C_5H_{11}NH)C_6H_4N$=N]$C_6H_4AsO_3H_2$, 3, 462
4-[4-(MeBuN)C_6H_4N=N]$C_6H_4AsO_3H_2$, 3, 462

$AsC_{17}H_{22}N_5O_5S_2$
4-[N=C(NHEt)N=C(NHEt)N=CO]C_6H_4-$As(SCH_2CO_2H)_2$, 3, 271

$AsC_{17}H_{23}O_6$ 5,2,7,2,7-$PhCH_2(Et)_2(Me)_2$-3,8-dioxo-1,4,6,9-tetraoxa-5-arsaspiro[4,4]nonane, 3, 581

$AsC_{17}H_{26}Br_2NO_4$ 4-[$BrCH_2CHBr(CH_2)_8CONH$]-$C_6H_4AsO_3H_2$, 3, 440

$AsC_{17}H_{26}IO_4$ [$MePh(EtO_2CCH_2CH_2)_2As$]I, 3, 321

$AsC_{17}H_{26}NO_8$ 4-[$EtO_2C(CH_2)_7CONH]C_6H_4$-$AsO_3H_2$, 3, 440
4-[$MeO_2C(CH_2)_8CONH]C_6H_4AsO_3H_2$, 3, 440

$AsC_{17}H_{28}N_3O_7S_2$ 3,4-$O_2N(MeO)C_6H_3As$-[$SCH_2CH(NH_2)CO_2Et$]$_2$, 3, 271

$AsC_{17}H_{27}N_2OS_4$ $MeOC_6H_4As[SC(S)NEt_2]_2$ isomers, 3, 283

$AsC_{17}H_{27}N_2O_5$ 4-[$EtNHOC(CH_2)_7CONH$]-$C_6H_4AsO_3H_2$, 3, 440

4-[$MeNHOC(CH_2)_8CONH$]$C_6H_4AsO_3H_2$, 3, 440

$AsC_{17}H_{27}N_6O_3$

4-[$\overline{N=C(NEt_2)N=C(NEt_2)N=C}NH$]$C_6H_4$-$AsO_3H_2$, 3, 449

$AsC_{17}H_{28}NO_4$ 4-(n-$C_{10}H_{21}CONH)C_6H_4$-AsO_3H_2, 3, 440

4-[Me(n-C_8H_{17})CHCONH]$C_6H_4AsO_3H_2$, 3, 440

$AsC_{17}H_{29}$ Et(n-C_9H_{19})PhAs, 3, 77

Et(n-C_8H_{17})(4-MeC_6H_4)As, 3, 77

$AsC_{17}H_{29}N_2O_{10}$ N-(Hydroxyethyl)-N-(N′-glucosyl-N′-methylcarbamoylmethyl)arsanilic acid, 3, 434

$AsC_{17}H_{29}S$ EtPh(C_9H_{19})AsS, 3, 365

$AsC_{17}H_{30}Br_2P$ 1,1,5,5-Et_4-2,3,4,5-tetrahydro-1H-1,5-benzophosphoniaarsepinium dibromide, 3, 644

$AsC_{17}H_{30}Cl$ [$Me_2(PhCH_2)$(n-C_8H_{17})As]Cl, 3, 321

$AsC_{17}H_{30}I$ [$Me_2(PhCH_2)$(n-C_8H_{17})As]I, 3, 321

$AsC_{17}H_{33}O_2S_4$ PrAs[$SC(S)OC_6H_{13}$]$_2$, 3, 280

$AsC_{17}H_{37}S_2$ $MeAs(SC_8H_{17})_2$, 3, 271

$PrAs(SC_7H_{15})_2$, 3, 271

$AsC_{17}H_{39}O_4$ $Pr_3As(OOCMe_3)_2$, 3, 361

$AsC_{18}H_6Br_9$ (2,4,6-$Br_3C_6H_2$)$_3$As, 3, 30

$AsC_{18}H_9Br_3N_3O_6$ [4,3-$Br(O_2N)C_6H_3$]$_3$As, 3, 30

$AsC_{18}H_9Br_3N_3O_7$ [4,3-$Br(O_2N)C_6H_3$]$_3$AsO, 3, 358

$AsC_{18}H_9Br_6$ (2,4,6-$Br_3C_6H_2$)$_2$AsPh, 3, 69

$AsC_{18}H_{11}Cl_4N_8O_2$ [4-($\overline{N=CClN=CClN=C}NH$)-$C_6H_4$]$_2AsO_2H$, 3, 386

$AsC_{18}H_{12}Br_3$ (4-BrC_6H_4)$_3$As, 3, 17

$AsC_{18}H_{12}Br_3O$ (4-BrC_6H_4)$_3$AsO, 3, 358

$AsC_{18}H_{21}ClNO_4$ 1,1-HO[o-$Cl(CH_2)_3C_6H_4$]-1,2,3,4-tetrahydroarsinolinium NO_3^-, 3, 592

$AsC_{18}H_{12}Cl_3$ (4-ClC_6H_4)$_3$As, 3, 17

$AsC_{18}H_{12}NO_3$ 3,5-O_2N(Ph)-dibenzarsole 5-oxide, 3, 564

$AsC_{18}H_{12}N_3O_6$ ($O_2NC_6H_4$)$_3$As isomers, 3, 30

$AsC_{18}H_{12}N_3O_7$ (3-$O_2NC_6H_4$)$_3$AsO, 3, 358

$AsC_{18}H_{13}$ Poly(4,4′-biphenylylenephenylarsine), 3, 113

5-Ph-dibenzarsole, 3, 565

$AsC_{18}H_{13}BrCl$ Ph(4-BrC_6H_4)(4-ClC_6H_4)As, 3, 77

$AsC_{18}H_{13}BrClO$ Ph(4-BrC_6H_4)(4-ClC_6H_4)-AsO, 3, 358

$AsC_{18}H_{13}Cl_2$ (4-ClC_6H_4)$_2$AsPh, 3, 69

$AsC_{18}H_{13}Cl_2N_8O_4$ {4-[$\overline{N=CClN=C(OH)N=C}NH$]-$C_6H_4$}$_2AsO_2H$, 3, 386

$AsC_{18}H_{13}N_2O_6$ PhAs($OC_6H_4NO_2$-p)$_2$, 3, 254

$AsC_{18}H_{13}O$ Poly{[oxybis(p-phenylene)]-phenylarsine}, 3, 113

5-Ph-dibenzarsole 5-oxide, 3, 565

10-Ph-phenoxarsine, 3, 610

$AsC_{18}H_{13}O_2$ 10-Ph-phenoxarsine 10-oxide, 3, 614

$AsC_{18}H_{14}Br$ (BrC_6H_4)$AsPh_2$ isomers, 3, 62, 69

$AsC_{18}H_{14}Cl$ 4-$ClC_6H_4AsPh_2$, 3, 69

Ph(4-PhC_6H_4)AsCl, 3, 175

$AsC_{18}H_{14}I_2NO_2S$ 4-(4-$PhC_6H_4NHO_2S$)C_6H_4-AsI_2, 3, 198

$AsC_{18}H_{14}N$ (MeC_6H_4)($C_{10}H_7$)AsCN isomers, 3, 205

3,5-H_2N(Ph)-dibenzarsole, 3, 565

10-Ph-5,10-dihydrophenarsazine, 3, 624

$AsC_{18}H_{14}NOS$ 10-(4-$H_2NC_6H_4S$)-phenoxarsine, 3, 615

$AsC_{18}H_{14}NO_2$ 4-$O_2NC_6H_4AsPh_2$, 3, 70

$AsC_{18}H_{14}NO_3S$ 4-($PhC_6H_4NHO_2S$)C_6H_4AsO, 3, 249

$AsC_{18}H_{14}NO_4$ Ph[2-(4-$O_2NC_6H_4$)C_6H_4]-AsO_2H, 3, 386

$AsC_{18}H_{15}$ Ph_3As, 3, 17

2-Ph-2,3-dihydro-1H-naphth[1,8-cd]-arsenin, 3, 599

$AsC_{18}H_{15}Br_2$ Ph_3AsBr_2, 3, 345

$AsC_{18}H_{15}Br_4$ Ph_3AsBr_4, 3, 345

$AsC_{18}H_{15}ClN$ Cl(Et)-7,12-dihydrobenzo-

phenarsazine isomers, 3, 628, 632
Cl(Me)$_2$-7,12-dihydrobenzophenarsazine isomers, 3, 628, 631, 633
$AsC_{18}H_{15}ClNO$ Cl(EtO)-7,12-dihydrobenzophenarsazine isomers, 3, 628
$AsC_{18}H_{15}Cl_2$ Ph_3AsCl_2, 3, 345
$AsC_{18}H_{15}Cl_2N_{10}O_2$
{4-[N=C(NH$_2$)N=CClN=CNH]C_6H_4}$_2AsO_2H$, 3, 386
$AsC_{18}H_{15}F_2$ Ph_3AsF_2, 3, 345
$AsC_{18}H_{15}I_2$ Ph_3AsI_2, 3, 345
$AsC_{18}H_{15}I_4$ Ph_3AsI_4, 3, 345
$AsC_{18}H_{15}N_2O_9S_2$ 2,2'-{1,1'-Diacetyl-2,2'-[5,5'-arsenobis(2-hydroxyphenyl)]hydrazo}-di-1-naphthol-4,8-disulfonic acid, 3, 135
$AsC_{18}H_{15}N_2O_{10}S_2$ 2-[1-Acetyl-2-(5-arsenoso-2-hydroxyphenyl)hydrazo]-1-naphthol-4,8-disulfonic acid, 3, 233
$AsC_{18}H_{15}N_4O_4$ 4-[4-(HOC_6H_4N=N)C_6H_4-N=N]$C_6H_4AsO_3H_2$, 3, 462
$AsC_{18}H_{15}N_4O_5$ 2,4,3,5-(HO)$_2$(PhN=N)$_2$-$C_6HAsO_3H_2$, 3, 504
$AsC_{18}H_{15}N_6O_5$ 5,2-O_2N[4-(PhN=N)C_6H_4-NHN=N]$C_6H_3AsO_3H_2$, 3, 499
$AsC_{18}H_{15}N_6O_8S$ O_2N[4-(4-$HO_3SC_6H_4$N=N)-C_6H_4NHN=N]$C_6H_3AsO_3H_2$ isomers, 3, 499
$AsC_{18}H_{15}N_8O_5$
{4-[N=C(OH)N=C(OH)N=CNH]C_6H_4}$_2$AsOH, 3, 208
$AsC_{18}H_{15}N_8O_6$
{4-[N=C(OH)N=C(OH)N=CNH]C_6H_4}$_2$-AsO_2H, 3, 386
$AsC_{18}H_{15}O$ Ph_2AsOPh, 3, 211
Ph_3AsO, 3, 352
$AsC_{18}H_{15}O_2$ {Ph_3AsO-O}$_n$, 3, 362
Ph(4-PhC_6H_4)AsO_2H, 3, 386
5,5-HO(Ph)-dibenzarsolium hydroxide, 3, 565
$AsC_{18}H_{15}O_3$ (3-HOC_6H_4)$_3$As, 3, 30
10-Ph-phenoxarsine 10,10-dihydroxide, 3, 615
$AsC_{18}H_{15}O_3S$ Ph_3AsSO_3, 3, 26
$AsC_{18}H_{15}O_4$ (HOC_6H_4)$_3$AsO isomers, 3, 358
$AsC_{18}H_{15}O_4S$ Ph_3AsOSO_3, 3, 26
$AsC_{18}H_{15}O_9S_3$ (3-$HO_3SC_6H_4$)$_3$As, 3, 30
$AsC_{18}H_{15}S$ Ph_2AsSPh, 3, 260
Ph_3AsS, 3, 365
$AsC_{18}H_{15}Se$ Ph_3AsSe, 3, 367
$AsC_{18}H_{15}S_2$ PhAs(SPh)$_2$, 3, 271
$AsC_{18}H_{16}BrO_2$ (4-BrC_6H_4)Ph_2As(OH)$_2$, 3, 350
$AsC_{18}H_{16}ClO$ Ph_3As(OH)Cl, 3, 346
$AsC_{18}H_{16}N$ Ph_3As=NH, 3, 369
2-Ph-1,2,3,4-tetrahydroarsenino[4,3-b]quinoline, 3, 591
6,12-Me$_2$-7,12-dihydrobenzo[b]phenarsazine, 3, 631
$AsC_{18}H_{16}NO$ 3,5-MeO(Me)-5,6-dihydroarsinolino[4,3-b]quinoline, 3, 593
10,12-MeO(Me)-7,12-dihydrobenzo[a]-phenarsazine, 3, 628
$AsC_{18}H_{16}NO_2$ 3,5-MeO(Me)-5,6-dihydroarsinolino[4,3-b]quinoline 5-oxide, 3, 593
HO(Me)$_2$-7,12-dihydrobenzophenarsazine 12-oxide isomers, 3, 628, 631
$AsC_{18}H_{16}NO_4$ Ph_3As(OH)NO_3, 3, 348
$AsC_{18}H_{16}NO_5S$ 4-(4-$PhC_6H_4NHO_2S$)C_6H_4-AsO_3H_2, 3, 418
$AsC_{18}H_{16}NO_7S$ 3,4-AcNH(2-$C_{10}H_7SO_3$)-$C_6H_3AsO_3H_2$, 3, 500
$AsC_{18}H_{16}N_3O_5$ 3,4-O_2N(4-$H_2NC_6H_4C_6H_4$NH)-$C_6H_3AsO_3H_2$, 3, 500
$AsC_{18}H_{17}ClN$ [Ph_3As=NH]Cl, 3, 368
2-Ph-1,2,3,4-tetrahydroarsenino-[4,3-b]quinoline·HCl, 3, 591
$AsC_{18}H_{17}N_2$ (4-$H_2NC_6H_4$)$_2$AsPh, 3, 70
$AsC_{18}H_{17}N_2O$ 2,3a,4,5,6,7-Hexahydro-2,5-diphenyl-3H-arsenino[4,3-c]-pyrazol-3-one, 3, 591
$AsC_{18}H_{17}N_2O_{10}S_2$ 4-{2-[1,4,8-HO(HO_3S)$_2$-2-naphthyl]-3-methyl-1-diaziridinyl]-$C_6H_4AsO_3H_2$, 3, 472
$AsC_{18}H_{17}N_{10}O_3$
{4-[N=C(NH$_2$)N=C(OH)N=CNH]C_6H_4}$_2$AsOH, 3, 208
$AsC_{18}H_{17}N_{10}O_4$
{4-[N=C(NH$_2$)N=C(OH)N=CNH]C_6H_4}$_2AsO_2H$, 3, 386
$AsC_{18}H_{17}O_2$ Ph_3As(OH)$_2$, 3, 350

$AsC_{18}H_{17}O_5S$ $Ph_3As(OH)SO_4H$, 3, 348

$AsC_{18}H_{18}NO$ 3,5,11-MeO(Me)$_2$-5,6-dihydro-11H-arsinolino[4,3-b]indole, 3, 593

$AsC_{18}H_{18}NOS_2$ 10-Phenoxarsinyl-SC(S)-$\overline{N(CH_2)_4C}H_2$, 3, 615

$AsC_{18}H_{18}NO_4$ 4-[$\overline{CH_2CH(Ph)CH_2COCH=C}NH$]-$C_6H_4AsO_3H_2$, 3, 443

$AsC_{18}H_{18}N_3$ $(H_2NC_6H_4)_3As$ isomers, 3, 31

$AsC_{18}H_{18}N_3O$ $(H_2NC_6H_4)_3AsO$ isomers, 3, 358

$AsC_{18}H_{18}N_3O_9$ 7,1,1-MeO(Me)$_2$-1,2,3,4-tetrahydro-4-oxoarsinolinium picrate, 3, 592

$AsC_{18}H_{18}N_7O_3$
4-[$\overline{N=C(NH_2)N=C(NHN=CHCH=CHPh)N=C}NH$]-$C_6H_4AsO_3H_2$, 3, 449*

$AsC_{18}H_{19}N_4O_7$ {$Me_2[NC(CH_2)_3]PhAs$}-$OC_6H_2(NO_2)_3$, 3, 317

$AsC_{18}H_{19}N_4O_8$ [$Me_2(NCCH_2CH_2)$(3-MeO-C_6H_4)As]$OC_6H_2(NO_2)_3$, 3, 317

$AsC_{18}H_{19}N_6O_4S$ 4-[4-(6-Methyl-2-pyrimidylsulfamoyl)phenylguanyl]arsanilic acid, 3, 433

$AsC_{18}H_{19}N_{12}O$
{4-[$\overline{N=C(NH_2)N=C(NH_2)N=C}NH$]$C_6H_4$}$_2$-AsOH, 3, 208

$AsC_{18}H_{19}N_{12}O_2$
{4-[$\overline{N=C(NH_2)N=C(NH_2)N=C}NH$]$C_6H_4$}$_2$-$AsO_2H$, 3, 386

$AsC_{18}H_{20}Br$ 3,3′,4,4′-Tetrahydro-1,1′-(2H,2′H)-spirobisarsinolinium Br^-, 3, 599
3,3′,4,4′-Tetrahydro-2,2′(1H,1′H)-spirobisisoarsinolinium Br^-, 3, 600

$AsC_{18}H_{20}I$ 3,3′,4,4′-Tetrahydro-1,1′-(2H,2′H)-spirobisarsinolinium I^-, 3, 599
3,3′,4,4′-Tetrahydro-2,2′(1H,1′H)-spirobisisoarsinolinium I^-, 3, 599

$AsC_{18}H_{21}$ 4-$(PhCH=CH)C_6H_4AsEt_2$, 3, 70
cyclo-$C_6H_{11}AsPh_2$, 3, 70

$AsC_{18}H_{21}Cl_2O$ 1-{2-[$Cl(CH_2)_3C_6H_4$]-1,2,3,4-tetrahydro-1-hydroxyarsinolinium}Cl^-, 3, 592

$AsC_{18}H_{21}N_2O_5$ 4-[$Me_2C_6H_3NHCOCH(Ac)NH$]-$C_6H_4AsO_3H_2$, 3, 440

$AsC_{18}H_{21}O$ (cyclo-C_6H_{11})Ph_2AsO, 3, 358
1-{2-[$HO(CH_2)_3$]C_6H_4}-1,2,3,4-tetrahydroarsinoline, 3, 592

$AsC_{18}H_{22}NO$ 10-($Et_2NCH_2CH_2$)-phenoxarsine, 3, 615

$AsC_{18}H_{22}NO_3S$ [$Me_2(NCCH_2CH_2)PhAs$]O_3S-C_6H_4Me, 3, 316

$AsC_{18}H_{22}NO_7$ 1-(4-Arsonophenyl)-2-methylpyrrole-3,4-$(CO_2Et)_2$, 3, 473*

$AsC_{18}H_{22}N_3O_3$ 4-[4-(cyclo-$C_6H_{11}NH$)-$C_6H_4N=N$]$C_6H_4AsO_3H_2$, 3, 462

$AsC_{18}H_{22}N_3O_7$ [$(CH_2=CHCH_2)_4As$]OC_6H_2-$(NO_2)_3$, 3, 317

$AsC_{18}H_{22}O_2PS_2$ 5-Dibenzarsolyl-SP(S)-$(OPr)_2$, 3, 565
5-Dibenzarsolyl-SP(S)$(OCHMe_2)_2$, 3, 565

$AsC_{18}H_{22}O_3PS_2$ 10-Phenoxarsinyl-SP(S)-$(OPr)_2$, 3, 615
10-Phenoxarsinyl-SP(S)$(OCHMe_2)_2$, 3, 615

$AsC_{18}H_{23}BrN$ [$\overline{CH_2CH_2As(Ph)CH_2CH_2N}$(Ph)-Et]Br, 3, 619

$AsC_{18}H_{23}N_2O_2S$ {Me[$H_2N(HN=)C$](HO_2CCH_2-CH_2)PhAs}SCH_2Ph, 3, 316*

$AsC_{18}H_{24}Br$ [$EtPrPh(PhCH_2)As$]Br, 3, 322
[$MeBuPh(PhCH_2)As$]Br, 3, 322

$AsC_{18}H_{24}Br_2$ 1,4,1,4-$Me_2(Ph)_2$-1,4-diarsenanium $2Br^-$, 3, 602

$AsC_{18}H_{24}Br_2N$ [$\overline{CH_2CH_2As(Ph)CH_2CH_2N}$(Ph)-Et]Br·HBr, 3, 619

$AsC_{18}H_{24}ClO_4$ [$MeBuPh(PhCH_2)As$]ClO_4, 3, 322

$AsC_{18}H_{24}IO$ [$Me_3(Ph_2CHOCH_2CH_2)As$]I, 3, 322

$AsC_{18}H_{24}NS$ $Ph_2AsSCH_2CH_2NEt_2$, 3, 260

$AsC_{18}H_{24}N_3O_3$ 4-[4-(n-$C_6H_{13}NH)C_6H_4N=N$]-$C_6H_4AsO_3H_2$, 3, 462
4-{4-[$Me_2CH(CH_2)_3NH$]$C_6H_4N=N$}C_6H_4-AsO_3H_2, 3, 462

$AsC_{18}H_{25}ClNS$ [$Ph_2AsSCH_2CH_2NHEt_2$]Cl, 3, 260

$AsC_{18}H_{25}IP$ [Ph_2AsPEt_3]I, 3, 291

$AsC_{18}H_{27}O_2$ PhAs[$OCH(Pr)CH=CH_2$]$_2$, 3, 254

$PhAs(OC_6H_{11})_2$, 3, 254
$AsC_{18}H_{27}O_2S_4$ $PhAs[SC(S)OC_5H_{11}]_2$, 3, 280
$PhAs[SC(S)OCH_2CH_2CHMe_2]_2$, 3, 280
$AsC_{18}H_{27}O_4P_2S_4$ $1\text{-}C_{10}H_7As[SP(S)\text{-}(OEt)_2]_2$, 3, 281
$AsC_{18}H_{28}NO_6$ $4\text{-}[EtO_2C(CH_2)_8CONH]C_6H_4\text{-}AsO_3H_2$, 3, 440
$AsC_{18}H_{28}NO_6S_2$ $3,4\text{-}H_2N(HOCH_2CH_2O)C_6H_3\text{-}As(SCH_2CO_2CHMe_2)_2$, 3, 271
$AsC_{18}H_{28}NO_{13}$ N-Lactosylarsanilic acid, 3, 440
N-Maltosylarsanilic acid, 3, 440
$AsC_{18}H_{29}N_2O_5$ $4\text{-}[Me_2NOC(CH_2)_8CONH]\text{-}C_6H_4AsO_3H_2$, 3, 440
$AsC_{18}H_{29}O_6$ 8-Butyl-7,9,17,18-tetraoxa-8-arsatrispiro[5.1.1.5.2.2]-nonadecane-16,19-dione, 3, 582
$AsC_{18}H_{31}S_2$ $PhAs(SC_6H_{13})_2$, 3, 271
$AsC_{18}H_{32}ClO_4P_2S_4$ $2\text{-}ClC_6H_4As[SP(S)\text{-}(OPr)_2]_2$, 3, 281
$AsC_{18}H_{32}IO$ $[Bu_2PhAs(OBu)]I$, 3, 346
$AsC_{18}H_{32}NO_4$ $[Bu_2PhAs(OBu)]NO_3$, 3, 350
$AsC_{18}H_{33}ClNO_3$ $3\text{-}ClC_6H_4AsO_3H_2\cdot H_2N\text{-}C_{12}H_{25}$, 3, 409
$4\text{-}ClC_6H_4AsO_3H_2\cdot H_2NC_{12}H_{25}$, 3, 409
$AsC_{18}H_{33}Cl_2$ $(C_6H_{11})_3AsCl_2$, 3, 345
$AsC_{18}H_{33}I_4$ $(C_6H_{11})_3AsI_4$, 3, 345
$AsC_{18}H_{33}N_2O_5$ $2\text{-}O_2NC_6H_4AsO_3H_2\cdot H_2N\text{-}C_{12}H_{25}$, 3, 419
$AsC_{18}H_{33}N_2O_6$ $4,3\text{-}HO(O_2N)C_6H_3AsO_3H_2\cdot\text{-}H_2NC_{12}H_{25}$, 3, 475
$AsC_{18}H_{33}N_2S_2$ $PhAs(SCH_2CH_2NEt_2)_2$, 3, 271
$AsC_{18}H_{33}O$ $(cyclo\text{-}C_6H_{11})_3AsO$, 3, 358
$AsC_{18}H_{33}S$ $(cyclo\text{-}C_6H_{11})_3AsS$, 3, 365
$AsC_{18}H_{33}Se$ $(cyclo\text{-}C_6H_{11})_3AsSe$, 3, 367
$AsC_{18}H_{34}Cl_2N_3O_2S_2$ $4\text{-}O_2NC_6H_4As(SCH_2\text{-}CH_2NHEt_2)_2]Cl_2$, 3, 271
$AsC_{18}H_{35}Cl_2N_2S_2$ $PhAs[(SCH_2CH_2NHEt_2)_2]Cl_2$, 3, 271
$AsC_{18}H_{35}O_5S$ $(cyclo\text{-}C_6H_{11})_3As(OH)SO_4H$, 3, 348
$AsC_{18}H_{37}$ $(C_8H_{17})_2AsCH{=}CH_2$, 3, 70
$AsC_{18}H_{37}Cl_2$ $C_{18}H_{37}AsCl_2$, 3, 198
$AsC_{18}H_{39}$ $(C_6H_{13})_3As$, 3, 27
$n\text{-}C_{16}H_{33}AsMe_2$, 3, 70
$AsC_{18}H_{39}S_2$ $EtAs(SC_8H_{17})_2$, 3, 271
$BuAs(SC_7H_{15})_2$, 3, 271
$AsC_{18}H_{40}I$ $[Et_3(n\text{-}C_{12}H_{25})As]I$, 3, 322
$AsC_{19}H_9Cl_3N_3S_4$ $N{=}CCH{=}C(Cl)CH{=}CAs[S\text{-}2\text{-}(6\text{-}Cl\text{-}benzothiazole)]_2$, 3, 272
$AsC_{19}H_{12}Br_3N_4O_9$ $\{Me[4,3\text{-}Br(O_2N)C_6H_3]_3\text{-}As\}NO_3$, 3, 322
$AsC_{19}H_{12}ClO_3$ 8,10-Cl(Ph)-2-phenoxarsine-CO_2H, 3, 615
$AsC_{19}H_{12}N_3S_4$ $N{=}CHCH{=}CHCH{=}CAs(S\text{-}2\text{-}benzothiazole)_2$, 3, 272
$AsC_{19}H_{13}ClN_3O_5$ N-[7,2-Cl(O_2N)-7-acridyl]-$NHC_6H_4AsO_3H_2$, 3, 449
$AsC_{19}H_{13}ClO_3$ p-(2-Cl-phenoxarsin-10-yl)-$C_6H_4CO_2H$, 3, 615
$AsC_{19}H_{13}ClO_4$ p-(2-Cl-phenoxarsin-10-yl)-$C_6H_4CO_2H$ 10-oxide, 3, 615
$AsC_{19}H_{13}O$ 5-Phenyl-10(5H)-acridarsinone, 3, 597
$AsC_{19}H_{13}O_2$ p-(5-Dibenzarsolyl)-C_6H_4-CO_2H, 3, 565
5-Ph-10(5H)-acridarsinone 5-oxide, 3, 597
$AsC_{19}H_{13}O_3$ p-(5-Dibenzarsolyl)-C_6H_4-CO_2H As-oxide, 3, 565
10-Ph-phenoxarsine-2-CO_2H, 3, 616
$AsC_{19}H_{13}O_4$ 10-Ph-phenoxarsine-2-CO_2H 10-oxide, 3, 616
$AsC_{19}H_{14}ClO$ 2,10-Cl(p-MeC_6H_4)-phenoxarsine, 3, 616
8,2,10-Cl(Me)(Ph)-phenoxarsine, 3, 616
$AsC_{19}H_{14}ClOS$ $Ph_2AsSC(O)C_6H_4Cl\text{-}p$, 3, 261
$AsC_{19}H_{14}NOS$ 10-(BzS)-5,10-dihydrophenarsazine, 3, 624
$AsC_{19}H_{14}NO_5S$ $Ph[3,4\text{-}O_2N(HO)C_6H_3]AsS\text{-}C_6H_4CO_2H$, 3, 260
$AsC_{19}H_{15}$ 5-p-Tolyldibenzarsole, 3, 566
5,10-Dihydro-5-phenylacridarsine, 3, 597
5,6-Dihydro-5-phenylarsanthridin, 3, 598
$AsC_{19}H_{15}Br_3I$ $[Me(4\text{-}BrC_6H_4)_3As]I$, 3, 322
$AsC_{19}H_{15}Br_3NO_3$ $[Me(4\text{-}BrC_6H_4)_3As]NO_3$,

3, 322
$AsC_{19}H_{15}ClN$ Cl(Et)(Me)-7,12-dihydrobenzophenarsazine isomers, 3, 628, 629, 633
13-Cl-7,13-dihydrobenzocyclopenteno[a,i]phenarsazine, 3, 635*
7-Cl-7,13-dihydrobenzocyclopenteno[c,i]phenarsazine, 3, 635*
$AsC_{19}H_{15}ClN_3O_6$ [Me(3-$O_2NC_6H_4$)$_3$As]Cl, 3, 322
$AsC_{19}H_{15}ClN_5O_{10}S_2$ N-{4,6-Cl[6,5,2,7-Cl(HO)(HO_3S)$_2$-4-naphthylamino]-s-triazin-2-yl}arsanilic acid, 3, 449
$AsC_{19}H_{15}Cl_2O_3$ $PhCH_2AsO(OC_6H_4Cl\text{-}p)_2$, 3, 546
$AsC_{19}H_{15}Cl_3I$ [$Ph_3As\cdot CCl_3I$], 3, 19
$AsC_{19}H_{15}N_2O_3$ N-9-Acridylarsanilic acid, 3, 449
N-(4-Benzo[h]quinolyl)arsanilic acid, 3, 449
N-(4-Benzo[f]quinolyl)arsanilic acid, 3, 449
4-(1-Isoquinolylamino)-$C_{10}H_6$-1-AsO_3H_2, 3, 510
$AsC_{19}H_{15}N_2O_4$ 3-(1-Benzo[f]quinolylamino)-4-hydroxybenzenearsonic acid, 3, 500
$AsC_{19}H_{15}N_2O_{12}S_2$ 4-{2-[1,4,8-HO(HO_3S)$_2$-2-naphthyl]-3,5-dioxo-1-pyrazolidyl}-$C_6H_4AsO_3H_2$, 3, 472
$AsC_{19}H_{15}N_4O_9$ [Me(3-$O_2NC_6H_4$)$_3$As]NO_3, 3, 322
$AsC_{19}H_{15}O$ 5-(p-$MeOC_6H_4$)-dibenzarsole, 3, 566
3,5-MeO(Ph)-dibenzarsole, 3, 566
$AsC_{19}H_{15}O_2$ 2-(Ph_2As)$C_6H_4CO_2H$, 3, 62
$AsC_{19}H_{15}O_2S$ $Ph_2AsSC_6H_4CO_2H$, 3, 260
$AsC_{19}H_{15}O_3$ ($HO_2CC_6H_4$)Ph_2AsO isomers, 3, 359
$AsC_{19}H_{15}O_3S$ Ph(4-HOC_6H_4)$AsSC_6H_4CO_2H$, 3, 260
$AsC_{19}H_{15}O_4$ 2-Benzyl-2,2'-spirobi-1,3,2-benzodioxarsole, 3, 581
$AsC_{19}H_{15}O_4S$ (4-HOC_6H_4)$_2AsSC_6H_4CO_2H$, 3, 260
$AsC_{19}H_{16}Br$ Ph(4-BrC_6H_4)(4-MeC_6H_4)As, 3, 77
$AsC_{19}H_{16}Cl$ Ph(4-ClC_6H_4)(4-MeC_6H_4)As, 3, 78
$AsC_{19}H_{16}ClO$ Ph(4-ClC_6H_4)(4-MeC_6H_4)AsO, 3, 359
$AsC_{19}H_{16}ClO_2$ (4-ClC_6H_4)(2-$PhC_6H_4CH_2$)-AsO_2H, 3, 387
$AsC_{19}H_{16}IO$ 10,10-Me(Ph)-phenoxarsinium iodide, 3, 610
$AsC_{19}H_{16}I_2NO_2S$ 4-[Ph($PhCH_2$)NO_2S]C_6H_4-AsI_2, 3, 198
$AsC_{19}H_{16}NO$ 3,5-MeO(Me)-5,6-dihydroarsinolino[4,3-b]quinoline-7-carboxylic acid, 3, 594
$AsC_{19}H_{16}NO_2$ Ph(4-$O_2NC_6H_4$)(4-MeC_6H_4)As, 3, 78
1,2,3,4-Tetrahydro-2-phenylarsenino[4,3-b]quinoline-10-carboxylic acid, 3, 591
5,6,8,13-Tetrahydro-3-methoxy-4H-benz[1,9]arsinolizino[2,3-b]indol-4-one, 3, 599
$AsC_{19}H_{16}NO_2S$ Ph(4-$H_2NC_6H_4$)$AsSC_6H_4CO_2H$, 3, 260
$AsC_{19}H_{16}NO_3S$ Ph[3,4-H_2N(HO)C_6H_3]AsS-$C_6H_4CO_2H$, 3, 260
$AsC_{19}H_{16}N_2O_3$ N-(1-Benzo[f]quinolyl)-m-arsanilic acid, 3, 425
N-(4-Benzo[h]quinolyl)-m-arsanilic acid, 3, 425
$AsC_{19}H_{16}N_3O_4$ 4-[$C_{10}H_7$NHCOCH(CN)NH]-$C_6H_4AsO_3H_2$, 3, 440
$AsC_{19}H_{17}$ (MeC_6H_4)$AsPh_2$ isomers, 3, 62, 70
Ph_3AsCH_2, 3, 309
$AsC_{19}H_{17}ClN$ Chlorotrimethyl-7,12-dihydrobenzophenarsazine isomers, 3, 629, 633
$AsC_{19}H_{17}INO_3$ [(ICH_2)Ph_3As]NO_3, 3, 322
$AsC_{19}H_{17}N_2O_4S$ [3,4-H_2N(HO)C_6H_3]$_2$AsS-$C_6H_4CO_2H$, 3, 260
$AsC_{19}H_{17}N_2O_6S$ 4-(4-$BzNHC_6H_4SO_2NH$)C_6H_4-AsO_3H_2, 3, 444
$AsC_{19}H_{17}N_4O_2S_4$ $PhCH_2As$[SC(S)NHN=CH-C=CHCH=CHO]$_2$, 3, 284*
$AsC_{19}H_{17}N_5O_9P$ Me(4-$O_2NC_6H_4$)As(O)OP-(O)($NHC_6H_4NO_2$-4)$_2$, 3, 388

$AsC_{19}H_{17}N_6O_8S$ 5,2-O_2N[4-(4-HO_3S-C_6H_4N=N)C_6H_4N(Me)N=N]$C_6H_3AsO_3H_2$, 3, 500

$AsC_{19}H_{17}N_8S_2$ 4-[N=C(NH_2)N=C(NH_2)N=CNH]C_6H_4As-(SC=CHCH=CHCH=N)$_2$, 3, 272

$AsC_{19}H_{17}O_2$ Ph(2-$PhC_6H_4CH_2$)AsO_2H, 3, 387

$AsC_{19}H_{17}O_2S$ Et($HO_2CC_6H_4$)($C_{10}H_7$)AsS isomers, 3, 365

$AsC_{19}H_{18}Br$ [$MePh_3As$]Br, 3, 323

$AsC_{19}H_{18}Cl$ [$MePh_3As$]Cl, 3, 323

$AsC_{19}H_{18}I$ [$MePh_3As$]I, 3, 323

$AsC_{19}H_{18}N$ 9,7-Et(Me)-7,12-dihydrobenzo[c]phenarsazine, 3, 633

Trimethyl-7,12-dihydrobenzophenarsazine isomers, 3, 629, 631, 633

$AsC_{19}H_{18}NO_2$ HO(Me)$_3$-7,12-dihydrobenzophenarsazine 12-oxide isomers, 3, 629, 633

$AsC_{19}H_{18}NO_3$ [$MePh_3As$]NO_3, 3, 323

$AsC_{19}H_{18}NO_4S$ 4-[Ph($PhCH_2$)NO_2S]C_6H_4-As$(OH)_2$, 3, 225

$AsC_{19}H_{18}NO_5S$ 4-[Ph($PhCH_2$)NO_2S]C_6H_4-AsO_3H_2, 3, 418

$AsC_{19}H_{18}N_3O_3$ 4-[4-($PhCH_2NH$)C_6H_4N=N]-$C_6H_4AsO_3H_2$, 3, 462

$AsC_{19}H_{18}N_5O_7$ [Me$(NCCH_2CH_2)_2$PhAs]O-$C_6H_2(NO_2)_3$, 3, 318

$AsC_{19}H_{19}$ Et($C_{10}H_7$)(MeC_6H_4)As isomers, 3, 78

$AsC_{19}H_{19}INO$ 3,5,5-MeO$(Me)_2$-5,6-dihydroarsinolino[4,3-b]quinolinium I^-, 3, 593

$AsC_{19}H_{19}O$ [$MePh_3As$]OH, 3, 323

$AsC_{19}H_{19}O_2S_2$ 10-Phenoxarsinyl-SC(S)O-CH$(CH_2)_4$CH$_2$, 3, 616

$AsC_{19}H_{20}NO_4S$ 10-Phenoxarsinyl-$SCMe_2$-CH(NHAc)CO_2H, 3, 616

$AsC_{19}H_{21}ClN_3$ [Me(3-$H_2NC_6H_4$)$_3$As]Cl, 3, 322, 327

$AsC_{19}H_{21}IN_3$ [Me(3-$H_2NC_6H_4$)$_3$As]I, 3, 323

$AsC_{19}H_{21}N_2O_2$ 7,1,3-MeO(Me)(4-Me_2N-C_6H_4N=)-1,2,3,4-tetrahydro-1-oxoarsinoline, 3, 592

$AsC_{19}H_{22}N_3O_9S_2$ 3-{[1,4,8-MeNH$(HO_3S)_2$-2-naphthyl]-N(Me)N(Me)}$C_6H_4AsO_3H_2$, 3, 454*

$AsC_{19}H_{22}I_2NO_2$ 3,5,5-MeO$(Me)_2$-5,6-dihydroarsinolino[4,3-b]quinolinium I^-·HI·H_2O, 3, 593

$AsC_{19}H_{22}NO_4$ 4-($PhCH_2$CHCH$_2$CH$_2$COCH$_2$CHNH)$C_6H_4AsO_3H_2$, 3, 443

$AsC_{19}H_{22}N_3O_7S_2$ 3,4-O_2N($PhCH_2O$)C_6H_3-As[$SCH_2CH(NH_2)CO_2H$]$_2$, 3, 272

$AsC_{19}H_{23}O$ 1-{2-[MeO$(CH_2)_3$]C_6H_4}-1,2,3,4-tetrahydroarsinoline, 3, 592

$AsC_{19}H_{23}O_5S$ 7,1,1-MeO$(Me)_2$-1,2,3,4-tetrahydro-4-oxoarsolinium p-toluenesulfonate, 3, 592

$AsC_{19}H_{24}I$ [Me(Ph_2(C_6H_{11})As]I, 3, 327

$AsC_{19}H_{24}N_3O_7$ [MeEtPr($PhCH_2$)As]OC_6H_2-$(NO_2)_3$, 3, 318*

$AsC_{19}H_{25}N_2O_3S$ {Me[H_2N(HN=)C](HO_2CCH_2-CH_2)(3-$MeOC_6H_4$)As}SCH_2Ph, 3, 317*

$AsC_{19}H_{25}N_2O_4$ 2-[2,4,3,5-HO(Me)$(Me_2CH)_2$-C_6HN=N]$C_6H_4AsO_3H_2$, 3, 462

$AsC_{19}H_{25}N_2O_5$ 3,4-$PhNHCOCH_2NH$(Me_2CH-CH_2CH_2O)$C_6H_3AsO_3H_2$, 3, 500

$AsC_{19}H_{26}Cl$ [EtBuPh($PhCH_2$)As]Cl, 3, 327

$AsC_{19}H_{26}N_3O_3$ 4-[4-(n-C_7H_{15}NH)C_6H_4-N=N]$C_6H_4AsO_3H_2$, 3, 462

$AsC_{19}H_{27}BrNO$ [Me_2(2-$Me_2NC_6H_4$)(2-MeO-$CH_2C_6H_4CH_2$)As]Br, 3, 327

$AsC_{19}H_{27}ClNO$ [Me_2(2-$Me_2NC_6H_4$)(2-MeO-$CH_2C_6H_4CH_2$)As]Cl, 3, 327

$AsC_{19}H_{27}ClNO_5$ [Me_2(2-$Me_2NC_6H_4$)(2-MeO-$CH_2C_6H_4CH_2$)As]ClO_4, 3, 327

$AsC_{19}H_{30}N_3O_4$ 6,8-MeO[$Et_2N(CH_2)_3$CH(Me)-NH]-5-quinolinearsonic acid, 3, 525

$AsC_{19}H_{41}S_2$ MeAs$(SC_9H_{19})_2$, 3, 272*

Pr$As(SC_8H_{17})_2$, 3, 272

$AsC_{19}H_{42}I$ [Me_3(n-$C_{16}H_{33}$)As]I, 3, 327

$AsC_{20}H_{12}ClN_2S_4$ 2-ClC_6H_4As(S-2-benzothiazole)$_2$, 3, 272

$AsC_{20}H_{12}N_3O_2S_4$ 4-$O_2NC_6H_4$As(S-2-benzothiazole)$_2$, 3, 272

$AsC_{20}H_{12}N_5O_7S_2$ N,N-Bis(6-nitro-2-benzothiazolyl)arsanilic acid, 3, 434

$AsC_{20}H_{13}$ Poly[(9,10-anthrylene)phenylarsine], 3, 114

$AsC_{20}H_{13}ClN$ 14-Cl-7,14-dihydrodibenzo[a,h]phenarsazine, **3**, 637
$AsC_{20}H_{13}ClN_3O_7$ 3,1-Cl(Ph)-arsindolinium picrate, **3**, 556
$AsC_{20}H_{13}N_2OS_4$ 4-HOC_6H_4As(S-2-benzothiazole)$_2$, **3**, 272
$AsC_{20}H_{13}N_2O_2$ 10-Phthalimino-5,10-dihydrophenarsazine, **3**, 624
$AsC_{20}H_{13}N_2O_9S_2$ 3,4,5-HO$(O_2N)_2C_6H_2$-As$(SC_6H_4CO_2H)_2$, **3**, 272
$AsC_{20}H_{14}Cl$ $(C_{10}H_7)_2AsCl$ isomers, **3**, 175
$AsC_{20}H_{14}NO_7S_2$ 3,4-$O_2N(HO)C_6H_3As$-$(SC_6H_4CO_2H)_2$, **3**, 272
$AsC_{20}H_{14}N_3O_3S_5$ 4-$HO_3SNHC_6H_4As$(S-2-benzothiazole)$_2$, **3**, 272
$AsC_{20}H_{14}N_3S_4$ 4-$H_2NC_6H_4As$(S-2-benzothiazole)$_2$, **3**, 272
$AsC_{20}H_{14}N_3S_4$ 4-$H_2NC_6H_4As$(S-2-benzothiazole)$_2$, **3**, 272
$AsC_{20}H_{15}$ $PhC{\equiv}CAsPh_2$, **3**, 70
$(1\text{-}C_{10}H_7)_2AsH$, **3**, 118
$AsC_{20}H_{15}ClNO_4S_2$ 3,4-Cl$(H_2N)C_6H_3As$-$(SC_6H_4CO_2H)_2$, **3**, 272
$AsC_{20}H_{15}ClN_3O_8$ 5,2,2-Cl(HO)(Ph)-isoarsindolinium picrate, **3**, 557
2,2-HO(4-ClC_6H_4)-isoarsindolinium picrate, **3**, 558
$AsC_{20}H_{15}N_2O_4$ 3-(2-$HOC_{10}H_6$-1-N=N)-$C_{10}H_6$-1-AsO_3H_2, **3**, 511
$AsC_{20}H_{15}N_2O_{10}S_2$ 2-[2,3,6-HO$(HO_3S)_2$-1-naphthylazo]naphthalenearsonic acid, **3**, 511
$AsC_{20}H_{15}O_2S_2$ PhAs(SCOPh)$_2$, **3**, 279
$AsC_{20}H_{15}O_3$ 4-(9-Anthryl)-$C_6H_4AsO_3H_2$, **3**, 511
p-(3-MeO-5-dibenzarsolyl)-$C_6H_4CO_2H$, **3**, 566
$AsC_{20}H_{16}BrO_2$ Ph(4-BrC_6H_4)(2-MeO_2C-C_6H_4)As, **3**, 78
$AsC_{20}H_{16}ClN_2O_4$ N-[Chloro(methoxy)-9-acridyl]arsanilic acid isomers, **3**, 450
$AsC_{20}H_{16}ClO_3$ (4-$HO_2CC_6H_4$)[2-(4-MeO-C_6H_4)C_6H_4]AsCl, **3**, 175
$AsC_{20}H_{16}Cl_3O_3$ [$Ph_3AsO\cdot CCl_3CO_2H$], **3**, 352
$AsC_{20}H_{16}NO$ 5,10-Ac(Ph)-5,10-dihydrophenarsazine, **3**, 624
$AsC_{20}H_{16}NO_2$ 5,6-Dihydro-3-methoxy-4H,8H-benz[i,j]arsinolizino[2,3-b]-quinolin-4-one, **3**, 599
$AsC_{20}H_{16}NO_5S_2$ 3,4-$H_2N(HO)C_6H_3As(SC_6H_4$-$CO_2H)_2$, **3**, 272
$AsC_{20}H_{16}N_3O_2$ 4-[NCC_6H_4C(=NPh)NH]-$C_6H_4As(OH)_2$, **3**, 225
$AsC_{20}H_{16}N_3O_7$ 2-Phenylisoarsindolinium picrate, **3**, 558
$AsC_{20}H_{16}N_3O_8$ 2-Hydroxy-2-phenylisoarsindolinium picrate, **3**, 558
$AsC_{20}H_{17}$ 4-$CH_2{=}CHC_6H_4AsPh_2$, **3**, 62
Poly[p-(diphenylarsino)styrene], **3**, 114
10,11-Dihydro-5-phenyl-5H-dibenz-[b,f]arsepin, **3**, 642
5,7-Dihydro-6-phenyl-6H-dibenz[c,e]-arsepin, **3**, 642*
$AsC_{20}H_{17}ClN$ Chlorohexahydrodibenzophenarsazine isomers, **3**, 637
$AsC_{20}H_{17}N_2O_3$ N-(2-Methyl-9-acridyl)-arsanilic acid, **3**, 450
$AsC_{20}H_{17}N_2O_4$ N-(2-Methoxy-9-acridyl)-arsanilic acid, **3**, 450
$AsC_{20}H_{17}N_2O_{11}S_2$ 2-[1,2-Diacetyl-2-(5-arsenoso-2-hydroxyphenyl)hydrazo]-1-naphthol-4,8-disulfonic acid, **3**, 233
$AsC_{20}H_{17}O$ 10,11-Dihydro-5-phenyl-5H-dibenz[b,f]arsepin 5-oxide, **3**, 642
$AsC_{20}H_{17}O_2$ 2-$(MeO_2C)C_6H_4AsPh_2$, **3**, 70
$AsC_{20}H_{17}O_2S$ $Ph_2AsSC_6H_4CO_2Me$, **3**, 260*
$AsC_{20}H_{17}O_5$ (4-$HO_2CC_6H_4$)[2-(4-$MeOC_6H_4$)-C_6H_4]AsO_2H, **3**, 387
$AsC_{20}H_{18}BrO_4$ Ph(4-BrC_6H_4)(2-MeO_2C-C_6H_4)As$(OH)_2$, **3**, 351
$AsC_{20}H_{18}Cl$ $(4\text{-}MeC_6H_4)_2AsC_6H_4Cl$-4, **3**, 70
$AsC_{20}H_{18}ClO_2$ [$Ph_3(HO_2CCH_2)As$]Cl, **3**, 327
$AsC_{20}H_{18}I$ 5,5-Me(Ph)-5,10-dihydroacridarsinium I^-, **3**, 597
5,5-Me(Ph)-5,6-dihydroarsanthridinium I^-, **3**, 598
$AsC_{20}H_{18}IO_2$ [$MePh_2(2\text{-}HO_2CC_6H_4)As$]I,

3, 327

$AsC_{20}H_{18}N$ 5-(p-$Me_2NC_6H_4$)-dibenzarsole, 3, 566

7,13-Dihydro-13-methylbenzocyclopenteno[a,i]phenarsazine, 3, 635

$AsC_{20}H_{18}NO_2$ Ph_3As=NOCOMe, 3, 369

$AsC_{20}H_{18}NS$ [$MePh_3As$]SCN, 3, 324

$AsC_{20}H_{18}N_3O_7S$ HO(4-$AcNHC_6H_4SO_2C_6H_4$-N=N)$C_6H_3AsO_3H_2$ isomers, 3, 500

$AsC_{20}H_{18}N_3O_{11}S$ 2-(2-Arsonophenylazo)-3-sulfo-1-hydroxy-6-naphthyliminodiacetic acid, 3, 463

$AsC_{20}H_{18}P$ 10,5-Et(Ph)-5,10-dihydrodibenzo[1,4]phosphenarsenin, 3, 640

$AsC_{20}H_{19}ClN$ Cl(Et)$(Me)_2$-7,12-dihydrobenzo[a]phenarsazine isomers, 3, 629

Cl(Me_2CH)(Me)-7,12-dihydrobenzophenarsazine isomers, 3, 629, 634

$AsC_{20}H_{19}N_2O_5$ 4-[$C_{10}H_7NHCOCH$(Ac)NH]-$C_6H_4AsO_3H_2$, 3, 440

$AsC_{20}H_{19}N_4O_2$ 4-[4-Guanyl-α-(phenylimino)benzylamino]benzenearsonous acid, 3, 226

$AsC_{20}H_{19}N_4O_6S$ H_2N(4-$AcNHC_6H_4SO_2C_6H_4$-N=N)$C_6H_3AsO_3H_2$ isomers, 3, 500

$AsC_{20}H_{19}O_2$ $(4\text{-}MeOC_6H_4)_2AsPh$, 3, 70

$AsC_{20}H_{19}S_2$ $PhAs(SCH_2Ph)_2$, 3, 272

$AsC_{20}H_{20}Br$ [$EtPh_3As$]Br, 3, 327

$AsC_{20}H_{20}ClO_4$ [$EtPh_3As$]ClO_4, 3, 327

$AsC_{20}H_{20}N$ Et$(Me)_2$-7,12-dihydrobenzophenarsazine isomers, 3, 629, 634

7,2-Dihydro-8,10,11,12-tetramethylbenzo[a]phenarsazine, 3, 629

$AsC_{20}H_{20}NO_3$ [$EtPh_3As$]NO_3, 3, 327

$AsC_{20}H_{20}NO_4$ [$Ph_3(HOCH_2CH_2)As$]NO_3, 3, 327

$AsC_{20}H_{20}NS_2$ S-(1,2,3,4-Tetrahydroarsinolin-1-yl)-N,N-pentamethylenedithiourethan, 3, 593

$AsC_{20}H_{20}N_3O_2$ Ph[4-(4-$Me_2NC_6H_4N_2$)-C_6H_4]AsO_2H, 3, 387

$AsC_{20}H_{21}N_2O_2S$ $Ph_2AsO_2H \cdot H_2NC$(=NH)SCH_2-Ph, 3, 376

$AsC_{20}H_{21}N_4O_5$ (4-$H_2NC_6H_4$){4-[3,5,2-EtO_2C(HO)(Me)-4-pyrrylazo]phenyl}-arsinic acid, 3, 377

$AsC_{20}H_{22}Br$ [MeEt($PhCH_2$)(1-$C_{10}H_7$)As]Br, 3, 328

$AsC_{20}H_{22}NOS_2$ 10-Phenoxarsinyl-SC(S)-$N(CH_2)_4CHEt$, 3, 616

$AsC_{20}H_{22}NO_2$ 3-(4-$Me_2NC_6H_4CH$=)-1,2,3,4-tetrahydro-7-methoxy-1-methyl-4-oxoarsinoline, 3, 593

$AsC_{20}H_{23}N_4O_7$ {Me_2[$NC(CH_2)_5$]PhAs}OC_6H_2-$(NO_2)_3$, 3, 319

$AsC_{20}H_{23}O_2S$ 10-Phenoxarsinyl-SC(O)-C_7H_{15}, 3, 616

$AsC_{20}H_{24}N_3O_9$ [$CH_2CH_2C(OMe)_2CH_2CH_2As$-MePh]$OC_6H_2(NO_2)_3$, 3, 590

$AsC_{20}H_{25}$ 4-(PhCH=CH)$C_6H_4As(CHMe_2)_2$, 3, 70

$AsC_{20}H_{26}BrO$ 1,1-Me{2-[$MeO(CH_2)_3$]C_6H_4}-1,2,3,4-tetrahydroarsinolinium Br^-, 3, 593

$AsC_{20}H_{26}NO_3S$ 10-Phenoxarsinyl-SCH_2-$CO_2H \cdot NEt_3$, 3, 617

$AsC_{20}H_{26}NO_{12}$ N-(Tetraacetatoglucosyl)-arsanilic acid, 3, 440

$AsC_{20}H_{26}O_3PS_2$ 10-Phenoxarsinyl-SP(S)-$(OCH_2CHMe_2)_2$, 3, 616

10-Phenoxarsinyl-SP(S)$(OBu)_2$, 3, 617

$AsC_{20}H_{27}N_2O_4$ 4-[2,5-HO($Me_3CCH_2CMe_2$)-C_6H_3N=N]$C_6H_4AsO_3H_2$, 3, 463

$AsC_{20}H_{28}Br$ [Et(n-C_5H_{11})Ph($PhCH_2$)As]Br, 3, 328

$AsC_{20}H_{28}Br_2P$ 5,5,12,12-$Et_2(Me)_2$-5,6,-11,12-tetrahydrodibenzo[b,f][1,4]-phosphoniaarsocinium 2 Br^-, 3, 648

$AsC_{20}H_{28}N_3O_3$ 4-[4-(n-C_8H_{17}NH)C_6H_4N=N]-$C_6H_4AsO_3H_2$, 3, 463

$AsC_{20}H_{30}N_5O_3$ 4-{4-[$Et_2N(CH_2)_3$N(Me)NH]-C_6H_4N=N}$C_6H_4AsO_3H_2$, 3, 463*

$AsC_{20}H_{31}O_2$ PhAs[OCH(Bu)CH=CH_2]$_2$, 3, 255

$AsC_{20}H_{31}O_2S_4$ PhAs[SC(S)OC_6H_{13}]$_2$, 3, 280

$AsC_{20}H_{32}N_3O_2S_4$ 4-$O_2NC_6H_4$As[SC(S)NPr_2]$_2$, 3, 284

$AsC_{20}H_{35}S_2$ $PhAs(SC_7H_{15})_2$, 3, 272

$AsC_{20}H_{37}ClNO_3$ 4-$ClC_6H_4AsO_3H_2 \cdot H_2N$-$C_{14}H_{29}$, 3, 409

$AsC_{20}H_{43}O_2$ $BuAs(OC_8H_{17})_2$, 3, 255

$AsC_{20}H_{43}S_2$ $EtAs(SC_9H_{19})_2$, 3, 272
$BuAs(SC_8H_{17})_2$, 3, 272
$AsC_{20}H_{45}O_4$ $Bu_3As(OOCMe_3)_2$, 3, 361
$AsC_{21}H_{12}F_9$ $(CF_3C_6H_4)_3As$ isomers, 3, 31
$AsC_{21}H_{14}N$ $(C_{10}H_7)_2AsCN$ isomers, 3, 205
$AsC_{21}H_{15}ClN$ 14,3-Cl(Me)-7,14-dihydrodibenzophenarsazine isomers, 3, 637
$AsC_{21}H_{15}N_2$ $Ph_3AsC(CN)_2$, 3, 309
$AsC_{21}H_{15}N_2S_4$ 4-MeC_6H_4As(S-2-benzothiazole)$_2$, 3, 272
$AsC_{21}H_{15}N_6O_8$ (N=CHCH=CHCH=C)$_3$As(OH)-$OC_6H_2(NO_2)_3$, 3, 348
$AsC_{21}H_{15}O_6S$ 2-[$(HO_2CC_6H_4)_2AsS$]C_6H_4-CO_2H isomers, 3, 260
$AsC_{21}H_{16}NO_7S_2$ 3,4-$O_2N(MeO)C_6H_3As$-$(SC_6H_4CO_2H)_2$, 3, 273
$AsC_{21}H_{17}ClN_3O_7$ 2,2-Me(4-ClC_6H_4)-isoarsindolinium picrate, 3, 558
$AsC_{21}H_{17}I_3N_3O_7$ β-{[3-(4-Arsonophenylazo)-4-hydroxy-5-iodophenoxy]-3,5-diiodophenyl}alanine, 3, 463
$AsC_{21}H_{17}N_2O_5S_2$ 4-$(H_2NCONH)C_6H_4As$-$(SC_6H_4CO_2H)_2$, 3, 273
$AsC_{21}H_{17}O_2$ 3,5-Me(p-$HO_2CC_6H_4$)-5,10-dihydroacridarsine, 3, 597
$AsC_{21}H_{17}O_3$ 3,5-Me(p-$HO_2CC_6H_4$)-5,10-dihydroacridarsine 5-oxide, 3, 598
$AsC_{21}H_{18}Br_2N_3O_5$ 4-[(4-$BrC_6H_4NHCO)_2$-CHNH]$C_6H_4AsO_3H_2$, 3, 440
$AsC_{21}H_{18}N_3O_7S$ N-(N-7-Sulfo-3-acridylglycyl)arsanilic acid, 3, 434
$AsC_{21}H_{18}N_3O_8$ 2,2-HO(p-MeC_6H_4)-isoarsindolinium picrate, 3, 559
2,2-HO(Ph)-1,2,3,4-tetrahydroisoarsinolinium picrate, 3, 595
$AsC_{21}H_{18}N_3O_9$ 2,2-HO(p-$MeOC_6H_4$)-isoarsindolinium picrate, 3, 559
$AsC_{21}H_{19}N_2O_3$ 9-(p-$Me_2NC_6H_4$)-2-acridinearsonic acid, 3, 525
$AsC_{21}H_{19}N_6OS_2$
4,3-HO[N=C(NH_2)N=C(NH_2)N=CNH]C_6H_3-$As(SPh)_2$, 3, 273
$AsC_{21}H_{19}N_6O_6S_4$
4-[N=C(NH_2)N=C(NH_2)N=CNH]C_6H_4As-$(SC_6H_4SO_3H)_2$, 3, 273
$AsC_{21}H_{19}O$ $Ph_2AsCH_2CH_2Bz$, 3, 63*
$Ph_3AsCHAc$, 3, 309
$AsC_{21}H_{19}O_2$ Ph(4-MeC_6H_4)(2-$MeO_2CC_6H_4$)As, 3, 78
$Ph_3AsCHCO_2Me$, 3, 309
$AsC_{21}H_{19}O_4$ (4-$HO_2CC_6H_4$)[2-(4-MeC_6H_4-CH_2)C_6H_4]AsO_2H, 3, 387
$AsC_{21}H_{20}Br$ [$Ph_3(CH_2$=$CHCH_2)As$]Br, 3, 328
$AsC_{21}H_{20}BrO$ [$Ph_3(AcCH_2)As$]Br, 3, 328
$AsC_{21}H_{20}BrO_2$ [$Ph_3(MeO_2CCH_2)As$]Br, 3, 328
$AsC_{21}H_{20}I$ [$Ph_3(CH_2$=$CHCH_2)As$]I, 3, 328
5,5-Me(Ph)-10,11-dihydrodibenz[b,f]-arsepinium I^-, 3, 642
$AsC_{21}H_{20}IO_2$ [$Ph_3(MeO_2CCH_2)As$]I, 3, 328
$AsC_{21}H_{20}N$ Hexahydromethyldibenzophenarsazine isomers, 3, 637
$AsC_{21}H_{20}NO_3$ [$Ph_3(CH_2$=$CHCH_2)As$]NO_3, 3, 328
$AsC_{21}H_{20}N_3O_7$ [$Me_2Ph(PhCH_2)As$]OC_6H_2-$(NO_2)_3$, 3, 319
[$Me_3(2$-$PhC_6H_4)As$]$OC_6H_2(NO_2)_3$, 3, 319
$AsC_{21}H_{20}N_5O_7$ $Ph_2AsCH_2CH_2C(=NH)NH_3$·O-$C_6H_2(NO_2)_3$, 3, 61
$AsC_{21}H_{21}$ $(PhCH_2)_3As$, 3, 27
$(MeC_6H_4)_3As$ isomers, 3, 27
Ph(4-EtC_6H_4)(4-MeC_6H_4)As, 3, 78
$AsC_{21}H_{21}ClN$ 12,9,8-Bu(Cl)(Me)-7,12-dihydrobenzo[a]phenarsazine, 3, 629
3,12,10-Me_3C(Cl)(Me)-7,12-dihydrobenzo[a]phenarsazine, 3, 629
Cl(MePrCH)-7,12-dihydrobenzophenarsazine isomers, 3, 629, 634
Cl(tert-C_5H_{11})-7,12-dihydrobenzophenarsazine isomers, 3, 629, 634
$AsC_{21}H_{21}ClNO$ 12,11,10,8-Cl(Me_2CH)-(MeO)(Me)-7,12-dihydrobenzo[b]-phenarsazine, 3, 631*
7,8,9,11-Cl(Me_2CH)(MeO)(Me)-7,12-dihydrobenzo[c]phenarsazine, 3, 634*
$AsC_{21}H_{21}N_4O_8S$ N-(4-$H_2NC_6H_4SO_2$)-3-(4-$H_2O_3AsC_6H_4N$=N)tyrosine, 3, 463
$AsC_{21}H_{21}N_6O_3$ [4-$(H_2NCONH)C_6H_4$]$_3$As,

3, 31
$AsC_{21}H_{21}O$ $(4\text{-}MeC_6H_4)_3AsO$, 3, 359
$AsC_{21}H_{21}O_3$ $(MeOC_6H_4)_3As$ isomers, 3, 28, 31
$AsC_{21}H_{21}O_4S$ 2,2-HO(Ph)-isoarsindolinium p-$MeC_6H_4SO_3^-$, 3, 558
$AsC_{21}H_{22}Br$ $[MePh(PhCH_2)_2As]Br$, 3, 328
$[Me_2Ph(4\text{-}PhC_6H_4CH_2)As]Br$, 3, 329
$AsC_{21}H_{22}ClO$ $(2\text{-}MeC_6H_4)_3As(OH)Cl$, 3, 346, 347
$AsC_{21}H_{22}N$ 12,9,10-$Me_2CH(Me)_2$-7,12-dihydrobenzo[a]phenarsazine, 3, 629
$AsC_{21}H_{23}N_2O_6$ $MeAs(O_2CC_6H_4CONMe_2)_2$, 3, 257
$AsC_{21}H_{23}O_2$ $(MeC_6H_4)_3As(OH)_2$ isomers, 3, 351
$AsC_{21}H_{23}O_5$ $(4\text{-}MeOC_6H_4)_3As(OH)_2$, 3, 351
$AsC_{21}H_{24}IO$ 2,2-Me(Ph)-2,3-dihydro-1H-naphth[1,8-cd]arseninium I^-·EtOH, 3, 599
$AsC_{21}H_{24}N$ 2-[4-$(Pr_2As)C_6H_4$]-quinoline, 3, 63
$AsC_{21}H_{25}O_2S$ 10-Phenoxarsinyl-SC(O)-C_8H_{17}, 3, 617
$AsC_{21}H_{25}O_2S_2$ 10-Phenoxarsinyl-SC(S)-OC_8H_{17}, 3, 617
$AsC_{21}H_{26}NOS_2$ 10-Phenoxarsinyl-SC(S)-NBu_2, 3, 616*
$AsC_{21}H_{27}O_6$ 8-Benzyl-7,9,17,18-tetraoxa-8-arsatrispiro[5.1.1.5.2.2]-nonadecane-16,19-dione, 3, 582
$AsC_{21}H_{31}N_6O_{10}S_2$
4-[$N{=}C(NH_2)N{=}C(NH_2)N{=}CNH]C_6H_4As$-(S-glucosyl)$_2$, 3, 273
$AsC_{21}H_{31}O_4$ $Me_3As(OOCMe_2Ph)_2$, 3, 361
$AsC_{21}H_{38}Cl$ $[Me_2(PhCH_2)(n\text{-}C_{12}H_{25})As]Cl$, 3, 329
$AsC_{21}H_{38}I$ $[Me_2(PhCH_2)(n\text{-}C_{12}H_{25})As]I$, 3, 329
$AsC_{21}H_{45}O_2$ $C_5H_{11}As(OC_8H_{17})_2$, 3, 255
$AsC_{21}H_{45}S_2$ $MeAs(SC_{10}H_{21})_2$, 3, 273
$PrAs(SC_9H_{19})_2$, 3, 273
$AsC_{22}H_{15}$ $(1\text{-}C_{10}H_7)_2AsC{\equiv}CH$, 3, 71
5-$(1\text{-}C_{10}H_7)$-dibenzarsole, 3, 566
$AsC_{22}H_{15}ClN$ Chloro-7,12-dihydrophenylbenzophenarsazine isomers, 3, 629, 634
Chloro-7,15-dihydro-H-benzindenophenarsazine isomers, 3, 635
$AsC_{22}H_{15}O$ 10-α-Naphthylphenoxarsine, 3, 617
10-α-Naphthylphenoxarsine, 3, 617

$AsC_{22}H_{16}Cl_2N_2O_7$ 2,2′-Spirobi(5-chloroisoarsindolinium) picrate, 3, 574
$AsC_{22}H_{16}N_3OS_4$ 4-$AcNHC_6H_4As$(S-2-benzothiazole)$_2$, 3, 273*
$AsC_{22}H_{16}N_3O_2S_4$ 2,2′-$(4\text{-}O_2NC_6H_4AsS_2)$-[6-methylbenzothiazole]$_2$, 3, 273
$AsC_{22}H_{16}N_5O_{13}S_2$ 2-[1,8,7,3,6-$(HO)_2$-$(o\text{-}O_2NC_6H_4N{=}N)(HO_3S)_2$-2-naphthylazo]-$C_6H_4AsO_3H_2$, 3, 464
$AsC_{22}H_{16}N_5O_{16}S_3$ 2-{1,8,7,3,6-$(HO)_2$-[4,2-$O_2N(HO_3S)C_6H_3N{=}N](HO_3S)_2$-2-naphthylazo}-$C_6H_4AsO_3H_2$, 3, 464
$AsC_{22}H_{17}$ $C_{10}H_7AsPh_2$ isomers, 3, 71
$AsC_{22}H_{17}N_4O_4$ 4-[4-$(1,4\text{-}HOC_{10}H_6N{=}N)$-$C_6H_4N{=}N]C_6H_4AsO_3H_2$, 3, 463
$AsC_{22}H_{17}N_4O_7S$ 4-{4-[1,2-$HO(HO_3S)C_{10}H_5$-4-$N{=}N]C_6H_4N{=}N\}C_6H_4AsO_3H_2$, 3, 463
$AsC_{22}H_{17}N_4O_{11}S_2$ 2-[1,8,7,3,6-$(HO)_2$-$(PhN{=}N)(HO_3S)_2$-2-naphthylazo]-$C_6H_4AsO_3H_2$, 3, 464
$AsC_{22}H_{17}N_4O_{12}S_2$ 2-[1,8,7,3,6-$(HO)_2$-$(2\text{-}HOC_6H_4N{=}N)(HO_3S)_2$-2-naphthylazo]-$C_6H_4AsO_3H_2$
$AsC_{22}H_{17}N_4O_{14}S_3$ 2-[1,8,3,6,7-$(HO)_2$-$(HO_3S)_2(3\text{-}HO_3SC_6H_4N{=}N)$-2-naphthylazo]-$C_6H_4AsO_3H_2$, 3, 464
2-[1,8,3,6,7-$(HO)_2(HO_3S)_2(2\text{-}HO_3S$-$C_6H_4N{=}N)$-2-naphthylazo]-$C_6H_4AsO_3H_2$, 3, 464
$AsC_{22}H_{17}O_2$ 5-$(1\text{-}C_{10}H_7)$-dibenzarsole 5,5-dihydroxide, 3, 566
$AsC_{22}H_{17}O_3$ 10-α-Naphthylphenoxarsine 10,10-dihydroxide, 3, 617
$AsC_{22}H_{18}N$ 7,14-Dihydro-3,14-dimethyldibenzo[-,-]phenarsazine isomers, 3, 637
$AsC_{22}H_{18}NO_2$ $Ph_3AsC(CN)CO_2Me$, 3, 309
$AsC_{22}H_{18}NO_6S_2$ 2,4-$HO(AcNH)C_6H_3As$-$(SC_6H_4CO_2H\text{-}2)_2$, 3, 273

$AsC_{22}H_{18}N_2O_7$ 2,2′-Spirobi(isoarsindolinium) picrate, 3, 574

$AsC_{22}H_{19}$ (1-$C_{10}H_7$)$_2$AsEt, 3, 71

$AsC_{22}H_{19}N_4O$ 4-(2-$HOC_{10}H_6N_2$)C_6H_4As(CH_2CH_2CN)$_2$, 3, 63

$AsC_{22}H_{19}O_2$ Ph_2(1-$C_{10}H_7$)As$(OH)_2$, 3, 351*

$AsC_{22}H_{19}O_2S_4$ PhAs[SC(S)OCH_2Ph]$_2$, 3, 281

$AsC_{22}H_{19}O_4S_2$ PhAs($SC_6H_4CO_2Me$)$_2$, 3, 273

$AsC_{22}H_{20}N_3O_7$ [Me_2(2-PhC_6H_4)(CH_2=CH)-As]$OC_6H_2(NO_2)_3$, 3, 320

$AsC_{22}H_{20}N_3O_8$ 2,2-Me(4-$MeOC_6H_4$)-isoarsindolinium picrate, 3, 559

$AsC_{22}H_{21}$ Benzyl(9-fluorenylidene)dimethylarsenic, 3, 308

$AsC_{22}H_{21}BrN_3O_7$ [Me_2($BrCH_2CH_2$)(2-Ph-C_6H_4)As]$OC_6H_2(NO_2)_3$, 3, 320

$AsC_{22}H_{21}ClN$ 12,10-Cl(cyclo-C_6H_{11})-7,12-dihydrobenzo[a]phenarsazine, 3, 630

7,9-Cl(cyclo-C_6H_{11})-7,12-dihydrobenzo[a]phenarsazine, 3, 634

$AsC_{22}H_{21}N_2S_4$ PhAs[SC(S)NMePh]$_2$, 3, 284*

$AsC_{22}H_{21}N_4O_8$ [CH_2CH_2As(Ph)CH_2CH_2N-(Ph)OH]$OC_6H_2(NO_2)_3$, 3, 619

$AsC_{22}H_{22}Br$ [Me_2($PhCH_2$)(9-fluorenyl)-As]Br, 3, 329

$AsC_{22}H_{22}BrO_2$ [Ph_3(EtO_2CCH_2)As]Br, 3, 329

$AsC_{22}H_{22}I$ [Me_2($PhCH_2$)(9-fluorenyl)-As]I, 3, 329

$AsC_{22}H_{24}Br$ [Me($PhCH_2$)(4-MeC_6H_4)$_2$As]-Br, 3, 329

$AsC_{22}H_{24}Cl$ [Me($PhCH_2$)$_3$As]Cl, 3, 329

$AsC_{22}H_{24}I$ [Me(MeC_6H_4)$_3$As]I isomers, 3, 329

$AsC_{22}H_{24}IO$ 6,6-Me(Ph)-5,7-dihydro-6H-dibenz[c,e]arsepinium I^-·MeOH, 3, 642

$AsC_{22}H_{24}IO_3$ [Me(3-$MeOC_6H_4$)$_3$As]I, 3, 329

$AsC_{22}H_{24}N$ 10,12-($EtMe_2C$)(Me)-7,12-dihydrobenzo[a]phenarsazine, 3, 630

12,10-Me(MePrCH)-7,12-dihydrobenzo-[a]phenarsazine, 3, 630

12,6,x-Bu$(Me)_2$-7,12-dihydrobenzo[b]-phenarsazine isomers, 3, 631

$AsC_{22}H_{26}N$ 2-[4-(Pr_2As)C_6H_4]-8-methylquinoline, 3, 63

$AsC_{22}H_{26}NO$ 2-[4-(Pr_2As)C_6H_4]-6-methoxyquinoline, 3, 63

$AsC_{22}H_{29}N_2O_5$ 4-[PhNHOC$(CH_2)_8$CONH]-$C_6H_4AsO_3H_2$, 3, 441

$AsC_{22}H_{32}Br_2P$ 5,5,12,12-Et_4-5,6,11,12-tetrahydrodibenzo[b,f][1,4]phosphoniaarsocinium $2Br^-$, 3, 648

$AsC_{22}H_{34}NO_5S_2$ 4-$AcNHC_6H_4$As(SCH_2CO_2-C_5H_{11})$_2$, 3, 273

$AsC_{22}H_{34}NO_6S_2$ 2,5-HO(AcNH)C_6H_3As-[SCH(Me)CO_2Bu]$_2$, 3, 273*

$AsC_{22}H_{35}O_2$ PhAs[OCH(Bu)CH=CHMe]$_2$, 3, 255

PhAs[OCH(C_5H_{11})CH=CH_2]$_2$, 3, 255

$AsC_{22}H_{36}NO_5S_2$ 3,4-H_2N(HO)C_6H_3As(SCH_2-$CO_2C_6H_{13}$)$_2$, 3, 273

$AsC_{22}H_{38}NO_4$ 4-(n-$C_{15}H_{31}$CONH)C_6H_4-AsO_3H_2, 3, 441

$AsC_{22}H_{38}N_3O_7$ [Bu_4As]$OC_6H_2(NO_2)_3$, 3, 321

$AsC_{22}H_{41}ClNO_3$ 4-$ClC_6H_4AsO_3H_2$·$H_2NC_{16}H_{33}$, 3, 409

$AsC_{22}H_{41}O_4P_2S_4$ PhAs[SP(S)$(OBu)_2$]$_2$, 3, 282

$AsC_{22}H_{47}S_2$ EtAs($SC_{10}H_{21}$)$_2$, 3, 273

BuAs(SC_9H_{19})$_2$, 3, 273

$AsC_{22}H_{48}I$ [Et_3(n-$C_{16}H_{33}$)As]I, 3, 330

$AsC_{23}H_{14}Cl_3N_4O_4S_2$

3,4-Cl[N=C(Cl)N=C(Cl)N=CNH]C_6H_3As-($SC_6H_4CO_2H$)$_2$, 3, 273

$AsC_{23}H_{17}N_4O_{13}S_2$ 2-[1,8,3,6,7-$(HO)_2$-$(HO_3S)_2$(3-$HO_2CC_6H_4$N=N)-2-naphthylazo]-$C_6H_4AsO_3H_2$, 3, 465

2-[1,8,3,6,7-$(HO)_2(HO_3S)_2$(2-$HO_2CC_6H_4$-N=N)-2-naphthylazo]-$C_6H_4AsO_3H_2$, 3, 465

$AsC_{23}H_{17}N_4O_{16}S_3$ 2-{1,8,3,6,7-$(HO)_2$-$(HO_3S)_2$[2,4-$HO_2C(HO_3S)C_6H_3$N=N]-2-naphthylazo}-$C_6H_4AsO_3H_2$, 3, 465*

$AsC_{23}H_{18}Cl_2N_5S_2$

3,4-AcNH[N=C(Cl)N=C(Cl)N=CNH]C_6H_4-

$As(SPh)_2$, 3, 273

$AsC_{23}Cl_8N$ 7,12-Dihydro-methylphenyl-benzophenarsazine isomers, 3, 630

$AsC_{23}H_{19}N_2O_4S_2$ (Benzo[f]quinolyl-4-NH)$C_6H_4As(SCH_2CO_2H)_2$ isomers, 3, 274

$AsC_{23}H_{19}N_2O_{11}S_2$ 4-[1,4,8-HO$(HO_3S)_2$-2-naphthyl-N(Bz)NH]$C_6H_4AsO_3H_2$, 3, 454

$AsC_{23}H_{19}N_4OS_2$ $\overline{N{=}CHCH{=}C(NHAc)CH{=}C}As$-[S-2-(7-methylbenzothiazole)]$_2$, 3, 274

$AsC_{23}H_{19}N_6O_4S_2$ 4-[$\overline{N{=}C(NH_2)N{=}C(NH_2)N{=}C}$-NH]$C_6H_4As(SC_6H_4CO_2H)_2$, 3, 274*

$AsC_{23}H_{19}O_5$ $Ph_3AsC(CO_2Me)C(O)CO_2H$, 3, 308

$AsC_{23}H_{20}Br$ [$Ph_3(\overline{CH{=}CHCH{=}CHCH})As$]Br, 3, 330

$AsC_{23}H_{20}I$ [$MePh_2(1\text{-}C_{10}H_7)As$]I, 3, 330

$AsC_{23}H_{20}N$ 3,14-Me(Et)-7,14-dihydro-dibenzophenarsazine isomers, 3, 637, 638

$AsC_{23}H_{20}NO$ 3,5,11-MeO(Me)(Ph)-5,6-dihydro-11H-arsinolino[4,3-b]indole, 3, 593

$AsC_{23}H_{21}BrClO$ 2,2-Ph(4-$ClC_6H_4COCH_2$)-1,2,3,4-tetrahydroisoarsinolinium Br^-, 3, 595

$AsC_{23}H_{21}ClIO$ 2,2-Ph(4-$ClC_6H_4COCH_2$)-1,2,3,4-tetrahydroisoarsinolinium I^-, 3, 595

$AsC_{23}H_{21}N_4O_7$ [Me($NCCHMeCH_2$)Ph_2As]-$OC_6H_2(NO_2)_3$, 3, 321

$AsC_{23}H_{23}ClN$ 12,3-Cl(1-Me-cyclohexyl)-7,12-dihydrobenzo[a]phenarsazine, 3, 630

$AsC_{23}H_{24}N_3O_5$ 4-[$(PhCH_2NHCO)_2CHNH$]-$C_6H_4AsO_3H_2$, 3, 441

4-[(4-$MeC_6H_4NHCO)_2CHNH$]$C_6H_4AsO_3H_2$, 3, 441

$AsC_{23}H_{24}N_7S_2$

4,2-Me_2N[$\overline{N{=}C(NH_2)N{=}C(NH_2)N{=}C}NH$]-$C_6H_3As(SPh)_2$, 3, 274

$AsC_{23}H_{25}N_8S_2$ 4-[$\overline{N{=}C(NH_2)N{=}C(NH_2)N{=}C}$-NH]$C_6H_4As$[$S\overline{C{=}CHC(Me){=}NC(Me){=}C}H$]$_2$, 3, 274

$AsC_{23}H_{29}O_3S$ 10-Phenoxarsinyl-S$(CH_2)_{10}$-CO_2H, 3, 617

$AsC_{23}H_{30}N_3O_7S_2$ 3,4-$O_2N(PhCH_2O)C_6H_3$-As[$SCH_2CH(NH_2)CO_2Et$]$_2$, 3, 274

$AsC_{23}H_{36}N_3O_3S_4$ Pip·$HO_3SC_6H_4As$[SC(S)-$\overline{N(CH_2)_4CH_2}$]$_2$, 3, 284

$AsC_{23}H_{38}NO_8$ 4-(n-$C_{11}H_{23}CO_2CH_2CHOHCH_2$-$O_2CCH_2NH)C_6H_4AsO_3H_2$, 3, 441

$AsC_{23}H_{43}O_4P_2S_4$ 4-MeC_6H_4As[SP(S)-$(OBu)_2$]$_2$, 3, 282

$AsC_{23}H_{49}S_2$ $MeAs(SC_{11}H_{23})_2$, 3, 274

$PrAs(SC_{10}H_{21})_2$, 3, 274

$AsC_{24}H_{10}Cl_{10}O_4PS$ $Ph_2AsO_2P(S)(OC_6Cl_5)_2$, 3, 377, 388

$AsC_{24}H_{14}ClN_2S_4$ 2-$ClC_{10}H_6As$(S-2-benzothiazole)$_2$, 3, 274

$AsC_{24}H_{15}$ $(PhC{\equiv}C)_3As$, 3, 28

$AsC_{24}H_{16}BrClN_3O_8$ Ph(4-BrC_6H_4)(4-Cl-C_6H_4)As(OH)$OC_6H_2(NO_2)_3$, 3, 348

$AsC_{24}H_{16}BrN_4O_8$ [$(3\text{-}O_2NC_6H_4)_4As$]Br, 3, 330

$AsC_{24}H_{16}Cl$ 5-[2-(2-$ClC_6H_4)C_6H_4$]-dibenzarsole, 3, 566

$AsC_{24}H_{16}ClN_4O_8$ [$(3\text{-}O_2NC_6H_4)_4As$]Cl, 3, 330

$AsC_{24}H_{16}IN_4O_8$ [$(3\text{-}O_2NC_6H_4)_4As$]I, 3, 330

$AsC_{24}H_{16}N_5O_{11}$ [$(3\text{-}O_2NC_6H_4)_4As$]$NO_3$, 3, 330

$AsC_{24}H_{17}$ 5-(2-Biphenylyl)-dibenzarsole, 3, 567

$AsC_{24}H_{17}ClNOS$ 5-(10-Phenoxarsinyl)-phenothiazinium chloride, 3, 611

$AsC_{24}H_{17}Cl_2$ 5-(2-Biphenylyl)dibenzarsole 5,5-dichloride, 3, 567

$AsC_{24}H_{17}Cl_2N_4O_4S_2$

3,4-Me[$\overline{N{=}C(Cl)N{=}C(Cl)N{=}C}NH$]$C_6H_3As$-$(SC_6H_4CO_2H)_2$, 3, 274

$AsC_{24}H_{17}N_2O_6$ [2-(4-$O_2NC_6H_4)C_6H_4$]$_2$-AsO_2H, 3, 387

$AsC_{24}H_{18}BrO$ 10,10-Ph_2-phenoxarsinium bromide, 3, 610

$AsC_{24}H_{18}Cl$ Ph(4-PhC_6H_4)(4-ClC_6H_4)As, 3, 78

$(2\text{-}PhC_6H_4)_2AsCl$, 3, 167

$AsC_{24}H_{18}ClO$ Ph(4-PhC_6H_4)(4-ClC_6H_4)-AsO, 3, 359

5,5'-Spirobi(dibenzarsolium)Cl^-·H_2O,

3, 571
$AsC_{24}H_{18}Cl_3$ $(2\text{-}PhC_6H_4)_2AsCl_3$, 3, 371
$AsC_{24}H_{18}N$ 7,15-Dihydro-7-methyl-13H-benz[h]indeno[1,2-b]phenarsazine, 3, 635
$AsC_{24}H_{18}N_3O_7$ $[Ph_3AsH]OC_6H_2(NO_2)_3$, 3, 342
$AsC_{24}H_{18}OP$ 5,10-Ph_2-5,10-dihydrodibenzo[1,4]phospharsenine 5 oxide, 3, 640
$AsC_{24}H_{18}O_2P$ 5,10-Ph_2-5,10-dihydrodibenzo[1,4]phospharsenin 5,10-dioxide, 3, 640
$AsC_{24}H_{18}P$ 5,10-Ph_2-5,10-dihydrodibenzo[1,4]phospharsenin, 3, 640
$AsC_{24}H_{19}N_4O_8$ 3,5-MeO(Me)-5,6-dihydroarsinolino[4,3-b]quinolinium picrate, 3, 593
$AsC_{24}H_{19}N_4O_9$ 3,5-MeO(Me)-5,6-dihydroarsinolino[4,3-b]quinoline 5-oxide picrate, 3, 594
$AsC_{24}H_{20}Br$ $[Ph_4As]Br$, 3, 330
$AsC_{24}H_{20}Br_3$ $[Ph_4As]Br_3$, 3, 330
$AsC_{24}H_{20}Cl$ $[Ph_4As]Cl$, 3, 330
$AsC_{24}H_{20}ClO_4$ $[Ph_4As]ClO_4$, 3, 332
$AsC_{24}H_{20}Cl_3$ $[Ph_4As]Cl_3$, 3, 331
$AsC_{24}H_{20}F_6P$ $[Ph_4As]PF_6$, 3, 333
$AsC_{24}H_{20}I$ $[Ph_4As]I$, 3, 331
$AsC_{24}H_{20}IO_4$ $[Ph_4As]IO_4$, 3, 332
$AsC_{24}H_{20}I_3$ $[Ph_4As]I_3$, 3, 331
$AsC_{24}H_{20}N$ 12,8-Et(Ph)-7,12-dihydrobenzo[a]phenarsazine, 3, 630
6,10,12-Me_2(Ph)-7,12-dihydrobenzo-[b]phenarsazine, 3, 631
$AsC_{24}H_{20}NO_3$ $[Ph_4As]NO_3$, 3, 332
$AsC_{24}H_{20}N_3O_{14}$ $[3,4\text{-}O_2N(AcO)C_6H_3]_3As(OH)_2$, 3, 351
$AsC_{24}H_{21}$ $(PhCH{=}CH)_3As$, 3, 28
$(4\text{-}CH_2{=}CHC_6H_4)_3As$, 3, 31
$AsC_{24}H_{21}ClN$ 14,1,13-Cl$(Et)_2$-7,14-dihydrodibenzo[a,j]phenarsazine, 3, 638
$AsC_{24}H_{21}O$ $[Ph_4As]OH$, 3, 332
$AsC_{24}H_{21}O_2$ $Ph_3As(OH)OPh$, 3, 348
$AsC_{24}H_{21}O_5$ $Ph_3AsC(CO_2Me)C(O)CO_2Me$, 3, 308
$AsC_{24}H_{22}NO_8S_2$ 2,4-HO(AcNH)C_6H_3As-$[SC_6H_3(OMe)CO_2H]_2$, 3, 274
$AsC_{24}H_{22}N_3O_7$ 3,3′,4,4′-Tetrahydro-1,1′(2H,2′H)-spirobisarsinolinium picrate, 3, 599
$AsC_{24}H_{23}N_2O_3S$ $[Ph_3AsNHO_2SC_6H_4NH_2]OH$, 3, 370
$AsC_{24}H_{23}N_{10}O_{14}$ $PhAs[CH_2CH_2C({=}NH)NH_3{\cdot}\text{-}OC_6H_2(NO_2)_3]_2$, 3, 50
$AsC_{24}H_{23}O_2$ $(BzCH_2CH_2)_2AsPh$, 3, 71
$PhAs[OCH(Ph)CH{=}CH_2]_2$, 3, 255
$AsC_{24}H_{24}BrN_4$ $[(3\text{-}H_2NC_6H_4)_4As]Br$, 3, 336
$AsC_{24}H_{24}ClN_4$ $[(3\text{-}H_2NC_6H_4)_4As]Cl$, 3, 336
$AsC_{24}H_{24}N_3O_3$ $(3\text{-}AcNHC_6H_4)_3As$, 3, 31
$AsC_{24}H_{24}N_3O_{10}$ $[4,3\text{-}EtO(O_2N)C_6H_3]_3AsO$, 3, 359
$AsC_{24}H_{25}ClN$ 12,10,3-Cl(Me)(1-Me-cyclohexyl)-7,12-dihydrobenzo[a]phenarsazine, 3, 630
$AsC_{24}H_{26}Br$ $[(PhCH_2)_3(CH_2{=}CHCH_2)As]Br$, 3, 336
$[Ph_3(cyclo\text{-}C_6H_{11})As]Br$, 3, 337
$AsC_{24}H_{26}ClN_{10}O_{15}$ $4\text{-}ClC_6H_4As[CH_2CH_2C\text{-}({=}NH)NH_3{\cdot}OC_6H_2(NO_2)_3]_2{\cdot}H_2O$, 3, 50
$AsC_{24}H_{26}N$ 12,3-Me(1-Me-cyclohexyl)-7,12-dihydrobenzo[a]phenarsazine, 3, 630
$AsC_{24}H_{26}N_3O_5$ $(3\text{-}AcNHC_6H_4)_3As(OH)_2$, 3, 351
$AsC_{24}H_{27}$ $(2,5\text{-}Me_2C_6H_3)_3As$, 3, 31
$AsC_{24}H_{27}O_3$ $(4\text{-}EtOC_6H_4)_3As$, 3, 31
$AsC_{24}H_{28}ClNO_4$ $[3,4\text{-}O_2N(EtO)C_6H_4]_3\text{-}As(OH)Cl$, 3, 347
$AsC_{24}H_{30}N_3O_3$ $[3,4\text{-}H_2N(EtO)C_6H_3]_3As$, 3, 31
$4\text{-}[2,1\text{-}(n\text{-}C_8H_{17}NH)C_{10}H_6N{=}N]C_6H_4\text{-}AsO_3H_2$, 3, 465
$AsC_{24}H_{30}N_3O_4$ $[3,4\text{-}H_2N(EtO)C_6H_3]_3AsO$, 3, 359
$AsC_{24}H_{31}O_3S$ 10-Phenoxarsinyl-$S(CH_2)_{11}\text{-}CO_2H$, 3, 617
$AsC_{24}H_{36}N_3O_3$ $4\text{-}[4\text{-}(n\text{-}C_{12}H_{25}NH)C_6H_4\text{-}N{=}N]C_6H_4AsO_3H_2$, 3, 465
$AsC_{24}H_{37}O_4$ $Et_3As(OOCMe_2Ph)_2$, 3, 361
$AsC_{24}H_{39}O_2S_4$ $PhAs[SC(S)OC_8H_{17}]_2$, 3, 281

$AsC_{24}H_{40}NO_5S_2$ 3,4-$H_2N(HO)C_6H_3As(SCH_2$-$CO_2CHEtBu)_2$, 3, 274
$AsC_{24}H_{40}NO_6$ 4-[$HO_2C(CH_2)_{16}CONH]C_6H_4$-$AsO_3H_2$, 3, 441
$AsC_{24}H_{41}N_2S_4$ PhAs[SC(S)$NBu_2]_2$, 3, 284
$AsC_{24}H_{42}NO_4$ 4-(n-$C_{17}H_{35}CONH)C_6H_4$-AsO_3H_2, 3, 441
$AsC_{24}H_{43}S_2$ PhAs$(SC_9H_{19})_2$, 3, 274
$AsC_{24}H_{45}ClNO_3$ 4-$ClC_6H_4AsO_3H_2\cdot H_2N$-$C_{18}H_{37}$, 3, 409
$AsC_{24}H_{51}$ (n-$C_8H_{17})_3$As, 3, 29
(EtBuCH$CH_2)_3$As, 3, 31
$AsC_{24}H_{51}S_2$ EtAs$(SC_{11}H_{23})_2$, 3, 274
BuAs$(SC_{10}H_{21})_2$, 3, 274
$AsC_{24}H_{51}Se$ $(C_8H_{17})_3$AsSe, 3, 367
$AsC_{25}H_{19}ClN_3O_8$ Ph(4-ClC_6H_4)(4-Me-C_6H_4)As(OH)$OC_6H_2(NO_2)_3$, 3, 348
$AsC_{25}H_{19}O$ 5,10-Dihydro-5,10-diphenyl-10-acridarsinol, 3, 598
$AsC_{25}H_{19}O_2$ 3-[Ph(4-PhC_6H_4)As]C_6H_4-CO_2H, 3, 73
$AsC_{25}H_{20}BrClI$ [Ph(4-BrC_6H_4)(4-Cl-C_6H_4)(4-MeC_6H_4)As]I, 3, 337
$AsC_{25}H_{20}NO_2$ Ph_3AsC(Ph)NO_2, 3, 309
$AsC_{25}H_{20}NO_2S$ 5,5-Ph(p-$MeC_6H_4SO_2N$=)-dibenzarsole, 3, 567
$AsC_{25}H_{20}NS$ [Ph_4As]SCN, 3, 334
$AsC_{25}H_{20}N_3O_7$ [MePh_3As]$OC_6H_2(NO_2)_3$, 3, 324
$AsC_{25}H_{21}$ Ph(4-PhC_6H_4)(3-MeC_6H_4)As, 3, 78
$AsC_{25}H_{21}BrNO_2$ [$Ph_3(O_2NC_6H_4CH_2)$As]Br isomers, 3, 337
$AsC_{25}H_{21}IP$ 5,10-Ph_2-5,10-dihydrodibenzo[1,4]phospharsenin methiodide, 3, 641
$AsC_{25}H_{21}N_8O_9$ 2,3,5,6-Tetrahydro-8-methoxy-1,7-bis(2,4-dinitrophenylhydrazono)-1H,7H-benz[i,j]arsinolizine, 3, 599
$AsC_{25}H_{21}O_3$ [Ph_4As]OCO_2H, 3, 332
$AsC_{25}H_{22}Br$ [$Ph_3(PhCH_2)$As]Br, 3, 337
$AsC_{25}H_{22}BrO$ $Ph_3As(OCH_2Ph)$Br, 3, 347
$AsC_{25}H_{22}Cl$ [$Ph_3(PhCH_2)$As]Cl, 3, 337
$AsC_{25}H_{22}I$ [$Ph_3(PhCH_2)$As]I, 3, 337
[$Ph_3(MeC_6H_4)$As]I isomers, 3, 337
$AsC_{25}H_{22}N$ 7,8,11-$PhCH_2(Me)_2$-7,12-dihydrobenzo[c]phenarsazine, 3, 634
$AsC_{25}H_{22}NO_2S$ $Ph_3As=NO_2SC_6H_4Me$-p, 3, 369
$AsC_{25}H_{22}NO_3$ [$Ph_3(PhCH_2)$As]NO_3, 3, 337, 338
$AsC_{25}H_{25}N_2O_{10}S_2$ 2,2′-[Arsenobis(p-phenylenedipropionylhydrazo)]di-1-naphthol-4,8-disulfonic acid dipropionate, 3, 135
$AsC_{25}H_{25}N_4O$ 2,3,5,6-Tetrahydro-8-methoxy-1,7-bis(phenylhydrazono)-1H,7H-benz[i,j]arsinolizine, 3, 599
$AsC_{25}H_{26}BrO_4$ {$Ph_3[(EtO_2C)_2CH]$As}Br, 3, 338
$AsC_{25}H_{26}IO_2$ [MePh$(BzCH_2CH_2)_2$As]I, 3, 338
$AsC_{25}H_{27}ClN_3O_3$ [Me(3-$AcNHC_6H_4)_3$As]Cl, 3, 338
$AsC_{25}H_{27}N_2O_{13}S_2$ 2-[2-(4-Arsonophenyl)-1,2-dipropionylhydrazo]-1-naphthol-4,8-disulfonic acid propionate, 3, 454
$AsC_{25}H_{27}N_{12}O_{15}$ 4-[$H_2NCONH)C_6H_4$As[CH_2-$CH_2C(=NH)NH_3\cdot OC_6H_2(NO_2)_3]_2$, 3, 57
$AsC_{25}H_{29}N_4O_8$ [$Me_2(2-Me_2NC_6H_4)$(2-MeO-$CH_2C_6H_4CH_2$)As]$OC_6H_2(NO_2)_3$, 3, 327
$AsC_{25}H_{31}O_3S$ [MeBuPh$(PhCH_2)$As]O_3S-C_6H_4Me, 3, 322
$AsC_{25}H_{35}N_4O_6$ 8-[$Et_2N(CH_2)_3NH$]-6-MeO-quinoline·3,4-AcNH(HO)$C_6H_3AsO_3H_2$ salt, 3, 477
$AsC_{25}H_{42}NO_6$ 4-[$MeO_2C(CH_2)_{16}CONH]C_6H_4$-$AsO_3H_2$, 3, 441
$AsC_{25}H_{42}NO_8$ 4-(n-$C_{13}H_{27}CO_2CH_2CHOHCH_2$-$O_2CCH_2NH)C_6H_4AsO_3H_2$, 3, 441
$AsC_{25}H_{46}Cl$ [$Me_2(PhCH_2)(n-C_{16}H_{33})$As]Cl, 3, 338
$AsC_{25}H_{46}I$ [$Me_2(PhCH_2)(n-C_{16}H_{33})$As]I, 3, 338
$AsC_{25}H_{53}S_2$ MeAs$(SC_{12}H_{25})_2$, 3, 275
PrAs$(SC_{11}H_{23})_2$, 3, 275
$AsC_{26}H_{15}ClN$ 16-Cl-7,16-dihydrobenzo-[a]fluoreno[1,9-hi]phenarsazine, 3, 636
16-Cl-7,16-dihydroacenaphtheno[1,2-l]-benzo[a]phenarsazine, 3, 636
$AsC_{26}H_{17}N_4O_3S_4$ 7,2-AcNH$(O_2N)C_{10}H_5$As-

(S-2-benzothiazole)$_2$, 3, 275

$AsC_{26}H_{19}$ (1-$C_{10}H_7$)$_2$AsPh, 3, 71

$AsC_{26}H_{19}Br_2D_2O$ [Ph_3(4-$BrC_6H_4COCD_2$)-As]Br, 3, 338

$AsC_{26}H_{19}N_4O_{12}S_2$ 2-[1,8,3,6,7-(HO)$_2$-(HO_3S)$_2$(2-HO-1-naphthylazo)-2-naphthylazo]-$C_6H_4AsO_3H_2$, 3, 465

$AsC_{26}H_{19}N_4O_{14}S_3$ 2-[1,8,3,6,7-(HO)$_2$-(HO_3S)(1-HO_3S-2-naphthylazo)-2-naphthylazo]-$C_6H_4AsO_3H_2$, 3, 465

$AsC_{26}H_{19}N_4O_{15}S_3$ 2-{1,8,3,6,7-(HO)$_2$-(HO_3S)$_2$[2,4-HO(HO_3S)-1-naphthylazo]-2-naphthylazo}-$C_6H_4AsO_3H_2$, 3, 466

2-{1,8,3,6,7-(HO)$_2$(HO_3S)$_2$[8,5-HO-(HO_3S)-1-naphthylazo]-2-naphthylazo}-$C_6H_4AsO_3H_2$, 3, 466

$AsC_{26}H_{19}N_4O_{18}S_4$ 2-{1,8,3,6,7-(HO)$_2$-(HO_3S)$_2$[1,3,8-HO(HO_3S)$_2$-2-naphthylazo]-2-naphthylazo}-$C_6H_4AsO_3H_2$, 3, 466

2-{1,8,3,6,7-(HO)$_2$(HO_3S)$_2$[8,3,5-HO-(HO_3S)$_2$-1-naphthylazo]-2-naphthylazo}-$C_6H_4AsO_3H_2$, 3, 466

2-{1,8,3,6,7-(HO)$_2$(HO_3S)$_2$[8,3,6-HO-(HO_3S)$_2$-1-naphthylazo]-2-naphthylazo}-$C_6H_4AsO_3H_2$, 3, 466

$AsC_{26}H_{20}BrD_2O$ [Ph_3($BzCD_2$)As]Br, 3, 338

$AsC_{26}H_{20}Cl$ 3,3'-Me_2-5,5'-spirobi(dibenzarsolium)Cl^-, 3, 571

$AsC_{26}H_{20}I$ 3,3'-Me_2-5,5'-spirobi(dibenzarsolium)I^-, 3, 571

$AsC_{26}H_{20}N_3O_7$ 5,5-Me(Ph)-5,10-dihydroacridarsinium picrate, 3, 597

5,5-Me(Ph)-5,6-dihydroacridarsinium picrate, 3, 598

$AsC_{26}H_{21}$ 4-(PhCH=CH)$C_6H_4AsPh_2$, 3, 71

$AsC_{26}H_{21}Br_2O$ [Ph_3(BzCHBr)As]Br, 3, 338

[Ph_3(4-$BrC_6H_4CH_2CO$)As]Br, 3, 338

[Ph_3(4-$BrC_6H_4COCH_2$)As]Br, 3, 338

$AsC_{26}H_{21}O$ $Ph_3AsCHBz$, 3, 309

$AsC_{26}H_{21}O_2$ Ph(1-$C_{10}H_7$)$_2$As(OH)$_2$, 3, 351

$AsC_{26}H_{21}O_4S$ $Ph_3\overset{+}{As}CH(Bz)SO_3^-$, 3, 309, 338

$AsC_{26}H_{22}BrClI$ [Ph(4-BrC_6H_4)(4-Cl-C_6H_4)(3,4-$Me_2C_6H_3$)As]I, 3, 338

$AsC_{26}H_{22}BrO$ [Ph_3($BzCH_2$)As]Br, 3, 338

$AsC_{26}H_{22}Cl$ [2,5-Ph(Me)C_6H_3]$_2$AsCl, 3, 167

$AsC_{26}H_{22}ClO$ [Ph_3($BzCH_2$)As]Cl, 3, 338

$AsC_{26}H_{22}N_3O_7$ [$EtPh_3As$]$OC_6H_2(NO_2)_3$, 3, 327

$AsC_{26}H_{23}ClI$ [Ph(4-ClC_6H_4)(3-MeC_6H_4)-(4-MeC_6H_4)As]I, 3, 338

$AsC_{26}H_{23}O_2$ [Ph_4As]OAc, 3, 332

[Ph_3($BzCH_2$)As]OH, 3, 338

$AsC_{26}H_{24}I$ [Ph_2(4-MeC_6H_4)$_2$As]I, 3, 339

[Ph_3(4-EtC_6H_4)As]I, 3, 339

$AsC_{26}H_{24}N_3O_7$ [MeEt($PhCH_2$)(1-$C_{10}H_7$)As]-$OC_6H_2(NO_2)_3$, 3, 328

$AsC_{26}H_{25}ClN$ 14,1,13-Cl(Pr)$_2$-7,14-dihydrodibenzo[a,j]phenarsazine, 3, 638

$AsC_{26}H_{25}N_2O_4S$ [$Ph_3AsNHO_2SC_6H_4NHAc$]OH, 3, 370

$AsC_{26}H_{26}Br$ [Et($PhCH_2$)(4-MeC_6H_4)(1-$C_{10}H_7$)As]Br, 3, 339

$AsC_{26}H_{27}N_8O_{12}$ PhAs[CH_2CH_2C(=NH)$NH_3\cdot$-$O_2CC_6H_3(NO_2)_2$-2,5]$_2$, 3, 50

$AsC_{26}H_{28}BrCl_2O_4S$ 2,2'-Spirobi(5-chloroisoarsindolinium)bromocamphorsulfonate, 3, 574

$AsC_{26}H_{29}Cl_2O_4S$ 2,2'-Spirobi(5-chloroisoarsindolinium)camphorsulfonate, 3, 574

$AsC_{26}H_{31}N_2O_5$ 4-[2,1-($C_9H_{19}CO_2$)$C_{10}H_6$-N=N]$C_6H_4AsO_3H_2$, 3, 467

$AsC_{26}H_{31}N_4O_6\cdot H_2O$ PhAs[CH_2CH_2C(=NH)-$NH_3\cdot O_2CC_6H_4OH$-2]$_2\cdot H_2O$, 3, 50

$AsC_{26}H_{31}O_2$ [$PhCH_2O(CH_2)_3$]$_2$AsPh, 3, 71

$AsC_{26}H_{32}N_3O_4$ 4-[2,1-($C_9H_{19}CONH$)$C_{10}H_6$-N=N]$C_6H_4AsO_3H_2$, 3, 467

$AsC_{26}H_{33}O_4$ $Ph_3As(OOCMe_3)_2$, 3, 361

$AsC_{26}H_{36}BrO_2$ {MePh($PhCH_2$)[2,5-Me_2CH-(Me)$C_6H_9OCOCH_2$]As}Br, 3, 339

$AsC_{26}H_{40}N_3O_3$ 4-[4-(n-$C_{14}H_{29}NH$)C_6H_4-N=N]$C_6H_4AsO_3H_2$, 3, 467

$AsC_{26}H_{44}NO_5S_2$ 3,4-H_2N(HO)C_6H_3As(SCH_2-$CO_2C_8H_{17}$)$_2$, 3, 275

3,4-H_2N(HO)C_6H_3As($SCH_2CO_2CHMeC_6H_{13}$)$_2$, 3, 275

$AsC_{26}H_{44}NO_6S_2$ 3,4-H_2N($HOCH_2CH_2O$)C_6H_3-

$As(SCH_2CO_2CHEtBu)_2$, 3, 275*
$AsC_{26}H_{44}N_4S_8$ {$-CH_2As[SC(S)-N(CH_2)_4CH_2]_2$}$_2$, 3, 284
$AsC_{26}H_{47}S_2$ $PhAs(SC_{10}H_{21})_2$, 3, 275
$AsC_{26}H_{55}S_2$ $EtAs(SC_{12}H_{25})_2$, 3, 275
$BuAs(SC_{11}H_{23})_2$, 3, 275
$AsC_{27}H_{19}ClN$ 3,14-Me(Ph)-7,14-dihydrodibenzophenarsazine isomers, 3, 638
$AsC_{27}H_{20}NO_7S_2$ 3,4-$O_2N(PhCH_2O)C_6H_3As$-$(SC_6H_4CO_2H)_2$, 3, 275
$AsC_{27}H_{21}Br_2N_8S_2$ 4-[$N=C(NH_2)N=C(NH_2)N=CNH$]C_6H_4As-[S-2-(6-bromoquinoline)]$_2$, 3, 275
$AsC_{27}H_{21}N_8S_2$ 4-[$N=C(NH_2)N=C(NH_2)N=C$-NH]C_6H_4As(S-2-quinoline)$_2$, 3, 275
4-[$N=C(NH_2)N=C(NH_2)N=CNH$]C_6H_4As-(S-8-quinoline)$_2$, 3, 275
$AsC_{27}H_{21}O_7$ $(o-HO_2CCH=CHC_6H_4)_3AsO$, 3, 355
$AsC_{27}H_{23}BrClO$ 6,6-($p-ClC_6H_4CH_2CO$)(Ph)-5,7-dihydro-6H-dibenz[c,e]arsepinium bromide, 3, 642
$AsC_{27}H_{23}N_2S_4$ $MeAs[SC(S)NPh_2]_2$, 3, 284
$AsC_{27}H_{23}O_2$ $Ph_3AsC(Ph)CO_2Me$, 3, 309
$AsC_{27}H_{24}BrO_2$ {$Ph_3[MeO_2CCH(Ph)]As$}Br, 3, 339
$AsC_{27}H_{24}N_3O_7$ [$Me_2Ph(2-PhC_6H_4CH_2)As$]-$OC_6H_2(NO_2)_3$, 3, 328
$AsC_{27}H_{24}N_3O_8$ $Ph(4-MeC_6H_4)(4-EtC_6H_4)$-$As(OH)OC_6H_2(NO_2)_3$, 3, 349
$AsC_{27}H_{25}BrP$ 5,10,5-$PhCH_2$(Et)(Ph)-5,-10-dihydrodibenzo[1,4]phosphoniaarsenin Br^-, 3, 640
$AsC_{27}H_{25}ClI$ [$(4-ClC_6H_4)(3-MeC_6H_4)$-$(4-MeC_6H_4)_2As$]I, 3, 339
$AsC_{27}H_{25}N_2O_2S$ 5,5-($p-Me_2NC_6H_4$)(p-Me-$C_6H_4SO_2N=$)-dibenzarsole, 3, 567
$AsC_{27}H_{25}N_4O$ 2,3,5,6-Tetrahydro-8-methoxy-1,7-bis(methylphenylhydrazono)-1H,7H-benz[i,j]arsinolizine, 3, 599
$AsC_{27}H_{26}I$ [$Ph(4-MeC_6H_4)_3As$]I, 3, 339
$AsC_{27}H_{29}N_2O_3S$ [$(MeC_6H_4)_3AsNHO_2SC_6H_4$-$NH_2$]OH isomers, 3, 370
$AsC_{27}H_{30}N_6O_{14}P$ 1,1,5,5-$Et_2(Me)_2$-2,3,-4,5-tetrahydro-1H-1,5-benzophosphoniaarsepinium dipicrate, 3, 644
$AsC_{27}H_{34}IO_2$ {$MePh[PhCH_2O(CH_2)_3]_2As$}I, 3, 339
$AsC_{27}H_{36}N_3$ [$4-(EtMeN)C_6H_4$]$_3$As, 3, 31
[$2,4-Me(Me_2N)C_6H_3$]$_3$As, 3, 31
$AsC_{27}H_{43}O_4$ $Pr_3As(OOCMe_2Ph)_2$, 3, 361
$AsC_{27}H_{51}$ $(3,3,5-Me_3C_6H_8)_3As$, 3, 31
$AsC_{27}H_{57}S_2$ $MeAs(SC_{13}H_{27})_2$, 3, 275
$PrAs(SC_{12}H_{25})_2$, 3, 275
$AsC_{28}H_{19}$ $(1-C_{10}H_7)_2AsC\equiv CPh$, 3, 71
$AsC_{28}H_{20}Cl$ 1-Chlorotetraphenylarsole, 3, 554
$AsC_{28}H_{20}N$ 8,12-Ph_2-7,12-dihydrobenzo-[a]phenarsazine, 3, 630
$AsC_{28}H_{21}N_6O_{11}S_2$ 2-{1,8,3,6,7-$(HO)_2$-$(HO_3S)_2$[$4-(PhN=N)C_6H_4N=N$]naphthylazo}-$C_6H_4AsO_3H_2$, 3, 467
$AsC_{28}H_{22}N$ 6,10,12-$Me_2(\alpha-C_{10}H_7)$-7,12-dihydrobenzo[b]phenarsazine, 3, 631
$AsC_{28}H_{25}$ [$4-(PhCH=CH)C_6H_4$]$As(CH_2Ph)_2$, 3, 71
$AsC_{28}H_{25}N_2S_4$ $EtAs[SC(S)NPh_2]_2$, 3, 284
$AsC_{28}H_{25}N_4O_{10}$ 5,6-Dihydro-3-methoxy-4-oxo-4H,8H-benz[i,j]arsinolizino-[2,3-b]quinolinium picrate·EtOH, 3, 599
$AsC_{28}H_{26}NS$ [$Ph(4-MeC_6H_4)_3As$]SCN, 3, 339
$AsC_{28}H_{27}N_2O_9$ Berberine·3,4-AcNH(HO)-$C_6H_3AsO_3H$ salt, 3, 477
Berberine·3,4-AcNH(HO)$C_6H_3AsO_3H$ salt, 3, 477
$AsC_{28}H_{28}I$ [$(4-MeC_6H_4)_4As$]I, 3, 339
[$(3-MeC_6H_4)_4As$]I, 3, 340
$AsC_{28}H_{28}NO_2S$ $(MeC_6H_4)_3As=NO_2SC_6H_4Me$-p isomers, 3, 369
$AsC_{28}H_{29}N_6O_2$ [$4-(4-Me_2NC_6H_4N_2)C_6H_4$]$_2$-$AsO_2H$, 3, 387
$AsC_{28}H_{30}N_9O_{13}$ 4-$AcNHC_6H_4As[CH_2CH_2C$-$(=NH)NH_3\cdot O_2CC_6H_3(NO_2)_2$-2,5]$_2$, 3, 58
$AsC_{28}H_{33}N_6O_6$ Bis[4-(3-ethoxycarbonyl-2,4-dimethyl-5-pyrrylazo)phenyl]-arsinic acid, 3, 377
$AsC_{28}H_{34}BrO_4S$ 3,3′,4,4′-Tetrahydro-1,1′(2H,2′H)-spirobisarsinolinium bromocamphorsulfonate, 3, 600
3,3′,4,4′-Tetrahydro-2,2′(1H,1′H)-

spirobisisoarsinolinium bromocamphorsulfonate, 3, 600
$AsC_{28}H_{40}P$ $Ph_2As(CH_2)_4P(C_6H_{11})_2$, 3, 63
$AsC_{28}H_{44}N_3O_3$ 4-[4-(n-$C_{18}H_{33}NH$)C_6H_4-N=N]$C_6H_4AsO_3H_2$, 3, 467
$AsC_{28}H_{47}O_2S_4$ $PhAs[SC(S)OC_{10}H_{21}]_2$, 3, 281
$AsC_{28}H_{48}NO_6S_2$ 3,4-$H_2N(HOCH_2CH_2O)C_6H_3$-$As(SCH_2CO_2C_8H_{17})_2$, 3, 275
3,4-$H_2N(HOCH_2CH_2O)C_6H_3As(SCH_2CO_2CH$-$MeC_6H_{13})_2$, 3, 275
$AsC_{28}H_{51}S_2$ $PhAs(SC_{11}H_{23})_2$, 3, 275
$AsC_{28}H_{59}S_2$ $EtAs(SC_{13}H_{27})_2$, 3, 275
$BuAs(SC_{12}H_{25})_2$, 3, 275
$AsC_{29}H_{22}NO_3$ 3-(Diphenylarsinomethyl)-cinchophen oxide, 3, 360
$AsC_{29}H_{22}N_3O_7$ [Ph_3(CH=CHCH=CHCH)As]-$OC_6H_2(NO_2)_3$, 3, 330
$AsC_{29}H_{23}ClN_3O_8$ 2-(4-Chlorophenacyl)-1,2,3,4-tetrahydro-2-phenylisoarsinolinium picrate, 3, 595
$AsC_{29}H_{24}I$ [$Ph_2(PhCH_2)(1$-$C_{10}H_7)As$]I, 3, 340
$AsC_{29}H_{25}N_8O_4S_2$
4-[N=C(NH_2)N=C(NH_2)N=CNH]C_6H_4As-($SCH_2CH_2NCOC_6H_4CO)_2$, 3, 276
$AsC_{29}H_{25}N_8S_2$
4-[N=C(NH_2)N=C(NH_2)N=CNH]C_6H_4As=[S-2-(4-methylquinoline)]$_2$, 3, 275
$AsC_{29}H_{25}O_8$ 1,1-Dihydro-1,1,1-triphenylarsole-2,3,4,5-tetracarboxylic acid tri-Me ester, 3, 554
$AsC_{29}H_{28}NS$ [(4-$MeC_6H_4)_4As$]SCN, 3, 339
$AsC_{29}H_{30}S_2$ $Ph_2As(S)(CH_2)_5As(S)Ph_2$, 3, 366
$AsC_{29}H_{31}N_2O_4S$ [(2-$MeC_6H_4)_3AsNHO_2S$-C_6H_4NHAc]OH, 3, 349, 370
$AsC_{29}H_{41}N_{12}O_{12}S_2$
4-[N=C(NH_2)N=C(NH_2)N=CNH]C_6H_4As-(S-glutathione ester)$_2$, 3, 276*
$AsC_{29}H_{41}N_{12}O_{13}S_2$
4,3-HO[N=C(NH_2)N=C(NH_2)N=CNH]C_6H_3-As(S-glutathione ester)$_2$, 3, 276*
$AsC_{29}H_{50}NO_8$ 4-(n-$C_{17}H_{35}CO_2CH_2CHOH$-$CH_2O_2CCH_2NH)C_6H_4AsO_3H_2$, 3, 441
$AsC_{29}H_{61}S_2$ $MeAs(SC_{14}H_{29})_2$, 3, 276
$PrAs(SC_{13}H_{27})_2$, 3, 276
$AsC_{30}H_{21}$ (1-$C_{10}H_7)_3As$, 3, 32
5-o-Terphenyldibenzarsole, 3, 567
5-Phenyl-5,5′-spirobi(dibenzarsole), 3, 572
17-Ph-17H-tetrabenz[b,d,f,h]arsonin, 3, 649
$AsC_{30}H_{21}Br_4$ 5,5-Ph[2-(2-$BrC_6H_4)C_6H_4$]-dibenzarsolium Br_3^-, 3, 567*
$AsC_{30}H_{21}ClN_3O_8$ Ph(4-PhC_6H_4)(4-ClC_6H_4)-$As(OH)OC_6H_2(NO_2)_3$, 3, 349
$AsC_{30}H_{22}I$ 5-(2-PhC_6H_4)-5-Ph-dibenzarsolium iodide, 3, 568
$AsC_{30}H_{22}N_3O_7$ [Ph_4As]$OC_6H_2(NO_2)_3$, 3, 333
$AsC_{30}H_{23}$ 2-(2-$PhC_6H_4)C_6H_4AsPh_2$, 3, 63
5,5-Dihydro-5,5,5-triphenyldibenzarsole, 3, 568
$AsC_{30}H_{24}I$ [Ph_3(2-PhC_6H_4)As]I, 3, 340
$AsC_{30}H_{25}$ Ph_5As, 3, 307
$AsC_{30}H_{25}N_4O_{12}S_2$ 4-[1,4,8-HO($HO_3S)_2$-2-naphthyl]-N(CONHPh)N(CONHPh)]-$C_6H_4AsO_3H_2$, 3, 454
$AsC_{30}H_{28}ClN_4O_8$ [$CH_2CH_2As(Ph)CH_2CH_2N$-(Ph)$CH_2COC_6H_4Cl$-p]$OC_6H_2(NO_2)_3$, 3, 619
[$CH_2CH_2As(Ph)CH_2CH_2N(Ph)CH_2COC_6H_4$-Cl-p]$OC_6H_2(NO_2)_3$, 3, 619
$AsC_{30}H_{27}O_8$ 1,1-Dihydro-1,1,1-triphenylarsole-2,3,4,5-tetracarboxylic acid tetra-Me ester, 3, 554
$AsC_{30}H_{29}N_{14}O_{21}$ 4-$H_2NC_6H_4As[CH_2CH_2C$-(=NH)$NH_2]_2[HOC_6H_2(NO_2)_3]_3$, 3, 51
$AsC_{30}H_{30}Br$ [$(PhCH_2)_2Et$(4-PhCH=CH-C_6H_4)As]Br, 3, 340
$AsC_{30}H_{36}BrO_2$ {Ph_3[2,5-$Me_2CH(Me)C_6H_9$-$OCOCH_2$]As}Br, 3, 340
$AsC_{30}H_{36}N_3O_7$ [3,4-AcNH(EtO)$C_6H_3]_3AsO$, 3, 359
$AsC_{30}H_{45}O_5$ 3,3′,4,4′-Tetrahydro-1,1′-(2H,2′H)-spirobisarsinolinium menthyloxyacetate·$2H_2O$, 3, 600
$AsC_{30}H_{48}NO_9S_2$ 5,2-$H_2N(HO)C_6H_3As$[SCH-($CO_2Bu)CH_2CO_2Bu]_2$, 3, 276
$AsC_{30}H_{48}N_3O_3$ 4-[4-(n-$C_{18}H_{37}NH)C_6H_4$-N=N]$C_6H_4AsO_3H_2$, 3, 467
$AsC_{30}H_{50}NO_6S_2$ 2,5-HO(AcNH)C_6H_3As-[SCH(Me)$CO_2C_8H_{17}]_2$, 3, 276

$AsC_{30}H_{52}NO_5S_2$ 3,4-$H_2N(HO)C_6H_3As$-($SCH_2CO_2C_{10}H_{21})_2$, 3, 276

$AsC_{30}H_{55}S_2$ $PhAs(SC_{12}H_{25})_2$, 3, 276

$AsC_{30}H_{63}S_2$ $EtAs(SC_{14}H_{29})_2$, 3, 276

$BuAs(SC_{13}H_{27})_2$, 3, 276

$AsC_{31}H_{23}$ 9-Fluorenylidenetriphenylarsenic, 3, 308

$AsC_{31}H_{23}N_3O_7P$ 5,10-Ph_2-5,10-dihydrodibenzo[1,4]phospharsenin methopicrate, 3, 641

$AsC_{31}H_{24}Br$ [Ph_3(9-fluorenyl)As]Br, 3, 340

$AsC_{31}H_{24}I$ 5-[2-(2-$MeC_6H_4)C_6H_4$]-5-Ph-dibenzarsolium iodide, 3, 569

$AsC_{31}H_{25}BrP$ 5,5,10-$PhCH_2(Ph)_2$-5,10-dihydrodibenzo[1,4]phosphoniaarsenin bromide, 3, 641

$AsC_{31}H_{25}ClI$ [Ph(4-$ClC_6H_4)(MeC_6H_4)$-(PhC_6H_4)As]I isomers, 3, 340

$AsC_{31}H_{25}O_4S_2$ $Ph_3AsC(O_2SPh)_2$, 3, 309

$AsC_{31}H_{65}S_2$ $MeAs(SC_{15}H_{31})_2$, 3, 276

$PrAs(SC_{14}H_{29})_2$, 3, 276

$AsC_{32}H_{20}I$ 7,7'-Spirobi(7H-benzo[b]naphth[1,2-d]arsolium)chloride, 3, 572

$AsC_{32}H_{22}Cl$ Bis(1-phenyl-2-naphthyl)-AsCl, 3, 167

$AsC_{32}H_{23}O_2$ (1-$PhC_{10}H_6)_2AsO_2H$, 3, 387

$AsC_{32}H_{25}$ 3,3',5-Me_2(Ph)-5,5'-spirobi(dibenzarsole), 3, 573

$AsC_{32}H_{25}N_2S_4$ $PhAs[SC(S)NPh_2]_2$, 3, 284

$AsC_{32}H_{26}N$ 5-(p-$Me_2NC_6H_4$)-5,5'-spirobi(dibenzarsole), 3, 573

17-(p-$Me_2NC_6H_4$)-17H-tetrabenz[b,d,f,h]arsonin, 3, 649

$AsC_{32}H_{27}$ (2-$PhC_6H_4CH_2)_2AsPh$, 3, 72

$AsC_{32}H_{27}ClI$ [Ph(4-ClC_6H_4)(4-EtC_6H_4)-(4-PhC_6H_4)As]I, 3, 340

$AsC_{32}H_{28}Cl_2NO_3$ 2,2'-Spirobi(5-chloroisoarsindolinium) N-(1-phenylethyl)phthalamate, 3, 574

$AsC_{32}H_{28}O_3PS$ 5,10-Ph_2-5,10-dihydrodibenzo[1,4]phospharsenin metho-p-toluenesulfonate, 3, 641

$AsC_{32}H_{32}ClN_4O_4$ [(3-$AcNHC_6H_4)_4As$]Cl, 3, 340

$AsC_{32}H_{32}N_6O_{14}P$ 5,5,12,12-$Et_2(Me)_2$-5,6,11,12-tetrahydrodibenzo[b,f]-[1,4]phosphoniaarsocinium dipicrate, 3, 648

$AsC_{32}H_{50}NO_{10}S_2$ HO(AcNH)$C_6H_3As[SCH(CO_2$-Bu)$CH_2CO_2Bu]_2$ isomers, 3, 276

$AsC_{32}H_{56}NO_6S_2$ 3,4-$H_2N(HOCH_2CH_2O)C_6H_3$-$As(SCH_2CO_2C_{10}H_{21})_2$, 3, 276

$AsC_{32}H_{59}S_2$ $PhAs(SC_{13}H_{27})_2$, 3, 276

$AsC_{32}H_{67}S_2$ $EtAs(SC_{15}H_{31})_2$, 3, 276

$BuAs(SC_{14}H_{29})_2$, 3, 276

$AsC_{33}H_{27}N_3O_7P$ 5,10,5-$PhCH_2$(Et)(Ph)-5,10-dihydrodibenzo[1,4]phosphoniaarsenin picrate, 3, 640,

$AsC_{33}H_{29}IN$ Trimethyl(p-5,5 -spirobi-[dibenzarsol]-5-ylphenyl)ammonium I^-, 3, 573

Trimethyl(p-17H-tetrabenz[b,d,f,h]-arsonin-17-ylphenyl)ammonium I^-, 3, 649*

$AsC_{33}H_{32}N_2O_{12}S_2$ 2,2'-$CH_2(C_{10}H_6$-1-SO_3-$H\cdot H_2NC_6H_4AsO_3H_2)_2$, 3, 428

$AsC_{33}H_{35}BrClO_5S$ 2-(4-Chlorophenacyl)-1,2,3,4-tetrahydro-2-phenylisoarsinolinium bromocamphorsulfonate, 3, 595

$AsC_{33}H_{36}ClN_4O_8$ 4-[6-Chloro-2-methoxy-9-acridinyl)amino]-α-(diethylamino)cresol 10-oxide·4-($HOCH_2CO$-NH)$C_6H_4AsO_3H_2$, 3, 429

$AsC_{33}H_{38}BrO_2$ {Ph($PhCH_2$)[$PhCH_2O(CH_2)_3]_2$-As}Br, 3, 341*

$AsC_{33}H_{51}O_{11}S_2$ 4-$MeO_2CCH_2OC_6H_4As$[SCH-($CO_2Bu)CH_2CO_2Bu]_2$, 3, 277

$AsC_{33}H_{69}S_2$ $MeAs(SC_{16}H_{33})_2$, 3, 277

$PrAs(SC_{15}H_{31})_2$, 3, 277

$AsC_{34}H_{25}$ Pentaphenylarsole, 3, 554

$AsC_{34}H_{25}O$ 1,1-Dihydro-1,2,3,4,5-pentaphenylarsole 1-oxide, 3, 554

$AsC_{34}H_{26}N_2O$ 7,7'-Oxybis(10-methyl-7,12-dihydrobenzo[c]phenarsazine, 3, 634

$AsC_{34}H_{31}N_4O_6S_2$ [4-($AcNHC_6H_4SO_2NH$)-$C_6H_4]_2AsPh$, 3, 71

$AsC_{34}H_{32}Br_2P$ 5,10,10,5-$(PhCH_2)_2$(Et)-(Ph)-5,10-dihydrodibenzo[1,4]phospharseninium dibromide, 3, 640

$AsC_{34}H_{32}I_2N$ 17-(p-$Me_2NC_6H_4$)-17H-tetra-

benz[b,d,f,h]arsonin dimethiodide, 3, 649

$AsC_{34}H_{33}O_8$ [MeEtPh($PhCH_2$)As]O_2CCH-(OBz)CH(OBz)CO_2H, 3, 321

$AsC_{34}H_{34}N_2O_{12}S_2$ 1,1′-MeCH(2-$C_{10}H_6SO_3$-$H\cdot H_2NC_6H_4AsO_3H_2$)$_2$, 3, 428

$AsC_{34}H_{36}ClO_5S$ 2-(4-Chlorophenacyl)-1,2,3,4-tetrahydro-2-phenylisoarsinolinium camphorsulfonate, 3, 595

$AsC_{34}H_{38}N_6O_{14}P$ 5,5,12,12-Et_4-5,6,11,12-tetrahydrodibenzo[b,f][1,4]-phosphoniaarsocinium dipicrate, 3, 648

$AsC_{34}H_{60}NO_5S_2$ 3,4-H_2N(HO)C_6H_3As(SCH_2-$CO_2C_{12}H_{25}$)$_2$, 3, 277

$AsC_{34}H_{63}S_2$ PhAs($SC_{14}H_{29}$)$_2$, 3, 277

$AsC_{34}H_{71}S_2$ EtAs($SC_{16}H_{33}$)$_2$, 3, 277
BuAs($SC_{15}H_{31}$)$_2$, 3, 277

$AsC_{35}H_{44}N_5O_4$ 7-[$C_8H_{17}NH(CH_2)_3NH$]benz-[c]acridine·4-(H_2NCONH)$C_6H_4AsO_3H_2$, 3, 429

$AsC_{35}H_{50}N_3O_7$ 3,4-AcNH(HO)$C_6H_3AsO_3H_2$·-vuzin salt, 3, 477

$AsC_{35}H_{73}S_2$ MeAs($SC_{17}H_{35}$)$_2$, 3, 277
PrAs($SC_{16}H_{33}$)$_2$, 3, 277

$AsC_{36}H_{27}$ (PhC_6H_4)$_3$As isomers, 3, 32

$AsC_{36}H_{27}ClI$ (2-PhC_6H_4)$_3$AsClI, 3, 345

$AsC_{36}H_{28}N$ N,N-Dimethyl-p-spiro[7H-benzo[b]naphth[1,2-d]arsole-7,5′-[5H]dibenzarsol]-7-ylaniline, 3, 573

$AsC_{36}H_{29}O_2$ (2-PhC_6H_4)$_3$As(OH)$_2$, 3, 351

$AsC_{36}H_{30}BrNP$ [(Ph_3As)(Ph_3P)N]Br, 3, 368

$AsC_{36}H_{36}N_3$ [4-(Me_2N)$C_{10}H_6$]$_3$As, 3, 32

$AsC_{36}H_{37}O_4$ Ph_3As($OOCMe_2Ph$)$_2$, 3, 361

$AsC_{36}H_{37}O_8$ [MeBuPh($PhCH_2$)As]O_2CCH-(OBz)CH(OBz)CO_2H, 3, 322

$AsC_{36}H_{45}N_4O_5$ 7-($C_8H_{17}NHCH_2CH_2CH_2NH$)-benz[c]acridine-3,4-AcNH(HO)C_6H_3-AsO_3H_2 salt, 3, 477

$AsC_{36}H_{67}S_2$ PhAs($SC_{15}H_{31}$)$_2$, 3, 277

$AsC_{36}H_{75}$ ($C_{12}H_{25}$)$_3$As, 3, 29

$AsC_{36}H_{75}S_2$ EtAs($SC_{17}H_{35}$)$_2$, 3, 277
BuAs($SC_{16}H_{33}$)$_2$, 3, 277

$AsC_{37}H_{25}N_2O_{11}S_2$ 2-[2-(p-Arsenosophenyl)-1,2-dibenzoylhydrazo]-1-naphthol-4,8-disulfonic acid benzoate, 3, 250

$AsC_{37}H_{27}N_2O_{13}S_2$ 2-[2-(4-Arsonophenyl)-1,2-dibenzoylhydrazo]-1-naphthol-4,8-disulfonic acid benzoate, 3, 454

$AsC_{37}H_{27}N_3O_7P$ 5,5,10-$PhCH_2$(Ph)$_2$-5,10-dihydrodibenzo[1,4]phosphoniaarsenin picrate, 3, 641

$AsC_{37}H_{30}I$ [Me(2-PhC_6H_4)$_3$As]I, 3, 341

$AsC_{37}H_{32}NO$ 5-(p-$Me_2NC_6H_4$)-5,5-dihydro-5-methoxy-5-(0-1-naphthylphenyl)dibenzarsole, 3, 569

$AsC_{37}H_{33}N_4O$ 2,3,5,6-Tetrahydro-8-methoxy-1,7-bis(diphenylhydrazono)-1H,7H-benz[i,j]arsinolizine, 3, 599

$AsC_{37}H_{77}S_2$ MeAs($SC_{18}H_{37}$)$_2$, 3, 277
PrAs($SC_{17}H_{35}$)$_2$, 3, 277

$AsC_{38}H_{32}Br_2P$ 5,10,5,10-($PhCH_2$)$_2$(Ph)$_2$-5,10-dihydrodibenzo[1,4]phospharseninium dibromide, 3, 641

$AsC_{38}H_{37}O_4$ Ph_3As(OO-tetrahydronaphthyl)$_2$, 3, 362

$AsC_{38}H_{71}S_2$ PhAs($SC_{16}H_{33}$)$_2$, 3, 277

$AsC_{38}H_{79}S_2$ EtAs($SC_{18}H_{37}$)$_2$, 3, 277
BuAs($SC_{17}H_{35}$)$_2$, 3, 277

$AsC_{39}H_{32}Br$ [(CH_2=$CHCH_2$)(PhC_6H_4)$_3$As]Br, 3, 341

$AsC_{39}H_{56}N_3O_3$ 4-[4-(Cholesterylamino)-C_6H_4N=N]$C_6H_4AsO_3H_2$, 3, 467*

$AsC_{39}H_{81}S_2$ PrAs($SC_{18}H_{37}$)$_2$, 3, 277

$AsC_{40}H_{34}NO_{12}$ N-(Tetrabenzoatoglucosyl)-arsanilic acid, 3, 441

$AsC_{40}H_{75}S_2$ PhAs($SC_{17}H_{35}$)$_2$, 3, 277

$AsC_{40}H_{83}S_2$ BuAs($SC_{18}H_{37}$)$_2$, 3, 277

$AsC_{41}H_{39}O_4$ Me_3As($OOCPh_3$)$_2$, 3, 362

$AsC_{42}H_{33}O_3$ 3,3′-Dimethyl-5,5′-spirobi(dibenzarsolium)dibenzoylhydrogentartrate, 3, 571

$AsC_{42}H_{62}O_8PS_2$ 5,5,12,12-Et_4-5,6,11,12-tetrahydrodibenzo[b,f][1,4]-phosphoniaarsocinium dicamphorsulfonate, 3, 648

$AsC_{42}H_{79}S_2$ PhAs($SC_{18}H_{37}$)$_2$, 3, 277

$AsC_{42}H_{87}$ $(C_{14}H_{29})_3As$, 3, 29
$AsC_{43}H_{28}N_8$ $[MePh_3As][(NC)_2CC_6H_4C(CN)_2]_2$, 3, 324
$AsC_{43}H_{29}Br_4O_5S$ $[Ph_4As]$bromophenol blue salt, 3, 332
$AsC_{44}H_{27}Br_4O_5$ $[Ph_4As]$eosinate, 3, 332
$AsC_{44}H_{28}N_8$ $[EtPh_3As][(NC)_2CC_6H_4C(CN)_2]_2$, 3, 327
$AsC_{44}H_{45}O_4$ $Et_3As(OOCPh_3)_2$, 3, 362
$AsC_{45}H_{30}N_8$ $[MePh_3As][(NC)_2CC_6H_3(Me)C(CN)_2]_2$, 3, 324
$AsC_{45}H_{60}$ $[2,4,6\text{-}Me(Me_3C)_2C_6H_8]_3As$, 3, 32
$AsC_{46}H_{36}N_6O_{14}P$ 5,10,10,5-$(PhCH_2)_2$-(Et)(Ph)-5,10-dihydrodibenzo[1,4]-phospharseninium dipicrate, 3, 640
$AsC_{50}H_{36}N_6O_{14}P$ 5,10,5,10-$(PhCH_2)_2$-$(Ph)_2$-5,10-dihydrodibenzo[1,4]-phospharseninium dipicrate, 3, 641
$AsC_{51}H_{47}Br_2O_5S$ $[Ph_4As]$bromothymol blue salt, 3, 332
$AsC_{69}H_{56}NO_{12}$ 5-(2-Biphenylyl)-5-(p-$Me_2NC_6H_4$)dibenzarsolium hydrogen tartrate, dibenzoate, methotartrate dibenzoate, 3, 570
$AsC_{74}H_{48}N_{12}P$ $\{[MePh_3As][MePh_3P]\}$-$[(NC)_2CC_6H_4C(CN)_2]_3$, 3, 324
$As_2CH_2Br_4$ $CH_2(AsBr_2)_2$, 3, 188
$As_2CH_2Cl_4$ $CH_2(AsCl_2)_2$, 3, 178
$As_2CH_2O_2$ $CH_2(AsO)_2$, 3, 227
$As_2CH_2S_2$ $CH_2(AsS)_2$, 3, 264
$As_2CH_6O_6$ $CH_2(AsO_3H_2)_2$, 3, 531
$As_2C_2H_4I_4$ $(\text{-}CH_2AsI_2)_2$, 3, 189
$AsC_2H_6S_2$ $[MeAs(S)=]_2$, 3, 551
$As_2C_3H_4O_7$ $[(HO_2C)_2(H_2O_3As)CAs]_n$, 3, 136
$As_2C_3H_6O_2$ $EtCH(AsO)_2$, 3, 234
$As_2C_3H_6O_{10}$ $(HO_2C)_2C(AsO_3H_2)_2$, 3, 531
$As_2C_3H_8Cl_2$ $CH_2(AsMeCl)_2$, 3, 168
$As_2C_3H_8S_2$ $MeAs(S)=As(S)Et$, 3, 551
$As_2C_4F_{12}$ $(CF_3)_2AsAs(CF_3)_2$, 3, 123
$AsC_4F_{12}O$ $[(CF_3)_2As]_2O$, 3, 216
$As_2C_4F_{12}S$ $[(CF_3)_2As]_2S$, 3, 262
$As_2C_4HF_{12}N$ $[(CF_3)_2As]_2NH$, 3, 288*
$As_2C_4H_4Cl_4O$ $[(ClCH=CH)AsCl]_2O$, 3, 305
$As_2C_4H_8Br_2$ $BrAs(CH_2CH_2)_2AsBr$, 3, 601
$As_2C_4H_8Cl_4O_2S$ $SO_2(CH_2CH_2AsCl_2)_2$, 3, 190
$As_2C_4H_8O_{10}$ $[HO_2CCH(AsO_3H_2)\text{-}]_2$, 3, 531
$As_2C_4H_9Br_3$ $[MeAs=CHBr]_2\cdot HBr$, 3, 305
$As_2C_4H_{10}S_2$ $[EtAs(S)=]_2$, 3, 551
$As_2C_4H_{12}$ $Me_2AsAsMe_2$, 3, 124
$As_2C_4H_{12}O$ $(Me_2As)_2O$, 3, 213
$As_2C_4H_{12}O_7$ $O(CH_2CH_2AsO_3H_2)_2$, 3, 395
$As_2C_4H_{12}O_8S$ $SO_2(CH_2CH_2AsO_3H_2)_2$, 3, 395
$As_2C_4H_{12}S$ $(Me_2As)_2S$, 3, 262
$As_2C_4H_{12}S_2$ $Me_2As(S)SAsMe_2$, 3, 262
$As_2C_5H_{12}S$ $Me_2AsC(S)AsMe_2$, 3, 79*
$As_2C_5H_{15}N$ $(Me_2As)_2NMe$, 3, 288
$As_2C_6H_3Cl_4NO_2$ 2,5-$(Cl_2As)_2C_6H_3NO_2$, 3, 191
$As_2C_6H_3Cl_5$ 2,1,4-$ClC_6H_3(AsCl_2)_2$, 3, 191
$As_2C_6H_3NO_4$ 2,1,4-$O_2NC_6H_3(AsO)_2$, 3, 234
$As_2C_6H_4Cl_2O$ 1,3-Dichloro-1,3-dihydro-2,1,3-benzoxadiarsole, 3, 575
$As_2C_6H_4Cl_4$ o-$C_6H_4(AsCl_2)_2$, 3, 192
$As_2C_6H_4O_2$ 1,4-$C_6H_4(AsO)_2$, 3, 236
$As_2C_6H_7NO_8$ $O_2NC_6H_3(AsO_3H_2)_2$ isomers, 3, 531
$As_2C_6H_7NO_9$ 4,5,3,1-$HO(O_2N)C_6H_2(AsO_3H_2)_2$, 3, 531
$As_2C_6H_8O_6$ $C_6H_4(AsO_3H_2)_2$ isomers, 3, 531
$As_2C_6H_8O_7$ $HOC_6H_3(AsO_3H_2)_2$ isomers, 3, 531
$As_2C_6H_{10}Cl_2O$ $[Me(ClCH=CH)As]_2O$, 3, 216
$As_2C_6H_{10}O_2S_2$ $[MeCOCH_2As(S)=]_2$, 3, 551
$As_2C_6H_{12}BrF_3$ $Me_2AsCF_2CBrFAsMe_2$, 3, 79
$As_2C_8H_{12}F_6S_2$ $Me_2AsSC(CF_3)=C(CF_3)S$-$AsMe_2$, 3, 259
$As_2C_6H_{14}$ 1,4-Me_2-1,4-diarsonane, 3, 601
$As_2C_6H_{14}S_2$ $[PrAs(S)=]_2$, 3, 551
$[Me_2CHAs(S)=]_2$, 3, 551
$As_2C_6H_{16}O$ $(MeEtAs)_2O$, 3, 216
$As_2C_6H_{16}O_3$ $[Me(HOCH_2CH_2)As]_2O$, 3, 216
$As_2C_6H_{18}N_2O_6$ $(\text{-}CH_2NHCH_2CH_2AsO_3H_2)_2$, 3, 531
$As_2C_7H_{12}F_6$ $Me_2AsCF_2CF(CF_3)AsMe_2$, 3, 79

$As_2C_8H_{12}F_6$ $Me_2AsC(CF_3)=C(CF_3)AsMe_2$, 3, 80

$As_2C_8H_{16}Cl_4O_3$ $O(CH_2CH_2OCH_2CH_2AsCl_2)_2$, 3, 195

$As_2C_8H_{20}$ $(Et_2As-)_2$, 3, 126

$As_2C_8H_{20}O$ $(Et_2As)_2O$, 3, 214

$As_2C_8H_{22}N_2$ $[Me_2AsN(Me)CH_2-]_2$, 3, 287

$As_2C_9H_{11}Br$ 1,4-Br(Me)-1,2,3,4-tetrahydro-1,4-benzodiarsenin, 3, 601

$As_2C_9H_{18}O_4$ 3,9-Et_2-2,4,8,10-tetraoxa-3,9-diarsaspiro[5,5]hendecane, 3, 618

$As_2C_{10}H_{12}$ 2,3-Dihydro-1,4-ethano-1,4-benzodiarsenin, 3, 601

$As_2C_{10}H_{14}$ 1,2,3,4-Tetrahydro-1,4-dimethyl-1,4-benzodiarsenin, 3, 601

$As_2C_{10}H_{14}Br_2$ 1-Bromo-1,2,3,4-tetrahydro-4-methyl-1,4-benzodiarsenin methobromide, 3, 601

$As_2C_{10}H_{14}I_2$ 1,2,3,4-Tetrahydro-1-iodo-4-methyl-1,4-benzodiarsenin methiodide, 3, 602

$As_2C_{10}H_{14}O_2$ 1,2,3,4-Tetrahydro-1,4-dimethyl-1,4-benzodiarsenin 1,4-dioxide, 3, 602

$As_2C_{10}H_{16}$ 2-$Me_2AsC_6H_4AsMe_2$, 3, 80
 4-$Me_2AsC_6H_4AsMe_2$, 3, 98

$As_2C_{10}H_{16}N_2O_7$ 1,3-Dihydro-1,1,3,3-tetramethyl-2,1,3-benzoxadiarsolium dinitrate, 3, 575*

$As_2C_{10}H_{20}$ $Et_2AsC{\equiv}CAsEt_2$, 3, 98

$As_2C_{10}H_{22}Cl_2$ $(-CH_2AsBuCl)_2$, 3, 171

$As_2C_{10}H_{24}N_2$ $Me_3C\overline{NAs(Me)N(CMe_3)A}sMe$, 3, 552

$As_2C_{10}H_{24}O$ $[Et(Me_2CH)As]_2O$, 3, 216
 $(MeBuAs)_2O$, 3, 216

$As_2C_{10}H_{24}O_4$ $[-CH_2(Bu)AsO_2H]_2$, 3, 379

$As_2C_{10}H_{24}S_6$ 2,2'-(Ethylenedithio)-bis[2,2-dihydro-2,2-dimethyl-1,3,2-dithiarsolane], 3, 585

$As_2C_{10}H_{28}O_6S$ $[Me_3(HOCH_2CH_2)As]_2SO_4$, 3, 312

$As_2C_{11}H_{12}N_2O_6$ 2-(p-Arsonoanilino)-5-pyridinearsonic acid, 3, 532

$As_2C_{11}H_{15}Br$ 2,3-Dihydro-1,4-ethano-1-methyl-1,4-benzodiarseninium Br^-, 3, 601

$As_2C_{11}H_{15}I$ 2,3-Dihydro-1,4-ethano-1-methyl-1,4-benzodiarseninium I^-, 3, 601

$As_2C_{11}H_{17}Br$ 1,2,3,4-Tetrahydro-1,1,4-trimethyl-1,4-benzodiarseninium Br^-, 3, 601

$As_2C_{11}H_{17}I$ 1,2,3,4-Tetrahydro-1,1,4-trimethyl-1,4-benzodiarseninium I^-, 3, 601

$As_2C_{11}H_{18}Br_2$ 2,3-Dihydro-1,1,3,3-tetramethyl-1H-1,3-benzodiarsolium $2Br^-$, 3, 560*

$As_2C_{11}H_{18}Cl_2O_8$ 2,3-Dihydro-1,1,3,3-tetramethyl-1H-1,3-benzodiarsolium diperchlorate, 3, 560*

$As_2C_{11}H_{18}I_2$ 2,3-Dihydro-1,1,3,3-tetramethyl-1H-1,3-benzodiarsolium $2I^-$, 3, 560*

$As_2C_{11}H_{19}Br$ $[2-(Me_2As)C_6H_4AsMe_3]Br$, 3, 317

$As_2C_{12}H_8Cl_2$ 5,10-Cl_2-5,10-dihydroarsanthrene, 3, 603

$As_2C_{12}H_8Cl_2S_2$ $[3-ClC_6H_4As(S)=]_2$, 3, 551

$As_2C_{12}H_8Cl_4N_2$ $(=NC_6H_4AsCl_2)_2$, 3, 197

$As_2C_{12}H_8Cl_4N_2O$ $Cl_2AsC_6H_4N=N(=O)C_6H_4-AsCl_2$, 3, 197

$As_2C_{12}H_8Cl_4S_2$ $(-SC_6H_4AsCl_2)_2$, 3, 197

$As_2C_{12}H_8O$ Poly[p-phenylene-(1,3-dihydro-2,1,3-benzoxadiarsolylene), 3, 576*

$As_2C_{12}H_8O_3$ $O(C_6H_4AsO)_2$, 3, 246

$As_2C_{12}H_9Cl_3$ 2-$(PhClAs)C_6H_4AsCl_2$, 3, 197

$As_2C_{12}H_{10}N_2O_{10}$ 3,3'-Dinitrobiphenyl-4,4'-diarsonic acid, 3, 532

$As_2C_{12}H_{10}O_2S_2$ $[4-HOC_6H_4As(S)=]_2$, 3, 551

$As_2C_{12}H_{10}S_2$ $[PhAs(S)=]_2$, 3, 551

$As_2C_{12}H_{11}NO$ 4,4'-$H_2N(HO)$-arsenobenzene, 3, 144

$As_2C_{12}H_{11}NO_8$ 3-Nitrobiphenyl-4,4'-diarsonic acid, 3, 532

$As_2C_{12}H_{12}F_{16}$ $Me_2As(CF_2)_8AsMe_2$, 3, 98

$As_2C_{12}H_{12}N_2O_6$ Azobenzenediarsonic acid isomers, 3, 532

$As_2C_{12}H_{12}N_2O_7$ Azoxybenzene-3,3'-diarsonic acid, 3, 532

$As_2C_{12}H_{12}O_6$ 4,4'-Biphenyldiarsonic acid, 3, 532

$As_2C_{12}H_{12}O_6S_2$ $(-SC_6H_4AsO_3H_2)_2$, 3, 417, 532

$As_2C_{12}H_{14}N_2O_6S_2$ 4,4'-Dithiodi-m-arsanilic acid, 3, 532

$As_2C_{12}H_{18}Br_2$ 2,3-Dihydro-1,4-dimethyl-1,4-ethano-1,4-benzodiarseninium $2Br^-$, 3, 602

$As_2C_{12}H_{18}I_2$ 2,3-Dihydro-1,4-ethano-1,4-dimethyl-1,4-benzodiarseninium $2I^-$, 3, 601

$As_2C_{12}H_{20}$ $2-Et_2AsC_6H_4AsMe_2$, 3, 98
  $[(CH_2{=}CHCH_2)_2As-]_2$, 3, 126

$As_2C_{12}H_{20}Br_2$ 1,1,4,4-Me_4-1,2,3,4-tetrahydro-1,4-benzodiarseninium $2Br^-$, 3, 601, 602

$As_2C_{12}H_{20}I_2$ 1,2,3,4-Tetrahydro-1,1,4,4-tetramethyl-1,4-benzodiarseninium $2I^-$, 3, 601

$As_2C_{12}H_{22}Br_2$ $[o-C_6H_4(AsMe_3)_2]Br_2$, 3, 317
  $[p-C_6H_4(AsMe_3)_2]Br_2$, 3, 318

$As_2C_{12}H_{22}I_2$ $[o-C_6H_4(AsMe_3)_2]I_2$, 3, 317

$As_2C_{12}H_{28}O$ $(Pr_2As)_2O$, 3, 216

$As_2C_{12}H_{31}IO_2$ $Et_3As(OH)OAs(I)Et_3$, 3, 362

$As_2C_{13}H_8N_2O_4$ 2,2',2'',2'''-Tetranitro-4,4'-diarsenobisphenylenemethane, 3, 134

$As_2C_{13}H_{10}N_2O_3$ 5,5'-Arseno-2,2'-dihydroxydiphenylurea, 3, 132*

$As_2C_{13}H_{10}N_2O_7$ α,α'-(2,2'-Dinitro-4,4'-arsenodi-p-phenylene)bis-(3-nitro-p-toluenearsonic acid), 3, 134

$As_2C_{13}H_{10}N_2O_{11}$ $4,4'-CO[3-O_2NC_6H_3-AsO_3H_2]_2$, 3, 533

$As_2C_{13}H_{10}O_4$ $[4-(4-H_2O_3AsC_6H_4CO)C_6H_4-As]_n$, 3, 141

$As_2C_{13}H_{10}O_7$ 9-Oxofluorene-2,7-diarsonic acid, 3, 533

$As_2C_{13}H_{11}Br$ 5-Br-5,10-dihydro-10-methylarsanthrene, 3, 603

$As_2C_{13}H_{11}I$ 5,10-Dihydro-5-iodo-10-methylarsanthrene, 3, 604

$As_2C_{13}H_{12}Cl_2$ $CH_2[As(Ph)Cl]_2$, 3, 174

$As_2C_{13}H_{12}Cl_2N_2O_5$ $3,3'-CO[NH-4-ClC_6H_3-As(OH)_2]_2$, 3, 224

$As_2C_{13}H_{12}Cl_2N_2O_7$ 5,5'-Ureylenebis(2-chlorobenzenearsonic acid), 3, 533

$As_2C_{13}H_{12}O_3$ $[4-(4-H_2O_3AsC_6H_4CH_2)C_6H_4-As]_n$, 3, 141

$As_2C_{13}H_{12}O_4$ $1,3-HO(p-MeC_6H_4)$-1,3-dihydro-2,1,3-benzoxadiarsole 1,3-dioxide, 3, 575

$As_2C_{13}H_{12}O_5$ $1,3-HO(p-MeOC_6H_4)$-1,3-dihydro-2,1,3-benzoxadiarsole 1,3-dioxide, 3, 576

$As_2C_{13}H_{12}O_7$ $4,4'-CO(C_6H_4AsO_3H_2)_2$, 3, 405, 533

$As_2C_{13}H_{14}N_2O_5$ $4,4'-CO[NHC_6H_4As(OH)_2]_2$, 3, 224

$As_2C_{13}H_{14}N_2O_6S$ $4,4'-CS(NHC_6H_4AsO_3H_2)_2$, 3, 533

$As_2C_{13}H_{14}N_2O_7$ $3,3'-CO[NH-4-(HO)C_6H_3-As(OH)_2]_2$, 3, 224
  $4,4'-CO(NHC_6H_4AsO_3H_2)_2$, 3, 533

$As_2C_{13}H_{14}N_2O_8$ 1-(p-Arsonophenylcarbamoylmethyl)-1,6-dihydro-6-oxo-3-pyridinearsonic acid, 3, 534

$As_2C_{13}H_{14}O_6$ $4,4'-CH_2(C_6H_4AsO_3H_2)_2$, 3, 406, 533

$As_2C_{13}H_{14}O_8$ $5,5'-CH_2[2-HOC_6H_3AsO_3H_2]_2$, 3, 534

$As_2C_{13}H_{22}Br_2$ 1,1,5,5-Me_4-2,3,4,5-tetrahydro-1H-1,5-benzodiarsepinium $2Br^-$, 3, 643

$As_2C_{14}H_8N_2$ $5,10-(NC)_2$-5,10-dihydroarsanthrene, 3, 604

$As_2C_{14}H_{10}Cl_4$ $(=CHC_6H_4AsCl_2)_2$, 3, 198

$As_2C_{14}H_{10}N_2O_4$ $(-NHCOC_6H_4AsO)_2$, 3, 248

$As_2C_{14}H_{11}N_3$ 4-Cyano-4'-guanylarsenobenzene, 3, 145

$As_2C_{14}H_{12}Cl_2O_6$ $4,4'-(Cl_2C{=}C)[C_6H_4-AsO_3H_2]_2$, 3, 406

$As_2C_{14}H_{13}Cl_3O_6$ $3,3'-(Cl_3CCH)[C_6H_4-AsO_3H_2]_2$, 3, 406, 534

$As_2C_{14}H_{14}$ 5,10-Me_2-5,10-dihydroarsanthrene, 3, 604

$As_2C_{14}H_{14}BrI$ 5,10-Br(Me)-5,10-dihydro-

arsanthrene methiodide, 3, 605

$As_2C_{14}H_{14}Br_2$ Arsanthrene dimethobromide, 3, 605

$As_2C_{14}H_{14}Cl_2$ $Ph(Cl)AsCH_2CH_2As(Cl)Ph$, 3, 166

$As_2C_{14}H_{14}I_2$ $[-CH_2As(Ph)I]_2$, 3, 175

Arsanthrene dimethiodide, 3, 605

$As_2C_{14}H_{14}N_2O_2$ $4\text{-}H_2NC_6H_4As\text{-}AsC_6H_4\text{-}(NHOCCH_2OH)\text{-}4]_n$, 3, 153

$As_2C_{14}H_{14}N_2O_3$ $[3,4\text{-}H_2N(HO)C_6H_3As\text{-}AsC_6H_4(NHOCCH_2OH)\text{-}4]_n$, 3, 153

$As_2C_{14}H_{14}N_2O_8$ $4,4'\text{-}(CONHC_6H_4AsO_3H_2)_2$, 3, 534

$As_2C_{14}H_{16}O_3$ $1,3\text{-}Ph_2\text{-}2,1,3$-oxadiarsolane 1,3-dioxide, 3, 575

$As_2C_{14}H_{14}S_2$ $[2\text{-}MeC_6H_4As(S)=]_2$, 3, 551

$[PhCH_2As(S)=]_2$, 3, 551

$As_2C_{14}H_{16}$ $(MePhAs\text{-})_2$, 3, 126

$As_2C_{14}H_{16}N_2O_8$ $5,10,5,10\text{-}(HO)_2(Me)_2$-5,10-dihydroarsanthrenium dinitrate, 3, 605

$As_2C_{14}H_{16}O$ $(MePhAs)_2O$, 3, 214

$1,3\text{-}Ph_2\text{-}2,1,3$-oxadiarsolane, 3, 575

$As_2C_{14}H_{16}O_4$ $[-CH_2(Ph)AsO_2H]_2$, 3, 385

$As_2C_{14}H_{16}O_8$ $4,4'\text{-}(\text{-}CH_2OC_6H_4AsO_3H_2)_2$, 3, 534

$As_2C_{14}H_{18}N_2O_2$ $\{Me[3,4\text{-}H_2N(HO)C_6H_3]\text{-}As\text{-}\}_2$, 3, 126

$As_2C_{14}H_{18}N_2O_6$ $4,4'\text{-}(\text{-}CH_2NHC_6H_4\text{-}AsO_3H_2)_2$, 3, 534

$As_2C_{14}H_{20}Cl_2N_2O_5$ $4,3\text{-}HO(HCl\cdot H_2N)C_6H_3\text{-}As(OH)OAs(OEt)C_6H_3(NH_2\cdot HCl)OH\text{-}3,4$, 3, 257

$As_2C_{14}H_{20}Cl_2N_2O_6$ $(2\text{-}ClC_6H_4AsO_3H_2\cdot H_2N\text{-}CH_2\text{-})_2$, 3, 409

$(4\text{-}ClC_6H_4AsO_3H_2\cdot H_2NCH_2\text{-})_2$, 3, 409

$As_2C_{14}H_{24}$ $2\text{-}Et_2AsC_6H_4AsEt_2$, 3, 98

$As_2C_{14}H_{32}O_2$ $(Pr_2AsOCH_2\text{-})_2$, 3, 211

$As_2C_{14}H_{32}S_6$ 2,2'-(Ethylenedithio)-bis[2,2-dihydro-2,2-diethyl-1,3,2-dithiaarsolane], 3, 587

$As_2C_{15}H_{13}N_3O_3S_2$ $[4\text{-}(\overline{SCH=CHN=C}NHO_2S)\text{-}C_6H_4As\text{-}AsC_6H_3(NH_2)OH\text{-}3,4]_n$, 3, 153

$As_2C_{15}H_{14}ClN_5O_6$ 6-Chloro-s-triazin-2,4-diyldiarsanilic acid, 3, 534

$As_2C_{15}H_{15}NO_4$ $[3,4\text{-}H_2N(HO)C_6H_3As\text{-}As\text{-}C_6H_3(Me)OCH_2CO_2H\text{-}3,4]_n$, 3, 153

$As_2C_{15}H_{15}N_3O_{15}$ 4,4'-(2-Hydroxytrimethylenedioxy)bis[3-nitrobenzenearsonic acid]nitrate, 3, 534

$As_2C_{15}H_{16}$ 5,10-Dihydro-2,5,10-trimethylarsanthrene, 3, 605

$5,7\text{-}Me_2$-6,7-dihydro-5H-dibenzo[d,f]-[1,3]diarsepin, 3, 643

$As_2C_{15}H_{16}Br_2$ $Ph(Br)As(CH_2)_3As(Br)Ph$, 3, 166

$As_2C_{15}H_{16}N_2O_3$ $[2,4\text{-}Me(HOCH_2CONH)C_6H_3\text{-}As\text{-}AsC_6H_3(NH_2)OH\text{-}3,4]_n$, 3, 153

$As_2C_{15}H_{16}N_2O_{13}$ 4,4'-(2-Hydroxytrimethylenedioxy)bis[3-nitrobenzenearsonic acid], 3, 535

$As_2C_{15}H_{16}N_6O_6$ $6\text{-}H_2N$-s-triazin-2,4-diyldiarsanilic acid, 3, 535

$As_2C_{15}H_{17}Br$ $5,10\text{-}Me_2$-5,10-dihydroarsanthrene methobromide, 3, 604

$As_2C_{15}H_{17}I$ $5,10\text{-}Me_2$-5,10-dihydroarsanthrene methiodide, 3, 604

$As_2C_{15}H_{18}$ $PhHAs(CH_2)_3AsHPh$, 3, 118

$As_2C_{15}H_{18}O_4$ $CH_2[CH_2(Ph)AsO_2H]_2$, 3, 385

$As_2C_{15}H_{18}O_8$ $4,4'\text{-}CH_2(CH_2OC_6H_4AsO_3H_2)_2$, 3, 535

$As_2C_{15}H_{18}O_9$ $4,4'\text{-}HOCH(CH_2OC_6H_4AsO_3H_2)_2$, 3, 535

$As_2C_{15}H_{20}N_2O_6$ $4,4'\text{-}CH_2(CH_2NHC_6H_4\text{-}AsO_3H_2)_2$, 3, 535

$As_2C_{15}H_{20}N_2O_7$ $4,4'\text{-}HOCH(CH_2NHC_6H_4\text{-}AsO_3H_2)_2$, 3, 535

$As_2C_{15}H_{20}N_2O_9$ 4,4'-(2-Hydroxytrimethylenedioxy)di-m-arsanilic acid, 3, 535

$As_2C_{15}H_{26}$ $4,1,2\text{-}MeC_6H_3(AsEt_2)_2$, 3, 99

$As_2C_{16}H_{12}O_4$ $[(\overline{OCH=CHCH=C})_2As\text{-}]_2$, 3, 126

$As_2C_{16}H_{12}O_5$ $[(\overline{OCH=CHCH=C})_2As]_2O$, 3, 216

$As_2C_{16}H_{14}Cl_2S$ $[Ph(ClCH=CH)As]_2S$, 3, 263

$As_2C_{16}H_{14}N_2O_4$ $(\text{-}CH_2NHOCC_6H_4AsO)_2$, 3, 248

$As_2C_{16}H_{16}N_2O_2$ 1,4-Bis(4-arsenosophenyl)piperazine, 3, 233

$As_2C_{16}H_{16}N_2O_4$ $[4\text{-}AcNHC_6H_4As\text{-}AsC_6H_3\text{-}$

$(NHOCCH_2OH)OH$-3,4]$_n$, 3, 153
[4-(H_2NCONH)C_6H_4As-AsC_6H_3(Me)OCH_2-CO_2H-2,4]$_n$, 3, 153

$As_2C_{16}H_{16}S$ Di-2-isoarsindolinyl sulfide, 3, 559*

$As_2C_{16}H_{18}$ 1,4-(Ph)$_2$-1,4-diarsenane, 3, 602

$As_2C_{16}H_{18}Br_2$ 5,10-Ethanoarsanthrene dimethobromide, 3, 606

$As_2C_{16}H_{18}I_2$ 5,10-Ethanoarsanthrene dimethiodide, 3, 606

$As_2C_{16}H_{18}N_2O_2$ [-CH_2N(Me)C_6H_4AsO]$_2$, 3, 249

$As_2C_{16}H_{18}N_2O_8$ 4,4'-(-$CH_2CONHC_6H_4$-AsO_3H_2)$_2$, 3, 535

$As_2C_{16}H_{18}N_2O_8S_2$ 4,4'-Dithiobis(3-$AcNHC_6H_3AsO_3H_2$), 3, 535

$As_2C_{16}H_{19}I$ 5,10-Dihydro-2,5,10-trimethylarsanthrene methiodide, 3, 606

$As_2C_{16}H_{20}$ ($MePhAsCH_2$-)$_2$, 3, 99
(2-$Me_2AsC_6H_4$-)$_2$, 3, 99
$PhHAs(CH_2)_4AsHPh$, 3, 118

$As_2C_{16}H_{20}Br_2$ 5,5,10,10-Me_4-5,10-dihydroarsanthrenium 2Br^-, 3, 604

$As_2C_{16}H_{20}Br_2O$ 5,5,7-$BrCH_2$(Me)$_2$-6,7-dihydro-5H-dibenzo[d,f][1,3]diarsepinium $Br^-\cdot H_2O$, 3, 643*

$As_2C_{16}H_{20}I_2$ 5,5,10,10-Me_4-5,10-dihydroarsanthrenium 2I^-, 3, 604

$As_2C_{16}H_{20}N_2O_6$ N,N'-Bis(4-arsonophenyl)piperazine, 3, 536

$As_2C_{16}H_{20}N_2O_8$ 1,4,1,4-(HO)$_2$(Ph)$_2$-1,4-diarsenanium dinitrate, 3, 602

$As_2C_{16}H_{20}O$ (EtPhAs)$_2$O, 3, 216
[Me(4-MeC_6H_4)As]$_2$O, 3, 216

$As_2C_{16}H_{20}O_4S$ S[CH_2CH_2(Ph)AsO_2H]$_2$, 3, 386

$As_2C_{16}H_{20}O_6S$ SO_2[CH_2CH_2(Ph)AsO_2H]$_2$, 3, 386

$As_2C_{16}H_{22}N_2O_6$ 4,4'-[-CH_2N(Me)C_6H_4-AsO_3H_2]$_2$, 3, 536

$As_2C_{16}H_{28}N_2S_4$ 1,4-Dimercapto-1,4-diarsenane diester with 1-piperidinecarbodithioic acid, 3, 602

$As_2C_{16}H_{36}O$ (Bu_2As)$_2$O, 3, 216
[(Me_2CHCH_2)$_2$As]$_2$O, 3, 216

$As_2C_{17}H_{13}NO_6$ $EtO_2CN(COC_6H_4AsO)_2$, 3, 249

$As_2C_{17}H_{17}NO_5$ [2,4-Me(HO_2CCH_2O)C_6H_3-As-AsC_6H_3(NHAc)OH-3,4]$_n$, 3, 154

$As_2C_{17}H_{17}N_3O_5$ [3,4-Me(HO_2CCH_2O)C_6H_3-As-AsC_6H_3(CH=$NNHCONH_2$)OH-4,2]$_n$, 3, 154

$As_2C_{17}H_{17}N_3O_8$ [HO(H_2NCONH)C_6H_3As-As-C_6H_3(NHAc)OCH_2CO_2H]$_n$ isomers, 3, 153, 154

$As_2C_{17}H_{18}N_2O_4$ [4,3-Me(H_2NCONH)C_6H_3-As-AsC_6H_3(Me)OCH_2CO_2H-2,4]$_n$, 3, 154
[2,4-Me($HOCH_2CONH$)C_6H_3As-AsC_6H_3-(NHAc)OH-3,4]$_n$, 3, 154

$As_2C_{17}H_{18}N_2O_7$ N-[5-(Arsonophenylimino)-4-hydroxy-1,3-pentadienyl]-arsanilic acid, 3, 536

$As_2C_{17}H_{18}O_4$ 3,9-Diphenyl-2,4,8,10-tetraoxa-3,9-diarsaspiro[5,5]hendecane, 3, 618

$As_2C_{17}H_{20}Br_2$ 5,10-Propanoarsanthrene dimethobromide, 3, 606

$As_2C_{17}H_{21}I$ 1,1,4-Me(Ph)$_2$-1,4-diarsenanium I^-, 3, 602

$As_2C_{17}H_{22}$ $PhHAs(CH_2)_5AsHPh$, 3, 118

$As_2C_{17}H_{22}Br_2$ 2,5,5,10,10-Me_5-5,10-dihydroarsanthrenium 2Br^-, 3, 606
5,5,7,7-Me_4-6,7-dihydro-5H-dibenzo-[d,f][1,3]diarsepinium 2Br^-, 3, 643

$As_2C_{17}H_{22}I_2$ 5,5,7,7-Me_4-6,7-dihydro-5H-dibenzo[d,f][1,3]diarsepinium 2I^-, 3, 644

$As_2C_{17}H_{24}O_2$ 5,5,7,7-Me_4-6,7-dihydro-5H-dibenzo[d,f][1,3]diarsepinium dihydroxide, 3, 644

$As_2C_{18}H_{12}$ 5,10-o-Benzenoarsanthrene, 3, 606
Poly(p-phenylene-5,10-arsanthrylene), 3, 606*

$As_2C_{18}H_{12}N_2O_2S_2$ 7,7'-Arsenobis(5-chloro-8-hydroxyquinoline)disulfide, 3, 551

$As_2C_{18}H_{14}Cl_2N_2O_8$ N,N'-(3,6-Dichloro-2,5-p-benzoquinonylene)diarsanilic acid, 3, 536*

$As_2C_{18}H_{14}O_8$ 4,4'-(2,5-p-benzoquinony-

lene)dibenzenearsonic acid, 3, 536

$As_2C_{18}H_{16}N_2O_3S$ [3,4-$H_2N(HO)C_6H_3As$-As-$C_6H_4(SO_2NHPh)$-4$]_n$, 3, 154

$As_2C_{18}H_{16}N_4O_6$ 4,4′-(p-Phenylenebisazo)dibenzenearsonic acid, 3, 537

$As_2C_{18}H_{16}N_4O_7$ 4,4′-(4-Hydroxy-m-phenylenebisazo)dibenzenearsonic acid, 3, 537

$As_2C_{18}H_{16}N_4O_8$ 4,4′-(3,4-Dihydroxy-p-phenylenebisazo)dibenzenearsonic acid, 3, 537

$As_2C_{18}H_{17}N_3O_2S$ [4-$H_2NC_6H_4As$-AsC_6H_4-$(NHO_2SC_6H_4NH_2)$-4$]_n$, 3, 154

$As_2C_{18}H_{17}N_3O_3S$ [3,4-$H_2N(HO)C_6H_3As$-$AsC_6H_4(NHO_2SC_6H_4NH_2)-4]_n$, 3, 154

$As_2C_{18}H_{17}N_3O_5$ 5-[3,4-AcNH(HO)C_6H_3-arseno]-1-methylbenzimidazol-2-yl-OCH_2CO_2H, 3, 145*

4-[8,3-AcNH(HO)-1,4-benzisoxazine-5-arseno]phenyl-$NHCH_2CO_2H$, 3, 145

$As_2C_{18}H_{18}N_2O_6$ [AcNH$(HO_2CCH_2O)C_6H_3As$-AsC_6H_3(NHAc)OH$]_n$ isomers, 3, 155

$As_2C_{18}H_{21}Br_2Cl$ 8,5,12-Cl$(Me)_2$-6,11-dihydro-5,12-ethanodibenzo[b,f]-[1,4]diarsocinium 2Br⁻, 3, 645

$As_2C_{18}H_{22}$ $(4\text{-}Me_2AsC_6H_4CH=)_2$, 3, 100

$As_2C_{18}H_{22}Br_2$ Ph(Br)As$(CH_2)_6$As(Br)Ph, 3, 167

5,12-Me_2-6,11-dihydro-5,12-ethanodibenzo[b,f][1,4]diarsocinium 2Br⁻, 3, 646

$As_2C_{18}H_{22}N_2O_8$ [4,4′-(-CH_2CH_2CONH)-$C_6H_4AsO_3H_2]_2$, 3, 537

$As_2C_{18}H_{23}Br_2Cl$ 8,5,5,12,2-Cl$(Me)_4$-5,6,11,12-tetrahydrodibenzo[b,f]-[1,4]diarsocinium 2Br⁻, 3, 646

$As_2C_{18}H_{24}$ $(EtPhAsCH_2\text{-})_2$, 3, 100

PhHAs$(CH_2)_6$AsHPh, 3, 118

$As_2C_{18}H_{24}Br_2$ 1,4,1,4-$Me_2(Ph)_2$-hexahydrodiarseninium 2Br⁻, 3, 603

1,1,4,4-Me_4-1,2,3,4-tetrahydrodibenzo[e,g][1,4]diarsocinium 2Br⁻, 3, 646

5,5,12,12-Me_4-5,6,11,12-tetrahydrodibenzo[b,f][1,4]diarsocinium 2Br⁻, 3, 647

$As_2C_{18}H_{24}I_2$ 1,1,4,4-Me_4-1,2,3,4-tetrahydrodibenzo[e,g][1,4]diarsocinium 2I⁻, 3, 647

$As_2C_{18}H_{24}N_2O_6S$ 3,4′-Dihydroxy-4-[bis-(2-hydroxyethyl)amino]-3′-[(N-methyl-N-sulfinomethyl)amino]arsenobenzene, 3, 146

$As_2C_{18}H_{24}O$ [Et$(PhCH_2)As]_2O$, 3, 216

$(PrPhAs)_2O$, 3, 217

[Et$(4\text{-}MeC_6H_4)As]_2O$, 3, 217

$As_2C_{18}H_{24}O_4$ [-$CH_2CH_2CH_2$(Ph)$AsO_2H]_2$, 3, 386

$As_2C_{18}H_{25}NO_8$ 3,3′-Diarsono-4,4′-dimethoxy-α,α′-dimethyldibenzylamine, 3, 537

$As_2C_{18}H_{26}Br_2$ [2-$Me_3AsC_6H_4$-$]_2Br_2$, 3, 322

$As_2C_{18}H_{26}I_2$ [$Me_2PhAsCH_2$-$]_2I_2$, 3, 322

[2-$Me_3AsC_6H_4$-$]_2I_2$, 3, 322

$As_2C_{18}H_{26}N_2O_6$ 4,4′-[-$CH_2N(Et)C_6H_4$-$AsO_3H_2]_2$, 3, 537

$As_2C_{18}H_{26}O_2$ 1,1,4,4-Me_4-1,2,3,4-tetrahydrodibenzo[e,g][1,4]diarsocinium dihydroxide, 3, 647

$As_2C_{18}H_{28}O_2$ [2-$Me_3AsC_6H_4$-$]_2(OH)_2$, 3, 322

$As_2C_{18}H_{40}$ $(Bu_2AsCH_2\text{-})_2$, 3, 100

$As_2C_{18}H_{40}S_6$ 2,2′-(Ethylenedithio)bis-[2,2-dihydro-2,2-dipropyl-1,3,2-dithiarsolane], 3, 588*

$As_2C_{19}H_{18}N_4O_7$ 4,4′-(6-Hydroxy-5-methyl-m-phenylenebisazo)dibenzenearsonic acid, 3, 538

$As_2C_{19}H_{19}NO_7$ [3,4-Me$(HO_2CCH_2O)C_6H_3As$-AsC_6H_3(NHAc)OCH_2CO_2H-3,4$]_n$, 3, 155

$As_2C_{19}H_{19}N_3O_5$ 5-[4-(Hydroxyacetamido)-2-methylphenylarseno]-1-methylbenzimidazol-2-yloxyacetic acid, 3, 147

{4-[MeNC(Me)=C(NH_2)C(=O)N]C_6H_4As-As-$C_6H_3(OCH_2CO_2H)$OH-4,3$\}_n$, 3, 155*

$As_2C_{19}H_{20}Br_2$ 5,5,9,9-Me_4-6,7,8,9-tetrahydro-5H-dibenzo[f,h][1,5]-diarsoninium 2Br⁻, 3, 649

$As_2C_{19}H_{20}I_2$ 5,5,9,9-Me_4-b,7,8,9-tetrahydro-5H-dibenzo[f,h][1,5]-diarsoninium 2I⁻, 3, 650

$As_2C_{19}H_{20}N_4O_3$ 4-(4-Amino-2,3-dimethyl-5-oxo-1-pyrazolinyl)-4′-(hydroxy-

acetamido)arsenobenzene, 3, 147

$As_2C_{19}H_{22}O_2$ 5,5,9,9-Me_4-6,7,8,9-tetrahydro-5H-dibenzo[f,h][1,5]-diarsoninium dihydroxide, 3, 650

$As_2C_{19}H_{22}O_4$ 3,9-Di-p-tolyl-2,4,8,10-tetraoxa-3,9-diarsaspiro[5,5]hendecane, 3, 618

$As_2C_{20}H_{18}N_2O_7S$ 2-Acetamido-4-[2-(2-carboxyethylthio)-5-benzoxazolylarseno]phenoxyacetic acid, 3, 149

$As_2C_{20}H_{18}N_2O_8$ $C_6H_4(CONHC_6H_4AsO_3H_2)_2$ isomers, 3, 538

$As_2C_{20}H_{20}$ 2-$Me_2AsC_6H_4AsPh_2$, 3, 100

$As_2C_{20}H_{22}N_4O_4$ 1,4-Bis[(4-arsenosophenylcarbamoyl)methyl]piperazine, 3, 250

1,4-Bis[N-(4-arsenosophenyl)glycyl]-piperazine, 3, 250

$As_2C_{20}H_{22}O_2S_4$ Tetrathio-4,4'-stilbenediarsonous acid bis[(hydroxymethyl)ethylene] ester, 3, 588

$As_2C_{20}H_{26}N_2O_6$ [-$CH_2N(CH_2CH{=}CH_2)C_6H_4$-$AsO_3H_2$-4]$_2$, 3, 538

$As_2C_{20}H_{26}N_4O_8$ 1,4-Bis[N-(p-arsonophenylglycyl]piperazine, 3, 539

4',4''-Diarsono-1,4-piperazinediacetanilide, 3, 539

$As_2C_{20}H_{28}$ (2-$Et_2AsC_6H_4$-)$_2$, 3, 100

$As_2C_{20}H_{28}Br_2$ 1,1,6,6-Me_4-1,2,3,4,5,6-hexahydrodibenzo[g,i][1,6]diarsecinium $2Br^-$, 3, 651

$As_2C_{20}H_{28}I_2$ [=$CHC_6H_4AsMe_3$]$_2I_2$, 3, 328

$As_2C_{20}H_{28}O$ $(BuPhAs)_2O$, 3, 217

$As_2C_{20}H_{30}I_2$ [-$CH_2CH_2AsPhMe_2$]$_2I_2$, 3, 328

$As_2C_{20}H_{30}N_4O_{10}$ [4-$(HOCH_2CONH)C_6H_4$-AsO_3H_2]$_2\cdot HN(CH_2CH_2)_2NH$, 3, 429

$As_2C_{20}H_{32}Br_3P$ [P(2-$Me_2AsC_6H_4AsMe_2$)-Br_3], 3, 92

$As_2C_{20}H_{32}I_3P$ [P(2-$Me_2AsC_6H_4AsMe_2$)-I_2]I, 3, 92

$As_2C_{21}H_{19}Br$ 5-Me-6,11-dihydro-5,12-o-benzenodibenzo[b,f][1,4]diarsocinium Br^-, 3, 647

$As_2C_{21}H_{19}I$ 5-Me-6,11-dihydro-5,12-o-benzenodibenzo[b,f][1,4]diarsocinium I^-, 3, 647

$As_2C_{21}H_{20}Cl_2N_2O_9S_4$ 3,3'-CO[NH-4-Cl-$C_6H_3As(SCH_2CO_2H)_2$]$_2$, 3, 273

$As_2C_{21}H_{21}N_5O_4$ 5-[4-(4-Amino-2,3-dimethyl-5-oxo-1-pyrazolinyl)phenylarseno]-1-methyl-2-benzimidazolyloxyacetic acid, 3, 150

$As_2C_{21}H_{21}N_5O_9$ Bis(4-arsonophenylazo)-tyrosine, 3, 539

$As_2C_{21}H_{22}N_2O_9S_4$ 4,4'-CO[NHC_6H_4As-$(SCH_2CO_2H)_2$]$_2$, 3, 273

$As_2C_{21}H_{22}N_2O_{11}S_4$ 3,3'-CO[NH-HOC_6H_3-$As(SCH_2CO_2H)_2$]$_2$, 3, 273

$As_2C_{21}H_{23}N_3O_7S$ 4-[2,3-Dimethyl-5-oxo-4-(sulfomethylamino)-1-pyrazolinyl]-3'-methylarsenobenzene-4'-oxyacetic acid, 3, 150

4-[2,4-Dimethyl-5-oxo-4-(sulfomethylamino)-1-pyrazolinyl]-2'-methylarsenobenzene-4'-oxyacetic acid, 3, 150

$As_2C_{21}H_{28}N_2O_8$ $CH_2[(CH_2)_3CONHC_6H_4$-AsO_3H_2-4]$_2$, 3, 539

$As_2C_{22}H_{14}Br_4N_4O_{14}S_2$ 2,2'[1,8,3,6-$(HO)_2(HO_3S)_2$-2,7-naphthylenebisazo]bis(3,5-dibromobenzenearsonic acid), 3, 527*

$As_2C_{22}H_{18}Cl_2$ 2,2'-o-Phenylenebis(5-chloroisoarsindoline), 3, 559*

$As_2C_{22}H_{18}N_4O_{11}S$ 2,2'-[2,3,6-$(HO)_2$-(HO_3S)-1,4-naphthylenebisazo]-[$C_6H_4AsO_3H_2$]$_2$, 3, 539

$As_2C_{22}H_{18}N_4O_{14}S_2$ 2,2'-[1,8,3,6-$(HO)_2$-$(HO_3S)_2$-2,7-naphthylenebisazo]-[$C_6H_4AsO_3H_2$]$_2$, 3, 527*

$As_2C_{22}H_{18}N_6O_{16}$ 2,3-Dihydro-1,4-dihydroxy-1,4-ethano-1,4-benzodiarsenium dipicrate, 3, 601

$As_2C_{22}H_{19}N_5O_{10}S$ 4,4'-[4,5,1-$H_2N(HO)$-(HO_3S)-2,6-naphthylenebisazo]-[$C_6H_4AsO_3H_2$]$_2$, 3, 539

$As_2C_{22}H_{20}$ 2,2'-o-Phenylenedi(isoarsindoline), 3, 560

$As_2C_{22}H_{20}O$ [Me(1-$C_{10}H_7$)As]$_2$O, 3, 217

$As_2C_{22}H_{22}$ 9,16-Me_2-9,10,15,16-tetrahydrotribenzo[b,d,h][1,6]diarsecin, 3, 651

$As_2C_{22}H_{22}BrN_3O_8$ 5,5,7-$BrCH_2(Me)_2$-6,7-

dihydro-5H-dibenzo[d,f][1,3]diarsepinium picrate·H_2O, 3, 643*
$As_2C_{22}H_{23}N_3O_5$ [2,4-Me(HO_2CCH_2O)C_6H_3-As-AsC_6H_4[NC(=O)C(NHAc)=C(Me)NMe)-4]$_n$, 3, 155
$As_2C_{22}H_{24}$ o-C_6H_4(C_6H_4AsMe_2)$_2$, 3, 101
$As_2C_{22}H_{24}N_2O_8S$ 3-Acetamido-4′-[2,3-dimethyl-5-oxo-4-(sulfomethylamino)-1-pyrazolinyl]arsenobenzene-4-oxyacetic acid, 3, 151
$As_2C_{22}H_{26}N_2$ [Ph($NCCH_2CH_2$)AsCH_2CH_2-]$_2$, 3, 101
$As_2C_{22}H_{26}N_2O_8S$ 3-{5-{3-[Bis(2,3-dihydroxypropyl)amino]-4-hydroxyphenylarseno}-2-benzoxazolylthio}propionic acid, 3, 151
$As_2C_{22}H_{28}O_4$ [Ph($HO_2CCH_2CH_2$)AsCH_2-CH_2-]$_2$, 3, 101
$As_2C_{22}H_{30}$ (4-Et_2AsC_6H_4CH=)$_2$, 3, 101
$As_2C_{22}H_{30}Br_2$ {-[Me_2(CH_2=CHCH_2)As]-C_6H_4-}$_2Br_2$, 3, 329
$As_2C_{22}H_{30}N_2O_8$ [-(CH_2)$_4$CONHC_6H_4AsO_3H_2-4]$_2$, 3, 539
$As_2C_{22}H_{32}$ (BuPhAsCH_2-)$_2$, 3, 101
$As_2C_{22}H_{32}S_2$ [BuPhAs(S)CH_2-]$_2$, 3, 365
$As_2C_{22}H_{34}Br_2$ [2-(Et_2MeAs)C_6H_4-]$_2Br_2$, 3, 329
[Et_2PhAsCH_2-]$_2Br_2$, 3, 329
$As_2C_{22}H_{34}I_2$ [Me_2PhAs$CH_2CH_2CH_2$-]$_2I_2$, 3, 330
$As_2C_{22}H_{36}Br_2N_2$ [Me_2(2-$Me_2NC_6H_4$)-AsCH_2-]$_2Br_2$, 3, 330
$As_2C_{22}H_{40}$ o-C_6H_4(AsBu_2)$_2$, 3, 101
$As_2C_{22}H_{48}S_6$ 2,2′-(Ethylenedithio)-bis[2,2-dihydro-2,2-di(sec-butyl)-1,3,2-dithiarsolane], 3, 588
$As_2C_{23}H_{23}I$ 2,2′-o-Phenylenedi(isoarsindoline)methiodide, 3, 560
$As_2C_{23}H_{23}N_5O_5$ 5-[4-(4-Acetamido-2,3-dimethyl-5-oxo-1-pyrazolinyl)-phenylarseno]-1-methyl-2-benzimidazolyloxyacetic acid, 3, 152*
$As_2C_{23}H_{24}N_4O_6$ [3,4-AcNH(HO_2CCH_2O)C_6H_3-As-AsC_6H_4[NC(=O)C(NHAc)=C(Me)NMe)-4]$_n$, 3, 155
$As_2C_{23}H_{34}$ (4-Et_2AsC_6H_4)$_2$CMe_2, 3, 102
$As_2C_{23}H_{38}Br_2N_2$ {[Me_2(2-$Me_2NC_6H_4$)-AsCH_2]$_2CH_2$}Br_2, 3, 330
$As_2C_{24}H_8Br_{12}$ [(2,4,6-$Br_3C_6H_2$)$_2$As-]$_2$, 3, 126
$As_2C_{24}H_8Br_{12}O$ [(2,4,6-$Br_3C_6H_2$)$_2$As]$_2$O, 3, 214
$As_2C_{24}H_{16}Br_4O$ [(4-BrC_6H_4)$_2$As]$_2$O, 3, 217
$As_2C_{24}H_{16}O_3$ 10,10′-Oxydiphenoxarsine, 3, 610
$As_2C_{24}H_{18}Br_2O$ [Ph(4-BrC_6H_4)As]$_2$O, 3, 214
$As_2C_{24}H_{18}Cl_2O$ [Ph(4-ClC_6H_4)As]$_2$O, 3, 217
$As_2C_{24}H_{18}N_2$ 10,0′(5H,5′H)-Biphenarsazine, 3, 625
$As_2C_{24}H_{18}N_2O$ 10,10′-Oxybis(5,10-dihydrophenarsazine), 3, 621
$As_2C_{24}H_{18}N_2O_5$ [Ph($O_2NC_6H_4$)As]$_2$O isomers, 3, 214, 217
$As_2C_{24}H_{20}$ Ph_2As-AsPh_2, 3, 125
$As_2C_{24}H_{20}N_4O_4$ [(4-$H_2NC_6H_4$)(4-$O_2NC_6H_4$)-As-]$_2$, 3, 126
$As_2C_{24}H_{20}N_4O_8$ 3,3′-(4,4′-Dihydroxy-3,3′-biphenylylenebisazo)dibenzenearsonic acid, 3, 540
$As_2C_{24}H_{20}O$ (Ph_2As)$_2$O, 3, 214
$As_2C_{24}H_{20}O_2$ [Ph(3-HOC_6H_4)As-]$_2$, 3, 126
$As_2C_{24}H_{20}S$ (Ph_2As)$_2$S, 3, 263
$As_2C_{24}H_{22}N_2$ [Ph($H_2NC_6H_4$)As-]$_2$ isomers, 3, 126
$As_2C_{24}H_{22}N_2O$ [Ph(2-$H_2NC_6H_4$)As]$_2$O, 3, 217
$As_2C_{24}H_{22}N_2O_5S_2$ [Ph(4-$H_2NO_2SC_6H_4$)-As]$_2$O, 3, 217
$As_2C_{24}H_{22}N_6O_{14}$ 2,3-Dihydro-1,4-dimethyl-1,4-ethano-1,4-benzodiarseninium dipicrate, 3, 602
$As_2C_{24}H_{24}Br_2$ 1,1″,2′,3,3′,3″-Hexahydrodispiro[isoarsindolinium-2,1′-[1,4]benzodiarsenin-4′,2″-isoarsindolinium] 2Br^-, 3, 574
$As_2C_{24}H_{24}N_4$ [(4-$H_2NC_6H_4$)$_2$As-]$_2$, 3, 127
$As_2C_{24}H_{24}N_4O_4$ {[3,4-H_2N(HO)C_6H_3]$_2$-As-}$_2$, 3, 127
$As_2C_{24}H_{24}N_4O_5$ {[3,4-H_2N(HO)C_6H_3]$_2$-As}$_2$O, 3, 217
$As_2C_{24}H_{24}N_6O_{14}$ 1,2,3,4-Tetrahydro-

1,1,4,4-tetramethyl-1,4-benzodiarseninium dipicrate, 3, 602

$As_2C_{24}H_{28}Br_2$ 9,9,16,16-Me_4-9,10,15,16-tetrahydrotribenzo[b,d,h][1,6]-diarsecinium $2Br^-$, 3, 651, 652

$As_2C_{24}H_{29}N_2O_9S$ 3-{5-{3-Acetamido-5-[bis(2,3-dihydroxypropyl)amino]-4-hydroxyphenylarseno}-2-benzoxazolylthio}propionic acid, 3, 152

$As_2C_{24}H_{30}I_2$ $[o\text{-}C_6H_4(2\text{-}C_6H_4AsMe_3)_2]I_2$, 3, 337

$As_2C_{24}H_{34}Br_2$ $[EtPh(CH_2{=}CHCH_2)AsCH_2\text{-}]_2\text{-}Br_2$, 3, 337

$As_2C_{25}H_{22}$ $(Ph_2As)_2CH_2$, 3, 102

$As_2C_{25}H_{22}O$ $Ph_2AsCH_2As(O)Ph_2$, 3, 355

$As_2C_{25}H_{22}O_2$ $CH_2[As(O)Ph_2]_2$, 3, 359

$As_2C_{25}H_{22}S$ $Ph_2AsCH_2As(S)Ph_2$, 3, 366

$As_2C_{25}H_{23}BrO_2$ $Ph_2As(O)CH_2AsPh_2(OH)Br$, 3, 347

$As_2C_{25}H_{24}Cl_2O_2$ $CH_2[AsPh_2(OH)Cl]_2$, 3, 347

$As_2C_{26}H_{16}Cl_2N_2O_4$ 2,2'-Cl_2-10,10'(5H,5'H)biphenarsazine-9,9'-$(CO_2H)_2$, 3, 625

$As_2C_{26}H_{16}O_4S_2$ 10,10'-(Oxalyldithio)-diphenoxarsine, 3, 611*

$As_2C_{26}H_{18}N_2O_4$ 10,10'(5,5'H)-Biphenarsazine-4,4'-$(CO_2H)_2$, 3, 625

$As_2C_{26}H_{20}$ $Ph_2AsC{\equiv}CAsPh_2$, 3, 102

$As_2C_{26}H_{20}Cl_2N_2$ 2,2',5,5'-$Cl_2(Me)_2$-10,10'(5H,5'H)-biphenarsazine, 3, 625

$As_2C_{26}H_{20}N_4O_4$ 1,1',2,2'-$Me_2(O_2N)_2$-10,10'(5H,5'H)biphenarsazine, 3, 625

$As_2C_{26}H_{20}N_6O_{16}$ 5,10,5,10-$(HO)_2(Me)_2$-5,10-dihydroarsanthrenium dipicrate, 3, 605

$As_2C_{26}H_{20}O_6$ 4,4'-(9,10-Anthrylene)-dibenzenearsonic acid, 3, 528

$As_2C_{26}H_{22}$ 5,10-Dihydro-5,10-di-p-tolylarsanthrene, 3, 606

$As_2C_{26}H_{22}Br_2$ 5,10-Di-p-tolylarsarthrenium $2Br^-$, 3, 607

$As_2C_{26}H_{22}Br_4$ 5,10-Dihydro-5,10-di-p-tolylarsanthrene $4Br^-$, 3, 607

$As_2C_{26}H_{22}N_2O_2$ 3,3'-$(MeO)_2$-10,10'-(5H,5'H)biphenarsazine, 3, 625

$As_2C_{26}H_{22}O_2$ 5,10-Dihydro-5,10-di-p-tolylarsanthrene 5,10-dioxide, 3, 607

$As_2C_{26}H_{22}S$ 5,10-Dihydro-5,10-di-p-tolylarsanthrene 5-sulfide, 3, 607

$As_2C_{26}H_{24}$ $(Ph_2AsCH_2\text{-})_2$, 3, 102

$As_2C_{26}H_{24}Br_4$ $[\text{-}CH_2AsPh_2Br_2]_2$, 3, 345

$As_2C_{26}H_{24}O$ $[Ph(4\text{-}MeC_6H_4)As]_2O$, 3, 218

$As_2C_{26}H_{24}O_2$ $[Ph(3\text{-}MeOC_6H_4)As\text{-}]_2$, 3, 127

$As_2C_{26}H_{24}S_2$ $[Ph_2As(S)CH_2\text{-}]_2$, 3, 366

$As_2C_{26}H_{26}O_4$ 5,10-Dihydro-5,10-di-p-tolylarsanthrene 5,5,10,10-tetrahydroxide, 3, 607

$As_2C_{26}H_{44}$ $(C_6H_{11})_2AsC{\equiv}CAs(C_6H_{11})_2$, 3, 103

$As_2C_{27}H_{21}N_3O_7$ 5-Me-6,11-dihydro-5,12-o-benzenodibenzo[b,f][1,4]diarsocinium picrate, 3, 647

$As_2C_{27}H_{26}$ $(Ph_2AsCH_2)_2CH_2$, 3, 103

$As_2C_{27}H_{26}S_2$ $[Ph_2As(S)CH_2]_2CH_2$, 3, 366

$As_2C_{27}H_{47}I$ $[(C_6H_{11})_2AsC{\equiv}CAsMe(C_6H_{11})_2]I$, 3, 339

$As_2C_{28}H_{16}F_{12}O$ $[(2\text{-}CF_3C_6H_4)_2As]_2O$, 3, 215

$As_2C_{28}H_{20}O_9$ $[(2\text{-}HO_2CC_6H_4)_2As]_2O$, 3, 218

$As_2C_{28}H_{22}N_2O_4$ 5,5'-Ac_2-10,10'(5H,5'H)-biphenarsazine-3,3'-diol, 3, 625

$As_2C_{28}H_{22}N_6O_{14}$ 5,10-Ethanoarsanthrene dimethopicrate, 3, 606

$As_2C_{28}H_{24}N_4O_6$ $[(4\text{-}AcNHC_6H_4)(4\text{-}O_2N\text{-}C_6H_4)As\text{-}]_2$, 3, 127

$As_2C_{28}H_{24}N_4O_7$ $[(4\text{-}AcNHC_6H_4)(4\text{-}O_2N\text{-}C_6H_4)As]O$, 3, 218

$As_2C_{28}H_{24}N_6O_{14}$ 5,10-Me_2-5,10-dihydro-arsanthrene dimethopicrate, 3, 605

$As_2C_{28}H_{28}$ $Ph_2As(CH_2)_4AsPh_2$, 3, 103

$As_2C_{28}H_{28}N_4O_2$ $[(4\text{-}AcNHC_6H_4)(4\text{-}H_2N\text{-}C_6H_4)As\text{-}]_2$, 3, 127

$As_2C_{28}H_{28}O$ $[(4\text{-}MeC_6H_4)_2As]_2O$, 3, 218

$As_2C_{28}H_{28}O_3$ $[Ph(2\text{-}EtOC_6H_4)As]_2O$, 3, 218

$As_2C_{28}H_{28}O_5$ $[(4\text{-}MeOC_6H_4)_2As]_2O$, 3, 218

$As_2C_{28}H_{28}S_2$ $[Ph_2As(S)CH_2CH_2\text{-}]_2$, 3, 366

$As_2C_{28}H_{30}I_2$ $[MePh_2AsCH_2-]_2I_2$, 3, 339*

$As_2C_{28}H_{40}Br_2N_2$ o-$C_6H_4[CH_2AsMe_2(C_6H_4NMe_2)]_2Br_2$, 3, 340

$As_2C_{28}H_{40}I_2O_2$ 2,5,2,5-$Me_2(Ph)_2$-1,2,3,4,5,6-hexahydro-2,5-benzodiarsocinium $2I^-$, 3, 647

$As_2C_{28}H_{52}$ $[(C_6H_{11})_2AsCH_2CH_2-]_2$, 3, 103

$As_2C_{29}H_{24}N_6O_{14}$ 5,10-Propanoarsanthrene dimethopicrate, 3, 606

$As_2C_{29}H_{26}N_6O_{14}$ 2,5,5,10,10-Me_5-5,10-dihydroarsanthrenium dipicrate, 3, 606

5,5,7,7-Me_4-6,7-dihydro-5H-dibenzo[d,f][1,3]diarsapinium dipicrate, 3, 644

$As_2C_{29}H_{30}$ $Ph_2As(CH_2)_5AsPh_2$, 3, 103

$As_2C_{30}H_{24}$ $Ph_2AsC_6H_4AsPh_2$ isomers, 3, 103

$As_2C_{30}H_{26}N_6O_{14}$ 5,12-Me_2-6,11-dihydro-5,12-ethanodibenzo[b,f][1,4]diarsocinium dipicrate, 3, 646

$As_2C_{30}H_{27}ClN_6O_{14}$ 8,5,5,12,12-$Cl(Me)_4$-5,6,11,12-tetrahydrodibenzo[b,f][1,4]diarsocinium dipicrate, 3, 646

$As_2C_{30}H_{28}N_6O_{14}$ 1,4,1,4-$Me_2(Ph)_2$-hexahydro-1,4-diarseninium dipicrate, 3, 603

5,5,12,12-Me_4-5,6,11,12-tetrahydrodibenzo[b,f][1,4]diarsocinium dipicrate, 3, 647

$As_2C_{30}H_{30}N_6O_{14}$ $[Me_2PhAsCH_2-]_2[OC_6H_2(NO_2)_3]_2$, 3, 322

$[2\text{-}Me_3AsC_6H_4-]_2[OC_6H_2(NO_2)_3]_2$, 3, 322

$As_2C_{30}H_{32}$ $Ph_2As(CH_2)_6AsPh_2$, 3, 104

$As_2C_{30}H_{32}Br_4$ $(-CH_2CH_2CH_2AsPh_2Br_2)_2$, 3, 345

$As_2C_{30}H_{32}S_2$ $Ph_2As(S)(CH_2)_6As(S)Ph_2$, 3, 366

$As_2C_{30}H_{34}Br_2$ $[Me_2(2\text{-}PhC_6H_4)AsCH_2-]_2Br_2$, 3, 340

$As_2C_{30}H_{34}O_6S_2$ 5,5,10,10-Me_4-5,10-dihydroarsanthrenium$[p\text{-}MeC_6H_4SO_3^-]_2$, 3, 604

$As_2C_{30}H_{46}N_2O_8$ $[-(CH_2)_8CONHC_6H_4AsO_3H_2\text{-}4]_2$, 3, 540

$As_2C_{31}H_{24}N_6O_{14}$ 5,5,9,9-Me_4-6,7,8,9-tetrahydro-5H-dibenzo[f,h][1,5]diarsoninium dipicrate, 3, 650

$As_2C_{31}H_{25}N_3O_8$ $Ph_2AsCH_2As(OH)Ph_2$-$[OC_6H_2(NO_2)_3]$, 3, 349

$As_2C_{31}H_{25}N_3O_9$ $Ph_2AsCH_2As(OH)Ph_2$-$[OC_6H(OH)(NO_2)_3]$, 3, 349

$As_2C_{31}H_{26}N_8O_7$ 4,6-Bis[4-(4-Arsonophenylazo)phenylazo]cresol, 3, 540

$As_2C_{32}H_{22}Br_2N_4O_{22}S_4$ 3,3'-[7,1,8,3,6-$Br(HO)_2(HO_3S)_2$-2-naphthylazo$]_2$biphenyl-4,4'-diarsonic acid, 3, 529

$As_2C_{32}H_{24}N_4O_{20}S_4$ [2,3,6-$HO(HO_3S)_2$-1-naphthylazo$]_2$biphenyldiarsonic acid isomers, 3, 529

$As_2C_{32}H_{24}N_4O_{22}S_4$ [1,8,3,6-$(HO)_2(HO_3S)_2$-naphthylazo$]_2$biphenyldiarsonic acid isomers, 3, 529, 530

$As_2C_{32}H_{24}O$ $[Ph(2\text{-}C_{10}H_7)As]_2O$, 3, 218

$As_2C_{32}H_{26}N_6O_6$ 4,4'-Diamino-2,2'-(4,4'-biphenylylenebiazo)dinaphthalenearsonic acid, 3, 540

$As_2C_{32}H_{38}Cl_2$ $[EtPh(PhCH_2)AsCH_2-]_2Cl_2$, 3, 340

$As_2C_{32}H_{38}O_6S_2$ 1,4,1,4-$Me_2(Ph)_2$-1,4-diarsenanium di-p-toluenesulfonate, 3, 602

$As_2C_{34}H_{38}N_6O_{14}$ $[Et_2PhAsCH_2-]_2[OC_6H_2(NO_2)_3]_2$, 3, 329

$As_2C_{34}H_{30}N_4O_{20}S_4$ 1,5-Bis(4-arsonophenyl)hexahydro-2,4-bis(1-hydroxy-4,8-disulfo-2-naphthyl)-s-tetrazine, 3, 530

$As_2C_{34}H_{40}Br_2N_2$ 4,4'-Ethylenebis(hexahydro-1,4-diphenyl-1,4-azarsinium Br^-), 3, 620

$As_2C_{34}H_{40}N_6O_{14}$ $[Me_2(2\text{-}Me_2NC_6H_4)AsCH_2-]_2$-$[OC_6H_2(NO_2)_3]_2$, 3, 330

$As_2C_{34}H_{46}ClN_5O_{10}$ Chloroquine·$[4\text{-}(HOCH_2CONH)C_6H_4AsO_3H_2]_2$, 3, 429

$As_2C_{35}H_{30}N_4O_6$ $[Ph_2AsCH_2As(OH)Ph_2]$-picrolonate, 3, 349

$As_2C_{35}H_{34}O_6$ 5,5,7,7-Me_4-6,7-dihydro-5H-dibenzo[d,f][1,3]diarsepinium dibenzoyltartrate, 3, 644

$As_2C_{36}H_{28}N_6O_{14}$ 1,1'',2',3,3',3''-Hexa-

hydrodispiro[isoarsindolinium-2,1′-[1,4]benzodiarsenin-4′,2″-isoarsindolinium]dipicrate, 3, 574

$As_2C_{36}H_{32}N_6O_{14}$ 9,9,16,16-Me_4-9,10,15,16-tetrahydrotribenzo[b,d,h][1,6]-diarsecinium dipicrate, 3, 652

$As_2C_{36}H_{32}O_6S$ $[Ph_3As(OH)]_2SO_4$, 3, 349

$As_2C_{36}H_{33}ClN_6O_{14}$ $[4,1,2\text{-}ClC_6H_3(CH_2\text{-}AsMe_2Ph)_2][OC_6H_2(NO_2)_3]_2$, 3, 341*

$As_2C_{36}H_{34}N_6O_{14}$ $[o\text{-}C_6H_4(CH_2AsMe_2Ph)_2]\text{-}[OC_6H_2(NO_2)_3]_2$, 3, 341*

$As_2C_{36}H_{36}O_6$ 1,1,4,4-Me_4-1,2,3,4-tetrahydrodibenzo[e,g][1,4]diarsocinium dibenzoyltartrate, 3, 647

$As_2C_{36}H_{42}N_6O_{14}$ $[MeBuPhAsCH_2\text{-}]_2[OC_6H_2\text{-}(NO_2)_3]_2$, 3, 337

$[MeBuPhAsCH_2\text{-}]_2[OC_6H_2(NO_2)_3]_2$, 3, 337

$As_2C_{38}H_{30}$ $(4\text{-}Ph_2AsC_6H_4CH{=})_2$, 3, 104

$As_2C_{38}H_{49}Br_2ClO_8S_2$ 8,5,12-$Cl(Me)_2$-6,11-dihydro-5,12-ethanodibenzo-[b,f][1,4]diarsocinium di-bromocamphorsulfonate, 3, 646

$As_2C_{38}H_{51}ClO_8S_2$ 8,5,12-$Cl(Me)_2$-6,11-dihydro-5,12-ethanodibenzo[b,f]-[1,4]diarsocinium di-camphorsulfonate, 3, 646

$As_2C_{38}H_{52}Br_2O_8S_2$ 1,1,4,4-Me_4-1,2,3,4-tetrahydrodibenzo[e,g][1,4]diarsocinium di-bromocamphorsulfonate, 3, 647

$As_2C_{38}H_{54}O_8S_2$ 1,1,4,4-Me_4-1,2,3,4-tetrahydrodibenzo[e,g][1,4]diarsocinium dicamphorsulfonate, 3, 647

$As_2C_{39}H_{30}Cl_6O_6S_3$ $[O_2SCCl(AsPh_3)SO_2CCl_2SO_2CCl(AsPh_3)]\text{-}Cl_2$, 3, 341*

$As_2C_{39}H_{48}Br_2O_8S_2$ 5,5,9,9-Me_4-6,7,8,9-tetrahydro-5H-dibenzo[f,h][1,5]-diarsoninium di-bromocamphorsulfonate, 3, 650

$As_2C_{40}H_{47}N_5O_{11}$ 12-$[HO(CH_2)_3N(Me)\text{-}(CH_2)_3NH]$-benz[b]acridine·$[3,4\text{-}AcNH(HO)C_6H_3AsO_3H_2]_2$, 3, 477

12-$[HO(CH_2)_2N(Et)(CH_2)_3NH]$-benz[b]-acridine·$[3,4\text{-}AcNH(HO)C_6H_3AsO_3H_2]_2$, 3, 477

$As_2C_{40}H_{47}N_5O_{12}$ 2,2′-[3-(12-Benz[b]-acridinylamino)propylimino]diethanol·$[4\text{-}(HOCH_2CONH)C_6H_4AsO_3H_2]_2$, 3, 429

$As_2C_{42}H_{28}$ $[(1\text{-}C_{10}H_7)_2AsC{\equiv}]_2$, 3, 104

$As_2C_{42}H_{28}O_2$ $[(C_{10}H_7)_2As(O)C{\equiv}]_2$, 3, 359

$As_2C_{42}H_{33}P$ $(2\text{-}Ph_2AsC_6H_4)_2PPh$, 3, 104

$As_2C_{42}H_{38}N_6O_{14}$ $[Me_2(2\text{-}PhC_6H_4)AsCH_2\text{-}]_2\text{-}[OC_6H_2(NO_2)_3]_2$, 3, 340

$As_2C_{42}H_{42}Br_2O$ $[(2\text{-}MeC_6H_4)_3AsBr]_2O$, 3, 362

$As_2C_{42}H_{43}BrO_2$ $(2\text{-}MeC_6H_4)_3As(OH)OAs\text{-}(Br)(C_6H_4Me\text{-}2)_3$, 3, 363

$As_2C_{42}H_{43}IO_2$ $(2\text{-}MeC_6H_4)_3As(OH)OAs(I)\text{-}(C_6H_4Me\text{-}2)_3$, 3, 363

$As_2C_{42}H_{50}N_5O_{10}$ 2-{Butyl[3-(12-benz[b]-acridinylamino)propylamino]}ethanol·$[4\text{-}(HOCH_2CONH)C_6H_4AsO_3H_2]_2$, 3, 429

$As_2C_{42}H_{53}N_7O_8$ 7-(3-Octylaminopropyl-amino)benz[c]acridine·$[4\text{-}(H_2NCO\text{-}NH)C_6H_4AsO_3H_2]_2$, 3, 429

$As_2C_{44}H_{55}N_5O_{10}$ 7-$[C_8H_{17}NH(CH_2)_3NH]$-benz[c]acridine·$[3,4\text{-}AcNH(HO)C_6H_3\text{-}AsO_3H_2]_2$, 3, 477

$As_2C_{44}H_{56}Br_2O_8S_2$ 9,9,16,16-Me_4-9,10,15,16-tetrahydrotribenzo[b,d,h]-[1,6]diarsecinium di-bromocamphorsulfonate, 3, 652

$As_2C_{44}H_{58}O_8S_2$ 9,9,16,16-Me_4-9,10,15,16-tetrahydrotribenzo[b,d,h][1,6]-diarsecinium di-camphorsulfonate, 3, 652

$As_2C_{45}H_{34}I_2N_{10}O_{20}S_4$ Bis{4-[7-amino-8-(4-arsonophenylazo)-1-hydroxy-3,6-disulfo-2-naphthylazo]-3-iodo-phenyl}methane, 3, 530

$As_2C_{46}H_{44}N_8O_{14}$ $[PhN(CH_2CH_2)_2As(Ph)\text{-}CH_2\text{-}]_2$ dipicrate, 3, 620

$As_2C_{48}H_{40}O_4S$ $[Ph_4As]_2SO_4$, 3, 336

$As_2C_{54}H_{46}N_6O_{15}$ $\{[(2\text{-}MeC_6H_4)_3]As[OC_6H_2\text{-}(NO_2)_3]\}_2O$, 3, 363

$As_2C_{56}H_{44}Br_2P_2$ 5,5′-(p-Phenylenedi-methylene)bis(5,10-dihydro-5,10-diphenyldibenzo[1,4]phosphoniaar-

senin $2Br^-$, 3, 641

$As_2C_{60}H_{60}O_{16}$ $[EtPh(CH_2{=}CHCH_2)AsCH_2\text{-}]_2[O_2CCH(OBz)CH(OBz)CO_2H]_2$, 3, 337

$As_2C_{67}H_{48}Br_4O_5S$ $[Ph_4As]_2$ bromophenol blue salt, 3, 334

$As_2C_{68}H_{46}Br_4O_5$ $[Ph_4As]_2$ eosinate, 3, 335

$As_2C_{68}H_{48}N_6O_{14}P_2$ 5,5'-(p-Phenylenedimethylene)bis(5,10-dihydro-5,10-diphenyldibenzo[1,4]phosphoniaarsenin dipicrate, 3, 641

$As_2C_{75}H_{66}Br_2O_5S$ $[Ph_4As]_2$ bromothymol blue salt, 3, 334

$As_3C_6H_{18}N$ $(Me_2As)_3N$, 3, 288

$As_3C_{10}H_{16}Br_3$ $[As(2\text{-}Me_2AsC_6H_4AsMe_2)\text{-}Br_3]$, 3, 80

$As_3C_{10}H_{16}ClI_2O_4 \cdot H_2O$ $[As(2\text{-}Me_2AsC_6H_4\text{-}AsMe_2)I_2]ClO_4 \cdot H_2O$, 3, 80

$As_3C_{10}H_{16}Cl_3$ $[As(2\text{-}Me_2AsC_6H_4AsMe_2)\text{-}Cl_3]$, 3, 80

$As_3C_{10}H_{16}I_3$ $[As(2\text{-}Me_2AsC_6H_4AsMe_2)I_2]\text{-}I$, 3, 80

$As_3C_{11}H_{27}$ $MeAs(CH_2CH_2CH_2AsMe_2)_2$, 3, 105

$MeC(CH_2AsMe_2)_3$, 3, 107

$As_3C_{15}H_{36}P$ $(Me_2AsCH_2CH_2CH_2)_3P$, 3, 107

$As_3C_{15}H_{36}PS_4$ $[Me_2As(S)CH_2CH_2CH_2]_3PS$, 3, 365

$As_3C_{24}H_{21}N_6O_{11}$ 2,4,6-Tris(p-arsonophenylazo)resorcinol, 3, 541

$As_3C_{24}H_{21}N_6O_{12}$ 2,4,6-Tris(p-arsonophenylazo)phloroglucinol, 3, 541

$As_3C_{27}H_{36}NO_{12}$ 3,3',3"-Triarsono-4,4',4"-trimethoxy-α,α',α"-trimethyltribenzylamine, 3, 541

$As_3C_{42}H_{33}$ $[2\text{-}(Ph_2As)C_6H_4]_2AsPh$, 3, 108

$As_3C_{42}H_{33}N_{12}O_{12}$ 2,4,6-Tris[4-(4-arsonophenylazo)phenylazo]phloroglucinol, 3, 541

$As_3C_{42}H_{56}ClN_6O_{15}$ Chloroquine·$[4\text{-}(HOCH_2CONH)C_6H_4AsO_3H_2]_3$, 3, 429

$As_3C_{54}H_{42}P$ $[2\text{-}Ph_2AsC_6H_4]_3P$, 3, 109

$As_4C_6H_4O_4$ $(\text{-}AsAsCH_2CO_2H)_2$, 3, 305

$As_4C_6H_4O_8$ s-Tetraarsin-3,3,6,6-tetracarboxylic acid, 3, 609

$As_4C_{15}H_{36}$ $(Me_2AsCH_2CH_2CH_2)_3As$, 3, 109

$As_4C_{21}H_{32}Br_4$ 1,1',3,3'-Tetrahydro-1,1,1',1',3,3,3',3'-octamethyl-2,2'-spirobi(2H-1,3-benzodiarsolium) $4Br^-$, 3, 574

$As_4C_{21}H_{32}Cl_4O_{16}$ 1,1',3,3'-Tetrahydro-1,1,1',1',3,3,3',3'-octamethyl-2,2'-spirobi(2H-1,3-benzodiarsolium)tetraperchlorate, 3, 574

$As_4C_{21}H_{32}I_4$ 1,1',3,3'-Tetrahydro-1,1,1',1',3,3,3',3'-octamethyl-2,2'-spirobi(2H-1,3-benzodiarsolium) $4I^-$, 3, 574

$As_4C_{30}H_{32}Br_2$ 5,5'-Ethylenebis(5,10-dihydro-10-methylarsanthrene)5,5'-dimethobromide, 3, 608

$As_4C_{30}H_{32}I_2$ 5,5'-Ethylenebis(5,10-dihydro-10-methylarsanthrene)5,5'-dimethiodide, 3, 608

$As_4C_{34}H_{44}Br_2$ $\{\{2\text{-}[2\text{-}(Me_2As)C_6H_4]C_6H_4\text{-}AsMe_2\}_2CHMe\}Br_2$, 3, 34

$As_4C_{36}H_{30}N_{10}O_{15}$ 3,3',5,5'-Tetrakis-(4-arsonophenylazo)-2,4,4'-trihydroxyazobenzene, 3, 541

$As_4C_{36}H_{48}N_6O_{14}$ $\{\{2\text{-}[2\text{-}(Me_2As)C_6H_4]\text{-}C_6H_4AsMe_2\}_2CHMe\}[OC_6H_2(NO_2)_3]_2$, 3, 341

$As_4C_{40}H_{32}Br_4$ 5,12,17,24-Tetrahydro-6,11,18,23-di-o-benzenotetrabenzo-[b,f,j,n][1,4,9,12]tetraarsacyclohexatetraenium $4Br^-$, 3, 608

$As_4C_{40}H_{32}I_4$ 5,12,17,24-Tetrahydro-6,11,18,23-di-o-benzenotetrabenzo-[b,f,j,n][1,4,9,12]tetraarsacyclohexatetraenium $4I^-$, 3, 608

$As_4C_{42}H_{36}N_6O_{14}$ 5,5'-Ethylenebis(5,10-dihydro-10-methylarsanthrene) 5,5'-dimethopicrate, 3, 608

$As_4C_{46}H_{48}I_4$ 5,12,17,24-Tetrahydro-6,8,11,18,21,23-hexamethyl-6,11,18,23-di-o-benzenotetrabenzo[b,f,j,n][1,4,9,12]-tetraarsacyclohexadecatetraenium $4I^-$, 3, 608*

$As_4C_{48}H_{38}N_{12}O_{16}$ 4,4'-(p-Biphenylenediazo)bis[2,6-bis(p-arsonophenylazo)resorcinol, 3, 541

$As_4C_{54}H_{42}$ $[2\text{-}(Ph_2As)C_6H_4]_3As$, 3, 110

$As_4C_{64}H_{40}N_{12}O_{28}$ 5,12,17,24-Tetrahydro-

6,11,18,23-di-o-benzenotetrabenzo[b,g,j,n][1,4,9,12]-tetraarsacyclohexatetraenium tetrapicrate, 3, 608

$As_4C_{70}H_{56}N_{12}O_{28}$ 5,12,17,24-Tetrahydro-6,8,11,18,21,23-hexamethyl-6,11,18,23-di-o-benzenotetrabenzo[b,f,j,n][1,4,9,12]-tetraarsacyclohexadecatetraenium tetrapicrate, 3, 608*

$As_4C_{72}H_{54}N_{20}O_{16}$ 4,4'-(p-Biphenylenebisazo)bis{2,4-bis[4-(4-arsonophenylazo)phenylazo]resorcinol}, 3, 542

$As_4C_{88}H_{108}Br_4O_{16}S_4$ 5,12,17,24-Tetrahydro-6,8,11,18,21,23-hexamethyl-6,11,18,23-di-o-benzenotetrabenzo[b,f,j,n][1,4,9,12]-tetraarsacyclohexadecatetraenium tetrakis(bromocamphorsulfonate), 3, 608*

$As_5C_5H_{15}$ Arsenomethane, 3, 130

$As_6C_{36}H_{30}$ Arsenobenzene, 3, 130

$AsAgC_2H_6O_2$ Me_2AsO_2Ag, 3, 374

$AsAgC_7H_8O_2$ $MePhAsO_2Ag$, 3, 375

$AsAgC_{14}H_{20}$ $[PhC{\equiv}CAgAsEt_3]$, 3, 11

$AsAg_2CH_3O_3$ $MeAsO_3Ag_2$, 3, 391

$AsAlC_{18}H_{15}Cl_3$ $[Al(AsPh_3)Cl_3]$, 3, 19

$AsAlC_{24}H_{20}BrCl_3$ $[Ph_4As]AlBrCl_3$, 3, 332

$AsAuC_3H_9Cl$ $[Au(AsMe_3)Cl]$, 3, 4

$AsAuC_3H_9I$ $[Au(AsMe_3)I]$, 3, 4

$AsAuC_4H_9NS$ $[Au(AsMe_3)SCN]$, 3, 4

$AsAuC_6H_5NS$ $[3,4\text{-}H_2N(AuS)C_6H_3As]_n$, 3, 136

$AsAuC_6H_7NO_3S$ $3,4\text{-}H_2N(AuS)C_6H_3AsO_3H_2$, 3, 475

$AsAuC_6H_{15}Cl$ $[Au(AsEt_3)Cl]$, 3, 8

$AsAuC_6H_{15}I$ $[Au(AsEt_3)I]$, 3, 8

$AsAuC_8H_{11}Br$ $[Au(Me_2AsPh)Br]$, 3, 40

$AsAuC_8H_{11}Cl$ $[Au(Me_2AsPh)Cl]$, 3, 40

$AsAuC_8H_{11}I$ $[Au(Me_2AsPh)I]$, 3, 40

$AsAuC_8H_{12}ClN$ $[Au(Me_2AsCH_2\overline{C{=}CHCH{=}CHCH{=}N})Cl]$, 3, 42

$AsAuC_9H_{11}N$ $[Au(Me_2AsPh)CN]_n$, 3, 40

$AsAuC_9H_{13}BrS$ $[Au(2\text{-}MeSC_6H_4AsMe_2)Br]$, 3, 45

$AsAuC_9H_{13}ClS$ $[Au(2\text{-}MeSC_6H_4AsMe_2)Cl]$, 3, 45

$AsAuC_9H_{13}IS$ $[Au(2\text{-}MeSC_6H_4AsMe_2)I]$, 3, 45

$AsAuC_9H_{21}Cl$ $[Au(AsPr_3)Cl]$, 3, 12

$AsAuC_9H_{21}I$ $[Au(AsPr_3)I]$, 3, 12

$AsAuC_{10}H_{21}NS$ $[Au(AsPr_3)SCN]$, 3, 12

$AsAuC_{13}H_{13}Br$ $[Au(Ph_2AsMe)Br]$, 3, 53

$AsAuC_{13}H_{13}I$ $[Au(Ph_2AsMe)I]$, 3, 53

$AsAuC_{14}H_{13}N$ $[Au(Ph_2AsMe)CN]$, 3, 53

$AsAuC_{14}H_{20}$ $[PhC{\equiv}CAuAsEt_3]$, 3, 8

$AsAuC_{23}H_{21}Cl_5O$ 2-(4-Chlorophenacyl)-1,2,3,4-tetrahydro-2-phenylisoarsinolinium $[AuCl_4]^-$, 3, 594

$AsBC_2H_8$ $(Me_2AsBH_2)_n$, 3, 297

$AsBC_3H_{12}$ $[Me_3AsBH_3]$, 3, 4

$AsBC_6H_{18}$ $[Et_3AsBH_3]$, 3, 7

$AsBC_{18}H_{18}$ $[Ph_3AsBH_3]$, 3, 19

$AsBC_{19}H_{17}F_3$ $Ph_3AsCH_2{\cdot}BF_3$, 3, 309

$AsBC_{23}H_{20}F_4$ $[Ph_3(\overline{CH{=}CHCH{=}CHCH})As]BF_4$, 3, 330

$AsBC_{23}H_{30}$ $[Me_4As][BMePh_3]$, 3, 312

$AsBC_{24}H_{18}Br_2$ $Ph_2AsB(C_6H_4Br\text{-}p)_2$, 3, 298

$AsBC_{24}H_{20}$ Ph_2AsBPh_2, 3, 298

$AsBC_{24}H_{20}F_4$ $[Ph_4As]BF_4$, 3, 332

$AsBC_{26}H_{24}$ $Ph_2AsB(C_6H_4Me\text{-}p)_2$, 3, 298

$AsBC_{37}H_{40}$ $[(-)\text{-}Me(CH_2{=}CHCH_2)PrPhAs]BPh_4$, 3, 318

$AsBC_{41}H_{42}$ $[(-)\text{-}MePrPh(PhCH_2)As]BPh_4$, 3, 321

$AsBC_{48}H_{32}$ 5,5'-Spirobi(dibenzarsolium) bi(2,2'-biphenylylene)borate, 3, 571

$AsBC_{48}H_{38}$ Triphenyl[2'-(5-phenyldibenzarsolio)-2-biphenylyl]borate, 3, 569

5,5'-Spirobi(dibenzarsolium)$[BPh_4]^-$, 3, 571

$AsBC_{48}H_{40}$ $[Ph_4As]BPh_4$, 3, 332

$AsBC_{50}H_{40}$ 3,3'-Dimethyl-5,5'-spirobi(dibenzarsolium)$[BPh_4]^-$, 3, 571

$AsBC_{54}H_{41}Br$ 5-(2'-Bromo-2-biphenylyl)-5-phenyldibenzarsolium $[BPh_4]^-$, 3, 569

$AsBC_{56}H_{40}$ 7,7'-Spirobi(7H-benzo[b]naphth[1,2-d]arsolium) $[BPh_4]^-$, 3, 572

$AsBC_{57}H_{49}N$ Trimethyl(p-5,5′-spirobi-[dibenzarsol]-5-ylphenylammonium $[BPh_4]^-$, 3, 573

$AsB_2C_{81}H_{70}N$ 5-(2-PhC_6H_4)-5-(p-Me_2N-C_6H_7)dibenzarsolium tetraphenylborate, methotetraphenylborate, 3, 570

$AsB_2C_{82}H_{72}N$ 5-[p-(Dimethylamino)-phenyl]-5-(2′-methyl-2-biphenylyl) dibenzarsolium tetraphenylborate, methotetraphenylborate, 3, 570

$AsB_{12}C_6H_{29}S$ $Me_3S[B_{12}H_{11}(AsMe_3)]$, 3, 4

$AsB_{12}C_7H_{32}N$ $Me_4N[B_{12}H_{11}(AsMe_3)]$, 3, 4

$AsBiC_9H_{14}I_4$ $[Me_3PhAs]BiI_4$, 3, 314

$AsBiC_{10}H_{16}I_4$ $[Me_3(4\text{-}MeC_6H_4)As]BiI_4$, 3, 316

$AsBiC_{14}H_{16}I_2O$ $[Me_2Ph_2As]BiOI_2$, 3, 319

$AsBiC_{14}H_{16}I_4$ $[Me_2Ph_2As]BiI_4$, 3, 319

$AsBiC_{16}H_{20}I_4$ $[Me_2(4\text{-}MeC_6H_4)_2As]BiI_4$, 3, 320

$AsBiC_{19}H_{18}I_4$ $[MePh_3As]BiI_4$, 3, 324

$AsBiC_{22}H_{24}I_4$ $[Me(4\text{-}MeC_6H_4)_3As]BiI_4$, 3, 329

$AsCdC_8H_{11}Br_2$ $[Cd(Me_2AsPh)Br_2]$, 3, 39

$AsCdC_8H_{11}Cl_2$ $[Cd(Me_2AsPh)Cl_2]$, 3, 39

$AsCdC_8H_{11}I_2$ $[Cd(Me_2AsPh)I_2]$, 3, 40

$AsCdC_9H_{13}I_2$ $[Cd(Me_2AsC_6H_4Me\text{-}2)]I_2$, 3, 43

$[Cd(Me_2AsC_6H_4Me\text{-}4)]I_2$, 3, 43

$AsCdC_{10}H_{16}Br_2N$ $\{Cd[4\text{-}(Me_2N)C_6H_4AsMe_2]Br_2\}$, 3, 49

$AsCdC_{10}H_{16}Cl_2N$ $\{Cd[4\text{-}(Me_2N)C_6H_4AsMe_2]Cl_2\}$, 3, 49

$AsCdC_{10}H_{16}I_2N$ $\{Cd[4\text{-}(Me_2N)C_6H_4AsMe_2]I_2\}$, 3, 49

$AsCdC_{13}H_{13}Br_2$ $[Cd(Ph_2AsMe)Br_2]$, 3, 52

$AsCoC_{11}H_{13}N_2S_3$ $[Co(2\text{-}MeSC_6H_4AsMe_2)(SCN)_2]$, 3, 44

$AsCoC_{18}H_{15}BrN_2O_2$ $[Co(AsPh_3)(NO)_2Br]$, 3, 19

$AsCoC_{18}H_{15}ClN_2O_2$ $[Co(AsPh_3)(NO)_2Cl]$, 3, 19

$AsCoC_{20}H_{12}Cl_3NO_3$ $\{Co[As(C_6H_4Cl\text{-}4)_3](CO)_2NO\}$, 3, 17

$AsCoC_{20}H_{15}NO_3$ $[Co(AsPh_3)(CO)_2(NO)]$, 3, 19

$AsCoC_{23}H_{15}F_5O_3$ $[C_2F_5Co(CO)_3(AsPh_3)]$, 1, 566, 3, 20

$AsCoC_{23}H_{21}NO_3$ $[CoAs(C_6H_4Me\text{-}p)_3(CO)_2(NO)]$, 3, 28

$AsCoC_{33}H_{30}S_4$ $[MePh_3As]$[bis(toluenedithiolato)cobaltate], 3, 323

$AsCoC_{37}H_{30}NO_2P$ $[Co(AsPh_3)(PPh_3)(CO)(NO)]$, 3, 20

$AsCoC_{42}H_{51}N_5S_4$ $Bu_4N\{Co[SC(CN)=C(CN)\text{-}S^-]_2(AsPh_3)\}$, 3, 20

$AsCo_2C_{25}H_{15}O_7$ $[Co_2(AsPh_3)(CO)_7]$, 3, 20

$AsCrC_{23}H_{15}O_5$ $[Cr(AsPh_3)(CO)_5]$, 3, 19

$AsCrC_{36}H_{20}F_{18}S_6$ $[Ph_4As]\{Cr[SC(CF_3)=C(CF_3)S]_3\}$, 3, 332*

$AsCsC_6H_9NO_4$ $4\text{-}H_2NC_6H_4As(O)(OH)OCs\cdot H_2O$, 3, 428

$AsCsC_6H_{13}NO_6$ $4\text{-}H_2NC_6H_4As(O)(OH)OCs\cdot 3H_2O$, 3, 428

$AsCuC_8H_{11}Br$ $[Cu(Me_2AsPh)Br]$, 3, 40

$AsCuC_8H_{11}I$ $[Cu(Me_2AsPh)I]$, 3, 40

$AsCuC_8H_{18}O_2$ Bu_2AsO_2Cu, 3, 375

$AsCuC_{10}H_{15}Br$ $[Cu(Et_2AsPh)Br]$, 3, 47

$AsCuC_{11}H_{17}Br$ $[EtPrPhAs\cdot CuBr]$, 3, 77

$AsCuC_{13}H_{13}NO_3$ $[Cu(Ph_2AsMe)NO_3]$, 3, 52

$AsCuC_{33}H_{30}S_4$ $[MePh_3As]$[bis(toluenedithiolato)cuprate], 3, 322

$AsCuC_{36}H_{30}BrO_3P$ $\{Cu(AsPh_3)[P(OPh)_3]Br\}$, 3, 20

$AsFeC_9H_5F_6O_3$ $C_5H_5Fe(CO)_2As(CF_3)_2$, 3, 303

$AsFeC_{18}H_{15}BrN_2O_2$ $[Fe(AsPh_3)(NO)_2Br]$, 3, 21

$AsFeC_{18}H_{15}ClN_2O_2$ $[Fe(AsPh_3)(NO)_2Cl]$, 3, 21

$AsFeC_{18}H_{15}IN_2O_2$ $[Fe(AsPh_3)(NO)_2I]$, 3, 21

$AsFeC_{19}H_{15}N_2O_3$ $[Fe(AsPh_3)(CO)(NO)_2]$, 3, 21

$AsFeC_{21}H_{15}I_2O_3$ $[Fe(AsPh_3)(CO)_3I_2]$, 3, 21

$AsFeC_{22}H_{15}O_4$ $[Fe(AsPh_3)(CO)_4]$, 3, 21

$AsFeC_{22}H_{21}N_2O_3$ $\{Fe[As(C_6H_4Me\text{-}4)_3](CO)(NO)_2\}$, 3, 28

$AsFeC_{24}H_{20}Cl_4$ $[Ph_4As]FeCl_4$, 3, 332
$AsFeC_{28}H_{23}O_2$ $[CH{=}CH(CH{=}CH)_2CH{=}CH]Fe(CO)_2(AsPh_3)$, 1, 501
$AsFeC_{33}H_{30}S_4$ $[MePh_3As]$[bis(toluenedithiolato)ferrate], 3, 323
$AsFeC_{37}H_{25}O_3$ Pentaphenylarsenole-$Fe(CO)_3$, 1, 496
$AsFe_2C_{16}H_{16}ClO_8$ $\{[C_5H_5Fe(CO)_2]_2AsMe_2\}ClO_4$, 1, 216, 3, 304
$AsFe_2C_{27}H_{25}O_5S_2$ $[Fe_2(AsPh_3)(CO)_5(SEt)_2]$, 3, 22
$AsFe_2C_{35}H_{25}O_5S_2$ $[Fe_2(AsPh_3)(CO)_5(SPh)_2]$, 3, 22
$AsGaC_6H_{18}$ $[Me_3AsGaMe_3]$, 3, 4
$AsHgC_4H_{12}I_3$ $[Me_4As]HgI_3$, 3, 312
$AsHgC_6H_3BrClO$ $4,2\text{-}Br(ClHg)C_6H_3AsO$, 3, 235
$AsHgC_6H_3Cl_2O$ $4,2\text{-}Cl(ClHg)C_6H_3AsO$, 3, 235
$AsHgC_6H_4ClO$ $4\text{-}ClHgC_6H_4AsO$, 3, 235
$AsHgC_7H_6ClO_4$ $2,4\text{-}HO_2C(ClHg)C_6H_3As(OH)_2$, 3, 222
$AsHgC_7H_{18}I_3$ $[Et_3MeAs]HgI_3$, 3, 313
$AsHgC_8H_5Cl_2O_3$ $2,4,6\text{-}Cl_2(AcOHg)C_6H_2AsO$, 3, 238
$AsHgC_8H_6BrO_3$ $4,2\text{-}Br(AcOHg)C_6H_3AsO$, 3, 238
$AsHgC_8H_6ClO_3$ $Cl(AcOHg)C_6H_3AsO$ isomers, 3, 238
$AsHgC_8H_6NO_4S$ $O_2N(AcOHg)C_6H_3AsS$ isomers, 3, 265
$AsHgC_8H_7O_2S$ $4\text{-}AcOHgC_6H_4AsS$, 3, 265
$AsHgC_8H_7O_3$ $4\text{-}AcOHgC_6H_4AsO$, 3, 239
$AsHgC_8H_8NO_6$ $O_2N(AcOHg)C_6H_3As(OH)_2$ isomers, 3, 223
$AsHgC_8H_{11}Br_2$ $[Hg(Me_2AsPh)Br_2]$, 3, 40
$AsHgC_8H_{11}Cl_2$ $[Hg(Me_2AsPh)Cl_2]$, 3, 40
$AsHgC_8H_{11}I_2$ $[Hg(Me_2AsPh)I_2]$, 3, 40
$AsHgC_8H_{20}I_3$ $[Et_4As]HgI_3$, 3, 314
$AsHgC_9H_9O_6$ $HO_2C(AcOHg)C_6H_3As(OH)_2$ isomers, 3, 223
$AsHgC_9H_{10}NO_6$ $2,4,5\text{-}Me(AcOHg)(O_2N)\text{-}C_6H_2As(OH)_2$, 3, 223
$AsHgC_9H_{14}I_3$ $[Me_3PhAs]HgI_3$, 3, 314
$AsHgC_{10}H_{11}O_3$ $2,4,6\text{-}Me_2(AcOHg)C_6H_2AsO$, 3, 243
$AsHgC_{10}H_{16}Br_2N$ $\{Hg[4\text{-}(Me_2N)C_6H_4AsMe_2]Br_2\}$, 3, 49
$AsHgC_{10}H_{16}Cl_2N$ $\{Hg[4\text{-}(Me_2N)C_6H_4AsMe_2]Cl_2\}$, 3, 49
$AsHgC_{12}H_{13}Cl_2$ $[Hg(1\text{-}C_{10}H_7AsMe_2)Cl_2]$, 3, 50
$AsHgC_{12}H_{13}Cl_2N_2$ $\{Hg[PhAs(CH_2CH_2CN)_2]Cl_2\}$, 3, 50
$AsHgC_{12}H_{20}I_3$ $[Et_3PhAs]HgI_3$, 3, 317
$AsHgC_{12}H_{21}Cl_2$ $[(CH_2{=}CMeCH_2)_3AsHgCl_2]$, 3, 15
$AsHgC_{13}H_{13}Br_2$ $[Hg(Ph_2AsMe)Br_2]$, 3, 54
$AsHgC_{13}H_{13}I_2$ $[Hg(Ph_2AsMe)I_2]$, 3, 54
$AsHgC_{13}H_{16}I_3$ $[Me_3(1\text{-}C_{10}H_7)As]HgI_3$, 3, 318
$AsHgC_{14}H_{12}Cl_3$ $[Hg(Ph_2AsCH{=}CHCl)Cl_2]$, 3, 58
$AsHgC_{14}H_{16}I_3$ $[Me_2Ph_2As]HgI_3$, 3, 319
$AsHgC_{16}H_{20}I_3$ $[Et_2Ph_2As]HgI_3$, 3, 320
$AsHgC_{18}H_{12}Cl_5$ $\{Hg[As(C_6H_4Cl\text{-}4)_3]Cl_2\}$, 3, 17
$\{Hg[As(C_6H_4Cl\text{-}4)_3]Cl_2\}$, 3, 17
$AsHgC_{19}H_{18}I_3$ $[MePh_3As]HgI_3$, 3, 323
$AsHgC_{24}H_{20}Cl_3$ $[Ph_4As]HgCl_3$, 3, 333
$AsHgC_{30}H_{22}Cl_3$ 5-(2-Biphenylyl)-5-phenyldibenzarsolium$[HgCl_3]^-$, 3 568
$AsHg_2C_3H_9Cl_4$ $[Me_3As\cdot 2HgCl_2]$, 3, 4
$AsHg_2C_5H_{13}O_3$ $MeAsO(OHgEt)_2$, 3, 547
$PrAsO(OHgMe)_2$, 3, 547
$AsHg_2C_6H_{15}O_3$ $EtAsO(OHgEt)_2$, 3, 547
$BuAsO(OHgMe)_2$, 3, 548
$AsHg_2C_{10}H_{14}ClO_3$ $4\text{-}ClC_6H_4AsO(OHgEt)_2$, 3, 548
$AsHg_2C_{13}H_{13}O_3$ $MeAsO(OHgPh)_2$, 3, 547
$AsHg_2C_{14}H_{15}O_3$ $EtAsO(OHgPh)_2$, 3, 547
$AsHg_2C_{15}H_{17}O_3$ $PrAsO(OHgPh)_2$, 3, 547
$AsHg_2C_{16}H_{19}O_3$ $EtAsO(OHgC_6H_4Me\text{-}p)_2$, 3, 547
$AsHg_2C_{18}H_{12}ClN_2O_7$ $o\text{-}ClC_6H_4AsO(OHgC_6H_4NO_2\text{-}p)_2$, 3, 409
$AsHg_2C_{19}H_{14}ClN_2O_7$ $2\text{-}ClC_6H_4CH_2AsO(OHgC_6H_4NO_2\text{-}p)_2$, 3, 549
$AsHg_2C_{19}H_{17}O_3$ $2\text{-}MeC_6H_4AsO(OHgPh)_2$, 3, 549
$AsHg_2C_{19}H_{18}I_5$ $[MePh_3As]Hg_2I_5$, 3, 323
$AsHg_2C_{20}H_{19}O_3$ $PhAsO(OHgC_6H_4Me\text{-}4)_2$, 3, 400, 549

$AsHg_2C_{30}H_{21}Cl_4$ 5-[2'-(chloromercuri)-2-biphenylyl]-5-phenyldibenzarsolium$[HgCl_3]^-$, 3, 568

$AsHg_2C_{32}H_{26}Cl_4N$ 5-[2-(2-$ClHgC_6H_4$)-C_6H_4]-5-(p-$Me_2NC_6H_4$)dibenzarsolium$[HgCl_3]^-$, 3, 569

$AsHg_4C_{14}H_{16}NO_{11}$ Tetrakis(acetoxymercuri)arsanilic acid, 3, 504

$AsInC_6H_{18}$ $[Me_3AsInMe_3]$, 3, 4

$AsIrC_{25}H_{20}I_4O$ $[Ph_4As][Ir(CO)I_4]$, 3, 333

$AsIrC_{25}H_{20}I_5O$ $[Ph_4As][Ir(CO)I_5]$, 3, 333

$AsIrC_{25}H_{23}I_4NO$ $[Ph_4As][Ir(CO)(NH_3)I_4]$, 3, 333

$AsIrC_{26}H_{20}I_3O_2$ $[Ph_4As][Ir(CO)_2I_3]$, 3, 333

$AsIrC_{26}H_{20}I_4O_2$ $[Ph_4As][Ir(CO)_2I_4]$, 3, 333

$AsIrC_{27}H_{27}I_4NO$ $[Ph_4As][Ir(CO)(EtNH_2)I_4]$, 3, 333

$AsIrC_{28}H_{29}I_2NO$ $[Ph_4As][Ir(CO)(Me_2CHNH_2)I_2]$, 3, 333

$AsIrC_{36}H_{50}Cl_4P_2$ $[Ph_4As][Ir(Et_3P)_2Cl_4]$, 3, 333

$AsIrC_{44}H_{50}Cl_4P_2$ $[Ph_4As][Ir(Et_2PhP)_2Cl_4]$, 3, 333

$AsKC_6H_6$ PhHAsK, 3, 296

$AsKC_8H_{10}$ EtPhAsK, 3, 296

$AsKC_{12}H_{16}$ $Ph(C_6H_{11})AsK$, 3, 296

$AsKC_{14}H_{14}$ $(4\text{-}MeC_6H_4)_2AsK$, 3, 296

$AsKC_{14}H_{22}O_4$ PhHAsK compd. with dioxane, 3, 296

$AsKC_{20}H_{14}$ $(1\text{-}C_{10}H_7)_2AsK$, 3, 296

$AsKC_{20}H_{26}O_4$ Ph_2AsK compd. with dioxane, 3, 296

$AsK_2C_{13}H_{13}N_6O_4S_2$ 4-$[N{=}C(NH_2)N{=}C(NH_2)N{=}CNH]C_6H_4As(SCH_2CO_2K)_2$, 3, 268

$AsK_2C_{14}H_{11}Cl_2N_4O_5S_2$ 3,4-MeO$(N{=}CClN{=}CClN{=}CNH)C_6H_3As(SCH_2CO_2K)_2$, 3, 269

$AsLiC_6H_5$ $PhAsLi_2$, 3, 294

$AsLiC_{10}H_{14}O_2$ PhHAsLi·p-dioxane, 3, 295

$AsLiC_{12}H_{18}O_2$ EtPhAsLi·p-dioxane, 3, 295

$AsLiC_{12}H_{22}$ $(C_6H_{11})_2AsLi$, 3, 295

$AsLiC_{16}H_{18}O_2$ Ph_2AsLi·p-dioxane, 3, 295

$AsLiC_{16}H_{24}O_2$ $Ph(C_6H_{11})AsLi$·p-dioxane, 3, 295

$AsLiC_{18}H_{22}O_2$ $(4\text{-}MeC_6H_4)_2AsLi$·p-dioxane, 3, 295

$AsLiC_{24}H_{24}O$ $(\alpha\text{-}C_{10}H_7)_2AsLi\cdot Et_2O$, 3, 295

$AsMgC_2H_6Br$ $Me_2AsMgBr$, 3, 297

$AsMg_2C_6H_5Br_2$ $PhAs(MgBr)_2$, 3, 297

$AsMnC_{14}H_{22}O_2$ $[(MeC_5H_4)Mn(AsEt_3)(CO)_2]$, 3, 8

$AsMnC_{15}H_{24}O_2$ $(EtC_5H_4)Mn(CO)_2(AsEt_3)$, 1, 170

$AsMnC_{17}H_{21}O_2$ $(C_5H_5)Mn(CO)_2[o\text{-}C_6H_4(AsMe_2)_2]$, 1, 170

$AsMnC_{17}H_{22}O_5$ Indenyl-$Mn(CO)_2[As(OEt)_3]$, 1, 170

$AsMnC_{18}H_{14}O_4$ $(4\text{-}MeC_6H_4)_2AsMn(CO)_4$, 3, 302

$AsMnC_{18}H_{15}N_2O_4$ $[Mn(AsPh_3)(NO)_2]$, 3, 22

$AsMnC_{18}H_{15}N_3O_3$ $[Mn(AsPh_3)(NO)_3]$, 3, 22

$AsMnC_{20}H_{39}O_7$ $(Bu_2C_5H_3)Mn(CO)[As(OMe)_3]_2$, 1, 170

$AsMnC_{22}H_{15}BrO_4$ $[Mn(AsPh_3)(CO)_4]Br$, 3, 22

$AsMnC_{22}H_{15}ClO_4$ $[Mn(AsPh_3)(CO)_4]Cl$, 3, 22

$AsMnC_{22}H_{15}IO_4$ $[Mn(AsPh_3)(CO)_4]I$, 3, 22

$AsMnC_{22}H_{15}O_4$ $[Mn(AsPh_3)(CO)_4]$, 3, 22

$AsMnC_{24}H_{20}O_4$ $[Ph_4As]MnO_4$, 3, 333

$AsMnC_{25}H_{20}O_2$ $(C_5H_5)Mn(CO)_2(AsPh_3)$, 1, 170, 3, 23

$AsMnC_{26}H_{22}O_2$ $[(MeC_5H_4)Mn(CO)_2(AsPh_3)]$, 1, 170, 3, 23

$AsMnC_{28}H_{26}O_2$ $[C_5H_5Mn(CO)_2As(CH_2Ph)_3]$, 1, 170, 3, 27

$AsMnC_{28}H_{50}O_2$ $[Me(n\text{-}C_8H_{17})C_5H_3]Mn(CO)_2(AsBu_3)$, 1, 170

$AsMnC_{38}H_{70}O_2$ $\{[Me(C_{18}H_{37})C_5H_3]Mn(AsBu_3)(CO)_2\}$, 3, 15

$AsMnC_{51}H_{48}O_2$ $\{(n\text{-}C_8H_{17}C_5H_4)Mn(CO)_2[As(C_6H_4Ph\text{-}p)_3]\}$, 1, 170

$AsMn_2C_{10}F_6IO_8$ $[(OC)_4MnAs(CF_3)_2Mn(CO)_4I]$, 3, 301*

$AsMoC_{10}H_5F_6O_3$ $[C_5H_5Mo(CO)_3As(CF_3)_2]$, 3, 300

$AsMoC_{11}H_{15}O_5$ [Mo(AsEt$_3$)(CO)$_5$], 3, 9

$AsMoC_{18}H_{15}Cl_4$ [Mo(AsPh$_3$)Cl$_4$], 3, 23

$AsMoC_{18}H_{15}Cl_5O$ [Mo(Ph$_3$AsO)Cl$_5$], 3, 354

$AsMoC_{21}H_{21}Cl_3O_3$ [MoO(Ph$_3$AsO)(Me$_2$CO)-Cl$_3$], 3, 354

$AsMoC_{23}H_{15}O_5$ [Mo(AsPh$_3$)(CO)$_5$], 3, 23

$AsNaC_2H_{12}O_3$ Me$_2$AsO$_2$Na·3H$_2$O, 3, 374

$AsNaC_6H_5NO_7$ 3,4-O$_2$N(NaO$_3$S)C$_6$H$_3$As(OH)$_2$, 3, 221

$AsNaC_6H_7NO_3$ 4-H$_2$NC$_6$H$_4$AsO(OH)ONa, 3, 428

$AsNaC_7H_7NO_4S$ [HO(NaO$_3$SCH$_2$NH)C$_6$H$_3$As]$_n$ isomers, 3, 136, 137
4,3-HO(NaO$_2$SCH$_2$NH)C$_6$H$_3$AsO, 3, 238

$AsNaC_7H_9NO_3S$ 3,4-HO(NaOSOCH$_2$NH)C$_6$H$_3$-AsH$_2$, 3, 120

$AsNaC_7H_9NO_5S$ 4-NaO$_3$SCH$_2$NHC$_6$H$_4$As(OH)$_2$, 3, 222

$AsNaC_8H_9NO_4$ 4-AcNHC$_6$H$_4$AsO(OH)ONa, 3, 429

$AsNaC_8H_{11}NO_3S$ 4,3-HO(NaOSOCH$_2$NMe)-C$_6$H$_3$AsH$_2$, 3, 121

$AsNaC_9H_{11}NO_5S$ [3,4-NaO$_3$SCH$_2$NH(HOCH$_2$-CH$_2$O)C$_6$H$_3$As]$_n$, 3, 139

$AsNaC_9H_{13}NO_7S$ 3,4-(NaO$_3$SCH$_2$NH)(HO-CH$_2$CH$_2$O)C$_6$H$_3$As(OH)$_2$, 3, 223

$AsNaC_{10}H_{13}NO_4S$ [3,4-NaO$_2$SCH$_2$NH(HO-CH$_2$CH$_2$CH$_2$O)C$_6$H$_3$As]$_n$, 3, 139
[3,4-NaO$_2$SCH$_2$NH(MeCHOHCH$_2$O)C$_6$H$_3$As]$_n$, 3, 139

$AsNaC_{10}H_{13}NO_5$ [3,4-NaO$_3$SCH$_2$NH(HOCH$_2$-CH$_2$CH$_2$O)C$_6$H$_3$As]$_n$, 3, 140
[3,4-NaO$_3$SCH$_2$NH(MeCHOHCH$_2$O)C$_6$H$_3$As]$_n$, 3, 140

$AsNaC_{12}H_{18}O_{12}S$ 4,3-HO[NaO$_3$S(CHOH)$_5$-CH$_2$O]C$_6$H$_3$As(OH)$_2$, 3, 224*

$AsNaC_{13}H_{19}N_3O_7S_3$ 3,4-(NaO$_3$SCH$_2$NH)-(HOCH$_2$CH$_2$O)C$_6$H$_3$As(SCH$_2$CONH$_2$)$_2$, 3, 268

$AsNaC_{14}H_{22}O_4$ PhHAsNa compd. with dioxane, 3, 295

$AsNaC_{16}H_{18}O_2$ Ph$_2$AsNa·dioxane, 3, 295

$AsNa_2CH_{15}O_9$ MeAsO$_3$Na$_2$·6H$_2$O, 3, 391

$AsNa_2C_9H_9N_6O_3$ 4-[N=C(NH$_2$)N=C(NH$_2$)N=C-NH]C$_6$H$_4$As(O)(ONa)$_2$, 3, 431

$AsNa_2C_{10}H_9N_4O_3$ 4-[CH=NC(NH$_2$)=NCH=CNH]-C$_6$H$_4$AsO(ONa)$_2$, 3, 432

$AsNa_2C_{23}H_{17}N_6O_4S_2$ 4-[N=C(NH$_2$)N=C(NH$_2$)N=CNH]C$_6$H$_4$As-(SC$_6$H$_4$CO$_2$Na)$_2$, 3, 274

$AsNa_3C_{13}H_{15}NO_9S_3$ 3,4-(NaO$_3$SCH$_2$NH)-(HOCH$_2$CH$_2$O)C$_6$H$_3$As(SCH$_2$CO$_2$Na)$_2$, 3, 268

$AsNa_3C_{16}H_{21}NO_{13}S_2$ 4,3-HO[NaO$_3$S(CHOH)$_5$-CH$_2$NH]C$_6$H$_3$As(SCH$_2$CO$_2$Na)$_2$, 3, 270

$AsNa_4C_{14}H_{10}NO_{11}S_2$ 3,4-O$_2$N(HO)C$_6$H$_3$-As[SCH(CO$_2$Na)CH$_2$CO$_2$Na]$_2$, 3, 269

$AsNiC_9H_{15}O_3$ [Ni(AsEt$_3$)(CO)$_3$], 3, 9

$AsNiC_{21}H_{15}O_3$ [Ni(AsPh$_3$)(CO)$_3$], 3, 24

$AsNiC_{33}H_{30}S_4$ [MePh$_3$As][bis(toluene-dithiolato)niccolate], 3, 323

$AsPaC_{24}H_{20}Cl_6$ [Ph$_4$As]PaCl$_6$, 3, 333

$AsPbC_{12}H_{22}NO_3$ p-H$_2$NC$_6$H$_4$AsO(OH)OPbEt$_3$, 2, 564

$AsPbC_{13}H_{14}ClO_3$ Ph$_2$Pb(HO$_3$AsMe)Cl, 2, 565

$AsPbC_{14}H_{16}ClO_2$ Me$_2$AsO$_2$PbClPh$_2$, 2, 566, 3, 374

$AsPbC_{15}H_{18}ClO_3$ Ph$_2$Pb(HO$_3$AsPr)Cl, 2, 566

$AsPbC_{20}H_{21}O_2$ Me$_2$AsO$_2$PbPh$_3$, 2, 565, 3, 374

$AsPb_2C_{25}H_{23}Cl_2O_3$ (Ph$_2$Pb)$_2$(O$_3$AsMe)Cl$_2$, 2, 565

$AsPdC_6H_{15}Br_2S$ [Pd(Me$_2$AsCH$_2$CH$_2$CH$_2$SMe)-Br$_2$], 3, 38

$AsPdC_6H_{15}Cl_2S$ [Pd(Me$_2$AsCH$_2$CH$_2$CH$_2$SMe)-Cl$_2$], 3, 38

$AsPdC_6H_{15}I_2S$ [Pd(Me$_2$SCH$_2$CH$_2$CH$_2$SMe)I$_2$], 3, 38

$AsPdC_7H_{11}Cl_2N_2$ {Pd[MeAs(CH$_2$CH$_2$CN)$_2$]-Cl$_2$}, 3, 38

$AsPdC_9H_{13}Br_2S$ [Pd(2-MeSC$_6$H$_4$AsMe$_2$)Br$_2$], 3, 45

$AsPdC_9H_{13}Cl_2S$ [Pd(2-MeSC$_6$H$_4$AsMe$_2$)Cl$_2$], 3, 45

$AsPdC_9H_{13}I_2S$ [Pd(2-MeSC$_6$H$_4$AsMe$_2$)I$_2$], 3, 45

$AsPdC_{10}H_{16}Br_2N$ {Pd[2-(Me$_2$N)C$_6$H$_4$AsMe$_2$]-Br$_2$}, 3, 48

$AsPdC_{10}H_{16}Cl_2N$ {Pd[2-(Me$_2$N)C$_6$H$_4$AsMe$_2$]-Cl$_2$}, 3, 48

$AsPdC_{11}H_{13}N_2S_3$ [Pd(2-MeSC$_6$H$_4$AsMe$_2$)-

$AsPdC_{11}H_{13}N_2S_3$

$(SCN)_2]$, 3, 45

$AsPdC_{11}H_{19}BrP$ $[MePd(o\text{-}Me_2AsC_6H_4PMe_2)\text{-}Br]$, 1, 664

$AsPdC_{11}H_{28}Cl_2N$ $[Pd(AsEt_3)(Pip)Cl_2]$, 3, 9

$AsPdC_{12}H_{13}Cl_2N_2$ $\{Pd[PhAs(CH_2CH_2CN)_2]\text{-}Cl_2\}$, 3, 50

$AsPdC_{12}H_{20}Br_2P$ $[Pd(2\text{-}Me_2AsC_6H_4PEt_2)\text{-}Br_2]$, 3, 51

$AsPdC_{12}H_{30}Cl_2N$ $[Pd(AsBu_3)(NH_3)Cl_2]$, 3, 16

$AsPdC_{13}H_{23}BrP$ $[MePd(2\text{-}Me_2AsC_6H_4PEt_2)\text{-}Br]$, 3, 51

$AsPdC_{14}H_{15}Cl_2$ $[Pd(p\text{-}PhC_6H_4AsMe_2)Cl_2]$, 3, 61

$AsPdC_{16}H_{16}Cl_2N$ $[Pd(Ph_2AsCH_2CHMeCN)\text{-}Cl_2]$, 3, 61

$AsPdC_{18}H_{15}Cl_2$ 2,3-Dihydro-2-phenyl-1H-naphth[1,8-cd]arsenin·$PdCl_2$, 3, 599

$AsPdC_{19}H_{15}Br_2$ 5,6-Dihydro-5-phenyl-arsanthridin·$PdBr_2$, 3, 598

$AsPdC_{19}H_{15}Cl_2$ 5,6-Dihydro-5-phenyl-arsanthridin·$PdCl_2$, 3, 598

$AsPdC_{24}H_{18}Br_2$ 5,10-Ph_2-5,10-dihydrodibenzo[1,4]phospharsenin-$PdBr_2$, 3, 641

$AsPtC_6H_{15}I_2S$ $[Pt(Me_2AsCH_2CH_2CH_2SMe)\text{-}I_2]$, 3, 38

$AsPtC_7H_{15}Br_2$ $\{Pt[Me_2As(CH_2)_3CH{=}CH_2]\text{-}Br_2\}$, 3, 39*

$AsPtC_7H_{15}Cl_2$ $\{Pt[Me_2As(CH_2)_3CH{=}CH_2]\text{-}Cl_2\}$, 3, 39

$AsPtC_7H_{15}I_2$ $\{Pt[Me_2As(CH_2)_3CH{=}CH_2]\text{-}I_2\}$, 3, 39

$AsPtC_9H_{13}I_2S$ $[Pt(2\text{-}MeSC_6H_4AsMe_2)I_2]$, 3, 45

$AsPtC_9H_{24}Cl_2N$ $[Pt(AsPr_3)(NH_3)Cl_2]$, 3, 13

$AsPtC_{14}H_{13}Br_2O$ $[Pt(Ph_2AsMe)(CO)Br_2]$, 3, 55

$AsPtC_{14}H_{13}Cl_2O$ $[Pt(Ph_2AsMe)(CO)Cl_2]$, 3, 55

$AsPtC_{14}H_{13}I_2O$ $[Pt(Ph_2AsMe)(CO)I_2]$, 3, 55

$AsPtC_{14}H_{24}Cl_2P$ $\{Pt[2\text{-}(Et_2As)C_6H_4\text{-}PEt_2]Cl_2\}$, 3, 60

$AsPtC_{14}H_{32}Cl_2N$ $[Pt(AsPr_3)(Pip)Cl_2]$, 3, 13

$AsPtC_{16}H_{30}Cl_2N$ $[Pt(AsPr_3)(p\text{-}H_2NC_6H_4\text{-}Me)Cl_2]$, 3, 14

$AsPtC_{18}H_{35}Br$ $[trans\text{-}PhPt(AsEt_3)_2Br]$, 1, 691

$AsPtC_{18}H_{42}Cl_2N$ $[Pt(AsPr_3)(NPr_3)Cl_2]$, 3, 13

$AsPtC_{19}H_{27}Cl_3NO$ $[Me_2(2\text{-}Me_2NC_6H_4)\text{-}(2\text{-}MeOCH_2C_6H_4CH_2)As]PtCl_3$, 3, 327

$AsPtC_{19}H_{29}Cl_2N_2$ $[Pt(AsPr_3)(BiPy)Cl_2]$, 3, 14

$AsPtC_{19}H_{37}Cl$ $[trans\text{-}PhCH_2Pt(AsEt_3)_2\text{-}Cl]$, 1, 691

$AsPtC_{24}H_{40}$ $[Ph_2Pt(AsEt_3)_2]$ isomers, 1, 694

$AsPtC_{25}H_{20}Br_3O$ $[Ph_4As][Pt(CO)Br_3]$, 3, 333

$AsPtC_{25}H_{20}Cl_3O$ $[Ph_4As][Pt(CO)Cl_3]$, 3, 333

$AsPtC_{25}H_{20}I_3O$ $[Ph_4As][Pt(CO)I_3]$, 3, 333

$AsPtC_{28}H_{40}$ $[trans\text{-}(PhC{\equiv}C)_2Pt(AsEt_3)_2]$, 1, 694

$AsPtC_{30}H_{52}$ $[trans\text{-}Ph_2Pt(AsPr_3)_2]$, 1, 694

$AsPtC_{40}H_{30}F_6$ $[(Ph_3As)_2Pt(CF_3C{\equiv}CCF_3)]$, 1, 730

$AsPt_2C_{42}H_{62}BrP_2$ $\{Ph_2As[PhPt(PPr_3)]_2\text{-}Br\}$, 3, 304

$AsRbC_6H_7NO_3$ $4\text{-}H_2NC_6H_4As(O)(OH)ORb$, 3, 428*

$AsReC_{18}H_{15}Cl_3$ $Re(AsPh_3)Cl_3$, 3, 25

$AsReC_{18}H_{15}Cl_3O$ $[Re(Ph_3AsO)Cl_3]$, 3, 355

$AsReC_{24}H_{20}O_4$ $[Ph_4As]ReO_4$, 3, 333

$AsRhC_{24}H_{22}O_2$ $[AcCH{=}C(Me)ORh(AsPh_3)\text{-}(CO)]$, 3, 26

$AsRhC_{25}H_{23}Cl$ [(Bicyclo[2.2.1]hepta-2,5-diene)$(Ph_3As)Rh]Cl$, 1, 625

$AsRuC_{24}H_{20}O_4$ $[Ph_4As]RuO_4$, 3, 333

$AsSbC_3H_9Cl_3$ $[Me_3AsSbCl_3]$, 3, 3

$AsSbC_{18}H_{15}Cl_5O$ $[Sb(Ph_3AsO)Cl_5]$, 3, 353

$AsSiC_3H_{12}I$ $[Me_3AsSiH_3I]$, 3, 5

$AsSiC_5H_{15}O$ $Me_2AsOSiMe_3$, 3, 209

$AsSiC_6H_{18}N$ $Me_2AsN(Me)SiMe_3$, 3, 286

$AsSiC_{20}H_{21}$ $Me_2AsSiPh_3$, 3, 298

$AsSiC_{21}H_{23}O_2$ $4\text{-}[Me(MeO)_2Si]C_6H_4AsPh_2$, 3, 71

$AsSiC_{22}H_{25}$ $Et_2AsSiPh_3$, 3, 299

$AsSiC_{22}H_{26}Br$ $[Ph_3(Me_3SiCH_2)As]Br$, 3, 329

$AsSiC_{28}H_{28}N_3O_7$ $[Ph_3(Me_3SiCH_2)As]O\text{-}C_6H_2(NO_2)_3$, 3, 329

$AsSi_2C_3H_9Cl_4O_3$ $MeAsO(OSiMeCl_2)_2$, 3, 544

$AsSi_2C_5H_{15}Cl_2O_3$ $MeAsO(OSiMe_2Cl)_2$, 3, 544

$AsSi_2C_8H_{24}N$ $Me_2AsN(SiMe_3)_2$, 3, 287

$AsSi_2C_9H_{15}Cl_4O_3$ $CH_2{=}CHCH_2AsO[OSiCl_2\text{-}(CH_2CH{=}CH_2)]_2$, 3, 545*

$AsSi_2C_{11}H_{17}Cl_2N_2O_3$ $MeAsO[OSiMe\text{-}(\text{-}NCH{=}CHCH{=}CH)Cl]_2$, 3, 544

$AsSi_2C_{13}H_{25}Cl_4O_3$ $MeAsO[OSiCl_2\text{-}(C_6H_{11})]_2$, 3, 544

$AsSi_2C_{17}H_{21}Cl_4O_3$ $MeAsO[OSiCl_2(CH_2\text{-}CH_2Ph)]_2$, 3, 545

$AsSi_2C_{18}H_{15}Cl_4O_3$ $PhAsO(OSiCl_2Ph)_2$, 3, 546

$AsSi_2C_{23}H_{25}Cl_4O_3$ $PhCH_2AsO[OSiCl_2\text{-}(CH_2CH_2Ph)]_2$, 3, 546

$AsSi_2C_{30}H_{25}Cl_2O_3$ $PhAsO(OSiClPh_2)_2$, 3, 546

$AsSi_2C_{37}H_{33}$ $MeAs(SiPh_3)_2$, 3, 299

$AsSi_2C_{42}H_{35}O_2$ $PhAs(OSiPh_3)_2$, 3, 255

$AsSi_2C_{42}H_{35}O_3$ $PhAsO(OSiPh_3)_2$, 3, 546

$AsSi_3C_{12}H_{33}$ $(Me_3SiCH_2)_3As$, 3, 16

$AsSi_3C_{12}H_{33}Br_2$ $(Me_3SiCH_2)_3AsBr_2$, 3, 344

$AsSi_3C_{12}H_{33}I_2$ $(Me_3SiCH_2)_3AsI_2$, 3, 344

$AsSi_3C_{13}H_{36}I$ $[Me(Me_3SiCH_2)_3As]I$, 3, 318

$AsSi_3C_{14}H_{38}I$ $[Et(Me_3SiCH_2)_3As]I$, 3, 319

$AsSi_3C_{21}H_{51}$ $(BuMe_2SiCH_2)_3As$, 3, 31

$AsSi_3C_{27}H_{57}$ $(cyclo\text{-}C_6H_{11}Me_2SiCH_2)_3As$, 3, 32

$AsSi_4C_9H_{27}O_6$ $\{As(Me)(O)O\}\{Si(Et)(H)\text{-}O\}_4$, 3, 545

$AsSi_4C_{16}H_{44}I$ $[(Me_3SiCH_2)_4As]I$, 3, 321

$AsSi_{12}C_{13}H_{39}O_{20}$ $\{As(Me)(O)O\}\{Si(Me)\text{-}O_{3/2}\}_{12}$, 3, 545

$AsSi_{12}C_{25}H_{75}O_{14}$ $\{As(Me)(O)O\}\{Si(Me)_2\text{-}O\}_{12}$, 3, 545

$AsSi_{12}C_{33}H_{78}NO_{15}$ 8-Hydroxy-5-quinoline-$As(O)O\{Si(Et)(H)O\}_{12}$, 3, 547

$AsSnC_3H_9O_3$ $\{As(Me)(O)OSn(Me)_2O\}_n$, 2, 349, 3, 547

$AsSnC_8H_{11}O_3$ $\{As(Ph)(O)OSn(Me)_2O\}_n$, 2, 349, 3, 549*

$AsSnC_{14}H_{18}O_6$ $Me_2Sn(HO_3AsPh)_2$, 2, 349

$AsSnC_{14}H_{22}NO_5$ $4\text{-}O_2NC_6H_4As(O)OSn(Bu)_2O$, 2, 354, 3, 548

$AsSnC_{14}H_{22}NO_6$ $3,4\text{-}O_2N(HO)C_6H_3As(O)\text{-}O_2SnBu_2$, 2, 353, 3, 475, 548

$AsSnC_{14}H_{23}N_2O_5$ $3,4\text{-}O_2N(H_2N)C_6H_3\text{-}As(O)OSn(Bu)_2O$, 2, 353, 3, 475, 549

$AsSnC_{14}H_{24}NO_3$ $4\text{-}H_2NC_6H_4As(O)OSn(Bu)_2O$, 2, 353, 3, 428, 549

$AsSnC_{15}H_{19}$ $Me_3SnAsPh_2$, 2, 484, 3, 299

$AsSnC_{15}H_{19}O_2$ $Ph_2AsO_2SnMe_3$, 3, 376

$AsSnC_{16}H_{19}O_3$ $BuAs(O)OSn(Ph)_2O$, 2, 356, 3, 548

$AsSnC_{18}H_{25}$ $Ph_2AsSnEt_3$, 3, 299

$AsSnC_{18}H_{25}O_2$ $Ph_2AsO_2SnEt_3$, 3, 376

$AsSnC_{21}H_{31}$ $Ph_2AsSnPr_3$, 3, 299

$AsSnC_{24}H_{20}NO_5$ $4\text{-}O_2NC_6H_4As(O)(OH)OSn\text{-}Ph_3$, 2, 355, 3, 548

$AsSnC_{24}H_{37}$ $Ph_2AsSnBu_3$, 3, 299

$AsSnC_{24}H_{37}O_2$ $Ph_2AsO_2SnBu_3$, 3, 377

$AsSnC_{30}H_{25}O_2$ $Ph_2AsO_2SnPh_3$, 3, 377

$AsSn_2C_{42}H_{35}O_3$ $PhAs(O)(OSnPh_3)_2$, 2, 355, 3, 549

$AsSn_3C_{54}H_{45}$ $(Ph_3Sn)_3As$, 2, 484

$AsTaC_{24}H_{20}F_6$ $[Ph_4As]TaF_6$, 3, 334

$AsTcC_{24}H_{20}O_4$ $[Ph_4As]TcO_4$, 3, 334

$AsTiC_{18}H_{15}Br_4$ $[TiBr_4(AsPh_3)]$, 3, 26

$AsTiC_{18}H_{15}Cl_4$ $[TiCl_4(AsPh_3)]$, 3, 26

$AsTlC_6H_{18}$ $[Me_3AsTlMe_3]$, 3, 6

$AsUC_6H_5O_5$ $UO_2[O_3AsPh]$, 3, 400

$AsUC_{24}H_{20}Cl_6$ $[Ph_4As]UCl_6$, 3, 334

$AsVC_{19}H_{18}Cl_4$ $[MePh_3As]VCl_4$, 3, 324

$AsWC_{18}H_{15}Cl_2N_2O_2$ $[W(AsPh_3)(NO)_2Cl_2]$, 3, 26

$AsWC_{24}H_{20}Cl_6$ $[Ph_4As]WCl_6$, 3, 334

$AsWC_{28}H_{21}N_4OS_4$ $[Ph_4As][W(OH)(SCN)_4]$, 3, 334

$AsWC_{36}H_{20}F_{18}S_6$ $[Ph_4As]\{W[SC(CF_3){=}C\text{-}(CF_3)S]_3\}$, 3, 334

$AsZnC_{13}H_{13}Cl_2$ $[Zn(Ph_2AsMe)Cl_2]$, 3, 57

$AsZrC_6H_5O_4$ $[ZrO][O_3AsPh]$, 3, 400

$As_2AgC_{18}H_{26}BrS_2$ $[Ag(2-MeSC_6H_4AsMe_2)_2Br]$, 3, 46

$As_2AgC_{18}H_{26}ClS_2$ $[Ag(2-MeSC_6H_4AsMe_2)_2Cl]$, 3, 46

$As_2AgC_{18}H_{26}IS_2$ $[Ag(2-MeSC_6H_4AsMe_2)_2I]$, 3, 46

$As_2AgC_{20}H_{32}IN_2$ $\{Ag[4-(Me_2N)C_6H_4AsMe_2]_2I\}$, 3, 49

$As_2AgC_{28}H_{48}IP_2$ $\{Ag[2-(Et_2As)C_6H_4PEt_2]_2\}I$, 3, 60

$As_2Ag_2C_{16}H_{24}Cl_2N_2$ $[Ag(Me_2AsCH_2C{=}CHCH{=}CHCH{=}N)_2][AgCl_2]$, 3, 42

$As_2Ag_2C_{28}H_{48}Cl_2P_2$ $\{Ag[2-(Et_2As)C_6H_4PEt_2]_2\}[AgCl_2]$, 3, 60

$As_2Ag_2C_{28}H_{48}I_2P_2$ $\{Ag[2-(Et_2As)C_6H_4PEt_2]_2\}[AgI_2]$, 3, 61

$As_2Al_2C_{28}H_{32}$ $(Ph_2AsAlMe_2)_2$, 3, 298

$As_2AuC_{28}H_{48}ClP_2$ $\{Au[2-(Et_2As)C_6H_4PEt_2]_2\}Cl$, 3, 59

$As_2AuC_{28}H_{48}IP_2$ $\{Au[2-(Et_2As)C_6H_4PEt_2]_2\}I$, 3, 59

$As_2AuC_{28}H_{49}O_4P_2S$ $\{Au[2-(Et_2As)C_6H_4PEt_2]_2\}OSO_3H$, 3, 60

$As_2AuC_{32}H_{53}O_6P_2$ $\{Au[2-(Et_2As)C_6H_4PEt_2]_2\}OC(O)CHOHCHOHCO_2H$, 3, 60

$As_2AuC_{34}H_{50}N_3O_7P_2$ $\{Au[2-(Et_2As)C_6H_4PEt_2]_2\}OC_6H_2(NO_2)_3$, 3, 59

$As_2AuC_{38}H_{62}BrO_4P_2S$ $\{Au[2-(Et_2As)C_6H_4PEt_2]_2\}$ (+)-bromocamphorsulfonate, 3, 60

$As_2AuC_{38}H_{62}NO_3P_2$ $\{Au[2-(Et_2As)C_6H_4PEt_2]_2\}$ (+)-camphornitronate, 3, 60

$As_2AuC_{38}H_{63}O_4P_2S$ $\{Au[2-(Et_2As)C_6H_4PEt_2]_2\}$ (+)-camphorsulfonate, 3, 60

$As_2AuC_{40}H_{69}O_3P_2$ $\{Au[2-(Et_2As)C_6H_4PEt_2]_2\}$ (-)-menthoxyacetate, 3, 60

$As_2AuC_{43}H_{61}NO_3P_2$ $\{Au[2-(Et_2As)C_6H_4PEt_2]_2\}$ (-)-N-1-phenylethylphthalamate, 3, 60

$As_2AuC_{46}H_{72}NO_3P_2$ $\{Au[2-(Et_2As)C_6H_4PEt_2]_2\}$ (-)-menthylphthalamate, 3, 60

$As_2Au_2C_{22}H_{24}Cl_2$ $\{Au_2[o-C_6H_4(C_6H_4AsMe_2)_2]Cl_2\}$, 3, 101

$As_2BC_6H_{20}F_6P$ $[H_2B(AsMe_3)_2]PF_6$, 3, 4

$As_2B_{10}C_{12}H_{42}$ $B_{10}H_{12}(AsEt_3)_2$, 3, 7

$As_2B_{10}C_{36}H_{42}$ $B_{10}H_{12}(AsPh_3)_2$, 3, 19

$As_2B_{12}C_6H_{28}$ $B_{12}H_{10}(AsMe_3)_2$, 3, 4

$As_2BaC_4H_{12}O_4$ $[Me_2AsO_2)_2Ba$, 3, 374

$As_2BaC_{12}H_{14}N_2O_6$ $[4-H_2NC_6H_4AsO(OH)O]_2Ba$, 3, 428

$As_2BaC_{12}H_{24}N_2O_{11}$ $[4-H_2NC_6H_4AsO(OH)O]_2Ba \cdot 5H_2O$, 3, 428

$As_2BiC_{10}H_{16}Br_3$ $[Bi(2-Me_2AsC_6H_4AsMe_2)Br_3]$, 3, 80

$As_2BiC_{10}H_{16}Cl_3$ $[Bi(2-Me_2AsC_6H_4AsMe_2)Cl_3]$, 3, 80

$As_2BiC_{10}H_{16}I_3$ $[Bi(2-Me_2AsC_6H_4AsMe_2)I_3]$, 3, 80

$As_2Bi_2C_{28}H_{32}I_2O_3$ $[Me_2Ph_2As]_2[Bi_2O_3I_2]$, 3, 319

$As_2CaC_4H_{12}O_4$ $(Me_2AsO_2)_2Ca$, 3, 374

$As_2CdC_4H_{12}O_4$ $(Me_2AsO_2)_2Cd$, 3, 374

$As_2CdC_{10}H_{16}Br_2$ $[Cd(2-Me_2AsC_6H_4AsMe_2)Br_2]$, 3, 81

$As_2CdC_{10}H_{16}Cl_2$ $[Cd(2-Me_2AsC_6H_4AsMe_2)Cl_2]$, 3, 81

$As_2CdC_{10}H_{16}I_2$ $[Cd(2-Me_2AsC_6H_4AsMe_2)I_2]$, 3, 81

$As_2CdC_{12}H_{30}I_2$ $[Cd(AsEt_3)_2I_2]$, 3, 7

$As_2CdC_{18}H_{28}I_4$ $[Me_3PhAs]_2CdI_4$, 3, 314

$As_2CdC_{18}H_{42}I_2$ $[Cd(AsPr_3)_2I_2]$, 3, 12

$As_2CdC_{20}H_{32}I_4$ $[Me_3(4-MeC_6H_4)As]_2CdI_4$, 3, 316

$As_2CdC_{28}H_{26}Br_2$ (1-Phenylarsindoline)$_2$·$CdBr_2$ complex, 3, 556

$As_2CdC_{28}H_{32}I_4$ $[Me_2Ph_2As]_2CdI_4$, 3, 319

$As_2CdC_{32}H_{40}I_4$ $[Me_2(4-MeC_6H_4)_2As]_2CdI_4$, 3, 320

$As_2CdC_{38}H_{36}I_4$ $[MePh_3As]_2CdI_4$, 3, 324

$As_2CdC_{44}H_{48}I_4$ $[Me(4-MeC_6H_4)_3As]_2CdI_4$, 3, 329

$As_2Cd_2C_{12}H_{30}Br_4$ $[Cd(AsEt_3)Br_2]_2$, 3, 7

$As_2Cd_2C_{12}H_{30}I_4$ $[Cd(AsEt_3)I_2]_2$, 3, 7

$As_2Cd_2C_{18}H_{42}I_4$ $[Cd(AsPr_3)I_2]_2$, 3, 12

$As_2CoC_{10}H_{24}S_2$ $Co[S(CH_2)_3AsMe_2]_2$, 3, 36

$As_2CoC_{12}H_{30}Br_2$ $[Co(AsEt_3)_2Br_2]$, 3, 7

$[Co(AsEt_3)_2Br_2]$, 3, 7

$As_2CoC_{12}H_{30}Cl_2$ $[Co(AsEt_3)_2Cl_2]$, 3, 7

$As_2CoC_{12}H_{30}I_2$ $[Co(AsEt_3)_2I_2]$, 3, 7

$As_2CoC_{18}H_{26}Cl_2O_8S_2$ $[Co(2-MeSC_6H_4-$

$AsMe_2)_2](ClO_4)_2$, 3, 44
$As_2CoC_{18}H_{26}I_2S_2$ $[Co(2-MeSC_6H_4AsMe_2)_2I_2]$, 3, 44
$As_2CoC_{22}H_{28}N_4S_4$ $[Me_3PhAs]_2[Co(SCN)_4]$, 3, 315
$As_2CoC_{24}H_{32}N_4S_4$ $[Me_3(4-MeC_6H_4)As]_2[Co(SCN)_4]$, 3, 316
$As_2CoC_{32}H_{32}N_4S_4$ $[Me_2Ph_2As]_2[Co(SCN)_4]$, 3, 319
$As_2CoC_{34}H_{52}N_5S_4$ $[Bu_4N]\{Co[SC(CN)=C(CN)S]_2(2-Me_2AsC_6H_4AsMe_2)\}$, 3, 83
$As_2CoC_{36}H_{30}Br_2O_2$ $[Co(Ph_3AsO)_2Br_2]$, 3, 353
$As_2CoC_{36}H_{30}Cl_2O_2$ $[Co(Ph_3AsO)_2Cl_2]$, 3, 353
$As_2CoC_{36}H_{30}I_2$ $Co(AsPh_3)_2I_2$, 3, 19
$As_2CoC_{36}H_{30}I_2O_2$ $[Co(Ph_3AsO)_2I_2]$, 3, 353
$As_2CoC_{36}H_{30}N_2O_8$ $[Co(Ph_3AsO)_2(NO_3)_2]$, 3, 353
$As_2CoC_{36}H_{40}N_4S_4$ $[Me_2(4-MeC_6H_4)_2As][Co(SCN)_4]$, 3, 320
$As_2CoC_{37}H_{30}NO_2$ $[Co(AsPh_3)_2(CO)(NO)]$, 3, 20
$As_2CoC_{38}H_{30}N_2O_2S_2$ $[Co(Ph_3AsO)_2(NCS)_2]$, 3, 353
$As_2CoC_{38}H_{36}Br_4$ $[MePh_3As]_2CoBr_4$, 3, 324
$As_2CoC_{38}H_{36}Cl_4$ $[MePh_3As]_2CoCl_4$, 3, 324
$As_2CoC_{38}H_{36}I_4$ $[MePh_3As]_2CoI_4$, 3, 324
$As_2CoC_{38}H_{36}N_4O_{12}$ $[MePh_3As]_2[Co(NO_3)_4]$, 3, 324
$As_2CoC_{40}H_{64}NS_4$ $[Bu_4N]$[bis(toluenedithiolato)$Co(2-Me_2AsC_6H_4AsMe_2)$], 3, 83
$As_2CoC_{42}H_{36}N_4S_4$ $[MePh_3As]_2[Co(SCN)_4]$, 3, 324
$As_2CoC_{48}H_{40}N_{12}$ $[Ph_4As]_2[Co(N_3)_4]$, 3, 335
$As_2CoC_{48}H_{48}N_4S_4$ $[Me(4-MeC_6H_4)_3As]_2[Co(SCN)_4]$, 3, 329
$As_2CoC_{52}H_{40}N_4S_4$ $[Ph_4As]_2[Co(SCN)_4]$, 3, 335
$As_2CoC_{52}H_{40}N_4Se_4$ $[Ph_4As]_2[Co(NCSe)_4]$, 3, 335
$As_2CoC_{56}H_{40}F_{12}O_8$ $[Ph_4As]_2[Co(O_2CCF_3)_4]$, 3, 335
$As_2CoC_{56}H_{40}N_4S_4$ $[Ph_4As]_2\{Co[SC(CN)=C(CN)S]_2\}$, 3, 335
$As_2Co_2C_{12}H_{30}Cl_4$ $[Co(AsEt_3)Cl_2]_2$, 3, 7
$As_2Co_2C_{12}H_{30}I_4$ $[Co(AsEt_3)I_2]_2$, 3, 8
$As_2Co_2C_{18}H_{26}Br_4S_2$ $[Co(2-MeSC_6H_4AsMe_2)_2][CoBr_4]$, 3, 44
$As_2Co_2C_{18}H_{26}Cl_4S_2$ $[Co(2-MeSC_6H_4AsMe_2)_2][CoCl_4]$, 3, 44
$As_2Co_2C_{18}H_{26}I_4S_2$ $[Co(2-MeSC_6H_4AsMe_2)_2][CoI_4]$, 3, 44
$As_2Co_2C_{40}H_{40}N_2O_2$ $[Co(AsPh_3)(NO)SEt]_2$, 3, 20
$As_2Co_2C_{42}H_{30}O_6$ $[Co(AsPh_3)(CO)_3]_2$, 3, 20
$As_2Co_2C_{43}H_{30}O_7$ $[Co(AsPh_3)_2(CO)_3][Co(CO)_4]$, 3, 20
$As_2Co_2C_{44}H_{48}N_2O_2S_2$ $[Co(AsPh_3)(NO)SCMe_3]_2$, 3, 20
$As_2Co_2C_{48}H_{40}N_2O_2S_2$ $[Co(AsPh_3)(NO)SPh]_2$, 3, 20
$As_2CrC_{10}H_{18}Br_3O$ $[Cr(2-Me_2AsC_6H_4AsMe_2)(H_2O)Br_3]$, 3, 81
$As_2CrC_{10}H_{18}Cl_3O$ $[Cr(2-Me_2AsC_6H_4AsMe_2)(H_2O)Cl_3]$, 3, 81
$As_2CrC_{12}H_{24}Cl_3O_2$ $[Cr(2-Me_2AsC_6H_4AsMe_2)(H_2O)(EtOH)Cl_2]Cl$, 3, 81
$As_2CrC_{14}H_{16}O_4$ $[Cr(2-Me_2AsC_6H_4AsMe_2)(CO)_4]$, 3, 82
$As_2CrC_{38}H_{36}O_6$ $[MePh_3As]_2CrO_6$, 3, 324
$As_2CrC_{60}H_{40}F_{18}S_6$ $[Ph_4As]_2\{Cr[SC(CF_3)=C(CF_3)S]_3\}$, 3, 334
$As_2Cr_2C_{12}H_{12}O_8$ $\{(OC)_4Cr \leftarrow As(Me)_2-As(Me)_2 \rightarrow Cr(CO)_4\}_n$, 3, 299
$As_2Cr_2C_{14}H_{12}O_{10}$ $(OC)_5Cr \leftarrow As(Me)_2-As(Me)_2 \rightarrow Cr(CO)_5$, 3, 300
$As_2Cr_2C_{38}H_{36}O_7$ $[MePh_3As]_2Cr_2O_7$, 3, 324
$As_2CuC_{12}H_{30}ClO_4S_2$ $[Cu(Me_2AsCH_2CH_2CH_2SMe)_2]ClO_4$, 1, 37
$As_2CuC_{16}H_{22}Cl$ $Cu(Me_2AsPh)_2Cl$, 3, 40
$As_2CuC_{16}H_{22}I$ $Cu(Me_2AsPh)_2I$, 3, 40
$As_2CuC_{16}H_{36}O_4$ $(Bu_2AsO)_2Cu$, 3, 375
$As_2CuC_{18}H_{26}BrS_2$ $[Cu(2-MeSC_6H_4AsMe_2)_2]Br$, 3, 44
$As_2CuC_{18}H_{26}ClO_4S_2$ $[Cu(2-MeSC_6H_4AsMe_2)_2]ClO_4$, 3, 44
$As_2CuC_{18}H_{26}ClS_2$ $[Cu(2-MeSC_6H_4AsMe_2)_2]$-

Cl, 3, 44
$As_2CuC_{18}H_{26}IS_2$ [Cu(2-MeSC$_6$H$_4$AsMe$_2$)$_2$]I 3, 44
$As_2CuC_{28}H_{48}BrP_2$ {Cu[2-(Et$_2$As)C$_6$H$_4$P-Et$_2$]$_2$}Br·1/2 EtCOMe, 3, 59
$As_2CuC_{28}H_{48}ClP_2$ {Cu[2-(Et$_2$As)C$_6$H$_4$P-Et$_2$]$_2$}Cl, 3, 59
$As_2CuC_{28}H_{48}IP_2$ {Cu[2-(Et$_2$As)C$_6$H$_4$P-Et$_2$]$_2$}I, 3, 59
$As_2CuC_{36}H_{30}Br_2O_2$ [Cu(Ph$_3$AsO)$_2$Br$_2$], 3, 353
$As_2CuC_{36}H_{30}Cl_2$ [Cu(Ph$_3$AsO)$_2$Cl$_2$], 3, 353
$As_2CuC_{38}H_{36}Br_4$ [MePh$_3$As]$_2$CuBr$_4$, 3, 325
$As_2CuC_{38}H_{36}Cl_4$ [MePh$_3$As]$_2$CuCl$_4$, 3, 325
$As_2CuC_{38}H_{36}N_4O_{12}$ [MePh$_3$As]$_2$[Cu(NO$_3$)$_4$], 3, 325
$As_2CuC_{38}H_{63}O_4P_2S$ {Cu[2-(Et$_2$As)C$_6$H$_4$-PEt$_2$]$_2$} (+)-camphorsulfonate, 3, 59
$As_2CuC_{42}H_{36}N_4S_4$ [MePh$_3$As]$_2$[Cu(SCN)$_4$], 3, 324
$As_2CuC_{48}H_{40}Cl_3$ [Ph$_4$As]$_2$CuCl$_3$, 3, 335
$As_2Cu_2C_{12}H_{30}Br_2S_2$ [Cu(Me$_2$AsCH$_2$CH$_2$CH$_2$-SMe)$_2$][CuBr$_2$], 3, 37
$As_2Cu_2C_{12}H_{30}Cl_2S_2$ [Cu(Me$_2$AsCH$_2$CH$_2$CH$_2$-SMe)$_2$][CuCl$_2$], 3, 37
$As_2Cu_2C_{12}H_{30}I_2S_2$ [Cu(Me$_2$AsCH$_2$CH$_2$CH$_2$-SMe)$_2$][CuI$_2$], 3, 37
$As_2Cu_2C_{16}H_{24}Cl_2N_2$ [Cu(Me$_2$AsCH$_2$-C=CHCH=CHCH=N)$_2$][CuCl$_2$], 3, 42
$As_2Cu_2C_{18}H_{26}Br_2S_2$ [Cu(2-MeSC$_6$H$_4$-AsMe$_2$)$_2$][CuBr$_2$], 3, 44
$As_2Cu_2C_{18}H_{26}Cl_2S_2$ [Cu(2-MeSC$_6$H$_4$-AsMe$_2$)$_2$][CuCl$_2$], 3, 44
$As_2Cu_2C_{18}H_{26}I_2S_2$ [Cu(2-MeSC$_6$H$_4$-AsMe$_2$)$_2$][CuI$_2$], 3, 44
$As_2Cu_2C_{20}H_{26}N_2S_4$ [Cu(2-MeSC$_6$H$_4$-AsMe$_2$)$_2$][Cu(SCN)$_2$], 3, 44
$As_2Cu_2C_{28}H_{48}Br_2P_2$ {Cu[2-(Et$_2$As)C$_6$H$_4$-PEt$_2$]$_2$}[CuBr$_2$], 3, 59
$As_2Cu_2C_{28}H_{48}Cl_2P_2$ {Cu[2-(Et$_2$As)C$_6$H$_4$-PEt$_2$]$_2$}[CuCl$_2$], 3, 59
$As_2Cu_3C_{28}H_{48}Cl_3P_2$ {Cu$_3$[2-(Et$_2$As)C$_6$H$_4$-PEt$_2$]$_2$Cl$_3$}, 3, 59
$As_2FeC_{12}H_{16}Br_2O_2$ [Fe(2-Me$_2$AsC$_6$H$_4$-AsMe$_2$)(CO)$_2$Br$_2$], 3, 85
$As_2FeC_{12}H_{16}IO_2$ [Fe(2-Me$_2$AsC$_6$H$_4$AsMe$_2$)-(CO)$_2$I], 3, 85
$As_2FeC_{12}H_{16}I_2O_2$ [Fe(2-Me$_2$AsC$_6$H$_4$AsMe$_2$)-(CO)$_2$I$_2$], 3, 85
$As_2FeC_{13}H_{16}O_3$ [Fe(2-Me$_2$AsC$_6$H$_4$AsMe$_2$)-(CO)$_3$], 3, 85
$As_2FeC_{29}H_{24}O_3$ [Fe(Ph$_2$AsCH$_2$-)$_2$(CO)$_3$], 3, 102
$As_2FeC_{36}H_{30}Cl_3$ [Fe(AsPh$_3$)$_2$Cl$_3$], 3, 21
$As_2FeC_{36}H_{30}Cl_3O_2$ [Fe(Ph$_3$AsO)$_2$Cl$_3$], 3, 353
$As_2FeC_{36}H_{30}N_2O_2$ [Fe(AsPh$_3$)$_2$(NO)$_2$], 3, 21
$As_2FeC_{36}H_{30}N_3O_9$ [Fe(AsPh$_3$)$_2$(NO$_3$)$_3$], 3, 21
$As_2FeC_{38}H_{30}Br_2O_2$ [Fe(AsPh$_3$)$_2$(CO)$_2$Br$_2$], 3, 21
$As_2FeC_{38}H_{30}Cl_2O_2$ [Fe(AsPh$_3$)$_2$(CO)$_2$Cl$_2$], 3, 21
$As_2FeC_{38}H_{36}Br_4$ [MePh$_3$As]$_2$FeBr$_4$, 3, 325
$As_2FeC_{38}H_{36}Cl_4$ [MePh$_3$As]$_2$FeCl$_4$, 3, 325
$As_2FeC_{38}H_{36}I_4$ [MePh$_3$As]$_2$FeI$_4$, 3, 325
$As_2FeC_{39}H_{30}O_3$ [Fe(AsPh$_3$)$_2$(CO)$_3$], 3, 21
$As_2Fe_2C_4F_{12}N_4O_4$ [Fe(NO)$_2$As(CF$_3$)$_2$]$_2$, 3, 302
$As_2Fe_2C_4H_{12}N_4O_4$ [Fe(NO)$_2$AsMe$_2$]$_2$, 3, 303
$As_2Fe_2C_{10}F_{12}O_6$ [Fe(CO)$_3$As(CF$_3$)$_2$]$_2$, 3, 303
$As_2Fe_2C_{10}H_{12}Br_2O_6$ [Fe(CO)$_3$(AsMe$_2$)Br]$_2$, 3, 303
$As_2Fe_2C_{10}H_{12}I_2O_6$ [Fe(CO)$_3$(AsMe$_2$)I]$_2$, 3, 303
$As_2Fe_2C_{10}H_{12}O_6$ [Fe(CO)$_3$AsMe$_2$]$_2$, 3, 303
$As_2Fe_2C_{16}H_{10}F_{12}O_2$ [C$_5$H$_5$Fe(CO)As(CF$_3$)$_2$]$_2$, 3, 304*
$As_2Fe_2C_{16}H_{22}O_2$ [C$_5$H$_5$Fe(CO)AsMe$_2$]$_2$, 3, 304
$As_2Fe_2C_{36}H_{30}O_2$ [C$_5$H$_5$Fe(CO)(AsPh$_2$)]$_2$, 1, 215
$As_2Ga_2C_{28}H_{32}$ (Ph$_2$AsGaMe$_2$)$_2$, 3, 298
$As_2GeC_{42}H_{34}N_2$ [As(Ph)$_2$=N-Ge(Ph)$_2$-N=As-(Ph)$_2$-C$_6$H$_4$]$_n$, 2, 128*, 3, 369
$As_2Ge_2C_{47}H_{88}N_2$ [Bu$_3$GeN=As(Et)$_2$-C$_6$H$_4$]$_2$-

CMe$_2$, 2, 128, 3, 369
$As_2HfC_{20}H_{32}Cl_4$ [Hf(2-Me$_2$AsC$_6$H$_4$As-Me$_2$)$_2$Cl$_4$], 3, 85
$As_2HgC_{10}H_{16}Br_2$ [Hg(2-Me$_2$AsC$_6$H$_4$AsMe$_2$)-Br$_2$], 3, 87
$As_2HgC_{10}H_{16}Cl_2$ [Hg(2-Me$_2$AsC$_6$H$_4$AsMe$_2$)-Cl$_2$], 3, 87
$As_2HgC_{10}H_{16}I_2$ [Hg(2-Me$_2$AsC$_6$H$_4$AsMe$_2$)-I$_2$], 3, 87
$As_2HgC_{12}H_{14}N_2O_6$ [4-H$_2$NC$_6$H$_4$As(O)(OH)-O]$_2$Hg, 3, 428
$As_2HgC_{14}H_{36}I_4$ [Et$_3$MeAs]$_2$HgI$_4$, 3, 313
$As_2HgC_{16}H_{22}Br_2$ Hg(Me$_2$AsPh)$_2$Br$_2$, 3, 41
$As_2HgC_{16}H_{22}I_2$ Hg(Me$_2$AsPh)$_2$I$_2$, 3, 41
$As_2HgC_{16}H_{40}I_4$ [Et$_4$As]$_2$HgI$_4$, 3, 314
$As_2HgC_{20}H_{32}Br_2N_2$ {Hg[4-(Me$_2$N)C$_6$H$_4$-AsMe$_2$]$_2$Br$_2$}, 3, 49
$As_2HgC_{20}H_{32}Cl_2N_2$ {Hg[4-(Me$_2$N)C$_6$H$_4$-AsMe$_2$]$_2$Cl$_2$}, 3, 49
$As_2HgC_{20}H_{32}I_2N_2$ {Hg[4-(Me$_2$N)C$_6$H$_4$-AsMe$_2$]$_2$I$_2$}, 3, 49
$As_2HgC_{26}H_{26}Br_2$ [Hg(Ph$_2$AsMe)$_2$Br$_2$], 3, 54
$As_2HgC_{26}H_{26}I_2$ [Hg(Ph$_2$AsMe)$_2$I$_2$], 3, 54
$As_2HgC_{36}H_{30}Br_2$ Hg(AsPh$_3$)$_2$Br$_2$, 3, 23
$As_2HgC_{36}H_{30}Cl_2$ Hg(AsPh$_3$)$_2$Cl$_2$, 3, 23
$As_2HgC_{36}H_{30}Cl_2O_2$ [Hg(Ph$_3$AsO)$_2$Cl$_2$], 3, 354
$As_2HgC_{36}H_{30}I_2$ Hg(AsPh$_3$)$_2$I$_2$, 3, 23
$As_2HgC_{38}H_{36}I_4$ [MePh$_3$As]$_2$HgI$_4$, 3, 326
$As_2HgC_{48}H_{40}Cl_4$ [Ph$_4$As]$_2$HgCl$_4$, 3, 335
$As_2Hg_2C_4H_{12}Cl_4O$ (Me$_2$As)$_2$O·2HgCl$_2$, 3, 213
$As_2Hg_2C_{12}H_{30}Cl_4$ [Hg(AsEt$_3$)Cl$_2$]$_2$, 3, 8
$As_2Hg_2C_{12}H_{30}I_4$ [Hg(AsEt$_3$)I$_2$], 3, 8
$As_2Hg_2C_{14}H_{30}Br_4$ {Hg[Me$_2$As(CH$_2$)$_3$-CH=CH$_2$]Br$_2$}$_2$, 3, 38
$As_2Hg_2C_{14}H_{30}I_4$ {Hg[Me$_2$As(CH$_2$)$_3$-CH=CH$_2$]I$_2$}$_2$, 3, 38
$As_2Hg_2C_{18}H_{42}Br_4$ [Hg(AsPr$_3$)Br$_2$]$_2$, 3, 12
$As_2Hg_2C_{18}H_{42}I_4$ [Hg(AsPr$_3$)I$_2$]$_2$, 3, 12
$As_2Hg_2C_{20}H_{32}I_4N_2$ {Hg[4-(Me$_2$N)C$_6$H$_4$-AsMe$_2$]I$_2$}$_2$, 3, 49
$As_2Hg_2C_{24}H_{54}Br_4$ [Hg(AsBu$_3$)Br$_2$]$_2$, 3, 15
$As_2Hg_2C_{24}H_{54}I_4$ [Hg(AsBu$_3$)I$_2$]$_2$, 3, 15
$As_2Hg_2C_{36}H_{28}Br_2Cl_4$ [Hg(Ph$_2$AsC$_6$H$_4$Br-4)-Cl$_2$]$_2$, 3, 62
$As_2Hg_2C_{36}H_{30}Br_4$ [Hg(AsPh$_3$)Br$_2$]$_2$, 3, 23
$As_2Hg_2C_{36}H_{30}Cl_4$ [Hg(AsPh$_3$)Cl$_2$]$_2$, 3, 23
$As_2Hg_2C_{36}H_{30}Cl_4O_2$ [Hg(Ph$_3$AsO)Cl$_2$]$_2$, 3, 354
$As_2Hg_2C_{38}H_{34}Cl_4$ [Hg(4-MeC$_6$H$_4$AsPh$_2$)-Cl$_2$]$_2$, 3, 62
$As_2Hg_2C_{42}H_{42}Cl_4O_6$ {Hg[As(C$_6$H$_4$OMe-3)$_3$]-Cl$_2$}$_2$, 3, 28
$As_2Hg_2C_{64}H_{54}Cl_4$ {Hg[(2-PhC$_6$H$_4$CH$_2$)$_2$-AsPh]Cl$_2$}$_2$, 3, 72
$As_2Hg_3C_{12}H_{30}I_6$ [(Et$_3$As)$_2$(HgI$_2$)$_3$], 3, 9
$As_2Hg_3C_{18}H_{42}Cl_6$ [(Pr$_3$As)$_2$(HgCl$_2$)$_3$], 3, 13
$As_2Hg_3C_{24}H_{54}Br_6$ [(Bu$_3$As)$_2$(HgBr$_2$)$_3$], 3, 15
$As_2Hg_3C_{24}H_{54}I_6$ [(Bu$_3$As)$_2$(HgI$_2$)$_3$], 3, 15
$As_2Hg_4C_{12}H_{30}Cl_8$ [(Et$_3$As)$_2$(HgCl$_2$)$_4$], 3, 9
$As_2Hg_4C_{32}H_{28}O_9$ [(4-AcOHgC$_6$H$_4$)$_2$As]$_2$O, 3, 218
$As_2In_2C_{28}H_{32}$ (Ph$_2$AsInMe$_2$)$_2$, 3, 298
$As_2IrC_{13}H_{30}Cl_3O$ [Ir(AsEt$_3$)$_2$(CO)Cl$_3$], 3, 8
$As_2IrC_{16}H_{23}Br_4$ H[Ir(Me$_2$AsPh)$_2$Br$_4$], 3, 40
$As_2IrC_{16}H_{23}Cl_4$ H[Ir(Me$_2$AsPh)$_2$Cl$_4$], 3, 40
$As_2IrC_{16}H_{26}Br_4N$ NH$_4$[Ir(Me$_2$AsPh)$_2$Br$_4$], 3, 40
$As_2IrC_{16}H_{26}Cl_4N$ NH$_4$[Ir(Me$_2$AsPh)$_2$Cl$_4$], 3, 40
$As_2IrC_{18}H_{26}Br_3S_2$ [Ir(2-MeSC$_6$H$_4$AsMe$_2$)$_2$-Br$_3$], 3, 45
$As_2IrC_{18}H_{26}Cl_3S_2$ [Ir(2-MeSC$_6$H$_4$AsMe$_2$)$_2$-Cl$_3$], 3, 45
$As_2IrC_{18}H_{26}I_3S_2$ [Ir(2-MeSC$_6$H$_4$AsMe$_2$)$_2$-I$_3$], 3, 45
$As_2IrC_{21}H_{28}Cl_4N$ Py·H[Ir(Me$_2$AsPh)$_2$Cl$_4$], 3, 40
$As_2IrC_{21}H_{30}Cl_3O$ [Ir(Et$_2$AsPh)$_2$(CO)Cl$_3$],

$As_2IrC_{21}H_{30}Cl_3O$
3, 47
$As_2IrC_{26}H_{27}Br_4$ $H[Ir(Ph_2AsMe)_2Br_4]$, 3, 53
$As_2IrC_{26}H_{27}Cl_4$ $H[Ir(Ph_2AsMe)_2Cl_4]$, 3, 53
$As_2IrC_{37}H_{30}I_3O$ $[Ir(AsPh_3)_2(CO)I_3]$, 3, 20
$As_2IrC_{37}H_{31}Cl_2O$ $[IrH(AsPh_3)_2(CO)Cl_2]$, 3, 20
$As_2IrC_{44}H_{42}BrO_2$ $\{Ir[As(C_6H_4Me\text{-}4)_3]_2(CO)_2Br\}$, 3, 28
$As_2IrC_{48}H_{40}Cl_6$ $[Ph_4As]_2IrCl_6$, 3, 335
$As_2IrC_{50}H_{40}Br_2O_2$ $[Ph_4As]_2[Ir(CO)_2Br_2]$, 3, 335
$As_2IrC_{52}H_{40}Br_3O_4$ $[Ph_4As]_2[Ir(CO)_4Br_3]$, 3, 335
$As_2Li_2C_{17}H_{20}$ $PhAs(Li)(CH_2)_5As(Li)Ph$, 3, 295
$As_2MgC_{12}H_{24}N_2O_{11}$ $[4\text{-}H_2NC_6H_4AsO(OH)O]_2Mg \cdot 5H_2O$, 3, 428
$As_2MnC_9H_{18}BrO_3$ $[Mn(AsMe_3)_2(CO)_3Br]$, 3, 4
$As_2MnC_{10}H_{16}Cl_3$ $[Mn(2\text{-}Me_2AsC_6H_4AsMe_2)Cl_3]$, 3, 86
$As_2MnC_{12}H_{16}Br_2O_2$ $[Mn(2\text{-}Me_2AsC_6H_4AsMe_2)(CO)_2Br_2]$, 3, 87
$As_2MnC_{12}H_{16}O_2$ $[Mn(2\text{-}Me_2AsC_6H_4AsMe_2)(CO)_2]$, 3, 86
$As_2MnC_{13}H_{16}BrO_3$ $[Mn(2\text{-}Me_2AsC_6H_4AsMe_2)(CO)_3Br]$, 3, 86
$As_2MnC_{13}H_{16}IO_3$ $[Mn(2\text{-}Me_2AsC_6H_4AsMe_2)(CO)_3I]$, 3, 86
$As_2MnC_{13}H_{16}O_3$ $[Mn(2\text{-}Me_2AsC_6H_4AsMe_2)(CO)_3]$, 3, 86
$As_2MnC_{18}H_{23}O_2$ $(MeC_5H_4)Mn(CO)_2[o\text{-}C_6H_4(AsMe_2)_2]$, 1, 170
$As_2MnC_{36}H_{30}Br_2$ $[Mn(AsPh_3)_2Br_2]$, 3, 22
$As_2MnC_{36}H_{30}Br_2O_2$ $[Mn(Ph_3AsO)_2Br_2]$, 3, 354
$As_2MnC_{36}H_{30}Cl_2$ $[Mn(AsPh_3)_2Cl_2]$, 3, 22
$As_2MnC_{36}H_{30}Cl_2O_2$ $[Mn(Ph_3AsO)_2Cl_2]$, 3, 353
$As_2MnC_{36}H_{30}IN_2O_2$ $[Mn(AsPh_3)_2(NO)_2I]$, 3, 22
$As_2MnC_{36}H_{30}I_2$ $[Mn(AsPh_3)_2I_2]$, 3, 22
$As_2MnC_{38}H_{36}Br_4$ $[MePh_3As]_2MnBr_4$, 3, 325
$As_2MnC_{38}H_{36}Cl_4$ $[MePh_3As]_2MnCl_4$, 3, 325
$As_2MnC_{38}H_{36}I_4$ $[MePh_3As]_2MnI_4$, 3, 326
$As_2MnC_{38}H_{36}N_4O_{12}$ $[MePh_3As]_2[Mn(NO_3)_4]$, 3, 326
$As_2MnC_{39}H_{39}BrO_3$ $[Mn(AsPh_3)_2(CO)_3]Br$, 3, 23
$As_2MnC_{39}H_{30}ClO_3$ $[Mn(AsPh_3)_2(CO)_3]Cl$, 3, 23
$As_2MnC_{39}H_{30}IO_3$ $[Mn(AsPh_3)_2(CO)_3]I$, 3, 23
$As_2MnC_{45}H_{42}IO_3$ $Mn[As(CH_2Ph)_3]_2(CO)_3I$, 3, 27
$As_2MnC_{48}H_{40}Br_2I_2$ $[Ph_4As]_2MnBr_2I_2$, 3, 335
$As_2MnC_{48}H_{40}Br_4$ $[Ph_4As]_2MnBr_4$, 3, 335
$As_2MnC_{48}H_{40}Cl_2I_2$ $[Ph_4As]_2MnCl_2I_2$, 3, 335
$As_2MnC_{48}H_{40}Cl_4$ $[Ph_4As]_2MnCl_4$, 3, 335
$As_2MnC_{48}H_{40}I_4$ $[Ph_4As]_2MnI_4$, 3, 335
$As_2MnC_{50}H_{87}O$ Fluorenyl-$Mn(CO)[As(C_6H_{13}\text{-}n)_3]_2$, 1, 170
$As_2MnC_{57}H_{66}IO_3$ $\{Mn[As(C_6H_2Me_3)_3]_2(CO)_3I\}$, 3, 29
$As_2MnC_{86}H_{171}O_7$ $(n\text{-}C_8H_{17}C_5H_4)Mn(CO)[As(OC_{12}H_{25})_3]_2$, 1, 170
$As_2Mn_2C_{12}F_{12}O_8$ $[Mn(CO)_4As(CF_3)_2]_2$, 3, 301
$As_2Mn_2C_{12}H_{12}O_8$ $[Mn(CO)_4AsMe_2]_2$, 3, 302
$As_2Mn_2C_{20}H_{30}O_8$ $[Mn(AsEt_3)(CO)_4]_2$, 3, 8
$As_2Mn_2C_{24}H_{26}O_4$ $\{[C_5H_5Mn(CO)_2]_2(2\text{-}Me_2AsC_6H_4AsMe_2)\}$, 3, 87
$As_2Mn_2C_{26}H_{30}O_4$ $\{[(MeC_5H_4)Mn(CO)_2]_2(2\text{-}Me_2AsC_6H_4AsMe_2)\}$, 3, 87
$As_2Mn_2C_{32}H_{20}O_8$ $[Mn(CO)_4AsPh_2]_2$, 3, 302
$As_2Mn_2C_{36}H_{28}O_8$ $[Mn(CO)_4As(C_6H_4Me\text{-}m)_2]_2$, 3, 302
$As_2MoC_{10}H_{16}Br_4$ $[Mo(2\text{-}Me_2AsC_6H_4AsMe_2)Br_4]$, 3, 87
$As_2MoC_{12}H_{16}I_2O_2$ $[Mo(2\text{-}Me_2AsC_6H_4AsMe_2)(CO)_2I_2]$, 3, 87
$As_2MoC_{13}H_{16}Br_2O_3$ $[Mo(2\text{-}Me_2AsC_6H_4AsMe_2)(CO)_3Br_2]$, 3, 88
$As_2MoC_{13}H_{16}I_2O_3$ $[Mo(2\text{-}Me_2AsC_6H_4AsMe_2)(CO)_3I_2]$, 3, 88
$As_2MoC_{14}H_{16}O_4$ $[Mo(2\text{-}Me_2AsC_6H_4AsMe_2)$-

$(CO)_4]$, 3, 87

$As_2MoC_{16}H_{30}O_4$ $[Mo(AsEt_3)_2(CO)_4]$, 3, 9

$As_2MoC_{36}H_{30}Cl_2N_2O_2$ $[Mo(AsPh_3)_2(NO)_2$-$Cl_2]$, 3, 24

$As_2MoC_{36}H_{30}Cl_2O_4$ $[MoO_2(Ph_3AsO)_2Cl_2]$, 3, 354

$As_2MoC_{36}H_{30}Cl_3O_3$ $[MoO(Ph_3AsO)_2Cl_3]$, 3, 354

$As_2MoC_{40}H_{30}O_4$ $Mo(AsPh_3)_2(CO)_4$, 3, 23

$As_2MoC_{48}H_{40}Cl_4N_2O_2$ $[Ph_4As]_2[Mo(NO)_2$-$Cl_4]$, 3, 336

$As_2MoC_{60}H_{40}F_{18}S_6$ $[Ph_2As]_2\{Mo[SC$-$(CF_3)=C(CF_3)S]_3\}$, 3, 336

$As_2Mo_2C_{12}H_{12}O_8$ $[(OC)_4MoAsMe_2]_2$, 3, 300

$As_2Mo_2C_{14}H_{12}O_{10}$ $(OC)_5Mo \leftarrow As(Me)_2$-$As(Me)_2 \rightarrow Mo(CO)_5$, 3, 300

$As_2Mo_2C_{18}H_{10}F_{12}O_4$ $[C_5H_5Mo(CO)_2As$-$(CF_3)_2]_2$, 3, 300

$As_2Mo_2C_{18}H_{22}O_4$ $[C_5H_5Mo(CO)_2(AsMe_2)]_2$, 1, 102, 3, 301

$As_2Mo_2C_{32}H_{20}O_8$ $[Mo(CO)_4AsPh_2]_2$, 3, 301

$As_2Mo_6C_{36}H_{30}Cl_{12}O_2$ $\{[Ph_3AsO]_2[(Mo_6$-$Cl_8)Cl_4]\}$, 3, 354

$As_2Mo_6C_{36}H_{32}Cl_{14}O_2$ $\{H_2[(Ph_3AsO)_2\cdot$-$(Mo_6Cl_8)Cl_6]\}$, 3, 354

$As_2NaC_{13}H_{13}N_2O_4S$ $[4,3\text{-}HO(NaO_2SCH_2$-$NH)C_6H_3AsAsC_6H_3(NH_2)OH\text{-}3,4]_n$, 3, 144

$As_2NaC_{16}H_{19}N_2O_7S$ $[4,3\text{-}HO(HOCH_2CHOH$-$CH_2NH)C_6H_3As\text{-}AsC_6H_3(NHCH_2SO_3Na)$-$OH\text{-}3,4]_n$, 3, 153

$As_2NaC_{17}H_{18}N_3O_7S$ 3-[8-Acetamido-3-hydroxy-1,4-benzisoxazine-6-arseno)-4-hydroxyanilinomethanesulfonic acid Na salt, 3, 145

$As_2NaC_{17}H_{21}N_2O_6S$ $\{3,4\text{-}HO[(HOCH_2CH_2)_2$-$N]C_6H_3As\text{-}AsC_6H_3(NHCH_2SO_2Na)OH$-$4,3]_n$, 3, 154*

$As_2NaC_{18}H_{23}NO_8S$ $\{3,4\text{-}HO[HOCH_2CH_2(HO$-$CH_2CHOHCH_2)N]C_6H_3As\text{-}AsC_6H_3(NHCH_2$-$SO_3Na)OH\text{-}3,4]_n$, 3, 155

$As_2NaC_{18}H_{23}N_2O_6S$ $\{4\text{-}[HOCH_2CH_2(HOCH_2$-$CHOHCH_2)N]C_6H_4As\text{-}AsC_6H_3(NHCH_2SO_2$-$Na)OH\text{-}3,4\}_n$, 3, 155*

$As_2NaC_{18}H_{23}N_2O_7S$ 4,4′-Dihydroxy-3-[(2-hydroxyethyl)(2,3-dihydroxypropyl)amino]-3′-(sulfinomethylamino)arsenobenzene Na salt, 3, 146*

$As_2NaC_{18}H_{23}N_2O_8S$ 3,4′-Dihydroxy-3′-[(2-hydroxyethyl)(2,3-dihydroxypropyl)amino]-4-(sulfomethylamino)-arsenobenzene Na salt, 3, 146*

$As_2NaC_{19}H_{17}N_4O_7S$ 8-Acetamido-8′-sulfinomethylamino-6,6′-arsenobis-(3-hydroxybenzisoxazine) Na salt, 3, 146

$As_2NaC_{19}H_{19}N_4O_8S$ 5-(8-Acetamido-3-hydroxybenzisoxazin-6-ylarseno)-3-acetamido-2-hydroxyanilinomethanesulfonic acid Na salt, 3, 147*

$As_2NaC_{19}H_{25}N_2O_6S$ 3-Amino-4,4′-di(2-hydroxypropoxy)-3′-(sulfinomethylamino)arsenobenzene Na salt, 3, 147

3-Amino-4,4′-di(3-hydroxypropoxy)-3′-(sulfinomethylamino)arsenobenzene Na salt, 3, 147

$As_2NaC_{19}H_{25}N_2O_7S$ 3-Amino-4,4′-bis(3-hydroxypropoxy)-3′-(sulfomethylamino)arsenobenzene Na salt, 3, 148*

3-Amino-4,4′-bis(2-hydroxypropoxy)-3′-(sulfomethylamino)arsenobenzene Na salt, 3, 148

3,4′-Dihydroxy-4-[(2-hydroxyethyl)-(2,3-dihydroxypropyl)amino]-3′-(N-methyl-N-sulfinomethylamino)arsenobenzene Na salt, 3, 148*

4,4′-Dihydroxy-3-[(2-hydroxyethyl)-(2,3-dihydroxypropyl)amino]-3′-[N-methyl-N-(sulfinomethyl)amino]-arsenobenzene Na salt, 3, 148

$As_2NaC_{19}H_{25}N_2O_8S$ 4,4′-Dihydroxy-3-[bis(2,3-dihydroxypropyl)amino]-3′-(sulfinomethylamino)arsenobenzene Na salt, 3, 149

$\{3,4\text{-}HO[(HOCH_2CHOHCH_2)_2N]C_6H_3As\text{-}As$-$C_6H_3(NHCH_2SO_2Na)OH\text{-}4,3]_n$, 3, 155

$As_2NaC_{19}H_{25}N_2O_9S$ 4,4′-Dihydroxy-3-[(2-hydroxyethyl)(2,3-dihydroxypropyl)amino]-5-methoxy-3′-(sulfomethylamino)arsenobenzene Na salt,

3, 149

$As_2NaC_{20}H_{27}N_2O_8S$ 3,4′-Dihydroxy-4-[bis(2,3-dihydroxypropyl)amino]-3′-[N-methyl-N-(sulfinomethyl)-amino]arsenobenzene Na salt, 3, 148

$As_2NaC_{20}H_{27}N_2O_9S$ {4,3-HO[($HOCH_2CHOH$-CH_2)$_2$N]C_6H_3As-AsC_6H_2($NHCH_2SO_2Na$)-(OH)OMe-3,4,5}$_n$, 3, 155

$As_2NaC_{21}H_{28}N_3O_9S$ 3-Acetamido-4,4′-dihydroxy-5-[bis(2,3-dihydroxypropyl)amino]-3′-(sulfinomethylamino)arsenobenzene Na salt, 3, 150

$As_2NaC_{21}H_{28}N_3O_{10}S$ {3,4,5-AcNH(HO)-[($HOCH_2CHOHCH_2$)$_2$N]C_6H_2As-AsC_6H_3-($NHCH_2SO_3Na$)OH-4,3}$_n$, 3, 155

{3,4,5-AcNH(HO)[($HOCH_2CHOHCH_2$)$_2$N]-C_6H_2As-AsC_6H_3($NHCH_2SO_3Na$)OH-3,4}$_n$, 3, 155

$As_2NaC_{22}H_{31}N_2O_{10}S$ 4,4′-Dihydroxy-N,N,N′-tris(2,3-dihydroxypropyl)-N′-sulfinomethyl-3,3′-diaminoarsenobenzene Na salt, 3, 151

$As_2NaC_{22}H_{31}N_2O_{11}S$ 4,4′-Dihydroxy-N,N,N′-tris(2,3-dihydroxypropyl)-N′-sulfomethyl-3,3′-diaminoarsenobenzene Na salt, 3, 151

$As_2NaC_{25}H_{23}N_2O_8S_3$ N‴-(Hydroxymethyl)-3′,3″-arsenobis(6-hydroxysulfanilylaniline)sulfoxylate Na salt, 3, 152

$As_2Na_3C_{15}H_{15}N_2O_{11}S_3$ {4,3-HO(NaO_3SCH_2-NH)C_6H_3As-AsC_6H_3[N(CH_2SO_3Na)$_2$]-OH-3,4}$_n$, 3, 153

$As_2NeC_{12}H_{10}O_6$ Ne[O_3AsPh]$_2$, 3, 400

$As_2NiC_{10}H_{16}Br_2$ [Ni(2-$Me_2AsC_6H_4AsMe_2$)-Br_2], 3, 88

$As_2NiC_{10}H_{16}Br_3$ [Ni(2-$Me_2AsC_6H_4AsMe_2$)-Br_3], 3, 89

$As_2NiC_{10}H_{16}I_2$ [Ni(2-$Me_2AsC_6H_4AsMe_2$)-I_2], 3, 88

$As_2NiC_{10}H_{24}S_2$ Ni[S(CH_2)$_3$AsMe_2]$_2$, 3, 36

$As_2NiC_{12}H_{16}O_2$ [Ni(2-$Me_2AsC_6H_4AsMe_2$)-(CO)$_2$], 3, 88

$As_2NiC_{14}H_{30}O_2$ [Ni(AsEt_3)$_2$(CO)$_2$], 3, 9

$As_2NiC_{18}H_{26}Br_2S_2$ [Ni(2-$MeSC_6H_4AsMe_2$)$_2$-Br_2], 3, 45

$As_2NiC_{18}H_{26}I_2S_2$ [Ni(2-$MeSC_6H_4AsMe_2$)$_2$-I_2], 3, 45

$As_2NiC_{20}H_{32}Br_2N_2$ {Ni[4-(Me_2N)C_6H_4As-Me_2]$_2Br_2$}, 3, 49

$As_2NiC_{20}H_{32}I_2N_2$ {Ni[4-(Me_2N)C_6H_4As-Me_2]$_2I_2$}, 3, 49

$As_2NiC_{28}H_{24}O_2$ [Ni(Ph_2AsCH_2-)$_2$(CO)$_2$], 3, 102

$As_2NiC_{32}H_{24}O_2$ [Ni(2-$Ph_2AsC_6H_4AsPh_2$)-(CO)$_2$], 3, 103

$As_2NiC_{36}H_{30}BrNO$ [Ni(AsPh_3)$_2$(NO)Br], 3, 24

$As_2NiC_{36}H_{30}Br_2O_2$ [Ni(Ph_3AsO)$_2Br_2$], 3, 354

$As_2NiC_{36}H_{30}Cl_2O_2$ [Ni(Ph_3AsO)$_2Cl_2$], 3, 354

$As_2NiC_{36}H_{30}Cl_2O_{10}$ [Ni(Ph_3AsO)$_2$(ClO_4)$_2$], 3, 355

$As_2NiC_{36}H_{30}INO$ [Ni(AsPh_3)$_2$(NO)I], 3, 24

$As_2NiC_{36}H_{30}I_2O_2$ [Ni(Ph_3AsO)$_2I_2$], 3, 355

$As_2NiC_{36}H_{30}N_2O_8$ [Ni(Ph_3AsO)$_2$(NO_3)$_2$], 3, 355

$As_2NiC_{38}H_{30}O_2$ [Ni(AsPh_3)$_2$(CO)$_2$], 3, 24

$As_2NiC_{38}H_{36}Br_4$ [MePh_3As]$_2$NiBr_4, 3, 326

$As_2NiC_{38}H_{36}Cl_4$ [MePh_3As]$_2$NiCl_4, 3, 326

$As_2NiC_{38}H_{36}I_4$ [MePh_3As]$_2$NiI_4, 3, 326

$As_2NiC_{38}H_{36}N_4O_{12}$ [MePh_3As]$_2$[Ni(NO_3)$_4$], 3, 326

$As_2NiC_{42}H_{36}O_4S_4$ [MePh_3As]$_2$[Ni(SOCC-OS)$_2$], 3, 326

$As_2NiC_{48}H_{40}Cl_4$ [Ph_4As]$_2$NiCl_4, 3, 335

$As_2Ni_2C_{14}H_{22}$ (C_5H_5NiAsMe_2)$_2$, 1, 649, 3, 304

$As_2OsC_{36}H_{30}Br_4$ [Os(AsPh_3)$_2Br_4$], 3, 24

$As_2PbC_{18}H_{22}O_6$ Ph_2Pb(HO_3AsCH_2CH=CH_2)$_2$, 2, 566

$As_2PbC_{24}H_{20}N_2O_{10}$ (m-$O_2NC_6H_4AsO_3H$)$_2$-PbPh_2, 2, 566

$As_2PbC_{24}H_{22}O_6$ Ph_2Pb(HO_3AsPh)$_2$, 2, 566

$As_2PbC_{26}H_{26}O_6$ Ph_2Pb(HO_3AsCH_2Ph)$_2$, 2, 566

$As_2Pb_2C_{28}H_{28}Cl_2O_5$ $(Ph_2PbClO_2AsMe)_2O$, 2, 566*

$As_2PdC_6H_{18}Br_2$ $[Pd(AsMe_3)_2Br_2]$, 3, 5

$As_2PdC_6H_{18}Cl_2$ $[Pd(AsMe_3)_2Cl_2]$, 3, 5

$As_2PdC_6H_{18}N_2O_4$ $[Pd(AsMe_3)_2(NO_2)_2]$, 3, 5

$As_2PdC_8H_{18}N_2S_2$ $[Pd(AsMe_3)_2(SCN)_2]$, 3, 5

$As_2PdC_{10}H_{24}S_2$ $Pd[S(CH_2)_3AsMe_2]_2$, 3, 36

$As_2PdC_{12}H_{22}$ $Me_2Pd(Me_2As)_2C_6H_4$, 1, 665

$As_2PdC_{12}H_{30}Cl_2$ $[Pd(AsEt_3)_2Cl_2]$, 3, 9

$As_2PdC_{14}H_{36}$ $[Me_2Pd(AsEt_3)_2]$, 1, 665, 3, 9

$As_2PdC_{16}H_{20}Br_2$ $[Pd(2\text{-}Me_2AsC_6H_4\text{-})_2Br_2]$, 3, 100

$As_2PdC_{16}H_{20}Br_4$ $[Pd(2\text{-}BrC_6H_4AsMe_2)_2Br_2]$, 3, 39

$As_2PdC_{16}H_{22}Cl_2$ $Pd(Me_2AsPh)_2Cl_2$, 3, 41

$As_2PdC_{16}H_{24}Cl_2N_2O_8$ $[Pd(Me_2AsCH_2\text{-}C{=}CHCH{=}CHCH{=}N)_2](ClO_4)_2$, 3, 42

$As_2PdC_{18}H_{26}Cl_2O_8S_2$ $[Pd(2\text{-}MeSC_6H_4\text{-}AsMe_2)_2](ClO_4)_2$, 3, 45

$As_2PdC_{18}H_{26}Cl_2S_2$ $[Pd(2\text{-}MeSC_6H_4\text{-}AsMe_2)_2Cl]Cl$, 3, 45

$As_2PdC_{18}H_{26}I_2S_2$ $[Pd(2\text{-}MeSC_6H_4AsMe_2)_2\text{-}I]I$, 3, 45

$As_2PdC_{18}H_{40}Cl_2$ $[Pd(Bu_2AsCH_2\text{-})_2Cl_2]$, 3, 100

$As_2PdC_{18}H_{42}Cl_2$ $[Pd(AsPr_3)_2Cl_2]$, 3, 13

$As_2PdC_{20}H_{32}Br_2$ $[Pd(2\text{-}Me_2AsC_6H_4AsMe_2)\text{-}Br_2]$, 3, 91

$As_2PdC_{20}H_{32}Br_2N_2$ $\{Pd[2\text{-}(Me_2N)C_6H_4\text{-}AsMe_2]_2Br\}Br$, 3, 48
$\{Pd[2\text{-}(Me_2N)C_6H_4AsMe_2]_2Br_2\}$, 3, 48

$As_2PdC_{20}H_{32}Cl_2$ $[Pd(2\text{-}Me_2AsC_6H_4\text{-}AsMe_2)Cl_2]$, 3, 91

$As_2PdC_{20}H_{32}Cl_2N_2$ $\{Pd[2\text{-}(Me_2N)C_6H_4\text{-}AsMe_2]_2Cl\}Cl$, 3, 48

$As_2PdC_{20}H_{32}I_2N_2$ $\{Pd[2\text{-}(Me_2N)C_6H_4\text{-}AsMe_2]_2I_2\}$, 3, 48

$As_2PdC_{20}H_{32}N_2O_4S \cdot 3H_2O$ $\{Pd[2\text{-}(Me_2N)\text{-}C_6H_4AsMe_2]_2\}SO_4 \cdot 3H_2O$, 3, 48

$As_2PdC_{20}H_{32}N_4O_4$ $\{Pd[2\text{-}(Me_2N)C_6H_4\text{-}AsMe_2]_2(NO_2)_2\}$, 3, 48

$As_2PdC_{20}H_{42}N_2S_2$ $[Pd(AsPr_3)_2(SCN)_2]$, 3, 13

$As_2PdC_{22}H_{20}Br_2$ 2,2'-o-Phenylenedi(isoarsindoline)-$PdBr_2$, 3, 560

$As_2PdC_{22}H_{24}Br_2$ $\{Pd[o\text{-}C_6H_4(C_6H_4AsMe_2)_2]\text{-}Br_2\}$, 3, 101

$As_2PdC_{22}H_{32}Cl_2$ $[Pd(BuPhAsCH_2\text{-})_2Cl_2]$, 3, 101

$As_2PdC_{22}H_{40}Cl_2$ $\{Pd[o\text{-}C_6H_4(AsBu_2)_2]Cl_2\}$, 3, 101

$As_2PdC_{24}H_{40}Br_2P_2 \cdot H_2O$ $[Pd(2\text{-}Me_2AsC_6H_4\text{-}PEt_2)_2]Br_2 \cdot H_2O$, 3, 51

$As_2PdC_{24}H_{40}P_2$ $\{Pd(2\text{-}Me_2AsC_6H_4AsMe_2)\text{-}[o\text{-}C_6H_4(PEt_2)_2]\}$, 3, 92

$As_2PdC_{24}H_{54}Cl_2$ $[Pd(AsBu_3)_2Cl_2]$, 3, 15

$As_2PdC_{24}H_{54}N_2O_4$ $[Pd(AsBu_3)_2(NO_2)_2]$, 3, 15

$As_2PdC_{26}H_{24}Br_2$ $[Pd(Ph_2AsCH_2\text{-})_2Br_2]$, 3, 102

$As_2PdC_{26}H_{24}Cl_2$ $[Pd(Ph_2AsCH_2\text{-})_2Cl_2]$, 3, 102

$As_2PdC_{26}H_{24}I_2$ $[Pd(Ph_2AsCH_2\text{-})_2I_2]$, 3, 102

$As_2PdC_{26}H_{26}Br_2$ $[Pd(Ph_2AsMe)_2Br_2]$, 3, 54

$As_2PdC_{26}H_{26}I_2$ $[Pd(Ph_2AsMe)_2I_2]$, 3, 54

$As_2PdC_{26}H_{34}BrN_5O_7$ $\{Pd[2\text{-}(Me_2N)C_6H_4\text{-}AsMe_2]_2Br\}OC_6H_2(NO_2)_3$, 3, 48

$As_2PdC_{26}H_{34}ClN_5O_7$ $\{Pd[2\text{-}(Me_2N)C_6H_4\text{-}AsMe_2]_2Cl\}OC_6H_2(NO_2)_3$, 3, 48

$As_2PdC_{26}H_{54}N_2S_2$ $[Pd(AsBu_3)_2(SCN)_2]$, 3, 15

$As_2PdC_{28}H_{30}Br_2$ $[Pd(2\text{-}PhC_6H_4AsMe_2)_2\text{-}Br_2]$, 3, 58

$As_2PdC_{28}H_{46}Cl_2$ $[Pd(Bu_2AsPh)_2Cl_2]$, 3, 58

$As_2PdC_{28}H_{48}P_2$ $[Pd(2\text{-}Et_2AsC_6H_4AsEt_2)\text{-}(2\text{-}Et_2PC_6H_4PEt_2)]$, 3, 99

$As_2PdC_{32}H_{36}N_8O_{14}$ $\{Pd[2\text{-}(Me_2N)C_6H_4\text{-}AsMe_2]_2\}[OC_6H_2(NO_2)_3]_2$, 3, 49

$As_2PdC_{36}H_{30}Cl_2$ $[Pd(AsPh_3)_2Cl_2]$, 3, 24

$As_2PdC_{36}H_{30}I_2$ $[Pd(AsPh_3)_2I_2]$, 3, 24

$As_2PdC_{38}H_{30}Br_2O_4$ $\{Pd[2\text{-}(Ph_2As)C_6H_4\text{-}CO_2H]_2Br_2\}$, 3, 62

$As_2PdC_{38}H_{30}N_2S_2$ $[Pd(AsPh_3)_2(SCN)_2]$, 3, 25

$As_2PdC_{38}H_{36}Cl_4$ $[MePh_3As]_2PdCl_4$, 3, 326

$As_2PdC_{64}H_{54}Cl_2$ $\{Pd[(2\text{-}PhC_6H_4CH_2)_2\text{-}AsPh]_2Cl_2\}\cdot 2.5\ H_2O$, 3, 72

$As_2Pd_2C_6H_{18}Br_2Cl_2$ $[Pd(AsMe_3)BrCl]_2$, 3, 5

$As_2Pd_2C_6H_{18}Br_4$ $[Pd(AsMe_3)Br_2]_2$, 3, 5

$As_2Pd_2C_6H_{18}Cl_2N_2O_4$ $[Pd(AsMe_3)Cl\text{-}(NO_2)]_2$, 3, 5

$As_2Pd_2C_6H_{18}Cl_4$ $[Pd(AsMe_3)Cl_2]_2$, 3, 5

$As_2Pd_2C_8H_{18}Br_2N_2S_2$ $[Pd(AsMe_3)Br\text{-}(SCN)]_2$, 3, 5

$As_2Pd_2C_8H_{18}Cl_2N_2S_2$ $[Pd(AsMe_3)Cl\text{-}(SCN)]_2$, 3, 5

$As_2Pd_2C_{10}H_{24}Br_2S_2$ $\{Pd[S(CH_2)_3AsMe_2]\text{-}Br\}_2$, 3, 36

$As_2Pd_2C_{10}H_{24}Cl_2S_2$ $\{Pd[S(CH_2)_3AsMe_2]\text{-}Cl\}_2$, 3, 36

$As_2Pd_2C_{10}H_{24}N_2O_4S_2$ $\{Pd[S(CH_2)_3AsMe_2]\text{-}NO_2\}_2$, 3, 36

$As_2Pd_2C_{12}H_{24}N_2S_4$ $\{Pd[S(CH_2)_3AsMe_2]\text{-}SCN\}_2$, 3, 36

$As_2Pd_2C_{12}H_{24}O_4S_2$ $\{Pd[S(CH_2)_3AsMe_2]\text{-}OOC\text{-}\}_2$, 3, 37

$As_2Pd_2C_{12}H_{30}Cl_4$ $[Pd(AsEt_3)Cl_2]_2$, 3, 9

$As_2Pd_2C_{18}H_{42}Cl_4$ $[Pd(AsPr_3)Cl_2]_2$, 3, 13

$As_2Pd_2C_{20}H_{32}Br_2I_2N_2$ $\{Pd[2\text{-}(Me_2N)C_6H_4\text{-}AsMe_2]BrI\}_2$, 3, 49

$As_2Pd_2C_{20}H_{32}I_4N_2$ $\{Pd[2\text{-}(Me_2N)C_6H_4\text{-}AsMe_2]I_2\}_2$, 3, 49

$As_2Pd_2C_{20}H_{42}Cl_2N_2S_2$ $[Pd(AsPr_3)(SCN)\text{-}Cl]_2$, 3, 13

$As_2Pd_2C_{22}H_{42}N_4S_4$ $[Pd(AsPr_3)(SCN)_2]_2$, 3, 13

$As_2Pd_2C_{24}H_{40}Br_4P_2$ $[Pd(2\text{-}Me_2AsC_6H_4\text{-}PEt_2)_2][PdBr_4]$, 3, 51

$As_2Pd_2C_{24}H_{54}Cl_2N_2O_4$ $[Pd(AsBu_3)(NO_2)\text{-}Cl]_2$, 3, 16

$As_2Pd_2C_{24}H_{54}Cl_4$ $[Pd(AsBu_3)Cl_2]_2$, 3, 16

$As_2Pd_2C_{26}H_{54}Cl_2N_2S_2$ $[Pd(AsBu_3)(SCN)\text{-}Cl]_2$, 3, 16

$As_2Pd_2C_{28}H_{46}Cl_4$ $[Pd(Bu_2AsPh)Cl_2]_2$, 3, 59

$As_2Pd_2C_{28}H_{54}N_4S_4$ $[Pd(AsBu_3)(SCN)_2]_2$, 3, 16

$As_2Pd_2C_{38}H_{36}Cl_6$ $[MePh_3As]_2[Pd_2Cl_6]$, 3, 326

$As_2PtC_4H_{12}Br_2O$ $\{Pt[(Me_2As)_2O]Br_2\}$, 3, 213

$As_2PtC_4H_{12}I_2O$ $\{Pt[(Me_2As)_2O]I_2\}$, 3, 214

$As_2PtC_4H_{14}Cl_2O_2$ $\{Pt[(Me_2As)_2O\cdot H_2O]\text{-}Cl_2\}$, 3, 213

$As_2PtC_{10}H_{24}S_2$ $Pt[S(CH_2)_3AsMe_2]_2$, 3, 37

$As_2PtC_{12}H_{18}Cl_2$ $\{cis\text{-}Pt[As(CH{=}CH_2)_3]_2\text{-}Cl_2\}$, 3, 6

$As_2PtC_{12}H_{30}Br_2$ $[Pt(AsEt_3)_2Br_2]$, 3, 9

$As_2PtC_{12}H_{30}Br_4$ $[Pt(AsEt_3)_2Br_4]$, 3, 10

$As_2PtC_{12}H_{30}Cl_2$ $[Pt(AsEt_3)_2Cl_2]$, 3, 9

$As_2PtC_{12}H_{30}Cl_4$ $[Pt(AsEt_3)_2Cl_4]$, 3, 10

$As_2PtC_{12}H_{30}I_2$ $[Pt(AsEt_3)_2I_2]$, 3, 10

$As_2PtC_{12}H_{30}I_4$ $[Pt(AsEt_3)_2I_4]$, 3, 10

$As_2PtC_{12}H_{30}N_2O_4$ $[Pt(AsEt_3)_2(NO_2)_2]$, 3, 10

$As_2PtC_{12}H_{31}Br$ $[Pt(AsEt_3)_2HBr]$, 3, 10

$As_2PtC_{12}H_{31}Cl$ $[Pt(AsEt_3)_2HCl]$, 3, 9

$As_2PtC_{12}H_{31}I$ $[Pt(AsEt_3)_2HI]$, 3, 10

$As_2PtC_{13}H_{31}NS$ $[Pt(AsEt_3)_2H(SCN)]$, 3, 10

$As_2PtC_{14}H_{38}Br_2N_4S_2$ $\{Pt(AsEt_3)_2[C(S)\text{-}(NH_2)_2]_2\}Br_2$, 3, 10

$As_2PtC_{14}H_{38}Cl_2N_4S_2$ $\{Pt(AsEt_3)_2[C(S)\text{-}(NH_2)_2]_2\}Cl_2$, 3, 10

$As_2PtC_{14}H_{38}I_2N_4S_2$ $\{Pt(AsEt_3)_2[C(S)\text{-}(NH_2)_2]_2\}I_2$, 3, 11

$As_2PtC_{16}H_{24}Cl_2N_2O_8$ $[Pt(Me_2AsCH_2\text{-}C{=}CHCH{=}CHCH{=}N)_2](ClO_4)_2$, 3, 42

$As_2PtC_{18}H_{26}BrClO_4S_2$ $[Pt(2\text{-}MeSC_6H_4\text{-}AsMe_2)_2Br]ClO_4$, 3, 46

$As_2PtC_{18}H_{26}Br_2S_2$ $[Pt(2\text{-}MeSC_6H_4AsMe_2)_2\text{-}Br]Br$, 3, 46

$As_2PtC_{18}H_{26}ClIO_4S_2$ $[Pt(2\text{-}MeSC_6H_4\text{-}AsMe_2)_2I]ClO_4$, 3, 46

$As_2PtC_{18}H_{26}Cl_2O_4S_2$ $[Pt(2\text{-}MeSC_6H_4\text{-}AsMe_2)_2Cl]ClO_4$, 3, 46

$As_2PtC_{18}H_{26}Cl_2O_8S_2$ $[Pt(2\text{-}MeSC_6H_4\text{-}AsMe_2)_2](ClO_4)_2$, 3, 45

$As_2PtC_{18}H_{26}Cl_2S_2$ $[Pt(2\text{-}MeSC_6H_4AsMe_2)_2\text{-}Cl]Cl$, 3, 46

$As_2PtC_{18}H_{26}I_2S_2$ $[Pt(2\text{-}MeSC_6H_4AsMe_2)_2\text{-}I]I$, 3, 46

$As_2PtC_{18}H_{42}Cl_2$ [Pt(AsPr$_3$)$_2$Cl$_2$], 3, 13

$As_2PtC_{20}H_{28}Br_2$ {Pt[Me$_2$As(CH$_2$)$_3$-CH=CH$_2$](Ph$_2$AsMe)Br$_2$}, 3, 39, 55

$As_2PtC_{24}H_{54}Cl_2$ [Pt(AsBu$_3$)$_2$Cl$_2$], 3, 16

$As_2PtC_{24}H_{54}N_2O_4$ [Pt(AsBu$_3$)$_2$(NO$_2$)$_2$], 3, 16

$As_2PtC_{26}H_{26}Br_2$ [Pt(Ph$_2$AsMe)$_2$Br$_2$], 3, 54

$As_2PtC_{26}H_{26}Br_2Cl_2$ [Pt(Ph$_2$AsMe)$_2$Br$_2$-Cl$_2$], 3, 55

$As_2PtC_{26}H_{26}Br_2I_2$ [Pt(Ph$_2$AsMe)$_2$Br$_2$-I$_2$], 3, 55

$As_2PtC_{26}H_{26}Br_4$ [Pt(Ph$_2$AsMe)$_2$Br$_4$], 3, 54

$As_2PtC_{26}H_{26}Cl_2$ [Pt(Ph$_2$AsMe)$_2$Cl$_2$], 3, 54

$As_2PtC_{26}H_{26}Cl_2I_2$ [Pt(Ph$_2$AsMe)$_2$Cl$_2$I$_2$], 3, 55

$As_2PtC_{26}H_{26}Cl_4$ [Pt(Ph$_2$AsMe)$_2$Cl$_4$], 3, 55

$As_2PtC_{26}H_{26}I_2$ [Pt(Ph$_2$AsMe)$_2$I$_2$], 3, 54

$As_2PtC_{26}H_{26}I_4$ [Pt(Ph$_2$AsMe)$_2$I$_4$], 3, 55

$As_2PtC_{28}H_{48}Cl_2P_2$ {Pt[2-(Et$_2$As)C$_6$H$_4$-PEt$_2$]$_2$}Cl$_2$, 3, 60

$As_2PtC_{36}H_{30}Cl_2$ [Pt(AsPh$_3$)$_2$Cl$_2$], 3, 25

$As_2PtC_{36}H_{30}Cl_4$ [Pt(AsPh$_3$)$_2$Cl$_4$], 3, 25

$As_2PtC_{38}H_{30}N_2S_2$ [Pt(AsPh$_3$)$_2$(NCS)$_2$], 3, 25

$As_2PtC_{38}H_{36}Cl_4$ [MePh$_3$As]$_2$PtCl$_4$, 3, 326

$As_2PtC_{46}H_{42}Cl_8O_2$ [2-(4-Chlorophenacyl)-1,2,3,4-tetrahydro-2-phenylisoarsinolinium]$_2$[PtCl$_6$], 3, 595

$As_2PtC_{48}H_{40}Br_6$ [Ph$_4$As]$_2$PtBr$_6$, 3, 336

$As_2PtC_{66}H_{76}Cl_6O_4$ {Ph(PhCH$_2$)[PhCH$_2$O-(CH$_2$)$_3$]$_2$As}$_2$PtCl$_4$, 3, 341*

$As_2Pt_2C_{10}H_{24}Cl_2S_2$ {Pt[S(CH$_2$)$_3$AsMe$_2$]-Cl}$_2$, 3, 37

$As_2Pt_2C_{12}H_{30}Br_4S_2$ [Pt(Me$_2$AsCH$_2$CH$_2$CH$_2$-SMe)$_2$][PtBr$_4$], 3, 38

$As_2Pt_2C_{12}H_{30}Br_6S_2$ [Pt(Me$_2$AsCH$_2$CH$_2$CH$_2$-SMe)$_2$][PtBr$_6$], 3, 38

$As_2Pt_2C_{12}H_{30}Cl_4$ [Pt(AsEt$_3$)Cl$_2$]$_2$, 3, 10

$As_2Pt_2C_{12}H_{30}Cl_4S_2$ [Pt(Me$_2$AsCH$_2$CH$_2$CH$_2$-SMe)$_2$][PtCl$_4$], 3, 38

$As_2Pt_2C_{12}H_{30}Cl_6S_2$ [Pt(Me$_2$AsCH$_2$CH$_2$CH$_2$-SMe)$_2$][PtCl$_6$], 3, 38

$As_2Pt_2C_{14}H_{38}Cl_4N_2$ [Pt(AsEt$_3$)$_2$(H$_2$N-CH$_2$-)$_2$][PtCl$_4$], 3, 10

$As_2Pt_2C_{18}H_{26}Br_4S_2$ [Pt(2-MeSC$_6$H$_4$-AsMe$_2$)$_2$][PtBr$_4$], 3, 46

$As_2Pt_2C_{18}H_{26}Br_6S_2$ [Pt(2-MeSC$_6$H$_4$-AsMe$_2$)$_2$][PtBr$_6$], 3, 46

$As_2Pt_2C_{18}H_{26}Cl_4S_2$ [Pt(2-MeSC$_6$H$_4$-AsMe$_2$)$_2$][PtCl$_4$], 3, 46

$As_2Pt_2C_{18}H_{26}Cl_6S_2$ [Pt(2-MeSC$_6$H$_4$-AsMe$_2$)$_2$][PtCl$_6$], 3, 46

$As_2Pt_2C_{18}H_{42}Cl_4$ [Pt(AsPr$_3$)Cl$_2$]$_2$, 3, 14

$As_2Pt_2C_{28}H_{48}Cl_4P_2$ {Pt[2-(Et$_2$As)C$_6$H$_4$-PEt$_2$]}[PtCl$_4$], 3, 60

$As_2Pt_2C_{42}H_{62}Cl_2P_2$ [(Pr$_3$P)Pt(AsPh$_2$)-Cl]$_2$, 3, 305

$As_2ReC_{16}H_{22}Cl_3O$ Re(O)(Me$_2$AsPh)$_2$Cl$_3$, 3, 41

$As_2ReC_{20}H_{30}Cl_3O$ [Re(O)(Et$_2$AsPh)$_2$Cl$_3$], 3, 47

$As_2ReC_{36}H_{30}Br_3O$ [Re(O)(AsPh$_3$)$_2$Br$_3$], 3, 25

$As_2ReC_{36}H_{30}Cl_3O$ [Re(O)(AsPh$_3$)$_2$Cl$_3$], 3, 25

$As_2ReC_{39}H_{30}ClO_3$ [Re(AsPh$_3$)$_2$(CO)$_3$Cl], 3, 25

$As_2ReC_{39}H_{30}IO_3$ [Re(AsPh$_3$)$_2$(CO)$_3$I], 3, 25

$As_2ReC_{48}H_{40}F_6$ [Ph$_4$As]$_2$ReF$_6$, 3, 336

$As_2Re_3C_{38}H_{36}Cl_{11}$ [MePh$_3$As]$_2$[Re$_3$Cl$_{11}$], 3, 326

$As_2Re_3C_{48}H_{40}Cl_{11}$ [Ph$_4$As]$_2$[Re$_3$Cl$_{11}$], 3, 336

$As_2RhC_{18}H_{26}I_7S_2$ [Rh(2-MeSC$_6$H$_4$AsMe$_2$)$_2$-I$_2$]I$_5$, 3, 46

$As_2RhC_{36}H_{30}Cl_2NO$ [Rh(AsPh$_3$)$_2$(NO)Cl$_2$], 3, 26

$As_2RhC_{37}H_{24}Cl_7O$ {Rh[As(C$_6$H$_4$Cl-4)$_3$]$_2$-(CO)Cl}, 3, 17

$As_2RhC_{37}H_{24}Cl_9O$ {Rh[As(C$_6$H$_4$Cl-4)$_3$]$_2$-(CO)Cl$_3$}, 3, 17

$As_2RhC_{37}H_{30}ClI_2O$ [Rh(AsPh$_3$)$_2$(CO)ClI$_2$], 3, 25

$As_2RhC_{37}H_{30}ClO$ [Rh(AsPh$_3$)$_2$(CO)Cl], 3, 25

$As_2RhC_{37}H_{30}Cl_3O$ [Rh(AsPh$_3$)$_2$(CO)Cl$_3$],

3, 25
$As_2RhC_{37}H_{30}I_3O$ [$Rh(AsPh_3)_2(CO)I_3$], 3, 26
$As_2RhC_{43}H_{42}ClO$ {$Rh[As(C_6H_4Me\text{-}4)_3]_2$-(CO)Cl}, 3, 28
$As_2RhC_{43}H_{42}ClO_7$ {$Rh[As(C_6H_4OMe\text{-}4)_3]_2$-(CO)Cl}, 3, 28
$As_2RhC_{43}H_{42}Cl_3O$ {$Rh[As(C_6H_4Me\text{-}4)_3]_2$-$(CO)Cl_3$}, 3, 28
$As_2RhC_{52}H_{44}ClN_2O_6$ [$(p\text{-}MeOC_6H_4NC)_2Rh$-$(AsPh_3)_2$]$ClO_4$, 1, 627
$As_2RuC_{12}H_{30}Cl_3NO$ [$Ru(AsEt_3)_2(NO)Cl_3$], 3, 11
$As_2RuC_{12}H_{30}I_3NO$ [$Ru(AsEt_3)_2(NO)I_3$], 3, 11
$As_2RuC_{16}H_{24}Cl_2N_2$ [$Ru(Me_2AsCH_2$-C=CHCH=CHCH=N$)_2Cl_2$], 3, 42
$As_2RuC_{16}H_{24}Cl_3N_2$ [$Ru(Me_2AsCH_2$-C=CHCH=CHCH=N$)_2Cl_2$]$Cl\cdot3H_2O$, 3, 42
$As_2RuC_{16}H_{24}Cl_3N_2O_4$ [$Ru(Me_2AsCH_2$-C=CHCH=CHCH=N$)_2Cl_2$]ClO_4, 3, 42
$As_2RuC_{22}H_{30}Cl_2O_2$ [$Ru(Et_2AsPh)_2(CO)_2$-Cl_2], 3, 47
$As_2RuC_{26}H_{26}Cl_3NO$ [$Ru(Ph_2AsMe)_2(NO)$-Cl_3], 3, 57
$As_2RuC_{36}H_{30}Br_3$ $Ru(AsPh_3)_2Br_3$, 3, 26
$As_2RuC_{36}H_{30}Cl_3$ $Ru(AsPh_3)_2Cl_3$, 3, 26
$As_2RuC_{36}H_{30}Cl_3NO$ [$Ru(AsPh_3)_2(NO)Cl_3$], 3, 26
$As_2RuC_{37}H_{34}Br_3O$ [$Ru(AsPh_3)_2(MeOH)$-Br_3], 3, 26
$As_2RuC_{38}H_{30}I_2O_2$ [$Ru(AsPh_3)_2(CO)_2I_2$], 3, 26
$As_2SbC_{10}H_{16}Br_3$ [$Sb(2\text{-}Me_2AsC_6H_4AsMe_2)$-$Br_3$], 3, 80
$As_2SbC_{10}H_{16}Cl_3$ [$Sb(2\text{-}Me_2AsC_6H_4AsMe_2)$-$Cl_3$], 3, 80
$As_2SbC_{10}H_{16}I_3$ [$Sb(2\text{-}Me_2AsC_6H_4AsMe_2)$-$I_3$], 3, 80
$As_2SbC_{36}H_{30}Cl_3O_2$ [$Sb(Ph_3AsO)_2Cl_3$], 3, 353
$As_2SiC_{38}H_{68}N_2O_2$ [$Bu_3As{=}N]_2Si(C_6H_4$-$OMe\text{-}p)_2$, 3, 369
$As_2Si_2C_{36}H_{30}O_4$ Hexaphenyl-1,3,5,7-tetraoxa-2,6-diarsa-4,8-disilacyclooctane, 3, 648
$As_2SnC_6H_{18}O_4$ $(Me_2AsO_2)_2SnMe_2$, 2, 349, 3, 374
$As_2SnC_{10}H_{16}Cl_4$ [$Sn(2\text{-}Me_2AsC_6H_4AsMe_2)$-$Cl_4$], 3, 96
$As_2SnC_{12}H_{30}Br_4$ [$(Et_3As)_2SnBr_4$], 3, 11
$As_2SnC_{12}H_{30}Cl_4$ [$(Et_3As)_2SnCl_4$], 3, 11
$As_2SnC_{14}H_{18}O_6$ $[PhAs(O)(OH)O]_2SnMe_2$, 3, 549
$As_2SnC_{20}H_{28}N_2O_{10}$ [$4\text{-}O_2NC_6H_4As(O)(OH)$-$O]_2SnBu_2$, 2, 353, 3, 548
$As_2SnC_{20}H_{28}N_2O_{12}$ [$3,4\text{-}O_2N(HO)C_6H_3$-$As(O)(OH)O]_2SnBu_2$, 2, 353, 3, 475, 548
$As_2SnC_{20}H_{30}N_4O_{10}$ [$3,4\text{-}O_2N(H_2N)C_6H_3$-$As(O)(OH)O]_2SnBu_2$, 2, 353, 3, 475, 549*
$As_2SnC_{20}H_{30}O_6$ $[BuAs(O)(OH)O]_2SnPh_2$, 2, 356, 3, 548
$As_2SnC_{36}H_{30}Br_4O_2$ [$Sn(Ph_3AsO)_2Br_4$], 3, 355
$As_2SnC_{36}H_{30}Cl_4O_2$ [$Sn(Ph_3AsO)_2Cl_4$], 3, 355
$As_2SrC_4H_{12}O_4$ $(Me_2AsO_2)_2Sr$, 3, 374
$As_2SrC_{12}H_{14}N_2O_6$ [$4\text{-}H_2NC_6H_4AsO(OH)O]_2$-Sr, 3, 428
$As_2SrC_{12}H_{24}N_2O_{11}$ [$4\text{-}H_2NC_6H_4AsO(OH)O]_2$-$Sr\cdot5H_2O$, 3, 428
$As_2TeC_{48}H_{40}Cl_6$ $[Ph_4As]_2TeCl_6$, 3, 336
$As_2ThC_{12}H_{10}O_6$ $Th(O_3AsPh)_2$, 3, 400
$As_2TiC_{10}H_{16}Br_4$ [$Ti(2\text{-}Me_2AsC_6H_4AsMe_2)$-$Br_4$], 3, 96
$As_2TiC_{10}H_{16}Cl_4$ [$Ti(2\text{-}Me_2AsC_6H_4AsMe_2)$-$Cl_4$], 3, 96
$As_2TiC_{10}H_{18}Br_3O$ [$Ti(2\text{-}Me_2AsC_6H_4AsMe_2)$-$(H_2O)Br_3$], 3, 96
$As_2TiC_{10}H_{18}Cl_3O$ [$Ti(2\text{-}Me_2AsC_6H_4AsMe_2)$-$(H_2O)Cl_3$], 3, 96
$As_2UC_{12}H_{10}O_6$ $U[O_3AsPh]_2$, 3, 400
$As_2UC_{16}H_{36}O_6$ $UO_2(O_2AsBu_2)_2$, 3, 375
$As_2VC_{60}H_{40}F_{18}S_6$ $[Ph_4As]_2\{V[SC(CF_3){=}C$-$(CF_3)S]_3\}$, 3, 336
$As_2WC_{13}H_{16}Br_2O_3$ [$W(2\text{-}Me_2AsC_6H_4AsMe_2)$-$(CO)_3Br_2$], 3, 97
$As_2WC_{13}H_{16}Br_3O_3$ [$W(2\text{-}Me_2AsC_6H_4AsMe_2)$-$(CO)_3Br_2$]Br, 3, 97
$As_2WC_{14}H_{16}I_2O_4$ [$W(2\text{-}Me_2AsC_6H_4AsMe_2)$-$(CO)_4I$]I, 3, 97
$As_2WC_{14}H_{16}I_4O_4$ [$W(2\text{-}Me_2AsC_6H_4AsMe_2)$-

$(CO)_4I]I_3$, 3, 97

$As_2WC_{14}H_{16}O_4$ $[W(2-Me_2AsC_6H_4AsMe_2)(CO)_4]$, 3, 97

$As_2WC_{46}H_{40}Cl_4N_2O_2$ $[Ph_4As]_2[W(NO)_2Cl_4]$, 3, 336

$As_2WC_{60}H_{40}F_{18}S_6$ $[Ph_4As]_2\{W[SC(CF_3)=C(CF_3)S]_3\}$, 3, 336

$As_2W_2C_{12}H_{12}O_8$ $[W(CO)_4AsMe_2]_2$, 3, 301

$As_2W_2C_{14}H_{12}O_{10}$ $(OC)_5W\leftarrow As(Me)_2-As(Me)_2\rightarrow W(CO)_5$, 3, 301

$As_2W_{12}C_{48}H_{41}O_{40}P$ $[Ph_4As]_2H\cdot PW_{12}O_{40}$, 3, 336

$As_2ZnC_4H_{12}O_4$ $(Me_2AsO_2)_2Zn$, 3, 374

$As_2ZnC_{10}H_{16}Br_2$ $[Zn(2-Me_2AsC_6H_4AsMe_2)Br_2]$, 3, 97

$As_2ZnC_{10}H_{16}I_2$ $[Zn(2-Me_2AsC_6H_4AsMe_2)I_2]$, 3, 97

$As_2ZnC_{12}H_{30}Br_2$ $[Zn(AsEt_3)_2Br_2]$, 3, 11

$As_2ZnC_{12}H_{30}Cl_2$ $[Zn(AsEt_3)_2Cl_2]$, 3, 11

$As_2ZnC_{12}H_{30}I_2$ $[Zn(AsEt_3)_2I_2]$, 3, 11

$As_2ZnC_{16}H_{22}Br_2$ $[Zn(Me_2AsPh)_2]Br_2$, 3, 42

$As_2ZnC_{16}H_{22}Cl_2$ $[Zn(Me_2AsPh)_2]Cl_2$, 3, 42

$As_2ZnC_{16}H_{36}O_4$ $(Bu_2AsO_2)_2Zn$, 3, 375

$As_2ZnC_{18}H_{26}I_2$ $[Zn(Me_2AsC_6H_4Me-o)_2]I_2$, 3, 43

$[Zn(Me_2AsC_6H_4Me-p)_2]I_2$, 3, 44

$As_2ZnC_{18}H_{34}$ $[Bu_2Zn(2-Me_2AsC_6H_4AsMe_2)]$, 3, 97

$As_2ZnC_{20}H_{32}I_2N_2$ $\{Zn[4-(Me_2N)C_6H_4AsMe_2]_2I_2\}$, 3, 49

$As_2ZnC_{22}H_{16}F_{10}$ $[(C_6F_5)_2Zn(2-Me_2AsC_6H_4AsMe_2)]$, 3, 98

$As_2ZnC_{22}H_{26}$ $[Ph_2Zn(2-Me_2AsC_6H_4AsMe_2)]$, 3, 98

$As_2ZnC_{36}H_{30}N_2O_8$ $[Zn(Ph_3AsO)_2(NO_3)_2]$, 3, 355

$As_2ZnC_{38}H_{36}Br_4$ $[MePh_3As]_2ZnBr_4$, 3, 326

$As_2ZnC_{38}H_{36}Cl_4$ $[MePh_3As]_2ZnCl_4$, 3, 326

$As_2ZnC_{38}H_{36}I_4$ $[MePh_3As]_2ZnI_4$, 3, 326

$As_2ZrC_{12}H_{10}O_6$ $Zr[O_3AsPh]_2$, 3, 400

$As_3AgC_{11}H_{27}Br$ $\{Ag[MeC(CH_2AsMe_2)_3]Br\}$, 3, 107

$As_3AuC_{27}H_{39}ClO_4S_3$ $[Au(2-MeSC_6H_4AsMe_2)_3]ClO_4$, 3, 45

$As_3Au_3C_{11}H_{27}Cl_3$ $\{(AuCl)_3[MeC(CH_2AsMe_2)_3]\}$, 3, 107

$As_3B_{13}C_9H_{40}$ $[H_2B(AsMe_3)_2][B_{12}H_{11}(AsMe_3)]$, 3, 4

$As_3BiC_{24}H_{54}O_6$ $(Bu_2AsO_2)_3Bi$, 3, 375

$As_3CoC_{11}H_{27}I_2$ $\{Co[MeAs(CH_2CH_2CH_2AsMe_2)_2]I_2\}$, 3, 105

$As_3CoC_{11}H_{27}I_3$ $\{Co[MeAs(CH_2CH_2CH_2AsMe_2)_2]I_3\}$, 3, 105

$As_3CrC_{14}H_{27}O_3$ $\{Cr[MeAs(CH_2CH_2CH_2AsMe_2)_2](CO)_3\}$, 3, 105

$As_3CrC_{27}H_{33}O_3$ $Cr(Me_2AsPh)_3(CO)_3$, 3, 40

$As_3CrC_{84}H_{60}N_6S_6$ $[Ph_4As]_3\{Cr[SC(CN)=C(CN)S]_3\}$, 3, 336

$As_3CuC_{11}H_{27}Br$ $\{Cu[MeC(CH_2AsMe_2)_3]Br\}$, 3, 107

$As_3CuC_{11}H_{27}I$ $\{Cu[MeAs(CH_2CH_2CH_2AsMe_2)_2]I\}$, 3, 105

$As_3CuC_{39}H_{39}Br$ $[Cu(Ph_2AsMe)_3Br]$, 3, 52

$As_3CuC_{39}H_{39}Cl$ $[Cu(Ph_2AsMe)_3Cl]$, 3, 52

$As_3CuC_{39}H_{39}I$ $[Cu(Ph_2AsMe)_3I]$, 3, 52

$As_3Cu_2C_{39}H_{39}Br_3$ $[Cu_2(Ph_2AsMe)_3Br_3]$, 3, 53

$As_3Cu_2C_{39}H_{39}Cl_3$ $[Cu_2(Ph_2AsMe)_3Cl_3]$, 3, 53

$As_3FeC_{14}H_{27}N_3S_3$ $\{Fe[MeAs(CH_2CH_2CH_2AsMe_2)_2](SCN)_3\}$, 3, 105

$As_3FeC_{63}H_{54}N_6S_6$ $[MePh_3As]_3[Fe(SCN)_6]$, 3, 326

$As_3Hg_2C_{18}H_{45}I_4$ $[(Et_3As)_3(HgI_2)_2]$, 3, 8

$As_3Hg_2C_{27}H_{63}I_4$ $[(Pr_3As)_3(HgI_2)_2]$, 3, 12

$As_3Hg_2C_{36}H_{81}I_4$ $[(Bu_3As)_3(HgI_2)_2]$, 3, 15

$As_3IrC_{18}H_{45}Cl_3$ $[trans-Ir(AsEt_3)_3Cl_3]$, 3, 8

$As_3IrC_{18}H_{46}Cl_2$ $[IrH(Et_3As)_3Cl_2]$, 3, 8

$As_3IrC_{24}H_{33}Br_3$ $Ir(Me_2AsPh)_3Br_3$, 3, 40

$As_3IrC_{24}H_{33}Cl_3$ $Ir(Me_2AsPh)_3Cl_3$, 3, 40

$As_3IrC_{24}H_{33}I_3$ $Ir(Me_2AsPh)_3I_3$, 3, 40

$As_3IrC_{30}H_{45}Cl_3$ $[Ir(Et_2AsPh)_3Cl_3]$, 3, 47

$As_3IrC_{39}H_{39}Br_3$ $[Ir(Ph_2AsMe)_3Br_3]$, 3, 53

$As_3IrC_{39}H_{39}Cl_3$ $[Ir(Ph_2AsMe)_3Cl_3]$, 3, 53

$As_3IrC_{39}H_{39}I_3$ $[Ir(Ph_2AsMe)_3I_3]$, 3, 53

$As_3IrC_{39}H_{40}Br_2$ $[IrH(Ph_2AsMe)_3Br_2]$, 3,

53
$As_3IrC_{54}H_{46}Cl_2$ $[IrH(AsPh_3)_3Cl_2]$, 3, 20
$As_3MnC_{13}H_{27}BrO_2$ $\{Mn[MeAs(CH_2CH_2CH_2AsMe_2)_2](CO)_2\}Br$, 3, 105
$As_3MnC_{13}H_{27}IO_2$ $\{Mn[MeAs(CH_2CH_2CH_2AsMe_2)_2](CO)_2\}I$, 3, 105
$\{Mn[MeAs(CH_2CH_2CH_2AsMe_2)_2](CO)_2\}I$, 3, 105
$As_3MoC_{14}H_{27}O_3$ $\{Mo[MeAs(CH_2CH_2CH_2AsMe_2)_2](CO)_3\}$, 3, 105
$As_3MoC_{21}H_{45}O_3$ $[Mo(AsEt_3)_3(CO)_3]$, 3, 9
$As_3MoC_{57}H_{45}O_3$ $Mo(AsPh_3)_3(CO)_3$, 3, 23
$As_3NiC_{11}H_{27}Br_2$ $\{Ni[MeAs(CH_2CH_2CH_2AsMe_2)_2]Br_2\}$, 3, 105
$As_3NiC_{11}H_{27}Br_3$ $\{Ni[MeAs(CH_2CH_2CH_2AsMe_2)_2]Br_3\}$, 3, 106
$As_3NiC_{11}H_{27}I_2$ $\{Ni[MeAs(CH_2CH_2CH_2AsMe_2)_2]I_2\}$, 3, 105
$As_3NiC_{15}H_{36}BrClO_4P$ $\{Ni[(Me_2AsCH_2CH_2CH_2)_3P]Br\}ClO_4$, 3, 108
$As_3NiC_{15}H_{36}Br_2P$ $\{Ni[(Me_2AsCH_2CH_2CH_2)_3P]Br_2\}$, 3, 108
$As_3NiC_{15}H_{36}ClIO_4P$ $\{Ni[(Me_2AsCH_2CH_2CH_2)_3P]I\}ClO_4$, 3, 108
$As_3NiC_{15}H_{36}ClNO_6P$ $\{Ni[(Me_2AsCH_2CH_2CH_2)_3P]NO_2\}ClO_4$, 3, 108
$As_3NiC_{15}H_{36}Cl_2O_4P$ $\{Ni[(Me_2AsCH_2CH_2CH_2)_3P]Cl\}ClO_4$, 3, 108
$As_3NiC_{15}H_{36}I_2P$ $\{Ni[(Me_2AsCH_2CH_2CH_2)_3P]I_2\}$, 3, 108
$As_3NiC_{15}H_{38}Cl_2O_9P$ $\{Ni[(Me_2AsCH_2CH_2CH_2)_3P]H_2O\}(ClO_4)_2$, 3, 108
$As_3NiC_{16}H_{36}ClNO_4P$ $\{Ni[(Me_2AsCH_2CH_2CH_2)_3P]CN\}ClO_4$, 3, 108
$As_3NiC_{16}H_{36}ClNO_4PS$ $\{Ni[(Me_2AsCH_2CH_2CH_2)_3P]NCS\}ClO_4$, 3, 108
$As_3NiC_{17}H_{41}ClO_4P$ $\{Ni[(Me_2AsCH_2CH_2CH_2)_3P]SEt\}ClO_4$, 3, 108
$As_3OsC_{24}H_{33}Br_3$ $Os(Me_2AsPh)_3Br_3$, 3, 41
$As_3OsC_{24}H_{33}Cl_3$ $Os(Me_2AsPh)_3Cl_3$, 3, 41
$As_3OsC_{39}H_{39}Br_3$ $[Os(Ph_2AsMe)_3Br_3]$, 3, 54
$As_3OsC_{39}H_{39}Cl_3$ $[Os(Ph_2AsMe)_3Cl_3]$, 3, 54
$As_3OsC_{54}H_{45}Br_3$ $[Os(AsPh_3)_3Br_3]$, 3, 24
$As_3OsC_{55}H_{46}BrO$ $[OsH(AsPh_3)_3(CO)Br]$, 3, 24
$As_3OsC_{55}H_{46}ClO$ $[OsH(AsPh_3)_3(CO)Cl]$, 3, 24
$As_3PdC_{11}H_{27}BrClO_4$ $\{Pd[MeAs(CH_2CH_2CH_2AsMe_2)_2]Br\}ClO_4$, 3, 106
$As_3PdC_{11}H_{27}Br_2$ $\{Pd[MeAs(CH_2CH_2CH_2AsMe_2)_2]Br\}Br$, 3, 106
$As_3PdC_{11}H_{27}ClIO_4$ $\{Pd[MeAs(CH_2CH_2CH_2AsMe_2)_2]I\}ClO_4$, 3, 106
$As_3PdC_{11}H_{27}Cl_2$ $\{Pd[MeAs(CH_2CH_2CH_2AsMe_2)_2]Cl\}Cl$, 3, 106
$As_3PdC_{11}H_{27}Cl_2O_4$ $\{Pd[MeAs(CH_2CH_2CH_2AsMe_2)_2]Cl\}ClO_4$, 3, 106
$As_3PdC_{11}H_{27}I_2$ $\{Pd[MeAs(CH_2CH_2CH_2AsMe_2)_2]I\}I$, 3, 106
$As_3PdC_{39}H_{39}Br_2$ $[Pd(Ph_2AsMe)_3Br]Br$, 3, 54
$As_3PdC_{42}H_{33}ClIO_4$ $\{Pd[(2\text{-}Ph_2AsC_6H_4)_2AsPh]I\}ClO_4$, 3, 108
$As_3PdC_{42}H_{33}I_2$ $\{Pd[(2\text{-}Ph_2AsC_6H_4)_2AsPh]I_2\}$, 3, 108
$As_3PtC_{11}H_{27}BrClO_4$ $\{Pt[MeAs(CH_2CH_2CH_2AsMe_2)_2]Br\}ClO_4$, 3, 106
$As_3PtC_{11}H_{27}Br_2$ $\{Pt[MeAs(CH_2CH_2CH_2AsMe_2)_2]Br\}Br$, 3, 106
$As_3PtC_{11}H_{27}Br_3ClO_4$ $\{Pt[MeAs(CH_2CH_2CH_2AsMe_2)_2]Br_3\}ClO_4$, 3, 106
$As_3PtC_{11}H_{27}Cl_2$ $\{Pt[MeAs(CH_2CH_2CH_2AsMe_2)_2]Cl\}Cl$, 3, 106
$As_3PtC_{11}H_{27}Cl_2O_4$ $\{Pt[MeAs(CH_2CH_2CH_2AsMe_2)_2]Cl\}ClO_4$, 3, 106
$As_3PtC_{11}H_{27}I_2$ $\{Pt[MeAs(CH_2CH_2CH_2AsMe_2)_2]I\}I$, 3, 106
$As_3PtC_{11}H_{27}I_4$ $\{Pt[MeAs(CH_2CH_2CH_2AsMe_2)_2]I_3\}I$, 3, 106
$As_3PtC_{42}H_{33}ClIO_4$ $\{Pt[(2\text{-}Ph_2AsC_6H_4)_2AsPh]I\}ClO_4$, 3, 109
$As_3PtC_{42}H_{33}I_2$ $\{Pt[(2\text{-}Ph_2AsC_6H_4)_2AsPh]I\}I$, 3, 109
$As_3Re_3C_{54}H_{45}Br_9$ $[Re(AsPh_3)Br_3]_3$, 3, 25
$As_3Re_3C_{57}H_{54}Cl_{12}$ $[MePh_3As]_3[Re_3Cl_{12}]$, 3, 326
$As_3RhC_{18}H_{45}Cl_3$ $[Rh(AsEt_3)_3Cl_3]$, 3, 11
$As_3RhC_{24}H_{33}Cl_3$ $Rh(Me_2AsPh)_3Cl_3$, 3, 41
$As_3RhC_{27}H_{39}Br_3$ $[Rh(Me_2AsC_6H_4Me\text{-}4)_3Br_3]$, 3, 43
$As_3RhC_{27}H_{39}Cl_3$ $[Rh(Me_2AsC_6H_4Me\text{-}4)_3$-

Cl_3], 3, 43
$As_3RhC_{27}H_{39}I_3$ [Rh($Me_2AsC_6H_4Me$-4)$_3I_3$], 3, 43
$As_3RhC_{36}H_{81}Cl_3$ [Rh($AsBu_3$)$_3Cl_3$], 3, 16
$As_3RhC_{39}H_{39}Br_3$ [Rh(Ph_2AsMe)$_3Br_3$], 3, 55
$As_3RhC_{39}H_{39}Cl_3$ [Rh(Ph_2AsMe)$_3Cl_3$], 3, 55
$As_3RhC_{39}H_{39}I_3$ [Rh(Ph_2AsMe)$_3I_3$], 3, 55
$As_3RhC_{39}H_{40}Br_2$ [RhH(Ph_2AsMe)$_3Br_2$], 3, 55
$As_3RhC_{39}H_{40}Cl_2$ [RhH(Ph_2AsMe)$_3Cl_2$], 3, 56
$As_3RhC_{39}H_{40}I_2$ [RhH(Ph_2AsMe)I_2], 3, 56
$As_3RhC_{42}H_{33}Cl_3$ {Rh[(2-$Ph_2AsC_6H_4$)$_2$-AsPh]Cl_3}, 3, 109
$As_3RhC_{54}H_{45}NO$ [Rh($AsPh_3$)$_3$(NO)], 3, 26
$As_3Rh_2C_{18}H_{45}Br_6$ [Rh_2($AsEt_3$)$_3Br_6$], 3, 11
$As_3Rh_2C_{18}H_{45}Cl_6$ [Rh_2($AsEt_3$)$_3Cl_6$], 3, 11
$As_3Rh_2C_{27}H_{39}Br_6S_3$ [Rh(2-$MeSC_6H_4$-$AsMe_2$)$_2Br_2$][Rh(2-$MeSC_6H_4AsMe_2$)Br_4], 3, 46
$As_3Rh_2C_{27}H_{39}Cl_6S_3$ [Rh(2-$MeSC_6H_4$-$AsMe_2$)$_2Cl_2$][Rh(2-$MeSC_6H_4AsMe_2$)Cl_4], 3, 46
$As_3Rh_2C_{27}H_{39}I_6S_3$ [Rh(2-$MeSC_6H_4$-$AsMe_2$)$_2I_2$][Rh(2-$MeSC_6H_4AsMe_2$)I_4], 3, 46
$As_3RuC_{30}H_{45}Cl_3$ [Ru(Et_2AsPh)$_3Cl_3$], 3, 47
$As_3RuC_{39}H_{39}Cl_3$ [Ru(Ph_2AsMe)$_3Cl_3$], 3, 57
$As_3RuC_{54}H_{45}Cl_3$ Ru($AsPh_3$)$_3Cl_3$, 3, 26
$As_3RuC_{55}H_{46}ClO$ [RuH($AsPh_3$)$_3$(CO)Cl], 3, 26
$As_3Ru_3C_{57}H_{45}Cl_6O_3$ [Ru($AsPh_3$)(CO)Cl_2]$_3$ 3, 26
$As_3TiC_{11}H_{27}Br_4$ {Ti[MeAs($CH_2CH_2CH_2$-$AsMe_2$)$_2$]Br_4}, 3, 107
$As_3TiC_{11}H_{27}I_4$ {Ti[MeAs($CH_2CH_2CH_2$-$AsMe_2$)$_2$]I_3}I, 3, 107
$As_3Ti_2C_{11}H_{27}Cl_8$ {($TiCl_4$)$_2$[MeAs(CH_2-$CH_2CH_2AsMe_2$)$_2$]}, 3, 107
$As_3Ti_2C_{11}H_{27}F_8$ {(TiF_4)$_2$[MeAs(CH_2CH_2-CH_2AsMe_2)$_2$]}, 3, 107
$As_3WC_{14}H_{27}O_3$ {W[MeAs($CH_2CH_2CH_2$-$AsMe_2$)$_2$](CO)$_3$}, 3, 107
$As_4AgC_{20}H_{32}ClO_4$ [Ag(2-$Me_2AsC_6H_4$-$AsMe_2$)$_2$]ClO_4, 3, 95
$As_4AgC_{20}H_{32}NO_3$ [Ag(2-$Me_2AsC_6H_4$-$AsMe_2$)$_2$]NO_3, 3, 95
$As_4Ag_2C_{20}H_{32}Br_2$ [Ag(2-$Me_2AsC_6H_4$-$AsMe_2$)$_2$][$AgBr_2$], 3, 95
$As_4Ag_2C_{20}H_{32}Cl_2$ [Ag(2-$Me_2AsC_6H_4$-$AsMe_2$)$_2$][$AgCl_2$], 3, 95
$As_4Ag_2C_{20}H_{32}I_2$ [Ag(2-$Me_2AsC_6H_4$-$AsMe_2$)$_2$][AgI_2], 3, 95
$As_4Ag_2C_{20}H_{32}N_2O_6$ [Ag(2-$Me_2AsC_6H_4$-$AsMe_2$)$_2$][Ag(NO_3)$_2$], 3, 95
$As_4Ag_4C_{24}H_{60}I_4$ [Ag($AsEt_3$)I]$_4$, 3, 11
$As_4Ag_4C_{36}H_{84}I_4$ [Ag($AsPr_3$)I]$_4$, 3, 14
$As_4AuC_{20}H_{32}Br$ [Au(2-$Me_2AsC_6H_4$-$AsMe_2$)$_2$]Br, 3, 84
$As_4AuC_{20}H_{32}Br_2ClO_4$ [Au(2-$Me_2AsC_6H_4$-$AsMe_2$)$_2Br_2$]ClO_4, 3, 85
$As_4AuC_{20}H_{32}ClI_2O_4$ [Au(2-$Me_2AsC_6H_4$-$AsMe_2$)$_2I_2$]ClO_4, 3, 85
$As_4AuC_{20}H_{32}ClO_4$ [Au(2-$Me_2AsC_6H_4$-$AsMe_2$)$_2$]ClO_4, 3, 84
$As_4AuC_{20}H_{32}Cl_2IO_8$ [Au(2-$Me_2AsC_6H_4$-$AsMe_2$)$_2$I](ClO_4)$_2$, 3, 84
$As_4AuC_{20}H_{32}Cl_3O_{12}$ [Au(2-$Me_2AsC_6H_4$-$AsMe_2$)$_2$](ClO_4)$_3$, 3, 84
$As_4AuC_{20}H_{32}I$ [Au(2-$Me_2AsC_6H_4AsMe_2$)$_2$]I, 3, 84
$As_4AuC_{20}H_{32}I_3$ [Au(2-$Me_2NC_6H_4AsMe_2$)$_2$-I_2]I, 3, 85
$As_4AuC_{20}H_{32}N_3O_9$ [Au(2-$Me_2AsC_6H_4$-$AsMe_2$)$_2NO_3$](NO_3)$_2$, 3, 85
$As_4AuC_{24}H_{40}I$ [Au(2-$Et_2AsC_6H_4AsMe_2$)$_2$]I, 3, 98
$As_4AuC_{26}H_{34}N_3O_7$ [Au(2-$Me_2AsC_6H_4$-$AsMe_2$)$_2$]OC_6H_2(NO_2)$_3$, 3, 84
$As_4AuC_{28}H_{48}Br$ [Au(2-$Et_2AsC_6H_4AsEt_2$)$_2$]-Br, 3, 99
$As_4AuC_{28}H_{48}I$ [Au(2-$Et_2AsC_6H_4AsEt_2$)$_2$]I, 3, 99
$As_4AuC_{34}H_{54}BrO_4S$ [Au(2-$Et_2AsC_6H_4$-$AsMe_2$)$_2$](+)-bromocamphorsulfonate, 3, 98
$As_4AuC_{34}H_{55}O_4S$ [Au(2-$Et_2AsC_6H_4$-$AsMe_2$)$_2$](+)-camphorsulfonate, 3, 98

$As_4Au_2C_{20}H_{32}I_2$ $[Au(2\text{-}Me_2AsC_6H_4AsMe_2)_2][AuI_2]$, 3, 84

$As_4Au_2C_{60}H_{100}O_8P_4 \cdot H_2O$ $\{Au[2\text{-}(Et_2As)C_6H_4PEt_2]_2\}_2[OC(O)CH(OH)\text{-}]_2 \cdot H_2O$, 3, 60

$As_4CdC_{20}H_{32}Cl_2O_8$ $[Cd(2\text{-}Me_2AsC_6H_4AsMe_2)_2](ClO_4)_2$, 3, 81

$As_4CoC_{15}H_{36}Br_2ClO_4$ $\{Co[(Me_2AsCH_2CH_2CH_2)_3As]Br_2\}ClO_4$, 3, 109

$As_4CoC_{15}H_{36}Br_3$ $\{Co[(Me_2AsCH_2CH_2CH_2)_3As]Br_2\}Br$, 3, 109

$As_4CoC_{15}H_{36}Cl_3O_4$ $\{Co[(Me_2AsCH_2CH_2CH_2)_3As]Cl_2\}ClO_4$, 3, 110

$As_4CoC_{15}H_{36}I_3$ $\{Co[(Me_2AsCH_2CH_2CH_2)_3As]I_2\}I$, 3, 110

$As_4CoC_{20}H_{32}Br_2$ $[Co(2\text{-}Me_2AsC_6H_4AsMe_2)_2]Br_2$, 3, 82

$As_4CoC_{20}H_{32}Br_2ClO_4$ $[Co(2\text{-}Me_2AsC_6H_4AsMe_2)_2Br_2]ClO_4$, 3, 82

$As_4CoC_{20}H_{32}Br_3$ $[Co(2\text{-}Me_2AsC_6H_4AsMe_2)_2Br_2]Br$, 3, 82

$As_4CoC_{20}H_{32}Cl_2$ $[Co(2\text{-}Me_2AsC_6H_4AsMe_2)_2]Cl_2$, 3, 82

$As_4CoC_{20}H_{32}Cl_3$ $[Co(2\text{-}Me_2AsC_6H_4AsMe_2)_2Cl_2]Cl$, 3, 83

$As_4CoC_{20}H_{32}Cl_3O_4$ $[Co(2\text{-}Me_2AsC_6H_4AsMe_2)_2Cl_2]ClO_4$, 3, 83

$As_4CoC_{20}H_{32}I_2$ $[Co(2\text{-}Me_2AsC_6H_4AsMe_2)_2]I_2$, 3, 82

$As_4CoC_{20}H_{32}I_3$ $[Co(2\text{-}Me_2AsC_6H_4AsMe_2)_2I_2]I$, 3, 83

$As_4CoC_{22}H_{32}ClN_2O_4S_2$ $[Co(2\text{-}Me_2AsC_6H_4AsMe_2)_2(SCN)_2]ClO_4$, 3, 83

$As_4CoC_{22}H_{32}N_2S_2$ $[Co(2\text{-}Me_2AsC_6H_4AsMe_2)_2](SCN)_2$, 3, 82

$As_4CoC_{24}H_{38}ClO_8$ $[Co(2\text{-}Me_2AsC_6H_4AsMe_2)_2(OAc)_2]ClO_4$, 3, 83

$As_4CoC_{72}H_{60}Cl_2O_{12}$ $[Co(Ph_3AsO)_4](ClO_4)_2$, 3, 353

$As_4CoC_{72}H_{60}I_2O_4$ $[Co(Ph_3AsO)_4]I_2$, 3, 353

$As_4CrC_{20}H_{32}Br_2ClO_4$ $[Cr(2\text{-}Me_2AsC_6H_4AsMe_2)_2Br_2]ClO_4$, 3, 81

$As_4CrC_{20}H_{32}ClI_2O_4$ $[Cr(2\text{-}Me_2AsC_6H_4AsMe_2)_2I_2]ClO_4$, 3, 81

$As_4CrC_{20}H_{32}Cl_3O_4$ $[Cr(2\text{-}Me_2AsC_6H_4AsMe_2)_2Cl_2]ClO_4$, 3, 81

$As_4CrC_{20}H_{32}I_5$ $[Cr(2\text{-}Me_2AsC_6H_4AsMe_2)_2I_2]I_3$, 3, 81

$As_4CrC_{22}H_{32}Br_2O_2$ $[Cr(2\text{-}Me_2AsC_6H_4AsMe_2)_2(CO)_2Br]Br$, 3, 82

$As_4CrC_{22}H_{32}Br_4O_2$ $[Cr(2\text{-}Me_2AsC_6H_4AsMe_2)_2(CO)_2Br]Br_3$, 3, 82

$As_4CrC_{22}H_{32}I_2O_2$ $[Cr(2\text{-}Me_2AsC_6H_4AsMe_2)_2(CO)_2I]I$, 3, 82

$As_4CrC_{22}H_{32}I_4O_2$ $[Cr(2\text{-}Me_2AsC_6H_4AsMe_2)_2(CO)_2I]I_3$, 3, 82

$As_4CrC_{22}H_{32}O_2$ $[Cr(2\text{-}Me_2AsC_6H_4AsMe_2)_2(CO)_2]$, 3, 82

$As_4CrC_{72}H_{60}Cl_3O_4$ $[Cr(Ph_3AsO)_4Cl_2]Cl$, 3, 353

$As_4CrC_{72}H_{60}Cl_3O_8$ $[Cr(Ph_3AsO)_4Cl_2]ClO_4$, 3, 353

$As_4Cr_2C_{20}H_{32}Br_6$ $[Cr(2\text{-}Me_2AsC_6H_4AsMe_2)Br_3]_2$, 3, 81

$As_4Cr_2C_{20}H_{32}I_6$ $[Cr(2\text{-}Me_2AsC_6H_4AsMe_2)I_3]_2$, 3, 81

$As_4CuC_{20}H_{32}Br$ $[Cu(2\text{-}Me_2AsC_6H_4AsMe_2)_2]Br$, 3, 84

$As_4CuC_{20}H_{32}ClO_4$ $[Cu(2\text{-}Me_2AsC_6H_4AsMe_2)_2]ClO_4$, 3, 84

$As_4CuC_{20}H_{32}I$ $[Cu(2\text{-}Me_2AsC_6H_4AsMe_2)_2]I$, 3, 84

$As_4CuC_{40}H_{40}I$ $[Cu(2\text{-}Me_2AsC_6H_4AsPh_2)_2]I$, 3, 100

$As_4CuC_{50}H_{54}BrO_4S$ $[Cu(2\text{-}Me_2AsC_6H_4AsPh_2)_2]$(+)-bromocamphorsulfonate, 3, 100

$As_4CuC_{52}H_{52}ClO_4$ $[Cu(Ph_2AsMe)_4]ClO_4$, 3, 52

$As_4CuC_{52}H_{52}I$ $[Cu(Ph_2AsMe)_4]I$, 3, 52

$As_4CuC_{52}H_{52}NO_3$ $[Cu(Ph_2AsMe)_4]NO_3$, 3, 52

$As_4CuC_{72}H_{60}N_2O_{10}$ $[Cu(Ph_3AsO)_4](NO_3)_2$, 3, 353

$As_4Cu_2C_{20}H_{32}Br_2$ $[Cu(2\text{-}Me_2AsC_6H_4AsMe_2)_2][CuBr_2]$, 3, 84

$As_4Cu_2C_{20}H_{32}Cl_2$ $[Cu(2\text{-}Me_2AsC_6H_4AsMe_2)_2][CuCl_2]$, 3, 84

$As_4Cu_2C_{20}H_{32}I_2$ $[Cu(2\text{-}Me_2AsC_6H_4AsMe_2)_2][CuI_2]$, 3, 84

$As_4Cu_2C_{30}H_{52}I_2$ $\{Cu[4,1,2\text{-}MeC_6H_3(AsEt_2)_2]_2\}[CuI_2]$, 3, 99

$As_4Cu_2C_{40}H_{40}I_2$ $[Cu(2\text{-}Me_2AsC_6H_4\text{-}$

$AsPh_2)_2][CuI_2]$, 3, 100

$As_4Cu_2C_{52}H_{52}Br_2$ $[Cu(Ph_2AsMe)_4][CuBr_2]$, 3, 52

$As_4Cu_2C_{52}H_{52}Cl_2$ $[Cu(Ph_2AsMe)_4][CuCl_2]$, 3, 52

$As_4Cu_4C_{12}H_{36}I_4$ $[Cu(AsMe_3)I]_4$, 3, 4

$As_4Cu_4C_{24}H_{60}Br_4$ $[Cu(AsEt_3)Br]_4$, 3, 8

$As_4Cu_4C_{24}H_{60}I_4$ $[Cu(AsEt_3)I]_4$, 3, 8

$As_4Cu_4C_{36}H_{84}I_4$ $[Cu(AsPr_3)I]_4$, 3, 12

$As_4Cu_4C_{48}H_{108}I_4$ $[Cu(AsBu_3)I]_4$, 3, 15

$As_4Cu_4C_{52}H_{52}Br_4$ $[Cu(Ph_2AsMe)Br]_4$, 3, 52

$As_4Cu_4C_{52}H_{52}Cl_4$ $[Cu(Ph_2AsMe)Cl]_4$, 3, 52

$As_4Cu_4C_{52}H_{52}I_4$ $[Cu(Ph_2AsMe)I]_4$, 3, 52

$As_4FeC_{17}H_{36}N_2S_2$ $\{Fe[(Me_2AsCH_2CH_2CH_2)_3As](SCN)_2\}$, 3, 110

$As_4FeC_{20}H_{32}Br_2$ $[Fe(2\text{-}Me_2AsC_6H_4AsMe_2)_2Br_2]$, 3, 85

$As_4FeC_{20}H_{32}Br_3$ $[Fe(2\text{-}Me_2AsC_6H_4AsMe_2)_2Br_2]Br$, 3, 86

$As_4FeC_{20}H_{32}Cl_3O_4$ $[Fe(2\text{-}Me_2AsC_6H_4AsMe_2)_2Cl_2]ClO_4$, 3, 86

$As_4FeC_{20}H_{32}Cl_4O_8$ $[Fe(2\text{-}Me_2AsC_6H_4AsMe_2)_2Cl_2](ClO_4)_2$, 3, 86

$As_4FeC_{20}H_{32}I_2$ $[Fe(2\text{-}Me_2AsC_6H_4AsMe_2)_2I_2]$, 3, 85

$As_4FeC_{21}H_{32}O$ $[Fe(2\text{-}Me_2AsC_6H_4AsMe_2)_2(CO)]$, 3, 85

$As_4FeC_{22}H_{32}N_2S_2$ $[Fe(2\text{-}Me_2AsC_6H_4AsMe_2)_2(SCN)_2]$, 3, 86

$As_4FeC_{72}H_{60}Cl_2O_{12}$ $[Fe(Ph_3AsO)_4](ClO_4)_2$, 3, 353

$As_4FeC_{72}H_{60}I_6$ $[Fe(AsPh_3)_4]I_6$, 3, 21

$As_4Fe_2C_{15}H_{36}Cl_6$ $\{Fe[(Me_2AsCH_2CH_2CH_2)_3As]Cl_2\}FeCl_4$, 3, 110

$As_4Fe_2C_{20}H_{32}Br_6$ $[Fe(2\text{-}Me_2AsC_6H_4AsMe_2)_2Br_2]FeBr_4$, 3, 86

$As_4Fe_2C_{20}H_{32}Cl_6$ $[Fe(2\text{-}Me_2AsC_6H_4AsMe_2)_2Cl_2]FeCl_4$, 3, 86

$As_4Fe_2C_{36}H_{52}Cl_6$ $[Fe(Me_2AsC_6H_4Me\text{-}2)_2Cl_3]_2$, 3, 43

$As_4Fe_2C_{52}H_{52}Cl_6$ $[Fe(Ph_2AsMe)_2Cl_3]_2$, 3, 53

$As_4Fe_2C_{72}H_{60}I_4$ $[Fe(AsPh_3)_4]FeI_4$, 3, 21

$As_4Fe_3C_{20}H_{32}Cl_{10}$ $[Fe(2\text{-}Me_2AsC_6H_4AsMe_2)_2Cl_2][FeCl_4]_2$, 3, 86

$As_4Ge_2C_{90}H_{74}N_4$ $p\text{-}C_6H_4[GePh(N{=}AsPh_3)_2]_2$, 2, 128, 3, 369

$As_4HgC_{20}H_{32}Cl_2O_8$ $[Hg(2\text{-}Me_2AsC_6H_4AsMe_2)_2](ClO_4)_2$, 3, 87

$As_4MnC_{20}H_{32}Br_2$ $[Mn(2\text{-}Me_2AsC_6H_4AsMe_2)_2Br_2]$, 3, 87

$As_4MnC_{20}H_{32}Cl_2$ $[Mn(2\text{-}Me_2AsC_6H_4AsMe_2)_2Cl_2]$, 3, 87

$As_4MnC_{20}H_{32}I_2$ $[Mn(2\text{-}Me_2AsC_6H_4AsMe_2)_2I_2]$, 3, 87

$As_4MnC_{20}H_{34}Br_2ClO_5$ $[Mn(2\text{-}Me_2AsC_6H_4AsMe_2)_2(H_2O)Br_2]ClO_4$, 3, 87

$As_4MnC_{72}H_{60}I_2O_4$ $[Mn(Ph_3AsO)_4I_2]$, 3, 354

$As_4MnC_{72}H_{60}Cl_2O_{12}$ $[Mn(Ph_3AsO)_4](ClO_4)_2$, 3, 354

$As_4MnC_{72}H_{60}I_8$ $[Mn(AsPh_3)_4]I_8$, 3, 22

$As_4Mn_2C_{26}H_{32}O_6$ $[Mn(2\text{-}Me_2AsC_6H_4AsMe_2)(CO)_3]_2$, 3, 86

$As_4MoC_{20}H_{32}Br_2$ $[Mo(2\text{-}Me_2AsC_6H_4AsMe_2)_2Br_2]$, 3, 88

$As_4MoC_{20}H_{32}Cl_2$ $[Mo(2\text{-}Me_2AsC_6H_4AsMe_2)_2Cl_2]$, 3, 88

$As_4MoC_{20}H_{32}I_2$ $[Mo(2\text{-}Me_2AsC_6H_4AsMe_2)_2I_2]$, 3, 88

$As_4MoC_{22}H_{32}Br_2O_2$ $[Mo(2\text{-}Me_2AsC_6H_4AsMe_2)_2(CO)_2Br]Br$, 3, 88

$As_4MoC_{22}H_{32}Br_4O_2$ $[Mo(2\text{-}Me_2AsC_6H_4AsMe_2)_2(CO)_2Br]Br_3$, 3, 88

$As_4MoC_{22}H_{32}I_2O_2$ $[Mo(2\text{-}Me_2AsC_6H_4AsMe_2)_2(CO)_2I]I$, 3, 88

$As_4MoC_{22}H_{32}O_2$ $[Mo(2\text{-}Me_2AsC_6H_4AsMe_2)_2(CO)_2]$, 3, 87

$As_4MoC_{72}H_{60}Cl_4O_4$ $[Mo(Ph_3AsO)_4Cl_4]$, 3, 354

$As_4Mo_6C_{72}H_{62}Cl_{14}O_4$ $\{(Ph_3AsO)_4\cdot H_2[(Mo_6Cl_8)Cl_6]\}$, 3, 354

$\{(Ph_3AsO)_4\cdot H_2[(Mo_6Cl_8)Cl_6]\}$, 3, 354

$As_4NiC_{15}H_{36}Br_2ClO_4$ $\{Ni[(Me_2AsCH_2CH_2CH_2)_3As]Br_2\}ClO_4$, 3, 110

$As_4NiC_{15}H_{36}Br_3$ $\{Ni[(Me_2AsCH_2CH_2CH_2)_3As]Br_2\}Br$, 3, 110

$As_4NiC_{15}H_{36}Cl_3O_4$ $\{Ni[(Me_2AsCH_2CH_2CH_2)_3As]Cl_2\}ClO_4$, 3, 110

$As_4NiC_{15}H_{36}I_2$ $\{Ni[(Me_2AsCH_2CH_2CH_2)_3As]I_2\}$, 3, 110

$As_4NiC_{15}H_{36}I_3$ $\{Ni[(Me_2AsCH_2CH_2CH_2)_3As]I_2\}I$, 3, 110

$As_4NiC_{20}H_{32}$ $[Ni(2-Me_2AsC_6H_4AsMe_2)_2]$, 3, 88

$As_4NiC_{20}H_{32}BrClO_4$ $[Ni(2-Me_2AsC_6H_4AsMe_2)_2Br]ClO_4$, 3, 89

$As_4NiC_{20}H_{32}Br_2$ $[Ni(2-Me_2AsC_6H_4AsMe_2)_2Br]Br$, 3, 89

$As_4NiC_{20}H_{32}Br_2Cl_2O_8$ $[Ni(2-Me_2AsC_6H_4AsMe_2)_2Br_2](ClO_4)_2$, 3, 90

$As_4NiC_{20}H_{32}Br_3$ $[Ni(2-Me_2AsC_6H_4AsMe_2)_2Br_2]Br$, 3, 90

$As_4NiC_{20}H_{32}Cl_2$ $[Ni(2-Me_2AsC_6H_4AsMe_2)_2Cl]Cl$, 3, 89

$As_4NiC_{20}H_{32}Cl_2O_4$ $[Ni(2-Me_2AsC_6H_4AsMe_2)_2Cl]ClO_4$, 3, 89

$As_4NiC_{20}H_{32}Cl_2O_8$ $[Ni(2-Me_2AsC_6H_4AsMe_2)_2](ClO_4)_2$, 3, 88

$As_4NiC_{20}H_{32}Cl_3$ $[Ni(2-Me_2AsC_6H_4AsMe_2)_2Cl_2]Cl$, 3, 90

$As_4NiC_{20}H_{32}Cl_3O_4$ $[Ni(2-Me_2AsC_6H_4AsMe_2)_2Cl_2]ClO_4$, 3, 90

$As_4NiC_{20}H_{32}Cl_4O_8$ $[Ni(2-Me_2AsC_6H_4AsMe_2)_2Cl_2](ClO_4)_2$, 3, 90

$As_4NiC_{20}H_{32}I_2$ $[Ni(2-Me_2AsC_6H_4AsMe_2)_2I]I$, 3, 89

$As_4NiC_{20}H_{32}I_3$ $[Ni(2-Me_2AsC_6H_4AsMe_2)_2I_2]I$, 3, 90

$As_4NiC_{20}H_{32}N_2O_4$ $[Ni(2-Me_2AsC_6H_4AsMe_2)_2NO_2]NO_2$, 3, 89

$As_4NiC_{20}H_{32}N_2O_6$ $[Ni(2-Me_2AsC_6H_4AsMe_2)_2NO_3]NO_3$, 3, 89

$As_4NiC_{22}H_{32}N_2S_2$ $[Ni(2-Me_2AsC_6H_4AsMe_2)_2(SCN)]SCN$, 3, 89

$As_4NiC_{23}H_{32}N_3S_3$ $[Ni(2-Me_2AsC_6H_4AsMe_2)_2(SCN)_2]SCN$, 3, 90

$As_4NiC_{72}H_{60}$ $[Ni(AsPh_3)_4]$, 3, 24

$As_4NiC_{72}H_{60}Cl_2O_{12}$ $[Ni(Ph_3AsO)_4](ClO_4)_2$, 3, 355

$As_4Ni\ C_{24}H_{32}N_4S_4$ $[Ni(2-Me_2AsC_6H_4AsMe_2)_2(SCN)_2]SCN$, 3, 90

$As_4OsC_{20}H_{32}BrI_2$ $[Os(2-Me_2AsC_6H_4AsMe_2)_2I_2]Br$, 3, 90

$As_4OsC_{20}H_{32}Br_2$ $[Os(2-Me_2AsC_6H_4AsMe_2)_2Br_2]$, 3, 90

$As_4OsC_{20}H_{32}Br_2ClO_4$ $[Os(2-Me_2AsC_6H_4AsMe_2)_2Br_2]ClO_4$, 3, 90

$As_4OsC_{20}H_{32}Br_2Cl_2O_8$ $[Os(2-Me_2AsC_6H_4AsMe_2)_2Br_2](ClO_4)_2$, 3, 91

$As_4OsC_{20}H_{32}Br_3$ $[Os(2-Me_2AsC_6H_4AsMe_2)_2Br_2]Br$, 3, 90

$As_4OsC_{20}H_{32}ClI_2O_4$ $[Os(2-Me_2AsC_6H_4AsMe_2)_2I_2]ClO_4$, 3, 91

$As_4OsC_{20}H_{32}Cl_2$ $[Os(2-Me_2AsC_6H_4AsMe_2)_2Cl_2]$, 3, 90

$As_4OsC_{20}H_{32}Cl_2I_2O_8$ $[Os(2-Me_2AsC_6H_4AsMe_2)_2I_2](ClO_4)_2$, 3, 91

$As_4OsC_{20}H_{32}Cl_3$ $[Os(2-Me_2AsC_6H_4AsMe_2)_2Cl_2]Cl$, 3, 90

$As_4OsC_{20}H_{32}Cl_3O_4$ $[Os(2-Me_2AsC_6H_4AsMe_2)_2Cl_2]ClO_4$, 3, 91

$As_4OsC_{20}H_{32}Cl_4O_8$ $[Os(2-Me_2AsC_6H_4AsMe_2)_2Cl_2](ClO_4)_2$, 3, 90

$As_4OsC_{20}H_{32}I_2$ $[Os(2-Me_2AsC_6H_4AsMe_2)_2I_2]$, 3, 90

$As_4OsC_{22}H_{32}ClN_2O_4S_2$ $[Os(2-Me_2AsC_6H_4AsMe_2)_2(SCN)_2]ClO_4$, 3, 91

$As_4OsC_{22}H_{32}N_2S_2$ $[Os(2-Me_2AsC_6H_4AsMe_2)_2(SCN)_2]$, 3, 90

$As_4OsC_{32}H_{44}Br_2$ $Os(Me_2AsPh)_4Br_2$, 3, 41

$As_4OsC_{32}H_{44}Cl_2$ $Os(Me_2AsPh)_4Cl_2$, 3, 41

$As_4PdC_{20}H_{32}$ $[Pd(2-Me_2AsC_6H_4AsMe_2)_2]$, 3, 91

$As_4PdC_{20}H_{32}BrClO_4$ $[Pd(2-Me_2AsC_6H_4AsMe_2)_2Br]ClO_4$, 3, 92

$As_4PdC_{20}H_{32}Br_2$ $[Pd(2-Me_2AsC_6H_4AsMe_2)_2Br]Br$, 3, 91

$As_4PdC_{20}H_{32}Br_2Cl_2O_8$ $[Pd(2-Me_2AsC_6H_4AsMe_2)_2Br_2](ClO_4)_2$, 3, 92

$As_4PdC_{20}H_{32}ClIO_4$ $[Pd(2-Me_2AsC_6H_4AsMe_2)_2I]ClO_4$, 3, 92

$As_4PdC_{20}H_{32}Cl_2$ $[Pd(2-Me_2AsC_6H_4AsMe_2)_2Cl]Cl$, 3, 91

$As_4PdC_{20}H_{32}Cl_2O_4$ $[Pd(2-Me_2AsC_6H_4AsMe_2)_2Cl]ClO_4$, 3, 92

$As_4PdC_{20}H_{32}Cl_2O_8$ $[Pd(2-Me_2AsC_6H_4AsMe_2)_2](ClO_4)_2$, 3, 91
 $[Pd(2-Me_2AsC_6H_4AsMe_2)_2](ClO_4)_2$, 3, 91

$As_4PdC_{20}H_{32}Cl_4O_8$ $[Pd(2-Me_2AsC_6H_4AsMe_2)_2Cl_2](ClO_4)_2$, 3, 92

$As_4PdC_{20}H_{32}I_2$ $[Pd(2-Me_2AsC_6H_4AsMe_2)_2I]I$, 3, 92

$As_4PdC_{20}H_{32}N_2O_4$ $[Pd(2-Me_2AsC_6H_4AsMe_2)$-

$NO_2]NO_2$, 3, 92
$As_4PdC_{22}H_{32}N_2S_2$ $[Pd(2\text{-}Me_2AsC_6H_4\text{-}AsMe_2)_2SCN]SCN$, 3, 92
$As_4PdC_{44}H_{80}Cl_2$ $\{Pd[o\text{-}C_6H_4(AsBu_2)_2]_2\}\text{-}Cl_2$, 3, 101
$As_4PdC_{52}H_{48}Cl_2O_8$ $\{Pd[(Ph_2AsCH_2\text{-})_2]_2\}\text{-}(ClO_4)_2$, 3, 103
$As_4PdC_{54}H_{42}Br_2$ $\{Pd[(2\text{-}Ph_2AsC_6H_4)_3As]\text{-}Br\}Br$, 3, 110
$As_4PdC_{54}H_{42}Cl_2$ $\{Pd[(2\text{-}Ph_2AsC_6H_4)_3As]\text{-}Cl\}Cl$, 3, 110
$As_4PdC_{54}H_{42}Cl_2O_4$ $\{Pd[(2\text{-}Ph_2AsC_6H_4)_3\text{-}As]Cl\}ClO_4$, 3, 110
$As_4PdC_{54}H_{42}I_2$ $\{Pd[(2\text{-}Ph_2AsC_6H_4)_3As]\text{-}I\}I$, 3, 111
$As_4PdC_{56}H_{42}N_2S_2$ $\{Pd[(2\text{-}Ph_2AsC_6H_4)_3\text{-}As](SCN)\}SCN$, 3, 111
$As_4PdC_{72}H_{60}$ $[Pd(AsPh_3)_4]$, 3, 24
$As_4Pd_2C_{44}H_{80}Cl_4$ $\{Pd[o\text{-}C_6H_4(AsBu_2)_2]\text{-}_2\}[PdCl_4]$, 3, 101
$As_4PtC_{20}H_{32}BrClO_4$ $[Pt(2\text{-}Me_2AsC_6H_4\text{-}AsMe_2)_2Br]ClO_4$, 3, 93
$As_4PtC_{20}H_{32}Br_2$ $[Pt(2\text{-}Me_2AsC_6H_4\text{-}AsMe_2)_2Br]Br$, 3, 93
$As_4PtC_{20}H_{32}Br_2Cl_2O_8$ $[Pt(2\text{-}Me_2AsC_6H_4\text{-}AsMe_2)_2Br_2](ClO_4)_2$, 3, 93
$As_4PtC_{20}H_{32}ClIO_4$ $[Pt(2\text{-}Me_2AsC_6H_4\text{-}AsMe_2)_2I]ClO_4$, 3, 93
$As_4PtC_{20}H_{32}Cl_2$ $[Pt(2\text{-}Me_2AsC_6H_4\text{-}AsMe_2)_2Cl]Cl$, 3, 93
$As_4PtC_{20}H_{32}Cl_2I_2O_8$ $[Pt(2\text{-}Me_2AsC_6H_4\text{-}AsMe_2)_2I_2](ClO_4)_2$, 3, 93
$As_4PtC_{20}H_{32}Cl_2O_4$ $[Pt(2\text{-}Me_2AsC_6H_4\text{-}AsMe_2)_2Cl]ClO_4$, 3, 93
$As_4PtC_{20}H_{32}Cl_2O_8$ $[Pt(2\text{-}Me_2AsC_6H_4\text{-}AsMe_2)_2](ClO_4)_2$, 3, 93
$As_4PtC_{20}H_{32}Cl_4O_8$ $[Pt(2\text{-}Me_2AsC_6H_4\text{-}AsMe_2)_2Cl_2](ClO_4)_2$, 3, 93
$As_4PtC_{20}H_{32}I_2$ $[Pt(2\text{-}Me_2AsC_6H_4AsMe_2)_2\text{-}I]I$, 3, 93
$As_4PtC_{20}H_{32}I_2N_2O_6$ $[Pt(2\text{-}Me_2AsC_6H_4\text{-}AsMe_2)_2I_2](NO_3)_2$, 3, 93
$As_4PtC_{20}H_{32}I_4$ $[Pt(2\text{-}Me_2AsC_6H_4\text{-}AsMe_2)_2]I_4$, 3, 93
$As_4PtC_{20}H_{32}N_2O_6$ $[Pt(2\text{-}Me_2AsC_6H_4\text{-}AsMe_2)_2NO_3]NO_3$, 3, 93
$As_4PtC_{22}H_{32}N_2S_2$ $[Pt(2\text{-}Me_2AsC_6H_4\text{-}AsMe_2)_2SCN]SCN$, 3, 93
$As_4PtC_{54}H_{42}Br_2$ $\{Pt[(2\text{-}Ph_2AsC_6H_4)_3As]\text{-}Br\}Br$, 3, 111
$As_4PtC_{54}H_{42}ClIO_4$ $\{Pt[(2\text{-}Ph_2AsC_6H_4)_3\text{-}As]I\}ClO_4$, 3, 111
$As_4PtC_{54}H_{42}Cl_2$ $\{Pt[(2\text{-}Ph_2AsC_6H_4)_3As]\text{-}Cl\}Cl$, 3, 111
$As_4PtC_{54}H_{42}Cl_2O_4$ $\{Pt[(2\text{-}Ph_2AsC_6H_4)_3\text{-}As]Cl\}ClO_4$, 3, 111
$As_4PtC_{54}H_{42}I_2$ $\{Pt[(2\text{-}Ph_2AsC_6H_4)_3As]\text{-}I\}I$, 3, 111
$As_4PtC_{56}H_{42}N_2S_2$ $\{Pt[(2\text{-}Ph_2AsC_6H_4)_3As]\text{-}(SCN)\}SCN$, 3, 111
$As_4PtC_{72}H_{60}$ $[Pt(AsPh_3)_4]$, 3, 25
$As_4PtC_{72}H_{48}Cl_{12}$ $\{Pt[As(C_6H_4Cl\text{-}4)_3]_4\}$, 3, 17
$As_4Pt_2C_{20}H_{32}Cl_4$ $[Pt(2\text{-}Me_2AsC_6H_4\text{-}AsMe_2)_2][PtCl_4]$, 3, 93
$As_4Pt_2C_{52}H_{52}Cl_6$ $[Pt(Ph_2AsMe)_4][PtCl_6]$, 3, 55
$As_4ReC_{20}H_{32}Br_2$ $[Re(2\text{-}Me_2AsC_6H_4AsMe_2)_2\text{-}Br_2]$, 3, 93
$As_4ReC_{20}H_{32}Br_2ClO_4$ $[Re(2\text{-}Me_2AsC_6H_4\text{-}AsMe_2)_2Br_2]ClO_4$, 3, 94
$As_4ReC_{20}H_{32}Br_7$ $[Re(2\text{-}Me_2AsC_6H_4AsMe_2)_2\text{-}Br_4]Br_3$, 3, 94
$As_4ReC_{20}H_{32}ClI_2O_4$ $[Re(2\text{-}Me_2AsC_6H_4\text{-}AsMe_2)_2I_2]ClO_4$, 3, 94
$As_4ReC_{20}H_{32}Cl_2$ $[Re(2\text{-}Me_2AsC_6H_4AsMe_2)_2\text{-}Cl_2]$, 3, 93
$As_4ReC_{20}H_{32}Cl_3O_4$ $[Re(2\text{-}Me_2AsC_6H_4\text{-}AsMe_2)_2Cl_2]ClO_4$, 3, 94
$As_4ReC_{20}H_{32}Cl_5O_4$ $[Re(2\text{-}Me_2AsC_6H_4\text{-}AsMe_2)_2Cl_4]ClO_4$, 3, 94
$As_4ReC_{20}H_{32}I_2$ $[Re(2\text{-}Me_2AsC_6H_4AsMe_2)_2\text{-}I_2]$, 3, 94
$As_4RhC_{20}H_{32}Br_3$ $[Rh(2\text{-}Me_2AsC_6H_4AsMe_2)_2\text{-}Br_2]Br$, 3, 94
$As_4RhC_{20}H_{32}Cl_3$ $[Rh(2\text{-}Me_2AsC_6H_4AsMe_2)_2\text{-}Cl_2]Cl$, 3, 94
$As_4RhC_{20}H_{32}I_3$ $[Rh(2\text{-}Me_2AsC_6H_4AsMe_2)_2\text{-}I_2]I$, 3, 94
$As_4RhC_{32}H_{44}Br_2$ $Rh(Me_2AsPh)_4Br_2$, 3, 41
$As_4RhC_{32}H_{44}Cl_2$ $Rh(Me_2AsPh)_4Cl_2$, 3, 41
$As_4RhC_{32}H_{44}I_2$ $Rh(Me_2AsPh)_4I_2$, 3, 41
$As_4RhC_{36}H_{52}Br_2$ $[Rh(Me_2AsC_6H_4Me\text{-}4)_4\text{-}Br_2]$, 3, 43

$As_4RhC_{38}H_{52}Cl_2$ $[Rh(Me_2AsC_6H_4Me\text{-}4)_4Cl_2]$, 3, 43
$As_4RhC_{38}H_{52}I_2$ $[Rh(Me_2AsC_6H_4Me\text{-}4)_4I_2]$, 3, 43
$As_4RhC_{54}H_{42}Br$ $\{Rh[(2\text{-}Ph_2AsC_6H_4)_3As]Br\}$, 3, 111
$As_4RhC_{54}H_{42}Br_3$ $\{Rh[(2\text{-}Ph_2AsC_6H_4)_3As]Br_2\}Br$, 3, 112
$As_4RhC_{54}H_{42}Cl$ $\{Rh[(2\text{-}Ph_2AsC_6H_4)_3As]Cl\}$, 3, 112
$As_4RhC_{54}H_{42}Cl_3$ $\{Rh[(2\text{-}Ph_2AsC_6H_4)_3As]Cl_2\}Cl$, 3, 112
$As_4RhC_{54}H_{42}Cl_3O_4$ $\{Rh[(2\text{-}Ph_2AsC_6H_4)_3As]Cl_2\}ClO_4$, 3, 112
$As_4RhC_{54}H_{42}I$ $\{Rh[(2\text{-}Ph_2AsC_6H_4)_3As]I\}$, 3, 112
$As_4RhC_{54}H_{42}I_3$ $\{Rh[(2\text{-}Ph_2AsC_6H_4)_3As]I_2\}I$, 3, 112
$As_4RhC_{55}H_{42}ClO$ $\{Rh[(2\text{-}Ph_2AsC_6H_4)_3As](CO)\}Cl$, 3, 112
$As_4Rh_2C_{24}H_{60}Cl_6$ $[Rh(AsEt_3)_2Cl_3]_2$, 3, 11
$As_4RuC_{20}H_{32}Br_2$ $[Ru(2\text{-}Me_2AsC_6H_4AsMe_2)_2Br_2]$, 3, 94
$As_4RuC_{20}H_{32}Br_3$ $[Ru(2\text{-}Me_2AsC_6H_4AsMe_2)_2Br_2]Br$, 3, 95
$As_4RuC_{20}H_{32}ClI_2O_4$ $[Ru(2\text{-}Me_2AsC_6H_4AsMe_2)_2I_2]ClO_4$, 3, 95
$As_4RuC_{20}H_{32}Cl_2$ $[Ru(2\text{-}Me_2AsC_6H_4AsMe_2)_2Cl_2]$, 3, 94
$As_4RuC_{20}H_{32}Cl_3$ $[Ru(2\text{-}Me_2AsC_6H_4AsMe_2)_2Cl_2]Cl$, 3, 95
$As_4RuC_{20}H_{32}Cl_3NO$ $[Ru(2\text{-}Me_2AsC_6H_4AsMe_2)_2(NO)Cl_3]$, 3, 95
$As_4RuC_{20}H_{32}Cl_3O_4$ $[Ru(2\text{-}Me_2AsC_6H_4AsMe_2)_2Cl_2]ClO_4$, 3, 95
$As_4RuC_{20}H_{32}I_2$ $[Ru(2\text{-}Me_2AsC_6H_4AsMe_2)_2I_2]$, 3, 94
$As_4RuC_{20}H_{33}Cl$ $[RuH(2\text{-}Me_2AsC_6H_4AsMe_2)_2Cl]$, 3, 94
$As_4RuC_{21}H_{35}Cl$ $[MeRu(2\text{-}Me_2AsC_6H_4AsMe_2)_2Cl]$, 3, 94
$As_4RuC_{22}H_{32}N_2S_2$ $[Ru(2\text{-}Me_2AsC_6H_4AsMe_2)_2(SCN)_2]$, 3, 95
$As_4RuC_{28}H_{48}Br_2$ $[Ru(2\text{-}Et_2AsC_6H_4AsEt_2)_2Br_2]$, 3, 99
$As_4RuC_{28}H_{48}Cl_2$ $[Ru(2\text{-}Et_2AsC_6H_4AsEt_2)_2Cl_2]$, 3, 99
$As_4RuC_{28}H_{48}I_2$ $[Ru(2\text{-}Et_2AsC_6H_4AsEt_2)_2I_2]\cdot 3H_2O$, 3, 99
$As_4RuC_{32}H_{44}Cl_4$ $Ru(Me_2AsPh)_4Cl_4$, 3, 42
$As_4RuC_{54}H_{42}Br_2$ $\{Ru[(2\text{-}Ph_2AsC_6H_4)_3As]Br_2\}$, 3, 112
$As_4RuC_{54}H_{42}Cl_2$ $\{Ru[(2\text{-}Ph_2AsC_6H_4)_3As]Cl_2\}$, 3, 112
$As_4RuC_{54}H_{42}I_2$ $\{Ru[(2\text{-}Ph_2AsC_6H_4)_3As]I_2\}$, 3, 112
$As_4RuC_{54}H_{42}N_2O_6$ $\{Ru[(2\text{-}Ph_2AsC_6H_4)_3As](NO_3)_2\}$, 3, 112
$As_4RuC_{56}H_{42}N_2S_2$ $\{Ru[(2\text{-}Ph_2AsC_6H_4)_3As](SCN)_2\}$, 3, 112
$As_4TcC_{20}H_{32}Br_2$ $[Tc(2\text{-}Me_2AsC_6H_4AsMe_2)_2Br_2]$, 3, 96
$As_4TcC_{20}H_{32}Br_3$ $[Tc(2\text{-}Me_2AsC_6H_4AsMe_2)_2Br_2]Br$, 3, 96
$As_4TcC_{20}H_{32}Cl_2$ $[Tc(2\text{-}Me_2AsC_6H_4AsMe_2)_2Cl_2]$, 3, 96
$As_4TcC_{20}H_{32}Cl_3$ $[Tc(2\text{-}Me_2AsC_6H_4AsMe_2)_2Cl_2]Cl$, 3, 96
$As_4TcC_{20}H_{32}Cl_5$ $[Tc(2\text{-}Me_2AsC_6H_4AsMe_2)_2Cl_4]Cl$, 3, 96
$As_4TcC_{20}H_{32}Cl_5O_4$ $[Tc(2\text{-}Me_2AsC_6H_4AsMe_2)_2Cl_4]ClO_4$, 3, 96
$As_4TcC_{20}H_{32}I_2$ $[Tc(2\text{-}Me_2AsC_6H_4AsMe_2)_2I_2]$, 3, 96
$As_4TcC_{20}H_{32}I_5$ $[Tc(2\text{-}Me_2AsC_6H_4AsMe_2)_2I_2]I_3$, 3, 96
$As_4TiC_{20}H_{32}Br_4$ $[Ti(2\text{-}Me_2AsC_6H_4AsMe_2)_2Br_4]$, 3, 97
$As_4TiC_{20}H_{32}Cl_4$ $[Ti(2\text{-}Me_2AsC_6H_4AsMe_2)_2Cl_4]$, 3, 97
$As_4VC_{20}H_{32}Cl_4$ $[V(2\text{-}Me_2AsC_6H_4AsMe_2)_2Cl_4]$, 3, 97
$As_4VC_{72}H_{60}Cl_2O_{13}$ $[VO(Ph_3AsO)_4](ClO_4)_2$, 3, 355
$As_4V_3C_{84}H_{60}O_{16}$ $[V(Ph_3AsO)_4][V(CO)_6]_2$, 3, 355
$As_4WC_{22}H_{32}Br_2O_2$ $[W(2\text{-}Me_2AsC_6H_4AsMe_2)_2(CO)_2Br]Br$, 3, 97
$As_4WC_{22}H_{32}Br_4O_2$ $[W(2\text{-}Me_2AsC_6H_4AsMe_2)_2(CO)_2Br]Br_3$, 3, 97
$As_4WC_{22}H_{32}I_2O_2$ $[W(2\text{-}Me_2AsC_6H_4AsMe_2)_2(CO)_2I]I$, 3, 97
$As_4WC_{22}H_{32}I_4O_2$ $[W(2\text{-}Me_2AsC_6H_4AsMe_2)_2$-

$(CO)_2I]I_3$, 3, 97
$AsWC_{22}H_{32}O_2$ $[W(2\text{-}Me_2AsC_6H_4AsMe_2)_2(CO)_2]$, 3, 97
$As_4ZnC_{20}H_{32}Cl_2O_4$ $[Zn(2\text{-}Me_2AsC_6H_4AsMe_2)_2](ClO_4)_2$, 3, 97
$As_4ZnC_{72}H_{60}Cl_2O_{12}$ $[Zn(Ph_3AsO)_4](ClO_4)_2$, 3, 355
$As_4ZrC_{20}H_{32}Cl_4$ $[Zr(2\text{-}Me_2AsC_6H_4AsMe_2)Cl_4]$, 3, 98
$As_6BiC_{30}H_{48}Cl_3O_{12}$ $[Bi(2\text{-}Me_2AsC_6H_4AsMe_2)_3](ClO_4)_3$, 3, 80
$As_6CoC_{22}H_{54}Cl_2O_8$ $\{Co[MeAs(CH_2CH_2CH_2AsMe_2)_2]_2\}(ClO_4)_2$, 3, 105
$As_6CoC_{30}H_{48}Cl_2O_8$ $[Co(2\text{-}Me_2AsC_6H_4AsMe_2)_3](ClO_4)_2$, 3, 83
$As_6CoC_{30}H_{48}Cl_3O_{12}$ $[Co(2\text{-}Me_2AsC_6H_4AsMe_2)_3](ClO_4)_3$, 3, 83
$As_6Cr_2C_{30}H_{48}Br_6$ $[Cr(2\text{-}Me_2AsC_6H_4AsMe_2)_2Br_2][Cr(2\text{-}Me_2AsC_6H_4AsMe_2)Br_4]$, 3, 81
$As_6Cr_2C_{30}H_{48}Cl_6$ $[Cr(2\text{-}Me_2AsC_6H_4AsMe_2)_2Cl_2][Cr(2\text{-}Me_2AsC_6H_4AsMe_2)Cl_4]$, 3, 81
$As_6Cr_2C_{30}H_{48}I_6$ $[Cr(2\text{-}Me_2AsC_6H_4AsMe_2)I_2][Cr(2\text{-}Me_2AsC_6H_4AsMe_2)I_4]$, 3, 82
$As_6Fe_4C_{54}H_{78}Cl_{12}$ $[(FeCl_3)_2(Me_2AsC_6H_4Me\text{-}4)_3]_2$, 3, 43
$As_6Fe_4C_{78}H_{78}Cl_{12}$ $[Fe_2(Ph_2AsMe)_3Cl_6]_2$, 3, 53
$As_6IrC_{54}H_{78}Br_2$ $[Ir(Me_2AsC_6H_4Me\text{-}4)_6Br_2]$, 3, 43
$As_6Ir_2C_{54}H_{78}Cl_4$ $[Ir(Me_2AsC_6H_4Me\text{-}4)_3Cl_2]_2$, 3, 43
$As_6Ir_2C_{54}H_{78}I_4$ $[Ir(Me_2AsC_6H_4Me\text{-}4)_3I_2]_2$, 3, 43
$As_6Ir_2C_{78}H_{78}Br_4$ $[Ir(Ph_2AsMe)_3Br_2]_2$, 3, 53
$As_6Ir_2C_{78}H_{78}Cl_4$ $[Ir(Ph_2AsMe)_3Cl_2]_2$, 3, 53
$As_6Ir_2C_{78}H_{78}I_4$ $[Ir(Ph_2AsMe)_3I_2]_2$, 3, 53
$As_6Mn_2C_{26}H_{54}O_4$ $\{Mn[MeAs(CH_2CH_2CH_2AsMe_2)_2](CO)_2\}_2$, 3, 105
$As_6NiC_{22}H_{54}Cl_2O_8$ $\{Ni[MeAs(CH_2CH_2CH_2AsMe_2)_2]_2\}(ClO_4)_2$, 3, 106
$As_6NiC_{30}H_{48}Cl_2O_8$ $[Ni(2\text{-}Me_2AsC_6H_4AsMe_2)_3](ClO_4)_2$, 3, 89
$As_6Ni_3C_{30}H_{72}I_6P_2$ $\{Ni[(Me_2AsCH_2CH_2CH_2)_3P]I\}_2[NiI_4]$ 3, 108
$As_6Os_2C_{48}H_{66}I_4$ $[Os(Me_2AsPh)_3I_2]_2$, 3, 41
$As_6Os_2C_{78}H_{78}I_4$ $[Os(Ph_2AsMe)_3I_2]_2$, 3, 54
$As_6PdC_{22}H_{54}Cl_2O_8$ $\{Pd[MeAs(CH_2CH_2CH_2AsMe_2)_2]_2\}(ClO_4)_2$, 3, 106
$As_6PtC_{22}H_{54}Cl_2O_8$ $\{Pt[MeAs(CH_2CH_2CH_2AsMe_2)_2]_2\}(ClO_4)_2$, 3, 106
$As_6Rh_2C_{48}H_{66}Cl_4$ $[Rh(Me_2AsPh)_3Cl_2]_2$, 3, 41
$As_6Rh_2C_{54}H_{78}Cl_4$ $[Rh(Me_2AsC_6H_4Me\text{-}4)_3Cl_2]_2$, 3, 43
$As_6Rh_2C_{54}H_{78}I_6$ $[Rh(Me_2AsC_6H_4Me\text{-}4)_6][RhI_6]$, 3, 43
$As_6Rh_2C_{78}H_{78}Br_6$ $[Rh(Ph_2AsMe)_6][RhBr_6]$, 3, 55
$As_6Rh_2C_{78}H_{78}Cl_6$ $[Rh(Ph_2AsMe)_6][RhCl_6]$, 3, 55
$As_6Rh_2C_{78}H_{78}I_6$ $[Rh(Ph_2AsMe)_6][RhI_6]$, 3, 55
$As_6Ru_2C_{30}H_{48}Br_6$ $[Ru(2\text{-}Me_2AsC_6H_4AsMe_2)_2Br_2][Ru(2\text{-}Me_2AsC_6H_4AsMe_2)Br_4]$, 3, 95
$As_6Ru_2C_{30}H_{48}Cl_6$ $[Ru(2\text{-}Me_2AsC_6H_4AsMe_2)_2Cl_2][Ru(2\text{-}Me_2AsC_6H_4AsMe_2)Cl_4]$, 3, 95
$As_7Rh_2C_{56}H_{77}I_5$ $[Rh(Me_2AsPh)_6][Rh(Me_2AsPh)I_5]$, 3, 41
$As_7Rh_2C_{63}H_{91}I_5$ $[Rh(Me_2AsC_6H_4Me\text{-}4)_6][Rh(Me_2AsC_6H_4Me\text{-}4)I_5]$, 3, 44
$As_8Co_3C_{40}H_{64}Cl_8$ $[Co(2\text{-}Me_2AsC_6H_4AsMe_2)_2Cl_2]_2[CoCl_4]$, 3, 83
$As_8Co_3C_{48}H_{64}N_8S_8$ $[Co(2\text{-}Me_2AsC_6H_4AsMe_2)_2(SCN)_2]_2[Co(SCN)_4]$, 3, 83
$As_8Mn_2C_{32}H_{72}O_2$ $\{Mn[(Me_2AsCH_2CH_2CH_2)_3As](CO)\}_2$, 3, 110
$As_8Ni_2PtC_{40}H_{64}Cl_{10}$ $[Ni(2\text{-}Me_2AsC_6H_4AsMe_2)_2Cl_2]_2[PtCl_6]$, 3, 90
$As_8Rh_2C_{64}H_{88}I_4$ $[Rh(Me_2AsPh)_6][Rh(Me_2AsPh)_2I_4]$, 3, 41
$AsBFe_2C_{40}H_{38}O_4$ $\{[C_5H_5Fe(CO)_2]_2AsMe_2\}BPh_4$, 1, 216, 3, 304
$AsGeMnC_{12}H_{14}I_2O_4$ $\{I_2EtGeMn[As(CH{=}CH_2)_3](CO)_4\}$, 3, 6
$AsHgSi_3C_{12}H_{33}Cl_2$ $[(Me_3SiCH_2)_3As\cdot Hg\text{-}$

$Cl_2]$, 3, 16

$AsMnPbC_{16}H_{14}Cl_2O_4$ $\{Cl_2PhPbMn[As(CH{=}CH_2)_3](CO)_4\}$, 3, 6

$AsMnPbC_{34}H_{42}O_4$ $\{Pr_3PbMn[As(CH_2Ph)_3](CO)_4\}$, 3, 27

$AsMnSnC_{10}H_{18}O_4$ $[Me_3SnMn(AsMe_3)(CO)_4]$, 3, 4

$AsMnSnC_{21}H_{21}ClO_4$ $\{Me_2Sn(Cl)Mn[As(C_5H_5)_3](CO)_4\}$, 3, 17

$AsMnSnC_{40}H_{30}O_4$ $[Ph_3SnMn(AsPh_3)(CO)_4]$, 2, 501, 3, 23

$As_2Co_2HgC_{42}H_{30}O_6$ $Hg[Co(AsPh_3)(CO)_3]_2$, 3, 20

$As_2Fe_2CdC_{40}H_{30}N_2O_6$ $Cd[Fe(AsPh_3)(NO)(CO)_2]_2$, 3, 22

$As_2Fe_2HgC_{40}H_{30}N_2O_6$ $Hg[Fe(AsPh_3)(NO)(CO)_2]_2$, 3, 22

$As_2Fe_2PtC_{50}H_{40}Cl_6O_4$ $[C_5H_5Fe(CO)_2(AsPh_3)]_2PtCl_6$, 1, 209

$As_2GeMn_2C_{58}H_{44}O_8$ $\{(PhCH_2)_2Ge[Mn(AsPh_3)(CO)_4]_2\}$, 3, 23

$As_2HgPdC_{18}H_{42}Br_4$ $[HgPd(AsPr_3)_2Br_4]$, 3, 13

$As_2Mn_2SiC_{36}H_{52}O_8$ $\{(Me_3CCH_2)_2Si\{Mn[As(CH_2CH{=}CH_2)_3](CO)_4\}_2\}$, 3, 11

$As_2Mn_2SiC_{120}H_{216}O_8$ $(2\text{-}EtC_6H_4)_2Si\{Mn(CO)_4[As(C_{16}H_{33})_3]\}_2$, 3, 29

$As_2Mn_2SnC_{30}H_{52}O_8$ $Et_2Sn[Mn(AsPr_3)(CO)_4]_2$, 3, 13

$As_2Mn_2SnC_{78}H_{84}O_8$ $(2,5\text{-}Me_2C_6H_3)_2Sn\{Mn[As(C_6H_2Me_3)_3](CO)_4\}_2$, 3, 29

$As_3Au_3Mn_3C_{26}H_{27}O_{15}$ $\{[MeC(CH_2AsMe_2)_3][AuMn(CO)_5]_3\}$, 3, 107

$As_3BNiC_{39}H_{56}BrP$ $\{Ni[(Me_2AsCH_2CH_2CH_2)_3P]Br\}BPh_4$, 3, 107

$As_3BNiC_{39}H_{56}ClP$ $\{Ni[(Me_2AsCH_2CH_2CH_2)_3P]Cl\}BPh_4$, 3, 107

$As_3BNiC_{39}H_{56}IP$ $\{Ni[(Me_2AsCH_2CH_2CH_2)_3P]I\}BPh_4$, 3, 108

$As_3CuMnC_{16}H_{27}O_5$ $\{Cu[MeC(CH_2AsMe_2)_3]Mn(CO)_5\}$, 3, 107

$As_3Fe_3SnC_{60}H_{45}ClN_3O_9$ $ClSn[Fe(AsPh_3)(NO)(CO)_2]_3$, 3, 22

$As_3GeMn_3C_{122}H_{228}O_{12}$ $CH_2{=}CHGe\{Mn[As(C_{12}H_{25})_3](CO)_4\}_3$, 3, 29

$As_3HgRhC_{39}H_{39}BrCl_2$ $[(Ph_2AsMe)_3Rh\text{-}Hg\text{-}BrCl_2]$, 3, 56

$As_3HgRhC_{39}H_{39}Cl_2F$ $[(Ph_2AsMe)_3Rh\text{-}HgFCl_2]$, 3, 56

$As_3HgRhC_{39}H_{39}Cl_2I$ $[(Ph_2AsMe)_3Rh\text{-}HgICl_2]$, 3, 56

$As_3HgRhC_{39}H_{39}Cl_3$ $[(Ph_2AsMe)_3Rh\text{-}HgCl_3]$, 3, 56

$As_3HgRhC_{41}H_{42}Cl_2O_2$ $[(Ph_2AsMe)_3Rh\text{-}Hg(OAc)Cl_2]$, 3, 56

$As_3RhSnC_{27}H_{39}Cl_5$ $[Rh(Me_2AsC_6H_4Me\text{-}4)_3SnCl_5]$, 3, 44

$As_4AgBC_{72}H_{60}F_4$ $[Ag(AsPh_3)_4]BF_4$, 3, 26

$As_4AuCuC_{20}H_{32}I_2$ $[Au(2\text{-}Me_2AsC_6H_4AsMe_2)_2][CuI_2]$, 3, 84

$As_4BPdC_{78}H_{62}Cl$ $\{Pd[(2\text{-}Ph_2AsC_6H_4)_3As]Cl\}BPh_4$, 3, 111

$As_4BPdC_{78}H_{62}I$ $\{Pd[(2\text{-}Ph_2AsC_6H_4)_3As]I\}BPh_4$, 3, 111

$As_4BPtC_{78}H_{62}Br$ $\{Pt[(2\text{-}Ph_2AsC_6H_4)_3As]Br\}BPh_4$, 3, 111

$As_4BPtC_{78}H_{62}Cl$ $\{Pt[(2\text{-}Ph_2AsC_6H_4)_4As]Cl\}BPh_4$, 3, 111

$As_4BPtC_{78}H_{62}I$ $\{Pt[(2\text{-}Ph_2AsC_6H_4)_3As]I\}BPh_4$, 3, 111

$As_4BPtC_{79}H_{62}NS$ $\{Pt[(2\text{-}Ph_2AsC_6H_4)_3As](SCN)\}BPh_4$, 3, 111

$As_4BRhC_{78}H_{62}Cl_2$ $\{Rh[(2\text{-}Ph_2AsC_6H_4)_3As]Cl_2\}BPh_4$, 3, 112

$As_4Cd_2Hg_2C_{36}H_{84}I_4$ $[HgCd(AsPr_3)_2I_4]_2$, 3, 13

$As_4Fe_4SnC_{80}H_{60}N_4O_{12}$ $Sn[Fe(AsPh_3)(NO)(CO)_2]_4$, 3, 22

$As_4GeMo_{12}C_{96}H_{80}O_{40}$ $[Ph_4As]_4[Ge(Mo_3O_{10})_4]$, 3, 336

$As_6Rh_2SnC_{78}H_{78}Cl_8$ $\{[Rh(Ph_2AsMe)_3Cl_3]_2SnCl_2\}$, 3, 56

$As_6Rh_2Sn_2C_{78}H_{78}Cl_8$ $[RhCl_2\cdot SnCl_2\cdot(Ph_2AsMe)_3]_2$, 3, 56

$AuAsC_{14}H_{20}$ $PhC{\equiv}CAu\cdot AsEt_3$, 3, 8

$AuGeC_{21}H_{24}P$ $Ph_3GeAu\cdot PMe_3$, 2, 139

$AuGeC_{36}H_{30}P$ $Ph_3GeAu\cdot PPh_3$, 2, 139

$AuGe_2C_{44}H_{50}N$ $Et_4N[(Ph_3Ge)_2Au]$, 2, 139

$AuSbC_{14}H_{20}$ $PhC{\equiv}CAu\cdot SbEt_3$, 3, 656

$AuGe_2LiC_{36}H_{30}$ $Li[(Ph_3Ge)_2Au]$, 2, 139

$BAsC_{23}H_{30}$ $[Me_4As][BMePh_3]$, 3, 312

$BAsC_{24}H_{18}Br_2$ $Ph_2AsB(C_6H_4Br\text{-}p)_2$, 3, 298

$BAsC_{24}H_{20}$ Ph_2AsBPh_2, 3, 298

$BAsC_{26}H_{24}$ $Ph_2AsB(C_6H_4Me\text{-}p)_2$, 3, 298

$BAsC_{37}H_{40}$ $[Me(CH_2{=}CHCH_2)PrPhAs]BPh_4$, 3, 318

$BAsC_{41}H_{42}$ $[MePrPh(PhCH_2)As]BPh_4$, 3, 321

$BAsC_{48}H_{32}$ 5,5′-Spirobi(dibenzarsolium)bi(2,2′-biphenylylene)borate, 3, 571

$BAsC_{48}H_{36}$ Triphenyl[2′-(5-phenyldibenzarsolio)-2-biphenylyl]borate, 3, 569

5,5′-Spirobi(dibenzoarsolium)-$[BPh_4]^-$, 3, 571

$BAsC_{48}H_{40}$ $[Ph_4As]BPh_4$, 3, 332

$BAsC_{50}H_{40}$ 3,3′-Dimethyl-5,5′-spirobi(dibenzarsolium)$[BPh_4]^-$, 3, 571

$BAsC_{54}H_{41}Br$ 5-(2′-Bromo-2-biphenylyl)-5-phenyldibenzarsolium$[BPh_4]^-$, 3, 569

$BAsC_{56}H_{40}$ 7,7′-Spirobi(7H-benzo[b]-naphth[1,2-d]arsolium)$[BPh_4]^-$, 3, 572

$BAsC_{57}H_{49}N$ Trimethyl(p-5,5′-spirobi-[dibenzarsol]-5-ylphenylammonium-$[BPh_4]^-$, 3, 573

$BBiC_{48}H_{40}$ $[Ph_4Bi][BPh_4]$, 3, 835

$BCrC_{34}H_{30}$ $[(C_5H_5)_2Cr]BPh_4$, 1, 68

$BCrC_{36}H_{30}D_2$ $[(PhD)_2Cr]BPh_4$, 1, 78

$BCrC_{36}H_{32}$ $[(C_6H_6)_2Cr]BPh_4$, 1, 78

$BCrC_{37}H_{34}O$ $[(C_6H_6)(PhCH_2OH)Cr]BPh_4$, 1, 86

$BCrC_{42}H_{35}D$ $[(PhD)(Ph_2)Cr]BPh_4$, 1, 86

$BCrC_{42}H_{36}$ $[(C_6H_6)(Ph_2)Cr]BPh_4$, 1, 87

$BCrC_{43}H_{36}O_2$ $[(BzOH)(Ph_2)Cr]BPh_4$, 1, 88

$BCrC_{44}H_{36}$ $[(C_{10}H_8)_2Cr]BPh_4$, 1, 81

$BCrC_{45}H_{42}$ $[(2\text{-}PhCH_2C_6H_4Me)(MePh)Cr]$-$BPh_4$, 1, 88

$BCrC_{48}H_{39}Br$ $[(Ph_2)(p\text{-}BrC_6H_4Ph)Cr]$-$BPh_4$, 1, 88

$BCrC_{48}H_{39}Cl$ $[(Ph_2)(p\text{-}ClC_6H_4Ph)Cr]$-$BPh_4$, 1, 88

$BCrC_{48}H_{39}I$ $[(Ph_2)(p\text{-}IC_6H_4Ph)Cr]BPh_4$, 1, 88

$BCrC_{48}H_{40}$ $[(Ph_2)_2Cr]BPh_4$, 1, 84

$BCrC_{50}H_{40}$ [Bis(fluorene)-Cr]BPh_4, 1, 85

$BCrC_{50}H_{42}N_4$ $[(C_6H_6)(BiPy)_2Cr]PhPh_4$, 1, 75

$BCrC_{52}H_{40}$ [Bis(phenanthrene)-Cr]BPh_4, 1, 85

$BCrC_{64}H_{48}$ [Bis(binaphthyl)-Cr]BPh_4, 1, 86

$BCrC_{72}H_{56}$ [Bis(quaterphenyl)-Cr]BPh_4, 1, 86

$BCrC_{120}H_{88}$ [Bis(octaphenyl)-Cr]BPh_4, 1, 85

$BFeC_{10}H_{10}BrO_2$ Br-Fcd-$B(OH)_2$, 1, 532

$BFeC_{10}H_{10}ClO_2$ Cl-Fcd-$B(OH)_2$, 1, 532

$BFeC_{10}H_{11}O_2$ $FcB(OH)_2$, 1, 531

$BFeC_{25}H_{20}F_4O_2P$ $[C_5H_5Fe(CO)_2(PPh_3)]BF_4$, 1, 209

$BFeC_{32}H_{25}O_3$ $[C_5H_5Fe(CO)_3]BPh_4$, 1, 208

$BFeC_{34}H_{30}$ $[FcH]BPh_4$, 1, 227

$BFeC_{35}H_{29}O_3$ [Bicyclo[5.1.0]octa-2,4-dienylium-Fe$(CO)_3$]BPh_4, 1, 499

$BFeC_{35}H_{31}$ $[C_6H_6FeC_5H_5]BPh_4$, 1, 218

$BFeC_{38}H_{37}$ $[(Me_3C_6H_3)FeC_5H_5]BPh_4$, 1, 218

$BFeC_{39}H_{37}$ [Tetralin-FeC_5H_5]BPh_4, 1, 218

$BFeC_{40}H_{41}$ $[(Me_3C_6H_3)Fe(C_5H_4Et)]BPh_4$, 1, 218

$BFeC_{45}H_{35}O_3$ $[(Ph_2CHC_5H_4)Fe(CO)_3]BPh_4$, 1, 209

$BFe_2C_{40}H_{36}O_4P$ $\{[C_5H_5Fe(CO)_2]_2PMe_2\}BPh_4$, 1, 216

$BFe_3C_{55}H_{47}$ $[Fc_3C]BPh_4$, 1, 431

$BGeC_4H_9F_3N$ $Me_3GeCN{\cdot}BF_3$, 2, 94

$BGeC_4H_{12}F_3O$ $Me_3GeOMe{\cdot}BF_3$, 2, 107

$BGeC_{40}H_{42}N$ $Me_4N[Ph_3GeBPh_3]$, 2, 139

$BGeC_{55}H_{48}P$ $MePh_3P[Ph_3GeBPh_3]$, 2, 139

$BGeC_{61}H_{52}P$ $[Ph_3PCH_2GePh_3]BPh_4$, 2, 45

$BGe_3C_9H_{27}O_3$ $(Me_3Ge)_3BO_3$, 2, 97

$BMnC_{12}H_8F_2O_5$ $[AcCH{=}C(OBF_2)C_5H_4]Mn$-$(CO)_3$, 1, 149

$BMnC_{18}H_{30}F_4N_6$ $[(EtNC)_6Mn]BF_4$, 1, 182

$BMnC_{33}H_{26}O_3$ $[(C_6H_6)Mn(CO)_3]BPh_4$, 1, 179

$BMnC_{36}H_{38}N_6$ $[(MeNC)_6Mn]BPh_4$, 1, 182

$BMnC_{48}H_{56}$ [Bis(hexamethylbenzene)-Mn]-BPh_4, 1, 131, 180

$BMnC_{72}H_{62}N_6O_6$ [(p-$MeOC_6H_4NC)_6Mn$]BPh_4, 1, 183

$BMoC_{32}H_{28}NO_3$ [$C_5H_5Mo(CO)_3NH_3$]BPh_4, 1, 99

$BMoC_{32}H_{29}N_2O_3$ [$C_5H_5Mo(CO)_3H_2NNH_2$]BPh_4, 1, 99

$BMoC_{34}H_{27}O_3$ [Cycloheptatrienium-Mo-$(CO)_3$]BPh_4, 1, 113

$BMoC_{36}H_{32}$ [$(C_6H_6)_2Mo$]BPh_4, 1, 108

$BNbC_{10}H_{14}Cl$ $(C_5H_5)_2NbClBH_4$, 1, 53

$BReC_{36}H_{32}$ [$(C_6H_6)_2Re$]BPh_4, 1, 190

$BReC_{42}H_{44}$ [$(Me_3C_6H_3)_2Re$]BPh_4, 1, 190

$BReC_{68}H_{57}INP_2$ [(p-$MeC_6H_4NC)(Ph_3P)_2$-ReI]BPh_4, 1, 191

$BReC_{72}H_{62}N_6$ [(p-$MeC_6H_4NC)_6Re$]BPh_4, 1, 192

$BSbC_{23}H_{30}$ [Me_4Sb][$BMePh_3$], 3, 714

$BSbC_{48}H_{40}$ [Ph_4Sb][BPh_4], 3, 718

$BSnC_8H_{21}ClN$ $Et_3SnB(NMe_2)Cl$, 2, 495

$BSnC_{10}H_{27}N_2$ $Et_3SnB(NMe_2)_2$, 2, 495

$BSn_2C_{14}H_{36}N$ $(Et_3Sn)_2BNMe_2$, 2, 495

$BSn_3C_{12}H_{33}$ $(Me_3SnCH_2)_3B$, 2, 253

$BTcC_{36}H_{32}$ [$(C_6H_6)_2Tc$]BPh_4, 1, 186

$BTiC_{10}H_{10}F_4$ $(C_5H_5)_2TiBF_4$, 1, 19

$BTiC_{10}H_{14}$ $(C_5H_5)_2TiBH_4$, 1, 19

$BVC_{36}H_{30}O_2$ [$(C_5H_5)_2V(CO)_2$]BPh_4, 1, 49

$BVC_{36}H_{30}O_4$ [(p-$Me_2C_6H_4)V(CO)_4$]BPh_4, 1, 50

$BWC_{10}H_7F_4O_3$ [Cycloheptatrienium-$W(CO)_3$]BF_4, 1, 127

$BWC_{18}H_{24}F_4$ [(1,3,5-$Me_3C_6H_3)_2W$]BF_4, 1, 126

$BWC_{32}H_{28}NO_3$ [$C_5H_5W(CO)_3NH_3$]BPh_4, 1, 122

$BWC_{32}H_{29}N_2O_3$ [$C_5H_5W(CO)_3H_2NNH_2$]BPh_4, 1, 122

$B_2AsC_{81}H_{70}N$ 5-(2-PhC_6H_4)-5-(p-Me_2N-C_6H_4)dibenzarsolium tetraphenylborate, methotetraphenylborate, 3, 570

$B_2AsC_{82}H_{72}N$ 5-(2'-Methyl-2-biphenylyl)-5-[p-(dimethylamino)phenyl]-dibenzarsolium tetraphenylborate, methotetraphenylborate, 3, 570

$B_2Cr_2C_{86}H_{70}BaO_4$ {[$(Ph_2)(BzO^-)Cr$]-BPh_4}$_2Ba^{++}$, 1, 88

$B_2FeC_{10}H_{12}O_4$ Fcd[$B(OH)_2$]$_2$, 1, 532

$(B_2FeC_{22}H_{18}N_4)_n$ Bis[(1,3-dihydro-1,3,2-benzodiazoborol-2-yl)cyclopentadienyl]-Fe polymer, 1, 532

$B_2FeC_{22}H_{20}N_4$ Fcd($BNHC_6H_4NH$)$_2$, 1, 532

$B_2SnC_{88}H_{80}P_2$ [$Me_2Sn(CH_2PPh_3)_2$][BPh_4]$_2$, 2, 243

$B_3Sn_3C_{42}H_{72}N_3$ cyclic$(Et_3SnCH_2CH_2NBPh)_3$, 2, 253

$B_3Sn_3C_{78}H_{72}N_3$ cyclic$(Ph_3SnCH_2CH_2NBPh)_3$, 2, 253

$BAsFe_2C_{40}H_{36}O_4$ {[$C_5H_5Fe(CO)_2$]$_2AsMe_2$}-BPh_4, 1, 216, 3, 304

$BAs_3NiC_{39}H_{56}BrP$ {Ni[$(Me_2AsCH_2CH_2CH_2)_3$-P]Br}BPh_4, 3, 107

$BAs_3NiC_{39}H_{56}ClP$ {Ni[$(Me_2AsCH_2CH_2CH_2)_3$-P]Cl}BPh_4, 3, 107

$BAs_3NiC_{39}H_{56}IP$ {Ni[$(Me_2AsCH_2CH_2CH_2)_3$-P]I}BPh_4, 3, 108

$BAs_4PdC_{78}H_{62}Cl$ {Pd[$(2-Ph_2AsC_6H_4)_3As$]-Cl}BPh_4, 3, 111

$BAs_4PdC_{78}H_{62}I$ {Pd[$(2-Ph_2AsC_6H_4)_3As$]I}-BPh_4, 3, 111

$BAs_4PtC_{78}H_{62}Br$ {Pt[$(2-Ph_2AsC_6H_4)_3As$]-Br}BPh_4, 3, 111

$BAs_4PtC_{78}H_{62}Cl$ {Pt[$(2-Ph_2AsC_6H_4)_3As$]-Cl}BPh_4, 3, 111

$BAs_4PtC_{78}H_{62}I$ {Pt[$(2-Ph_2AsC_6H_4)_3As$]I}-BPh_4, 3, 111

$BAs_4PtC_{79}H_{62}NS$ {Pt[$(2-Ph_2AsC_6H_4)_3As$]-(SCN)}BPh_4, 3, 111

$BAs_4RhC_{78}H_{62}Cl_2$ {Ph[$(2-Ph_2AsC_6H_4)_3As$]-Cl_2}BPh_4, 3, 112

$BGeLiC_{36}H_{30}$ Li[Ph_3GeBPh_3], 2, 139

$BaMnC_8H_4$ Ba[$Mn(C{\equiv}CH)_4$], 1, 133

$BaMnC_{32}H_{20}$ Ba[$Mn(C{\equiv}CPh)_4$], 1, 133

$BiCF_3I_2$ CF_3BiI_2, 3, 831

$BiCH_2ClO$ $ClCH_2BiO$, 3, 832

$BiCH_2Cl_3$ $ClCH_2BiCl_2$, 3, 831

$BiCH_3Br_2$ $MeBiBr_2$, 3, 831

$BiCH_3Cl_2$ $MeBiCl_2$, 3, 831

$BiCH_5$ $MeBiH_2$, 3, 827

BiC_2F_6I $(CF_3)_2BiI$, 3, 829

$BiC_2H_3Cl_2$ $CH_2{=}CHBiCl_2$, 3, 831

$BiC_2H_3F_3I$ $Me(CF_3)BiI$, 3, 829

BiC_2H_6Br Me_2BiBr, 3, 829

BiC_2H_6Cl Me_2BiCl, 3, 829

BiC_2H_7 Me_2BiH, 3, 827

BiC_3F_9 $(CF_3)_3Bi$, 3, 821
$BiC_3H_3F_6$ $Me(CF_3)_2Bi$, 3, 824
$BiC_3H_5Br_2$ $CH_2{=}C(Me)BiBr_2$, 3, 831
$BiC_3H_6F_3$ $Me_2(CF_3)Bi$, 3, 824
BiC_3H_9 Me_3Bi, 3, 816
BiC_3N_3 $(NC)_3Bi$, 3, 821
$BiC_4H_5F_6$ $(CF_3)_2EtBi$, 3, 825
$BiC_4H_6F_5$ $Me_2(C_2F_5)Bi$, 3, 825
$BiC_5H_3F_{10}$ $(C_2F_5)_2MeBi$, 3, 825
$BiC_5H_6F_7$ $Me_2(CF_3CF_2CF_2)Bi$, 3, 825
$BiC_5H_{10}F_3$ $Et_2(CF_3)Bi$, 3, 825
BiC_6F_9 $(CF_2{=}CF)_3Bi$, 3, 821
BiC_6H_5 $(PhBi)_n$, 3, 828
$BiC_6H_5Br_2$ $PhBiBr_2$, 3, 831
$BiC_6H_5Cl_2$ $PhBiCl_2$, 3, 831
$BiC_6H_5I_2$ $PhBiI_2$, 3, 831
BiC_6H_5O $PhBiO$, 3, 832
BiC_6H_9 $(CH_2{=}CH)_3Bi$, 3, 821
BiC_6H_{15} Et_3Bi, 3, 816
$BiC_6H_{15}O$ Et_2BiOEt, 3, 832
$BiC_7H_3F_{14}$ $(CF_3CF_2CF_2)_2MeBi$, 3, 825
$BiC_7H_7Br_2$ 2-$MeC_6H_4BiBr_2$, 3, 831
$BiC_8H_7Cl_2O_2$ 2-$MeO_2CC_6H_4BiCl_2$, 3, 831
BiC_9H_{15} $(CH_2{=}CHCH_2)_3Bi$, 3, 821
  $[CH_2{=}C(Me)]_3Bi$, 3, 821
  $(MeCH{=}CH)_3Bi$ isomers, 3, 821
$BiC_9H_{15}Br_2$ $(MeCH{=}CH)_3BiBr_2$ isomers, 3, 837
BiC_9H_{21} Pr_3Bi, 3, 821
$BiC_{10}H_7Br_2$ 1-$C_{10}H_7BiBr_2$, 3, 831
$BiC_{10}H_{10}Cl$ $(C_5H_5)_2BiCl$, 3, 829
$BiC_{12}H_8Br_2Cl$ (4-$BrC_6H_4)_2BiCl$, 3, 829
$BiC_{12}H_8Cl_2I$ (4-$ClC_6H_4)_2BiI$, 3, 829
$BiC_{12}H_8Cl_3$ (4-$ClC_6H_4)_2BiCl$, 3, 829
$BiC_{12}H_9S_3$ $(SCH{=}CHCH{=}C)_3Bi$, 3, 821
$BiC_{12}H_{10}Br$ Ph_2BiBr, 3, 829
$BiC_{12}H_{10}Cl$ Ph_2BiCl, 3, 828
$BiC_{12}H_{10}I$ Ph_2BiI, 3, 830
$BiC_{12}H_{21}$ $(EtCH{=}CH)_3Bi$, 3, 821
$BiC_{12}H_{27}$ Bu_3Bi, 3, 821
  $(Me_2CHCH_2)_3Bi$, 3, 821
$BiC_{13}H_9O_2S$ 4-Oxo-2-phenyl-3,1,2-benzoxathiabismuthine, 3, 841
$BiC_{13}H_{10}N$ Ph_2BiCN, 3, 832
$BiC_{14}H_{12}BrO_2$ $Ph_2BiO_2CCH_2Br$, 3, 832
$BiC_{14}H_{12}IO_2$ $Ph_2BiO_2CCH_2I$, 3, 832
$BiC_{14}H_{14}Cl$ (4-$MeC_6H_4)_2BiCl$, 3, 830
$BiC_{14}H_{14}ClO_2$ (4-$MeOC_6H_4)_2BiCl$, 3, 830
$BiC_{14}H_{14}I$ (4-$MeC_6H_4)_2BiI$, 3, 830
$BiC_{15}H_{15}$ $(C_5H_5)_3Bi$, 3, 817
$BiC_{15}H_{15}O_2$ Ph_2BiO_2CEt, 3, 832
$BiC_{16}H_{14}ClO_4$ (2-$MeO_2CC_6H_4)_2BiCl$, 3, 830
$BiC_{16}H_{15}$ (4-$MeC_6H_4)_2(HC{\equiv}C)Bi$, 3, 825
$BiC_{16}H_{20}ClN_2$ [4-$(Me_2N)C_6H_4]_2BiCl$, 3, 830
$BiC_{18}H_{12}Br_3$ (4-$BrC_6H_4)_3Bi$, 3, 817
$BiC_{18}H_{12}Cl_2N_3O_6$ $(O_2NC_6H_4)_3BiCl_2$ isomers, 3, 837
$BiC_{18}H_{12}Cl_3$ (4-$ClC_6H_4)_3Bi$, 3, 817
  (2-$ClC_6H_4)_3Bi$, 3, 822
$BiC_{18}H_{12}Cl_5$ (4-$ClC_6H_4)_3BiCl_2$, 3, 837
$BiC_{18}H_{12}F_3$ (4-$FC_6H_4)_3Bi$, 3, 822
$BiC_{18}H_{12}I_3$ (4-$IC_6H_4)_3Bi$, 3, 822
$BiC_{18}H_{12}N_3O_6$ (4-$O_2NC_6H_4)_3Bi$, 3, 822
$BiC_{18}H_{13}$ 5-Phenyldibenzobismole, 3, 841
$BiC_{18}H_{14}Cl$ Ph_2(4-$ClC_6H_4)Bi$, 3, 825
$BiC_{18}H_{14}ClS_2$ 4-$ClC_6H_4Bi(SPh)_2$, 3, 833
$BiC_{18}H_{15}$ Ph_3Bi, 3, 818
$BiC_{18}H_{15}Br_2$ Ph_3BiBr_2, 3, 837
$BiC_{18}H_{15}Cl_2$ Ph_3BiCl_2, 3, 836
$BiC_{18}H_{15}F_2$ Ph_3BiF_2, 3, 837
$BiC_{18}H_{15}N_2O_6$ $Ph_3Bi(NO_3)_2$, 3, 840
$BiC_{18}H_{15}O$ Ph_2(4-$HOC_6H_4)Bi$, 3, 825
$BiC_{18}H_{15}O_3S$ Ph_3BiSO_3, 3, 820
$BiC_{18}H_{15}S$ Ph_2BiSPh, 3, 833
$BiC_{18}H_{15}S_2$ $PhBi(SPh)_2$, 3, 833
$BiC_{18}H_{16}ClO$ $Ph_3Bi(OH)Cl$, 3, 839
$BiC_{18}H_{18}ClO_2$ [2,4-$MeO(CH_2{=}CH)C_6H_3]_2$-$BiCl$, 3, 830
$BiC_{18}H_{18}ClO_4$ (2-$EtO_2CC_6H_4)_2BiCl$, 3, 830
$BiC_{18}H_{18}N_3O_6S_3$ (4-$H_2NO_2SC_6H_4)_3Bi$, 3, 822
$BiC_{18}H_{21}$ $(MeC_5H_4)_3Bi$, 3, 820
$BiC_{19}H_{15}Br_2Cl_2$ (4-$ClC_6H_4)_2$(2-MeC_6H_4)-$BiBr_2$, 3, 837
$BiC_{19}H_{15}Cl_2$ (4-$ClC_6H_4)_2$(2-$MeC_6H_4)Bi$, 3, 825
$BiC_{19}H_{15}Cl_4$ (4-$ClC_6H_4)_2$(2-MeC_6H_4)-$BiCl_2$, 3, 837
$BiC_{19}H_{17}$ Ph_2(4-$MeC_6H_4)Bi$, 3, 825
$BiC_{19}H_{17}Cl_2$ Ph_2(4-$MeC_6H_4)BiCl_2$, 3,

837
$BiC_{19}H_{17}S_2$ $4\text{-}MeC_6H_4Bi(SPh)_2$, 3, 833
$BiC_{20}H_{13}Cl_2$ $(4\text{-}ClC_6H_4)_2(PhC{\equiv}C)Bi$, 3, 825
$BiC_{20}H_{15}$ $Ph_2(PhC{\equiv}C)Bi$, 3, 825
$BiC_{20}H_{15}O_4$ $PhBi(OBz)_2$, 3, 832
$BiC_{20}H_{17}$ $Ph_2(4\text{-}CH_2{=}CHC_6H_4)Bi$, 3, 826
$(Ph_2BiC_6H_4CHCH_2)_n$, 3, 827
$BiC_{20}H_{18}Cl$ $(4\text{-}MeC_6H_4)_2(4\text{-}ClC_6H_4)Bi$, 3, 826
$BiC_{20}H_{19}Cl_2$ $(4\text{-}MeC_6H_4)_2PhBiCl_2$, 3, 837
$BiC_{21}H_{18}Br_3$ $(2\text{-}BrC_6H_4CH_2)_3Bi$, 3, 822
$BiC_{21}H_{18}Cl_3$ $(2\text{-}ClC_6H_4CH_2)_3Bi$, 3, 822
$BiC_{21}H_{21}$ $(PhCH_2)_3Bi$, 3, 822
$(MeC_6H_4)_3Bi$ isomers, 3, 822
$BiC_{21}H_{21}Br_2$ $(2\text{-}MeC_6H_4)_3BiBr_2$, 3, 837
$BiC_{21}H_{21}Cl_2$ $(MeC_6H_4)_3BiCl_2$ isomers, 3, 837
$BiC_{21}H_{21}O_3$ $(MeOC_6H_4)_3Bi$ isomers, 3, 823
$BiC_{22}H_{15}Cl_2$ $(4\text{-}ClC_6H_4)_2(1\text{-}C_{10}H_7)Bi$, 3, 826
$BiC_{22}H_{15}Br_2Cl_2$ $(4\text{-}ClC_6H_4)_2(1\text{-}C_{10}H_7)\text{-}BiBr_2$, 3, 838
$BiC_{22}H_{15}Cl_4$ $(4\text{-}ClC_6H_4)_2(1\text{-}C_{10}H_7)Bi\text{-}Cl_2$, 3, 838
$BiC_{22}H_{17}$ $Ph_2(1\text{-}C_{10}H_7)Bi$, 3, 826
$BiC_{22}H_{17}Cl_2$ $Ph_2(1\text{-}C_{10}H_7)BiCl_2$, 3, 838
$BiC_{22}H_{17}S_2$ $1\text{-}C_{10}H_7Bi(SPh)_2$, 3, 833
$BiC_{22}H_{19}$ $(4\text{-}MeC_6H_4)_2(PhC{\equiv}C)Bi$, 3, 826
$BiC_{22}H_{19}Cl_2O_4$ $Ph_3Bi(O_2CCH_2Cl)_2$, 3, 840
$BiC_{22}H_{19}O_4S_2$ $PhBi(SC_6H_4CO_2Me\text{-}2)_2$, 3, 833
$BiC_{22}H_{21}O_4$ $Ph_3Bi(OAc)_2$, 3, 840
$BiC_{24}H_{18}Cl$ $(4\text{-}PhC_6H_4)_2BiCl$, 3, 830
$BiC_{24}H_{19}O_4$ $PhBi(O_2CCH{=}CHPh)_2$, 3, 832
$BiC_{24}H_{20}Br$ $[Ph_4Bi]Br$, 3, 835
$BiC_{24}H_{20}Br_3$ $[Ph_4Bi]Br_3$, 3, 835
$BiC_{24}H_{20}Cl$ $[Ph_4Bi]Cl$, 3, 835
$BiC_{24}H_{20}ClO_4$ $[Ph_4Bi]ClO_4$, 3, 835
$BiC_{24}H_{20}NO_3$ $[Ph_4Bi]NO_3$, 3, 835
$BiC_{24}H_{21}$ $(3\text{-}CH_2{=}CHC_6H_4)_3Bi$, 3, 823
$(MeC_6H_4)_2(C_{10}H_7)Bi$ isomers, 3, 826
$BiC_{24}H_{21}Br_2$ $(MeC_6H_4)_2(1\text{-}C_{10}H_7)BiBr_2$ isomers, 3, 838
$BiC_{24}H_{21}Cl_2$ $(MeC_6H_4)_2(1\text{-}C_{10}H_7)BiCl_2$ isomers, 3, 838
$BiC_{24}H_{21}O_2$ $(4\text{-}MeOC_6H_4)_2(1\text{-}C_{10}H_7)Bi$, 3, 826
$BiC_{24}H_{27}$ $(2,4\text{-}Me_2C_6H_3)_3Bi$, 3, 823
$BiC_{24}H_{27}Cl_2$ $(2,4\text{-}Me_2C_6H_3)_3BiCl_2$, 3, 838
$BiC_{24}H_{27}O_3$ $(EtOC_6H_4)_3Bi$ isomers, 3, 823
$BiC_{24}H_{30}N_3$ $(4\text{-}Me_2NC_6H_4)_3Bi$, 3, 823
$BiC_{25}H_{22}NO_2S$ $Ph_3Bi{=}NO_2SC_6H_4Me\text{-}p$, 3, 841
$BiC_{26}H_{19}O_4$ 3,3-Dihydro-1,5-dioxo-3,3,3-triphenyl-2,4,3-benzodioxabismepin, 3, 842
$BiC_{26}H_{25}O_2$ $(4\text{-}EtOC_6H_4)_2(1\text{-}C_{10}H_7)Bi$, 3, 826
$BiC_{27}H_{27}Cl_2O_6$ $(4\text{-}EtO_2CC_6H_4)_3BiCl_2$, 3, 838
$BiC_{27}H_{33}$ $(2,4,6\text{-}Me_3C_6H_2)_3Bi$, 3, 823
$BiC_{27}H_{33}Br_2$ $(2,4,6\text{-}Me_3C_6H_2)_3BiBr_2$, 3, 838
$BiC_{27}H_{33}Cl_2$ $(2,4,6\text{-}Me_3C_6H_2)_3BiCl_2$, 3, 838
$BiC_{28}H_{29}$ $(2,4,6\text{-}Me_3C_6H_2)_2(1\text{-}C_{10}H_7)Bi$, 3, 826
$BiC_{30}H_{21}$ $(1\text{-}C_{10}H_7)_3Bi$, 3, 820
$BiC_{30}H_{25}$ Ph_5Bi, 3, 834
$BiC_{30}H_{25}S_2$ $Ph_3Bi(SPh)_2$, 3, 841
$BiC_{30}H_{39}$ $[2,5\text{-}Me(Me_2CH)C_6H_3]_3Bi$, 3, 823
$BiC_{30}H_{39}Br_2$ $[2,5\text{-}Me(Me_2CH)C_6H_3]_3Bi\text{-}Br_2$, 3, 838
$BiC_{30}H_{39}Cl_2$ $[2,5\text{-}Me(Me_2CH)C_6H_3]_3Bi\text{-}Cl_2$, 3, 838
$BiC_{31}H_{25}$ $Ph_2(Ph_3C)Bi$, 3, 826
$BiC_{32}H_{22}Cl_3O_6$ $(4\text{-}ClC_6H_4)_3Bi(O_2CC_6H_4\text{-}OH\text{-}2)_2$, 3, 840
$BiC_{32}H_{25}O_6$ $Ph_3Bi(O_2CC_6H_4OH)_2$ isomers, 3, 840
$BiC_{32}H_{27}N_2O_4$ $Ph_3Bi(O_2CC_6H_4NH_2)_2$ isomers, 3, 840
$BiC_{34}H_{25}O_8$ $Ph_3Bi(O_2CC_6H_4CO_2H\text{-}o)_2$, 3, 840
$BiC_{35}H_{31}O_4$ $(4\text{-}MeC_6H_4)_3Bi(OBz)_2$, 3, 840

$BiC_{35}H_{31}O_8$ $(MeC_6H_4)_3Bi(O_2CC_6H_4OH\text{-}o)_2$ isomers, 3, 840
$BiC_{36}H_{27}$ $(4\text{-}PhC_6H_4)_3Bi$, 3, 823
$BiC_{36}H_{27}Br_2$ $(4\text{-}PhC_6H_4)_3BiBr_2$, 3, 838
$BiC_{36}H_{27}Cl_2$ $(4\text{-}PhC_6H_4)_3BiCl_2$, 3, 838
$BiC_{36}H_{27}N_2O_6$ $(4\text{-}PhC_6H_4)_3Bi(NO_3)_2$, 3, 840
$BiC_{36}H_{27}O_4$ $Ph_2(2\text{-}C_{10}H_7)Bi(OBz)_2$, 3, 840
$BiC_{36}H_{29}O_4$ $Ph_3Bi(O_2CCH{=}CHPh)_2$, 3, 840
$BiC_{38}H_{29}O_4$ $Ph_2(4\text{-}PhC_6H_4)Bi(OBz)_2$, 3, 840
$Bi_2C_{26}H_{16}Cl_4$ $(4\text{-}ClC_6H_4)_2BiC{\equiv}CBi(C_6H_4Cl\text{-}4)_2$, 3, 826
$Bi_2C_{26}H_{20}$ $Ph_2BiC{\equiv}CBiPh_2$, 3, 827
$Bi_2C_{30}H_{28}$ $(4\text{-}MeC_6H_4)_2BiC{\equiv}CBi(C_6H_4\text{-}Me\text{-}4)_2$, 3, 827
$Bi_2C_{39}H_{30}Cl_6O_6S_3$ (2,4,6,6-Cl_4-s-trithiane-2,4-diyl)bis[triphenylarsonium chloride]S,S,S′,S″,S″-hexaoxide, 3, 835
$BiAgC_{18}H_{15}ClO_4$ $[AgBiPh_3]ClO_4$, 3, 820
$BiBC_{48}H_{40}$ $[Ph_4Bi][BPh_4]$, 3, 835
$BiMoC_{11}H_{15}O_5$ $[Mo(BiEt_3)(CO)_5]$, 3, 816
$BiMoC_{23}H_{15}O_5$ $[Mo(BiPh_3)(CO)_5]$, 3, 820
$BiNiC_9H_{15}O_3$ $[Ni(BiEt_3)(CO)_3]$, 3, 816
$BiRaC_{30}H_{21}Cl_2$ $(1\text{-}C_{10}H_7)_3Bi(RaE)Cl_2$, 3, 838
$BiReC_{17}H_{10}O_5$ $Ph_2BiRe(CO)_5$, 3, 833*
$BiSi_3C_{12}H_{33}$ $(Me_3SiCH_2)_3Bi$, 3, 822
$BiSn_3C_{54}H_{45}$ $(Ph_3Sn)_3Bi$, 2, 484
$BiTlC_{24}H_{30}$ $[Et_3Bi\cdot TlPh_3]$, 3, 816
$Bi_2Mn_2C_{44}H_{30}O_8$ $[Mn(BiPh_3)(CO)_4]_2$, 3, 820
$Bi_2MoC_{16}H_{30}O_4$ $[Mo(BiEt_3)_2(CO)_4]$, 3, 816
$CdGe_2C_{12}H_{30}$ $(Et_3Ge)_2Cd$, 2, 134
$CdMn_2C_{96}H_{84}Br_4N_{12}$ $[(p\text{-}MeC_6H_4NC)_6Mn]_2CdBr_4$, 1, 182
$CdMn_2C_{96}H_{84}Br_4N_{12}O_{12}$ $[(p\text{-}MeOC_6H_4NC)_6Mn]_2CdBr_4$, 1, 183
$CeC_{15}H_{15}$ $(C_5H_5)_3Ce$, 1, 4
$CoC_5F_3O_4$ $CF_3Co(CO)_4$, 1, 565
$CoC_5H_3O_4$ $MeCo(CO)_4$, 1, 565
$CoC_5H_5Cl_2$ $C_5H_5CoCl_2$, 1, 584
$CoC_6F_3O_5$ $CF_3COCo(CO)_4$, 1, 567
$CoC_6F_5O_4$ $C_2F_5Co(CO)_4$, 1, 565
$CoC_6H_3O_4$ π-$CH_2{=}CHCOCo(CO)_3$, 1, 592
$CoC_6H_3O_5$ $AcCo(CO)_4$, 1, 567
$CoC_6H_4BrO_3$ π-$CH_2{=}CBrCH_2Co(CO)_3$, 1, 592
$CoC_6H_4ClO_3$ π-$CH_2{=}CHCHClCo(CO)_3$, 1, 592
π-$CH_2{=}CClCH_2Co(CO)_3$, 1, 592
$CoC_6H_5O_3$ π-$CH_2{=}CHCH_2Co(CO)_3$, 1, 592
$CoC_6H_5O_4$ $EtCo(CO)_4$, 1, 565
CoC_6H_6ClNO $C_6H_6Co(NO)Cl$, 1, 597
$CoC_7F_5O_5$ $C_2F_5COCo(CO)_4$, 1, 567
$CoC_7F_7O_4$ n-$C_3F_7Co(CO)_4$, 1, 565
$CoC_7H_5F_3IO$ $C_5H_5CoCF_3(CO)I$, 1, 577
$CoC_7H_5O_2$ $C_5H_5Co(CO)_2$, 1, 576
$CoC_7H_5O_4$ π-$MeCH{=}CHCOCo(CO)_3$, 1, 571
$CoC_7H_5O_5$ $EtCOCo(CO)_4$, 1, 567
$CoC_7H_5O_6$ $MeOCH_2COCo(CO)_4$, 1, 567
$CoC_7H_7O_3$ π-$CH_2{=}CHCH(Me)Co(CO)_3$, 1, 591
π-$CH_2{=}C(Me)CH_2Co(CO)_3$, 1, 592
$CoC_7H_7O_4$ π-$HOCH_2CH{=}CHCH_2Co(CO)_3$, 1, 592
CoC_7H_9I $MeCoI\cdot C_6H_6$, 1, 576
$CoC_7H_9N_3O$ $(MeNC)_3CoCO$, 1, 611
$CoC_8H_5F_5IO$ $C_5H_5Co(C_2F_5)(CO)I$, 1, 577
$CoC_8H_5O_5$ $MeCH{=}CHCOCo(CO)_4$, 1, 567
$CoC_8H_6ClO_5$ $Cl(CH_2)_3COCo(CO)_4$, 1, 567
$CoC_8H_7O_2$ $(MeC_5H_4)Co(CO)_2$, 1, 577
$CoC_8H_7O_4$ π-$CH_2{=}CHCH_2CH_2COCo(CO)_3$, 1, 571
$CoC_8H_7O_5$ $Me_2CHCOCo(CO)_4$, 1, 567
π-$CH_2{=}CHOCH(Me)COCo(CO)_3$, 1, 571
π-$MeO_2CCH{=}CHCH_2Co(CO)_3$, 1, 592
CoC_8H_8 Cyclooctatetraene-Co, 1, 602
$CoC_8H_9O_3$ π$MeCH{=}CHCH(Me)Co(CO)_3$, 1, 592
π-$EtCH{=}CHCH_2Co(CO)_3$, 1, 592
π-$Me_2C{=}CHCH_2Co(CO)_3$, 1, 592
$CoC_8H_{12}Br_2N_4$ $(MeNC)_4CoBr_2$, 1, 612
$CoC_8H_{12}Cl_2N_4$ $(MeNC)_4CoCl_2$, 1, 612
$CoC_8H_{12}I_2N_4$ $(MeNC)_4CoI_2$, 1, 612
$CoC_8H_{12}O_7P$ $AcCo(CO)_3[P(OMe)_3]$, 1, 568
$CoC_9H_5F_6S_2$ $C_5H_5\overline{CoSC(CF_3){=}C(CF_3)S}$, 1,

584

$CoC_9H_5F_7IO$ $C_5H_5Co(C_3F_7)(CO)I$, 1, 577

$CoC_9H_7O_6$ $CH_2=CHCH_2OCH_2COCo(CO)_4$, 1, 567

$HOCH_2CH=CHCH_2COCo(CO)_4$, 1, 567

$CoC_9H_7O_7$ $EtO_2CCH_2COCo(CO)_4$, 1, 567

$CoC_9H_9O_2$ $(EtC_5H_4)Co(CO)_2$, 1, 577

$CoC_9H_9O_4$ π-$AcCH_2CH=CHCH_2Co(CO)_3$, 1, 593

$CoC_9H_9O_5$ $BuCOCo(CO)_4$, 1, 567

$Me_2CHCH_2COCo(CO)_4$, 1, 567

$Me_3CCOCo(CO)_4$, 1, 567

$CoC_9H_{10}BrO$ $[\pi$-$CH_2=CHCH_2CoC_5H_5(CO)Br]$, 1, 578

$[\pi$-$CH_2=CHCH_2CoC_5H_5(CO)]Br$, 1, 578

$CoC_9H_{10}F_6OP$ $[\pi$-$CH_2=CHCH_2CoC_5H_5(CO)]$-PF_6, 1, 578

$CoC_9H_{10}IO$ $[\pi$-$CH_2=CHCH_2CoC_5H_5(CO)I]$, 1, 578

$[\pi$-$CH_2=CHCH_2CoC_5H_5(CO)]I$, 1, 578

$CoC_9H_{11}O_3$ π-$Me_2C=C(Me)CH_2Co(CO)_3$, 1, 592

$CoC_9H_{13}I_3$ $PrCoI_3 \cdot C_6H_6$, 1, 576

$Me_2CHCoI_3 \cdot C_6H_6$, 1, 576

CoC_9H_{15} $(\pi$-$CH_2=CHCH_2)_3Co$, 1, 597

$CoC_9H_{21}O_8P_2$ $MeCo(CO)_2[P(OMe)_3]_2$, 1, 566

$CoC_{10}H_5F_8O$ $C_5H_5CoCF_2CF_2CF_2CF_2(CO)$, 1, 575*

$CoC_{10}H_7I_3$ $C_{10}H_7CoI_3$ isomers, 1, 573

$CoC_{10}H_7O_3$ Cycloheptatrienyl-$Co(CO)_3$, 1, 592

$CoC_{10}H_9O_4$ π-$CH_2=CHCOCH_2CH=CHCH_2Co$-$(CO)_3$, 1, 593

π-$AcCHCH_2CH=CHCHCo(CO)_3$, 1, 594

$CoC_{10}H_9O_5$ $CH_2=CH(CH_2)_3COCo(CO)_4$, 1, 567

$CoC_{10}H_{10}$ $(C_5H_5)_2Co$, 1, 584, 586

$CoC_{10}H_{10}Br$ $[(C_5H_5)_2Co]Br$, 1, 586

$CoC_{10}H_{10}Br_2$ Dibromocyclopentenyl-CoC_5H_5, 1, 584

$CoC_{10}H_{10}Br_3$ $[(C_5H_5)_2Co]Br_3$, 1, 586, 587

$CoC_{10}H_{10}Cl$ $[(C_5H_5)_2Co]Cl$, 1, 586

$CoC_{10}H_{10}ClO_4$ $[(C_5H_5)_2Co]ClO_4$, 1, 587

$CoC_{10}H_{10}Cl_2$ Dichlorocyclopentenyl-CoC_5H_5, 1, 584

$CoC_{10}H_{10}D$ $(DC_5H_5)CoC_5H_5$, 1, 580

$CoC_{10}H_{10}F_6P$ $[(C_5H_5)_2Co]PF_6$, 1, 588

$CoC_{10}H_{10}I_3$ $[(C_5H_5)_2Co]I_3$, 1, 587

$CoC_{10}H_{10}NO_3$ $[(C_5H_5)_2Co]NO_3$, 1, 587

$CoC_{10}H_{11}$ $C_5H_5CoC_5H_6$, 1, 579

$CoC_{10}H_{11}O$ $[(C_5H_5)_2Co]OH$, 1, 586

$CoC_{10}H_{11}O_4$ π-$AcCH_2CH=C(Me)CH_2Co(CO)_3$, 1, 593

π-$AcCH_2CH=CHCH(Me)Co(CO)_3$, 1, 593

π-$AcCH(Me)CH=CHCH_2Co(CO)_3$, 1, 593

π-$EtCOCH_2CH=CHCH_2Co(CO)_3$, 1, 593

$CoC_{10}H_{11}O_5$ n-$C_5H_{11}COCo(CO)_4$, 1, 567

$Et_2CHCOCo(CO)_4$, 1, 567

π-$AcCH_2CH=CHCH(OMe)Co(CO)_3$, 1, 593

π-$MeOCH_2COCH_2CH=CHCH_2Co(CO)_3$, 1, 593

$CoC_{10}H_{14}O_6P$ $MeCo(CO)_3$-$[P(OCH_2CH_2)_2CHCH_2O]$, 1, 566

$CoC_{10}H_{15}ClN_5O_4$ $[(MeNC)_5Co]ClO_4$, 1, 614

$CoC_{10}H_{15}Cl_2N_5O_8$ $[(MeNC)_5Co](ClO_4)_2$, 1, 616

$CoC_{10}H_{21}O_9P_2$ $MeCOCo(CO)_2[P(OMe)_3]_2$, 1, 571

$CoC_{10}H_{28}N_6$ $[(C_5H_5)_2Co(NH_3)_6]$, 1, 585

$CoC_{11}F_{15}O_4$ n-$C_7F_{15}Co(CO)_4$, 1, 566

$CoC_{11}H_4NO_7$ p-$O_2NC_6H_4COCo(CO)_4$, 1, 568

$CoC_{11}H_5O_5$ $PhCOCo(CO)_4$, 1, 568

$CoC_{11}H_7O_4$ $PhCH_2Co(CO)_4$, 1, 566

$CoC_{11}H_{10}Cl_3$ $(CCl_3C_5H_5)CoC_5H_5$, 1, 580

$CoC_{11}H_{10}F_3$ $(CF_3C_5H_5)CoC_5H_5$, 1, 580

$CoC_{11}H_{11}$ $C_5H_5CoC_6H_6$, 1, 582

$CoC_{11}H_{11}Cl_2$ $(Cl_2CHC_5H_5)CoC_5H_5$, 1, 580

$CoC_{11}H_{11}O_4$ π-$AcCHCH_2CH_2CH=CHCHCo(CO)_3$, 1, 594*

$CoC_{11}H_{11}O_5$ $MeCH=CH(CH_2)_3COCo(CO)_4$, 1, 568

$CoC_{11}H_{11}O_6$ π-$MeO_2CCH_2COCH_2CH=CHCH_2$-$Co(CO)_3$, 1, 593

$CoC_{11}H_{13}$ $(MeC_5H_5)CoC_5H_5$, 1, 580

$CoC_{11}H_{13}O_2$ $(Me_2CHCH_2C_5H_4)Co(CO)_2$, 1, 577

$CoC_{11}H_{13}O_4$ π-$AcCH_2C(Me)=C(Me)CH_2Co$-$(CO)_3$, 1, 593

π-$AcCH_2CH=CHCMe_2Co(CO)_3$, 1, 593

π-$EtCOCH_2CH=C(Me)CH_2Co(CO)_3$, 1, 593

$CoC_{11}H_{13}O_6$ π-$MeOCH_2COCH_2CH=CHCH(OMe)$-$Co(CO)_3$, 1, 593

$CoC_{11}H_{14}O_7P$ $AcCo(CO)_3$-

[$P(OCH_2CH_2)_2CHCH_2O$], 1, 568
$CoC_{11}H_{18}N_3O_2$ $(Me_3CNC)_2Co(CO)NO$, 1, 611
$CoC_{12}H_7O_5$ $PhCH_2COCo(CO)_4$, 1, 568
$MeC_6H_4COCo(CO)_4$ isomers, 1, 568
$CoC_{12}H_9O_3$ π-$CH_2C(Ph)CH_2Co(CO)_3$, 1, 592
$CoC_{12}H_{10}\cdot(THF)_n$ Ph_2Co complex with THF, 1, 572
$CoC_{12}H_{10}BrO_4$ $[(HO_2CC_5H_4)_2Co]Br$, 1, 589
$CoC_{12}H_{10}ClO_8$ $[(HO_2CC_5H_4)_2Co]ClO_4$, 1, 589
$CoC_{12}H_{11}Br_3$ $PhCoBr_3\cdot C_6H_6$, 1, 573
$CoC_{12}H_{12}ClI_3$ $EtCoI_3\cdot C_{10}H_7Cl$, 1, 576
$CoC_{12}H_{13}$ Bicyclo[2.2.1]heptadiene-CoC_5H_5, 1, 583
$CoC_{12}H_{13}O_4$ π-$MeCH{=}CHCH_2COCH_2CH{=}CH$-$CH_2Co(CO)_3$, 1, 594
π-$EtCOCHCH_2CH_2CH{=}CHCHCo(CO)_3$, 1, 594*
$CoC_{12}H_{13}O_5$ π-$OC(O)C(Et)C(Et)C(Me)$-$Co(CO)_3$, 1, 595
$CoC_{12}H_{13}O_6$ π-$AcCH(Me)CHCHCH(CO_2Me)$-$Co(CO)_3$, 1, 594
$CoC_{12}H_{14}$ $(MeC_5H_4)_2Co$, 1, 588
$CoC_{12}H_{14}Br_3$ $[(MeC_5H_4)_2Co]Br_3$, 1, 588
$CoC_{12}H_{15}O_4$ π-$Me_3CCOCH_2CH{=}CHCH_2Co$-$(CO)_3$, 1, 594
$CoC_{12}H_{18}Cl_2N_6$ $(MeNC)_6CoCl_2$, 1, 617
$CoC_{12}H_{30}O_{11}P_3$ $MeCOCo(CO)[P(OMe)_3]_3$, 1, 571
$CoC_{13}H_{13}$ Cyclooctatetraene-CoC_5H_5, 1, 582
$CoC_{13}H_{15}$ 1,3,5-Cyclooctatriene-CoC_5H_5, 1, 583
Bicyclo[4.2.0]octadiene-CoC_5H_5, 1, 583
$CoC_{13}H_{15}O_7$ π-$EtO_2CCH_2COCH_2CHCHCH$-$(OMe)Co(CO)_3$, 1, 594
$CoC_{13}H_{17}$ 1,5-Cyclooctadiene-CoC_5H_5, 1, 583
$CoC_{13}H_{17}O$
$C_5H_5Co[C(Me){=}C(Me)C(Me){=}C(Me)O]$, 1, 581
$CoC_{13}H_{17}O_5$ n-$C_8H_{17}COCo(CO)_4$, 1, 568
$CoC_{14}H_5F_{12}O$ C_5H_5Co-[$C(CF_3){=}C(CF_3)C(CF_3){=}C(CF_3)CO$], 1, 581
$C_5H_5Co[C(CF_3){=}C(CF_3)C(CF_3){=}C(CF_3)CO]$, 1, 581
$CoC_{14}H_9O_3$ $(BzC_5H_4)Co(CO)_2$, 1, 577
$CoC_{14}H_{11}O_4$ π-$BzCH_2CH{=}CHCH_2Co(CO)_3$, 1, 594
$CoC_{14}H_{11}O_5$ 2,4,6-$Me_3C_6H_2COCo(CO)_4$, 1, 568
$CoC_{15}H_{17}O_2$ Duroquinone-CoC_5H_5, 1, 582
$CoC_{15}H_{19}O_4$ π-$AcCH_2CH{=}C(CH_2CH{=}CHCHMe_2)$-$CH_2Co(CO)_3$, 1, 594
$CoC_{15}H_{19}O_5$ 2-$HOC_6H_{10}COCH_2CH{=}C(Me)CH_2$-$Co(CO)_3$, 1, 594
$CoC_{15}H_{20}BrO_3$ $\{[Me_3CC{=}CHCH{=}C(CMe_3)CO]$-$Co(CO)_2\}Br$, 1, 600
$CoC_{15}H_{20}IO_3$ $\{[Me_3CC{=}CHCH{=}C(CMe_3)CO]$-$Co(CO)_2\}I$, 1, 600
$CoC_{15}H_{25}O_8P_2$ $MeCo(CO)_2$-$[P(OCH_2CH_2)_2CHCH_2O]_2$, 1, 566
$CoC_{16}H_{10}Cl_4O_2$ $(C_5H_5)_2Co\cdot$chloranil, 1, 585
$CoC_{16}H_{12}N_3O_7$ $[(C_5H_5)_2Co][O(C_6H_2)(NO_2)_3]$, 1, 588
$CoC_{16}H_{13}N_2O_5$ $[(C_5H_5)_2Co][OC_6H_3(NO_2)_2]$, 1, 588
$CoC_{16}H_{13}N_2O_7S$ $[(C_5H_5)_2Co][O_3SC_6H_3$-$(NO_2)_2]$, 1, 588*
$CoC_{16}H_{13}N_3O_6$ $[(C_5H_5)_2Co\cdot C_6H_3(NO_2)_3]$, 1, 585
$CoC_{16}H_{14}F_6P$ $[C_5H_5Co(C_5H_4Ph)]PF_6$, 1, 588
$CoC_{16}H_{15}$ $(PhC_5H_5)CoC_5H_5$, 1, 580
$CoC_{16}H_{16}$ Bis(cyclooctatetraene)-Co, 1, 602
$CoC_{16}H_{22}ClO_4$ $[(Me_2CHC_5H_4)_2Co]ClO_4$, 1, 589
$CoC_{16}H_{25}O_9P_2$ $MeCOCo(CO)_2$-$[P(OCH_2CH_2)_2CHCH_2O]_2$, 1, 571*
$CoC_{16}H_{27}O$ $(C_{10}H_{21}C_5H_4)Co(CO)H_2$, 1, 584
$CoC_{17}H_{14}N_3O_2$ $(p$-$MeC_6H_4NC)_2Co(CO)NO$, 1, 611
$CoC_{17}H_{15}O$ $(BzC_5H_5)Co(C_5H_5)$, 1, 580
$CoC_{17}H_{15}O_5$ π-$OC(O)C(Et){=}C(Et)C(Ph)$-$Co(CO)_3$, 1, 595
$CoC_{18}H_{14}$ Bis(indenyl)-Co, 1, 590

$CoC_{18}H_{14}Br$ [Bis(indenyl)-Co]Br, 1, 590
$CoC_{18}H_{14}Br_3$ [Bis(indenyl)-Co]Br_3, 1, 590
$CoC_{18}H_{14}ClO_4$ [Bis(indenyl)-Co]ClO_4, 1, 590
$CoC_{18}H_{14}F_6P$ [Bis(indenyl)-Co]PF_6, 1, 590
$CoC_{18}H_{14}I$ [Bis(indenyl)-Co]I, 1, 590
$CoC_{18}H_{14}I_3$ [Bis(indenyl)-Co]I_3, 1, 590
$CoC_{18}H_{14}NO_3$ [Bis(indenyl)-Co]NO_3, 1, 590
$CoC_{18}H_{16}Br_2$ $Ph_2CoBr_2 \cdot C_6H_6$, 1, 573
$CoC_{18}H_{22} \cdot (THF)_n$ $(2,4,6\text{-}Me_3C_6H_2)_2Co \cdot (THF)_n$, 1, 518
$CoC_{18}H_{32}N_6$ Bis(indenyl)-Co·$(NH_3)_6$, 1, 590
$CoC_{19}H_{17}$ (Indenylcyclopentadiene)-CoC_5H_5, 1, 580
$CoC_{19}H_{19}F_6$ [2,3,5,6-Tetramethyl-7,8-bis(trifluoromethyl)bicyclo[2.2.2]octatriene]-CoC_5H_5, 1, 583
$CoC_{20}H_{14}Br_2$ $(\beta\text{-}C_{10}H_7)_2CoBr_2$, 1, 573
$CoC_{20}H_{14}I_2$ $(C_{10}H_7)_2CoI_2$ isomers, 1, 573
$CoC_{20}H_{15}F_5N_2O_2P$ $C_2F_5Co(NO)_2(PPh_3)$, 1, 566
$CoC_{20}H_{36}O_{10}P_3$ MeCo(CO)-$[P(OCH_2CH_2)_2CHCH_2O]_3$, 1, 566
$CoC_{21}H_{12}Cl_3N_4O$ $(p\text{-}ClC_6H_4NC)_3CoNO$, 1, 611
$CoC_{21}H_{15}N_4O$ $(PhNC)_3CoNO$, 1, 611
$CoC_{21}H_{36}O_{11}P_3$ MeCOCo(CO)-$[P(OCH_2CH_2)_2CHCH_2O]_3$, 1, 571
$CoC_{22}H_{10}Cl_8O_4$ $\{(C_5H_5)_2Co[OC(CCl{=}CCl)_2CO]_2\}$, 1, 585
$CoC_{22}H_{15}F_3O_3P$ $CF_3Co(CO)_3(PPh_3)$, 1, 566
$CoC_{22}H_{18}ClN_4$ $[(PhN_2C_5H_4)_2Co]Cl$, 1, 589
$CoC_{22}H_{18}O_3P$ $MeCo(CO)_3(PPh_3)$, 1, 566
$CoC_{22}H_{18}O_6P$ $MeCo(CO)_3[P(OPh)_3]$, 1, 566
$CoC_{22}H_{24}N_3O_7$ $[(Me_2CHC_5H_4)_2Co]OC_6H_2(NO_2)_3$, 1, 589
$CoC_{23}H_{15}F_3O_4P$ $CF_3COCo(CO)_3(PPh_3)$, 1, 568
$CoC_{23}H_{15}F_5O_3P$ $C_2F_5Co(CO)_3(PPh_3)$, 1, 566
$CoC_{23}H_{17}ClO_4P$ $ClCH_2COCo(CO)_3(PPh_3)$, 1, 568
$CoC_{23}H_{18}O_3P$ $\pi\text{-}CH_2{=}CHCOCo(CO)_2(PPh_3)$, 1, 595
$CoC_{23}H_{18}O_4P$ $AcCo(CO)_3(PPh_3)$, 1, 568
$CoC_{23}H_{20}O_2P$ $\pi\text{-}CH_2{=}CHCH_2Co(CO)_2(PPh_3)$, 1, 595
$CoC_{24}H_{15}F_5O_4P$ $C_2F_5COCo(CO)_3(PPh_3)$, 1, 568
$CoC_{24}H_{17}NO_4P$ $NCCH_2COCo(CO)_3(PPh_3)$, 1, 568
$CoC_{24}H_{20}O_4P$ $EtCOCo(CO)_3(PPh_3)$, 1, 568
$CoC_{24}H_{20}O_5P$ $HOCH_2CH_2COCo(CO)_3(PPh_3)$, 1, 569
$MeOCH_2COCo(CO)_3(PPh_3)$, 1, 569
$CoC_{24}H_{21}N_4O$ $(p\text{-}MeC_6H_4NC)_3CoNO$, 1, 611
$CoC_{24}H_{22}O_2P$ $\pi\text{-}MeCH{=}CHCH_2Co(CO)_2(PPh_3)$, 1, 595
$CoC_{24}H_{36}$ $(C_6Me_6)_2Co$, 1, 597
$CoC_{24}H_{36}Cl$ $[(C_6Me_6)_2Co]Cl$, 1, 598
$CoC_{24}H_{36}F_6P$ $[(C_6Me_6)_2Co]PF_6$, 1, 598
$CoC_{24}H_{36}F_{12}P_2$ $[(C_6Me_6)_2Co](PF_6)_2$, 1, 598
$CoC_{25}H_{15}O_4$ $PhC{=}C(Ph)C(Ph)COCo(CO)_3$, 1, 571*
$CoC_{25}H_{20}O_3P$ $\pi\text{-}CH_2C(O)CH{=}CHCHCo(CO)_2\text{-}(PPh_3)$, 1, 596
$CoC_{25}H_{20}O_4P$ $CH_2{=}CHCH_2COCo(CO)_3(PPh_3)$, 1, 569
$MeCH{=}CHCOCo(CO)_3(PPh_3)$, 1, 569
$CoC_{25}H_{21}ClO_4P$ $Cl(CH_2)_3COCo(CO)_3(PPh_3)$, 1, 569
$CoC_{25}H_{22}O_4P$ $Me_2CHCOCo(CO)_3(PPh_3)$, 1, 569
$\pi\text{-}MeO_2CCH{=}CHCH_2Co(CO)_2(PPh_3)$, 1, 595
$CoC_{25}H_{22}O_5P$ $HO(CH_2)_3COCo(CO)_3(PPh_3)$, 1, 569
$MeCH(OH)CH_2COCo(CO)_3(PPh_3)$, 1, 569
$MeCH(OH)CH_2COCo(CO)_3(PPh_3)$, 1, 569
$CoC_{25}H_{27}O_6P_2$ $MeCOCo(CO)_2[P(OMe)_3\text{-}(PPh_3)]$, 1, 571
$CoC_{26}H_{20}O_4P$ $CH_2{=}CHCH{=}CHCOCo(CO)_3\text{-}(PPh_3)$, 1, 569
$CoC_{26}H_{22}O_3P$ $\pi\text{-}MeCHC(O)CH{=}CHCHCo(CO)_2\text{-}$

(PPh_3), 1, 596
$CoC_{26}H_{22}O_4P$ $MeCH{=}CHCH_2COCo(CO)_3(PPh_3)$, 1, 569
$CH_2{=}CHCH_2CH_2COCo(CO)_3(PPh_3)$, 1, 569
$CoC_{26}H_{22}O_6P$ $MeO_2CCH(Me)COCo(CO)_3$-(PPh_3), 1, 569
$MeO_2CCH_2CH_2COCo(CO)_3(PPh_3)$, 1, 569
$EtO_2CCH_2COCo(CO)_3(PPh_3)$, 1, 570
$CoC_{26}H_{24}O_3P$ π-$AcCH_2CH{=}CHCH_2Co(CO)_2$-$(PPh_3)$, 1, 596
$CoC_{26}H_{24}O_5P$ $HOCH_2CMe_2COCo(CO)_3(PPh_3)$, 1, 569
$CoC_{26}H_{24}O_7P$ $AcCo(CO)_3[P(C_6H_4OMe\text{-}p)_3]$, 1, 568
$CoC_{26}H_{29}O_7P_2$ $MeOCH_2COCo(CO)_2[P(OMe)_3$-$(PPh_3)]$, 1, 571
$CoC_{27}H_{20}O_3P$ $PhCo(CO)_3(PPh_3)$, 1, 566
$CoC_{27}H_{22}O_4P$ $Me(CH{=}CH)_2COCo(CO)_3(PPh_3)$, 1, 570
$CoC_{27}H_{22}O_6P$ $MeO_2CCH{=}CHCH_2COCo(CO)_3$-$(PPh_3)$, 1, 570
$CoC_{27}H_{24}O_3P$ π-$AcCHCH_2CH{=}CHCHCo(CO)_2$-$(PPh_3)$, 1, 596
$CoC_{27}H_{24}O_4P$ $CH_2{=}CH(CH_2)_3COCo(CO)_3$-$(PPh_3)$, 1, 570
$CoC_{27}H_{26}O_3P$ π-$AcCH(Me)CH{=}CHCH_2Co$-$(CO)_2(PPh_3)$, 1, 596
π-$AcCH_2CH{=}C(Me)CH_2Co(CO)_2(PPh_3)$, 1, 596
$CoC_{27}H_{26}O_4P$ n-$C_5H_{11}COCo(CO)_3(PPh_3)$, 1, 570
π-$AcCH_2CH{=}CHCH(OMe)Co(CO)_2(PPh_3)$, 1, 596
$CoC_{27}H_{30}N_7O$ $(p\text{-}Me_2NC_6H_4NC)_3CoNO$, 1, 611
$CoC_{28}H_{12}Cl_8I_2N_4$ $(2,5\text{-}Cl_2C_6H_3NC)_4CoI_2$, 1, 612
$CoC_{28}H_{16}Cl_4I_2N_4$ $(ClC_6H_4NC)_4CoI_2$ isomers, 1, 612
$CoC_{28}H_{20}Br_2N_4$ $(PhNC)_4CoBr_2$, 1, 613
$CoC_{28}H_{20}Br_3N_4$ $(PhNC)_4CoBr_3$, 1, 614
$CoC_{28}H_{20}Cl_2N_4$ $(PhNC)_4CoCl_2$, 1, 613
$CoC_{28}H_{20}I_2N_4$ $(PhNC)_4CoI_2$, 1, 613
$CoC_{28}H_{20}O_4P$ $PhCOCo(CO)_3(PPh_3)$, 1, 570
$CoC_{28}H_{24}O_4P$ $CH_2{=}CHCH{=}CH(CH_2)_2COCo$-$(CO)_3(PPh_3)$, 1, 570
$CoC_{28}H_{26}O_4P$ $CH_2{=}CH(CH_2)_4COCo(CO)_3$-$(PPh_3)$, 1, 570
$Me_2C{=}C(Me)CH_2COCo(CO)_3(PPh_3)$, 1, 570
$CoC_{28}H_{26}O_5P$ $CH_2(CH_2)_3CH(OH)CHCOCo$-$(CO)_3(PPh_3)$, 1, 570
$CoC_{28}H_{28}O_3P$ π-$EtCOCH_2CH{=}C(Me)CH_2Co$-$(CO)_2(PPh_3)$, 1, 596
π-$AcCH_2C(Me){=}C(Me)CH_2Co(CO)_2(PPh_3)$, 1, 596
π-$AcCH_2CH{=}CHC(Me)_2Co(CO)_2(PPh_3)$, 1, 596
π-$OC(Et){=}C(Et)C(Me)Co(CO)_2(PPh_3)$, 1, 596
$CoC_{29}H_{22}O_5P$ $p\text{-}MeOC_6H_4COCo(CO)_3(PPh_3)$, 1, 570
$CoC_{29}H_{24}O_2P$ π-$CH_2{=}C(Ph)CH_2Co(CO)_2(PPh_3)$, 1, 595
$CoC_{29}H_{28}O_5P$ π-$AcCH(Me)CH{=}CHCH(CO_2Me)$-$Co(CO)_2(PPh_3)$, 1, 596
$CoC_{30}H_{20}BrO_2$ $\{[PhC{=}C(Ph)C(Ph){=}CPh]Co$-$(CO)_2\}Br$, 1, 599
$CoC_{30}H_{24}O_5P$ $PhCH(OH)CH_2COCo(CO)_3(PPh_3)$, 1, 570
$CoC_{30}H_{38}N_3O_7$ $[(C_6Me_6)_2Co]OC_6H_2(NO_2)_3$, 1, 598
$CoC_{30}H_{52}P_2$ $(2,4,6\text{-}Me_3C_6H_2)_2Co(PEt_3)_2$, 1, 572
$CoC_{31}H_{26}O_4P$ $2,4,6\text{-}Me_3C_6H_2COCo(CO)_3$-$(PPh_3)$, 1, 570
$CoC_{32}H_{24}Cl_4I_2N_4$ $[3,2\text{-}Cl(Me)C_6H_3NC]_4$-$CoI_2$, 1, 613
$[4,2\text{-}Cl(Me)C_6H_3NC]_4CoI_2$, 1, 613
$CoC_{32}H_{24}Cl_5I_2N_4O_4$ $\{[2,6\text{-}Cl(Me)C_6H_3$-$NC]_4CoI_2\}ClO_4$, 1, 614
$CoC_{32}H_{28}ClI_2N_4O_4$ $[(p\text{-}MeC_6H_4NC)_4CoI_2]$-$ClO_4$, 1, 614
$CoC_{32}H_{28}I_2N_4$ $(p\text{-}MeC_6H_4NC)_4CoI_2$, 1, 613
$CoC_{32}H_{28}I_2N_4O_4$ $(p\text{-}MeOC_6H_4NC)_4CoI_2$, 1, 613
$CoC_{32}H_{28}I_3N_4$ $(p\text{-}MeC_6H_4NC)_4CoI_3$, 1, 614
$CoC_{32}H_{28}I_3N_4O_4$ $(p\text{-}MeOC_6H_4NC)_4CoI_3$, 1, 614
$CoC_{32}H_{30}Cl_{10}P_2$ $(C_6Cl_5)_2Co(PEt_2Ph)_2$, 1, 572
$CoC_{33}H_{25}$ $C_5H_5Co[PhC{=}C(Ph)C(Ph){=}CPh]$, 1, 579

$CoC_{33}H_{27}$ $C_5H_5Co(PhCH{=}CPh\text{-})_2$, 1, 578
$CoC_{34}H_{18}N_8$ $[(C_5H_5)_2Co][(NC)_2\dot{C}C_6H_4\bar{C}\text{-}(CN)_2\cdot(NC)_2C{=}C_6H_4{=}C(CN)_2]$, 1, 588
$CoC_{34}H_{25}O$ $C_5H_5Co\text{-}[\overline{PhC{=}C(Ph)C(Ph){=}C(Ph)C}O]$, 1, 581
$CoC_{35}H_{20}Cl_5IN_5$ $[(p\text{-}ClC_6H_4NC)_5Co]I$, 1, 615
$CoC_{35}H_{20}Cl_5N_6O_3$ $[(p\text{-}ClC_6H_4NC)_5Co]NO_3$, 1, 615
$CoC_{35}H_{20}Cl_6N_5O_4$ $[(p\text{-}ClC_6H_4NC)_5Co]\text{-}ClO_4$, 1, 615
$CoC_{35}H_{25}BrClN_5O_4$ $[(PhNC)_5Co]Br(ClO_4)$, 1, 616
$CoC_{35}H_{25}ClIN_5O_4$ $[(PhNC)_5Co]I(ClO_4)$, 1, 616
$CoC_{35}H_{25}ClN_5O_4$ $[(PhNC)_5Co]ClO_4$, 1, 615
$CoC_{35}H_{25}Cl_2N_5O_8$ $[(PhNC)_5Co](ClO_4)_2$, 1, 616
$CoC_{35}H_{25}IN_5$ $[(PhNC)_5Co]I$, 1, 615
$CoC_{35}H_{25}N_6O_3$ $[(PhNC)_5Co]NO_3$, 1, 615
$CoC_{36}H_{36}I_2N_4$ $(2,4\text{-}Me_2C_6H_3NC)_4CoI_2$, 1, 614
$CoC_{38}H_{52}P_2$ $(2,4,6\text{-}Me_3C_6H_2)_2Co(PEt_2\text{-}Ph)_2$, 1, 572
$CoC_{40}H_{30}Cl_6N_5O_4$ $\{[3,2\text{-}Cl(Me)C_6H_3\text{-}NC]_5Co\}ClO_4$, 1, 615
$CoC_{40}H_{35}ClIN_5O_4$ $[(p\text{-}MeC_6H_4NC)_5Co]I\text{-}(ClO_4)$, 1, 617
$CoC_{40}H_{35}ClIN_5O_9$ $[(p\text{-}MeOC_6H_4NC)_5Co]I\text{-}(ClO_4)$, 1, 617
$CoC_{40}H_{35}ClN_5O_4$ $[(p\text{-}MeC_6H_4NC)_5Co]ClO_4$, 1, 615
$CoC_{40}H_{35}Cl_2IN_5O_8$ $[(MeC_6H_4NC)_5CoI]\text{-}(ClO_4)_2$ isomers, 1, 617
$[(PhCH_2NC)_5CoI](ClO_4)_2$, 1, 617
$CoC_{40}H_{35}Cl_2N_5O_8$ $[(MeC_6H_4NC)_5Co]\text{-}(ClO_4)_2$ isomers, 1, 616, 617
$CoC_{40}H_{35}IN_5$ $[(p\text{-}MeC_6H_4NC)_5Co]I$, 1, 615
$CoC_{40}H_{36}N_5O$ $[(p\text{-}MeC_6H_4NC)_5Co]OH$, 1, 615
$CoC_{42}H_{48}P_2$ $(2\text{-}MeC_{10}H_6)_2Co(PEt_2Ph)_2$, 1, 572
$CoC_{44}H_{28}I_2N_4$ $(2\text{-}C_{10}H_7NC)_4CoI_2$, 1, 614
$CoC_{44}H_{48}P_2$ $(2\text{-}PhC_6H_4)_2Co(PEt_2Ph)_2$, 1, 572
$CoC_{58}H_{41}O_2$ (Hydroxytetraphenylcyclopentadienyl)(tetraphenylcyclopentadienone)-Co, 1, 582
$CoC_{59}H_{43}O_2$ (Methoxytetraphenylcyclopentadienyl)(tetraphenylcyclopentadienone)-Co, 1, 582
$CoC_{60}H_{43}O_3$ (Acetoxytetraphenylcyclopentadienyl)(tetraphenylcyclopentadienone)-Co, 1, 582
$CoC_{60}H_{51}BrN_3P_2$ $[(p\text{-}MeC_6H_4NC)_3(Ph_3P)_2\text{-}Co]Br$, 1, 611
$CoC_{60}H_{51}ClN_3O_4P_2$ $[(p\text{-}MeC_6H_4NC)_3(Ph_3P)_2\text{-}Co]ClO_4$, 1, 611
$CoC_{60}H_{51}ClN_3P_2$ $[(p\text{-}MeC_6H_4NC)_3(Ph_3P)_2\text{-}Co]Cl$, 1, 611
$CoC_{60}H_{51}IN_3P_2$ $[(p\text{-}MeC_6H_4NC)_3(Ph_3P)_2\text{-}Co]I$, 1, 611
$Co_2C_8H_2O_6$ $(HC{\equiv}CH)Co_2(CO)_6$, 1, 604
$Co_2C_8H_{12}Br_4N_4$ $[(MeNC)_4Co]CoBr_4$, 1, 612
$Co_2C_8N_{10}O_4$ $[(NC)_3CCo(NO)_2]_2$, 1, 575
$Co_2C_9H_3BrO_6$ $(HC{\equiv}CCH_2Br)Co_2(CO)_6$, 1, 604
$Co_2C_9H_3ClO_6$ $(HC{\equiv}CCH_2Cl)Co_2(CO)_6$, 1, 604
$Co_2C_9H_4O_7$ $(HC{\equiv}CCH_2OH)Co_2(CO)_6$, 1, 604
$Co_2C_{10}F_4O_8$ $[\text{-}CF_2Co(CO)_4]_2$, 1, 566
$Co_2C_{10}F_6O_6$ $(F_3CC{\equiv}CCF_3)Co_2(CO)_6$, 1, 604
$Co_2C_{10}H_2O_{10}$ $(HO_2CC{\equiv}CCO_2H)Co_2(CO)_6$, 1, 604
$Co_2C_{10}H_4Br_2O_6$ $(BrCH_2C{\equiv}CCH_2Br)Co_2(CO)_6$, 1, 604
$Co_2C_{10}H_6O_6$ $(MeC{\equiv}CMe)Co_2(CO)_6$, 1, 604
$Co_2C_{10}H_6O_7$ $(HC{\equiv}CCH_2OMe)Co_2(CO)_6$, 1, 604
$Co_2C_{10}H_6O_8$ $(HOCH_2C{\equiv}CCH_2OH)Co_2(CO)_6$, 1, 604
$Co_2C_{11}H_2O_9$ $\overline{OC({=}O)CH{=}CHC}Co_2(CO)_7$, 1, 746
$Co_2C_{11}H_3Cl_3O_8$ $(HC{\equiv}CCH_2O_2CCCl_3)Co_2(CO)_6$, 1, 605
$Co_2C_{11}H_3F_3O_8$ $(HC{\equiv}CCH_2O_2CCF_3)Co_2(CO)_6$, 1, 605
$Co_2C_{11}H_6O_6$ $[HC{\equiv}CC(Me){=}CH_2]Co_2(CO)_6$, 1, 605*
$Co_2C_{11}H_6O_8$ $(HC{\equiv}CCH_2OAc)Co_2(CO)_6$, 1, 605

$Co_2C_{11}H_7ClO_6$ $(HC{\equiv}CCMe_2Cl)Co_2(CO)_6$, 1, 605
$Co_2C_{11}H_8O_6$ $(HC{\equiv}CPr)Co_2(CO)_6$, 1, 605
$Co_2C_{11}H_8O_7$ $(HC{\equiv}CCMe_2OH)Co_2(CO)_6$, 1, 605
$Co_2C_{12}H_3Cl_3O_8$ $(HC{\equiv}CCH_2O_2CCCl{=}CCl_2)$-$Co_2(CO)_6$, 1, 605
$Co_2C_{12}H_6O_{10}$ $(MeO_2CC{\equiv}CCO_2Me)Co_2(CO)_6$, 1, 605
$Co_2C_{12}H_8O_8$ $[Me(CH_2)_2C{\equiv}CCO_2H]Co_2(CO)_6$, 1, 605
$Co_2C_{12}H_9DO_6$ $(DC{\equiv}CBu)Co_2(CO)_6$, 1, 605
$Co_2C_{12}H_9IO_6$ $[HC{\equiv}C(CH_2)_4I]Co_2(CO)_6$, 1, 605
$Co_2C_{12}H_{10}O_6$ $[-C(Me){=}CH_2Co(CO)_3]_2$, 1, 599
$(HC{\equiv}CBu)Co_2(CO)_6$, 1, 605
$(HC{\equiv}CCMe_3)Co_2(CO)_6$, 1, 605
$(MeC{\equiv}CPr)Co_2(CO)_6$, 1, 605
$(EtC{\equiv}CEt)Co_2(CO)_6$, 1, 605
$Co_2C_{12}H_{10}O_7$ $(HC{\equiv}COBu)Co_2(CO)_6$, 1, 605
$Co_2C_{12}H_{12}N_8S_4$ $[(MeNC)_4Co][Co(CNS)_4]$, 1, 612
$Co_2C_{12}H_{12}O_4$ $[CH_2{=}CHCH{=}CH_2Co(CO)_2]_2$, 1, 599
$Co_2C_{12}H_{16}S_2$ $(C_5H_5CoSMe)_2$, 1, 584
$Co_2C_{13}H_6O_9$ $\overline{OC({=}O)C(Me){=}C(Me)C}Co_2(CO)_7$, 1, 749
$Co_2C_{13}H_8O_6$ Bicyclo[2.2.1]hepta-2,5-diene-$[Co(CO)_3]_2$, 1, 602*
$Co_2C_{13}H_{10}O_8$ $(BuC{\equiv}CCO_2H)Co_2(CO)_6$, 1, 605
$Co_2C_{13}H_{11}BrO_6$ $[HC{\equiv}C(CH_2)_5Br]Co_2(CO)_6$, 1, 605
$(BuC{\equiv}CCH_2Br)Co_2(CO)_6$, 1, 605
$Co_2C_{13}H_{11}ClO_6$ $(BuC{\equiv}CCH_2Cl)Co_2(CO)_6$, 1, 605
$Co_2C_{13}H_{11}IO_6$ $(BuC{\equiv}CCH_2I)Co_2(CO)_6$, 1, 605
$Co_2C_{13}H_{12}O_7$ $(BuC{\equiv}CCH_2OH)Co_2(CO)_6$, 1, 605
$(Me_3CC{\equiv}CCH_2OH)Co_2(CO)_6$, 1, 605
$Co_2C_{14}H_5BrO_6$ $(p\text{-}BrC_6H_4C{\equiv}CH)Co_2(CO)_6$, 1, 605
$Co_2C_{14}H_5ClO_6$ $(ClC_6H_4C{\equiv}CH)Co_2(CO)_6$ isomers, 1, 605
$Co_2C_{14}H_5FO_6$ $(p\text{-}FC_6H_4C{\equiv}CH)Co_2(CO)_6$, 1, 605
$Co_2C_{14}H_6O_6$ $(PhC{\equiv}CH)Co_2(CO)_6$, 1, 606
$Co_2C_{14}H_{10}O_4$ $[(C_5H_5)_2Co][Co(CO)_4]$, 1, 587
$Co_2C_{14}H_{10}O_6$ $[HC{\equiv}C(CH_2)_4C{\equiv}CH]Co_2(CO)_6$, 1, 606
$Co_2C_{14}H_{10}O_{10}$ $(EtO_2CC{\equiv}CCO_2Et)Co_2(CO)_6$, 1, 606
$Co_2C_{14}H_{12}O_8$ $[EtC{\equiv}C(CH_2)_3CO_2H]Co_2(CO)_6$, 1, 606
$[Me(CH_2)_4C{\equiv}CCO_2H]Co_2(CO)_6$, 1, 606
$[\overline{OCH_2(CH_2)_3C}HOCH_2C{\equiv}CH]Co_2(CO)_6$, 1, 606
$Co_2C_{14}H_{15}N_5O_4$ $[(MeNC)_5Co][Co(CO)_4]$, 1, 615
$Co_2C_{15}H_6O_8$ $(PhC{\equiv}CCO_2H)Co_2(CO)_6$, 1, 606
$Co_2C_{15}H_8O_6$ $(MeC_6H_4C{\equiv}CH)Co_2(CO)_6$ isomers, 1, 606
$Co_2C_{15}H_{14}O_8$ $[HC{\equiv}C(CH_2)_5CO_2Me]Co_2(CO)_6$, 1, 606
$[MeC{\equiv}C(CH_2)_4CO_2Me]Co_2(CO)_6$, 1, 606
$[EtC{\equiv}C(CH_2)_3CO_2Me]Co_2(CO)_6$, 1, 606
$[n\text{-}C_5H_{11}C{\equiv}CCO_2Me)Co_2(CO)_6$, 1, 606
$Co_2C_{16}H_6N_2O_{12}$ $[3,5\text{-}(O_2N)_2C_6H_3CO_2CH_2$-$C{\equiv}CH]Co_2(CO)_6$, 1, 606
$Co_2C_{16}H_7NO_{10}$ $(p\text{-}O_2NC_6H_4CO_2CH_2C{\equiv}CH)Co_2$-$(CO)_6$, 1, 606
$Co_2C_{16}H_8O_8$ $(PhC{\equiv}CCO_2Me)Co_2(CO)_6$, 1, 606
$(BzOCH_2C{\equiv}CH)Co_2(CO)_6$, 1, 606
$Co_2C_{16}H_{10}O_6$ $(Me_2C_6H_3C{\equiv}CH)Co_2(CO)_6$ isomers, 1, 606
$Co_2C_{16}H_{14}O_8$ $[Me_2C(OH)C{\equiv}CC{\equiv}CC(OH)Me_2]$-$Co_2(CO)_6$, 1, 606
$Co_2C_{16}H_{16}O_4$ [1,3-Cyclohexadiene-Co-$(CO)_2]_2$, 1, 602
$Co_2C_{16}H_{16}O_6$ $[\overline{CH_2C{\equiv}C(CH_2)_6C}H_2]Co_2(CO)_6$, 1, 606
$Co_2C_{16}H_{18}O_6$ $(Me_3CC{\equiv}CCMe_3)Co_2(CO)_6$, 1, 606
$Co_2C_{17}H_{10}O_8$ $(PhC{\equiv}CCO_2Et)Co_2(CO)_6$, 1, 607
$Co_2C_{17}H_{12}O_6$ $(2,4,6\text{-}Me_3C_6H_2C{\equiv}CH)Co_2$-$(CO)_6$, 1, 606
$Co_2C_{18}H_{16}O_4$ Bis{bicyclo[2.2.1]hepta-2,5-diene-$Co(CO)_2$}, 1, 602
$Co_2C_{18}H_{18}$ Cyclooctatetraene$[CoC_5H_5]_2$,

1, 583

$Co_2C_{18}H_{18}O_{10}$ $(BuO_2CC\equiv CCO_2Bu)Co_2(CO)_6$, 1, 607

$Co_2C_{20}H_8N_2O_{10}$ $(p\text{-}O_2NC_6H_4C\equiv CC_6H_4NO_2\text{-}p)\text{-}Co_2(CO)_6$, 1, 607

$Co_2C_{20}H_{10}O_6$ $(PhC\equiv CPh)Co_2(CO)_6$, 1, 607

$Co_2C_{20}H_{18}O_6$ $[2,4,6\text{-}Me_2(Me_3C)C_6H_2\text{-}C\equiv CH]Co_2(CO)_6$, 1, 607

$Co_2C_{20}H_{20}O_4$ Bis[1,3,6-cyclooctatriene-$Co(CO)_2$], 1, 602

$Co_2C_{20}H_{30}Cl_2IN_{10}O_8$ $[(MeNC)_{10}Co_2I]\text{-}(ClO_4)_2$, 1, 617

$Co_2C_{22}H_{10}O_6$ $(PhC\equiv CC\equiv CPh)Co_2(CO)_6$, 1, 607

$Co_2C_{22}H_{18}O_{18}$ $(MeO_2CC\equiv CCO_2Me)_3Co_2\text{-}(CO)_4$, 1, 609

$Co_2C_{22}H_{20}F_4$ $[(\text{-}CF_2C_5H_5)CoC_5H_5]_2$, 1, 580

$Co_2C_{22}H_{30}O_4$ $(HC\equiv CCMe_3)_3Co_2(CO)_4$, 1, 609

$Co_2C_{26}H_{26}$ $MeC_5H_3(C_5H_5CoC_5H_5)_2$, 1, 581

$Co_2C_{28}H_{18}O_7$ $(Ph_3COCH_2C\equiv CH)Co_2(CO)_6$, 1, 607

$Co_2C_{28}H_{20}Br_4N_4$ $[(PhNC)_4Co]CoBr_4$, 1, 612

$Co_2C_{30}H_{18}O_6$ $(PhC\equiv CH)_3Co_2(CO)_6$, 1, 610

$Co_2C_{30}H_{21}Br_5$ $(C_{10}H_7)_3Co_2Br_5$ isomers, 1, 573

$Co_2C_{30}H_{21}I_5$ $(\beta\text{-}C_{10}H_7)_3Co_2I_5$, 1, 573

$Co_2C_{30}H_{40}O_6$ $\{[Me_3C\overline{C=CHCH=C(CMe_3)C}O]\text{-}Co(CO)_2\}_2$, 1, 601

$Co_2C_{31}H_{24}O_4$ $(PhC\equiv CMe)_3Co_2(CO)_4$, 1, 609

$Co_2C_{34}H_{24}O_{10}$ $(PhC\equiv CCO_2Me)_3Co_2(CO)_4$, 1, 609

$Co_2C_{37}H_{25}O_5P$ $(PhC\equiv CPh)Co_2(CO)_5(PPh_3)$, 1, 607

$Co_2C_{37}H_{30}O_{10}$ $(PhC\equiv CCO_2Et)_3Co_2(CO)_4$, 1, 609

$Co_2C_{38}H_{24}O_6$ $\{[Ph\overline{C=CHCH=C(Ph)C}O]Co\text{-}(CO)_2\}_2$, 1, 601

$Co_2C_{38}H_{25}N_5O_3$ $[(PhNC)_5Co][Co(CO)_3]$, 1, 617*

$Co_2C_{39}H_{20}Cl_5N_5O_4$ $[(p\text{-}ClC_6H_4NC)_5Co]\text{-}[Co(CO)_4]$, 1, 615

$Co_2C_{39}H_{45}O_3P$ $(HC\equiv CCMe_3)_3Co_2(CO)_3\text{-}(PPh_3)$, 1, 609

$Co_2C_{42}H_{32}O_4P_2$ $(HC\equiv CH)Co_2(CO)_4(PPh_3)_2$, 1, 607

$Co_2C_{44}H_{35}N_5O_4$ $[(p\text{-}MeC_6H_4NC)_5Co][Co\text{-}(CO)_4]$, 1, 615

$Co_2C_{45}H_{32}O_7P_2$ $\overline{OC(=O)CH=CHC}Co_2(CO)_5\text{-}(PPh_3)_2$, 1, 746

$Co_2C_{46}H_{30}O_4$ $(PhC\equiv CPh)_3Co_2(CO)_4$, 1, 609

$Co_2C_{50}H_{34}O_8P_2$ $o\text{-}C_6H_4[COCo(CO)_3(PPh_3)]_2$, 1, 570

$Co_2C_{54}H_{40}O_4P_2$ $(PhC\equiv CPh)Co_2(CO)_4(PPh_3)_2$, 1, 607

$Co_3C_{10}BrO_9$ $BrC[Co(CO)_3]_3$, 1, 574

$Co_3C_{10}ClO_9$ $ClC[Co(CO)_3]_3$, 1, 574

$Co_3C_{10}FO_9$ $FC[Co(CO)_3]_3$, 1, 574

$Co_3C_{10}HO_9$ $HC[Co(CO)_3]_3$, 1, 574

$Co_3C_{10}IO_9$ $IC[Co(CO)_3]_3$, 1, 574

$Co_3C_{11}F_3O_9$ $CF_3C[Co(CO)_3]_3$, 1, 574

$Co_3C_{11}HO_{11}$ $HO_2CC[Co(CO)_3]_3$, 1, 574

$Co_3C_{11}H_3O_9$ $MeC[Co(CO)_3]_3$, 1, 574

$Co_3C_{12}H_3O_{11}$ $MeO_2CC[Co(CO)_3]_3$, 1, 574

$Co_3C_{12}H_5O_9$ $EtC[Co(CO)_3]_3$, 1, 574

$Co_3C_{13}H_5O_{11}$ $EtO_2CC[Co(CO)_3]_3$, 1, 574

$Co_3C_{14}H_9O_9$ $BuC[Co(CO)_3]_3$, 1, 574

$Co_3C_{15}H_7O_{13}$ $(AcO)_2CHC[Co(CO)_3]_3$, 1, 574

$Co_3C_{16}H_5O_9$ $PhC[Co(CO)_3]_3$, 1, 574

$Co_3C_{17}H_7O_9$ $PhCH_2C[Co(CO)_3]_3$, 1, 574

$Co_3C_{20}H_{18}BrO_2$ $[(C_6H_6Co)_3(CO)_2]Br$, 1, 598

$Co_3C_{20}H_{18}ClO_6$ $[(C_6H_6Co)_3(CO)_2]ClO_4$, 1, 598

$Co_3C_{20}H_{18}IO_2$ $[(C_6H_6Co)_3(CO)_2]I$, 1, 598

$Co_3C_{20}H_{20}Br_3Cl$ $[(C_5H_5)_2Co]_2[CoCl_{1.4}\text{-}Br_{2.6}]$, 1, 587

$Co_3C_{20}H_{20}Br_4$ $[(C_5H_5)_2Co]_2CoBr_4$, 1, 587

$Co_3C_{20}H_{20}Cl_4$ $[(C_5H_5)_2Co]_2CoCl_4$, 1, 587

$Co_3C_{23}H_{24}BrO_2$ $\{[(MePh)Co]_3(CO)_2\}Br$, 1, 598

$Co_3C_{23}H_{24}IO_2$ $\{[(MePh)Co]_3(CO)_2\}I$, 1, 598

$Co_3C_{26}H_{20}N_3O_9$ $[(C_6H_6Co)_3(CO)_2]OC_6H_2\text{-}(NO_2)_3$, 1, 598

$Co_3C_{32}H_{16}Br_2N_4O_8$ $[Co_2(O_2CC_6H_4NC)_4Co]\text{-}Br_2$, 1, 614

$Co_3C_{32}H_{16}I_2N_4O_8$ $[Co_2(O_2CC_6H_4NC)_4Co]I_2$,

1, 614*

$Co_3C_{36}H_{28}Br_4$ [Bis(indenyl)-Co]$_2$$CoBr_4$, 1, 590

$Co_3C_{116}H_{80}O_4$ Tetrakis(tetraphenylcyclopentadienone)-Co_3, 1, 602

Co{[PhC=C(Ph)C(Ph)=C(Ph)CO]$_2$Co}$_2$, 1, 602

$Co_3C_{120}H_{102}Cl_4N_6P_4$ [(p-$MeC_6H_4NC)_3$-($Ph_3P)_2Co]_2CoCl_4$, 1, 611

$Co_4C_{12}H_2O_{10}$ (HC≡CH)$Co_4(CO)_{10}$, 1, 608

$Co_4C_{16}H_{10}O_{10}$ (EtC≡CEt)$Co_4(CO)_{10}$, 1, 608

$Co_4C_{18}H_6O_{10}$ (HC≡CPh)$Co_4(CO)_{10}$, 1, 608

$Co_4C_{20}H_8O_{12}$ ($PhC≡CCO_2Me)Co_4(CO)_{10}$, 1, 608

$Co_4C_{22}H_{12}Cl_2O_{12}$ [⟨$CH_2CH_2CH_2C≡CCl)Co_2$-$(CO)_6]_2$, 1, 607

$Co_4C_{24}H_{10}O_{10}$ (PhC≡CPh)$Co_4(CO)_{10}$, 1, 608

$Co_4C_{24}H_{18}O_6$ [($C_6H_6Co)_3(CO)_2$][$Co(CO)_4$], 1, 598

$Co_4C_{28}H_{10}O_{12}$ [⟨C≡CPh)$Co_2(CO)_6]_2$, 1, 607

$Co_6C_{20}O_{18}$ {-C[$Co(CO)_3]_3$}$_2$, 1, 574

$Co_6C_{30}H_6O_{18}$ 1,3,5-C_6H_3[⟨C≡CH)Co_2-$(CO)_6]_3$, 1, 607

1,2,4-C_6H_3[⟨C≡CH)$Co_2(CO)_6]_3$, 1, 607

$CoAgC_{40}H_{35}I_2N_5$ [(p-$MeC_6H_4NC)_5Co]AgI_2$, 1, 616

$CoAlC_{12}H_{12}Br_3$ ($C_6H_6)_2Co\cdot AlBr_3$, 1, 597

$CoAlC_{14}H_{16}Cl_3$ (MePh)$_2Co\cdot AlCl_3$, 1, 597

$CoAlC_{18}H_{24}Cl_3$ ($Me_3C_6H_3)_2Co\cdot AlCl_3$, 1, 597

$CoAlC_{36}H_{60}Cl_3$ ($C_6Et_6)_2Co\cdot AlCl_3$, 1, 598

$(CoAl_2C_{15}H_{13}Cl_5)_n$ ($Ph_2Al_2CoCl_5\cdot 1/2$ $C_6H_6)_n$, 1, 618

$CoAsC_{20}H_{15}NO_3$ [Co($AsPh_3)(CO)_2$(NO)], 3, 19

$CoAsC_{23}H_{15}F_5O_3$ $C_2F_5Co(CO)_3(AsPh_3)$, 1, 566, 3, 20

$CoAsC_{23}H_{21}NO_3$ Co[As(C_6H_4Me-4)$_3(CO)_2$-(NO)], 3, 28

$CoAsC_{37}H_{30}NO_2P$ [Co($AsPh_3)(PPh_3)$(CO)-(NO)], 3, 20

$CoAs_2C_{37}H_{30}NO_2$ [Co($AsPh_3)_2$(CO)(NO)], 3, 20

$CoBC_{26}H_{21}F_4O_3P$ [(MeC=CHCH=CHCO)Co-$(CO)_2(PPh_3)]BF_4$, 1, 600

$CoBC_{34}H_{30}$ [($C_5H_5)_2Co]BPh_4$, 1, 587

$CoBC_{38}H_{34}O_2$ [($AcC_5H_4)_2Co]BPh_4$, 1, 589

$CoBC_{40}H_{34}$ [$C_5H_5Co(C_5H_4Ph)]BPh_4$, 1, 588

$CoBC_{42}H_{34}$ [Bis(indenyl)-Co]BPh_4, 1, 590

$CoBC_{42}H_{48}N_2$ {[$Me_2NCH(Me)C_5H_4]_2Co$}BPh_4, 1, 589

$CoBC_{46}H_{38}N_4$ [($PhN_2C_5H_4)_2Co]BPh_4$, 1, 589

$CoBC_{48}H_{56}$ [($C_6Me_6)_2Co]BPh_4$, 1, 598

$CoB_2C_{11}H_{11}F_8$ [($C_6H_6)(C_5H_5)Co][BF_4]_2$, 1, 578

$CoB_2C_{58}H_{55}N_5$ [(MeNC)$_5$Co][$BPh_4]_2$, 1, 616

$CoBi_3C_{10}H_{10}I_{10}$ [($C_5H_5)_2Co][BiI_4\cdot 2BiI_3$], 1, 587

$CoCdC_8H_{12}Br_4N_4$ [(MeNC)$_4$Co]$CdBr_4$, 1, 612

$CoCdC_8H_{12}I_4N_4$ [(MeNC)$_4$Co]CdI_4, 1, 612

$CoCrC_{14}H_{16}N_6S_4$ [($C_5H_5)_2Co][Cr(NH_3)_2$-$(SCN)_4$], 1, 587

$CoCrC_{16}H_{20}N_6S_4$ [($MeC_5H_4)_2Co][Cr(NH_3)_2$-$(SCN)_4$], 1, 588

$CoCrC_{16}H_{22}N_6OS_4$ {[1,3-HO(Me)$C_5H_3]_2$Co}-reineckate$\cdot H_2O$, 1, 589

$CoCrC_{18}H_{15}O_3$ [($C_5H_5)_2Co][C_5H_5Cr(CO)_3$], 1, 587

$CoCrC_{18}H_{20}N_6S_4$ [($AcC_5H_4)_2Co][Cr(NH_3)_2$-$(SCN)_4$], 1, 589

$CoCrC_{22}H_{16}O_4$ [($C_6H_6)(Ph_2)Cr][Co(CO)_4$], 1, 87

$CoCrC_{22}H_{20}N_6S_4$ [Bis(indenyl)-Co][Cr-$(NH_3)_2(SCN)_4$], 1, 590

$CoCrC_{22}H_{34}N_6S_4$ {[$Me_2NCH(Me)C_5H_4]_2$Co}-[$Cr(NH_3)_2(SCN)_4$], 1, 589

$CoCrC_{26}H_{24}N_{10}S_4$ [($PhN_2C_5H_4)_2$Co][Cr-$(NH_3)_2(SCN)_4$], 1, 589

$CoCrC_{28}H_{20}O_4$ [($Ph_2)_2Cr][Co(CO)_4$], 1, 83

$CoFeC_{11}H_5O_6$ $C_5H_5Fe(CO)$-μ-$(CO)_2$-Co-$(CO)_3$, 1, 547

$CoFeC_{14}H_{11}O_4$ [($C_5H_5)_2Co][HFe(CO)_4$], 1, 587

$CoFeC_{17}H_{17}O_3$ ($Me_3C_6H_3)Co(CO)(C_5H_5)Fe$-$(CO)_2$, 1, 547

$CoFeC_{18}H_{15}O_2$ $FcCH_2C_5H_4Co(CO)_2$, 1, 547

$CoFe_2C_{44}H_{38}Cl$ $\{[FcCH(Ph)C_5H_4]_2Co\}Cl$, 1, 547

$CoFe_2C_{44}H_{38}ClO_4$ $\{[FcCH(Ph)C_5H_4]_2Co\}$-ClO_4, 1, 547

$CoFe_2C_{50}H_{40}N_3O_7$ $\{[FcCH(Ph)C_5H_4]_2Co\}$-$OC_6H_2(NO_2)_3$, 1, 547

$CoGeC_{27}H_{26}O$ $(PhCH_2)_3Ge[C_5H_5Co(CO)]$, 2, 142

$CoHgC_{35}H_{25}I_3N_5$ $[(PhNC)_5Co]HgI_3$, 1, 615

$CoHgC_{40}H_{35}Br_3N_5$ $[(p\text{-}MeC_6H_4NC)_5Co]Hg$-$Br_3$, 1, 616

$CoHgC_{40}H_{35}Cl_3N_5O_4$ $[(p\text{-}MeC_6H_4NC)_5Co]$-$ClO_4 \cdot HgCl_2$, 1, 616

$CoHgC_{40}H_{35}I_3N_5$ $[(p\text{-}MeC_6H_4NC)_5Co]HgI_3$, 1, 616

$CoHg_2C_{40}H_{35}I_4N_5$ $[(p\text{-}MeC_6H_4NC)_5Co]$-$Hg_2I_4$, 1, 616

$CoK_4C_{12}H_6 \cdot (NH_3)_n$ $K_4[Co(C{\equiv}CH)_6] \cdot (NH_3)_n$ 1, 575

$CoK_4C_{28}H_{16}Br_2N_4O_{12}S_4$ $(p\text{-}KO_3SC_6H_4NC)_4$-$CoBr_2$, 1, 614

$CoK_4C_{28}H_{16}I_2N_4O_{12}S_4$ $(p\text{-}KO_3SC_6H_4NC)_4$-$CoI_2$, 1, 614

$CoMnC_{15}H_{10}O_5$ $[(C_5H_5)_2Co][Mn(CO)_5]$, 1, 587

$CoMoC_{18}H_{15}O_3$ $[(C_5H_5)_2Co][C_5H_5Mo(CO)_3]$ 1, 118

$CoNa_3C_6H_3N_5$ $Na_3[MeCo(CN)_5]$, 1, 575

$CoNa_3C_7H_5N_5$ $Na_3[EtCo(CN)_5]$, 1, 575

$CoNa_3C_{12}H_{11}N_5O_2$ $Na_3[PhCH_2Co(CN)_5] \cdot$-$2H_2O$, 1, 575

$CoNa_4C_{18}H_{18}$ $Na_4[Co(C{\equiv}CMe)_6]$, 1, 575

$CoPbC_{22}H_{15}O_4$ $Ph_3PbCo(CO)_4$, 2, 595

$CoPtC_{13}H_{24}Cl_5N_4$ $[Co(H_2NCH_2CH_2NH_2)_2$-$Cl_2][Pt(indene)Cl_3]$, 1, 724

$CoPtC_{24}H_{36}Cl_6$ $[(C_6Me_6)_2Co]PtCl_6$, 1, 598

$CoSbC_{20}H_{12}Cl_3NO_3$ $\{Co[Sb(C_6H_4Cl\text{-}p)_3]$-$(NO)(CO)_2\}$, 3, 660

$CoSbC_{20}H_{15}NO_3$ $[Co(SbPh_3)(CO)_2(NO)]$, 3, 662

$CoSbC_{23}H_{15}F_5O_3$ $C_2F_5Co(CO)_3(SbPh_3)$, 1, 566, 3, 662

$CoSbC_{23}H_{21}NO_3$ $\{Co[Sb(C_6H_4Me)_3](CO)_2$-$(NO)\}$, 3, 667

$CoSb_2C_{37}H_{30}NO_2$ $[Co(SbPh_3)_2(CO)(NO)]$, 3, 662

$CoSi_2C_{46}H_{38}$ $(Ph_3SiC_5H_4)_2Co$, 1, 618

$CoSnC_9H_{12}ClO$ $(MeC_5H_4)Co(CO)SnEtCl$, 1, 618, 2, 502

$CoSnC_{16}H_{27}O_4$ $Bu_3SnCo(CO)_4$, 2, 502

$CoSnC_{17}H_{16}IO$ $(Me_2C_5H_3)Co(CO)Sn(in$-$denyl)I$, 1, 618, 2, 502

$CoSnC_{30}H_{46}ClO_2$ $(C_{16}H_{33}C_5H_4)Co(CO)Sn$-$(C_6H_4OEt)Cl$, 1, 618*, 2, 502

$CoSnC_{33}H_{42}O_3P$ $Bu_3SnCo(CO)_3PPh_3$, 2, 502

$CoWC_{16}H_{15}O$ $C_5H_5W(CO)Co(C_5H_5)_2$, 1, 130

$CoWC_{18}H_{15}O_3$ $[(C_5H_5)_2Co][C_5H_5W(CO)_3]$, 1, 588

$Co_2Ag_3C_{24}H_{16}Cl_2O_{12}$ $Ag[(O_2CC_5H_4)_2Co] \cdot$ $1/2AgClO_4$, 1, 589

$Co_2Al_7C_{24}H_{24}Cl_{21}$ $(C_6H_6)_2Co(AlCl_3)_{3.5}$, 1, 597

$Co_2AsC_{25}H_{15}O_7$ $[Co_2(AsPh_3)(CO)_7]$, 3, 20

$Co_2As_2C_{42}H_{30}O_6$ $[Co(AsPh_3)(CO)_3]_2$, 3, 20

$Co_2AsC_{43}H_{30}O_7$ $[Co(AsPh_3)_2(CO)_3][Co$-$(CO)_4]$, 3, 20

$Co_2CdC_{80}H_{70}I_4N_{10}$ $[(p\text{-}MeC_6H_4NC)_5Co]_2$-$CdI_4$, 1, 616

$Co_2Fe_2C_{28}H_{20}O_8$ $[(C_5H_5)_2Co]_2[Fe_2(CO)_8]$, 1, 587

$Co_2Fe_4C_{33}H_{20}O_{13}$ $[(C_5H_5)_2Co]_2[Fe_4$-$(CO)_{13}]$, 1, 587

$Co_2HgC_{20}H_{12}O_4$ $\{Co(CO)_2$-$[PhC{=}C(Ph)CH{=}CH]Co(CO)_2Hg\}_n$, 1, 600

$Co_2HgC_{30}H_{40}O_6$ $\{[Me_3CC{=}CHCH{=}C(CMe_3)CO]$-$Co(CO)_2\}_2Hg$, 1, 601

$Co_2HgC_{62}H_{40}O_6$ $\{[PhC{=}C(Ph)C(Ph){=}C(Ph)CO]$-$Co(CO)_2\}_2Hg$, 1, 601

$Co_2Hg_2C_{80}H_{70}I_4N_{10}$ $[(p\text{-}MeC_6H_4NC)_5Co]_2$-$Hg_2I_4$, 1, 616

$Co_2Hg_3C_{20}H_{20}I_8$ $[(C_5H_5)_2Co]_2[HgI_4 \cdot 2Hg$-$I_2]$, 1, 587

$Co_2K_2C_{32}H_{16}Br_2N_4O_8$ $[Co(O_2CC_6H_4NC)_2$-$(KO_2CC_6H_4NC)_2Co]Br_2$, 1, 614

$Co_2K_2C_{32}H_{16}I_2N_4O_8$ $[Co(O_2CC_6H_4NC)_2$-$(KO_2CC_6H_4NC)_2Co]I_2$, 1, 614

$Co_2K_6C_{12}H_{10}N_{10}O_4$ $K_6[{=}CHCo(CN)_5]_2 \cdot 4H_2O$, 1, 575

$Co_2MnC_{28}H_{28}O_{14}$ $[Mn_{0.9}Co_{0.1}(H_2O)_6]$-$[(O_2CC_5H_3CO_2H)_2Co]_2$, 1, 589

$Co_2Ni_2C_{32}H_{20}O_6$ $\{(C_5H_5Ni)_2(PhC{\equiv}CC{\equiv}CPh)$-$[Co(CO)_3]_2\}$, 1, 618

$Co_2PtC_{20}H_{20}Cl_6$ $[(C_5H_5)_2Co]_2PtCl_6$, 1, 588

$Co_2Sb_2C_{42}H_{30}O_6$ $[Co(SbPh_3)(CO)_3]_2$, 3, 662

$Co_2Sb_2C_{43}H_{30}O_7$ $[Co(SbPh_3)_2(CO)_3][Co$-$(CO)_4]$, 3, 662

$Co_2SiC_{11}H_{10}O_6$ $(HC{\equiv}CSiMe_3)Co_2(CO)_6$, 1, 605

$Co_2SiC_{17}H_{14}O_6$ $(PhC{\equiv}CSiMe_3)Co_2(CO)_6$, 1, 606

$Co_2Si_2C_{14}H_{18}O_6$ $(Me_3SiC{\equiv}CSiMe_3)Co_2$-$(CO)_6$, 1, 606

$Co_2Si_2C_{26}H_{40}O_6$ $\{[Me_3Si$-$C{=}CHCH{=}C(SiMe_3)CO]Co(CO)_2\}_2$, 1, 601

$Co_2Si_3C_{19}H_{30}O_4$ $(HC{\equiv}CSiMe_3)_3Co_2(CO)_4$, 1, 609

$Co_2SnC_{10}H_6O_8$ $Me_2Sn[Co(CO)_4]_2$, 2, 502

$Co_2SnC_{16}H_{18}O_8$ $Bu_2Sn[Co(CO)_4]_2$, 2, 502

$Co_2SnC_{44}H_{36}O_6P_2$ $Me_2Sn[Co(CO)_3PPh_3]_2$, 2, 502

$Co_2SnC_{50}H_{48}O_6P_2$ $Bu_2Sn[Co(CO)_3PPh_3]_2$, 2, 502

$Co_3BC_{44}H_{38}O_2$ $[(C_6H_6Co)_3(CO)_2]BPh_4$, 1, 598

$Co_3CrC_{24}H_{24}N_6O_2S_4$ $[(C_6H_6Co)_3(CO)_2]$-$[Cr(NH_3)_2(SCN)_4]$, 1, 598

$Co_3CrC_{27}H_{30}N_6O_2S_4$ $\{[(MePh)Co]_3(CO)_2\}$-$[Cr(NH_3)_2(SCN)_4]$, 1, 598

$Co_4Hg_2C_{24}H_{20}O_{12}$ $\{[Me_3CC{=}C(CMe_3)CH{=}CH]$ $[Co(CO)_2HgCo(CO)_4]_2\}$, 1, 599

$Co_4Hg_2C_{28}H_{12}O_{12}$ $\{[PhC{=}C(Ph)CH{=}CH]$-$[Co(CO)_2HgCo(CO)_4]_2\}$, 1, 599

$Co_4SiC_{15}H_{10}O_{10}$ $(HC{\equiv}CSiMe_3)Co_4(CO)_{10}$, 1, 608*

$Co_4SiC_{21}H_{14}O_{10}$ $(PhC{\equiv}CSiMe_3)Co_4(CO)_{10}$, 1, 608

$Co_6Hg_3C_{48}H_{24}O_{16}$ $Hg\{Co(CO)_2$-$[PhC{=}C(Ph)CH{=}CH]Co(CO)_2HgCo(CO)_4\}_2$ 1, 600

$CoCrFe_2C_{48}H_{44}N_6S_4$ $\{[FcCH(Ph)C_5H_4]_2$-$Co\}[Cr(NH_3)_2(SCN)_4]$, 1, 547

$CoFeHgC_{11}H_5O_6$ $C_5H_5Fe(CO)_2HgCo(CO)_4$, 1, 529

$Co_2As_2HgC_{42}H_{30}O_6$ $Hg[Co(AsPh_3)(CO)_3]_2$, 3, 20

$Co_2HgSb_2C_{42}H_{30}O_6$ $Hg[Co(SbPh_3)(CO)_3]_2$, 3, 662

$Co_2HgSi_4C_{26}H_{40}O_6$ $\{[Me_3SiC{=}CHCH{=}C(SiMe_3)CO]Co(CO)_2\}_2$-$Hg$, 1, 601

$CrCH_{11}Cl_4O_{13}$ $[Cl_2CHCr(H_2O)_5](ClO_4)_2$, 1, 56

CrC_2H_6 Me_2Cr, 1, 55

CrC_3H_9 Me_3Cr, 1, 55

$CrC_5H_5BrN_2O_2$ $C_5H_5Cr(NO)_2Br$, 1, 64

$CrC_5H_5ClN_2O_2$ $C_5H_5Cr(NO)_2Cl$, 1, 63

$CrC_5H_5Cl_2$ $C_5H_5CrCl_2$, 1, 59

$CrC_5H_5IN_2O_2$ $C_5H_5Cr(NO)_2I$, 1, 64

$CrC_5H_9Br_3N$ $NH_4[C_5H_5CrBr_3]$, 1, 60

$CrC_5H_9Cl_3N$ $NH_4[C_5H_5CrCl_3]$, 1, 60

$CrC_5H_{11}Cl_2O$ $MeCrCl_2{\cdot}THF$, 1, 56

$CrC_6H_5N_3O_2S$ $C_5H_5Cr(NO)_2SCN$, 1, 64

$CrC_6H_7ClN_2O_2$ $C_5H_5Cr(NO)_2CH_2Cl$, 1, 64

$CrC_6H_8N_2O_2$ $C_5H_5Cr(NO)_2Me$, 1, 64

CrC_6H_{15} Et_3Cr, 1, 55

$CrC_7H_4O_3S$ Thiophene-$Cr(CO)_3$, 1, 89*

$CrC_7H_5NO_3$ $C_5H_5Cr(CO)_2NO$, 1, 62

$CrC_7H_{10}N_2O_2$ $C_5H_5Cr(NO)_2Et$, 1, 64

$CrC_7H_{17}Cl_2O_{13}$ $[PhCH_2Cr(H_2O)_5](ClO_4)_2$, 1, 56

$CrC_8H_6O_3$ $C_5H_5Cr(CO)_3H$, 1, 61

$CrC_9H_4Cl_2O_3$ $(p\text{-}Cl_2C_6H_4)Cr(CO)_3$, 1, 736

$CrC_9H_4F_2O_3$ $(p\text{-}F_2C_6H_4)Cr(CO)_3$, 1, 736

$CrC_9H_5ClO_3$ $(PhCl)Cr(CO)_3$, 1, 70

$CrC_9H_5FO_3$ $(PhF)Cr(CO)_3$, 1, 736

$CrC_9H_6ClO_3$ $[C_6H_6Cr(CO)_3]Cl$, 1, 70

$CrC_9H_6O_3$ $C_6H_6Cr(CO)_3$, 1, 69

$CrC_9H_6O_4$ $(PhOH)Cr(CO)_3$, 1, 71

$CrC_9H_7NO_3$ $(PhNH_2)Cr(CO)_3$, 1, 71
(Methylpyridine)-$Cr(CO)_3$, 1, 90

$CrC_9H_7NO_4$ $(AcC_5H_4)Cr(CO)_2NO$, 1, 63

$CrC_9H_8ClNO_3$ $(PhNH_2{\cdot}HCl)Cr(CO)_3$, 1, 736

$CrC_9H_8O_3$ $C_5H_5Cr(CO)_3Me$, 1, 62

$CrC_9H_9N_3O_3$ $(MeNC)_3Cr(CO)_3$, 1, 92

CrC_9H_{15} $(CH_2{=}CHCH_2)_3Cr$, 1, 69

$CrC_9H_{13}Br_2O$ $C_5H_5CrBr_2{\cdot}THF$, 1, 60

$CrC_9H_{13}Cl_2O$ $C_5H_5CrCl_2{\cdot}THF$, 1, 59

$CrC_9H_{13}I_2O$ $C_5H_5CrI_2{\cdot}THF$, 1, 59

$CrC_9H_{17}Cl_3NO$ [$Me_2C(OH)CH=C=CHCH_2$-NMe_2]$CrCl_3$, 1, 89
$CrC_{10}H_6O_4$ (BzH)$Cr(CO)_3$, 1, 736
$CrC_{10}H_6O_5$ (BzOH)$Cr(CO)_3$, 1, 71
$CrC_{10}H_7ClO_3$ (ClC_6H_4Me)$Cr(CO)_3$ isomers, 1, 736
$CrC_{10}H_7ClO_4$ (p-ClC_6H_4OMe)$Cr(CO)_3$, 1, 736
$CrC_{10}H_7ClO_7$ [Cycloheptatrienium-Cr-$(CO)_3$]ClO_4, 1, 91
$CrC_{10}H_7FO_3$ (FC_6H_4Me)$Cr(CO)_3$ isomers, 1, 736
$CrC_{10}H_7I_3O_3$ [Cycloheptatrienium-Cr-$(CO)_3$]I_3, 1, 90
$CrC_{10}H_7NO_6$ ($Me_2CHCONC$)$Cr(CO)_5$, 1, 92
$CrC_{10}H_8O_3$ (MePh)$Cr(CO)_3$, 1, 71
Cycloheptatriene-$Cr(CO)_3$, 1, 743
$CrC_{10}H_8O_3S$ (1-Mercaptocycloheptatriene)-$Cr(CO)_3$, 1, 744
$CrC_{10}H_8O_4$ (PhOMe)$Cr(CO)_3$, 1, 72
($PhCH_2OH$)$Cr(CO)_3$, 1, 736
$CrC_{10}H_9NO_3$ (MeNHPh)$Cr(CO)_3$, 1, 736
($MeC_6H_4NH_2$)$Cr(CO)_3$ isomers, 1, 736
$CrC_{10}H_{10}$ $(C_5H_5)_2Cr$, 1, 67
$CrC_{10}H_{10}Br$ $(C_5H_5)_2CrBr$, 1, 68
$CrC_{10}H_{10}Br_2N$ $C_5H_5CrBr_2 \cdot Py$, 1, 60
$CrC_{10}H_{10}I$ $(C_5H_5)_2CrI$, 1, 68
$CrC_{10}H_{10}I_2N$ $C_5H_5CrI_2 \cdot Py$, 1, 60
$CrC_{10}H_{10}N_2O_2$ $(\pi$-$C_5H_5)(\sigma$-$C_5H_5)Cr(NO)_2$, 1, 64
$CrC_{10}H_{12}BrO_2$ (C_5H_5)[AcCH=C(Me)O]CrBr, 1, 61
$CrC_{10}H_{12}O_3S$ [$(C_6H_6)Cr(CO)_2(Me_2SO)$], 1, 742
$CrC_{11}H_6O_3S$ (Thianaphthene)-$Cr(CO)_3$, 1, 736
$CrC_{11}H_7NO_3$ Cycloheptatrienylcarbonitrile-$Cr(CO)_3$, 1, 744
$CrC_{11}H_8O_4$ Bicyclo[2.2.1]hepta-2,5-diene-$Cr(CO)_4$, 1, 91
(AcPh)$Cr(CO)_3$, 1, 737
$CrC_{11}H_8O_5$ (BzOMe)$Cr(CO)_3$, 1, 72
(PhOAc)$Cr(CO)_3$, 1, 737
($PhCH_2CO_2H$)$Cr(CO)_3$, 1, 737
$CrC_{11}H_8O_6$ (m-$MeOC_6H_4CO_2H$)$Cr(CO)_3$, 1, 737
$CrC_{11}H_9ClO_7$ [Methylcycloheptatrienium-$Cr(CO)_3$]ClO_4, 1, 91
$CrC_{11}H_9I_3O_3$ [Methylcycloheptatrienium-$Cr(CO)_3$]I_3, 1, 91
$CrC_{11}H_9NO_4$ (PhNHAc)$Cr(CO)_3$, 1, 737
$CrC_{11}H_{10}N_2O_2$ $C_5H_5Cr(NO)_2Ph$, 1, 64
$CrC_{11}H_{10}O_3$ ($Me_2C_6H_4$)$Cr(CO)_3$ isomers, 1, 72, 73
Cyclo-1,3,5-octatriene-$Cr(CO)_3$, 1, 91
(1-Methylcycloheptatriene)-$Cr(CO)_3$, 1, 744
$CrC_{11}H_{10}O_4$ (MeC_6H_4OMe)$Cr(CO)_3$ isomers, 1, 737
(1-Methoxycycloheptatriene)-$Cr(CO)_3$, 1, 744
$CrC_{11}H_{10}O_5$ [p-$(MeO)_2C_6H_4$]$Cr(CO)_3$, 1, 737
$CrC_{11}H_{11}$ $C_6H_6CrC_5H_5$, 1, 65
$CrC_{11}H_{11}NO_3$ ($PhNMe_2$)$Cr(CO)_3$, 1, 73
6-(Dimethylamino)fulvene-$Cr(CO)_3$, 1, 62
$CrC_{11}H_{12}O_2$ $C_5H_5Cr(CO)_2CH_2CH$=CHMe, 1, 62
$CrC_{11}H_{12}O_3$ (PrC_5H_4)$Cr(CO)_3H$, 1, 62
$CrC_{11}H_{15}Cl_2O$ $PhCH_2CrCl_2 \cdot THF$, 1, 57
$CrC_{12}H_8O_3$ Indene-$Cr(CO)_3$, 1, 737
$CrC_{12}H_{10}D_2$ $(PhD)_2Cr$, 1, 77
$CrC_{12}H_{10}O_3$ 5,6-Bis(methylene)bicyclo-[2.2.1]hept-2-ene-$Cr(CO)_3$, 1, 89
$CrC_{12}H_{10}O_4$ (MeC_6H_4Ac)$Cr(CO)_3$ isomers, 1, 738
$CrC_{12}H_{10}O_5$ (BzOEt)$Cr(CO)_3$, 1, 737
($PhCH_2OAc$)$Cr(CO)_3$, 1, 737
$CrC_{12}H_{10}O_6$ (m-$MeOC_6H_4CO_2Me$)$Cr(CO)_3$, 1, 738
$CrC_{12}H_{12}$ $(C_6H_6)_2Cr$, 1, 76
π-Cycloheptatrienyl-$Cr(C_5H_5)$, 1, 65
$CrC_{12}H_{12}ClO_4$ [$(C_6H_6)_2Cr$]ClO_4, 1, 78
$CrC_{12}H_{12}F_6P$ [π-Cycloheptatrienyl-$(C_5H_5)Cr$]PF_6, 1, 65
$CrC_{12}H_{12}I$ [$(C_6H_6)_2Cr$]I, 1, 78
$CrC_{12}H_{12}O_2$ Cyclopentadienylcyclopentenyl-$Cr(CO)_2$, 1, 62
$CrC_{12}H_{12}O_3$ (1,3,5-$Me_3C_6H_3$)$Cr(CO)_3$, 1, 73
(Me_2CHPh)$Cr(CO)_3$, 1, 738
(1,2,3-$Me_3C_6H_3$)$Cr(CO)_3$, 1, 738
(1,2,4-$Me_3C_6H_3$)$Cr(CO)_3$, 1, 738

$CrC_{12}H_{12}O_4$ 1,5-Cyclooctadiene-$Cr(CO)_4$, 1, 92
(p-$EtOC_6H_4Me$)$Cr(CO)_3$, 1, 738
$CrC_{12}H_{13}$ π-Cycloheptatriene-$(C_5H_5)Cr$, 1, 65
$CrC_{12}H_{13}NO_3$ (o-$MeC_6H_4NMe_2$)$Cr(CO)_3$, 1, 738
$CrC_{13}H_5NO_6$ (BzNC)$Cr(CO)_5$, 1, 92
$CrC_{13}H_7NO_6$ (p-$MeOC_6H_4NC$)$Cr(CO)_5$, 1, 92
$CrC_{13}H_8O_3$ Naphthalene-$Cr(CO)_3$, 1, 74
$CrC_{13}H_9NO_4$ [ON=C(Me)CH=CPh]$Cr(CO)_3$, 1, 738*
$CrC_{13}H_{10}N_2O_3$ [CH=C(Me)NHN=CPh]$Cr(CO)_3$ 1, 738
$CrC_{13}H_{10}O_3$ (CH_2=CHCH=CHPh)$Cr(CO)_3$, 1, 738
$CrC_{13}H_{10}O_5$ ($AcCH_2Bz$)$Cr(CO)_3$, 1, 738
$CrC_{13}H_{10}O_7$ [p-$(MeO_2C)_2C_6H_4$]$Cr(CO)_3$, 1, 74
$CrC_{13}H_{11}NO_2$ [$(C_6H_6)Cr(CO)_2Py$], 1, 742
$CrC_{13}H_{12}O_2$ $(C_6H_6)Cr(BzOH)$, 1, 86
$CrC_{13}H_{12}O_3$ Tetralin-$Cr(CO)_3$, 1, 739
$CrC_{13}H_{12}O_4$ ($PhCH_2COEt$)$Cr(CO)_3$, 1, 739
(EtC_6H_4Ac)$Cr(CO)_3$, 1, 739
$CrC_{13}H_{14}$ (C_5H_5)(methylcycloheptatrienyl)-Cr, 1, 66
$CrC_{13}H_{14}O_3$ (BuPh)$Cr(CO)_3$, 1, 739
(Me_3CPh)$Cr(CO)_3$, 1, 739
(1,2,4,5-$Me_4C_6H_2$)$Cr(CO)_3$, 1, 739
$CrC_{13}H_{15}$ 1,3,5-Cyclooctatriene-Cr-C_5H_5, 1, 67
$CrC_{13}H_{16}N_2O_3$ 6,6-Bis(dimethylamino)-fulvene-$Cr(CO)_3$, 1, 62
$CrC_{13}H_{16}O_2$ [(1,3,5-$Me_3C_6H_3$)$Cr(CO)_2$-(C_2H_4)], 1, 742
$CrC_{13}H_{17}NO_2$ [$(C_6H_6)Cr(CO)_2Pip$], 1, 742
$CrC_{13}H_{18}O_3S$ [(1,3,5-$Me_3C_6H_3$)$Cr(CO)_2$-(Me_2SO)], 1, 742
$CrC_{14}H_{13}NO_6$ {[p-$(MeO_2C)_2C_6H_4$]$Cr(CO)_2$-(MeCN)}, 1, 742
$CrC_{14}H_{14}Cl$ $(PhCH_2)_2CrCl$, 1, 55
$CrC_{14}H_{14}I$ [$(-CH_2Ph)_2Cr$]I, 1, 80
$CrC_{14}H_{14}O_2$ (o-$MeOC_6H_4$)$_2Cr$, 1, 57
$CrC_{14}H_{16}$ $(MePh)_2Cr$, 1, 79
$CrC_{14}H_{16}O_5$ [$PhCH(OEt)_2$]$Cr(CO)_3$, 1, 739
(1,3-Diethoxycycloheptatriene)-Cr-$(CO)_3$, 1, 744
$CrC_{14}H_{16}O_7S$ {[p-$(MeO_2C)_2C_6H_4$]$Cr(CO)_2$-(Me_2SO)}, 1, 743
$CrC_{15}H_8O_3$ Acenaphthylene-$Cr(CO)_3$, 1, 739
$CrC_{15}H_{10}O_3$ (Ph_2)$Cr(CO)_3$, 1, 739
$CrC_{15}H_{11}NO_2$ [$(C_6H_6)Cr(CO)_2(PhCN)$], 1, 742
$CrC_{15}H_{12}O_3$ (Cyclopentadienylcycloheptatriene)-$Cr(CO)_3$, 1, 744
$CrC_{15}H_{14}$ Azulenium-$(C_5H_5)Cr$, 1, 66
$CrC_{15}H_{14}F_6P$ [Azulenium-$(C_5H_5)Cr$]PF_6, 1, 66
$CrC_{15}H_{14}O_9$ [p-$(HOCH_2CH_2O_2C)_2C_6H_4$]Cr-$(CO)_3$, 1, 739
$CrC_{15}H_{16}$ (C_5H_5)(1,2-dihydroazulenium)-Cr, 1, 66
$CrC_{15}H_{16}O_3$ ($C_6H_{11}Ph$)$Cr(CO)_3$, 1, 739
$CrC_{15}H_{17}NO_2$ [$(C_6H_6)Cr(CO)_2(C_6H_{11}NC)$], 1, 742
$CrC_{15}H_{18}O_3$ (C_6Me_6)$Cr(CO)_3$, 1, 74
$CrC_{15}H_{33}N_6$ $(C_5H_5)_3Cr(NH_3)_6$, 1, 69
$CrC_{16}H_{10}O_3$ Fluorene-$Cr(CO)_3$, 1, 739
$CrC_{16}H_{12}O_3$ ($PhCH_2Ph$)$Cr(CO)_3$, 1, 739
(1-Phenylcycloheptatriene)-$Cr(CO)_3$, 1, 744
$CrC_{16}H_{13}NO_3$ ($PhCH_2NHPh$)$Cr(CO)_3$, 1, 740
$CrC_{16}H_{16}O_4$ $(BzOMe)_2Cr$, 1, 80
$CrC_{16}H_{16}O_7$ [1,1-Bis(ethoxycarbonyl)-cycloheptatriene]-$Cr(CO)_3$, 1, 744
$CrC_{16}H_{17}NO_2$ [(1,3,5-$Me_3C_6H_3$)$Cr(CO)_2$-Py], 1, 742
$CrC_{16}H_{20}$ $(Me_2C_6H_4)_2Cr$ isomers, 1, 80
$(EtPh)_2Cr$, 1, 80
$CrC_{16}H_{20}I$ [(1,3-$Me_2C_6H_4$)$_2Cr$]I, 1, 80
$CrC_{16}H_{23}NO_2$ [(1,3,5-$Me_3C_6H_3$)$Cr(CO)_2$-Pip], 1, 742
$CrC_{16}H_{24}O_3S$ [$(C_6Me_6)Cr(CO)_2(Me_2SO)$], 1, 743
$CrC_{17}H_{10}O_3$ Anthracene-$Cr(CO)_3$, 1, 74
Phenanthrene-$Cr(CO)_3$, 1, 740
$CrC_{17}H_{12}O_3$ Stilbene-$Cr(CO)_3$, 1, 74
9,10-Dihydroanthracene-$Cr(CO)_3$, 1, 740
$CrC_{17}H_{13}NO_2$ [$(C_6H_6)Cr(CO)_2$(quinoline)],

1, 742
$CrC_{17}H_{14}O_3$ $(PhCH_2-)_2Cr(CO)_3$, 1, 740
$(Ph_2CHMe)Cr(CO)_3$, 1, 740
(Bicycloheptatrienyl)-$Cr(CO)_3$, 1, 744
$CrC_{17}H_{15}NO_6$ $\{[p-(MeO_2C)_2C_6H_4]Cr(CO)_2-Py\}$, 1, 742
$CrC_{17}H_{18}O_7$ $[(EtO_2C)_2CH-CHCH=CHCH=CHCH=CH]Cr(CO)_3$, 1, 744
$CrC_{17}H_{21}NO_6$ $\{[p-(MeO_2C)_2C_6H_4]Cr(CO)_2-Pip\}$, 1, 742
$CrC_{18}H_{14}$ Bis(indenyl)-Cr, 1, 69
$CrC_{18}H_{14}N_3O_7$ $[(C_6H_6)_2Cr]$picrate, 1, 79
$CrC_{18}H_{16}$ $(C_6H_6)(Ph_2)Cr$, 1, 86
(C_5H_5)(phenylcycloheptatrienyl)-Cr, 1, 66
$CrC_{18}H_{16}I$ $[(C_6H_6)(Ph_2)Cr]I$, 1, 87
$CrC_{18}H_{16}O_3$ $[CH_2(CH_2Ph)_2]Cr(CO)_3$, 1, 740
$CrC_{18}H_{17}NO_2$ $[(1,3,5-Me_3C_6H_3)Cr(CO)_2-(PhCN)]$, 1, 742
$CrC_{18}H_{17}NO_6$ $\{[p-(MeO_2C)_2C_6H_4]Cr(CO)_2-(PhNH_2)\}$, 1, 742
$CrC_{18}H_{17}O$ $[(C_6H_6)(Ph_2)Cr]OH$, 1, 87
$CrC_{18}H_{18}O_3$ $[p-(CH_2)_4C_6H_4(CH_2)_3C\equiv C]-Cr(CO)_3$, 1, 740
$CrC_{18}H_{20}$ $(CH_2=CHCH_2Ph)_2Cr$, 1, 81
Bis(indane)-Cr, 1, 81
(C_5H_5)(4,6,8-trimethylazulenium)-Cr, 1, 66
$CrC_{18}H_{20}O_7$ $[(EtO_2C)_2CHCH_2-CHCH=CHCH=CHCH=CH]-Cr(CO)_3$, 1, 744
$CrC_{18}H_{22}$ Dimesityl-Cr, 1, 55*
$CrC_{18}H_{23}NO_2$ $[(1,3,5-Me_3C_6H_3)Cr(CO)_2-(C_6H_{11}NC)]$, 1, 742
$CrC_{18}H_{24}$ $(Me_3C_6H_3)_2Cr$ isomers, 1, 81
$CrC_{18}H_{24}O_3$ $[1,3,5-(Me_2CH)_3C_6H_3]Cr-(CO)_3$, 1, 739
$CrC_{19}H_{11}NO_5$ $(Ph_2CHNC)Cr(CO)_5$, 1, 92
$CrC_{19}H_{14}O_3$ $(PhCH=CH-)_2Cr(CO)_3$, 1, 740
$CrC_{19}H_{15}NO_6$ $\{[p-(MeO_2C)_2C_6H_4]Cr-(CO)_2(PhCN)\}$, 1, 743
$CrC_{19}H_{16}O$ $(BzPh)(C_6H_6)Cr$, 1, 88
$CrC_{19}H_{16}O_3$ [2.2]-Paracyclophane-$Cr(CO)_3$, 1, 741
$CrC_{19}H_{18}O_3$ $(PhCH_2CH_2-)_2Cr(CO)_3$, 1, 740
$CrC_{19}H_{20}O_3$ [1-(tert-Butylcyclopentadienyl)cycloheptatriene]-$Cr(CO)_3$, 1, 744
$CrC_{19}H_{21}NO_6$ $\{[(p-MeO_2C)_2C_6H_4]Cr(CO)_2-(NCC_6H_{11})\}$, 1, 742
$CrC_{19}H_{23}NO_2$ $[(C_6Me_6)Cr(CO)_2Py]$, 1, 743
$CrC_{19}H_{26}O_3$ Pentaethylbenzene-$Cr(CO)_3$, 1, 740
$CrC_{19}H_{29}NO_2$ $[(C_6Me_6)Cr(CO)_2Pip]$, 1, 743
$CrC_{20}H_{18}$ Azulene(1,2-dihydroazulene)-Cr, 1, 67
$CrC_{20}H_{18}O_3$ [2.3]-Paracyclophane-Br-$(CO)_3$, 1, 741
$CrC_{20}H_{19}NO_2$ $[(1,3,5-Me_3C_6H_3)Cr(CO)_2-$(quinoline)], 1, 742
$CrC_{20}H_{24}$ (1,2-Dihydroazulene)(3a,4,5,6,7,8-hexahydroazulene)-Cr, 1, 67
Bis(tetrahydronaphthalene)-Cr, 1, 81
$CrC_{20}H_{25}F_6P$ [(1,2-Dihydroazulene)-(3a,4,5,6,7,8-hexahydroazulene)-Cr]PF_6, 1, 67
$CrC_{20}H_{28}$ Bis(cumene)-Cr, 1, 82
$CrC_{21}H_{12}O_3$ Chrysene-$Cr(CO)_3$, 1, 740
$CrC_{21}H_{14}O_3$ $(CH=CHCH=CHC=CPh_2)Cr(CO)_3$, 1, 740
5,10-Dihydrotetracene-$Cr(CO)_3$, 1, 740
$CrC_{21}H_{17}NO_6$ $\{[p-(MeO_2C)_2C_6H_4]Cr(CO)_2-$(quinoline)}, 1, 742
$CrC_{21}H_{23}NO_2$ $[(C_6Me_6)Cr(CO)_2(PhCN)]$, 1, 743
$CrC_{21}H_{24}O_3$ Dodecahydrotriphenylene-$Cr(CO)_3$, 1, 740
$CrC_{21}H_{27}F_9O_3$ $(BuCF=CF_2)_3Cr(CO)_3$, 1, 89
$CrC_{21}H_{29}NO_2$ $[(C_6Me_6)Cr(CO)_2(C_6H_{11}NC)]$, 1, 743
$CrC_{21}H_{30}O_3$ Hexaethylbenzene-$Cr(CO)_3$, 1, 740
$CrC_{22}H_{22}O_3$ [3.4]-Paracyclophane-Cr-$(CO)_3$, 1, 741
$CrC_{22}H_{30}O$ $(2,4,6-Me_3C_6H_2)_2Cr\cdot THF$, 1, 57
$CrC_{23}H_{20}Br_2P$ $C_5H_5CrBr_2(PPh_3)$, 1, 60
$CrC_{23}H_{20}I_2P$ $C_5H_5CrI_2(PPh_3)$, 1, 60
$CrC_{23}H_{24}O_3$ [4.4]-Paracyclophane-Cr-

(CO)$_3$, 1, 741

$CrC_{23}H_{25}NO_2$ [(C_6Me_6)Cr(CO)$_2$(quinoline)], 1, 743

$CrC_{24}H_{16}Cl_4O_2$ (C_6H_6)(Ph_2)Cr-chloranil adduct, 1, 87

$CrC_{24}H_{18}Cl_2$ (p-ClC_6H_4Ph)$_2$Cr, 1, 81

$CrC_{24}H_{19}FI$ [(Ph_2)(p-FC_6H_4Ph)Cr]I, 1, 88

$CrC_{24}H_{20}$ (Ph_2)$_2$Cr, 1, 81

$CrC_{24}H_{20}Cl$ [(Ph_2)$_2$Cr]Cl, 1, 83

$CrC_{24}H_{20}ClO_4$ [(Ph_2)$_2$Cr]ClO_4, 1, 83

$CrC_{24}H_{20}I$ [(Ph_2)$_2$Cr]I, 1, 83

$CrC_{24}H_{21}O$ [(Ph_2)$_2$Cr]OH, 1, 82

$CrC_{24}H_{26}O_3$ [4.5]-Paracyclophane-Cr-(CO)$_3$, 1, 741

[1.8]-Paracyclophane-Cr(CO)$_3$, 1, 741

[9.0]-Paracyclophane-Cr(CO)$_3$, 1, 741

$CrC_{24}H_{30}N_3$ [o-(Me_2N)C_6H_4]$_3$Cr, 1, 57

$CrC_{24}H_{32}$ (cyclo-$C_6H_{11}Ph$)$_2$Cr, 1, 84

$CrC_{24}H_{36}$ ($Et_3C_6H_3$)$_2$Cr, 1, 84

Bis(hexamethylbenzene)-Cr, 1, 84

$CrC_{25}H_{15}NO_5$ (Ph_3CNC)Cr(CO)$_5$, 1, 92

$CrC_{25}H_{22}NO_2$ [(C_6H_6)(Ph_2)Cr]$O_2CC_6H_4$-NH_2-o, 1, 87

$CrC_{25}H_{28}O_3$ [4.6]-Paracyclophane-Cr-(CO)$_3$, 1, 741

[5.5]-Paracyclophane-Cr(CO)$_3$, 1, 741

[1.9]-Paracyclophane-Cr(CO)$_3$, 1, 741

[10.0]-Paracyclophane-Cr(CO)$_3$, 1, 741

$CrC_{26}H_{19}O_3P$ ($Ph_3PC_5H_4$)Cr(CO)$_3$, 1, 89

$CrC_{26}H_{21}O_2P$ [(C_6H_6)Cr(CO)$_2$(Ph_3P)], 1, 742

$CrC_{26}H_{22}IN_4$ Benzenebis(bipyridyl)-CrI, 1, 75

$CrC_{26}H_{30}O_3$ [5.6]-Paracyclophane-Cr-(CO)$_3$, 1, 741

[1.10]-Paracyclophane-Cr(CO)$_3$, 1, 741

$CrC_{27}H_{21}N_3O_3$ (p-MeC_6H_4NC)$_3$Cr(CO)$_3$, 1, 92

$CrC_{27}H_{24}O_3$ {$CH_2C_6H_4CH_2$}$_3$Cr(CO)$_3$, 1, 740

$CrC_{27}H_{32}O_3$ [6.6]-Paracyclophane-Cr-(CO)$_3$, 1, 741

[1.11]-Paracyclophane-Cr(CO)$_3$, 1, 741

[12.0]-Paracyclophane-Cr(CO)$_3$, 1, 741

$CrC_{27}H_{33}$ Trimesityl-Cr, 1, 55

$CrC_{28}H_{24}$ Bis(9,10-dihydrophenanthrene)-Cr, 1, 85

$CrC_{28}H_{26}I$ [1,2,3,4-Tetraphenylbutane-Cr]I, 1, 85

$CrC_{28}H_{34}O_3$ [1.12]-Paracyclophane-Cr-(CO)$_3$, 1, 741

$CrC_{29}H_{27}O_2P$ [(1,3,5-$Me_3C_6H_3$)Cr(CO)$_2$-(Ph_3P)], 1, 742

$CrC_{30}H_{21}$ Trinaphthyl-Cr isomers, 1, 56

$CrC_{30}H_{22}N_3O_7$ [(Ph_2)$_2$Cr]OC_6H_2(NO_2)$_3$, 1, 84

$CrC_{30}H_{25}O$ [(Ph_2)$_2$Cr]OPh, 1, 84

$CrC_{30}H_{25}O_6P$ [p-(MeO_2C)$_2C_6H_4$]Cr(CO)$_2$-(PPh_3), 1, 743

$CrC_{30}H_{32}$ (2,4,6-$Me_3C_6H_2Ph$)$_2$Cr, 1, 85

$CrC_{30}H_{36}Br_3O_3$ (p-BrC_6H_4)$_3$Cr·3THF, 1, 57

$CrC_{30}H_{36}Cl_3O_3$ (m-ClC_6H_4)$_3$Cr·3THF, 1, 57

$CrC_{30}H_{39}O_3$ Ph_3Cr·3THF, 1, 56

$CrC_{30}H_{42}N_3$ [o-(Et_2N)C_6H_4]$_3$Cr, 1, 57

$CrC_{31}H_{26}NO_2$ [(Ph_2)$_2$Cr]$O_2CC_6H_4NH_2$-o, 1, 84

$CrC_{32}H_{20}O_4$ [OCC(Ph)=C(Ph)C(Ph)=CPh]Cr-(CO)$_3$, 1, 740

$CrC_{32}H_{22}O_4$ [OCC(Ph)=C(Ph)CH(Ph)CHPh]-Cr(CO)$_3$, 1, 740

$CrC_{32}H_{26}IN_4$ Biphenylbis(bipyridyl)-CrI, 1, 76

$CrC_{32}H_{33}O_2P$ [(C_6Me_6)Cr(CO)$_2$(Ph_3P], 1, 743

$CrC_{32}H_{48}O_4$ Pentabutylcyclooctatetraene-Cr(CO)$_4$, 1, 92

$CrC_{33}H_{27}Cl_3N_3$ (m-ClC_6H_4)$_3$Cr·3Py, 1, 57

$CrC_{36}H_{28}$ (p-PhC_6H_4Ph)$_2$Cr, 1, 85

$CrC_{36}H_{60}$ Bis(hexaethylbenzene)-Cr, 1, 85

$CrC_{42}H_{18}Cl_{12}N_6$ (2,5-$Cl_2C_6H_3NC$)$_6$Cr, 1, 93

$CrC_{42}H_{24}Cl_6N_6$ (p-ClC_6H_4NC)$_6$Cr, 1, 93

$CrC_{42}H_{30}N_6$ (PhNC)$_6$Cr, 1, 93

$CrC_{48}H_{36}Cl_6N_6$ [Cl(Me)C_6H_3NC]$_6$Cr isomers, 1, 93

$CrC_{48}H_{42}N_6$ (p-MeC_6H_4NC)$_6$Cr, 1, 93

$CrC_{48}H_{42}N_6O_6$ (p-$MeOC_6H_4NC$)$_6$Cr, 1, 93

$CrC_{54}H_{54}N_6$ (2,4-$Me_2C_6H_3NC$)$_6$Cr, 1, 93

$Cr_2C_{10}H_{10}N_4O_4$ [C_5H_5Cr(NO)$_2$]$_2$, 1, 63

$Cr_2C_{12}H_{13}O_4$ [(C_6H_6)$_2$Cr][CrO_4H], 1, 79

$Cr_2C_{13}H_{19}S_3$ $C_5H_5Cr(MeS)_3CrC_5H_5$, 1, 63

$Cr_2C_{14}H_{13}N_6O_3S_4$ [Cycloheptatrienium-$Cr(CO)_3$][$Cr(NH_3)_2(SCN)_4$], 1, 91

$Cr_2C_{14}H_{18}Cl_4O_2$ $(C_5H_5CrCl_2)_2$·Dioxane, 1, 60

$Cr_2C_{15}H_{15}Cl_3$ [$(C_5H_5)_2Cr$][$C_5H_5CrCl_3$], 1, 68

$Cr_2C_{16}H_{10}O_6$ [$C_5H_5Cr(CO)_3$]$_2$, 1, 61

$Cr_2C_{16}H_{12}N_2O_{12}$ [$(HNC)Cr(CO)_5$]$_2$·MeO-CH_2CH_2OMe, 1, 42

$Cr_2C_{16}H_{18}N_6S_4$ [$(C_6H_6)_2Cr$][$Cr(SCN)_4$-$(NH_3)_2$], 1, 79

$Cr_2C_{18}H_{10}F_{12}$ [$C_5H_5Cr(CF_3C{=}CCF_3)$]$_2$, 1, 63

$Cr_2C_{18}H_{11}NO_6$ [Ph_2NH][$Cr(CO)_3$]$_2$, 1, 75

$Cr_2C_{18}H_{12}N_2O_6$ $(p\text{-}H_2NC_6H_4\text{-})_2[Cr(CO)_3]_2$ 1, 75

$Cr_2C_{18}H_{15}O_3$ [$(C_5H_5)_2Cr$][$C_5H_5Cr(CO)_3$], 1, 68

$Cr_2C_{18}H_{16}Cl_3$ [$(C_6H_6)(Ph_2)Cr$]$CrCl_3$, 1, 87

$Cr_2C_{19}H_{12}O_6$ $(Ph_2CH_2)[Cr(CO)_3]_2$, 1, 75

$Cr_2C_{19}H_{18}N_2O_{12}$ [$(HNC)Cr(CO)_5$]$_2$·EtO-$(CH_2)_3OEt$, 1, 92

$Cr_2C_{20}H_{12}O_6$ $(PhCH{=}CHPh)[Cr(CO)_3]_2$, 1, 75

$Cr_2C_{20}H_{14}O_6$ $(PhCH_2CH_2Ph)[Cr(CO)_3]_2$, 1, 75

Bicycloheptatrienyl$[Cr(CO)_3]_2$, 1, 90

$Cr_2C_{20}H_{14}O_6S_2$ Bis(cycloheptatrienylsulfide)-$Cr_2(CO)_6$, 1, 90

$Cr_2C_{20}H_{16}N_2O_6$ [3,4-$Me(H_2N)C_6H_3$-]$_2$-$[Cr(CO)_3]_2$, 1, 75

$Cr_2C_{20}H_{17}O_3$ [$(C_6H_6)_2Cr$][$C_5H_5Cr(CO)_3$], 1, 79*

$Cr_2C_{20}H_{20}N_2O_{12}$ [$(HNC)Cr(CO)_5$]$_2$·PrO-CH_2CH_2OPr, 1, 92

$Cr_2C_{22}H_{14}O_6$ $(PhCH{=}CH\text{-})_2[Cr(CO)_3]_2$, 1, 75

$Cr_2C_{22}H_{22}N_6S_2$ [$(C_6H_6)(Ph_2)Cr$][Cr-$(NH_3)_2(SCN)_4$], 1, 87

$Cr_2C_{24}H_{20}Cl_3$ [$(Ph_2)_2Cr$]$CrCl_3$, 1, 84

$Cr_2C_{27}H_{26}O_6$ Tricyclo[13.2.2.2^{6,9}]-heneicosa-6,8,15,17,18,20-hexaene-$[Cr(CO)_3]_2$, 1, 75

$Cr_2C_{28}H_{26}N_6S_4$ [$(Ph_2)_2Cr$][$Cr(NH_3)_2$-$(SCN)_4$], 1, 83

$Cr_2C_{30}H_{32}O_6$ Tricyclo[16.2.2.2^{8,11}]-tetracosa-8,10,18,20,21,23-hexaene-$[Cr(CO)_3]_2$, 1, 75

$Cr_2C_{36}H_{42}N_6S_4$ Bis(3,3′,4,4′-tetramethylbiphenyl)chromium reineckate, 1, 85

$Cr_3C_{22}H_{13}O_{10}$ [$(C_6H_6)_2Cr$][$Cr_2(CO)_{10}H$], 1, 79

$Cr_3C_{116}H_{80}O_4$ Tetrakis(tetraphenylcyclopentadienone)-Cr_3, 1, 89

$Cr_4C_{20}H_{20}O_4$ $(C_5H_5CrO)_4$, 1, 63

$CrAlC_{12}H_{12}Cl_4$ [$(C_6H_6)_2Cr$]$AlCl_4$, 1, 78

$CrAlC_{12}H_{15}O_3$ $Et_2Al(C_5H_5)Cr(CO)_3$, 1, 93

$CrAlC_{18}H_{16}Cl_4$ [$(C_6H_6)(Ph_2)Cr$]$AlCl_4$, 1, 87

$CrAlC_{24}H_{20}Br_4$ [$(Ph_2)_2Cr$]$AlBr_4$, 1, 83

$CrAlC_{24}H_{20}Cl_4$ [$(Ph_2)_2Cr$]$AlCl_4$, 1, 83

$CrAl_3C_6H_6Cl_9$ $C_6H_6Cr(AlCl_3)_3$, 1, 93

$CrAsC_{23}H_{15}O_5$ $Cr(AsPh_3)(CO)_5$, 3, 19

$CrAs_2C_{14}H_{16}O_4$ $Cr(2\text{-}Me_2AsC_6H_4AsMe_2)$-$(CO)_4$, 3, 82

$CrAs_3C_{14}H_{27}O_3$ Cr[$MeAs(CH_2CH_2CH_2$-$AsMe_2)_2$]$(CO)_3$, 3, 105

$CrAs_3C_{27}H_{33}O_3$ $Cr(Me_2AsPh)_3(CO)_3$, 3, 40

$CrAs_4C_{22}H_{32}Br_2O_2$ [Cr(2-$Me_2AsC_6H_4$-$AsMe_2)_2(CO)_2Br$]Br, 3, 82

$CrAs_4C_{22}H_{32}Br_4O_2$ [Cr(2-$Me_2AsC_6H_4$-$AsMe_2)_2(CO)_2Br$]Br_3, 3, 82

$CrAs_4C_{22}H_{32}I_2O_2$ [Cr(2-$Me_2AsC_6H_4$-$AsMe_2)_2(CO)_2I$]I, 3, 82

$CrAs_4C_{22}H_{32}I_4O_2$ [Cr(2-$Me_2AsC_6H_4$-$AsMe_2)_2(CO)_2I$]I_3, 3, 82

$CrAs_4C_{22}H_{32}O_2$ $Cr(2\text{-}Me_2AsC_6H_4AsMe_2)_2$-$(CO)_2$, 3, 82

$CrBC_9H_5F_4O_4$ [$C_5H_5Cr(CO)_4$]BF_4, 1, 62

$CrBC_{10}H_7F_4O_3$ [Cycloheptatrienium-Cr-$(CO)_3$]BF_4, 1, 90

$CrBC_{24}H_{20}F_4$ [$(Ph_2)_2Cr$]BF_4, 1, 83

$CrBC_{34}H_{30}$ [$(C_5H_5)_2Cr$]BPh_4, 1, 68

$CrBC_{36}H_{30}D_2$ [$(PhD)_2Cr$]BPh_4, 1, 78

$CrBC_{36}H_{32}$ [$(C_6H_6)_2Cr$]BPh_4, 1, 78

$CrBC_{37}H_{34}O$ [$(C_6H_6)(PhCH_2OH)Cr$]BPh_4, 1, 86

[$(C_6H_6)(PhCH_2OH)Cr$]BPh_4, 1, 86

$CrBC_{42}H_{35}D$ [$(PhD)(Ph_2)Cr$]BPh_4, 1, 86

$CrBC_{42}H_{36}$ [$(C_6H_6)(Ph_2)Cr$]BPh_4, 1, 87

$CrBC_{43}H_{36}O_2$ [(BzOH)(Ph_2)Cr]BPh_4, 1, 88

$CrBC_{44}H_{36}$ Bis(naphthalene)chromium-BPh_4, 1, 81

$CrBC_{45}H_{42}$ [(2-$PhCH_2C_6H_4Me$)(MePh)Cr]-BPh_4, 1, 88

$CrBC_{48}H_{39}Br$ [(Ph_2)(p-BrC_6H_4Ph)Cr]-BPh_4, 1, 88

$CrBC_{48}H_{39}Cl$ [(Ph_2)(p-ClC_6H_4Ph)Cr]-BPh_4, 1, 88

$CrBC_{48}H_{39}I$ [(Ph_2)(p-IC_6H_4Ph)Cr]BPh_4, 1, 88

$CrBC_{48}H_{40}$ [$(Ph_2)_2$Cr]BPh_4, 1, 84

$CrBC_{50}H_{40}$ [Bis(fluorene)-Cr]BPh_4, 1, 85

$CrBC_{50}H_{42}N_4$ Benzenebis(bipyridyl)-chromium-BPh_4, 1, 75

$CrBC_{52}H_{40}$ [Bis(phenanthrene)-Cr]BPh_4, 1, 85

[Bis(anthracene-Cr]BPh_4, 1, 85

$CrBC_{64}H_{48}$ [Bis(binaphthyl)-Cr]BPh_4, 1, 86

$CrBC_{72}H_{56}$ [Bis(quaterphenyl)-Cr]BPh_4, 1, 86

$CrBC_{120}H_{88}$ [Bis(octaphenyl)-Cr]BPh_4, 1, 86

$CrCoC_{18}H_{15}O_3$ [$(C_5H_5)_2$Co][$C_5H_5Cr(CO)_3$], 1, 587

$CrCoC_{22}H_{16}O_4$ [(C_6H_6)(Ph_2)Cr][$Co(CO)_4$] 1, 87

$CrCoC_{28}H_{20}O_4$ [$(Ph_2)_2$Cr][$Co(CO)_4$], 1, 83

$CrFeC_{16}H_{10}O_6$ $Fe(CO)_3$(CH_2=CHCH=CHPh)-$Cr(CO)_3$, 1, 94

$CrFeC_{20}H_{16}O_3$ ($FcCH_2Ph$)$Cr(CO)_3$, 1, 94

$CrFeC_{22}H_{14}O_6$ $Fe(CO)_3$(PhCH=CHCH=CHPh)-$Cr(CO)_3$, 1, 94

$CrFe_4C_{37}H_{20}O_{13}$ [$(Ph_2)_2$Cr]$Fe_4(CO)_{13}$, 1, 84

$CrGeC_{12}H_{14}O_3$ (Me_3GePh)$Cr(CO)_3$, 1, 93, 2, 140

$CrGeC_{24}H_{34}O_3$ $(Me_2CHCH_2)_3$Ge[indenyl-$Cr(CO)_3$], 2, 140

$CrHgC_{18}H_{16}I_3$ [(C_6H_6)(Ph_2)Cr]I·HgI_2, 1, 87

$CrK_3C_{12}H_6$ $K_3[Cr(C{\equiv}CH)_6]$, 1, 57

$CrLiC_{14}H_{14}O_3$ Li[$Cr(C_6H_4OMe\text{-}o)_2O$], 1, 58

$CrLiC_{23}H_{20}$ Li[$Cr(C_5H_5)Ph_3$], 1, 58

$CrLiC_{24}H_{30}N_3$ Li[$Cr(C_6H_4NMe_2\text{-}p)_3$], 1, 58

$CrLiC_{27}H_{22}$ Li[indenyltriphenyl-Cr], 1, 58

$CrLiC_{28}H_{28}O_4$ Li[$Cr(C_6H_4OMe\text{-}o)_4$], 1, 58

$CrLi_2C_{30}H_{25}$ $Li_2[CrPh_5]$, 1, 58

$CrLi_3C_6H_{18}$ $Li_3[CrMe_6]$ complex with dioxan, 1, 58

$CrLi_3C_{12}H_8BrO_2$ Li[$Cr(C_6H_4OLi\text{-}o)_2Br$], 1, 58*

$CrLi_3C_{24}H_{38}BrO_5$ Li[$Cr(C_6H_4OLi\text{-}o)_2Br$]·-$3Et_2O$, 1, 58

$CrLi_3C_{30}H_{26}$ $Li_3[CrHPh_5]$, 1, 58

$CrLi_3C_{36}H_{24}Br_6$ $Li_3[Cr(C_6H_4Br\text{-}p)_6]$, 1, 59

$CrLi_3C_{36}H_{24}Cl_6$ $Li_3[Cr(C_6H_4Cl\text{-}m)_6]$, 1, 59

$CrLi_3C_{36}H_{27}Cl_3$ $Li_3[Cr(C_6H_4Cl\text{-}m)_3Ph_3]$, 1, 58

$CrLi_3C_{36}H_{30}$ $Li_3[CrPh_6]$, 1, 58

$CrLi_3C_{42}H_{42}$ $Li_3[Cr(C_6H_4Me\text{-}p)_6]$, 1, 59

$CrLi_3C_{42}H_{42}O_6$ $Li_3[Cr(C_6H_4OMe\text{-}p)_6]$, 1, 59

$CrLi_3C_{48}H_{60}N_6$ $Li_3[Cr(C_6H_4NMe_2\text{-}p)_6]$, 1, 59

$CrLi_3C_{72}H_{54}$ $Li_3[Cr(C_6H_4Ph\text{-}p)_6]$, 1, 59

$CrLi_4C_{36}H_{24}BrCl_6$ $Li_4[Cr(C_6H_4Cl\text{-}m)_6Br]$, 1, 59

$CrLi_4C_{36}H_{24}Br_7$ $Li_4[Cr(C_6H_4Br\text{-}p)_6Br]$, 1, 59

$CrMnC_{13}H_{12}N_6O_3S_4$ [(C_6H_6)$Mn(CO)_3$][Cr-$(NH_3)_2(SCN)_4$], 1, 179

$CrMnC_{17}H_{12}O_5$ [$(C_6H_6)_2$Cr][$Mn(CO)_5$], 1, 79, 94*

$CrMnC_{19}H_{24}N_6O_3S_4$ [(Hexamethylbenzene)-$Mn(CO)_3$][$Cr(NH_3)_2(SCN)_4$], 1, 179

[(Hexamethylbenzene)$Mn(CO)_3$][$Cr(NH_3)_2$-$(SCN)_4$], 1, 179

$CrMoC_{12}H_{14}N_7O_3S_4$ [$C_5H_5Mo(CO)_3NH_3$][Cr-$(NH_3)_2(SCN)_4$], 1, 99

$CrMoC_{12}H_{15}N_8O_3S_4$ [$C_5H_5Mo(CO)_3H_2NNH_2$]-[$Cr(NH_3)_2(SCN)_4$], 1, 99

$CrMoC_{14}H_{13}N_6O_3S_4$ [Cycloheptatrienium-$Mo(CO)_3$][$Cr(NH_3)_2(SCN)_4$], 1, 113

$CrMoC_{14}H_{16}ClN_6S_4$ $(C_5H_5)_2$MoCl[$Cr(NH_3)_2$-$(SCN)_4$], 1, 118

$CrMoC_{20}H_{14}O_6$ (Bicyclohepta-2,4,6-trienyl)[$Cr(CO)_3$][$Mo(CO)_3$], 1, 94
$CrNi_4C_{21}H_{13}O_9$ [$(C_6H_6)_2Cr$][$Ni_4(CO)_9H$], 1, 95*
$CrReC_{19}H_{24}N_6O_3S_4$ [(Hexamethylbenzene)-$Re(CO)_3$][$Cr(NH_3)_2(SCN)_4$], 1, 190
$CrReC_{22}H_{30}N_6S_4$ [$(Me_3C_6H_3)_2Re$][Cr-$(NH_3)_2(SCN)_4$], 1, 190
$CrSbC_{23}H_{15}O_5$ [$Cr(SbPh_3)(CO)_5$], 3, 662
$CrSiC_{12}H_{14}O_3$ $(Me_3SiPh)Cr(CO)_3$, 1, 93, 738
$CrSnC_{12}H_{14}O_3$ $(Me_3SnPh)Cr(CO)_3$, 1, 94, 2, 500
$CrWC_{12}H_{14}N_7O_3S_4$ [$C_5H_5W(CO)_3NH_3$][Cr-$(NH_3)_2(SCN)_4$], 1, 122
$CrWC_{12}H_{15}N_8O_3S_4$ [$C_5H_5W(CO)_3H_2NNH_2$]-[$Cr(NH_3)_2(SCN)_4$], 1, 122
$(Cr_2As_2C_{12}H_{12}O_8)_n$ [$(OC)_4CrAs(Me)_2$-$As(Me)_2Cr(CO)_4$]$_n$, 3, 299
$Cr_2As_2C_{14}H_{12}O_{10}$ $(OC)_5CrAs(Me)_2$-$As(Me_2)Cr(CO)_5$, 3, 300
$Cr_2FeC_{25}H_{14}O_9$ $(OC)_3Fe$[-CH=CHPhCr-$(CO)_3$]$_2$, 1, 94
$Cr_2GeC_{44}H_{64}O_6$ (cyclo-C_6H_{11})$_2$Ge[(n-$C_8H_{17}C_5H_4)Cr(CO)_3$]$_2$, 1, 93, 2, 140
$Cr_2HgC_{16}H_{10}O_6$ Hg[$C_5H_5Cr(CO)_3$]$_2$, 1, 61
$Cr_2Li_3C_{36}H_{33}$ $Li_3[Cr_2H_3Ph_6]$ complex with dioxan, 1, 59
$Cr_2NiC_{21}H_{14}O_9$ [$(MeC_5H_4)Cr(CO)_3$]$_2$Ni-$(CO)_3$, 1, 94
$Cr_2PbC_{32}H_{32}O_6$ $(C_5H_5)_2$Pb[$Me_2CHC_5H_4$Cr-$(CO)_3$]$_2$, 2, 594
$Cr_2SnC_{36}H_{40}O_6$ $(EtPrC_5H_3)_2$Sn[$Cr(C_5H_5)$-$(CO)_3$]$_2$, 1, 94, 2, 500
$Cr_3AlC_{24}H_{15}O_9$ Al[$(C_5H_5)Cr(CO)_3$]$_3$, 1, 93
$Cr_3HgC_{72}H_{60}Cl_5$ {[$(Ph_2)_2Cr$]Cl}$_3\cdot HgCl_2$, 1, 83
$CrGeSiC_6H_{18}O_4$ $Me_3GeCrO_4SiMe_3$, 2, 97
$Cr_2B_2BaC_{88}H_{70}O_4$ [$(Ph_2)(BzOBa_{1/2})Cr$]-BPh_4, 1, 88
$CuGeC_{72}H_{60}P_3$ $Ph_3GeCu\cdot 3PPh_3$, 2, 139
$DyC_{10}H_{10}Cl$ $(C_5H_5)_2DyCl$, 1, 2
$DyC_{11}H_{13}O$ $(C_5H_5)_2DyOMe$, 1, 3
$DyC_{15}H_{15}$ $(C_5H_5)_3Dy$, 1, 4
$DyC_{17}H_{29}Cl_2O_3$ $C_5H_5DyCl_2\cdot 3THF$, 1, 1
$ErC_{10}H_{10}Cl$ $(C_5H_5)_2ErCl$, 1, 2
$ErC_{10}H_{10}I$ $(C_5H_5)_2ErI$, 1, 2
$ErC_{10}H_{12}N$ $(C_5H_5)_2ErNH_2$, 1, 3
$ErC_{11}H_{11}O_2$ $(C_5H_5)_2ErOOCH$, 1, 3
$ErC_{11}H_{13}O$ $(C_5H_5)_2ErOMe$, 1, 3
$ErC_{12}H_{13}O_2$ $(C_5H_5)_2ErOAc$, 1, 3
$ErC_{12}H_{14}Cl$ $(MeC_5H_4)_2ErCl$, 1, 2
$ErC_{14}H_{17}O_2$ $(MeC_5H_4)_2ErOAc$, 1, 3
$ErC_{15}H_{15}$ $(C_5H_5)_3Cr$, 1, 4
$ErC_{17}H_{29}Cl_2O_3$ $C_5H_5ErCl_2\cdot 3THF$, 1, 1
$EuC_{10}H_{10}$ $(C_5H_5)_2Eu$, 1, 2
$EuC_{17}H_{29}Cl_2O_3$ $C_5H_5EuCl_2\cdot 3THF$, 1, 1
$FeC_4H_6N_4O_2$ $(MeNC)_2Fe(NO)_2$, 1, 525
$FeC_5F_3IO_4$ [$CF_3Fe(CO)_4$]I, 1, 194
$FeC_5H_3I_2NO_3$ $(MeNC)Fe(CO)_3I_2$, 1, 524
$FeC_5H_5Cl_2$ $C_5H_5FeCl_2$, 1, 219
$FeC_6F_5IO_4$ [$C_2F_5Fe(CO)_4$]I, 1, 194
$FeC_6H_3NO_4$ $(MeNC)Fe(CO)_4$, 1, 519
$FeC_6H_4Br_2O_3$ [π-CH_2=$CBrCH_2$-$Fe(CO)_3$]Br, 1, 471
[π-BrCH=$CHCH_2$-$Fe(CO)_3$]Br, 1, 471
$FeC_6H_4O_2S$ Thiophene-$Fe(CO)_2$, 1, 496
$FeC_6H_4O_4$ $(CH_2$=$CH_2)Fe(CO)_4$, 1, 476
$FeC_6H_5BrO_3$ [π-Allyl-$Fe(CO)_3$]Br, 1, 471
$FeC_6H_5ClO_3$ [π-Allyl-$Fe(CO)_3$]Cl, 1, 471
$FeC_6H_5Cl_2$ $PhFeCl_2$, 1, 193
$FeC_6H_5IO_3$ [π-Allyl-$Fe(CO)_3$]I, 1, 471
$FeC_6H_6I_2N_2O_2$ $(MeNC)_2Fe(CO)_2I_2$, 1, 524
$FeC_6H_{10}N_4O_2$ $(EtNC)_2Fe(NO)_2$, 1, 525
$FeC_7Cl_2F_6O_3$ $(CClF$=$CF_2)_2Fe(CO)_3$, 1, 481
$FeC_7F_7IO_4$ [$C_3F_7Fe(CO)_4$]I, 1, 195
$FeC_7F_8O_3$ $(CF_2$=$CF_2)_2Fe(CO)_3$, 1, 481
$FeC_7H_3NO_4$ $(CH_2$=$CHCN)Fe(CO)_4$, 1, 476
$FeC_7H_4O_5$ $(CH_2$=$CHCHO)Fe(CO)_4$, 1, 476
$FeC_7H_4O_6$ $(CH_2$=$CHCO_2H)Fe(CO)_4$, 1, 476
$FeC_7H_5BrO_2$ $C_5H_5Fe(CO)_2Br$, 1, 199
$FeC_7H_5BrO_3$ $HOC_5H_4Fe(CO)_2Br$, 1, 200
$FeC_7H_5ClO_2$ $C_5H_5Fe(CO)_2Cl$, 1, 198
$FeC_7H_5ClO_3$ $HOC_5H_4Fe(CO)_2Cl$, 1, 200
$FeC_7H_5DO_2$ $C_5H_5Fe(CO)_2D$, 1, 198
$FeC_7H_5IO_2$ $C_5H_5Fe(CO)_2I$, 1, 199
$FeC_7H_5NO_4$ $(EtNC)Fe(CO)_4$, 1, 524
$FeC_7H_6N_2O_3$ $(MeNC)_2Fe(CO)_3$, 1, 524
$FeC_7H_6O_2$ $C_5H_5Fe(CO)_2H$, 1, 198
$FeC_7H_6O_3$ $(CH_2$=$CHCH$=$CH_2)Fe(CO)_3$, 1, 478
$FeC_7H_7BrO_3$ [π-MeCH=$CHCH_2$-$Fe(CO)_3$]Br, 1, 471

$FeC_7H_7ClO_3$ [π-MeCH=CHCH$_2$-Fe(CO)$_3$]Cl, 1, 471
[π-CH$_2$=C(Me)CH$_2$-Fe(CO)$_3$]Cl, 1, 472
$FeC_7H_7ClO_7$ [π-MeCH=CHCH$_2$-Fe(CO)$_3$]-ClO$_4$, 1, 471
[π-CH$_2$=C(Me)CH$_2$-Fe(CO)$_3$]ClO$_4$, 1, 472
$FeC_7H_7Cl_4$ Tropenium tetrachloroferrate, 1, 198
$FeC_7H_7IO_3$ [π-MeCH=CHCH$_2$-Fe(CO)$_3$]I, 1, 471
[π-CH$_2$=C(Me)CH$_2$-Fe(CO)$_3$]I, 1, 472
$FeC_8F_8O_4$ CF$_2$CF$_2$CF$_2$CF$_2$Fe(CO)$_4$, 1, 196
$FeC_8F_{10}O_4$ (C$_2$F$_5$)$_2$Fe(CO)$_4$, 1, 196
$FeC_8H_2O_6$ COCH=CHCOFe(CO)$_4$, 1, 735
$FeC_8H_2O_7$ (COCH=CHCOO)Fe(CO)$_4$, 1, 476
$FeC_8H_3NO_6$ (COCH=CHCONH)Fe(CO)$_4$, 1, 476
$FeC_8H_4O_4$ Cyclopentadienone-Fe(CO)$_3$, 1, 483
$FeC_8H_4O_8$ (=CHCO$_2$H)$_2$Fe(CO)$_4$ isomers, 1, 476
$FeC_8H_5BrO_4$ (CH=CHCH=CHCO)-Fe(CO)$_3$·HBr, 1, 483
$FeC_8H_5ClO_4$ (CH=CHCH=CHCO)-Fe(CO)$_3$·HCl, 1, 483
$FeC_8H_5F_3O_2$ C$_5$H$_5$Fe(CO)$_2$CF$_3$, 1, 205
$FeC_8H_5F_6O_3P$ [C$_5$H$_5$Fe(CO)$_3$]PF$_6$, 1, 208
$FeC_8H_5IO_4$ (CH=CHCH=CHCO)-Fe(CO)$_3$·HI, 1, 483
$FeC_8H_5NO_2$ C$_5$H$_5$Fe(CO)$_2$CN, 1, 201
$FeC_8H_6N_2O$ H[C$_5$H$_5$Fe(CO)(CN)$_2$], 1, 207
$FeC_8H_6O_4$ (CH$_2$=CHCH=CH$_2$)Fe(CO)$_4$, 1, 476
$FeC_8H_6O_5$ π-CH$_2$=CHCHCH$_2$OCOFe(CO)$_3$, 1, 473
π-CH$_2$=CCH$_2$Fe(CO)$_3$C(O)OCH$_2$, 1, 473
$FeC_8H_6O_6$ (CH$_2$=CHOAc)Fe(CO)$_4$, 1, 476
[CH$_2$=C(Me)CO$_2$H]Fe(CO)$_4$, 1, 476
(CH$_2$=CHCO$_2$Me)Fe(CO)$_4$, 1, 476
(MeO$_2$CC≡CCO$_2$Me)$_3$Fe(CO)$_2$, 1, 517
$FeC_8H_7BrO_5$ [π-CH$_2$=CHCH(OAc)-Fe(CO)$_3$]-Br, 1, 472
[π-CH$_2$=CHCH(CO$_2$Me)-Fe(CO)$_3$]Br, 1, 472
$FeC_8H_7ClO_2$ (MeC$_5$H$_4$)Fe(CO)$_2$Cl, 1, 200
$FeC_8H_7ClO_3$ (MeOC$_5$H$_4$)Fe(CO)$_2$Cl, 1, 200
$FeC_8H_7ClO_4$ [π-MeCH=CHCH$_2$-Fe(CO)$_4$]Cl, 1, 473
$FeC_8H_7ClO_7$ [(CH$_2$CH=CHCH=CH$_2$)Fe(CO)$_3$]$^+$-[ClO$_4$]$^-$, 1, 497
$FeC_8H_8N_2O_3$ (EtNC)(MeNC)Fe(CO)$_3$, 1, 524
$FeC_8H_8O_2$ C$_5$H$_5$Fe(CO)$_2$Me, 1, 202
$FeC_8H_8O_2S$ C$_5$H$_5$Fe(CO)$_2$SMe, 1, 201
$FeC_8H_8O_3$ Isoprene-Fe(CO)$_3$, 1, 478
1,3-Pentadiene-Fe(CO)$_3$, 1, 478
$FeC_8H_8O_4$ (HOCH$_2$CH=CHCH=CH$_2$)Fe(CO)$_3$, 1, 478
$FeC_8H_9ClO_3$ [π-MeCH=CHCH(Me)-Fe(CO)$_3$]Cl, 1, 472
$FeC_8H_9N_3O$ NH$_4$[C$_5$H$_5$Fe(CO)(CN)$_2$], 1, 207
FeC_8N_6 Fe[C(CN)$_3$]$_2$, 1, 193
$FeC_9F_8O_3$ Octafluoro-1,3-cyclohexadiene-Fe(CO)$_3$, 1, 487
$FeC_9H_4O_6$ COCH=C(Me)COFe(CO)$_4$, 1, 735
$FeC_9H_4O_7$ (Citraconic anhydride)Fe(CO)$_4$, 1, 477
$FeC_9H_5F_3O_3$ C$_5$H$_5$Fe(CO)$_2$COCF$_3$, 1, 206
$FeC_9H_5F_5O_2$ C$_5$H$_5$Fe(CO)$_2$C$_2$F$_5$, 1, 205
$FeC_9H_6O_4$ 2,4-Cyclohexadienone-Fe(CO)$_3$, 1, 487
$FeC_9H_6O_7$ (-CH=CHCO$_2$H)$_2$Fe(CO)$_3$, 1, 479
$FeC_9H_6O_8$ (MeO$_2$CCH=CHCO$_2$H)Fe(CO)$_4$, 1, 477
[HO$_2$CC(Me)=CHCO$_2$H]Fe(CO)$_4$, 1, 477
$FeC_9H_7ClO_4$ (AcOC$_5$H$_4$)Fe(CO)$_2$Cl, 1, 200
$FeC_9H_7IO_2$ [Cycloheptatrienylium-Fe-(CO)$_2$]I, 1, 500
$FeC_9H_7NO_2$ C$_5$H$_5$Fe(CO)$_2$CH$_2$CN, 1, 202
$FeC_9H_7NO_3$ (MeCH=CHCH=CHCN)Fe(CO)$_3$, 1, 481
$FeC_9H_8ClNO_2$ [C$_5$H$_5$Fe(CO)$_2$(HN=C=CH$_2$)]Cl, 1, 210
$FeC_9H_8ClNO_6$ [C$_5$H$_5$Fe(CO)$_2$(HN=C=CH$_2$)]-ClO$_4$, 1, 210
$FeC_9H_8O_2$ C$_5$H$_5$Fe(CO)$_2$CH=CH$_2$, 1, 205
$FeC_9H_8O_3$ C$_5$H$_5$Fe(CO)$_2$CH$_2$CHO, 1, 202
C$_5$H$_5$Fe(CO)$_2$Ac, 1, 206
(CH$_2$=CHCH=CHCH=CH$_2$)Fe(CO)$_3$, 1, 481
1,3-Cyclohexadiene-Fe(CO)$_3$, 1, 488
$FeC_9H_8O_4$ [MeCH=CHC(Me)=CO]Fe(CO)$_3$, 1, 479*
[MeCH=C(Me)CH=CO]Fe(CO)$_3$, 1, 479*
[Me(CH=CH)$_2$CHO]Fe(CO)$_3$, 1, 481

Hydroxycyclohexadiene-Fe$(CO)_3$, 1, 488

$FeC_9H_8O_5$ [Me(CH=CH)$_2$CO$_2$H]Fe$(CO)_3$, 1, 479

$FeC_9H_8O_5S$ [CH=C(Me)C(Me)=CHSO$_2$]-Fe$(CO)_3$, 1, 496

$FeC_9H_8O_6$ [CH$_2$=C(Me)CO$_2$Me]Fe$(CO)_4$, 1, 477

(CH$_2$=CHCO$_2$Et)Fe$(CO)_4$, 1, 477

$FeC_9H_9BrO_3$ [C$_5$H$_5$Fe(CO)$_2$(CH$_2$=CHOH)]Br, 1, 209

$FeC_9H_9ClO_6$ [C$_5$H$_5$Fe(CO)$_2$(C$_2$H$_4$)]ClO$_4$, 1, 209

$FeC_9H_9ClO_7$ [1-Methylpentadienylium-Fe$(CO)_3$]ClO$_4$, 1, 497, 498

$FeC_9H_9F_6O_2P$ [C$_5$H$_5$Fe(CO)$_2$(C$_2$H$_4$)]PF$_6$, 1, 209

$FeC_9H_9F_6O_3P$ [1-Methylpentadienylium-Fe$(CO)_3$]PF$_6$, 1, 498

$FeC_9H_9IO_2$ [C$_5$H$_5$Fe(CO)$_2$(C$_2$H$_4$)]I, 1, 209

FeC_9H_9N Cyclopentadienylpyrrolyl-Fe, 1, 462

$FeC_9H_9NO_4$ [Me(CH=CH)$_2$CONH$_2$]Fe$(CO)_3$, 1, 481

(Me$_3$CNC)Fe$(CO)_4$, 1, 524

$FeC_9H_{10}N_2O_3$ (EtNC)$_2$Fe$(CO)_3$, 1, 525

$FeC_9H_{10}O$ Allyl(C$_5$H$_5$)Fe(CO), 1, 217

$FeC_9H_{10}O_2$ C$_5$H$_5$Fe(CO)$_2$Et, 1, 202

$FeC_9H_{10}O_3$ (EtCH=CHCH=CH$_2$)Fe$(CO)_3$, 1, 481

(MeCH=CHCH=CHMe)Fe$(CO)_3$, 1, 481

$FeC_9H_{10}O_4$ [MeCH(OH)CH=CHCH=CH$_2$]Fe$(CO)_3$, 1, 479

(MeCH=CHCH=CHCH$_2$OH)Fe$(CO)_3$, 1, 479

$FeC_9H_{11}IO_2S$ [C$_5$H$_5$Fe(CO)$_2$(SMe$_2$)]I, 1, 219

FeC_9H_{15} Tris-π-allyl-Fe, 1, 470

$FeC_9H_{19}Cl_3NO_2$ [Me$_2$C(OH)CH=C=CHCH$_2$-NMe$_2$]FeCl$_3$·H$_2$O, 1, 514

$FeC_{10}D_{10}$ (C$_5$D$_5$)$_2$Fe, 1, 226

$FeC_{10}F_5IO_4$ [C$_6$F$_5$Fe(CO)$_4$]I, 1, 195

$FeC_{10}F_{14}O_4$ (C$_3$F$_7$)$_2$Fe$(CO)_4$, 1, 196

(C$_3$F$_7$)$_2$Fe$(CO)_4$, 1, 196

$FeC_{10}H_5F_5O_2$ C$_5$H$_5$Fe(CO)$_2$CF$_2$CF=CF$_2$, 1, 203

$FeC_{10}H_5F_5O_3$ C$_5$H$_5$Fe(CO)$_2$COC$_2$F$_5$, 1, 206

$FeC_{10}H_5F_7O_2$ C$_5$H$_5$Fe(CO)$_2$CF$_2$CF$_2$CF$_3$, 1, 205

$FeC_{10}H_6O_4$ Tropone-Fe$(CO)_3$, 1, 488

[(C$_2$H$_2$)$_3$Fe(CO)$_4$]$_n$, 1, 516

$FeC_{10}H_8O_6$ COC(Me)=C(Me)COFe$(CO)_4$, 1, 735

COCH=C(Et)COFe$(CO)_4$, 1, 735

$(FeC_{10}H_8)_n$ (-Fcd-)$_n$, 1, 463

$FeC_{10}H_8BrI$ Br-Fcd-I, 1, 396

$FeC_{10}H_8Br_2$ FcdBr$_2$, 1, 361

$FeC_{10}H_8ClI$ Cl-Fcd-I, 1, 396

$FeC_{10}H_8Cl_2$ FcdCl$_2$, 1, 361

$FeC_{10}H_8Cl_2O_4S_2$ Fcd(SO$_2$Cl)$_2$, 1, 377

$FeC_{10}H_8I_2$ FcdI$_2$, 1, 361

$FeC_{10}H_8N_6$ Fcd(N$_3$)$_2$, 1, 363

$FeC_{10}H_8O_2$ C$_5$H$_5$Fe(CO)$_2$CH$_2$C≡CH, 1, 203

$FeC_{10}H_8O_3$ C$_5$H$_5$Fe(CO)$_2$COCH=CH$_2$, 1, 206

Cycloheptatriene-Fe$(CO)_3$, 1, 489

Bicyclo[2.2.1]hepta-2,5-diene-Fe$(CO)_3$, 1, 493

$FeC_{10}H_8O_4$ (CH$_2$=CHCH=CHCH=CH$_2$)Fe$(CO)_4$, 1, 477

2,4-Cycloheptadienone-Fe$(CO)_3$, 1, 489*

3,5-Cycloheptadienone-Fe$(CO)_3$, 1, 489

2,4-Cycloheptadienone-Fe$(CO)_3$ oligomers, 1, 508

$FeC_{10}H_8O_5$ Hydroxycycloheptadienone-Fe$(CO)_3$, 1, 490

$FeC_{10}H_8O_8$ (=CHCO$_2$Me)$_2$Fe$(CO)_4$ isomers, 1, 477

$FeC_{10}H_9Br$ FcBr, 1, 256

$FeC_{10}H_9Cl$ FcCl, 1, 255

$FeC_{10}H_9ClO_2S$ FcSO$_2$Cl, 1, 317

$FeC_{10}H_9Cl_2P$ FcPCl$_2$, 1, 334

$FeC_{10}H_9F_6O_3P$ [Cycloheptadienylium-Fe$(CO)_3$]PF$_6$, 1, 498

$FeC_{10}H_9I$ FcI, 1, 256

$FeC_{10}H_9NO_2$ C$_5$H$_5$Fe(CO)$_2$CHMeCN, 1, 203

C$_5$H$_5$Fe(CO)$_2$CH$_2$CH$_2$CN, 1, 205

FcNO$_2$, 1, 332

$FeC_{10}H_9N_3$ FcN$_3$, 1, 264

$FeC_{10}H_{10}$ FcH, 1, 220

$FeC_{10}H_{10}BrDO_3$ [C$_5$H$_5$Fe(CO)$_2$(CH$_2$=CODMe)]-Br, 1, 210

$FeC_{10}H_{10}Br_2$ C$_5$H$_5$Fe(C$_5$H$_5$Br$_2$), 1, 219

$FeC_{10}H_{10}Br_4$ C$_5$H$_5$Fe(C$_5$H$_5$Br$_4$), 1, 219

$FeC_{10}H_{10}Cl$ [FcH]Cl, 1, 226

$FeC_{10}H_{10}ClN$ Cl-Fcd-NH_2, 1, 386

$FeC_{10}H_{10}ClNO_6$ [$C_5H_5Fe(CO)_2$(HN=C=CH-Me)]ClO_4, 1, 210

$FeC_{10}H_{10}ClO_4$ [FcH]ClO_4, 1, 227

$FeC_{10}H_{10}Cl_2$ $C_5H_5Fe(C_5H_5Cl_2)$, 1, 219

$FeC_{10}H_{10}I_2$ $C_5H_5Fe(C_5H_5I_2)$, 1, 219

$FeC_{10}H_{10}I_3$ [FcH]I_3, 1, 226

$FeC_{10}H_{10}I_4$ $C_5H_5Fe(C_5H_5I_4)$, 1, 219

$FeC_{10}H_{10}NO_3$ [FcH]NO_3, 1, 227

$FeC_{10}H_{10}O$ FcOH, 1, 265

$FeC_{10}H_{10}O_2$ $C_5H_5Fe(CO)_2CH_2CH$=CH_2, 1, 203

Fcd$(OH)_2$, 1, 363

$FeC_{10}H_{10}O_2S$ $FcSO_2H$, 1, 316

$FeC_{10}H_{10}O_3$ $C_5H_5Fe(CO)_2CH_2Ac$, 1, 205

$C_5H_5Fe(CO)_2COEt$, 1, 206

1,3-Cycloheptadiene-$Fe(CO)_3$, 1, 490

$FeC_{10}H_{10}O_3S$ $FcSO_3H$, 1, 317

$FeC_{10}H_{10}O_4$ (MeCH=CHCH=CHAc)$Fe(CO)_3$, 1, 481

Methoxycyclohexadiene-$Fe(CO)_3$, 1, 488

$FeC_{10}H_{10}O_4S_2$ Fcd$(SO_2H)_2$, 1, 377

$FeC_{10}H_{10}O_5$ π-CH_2=C(Me)C(Me)CH_2OCOFe-$(CO)_3$, 1, 473

π-MeCH=CHCHCH(Me)OCOFe$(CO)_3$, 1, 473

[Me(CH=CH)$_2CO_2$Me]$Fe(CO)_3$, 1, 481

$FeC_{10}H_{10}O_6S_2$ Fcd$(SO_3H)_2$, 1, 377

$FeC_{10}H_{10}S$ FcSH, 1, 316

$FeC_{10}H_{11}BrO_3$ [$C_5H_5Fe(CO)_2$(CH_2=COH-Me)]Br, 1, 210

$FeC_{10}H_{11}ClO_3$ [$C_5H_5Fe(CO)_2$(CH_2=COH-Me)]Cl, 1, 210

$FeC_{10}H_{11}F_6O_2P$ [$C_5H_5Fe(CO)_2$(CH_2=CH-Me)]PF_6, 1, 210

$FeC_{10}H_{11}IN_2O$ [$C_5H_5Fe(CO)(CNMe)_2$]I, 1, 211

$FeC_{10}H_{11}N$ $FcNH_2$, 1, 319

$FeC_{10}H_{11}NO_2S$ $FcSO_2NH_2$, 1, 317

$FeC_{10}H_{11}NO_3$ $C_5H_5Fe(CO)_2CONMe_2$, 1, 206

$FeC_{10}H_{11}O_2P$ FcP(O)H(OH), 1, 334

$FeC_{10}H_{12}N_2$ Fcd$(NH_2)_2$, 1, 378

$FeC_{10}H_{12}N_6$ $(MeNC)_4Fe(CN)_2$, 1, 525

$FeC_{10}H_{12}O$ 1-Methallyl(C_5H_5)Fe(CO), 1, 217

$FeC_{10}H_{12}O_2$ $C_5H_5Fe(CO)_2Pr$, 1, 203

$C_5H_5Fe(CO)_2CHMe_2$, 1, 203

$FeC_{10}H_{12}O_2S$ $C_5H_5Fe(CO)_2CH_2CH_2SMe$, 1, 203

$C_5H_5Fe(CO)COCH_2CH_2SMe$, 1, 219

$FeC_{10}H_{12}O_4$ [MeCH=CHCH=CHCH(OH)Me]Fe-$(CO)_3$, 1, 479

[MeCH(OMe)CH=CHCH=CH_2]$Fe(CO)_3$, 1, 481

$FeC_{10}H_{13}NO_3S$ NH_4O_3SFc, 1, 318

$FeC_{11}H_5F_7O_3$ $C_5H_5Fe(CO)_2COCF_2CF_2CF_3$, 1, 206

$FeC_{11}H_5NO_4$ (PhNC)$Fe(CO)_4$, 1, 520

$FeC_{11}H_5N_3O_2$ $C_5H_5Fe(CO)_2C(CN)_3$, 1, 205

$FeC_{11}H_7ClO_3$ Chlorocyclooctatetraene-$Fe(CO)_3$, 1, 490

$FeC_{11}H_8BrN$ Br-Fcd-CN, 1, 396

$FeC_{11}H_8Cl_2O_3S$ ClO_2S-Fcd-COCl, 1, 386

$FeC_{11}H_8N_2O_2$ O_2N-Fcd-CN, 1, 396

$FeC_{11}H_8O_3$ Cyclooctatetraene-$Fe(CO)_3$, 1, 491

Dihydroindenone-$Fe(CO)_2$, 1, 500*

$FeC_{11}H_8O_4$ Cyclooctatrienone-$Fe(CO)_3$, 1, 491

Bicyclo[5.1.0]octa-2,4-dien-6-one-$Fe(CO)_3$, 1, 494

$FeC_{11}H_8O_5$ (Bicyclo[2.2.1]hepta-2.5-diene-2-carboxylic acid)-$Fe(CO)_3$, 1, 494*

$FeC_{11}H_8O_7$ C(CO_2Me)=CHCH=C(CO_2Me)Fe-$(CO)_3$, 1, 735

$FeC_{11}H_9BrO_2$ Br-Fcd-CO_2H, 1, 386

$FeC_{11}H_9ClO$ FcCOCl, 1, 298

$FeC_{11}H_9ClO_2$ Carboxychloroferrocene, 1, 406

$FeC_{11}H_9ClO_3$ [Bicyclo[5.1.0]octa-2,4-dienylium-$Fe(CO)_3$]Cl, 1, 499

$FeC_{11}H_9ClO_7$ [Bicyclo[5.1.0]octa-2,4-dienylium-$Fe(CO)_3$]ClO_4, 1, 499

$FeC_{11}H_9FO_4S$ FO_2S-Fcd-CO_2H, 1, 386

$FeC_{11}H_9N$ FcCN, 1, 262

Poly(cyanoferrocenylene), 1, 463

$FeC_{11}H_9NO$ FcNCO, 1, 264

$FeC_{11}H_9N_3O$ $FcCON_3$, 1, 300*

$FeC_{11}H_{10}$ Poly(methanoferrocene), 1, 463

$FeC_{11}H_{10}ClNO_3S$ ClO_2S-Fcd-$CONH_2$, 1, 386

$FeC_{11}H_{10}N_2O$ [$C_5H_5Fe(CO)(CNCH_2CH$=CH_2)]-CN, 1, 207

$FeC_{11}H_{10}O$ FcCHO, 1, 280
$FeC_{11}H_{10}O_2$ $FcCO_2H$, 1, 297
$FeC_{11}H_{10}O_3$ 1,3,5-Cyclooctatriene-Fe-$(CO)_3$, 1, 492
Bicyclo[4.2.0]octa-2,4-diene-Fe$(CO)_3$, 1, 494
Bicyclo[5.1.0]octa-2,4-diene-Fe$(CO)_3$, 1, 495
$FeC_{11}H_{10}O_4$ 6-Hydroxybicyclo[5.1.0]-octa-2,4-diene-Fe$(CO)_3$, 1, 494
$FeC_{11}H_{10}O_5S$ HO_3S-Fcd-CO_2H, 1, 386
$FeC_{11}H_{10}O_7$ $(-CH{=}CHCO_2Me)_2Fe(CO)_3$, 1, 480
$FeC_{11}H_{11}Br$ $[C_6H_6FeC_5H_5]Br$, 1, 218
$FeC_{11}H_{11}D$ Cyclohexadienyl-d_1-(C_5H_5)Fe, 1, 217
$FeC_{11}H_{11}F_6O_2P$ $[C_5H_5Fe(CO)_2(CH_2{=}CHCH{=}CH_2)]PF_6$, 1, 210
$FeC_{11}H_{11}I$ $[C_6H_6FeC_5H_5]I$, 1, 218
$FeC_{11}H_{11}I_3$ $[C_6H_6FeC_5H_5]I_3$, 1, 218
$FeC_{11}H_{11}I_n$ $[C_6H_6FeC_5H_5]I_n$, 1, 218
$FeC_{11}H_{11}NO$ FcCH=NOH, 1, 281
$FcCONH_2$, 1, 300
$FeC_{11}H_{11}NO_2$ $(PrC_5H_4)Fe(CO)_2CN$, 1, 201
$FeC_{11}H_{11}NO_4S$ H_2NO_2S-Fcd-CO_2H, 1, 386
$FeC_{11}H_{12}$ Cyclohexadienyl(C_5H_5)Fe, 1, 217
FcMe, 1, 228
$FeC_{11}H_{12}N_2O$ $FcCONHNH_2$, 1, 300
$FcNHCONH_2$, 1, 319
$FeC_{11}H_{12}O$ $FcCH_2OH$, 1, 265
FcOMe, 1, 275
$FeC_{11}H_{12}O_2$ $C_5H_5Fe(CO)_2CH_2CH{=}CHMe$, 1, 203
$FeC_{11}H_{12}O_3$ 1,5-Cyclooctadiene-Fe$(CO)_3$, 1, 492
$FeC_{11}H_{12}O_3S$ $FcSO_3Me$, 1, 318
$FcCH_2SO_3H$ 1, 318
$FeC_{11}H_{12}O_5$ $[Me(CH{=}CH)_2CO_2Et]Fe(CO)_3$, 1, 481, 482
$FeC_{11}H_{12}S$ FcSMe, 1, 316
$FeC_{11}H_{13}ClO_2$ $[C_5H_5Fe(CO)_2(CH_2{=}CHEt)]$-Cl, 1, 210
$FeC_{11}H_{13}ClO_6$ $[C_5H_5Fe(CO)_2(CH_2{=}CHEt)]$-$ClO_4$, 1, 210
$FeC_{11}H_{13}F_6O_2P$ $[C_5H_5Fe(CO)_2(CH_2{=}CHEt)]$-$PF_6$, 1, 210
$[C_5H_5Fe(CO)_2(MeCH{=}CHMe)]PF_6$, 1, 210
$FeC_{11}H_{13}N$ FcNHMe, 1, 320
$FcCH_2NH_2$, 1, 321
Cyclopentadienyl(dimethylpyrrolyl)-Fe, 1, 462
$FeC_{11}H_{14}ClN$ $[FcCH_2NH_2]HCl$, 1, 321
$FeC_{11}H_{14}O_4$ (Ethyl hexadienyl ether)-Fe$(CO)_3$, 1, 480
$FeC_{11}H_{17}IN_6O$ $[(MeNC)_5Fe(CN)]I \cdot H_2O$, 1, 526
$FeC_{12}F_{12}O_4$ Tetrakis(trifluoromethyl)-cyclopentadienone-Fe$(CO)_3$, 1, 484
$FeC_{12}H_7NO_5$ $(p\text{-}MeOC_6H_4NC)Fe(CO)_4$, 1, 524
$(p\text{-}MeOC_6H_4NC)Fe(CO)_4$, 1, 524

$FeC_{12}H_8Cl_2O_2$ $Fcd(COCl)_2$, 1, 373
$FeC_{12}H_8N_2$ $Fcd(CN)_2$, 1, 362
$FeC_{12}H_8N_2O_2$ $Fcd(NCO)_2$, 1, 363
$FeC_{12}H_8N_6O_2$ $Fcd(CON_3)_2$, 1, 374
$FeC_{12}H_8O_3$ CO-Fcd-COO, 1, 436
$C_5H_5FeC_5H_3(CO)_2O$, 1, 454
$FeC_{12}H_8O_4$ Cinnamaldehyde-Fe$(CO)_3$, 1, 480
$FeC_{12}H_9Cl_3O$ $FcCOCCl_3$, 1, 257
$FeC_{12}H_9NO_2$ NC-Fcd-CO_2H, 1, 386
$FeC_{12}H_{10}$ FcC≡CH, 1, 253
Poly(ferrocenylvinylene), 1, 465
$FeC_{12}H_{10}Cl_2$ $FcCH{=}CCl_2$, 1, 257
$FeC_{12}H_{10}Cl_2O$ $FcCOCHCl_2$, 1, 257
$FeC_{12}H_{10}F_3NO_4$ $[C_5H_5Fe(CO)_2(HN{=}C{=}CH\text{-}Me)]O_2CCF_3$, 1, 210
$FeC_{12}H_{10}O_2$ $(C_5H_5)(\sigma\text{-}C_5H_5)Fe(CO)_2$, 1, 204
$Fcd(CHO)_2$, 1, 366
$FeC_{12}H_{10}O_3$ Methylcyclooctatetraene-Fe-$(CO)_3$, 1, 492
3a,7a-Dihydroindene-Fe$(CO)_3$, 1, 493
$FeC_{12}H_{10}O_4$ 1,2-$(HO_2C)_2C_5H_3FeC_5H_5$, 1, 336
$Fcd(CO_2H)_2$, 1, 373
$FeC_{12}H_{10}O_5$ 2-Methoxycarbonylbicyclo-[2.2.1]hepta-2,5-diene-Fe$(CO)_3$, 1, 494
7-Acetoxybicyclo[2.2.1]hepta-2,5-diene-Fe$(CO)_3$, 1, 494
$FeC_{12}H_{10}O_6$ COC(Et)=C(Et)COFe$(CO)_4$, 1,

735
COCH=C(Bu)COFe$(CO)_4$, 1, 735
$FeC_{12}H_{11}Br$ FcCH=CHBr, 1, 258
$FeC_{12}H_{11}BrO$ Ac-Fcd-Br, 1, 387
$FeC_{12}H_{11}BrO_2$ Br-Fcd-CO_2Me, 1, 387
$FeC_{12}H_{11}Br_3O$ FcCH(OH)CBr_3, 1, 258
$FeC_{12}H_{11}Cl$ FcCCl=CH_2, 1, 257
FcCH=CHCl, 1, 257
$FeC_{12}H_{11}ClO$ FcCOCH_2Cl, 1, 258
FcCH_2COCl, 1, 304
AcFcd-Cl, 1, 387
$FeC_{12}H_{11}ClO_4S$ ClO_2S-Fcd-CO_2Me, 1, 387
$FeC_{12}H_{11}ClO_6$ [C_5H_5Fe$(CO)_2$-(CH=CHCH=CHCH_2)]ClO_4, 1, 210
$FeC_{12}H_{11}Cl_3O$ FcCH(OH)CCl_3, 1, 258
$FeC_{12}H_{11}FO_4S$ FO_2S-Fcd-CO_2Me, 1, 387
$FeC_{12}H_{11}F_6O_2P$ [C_5H_5Fe$(CO)_2$-(CH=CHCH=CHCH_2)]PF_6, 1, 210
$FeC_{12}H_{11}I_3O$ FcCH(OH)CI_3, 1, 258
$FeC_{12}H_{11}N$ FcCH_2CN, 1, 263
NC(Me)C_5H_3FeC_5H_5, 1, 336
Me-Fcd-CN, 1, 396
1-Cyano-x-methylferrocene, 1, 406
$FeC_{12}H_{11}NO$ FcCH(OH)CN, 1, 263
FcCH_2NCO, 1, 264
$FeC_{12}H_{11}NO_3$ H_2NCO-Fcd-CO_2H, 1, 387
$FeC_{12}H_{11}NS$ FcCH_2SCN, 1, 265
$FeC_{12}H_{12}$ FcCH=CH_2, 1, 248
$FeC_{12}H_{12}BrNO$ MeC(=NOH)-Fcd-Br, 1, 387
$FeC_{12}H_{12}Br_2$ FcCHBrCH_2Br, 1, 259
$FeC_{12}H_{12}ClNO$ Cl-Fcd-NHAc, 1, 396
$FeC_{12}H_{12}Cl_2$ FcCHClCH_2Cl, 1, 258
FcCH_2CHCl_2, 1, 259
FcCCl_2Me, 1, 259
$FeC_{12}H_{12}Cl_2O$ FcCH(OH)CHCl_2, 1, 259
$FeC_{12}H_{12}N_2$ (p-H_2NC_6H_4)$_2$Fe, 1, 193
$FeC_{12}H_{12}N_2O_2$ Fcd(CH=NOH)$_2$, 1, 366
Fcd(CONH_2)$_2$, 1, 374
$FeC_{12}H_{12}O$ FcCH_2CHO, 1, 283
FcAc, 1, 285
[HO(CH_2=CH)C_5H_3]FeC_5H_5, 1, 336
[Me(CHO)C_5H_3]FeC_5H_5, 1, 336
Me-Fcd-CHO, 1, 387
$FeC_{12}H_{12}OS$ FcSAc, 1, 316
$FeC_{12}H_{12}O_2$ FcOAc, 1, 274
FcCO_2Me, 1, 298
FcCH_2CO_2H, 1, 303
[HO_2C(Me)C_5H_3]FeC_5H_5, 1, 336
Ac-Fcd-OH, 1, 388
HOCH_2-Fcd-CHO, 1, 388
Me-Fcd-CO_2H, 1, 388
$FeC_{12}H_{12}O_2S$ FcSCH_2CO_2H, 1, 305
$FeC_{12}H_{12}O_3$ FcOCH_2CO_2H, 1, 305
$FeC_{12}H_{12}O_4S$ Ac-Fcd-SO_3H, 1, 388
$FeC_{12}H_{12}O_5S$ HO_3S-Fcd-CO_2Me, 1, 388
$FeC_{12}H_{12}O_8$ trans-(=CHCO_2Et)$_2$Fe$(CO)_4$, 1, 477
$FeC_{12}H_{13}Br$ FcCHBrMe, 1, 260
Et-Fcd-Br, 1, 388
$FeC_{12}H_{13}Cl$ FcCHClMe, 1, 259
FcCH_2CH_2Cl, 1, 259
$FeC_{12}H_{13}ClO$ FcCH(OH)CH_2Cl, 1, 260
$FeC_{12}H_{13}N$ FcCH_2N=CH_2, 1, 322
$FeC_{12}H_{13}NO$ FcC(=NOH)Me, 1, 287
FcCONHMe, 1, 300
FcNHAc, 1, 319
[Me(H_2NCO)C_5H_3]FeC_5H_5, 1, 337
Me-Fcd-CONH_2, 1, 388
$FeC_{12}H_{13}NO_2$ FcNHCO_2Me, 1, 321
$FeC_{12}H_{13}NO_4S$ H_2NO_2S-Fcd-CO_2Me, 1, 388
$FeC_{12}H_{13}N_3$ FcCH(Me)N_3, 1, 264
$FeC_{12}H_{13}N_3O$ FcCH=NNHCONH_2, 1, 282
$FeC_{12}H_{13}N_3S$ FcCH=NNHCSNH_2, 1, 282
$FeC_{12}H_{14}$ FcEt, 1, 229
$FeC_{12}H_{14}$ ($Me_2C_5H_3$)FeC_5H_5, 1, 337
FcdMe_2, 1, 350
1,x-Dimethylferrocene, 1, 379
$FeC_{12}H_{14}N_4$ Fcd(N=NMe)$_2$, 1, 379
$FeC_{12}H_{14}O$ FcCH(OH)Me, 1, 266
FcCH_2CH_2OH, 1, 267
FcCH_2OMe, 1, 276
Me(HOCH_2)C_5H_3FeC_5H_5, 1, 337
Et(HO)C_5H_3FeC_5H_5, 1, 337
Et-Fcd-OH, 1, 396
$FeC_{12}H_{14}O_2$ FcCH(OH)CH_2OH, 1, 267
$FeC_{12}H_{14}O_2$ FcCH(OH)CH_2OH, 1, 267
Fcd(CH_2OH)$_2$, 1, 363
Fcd(OMe)$_2$, 1, 366
[HO(Me)C_5H_3]$_2$Fe, 1, 380
$FeC_{12}H_{14}O_3$ 1,2-Dihydromesitylene-Fe$(CO)_3$, 1, 488
$FeC_{12}H_{14}O_6S_2$ Fcd(SO_3Me)$_2$, 1, 377
$FeC_{12}H_{14}S_2$ Fcd(SMe)$_2$, 1, 377
$FeC_{12}H_{15}N$ FcNHEt, 1, 320

$FeC_{12}H_{15}N$

$FcCH(Me)NH_2$, 1, 325
$FcCH_2CH_2NH_2$, 1, 327
$Me(H_2NCH_2)C_5H_3FeC_5H_5$, 1, 377
H_2N-Fcd-Et, 1, 389
$FeC_{12}H_{15}NO$ FcCH(Me)NHOH, 1, 326
$FeC_{12}H_{15}NO_3$ $C_5H_5Fe(CO)_2CONEt_2$, 1, 206
$FeC_{12}H_{16}ClN$ $FcCH(Me)NH_2\cdot HCl$, 1, 326
$FeC_{12}H_{18}Br$ 1,5,9-Cyclododecatriene-FeBr, 1, 514
$FeC_{12}H_{18}Cl$ 1,5,9-Cyclododecatriene-FeCl, 1, 514
$FeC_{12}H_{18}I_2N_6$ $[(MeNC)_6Fe]I_2$, 1, 528
$FeC_{12}H_{21}O_5P$ $(CH_2{=}CHCH{=}CH_2)Fe(CO)_2$-$[P(OEt)_3]$, 1, 501
$FeC_{12}H_{22}Br_2N_6O_2$ $[(MeNC)_6Fe]Br_2\cdot 2H_2O$, 1, 528
$FeC_{12}H_{22}Cl_2N_2$ $(n\text{-}C_5H_{11}NC)_2FeCl_2$, 1, 524
$(iso\text{-}C_5H_{11}NC)_2FeCl_2$, 1, 524
$FeC_{12}H_{22}N_2O_4S$ $(iso\text{-}C_5H_{11}NC)_2FeSO_4$, 1, 524
$FeC_{12}H_{24}Cl_2N_6O_3$ $[(MeNC)_6Fe]Cl_2\cdot 3H_2O$, 1, 522
$FeC_{13}H_5F_5O_2$ $C_5H_5Fe(CO)_2C_6F_5$, 1, 205
$FeC_{13}H_6F_4O_2$ $(2,3,5,6\text{-}F_4C_6H)(C_5H_5)Fe(CO)_2$, 1, 205
$FeC_{13}H_7F_3O_2$ $(2,3,6\text{-}F_3C_6H_2)(C_5H_5)Fe(CO)_2$, 1, 205
$FeC_{13}H_8O_3$ Naphthelene-$Fe(CO)_3$, 1, 493
$FeC_{13}H_8O_5$ trans-$(PhCH{=}CHCHO)Fe(CO)_4$, 1, 477
$FeC_{13}H_8O_6$ trans-$(PhCH{=}CHCO_2H)Fe(CO)_4$, 1, 477
$FeC_{13}H_{10}O$ $FcCOC{\equiv}CH$, 1, 287
$FeC_{13}H_{10}O_2$ $C_5H_5Fe(CO)_2Ph$, 1, 204
$FcC{\equiv}CCO_2H$, 1, 305
$FeC_{13}H_{10}O_3$ $(PhCH{=}CHCH{=}CH_2)Fe(CO)_3$, 1, 482
$FeC_{13}H_{10}O_5$ $(p\text{-}MeOC_6H_4CH{=}CHCH{=}O)Fe(CO)_3$, 1, 482
$FeC_{13}H_{11}ClO$ FcCOCH=CHCl, 1, 260
$FeC_{13}H_{11}ClO_3$ MeO_2C-Fcd-COCl, 1, 396
$FeC_{13}H_{11}NO$ Isoxazolyl-Fc, 1, 238
$FeC_{13}H_{12}$ $FcC{\equiv}CMe$, 1, 254
$FcCH_2C{\equiv}CH$, 1, 254
$FcCH{=}C{=}CH_2$, 1, 249
$CH{=}CH$-Fcd-CH_2, 1, 437
$FeC_{13}H_{12}Cl_2$ $FcCHCH_2CCl_2$, 1, 260
$FeC_{13}H_{12}F_9NO_3$ Me_4N^+[nonafluorocyclohexenyl-$Fe(CO)_3]^-$, 1, 473
$FeC_{13}H_{12}N_2$ 3-Pyrazolyl-Fc, 1, 238
$FeC_{13}H_{12}N_2O_2$ 2,5-Dioxoimidazolidin-4-yl-Fc, 1, 238
$FeC_{13}H_{12}O$ $FcCH(OH)C{\equiv}CH$, 1, 267
FcCH=CHCHO, 1, 283
$FcCOCH{=}CH_2$, 1, 287, 295
CO-Fcd-CH_2CH_2, 1, 437
CH_2-Fcd-CH_2CO, 1, 437
Cyclopentadienyl(tetrahydro-4-oxopentalenyl)iron, 1, 454
$FeC_{13}H_{12}O_2$ $FcCOCH_2CHO$, 1, 283
$FcCH{=}CHCO_2H$, 1, 305
Ac-Fcd-CHO, 1, 389
$FeC_{13}H_{12}O_3$ $FcCOCO_2Me$, 1, 304
Ac-Fcd-CO_2H, 1, 389
$FeC_{13}H_{12}O_4$ $[HO_2C(MeO_2C)C_5H_3]FeC_5H_5$, 1, 338
AcO-Fcd-CO_2H, 1, 389
MeO_2C-Fcd-CO_2H, 1, 389, 396
$FeC_{13}H_{12}O_5$ Duroquinone-$Fe(CO)_3$, 1, 488
$FeC_{13}H_{13}F_6O_2P$ $[C_5H_5Fe(CO)_2$-$(CH{=}CHCH{=}CHCH_2CH_2)]PF_6$, 1, 210
$FeC_{13}H_{13}N$ FcCH(CN)Me, 1, 263
$[Me(NCCH_2)C_5H_3]FeC_5H_5$, 1, 338
$[NC(Et)C_5H_3]FeC_5H_5$, 1, 338
Et-Fcd-CN, 1, 390
1-Cyano-x-ethylferrocene, 1, 406
Cyclopentadienyl(4,5-dihydropyrindenyl)-Fe, 1, 458
$FeC_{13}H_{13}NO$ Ac-Fcd-$CONH_2$, 1, 390
$FeC_{13}H_{13}NO_2$ FcCH=NOAc, 1, 282
$FcCH{=}C(NO_2)Me$, 1, 332
$FeC_{13}H_{13}NO_3$ $FcCONHCH_2CO_2H$, 1, 301
MeO_2C-Fcd-$CONH_2$, 1, 396
$FeC_{13}H_{14}$ $FcCHCH_2CH_2$, 1, 230
$FcC(Me){=}CH_2$, 1, 249
FcCH=CHMe, 1, 250
$FcCH_2CH{=}CH_2$, 1, 250
$[Me(CH_2{=}CH)C_5H_3]FeC_5H_5$, 1, 338
CH_2-Fcd-CH_2CH_2, 1, 437
$FeC_{13}H_{14}N_2O$ FcCH=NNHAc, 1, 282
$FeC_{13}H_{14}O$ $FcCH(OH)CH{=}CH_2$, 1, 268
$FcCH{=}CHCH_2OH$, 1, 268
$FcC(OH)(Me)C{\equiv}CH$, 1, 269

$FcOCH_2CH{=}CH_2$, 1, 279
FcCOEt, 1, 287
$[Ac(Me)C_5H_3]FeC_5H_5$, 1, 338
$[Et(HCO)C_5H_3]FeC_5H_5$, 1, 338, 339
Ac-Fcd-Me, 1, 390
Et-Fcd-CHO, 1, 396
1-Acetyl-x-methylferrocene, 1, 406
$MeCH\text{-}Fcd\text{-}CH_2O$, 1, 436
$CH_2CH_2\text{-}Fcd\text{-}CHOH$, 1, 438
$FeC_{13}H_{14}O_2$ $FcCHOCH_2CH_2O$, 1, 244
FcOOCEt, 1, 274
$FcCH_2OAc$, 1, 274
$FcCO_2Et$, 1, 299
$FcCH_2CO_2Me$, 1, 304
$FcCH_2CH_2CO_2H$, 1, 306
$[Me(MeO_2C)C_5H_3]FeC_5H_5$, 1, 339
$[Me(HO_2C)C_5H_3]Fe(C_5H_4Me)$, 1, 399, 400
1-Carboxy-x-ethylferrocene, 1, 407
$Et\text{-}Fcd\text{-}CO_2H$, 1, 390
$Me\text{-}Fcd\text{-}CO_2Me$, 1, 390
$FeC_{13}H_{14}O_3$ $FcCH(OMe)CO_2H$, 1, 304
1,1'-Bicyclopentenyl-$Fe(CO)_3$, 1, 492
$FeC_{13}H_{14}O_4S$ $Ac\text{-}Fcd\text{-}SO_3Me$, 1, 390
$FeC_{13}H_{14}O_5S$ $MeO_3S\text{-}Fcd\text{-}CO_2Me$, 1, 390
$FeC_{13}H_{14}O_7$ $(\text{-}CH{=}CHCO_2Et)_2Fe(CO)_3$, 1, 480
$FeC_{13}H_{14}S$ $FcSCH_2CH{=}CH_2$, 1, 316
$FeC_{13}H_{15}Cl$ $FcCMe_2Cl$, 1, 260
$FeC_{13}H_{15}ClO$ $FcCH_2CH(OH)CH_2Cl$, 1, 261
$FeC_{13}H_{15}F_6O_2P$ $[C_5H_5Fe(CO)_2(cyclohexene)]PF_6$, 1, 210
$FeC_{13}H_{15}N$ $FcCH_2CH_2N{=}CH_2$, 1, 327
Cyclopentadienyl(4,5,6,7-tetrahydropyrindenyl)-Fe, 1, 458
$FeC_{13}H_{15}NO$ $FcC({=}NOH)Et$, 1, 287
FcCONHEt, 1, 300
$FcCH_2CH_2NHCHO$, 1, 327
$[Et(H_2NCO)C_5H_3]FeC_5H_5$, 1, 339
$Et\text{-}Fcd\text{-}CONH_2$, 1, 396
$FeC_{13}H_{15}NO_2$ $FcCH_2OOCNHMe$, 1, 274
$FcCH_2CH(NH_2)CO_2H$, 1, 307
$FcNHCO_2Et$, 1, 321
$FeC_{13}H_{15}N_3$ $FcCMe_2N_3$, 1, 265
$FeC_{13}H_{15}N_3O$ $FcC(Me){=}NNHCONH_2$, 1, 287
$[Me(H_2NCONHN{=}CH)C_5H_3]FeC_5H_5$, 1, 336
$FeC_{13}H_{15}N_3O_2$ $HOCH_2\text{-}Fcd\text{-}CH{=}NNHCONH_2$, 1, 388
$FeC_{13}H_{15}N_3S$ $FcC(Me){=}NNHCSNH_2$, 1, 287
$FeC_{13}H_{16}$ FcPr, 1, 230
$FcCHMe_2$, 1, 230
$[Et(Me)C_5H_3]FeC_5H_5$, 1, 339
$(1,2\text{-}Me_2C_5H_3)Fe(C_5H_4Me)$, 1, 405
$(1,3\text{-}Me_2C_5H_3)Fe(C_5H_4Me)$, 1, 405
Et-Fcd-Me, 1, 390
$FeC_{13}H_{16}ClN$ $[Cl(Me_2NCH_2)C_5H_3]FeC_5H_5$, 1, 339
$Cl\text{-}Fcd\text{-}CH_2NMe_2$, 1, 396
$FeC_{13}H_{16}O$ $FcCH_2CH_2CH_2OH$, 1, 268
$FcC(OH)Me_2$, 1, 268
FcCH(OH)Et, 1, 268
$FcCH_2CH(OH)Me$, 1, 268
$FcCH_2OEt$, 1, 276
FcCH(OMe)Me, 1, 276
$Me(MeCHOH)C_5H_3FeC_5H_5$, 1, 339
$FeC_{13}H_{16}O_2$ $Et(HO_2CH_2)C_5H_3FeC_5H_5$, 1, 339, 340
$MeCH(OH)\text{-}Fcd\text{-}CH_2OH$, 1, 390
$FeC_{13}H_{16}O_3$ $(Me_2C{=}CHCH{=}CHCH{=}CMe_2)Fe(CO)_3$, 1, 482
α-Terpinene-$Fe(CO)_3$, 1, 488
$FeC_{13}H_{17}N$ $FcCH_2NMe_2$, 1, 323
$FcCH_2CH(Me)NH_2$, 1, 328
$FcCH_2CH_2NHMe$, 1, 328
$Me(H_2NCH_2CH_2)C_5H_3FeC_5H_5$, 1, 340
$FeC_{13}H_{17}NO$ $FcCH_2NHCH_2CH_2OH$, 1, 322
$FeC_{13}H_{18}N_2O_3$ $(Me_3CNC)_2Fe(CO)_3$, 1, 525
$FeC_{13}H_{23}IN_3O_2P$ $\{C_5H_5Fe(CO)_2[P(NMe_2)_3]\}I$, 1, 209
$FeC_{14}H_5F_7O_2$ $(p\text{-}CF_3C_6F_4)(C_5H_5)Fe(CO)_2$, 1, 205
$FeC_{14}H_6O_6$ $COCH{=}C(Ph)COFe(CO)_4$, 1, 735
$FeC_{14}H_7N_3O_{11}$ $(CH{=}CHCH{=}CHCO)\text{-}Fe(CO)_3\cdot\text{-}HOC_6H_2(NO_2)_3$, 1, 484
$FeC_{14}H_9BrO_5$ $(CH{=}CHCH{=}CHCO)\text{-}Fe(CO)_3\cdot\text{-}p\text{-}HOC_6H_4Br$, 1, 484
$FeC_{14}H_9ClO_4$ $(BzOC_5H_4)Fe(CO)_2Cl$
$FeC_{14}H_{10}$ Poly(ferrocenylenedivinylene), 1, 465
$FeC_{14}H_{10}Cl_2N_2$ $(PhNC)_2FeCl_2$, 1, 524
$FeC_{14}H_{10}N_4O_2$ $(PhNC)_2Fe(NO)_2$, 1, 520
$FeC_{14}H_{10}O_2$ Cyclopentadienyl(4,7-dihydro-4,7-dioxoindenyl)-Fe, 1, 455
$FeC_{14}H_{10}O_3$ $C_5H_5Fe(CO)_2Bz$, 1, 206

$FeC_{14}H_{10}O_3S$ $C_5H_5Fe(CO)_2SBz$, 1, 201
$(FeC_{14}H_{10}O_4)_n$ $[(C_2H_2)_5Fe(CO)_4]_n$, 1, 516
$FeC_{14}H_{10}O_6$ trans-$(PhCH{=}CHCO_2Me)Fe(CO)_4$, 1, 477
trans-$(p\text{-}MeOC_6H_4CH{=}CHCHO)Fe(CO)_4$, 1, 477
$(CH{=}CHCH{=}CHCO)\cdot Fe(CO)_3\cdot p\text{-}C_6H_4(OH)_2$, 1, 483
$FeC_{14}H_{11}BrO_2$ $(PhCH_2C_5H_4)Fe(CO)_2Br$, 1, 200
$FeC_{14}H_{11}F_3O_2$ $FcCOCH_2COCF_3$, 1, 261
$FeC_{14}H_{11}NOS_2$ $FcCH{=}CSC(S)NHCO$, 1, 243, 282
$FeC_{14}H_{11}NO_2$ $FcCH{=}C(CN)CO_2H$, 1, 305
$FeC_{14}H_{12}$ $FcCH{=}CHC{\equiv}CH$, 1, 250
(C_5H_5)-indenyl-Fe, 1, 454
$FeC_{14}H_{12}Cl_6O_2$ $Fcd[CH(OH)CCl_3]_2$, 1, 361
$FeC_{14}H_{12}N_2O$ $FcC{=}NC(O)NHCH{=}CH$, 1, 240
$FeC_{14}H_{12}N_2OS$ $FcCONHC{=}NCH{=}CHS$, 1, 301
$FeC_{14}H_{12}N_2S$ $FcCH{=}NC{=}NCH{=}CHS$, 1, 282
$FeC_{14}H_{12}O_2$ $(C_5H_5)Fe(CO)_2CH_2Ph$, 1, 205
Cyclopentadienyl(4,5,6,7-tetrahydro-4,7-dioxoindenyl)-Fe, 1, 456
$FeC_{14}H_{12}S$ $FcC{=}CHCH{=}CHS$, 1, 247
$FeC_{14}H_{13}ClO_4S$ 1,x-$Ac(ClO_2S)C_5H_3FeC_5H_4Ac$, 1, 408
$FeC_{14}H_{13}ClO_6S$ 1,x-$ClO_2S(MeO_2C)C_5H_3FeC_5H_4CO_2Me$, 1, 408
$FeC_{14}H_{13}F_3O$ $FcCH_2CH_2COCF_3$, 1, 287
$FeC_{14}H_{13}N$ N-Pyrryl-Fc, 1, 239
$[NCCH_2(CH_2{=}CH)C_5H_3]FeC_5H_5$, 1, 340
$FeC_{14}H_{13}N_3O$ $FcCH{=}NNHOCCH_2CN$, 1, 282
$FeC_{14}H_{14}$ $(PhCH_2)_2Fe$, 1, 193
$(p\text{-}MeC_6H_4)_2Fe$, 1, 194
$FcCH{=}CHCH{=}CH_2$, 1, 250
$FcCH(Me)C{\equiv}CH$, 1, 254
Cyclopentadienyl(4,5-dihydroindenyl)-Fe, 1, 454
Poly(ferrocenylenediethylene), 1, 465
$FeC_{14}H_{14}N_2$ 5-Methylpyrazolyl-Fc, 1, 239
$FeC_{14}H_{14}N_2O_2$ $FcCH_2CHCONHCONH$, 1, 241
$(\text{-}Fcd\text{-}CONHCH_2CH_2NHCO\text{-})_n$, 1, 469
$FeC_{14}H_{14}N_4$ Cyclopentadienyl(4H-5,6-dihydrotetrazoloazepino[7,8-a]cyclopentadienyl)-Fe, 1, 460
$FeC_{14}H_{14}O$ $FcCH{=}CHAc$, 1, 294
$CH_2\text{-}Fcd\text{-}COCHMe$, 1, 438
3-$(HCO)C_5H_3FeC_5H_4CH_2CH_2CH_2$, 1, 440
$CH_2CH_2\text{-}Fcd\text{-}CH_2CO$, 1, 443
Cyclopentadienyl(4,5,6,7-tetrahydro-4-oxoindenyl)-Fe, 1, 455
$FeC_{14}H_{14}O_2$ $(p\text{-}MeOC_6H_4)_2Fe$, 1, 194
$FcCHCH_2CH_2C(O)O$, 1, 245
$FcC(OAc){=}CH_2$, 1, 274
$FcCOCH_2Ac$, 1, 288
$FcCH{=}CHCO_2Me$, 1, 305
$(Ac_2C_5H_3)FeC_5H_5$, 1, 340
$FcdAc_2$, 1, 367
$COCH_2\text{-}Fcd\text{-}CH_2CHOH$, 1, 443
$FeC_{14}H_{14}O_3$ $FcCOCH_2CH_2CO_2H$, 1, 308
$Ac\text{-}Fcd\text{-}CO_2Me$, 1, 391
$EtCO\text{-}Fcd\text{-}CO_2H$, 1, 391
$[CH_2(CH_2)_4C{=}CCH{=}CHCH{=}CH]\text{-}Fe(CO)_3$, 1, 487
$FeC_{14}H_{14}O_4$ $FcCH_2CH(CO_2H)_2$, 1, 314
$Fcd(OAc)_2$, 1, 365
$Fcd(CO_2Me)_2$, 1, 374
$Fcd(CH_2CO_2H)_2$, 1, 375
$[HO_2C(Me)C_5H_3]_2Fe$, 1, 380
$EtO_2C\text{-}Fcd\text{-}CO_2H$, 1, 391
$AcO\text{-}Fcd\text{-}CO_2Me$, 1, 396
$FeC_{14}H_{14}O_6$ $Fcd(OCH_2CO_2H)_2$, 1, 366
$FeC_{14}H_{14}O_7S$ 1,x-$MeO_2C(HO_3S)C_5H_3FeC_5H_4CO_2Me$, 1, 408
$FeC_{14}H_{15}BrN_2O$ $[C_5H_5Fe(CO)(CNCH_2CH{=}CH_2)_2]Br$, 1, 211
$FeC_{14}H_{15}ClO_2$ $MeCH(OAc)\text{-}Fcd\text{-}Cl$, 1, 391
$FeC_{14}H_{15}N$ $FcCMe_2CN$, 1, 263
4,5-Dihydromethylpyrindenyl-FeC_5H_5 isomers, 1, 458
3H-4,5-Dihydrocyclopent[c]azepinyl-FeC_5H_5, 1, 459
$FeC_{14}H_{15}NO$ 4,5,6,7-Tetrahydro-4-hydroximinoindenyl-FeC_5H_5, 1, 456
3H-1,2,4,5-Tetrahydro-1-oxocyclopent[c]azepinyl-FeC_5H_5, 1, 459
$FeC_{14}H_{15}NO_2$ $AcNH(Ac)C_5H_3FeC_5H_5$, 1, 340
$AcNH\text{-}Fcd\text{-}Ac$, 1, 396
$FeC_{14}H_{15}NO_3$ $FcCH_2CH(NHCHO)CO_2H$, 1, 307
$Ac(MeO_2CNH)C_5H_3FeC_5H_5$, 1, 340

FeC$_{14}$H$_{15}$NO$_3$ FcCH$_2$CH(NHCHO)CO$_2$H, 1, 307
Ac(MeO$_2$CNH)C$_5$H$_3$FeC$_5$H$_5$, 1, 340
MeO$_2$CNH-Fcd-Ac, 1, 397
FeC$_{14}$H$_{16}$ FcCH$_2$CH$_2$CH=CH$_2$, 1, 250
FcC(Me)=CHMe, 1, 250
[Et(CH$_2$=CH)C$_5$H$_3$]FeC$_5$H$_5$, 1, 340
CH$_2$CH$_2$-Fcd-CH$_2$CH$_2$, 1, 443
4,5,6,7-Tetrahydroindenyl-FeC$_5$H$_5$, 1, 455
Polymers, 1, 466
FeC$_{14}$H$_{16}$N$_2$ FcCH(CN)NMe$_2$, 1, 325
FeC$_{14}$H$_{16}$N$_2$O$_2$ Fcd[C(=NOH)Me]$_2$, 1, 368
FeC$_{14}$H$_{16}$N$_2$O$_4$ Fcd(NHCO$_2$Me)$_2$, 1, 378
FeC$_{14}$H$_{16}$O FcCH(OH)CH$_2$CH=CH$_2$, 1, 269
FcC(OH)(Me)CH=CH$_2$, 1, 269
FcCH$_2$OCH$_2$CH=CH$_2$, 1, 276
FcCH$_2$CH$_2$Ac, 1, 288
FcCOPr, 1, 293
FcCOCHMe$_2$, 1, 293
[Ac(Et)C$_5$H$_3$]FeC$_5$H$_5$, 1, 341
[HCO(Me$_2$CH)C$_5$H$_3$]FeC$_5$H$_5$, 1, 341
Ac-Fcd-Et, 1, 391
[Ac(Me)C$_5$H$_3$]Fe(C$_5$H$_4$Me), 1, 400
1,x-Ac(Et)-ferrocene, 1, 407
MeCH-Fcd-CH(Me)O, 1, 436
CH$_2$CH$_2$-Fcd-CHOMe, 1, 438
4,5,6,7-Tetrahydro-4-hydroxyindenyl-FeC$_5$H$_5$, 1, 456
FeC$_{14}$H$_{16}$O$_2$ [(C$_6$H$_6$)Fe(C$_5$H$_4$Me)]OAc, 1, 218
FcCH(OAc)Me, 1, 274
FcCH$_2$CH$_2$CO$_2$Me, 1, 306
Et(HO$_2$CCH$_2$)C$_5$H$_3$FeC$_5$H$_5$, 1, 341
Et(MeO$_2$C)C$_5$H$_3$FeC$_5$H$_5$, 1, 341
HO$_2$C(Me$_2$CH)C$_5$H$_3$FeC$_5$H$_5$, 1, 341, 342
FcCH$_2$CH(Me)CO$_2$H, 1, 306
Fc(CH$_2$)$_3$CO$_2$H, 1, 308
Et-Fcd-CO$_2$Me, 1, 391
Pr-Fcd-CO$_2$H, 1, 392
Et-Fcd-OAc, 1, 397
Me$_2$CH-Fcd-CO$_2$H, 1, 397
FeC$_{14}$H$_{16}$O$_3$ MeO$_2$C-Fcd-CH(OH)Me, 1, 397
FeC$_{14}$H$_{17}$Br$_3$ [(Me$_3$C$_6$H$_3$)FeC$_5$H$_5$]Br$_3$, 1, 218
FeC$_{14}$H$_{17}$I [(Me$_3$C$_6$H$_3$)FeC$_5$H$_5$]I, 1, 218
FeC$_{14}$H$_{17}$N 4-H$_2$N-4,5,6,7-tetrahydroindenyl-FeC$_5$H$_5$, 1, 455
3H-1,2,4,5-Tetrahydrocyclopent[c]azepinyl-FeC$_5$H$_5$, 1, 459
4,5,6,7-Tetrahydromethylpyrindenyl-FeC$_5$H$_5$ isomers, 1, 458, 459
FeC$_{14}$H$_{17}$NO FcCH$_2$CH$_2$NHAc, 1, 328
[H$_2$NCO(Me$_2$CH)C$_5$H$_3$]FeC$_5$H$_5$, 1, 342
AcNH-Fcd-Et, 1, 389
Me$_2$CH-Fcd-CONH$_2$, 1, 392
FeC$_{14}$H$_{17}$NS FcCH$_2$CSNMe$_2$, 1, 304
FeC$_{14}$H$_{18}$ FcBu, 1, 231
FcCH$_2$CHMe$_2$, 1, 231
FcCHMeEt, 1, 231
FcCMe$_3$, 1, 231
Et$_2$C$_5$H$_3$FeC$_5$H$_5$, 1, 342
FcdEt$_2$, 1, 351
1,2-Et(Me)C$_5$H$_3$FeC$_5$H$_4$Me, 1, 405
1,3-Et(Me)C$_5$H$_3$FeC$_5$H$_4$Me, 1, 405
1,x-Et$_2$-ferrocene, 1, 407
(1,3-Cycloheptadiene)(1,3,5-cycloheptatriene)-Fe, 1, 513
FeC$_{14}$H$_{18}$O FcC(OH)MeEt, 1, 269
FcCH$_2$CH$_2$CH(OH)Me, 1, 269
FcC(OMe)Me$_2$, 1, 277
FcCH(OEt)Me, 1, 279
[Me(MeCHOH)C$_5$H$_3$]Fe(C$_5$H$_4$Me), 1, 400
FeC$_{14}$H$_{18}$O$_2$ (MeCHOH)$_2$C$_5$H$_3$FeC$_5$H$_5$, 1, 342
HOCH$_2$[HO(CH$_2$)$_3$]C$_5$H$_3$FeC$_5$H$_5$, 1, 342
Fcd[CH(OH)Me]$_2$, 1, 363
[MeO(Me)C$_5$H$_3$]$_2$Fe, 1, 380
FeC$_{14}$H$_{19}$N FcNEt$_2$, 1, 320
FcCH(Me)NMe$_2$, 1, 326
FcCH$_2$CH$_2$NMe$_2$, 1, 328
Me(Me$_2$NCH$_2$)C$_5$H$_3$FeC$_5$H$_5$, 1, 342
FeC$_{14}$H$_{19}$NO$_2$S FcSO$_2$NEt$_2$, 1, 317
FeC$_{14}$H$_{20}$IN [FcNEt$_2$H]I, 1, 320
[FcCH$_2$NMe$_3$]I, 1, 323
FeC$_{14}$H$_{20}$N$_2$ Fcd[CH(Me)NH$_2$]$_2$, 1, 378
FeC$_{14}$H$_{20}$N$_6$ (EtNC)$_4$Fe(CN)$_2$, 1, 525
FeC$_{14}$H$_{21}$NO$_3$S [Et$_2$NH$_2$]O$_3$SFc, 1, 318
FeC$_{14}$H$_{23}$O$_5$P (CH$_2$=CHCH=CH$_2$)Fe(CO)$_2$-[EtOCH=CHP(OEt)$_2$], 1, 501
FeC$_{15}$H$_8$F$_7$IN$_2$O$_2$ [C$_3$F$_7$Fe(CO)$_2$(BiPy)]I, 1, 195
FeC$_{15}$H$_{10}$F$_7$IN$_2$O$_2$ [C$_3$F$_7$Fe(CO)$_2$Py$_2$]I, 1, 195
FeC$_{15}$H$_{11}$ClN$_2$O [C$_5$H$_5$Fe(CO)(CNCH$_2$C$_6$H$_4$-

Cl)]CN, 1, 207
$FeC_{15}H_{12}N_2O$ [C_5H_5Fe(CO)(CNCH_2Ph)]CN, 1, 207
$FeC_{15}H_{12}N_2O_3$ FcCH=CCONHCONHCO, 1, 294
$FeC_{15}H_{12}OS$ FcCOC=CHCH=CHS, 1, 247*
$FeC_{15}H_{12}O_2$ FcCOC=CHCH=CHO, 1, 246
$FeC_{15}H_{12}O_4$ (p-AcC_6H_4CH=CHCH=CH_2)Fe-(CO)$_3$, 1, 480
$FeC_{15}H_{13}N$ 3-Pyridylferrocene, 1, 240
2-Pyridylferrocene, 1, 745
$FeC_{15}H_{13}NO$ FcCOC=CHCH=CHNH, 1, 295
$FeC_{15}H_{13}NO_2$ FcNHCOC=CHCH=CHO, 1, 320
$FeC_{15}H_{13}NS$ FcN=CHC=CHCH=CHS, 1, 248
$FeC_{15}H_{14}O$ Fc(CH=CH)$_2$CHO, 1, 284
$FeC_{15}H_{14}OS$ FcCH(OH)C=CHCH=CHS, 1, 247
$FeC_{15}H_{14}O_2$ (PhCH_2C$H_2$$C_5H_4$)Fe(CO)$_2$H, 1, 198
CH_2-Fcd-C(OAc)=CH, 1, 438
CH_2CH_2CO-Fcd-COCH_2, 1, 444
$FeC_{15}H_{14}O_3$ FcCOCH=CHCO_2Me, 1, 308
Ac-Fcd-CH=CHCO_2H, 1, 392
COCH_2-Fcd-CHCO_2Me, 1, 438
$FeC_{15}H_{14}O_6$ Dimethylferrocenetricarboxylic acid, 1, 410
$FeC_{15}H_{14}S$ FcCH_2C=CHCH=CHS, 1, 247
$FeC_{15}H_{15}NO_5$ FcCH_2C(NHCHO)(CO_2H)$_2$, 1, 314
$FeC_{15}H_{16}$ Cyclopentenyl-Fc, 1, 250
CH-Fcd-CHCH_2CH_2CH_2, 1, 439
6,7-Dihydro-4-methylindenyl-FeC_5H_5, 1, 455
$FeC_{15}H_{16}N_2O$ FcCH=NNHCOCHCH_2CH_2, 1, 282
$FeC_{15}H_{16}O$ CO-Fcd-CH_2CMe_2, 1, 439
CH_2CH_2C$H_2$$C_5H_4FeC_5H_3$Ac, 1, 440
Tetrahydro-5,5-dimethyl-4-oxopentalenyl-FeC_5H_5, 1, 454
5,6,7,8-Tetrahydro-4-oxoazulenyl-FeC_5H_5, 1, 459
$FeC_{15}H_{16}O_2$ FcCH(OH)C≡CCH(OH)Me, 1, 269
FcCOCH_2COEt, 1, 288
$FeC_{15}H_{16}O_3$ FcCOCH_2CO_2Et, 1, 306
FcCOCH_2CH_2CO_2Me, 1, 308
FcCOCH(Me)CH_2CO_2H, 1, 309
FcCOCH_2CH(Me)CO_2H, 1, 309
FcCO(CH_2)$_3$CO_2H, 1, 310
HCO(MeO_2CCH_2CH_2)C_5H_3FeC_5H_5, 1, 342
CH_2CH_2-Fcd-CHOCO_2Me, 1, 439
PrCO-Fcd-CO_2H, 1, 392
Ac-Fcd-CH_2CH_2CO_2H, 1, 397
EtCO-Fcd-CO_2Me, 1, 397
(CH_2=CHCH=CHCH=CH_2)$_2$Fe(CO)$_3$, 1, 481
$FeC_{15}H_{16}O_4$ FcCO_2CHCH_2OCH_2OCH_2, 1, 299
FcCO_2CH_2CHCH_2OCH_2O, 1, 299
FcCH_2C(Me)(CO_2H)$_2$, 1, 314
[Me(MeO_2C)C_5H_3]Fe(C_5H_4CO_2Me), 1, 400
$FeC_{15}H_{17}I$ [Tetralin-FeC_5H_5]I, 1, 218
$FeC_{15}H_{17}N$ [(M$e_3$$C_6H_3$)Fe$C_5H_5$]CN, 1, 218
$FeC_{15}H_{17}NO_2$ FcCON(CH_2CH_2)$_2$O, 1, 302
$FeC_{15}H_{17}NO_3$ FcCONHCH_2CO_2Et, 1, 301
Ac(EtO_2CNH)C_5H_3FeC_5H_5, 1, 343
EtO_2CNH-Fcd-Ac, 1, 397
$FeC_{15}H_{18}$ FcCHCH_2CMe_2, 1, 232
FcCH(CH_2)$_3$CH_2, 1, 232
CH_2-Fcd-CH_2CMe_2, 1, 439
CH_2(CH_2)$_3$-Fcd-CH_2, 1, 444
Polymers, 1, 466
$FeC_{15}H_{18}O$ FcC(OH)(CH_2)$_3$CH_2, 1, 269
FcCOMe_3, 1, 288
Ac-Fcd-CHMe_2, 1, 392
FcCOBu, 1, 293
Ac(Me_2CH)C_5H_3FeC_5H_5, 1, 343
[Et(HCO)C_5H_3]Fe(C_5H_4Et), 1, 401
CH_2CH_2-Fcd-CHOEt, 1, 439
2-(MeCHOH)C_5H_3FeC_5H_4CH_2CH_2CH_2, 1, 440
$FeC_{15}H_{18}O_2$ FcCH_2CMe_2CO_2H, 1, 307
Fc(CH_2)$_4$CO_2H, 1, 310
FcCH_2CH_2CO_2Et, 1, 306
Me_2CH(MeO_2C)C_5H_3FeC_5H_5, 1, 343
{Me[MeCH(OAc)]C_5H_3}FeC_5H_5, 1, 343
Me_2CH-Fcd-CO_2Me, 1, 392
Bu-Fcd-CO_2H, 1, 392
Et-Fcd-CH_2CH_2CO_2H, 1, 393
$FeC_{15}H_{18}O_3$ FcCOCH_2CH(OMe)$_2$, 1, 283, 288
1,1'-Bicyclohexenyl-Fe(CO)$_3$, 1, 493
$FeC_{15}H_{19}F_6O_2P$ [C_5H_5Fe(CO)$_2$(cyclooctene)]PF_6, 1, 210
$FeC_{15}H_{19}N$ FcCH_2N(CH_2)$_3$CH_2, 1, 241
[CH_2=CH(Me_2NCH_2)C_5H_3]FeC_5H_5, 1, 343
FcCH(CH_2)$_4$NH, 1, 746
$FeC_{15}H_{19}NO$ FcCOCH_2CH_2NMe_2, 1, 288
FcCH_2N(CH_2CH_2)$_2$O, 1, 325

$FeC_{15}H_{19}NO_2$ $FcCH_2NHCH_2CO_2Et$, 1, 308

$FeC_{15}H_{20}$ FcC_5H_{11}, 1, 232
- $FcCMe_2Et$, 1, 232
- (Pentamethylcyclopentadienyl)-Fe-C_5H_5, 1, 349

$FeC_{15}H_{20}IN$ 4,5,6,7-Tetrahydromethylpyrindenyl-FeC_5H_5 methiodide isomers, 1, 459

$FeC_{15}H_{20}O$ $FcCH(OH)Bu$, 1, 269
- $FcCH_2OBu$, 1, 279
- $FcCH_2OCMe_3$, 1, 279

$FeC_{15}H_{21}N$ $FcCH_2NEt_2$, 1, 325
- $Fc(CH_2)_3NMe_2$, 1, 329
- $Et(Me_2NCH_2)C_5H_3FeC_5H_5$, 1, 343

$FeC_{15}H_{21}NO$ $FcCHOHCH_2CH_2NMe_2$, 1, 269

$FeC_{15}H_{21}NS_2$ 1,2-$MeS(Me_2NCH_2)C_5H_3Fe$-C_5H_4SMe, 1, 405
- 1,3-$MeS(Me_2NCH_2)C_5H_3FeC_5H_4SMe$, 1, 405

$FeC_{15}H_{22}BrN$ $[FcCH_2NMe_2Et]Br$, 1, 324

$FeC_{15}H_{22}ClN$ $[FcCH_2NMe_2Et]Cl$, 1, 324

$FeC_{15}H_{22}IN$ $[FcCH_2NMe_2Et]I$, 1, 324
- $[FcCH(Me)NMe_3]I$, 1, 326
- $[FcCH_2CH_2NMe_3]I$, 1, 328
- $[Me(Me_3NCH_2)C_5H_3FeC_5H_5]I$, 1, 342

$FeC_{15}H_{22}N_2$ $FcCH_2NHCH_2CH_2NMe_2$, 1, 322

$FeC_{16}H_{11}N_3O_8S$ $FcSO_2C_6H_2(NO_2)_3$, 1, 317

$FeC_{16}H_{12}$ Polymer, 1, 648

$FeC_{16}H_{12}N_2O_3$ $[C_5H_5Fe(CO)(CNCH_2C_6H_4$-$CO_2H)]CN$, 1, 207

$FeC_{16}H_{12}N_3O_7$ $[FcH]OC_6H_2(NO_2)_3$, 1, 228

$FeC_{16}H_{12}O_3$ $C_5H_5Fe(CO)_2COCH{=}CHPh$, 1, 206

$FeC_{16}H_{13}Br$ FcC_6H_4Br, 1, 261

$FeC_{16}H_{13}BrO$ $FcOC_6H_4Br$, 1, 279

$FeC_{16}H_{13}Cl$ FcC_6H_4Cl, 1, 261

$FeC_{16}H_{13}F$ FcC_6H_4F, 1, 261

$FeC_{16}H_{13}I$ FcC_6H_4I, 1, 262

$FeC_{16}H_{13}NO_2$ $FcC_6H_4NO_2$, 1, 332, 333

$FeC_{16}H_{14}$ FcPh, 1, 236
- $FcCH{=}CCH{=}CHCH{=}CH$, 1, 250
- Bis(1,3a-dihydropentalenyl)-Fe, 1, 460

$FeC_{16}H_{14}N_2$ $FcN{=}NPh$, 1, 330

$FeC_{16}H_{14}N_2O$ $PhN({=}O)NFc$, 1, 331

$FeC_{16}H_{14}N_4$ $Fcd(C{=}NNHCH{=}CH)_2$, 1, 357

$FeC_{16}H_{14}N_4O_2$ $(p\text{-}MeC_6H_4NC)_2Fe(NO)_2$, 1, 525

$FeC_{16}H_{14}N_4O_4$ $(p\text{-}MeOC_6H_4NC)_2Fe(NO)_2$, 1, 525

$FeC_{16}H_{14}O$ FcC_6H_4OH, 1, 270
- FcOPh, 1, 279

$FeC_{16}H_{14}O_2$ $Fcd(COCH{=}CH_2)_2$, 1, 368
- 1,1';3,3'-Bis(α-oxotrimethylene)-ferrocene, 1, 445

$FeC_{16}H_{14}O_2S$ $FcSO_2Ph$, 1, 317

$FeC_{16}H_{14}O_3S$ FcO_3SPh, 1, 274
- $FcC_6H_4SO_3H$, 1, 318
- Ph-Fcd-SO_3H, 1, 392

$FeC_{16}H_{14}S$ $FcCH{=}CHC{=}CHCH{=}CHS$, 1, 253
- FcSPh, 1, 316

$FeC_{16}H_{15}N$ FcNHPh, 1, 320
- $FcC_6H_4NH_2$, 1, 329
- $Me(N{=}CHCH{=}CHCH{=}C)C_5H_3FeC_5H_5$, 1, 344
- Me-Fcd-$C{=}NCH{=}CHCH{=}CH$, 1, 392

$FeC_{16}H_{15}NO_2$ $FcCH{=}C(CN)CO_2Et$, 1, 305

$FeC_{16}H_{15}N_3$ $PhNHN{=}NFc$, 1, 332

$FeC_{16}H_{16}$ $FcCH_2C_5H_5$, 1, 232

$FeC_{16}H_{16}Cl_2O_2$ $Fcd(COCH_2CH_2Cl)_2$, 1, 361

$FeC_{16}H_{16}N_2$ 2-Pyridyl-$NHCH_2Fc$, 1, 322
- FcNHNHPh, 1, 331

$(FeC_{16}H_{16}N_2O_2)_n$ $\{Fcd\text{-}CON(CH_2CH_2)_2NCO\}_n$, 1, 470

$FeC_{16}H_{16}O$ 4-[3-(Cyclopentadienyl)propyl]dihydro-1(2H)-pentalenone-Fe, 1, 440*
- 1,1'-(α-Oxotrimethylene)-2,2'-trimethyleneferrocene, 1, 445
- 1,3'-(α-Oxotrimethylene)-3,3'-trimethyleneferrocene, 1, 446

$FeC_{16}H_{16}O_2$ $FcCH_2OCH_2C{=}CHCH{=}CHO$, 1, 277
- $Fc(CH{=}CH)_2CO_2Me$, 1, 310
- 3-$(HO_2CCH{=}CH)C_5H_3FeC_5H_4CH_2CH_2CH_2$, 1, 440
- CH_2-Fcd-$CH_2CH{=}COAc$, 1, 443
- $MeCHCH_2CO$-Fcd-$COCH_2$, 1, 444

$FeC_{16}H_{16}O_3$ 1,x-$Ac_2C_5H_3FeC_5H_4Ac$, 1, 408
- 3-$(HO_2CCH_2CH_2)C_5H_3FeC_5H_4C(O)CH_2CH_2$, 1, 441
- 3-$(HO_2CCH_2CH_2)C_5H_3FeC_5H_4CH_2CH_2CO$, 1, 441

$FeC_{16}H_{16}O_4$ $FcCH{=}C(CO_2Me)_2$, 1, 314
- $HO_2CCH{=}CH(HO_2CCH_2CH_2)C_5H_3FeC_5H_5$, 1, 344

FeC$_{16}$H$_{18}$ Fcd(CH$_2$CH=CH$_2$)$_2$, 1, 360
Fcd[C(Me)=CH$_2$]$_2$, 1, 360
1,1';2,2'-Bis(trimethylene)ferrocene, 1, 446
1,1';3,3'-Bis(trimethylene)ferrocene, 1, 446

FeC$_{16}$H$_{18}$N$_2$O$_2$ 5-Ethyl-3-methyl-5-hydantoinyl-Fc, 1, 239
{Fcd-CONH(CH$_2$)$_4$NHCO}$_n$, 1, 470

FeC$_{16}$H$_{18}$O Ph$_2$Fe·THF, 1, 193
FcCH$_2$CH(CH$_2$)$_3$CO, 1, 246

FeC$_{16}$H$_{18}$O$_2$ Fcd(CH$_2$CHCH$_2$O)$_2$, 1, 358
Fcd(COEt)$_2$, 1, 368
[Ac(Me)C$_5$H$_3$]$_2$Fe, 1, 380
Diacetylethylferrocene, 1, 408
3-(HO$_2$CCH$_2$CH$_2$)C$_5$H$_3$FeC$_5$H$_4$CH$_2$CH$_2$CH$_2$, 1, 441
COCH$_2$CH$_2$-Fcd-CH$_2$CH$_2$CHOH, 1, 444
Cyclopentadienyl(4-acetoxy-4,5,6,7-tetrahydroindenyl)-Fe, 1, 456

FeC$_{16}$H$_{18}$O$_3$ FcCO(CH$_2$)$_4$CO$_2$H, 1, 310
Et(HO$_2$CCH$_2$CH$_2$CO)C$_5$H$_3$FeC$_5$H$_5$, 1, 344
PrCO-Fcd-CO$_2$Me, 1, 393
Et-Fcd-COCH$_2$CH$_2$CO$_2$H, 1, 393
CH$_2$CH$_2$-Fcd-CHOCO$_2$Et, 1, 439

FeC$_{16}$H$_{18}$O$_4$ [(HO$_2$CCH$_2$CH$_2$)$_2$C$_5$H$_3$]FeC$_5$H$_5$, 1, 344
Fcd(CO$_2$Et)$_2$, 1, 374
Fcd(CH$_2$CH$_2$CO$_2$H)$_2$, 1, 375
Fcd(CH$_2$CO$_2$Me)$_2$, 1, 375
1,2-Et(MeO$_2$C)C$_5$H$_3$FeC$_5$H$_4$CO$_2$Me, 1, 405
1,3-Et(MeO$_2$C)C$_5$H$_3$FeC$_5$H$_4$CO$_2$Me, 1, 405

FeC$_{16}$H$_{18}$O$_6$ [HO$_2$CCH$_2$O(Me)C$_5$H$_3$]$_2$Fe, 1, 381

FeC$_{16}$H$_{19}$ClO [(Me$_3$C$_6$H$_3$)Fe(C$_5$H$_4$Ac)]Cl, 1, 218

FeC$_{16}$H$_{19}$NOS FcCH$_2$CSN(CH$_2$CH$_2$)$_2$O, 1, 304

FeC$_{16}$H$_{19}$NO$_2$ FcCH$_2$CON(CH$_2$CH$_2$)$_2$O, 1, 304

FeC$_{16}$H$_{20}$ cyclo-C$_6$H$_{11}$Fc, 1, 233
Me$_2$C-Fcd-CMe$_2$, 1, 434

FeC$_{16}$H$_{20}$N$_2$O$_4$ Fcd(NHCO$_2$Et)$_2$, 1, 378

FeC$_{16}$H$_{20}$O FcCOC$_5$H$_{11}$, 1, 293
FcCOCHEt$_2$, 1, 293
[Ac(Me$_3$C)C$_5$H$_3$]FeC$_5$H$_5$, 1, 344
Me$_3$C-Fcd-Ac, 1, 397
[Ac(Et)C$_5$H$_3$]Fe(C$_5$H$_4$Et), 1, 401

FeC$_{16}$H$_{20}$O$_2$ Fc(CH$_2$)$_5$CO$_2$H, 1, 310
Me$_3$C(MeO$_2$C)C$_5$H$_3$FeC$_5$H$_5$, 1, 344
Bu-Fcd-CO$_2$Me, 1, 393
[Me$_3$C(MeO$_2$C)C$_5$H$_3$FeC$_5$H$_5$, 1, 344

FeC$_{16}$H$_{21}$N FcCH$_2$N(CH$_2$)$_4$CH$_2$, 1, 241, 325
[Me$_2$NCH$_2$CH$_2$(CH$_2$=CH)C$_5$H$_3$]FeC$_5$H$_5$, 1, 345

FeC$_{16}$H$_{21}$NO FcCOCH(Me)CH$_2$NMe$_2$, 1, 289

FeC$_{16}$H$_{21}$O$_4$P (CH$_2$=CHCH=CH$_2$)Fe(CO)$_2$-[PhP(OEt)$_2$], 1, 501

FeC$_{16}$H$_{22}$ FcCH(Me)CMe$_3$, 1, 233
n-C$_6$H$_{13}$Fc, 1, 235
FcCH$_2$CHEt$_2$, 1, 235
FcdPr$_2$, 1, 352
Fcd(CHMe$_2$)$_2$, 1, 352
[Me(Et)C$_5$H$_3$]$_2$Fe, 1, 381
1,2-Et$_2$C$_5$H$_3$FeC$_5$H$_4$Et, 1, 405
1,3-Et$_2$C$_5$H$_3$FeC$_5$H$_4$Et, 1, 405
Diisopropylferrocene, 1, 407
Triethylferrocene, 1, 408
(1,5-Cyclooctadiene)(1,3,5-cyclooctatriene)-Fe, 1, 514*

FeC$_{16}$H$_{22}$BrN [FcCH$_2$NMe$_2$CH$_2$CH=CH$_2$]Br, 1, 324

FeC$_{16}$H$_{22}$IN [FcCH$_2$NMe$_2$CH$_2$CH=CH$_2$]I, 1, 324

FeC$_{16}$H$_{22}$O$_2$ FcCH$_2$CH(OEt)$_2$, 1, 283
Fcd[C(OH)Me$_2$]$_2$, 1, 364

FeC$_{16}$H$_{23}$N Et(Me$_2$NCH$_2$CH$_2$)C$_5$H$_3$FeC$_5$H$_5$, 1, 345

FeC$_{16}$H$_{24}$BrN [FcCH$_2$NMe$_2$Pr]Br, 1, 324
[FcCH$_2$NMe$_2$(CHMe$_2$)]Br, 1, 324

FeC$_{16}$H$_{24}$IN [FcNEt$_3$]I, 1, 320
[FcCH$_2$NMe$_2$Pr]I, 1, 324
[FcCH$_2$NEt$_2$Me]I, 1, 325
[Fc(CH$_2$)$_3$NMe$_3$]I, 1, 329

FeC$_{16}$H$_{24}$N$_2$ Fcd(CH$_2$NMe$_2$)$_2$, 1, 378

FeC$_{17}$H$_{10}$N$_4$O$_3$ Bicyclo[4.2.2]deca-2,4,9-triene-7,7,8,8-tetracarbonitrile-Fe(CO)$_3$, 1, 495

FeC$_{17}$H$_{12}$N$_2$O$_2$ O$_2$NC$_6$H$_4$-Fcd-CN, 1, 397

FeC$_{17}$H$_{12}$O 1,8-Dihydro-8-oxocyclopent[a]indenyl-FeC$_5$H$_5$, 1, 746

FeC$_{17}$H$_{12}$O$_3$ Phenylcyclooctatetraene-Fe(CO)$_3$, 1, 492

FeC$_{17}$H$_{13}$F$_3$ FcC$_6$H$_4$CF$_3$, 1, 262

$FeC_{17}H_{13}N$ FcC_6H_4CN, 1, 263
  $Ph(NC)C_5H_3FeC_5H_5$, 1, 345
$FeC_{17}H_{14}BrNO$ $FcCONHC_6H_4Br$, 1, 301
$FeC_{17}H_{14}F_6O_3$ 2,3,5,6-Tetramethyl-7,8-bis(trifluoromethyl)bicyclo[2.2.2]octatriene-$Fe(CO)_3$, 1, 495
$FeC_{17}H_{14}N_2O$ 3-(2-Furyl)pyrazol-5-yl-Fc, 1, 239
$FeC_{17}H_{14}N_2O_3$ $FcCOCH{=}CHC{=}CHC(NO_2){=}CHNH$, 1, 295
  $FcCOCH{=}CHC{=}CHCH{=}C(NO_2)NH$, 1, 295
$FeC_{17}H_{14}N_4O_4$ $FcCH{=}NNHC_6H_3(NO_2)_2$, 1, 282
$FeC_{17}H_{14}O$ FcBz, 1, 289
$FeC_{17}H_{14}OS$ $FcC(CHO){=}CHC{=}CHCH{=}CHS$, 1, 284
  $FcCH{=}CHCOC{=}CHCH{=}CHS$, 1, 294
  $FcCOCH{=}CHC{=}CHCH{=}CHS$, 1, 295
  FcSBz, 1, 316
$FeC_{17}H_{14}O_2$ FcOBz, 1, 274
  $FcCOC_6H_4OH$, 1, 290
  $FcCOCH{=}CHC{=}CHCH{=}CHO$, 1, 295
  $FcCO_2Ph$, 1, 299
  $FcC_6H_4CO_2H$, 1, 310, 311
$FeC_{17}H_{14}O_2S$ $FcCOC{=}CHCH{=}C(Ac)S$, 1, 247
  $FcCOCH_2COC{=}CHCH{=}CHS$, 1, 290
$FeC_{17}H_{14}O_3$ $FcC_6H_3(CO_2H)OH$, 1, 270
  $FcCOC_6H_3(OH)_2$, 1, 290
  $FcCOCH_2COC{=}CHCH{=}CHO$, 1, 290
$FeC_{17}H_{15}N$ $FcCH{=}NPh$, 1, 282
$FeC_{17}H_{15}NO$ $FcC({=}NOH)Ph$, 1, 289
  $FcCH{=}CHCOC{=}CHCH{=}CHNH$, 1, 294*
  $FcCOCH{=}CHC{=}CHCH{=}CHNH$, 1, 295
  FcCONHPh, 1, 301
  FcNHBz, 1, 320
  $Ph(H_2NCO)C_5H_3FeC_5H_5$, 1, 345
  Ph-Fcd-$CONH_2$, 1, 397
$FeC_{17}H_{15}NO_2$ $FcC_6H_3(Me)NO_2$, 1, 333
$FeC_{17}H_{15}N_3$ $FcCH(Ph)N_3$, 1, 264
$FeC_{17}H_{15}N_3O$ $FcCH{=}NNHCOC{=}CHCH{=}NCH{=}CH$, 1, 282
$FeC_{17}H_{16}$ Phenylcyclohexadienyl-(C_5H_5)-Fe, 1, 217
  $FcCH_2Ph$, 1, 233
  FcC_6H_4Me, 1, 237
  $FcC(Me){=}CCH{=}CHCH{=}CH$, 1, 251
  Me-Fcd-Ph, 1, 393
$FeC_{17}H_{16}ClN$ $FcCH_2NHC_6H_4Cl$, 1, 322
$FeC_{17}H_{16}N_2O$ FcNHCONHPh, 1, 320
$FeC_{17}H_{16}N_2O_3S$ $FcCONHNHO_2SPh$, 1, 303
$FeC_{17}H_{16}N_4O_7$ $[FcCH_2NH_3]OC_6H_2(NO_2)_3$, 1, 321
$FeC_{17}H_{16}O$ FcCH(OH)Ph, 1, 270
  $FcC_6H_4CH_2OH$, 1, 270
  $FcCH_2C_6H_4OH$, 1, 270
  FcC_6H_4OMe, 1, 277, 279
  $FcCH_2OPh$, 1, 279
  $Fc(CH{=}CH)_3CHO$, 1, 284
$FeC_{17}H_{16}OS$ $FcCH_2C{=}CHCH{=}C(Ac)S$, 1, 247
$FeC_{17}H_{16}O_2$ $FcCH(OH)C_6H_4OH$, 1, 270
$FeC_{17}H_{16}O_2S$ $FcSO_2CH_2Ph$, 1, 317
$FeC_{17}H_{16}O_4$ MeO_2CCH-Fcd-$CH{=}COAc$, 1, 439
$FeC_{17}H_{17}N$ $FcCH_2NHPh$, 1, 322
  $FcCH(Ph)NH_2$, 1, 327
$FeC_{17}H_{18}ClNO_4$ $[FcCH{=}CC(Me){=}CHC(Me){=}NH]$-$ClO_4$, 1, 241
$FeC_{17}H_{18}O$ 4-Formyl-1,1';3,3'-bis(trimethylene)ferrocene, 1, 447
$FeC_{17}H_{18}O_2$ $CH_2[CH_2(Ac)C_5H_3]_2Fe$ isomers, 1, 441
  $EtCHCH_2CO$-Fcd-$COCH_2$, 1, 444
$FeC_{17}H_{18}S$ $FcCH_2C{=}CHCH{=}C(Et)S$, 1, 248
$FeC_{17}H_{20}O$ $FcCH_2CH(CH_2)_4CO$, 1, 246
  cyclo-$C_6H_{11}COFc$, 1, 293
  $EtC_5H_3FeC_5H_3(Et)COCH_2CH_2$, 1, 442
$FeC_{17}H_{20}O_2$ $Et(HO_2CCH{=}CH)C_5H_3FeC_5H_4Et$, 1, 401
  2-$[MeCH(OAc)]C_5H_3FeC_5H_4CH_2CH_2CH_2$, 1, 441
$FeC_{17}H_{20}O_3$ $FcCH_2CH(Ac)CO_2Et$, 1, 307
$FeC_{17}H_{20}O_4$ 1,2-$Me_2CH(MeO_2C)C_5H_3Fe$-$C_5H_4CO_2Me$, 1, 405
  1,3-$Me_2CH(MeO_2C)C_5H_3FeC_5H_4CO_2Me$, 1, 405
$FeC_{17}H_{21}NO_2$ $FcCOCH_2CH_2N(CH_2CH_2)_2O$, 1, 290
$FeC_{17}H_{22}$ cyclo-$C_6H_{11}CH_2Fc$, 1, 235
  $CH_2[CH_2(Et)C_5H_3]_2Fe$ isomers, 1, 442
$FeC_{17}H_{22}N_2O_3$ $FcCH_2NHCONHCH(CHMe_2)CO_2H$, 1, 322
$FeC_{17}H_{22}O$ Me_3CCH_2-Fcd-Ac, 1, 397
  $HCO(Me_2CH)C_5H_3FeC_5H_4CHMe_2$, 1, 402
$FeC_{17}H_{22}O_2$ $FcCH_2CMe_2CO_2Et$, 1, 307
  $[Et(HO_2CCH_2CH_2)C_5H_3]FeC_5H_4Et$, 1, 402

$FeC_{17}H_{22}O_3$ $FcCH(OH)CMe_2CO_2Et$, 1, 307
1,1'-Bicycloheptenyl-$Fe(CO)_3$, 1, 493
$FeC_{17}H_{22}O_6$
$FcCH_2OCH[CH(OH)]_3CH(CH_2OH)O$, 1, 277
$FeC_{17}H_{23}NO$ $FcCOCH_2CH_2NEt_2$, 1, 289
$FeC_{17}H_{24}$ n-$C_5H_{11}CH(Me)Fc$, 1, 235
$FeC_{17}H_{24}ClNO_2$ $[FcCH_2NMe_2(CH_2CO_2Et)]$-Cl, 1, 325
$FeC_{17}H_{24}IN$ $[Me_3NCH_2CH_2(CH_2{=}CH)C_5H_3$-$FeC_5H_5]I$, 1, 345
$FeC_{17}H_{24}N_2O$ $Me(BuNHCONHCH_2)C_5H_3Fe$-C_5H_5, 1, 345
$FeC_{17}H_{24}O$ $Fc(CH_2)_5CH(OH)Me$, 1, 271
$FeC_{17}H_{26}IN$ $[FcCH_2NMe_2Bu]I$, 1, 325
$[Et(Me_3NCH_2CH_2)C_5H_3FeC_5H_5]I$, 1, 345
$FeC_{17}H_{26}N_2$ $(Me_2NCH_2)_2$Me-ferrocene, 1, 408
$FeC_{18}H_8F_{14}N_2O_2$ $(C_3F_7)_2Fe(CO)_2(BiPy)$, 1, 196
$FeC_{18}H_{10}F_{14}N_2O_2$ $(C_3F_7)_2Fe(CO)_2(Py)_2$, 1, 196
$FeC_{18}H_{12}ClNO_2$ Cl-Fcd-$NCOC_6H_4CO$, 1, 397
$FeC_{18}H_{13}Cl_2O_3P$ $FcCOC_6H_4CO_2PCl_2$, 1, 312
$FeC_{18}H_{13}NO_2$ $FcNCOC_6H_4CO$, 1, 321
$FeC_{18}H_{14}$ $FcC{\equiv}CPh$, 1, 254
Diindenyl-Fe, 1, 460
$(FeC_{18}H_{14}N_2O_2)_n$ $\leftarrow$Fcd-$CONHC_6H_4NHCO\rightarrow_n$, 1, 470
$FeC_{18}H_{14}O_2$ 3-Phthalidyl-Fc, 1, 245
$FeC_{18}H_{14}O_3$ $FcCOC_6H_4CO_2H$, 1, 311
Bz-Fcd-CO_2H, 1, 393
$PhCH_2$-cyclooctatetraene-$Fe(CO)_3$, 1, 492
$FeC_{18}H_{15}N$ N-Indolyl-Fc, 1, 239
$FeC_{18}H_{15}NO$ $FcCOCH{=}CHC{=}CHCH{=}CHCH{=}N$, 1, 295
$FcCOCH{=}CHC{=}CHCH{=}CHN{=}CH$, 1, 295
$FcCOCH{=}CHC{=}CHCH{=}NCH{=}CH$, 1, 295
$FcCH{=}CHCOC{=}NCH{=}CHCH{=}CH$, 1, 294
$FeC_{18}H_{15}NO_2$ $FcCOCH_2COC{=}CHCH{=}NCH{=}CH$, 1, 290
$FcCOCH_2COC{=}CHN{=}CHCH{=}CH$, 1, 291
$FcCOCH_2COC{=}NCH{=}CHCH{=}CH$, 1, 291
$FeC_{18}H_{15}NO_3$ (7-Oxabicyclo[2.2.1]hept-5-ene-2,3-dicarboximido)-Fc, 1, 240
Ac-Fcd-$C_6H_4NO_2$, 1, 397
1,x-Ac(p-$O_2NC_6H_4$)-ferrocene, 1, 407
$FeC_{18}H_{15}NO_4$ MeO_2C-Fcd-$C_6H_4NO_2$, 1, 397
$FeC_{18}H_{15}NS$ (Methylbenzothiazolyl)-Fc, 1, 243, 244
$FeC_{18}H_{15}N_3$ (3-Pyridylpyrazol-5-yl)-Fc, 1, 239
$FeC_{18}H_{16}$ $FcCH{=}CHPh$, 1, 251, 253
Poly(ferrocenylstyrene), 1, 466
$FeC_{18}H_{16}N_2O$ $FcCH{=}NNHBz$, 1, 282
$FcCONHN{=}CHPh$, 1, 303
$FeC_{18}H_{16}N_2O_2$ $FcCH{=}NNHCOC_6H_4OH$, 1, 282
$Fcd(COCH_2CH_2CN)_2$, 1, 362
$FeC_{18}H_{16}N_2O_3$ $[C_5H_5Fe(CO)(CNCH_2C_6H_4$-$CO_2Et)]CN$, 1, 207
$FeC_{18}H_{16}O$ $FcCOCH_2Ph$, 1, 291
$FcCH_2Bz$, 1, 291
FcC_6H_4Ac, 1, 291
$FcCOC_6H_4Me$, 1, 293
$Ac(Ph)C_5H_3FeC_5H_5$, 1, 345
Ac-Fcd-Ph, 1, 393
$FeC_{18}H_{16}O_2$ $FcCH_2OBz$, 1, 275
$FcCOCH(OH)Ph$, 1, 291
$FcCOC_6H_4OMe$ isomers, 1, 291, 293*
$FcC_6H_4CO_2Me$, 1, 310, 311
$FcCH_2C_6H_4CO_2H$, 1, 312
$PhCH_2$-Fcd-CO_2H, 1, 393
$FeC_{18}H_{16}O_3$ $FcCOC_6H_3(OH)OMe$, 1, 292
$PhCH(OH)$-Fcd-CO_2H, 1, 394
$FeC_{18}H_{16}O_6$ $Fcd(CO_2CHCHCH_2O)_2$, 1, 374
$FeC_{18}H_{16}S$ 2-Thienyl-$CH{=}CHCH{=}CHFc$, 1, 253
$FeC_{18}H_{17}N$ $FcCH{=}NC_6H_4Me$, 1, 282
$FeC_{18}H_{17}NO$ $FcCH{=}NC_6H_4OMe$, 1, 282
FcC_6H_4NHAc, 1, 329
$FeC_{18}H_{17}NO_2$ $FcCH_2O_2CC_6H_4NH_2$, 1, 275
$FcCH_2O_2CNHPh$, 1, 275
$FcNHCO_2CH_2Ph$, 1, 321
$FcC_6H_2(NO_2)Me_2$, 1, 333
$FeC_{18}H_{18}$ $FcCH_2CH_2Ph$, 1, 234
$FcCH(Me)Ph$, 1, 234
Et-Fcd-Ph, 1, 397
$FeC_{18}H_{18}ClNO_4$ $[FcCH{=}N(Me)Ph]ClO_4$, 1, 282
$FeC_{18}H_{18}IN$ $[FcCH{=}CHC{=}CHCH{=}CHCH{=}NMe]I$,

1, 243
$FeC_{18}H_{18}N_4$ Fcd[C=CHC(Me)=NNH]$_2$, 1, 357
$FeC_{18}H_{18}O$ $FcCH_2CH(OH)Ph$, 1, 271
FcC(OH)(Me)Ph, 1, 271
$FcCH(OH)CH_2Ph$, 1, 271
$FcCH_2OCH_2Ph$, 1, 277
FcCH(OMe)Ph, 1, 278
FcCH(OPh)Me, 1, 278
$FcCH_2C_6H_4OMe$, 1, 279
FcC_6H_4OEt, 1, 279
$FeC_{18}H_{18}OS$ (4,5,6,7-Tetrahydro-4-hydroxy-4-thienylindenyl)-FeC_5H_5, 1, 456
$FeC_{18}H_{18}O_2$ $FcCH(OMe)C_6H_4OH$, 1, 271
$Fc(CH{=}CH)_3CO_2Me$, 1, 311
Bis(4,5,6,7-tetrahydro-4-oxoindenyl)-Fe, 1, 461
$FeC_{18}H_{18}O_4$ FcCH=CHCH=C(CO_2Me)$_2$, 1, 315
Fcd($CHCH_2CH_2COO$)$_2$, 1, 359
Fcd[C(OAc)=CH_2]$_2$, 1, 365
Fcd($COCH_2COMe$)$_2$, 1, 369
Tetraacetylferrocene, 1, 409
$FeC_{18}H_{18}O_6$ Fcd[CO(CH_2)$_2CO_2H$]$_2$, 1, 371
$FeC_{18}H_{19}N$ $FcCH_2NMePh$, 1, 325
$FeC_{18}H_{20}N_2O_3$ $FcCH_2C(CN)(NHAc)CO_2Et$, 1, 308
$FeC_{18}H_{20}O$ Acetyl-1,1';3,3'-bis(trimethylene)ferrocene isomers, 1, 447
Acetyl-1,1';2,2'-bis(trimethylene)-ferrocene isomers, 1, 447
$FeC_{18}H_{20}O_2$ Pr$CHCH_2CO$-Fcd-$COCH_2$, 1, 444
$Me_2CHCHCH_2CO$-Fcd-$COCH_2$, 1, 444
CO(CH_2)$_3$-Fcd-(CH_2)$_3$CO, 1, 444
Bis(6,6-dimethylfulvene)-Fe(CO)$_2$, 1, 500
$FeC_{18}H_{20}O_3$ $FcCOC_6H_{10}CO_2H$, 1, 312
Triacetyl-1,1'-dimethylferrocene, 1, 410
3-(EtO_2CCH_2CO)$C_5H_3FeC_5H_4CH_2CH_2CH_2$, 1, 442
$FeC_{18}H_{20}O_5$ $HO_2C(CH_2)_3$-Fcd-CO(CH_2)$_2$-CO_2H, 1, 397
$FeC_{18}H_{21}ON$ Et($NCCH_2CH_2CO$)$C_5H_3FeC_5H_4Et$, 1, 402
$FeC_{18}H_{22}$ Fcd[C(Me)=CHMe]$_2$, 1, 360
Bis(4,5,6,7-tetrahydroindenyl)-Fe, 1, 460
$(FeC_{18}H_{22}N_2O_2)_n$ {Fcd-CONH(CH_2)$_6$NHCO}$_n$, 1, 470
$FeC_{18}H_{22}N_2O_5$ Ac(EtO_2CNH)$C_5H_3FeC_5H_4$-$NHCO_2Et$, 1, 402
$FeC_{18}H_{22}O$ cyclo-$C_6H_{11}CH_2COFc$, 1, 293
(Ethyltetrahydrooxoindenyl)-Fe(C_5H_4Et) isomers, 1, 457, 458
$FeC_{18}H_{22}O_2$ $FcCH_2C_6H_{10}CO_2H$, 1, 312
Fcd(COPr)$_2$, 1, 369
Fcd($COCHMe_2$)$_2$, 1, 369
3-($EtO_2CCH_2CH_2$)$C_5H_3FeC_5H_4CH_2CH_2CH_2$, 1, 442
$FeC_{18}H_{22}O_3$ Et($HO_2CCH_2CH_2CO$)C_5H_3Fe-C_5H_4Et, 1, 402, 403
$FeC_{18}H_{22}O_4$ $FcCH_2CH(CO_2Et)_2$, 1, 314
Fcd[(CH_2)$_3CO_2H$]$_2$, 1, 376
1,2-Me_3C(MeO_2C)$C_5H_3FeC_5H_4CO_2Me$, 1, 405
1,3-Me_3C(MeO_2C)$C_5H_3FeC_5H_4CO_2Me$, 1, 405
$FeC_{18}H_{23}NO_3S$ $FcCONHCH(CO_2Et)CH_2CH_2SMe$, 1, 302
$FeC_{18}H_{24}$ cyclo-$C_6H_{11}CH_2CH_2Fc$, 1, 235
(Ethyltetrahydroindenyl)-Fe(C_5H_4Et) isomers, 1, 457
Poly(diisobutylferrocene), 1, 466
$FeC_{18}H_{24}O$ $FcCOC_7H_{15}$, 1, 293
$FcCOCEt_3$, 1, 293
[$Me_3CCO(Pr)C_5H_3$]FeC_5H_5, 1, 345
Me_3CCH_2-Fcd-COEt, 1, 394
Me_3CCO-Fcd-Pr, 1, 397
1,2-Ac(Me_2CH)$C_5H_3FeC_5H_4CHMe_2$, 1, 405
1,3-Ac(Me_2CH)$C_5H_3FeC_5H_4CHMe_2$, 1, 405
$FeC_{18}H_{24}O_2$ Et[$HO_2C(CH_2)_3$]$C_5H_3FeC_5H_4$-Et, 1, 403
Et($MeO_2CCH_2CH_2$)$C_5H_3FeC_5H_4Et$, 1, 403
$FeC_{18}H_{25}O_4P$ (CH_2=CHCH=CH_2)Fe(CO)$_2$-[PhP(OPr)$_2$], 1, 501
$FeC_{18}H_{26}$ $FcCMe_2CH_2CMe_3$, 1, 234
n-$C_8H_{17}Fc$, 1, 235
Et_3CCH_2Fc, 1, 235
(Me_3C)$_2C_5H_3FeC_5H_5$, 1, 346

$FcdBu_2$, 1, 352
$Fcd(CH_2CHMe_2)_2$, 1, 352
$Fcd(CMe_3)_2$, 1, 353
$Fcd(CHMeEt)_2$, 1, 355
$(Et_2C_5H_3)_2Fe$, 1, 381
Me_3CCH_2-Fcd-Pr, 1, 397
Di-t-butylferrocene, 1, 407
$FeC_{18}H_{26}O_2$ Fcd[C(OH)EtMe]$_2$, 1, 364
$FeC_{18}H_{28}N_2$ Fcd[CH(Me)NMe$_2$]$_2$, 1, 378
Fcd[$(CH_2)_4NH_2$]$_2$, 1, 378
$FeC_{18}H_{28}N_2O_4S_2$ $Fcd(SO_2NEt_2)_2$, 1, 377
$FeC_{19}H_{14}N_2O_5$ $(p\text{-}MeOC_6H_4NC)_2Fe(CO)_3$, 1, 525
$FeC_{19}H_{14}O$ FcCOC≡CPh, 1, 292
$FeC_{19}H_{14}O_2$ (1-Oxo-1H-2-benzopyran-3-yl)-Fc, 1, 245
1,3-Dioxo-2-indanyl-Fc, 1, 292
$FeC_{19}H_{14}O_3$ $(PhCH{=}CH\text{-})_2Fe(CO)_3$, 1, 480
$FeC_{19}H_{15}ClO$ $FcCOCH{=}CHC_6H_4Cl$, 1, 296
$FeC_{19}H_{15}N$ α-Quinolyl-Fc, 1, 241
FcC(=CHPh)CN, 1, 263
$FeC_{19}H_{15}NO_2$ $FcCH_2NCOC_6H_4CO$, 1, 325
$FeC_{19}H_{15}NO_3$ $FcCH{=}CHCOC_6H_4NO_2$, 1, 294
$FcCOCH{=}CHC_6H_4NO_2$, 1, 296
$FeC_{19}H_{15}O_2$ (4-Oxo-2-chromanyl)-Fc, 1, 245
$FeC_{19}H_{16}$ FcCH=C=CHPh, 1, 251
$FcCH_2C{\equiv}CPh$, 1, 254
$FcC{\equiv}CCH_2Ph$, 1, 254
$FeC_{19}H_{16}N_2$ FcC=NNHC(Ph)=CH, 1, 239
$FeC_{19}H_{16}N_2O$ (1-Phenyl-5-oxopyrazol-3-yl)-Fc, 1, 239
$FeC_{19}H_{16}O$ FcC(OH)(Ph)C≡CH, 1, 271
FcCH(OH)C≡CPh, 1, 271
FcCH=C(Ph)CHO, 1, 283
FcC(CHO)=CHPh, 1, 284
FcCH=CHBz, 1, 294
FcCOCH=CHPh, 1, 294
CH_2-Fcd-COCHPh, 1, 439
$FeC_{19}H_{16}O_2$ $(PhCH_2C_5H_4)(\sigma\text{-}C_5H_5)Fe(CO)_2$ 1, 204
$(C_5H_5)Fe(CO)_2(\sigma\text{-}C_5H_4CH_2Ph)$, 1, 205
$FcCOCH_2Bz$, 1, 293
$FcCH{=}CHCOC_6H_4OH$, 1, 294
Ac-Fcd-Bz, 1, 398
$FeC_{19}H_{16}O_3$ (1-Methoxy-3-oxo-1-phthalanyl)-Fc, 1, 245
$FcCOC_6H_4OAc$, 1, 292
$FcCOC_6H_4CO_2Me$, 1, 312
$FcCOC_6H_4CH_2CO_2H$, 1, 312
$FcCOCH_2C_6H_4CO_2H$, 1, 313
Ac-Fcd-OBz, 1, 398
MeO_2C-Fcd-Bz, 1, 398
$OCH{=}CHCH{=}CCHCH_2CO$-Fcd-$COCH_2$, 1, 444
(Cyclooctatetraene dimer)-$Fe(CO)_3$, 1, 495
$FeC_{19}H_{17}NO$ FcCOCH=CHNHPh, 1, 296
$FeC_{19}H_{17}NO_2$ (5-Norbornene-2,3-dicarboximido)-Fc, 1, 240
$FeC_{19}H_{17}NO_3$ (1-Methyl-7-oxabicyclo-[2.2.1]hept-5-ene-2,3-dicarboximido)-Fc, 1, 240*
$FeC_{19}H_{18}$ $FcCH{=}CHC_6H_4Me$, 1, 251
FcC(Ph)=CHMe, 1, 252
$FeC_{19}H_{18}ClNO$ $FcCH_2N(OCCH_2Cl)Ph$, 1, 323
$FeC_{19}H_{18}N_2$ $FcC{=}NN(Ph)CH_2CH_2$, 1, 240
$FeC_{19}H_{18}N_2O$ $FcCH{=}NNHCOC_6H_4Me$, 1, 282
$FeC_{19}H_{18}O$ FcCH(OH)CH=CHPh, 1, 272
$FcC({=}CHPh)CH_2OH$, 1, 272
$Fc(CH{=}CH)_4CHO$, 1, 284
$FcCOCH_2CH_2Ph$, 1, 292
$FcCH_2CH_2Bz$, 1, 292
$PhCH_2$-Fcd-Ac, 1, 398
$FeC_{19}H_{18}O_2$ $FcCH_2CH_2OBz$, 1, 275
$FcCH_2CH(Ph)CO_2H$, 1, 306
$FcC_6H_4CO_2Et$, 1, 310, 311
$FcCH_2CH_2C_6H_4CO_2H$, 1, 313
$FcCH_2C_6H_4CH_2CO_2H$, 1, 313
$Ac(MeOC_6H_4)C_5H_3FeC_5H_5$, 1, 346
$MeOC_6H_4$-Fcd-Ac, 1, 398
Et-Fcd-OBz, 1, 398
$FeC_{19}H_{18}O_2S$ Ac-Fcd-$CH_2C{=}CHCH{=}C(Ac)S$, 1, 394
$FeC_{19}H_{18}O_3$ $FcCOC_6H_3(OMe)_2$, 1, 293
$FeC_{19}H_{19}NO$ $FcCH_2CONMePh$, 1, 304
$FeC_{19}H_{19}NO_2$ $FcC_6H_4CH_2CH(NH_2)CO_2H$, 1, 313
$FeC_{19}H_{20}$ $Fc(CH_2)_3Ph$, 1, 234
$FcCH(Me)CH_2Ph$, 1, 235
FcCH(Et)Ph, 1, 235
$FeC_{19}H_{20}N_2O$ $FcCH_2NHCONHCH_2Ph$, 1, 322
FcCH(Me)NHCONHPh, 1, 326
$Me(PhNHCONHCH_2)C_5H_3FeC_5H_5$, 1, 346

$FeC_{19}H_{20}O$ FcCH(OH)CH_2CH_2Ph, 1, 272
FcCH(CH_2OH)CH_2Ph, 1, 272
FcC(OH)(Me)CH_2Ph, 1, 272
FcC(OH)(Ph)Et, 1, 272
FcCH(OMe)CH_2Ph, 1, 279
3,4-(γ-Oxotrimethylene)-1,1′;2,2′-bis(trimethylene)ferrocene, 1, 448
4,5-(α-Oxotrimethylene)-1,1′;3,3′-bis(trimethylene)ferrocene, 1, 449
3,3′-(α-Oxotrimethylene)-1,1′;2,2′-bis(trimethylene)ferrocene, 1,451*
2,2′-(α-Oxotrimethylene)-1,1′;3,3′-bis(trimethylene)ferrocene, 1, 451
4,4′-(α-Oxotrimethylene)-1,1′;3,3′-bis(trimethylene)ferrocene, 1, 451
4,4′-(α-Oxotrimethylene)-1,1′;2,2′-bis(trimethylene)ferrocene, 1, 452
$FeC_{19}H_{20}O_2$ FcCH(OMe)C_6H_4OMe, 1, 279
FcCH(OEt)C_6H_4OH, 1, 272
x-(2-Carboxyvinyl)-1,1′;3,3′-bis-(trimethylene)ferrocene, 1, 447
$FeC_{19}H_{20}O_3$ 4-(2-Carboxyethyl)-3,3′-(γ-oxotrimethylene)-1,1′-trimethylene-ferrocene, 1, 448*
(Cyclooctatriene dimer)-Fe$(CO)_3$, 1, 495
$FeC_{19}H_{20}O_6$ $HO_2C(CH_2)_2$CO-Fcd-CO$(CH_2)_3$-CO_2H, 1, 396
$FeC_{19}H_{21}N$ Ph(Me_2NCH_2)$C_5H_3FeC_5H_5$, 1, 346
Me_2NCH_2-Fcd-Ph, 1, 394
$FeC_{19}H_{21}NO_2S$ FcN(Et)$O_2SC_6H_4$Me, 1, 321
$FeC_{19}H_{22}$ $CH_2[CH_2(3\text{-}HO_2CCH_2CH_2)C_5H_3]_2$-Fe, 1, 442
1,1′;3,3′;4,5-Tris(trimethylene)-ferrocene, 1, 449
1,1′;2,2′;3,3′-Tris(trimethylene)-ferrocene, 1, 452
1,1′;2,2′;4,4′-Tris(trimethylene)-ferrocene, 1, 452
$FeC_{19}H_{22}ClNO_4$
[FcCH=CC(Me)=C(Et)C(Me)=NH]ClO_4, 1, 242
$FeC_{19}H_{22}O_2$ $Me_2CHCH_2CHCH_2$CO-Fcd-COCH_2, 1, 444
x-(2-Carboxyethyl)-1,1′;3,3′-bis-(trimethylene)ferrocene, 1, 448
$FeC_{19}H_{23}NO_5$ $FcCH_2$C(NHCHO)$(CO_2Et)_2$, 1, 314
$FeC_{19}H_{24}N_2O_4$ $FcCONHCH_2$CONHCH(CH_2CH-Me_2)CO_2H, 1, 301
$FeC_{19}H_{24}O_2$ $CO(CH_2)_3$-Fcd-$(CH_2)_4$CHOH, 1, 445
$FeC_{19}H_{24}O_3$ 1,x-Et(CHO)$C_5H_3FeC_5H_3$(CH_2-CH_2CO_2Me)Et-1,x, 1, 409
$FeC_{19}H_{24}O_4$ $FcCH_2$C(Me)$(CO_2Et)_2$, 1, 314
$HO_2C(CH_2)_4$-Fcd-$(CH_2)_3CO_2$H, 1, 399
$FeC_{19}H_{25}NO_3$ FcCONHCH(CH_2CHMe_2)CO_2Et, 1, 301
$FeC_{19}H_{26}$ cyclo-$C_6H_{11}CH_2$CH(Me)Fc, 1, 235
$FeC_{19}H_{26}O$ FcCOCH(Bu)$CHMe_2$, 1, 293
Me_2CHCO(Me_3CCH_2)$C_5H_3FeC_5H_5$, 1, 346
Me_3CCH_2-Fcd-$COCHMe_2$, 1, 398
Me_2CHCH_2-Fcd-$COCMe_3$, 1, 398
$FeC_{19}H_{26}O_2$ Et($EtO_2CCH_2CH_2$)C_5H_3Fe-C_5H_4Et, 1, 403
$FeC_{19}H_{28}$ $FcCH_2$CH(Bu)$CHMe_2$, 1, 235
[Me_2CHCH_2(Me_3CCH_2)C_5H_3]FeC_5H_5, 1, 346
Me_2CHCH_2-Fcd-CH_2CMe_3, 1, 398
$FeC_{20}H_{10}Br_2O_4$ Bis(p-bromophenyl)cyclopentadienone-Fe$(CO)_3$, 1, 484
$FeC_{20}H_{10}O_6$ COC(Ph)=C(Ph)COFe$(CO)_4$, 1, 735
$FeC_{20}H_{12}O_4$ [PhC=CHCH=C(Ph)CO]-Fe$(CO)_3$, 1, 484
[CH=C(Ph)C(Ph)=CHCO]-Fe$(CO)_3$, 1, 485
$FeC_{20}H_{14}$ FcC≡CC≡CPh, 1, 255
(8H-Cyclopent[a]acenaphthylenyl)-cyclopentadienyl-Fe, 1, 460
Polymer, 1, 468
$FeC_{20}H_{14}O_2$ [(1,3-Dioxo-2-indanylidene)methyl]-Fc, 1, 246
FcCH=$CCOC_6H_4$CO, 1, 252, 294
$FeC_{20}H_{14}O_2S_2$ Fcd(COC=CHCH=CHS$)_2$, 1, 359
$FeC_{20}H_{14}O_3$ FcCH=CC_6H_4C(O)OCO, 1, 246, 294
$FeC_{20}H_{15}ClO_2$ ($Ph_2CHC_5H_4$)Fe$(CO)_2$Cl, 1, 200
$FeC_{20}H_{15}IO_2$ ($Ph_2CHC_5H_4$)Fe$(CO)_2$I, 1, 200
$FeC_{20}H_{15}NO$ FcCOCH=CHC_6H_4CN, 1, 296

$FeC_{20}H_{15}NO_2$ FcC=NC(=CHPh)C(O)O, 1, 240
 $FcCH=CC_6H_4CONHCO$, 1, 241, 295
 FcCH=CN=C(Ph)OCO, 1, 242
$FeC_{20}H_{16}$ Naphthyl-Fc, 1, 237
 Bis(azulene)-Fe, 1, 514
$FeC_{20}H_{16}N_2$ Fcd(C=CHCH=CHCH=N)$_2$, 1, 357
 Fcd(C=CHN=CHCH=CH)$_2$, 1, 357
$FeC_{20}H_{16}N_2O$ $FcN=NC_{10}H_6OH$, 1, 330
$FeC_{20}H_{16}O$ β-Naphthyl-OFc, 1, 279
$FeC_{20}H_{16}O_2$ $FcC=C(Ph)CH_2C(O)O$, 1, 245
 FcC=CHCH(Ph)C(O)O, 1, 246
$FeC_{20}H_{16}O_3$ [(1,3,3a,4,7,7a-Hexahydro-1,3-dioxo-4,7-methanoisobenzofuran-8-ylidene)methyl]-Fc, 1, 246
 $FcCOCH=CHC_6H_3OCH_2O$, 1, 297
$FeC_{20}H_{16}O_4$ (1-Acetoxy-3-oxo-1-phthalanyl)-Fc, 1, 245
$FeC_{20}H_{17}NO_2$ p-$O_2NC_6H_4CH=CHCH=CHFc$, 1, 253
 FcCH(Me)$NCOC_6H_4CO$, 1, 327
 Et-Fcd-$NCOC_6H_4CO$, 1, 394
$FeC_{20}H_{17}NO_3$ FcCH=C(NHCOPh)CO_2H, 1, 305
$FeC_{20}H_{17}NO_6$ 1,x-MeO_2C(p-$O_2NC_6H_4$)C_5H_3-$FeC_5H_4CO_2Me$, 1, 408
$FeC_{20}H_{18}$ FcCH=CHCH=CHPh, 1, 252, 253
 (4,5-Dihydro-7-phenylindenyl)-Fe-C_5H_5, 1, 455
 Bis(1-methylindenyl)-Fe, 1, 461
 Bis(3a,4-dihydroazulenyl)-Fe, 1, 462
$FeC_{20}H_{18}IN$ α-Quinolyl-Fc methiodide, 1, 241
$FeC_{20}H_{18}N_2O_2$ $FcCH_2NCONHCH(Ph)CO$, 1, 242
$FeC_{20}H_{18}O$ (4,5,6,7-Tetrahydro-4-oxo-5-phenylindenyl)-FeC_5H_5, 1, 457
 (4,5,6,7-Tetrahydro-4-oxo-6-phenylindenyl)-FeC_5H_5, 1, 457
$FeC_{20}H_{18}O_2$ FcCH(OH)C≡CCH(OH)Ph, 1, 273*
 $FcCOCH=CHC_6H_4OMe$, 1, 296
$FeC_{20}H_{18}O_2S_2$ Fcd[CH(OH)C=CHCH=CHS]$_2$, 1, 259
$FeC_{20}H_{18}O_3$ (1-Ethoxy-3-oxo-1-phthalanyl)-Fc, 1, 245
 FcCOCH(Ph)CH_2CO_2H, 1, 309
 $FcCOCH_2CH(Ph)CO_2H$, 1, 309
$FeC_{20}H_{18}O_6$ Fcd($COCH=CHCO_2Me$)$_2$, 1, 369
$FeC_{20}H_{18}O_{14}$ ($MeO_2CC≡CCO_2Me$)$_3$Fe(CO)$_2$, 1, 518
$FeC_{20}H_{18}S$ Fc(CH=CH)$_3$C=CHCH=CHS, 1, 253
$FeC_{20}H_{18}S_2$ Fcd($CH_2C=CHCH=CHS$)$_2$, 1, 359
$FeC_{20}H_{19}NO_3$ FcCONHCH(CH_2Ph)CO_2H, 1, 302
 $FcCH_2CH(NHCOPh)CO_2H$, 1, 307
 $FcC_6H_4CH_2CH(NHCHO)CO_2H$, 1, 313
$FeC_{20}H_{20}$ (4,5,6,7-Tetrahydro-4-oxo-phenylindenyl)-FeC_5H_5, 1, 457
$FeC_{20}H_{20}N_2O_2S$ CH_2CH_2-Fcd-C=$NNHO_2S$-C_6H_4Me-p, 1, 439
$FeC_{20}H_{20}N_4O_7$ (4,5,6,7-Tetrahydro-4-methylpyrindenyl)-FeC_5H_5 picrate, 1, 459
 (4,5,6,7-Tetrahydro-6-methylpyrindenyl)-FeC_5H_5 picrate, 1, 459
$FeC_{20}H_{20}O$ $FcCOCH_2C_6H_3Me_2$, 1, 293
 (4,5,6,7-Tetrahydro-4-endo-hydroxy-4-phenylindenyl)-FeC_5H_5, 1, 457
$FeC_{20}H_{20}O_2$ $FcCH_2CH(Ph)OAc$, 1, 275
 $FcCH_2CH(Ph)CH_2CO_2H$, 1, 309
 $FcCH_2CH_2CH(Ph)CO_2H$, 1, 310
 Fc(CH=CH)$_4CO_2Me$, 1, 313
$FeC_{20}H_{20}O_3$ [(2,4-Diethylphenyl)cycloheptatriene]-Fe(CO)$_3$, 1, 490
$FeC_{20}H_{20}O_4$ Fc(CH=CH)$_2$CH=C(CO_2Me)$_2$, 1, 315
$FeC_{20}H_{21}F_3O$ 2-(4,4,4-Trifluoro-3-oxobutyl)-1,1′;3,3′-bis(trimethylene)-ferrocene, 1, 448
$FeC_{20}H_{21}N$ $FcCH=NC_6H_2Me_3$, 1, 282
$FeC_{20}H_{21}N_3O_5$ [5,6-Di(methoxycarbonyl)-3a,4,5,6,7,7a-hexahydro-4,7-epoxy-1H-benzotriazol-1-yl]-Fc, 1, 240
$FeC_{20}H_{22}$ Fc(CH_2)$_4$Ph, 1, 235
 3,4-$Me_2C_6H_3CH_2CH_2Fc$, 1, 236
 Fcd($C=CHCH_2CH_2CH_2$)$_2$, 1, 360
 Bis(tricyclo[5.2.1.0^{2,6}]deca-2,5-dienyl)-Fe, 1, 461
$FeC_{20}H_{22}ClNO_6$
 [FcCH=CC(Me)=C(CO_2Et)C(Me)=NH]ClO_4, 1, 242
$FeC_{20}H_{22}N_4$ Fcd[C=NNHC(Et)=CH]$_2$, 1, 357

$FeC_{20}H_{22}O$ FcC(OH)(Ph)Pr, 1, 273
FcCH(OCHMe$_2$)Ph, 1, 279
4′-Formyl-1,1′;3,3′;4,5-tris(trimethylene)ferrocene, 1, 449
2-Formyl-1,1′;3,3′;4,5-tris(trimethylene)ferrocene, 1, 449*
2-Formyl-1,1′;3,3′;4,4′-tris(trimethylene)ferrocene, 1, 452
$FeC_{20}H_{22}O_2$ FcCH(OCHMe$_2$)C$_6$H$_4$OH, 1, 273
$FeC_{20}H_{22}O_4$ Fcd[C(OOCEt)=CH$_2$]$_2$, 1, 365
Fcd(COCH$_2$COEt)$_2$, 1, 369
$FeC_{20}H_{22}O_6$ Fcd(COCH$_2$CO$_2$Et)$_2$, 1, 371
Fcd[CO(CH$_2$)$_2$CO$_2$Me]$_2$, 1, 371
Fcd[CO(CH$_2$)$_3$CO$_2$H]$_2$, 1, 372
$FeC_{20}H_{23}N$ FcCH(CH$_2$Ph)NMe$_2$, 1, 327
FcCH$_2$CH(Ph)NMe$_2$, 1, 328
Me$_2$NCH$_2$(MeC$_6$H$_4$)C$_5$H$_3$FeC$_5$H$_5$, 1, 346
Me$_2$NCH$_2$-Fcd-C$_6$H$_4$Me, 1, 394
$FeC_{20}H_{24}ClN$ [FcCH$_2$NMe$_2$(CH$_2$Ph)]Cl, 1, 325
$FeC_{20}H_{24}IN$ [FcCH$_2$NMe$_2$(CH$_2$Ph)]I, 1, 325
$FeC_{20}H_{24}O$ 4,4′-Diethyl-1,1′-(α-oxotrimethylene)-2,2′-trimethyleneferrocene, 1, 448
$FeC_{20}H_{24}OS$ BuCH(OH)-Fcd-CH$_2$C=CHCH=CHS, 1, 394
$FeC_{20}H_{24}O_3$ 3,4-Et(HO$_2$CCH$_2$CH$_2$)-C$_5$H$_2$FeC$_5$H$_3$(3-Et)CH$_2$CH$_2$CO, 1, 443
$FeC_{20}H_{24}O_4$ 1,x-Et(HO$_2$CCH=CH)C$_5$H$_3$Fe-C$_5$H$_3$(CH$_2$CH$_2$CO$_2$H)Et-1,x, 1, 409
$FeC_{20}H_{25}NO_3$ FcCH$_2$CH(NHCOC$_6$H$_{11}$)CO$_2$H, 1, 308
$FeC_{20}H_{26}$ 4,4′-Diethyl-1,1′;2,2′-bis-(trimethylene)ferrocene, 1, 448
Bis(3a,4,5,6,7,8-hexahydroazulenyl)-Fe, 1, 462
(3a,4,5,6,7,8-Hexahydroazulene)(1,2,3,3a-tetrahydroazulene)-Fe, 1, 514
$FeC_{20}H_{26}O_2$ Fcd(COBu)$_2$, 1, 369
Fcd(COCMe$_3$)$_2$, 1, 369
CO(CH$_2$)$_4$-Fcd-(CH$_2$)$_4$CHOH, 1, 445
$FeC_{20}H_{26}O_4$ Fcd(CH$_2$CH$_2$CO$_2$Et)$_2$, 1, 376
Fcd[(CH$_2$)$_3$CO$_2$Me]$_2$, 1, 376
Fcd[(CH$_2$)$_4$CO$_2$H]$_2$, 1, 376
[Et(HO$_2$CCH$_2$CH$_2$)C$_5$H$_3$]$_2$Fe, 1, 381
$FeC_{20}H_{28}$ Me(Me$_2$CH)C-Fcd-C(CHMe$_2$)Me, 1, 434
$FeC_{20}H_{28}N_2$ Fcd[CH(CH$_2$)$_4$NH]$_2$, 1, 357
$FeC_{20}H_{28}N_2O_2$ Fcd(COCH$_2$CH$_2$NMe$_2$)$_2$, 1, 370
$FeC_{20}H_{28}O$ FcCOC$_9$H$_{19}$, 1, 293
[Me$_3$CCH$_2$(Me$_3$CCO)C$_5$H$_3$]FeC$_5$H$_5$, 1, 346
Me$_3$CCH$_2$-Fcd-COCMe$_3$, 1, 398
$FeC_{20}H_{29}N_2P$ FcP[N(CH$_2$)$_4$CH$_2$]$_2$, 1, 334
$FeC_{20}H_{29}O_4P$ (CH$_2$=CHCH=CH$_2$)Fe(CO)$_2$-[PhP(OBu)$_2$], 1, 501
$FeC_{20}H_{30}$ n-C$_{10}$H$_{21}$Fc, 1, 236
(Me$_3$CCH$_2$)$_2$C$_5$H$_3$FeC$_5$H$_5$, 1, 347
(EtMe$_2$C)$_2$C$_5$H$_3$FeC$_5$H$_5$, 1, 347
Fcd(C$_5$H$_{11}$)$_2$, 1, 353
Fcd(CMe$_2$Et)$_2$, 1, 353, 355*
Fcd(CH$_2$CMe$_3$)$_2$, 1, 353
Fcd(CH$_2$CH$_2$CHMe$_2$)$_2$, 1, 355
Fcd(CHEt$_2$)$_2$, 1, 355
Di-t-pentylferrocene, 1, 407
$FeC_{20}H_{30}INO$ [Et(Me$_3$NCH$_2$CH$_2$CO)C$_5$H$_3$Fe-C$_5$H$_4$Et]I, 1, 403, 404
$FeC_{20}H_{30}O$ FcCH$_2$OC$_9$H$_{19}$, 1, 279
$FeC_{21}H_{12}Cl_2O_3$ (p-ClC$_6$H$_4$)$_2$C=CCH=CHCH=CH]-Fe(CO)$_3$, 1, 487
$FeC_{21}H_{14}O_3$ [Ph$_2$C=CCH=CHCH=CH]-Fe(CO)$_3$, 1, 487
$FeC_{21}H_{15}ClO_3$ [(Ph$_2$CHC$_5$H$_4$)Fe(CO)$_3$]Cl, 1, 209
$FeC_{21}H_{15}Cl_4$ [PhC=C(Ph)C(Ph)]FeCl$_4$, 1, 196
$FeC_{21}H_{15}F_6O_3P$ [(Ph$_2$CHC$_5$H$_4$)Fe(CO)$_3$]PF$_6$, 1, 209
$FeC_{21}H_{16}OS$ FcCH=CHCO-thianaphthene, 1, 294
$FeC_{21}H_{16}O_2$ FcC$_{10}$H$_6$CO$_2$H, 1, 313
$FeC_{21}H_{17}ClO_2$ o-ClC$_6$H$_4$CHCH$_2$CO-Fcd-COCH$_2$, 1, 444
$FeC_{21}H_{17}N$ FcCH=NC$_{10}$H$_7$, 1, 282
$FeC_{21}H_{17}NO$ FcCONHC$_{10}$H$_7$, 1, 302
$FeC_{21}H_{17}NO_2$ FcCH=CC$_6$H$_4$CON(Me)CO, 1, 242
FcCH=CCON(Me)COC$_6$H$_4$, 1, 294
$FeC_{21}H_{17}NO_4$ Ac-Fcd-COCH=CHC$_6$H$_4$NO$_2$, 1, 398
m-O$_2$NC$_6$H$_4$CHCH$_2$CO-Fcd-COCH$_2$, 1, 444
$FeC_{21}H_{18}$ FcC$_5$H$_4$Ph, 1, 234
α-Naphthyl-CH$_2$Fc, 1, 236

$FeC_{21}H_{18}O$ FcCOCH=CHCH=CHPh, 1, 296
$FeC_{21}H_{18}O_2$ PhCH=CHCO-Fcd-Ac, 1, 395
PhCHCH$_2$CO-Fcd-COCH$_2$, 1, 444
$FeC_{21}H_{18}O_3$ [1-(1,3,3a,4,7,7a-Hexahydro-1,3-dioxo-4,7-methanoisobenzofuran-8-ylidene)ethyl]-Fc, 1, 247
$FeC_{21}H_{19}NO_2$ [(3a,4,7,7a-Tetrahydro-2-methyl-1,3-dioxo-4,7-methanoisoindolin-8-ylidene)methyl]-Fc, 1, 243
$FeC_{21}H_{20}INS$ (2-Benzothiazolylvinyl)-Fc ethiodide, 1, 244
$FeC_{21}H_{20}O_3S$ Ac[SC(Ac)=CHCH=CCH$_2$]-C$_5$H$_3$FeC$_5$H$_4$Ac, 1, 404
$FeC_{21}H_{22}$ 1-Cyclopentyl-x-phenylferrocene, 1, 407
$FeC_{21}H_{24}BrNO$ [FcCH$_2$NMe$_2$(CH$_2$Bz)]Br, 1, 325
$FeC_{21}H_{24}N_4O_7$ [FcCH$_2$NMe$_2$Et]OC$_6$H$_2$(NO$_2$)$_3$ 1, 324
[Fc(CH$_2$)$_3$NHMe$_2$]OC$_6$H$_2$(NO$_2$)$_3$, 1, 329
$FeC_{21}H_{24}O$ 4'-Acetyl-1,1';3,3';4,5-tris(trimethylene)ferrocene, 1, 450
2-Acetyl-1,1';3,3';4,4'-tris(trimethylene)ferrocene, 1, 453
$FeC_{21}H_{24}O_6$ MeO$_2$C(CH$_2$)$_2$CO-Fcd-CO-(CH$_2$)$_3$CO$_2$Me, 1, 399
$FeC_{21}H_{30}Br_3O_{12}$ Fe[BrC(CO$_2$Et)$_2$]$_3$, 1, 193
$FeC_{21}H_{32}O$ FcCH$_2$OC$_{10}$H$_{21}$, 1, 279*
$FeC_{22}H_{16}Br_2$ Fcd(C$_6$H$_4$Br)$_2$, 1, 362
$FeC_{22}H_{16}Cl_2$ Fcd(C$_6$H$_4$Cl)$_2$, 1, 361
$FeC_{22}H_{16}N_2O_4$ Fcd(C$_6$H$_4$NO$_2$)$_2$, 1, 379
$FeC_{22}H_{16}O_4$ [MeC=C(Ph)C(Ph)=C(Me)CO]-Fe(CO)$_3$, 1, 485
[PhC=C(Me)C(Me)=C(Ph)CO]-Fe(CO)$_3$, 1, 485
$FeC_{22}H_{17}BrN_2$ FcC$_6$H$_4$N=NC$_6$H$_4$Br, 1, 331
$FeC_{22}H_{17}ClN_2$ FcC$_6$H$_4$N=NC$_6$H$_4$Cl, 1, 331
$FeC_{22}H_{17}IN_2$ FcC$_6$H$_4$N=NC$_6$H$_4$I, 1, 331
$FeC_{22}H_{17}NO$ 2-(FcCOCH=CH)-Quinoline, 1, 296
$FeC_{22}H_{17}N_3O_2$ FcC$_6$H$_4$N=NC$_6$H$_4$NO$_2$, 1, 331
$FeC_{22}H_{18}$ FcC$_6$H$_4$Ph, 1, 238
Ph$_2$C$_5$H$_3$FeC$_5$H$_5$, 1, 347
FcdPh$_2$, 1, 356
$FeC_{22}H_{18}N_2$ FcC$_6$H$_4$N=NPh, 1, 330
$FeC_{22}H_{18}N_4$ Fcd(N=NPh)$_2$, 1, 379
$FeC_{22}H_{18}O$ FcOC$_6$H$_4$Ph, 1, 279
$FeC_{22}H_{18}O_4$ (1-Benzhydrylcyclopentadienylcarbonyl)dicarbonylmethoxy-Fe, 1, 500
$FeC_{22}H_{18}O_6S_2$ Fcd(O$_3$SPh)$_2$, 1, 365
$FeC_{22}H_{19}BrN_2O$ [C$_5$H$_5$Fe(CO)(CNCH$_2$Ph)$_2$]-Br, 1, 211
$FeC_{22}H_{19}N$ FcNPh$_2$, 1, 321
$FeC_{22}H_{19}NO_2$ FcCH=CCON(Et)COC$_6$H$_4$, 1, 294
$FeC_{22}H_{19}OP$ FcP(O)Ph$_2$, 1, 334
$FeC_{22}H_{18}O_4$ 3,4-OCH$_2$OC$_6$H$_4$-CHCH$_2$CO-Fcd-COCH$_2$, 1, 444
$FeC_{22}H_{19}P$ FcPPh$_2$, 1, 334
$FeC_{22}H_{20}$ Fc(CH=CH)$_3$Ph, 1, 253
$FeC_{22}H_{20}N_2$ Fcd(C$_6$H$_4$NH$_2$)$_2$, 1, 379
$FeC_{22}H_{20}N_6$ Fcd(N=NNHPh)$_2$, 1, 379
$FeC_{22}H_{20}O$ FcCHCH$_2$C(Ph)=CHC(O)CH$_2$, 1, 246*
$FeC_{22}H_{20}O_2$ PhCH$_2$CHCH$_2$CO-Fcd-COCH$_2$, 1, 444
$FeC_{22}H_{21}NO_2$ (7-Isopropylidene-5-norbornene-2,3-dicarboximido)-Fc, 1, 241
[(2-Ethyl-3a,4,7,7a-tetrahydro-1,3-dioxo-4,7-methanoisoindolin-8-ylidene)methyl]-Fc, 1, 243
[1-(3a,4,7,7a-Tetrahydro-2-methyl-1,3-dioxo-4,7-methanoisoindolin-8-ylidene)ethyl]-Fc, 1, 243
$FeC_{22}H_{22}$ Bis(ethylindenyl)-Fe, 1, 461
$FeC_{22}H_{22}N_4$ Fcd(NHNHPh)$_2$, 1, 379
$FeC_{22}H_{22}O$ FcCOCH=CHC$_6$H$_2$Me$_3$, 1, 296
$FeC_{22}H_{22}O_2$ Fc(CH=CH)$_5$CO$_2$Me, 1, 313
$FeC_{22}H_{22}O_4$ Fc(CH=CH)$_3$CH=C(CO$_2$Me)$_2$, 1, 315
$FeC_{22}H_{23}ClO$ Me$_3$CCH$_2$-Fcd-COC$_6$H$_4$Cl, 1, 398
$FeC_{22}H_{23}N$ p-(Me$_2$N)C$_6$H$_4$CH=CHCH=CHFc, 1, 253
$FeC_{22}H_{23}NO_3$ FcCONHCH(CH$_2$Ph)CO$_2$Et, 1, 302
$FeC_{22}H_{24}N_2O_3$ FcCH$_2$NHCONHCH(Ph)CO$_2$Et, 1, 323

$FeC_{22}H_{24}N_4O_7$ $[FcCH_2NMe_2CH_2CH{=}CH_2]\text{-}OC_6H_2(NO_2)_3$, 1, 324
$FeC_{22}H_{24}O$ $FcCOCH_2CH_2C_6H_2Me_3$, 1, 292
Me_3CCH_2-Fcd-Bz, 1, 398
4,5-(α-Oxotrimethylene)-1,1′;3,3′;4′,5′-tris(trimethylene)ferrocene, 1, 450*
2,2′-(α-Oxotrimethylene)-1,1′;3,3′;4,4′-tris(trimethylene)ferrocene, 1, 453
$FeC_{22}H_{24}O_2$ 2-(2-Carboxyvinyl)-1,1′;3,3′;4,4′-tris(trimethylene)ferrocene, 1, 453
$FeC_{22}H_{25}Cl$ Me_3CCH_2-Fcd-$CH_2C_6H_4Cl$, 1, 398
$FeC_{22}H_{26}$ Fcd$[\overline{C{=}CH(CH_2)_3C}H_2]_2$, 1, 360
1,1′;3,3′;4,5;4′,5′-Tetrakis(trimethylene)ferrocene, 1, 450
1,1′;2,2′;3,3′;4,4′-Tetrakis(trimethylene)ferrocene, 1, 453
$FeC_{22}H_{26}N_4O_7$ $[Et(Me_2HNCH_2CH_2)C_5H_3Fe\text{-}C_5H_5]OC_6H_2(NO_2)_3$, 1, 345
$FeC_{22}H_{26}O$ $FcC(OH)(Ph)C_5H_{11}$, 1, 273
$FeC_{22}H_{26}O_2$ 4′-(2-Carboxyethyl)-1,1′;3,3′;4,5-tris(trimethylene)ferrocene, 1, 450
2-(2-Carboxyethyl)-1,1′;3,3′;4,4′-tris(trimethylene)ferrocene, 1, 453
$FeC_{22}H_{26}O_6$ Fcd$[CO(CH_2)_2CO_2Et]_2$, 1, 372
Fcd$[CO(CH_2)_3CO_2Me]_2$, 1, 372
$FeC_{22}H_{27}BrN_7O$ $[(CH_2{=}CHCH_2NC)_5Fe\text{-}(CN)_2]Br\cdot H_2O$, 1, 526
$FeC_{22}H_{28}N_2O_2S_2$ Fcd$[CH_2CSN(CH_2CH_2)_2O]_2$ 1, 375
$FeC_{22}H_{30}$ $(cyclo\text{-}C_6H_{11})_2C_5H_3FeC_5H_5$, 1, 347
$(cyclo\text{-}C_6H_{11})_2$Fcd, 1, 353
$FeC_{22}H_{30}O$ $(Me_3C_6H_2)_2Fe\cdot THF$, 1, 194
$[cyclo\text{-}C_6H_{11}CO(Me_3CCH_2)C_5H_3]FeC_5H_5$, 1, 347
$cyclo\text{-}C_6H_{11}CO$-Fcd-CH_2CMe_3, 1, 395
$FeC_{22}H_{30}O_2$ Fcd$(COC_5H_{11})_2$, 1, 372
Fcd$(COCH_2CMe_3)_2$, 1, 372
$FeC_{22}H_{30}O_4$ Fcd$[(CH_2)_3CO_2Et]_2$, 1, 376
Fcd$[(CH_2)_4CO_2Me]_2$, 1, 376
$FeC_{22}H_{32}$ $(cyclo\text{-}C_6H_{11}CH_2(Me_3CCH_2)C_5H_3]\text{-}FeC_5H_5$, 1, 347
Me_3CCH_2-Fcd-$CH_2C_6H_{11}$, 1, 398
$Me(Me_2CHCH_2)\overline{C\text{-Fcd-}C}(CH_2CHMe_2)Me$, 1, 435
$FeC_{22}H_{32}O$ $FcCOC_{11}H_{23}$, 1, 293
$FeC_{22}H_{34}$ $C_{12}H_{25}Fc$, 1, 236
Fcd$(C_6H_{13})_2$, 1, 355
Fcd$(CMeEt_2)_2$, 1, 355
Fcd$(CPrMe_2)_2$, 1, 355
Fcd$(CH_2CH_2CMe_3)_2$, 1, 355
$(Et_3C_5H_2)_2Fe$, 1, 384
1,3-$(Me_3C)_2C_5H_3FeC_5H_4CMe_3$, 1, 406
$FeC_{22}H_{36}BrN$ $[FcCH_2NMe_2C_9H_{19}]Br$, 1, 325
$FeC_{22}H_{36}IN$ $[FcCH_2NMe_2C_9H_{19}]I$, 1, 325
$FeC_{22}H_{36}N_2$ Fcd$(CHMeCMe_2CH_2NH_2)_2$, 1, 379
Fcd$(CMe_2CH_2CHMeNH_2)_2$, 1, 379
$FeC_{23}H_{16}O$ (1,8-Dihydro-8-oxocyclopent[a]indenyl)(phenylcyclopentadienyl)-Fe, 1, 746
$FeC_{23}H_{17}F_3N_2$ $FcC_6H_4N{=}NC_6H_4CF_3$, 1, 331
$FeC_{23}H_{17}NOS$ Phenothiazin-10-yl-COFc, 1, 302
$FeC_{23}H_{17}NO_2S$ $PhSO_2C_6H_4$-Fcd-CN, 1, 397
$FeC_{23}H_{19}N$ $FcC_6H_4N{=}CHPh$, 1, 329
$FeC_{23}H_{19}NO$ $FcCONHC_6H_4Ph$, 1, 302
$FcCONPh_2$, 1, 302
FcNPhBz, 1, 320
FcC_6H_4NHBz, 1, 329, 330
$FeC_{23}H_{19}NO_3$ $[MeO(4\text{-}O_2NC_6H_4)C_6H_3]Fc$, 1, 278
$[(MeOC_6H_4)(O_2NC_6H_4)C_5H_3]FeC_5H_5$, 1, 347
$MeOC_6H_4$-Fcd-$C_6H_4NO_2$, 1, 395
$FeC_{23}H_{19}N_3$ $FcCH{=}NC_6H_4N_2Ph$, 1, 282
$FeC_{23}H_{20}BrO_2P$ $[\pi\text{-}CH_2{=}CHCH_2\text{-}Fe(CO)_2\text{-}(PPh_3)]Br$, 1, 472
$FeC_{23}H_{20}ClO_2P$ $[\pi\text{-}CH_2{=}CHCH_2\text{-}Fe(CO)_2\text{-}(PPh_3)]Cl$, 1, 472
$FeC_{23}H_{20}IO_2P$ $[\pi\text{-}CH_2{=}CHCH_2\text{-}Fe(CO)_2\text{-}(PPh_3)]I$, 1, 472
$FeC_{23}H_{20}N_2$ $FcC_6H_4N{=}NC_6H_4Me$, 1, 331
$Fe_{23}H_{20}N_2O$ $FcCH_2OC_6H_4N{=}NPh$, 1, 279
$FcC_6H_4N{=}NC_6H_4OMe$, 1, 331
$FeC_{23}H_{20}O$ $FcCPh_2OH$, 1, 273
$FeC_{23}H_{21}NO_2$ [(2-Allyl-3a,4,7,7a-tetra-

hydro-1,3-dioxo-4,7-methanoisoindolin-8-ylidene)methyl]-Fc, 1, 243
$FeC_{23}H_{21}P$ $FcCH_2PPh_2$, 1, 334
$FeC_{23}H_{22}IP$ $[FcPPh_2Me]I$, 1, 334
$FeC_{23}H_{22}O_3$ 1,3-$Ac_2C_5H_3FeC_5H_4CH_2C_6H_4$-Ac-p, 1, 405
$FeC_{23}H_{23}NO_2$ [1-(2-Ethyl-3a,4,7,7a-tetrahydro-1,3-dioxo-4,7-methanoisoindolin-8-ylidene)ethyl]-Fc, 1, 243
[(3a,4,7,7a-Tetrahydro-2-isopropyl-1,3-dioxo-4,7-methanoisoindolin-8-ylidene)methyl]-Fc, 1, 243
$FeC_{23}H_{24}O_2$ 4-Phenylcyclohexyl-O_2CFc, 1, 299
$FeC_{23}H_{25}Cl_2N_3O$ 2,4-Me[$(ClCH_2CH_2)_2N$]-C_6H_3CH=NNHCOFc, 1, 303
$FeC_{23}H_{26}N_2O_3$ Et($O_2NC_6H_4NHCOCH_2CH_2$)-$C_5H_3FeC_5H_4Et$, 1, 404
$FeC_{23}H_{26}O$ [$Me_3CCH_2(PhCH_2CO)C_5H_3$]Fe-$C_5H_5$, 1, 347
[$PhCH_2CH_2(Me_3CCO)C_5H_3$]$FeC_5H_5$, 1, 348
Me_3CCH_2-Fcd-$COCH_2Ph$, 1, 398
Me_3CCO-Fcd-CH_2CH_2Ph, 1, 398
$FeC_{23}H_{26}O_2$ 2,x'-Diacetyl-1,1';3,3';4,5-tris(trimethylene)ferrocene, 1, 450
$FeC_{23}H_{28}$ Me_3CCH_2-Fcd-CH_2CH_2Ph, 1, 398
$FeC_{23}H_{36}O_3$ [PrCH=C(Bu)CH=CHC(Bu)=CH-Pr]$Fe(CO)_3$, 1, 482*
$FeC_{24}H_{15}F_7IO_3P$ [$C_3F_7Fe(CO)_3(PPh_3)$]I, 1, 195
$FeC_{24}H_{15}NO_4$ Triphenyl-2-cyclopropen-1-ylcarbonyl-$Fe(CO)_2(NO)$, 1, 194
$FeC_{24}H_{16}Cl_2O_2$ Fcd$(COC_6H_4Cl)_2$, 1, 362
$FeC_{24}H_{16}F_2O_2$ Fcd$(COC_6H_4F)_2$, 1, 362
$FeC_{24}H_{18}Br_2O$ 1,2-Ac(p-BrC_6H_4)C_5H_3-$FeC_5H_4C_6H_4Br$-p, 1, 406
1,3-Ac(p-BrC_6H_4)$C_5H_3FeC_5H_4C_6H_4Br$-p, 1, 406
$FeC_{24}H_{18}O_2$ $FcdBz_2$, 1, 370
[1-Carboxy-2(hydroxydiphenylmethyl)-cyclopentadienyl]cyclopentadienyliron γ-lactone, 1, 454
$FeC_{24}H_{18}O_4$ Fcd(COCH=CHC=CHCH=CHO)$_2$, 1, 359
Fcd$(OBz)_2$, 1, 365
Fcd$(C_6H_4CO_2H)_2$, 1, 376
$FeC_{24}H_{18}O_6$ [$HO(HO_2C)C_6H_3$]$_2$Fcd, 1, 364
$FeC_{24}H_{20}N_2O_2$ Fcd$(CONHPh)_2$, 1, 374
Fcd[C(=NOH)Ph]$_2$, 1, 370
$FeC_{24}H_{20}O$ $PhCH_2$-Fcd-Bz, 1, 398
Ph-Fcd-C_6H_4Ac, 1, 398
Ac(Ph)$C_5H_3FeC_5H_4Ph$, 1, 404
PhCH-Fcd-CH(Ph)O, 1, 436
$FeC_{24}H_{20}O_2$ $FcCO_2CHPh_2$, 1, 300
$FeC_{24}H_{20}O_3$ $HO_2C(Ph_2COH)C_5H_3FeC_5H_5$, 1, 348
[2-p-Tolyl-1,1'-bicyclohepta-2,4,6-trienyl]-$Fe(CO)_3$, 1, 490*
$FeC_{24}H_{21}NO$ FcCH(Ph)NHBz, 1, 327
$FeC_{24}H_{22}$ $FcCH_2CHPh_2$, 1, 236
Fc$(CH=CH)_4$Ph, 1, 253
Fcd[C(Me)=CCH=CHCH=CH]$_2$, 1, 360
Dicycloheptatrienyl-Fcd, 1, 353
Fcd$(CH_2Ph)_2$, 1, 354
Fcd$(C_6H_4Me)_2$, 1, 356
1,3-Et(Ph)$C_5H_3FeC_5H_4Ph$, 1, 406
Bis(allylindenyl)-Fe, 1, 461
$FeC_{24}H_{22}N_2O_3S$ $FcC_6H_4NHO_2SC_6H_4NHAc$, 1, 330
$FeC_{24}H_{22}O$ FcC(Ph)(C_6H_4Me)OH, 1, 273
Me$(Ph_2COH)C_5H_3FeC_5H_5$, 1, 348
$FeC_{24}H_{22}O_2$ $(MeOC_6H_4)_2C_5H_3FeC_5H_5$, 1, 348
$HOCH_2(Ph_2COH)C_5H_3FeC_5H_5$, 1, 348
Fcd[CH(OH)Ph]$_2$, 1, 364
Fcd$(C_6H_4OMe)_2$, 1, 366
1,x-(p-$MeOC_6H_4$)-ferrocene, 1, 407
$FeC_{24}H_{22}O_2S_2$ Fcd[CH_2C=CHCH=C(Ac)S]$_2$, 1, 359
$FeC_{24}H_{22}O_6S_2$ [$PhSO_3(Me)C_5H_3$]$_2$Fe, 1, 381
$FeC_{24}H_{23}NO_2$ [1-(2-Allyl-3a,4,7,7a-tetrahydro-1,3-dioxo-4,7-methanoisoindolin-8-ylidene)ethyl]-Fc, 1, 243
$FeC_{24}H_{24}IP$ [$FcCH_2PMePh_2$]I, 1, 334
$FeC_{24}H_{24}O_4$ Fc$(CH=CH)_4$CH=C$(CO_2Me)_2$, 1, 315
$FeC_{24}H_{25}NO_2$ [(2-Butyl-3a,4,7,7a-tetrahydro-1,3-dioxo-4,7-methanoisoindolin-8-ylidene)methyl]-Fc, 1, 243

[1-(3a,4,7,7a-Tetrahydro-2-isopropyl-1,3-dioxo-4,7-methanoisoindolin-8-ylidene)ethyl]-Fc, 1, 243

$FeC_{24}H_{26}O$ Me_3CCH_2-Fcd-COCH=CHPh, 1, 398

$FeC_{24}H_{26}O_4$ $FcCH_2C(Ph)(CO_2Et)_2$, 1, 314

$FeC_{24}H_{28}N_2O$ $FcCH_2N(Ph)OCCH_2N(CH_2)_4CH_2$ 1, 327

$FeC_{24}H_{28}O_2$ Bis(6,6-pentamethylenefulvene)-$Fe(CO)_2$, 1, 500

$FeC_{24}H_{30}O_2$ $Fcd(COC_6H_{11})_2$, 1, 370

$FeC_{24}H_{32}N_2O_2$ $FcCON(C_6H_{11})CONHC_6H_{11}$, 1, 303

$FeC_{24}H_{34}$ Bis(1-methylcyclohexyl)-Fcd, 1, 354

Bis(cyclohexylmethyl)-Fcd, 1, 355

$FeC_{24}H_{36}O$ Me_3CCH_2-Fcd-$COCH_2CH(Me)CH_2$-CMe_3, 1, 399

$FeC_{24}H_{38}$ $Me_3CCH_2(Me_3CCH_2CHMeCH_2CH_2)$-$C_5H_3FeC_5H_5$, 1, 348

$Me(Et_2NCH_2)C$-Fcd-$C(CH_2NEt_2)Me$, 1, 435

$FeC_{25}H_{17}Br_2N_5OS_5$ $[(CH{=}CHSCH{=}CNC)_5Fe]$-$Br_2 \cdot H_2O$, 1, 526

$FeC_{25}H_{17}O_6P$ $(OCOCH{=}CHCO)Fe(CO)_3(PPh_3)$ 1, 500

$(OCOCH{=}CHCO)Fe(CO)_3(PPh_3)$, 1, 500

$FeC_{25}H_{19}O_3P$ $(CH{=}CHCH{=}CHCO)Fe(CO)_2$-$(PPh_3)$, 1, 513

$FeC_{25}H_{20}Br_3O_2P$ $[C_5H_5Fe(CO)_2(PPh_3)]Br_3$ 1, 209

$FeC_{25}H_{20}ClO_2P$ $[C_5H_5Fe(CO)_2(PPh_3)]Cl$, 1, 209

$FeC_{25}H_{21}I_2N_3O_4$ $[(p\text{-}MeOC_6H_4NC)_3Fe$-$(CO)]I_2$, 1, 525

$FeC_{25}H_{21}NO_2$ Ac-Fcd-$CONPh_2$, 1, 395

$FeC_{25}H_{21}O_2P$ $(CH{=}CHCH{=}CHCH_2)Fe(CO)_2$-$(PPh_3)$, 1, 501

$FeC_{25}H_{22}$ $FcCHCH_2CPh_2$, 1, 234

$FeC_{25}H_{22}N_2O_2$ $FcCH{=}NNHOCC(OH)Ph_2$, 1, 282

$FeC_{25}H_{22}O_5$ $BzCH_2COCH_2CO$-Fcd-$COCH_2Ac$, 1, 395

$FeC_{25}H_{24}O_3$ $FcC{=}CHCOCH(CO_2Et)CH(Ph)CH_2$ 1, 246

$FeC_{25}H_{27}NO_2$ [1-(2-Butyl-3a,4,7,7a-tetrahydro-1,3-dioxo-4,7-methanoisoindolin-8-ylidene)ethyl]-Fc, 1, 243

$FeC_{25}H_{27}NO_5$ $FcC_6H_4CH_2C(NHCHO)(CO_2Et)_2$, 1, 315

$FeC_{25}H_{30}O_{10}$ $FcCH_2OCH[CH(OAc)]_3CH(CH_2OAc)O$, 1, 278

$FeC_{25}H_{31}NO_3$ $[(Ph_2)Fe(C_5H_4C_8H_{17})]NO_3$, 1, 218

$FeC_{26}H_{10}N_8$ $Fe[C(CN)_2C(Ph){=}C(CN)_2]_2$, 1, 197

$FeC_{26}H_{18}O_6$ $Fcd(COC_6H_4CO_2H)_2$, 1, 371

$FeC_{26}H_{19}NO_2$ $FcCH{=}CC_6H_4CON(Ph)CO$, 1, 242, 294

$FeC_{26}H_{20}$ (2-Anthrylvinyl)-Fc, 1, 252

$FeC_{26}H_{20}N_2S_2$ Bis(2-methyl-5-benzothiazolyl)-Fcd, 1, 357

Bis(2-methyl-6-benzothiazolyl)-Fcd, 1, 358

$FeC_{26}H_{21}NO_2$ [(3a,4,7,7a-Tetrahydro-1,3-dioxo-2-phenyl-4,7-methanoisoindolin-8-ylidene)methyl]-Fc, 1, 243

$FeC_{26}H_{22}$ $FcCH{=}CHC_6H_4CH{=}CHPh$, 1, 252

$FeC_{26}H_{22}N_4O_6$ $Fcd[CON(Me)C_6H_4NO_2]_2$, 1, 375

$FeC_{26}H_{22}O_2$ $(AcC_6H_4)_2C_5H_3FeC_5H_5$, 1, 348, 349

$Fcd(COCH_2Ph)_2$, 1, 370

$Fcd(C_6H_4Ac)_2$, 1, 372

$FeC_{26}H_{22}O_3$ Ac-Fcd-CO_2CHPh_2, 1, 399

$FeC_{26}H_{22}O_4$ $[BzO(Me)C_5H_3]_2Fe$, 1, 382

$FeC_{26}H_{24}$ $Fc(CH{=}CH)_5Ph$, 1, 253

MePhC-Fcd-CPhMe, 1, 435

$FeC_{26}H_{24}N_2OS_3$ {3-Ethyl-2-[(3-ethyl-4-oxo-2-thioxothiazolidin-5-yliden)ethylidene]benzothiazolinyl}-Fc, 1, 244

$FeC_{26}H_{24}N_2O_4$ $Fcd(NHCO_2CH_2Ph)_2$, 1, 378

$FeC_{26}H_{24}O$ $FcCOCH(CH_2Ph)_2$, 1, 292

$FeC_{26}H_{24}O_2$ Et-Fcd-CO_2CHPh_2, 1, 399

$FeC_{26}H_{24}O_3$ 1,2-Ac(p-$MeOC_6H_4$)C_5H_3Fe-$C_5H_4C_6H_4OMe$-p, 1, 406

1,3-Ac(p-$MeOC_6H_4$)$C_5H_3FeC_5H_4C_6H_4OMe$-p, 1, 406

$FeC_{26}H_{26}$ $Fcd(CHMePh)_2$, 1, 355

$Fcd(CH_2CH_2Ph)_2$, 1, 355

$FeC_{26}H_{26}N_2$ Me(m-$H_2NC_6H_4$)C-Fcd-C(C_6H_4-NH_2-m)Me, 1, 435

$FeC_{26}H_{26}N_4O_2$ Fcd[CON(Me)$C_6H_4NH_2$]$_2$, 1, 375

$FeC_{26}H_{26}O_2$ Fcd[CH(OH)CH_2Ph]$_2$, 1, 364

$FeC_{26}H_{28}O_5$ FcCOCH[CH(CO_2Et)$_2CH_2$Ph, 1, 315

FcCOCH(CH_2CO_2Et)CH(Ph)CO_2Et, 1, 315

$FeC_{26}H_{30}$ Bis(butylindenyl)-Fe, 1, 461

$FeC_{26}H_{32}O_{10}$ FcCH_2CH_2O-CH[CH(OAc)]$_3$CH(CH_2OAc)O, 1, 278

$FeC_{26}H_{38}O_2$ Fcd(COC_7H_{15})$_2$, 1, 370

$FeC_{26}H_{40}$ BuEtCHCH-Fcd-CHCHEtBu, 1, 435

$FeC_{26}H_{40}O$ $FcCOC_{15}H_{31}$, 1, 293

$FeC_{26}H_{42}$ $FcC_{16}H_{33}$, 1, 234

[Me(CH_2)$_4$$CMe_2$]$_2C_5H_3FeC_5H_5$, 1, 349

Fcd(C_8H_{17})$_2$, 1, 355

Fcd($CMe_2CH_2CMe_3$)$_2$, 1, 355

Fcd(CMeEtBu)$_2$, 1, 355

[(Me_3C)$_2C_5H_3$]$_2$Fe, 1, 382

(Et_4C_5H)$_2$Fe, 1, 384

(Me_3C)$_4$-ferrocene, 1, 409

$FeC_{27}H_{20}N_2O$ [C_5H_5Fe(CO)($CNCPh_3$)]CN, 1, 207

$FeC_{27}H_{21}NO$ FcCON(Ph)$C_{10}H_7$, 1, 303

$FeC_{27}H_{21}O_3$ [(PhCH=CH-)$_2$]$_{1.5}$Fe(CO)$_3$, 1, 480

$FeC_{27}H_{21}O_3P$ (CH=CHCH=CHCH=CHCO)Fe-(CO)$_2$(PPh_3), 1, 501

$FeC_{27}H_{23}NO_2$ [1-(3a,4,7,7a-tetrahydro-1,3-dioxo-2-phenyl-4,7-methanoisoindolin-8-ylidene)ethyl]-Fc, 1, 243

$FeC_{27}H_{23}O_3P$ (CH_2CH_2CH=CHCH=CHCO)Fe-(CO)$_2$(PPh_3), 1, 501

$FeC_{28}H_{15}Br_3O_4$ (p-BrC_6H_4C≡CH)$_3$Fe(CO)$_4$, 1, 518

$FeC_{28}H_{18}O_4$ 1,4,6-Triphenylcycloheptadienone-Fe(CO)$_3$, 1, 490

(PhC≡CH)$_3$Fe(CO)$_4$, 1, 518

$FeC_{28}H_{20}N_2$ Di-α-quinolyl-Fcd, 1, 357

$FeC_{28}H_{22}$ Ph_3-ferrocene, 1, 409

$FeC_{28}H_{22}N_4$ Fcd[C=NNHC(Ph)=CH]$_2$, 1, 357

$FeC_{28}H_{22}O_2$ Fcd(COCH=CHPh)$_2$, 1, 372

$FeC_{28}H_{22}O_3$ $FcOC_6H_4OC_6H_4OPh$, 1, 279

$FeC_{28}H_{22}O_4$ Fcd[C(OBz)=CH_2]$_2$, 1, 365

Fcd($COCH_2Bz$)$_2$, 1, 371

$FeC_{28}H_{22}O_6$ Fcd($COC_6H_4CO_2Me$)$_2$, 1, 372

$FeC_{28}H_{24}N_2O_6$ [Me($H_2NOCC_6H_4CO_2$)C_5H_3]$_2$-Fe, 1, 382

$FeC_{28}H_{26}O_2$ Fcd($COCH_2CH_2Ph$)$_2$, 1, 372

$FeC_{28}H_{30}$ Fcd(CMe_2Ph)$_2$, 1, 354

Fcd[CH_2CH(Ph)Me]$_2$, 1, 354

$FeC_{28}H_{40}$ (C_6H_{11})$_2C_5H_3FeC_5H_4C_6H_{11}$, 1, 404

$FeC_{28}H_{42}O_2$ [Me_3C(Me_3CCO)C_5H_3]$_2$Fe, 1, 282

$FeC_{28}H_{46}$ Fcd($CH_2CH_2CHMeCH_2CMe_3$)$_2$, 1, 355

[Me_3C(Me_3CCH_2)C_5H_3]$_2$Fe, 1, 383

$FeC_{29}H_{15}F_{12}O_3P$ [CF_3C=C(CF_3)C(CF_3)=C(CF_3)CO]Fe(CO)$_2$-(PPh_3), 1, 501

$FeC_{29}H_{22}N_2O_5$ 1,x-(p-$O_2NC_6H_4$)$_2C_5H_3$Fe-$C_5H_4C_6H_4$OMe-p, 1, 409

$FeC_{29}H_{24}$ $FcCPh_3$, 1, 235

$FeC_{29}H_{24}O$ FcC_6H_4C(OH)Ph_2, 1, 273

$FeC_{29}H_{24}O_2S$ $FcSO_2CPh_3$, 1, 317

$FeC_{29}H_{25}P$ FcCH=PPh_3, 1, 334

$FeC_{29}H_{26}IP$ [$FcCH_2PPh_3$]I, 1, 334

$FeC_{29}H_{47}NO$ $FcCONHC_{18}H_{37}$, 1, 303

$FeC_{30}H_{20}O_3$ (Ph_2C=C=CPh_2)Fe(CO)$_3$, 1, 482

$FeC_{30}H_{22}$ Bis(phenylindenyl)-Fe, 1, 461

$FeC_{30}H_{22}N_4O_2$ Fcd(N=N$C_{10}H_6$OH)$_2$, 1, 379

$FeC_{30}H_{23}NO_2$ Bz-Fcd-$CONPh_2$, 1, 395

$FeC_{30}H_{24}O_2$ $FcC_6H_4CO_2CHPh_2$, 1, 311

$FeC_{30}H_{25}NO_2$ PhCH(OH)-Fcd-$CONPh_2$, 1, 395

$FeC_{30}H_{28}$ Fc(CH=CH)$_7$Ph, 1, 253

$FeC_{30}H_{28}ClP$ [FcCH(Me)PPh_3]Cl, 1, 335

$FeC_{30}H_{34}O_2$ Fcd[CH($OCHMe_2$)Ph]$_2$, 1, 366

$FeC_{30}H_{46}O_2$ Fcd(COC_9H_{19})$_2$, 1, 372

$FeC_{30}H_{50}$ Fcd($C_{10}H_{21}$), 1, 355

[($EtMe_2C$)$_2C_5H_3$]$_2$Fe, 1, 383

Decaethylferrocene, 1, 384

Tetrakis-t-pentylferrocene, 1, 410

Bu_5-ferrocene, 1, 410

$FeC_{31}H_{20}O_3$ (Tetraphenylcyclobutadiene)-Fe(CO)$_3$, 1, 482

$FeC_{31}H_{20}O_5S$
[$PhC=C(Ph)C(Ph)=C(Ph)SO_2$]-$Fe(CO)_3$, 1, 497
$FeC_{31}H_{22}O_3$ [$PhCH=C(Ph)$-]$_2Fe(CO)_3$, 1, 480
$FeC_{31}H_{22}O_6$ (o-$HO_2CC_6H_4$)$_3$-ferrocene, 1, 409
$FeC_{31}H_{24}F_7IO_2P_2$ [$C_3F_7Fe(CO)_2(Ph_2PCH_2CH_2PPh_2)$]I, 1, 195
$FeC_{31}H_{24}O_3$ Bz-Fcd-CO_2CHPh_2, 1, 399
$FeC_{31}H_{25}NO_5$ $Ph_2COH(O_2NC_6H_4CH_2O_2C)$-$C_5H_3FeC_5H_5$, 1, 349
$FeC_{31}H_{26}O_2$ $PhCH_2$-Fcd-CO_2CHPh_2, 1, 399
$FeC_{31}H_{26}O_3$ $PhCH(OH)$-Fcd-CO_2CHPh_2, 1, 395, 399
$FeC_{32}H_{16}Cl_4O_4$ (p-ClC_6H_4)$_4$-cyclopentadienone-$Fe(CO)_3$, 1, 485
$FeC_{32}H_{18}Cl_2O_4$ 2,5,3,4-Ph_2(p-ClC_6H_4)$_2$-cyclopentadienone-$Fe(CO)_3$, 1, 486
$FeC_{32}H_{18}O_4$ 1,3-Ph_2-2H-cyclopenta[1]-phenanthrenone-$Fe(CO)_3$, 1, 493
$FeC_{32}H_{20}O_4$ Ph_4-cyclopentadienone-$Fe(CO)_3$, 1, 485
$FeC_{32}H_{24}F_8O_2P_2$ [$CF_2CF_2CF_2CF_2Fe(CO)_2$-($Ph_2PCH_2CH_2PPh_2$), 1, 196
$FeC_{32}H_{26}$ Bis(benzylindenyl)-Fe, 1, 461
Bis(3a,4-dihydro-4-phenylazulenyl)-Fe, 1, 462
$FeC_{32}H_{26}O_6$ Fcd-bis(7-ethylidene-5-norbornene-2,3-dicarboxylic anhydride), 1, 359
$FeC_{32}H_{30}Cl_{10}P_2$ $(C_6Cl_5)_2Fe(PEt_2Ph)_2$, 1, 194
$FeC_{33}H_{46}O_5$ Acetylergosterol-$Fe(CO)_3$, 1, 493
$FeC_{34}H_{18}N_8$ $[FcH]^+[(NC)_2\bar{C}C_6H_4\dot{C}(CN)_2\cdot$-$(NC)_2C=C_6H_4=C(CN)_2]$, 1, 228
$FeC_{34}H_{24}N_{12}O_{13}$ $(O_2NC_6H_4CH_2NC)_2$-$[(O_2N)_2C_6H_3CH_2NC]_2Fe(CN)_2\cdot H_2O$, 1, 526
$FeC_{34}H_{26}$ $Ph_4C_5HFeC_5H_5$, 1, 349
$(Ph_2C_5H_3)_2Fe$, 1, 383
$FeC_{34}H_{26}O_2$ $Fcd(C_6H_4OPh)_2$, 1, 366
$FeC_{34}H_{27}N_7O_2$ $(O_2NC_6H_4CH_2NC)(PhCH_2NC)_3Fe(CN)_2$, 1, 526
$FeC_{34}H_{28}N_6$ $(PhCH_2NC)_4Fe(CN)_2$, 1, 520
$FeC_{34}H_{30}N_6O_{16}S_5$ (p-$HO_3SC_6H_4CH_2NC)_4Fe$-$(CN)_2\cdot H_2SO_4$, 1, 525
$FeC_{34}H_{32}N_2O_4$ Fcd-bis(7-ethylidene-N-methyl-5-norbornene-2,3-dicarboximide), 1, 358
$FeC_{34}H_{38}$ Fcd[$C(Ph)CH_2CH_2CH_2CH_2CH_2$]$_2$, 1, 355
$FeC_{34}H_{54}O_2$ $Fcd(COC_{11}H_{23})_2$, 1, 372
$FeC_{34}H_{58}$ $(C_{12}H_{25})_2C_5H_3FeC_5H_5$, 1, 349
$Fcd(C_{12}H_{25})_2$, 1, 355
$(C_8H_{17})_2C_5H_3FeC_5H_4C_8H_{17}$, 1, 404
$FeC_{35}H_{24}O_3$ $(PhC\equiv CH)_4Fe(CO)_3$, 1, 517, 518
$FeC_{36}H_{20}O_4$ [$Ph_2(PhC\equiv C)_2$-cyclopentadienone]-$Fe(CO)_3$ isomers, 1, 486
$FeC_{36}H_{22}N_8$ $[FcdMe_2]^+[(NC)_2CC_6H_4C$-$(CN)_2]_2^{\cdot}$, 1, 351
$FeC_{36}H_{28}N_2O_2$ $Fcd(CONPh_2)_2$, 1, 375
$FeC_{36}H_{29}OP$ $FcC(Bz)=PPh_3$, 1, 335
$FeC_{36}H_{30}$ $Fcd(CHPh_2)_2$, 1, 354
$FeC_{36}H_{30}IP$ $[FcCH(Bz)PPh_3]I$, 1, 335
$FeC_{36}H_{30}O_2$ $(Ph_2COH)_2C_5H_3FeC_5H_5$, 1, 349
$Fcd[C(OH)Ph_2]_2$, 1, 364
$FeC_{36}H_{36}N_2O_4$ Fcd-bis(7-ethylidene-N-ethyl-5-norbornene-2,3-dicarboximide), 1, 358
$FeC_{36}H_{58}O_2$ $Fcd(COC_{12}H_{25})_2$, 1, 372
$FeC_{36}H_{62}$ $Fcd(C_{13}H_{27})_2$, 1, 355
$FeC_{37}H_{25}O_3P$ Pentaphenylphosphole-$Fe(CO)_3$, 1, 496
$FeC_{37}H_{25}O_3PS$ (Pentaphenylphosphole sulfide)-$Fe(CO)_3$, 1, 496
$FeC_{37}H_{25}O_4P$ (Pentaphenylphosphole oxide)-$Fe(CO)_3$, 1, 496
$FeC_{38}H_{25}O_4P$ Pentaphenylphosphole-Fe-$(CO)_4$, 1, 496
$FeC_{38}H_{26}Cl_4$ $Fcd[C(C_6H_4Cl)=CHC_6H_4Cl]_2$, 1, 362
$FeC_{38}H_{26}O_3$ Pentaphenylcyclopentadiene-$Fe(CO)_3$, 1, 487
$FeC_{38}H_{26}O_{12}$ $\{[HO(HO_2C)C_6H_3]_2C_5H_3\}_2Fe$, 1, 383
$FeC_{38}H_{30}$ $Fcd[C(Ph)=CHPh]_2$, 1, 360
$FeC_{38}H_{30}Br_4$ $[(BrC_6H_4CH_2)_2C_5H_3]_2Fe$, 1, 383, 384
$FeC_{38}H_{30}Cl_4$ $[(ClC_6H_4CH_2)_2C_5H_3]_2Fe$,

1, 383
$FeC_{38}H_{30}Cl_4O_2$ Fcd[C(OH)(C_6H_4Cl)-CH_2-C_6H_4Cl]$_2$, 1, 362
$FeC_{38}H_{30}F_4$ [($FC_6H_4CH_2$)$_2C_5H_3$]$_2$Fe, 1, 383
$FeC_{38}H_{34}$ Fcd[CH(Ph)CH_2Ph]$_2$, 1, 356
Fcd(CH_2CHPh_2)$_2$, 1, 356
[($PhCH_2$)$_2C_5H_3$]$_2$Fe, 1, 384
$FeC_{38}H_{34}O_2$ Fcd[C(OH)(Ph)CH_2Ph]$_2$, 1, 365
$FeC_{38}H_{36}N_2O_4$ Fcd-bis(7-ethylidene-N-allyl-5-norbornene-2,3-dicarboximide), 1, 358
$FeC_{38}H_{38}N_6O$ (o-$MeC_6H_4CH_2NC$)$_4$Fe-(CN)$_2$·H_2O, 1, 526
$FeC_{38}H_{40}N_2O_4$ Fcd-bis(7-ethylidene-N-isopropyl-5-norbornene-2,3-dicarboximide), 1, 358
$FeC_{38}H_{54}N_4O_4$ Fcd[CON(C_6H_{11})CONH-C_6H_{11}]$_2$, 1, 375
$FeC_{40}H_{30}$ Ph_5-ferrocene, 1, 410
$FeC_{40}H_{30}N_2O_3$
[PhC=C(Ph)C(Ph)=C(Ph)C=$NC_6H_4NMe_2$]-Fe(CO)$_3$, 1, 486
$FeC_{40}H_{35}Br_2N_9O_{10}$ [(p-$O_2NC_6H_4CH_2NC$)$_4$-($PhCH_2NC$)Fe]Br_2·2H_2O, 1, 528
$FeC_{40}H_{44}N_2O_4$ Fcd-bis(7-ethylidene-N-butyl-5-norbornene-2,3-dicarboximide), 1, 358
$FeC_{41}H_{30}BrCl_5N_6$ [(p-$ClC_6H_4CH_2NC$)$_5$-Fe(CN)]Br, 1, 520
$FeC_{41}H_{35}BrN_6$ [($PhCH_2NC$)$_5$Fe(CN)]Br, 1, 521
$FeC_{41}H_{35}BrN_8O_5$ [(o-$O_2NC_6H_4CH_2NC$)$_2$-($PhCH_2NC$)$_3$Fe(CN)]Br·H_2O, 1, 527
[(p-$O_2NC_6H_4CH_2NC$)$_2$($PhCH_2NC$)$_3$Fe(CN)]-Br·H_2O, 1, 528
$FeC_{41}H_{35}ClN_6$ [($PhCH_2NC$)$_5$Fe(CN)]Cl, 1, 525
$FeC_{41}H_{35}ClN_6O_4$ [($PhCH_2NC$)$_5$Fe(CN)]ClO_4 1, 526
$FeC_{41}H_{35}N_7O_6S$ [($O_2NC_6H_4CH_2NC$)($PhCH_2$-NC)$_4$Fe(CN)]OSO_3H, 1, 522
$FeC_{41}H_{36}Br_2N_6O$ [(p-$BrC_6H_4CH_2NC$)(Ph-CH_2NC)$_4$Fe(CN)]Br·H_2O, 1, 528
$FeC_{41}H_{36}N_6O_4S$ [($PhCH_2NC$)$_5$Fe(CN)]O-SO_3H, 1, 422
$FeC_{41}H_{37}BrN_6O$ [($PhCH_2NC$)$_5$Fe(CN)]Br·H_2O, 1, 521
$FeC_{41}H_{37}N_7O_4$ [($PhCH_2NC$)$_5$Fe(CN)]NO_3·H_2O, 1, 527
$FeC_{41}H_{42}N_6O_{22}S_6$ [($HO_3SC_6H_4CH_2NC$)$_5$Fe-(CN)]OSO_3H·3H_2O, 1, 521
$FeC_{42}H_{35}N_7S$ [($PhCH_2NC$)$_5$Fe(CN)]SCN, 1, 526
$FeC_{42}H_{38}O_2$ Fcd[COCH(CH_2Ph)$_2$]$_2$, 1, 371
$FeC_{42}H_{42}$ [($MeC_6H_4CH_2$)$_2C_5H_3$]$_2$Fe, 1, 384
$FeC_{42}H_{70}O_2$ Fcd($COC_{15}H_{31}$)$_2$, 1, 372
$FeC_{42}H_{74}$ Fcd($C_{16}H_{33}$)$_2$, 1, 356
$FeC_{43}H_{28}O_6$ Tris(8-carboxy-1-naphthyl)-ferrocene, 1, 409
$FeC_{43}H_{35}F_3N_6O_2$ [($PhCH_2NC$)$_5$Fe(CN)]O_2-CCF_3, 1, 526
($FeC_{43}H_{43}Cl_3N_6O_8S$)$_n$ {[-CH(CCl_3)C_6H_3-CH_2NC]($PhCH_2NC$)$_4$Fe(CN)OSO_3H·4H_2O}$_n$, 1, 523
$FeC_{44}H_{25}Br_5O_4$ (p-$BrC_6H_4C≡CH$)$_5$Fe(CO)$_4$, 1, 518
$FeC_{44}H_{30}O_4$ (PhC≡CH)$_5$Fe(CO)$_4$, 1, 517, 518
$FeC_{44}H_{36}N_2O_4$ Fcd-bis(7-ethylidene-N-phenyl-5-norbornene-2,3-dicarboximide), 1, 358*
$FeC_{44}H_{38}O_8$ [Me($PhC_6H_4COCH_2O_2CCH_2O$)-C_5H_3]$_2$Fe, 1, 384
$FeC_{45}H_{30}O_{15}$ [3,4-HO(HO_2C)C_6H_3]$_5$-ferrocene, 1, 410
$FeC_{45}H_{35}Br_2N_5O_5$ [($PhCOCH_2NC$)$_5$Fe]Br_2, 1, 527
$FeC_{45}H_{43}BrN_6$ [(p-$MeC_6H_4CH_2NC$)$_4$($PhCH_2$-NC)Fe(CN)]Br, 1, 528
$FeC_{46}H_{34}$ ($Ph_3C_5H_2$)$_2$Fe, 1, 384
(p-PhC_6H_4)$_3$-ferrocene, 1, 409
$FeC_{46}H_{42}N_6O_{16}S$ [(-$COC_6H_4CH_2NC$)$_2$(-CO-$C_6H_3CH_2NC$)$_3$Fe(CN)OSO_3H·7H_2O]$_n$, 1, 523
$FeC_{46}H_{45}BrN_6$ [(o-$MeC_6H_4CH_2NC$)$_5$Fe(CN)]-Br, 1, 527
[(p-$MeC_6H_4CH_2NC$)$_5$Fe(CN)]Br, 1, 527
$FeC_{46}H_{47}BrN_6O$ [(m-$MeC_6H_4CH_2NC$)$_5$Fe-(CN)]Br·H_2O, 1, 527
[(p-$MeC_6H_4CH_2NC$)$_5$Fe(CN)]Br·H_2O, 1, 527
($FeC_{46}H_{51}N_6O_9S$)$_n$, 1, 523*

$FeC_{46}H_{56}N_6O_{24}S$ $[(HO_2CC_6H_4CH_2NC)_2$-$(-COC_6H_3CH_2NC)_3Fe(CN)OSO_3H\cdot 13H_2O]_n$ 1, 523*

$FeC_{47}H_{41}BrN_6O_8$ $[(p$-$MeO_2CC_6H_4CH_2NC)_3$-$(PhCH_2NC)_2Fe(CN)]Br$, 1, 528

$FeC_{48}H_{38}$ $Fcd(CPh_3)_2$, 1, 354
Bis[1-phenyl-3-(3-phenylallyl)indenyl]iron, 1, 384

$FeC_{48}H_{41}N_7O_9S_2$ $[(HO_3SC_6H_4CH_2NC)(O_2N$-$C_6H_4CH_2NC)(PhCH_2NC)_4Fe]SO_4$, 1, 528

$FeC_{48}H_{42}Br_2N_6$ $[(PhCH_2NC)_6Fe]Br_2$, 1, 522*

$FeC_{48}H_{42}Cl_2N_6O_8$ $[(PhCH_2NC)_6Fe](ClO_4)_2$ 1, 528

$FeC_{48}H_{42}N_6O_3S$ $[(PhCH_2NC)_5Fe(CN)]O_3S$-C_6H_4Me-p, 1, 527

$FeC_{48}H_{42}N_6O_4S$ $[(PhCH_2NC)_6Fe]SO_4$, 1, 523

$FeC_{48}H_{44}Br_2N_6O$ $[(PhCH_2NC)_6Fe]Br_2\cdot H_2O$, 1, 528

$FeC_{48}H_{47}N_7O_9S$ $[(O_2NC_6H_4CH_2NC)(PhCH_2$-$NC)_5Fe]SO_4\cdot 3H_2O$, 1, 528

$FeC_{49}H_{35}O_3P$ $[PhC{=}C(Ph)C(Ph){=}C(Ph)CO]$-$Fe(CO)_2(PPh_3)$, 1, 501

$FeC_{49}H_{35}O_6P$ $[PhC{=}C(Ph)C(Ph){=}C(Ph)CO]$-$Fe(CO)_2[P(OPh)_3]$, 1, 501

$FeC_{49}H_{39}BrN_6$ $[(\beta$-$C_{10}H_7CH_2NC)_2(PhCH_2$-$NC)_3Fe(CN)]Br$, 1, 527

$FeC_{49}H_{43}BrN_6O_8$ $[(p$-$MeO_2CC_6H_4CH_2NC)_4$-$(PhCH_2NC)Fe(CN)]Br$, 1, 528

$FeC_{49}H_{45}BrN_6O_9$ $[(p$-$MeO_2CC_6H_4CH_2NC)_4$-$(PhCH_2NC)Fe(CN)]Br\cdot H_2O$, 1, 528

$FeC_{51}H_{45}BrN_6O_{10}$ $[(p$-$MeO_2CC_6H_4CH_2NC)_5$-$Fe(CN)]Br$, 1, 527

$FeC_{51}H_{47}BrN_6O_{11}$ $[(p$-$AcOC_6H_4CH_2NC)_5$-$Fe(CN)]Br\cdot H_2O$, 1, 527
$[(p$-$MeO_2CC_6H_4CH_2NC)_5Fe(CN)]Br\cdot H_2O$, 1, 527

$FeC_{51}H_{57}BrN_6O_6$ $\{[PhO(CH_2)_3NC]_5Fe$-$(CN)\}Br\cdot H_2O$, 1, 527

$FeC_{57}H_{44}BrN_5$ $(Ph_2CHNC)_4Fe(CN)Br$, 1, 526

$FeC_{57}H_{45}BrN_6O$ $[(\beta$-$C_{10}H_7CH_2NC)_4(Ph$-$CH_2NC)Fe(CN)]Br\cdot H_2O$, 1, 527

$FeC_{58}H_{42}$ 1,1',2,2',3,3',4,4'-Ph_8-ferrocene, 1, 384

$FeC_{58}H_{44}N_6$ $(Ph_2CHNC)_4Fe(CN)_2$, 1, 526

$FeC_{58}H_{78}N_6O$ $[(Me_2CH)_2C_6H_3CH_2NC]_4Fe$-$(CN)_2\cdot H_2O$, 1, 526

$FeC_{62}H_{78}N_6O_4S$ $\{(Me_2CHC_6H_4CH_2NC)_3$-$[(Me_2CH)_2C_6H_3CH_2NC]_2Fe(CN)\}OSO_3H$, 1, 522

$FeC_{65}H_{121}Br_2N_5O_{13}$ $\{[HO_2C(CH_2)_{10}CH_2$-$NC]_5Fe\}Br_2\cdot 3H_2O$, 1, 527

$FeC_{66}H_{115}O_6$ Cyclohexeneiron(III) stearate, 1, 514

$FeC_{77}H_{139}IN_6$ $\{[Me(CH_2)_{14}CH_2NC]_4(Ph$-$CH_2NC)Fe(CN)\}I$, 1, 527

$FeC_{82}H_{60}BrN_6$ $[(Ph_3CNC)_4Fe(CN)_2]Br$, 1, 526

$FeC_{86}H_{165}IN_6$ $\{[Me(CH_2)_{14}CH_2NC]_5Fe$-$(CN)\}I$, 1, 527

$FeC_{101}H_{75}BrN_6$ $[(Ph_3CNC)_5Fe(CN)]Br$, 1, 527

$FeC_{101}H_{155}BrN_6$ $\{[p$-$Me(CH_2)_{11}C_6H_4CH_2$-$NC]_5Fe(CN)\}Br$, 1, 527

$Fe_2C_6H_{15}Br_3$ $Et_2FeBr\cdot EtFeBr_2$, 1, 193

$Fe_2C_9H_6O_6S$ (Methanethiolato)vinyl-$[Fe(CO)_3]_2$, 1, 510

$Fe_2C_{10}F_{10}I_2O_6$ $[C_2F_5Fe(CO)_3I]_2$, 1, 194

$Fe_2C_{10}H_4O_6$ $[CH{=}CHCH{=}CHFe(CO)_3]Fe(CO)_3$, 1, 503

$Fe_2C_{10}H_6O_6S$ (Ethenethiolato)vinyl-$[Fe(CO)_3]_2$, 1, 510

$Fe_2C_{10}H_8O_6S$ (Ethanethiolato)vinyl-$[Fe(CO)_3]_2$, 1, 510

$Fe_2C_{10}H_{10}Br_4$ $[FcH]FeBr_4$, 1, 227

$Fe_2C_{10}H_{10}Cl_3$ $[FcH]FeCl_3$, 1, 227

$Fe_2C_{10}H_{10}Cl_4$ $[FcH]FeCl_4$, 1, 227

$Fe_2C_{11}H_6O_8$ $[HOC{=}C(Me)CH{=}C(OH)Fe(CO)_3]$-$Fe(CO)_3$, 1, 505

$Fe_2C_{11}H_8O_9S$ $[MeOC{=}CHCH{=}C(OMe)Fe(CO)_3]$-$Fe(CO)_2(SO_2)$, 1, 506

$Fe_2C_{11}H_{10}O_6S$ (2-Propanethiolato)vinyl-$[Fe(CO)_3]_2$, 1, 510

$Fe_2C_{12}F_{14}I_2O_6$ $[C_3F_7Fe(CO)_3I]_2$, 1, 194

$Fe_2C_{12}H_6O_6$ $[CH{=}CHC({=}CH_2)CH{=}CHFe(CO)_3]$-$Fe(CO)_3$, 1, 507
Fulvene-$Fe_2(CO)_6$, 1, 216, 745

$Fe_2C_{12}H_6O_8$ $(CH_2{=}CHCH{=}CH_2)[Fe(CO)_4]_2$, 1, 502

$Fe_2C_{12}H_8O_8$ $[HOC{=}C(Me)C(Me){=}C(OH)Fe$-$(CO)_3]Fe(CO)_3$, 1, 505
$[HOC{=}C(Et)CH{=}C(OH)Fe(CO)_3]Fe(CO)_3$,

1, 505

$[MeOC{=}CHCH{=}C(OMe)Fe(CO)_3]Fe(CO)_3$, 1, 505

$Fe_2C_{13}H_6O_7$ $(C_2H_2)_3Fe_2(CO)_7$, 1, 516

$Fe_2C_{13}H_8O_6$ 1,3,5-Cycloheptatriene-$[Fe(CO)_3]_2$, 1, 508

$Fe_2C_{13}H_8O_7$ [1,3,5-Hexatriene]$Fe_2(CO)_7$, 1, 502

$Fe_2C_{13}H_8O_9$ $[AcOC{=}CHC(Me){=}C(OH)Fe(CO)_3]Fe(CO)_3$, 1, 505

$Fe_2C_{13}H_9O_9$ $[AcOCH{=}C(Me)CH{=}C(OH)Fe(CO)_3]Fe(CO)_3$, 1, 505

$Fe_2C_{13}H_{10}O_5$ 6,6-Dimethylfulvene-$Fe_2(CO)_5$, 1, 511

$Fe_2C_{14}F_{20}O_6S_2$ Bis(trifluoromethylthiolato)-$[C_3F_7Fe(CO)_3]_2$, 1, 197

$Fe_2C_{14}H_6O_6S$ Thianaphthene-$[Fe(CO)_3]_2$, 1, 511

$Fe_2C_{14}H_8O_6$ Cyclooctatetraene-$[Fe(CO)_3]_2$, 1, 508

$[(CH{=}CHCH{=}CHCO)\text{-}Fe(CO)_2]_2$, 1, 512

$Fe_2C_{14}H_8O_7$ 2,4,6-Cyclooctatrienone-$[Fe(CO)_3]_2$, 1, 509*

$Fe_2C_{14}H_8O_8$ (1,3,5-Hexatriene)$[Fe(CO)_4]_2$, 1, 502

$Fe_2C_{14}H_9F_6O_6P$ {Cyclooctatrienylium-$[Fe(CO)_3]_2$}PF_6,

$Fe_2C_{14}H_{10}O_4$ $[C_5H_5Fe(CO)_2]_2$, 1, 211

$Fe_2C_{14}H_{10}O_6$ 6,6-Dimethylfulvene-$Fe_2(CO)_6$, 1, 216

1,3,5-Cyclooctatriene-$[Fe(CO)_3]_2$, 1, 509

$Fe_2C_{14}H_{10}O_7$ 7-Methoxy-1,3,5-cycloheptatriene-$[Fe(CO)_3]_2$, 1, 508

$Fe_2C_{14}H_{11}F_6O_4P$ $[C_5H_5Fe(CO)_2]_2H\cdot PF_6$, 1, 212

$Fe_2C_{14}H_{12}O_8$ $[HOC{=}C(CMe_3)CH{=}C(OH)Fe(CO)_3]Fe(CO)_3$, 1, 606

$Fe_2C_{14}H_{16}O_2S_2$ $[C_5H_5Fe(CO)SMe]_2$, 1, 215

$Fe_2C_{14}H_{17}O_2P$ $C_5H_5Fe(CO)H(PMe_2)Fe(CO)\text{-}C_5H_5$, 1, 215

$Fe_2C_{15}H_6D_2O_5$ Azulene-1,3-d_2-$Fe_2(CO)_5$, 1, 511

$Fe_2C_{15}H_7BrO_5$ 1-Bromoazulene-$Fe_2(CO)_5$, 1, 512

$Fe_2C_{15}H_8O_5$ Azulene-$Fe_2(CO)_5$, 1, 511

$Fe_2C_{15}H_8O_7$ (Cyclooctatetraene)$Fe_2(CO)_7$, 1, 502

$Fe_2C_{15}H_{10}O_6$ Vinylcycloheptatriene-$[Fe(CO)_3]_2$, 1, 508

$Fe_2C_{15}H_{33}Br_3$ $(n\text{-}C_5H_{11})_2FeBr\cdot n\text{-}C_5H_{11}FeBr_2$, 1, 193

$Fe_2C_{16}H_4N_2O_8$ p-$C_6H_4[NCFe(CO)_4]_2$, 1, 524

$Fe_2C_{16}H_{10}F_4O_4$ $[\text{-}CF_2Fe(CO)_2(C_5H_5)]_2$, 1, 205

$Fe_2C_{16}H_{10}O_5$ 4-Methylazulene-$Fe_2(CO)_5$, 1, 512

$Fe_2C_{16}H_{10}O_6$ [1,3-$(CH_2{=}CH)_2C_6H_4$][$Fe(CO)_3]_2$, 1, 507

[1,4-$(CH_2{=}CH)_2C_6H_4$][$Fe(CO)_3]_2$, 1, 508

9,10-Dihydronaphthalene-$[Fe(CO)_3]_2$, 1, 509

Bicyclo[4.2.2]deca-2,4,7,9-tetraene-$[Fe(CO)_3]_2$, 1, 509*

$Fe_2C_{16}H_{12}Cl_2O_4$ $\{CH_2C_5H_4Fe(CO)_2Cl]_2$, 1, 735

$Fe_2C_{16}H_{12}O_4$ $\{CH_2C_5H_4Fe(CO)_2]_2$, 1, 735

$Fe_2C_{16}H_{14}O_4$ $[(MeC_5H_4)Fe(CO)_2]_2$, 1, 213

$Fe_2C_{16}H_{22}O_2P_2$ $[C_5H_5Fe(CO)(PMe_2)]_2$, 1, 215*

$Fe_2C_{17}H_{10}ClN_3O_4$ 6-Cl-s-triazine-2,4-diyl-$[Fe(CO)_2C_5H_5]_2$, 1, 204*

$Fe_2C_{17}H_{14}O_6$ 6,6-Pentamethylenefulvene-$Fe_2(CO)_6$, 1, 216

$Fe_2C_{17}H_{16}O_4$ $CH_2[CH_2Fe(CO)_2(C_5H_5)]_2$, 1, 205

$Fe_2C_{18}H_8O_6$ Acenaphthylene-$[Fe(CO)_3]_2$, 1, 510

$Fe_2C_{18}H_{12}O_{14}$ $(MeO_2CC{\equiv}CCO_2Me)_2Fe_2(CO)_6$, 1, 518

$Fe_2C_{18}H_{14}O_4$ $[{\equiv}CCH_2Fe(CO)_2(C_5H_5)]_2$, 1, 205

$Fe_2C_{18}H_{14}O_5$ 4,6,8-Trimethylazulene-$Fe_2(CO)_5$, 1, 512

$Fe_2C_{18}H_{15}Br_3$ $Ph_2FeBr\cdot PhFeBr_2$, 1, 193

$Fe_2C_{18}H_{16}O_5$ $C_5H_5Fe(CO)_2(CH_2)_3COFe(CO)_2C_5H_5$, 1, 204

$Fe_2C_{18}H_{18}O_4$ $\{CH_2CH_2Fe(CO)_2(C_5H_5)]_2$, 1, 205

$[(EtC_5H_4)Fe(CO)_2]_2$, 1, 213

$Fe_2C_{18}H_{18}O_6$ $[\text{-}CH(Me)CH{=}CHCH{=}CH_2]_2\text{-}$

$[Fe(CO)_3]_2$, 1, 507
$Fe_2C_{18}H_{18}O_7$ $O[CH(Me)CH{=}CHCH{=}CH_2]_2$-$[Fe(CO)_3]_2$, 1, 507
$Fe_2C_{18}H_{20}O_6$ $(EtC{\equiv}CEt)_2Fe_2(CO)_6$, 1, 518
$Fe_2C_{19}H_{10}F_6O_6$ $CF_2[CF_2COFe(CO)_2$-$(C_5H_5)]_2$, 1, 206
$Fe_2C_{19}H_{16}O_6$ $CH_2[CH_2COFe(CO)_2(C_5H_5)]_2$, 1, 206
$Fe_2C_{19}H_{20}O_4$ $CH_2[CH_2CH_2Fe(CO)_2(C_5H_5)]_2$ 1, 205
$Fe_2C_{19}H_{20}O_7$ $(EtC{\equiv}CEt)_2Fe_2(CO)_7$, 1, 518
$Fe_2C_{20}H_{10}O_6$ $(PhC{\equiv}CPh)[Fe(CO)_3]_2$, 1, 516
$Fe_2C_{20}H_{12}O_6$ Phenylcyclooctatetraene-$[Fe(CO)_3]_2$, 1, 509
$Fe_2C_{20}H_{14}O_8$ $[(CH{=}CHCH{=}CHCO)Fe(CO)_2]_2 \cdot$ p-$C_6H_4(OH)_2$, 1, 513
$Fe_2C_{20}H_{16}Cl_2$ $(ClFcd-)_2$, 1, 412
$Fe_2C_{20}H_{16}Cl_2N_2$ Cl-Fcd-N=N-Fcd-Cl, 1, 426
$Fe_2C_{20}H_{18}$ Fc_2, 1, 411
$Fe_2C_{20}H_{18}N_2$ FcN=NFc, 1, 425
$Fe_2C_{20}H_{18}O_2S$ $FcSO_2Fc$, 1, 430
$Fe_2C_{20}H_{18}O_5$ 7-Isopropyl-1,4-dimethylazulene-$Fe_2(CO)_5$, 1, 512
$Fe_2C_{20}H_{18}O_6$ $[-CH_2CH_2COFe(CO)_2$-$(C_5H_5)]_2$, 1, 206
$Fe_2C_{20}H_{18}O_{10}S$ [Cycloheptadienylium-$Fe(CO)_3]_2SO_4$, 1, 498
$Fe_2C_{20}H_{18}S$ Fc_2S, 1, 430
$Fe_2C_{20}H_{18}S_2$ FcSSFc, 1, 430
$Fe_2C_{20}H_{18}Se$ Fc_2Se, 1, 431
$Fe_2C_{20}H_{19}N_3$ FcN=NNHFc, 1, 426
$Fe_2C_{20}H_{19}O_2P$ $Fc_2P(O)OH$, 1, 335, 428
$Fe_2C_{20}H_{20}O_4$ Tetramethylethylene-$\{C_5H_4Fe(CO)_2]_2$, 1, 214
$Fe_2C_{20}H_{20}O_4S$ $[FcH]_2SO_4$, 1, 227
$Fe_2C_{20}H_{22}N_2O_4$ 1,2-Bis(dimethylamino)-ethylene$\{C_5H_4Fe(CO)_2]_2$, 1, 214
$Fe_2C_{20}H_{22}O_4$ $\{-CH_2CH_2CH_2Fe(CO)_2$-$(C_5H_5)]_2$, 1, 205
$[(Me_2CHC_5H_4)Fe(CO)_2]_2$, 1, 213
$Fe_2C_{20}H_{24}Cl_4N_2O_4$ 1,2-Bis(dimethylamino)ethylene$\{C_5H_4Fe(CO)_2Cl]_2 \cdot$ 2HCl, 1, 201
$Fe_2C_{21}H_{18}O$ FcCOFc, 1, 413
$Fe_2C_{21}H_{19}ClO_4$ $[Fc_2CH]ClO_4$, 1, 414
$Fe_2C_{21}H_{19}NO$ FcCONHFc, 1, 426
$Fe_2C_{21}H_{20}$ Fc_2CH_2, 1, 414
$Fe_2C_{21}H_{20}N_2O$ $(FcNH)_2CO$, 1, 426
$Fe_2C_{21}H_{20}O$ Fc_2CHOH, 1, 414
$Fe_2C_{21}H_{20}O_2S$ $FcSO_2CH_2Fc$, 1, 431
$Fe_2C_{22}H_{10}Br_2O_6$ $(p\text{-}BrC_6H_4C{\equiv}CH)_2Fe_2(CO)_6$, 1, 518
$Fe_2C_{22}H_{12}O_6$ [Diphenyl-1,3-butadienylene-$Fe(CO)_3]Fe(CO)_3$, 1, 503
$Fe_2C_{22}H_{14}O_4$ [Indenyl-$Fe(CO)_2]_2$, 1, 214
$Fe_2C_{22}H_{16}O_6$ Pentacyclo$[8.3.3.0^{2,9}.0^{3,8}.0^{13,14}]$hexadeca-4,6,11,15-tetraene-$[Fe(CO)_3]_2$, 1, 510
$FeC_{22}H_{18}$ FcC≡CFc, 1, 416
$Fe_2C_{22}H_{18}O_2$ $(FcCO-)_2$, 1, 416
$Fe_2C_{22}H_{18}O_3$ $(FcCO)_2O$, 1, 429
$Fe_2C_{22}H_{18}O_6$ Bicycloocta-2,4,6-trienyl-$[Fe(CO)_3]_2$, 1, 509
$Fe_2C_{22}H_{20}$ $Fc_2C{=}CH_2$, 1, 414
FcCH=CHFc, 1, 417
$Fe_2C_{22}H_{20}N_2$ FcCH=NN=CHFc, 1, 283, 427
$Fe_2C_{22}H_{20}O$ $FcC_5H_3(Ac)FeC_5H_5$, 1, 412
Ac-Fcd-Fc, 1, 412
$FcCH_2COFc$, 1, 417
$Fe_2C_{22}H_{22}$ Fc_2CHMe, 1, 414
$FcCH_2CH_2Fc$, 1, 417
$Fe_2C_{22}H_{22}O$ $(FcCH_2)_2O$, 1, 429
$FcCH_2CH(OH)Fc$, 1, 418
$Fe_2C_{22}H_{22}O_4$ [Tetrahydroindenyl-Fe-$(CO)_2]_2$, 1, 214
$Fe_2C_{22}H_{22}S$ $(FcCH_2)_2S$, 1, 431
$Fe_2C_{22}H_{28}N_2O_4$ 1,2-Bis(dimethylamino)-ethylene$\{C_5H_4Fe(CO)_2Me]_2$, 1, 204
$Fe_2C_{23}H_{14}O_5$ 6,6-Diphenylfulvene-$Fe_2(CO)_5$, 1, 217
$Fe_2C_{23}H_{20}N_2$ FcC=NNHC(Fc)=CH, 1, 421
$Fe_2C_{23}H_{20}O$ FcCH=CHCOFc, 1, 294, 420
$Fe_2C_{23}H_{20}O_2$ $(FcCO)_2CH_2$, 1, 420
$Fe_2C_{23}H_{22}$ $FcCH{=}CHCH_2Fc$, 1, 420
$Fe_2C_{23}H_{22}O$ $FcCH_2CH_2COFc$, 1, 420
$Fe_2C_{23}H_{24}$ $Fc(CH_2)_3Fc$, 1, 420
$Fe_2C_{23}H_{24}O$ Fc_2CHOEt, 1, 415
$Fc_2C(OH)Et$, 1, 415
$Fe_2C_{24}H_{12}O_{10}$ [BzOC=CHCH=C(OBz)Fe-$(CO)_3]Fe(CO)_3$, 1, 505

$Fe_2C_{24}H_{16}$ Bis(as-indacenyleneiron), 1, 425

$Fe_2C_{24}H_{16}O_6$ $(PhC{\equiv}CMe)_2Fe_2(CO)_6$, 1, 518

$Fe_2C_{24}H_{18}$ $FcC{\equiv}CC{\equiv}CFc$, 1, 421

$Fe_2C_{24}H_{18}O_2$ $FcCOC{\equiv}CCOFc$, 1, 421

$Fe_2C_{24}H_{21}O_2P$ $C_5H_5Fe(CO)H(PPh_2)Fe(CO)C_5H_5$, 1, 215

$Fe_2C_{24}H_{22}$ $FcCH_2C{\equiv}CCH_2Fc$, 1, 421

$FcCH{=}CHCH{=}CHFc$, 1, 422

$Fe_2C_{24}H_{22}O$ $FcCOCH{=}C(Me)Fc$, 1, 420

$Fe_2C_{24}H_{22}O_2$ (3-Acetylbicyclopentadienylene)(acetylcyclopentadienyliron)(cyclopentadienyliron), 1, 412

$(AcFcd-)_2$, 1, 412

$[C_5H_5FeC_5H_3(Ac)-]_2$, 1, 413

$Fe_2CCH_2CH_2COO$, 1, 415

$FcCH(OH)C{\equiv}CCH(OH)Fc$, 1, 421

$(FcCOCH_2-)_2$, 1, 422

$Fe_2C_{24}H_{22}O_4$ $(MeO_2CFcd-)_2$, 1, 413

$Fe_2C_{24}H_{23}NO_3$ $FcCO_2CH_2CH_2NHCOFc$, 1, 427

$Fe_2C_{24}H_{24}Cl_2O$ $[FcCH(CH_2Cl)]_2O$, 1, 429

$Fe_2C_{24}H_{24}N_2$ $[FcC(Me){=}N-]_2$, 1, 427

$Fe_2C_{24}H_{24}O$ $Fc_2C(Me)Ac$, 1, 415

$FcCOCH_2CH(Me)Fc$, 1, 421

$FcCHCH_2CH_2CH(Fc)O$, 1, 429

$Fe_2C_{24}H_{26}$ $(EtFcd-)_2$, 1, 413

$[FcCH(Me)-]_2$, 1, 418

$Fc(CH_2)_4Fc$, 1, 422

$Fe_2C_{24}H_{26}O$ $FcCH(OH)CH_2CH(Me)Fc$, 1, 421

$[FcCH(Me)-]_2O$, 1, 429

$Fe_2C_{24}H_{26}O_2$ $[FcCH(OH)CH_2-]_2$, 1, 422

$Fe_2C_{24}H_{26}O_4$ [(Cyclopentylcyclopentadienyl)$Fe(CO)_2]_2$, 1, 213

$Fe_2C_{24}H_{28}BrN$ $[(FcCH_2)_2NMe_2]Br$, 1, 426

$Fe_2C_{25}H_{16}O_7$ $(PhC{\equiv}CMe)_2Fe_2(CO)_7$, 1, 518

$Fe_2C_{25}H_{22}O$ $(FcCH{=}CH)_2CO$, 1, 422

$Fe_2C_{25}H_{23}NO_2$ $Fc_2C(CN)CO_2Et$, 1, 415

$Fe_2C_{25}H_{24}O_2$ $(FcCO)_2CMe_2$, 1, 421

$FcCOCH_2CH(Me)COFc$, 1, 422

$Fe_2C_{25}H_{25}NO_3$ $Fc_2C(CONH_2)CO_2Et$, 1, 415

$Fe_2C_{25}H_{28}O$ $Fc_2C(OH)Bu$, 1, 415

$Fe_2C_{26}H_{12}Cl_2O_8$ [6,6-Di(p-chlorophenyl)-fulvene]$[Fe(CO)_4]_2$, 1, 502

$Fe_2C_{26}H_{14}O_8$ (6,6-Diphenylfulvene)-$[Fe(CO)_4]_2$, 1, 502

$Fe_2C_{26}H_{18}O_2$ $(FcCOC{\equiv}C-)_2$, 1, 423

$Fe_2C_{26}H_{18}O_9$ [Tropone-$Fe(CO)_3]_2{\cdot}HOC_6H_4$-OH, 1, 489

$Fe_2C_{26}H_{20}$ $FcCH{=}C{=}C{=}C{=}C{=}CHFc$, 1, 423

$Fe_2C_{26}H_{22}$ 1,3-$Fc(Ph)C_5H_3FeC_5H_5$, 1, 413

$FcCH{=}CHC{\equiv}CCH{=}CHFc$, 1, 423

m-$C_6H_4Fc_2$, 1, 425

$Fe_2C_{26}H_{22}Br_4O_4$ $\{[BzO(Me)C_5H_3]_2Fe\}FeBr_4$, 1, 382

$Fe_2C_{26}H_{22}O_2$ $[FcCH(OH)C{\equiv}C-]_2$, 1, 423

$Fe_2C_{26}H_{23}OP$ $Fc_2P(Ph)O$, 1, 428

$Fe_2C_{26}H_{23}P$ Fc_2PPh, 1, 335, 428

$Fe_2C_{26}H_{24}$ $Fc(CH{=}CH)_3Fc$, 1, 423

$Fe_2C_{26}H_{24}N_2$ $(FcCH{=}CHCH{=}N-)_2$, 1, 427

$Fe_2C_{26}H_{24}O_3$ Triacetylbiferrocenyl, 1, 413

$Fe_2C_{26}H_{26}$ $(FcCH{=}CHCH_2-)_2$, 1, 423

$Fe_2C_{26}H_{26}O$ $(CH_2CH_2\text{-}Fcd\text{-}CH)_2O$, 1, 437

$Fe_2C_{26}H_{26}O_2$ $[FcCH(OH)CH{=}CH-]_2$, 1, 424

$Fe_2C_{26}H_{30}$ $(FcCMe_2-)_2$, 1, 418

$Fc(CH_2)_6Fc$, 1, 424

$Fe_2C_{26}H_{30}O$ $(FcCHEt)_2O$, 1, 430

$Fe_2C_{26}H_{30}O_2$ $[FcCH(OH)CH_2CH_2-]_2$, 1, 424

$Fe_2C_{26}H_{30}O_4$ [(cyclo-$C_6H_{11}C_5H_4)Fe(CO)_2]_2$, 1, 213

$Fe_2C_{27}H_{19}O_5P$ $(CH{=}CHCH{=}CHFe)Fe(CO)_5(PPh_3)$, 1, 506

$Fe_2C_{27}H_{24}$ Fc_2CHPh, 1, 415

$Fe_2C_{27}H_{24}O$ $Fc_2C(OH)Ph$, 1, 415

$Fe_2C_{27}H_{26}IP$ $[Fc_2PPhMe]I$, 1, 428

$Fe_2C_{27}H_{34}N_2$ $[FcCH_2CH_2N(Me)]_2CH_2$, 1, 427

$Fe_2C_{28}H_{18}O_8$ $[(BzOC_5H_4)Fe(CO)_2]_2$, 1, 213

$Fe_2C_{28}H_{22}$ $(FcCH{=}CHC{\equiv}C-)_2$, 1, 424

$Fe_2C_{28}H_{22}O_4$ $[(PhCH_2C_5H_4)Fe(CO)_2]_2$, 1, 213

$Fe_2C_{28}H_{26}$ $Fc(CH{=}CH)_4Fc$, 1, 424

$Fe_2C_{28}H_{26}ClN$ $(FcCH_2)_2NC_6H_4Cl$-p, 1, 426

$Fe_2C_{28}H_{26}O$ $Fc_2C(Ph)OMe$, 1, 416

$Fe_2C_{28}H_{26}O_2$ $[FcC(OH)(Me)C{\equiv}C-]_2$, 1, 424

$Fe_2C_{28}H_{27}N$ $(FcCH_2)_2NPh$, 1, 426

$Fe_2C_{28}H_{32}$
CH_2CH_2-Fcd-$(CH_2)_4$-Fcd-CH_2CH_2, 1, 422
$Fe_2C_{28}H_{34}$ [FcCH(Me)CH_2CH_2-$]_2$, 1, 424
Fc$(CH_2)_8$Fc, 1, 424
$Fe_2C_{29}H_{21}O_5P$ Fulvene-$Fe_2(CO)_5(PPh_3)$, 1, 216
$Fe_2C_{29}H_{32}N_4O_2$ $(FcCH{=}NNHCOCH_2CH_2)_2CH_2$, 1, 427
$Fe_2C_{30}H_{15}Br_3O_6$ $(p\text{-}BrC_6H_4C{\equiv}CH)_3Fe_2(CO)_6$, 1, 518
$Fe_2C_{30}H_{17}O_6$ Pentacarbonyl(dihydro-6-oxo-2,4,7-triphenyl-1-ferripentalene)iron, 1, 507
$Fe_2C_{30}H_{18}O_6$ $(PhC{\equiv}CH)_3Fe_2(CO)_6$, 1, 518
$Fe_2C_{30}H_{25}NO_2$ N-Ferrocenyl-7-(ferrocenylmethylene)-5-norbonene-2,3-dicarboximide, 1, 427
$Fe_2C_{30}H_{26}$ p-$(FcCH{=}CH)_2C_6H_4$, 1, 425
$Fe_2C_{30}H_{28}$ Fc$(CH{=}CH)_5$Fc, 1, 425
$Fe_2C_{30}H_{31}N$ $[FcCH(Me)]_2NPh$, 1, 427
$Fe_2C_{30}H_{34}$ $CH_2CH_2CH_2CH_2C$(Fc)-C(Fc)$CH_2CH_2CH_2CH_2$, 1, 418
$Fe_2C_{30}H_{38}O$ $(FcCHBu)_2O$, 1, 430
$Fe_2C_{31}H_{27}NO_2$ N-Ferrocenyl-7-(1-ferrocenylethylidene)-5-norbornene-2,3-dicarboximide, 1, 427
$Fe_2C_{32}H_{26}$ $(m\text{-}FcC_6H_4\text{-})_2$, 1, 425
$(p\text{-}FcC_6H_4\text{-})_2$, 1, 425
$Fe_2C_{33}H_{20}O_5$ $(Ph_2C{=}C{=}C{=}CPh_2)Fe_2(CO)_5$, 1, 512
$Fe_2C_{33}H_{24}O_6$ $(PhC{\equiv}CMe)_3Fe_2(CO)_6$, 1, 518
$Fe_2C_{34}H_{16}Cl_4O_6$ $(ClC_6H_4C{\equiv}CC_6H_4Cl)_2$-$Fe_2(CO)_6$, 1, 517
$Fe_2C_{34}H_{20}O_6$ [PhC=C(Ph)C(Ph)=C(Ph)Fe-$(CO)_3$]$Fe(CO)_3$, 1, 503
$Fe_2C_{34}H_{28}$ FcC(Ph)=C(Ph)Fc, 1, 419
$Fe_2C_{34}H_{28}O$ Fc_2C(Ph)Bz; 1, 416
$Fe_2C_{34}H_{30}$ $Fc_2C(Ph)CH_2Ph$, 1, 416
[FcCH(Ph)-$]_2$, 1, 419
$Fe_2C_{34}H_{30}O$ FcCH(Ph)-Fcd-CH(Ph)OH, 1, 416
[FcCH(Ph)-$]_2$O, 1, 430
$Fe_2C_{34}H_{30}O_2$ [FcC(OH)(Ph)-$]_2$, 1, 419
$Fe_2C_{35}H_{16}Cl_4O_7$ $(ClC_6H_4C{\equiv}CC_6H_4Cl)_2$-$Fe_2(CO)_7$, 1, 517
$Fe_2C_{35}H_{20}O_7$ [PhC=C(Ph)C(Ph)=C(Ph)CO]-$Fe_2(CO)_6$, 1, 506
[PhC=C(Ph)C(O)C(Ph)=C(Ph)Fe$(CO)_3$]Fe-$(CO)_3$, 1, 506
$Fe_2C_{36}H_{22}O_6$ 2,3,4,5-Tetraphenylfulvene-$Fe_2(CO)_6$, 1, 217
$Fe_2C_{36}H_{26}O_4$ (3,3′-Biphthalid-3,3′-diylbiscyclopentadienylene)bis-(cyclopentadienyliron), 1, 419
$Fe_2C_{36}H_{30}O_2P_2$ $[C_5H_5Fe(CO)(PPh_2)]_2$, 1, 214
$Fe_2C_{38}H_{20}O_6$ [Diphenylbis(phenylethynyl)-1,3-butadienylene-$Fe(CO)_3$]-$Fe(CO)_3$ isomers, 1, 504
$Fe_2C_{39}H_{20}O_7$ [3-Oxodiphenylbis(phenylethynyl)-1,4-pentadienylene-Fe-$(CO)_3$]$Fe(CO)_3$, 1, 507
$Fe_2C_{40}H_{25}O_6P$ Pentaphenylphosphole-$Fe_2(CO)_6$, 1, 496
$Fe_2C_{40}H_{30}O_4$ $[(Ph_2CHC_5H_4)Fe(CO)_2]_2$, 1, 213
$Fe_2C_{40}H_{32}O_8$ $\{COC_6H_4CO_2C_5H_3(Me)FeC_5H_3(Me)O\}_2$, 1, 430
$Fe_2C_{40}H_{33}OP$ FcCOC(Fc)=PPh_3, 1, 335, 417
$Fe_2C_{40}H_{34}IOP$ [FcCOCH(Fc)PPh_3]I, 1, 335, 417
$Fe_2C_{41}H_{37}IN_2S_2$

1, 428

1, 428

$Fe_2C_{54}H_{44}O_8P_2$ [(OCH=CHCH=CH)$Fe(CO)_2$-$(PPh_3)]_2$·p-$C_6H_4(OH)_2$, 1, 501
$Fe_2C_{58}H_{40}O_2$ (PhC=C(Ph)C(Ph)=C(Ph)CO$)_2$-Fe_2, 1, 514
$Fe_3C_{14}H_8O_8$ $(C_2H_2)_3Fe_3(CO)_8$, 1, 516
$Fe_3C_{19}H_8O_{10}$ $(PhC{\equiv}CMe)Fe_3(CO)_{10}$, 1, 518
$Fe_3C_{20}H_{10}O_{10}$ $(C_2H_2)_5Fe_3(CO)_{10}$, 1, 517
$Fe_3C_{23}H_{16}O_9$ Bis(cycloheptatriene)-$[Fe(CO)_3]_3$, 1, 513

$Fe_3C_{25}H_{16}O_9$ (Cyclooctatetraene dimer)-[Fe(CO)$_3$]$_3$, 1, 513
$Fe_3C_{30}H_{26}$ FcdFc$_2$, 1, 431
$Fe_3C_{30}H_{27}OP$ Fc$_3$PO, 1, 433
$Fe_3C_{30}H_{27}PS$ Fc$_3$PS, 1, 434
$Fe_3C_{31}H_{27}Cl$ Fc$_3$CCl, 1, 431
$Fe_3C_{31}H_{28}$ Fc$_3$CH, 1, 432
$Fe_3C_{31}H_{28}O$ Fc$_3$COH, 1, 432
$Fe_3C_{31}H_{30}IP$ [Fc$_3$PMe]I, 1, 434
$Fe_3C_{32}H_{30}$ Fcd(CH$_2$Fc)$_2$, 1, 433
$Fe_3C_{32}H_{30}O$ Fc$_3$COMe, 1, 342
$Fe_3C_{34}H_{30}$ Fcd(CH=CHFc)$_2$, 1, 433
$Fe_3C_{35}H_{24}O_8$ (PhC≡CMe)$_3$Fe$_3$(CO)$_8$, 1, 518
$Fe_3C_{35}H_{32}O_2$ FcCH(CH$_2$COFc)$_2$, 1, 433
$Fe_3C_{36}H_{16}Cl_4O_8$ (p-ClC$_6$H$_4$C≡CC$_6$H$_4$Cl-p)$_2$Fe$_3$(CO)$_8$, 1, 518
$Fe_3C_{36}H_{20}O_8$ (PhC≡CPh)$_2$Fe$_3$(CO)$_8$, 1, 517
$Fe_3C_{39}H_{30}O_3$ 1,3,5-(FcCO)$_3$C$_6$H$_3$, 1, 433
$Fe_3C_{39}H_{36}$ 1,3,5-(FcCH$_2$)$_3$C$_6$H$_3$, 1, 433
$Fe_3C_{42}H_{34}$ [FcC$_5$H$_3$(Ph)]$_2$Fe, 1, 431
$Fe_3C_{44}H_{38}$ Fcd[CH(Ph)Fc]$_2$, 1, 433
$Fe_4C_{42}H_{38}O$ (Fc$_2$CH)$_2$O, 1, 434
$Fe_4C_{129}H_{115}N_{24}O_5$ [(PhCH$_2$NC)$_5$Fe(CN)]$_3$-[Fe(CN)$_6$]·5H$_2$O, 1, 526
$Fe_5C_{24}H_{12}Cl_6O_{12}$ [(CH=CHCH=CHCO)-Fe-(CO)$_3$]$_3$[FeCl$_3$]$_2$, 1, 484
$Fe_5C_{32}H_{16}O_{12}$ Bis(azulene)-Fe$_5$(CO)$_{12}$, 1, 513
$Fe_5C_{33}H_{16}O_{13}$ Bis(azulene)-Fe$_5$(CO)$_{13}$, 1, 513
$Fe_5C_{39}H_{36}O_{15}$ (C$_4$H$_6$)$_6$[Fe(CO)$_3$]$_5$, 1, 513
$FeAlC_{10}H_{11}Cl_4$ FcH·HAlCl$_4$, 1, 226
$FeAsC_9H_5F_6O_2$ C$_5$H$_5$Fe(CO)$_2$As(CF$_3$)$_2$, 3, 303
$FeAsC_{21}H_{15}I_2O_3$ Fe(AsPh$_3$)(CO)$_3$I$_2$, 3, 21
$FeAsC_{22}H_{15}O_4$ Fe(AsPh$_3$)(CO)$_4$, 3, 21
$FeAsC_{22}H_{21}N_2O_3$ Fe[As(C$_6$H$_4$Me-4)$_3$](CO)-(NO)$_2$, 3, 28
$FeAsC_{28}H_{23}O_2$ [CH=CH(CH=CH)$_2$CH=CH]Fe-(CO)$_2$(AsPh$_3$), 1, 501
$FeAsC_{37}H_{25}O_3$ Pentaphenylarsenole-Fe-(CO)$_3$, 1, 496
$FeAs_2C_{12}H_{16}Br_2O_2$ [Fe(2-Me$_2$AsC$_6$H$_4$-AsMe$_2$)(CO)$_2$Br$_2$], 3, 85
$FeAs_2C_{12}H_{16}IO_2$ [Fe(2-Me$_2$AsC$_6$H$_4$AsMe$_2$)-(CO)$_2$I], 3, 85
$FeAs_2C_{12}H_{16}I_2O_2$ [Fe(2-Me$_2$AsC$_6$H$_4$AsMe$_2$)-(CO)$_2$I$_2$], 3, 85
$FeAs_2C_{13}H_{16}O_3$ Fe(2-Me$_2$AsC$_6$H$_4$AsMe$_2$)-(CO)$_3$, 3, 85
$FeAs_2C_{29}H_{24}O_3$ [Fe(Ph$_2$AsCH$_2$-)$_2$(CO)$_3$], 3, 102
$FeAs_2C_{38}H_{30}Br_2O_2$ [Fe(AsPh$_3$)$_2$(CO)$_2$Br$_2$], 3, 21
$FeAs_2C_{38}H_{30}Cl_2O_2$ [Fe(AsPh$_3$)$_2$(CO)$_2$Cl$_2$], 3, 21
$FeAs_2C_{39}H_{30}O_3$ Fe(AsPh$_3$)$_2$(CO)$_3$, 3, 21
$FeAs_4C_{21}H_{32}O$ [Fe(2-Me$_2$AsC$_6$H$_4$AsMe$_2$)$_2$-(CO)], 3, 85
$FeBC_7H_7F_4O_3$ [π-MeCH=CHCH$_2$-Fe(CO)$_3$]BF$_4$, 1, 471
$FeBC_8H_9F_4O_3$ [π-MeCH=CHCH(Me)-Fe(CO)$_3$]-BF$_4$, 1, 472
[π-CH$_2$=C(Me)CH(Me)-Fe(CO)$_3$]BF$_4$, 1, 472
$FeBC_9H_7F_4O_3$ [Cyclohexadienylium-Fe-(CO)$_3$]BF$_4$, 1, 498
$FeBC_9H_9F_4O_2$ [C$_5$H$_5$Fe(CO)$_2$(C$_2$H$_4$)]BF$_4$, 1, 209
$FeBC_9H_9F_4O_3$ [1-Methylpentadienylium-Fe(CO)$_3$]BF$_4$, 1, 497, 498
$FeBC_{10}H_7F_4O_3$ [Cycloheptatrienylium-Fe(CO)$_3$]BF$_4$, 1, 498
$FeBC_{10}H_9F_4O_3$ [Cycloheptadienylium-Fe-(CO)$_3$]BF$_4$, 1, 498
$FeBC_{10}H_9F_4O_4$ [1-Methoxycyclohexadienylium-Fe(CO)$_3$]BF$_4$, 1, 498
$FeBC_{10}H_{10}BrO_2$ Br-Fcd-B(OH)$_2$, 1, 532
$FeBC_{10}H_{10}ClO_2$ Cl-Fcd-B(OH)$_2$, 1, 532
$FeBC_{10}H_{10}F_4$ [FcH]BF$_4$, 1, 227
$FeBC_{10}H_{11}F_4O_2$ [C$_5$H$_5$Fe(CO)$_2$(CH$_2$=CHMe)]-BF$_4$, 1, 210
$FeBC_{10}H_{11}F_4O_3$ {[MeCH=CHCH=CH$\overset{+}{C}$H(Me)]-Fe(CO)$_3$}$\bar{B}$F$_4$, 1, 497, 498*
$FeBC_{10}H_{11}O_2$ FcB(OH)$_2$, 1, 531
$FeBC_{11}H_9F_4O_3$ [Bicyclo[5.1.0]octa-2,4-dienylium-Fe(CO)$_3$]BF$_4$, 1, 499
$FeBC_{11}H_{11}F_4O_3$ [Cyclooctadienylium-Fe(CO)$_3$]BF$_4$, 1, 499
$FeBC_{13}H_{11}F_4O_3$ [π-MeCH=CHCH(Ph)-Fe-

$(CO)_3]BF_4$, 1, 472
$FeBC_{25}H_{20}F_4O_2P$ $[C_5H_5Fe(CO)_2(PPh_3)]$-BF_4, 1, 209
$FeBC_{29}H_{23}F_4O_3$ [Tritylcycloheptadienylium-$Fe(CO)_3]BF_4$, 1, 499
$FeBC_{32}H_{25}O_3$ $[C_5H_5Fe(CO)_3]BPh_4$, 1, 208
$FeBC_{34}H_{30}$ $[FcH]BPh_4$, 1, 227
$FeBC_{35}H_{29}O_3$ [Bicyclo[5.1.0]octa-2,4-dienylium-$Fe(CO)_3]BPh_4$, 1, 499
$FeBC_{35}H_{31}$ $[C_6H_6FeC_5H_5]BPh_4$, 1, 218
$FeBC_{38}H_{37}$ $[(Me_3C_6H_3)FeC_5H_5]BPh_4$, 1, 218
$FeBC_{39}H_{37}$ [Tetralin-$FeC_5H_5]BPh_4$, 1, 218
$FeBC_{40}H_{41}$ $[(Me_3C_6H_3)Fe(C_5H_4Et)]BPh_4$, 1, 218
$FeBC_{45}H_{35}O_3$ $[(Ph_2CHC_5H_4)Fe(CO)_3]BPh_4$, 1, 209
$FeB_2C_{10}H_{12}O_4$ $Fcd[B(OH)_2]_2$, 1, 532
$(FeB_2C_{22}H_{18}N_4)_n$ Bis[(1,3-dihydro-1,3,2-denzodiazoborol-2-yl)cyclopentadienyl]-Fe polymer, 1, 532
$FeB_2C_{22}H_{20}N_4$ $Fcd(BNHC_6H_4NH)_2$, 1, 532
$FeCoC_{11}H_5O_6$ Di-μ-carbonyl(carbonylcyclopentadienyliron)(tricarbonylcobalt), 1, 547
$FeCoC_{14}H_{11}O_4$ $[(C_5H_5)_2Co][HFe(CO)_4]$, 1, 587
$FeCoC_{17}H_{17}O_3$ $(Me_3C_6H_3)Co(CO)(C_5H_5)$-$Fe(CO)_2$, 1, 547
$FeCoC_{18}H_{15}O_2$ $FcCH_2C_5H_4Co(CO)_2$, 1, 547
$FeCrC_{15}H_{15}N_6O_3S_4$ [Bicyclo[5.1.0]octa-2,4-dienylium-$Fe(CO)_3][Cr(NH_3)_2$-$(SCN)_4]$, 1, 499
$FeCrC_{16}H_{10}O_6$ $Fe(CO)_3(CH_2{=}CHCH{=}CHPh)$-$Cr(CO)_3$, 1, 94
$FeCrC_{20}H_{16}O_3$ $(FcCH_2Ph)Cr(CO)_3$, 1, 94
$FeCrC_{22}H_{14}O_6$ $Fe(CO)_3(PhCH{=}CHCH{=}CHPh)$-$Cr(CO)_3$, 1, 94
$FeCrC_{29}H_{26}N_6O_2PS_4$ $[C_5H_5Fe(CO)_2$-$(PPh_3)][Cr(NH_3)_2(SCN)_4]$, 1, 209
$FeCr_2C_{25}H_{14}O_9$ $(OC)_3Fe[$-$CH{=}CHPhCr$-$(CO)_3]_2$, 1, 94
$FeGaC_{10}H_{10}Cl_4$ $[FcH]GaCl_4$, 1, 228
$FeGeC_8H_9NO_4$ $(Me_3GeNC)Fe(CO)_4$, 1, 524, 2, 141
$FeGeC_9H_{11}FO_2$ $C_5H_5Fe(CO)_2GeMe_2F$, 1, 533, 2, 141*
$FeGeC_{25}H_{20}O_2$ $C_5H_5Fe(CO)_2GePh_3$, 1, 533, 2, 141*
$FeGeC_{28}H_{24}$ $FcGePh_3$, 1, 537, 2, 141
$FeGeC_{30}H_{25}BrO_2$ $(EtMeC_5H_3)Fe(CO)_2Ge$-$(C_{10}H_7)_2Br$, 1, 534, 2, 141
$FeGeC_{33}H_{43}ClO_2$ $(n\text{-}C_{12}H_{25}C_5H_4)Fe(CO)_2$-$Ge(CH_2Ph)_2Cl$, 1, 534, 2, 141
$FeGe_2C_{46}H_{38}$ $Fcd(GePh_3)_2$, 1, 540, 2, 141
$FeHgC_7H_5ClO_2$ $C_5H_5Fe(CO)_2HgCl$, 1, 207
$(FeHgC_{10}H_8)_n$ $(\text{-}Fcd\text{-}Hg\text{-})_n$, 1, 531
$FeHgC_{10}H_8Br_2$ Br-Fcd-HgBr, 1, 530
$FeHgC_{10}H_8Cl_2$ Cl-Fcd-HgCl, 1, 529
$FeHgC_{10}H_9Cl$ FcHgCl, 1, 529
$FeHgC_{12}H_{12}O_2$ FcHgOAc, 1, 529
$FeHg_2C_{10}H_8Cl_2$ $Fcd(HgCl)_2$, 1, 530
$FeHg_2C_{14}H_{14}O_4$ $Fcd(HgOAc)_2$, 1, 530
$FeHg_2C_{16}H_{11}Cl_2NO_2$ Bis(chloromercuri)-p-nitrophenylferrocene, 1, 530
$FeKC_8H_5N_2O$ $K[C_5H_5Fe(CO)(CN)_2]$, 1, 207
$FeK_3C_{12}H_6$ $K_3[Fe(C{\equiv}CH)_6]$, 1, 197
$FeK_4C_7H_4N_6$ $K_4[Fe(CN)_5C{\equiv}CH]\cdot NH_3$, 1, 197
$FeK_4C_8H_3N_5$ $K_4[Fe(CN)_5C{\equiv}CMe]$, 1, 197
$FeK_4C_{12}H_6$ $K_4[Fe(C{\equiv}CH)_6]$, 1, 197
$FeK_4C_{13}H_5N_5$ $K_4[Fe(CN)_5C{\equiv}CPh]$, 1, 197
$FeK_4C_{18}H_{18}$ $K_4[Fe(C{\equiv}CMe)_6]$, 1, 197
$FeK_4C_{28}H_{16}Br_2N_4O_{12}S_4$ $[(p\text{-}KO_3SC_6H_4NC)_4$-$Fe]Br_2$, 1, 525
$FeK_4C_{28}H_{16}Cl_2N_4O_{12}S_4$ $[(p\text{-}KO_3SC_6H_4NC)_4$-$Fe]Cl_2$, 1, 525
$FeK_4C_{28}H_{16}I_2N_4O_{12}S_4$ $[(p\text{-}KO_3SC_6H_4NC)_4$-$Fe]I_2$, 1, 525
$FeK_4C_{48}H_{30}$ $K_4[Fe(C{\equiv}CPh)_6]$, 1, 197
$FeMnC_{12}H_5DF_6O_7P$ $[C_5H_5Fe(CO)_2Mn(CO)_5]$-$D\cdot PF_6$, 1, 184
$FeMnC_{12}H_5O_7$ $(C_5H_5)Fe(CO)_2Mn(CO)_5$, 1, 184
$FeMnC_{12}H_6F_6O_7P$ $[C_5H_5Fe(CO)_2Mn(CO)_5]$-$H\cdot PF_6$, 1, 184
$FeMnC_{15}H_{11}O_4$ $C_6H_6Mn(CO)_2Fe(C_5H_5)(CO)_2$, 1, 184
$FeMnC_{19}H_{13}O_4$ $FcCOC_5H_4Mn(CO)_3$, 1, 184
$FeMnC_{19}H_{15}O_3$ $FcCH_2C_5H_4Mn(CO)_3$, 1, 185
$FeMn_2C_{28}H_{16}O_8$ $Fcd[COC_5H_4Mn(CO)_3]_2$, 1, 185
$FeMn_2C_{30}H_{20}O_8$ $Fcd[CO(Me)C_5H_3Mn(CO)_3]_2$,

1, 185
$FeMoC_{15}H_{10}O_5$ $(C_5H_5)Fe(CO)_2Mo(C_5H_5)(CO)_3$, 1, 118
$FeMoC_{16}H_7BrO_6$ 1-Bromozaulene-[Fe(CO)$_3$][Mo(CO)$_3$], 1, 118
$FeNa_2C_{11}H_{10}O_4S$ $FcCHO \cdot Na_2SO_3$, 1, 283
$FeNa_2C_{16}H_{12}O_4$ $Fcd(COCH{=}CHONa)_2$, 1, 366
$FeNa_4C_{12}H_6$ $Na_4[Fe(C{\equiv}CH)_6]_2$, 1, 197
$FeNa_4C_{18}H_{18}$ $Na_4[Fe(C{\equiv}CMe)_6]$, 1, 197
$FeNiC_{13}H_{10}O_3$ $C_5H_5Ni(CO)_2Fe(CO)C_5H_5$, 1, 547
$FeNi_2C_{21}H_{16}O_3$ $[(C_5H_5Ni)_2(PhC{\equiv}CH)]Fe(CO)_3$, 1, 547
$FeNi_2C_{27}H_{20}O_3$ $[(C_5H_5Ni)_2(PhC{\equiv}CPh)]Fe(CO)_3$, 1, 547
$FeNi_2C_{29}H_{20}O_3$ $(C_5H_5Ni)_2(PhC{\equiv}CC{\equiv}CPh)Fe(CO)_3$, 1, 548
$FeNi_2C_{29}H_{22}O_3$ $[(C_5H_5Ni(PhC{\equiv}CH)]_2Fe(CO)_3$, 1, 548
$FePb_2C_{40}H_{30}O_4$ $(Ph_3Pb)_2Fe(CO)_4$, 2, 595
$FePb_2C_{40}H_{66}O_4$ (Cyclohexyl$_3$Pb)$_2$Fe(CO)$_4$, 2, 595
$FeReC_{12}H_5O_7$ $C_5H_5Fe(CO)_2Re(CO)_5$, 1, 192
$FeRuC_{21}H_{18}O$ $FcCOC_5H_4RuC_5H_5$, 1, 546
$FeSbC_7H_7Cl_6O_3$ [π-MeCH=CHCH$_2$-Fe(CO)$_3$]SbCl$_6$, 1, 471
$FeSbC_9H_7Cl_6DNO_2$ [C$_5$H$_5$Fe(CO)$_2$(DN=C=CH$_2$)]SbCl$_6$, 1, 210
$FeSbC_9H_8Cl_6NO_2$ [C$_5$H$_5$Fe(CO)$_2$(HN=C=CH$_2$)]SbCl$_6$, 1, 210
$FeSbC_9H_9Cl_6O_3$ [1-Methylpentadienylium-Fe(CO)$_3$]SbCl$_6$, 1, 498
$FeSbC_{10}H_9Cl_6DNO_2$ [C$_5$H$_5$Fe(CO)$_2$(DN=C=CHMe)]SbCl$_6$, 1, 210
$FeSbC_{10}H_9Cl_6O_2$ [C$_5$H$_5$Fe(CO)$_2$(CH$_2$=C=CH$_2$)]SbCl$_6$, 1, 209
$FeSbC_{10}H_{10}Cl_6NO_2$ [C$_5$H$_5$Fe(CO)$_2$(HN=C=CHMe)]SbCl$_6$, 1, 210
$FeSbC_{11}H_9Cl_6O_3$ [Bicyclo[5.1.0]octa-2,4-dienylium-Fe(CO)$_3$]SbCl$_6$, 1, 499
$FeSbC_{12}H_{11}Cl_6O_2$ [C$_5$H$_5$Fe(C$_5$H$_6$)(CO)$_2$]SbCl$_6$, 1, 210
$FeSbC_{19}H_{15}N_2O_3$ [Fe(SbPh$_3$)(CO)(NO)$_2$], 3, 662
$FeSbC_{21}H_{15}I_2O_3$ [Fe(SbPh$_3$)(CO)$_3$I$_2$], 3, 662
$FeSbC_{22}H_{15}O_4$ [Fe(SbPh$_3$)(CO)$_4$], 3, 662
$FeSbC_{22}H_{21}N_2O_3$ {Fe[Sb(C$_6$H$_4$Me)$_3$](CO)(NO)$_2$}, 3, 667
$FeSbC_{28}H_{23}O_2$ [CH=CH(CH=CH)$_2$CH=CH]Fe(CO)$_2$(SbPh$_3$), 1, 501, 3, 662
$FeSbC_{39}H_{30}O_3P$ [Fe(SbPh$_3$)(PPh$_3$)(CO)$_3$], 3, 662
$FeSb_2C_{39}H_{30}O_3$ [Fe(SbPh$_3$)$_2$(CO)$_3$], 3, 662
$FeSiC_8H_9NO_4$ (Me$_3$SiNC)Fe(CO)$_4$, 1, 524
$FeSiC_{10}H_{14}O_2$ C$_5$H$_5$Fe(CO)$_2$SiMe$_3$, 1, 533
$FeSiC_{13}H_{18}$ FcSiMe$_3$, 1, 535
FcSiHEtMe, 1, 535
$(FeSiC_{14}H_{16}O_2)_n$ [FcCOCH$_2$CH$_2$Si(Me)O]$_n$, 1, 546
$FeSiC_{14}H_{20}$ FcSiHEt$_2$, 1, 535
FcCH$_2$SiMe$_3$, 1, 536
$FeSiC_{14}H_{20}O$ FcSi(OEt)Me$_2$, 1, 535
$FeSiC_{15}H_{20}$ FcSi(CH$_2$CH=CH$_2$)Me$_2$, 1, 535
$FeSiC_{15}H_{20}O$ FcCOCH$_2$SiMe$_3$, 1, 536
$FeSiC_{15}H_{22}$ FcCH(Me)SiMe$_3$, 1, 536
$FeSiC_{16}H_{22}O$ FcCOCH$_2$CH$_2$SiMe$_3$, 1, 536
$FeSiC_{16}H_{24}$ FcSiEt$_3$, 1, 535
Fc(CH$_2$)$_3$SiMe$_3$, 1, 536
$FeSiC_{16}H_{24}O$ Fc(CH$_2$)$_4$Si(OH)Me$_2$, 1, 536
$FeSiC_{17}H_{22}O_3$ MeO$_2$C-Fcd-COCH$_2$SiMe$_3$, 1, 541
$FeSiC_{17}H_{24}O$ FcCO(CH$_2$)$_3$SiMe$_3$, 1, 536
$FeSiC_{17}H_{26}$ Fc(CH$_2$)$_4$SiMe$_3$, 1, 536
Me-Fcd-SiEt$_3$, 1, 541
$FeSiC_{17}H_{26}O$ FcCH(OH)(CH$_2$)$_3$SiMe$_3$, 1, 536
$FeSiC_{18}H_{20}$ FcSiPhMe$_2$, 1, 535
$FeSiC_{18}H_{28}O$ FcC(OH)(Me)(CH$_2$)$_3$SiMe$_3$, 1, 536
$FeSiC_{19}H_{19}F_3$ FcSi(C$_6$H$_4$CF$_3$-m)Me$_2$, 1, 535
$FeSiC_{19}H_{21}F_9$ FcSi(CH$_2$CH$_2$CF$_3$)$_3$, 1, 535
$FeSiC_{19}H_{30}$ Me$_2$CH-Fcd-SiEt$_3$, 1, 541
FcSi(C$_6$H$_{13}$)EtMe, 1, 533
$FeSiC_{20}H_{32}$ Me$_3$C-Fcd-SiEt$_3$, 1, 541
$FeSiC_{23}H_{30}O$ FcC(OH)(Ph)(CH$_2$)$_3$SiMe$_3$, 1, 536
$FeSiC_{28}H_{24}$ FcSiPh$_3$, 1, 535

$FeSiC_{28}H_{48}$ $FcSi(C_6H_{13}\text{-}n)_3$, 1, 535
$FeSi_2C_{12}H_{16}O_3$ $Me\overline{Si(OH)\text{-}Fcd\text{-}SiMe(OH)O}$, 1, 544
$FeSi_2C_{12}H_{18}O_4$ $Fcd[SiMe(OH)_2]_2$, 1, 538
$FeSi_2C_{14}H_{20}Cl_2$ $Fcd(SiMe_2Cl)_2$, 1, 538
$FeSi_2C_{14}H_{20}O$ $Me_2\overline{Si\text{-}Fcd\text{-}SiMe_2O}$, 1, 544
$FeSi_2C_{14}H_{20}O_4$ Bis(trimethylsilyl)-cyclopentadienone-$Fe(CO)_3$, 1, 484, 546
$FeSi_2C_{16}H_{24}Cl_2$ $FcSiMe_2(CH_2)_3SiCl_2Me$, 1, 535
$Fcd(SiMe_2CH_2Cl)_2$, 1, 538
$FeSi_2C_{16}H_{24}O_3$ $Me(EtO)\text{-}\overline{Si\text{-}Fcd\text{-}Si(OEt)MeO}$, 1, 544
$FeSi_2C_{16}H_{26}$ $Fcd(SiMe_3)_2$, 1, 538
$Fcd(SiHMeEt)_2$, 1, 538
$FeSi_2C_{16}H_{28}N_2$ $Fcd[Si(CH_2NH_2)Me_2]_2$, 1, 538
$FeSi_2C_{18}H_{29}Cl$ $FcSiMe_2(CH_2)_3SiClEtMe$, 1, 536
$FeSi_2C_{18}H_{30}$ $Fcd(SiMe_2Et)_2$, 1, 538
$Fcd(SiHEt_2)_2$, 1, 538
$FeSi_2C_{18}H_{30}O_2$ $Fcd[Si(OEt)Me_2]_2$, 1, 538
$FeSi_2C_{19}H_{32}$ $Me_3Si\text{-}Fcd\text{-}SiEt_3$, 1, 541
$FeSi_2C_{20}H_{26}O$ $FcSiMe_2OSiPhMe_2$, 1, 535
$(FeSi_2C_{20}H_{28}O_3)_n$ $\{SiMe_2CH_2CH_2CO\text{-}Fcd\text{-}COCH_2CH_2SiMe_2O\}_n$, 1, 545
$(FeSi_2C_{20}H_{32}O)_n$ $\{SiMe_2(CH_2)_3\text{-}Fcd\text{-}(CH_2)_3SiMe_2O\}_n$, 1, 545
$FeSi_2C_{20}H_{34}$ $FcSiMe_2(CH_2)_3SiEt_2Me$, 1, 536
$FeSi_2C_{20}H_{34}O_4$ $Fcd[SiMe(OEt)_2]_2$, 1, 538
$FeSi_2C_{21}H_{36}$ $FcSiMe_2(CH_2)_3SiEt_3$, 1, 536
$(FeSi_2C_{22}H_{32}O_3)_n$ $\{SiMe_2(CH_2)_3CO\text{-}Fcd\text{-}CO(CH_2)_3SiMe_2O\}_n$, 1, 545
$FeSi_2C_{22}H_{34}O_2$ $Fcd(COCH_2CH_2SiMe_3)_2$, 1, 538
$(FeSi_2C_{22}H_{36}O)_n$ $\{SiMe_2(CH_2)_4\text{-}Fcd\text{-}(CH_2)_4SiMe_2O\}_n$, 1, 546
$FeSi_2C_{22}H_{38}$ $Fcd(SiEt_3)_2$, 1, 538
$Fcd[(CH_2)_3SiMe_3]_2$, 1, 538
$FeSi_2C_{23}H_{40}O$ $FcC(OH)[(CH_2)_3SiMe_3]_2$, 1, 536
$FeSi_2C_{24}H_{30}$ $Fcd(SiMe_2C_5H_5)_2$, 1, 539
$FeSi_2C_{24}H_{38}O_2$ $Fcd[CO(CH_2)_3SiMe_3]_2$, 1, 538
$FeSi_2C_{24}H_{40}N_2$ $Fcd[SiMe_2\overline{N(CH_2)_4CH_2}]_2$, 1, 539
$FeSi_2C_{24}H_{42}$ $FcSiMe_2(CH_2)_3SiBu_2Me$, 1, 536
$Fcd[(CH_2)_4SiMe_3]_2$, 1, 539
$FeSi_2C_{26}H_{28}O_4$ Diphenylbis(trimethylsilyl)cyclopentadienone-$Fe(CO)_3$, 1, 485, 546
$FeSi_2C_{28}H_{28}F_6$ $Fcd[Si(C_6H_4CF_3\text{-}m)Me_2]_2$, 1, 539
$FeSi_2C_{28}H_{32}F_{18}$ $Fcd[Si(CH_2CH_2CF_3)_3]_2$, 1, 746
$FeSi_2C_{28}H_{50}$ $Fcd(SiPr_3)_2$, 1, 539
$FeSi_2C_{32}H_{32}N_2O_4$ $Fcd[SiMe_2CH_2N(CO)_2\text{-}C_6H_4]_2$, 1, 539
$FeSi_2C_{46}H_{38}$ $Fcd(SiPh_3)_2$, 1, 540
$FeSi_2C_{46}H_{86}$ $Fcd[Si(C_6H_{13})_3]_2$, 1, 540
$FeSi_2C_{82}H_{158}$ $Fcd[Si(C_{12}H_{25})_3]_2$, 1, 540
$FeSi_3C_{16}H_{26}O_2$ $Me_2\overline{Si\text{-}Fcd\text{-}SiMe_2OSiMe_2O}$, 1, 545
$FeSi_3C_{23}H_{34}O$ $Me_3Si\text{-}Fcd\text{-}SiMe_2OSiPhMe_2$, 1, 541
$FeSi_3C_{27}H_{41}NO$ $Me_2[\overline{CH_2(CH_2)_4N}]Si\text{-}Fcd\text{-}SiMe_2OSiMe_2Ph$, 1, 541
$FeSi_3C_{28}H_{36}O$ $Me_2PhSi\text{-}Fcd\text{-}SiMe_2OSi\text{-}PhMe_2$, 1, 541
$FeSi_3C_{28}H_{52}$ Tris(triethylsilyl)ferrocene, 1, 542
$FeSi_4C_{20}H_{38}O_2$ $Fcd(SiMe_2OSiMe_3)_2$, 1, 538
$FeSi_4C_{30}H_{38}Cl_4O_2$ $Fcd(SiMe_2OSiMe_2C_6H_3\text{-}Cl_2)_2$, 1, 539
$FeSi_4C_{30}H_{42}O_2$ $Fcd(SiMe_2OSiMe_2Ph)_2$, 1, 539
$FeSi_4C_{32}H_{46}O_2$ $Fe[C_5H_3(Me)SiMe_2OSiMe_2\text{-}Ph]_2$, 1, 539
$FeSi_4C_{36}H_{70}O_2$ $Fcd\{C(OH)[(CH_2)_3Si\text{-}Me_3]_2\}_2$, 1, 539
$FeSi_4C_{50}H_{50}$ $Fcd(SiMe_2OSiPh_3)_2$, 1, 540
$FeSi_6C_{26}H_{54}O_6$ $Fcd(SiMe_2OSiMe_2OSiMe_2O\text{-}Et)_2$, 1, 539
$FeSi_6C_{34}H_{54}O_4$ $Fcd(SiMe_2OSiMe_2OSiMe_2\text{-}Ph)_2$, 1, 539*
$FeSi_8C_{38}H_{66}O_6$ $Fcd[SiMe_2(OSiMe_2)_2OSi\text{-}$

$Me_2Ph]_2$, 1, 540
$(FeSi_{13}C_{41}H_{98}O_{12})_n$ $\{[SiMe_2(CH_2)_3$-Fcd-$(CH_2)_2SiMe_2O][SiMe_2O]_{11}\}_n$, 1, 546
$FeSnC_7H_5Cl_3O_2$ $C_5H_5Fe(CO)_2SnCl_3$, 1, 533
$FeSnC_{12}H_{18}O_4$ $Bu_2SnFe(CO)_4$, 2, 502
$FeSnC_{25}H_{20}O_2$ $C_5H_5Fe(CO)_2SnPh_3$, 1, 533, 2, 502
$FeSnC_{29}H_{33}O_3P$ $Bu_2SnFe(CO)_3PPh_3$, 2, 502
$FeSn_2C_{28}H_{54}O_4$ $(Bu_3Sn)_2Fe(CO)_4$, 2, 502
$FeSn_2C_{40}H_{30}O_4$ $(Ph_3Sn)_2Fe(CO)_4$, 2, 502
$FeSn_3C_{12}H_{18}N_6O$ $(Me_2Sn)_2Fe(CN)_6 \cdot Me_2SnO$, 2, 345
$FeSn_3C_{60}H_{45}N_6$ $(Ph_3Sn)_3Fe(CN)_6$, 2, 346
$FeSn_4C_{78}H_{60}N_6$ $(Ph_3Sn)_4Fe(CN)_6$, 2, 346
$Fe_2AsC_{16}H_{16}ClO_8$ $\{[C_5H_5Fe(CO)_2]_2AsMe_2\}ClO_4$, 1, 216, 3, 304
$Fe_2AsC_{27}H_{25}O_5S_2$ $[Fe_2(AsPh_3)(CO)_5(SEt)_2]$, 3, 22
$Fe_2AsC_{35}H_{25}O_5S_2$ $[Fe_2(AsPh_3)(CO)_5(SPh)_2]$, 3, 22
$Fe_2As_2C_4F_{12}N_4O_4$ $[Fe(NO)_2As(CF_3)_2]_2$, 3, 302
$Fe_2As_2C_4H_{12}N_4O_4$ $[Fe(NO)_2AsMe_2]_2$, 3, 303
$Fe_2As_2C_{10}F_{12}O_6$ $[Fe(CO)_3As(CF_3)_2]_2$, 3, 303
$Fe_2As_2C_{10}H_{12}Br_2O_6$ $[Fe(CO)_3(AsMe_2)Br]_2$, 3, 303
$Fe_2As_2C_{10}H_{12}I_2O_6$ $[Fe(CO)_3(AsMe_2)I]_2$, 3, 303
$Fe_2As_2C_{10}H_{12}O_6$ $[Fe(CO)_3AsMe_2]_2$, 3, 303
$Fe_2As_2C_{16}H_{10}F_{12}O_2$ $[C_5H_5Fe(CO)As(CF_3)_2]_2$, 3, 304
$Fe_2As_2C_{16}H_{22}O_2$ $[C_5H_5Fe(CO)AsMe_2]_2$, 3, 304
$Fe_2As_2C_{36}H_{30}O_2$ $[C_5H_5Fe(CO)(AsPh_2)]_2$, 1, 215
$Fe_2BC_{13}H_7F_4O_6$ {Cycloheptatrienylium-$[Fe(CO)_3]_2\}BF_4$, 1, 511
$Fe_2BC_{40}H_{36}O_4P$ $\{[C_5H_5Fe(CO)_2]_2PMe_2\}BPh_4$, 1, 216
$Fe_2CoC_{44}H_{38}Cl$ $\{[FcCH(Ph)C_5H_4]_2Co\}Cl$, 1, 547
$Fe_2CoC_{44}H_{38}ClO_4$ $\{[FcCH(Ph)C_5H_4]_2Co\}ClO_4$, 1, 547
$Fe_2CoC_{50}H_{40}N_3O_7$ $\{[FcCH(Ph)C_5H_4]_2Co\}OC_6H_2(NO_2)_3$, 1, 547
$Fe_2Co_2C_{28}H_{20}O_8$ $[(C_5H_5)_2Co]_2[Fe_2(CO)_8]$, 1, 587
$Fe_2HgC_{14}H_{10}O_4$ $Hg[C_5H_5Fe(CO)_2]_2$, 1, 207
$Fe_2HgC_{20}H_{16}Br_2$ $(BrFcd)_2Hg$, 1, 531
$Fe_2HgC_{20}H_{16}Cl_2$ $(ClFcd)_2Hg$, 1, 531
$Fe_2HgC_{20}H_{18}$ Fc_2Hg, 1, 530
$Fe_2K_8C_{14}H_6N_{12}$ $K_8[-C{\equiv}CFe(CN)_5]_2 \cdot 2NH_3$, 1, 197
$Fe_2NiC_{17}H_{10}O_7$ $[C_5H_5Fe(CO)_2]_2Ni(CO)_3$, 1, 548
$Fe_2Ni_2C_{24}H_{16}O_6$ $[(C_5H_5Ni)_2(PhC{\equiv}CH)][Fe(CO)_3]_2$, 1, 548
$Fe_2Ni_2C_{30}H_{20}O_6$ $[(C_5H_5Ni)_2(PhC{\equiv}CPh)][Fe(CO)_3]_2$, 1, 548
$Fe_2PbC_{16}H_{16}O_4$ $[C_5H_5Fe(CO)_2]_2PbMe_2$, 1, 534, 2, 595
$Fe_2PbC_{20}H_{24}O_9S_2$ $(FcSO_3)_2Pb \cdot 3H_2O$, 1, 318
$Fe_2Pb_2C_{12}H_{12}O_8$ $[Me_2PbFe(CO)_4]_2$, 2, 595
$Fe_2Pb_2C_{16}H_{20}O_8$ $[Et_2PbFe(CO)_4]_2$, 2, 595
$Fe_2Pb_2C_{20}H_{28}O_8$ $[Pr_2PbFe(CO)_4]_2$, 2, 595
$Fe_2Pb_2C_{24}H_{36}O_8$ $[Bu_2PbFe(CO)_4]_2$, 2, 595
$Fe_2Pb_2C_{32}H_{20}O_8$ $[Ph_2PbFe(CO)_4]_2$, 2, 595
$Fe_2PtC_{18}H_{16}Cl_6N_2O_4$ $[C_5H_5Fe(CO)_2(HN{=}C{=}CH_2)]_2PtCl_6$, 1, 210
$Fe_2PtC_{20}H_{20}Cl_6$ $[FcH]_2PtCl_6$, 1, 228
$Fe_2PtC_{20}H_{20}Cl_6N_2O_4$ $[C_5H_5Fe(CO)_2(HN{=}C{=}CHMe)]_2PtCl_6$, 1, 210
$Fe_2PtC_{22}H_{26}Cl_6O_4$ $[C_5H_5Fe(CO)_2(CH_2{=}CHEt)]_2PtCl_6$, 1, 210
$Fe_2PtC_{50}H_{40}Cl_6O_4P_2$ $[C_5H_5Fe(CO)_2(PPh_3)]_2PtCl_6$, 1, 209
$Fe_2SbC_{35}H_{25}O_5S_2$ $[Fe_2(Ph_3Sb)(CO)_5(SPh)_2]$, 3, 663

$Fe_2Si_2C_{16}H_{20}O_6$ $(Me_3SiC{\equiv}CH)_2Fe_2(CO)_6$, 1, 518

$Fe_2Si_2C_{28}H_{28}O_6$ $(PhC{\equiv}CSiMe_3)_2Fe_2(CO)_6$, 1, 518*

$Fe_2Si_2C_{29}H_{28}O_7$ $(PhC{\equiv}CSiMe_3)_2Fe_2(CO)_7$, 1, 518

$Fe_2Si_2C_{30}H_{38}O_3$ $(FcCOCH_2CH_2SiMe_2)_2O$, 1, 543

$Fe_2Si_2C_{30}H_{42}O$ $[Fc(CH_2)_3SiMe_2]_2O$, 1, 543

$Fe_2Si_2C_{32}H_{42}O_3$ $[FcCO(CH_2)_3SiMe_2]_2O$, 1, 543

$Fe_2Si_2C_{32}H_{46}O$ $[Fc(CH_2)_4SiMe_2]_2O$, 1, 543

$Fe_2Si_2C_{40}H_{46}O_4$ p-$(Me_3SiCH_2CH_2CO$-Fcd-$CO)_2C_6H_4$, 1, 543

$Fe_2Si_4C_{24}H_{28}O_4$ $\{OSi(Me)$-Fcd-$Si(Me)$-$(O)\}_2$, 1, 543

$Fe_2Si_4C_{40}H_{50}O$ $(Me_2PhSi$-Fcd-$SiMe_2)_2O$, 1, 543

$Fe_2Si_5C_{32}H_{52}O_2$ $(Me_3Si$-Fcd-$SiMe_2O)_2$-$SiMe_2$, 1, 544

$Fe_2Si_5C_{42}H_{56}O_2$ $(Me_2PhSi$-Fcd-$SiMe_2O)_2$-$SiMe_2$, 1, 544

$Fe_2Si_6C_{44}H_{62}O_3$ $(Me_2PhSiOSiMe_2$-Fcd-$SiMe_2)_2O$, 1, 543

$Fe_2SnC_{14}H_{10}Cl_2O_4$ $[C_5H_5Fe(CO)_2]_2SnCl_2$, 1, 534

$Fe_2SnC_{16}H_{16}O_4$ $[C_5H_5Fe(CO)_2]_2SnMe_2$, 1, 533, 2, 502

$Fe_2SnC_{18}H_{20}O_4$ $[C_5H_5Fe(CO)_2]_2SnEt_2$, 1, 533, 2, 502

$Fe_2SnC_{20}H_{18}S_2$ $(FcS)_2Sn$, 1, 544

$Fe_2SnC_{26}H_{20}O_4$ $[C_5H_5Fe(CO)_2]_2SnPh_2$, 1, 534

$Fe_2Sn_2C_{12}H_{12}O_8$ $[Me_2SnFe(CO)_4]_2$, 2, 502

$Fe_2Sn_2C_{16}H_{20}O_8$ $[Et_2SnFe(CO)_4]_2$, 2, 502

$Fe_2Sn_2C_{24}H_{36}O_8$ $[Bu_2SnFe(CO)_4]_2$, 2, 502

$Fe_2Sn_2C_{36}H_{28}O_8$ $[(PhCH_2)_2SnFe(CO)_4]_2$, 2, 502

$Fe_2Sn_2C_{36}H_{30}O_4$ $[Ph_2SnC_5H_5Fe(CO)_2]_2$, 2, 502

$Fe_2Sn_2C_{46}H_{34}O_4$ $[Ph_2SnC_9H_7Fe(CO)_2]_2$, 2, 502

$Fe_2Sn_3C_{18}H_{18}N_{12}$ $(Me_2Sn)_3[Fe(CN)_6]_2$, 2, 345*

$Fe_3AlC_{31}H_{27}Cl_4$ $[Fc_3C]AlCl_4$, 1, 431

$Fe_3BC_{55}H_{47}$ $[Fc_3C]BPh_4$, 1, 341

$Fe_3GeC_{49}H_{36}O_6$ BuGe[9-fluorenyl-Fe-$(CO)_2]_3$, 2, 141

$Fe_3Si_2C_{30}H_{28}O_8$ $(PhC{\equiv}CSiMe_3)_2Fe_3(CO)_8$, 1, 518

$Fe_3Si_6C_{54}H_{70}O_2$ $(Me_2PhSi$-Fcd-$SiMe_2O$-$SiMe_2)_2$Fcd, 1, 544

$Fe_3Si_7C_{61}H_{78}O_3$ $(Me_2PhSi$-Fcd-$SiMe_2O)_3$-SiMe, 1, 544

$Fe_4Co_2C_{33}H_{20}O_{13}$ $[(C_5H_5)_2Co]_2[Fe_4$-$(CO)_{13}]$, 1, 587

$Fe_4CrC_{37}H_{20}O_{13}$ $[(Ph_2)_2Cr]Fe_4(CO)_{13}$, 1, 84

$Fe_4Si_9C_{80}H_{100}O_4$ $(Me_2PhSi$-Fcd-$SiMe_2O)_4$-Si, 1, 544

$FeCoHgC_{11}H_5O_6$ $C_5H_5Fe(CO)_2HgCo(CO)_4$, 1, 529

$Fe_2AsBC_{40}H_{38}O_4$ $\{[C_5H_5Fe(CO)_2]_2AsMe_2\}$-$BPh_4$, 1, 216, 3, 304

$Fe_2As_2CdC_{40}H_{30}N_2O_6$ $Cd[Fe(AsPh_3)(NO)$-$(CO)_2]_2$, 3, 22

$Fe_2As_2HgC_{40}H_{30}N_2O_6$ $Hg[Fe(AsPh_3)(NO)$-$(CO)_2]_2$, 3, 22

$Fe_2As_2PtC_{50}H_{40}Cl_6O_4$ $[C_5H_5Fe(CO)_2$-$(AsPh_3)]_2PtCl_6$, 1, 209

$Fe_2CoCrC_{48}H_{44}N_6S_4$ $\{[FcCH(Ph)C_5H_4]_2$-$Co\}[Cr(NH_3)_2(SCN)_4]$, 1, 547

$Fe_2HgSb_2C_{40}H_{30}N_2O_6$ $Hg[Fe(SbPh_3)(CO)_2$-$(NO)]_2$, 3, 663

$Fe_2PtSb_2C_{50}H_{40}Cl_6O_4$ $[C_5H_5Fe(CO)_2$-$(SbPh_3)]_2PtCl_6$, 1, 209

$Fe_3As_3SnC_{60}H_{45}ClN_3O_9$ $ClSn[Fe(AsPh_3)$-$(NO)(CO)_2]_3$, 3, 22

$Fe_4As_4SnC_{80}H_{60}N_4O_{12}$ $Sn[Fe(AsPh_3)(NO)$-$(CO)_2]_4$, 3, 22

$GaAsC_6H_{18}$ $Me_3AsGaMe_3$, 3, 4

$GaSbC_6H_{18}$ $[Me_3SbGaMe_3]$, 3, 655

$Ga_2As_2C_{28}H_{32}$ $(Ph_2AsGaMe_2)_2$, 3, 298

$GdC_{10}H_{10}Cl$ $(C_5H_5)_2GdCl$, 1, 2

$GdC_{12}H_{14}Cl$ $(MeC_5H_4)_2GdCl$, 1, 2

$GdC_{14}H_{17}O_2$ $(MeC_5H_4)_2GdOAc$, 1, 3

$GdC_{15}H_{15}$ $(C_5H_5)_3Gd$, 1, 4

$GdC_{17}H_{29}Cl_2O_3$ $C_5H_5GdCl_2 \cdot 3THF$, 1, 1

$GeCBr_3D_3$ CD_3GeBr_3, 2, 86

$GeCCl_3F_3$ CF_3GeCl_3, 2, 90
$GeCCl_6$ Cl_3CGeCl_3, 2, 90
$GeCF_3I_3$ CF_3GeI_3, 2, 89
$GeCF_6$ CF_3GeF_3, 2, 90
$GeCH_2Br_4$ $BrCH_2GeBr_3$, 2, 90
$GeCH_2Cl_4$ $ClCH_2GeCl_3$, 2, 90
$GeCH_2I_4$ ICH_2GeI_3, 2, 90
$GeCH_3Br_3$ $MeGeBr_3$, 2, 84
$GeCH_3Cl_3$ $MeGeCl_3$, 2, 83
$GeCH_3D_3$ $MeGeD_3$, 2, 61
  CD_3GeH_3, 2, 61
$GeCH_3F_3$ $MeGeF_3$, 2, 86
$GeCH_3I_3$ $MeGeI_3$, 2, 86
$GeCH_4Cl_2$ $MeHGeCl_2$, 2, 64
$GeCH_5Br$ MeH_2GeBr, 2, 64
$GeCH_5Cl$ MeH_2GeCl, 2, 64
$GeCH_6$ $MeGeH_3$, 2, 53
$GeC_2Br_2D_6$ $(CD_3)_2GeBr_2$, 2, 80
$GeC_2F_6I_2$ $(CF_3)_2GeI_2$, 2, 82
$GeC_2HCl_3F_2$ $(C_2F_2H)GeCl_3$, 2, 90
GeC_2HCl_7 $Cl_2CHCCl_2GeCl_3$, 2, 90
$GeC_2H_2Cl_4$ $ClCH{=}CHGeCl_3$, 2, 90
  $CH_2{=}CClGeCl_3$, 2, 90
$GeC_2H_2Cl_6$ $Cl_2CHCHClGeCl_3$, 2, 90
$GeC_2H_2D_6$ $(CD_3)_2GeH_2$, 2, 61
$GeC_2H_3Br_3$ $CH_2{=}CHGeBr_3$, 2, 88
$GeC_2H_3Cl_3$ $CH_2{=}CHGeCl_3$, 2, 85
$GeC_2H_3Cl_5$ $Cl_2CHCH_2GeCl_3$, 2, 90
  $ClCH_2CHClGeCl_3$, 2, 90
$GeC_2H_4Cl_4$ $(CH_2Cl)_2GeCl_2$, 2, 82
  $ClCH_2CH_2GeCl_3$, 2, 89
  $MeCHClGeCl_3$, 2, 90
$GeC_2H_5Br_3$ $EtGeBr_3$, 2, 86
$GeC_2H_5Cl_3$ $Me(CH_2Cl)GeCl_2$, 2, 82
  $EtGeCl_3$, 2, 84
$GeC_2H_5D_3$ $EtGeD_3$, 2, 61
$GeC_2H_5I_3$ $EtGeI_3$, 2, 84
$GeC_2H_5I_3O$ $MeOCH_2GeI_3$, 2, 91
GeC_2H_6 $CH_2{=}CHGeH_3$, 2, 56
$(GeC_2H_6)_n$ $(Me_2Ge)_n$, 2, 131
$GeC_2H_6Br_2$ Me_2GeBr_2, 2, 78
$GeC_2H_6Cl_2$ Me_2GeCl_2, 2, 77, 157
  $EtHGeCl_2$, 2, 64
$GeC_2H_6D_2$ Me_2GeD_2, 2, 61
$GeC_2H_6F_2$ Me_2GeF_2, 2, 80
$GeC_2H_6I_2$ Me_2GeI_2, 2, 80
$GeC_2H_6N_6$ $Me_2Ge(N_3)_2$, 2, 94
$(GeC_2H_6O)_n$ $(Me_2GeO)_n$, 2, 102
$GeC_2H_6O_4S$ Me_2GeSO_4, 2, 97
$(GeC_2H_6Se)_n$ $(Me_2GeSe)_n$, 2, 121
GeC_2H_7Cl Me_2HGeCl, 2, 64
GeC_2H_8 Me_2GeH_2, 2, 53
  $EtGeH_3$, 2, 56
$GeC_3H_4Cl_3F_3$ $CF_3CH_2CH_2GeCl_3$, 2, 90
$GeC_3H_4Cl_3N$ $NCCH_2CH_2GeCl_3$, 2, 91
$GeC_3H_5Br_3$ $CH_2{=}CHCH_2GeBr_3$, 2, 88
$GeC_3H_5Cl_3$ $CH_2{=}CHCH_2GeCl_3$, 2, 85
$GeC_3H_5D_3$ $CH_2{=}CHCH_2GeD_3$, 2, 61, 156
$GeC_3H_5F_3$ $CH_2{=}CHCH_2GeF_3$, 2, 88
$GeC_3H_6Cl_4$ $Cl(CH_2)_3GeCl_3$, 2, 90
  $ClCH_2CHMeGeCl_3$, 2, 91
  $MeCHClCH_2GeCl_3$, 2, 91
$(GeC_3H_7)_n$ $(PrGe)_n$ (?), 2, 132
  $(i\text{-}PrGe)_n$, 2, 132
$GeC_3H_7Br_3$ $PrGeBr_3$, 2, 86
  $i\text{-}PrGeBr_3$, 2, 86
$GeC_3H_7Cl_3$ $Me(CH_2Cl)_2GeCl$, 2, 75
  $PrGeCl_3$, 2, 84
  $i\text{-}PrGeCl_3$, 2, 86
$GeC_3H_7D_3$ $PrGeD_3$, 2, 61
$GeC_3H_7F_3$ $CF_3CH_2CH_2GeH_3$, 2, 62
GeC_3H_8 $CH_2{=}CHCH_2GeH_3$, 2, 57, 156
$GeC_3H_8Cl_2$ $MeEtGeCl_2$, 2, 80
GeC_3H_9Br Me_3GeBr, 2, 66
GeC_3H_9Cl Me_3GeCl, 2, 66
  PrH_2GeCl, 2, 64
$GeC_3H_9ClO_4$ Me_3GeClO_4, 2, 97
$GeC_3H_9Cl_2O_2P$ $Me_3GePO_2Cl_2$, 2, 97
GeC_3H_9D Me_3GeD, 2, 61
GeC_3H_9F Me_3GeF, 2, 71
GeC_3H_9I Me_3GeI, 2, 67
$GeC_3H_9NO_3$ Me_3GeNO_3, 2, 97
$GeC_3H_9N_3$ Me_3GeN_3, 2, 94
GeC_3H_{10} Me_3GeH, 2, 53
  $PrGeH_3$, 2, 56
$GeC_3H_{10}O$ Me_3GeOH, 2, 101
$GeC_4H_6Cl_2$ $\overline{CH_2CH{=}CHCH_2Ge}Cl_2$, 2, 152
$GeC_4H_6F_2$ $(CH_2{=}CH)_2GeF_2$, 2, 80
$GeC_4H_6N_2S_2$ $Me_2Ge(SCN)_2$, 2, 94
$GeC_4H_7Cl_3$ $MeCH{=}CHCH_2GeCl_3$, 2, 88
  $CH_2{=}CMeCH_2GeCl_3$, 2, 85
$GeC_4H_7Cl_3O_2$ $AcOCH_2CH_2GeCl_3$, 2, 91
$GeC_4H_7I_3O_2$ $EtO_2CCH_2GeI_3$, 2, 91
$GeC_4H_8Br_2$ $Me_2(BrCH{=}CH)GeBr$, 2, 75

$CH_2(CH_2)_3GeBr_2$, 2, 151
$GeC_4H_8Br_4$ $Me_2CBrCH_2GeBr_3$, 2, 91
$GeC_4H_8Cl_2$ $CH_2(CH_2)_3GeCl_2$, 2, 151
$GeC_4H_8Cl_4$ $ClCH_2CHMeCH_2GeCl_3$, 2, 91
$GeC_4H_8I_2$ $CH_2(CH_2)_3GeI_2$, 2, 151
GeC_4H_8O $CH_2(CH_2)_3GeO$, 2, 154
$GeC_4H_8O_2S$ $Me_2GeO_2CCH_2S$, 2, 117
$GeC_4H_9Br_3$ $BuGeBr_3$, 2, 86
$GeC_4H_9Cl_3$ $BuGeCl_3$, 2, 86
i-$BuGeCl_3$, 2, 86
$GeC_4H_9I_3$ $BuGeI_3$, 2, 86
GeC_4H_9I $CH_2(CH_2)_3GeHI$, 2, 149
GeC_4H_9N Me_3GeCN, 2, 94
GeC_4H_9NS Me_3GeSCN, 2, 94
GeC_4H_{10} $CH_2(CH_2)_3GeH_2$, 2, 149
$GeC_4H_{10}Br_2$ Et_2GeBr_2, 2, 80
$GeC_4H_{10}Cl_2$ $Me_3GeCHCl_2$, 2, 28
Et_2GeCl_2, 2, 78
$GeC_4H_{10}D_2$ Et_2GeD_2, 2, 61
$GeC_4H_{10}I_2$ Et_2GeI_2, 2, 80
$(GeC_4H_{10}O)_n$ $(Et_2GeO)_n$, 2, 102
$GeC_4H_{10}OS$ $Me_2GeOCH_2CH_2S$, 2, 110
$GeC_4H_{10}O_2$ $Me_2GeOCH_2CH_2O$, 2, 107
$(GeC_4H_{10}O_2)_n$ $(Me_2GeOCH_2CH_2O)_n$, 2, 107
$GeC_4H_{10}O_4S$ Et_2GeSO_4, 2, 98
$GeC_4H_{10}S_2$ $Me_2GeSCH_2CH_2S$, 2, 122
$GeC_4H_{11}Br$ Me_3GeCH_2Br, 2, 28
$Me_2EtGeBr$, 2, 74
Et_2HGeBr, 2, 64
BuH_2GeBr, 2, 64
$GeC_4H_{11}D$ Me_2EtGeD, 2, 61
$GeC_4H_{11}Cl$ Me_3GeCH_2Cl, 2, 27
$Me_2EtGeCl$, 2, 74
Et_2HGeCl, 2, 63
BuH_2GeCl, 2, 64
$GeC_4H_{11}I$ Me_3GeCH_2I, 2, 28
Et_2HGeI, 2, 64
BuH_2GeI, 2, 64
GeC_4H_{12} Me_4Ge, 2, 1
Me_2EtGeH, 2, 60
Et_2GeH_2, 2, 56
$BuGeH_3$, 2, 57
$GeC_4H_{12}O$ Me_3GeCH_2OH, 2, 33
Me_3GeOMe, 2, 107
$GeC_4H_{12}O_2$ $Me_2Ge(OMe)_2$, 2, 107
$GeC_4H_{12}O_3$ $MeGe(OMe)_3$, 2, 107
$GeC_4H_{13}O_3P$ $Me_3GeCH_2PO_3H_2$, 2, 44
GeC_4N_4 $Ge(CN)_4$, 2, 96
$GeC_5H_5N_3$ $EtGe(CN)_3$ (?), 2, 95
$GeC_5H_5N_3O_3$ $EtGe(OCN)_3$, 2, 95
$GeC_5H_7Cl_3$ Cyclopentenyl-$GeCl_3$, 2, 88
$GeC_5H_7Cl_4F_3O$ $ClFCHCF_2O(CH_2)_3GeCl_3$, 2, 91
$GeC_5H_8O_6$ $EtGe(O_2CH)_3$, 2, 114
$GeC_5H_9Cl_3$ Cyclopentyl-$GeCl_3$, 2, 86
$MeCH{=}CHCHMeGeCl_3$, 2, 88
$GeC_5H_9Cl_3O_2$ $AcO(CH_2)_3GeCl_3$, 2, 91
GeC_5H_{10} $Me_3GeC{\equiv}CH$, 2, 22
$GeC_5H_{10}Br_2$ $CH_2(CH_2)_4GeBr_2$, 2, 152
$GeC_5H_{10}Cl_2$ $CH_2(CH_2)_4GeCl_2$, 2, 152
$GeC_5H_{10}Cl_4$ $MeCHCl(CH_2)_3GeCl_3$, 2, 91
$GeC_5H_{10}I_2$ $CH_2(CH_2)_4GeI_2$, 2, 152
$GeC_5H_{10}O$ $CH_2(CH_2)_4GeO$, 2, 154
$GeC_5H_{11}Cl$ $Me_3GeCH{=}CHCl$, 2, 20
$Me_3GeCCl{=}CH_2$, 2, 20
$GeC_5H_{11}Cl_3$ n-$C_5H_{11}GeCl_3$, 2, 86
$GeC_5H_{11}I$ $CH_2(CH_2)_4GeHI$, 2, 149
$GeC_5H_{11}NS$ Me_3GeCH_2SCN, 2, 31
GeC_5H_{12} $Me_3GeCH{=}CH_2$, 2, 13, 156
$CH_2(CH_2)_4GeH_2$, 2, 149
$GeC_5H_{12}O_2$ $Me_3GeCH_2CO_2H$, 2, 39
Me_3GeOAc, 2, 113
$GeC_5H_{12}S_2$ $Me_2GeS(CH_2)_3S$, 2, 122
$GeC_5H_{13}Cl$ $Me_3GeCHMeCl$, 2, 28
GeC_5H_{14} Me_3GeEt, 2, 7
n-$C_5H_{11}GeH_3$, 2, 57
$GeC_5H_{14}O$ Me_3GeCH_2OMe, 2, 33
$Me_3GeCH_2CH_2OH$, 2, 33
$GeC_5H_{14}O_3$ $EtGe(OMe)_3$, 2, 109
$GeC_6H_4Cl_4$ p-$ClC_6H_4GeCl_3$, 2, 91
$GeC_6H_5Cl_3$ $PhGeCl_3$, 2, 85
$GeC_6H_5I_3$ $PhGeI_3$, 2, 87
$GeC_6H_6F_6$ $Me_2Ge(CF{=}CF_2)_2$, 2, 20
GeC_6H_8 $PhGeH_3$, 2, 58
GeC_6H_9Br $(CH_2{=}CH)_3GeBr$, 2, 71
GeC_6H_9I $(CH_2{=}CH)_3GeI$, 2, 71
GeC_6H_{10} $CH{=}CHCH{=}CHGeMe_2$, 2, 146
$GeC_6H_{10}Cl_2$ 1,1-Dichlorodimethylgermacyclopent-3-ene, 2, 152
$GeC_6H_{10}N_2O_2$ $Et_2Ge(OCN)_2$, 2, 94
$GeC_6H_{10}N_2S_2$ $Et_2Ge(SCN)_2$, 2, 94
$GeC_6H_{11}Cl_3$ Cyclohexyl-$GeCl_3$, 2, 86
$GeC_6H_{11}Cl_3F_2$ (n-$C_6H_{11}F_2$)$GeCl_3$, 2, 91
GeC_6H_{12} $CH_2CH{=}CHCH_2GeMe_2$, 2, 146, 156

$GeC_6H_{12}Cl_4$ $(n\text{-}C_6H_{12}Cl)GeCl_3$, 2, 91
$GeC_6H_{12}O_4$ $Me_2Ge(OAc)_2$, 2, 113
$Et_2Ge(O_2CH)_2$, 2, 114
$GeC_6H_{13}Br$ $CH_2(CH_2)_3GeEtBr$, 2, 151
$GeC_6H_{13}Cl_3$ $n\text{-}C_6H_{13}GeCl_3$, 2, 86
$GeC_6H_{13}I$ $CH_2(CH_2)_3GeEtI$, 2, 151
$GeC_6H_{13}NS$ $Me_3GeCHMeSCN$, 2, 31
$GeC_6H_{13}NS$ $Me_3GeCHMeSCN$, 2, 31
GeC_6H_{14} Cyclopropyl-$GeMe_3$, 2, 7
$Me_3GeCH_2CH{=}CH_2$, 2, 14, 156
$Me_3GeCH{=}CHMe$, 2, 14
$CH_2(CH_2)_3GeMe_2$, 2, 145
$GeC_6H_{14}Br_2$ Pr_2GeBr_2, 2, 80
$i\text{-}Pr_2GeBr_2$, 2, 80
$GeC_6H_{14}Cl_2$ Pr_2GeCl_2, 2, 80
$i\text{-}Pr_2GeCl_2$, 2, 80
$GeC_6H_{14}F_2$ Pr_2GeF_2, 2, 80
$i\text{-}Pr_2GeF_2$, 2, 80
$GeC_6H_{14}I_2$ Pr_2GeI_2, 2, 80
$i\text{-}Pr_2GeI_2$, 2, 80
$(GeC_6H_{14}O)_n$ $(Pr_2GeO)_n$, 2, 103
$GeC_6H_{14}O_2$ Me_3GeCH_2OAc, 2, 33
$GeC_6H_{14}O_4S$ Pr_2GeSO_4, 2, 98
$i\text{-}Pr_2GeSO_4$, 2, 98
$GeC_6H_{15}Br$ Et_3GeBr, 2, 67
$GeC_6H_{15}Cl$ $Me_3Ge(CH_2)_3Cl$, 2, 29
Et_3GeCl, 2, 67
$GeC_6H_{15}D$ Et_3GeD, 2, 61
$GeC_6H_{15}F$ Et_3GeF, 2, 71
$GeC_6H_{15}I$ Et_3GeI, 2, 71
GeC_6H_{16} Me_3GePr, 2, 7
Me_2GeEt_2, 2, 7
Et_3GeH, 2, 53
Pr_2GeH_2, 2, 56
$i\text{-}Pr_2GeH_2$, 2, 57
$n\text{-}C_6H_{13}GeH_3$, 2, 57
$GeC_6H_{16}O$ $Me_3Ge(CH_2)_3OH$, 2, 33
Et_3GeOH, 2, 101
$GeC_6H_{16}O_2$ $Me_2Ge(OEt)_2$, 2, 107
$GeC_6H_{16}S$ $Me_3Ge(CH_2)_3SH$, 2, 43
$GeC_7H_7Cl_3$ $Ph(CH_2Cl)GeCl_2$, 2, 82
$PhCH_2GeCl_3$, 2, 86
$GeC_7H_8Br_2$ $MePhGeBr_2$, 2, 80
$GeC_7H_8O_2$ $m\text{-}MeC_6H_4GeO_2H$, 2, 104
$GeC_7H_{14}O_2$ $Me_3GeCH_2O_2CCH{=}CH_2$, 2, 33
$Me_3GeO_2CCMe{=}CH_2$, 2, 156
$GeC_7H_{15}Br$ $CH_2(CH_2)_3GePrBr$, 2, 151
$GeC_7H_{15}Cl_3$ $n\text{-}C_7H_{15}GeCl_3$, 2, 86
$GeC_7H_{15}N$ Et_3GeCN, 2, 94
$GeC_7H_{15}NO$ Et_3GeOCN, 2, 94
$GeC_7H_{15}NS$ Et_3GeSCN, 2, 94
GeC_7H_{16} $Me_3GeCH_2CHCH_2CH_2$, 2, 7
$Me_3GeCH_2CH_2CH{=}CH_2$, 2, 16
$Me_3GeCH_2CMe{=}CH_2$, 2, 16
$Me_3GeCH_2{=}CHMe$, 2, 16
$GeC_7H_{16}Cl_2$ $Cl(CH_2)_3GeEt_2Cl$, 2, 75
$GeC_7H_{16}O$ $Me_3GeCH_2CHOHCH{=}CH_2$, 2, 20
$GeC_7H_{16}O_2$ Et_3GeCO_2H, 2, 39
Et_3GeO_2CH, 2, 113
GeC_7H_{18} Me_3GeBu, 2, 7
$MeGeEt_3$, 2, 8
$Et(i\text{-}C_5H_{11})GeH_2$, 2, 60
$n\text{-}C_7H_{15}GeH_3$, 2, 57
$GeC_7H_{18}O$ Et_3GeOMe, 2, 107
$GeC_7H_{18}O_3$ $MeGe(OEt)_3$, 2, 107
$GeC_7H_{18}O_3S$ Et_3GeO_3SMe, 2, 97
GeC_8F_{12} $(CF_2{=}CF)_4Ge$, 2, 5
$GeC_8H_7Cl_3$ $PhCH{=}CHGeCl_3$, 2, 88
$GeC_8H_9Cl_3$ $Ph(ClCH_2)_2GeCl$, 2, 76
$PhCH_2CH_2GeCl_3$, 2, 86
$GeC_8H_9F_3$ $2,4\text{-}Me_2C_6H_3GeF_3$, 2, 87
$GeC_8H_9I_3$ $EtC_6H_4GeI_3$, 2, 87
$GeC_8H_{10}Cl_2$ $Me_2(p\text{-}ClC_6H_4)GeCl$, 2, 75
$GeC_8H_{11}NO$ $o\text{-}OC_6H_4NHGeMe_2$, 2, 110*
$GeC_8H_{11}NS$ $o\text{-}SC_6H_4NHGeMe_2$, 2, 122
GeC_8H_{12} $(CH_2{=}CH)_4Ge$, 2, 5
$GeC_8H_{12}N_2$ $o\text{-}HNC_6H_4NHGeMe_2$, 2, 126
GeC_8H_{14} $EtGe(CH{=}CH_2)_3$, 2, 18
$GeC_8H_{14}N_2$ $i\text{-}Pr_2Ge(CN)_2$, 2, 95
$GeC_8H_{14}N_2O_2$ $i\text{-}Pr_2Ge(OCN)_2$, 2, 95
$GeC_8H_{14}N_2S_2$ $i\text{-}Pr_2Ge(SCN)_2$, 2, 95
$GeC_8H_{14}O_6$ $EtGe(OAc)_3$, 2, 114
$GeC_8H_{15}Cl_3O_2$ $Et_3GeO_2CCCl_3$, 2, 117
$GeC_8H_{15}F_3$ $Et_3GeCF{=}CF_2$, 2, 20
$GeC_8H_{15}F_3O_2$ $Et_3GeO_2CCF_3$, 2, 117
GeC_8H_{16} Cyclopentenyl-$GeMe_3$, 2, 16
$Me_2Ge(CH_2CH{=}CH_2)_2$, 2, 16, 156, 157
$Et_2Ge(CH{=}CH_2)_2$, 2, 17, 156, 157
$Et_3GeC{\equiv}CH$, 2, 22
5-Germaspiro[4.4]nonane, 2, 144
3-Tetramethylgermacyclopentene, 2, 146
$GeC_8H_{16}ClN$ $NC(CH_2)_3GeEt_2Cl$, 2, 75
$GeC_8H_{16}Cl_2O_2$ $Et_3GeO_2CCHCl_2$, 2, 117

$GeC_8H_{16}O$ $Me_3GeC{\equiv}CCMe_2OH$, 2, 24
$GeC_8H_{16}O_2$ $Me_3GeCH_2O_2CCMe{=}CH_2$, 2, 33
$GeC_8H_{16}O_4$ $Et_2Ge(OAc)_2$, 2, 114
$GeC_8H_{17}Br$ $CH_2(CH_2)_3GeBuBr$, 2, 151
$GeC_8H_{17}BrO_2$ $Et_3GeO_2CCH_2Br$, 2, 117
$GeC_8H_{17}ClO$ $Et_2(AcCH_2CH_2)GeCl$, 2, 75
$GeC_8H_{17}ClO_2$ $Et_2(AcOCH_2CH_2)GeCl$, 2, 75
$Et_3GeO_2CCH_2Cl$, 2, 117
$GeC_8H_{17}Cl_3$ n-$C_8H_{17}GeCl_3$, 2, 86
i-$C_8H_{17}GeCl_3$, 2, 87
$GeC_8H_{17}IO_2$ $Et_3GeO_2CCH_2I$, 2, 117
$GeC_8H_{17}I_3$ n-$C_8H_{17}GeI_3$, 2, 87
GeC_8H_{18} Cyclopentyl-$GeMe_3$, 2, 7
i-$C_5H_9GeMe_3$ (isomer mixture), 2, 16
$Et_3GeCH{=}CH_2$, 2, 14
$CH_2(CH_2)_3GeEt_2$, 2, 145
$CH_2(CH_2)_3GeHBu$, 2, 149
$GeC_8H_{18}Br_2$ Bu_2GeBr_2, 2, 81
$GeC_8H_{18}Cl_2$ Bu_2GeCl_2, 2, 81
$GeC_8H_{18}F_2$ Bu_2GeF_2, 2, 80
$GeC_8H_{18}I_2$ Bu_2GeI_2, 2, 81
$GeC_8H_{18}N_6$ $Bu_2Ge(N_3)_2$, 2, 96
$(GeC_8H_{18}O)_n$ $(Bu_2GeO)_n$, 2, 103
$GeC_8H_{18}OS$ $Me_3Ge(CH_2)_3SAc$, 2, 43
$GeC_8H_{18}O_2$ $Me_3GeCH_2CH_2CO_2Et$, 2, 39
Et_3GeOAc, 2, 113
$GeC_8H_{19}Br$ $Et_2BuGeBr$, 2, 74
Bu_2HGeBr, 2, 64
$GeC_8H_{19}Cl$ $Et_3GeCHMeCl$, 2, 29
Bu_2HGeCl, 2, 63
$GeC_8H_{19}I$ Bu_2HGeI, 2, 64
GeC_8H_{20} Et_4Ge, 2, 2
Me_2GePr_2, 2, 7
Bu_2GeH_2, 2, 55, 156
n-$C_8H_{17}GeH_3$, 2, 57
$GeC_8H_{20}O$ $Et_3GeCH_2CH_2OH$, 2, 33
Et_3GeOEt, 2, 108
$GeC_8H_{20}O_3$ $EtGe(OEt)_3$, 2, 109
$GeC_8H_{20}O_3S$ Et_3GeO_3SEt, 2, 98
$GeC_8H_{20}O_6S_2$ i-$Pr_2Ge(O_3SMe)_2$, 2, 98
$GeC_8H_{21}N$ $Me_3GeCH_2NEt_2$, 2, 44
$Et_2HGe(CH_2)_4NH_2$, 2, 62
$GeC_8H_{21}O_3P$ $Me_3GeCH_2PO_3Et_2$, 2, 44
$GeC_8H_{23}N_3$ $EtGe(NMe_2)_3$, 2, 126
$GeC_8H_{24}N_4$ $(Me_2N)_4Ge$, 2, 126
$GeC_9H_{10}O_2S$ o-$SC_6H_4CO_2GeMe_2$, 2, 117
$GeC_9H_{10}O_3$ o-$OC_6H_4CO_2GeMe_2$, 2, 117
$GeC_9H_{12}O_2$ o-$OC_6H_4CH_2OGeMe_2$, 2, 107
$GeC_9H_{13}Cl$ p-$Me_3GeC_6H_4Cl$, 2, 29
$GeC_9H_{13}Br$ p-$BrC_6H_4GeMe_3$, 2, 29
GeC_9H_{14} Me_3GePh, 2, 4
$GeC_9H_{15}F_5O_2$ $Et_3GeO_2CC_2F_5$, 2, 117
GeC_9H_{16} $Me_2EtGeC_5H_5$, 2, 16
$GeC_9H_{17}NO_2$ $Et_3GeO_2CCH_2CN$, 2, 117
GeC_9H_{18} $Et_3GeCH_2C{\equiv}CH$, 2, 22
5-Germaspiro[4.5]decane, 2, 144
$GeC_9H_{18}N_2$ N-Imidazolyl-$GeEt_3$, 2, 126
$GeC_9H_{18}O$ $Me_3GeC{\equiv}CCMeEtOH$, 2, 24
$Et_3GeC{\equiv}CCH_2OH$, 2, 25
$Et_3GeOCH_2C{\equiv}CH$, 2, 108
$GeC_9H_{18}O_2$ $Et_3GeO_2CCH{=}CH_2$, 2, 113, 156
$GeC_9H_{18}O_4$ $Me_3GeCH_2CH(CO_2Me)_2$, 2, 39
$GeC_9H_{19}BrO_2$ $Et_3GeO_2CCH_2CH_2Br$, 2, 117
$Et_3GeO_2CCHBrMe$, 2, 117
$GeC_9H_{19}ClO_2$ $Et_2(AcOCHMeCH_2)GeCl$, 2, 75
$Et_2(EtO_2CCH_2CH_2)GeCl$, 2, 75
$Et_3GeO_2CCH_2CH_2Cl$, 2, 117
$Et_3GeO_2CCHClMe$, 2, 117
$GeC_9H_{19}IO_2$ $Et_3GeO_2CCH_2CH_2I$, 2, 117
$GeC_9H_{19}N$ $Et_3GeCH_2CH_2CN$, 2, 31
$Pr_2(NCCH_2CH_2)GeH$, 2, 62
Bu_2HGeCN, 2, 65
$GeC_9H_{19}NS$ $Me_3Ge(CH_2)_3S(CH_2)_2CN$, 2, 43
$Bu_2HGeNCS$, 2, 65
GeC_9H_{20} $Et_3GeCH_2CH{=}CH_2$, 2, 14, 156
$CH_2(CH_2)_4GeEt_2$, 2, 147
$GeC_9H_{20}O$ Et_3GeCH_2Ac, 2, 37
$Et_3GeCH{=}CHCH_2OH$, 2, 20
$Et_3GeCH_2CH_2CHO$, 2, 37
$Et_3GeOCH_2CH{=}CH_2$, 2, 108
$GeC_9H_{20}O_2$ $Et_3GeCH_2CH_2CO_2H$, 2, 39
$Pr_2(HO_2CCH_2CH_2)GeH$, 2, 62
Et_3GeO_2CEt, 2, 113
$GeC_9H_{21}Br$ Pr_3GeBr, 2, 71
i-Pr_3GeBr, 2, 72
$GeC_9H_{21}Cl$ $Et_3Ge(CH_2)_3Cl$, 2, 29
Pr_3GeCl, 2, 71
i-Pr_3GeCl, 2, 72
$GeC_9H_{21}F$ Pr_3GeF, 2, 71
$GeC_9H_{21}I$ Pr_3GeI, 2, 71
i-Pr_3GeI, 2, 72
GeC_9H_{22} Et_3GePr, 2, 9
Pr_3GeH, 2, 54

i-Pr_3GeH, 2, 57
Et_2(i-C_5H_{11})GeH, 2, 60
$GeC_9H_{22}O$ $Et_3Ge(CH_2)_3OH$, 2, 33
i-Pr_3GeOH, 2, 101
Et_3GeOPr, 2, 108
$Et_3GeOCHMe_2$, 2, 108
$GeC_9H_{22}O_2$ $Et_3GeCH_2CHOHCH_2OH$, 2, 34
$GeC_9H_{24}IN$ $Me_3GeCH_2NEt_2 \cdot MeI$, 2, 44
$GeC_{10}H_{13}Br$ $CH_2(CH_2)_3GePhBr$, 2, 151
$GeC_{10}H_{14}$ $CH_2(CH_2)_3GeHPh$, 2, 149
$GeC_{10}H_{14}O_2$ p-$Me_3GeC_6H_4CO_2H$, 2, 39
$GeC_{10}H_{15}Cl$ m-$ClC_6H_4GeMe_3$, 2, 28
p-$ClC_6H_4GeMe_3$, 2, 28
$GeC_{10}H_{15}F_7O_2$ $Et_3GeO_2CC_3F_7$, 2, 117
$GeC_{10}H_{16}$ Me_3GeCH_2Ph, 2, 7
$GeC_{10}H_{18}$ $MeGe(CH_2CH{=}CH_2)_3$, 2, 17, 156
1,1-Dimethyldivinylgermacyclopentane, 2, 146
$GeC_{10}H_{18}Cl_2O_4$ $Pr_2Ge(O_2CCH_2Cl)_2$, 2, 118
i-$Pr_2Ge(O_2CCH_2Cl)_2$, 2, 119
$GeC_{10}H_{18}N_2O_2$ $Bu_2Ge(OCN)_2$, 2, 96
$GeC_{10}H_{18}N_2S_2$ $Bu_2Ge(SCN)_2$, 2, 96
$GeC_{10}H_{20}$ $Et_2Ge(CH_2CH{=}CH_2)_2$, 2, 18, 156
6-Germaspiro[5.5]undecane, 2, 144
$CH_2(CH_2)_3Ge(Pr)CH_2CH{=}CH_2$, 2, 145
$GeC_{10}H_{20}O_2$ $Me_3GeCH_2CHCH_2CHCO_2Et$, 2, 39
$Et_3GeO_2CCMe{=}CH_2$, 2, 113, 156
$GeC_{10}H_{20}O_4$ $Pr_2Ge(OAc)_2$, 2, 114
i-$Pr_2Ge(OAc)_2$, 2, 115
$GeC_{10}H_{21}Cl_3$ n-$C_{10}H_{21}GeCl_3$, 2, 87
$GeC_{10}H_{21}N$ Pr_3GeCN, 2, 95
i-Pr_3GeCN, 2, 95
$GeC_{10}H_{21}NS$ Pr_3GeSCN, 2, 95
i-Pr_3GeSCN, 2, 95
$GeC_{10}H_{22}I_2$ (n-$C_5H_{11})_2GeI_2$, 2, 81
$GeC_{10}H_{22}O$ $Et_3GeCH_2CH_2Ac$, 2, 37
$Et_3GeOCH_2CH{=}CHMe$, 2, 108
$GeC_{10}H_{22}O_2$ $Et_3GeCH_2CO_2Et$, 2, 39
Et_3GeO_2CPr, 2, 113
Pr_3GeO_2CH, 2, 114
$GeC_{10}H_{22}O_2S$ $Et_3GeCH_2CH_2SCH_2CO_2H$, 2, 43
$GeC_{10}H_{23}Br$ $Et_3Ge(CH_2)_4Br$, 2, 29
$GeC_{10}H_{23}Cl$ $Et_3GeCH_2CHMeCH_2Cl$, 2, 29
$GeC_{10}H_{24}$ $MeGePr_3$, 2, 8
Et_3GeBu, 2, 9
(n-$C_5H_{11})_2GeH_2$, 2, 57
$GeC_{10}H_{24}O$ $Et_3Ge(CH_2)_4OH$, 2, 34
Et_3GeOBu, 2, 108
$Et_3GeOCMe_3$, 2, 108
$GeC_{10}H_{24}O_2$ $Et_3GeO_2CMe_3$, 2, 105
$GeC_{10}H_{24}O_6S_2$ $Bu_2Ge(O_3SMe)_2$, 2, 98
$GeC_{10}H_{24}S$ Et_3GeSBu, 2, 122
$GeC_{11}H_{15}Br$ $CH_2(CH_2)_4GePhBr$, 2, 152
$GeC_{11}H_{16}$ p-$Me_3GeC_6H_4CH{=}CH_2$, 2, 16, 156
$CH_2(CH_2)_4GeHPh$, 2, 149
$GeC_{11}H_{16}O_2$ p-$Me_3GeCH_2C_6H_4CO_2H$, 2, 39
$GeC_{11}H_{20}$ $Et_3GeC{\equiv}CCH_2CH{=}CH_2$, 2, 22
$Et_3GeC{\equiv}CCMe{=}CH_2$, 2, 22
$Et_3GeC_5H_5$, 2, 15
$GeC_{11}H_{20}OS$ 2-Furfurylthio-$GeEt_3$, 2, 122
$GeC_{11}H_{20}O_2$ 2-Furfuryl-$GeEt_3$, 2, 108
$GeC_{11}H_{20}O_6$ $EtGe(O_2CEt)_3$, 2, 114
$GeC_{11}H_{21}Cl$ $Et_3GeC{\equiv}CCClMe_2$, 2, 25
$GeC_{11}H_{21}Cl_3O_2$ i-$Pr_3GeO_2CCCl_3$, 2, 118
$GeC_{11}H_{21}F_3O_2$ $Pr_3GeO_2CCF_3$, 2, 118
i-$Pr_3GeO_2CF_3$, 2, 118
$GeC_{11}H_{22}ClN$ $Bu_2(NCCH_2CH_2)GeCl$, 2, 76
$GeC_{11}H_{22}Cl_2O_2$ $Pr_3GeO_2CCHCl_2$, 2, 118
i-$Pr_3GeO_2CCHCl_2$, 2, 118
$GeC_{11}H_{22}O$ $Et_3GeC{\equiv}CCMe_2OH$, 2, 26
$GeC_{11}H_{23}BrO_2$ $Pr_3GeO_2CCH_2Br$, 2, 118
i-$Pr_3GeO_2CCH_2Br$, 2, 118
$GeC_{11}H_{23}ClO_2$ $Pr_3GeO_2CCH_2Cl$, 2, 118
i-$Pr_3GeO_2CCH_2Cl$, 2, 118
$GeC_{11}H_{23}IO_2$ i-$Pr_3GeO_2CCH_2I$, 2, 118
$GeC_{11}H_{23}N$ $Bu_2(NCCH_2CH_2)GeH$, 2, 62
$GeC_{11}H_{24}$ Cyclopentyl-$GeEt_3$, 2, 9
$Bu_2(CH_2{=}CHCH_2)GeH$, 2, 60
$GeC_{11}H_{24}Cl_2$ $Bu_2(ClCH_2CH_2CH_2)GeCl$, 2, 76
$GeC_{11}H_{24}O$ $Pr_3GeOCH{=}CH_2$, 2, 109
$GeC_{11}H_{24}O_2$ $Et_3Ge(CH_2)_3OAc$, 2, 34
$Et_3GeCH_2CH_2CO_2Et$, 2, 39
$Et_3Ge(CH_2)_4CO_2H$, 2, 39
Et_3GeO_2CBu, 2, 113
Pr_3GeOAc, 2, 114
i-Pr_3GeOAc, 2, 114
$GeC_{11}H_{24}O_2S$ $Et_3Ge(CH_2)_3SCH_2CO_2H$, 2,

43
$GeC_{11}H_{25}Br$ $Et_3Ge(CH_2)_5Br$, 2, 29
$GeC_{11}H_{25}ClO$ $Bu_2(HOCH_2CH_2CH_2)GeCl$, 2, 76
$GeC_{11}H_{26}$ $n\text{-}C_5H_{11}GeEt_3$, 2, 9
$GeC_{11}H_{26}O$ $Et_3Ge(CH_2)_5OH$, 2, 34
$Pr_3GeCH_2CH_2OH$, 2, 34
$Bu_2(HOCH_2CH_2CH_2)GeH$, 2, 62
$GeC_{11}H_{26}OS$ $Et_3Ge(CH_2)_3SCH_2CH_2OH$, 2, 43
$GeC_{11}H_{27}N$ $Bu_2(H_2NCH_2CH_2CH_2)GeH$, 2, 62
$GeC_{12}H_8Br_2I_2$ $(o\text{-}BrC_6H_4)_2GeI_2$, 2, 82
$GeC_{12}H_8Br_2O$ $(p\text{-}BrC_6H_4)_2GeO$, 2, 103
$GeC_{12}H_8Cl_2I_2$ $(o\text{-}ClC_6H_4)_2GeI_2$, 2, 82
$(m\text{-}ClC_6H_4)_2GeI_2$, 2, 82
$(p\text{-}ClC_6H_4)_2GeI_2$, 2, 82
$GeC_{12}H_8Cl_2O$ $(p\text{-}ClC_6H_4)_2GeO$, 2, 103
$(GeC_{12}H_{10})_n$ $(Ph_2Ge)_n$, 2, 131
$GeC_{12}H_{10}Br_2$ Ph_2GeBr_2, 2, 79
$GeC_{12}H_{10}Cl_2$ Ph_2GeCl_2, 2, 79
$GeC_{12}H_{10}I_2$ Ph_2GeI_2, 2, 81
$GeC_{12}H_{10}N_6$ $Ph_2Ge(N_3)_2$, 2, 96
$(GeC_{12}H_{10}O)_n$ $(Ph_2GeO)_n$, 2, 103
$GeC_{12}H_{12}$ $EtPhGe(C{\equiv}CH)_2$, 2, 23
Ph_2GeH_2, 2, 56, 156
$GeC_{12}H_{14}Cl_2O_2$ $CH_2(CH_2)_3GePhO_2CCHCl_2$, 2, 154
$GeC_{12}H_{14}F_2$ $(MeC_5H_4)_2GeF_2$, 2, 81
$GeC_{12}H_{16}$ $Me_2Ge(C_5H_5)_2$, 2, 16
$GeC_{12}H_{18}$ $p\text{-}Me_3GeC_6H_4CMe{=}CH_2$, 2, 16
$CH_2(CH_2)_3GeEtPh$, 2, 145
$GeC_{12}H_{19}Br$ $m\text{-}BrC_6H_4GeEt_3$, 2, 30
$p\text{-}BrC_6H_4GeEt_3$, 2, 30
$GeC_{12}H_{19}Cl$ $m\text{-}ClC_6H_4GeEt_3$, 2, 30
$p\text{-}ClC_6H_4GeEt_3$, 2, 30
$GeC_{12}H_{19}ClO_2S$ $p\text{-}Et_3GeC_6H_4SO_2Cl$, 2, 43
$GeC_{12}H_{19}F$ $m\text{-}FC_6H_4GeEt_3$, 2, 30
$p\text{-}FC_6H_4GeEt_3$, 2, 30
$GeC_{12}H_{19}I$ $p\text{-}IC_6H_4GeEt_3$, 2, 30
$GeC_{12}H_{19}NO_2$ $m\text{-}Et_3GeC_6H_4NO_2$, 2, 44
$p\text{-}Et_3GeC_6H_4NO_2$, 2, 44
$GeC_{12}H_{20}$ $(CH_2{=}CHCH_2)_4Ge$, 2, 3, 156
Et_3GePh, 2, 9
$Et_3GeC{\equiv}CC{\equiv}CEt$, 2, 22
$GeC_{12}H_{20}O$ Et_3GeOPh, 2, 108
$p\text{-}Et_3GeC_6H_4OH$, 2, 34
$GeC_{12}H_{20}O_2$ $Me_2Ge(C{\equiv}CCMe_2OH)_2$, 2, 24
$GeC_{12}H_{20}O_3S$ Et_3GeO_3SPh, 2, 98
$GeC_{12}H_{20}S$ Et_3GeSPh, 2, 122
$GeC_{12}H_{21}NO_2S$ $p\text{-}Et_3GeC_6H_4SO_2NH_2$, 2, 43
$GeC_{12}H_{21}NS$ $o\text{-}Et_3GeSC_6H_4NH_2$, 2, 122
$GeC_{12}H_{22}$ $Et_3GeCH_2C{\equiv}CCH_2CH{=}CH_2$, 2, 22
$GeC_{12}H_{22}Br_2$ $Cyclohexyl_2GeBr_2$, 2, 81
$GeC_{12}H_{22}Cl_2O_4$ $i\text{-}Pr_2Ge(O_2CCH_2CH_2Cl)_2$, 2, 119
$GeC_{12}H_{22}N_2$ $Pr_2Ge(CH_2CH_2CN)_2$, 2, 31
$GeC_{12}H_{22}O_4S$ $p\text{-}Et_3GeC_6H_4SO_3H{\cdot}H_2O$, 2, 43
$GeC_{12}H_{24}$ 2,5-Hexadienyl-$GeEt_3$, 2, 17
$n\text{-}BuGePh_3$, 2, 11
$i\text{-}Bu_3GePh_3$, 2, 11
$GeC_{12}H_{24}ClO$ $Bu_2(OHCCHMeCH_2)GeCl$, 2, 76*
$GeC_{12}H_{24}O$ $Et_3GeC(CH_2CH_2OH){=}CHCH{=}CH_2$, 2, 20
$Et_3GeC(CHOHMe){=}CHCH{=}CH_2$, 2, 20
$Et_3GeC{\equiv}CCHPrOH$, 2, 26
$Et_3GeC{\equiv}CCMeEtOH$, 2, 26
$GeC_{12}H_{24}O_2S$ $CH_2(CH_2)_3GePr(CH_2)_3SCH_2CO_2H$, 2, 145
$GeC_{12}H_{24}O_4$ $i\text{-}Pr_2Ge(O_2CEt)_2$, 2, 115
$GeC_{12}H_{25}BrO_2$ $Et_3GeCHBrCH_2CH_2CO_2Et$, 2, 29
$GeC_{12}H_{25}ClO_2$ $Bu_2(MeO_2CCH_2CH_2)GeCl$, 2, 76
$i\text{-}Pr_3GeO_2CCH_2CH_2Cl$, 2, 118
$i\text{-}Pr_3GeO_2CCHMeCl$, 2, 118
$GeC_{12}H_{25}N$ $Pr_3GeCH_2CH_2CN$, 2, 31
$GeC_{12}H_{26}$ $Et_3GeCH{=}CHBu$, 2, 17
$GeC_{12}H_{26}Cl_2$ $Bu_2(ClCH_2CHMeCH_2)GeCl$, 2, 76
$GeC_{12}H_{26}O$ $Et_3Ge(CH_2)_3COEt$, 2, 37
$Bu_2(AcCH_2CH_2)GeH$, 2, 62
Cyclohexyloxy-$GeEt_3$, 2, 108
$GeC_{12}H_{26}O_2$ $Et_3Ge(CH_2)_3CO_2Et$, 2, 39
$Pr_3GeCH_2CO_2Me$, 2, 40
$Bu_2(EtO_2CCH_2)GeH$, 2, 62
$Bu_2(MeO_2CCH_2CH_2)GeH$, 2, 62
$Bu_2GeO(CH_2)_4O$, 2, 110
Pr_3GeO_2CEt, 2, 114
$i\text{-}Pr_3GeO_2CEt$, 2, 114
$GeC_{12}H_{27}Br$ Bu_3GeBr, 2, 72
$GeC_{12}H_{27}BrO$ $Pr_3GeOCH_2CHMeBr$, 2, 109
$GeC_{12}H_{27}Cl$ Bu_3GeCl, 2, 68

$GeC_{12}H_{27}F$ Bu_3GeF, 2, 72
$GeC_{12}H_{27}I$ Bu_3GeI, 2, 72
i-Bu_3GeI, 2, 72
$GeC_{12}H_{28}$ Pr_4Ge, 2, 2
Et_2GeBu_2, 2, 10
Bu_3GeH, 2, 54
(n-$C_6H_{13})_2GeH_2$, 2, 57
$GeC_{12}H_{28}O$ $Et_3Ge(CH_2)_5OMe$, 2, 34
$Et_3Ge(CH_2)_6OH$, 2, 34
$Pr_3Ge(CH_2)_3OH$, 2, 34
$Bu_2(HOCH_2CHMeCH_2)GeH$, 2, 62
$GeC_{12}H_{28}S$ n-$C_6H_{13}SGeEt_3$, 2, 122
$GeC_{12}H_{30}N_2$ $Bu_2Ge(NMe_2)_2$, 2, 127
$GeC_{13}H_{12}Cl_2$ $Ph_2(ClCH_2)GeCl$, 2, 76
$GeC_{13}H_{14}$ $MePh_2GeH$, 2, 60
$GeC_{13}H_{18}$ $\overline{CH_2(CH_2)_3Ge}PhCH_2CH{=}CH_2$, 2, 145
$GeC_{13}H_{20}$ $\overline{CH_2(CH_2)_3Ge}PrPh$, 2, 145
$GeC_{13}H_{20}O_2$ m-$Et_3GeC_6H_4CO_2H$, 2, 40
p-$Et_3GeC_6H_4CO_2H$, 2, 40
Et_3GeOBz, 2, 113
$GeC_{13}H_{19}F_3$ m-$Et_3GeC_6H_4CF_3$, 2, 30
p-$Et_3GeC_6H_4CF_3$, 2, 30
$GeC_{13}H_{21}Br$ m-$BrC_6H_4CH_2GeEt_3$, 2, 29
p-$BrC_6H_4CH_2GeEt_3$, 2, 29
$GeC_{13}H_{21}Cl$ o-$ClC_6H_4CH_2GeEt_3$, 2, 29
m-$ClC_6H_4CH_2GeEt_3$, 2, 29
p-$ClC_6H_4CH_2GeEt_3$, 2, 29
$GeC_{13}H_{21}F$ p-$FC_6H_4CH_2GeEt_3$, 2, 29
$GeC_{13}H_{21}NO_2$ p-$H_2NC_6H_4CO_2GeEt_3$, 2, 118
$GeC_{13}H_{22}$ $PhCH_2GeEt_3$, 2, 8
o-$MeC_6H_4GeEt_3$, 2, 9
m-$MeC_6H_4GeEt_3$, 2, 10
p-$MeC_6H_4GeEt_3$, 2, 10
$GeC_{13}H_{22}O$ o-$Et_3GeC_6H_4OMe$, 2, 34
m-$Et_3GeC_6H_4OMe$, 2, 34
p-$Et_3GeC_6H_4OMe$, 2, 34
Et_3GeOCH_2Ph, 2, 107
$GeC_{13}H_{22}S$ Et_3GeSCH_2Ph, 2, 122
o-$Et_3GeSC_6H_4Me$, 2, 122
m-$Et_3GeSC_6H_4Me$, 2, 122
$GeC_{13}H_{24}$ 3-(2,1,3,5-Methylhexatrienyl)$GeEt_3$, 2, 17
$GeC_{13}H_{26}$ $Et_3GeCH_2C{\equiv}CBu$, 2, 22
$GeC_{13}H_{26}O$ $Et_3GeC(CMe_2OH){=}CHCH{=}CH_2$, 2, 20
$Et_3GeC{\equiv}CCEt_2OH$, 2, 26
$GeC_{13}H_{26}O_2$ 9-Decalinperoxy-$GeMe_3$, 2, 105
$GeC_{13}H_{27}ClO_2$ $Bu_2(MeO_2CCHMeCH_2)GeCl$, 2, 76
$Bu_2(AcOCH_2CH_2CH_2)GeCl$, 2, 76
$GeC_{13}H_{27}NO$ Bu_3GeOCN, 2, 95
$GeC_{13}H_{27}NS$ Bu_3GeSCN, 2, 95
Bu_3GeSCN, 2, 95
$GeC_{13}H_{28}O_2$ $Pr_3GeCH_2CO_2Et$, 2, 40
$Pr_3GeCH_2CH_2CO_2Me$, 2, 40
i-Pr_3GeO_2CPr, 2, 115
Bu_3GeO_2CH, 2, 115
$GeC_{13}H_{28}O_4$ $\overline{CH_2(CH_2)_4Ge}(O_2CMe_3)_2$, 2, 154
$GeC_{13}H_{30}$ $MeGeBu_3$, 2, 8
$GeC_{13}H_{30}O$ $Et_3Ge(CH_2)_4CMe_2OH$, 2, 34
$Et_3GeCH_2CH_2CEt_2OH$, 2, 34
n-$C_7H_{15}OGeEt_3$, 2, 108
Bu_3GeOMe, 2, 109
$GeC_{13}H_{30}O_2$ $Pr_3GeO_2CMe_3$, 2, 105
$GeC_{13}H_{30}O_3S$ Bu_3GeO_3SMe, 2, 98
$GeC_{13}H_{30}S$ n-$C_7H_{15}SGeEt_3$, 2, 122
$GeC_{14}H_{14}Cl_2$ (p-$ClC_6H_4)_2GeMe_2$, 2, 29
$(PhCH_2)_2GeCl_2$, 2, 80
$GeC_{14}H_{14}I_2$ (o-$MeC_6H_4)_2GeI_2$, 2, 81
$GeC_{14}H_{14}O$ (p-$MeC_6H_4)_2GeO$, 2, 103
$GeC_{14}H_{16}$ Me_2GePh_2, 2, 8
$GeC_{14}H_{19}Br$ o-$BrC_6H_4C{\equiv}CGeEt_3$, 2, 25
m-$BrC_6H_4C{\equiv}CGeEt_3$, 2, 25
p-$BrC_6H_4C{\equiv}CGeEt_3$, 2, 25
$GeC_{14}H_{19}Cl$ o-$ClC_6H_4C{\equiv}CGeEt_3$, 2, 25
m-$ClC_6H_4C{\equiv}CGeEt_3$, 2, 25
p-$ClC_6H_4C{\equiv}CGeEt_3$, 2, 25
$GeC_{14}H_{19}F$ p-$FC_6H_4C{\equiv}CGeEt_3$, 2, 25
$GeC_{14}H_{19}I$ p-$IC_6H_4C{\equiv}CGeEt_3$, 2, 25
$GeC_{14}H_{20}$ $Et_3GeC{\equiv}CPh$, 2, 22
$Et_2Ge(C{\equiv}CCMe{=}CH_2)_2$, 2, 22
$GeC_{14}H_{22}$ p-$Et_3GeC_6H_4CH{=}CH_2$, 2, 17, 156
$GeC_{14}H_{22}O$ Et_3GeCH_2Bz, 2, 37
$(GeC_{14}H_{22}O_2)_n$ (p-$Bu_2GeOC_6H_4O)_n$, 2, 110
$GeC_{14}H_{24}$ p-$Et_3GeCH_2C_6H_4Me$, 2, 9
p-$Et_3GeC_6H_4Et$, 2, 10
Bu_2PhGeH, 2, 60
$GeC_{14}H_{24}O$ p-$Et_3GeC_6H_4CH_2CH_2OH$, 2, 34
p-$Et_3GeC_6H_4CHOHMe$, 2, 34
$GeC_{14}H_{24}O_2$ $Me_2Ge(C{\equiv}CCMeEtOH)_2$, 2, 24

$Et_2Ge(C{\equiv}CCMe_2OH)_2$, 2, 26
$GeC_{14}H_{25}N$ o-$Et_3GeC_6H_4NMe_2$, 2, 44
p-$Et_3GeC_6H_4NMe_2$, 2, 44
$GeC_{14}H_{26}$ 4-(5,1,3,5-Methylheptatrienyl)$GeEt_3$, 2, 17
$GeC_{14}H_{26}O_6$ $EtGe(O_2CPr)_3$, 2, 114
$GeC_{14}H_{27}F_3O_2$ $Bu_3GeO_2CCF_3$, 2, 119*
$GeC_{14}H_{28}$ $Bu_3GeC{\equiv}CH$, 2, 23
$Bu_2Ge(CH_2CH{=}CH_2)_2$, 2, 18
$GeC_{14}H_{28}O$ $Et_3GeC(CHPrOH){=}CHCH{=}CH_2$, 2, 20
$Et_3GeC(CMeEtOH){=}CHCH{=}CH_2$, 2, 20
$GeC_{14}H_{28}O_4$ i-$Pr_2Ge(O_2CPr)_2$, 2, 115
$GeC_{14}H_{29}Cl$ $Bu_2(C_6H_{11})GeCl$, 2, 74
$Bu_2(BuCH{=}CH)GeCl$, 2, 74
$GeC_{14}H_{29}N_3$ N-1,2,4-Triazolyl-$GeBu_3$, 2, 127
$GeC_{14}H_{30}$ $Bu_3GeCH{=}CH_2$, 2, 15
$GeC_{14}H_{30}Br_2$ $Bu_3GeCHBrCH_2Br$, 2, 30
$GeC_{14}H_{30}O$ (n-$C_5H_{11})_2(OHCCH_2CHMe)GeH$, 2, 62
$Bu_3GeOCH{=}CH_2$, 2, 109
$GeC_{14}H_{30}O_2$ $Et_3GeC(CMe_2OH){=}CHCMe_2OH$, 2, 21
$Pr_3GeCH_2CO_2Pr$, 2, 40
i-Pr_3GeO_2CBu, 2, 115
Bu_3GeOAc, 2, 115
i-Bu_3GeOAc, 2, 115
$GeC_{14}H_{31}ClO$ $Bu_2(BuOCH_2CH_2)GeCl$, 2, 76
$GeC_{14}H_{32}$ (n-$C_7H_{15})_2GeH_2$, 2, 57
$GeC_{14}H_{32}O$ $Et_3Ge(CH_2)_4CHPrOH$, 2, 34
Bu_3GeCH_2OMe, 2, 34
Bu_3GeOEt, 2, 109
$GeC_{14}H_{32}O_3S$ Bu_3GeO_3SEt, 2, 98
$GeC_{14}H_{35}N_3$ $EtGe(NEt_2)_3$, 2, 126
$GeC_{14}H_{36}N_2O_6P_3$ $Bu_2Ge[N{=}P(OMe)_3]_2$, 2, 128*
$GeC_{15}H_{19}F_3$ m-$Et_3GeC{\equiv}CC_6H_4CF_3$, 2, 25
$GeC_{15}H_{22}$ o-$MeC_6H_4C{\equiv}CGeEt_3$, 2, 22
m-$MeC_6H_4C{\equiv}CGeEt_3$, 2, 22
p-$MeC_6H_4C{\equiv}CGeEt_3$, 2, 22
Indenyl-$GeEt_3$, 2, 10
$GeC_{15}H_{22}O$ o-$MeOC_6H_4C{\equiv}CGeEt_3$, 2, 25
m-$MeOC_6H_4C{\equiv}CGeEt_3$, 2, 25
p-$MeOC_6H_4C{\equiv}CGeEt_3$, 2, 25
$GeC_{15}H_{24}$ p-$Et_3GeC_6H_4CMe{=}CH_2$, 2, 17
$GeC_{15}H_{24}O_2$ $Et_3GeCH_2CH_2OBz$, 2, 33
$GeC_{15}H_{26}$ Mesityl-$GeEt_3$, 2, 10
p-$Me_2CHC_6H_4GeEt_3$, 2, 10
1-(1-Cyclopentenyl)butadienyl-$GeEt_3$, 2, 17
$PhCH_2Bu_2GeH$, 2, 60
$GeC_{15}H_{28}IN$ m-$Et_3GeC_6H_4NMe_2{\cdot}MeI$, 2, 44
p-$Et_3GeC_6H_4NMe_2{\cdot}MeI$, 2, 44
$GeC_{15}H_{28}O$ 1-(1-HO-Cyclopentyl)butadienyl-$GeEt_3$, 2, 20
$GeC_{15}H_{30}N_2$ N-Imidazolyl-$GeBu_3$, 2, 127
N-Pyrazolyl-$GeBu_3$, 2, 127
$GeC_{15}H_{30}O$ $Bu_3GeOCH_2C{\equiv}CH$, 2, 109
$GeC_{15}H_{30}O_2$ $Me_3GeC{\equiv}CCMeEtOCHMeOBu$, 2, 24
$Et_3GeC{\equiv}CCH_2OCHMeOBu$, 2, 25
$Bu_3GeO_2CCH{=}CH_2$, 2, 115, 156
$GeC_{15}H_{31}Cl_3$ $Bu_3GeCH_2CH_2CCl_3$, 2, 30
$GeC_{15}H_{31}N$ $Bu_3GeCH_2CH_2CN$, 2, 31
$GeC_{15}H_{32}$ $Bu_3GeCH_2CH{=}CH_2$, 2, 15
$GeC_{15}H_{32}O$ $Bu_3GeCH{=}CHCH_2OH$, 2, 21
$Bu_3GeCH_2CH_2CHO$, 2, 37
$GeC_{15}H_{32}O_2$ $Bu_3GeCH_2CO_2Me$, 2, 40
$GeC_{15}H_{33}Cl$ $Bu_3Ge(CH_2)_3Cl$, 2, 30
(n-$C_5H_{11})_3GeCl$, 2, 72
$GeC_{15}H_{33}ClO_2$ $Bu_2[(EtO)_2CHCH_2CH_2]GeCl$, 2, 76
$GeC_{15}H_{33}I$ (n-$C_5H_{11})_3GeI$, 2, 72
(i-$C_5H_{11})_3GeI$, 2, 72*
$GeC_{15}H_{34}$ (n-$C_5H_{11})_3GeH$, 2, 57
(i-$C_5H_{11})_3GeH$, 2, 57
$GeC_{15}H_{34}O$ $Bu_3Ge(CH_2)_3OH$, 2, 35
$Bu_2HGeOC_7H_{15}$, 2, 65
$GeC_{15}H_{34}S$ $Bu_3Ge(CH_2)_3SH$, 2, 43
$GeC_{15}H_{35}N$ $Bu_3Ge(CH_2)_3NH_2$, 2, 44
$GeC_{16}H_{12}O_4$ 2-$Furyl_4Ge$, 2, 6
$GeC_{16}H_{16}$ $(CH{=}CH_2)_2GePh_2$, 2, 15
$GeC_{16}H_{18}$ 9-Fluorenyl-$GeMe_3$, 2, 16
$CH_2(CH_2)_3GePh_2$, 2, 145
$GeC_{16}H_{18}I_2O_2$ (o-$EtOC_6H_4)_2GeI_2$, 2, 82
$GeC_{16}H_{20}$ $Me_3GeCHPh_2$, 2, 7
Et_2GePh_2, 2, 10
$GeC_{16}H_{22}$ (1-$C_{10}H_7)GeEt_3$, 2, 10
(2-$C_{10}H_7)GeEt_3$, 2, 10
$GeC_{16}H_{22}S$ 2-$C_{10}H_7SGeEt_3$, 2, 122
$GeC_{16}H_{24}$ 2,3-$Me_2C_6H_4C{\equiv}CGeEt_3$, 2, 22
$GeC_{16}H_{24}O_3$ $MeGe(C{\equiv}CCMe_2OH)_3$, 2, 25
$GeC_{16}H_{24}O_4$ $Me_2Ge(C{\equiv}CCMe_2OAc)_2$, 2, 24

$GeC_{16}H_{25}Cl$ $Bu_2(PhCH{=}CH)GeCl$, 2, 74
$GeC_{16}H_{26}$ $Bu_2(PhCH{=}CH)GeH$, 2, 60
$GeC_{16}H_{26}N_2$ (N-Pyrrolyl)$_2$GeBu$_2$, 2, 127
$GeC_{16}H_{26}O_2$ i-Pr$_3$GeOBz, 2, 115
$GeC_{16}H_{28}$ m-Et$_3$GeC$_6$H$_4$CMe$_3$, 2, 10
p-Et$_3$GeC$_6$H$_4$CMe$_3$, 2, 10
1-(1-Cyclohexenyl)butadienyl-GeEt$_3$, 2, 17
$GeC_{16}H_{28}O_2$ Et$_2$Ge(C≡CCMeEtOH)$_2$, 2, 26
$GeC_{16}H_{30}O$ 1-(1-Hydroxycyclohexyl)-butadienyl-GeEt$_3$, 2, 20
$GeC_{16}H_{31}N$ N-Bu$_3$GeC$_5$H$_5$N, 2, 126
$GeC_{16}H_{31}NO_2$ N-Succinimidyl-GeBu$_3$, 2, 127
$GeC_{16}H_{32}O$ Bu$_2$Ge(CH$_2$CH$_2$Ac)$_2$, 2, 37
Bu$_3$GeO$_2$CCMe=CH$_2$, 2, 115, 156
$GeC_{16}H_{32}O_4$ Bu$_2$Ge(CH$_2$CO$_2$Et)$_2$, 2, 41
Bu$_2$Ge(CH$_2$CH$_2$CO$_2$Me)$_2$, 2, 41
$GeC_{16}H_{34}O$ Bu$_3$GeOCH$_2$CMe=CH$_2$, 2, 109
$GeC_{16}H_{34}O_2$ Et$_3$GeC(CMe$_2$OH)=CHCEt$_2$OH, 2, 21
Bu$_3$GeCH$_2$CO$_2$Et, 2, 40
Bu$_3$GeCH$_2$CH$_2$CO$_2$Me, 2, 40
$GeC_{16}H_{35}Br$ Bu$_2$(n-C$_8$H$_{17}$)GeBr, 2, 74
$GeC_{16}H_{35}Cl$ Bu$_2$(n-C$_8$H$_{17}$)GeCl, 2, 74
$GeC_{16}H_{35}I$ Bu$_2$(n-C$_8$H$_{17}$)GeI, 2, 74
$GeC_{16}H_{35}N$ N-Pyrrolidinyl-GeBu$_3$, 2, 127
$GeC_{16}H_{36}$ Bu$_4$Ge, 2, 3
i-Bu$_4$Ge, 2, 5
Bu$_2$(n-C$_8$H$_{17}$)GeH, 2, 60
(n-C$_8$H$_{17}$)$_2$GeH$_2$, 2, 57
$GeC_{16}H_{36}O$ Bu$_3$GeCH$_2$CHMeCH$_2$OH, 2, 35
$GeC_{16}H_{37}N$ Bu$_3$GeNEt$_2$, 2, 126
$GeC_{16}H_{40}N_4$ (Et$_2$N)$_4$Ge, 2, 126
$GeC_{17}H_{15}Br$ MePh(1-C$_{10}$H$_7$)GeBr, 2, 74
$GeC_{17}H_{15}Cl$ MePh(1-C$_{10}$H$_7$)GeCl, 2, 74
$GeC_{17}H_{18}$ MePh(1-C$_{10}$H$_7$)GeH, 2, 58
$GeC_{17}H_{20}$ $\overline{CH_2(CH_2)_4Ge}Ph_2$, 2, 147*
$GeC_{17}H_{26}$ Mesityl-C≡CGeEt$_3$, 2, 22
Pr$_3$GeC≡CPh, 2, 23
$GeC_{17}H_{26}O$ Et$_3$GeC(CHPhOH)=CHCH=CH$_2$, 2, 20
$GeC_{17}H_{26}O_3$ (Me$_2$COHC≡C)$_3$GeEt, 2, 26
$GeC_{17}H_{32}$ Bu$_3$GeC$_5$H$_5$, 2, 18
Bu$_3$GeC≡CCH$_2$CH=CH$_2$, 2, 23
$GeC_{17}H_{32}O_6$ EtGe(O$_2$CBu)$_3$, 2, 114
$GeC_{17}H_{34}$ Bu$_3$GeC≡CPr, 2, 23
Bu$_3$GeCH=CHCH$_2$CH=CH$_2$, 2, 18
Bu$_3$GeCH=CHCMe=CH$_2$, 2, 18
$GeC_{17}H_{34}O_2$ Et$_3$GeC≡CCMe$_2$COCHMeOBu, 2, 26
$GeC_{17}H_{36}O$ Bu$_3$GeCH=CHCMe$_2$OH, 2, 21
$GeC_{17}H_{36}O_2$ Bu$_3$GeCH$_2$CO$_2$Pr, 2, 40
Bu$_3$GeCH$_2$CH$_2$CO$_2$Et, 2, 40
(n-C$_5$H$_{11}$)$_3$GeOAc, 2, 115
(i-C$_5$H$_{11}$)$_3$GeOAc, 2, 115
$GeC_{17}H_{36}O_3$ Pr$_2$Ge(OPr)O$_2$CC$_7$H$_{15}$, 2, 116*
$GeC_{17}H_{38}$ (n-C$_5$H$_{11}$)$_3$GeEt, 2, 10
$GeC_{17}H_{38}O$ Bu$_3$Ge(CH$_2$)$_5$OH, 2, 35
$GeC_{18}H_{12}BrI$ (p-BrC$_6$H$_4$)$_3$GeI, 2, 75
$GeC_{18}H_{12}ClI$ (p-ClC$_6$H$_4$)$_3$GeI, 2, 75
$GeC_{18}H_{15}Br$ Ph$_3$GeBr, 2, 69
$GeC_{18}H_{15}Cl$ Ph$_3$GeCl, 2, 69
$GeC_{18}H_{15}F$ Ph$_3$GeF, 2, 73
$GeC_{18}H_{15}I$ Ph$_3$GeI, 2, 73
$GeC_{18}H_{15}N_3$ Ph$_3$GeN$_3$, 2, 96
$GeC_{18}H_{16}$ Ph$_3$GeH, 2, 55
$GeC_{18}H_{16}O$ Ph$_3$GeOH, 2, 100
$GeC_{18}H_{16}O_2$ MePh(1-C$_{10}$H$_7$)GeCO$_2$H, 2, 39
$GeC_{18}H_{16}S$ Ph$_3$GeSH, 2, 123
$GeC_{18}H_{17}N$ Ph$_3$GeNH$_2$, 2, 127
$GeC_{18}H_{18}$ EtPh(1-C$_{10}$H$_7$)GeH, 2, 59
$GeC_{18}H_{18}O$ MePh(1-C$_{10}$H$_7$)GeOMe, 2, 110
$GeC_{18}H_{20}$ Ph$_2$Ge(CH$_2$CH=CH$_2$)$_2$, 2, 19
$GeC_{18}H_{22}$ 1,1-Dimethyldiphenylgermacyclopentane, 2, 146
$GeC_{18}H_{24}$ o-PhC$_6$H$_4$GeEt$_3$, 2, 10
p-PhC$_6$H$_4$GeEt$_3$, 2, 10
$GeC_{18}H_{24}O$ p-Et$_3$GeC$_6$H$_4$OPh, 2, 34
$GeC_{18}H_{25}O_2P$ Et$_3$GeO$_2$PPh$_2$, 2, 97
$GeC_{18}H_{25}P$ Et$_3$GePPh$_2$, 2, 127
$GeC_{18}H_{28}$ p-(Et$_3$GeC≡C)C$_6$H$_4$CMe$_3$, 2, 22
$GeC_{18}H_{28}O_4$ Me$_2$Ge(C≡CCMeEtOAc)$_2$, 2, 24
Et$_2$Ge(C≡CCMe$_2$OAc)$_2$, 2, 26
$GeC_{18}H_{33}Br$ Cyclohexyl$_3$GeBr, 2, 72
$GeC_{18}H_{33}Cl$ Cyclohexyl$_3$GeCl, 2, 72
$GeC_{18}H_{33}I$ Cyclohexyl$_3$GeI, 2, 72
$GeC_{18}H_{34}$ Cyclohexyl$_3$GeH, 2, 57
$GeC_{18}H_{34}O$ Cyclohexyl$_3$GeOH, 2, 101
$GeC_{18}H_{35}N$ N-(2,4-Dimethylpyrrolyl)-GeBu$_3$, 2, 126
$GeC_{18}H_{36}$ Bu$_3$GeC≡CBu, 2, 23
$GeC_{18}H_{36}O_2$ Et$_3$GeC≡CCHPrOCHMeOBu, 2, 26

$Et_3GeC{\equiv}CCMeEtOCHMeOBu$, 2, 26
$GeC_{18}H_{37}N$ $(n\text{-}C_5H_{11})_3GeCH_2CH_2CN$, 2, 31
$GeC_{18}H_{38}$ $Bu_3GeCH{=}CHBu$, 2, 18
$GeC_{18}H_{38}O$ Cyclohexyloxy-$GeBu_3$, 2, 109
$GeC_{18}H_{39}Br$ $(C_6H_{13})_3GeBr$, 2, 72
$GeC_{18}H_{39}Cl$ $(C_6H_{13})_3GeCl$, 2, 72
$GeC_{18}H_{39}I$ $(C_6H_{13})_3GeI$, 2, 72
$GeC_{18}H_{40}$ $(n\text{-}C_6H_{13})_3GeH$, 2, 57
$GeC_{18}H_{40}O$ $Bu_3GeCH_2CH_2OBu$, 2, 35
$GeC_{18}H_{40}S$ $n\text{-}C_{12}H_{25}SGeEt_3$, 2, 122
$GeC_{19}H_{15}NS$ Ph_3GSCN, 2, 96
$GeC_{19}H_{16}O_2$ Ph_3GeCO_2H, 2, 41
$GeC_{19}H_{17}Cl$ Ph_3GeCH_2Cl, 2, 30
$GeC_{19}H_{18}$ $MeGePh_3$, 2, 8
$GeC_{19}H_{18}O$ Ph_3GeCH_2OH, 2, 35
Ph_3GeOMe, 2, 110
$GeC_{19}H_{18}O_2$ $MePh(1\text{-}C_{10}H_7)GeCO_2Me$, 2, 39
$GeC_{19}H_{18}S$ Ph_3GeSMe, 2, 123
$GeC_{19}H_{24}$ 9-Fluorenyl-$GeEt_3$, 2, 10
$GeC_{19}H_{26}$ $Ph_2CHGeEt_3$, 2, 9
$GeC_{19}H_{30}O_3$ $(EtMeCOHC{\equiv}C)_3GeMe$, 2, 25
$GeC_{19}H_{36}$ Cyclohexyl$_3$GeMe, 2, 8
$GeC_{19}H_{38}O_2$ $Et_3GeC{\equiv}CCEt_2OCHMeOBu$, 2, 26
9-Decacalinperoxy-$GePr_3$, 2, 105
$GeC_{19}H_{39}N$ $(i\text{-}C_5H_{11})_3Ge(CH_2)_3CN$, 2, 31
$GeC_{19}H_{40}O$ $(n\text{-}C_5H_{11})_3GeCH_2CO_2Et$, 2, 41
$Bu_3GeO_2CC_6H_{13}$, 2, 115
$GeC_{20}H_{14}Br_2$ $(1\text{-}C_{10}H_7)_2GeBr_2$, 2, 81
$GeC_{20}H_{14}O$ $(1\text{-}C_{10}H_7)_2GeO$, 2, 103
$GeC_{20}H_{15}F_3$ $Ph_3GeCF{=}CF_3$, 2, 21
$GeC_{20}H_{18}$ $Ph_3GeCH{=}CH_2$, 2, 18
$GeC_{20}H_{18}O_2$ Ph_3GeCO_2Me, 2, 41
Ph_3GeOAc, 2, 116
$GeC_{20}H_{20}$ $Et_2Ge(C{\equiv}CPh)_2$, 2, 22
$GeC_{20}H_{20}S_2$ Ph_3GeSCH_2SMe, 2, 123
$GeC_{22}H_{22}O$ $Ph_3GeCH_2CH_2OAc$, 2, 37
$GeC_{22}H_{22}O_2$ $Ph_3GeCH_2CH_2OAc$, 2, 35
$Ph_3GeCH_2CH_2CO_2Me$, 2, 41
$(p\text{-}MeC_6H_4)_3GeCO_2H$, 2, 41
$GeC_{22}H_{24}O$ $Ph_3Ge(CH_2)_4OH$, 2, 35
$GeC_{20}H_{24}O_4$ $i\text{-}Pr_2Ge(OBz)_2$, 2, 115
$GeC_{20}H_{26}$ 1,1-Dimethylditolylgermacyclopentane, 2, 146
$GeC_{20}H_{28}$ $n\text{-}Bu_2GePh_2$, 2, 11
$GeC_{20}H_{31}NO_2$ N-Phthalimidyl-$GeBu_3$, 2, 127
$GeC_{20}H_{32}$ $Bu_3GeC{\equiv}CPh$, 2, 23
$GeC_{20}H_{33}N$ N-Indolyl-$GeBu_3$, 2, 127
$GeC_{20}H_{34}$ $Bu_3GeCH{=}CHPh$, 2, 18
$GeC_{20}H_{36}O_2$ Cyclohexyl$_3$GeOAc, 2, 116
$GeC_{20}H_{38}$ Cyclohexyl$_3$GeEt, 2, 10
$Bu_2Ge(CH{=}CHBu)C{\equiv}CBu$, 2, 18
$GeC_{20}H_{40}O$ $Bu_3GeOCH_2CH_2C{\equiv}CBu$, 2, 109
$GeC_{20}H_{40}O_2$ Cyclohexyloxy$_2$GeBu$_2$, 2, 110
$GeC_{20}H_{42}O_2$ $(C_6H_{13})_3GeOAc$, 2, 116
$GeC_{20}H_{44}$ $(n\text{-}C_5H_{11})_4Ge$, 2, 5
$GeC_{20}H_{44}O$ $Bu_3GeOCH_2CHEtBu$, 2, 109*
$n\text{-}C_6H_{13}CHMeOGeBu_3$, 2, 109
$GeC_{20}H_{45}N$ $Bu_3GeNHC_8H_{17}$, 2, 126
$GeC_{21}H_{18}O_2$ $Ph_3GeO_2CCH{=}CH_2$, 2, 116, 156
$GeC_{21}H_{19}N$ $Ph_3GeCH_2CH_2CN$, 2, 31
$GeC_{21}H_{20}$ $Ph_3GeCH_2CH{=}CH_2$, 2, 18
$GeC_{21}H_{21}Br$ $(o\text{-}MeC_6H_4)_3GeBr$, 2, 73
$(p\text{-}MeC_6H_4)_3GeBr$, 2, 73
$GeC_{21}H_{21}Cl$ $(p\text{-}MeC_6H_4)_3GeCl$, 2, 73
$GeC_{21}H_{21}F$ $(p\text{-}MeC_6H_4)_3GeF$, 2, 73
$GeC_{21}H_{21}I$ $(m\text{-}MeC_6H_4)_3GeI$, 2, 73
$GeC_{21}H_{21}IO_3$ $(p\text{-}MeOC_6H_4)_3GeI$, 2, 75
$GeC_{21}H_{21}NO$ $Ph_3GeCH_2CH_2CONH_2$, 2, 41
$GeC_{21}H_{22}$ $i\text{-}PrGePh_3$, 2, 11
$(o\text{-}MeC_6H_4)_3GeH$, 2, 58
$(m\text{-}MeC_6H_4)_3GeH$, 2, 58
$(p\text{-}MeC_6H_4)_3GeH$, 2, 58
$GeC_{21}H_{30}$ $(PhCH_2)_2C_7H_{15}GeH$, 2, 60*
$GeC_{21}H_{40}$ Cyclohexyl$_3$GePr, 2, 11
$GeC_{21}H_{45}Cl$ $(n\text{-}C_7H_{15})_3GeCl$, 2, 73
$GeC_{21}H_{46}$ $(n\text{-}C_7H_{15})_3GeH$, 2, 57
$GeC_{22}H_{19}N$ N-$(Ph_3Ge)C_4H_4N$, 2, 127
$GeC_{22}H_{20}O_2$ $Ph_3GeO_2CCMe{=}CH_2$, 2, 116, 156
$GeC_{22}H_{23}BrO_2$ $Ph_3GeOCH_2CHBrCH_2OMe$, 2, 110
$GeC_{22}H_{24}$ Me_3GeCPh_3, 2, 7
$i\text{-}Pr_2Ge(C{\equiv}CPh)_2$, 2, 23
$GeC_{22}H_{24}O_2$ $Ph_3GeO_2CMe_3$, 2, 105
$GeC_{22}H_{24}S$ Ph_3GeSBu, 2, 123
$GeC_{22}H_{36}O_3$ $(Et_2COHC{\equiv}C)_3GeMe$, 2, 25
$GeC_{22}H_{38}O_2$ Cyclohexyl$_3$GeO$_2$CCMe=CH$_2$, 2, 116, 156

$GeC_{22}H_{42}$ Cyclohexyl$_3$GeBu, 2, 11
$GeC_{23}H_{20}$ $Ph_3GeC_5H_5$, 2, 19
$GeC_{23}H_{20}O_2$ m-$Ph_3GeC_6H_4CO_2H$, 2, 41
p-$Ph_3GeC_6H_4CO_2H$, 2, 41
$GeC_{23}H_{24}O$ $Ph_3GeCH{=}CHCOHMe_2$, 2, 21
$GeC_{23}H_{26}$ i-$C_5H_{11}GePh_3$, 2, 11
$GeC_{23}H_{26}O$ $Ph_3Ge(CH_2)_5OH$, 2, 35
$GeC_{23}H_{38}O_3$ $(Et_2COHC{\equiv}C)_3GeEt$, 2, 26
$GeC_{23}H_{44}$ Cyclohexyl$_3GeC_5H_{11}$, 2, 11
$GeC_{24}H_{16}$ 5,5'-Spirobi[dibenzogermole], 2, 144*, 609*
$GeC_{24}H_{16}F_4$ $(m\text{-}FC_6H_4)_4Ge$, 2, 6
$GeC_{24}H_{19}Br$ p-$BrC_6H_4GePh_3$, 2, 30
$GeC_{24}H_{19}Cl$ p-$ClC_6H_4GePh_3$, 2, 30
$GeC_{24}H_{19}F$ p-$FC_6H_4GePh_3$, 2, 30
$GeC_{24}H_{19}I$ p-$IC_6H_4GePh_3$, 2, 30
$GeC_{24}H_{20}$ Ph_4Ge, 2, 3
$GeC_{24}H_{21}N$ p-$Ph_3GeC_6H_4NH_2$, 2, 44
$GeC_{24}H_{26}$ Cyclohexyl-$GePh_3$, 2, 12
$GeC_{24}H_{27}IO_3$ $(p\text{-}EtOC_6H_4)_3GeI$, 2, 75
$GeC_{24}H_{28}$ n-$C_6H_{13}GePh_3$, 2, 11
$GeC_{24}H_{37}Cl$ p-ClC_6H_4Ge-cyclohexyl$_3$, 2, 30
$GeC_{24}H_{38}$ Cyclohexyl$_3$GePh, 2, 11
$GeC_{24}H_{39}BrO_6$ $(AcOC_6H_{10})_3GeBr$, 2, 75
$GeC_{24}H_{44}$ $(n\text{-}C_6H_{13})_3GePh$, 2, 11
$GeC_{24}H_{52}$ $(n\text{-}C_6H_{13})_4Ge$, 2, 5
$(n\text{-}C_8H_{17})_3GeH$, 2, 57
$GeC_{25}H_{19}ClO$ p-$Ph_3GeC_6H_4COCl$, 2, 41
$GeC_{25}H_{19}NO_3S$ p-$O_2NC_6H_4COSGePh_3$, 2, 123
$GeC_{25}H_{20}Br_2$ $Ph_3GeCPhBr_2$, 2, 30
$GeC_{25}H_{20}O$ Ph_3GeBz, 2, 37
$GeC_{25}H_{20}OS$ Ph_3GeSBz, 2, 123
$GeC_{25}H_{21}Br$ p-$BrCH_2C_6H_4GePh_3$, 2, 30
$GeC_{25}H_{21}NO$ m-$Ph_3GeC_6H_4CONH_2$, 2, 41
p-$Ph_3GeC_6H_4CONH_2$, 2, 41
$GeC_{25}H_{22}$ $PhCH_2GePh_3$, 2, 9
o-$MeC_6H_4GePh_3$, 2, 12
m-$MeC_6H_4GePh_3$, 2, 12
p-$MeC_6H_4GePh_3$, 2, 13
$GeC_{25}H_{22}O$ $Ph_3GeCHOHPh$, 2, 35
p-$Ph_3GeC_6H_4CH_2OH$, 2, 35
o-$Ph_3GeC_6H_4OMe$, 2, 35
p-$Ph_3GeC_6H_4OMe$, 2, 35
$GeC_{25}H_{22}S$ Ph_3GeSCH_2Ph, 2, 123
$GeC_{25}H_{30}$ Ph_3CGeEt_3, 2, 9
n-$C_7H_{15}GePh_3$, 2, 12
$GeC_{25}H_{40}$ Cyclohexyl$_3GeCH_2Ph$, 2, 8
m-MeC_6H_4Ge-cyclohexyl$_3$, 2, 11
p-MeC_6H_4Ge-cyclohexyl$_3$, 2, 11
$GeC_{25}H_{40}O$ p-$MeOC_6H_4Ge$-cyclohexyl$_3$, 2, 35
$GeC_{25}H_{54}O_3$ n-$C_7H_{15}Ge(CH_2CH_2OBu)_3$, 2, 35
$GeC_{26}H_{20}$ $EtGe(C{\equiv}CPh)_3$, 2, 23
$Ph_3GeC{\equiv}CPh$, 2, 23
$GeC_{26}H_{22}$ $Ph_3GeCH{=}CHPh$, 2, 18
p-$Ph_3GeC_6H_4CH{=}CH_2$, 2, 15
$GeC_{26}H_{22}O$ p-$Ph_3GeC_6H_4Ac$, 2, 37
$GeC_{26}H_{23}N$ o,o'-$C_6H_4NEtC_6H_4GePh_2$, 2, 148*, 638*
$GeC_{26}H_{24}$ $PhCH_2CH_2GePh_3$, 2, 11
$GeC_{26}H_{24}O$ p-$Ph_3GeC_6H_4OEt$, 2, 35
$GeC_{26}H_{25}N$ m-$Ph_3GeC_6H_4NMe_2$, 2, 44
p-$Ph_3GeC_6H_4NMe_2$, 2, 44
$GeC_{26}H_{26}O$ $EtPh(1\text{-}C_{10}H_7)GeOCHMePh$, 2, 110
$GeC_{26}H_{32}$ n-$C_8H_{17}GePh_3$, 2, 12
$GeC_{26}H_{38}$ Cyclohexyl$_3GeC{\equiv}CPh$, 2, 23
$GeC_{26}H_{42}$ p-EtC_6H_4Ge-cyclohexyl$_3$, 2, 11
$GeC_{26}H_{43}N$ p-$Me_2NC_6H_4Ge$-cyclohexyl$_3$, 2, 44
$GeC_{26}H_{48}O_4$ 9-Decalinperoxy$_2GePr_2$, 2, 105
$GeC_{27}H_{22}$ Indenyl-$GePh_3$, 2, 19
$GeC_{27}H_{24}O_3$ Isochromylperoxy-$GePh_3$, 2, 105
$GeC_{27}H_{26}O_2$ $Ph_3GeO_2CMe_2Ph$, 2, 105
$GeC_{27}H_{34}$ $C_9H_{19}GePh_3$, 2, 12
$GeC_{27}H_{34}O$ (-)Mentholoxy-$GeMePh(1\text{-}C_{10}H_7)$, 2, 110
$GeC_{27}H_{44}$ p-(Cyclohexyl$_3$Ge)$C_6H_4CHMe_2$, 2, 11
$GeC_{28}H_{20}$ $Ph_2Ge(C{\equiv}CPh)_2$, 2, 23
$GeC_{28}H_{24}$ $(p\text{-}CH_2{=}CHC_6H_4)_2GePh_2$, 2, 15, 157
$GeC_{28}H_{26}N_2$ (o,o'-$C_6H_4NEtC_6H_4)_2Ge$, 2, 144*, 638*
$GeC_{28}H_{28}$ $(PhCH_2)_4Ge$, 2, 5
$(m\text{-}MeC_6H_4)_4Ge$, 2, 6
$(p\text{-}MeC_6H_4)_4Ge$, 2, 6
$GeC_{28}H_{32}O_2$ 9-Decalinperoxy-$GePh_3$, 2, 105

$GeC_{28}H_{36}$ n-$C_7H_{15}Ge(CH_2Ph)_3$, 2, 8
$C_{10}H_{21}GePh_3$, 2, 12
$GeC_{28}H_{36}O$ (-)Mentholoxy-GeEtPh-($1\text{-}C_{10}H_7$), 2, 110
$GeC_{28}H_{46}$ p-(Cyclohexyl$_3$Ge)$C_6H_4CMe_3$, 2, 11
$GeC_{28}H_{54}$ $C_7H_{15}Ge(CH{=}CHC_5H_{11})_3$, 2, 18
$GeC_{28}H_{60}$ (n-C_7H_{15})$_4$Ge, 2, 6
$GeC_{30}H_{20}$ $PhGe(C{\equiv}CPh)_3$, 2, 23
$GeC_{30}H_{21}Br$ ($1\text{-}C_{10}H_7$)$_3$GeBr, 2, 73
$GeC_{30}H_{21}I$ ($2\text{-}C_{10}H_7$)$_3$GeI, 2, 73
$GeC_{30}H_{22}$ ($1\text{-}C_{10}H_7$)$_3$GeH, 2, 58
$GeC_{30}H_{22}O$ ($1\text{-}C_{10}H_7$)$_3$GeOH, 2, 101
$GeC_{30}H_{24}$ Indenyl$_2$GePh$_2$, 2, 19
$GeC_{30}H_{25}N$ Ph_3GeNPh_2, 2, 127
$GeC_{30}H_{26}$ $CPh{=}CPhCPh{=}CPhGeMe_2$, 2, 146
$GeC_{30}H_{26}Cl_2$ $Me_2(ClCPh{=}CPhCPh{=}CPh)GeCl$, 2, 75
$GeC_{30}H_{26}N_2$ N,N'-Diphenylhydrazinyl-GePh$_3$, 2, 127
$GeC_{31}H_{24}$ 9-Fluorenyl-GePh$_3$, 2, 19
$GeC_{31}H_{26}O$ $Ph_3GeCOHPh_2$, 2, 35
$GeC_{32}H_{20}$ $(PhC{\equiv}C)_4Ge$, 2, 5
$GeC_{32}H_{28}$ $(PhCH{=}CH)_4Ge$, 2, 5
(p-$CH_2{=}CHC_6H_4$)$_4$Ge, 2, 6
$Ph_2CHCH_2GePh_3$, 2, 11
$GeC_{32}H_{36}$ $(PhCH_2CH_2)_4Ge$, 2, 5
$GeC_{32}H_{44}O_4$ 9-Decalinperoxy$_2$GePh$_2$, 2, 105
$GeC_{32}H_{60}$ (Cyclohexyl-CH_2CH_2)$_4$Ge, 2, 5
$GeC_{32}H_{68}$ (n-C_8H_{17})$_4$Ge, 2, 6
$(BuEtCHCH_2)_4Ge$, 2, 6
$GeC_{33}H_{28}O$ $Ph_3GeCHPhCH_2Bz$, 2, 37
$GeC_{36}H_{28}$ Indenyl$_4$Ge, 2, 6
$GeC_{36}H_{30}NP$ $Ph_3GeN{=}PPh_3$, 2, 128
$GeC_{36}H_{44}$ $(PhCH_2CH_2CH_2)_4Ge$, 2, 5
$GeC_{36}H_{52}$ n-$C_{18}H_{37}GePh_3$, 2, 12*
$GeC_{37}H_{30}$ Ph_3GeCPh_3, 2, 9
p-$Ph_3GeC_6H_4CHPh_2$, 2, 13
$GeC_{37}H_{32}BrP$ [$Ph_3PCH_2GePh_3$]Br, 2, 45
$GeC_{38}H_{56}$ $PhCH_2CH_2$(n-$C_{18}H_{37}$)$GePh_2$, 2, 10
$GeC_{40}H_{26}O$ MePh($1\text{-}C_{10}H_7$)$GeCPh_2OH$, 2, 33
$GeC_{40}H_{28}$ ($1\text{-}C_{10}H_7$)$_4$Ge, 2, 6
$GeC_{40}H_{84}$ (n-$C_{10}H_{21}$)$_4$Ge, 2, 6
$GeC_{48}H_{100}$ (n-$C_{12}H_{25}$)$_4$Ge, 2, 6
$GeC_{56}H_{40}$ 5,5'-Spiro[tetraphenylgermole], 2, 144
$GeC_{56}H_{116}$ (n-$C_{14}H_{29}$)$_4$Ge, 2, 6
$GeC_{64}H_{132}$ (n-$C_{16}H_{33}$)$_4$Ge, 2, 6
$GeC_{72}H_{148}$ (n-$C_{18}H_{37}$)$_4$Ge, 2, 6
$Ge_2CH_4O_4$ $CH_2(GeO_2H)_2$, 2, 104
$Ge_2C_2H_2Cl_6$ $(Cl_3GeCH{=})_2$, 2, 92
$Ge_2C_2H_4Cl_6$ $(Cl_3GeCH_2)_2$, 2, 92
$(Ge_2C_2H_6O_3)_n$ $(Me_2Ge_2O_3)_n$, 2, 104
$Ge_2C_3H_6Cl_6$ $(Cl_3GeCH_2)_2CH_2$, 2, 92
$Ge_2C_4H_4Cl_4$ $CH{=}CHGeCl_2CH{=}CHGeCl_2$, 2, 152
$Ge_2C_4H_4I_4$ $CH{=}CHGeI_2CH{=}CHGeI_2$, 2, 153
$Ge_2C_4H_8Cl_6$ $(Cl_3GeCH_2)_2CHMe$, 2, 92
$Ge_2C_4H_8O_4$ $CH{=}CHGe(OH)_2CH{=}CHGe(OH)_2$, 2, 154
$Ge_2C_5H_{11}Cl_3$ $Me_3GeCH{=}CHGeCl_3$, 2, 46
$(Ge_2C_6H_{14}O_3)_n$ $(Pr_2Ge_2O_3)_n$, 2, 104
$Ge_2C_6H_{16}Cl_2O$ $(ClCH_2GeMe_2)_2O$, 2, 101
$Ge_2C_6H_{18}$ Me_6Ge_2, 2, 130
$Ge_2C_6H_{18}O$ $(Me_3Ge)_2O$, 2, 99
$Ge_2C_6H_{18}O_4S$ $(Me_3Ge)_2SO_4$, 2, 97
$Ge_2C_6H_{18}O_4Se$ $(Me_3Ge)_2SeO_4$, 2, 97
$Ge_2C_6H_{18}S$ $(Me_3Ge)_2S$, 2, 120
$Ge_2C_6H_{18}Se$ $(Me_3Ge)_2Se$, 2, 121
$Ge_2C_6H_{19}N$ $(Me_3Ge)_2NH$, 2, 126
$Ge_2C_7H_{20}$ $(Me_3Ge)_2CH_2$, 2, 46
$Ge_2C_7H_{21}N$ $(Me_3Ge)_2NMe$, 2, 126
$(Me_3Ge)_2NMe$, 2, 126
$Ge_2C_8H_{13}Cl_5$ $H(CH_2CH{=}CHCH_2GeCl_2)_2Cl$, 2, 92
$Ge_2C_8H_{16}$ $CH{=}CHGeMe_2CH{=}CHGeMe_2$, 2, 147
$Ge_2C_8H_{18}$ $(Me_3GeC{\equiv})_2$, 2, 46
$(Ge_2C_8H_{18}O_3)_n$ $(Bu_2Ge_2O_3)_n$, 2, 104
$Ge_2C_8H_{20}$ $(Me_3GeCH{=})_2$, 2, 47
$(Me_3Ge)_2C{=}CH_2$, 2, 46
$CH_2CH_2GeMe_2CH_2CH_2GeMe_2$, 2, 147
$Ge_2C_8H_{22}$ $(Me_3GeCH_2)_2$, 2, 46
$Ge_2C_9H_{24}$ $(Me_3GeCH_2)_2CH_2$, 2, 46
$Ge_2C_{10}H_{16}Cl_2$ p-$(Me_2GeCl)_2C_6H_4$, 2, 76
$Ge_2C_{10}H_{24}Cl_2$ $(Et_2GeClCH_2)_2$, 2, 76
$(Ge_2C_{12}H_8Br_2O_3)_n$ [(p-BrC_6H_4)$_2Ge_2O_3$]$_n$, 2, 104
$(Ge_2C_{12}H_8Cl_2O_3)_n$ [(p-ClC_6H_4)$_2Ge_2O_3$]$_n$, 2, 104
$Ge_2C_{12}H_{10}Br_4$ $(PhGeBr_2)_2$, 2, 133

$(Ge_2C_{12}H_{10}O_3)_n$ $(Ph_2Ge_2O_3)_n$, 2, 104
$Ge_2C_{12}H_{18}$ $(CH_2{=}CH)_6Ge_2$, 2, 130
$Ge_2C_{12}H_{22}$ p-$(Me_3Ge)_2C_6H_4$, 2, 46
$Ge_2C_{12}H_{30}$ Et_6Ge_2, 2, 129
$Ge_2C_{12}H_{30}O$ $(Et_3Ge)_2O$, 2, 99
$Ge_2C_{12}H_{30}O_2$ $(Et_3GeO)_2$, 2, 105
$Ge_2C_{12}H_{30}O_4S$ $(Et_3Ge)_2SO_4$, 2, 97
$Ge_2C_{12}H_{30}S$ $(Et_3Ge)_2S$, 2, 120
$Ge_2C_{13}H_{20}$ $Me(Me_2GeCH_2CH{=}CHCH_2)_2H$, 2, 46
$(Ge_2C_{14}H_{10}O_7)_n$ $[(p\text{-}HO_2CC_6H_4)_2GeO_3]_n$, 2, 104
$(Ge_2C_{14}H_{14}O_3)_n$ $[(p\text{-}MeC_6H_4)_2Ge_2O_3]_n$, 2, 104
$(Ge_2C_{14}H_{14}O_5)_n$ $[(p\text{-}MeOC_6H_4)_2Ge_2O_3]_n$, 2, 104
$Ge_2C_{14}H_{30}$ $(Et_3GeC{\equiv})_2$, 2, 47
$Ge_2C_{14}H_{32}O_2S$ $Et_3GeO_2CCH_2SGeEt_3$, 2, 118
$Ge_2C_{14}H_{34}$ $(Et_3GeCH_2)_2$, 2, 47
$Ge_2C_{14}H_{34}O_2$ $(Et_3GeOCH_2)_2$, 2, 108
$Ge_2C_{15}H_{32}O$ $Et_3GeC{\equiv}CCH_2OGeEt_3$, 2, 25
$Ge_2C_{15}H_{34}O$ $Et_3GeCH{=}CHCH_2OGeEt_3$, 2, 20
$Ge_2C_{15}H_{36}$ $(Et_3GeCH_2)_2CH_2$, 2, 47
$Ge_2C_{15}H_{36}O$ $(Et_3Ge(CH_2)_3OGeEt_3$, 2, 33
$(Ge_2C_{16}H_{18}N_4O_7)_n$ $[(3,4\text{-}O_2NMe_2NC_6H_3)_2\text{-}Ge_2O_3]_n$, 2, 104
$(Ge_2C_{16}H_{18}O_5)_n$ $[(p\text{-}EtOC_6H_4)_2Ge_2O_3]_n$, 2, 104
$(Ge_2C_{16}H_{20}N_2O_3)_n$ $[(p\text{-}Me_2NC_6H_4)_2\text{-}Ge_2O_3]_n$, 2, 104
$Ge_2C_{16}H_{34}O$ $[CH_2(CH_2)_3GeBu]_2O$, 2, 154
$Ge_2C_{16}H_{34}O_2$ $(Et_3GeOCH_2C{\equiv})_2$, 2, 108
$Ge_2C_{16}H_{36}$ $CH_2(CH_2)_3GeBuCH_2CH_2GeEt_3$, 2, 145
$CH_2CH_2GeEt_2CH_2CH_2GeBu_2$, 2, 147
$Ge_2C_{16}H_{36}I_2$ $(Bu_2GeI)_2$, 2, 133
$Ge_2C_{16}H_{36}O_2$ $(Et_3GeOCH_2CH{=})_2$, 2, 108
$Ge_2C_{16}H_{38}$ $(Et_3GeCH_2CH_2)_2$, 2, 47
$Ge_2C_{16}H_{38}O_2$ $(Et_3GeOCH_2CH_2)_2$, 2, 108
$Ge_2C_{18}H_{14}Cl_4$ p-$(PhGeCl_2)_2C_6H_4$, 2, 82
$Ge_2C_{18}H_{34}$ m-$(Et_3Ge)_2C_6H_4$, 2, 47*
p-$(Et_3Ge)_2C_6H_4$, 2, 47*
$Ge_2C_{18}H_{34}O_2$ $(p\text{-}Et_3GeO)_2C_6H_4$, 2, 108
$Ge_2C_{18}H_{34}O_3S$ p-$Et_3GeC_6H_4SO_3GeEt_3$, 2, 43
$Ge_2C_{18}H_{38}O_2$ $(Et_3GeOCHMeC{\equiv})_2$, 2, 108
$Ge_2C_{18}H_{40}O$ $Et_3GeCH_2CH{=}CHCH(CH_2CH_2OH)\text{-}GeEt_3$, 2, 47
$Et_3GeCH_2CH{=}CHCH(CHMeOH)GeEt_3$, 2, 47
$Ge_2C_{18}H_{42}O$ $(Pr_3Ge)_2O$, 2, 101
$(i\text{-}Pr_3Ge)_2O$, 2, 101
$Ge_2C_{18}H_{42}O_2$ $(Pr_3GeO)_2$, 2, 105
$Ge_2C_{18}H_{42}O_4S$ $(Pr_3Ge)_2SO_4$, 2, 98
$(i\text{-}Pr_3Ge)_2SO_4$, 2, 98
$Ge_2C_{20}H_{26}$ Bi[1,1-Phenylgermacyclopentane], 2, 155
$Ge_2C_{20}H_{42}$ $Et_3GeC{\equiv}CGeBu_3$, 2, 47
$Ge_2C_{20}H_{44}O$ $Et_3GeCH_2CH{=}CHCHGeEt_3CHOH\text{-}Pr$, 2, 47*
$Ge_2C_{23}H_{42}O$ $Et_3GeCH_2CH{=}CHCH(GeEt_3)\text{-}CHOHPh$, 2, 47*
$Ge_2C_{24}H_{20}Br_2$ $(Ph_2GeBr)_2$, 2, 133
$Ge_2C_{24}H_{54}O$ $(Bu_3Ge)_2O$, 2, 101
$Ge_2C_{24}H_{55}N$ $(Bu_3Ge)_2NH$, 2, 126
$Ge_2C_{25}H_{20}O_2$ $Ph_3GeCO_2GePh_3$, 2, 41
$Ge_2C_{25}H_{54}N_2$ $(Bu_3Ge)_2N_2C$, 2, 126
$(Ge_2C_{26}H_{24})_n$ $(p\text{-}Me_2GeC_6H_4GePh_2C_6H_4)_n$, 2, 46
$Ge_2C_{26}H_{56}$ $(Bu_3GeCH{=})_2$, 2, 47
$Ge_2C_{28}H_{20}Br_4$ $CPh{=}CPhGeBr_2CPh{=}CPhGeBr_2$, 2, 153
$Ge_2C_{28}H_{20}Cl_4$ $CPh{=}CPhGeCl_2CPh{=}CPhGeCl_2$, 2, 153
$Ge_2C_{28}H_{20}I_4$ $CPh{=}CPhGeI_2CPh{=}CPhGeI_2$, 2, 153
$Ge_2C_{28}H_{24}O_4$ $CPh{=}CPhGe(OH)_2CPh{=}CPhGe\text{-}(OH)_2$, 2, 154
$Ge_2C_{30}H_{66}O$ $(n\text{-}C_5H_{11})_6Ge_2O$, 2, 101
$Ge_2C_{32}H_{32}$ $CPh{=}CPhGeMe_2CPh{=}CPhGeMe_2$, 2, 147
$Ge_2C_{36}H_{30}$ Ph_6Ge_2, 2, 129
$Ge_2C_{36}H_{30}O$ $(Ph_3Ge)_2O$, 2, 100
$Ge_2C_{36}H_{30}O_2$ $(Ph_3GeO)_2$, 2, 105
$Ge_2C_{36}H_{30}S$ $(Ph_3Ge)_2S$, 2, 120
$Ge_2C_{36}H_{30}S_2$ $(Ph_3GeS)_2$, 2, 120
$Ge_2C_{36}H_{40}$ $CPh{=}CPhGeEt_2CPh{=}CPhGeEt_2$, 2, 148*
$Ge_2C_{36}H_{66}$ $Cyclohexyl_6Ge_2$, 2, 130
$Ge_2C_{36}H_{66}S_2$ $(Cyclohexyl_3GeS)_2$, 2, 120
$Ge_2C_{36}H_{78}O$ $(C_6H_{13})_6Ge_2O$, 2, 101
$Ge_2C_{38}H_{30}$ $(Ph_3GeC{\equiv})_2$, 2, 48
$Ge_2C_{38}H_{68}$ $(Cyclohexyl_3GeC{\equiv})_2$, 2, 47
$Ge_2C_{39}H_{36}$ $(Ph_3GeCH_2)_2CH_2$, 2, 48

$Ge_2C_{40}H_{30}$ $(Ph_3GeC{\equiv}C)_2$, 2, 48
$Ge_2C_{41}H_{40}$ $(Ph_3GeCH_2CH_2)_2CH_2$, 2, 48
$Ge_2C_{42}H_{34}$ p-$(Ph_3Ge)_2C_6H_4$, 2, 48
$Ge_2C_{42}H_{42}$ (o-$MeC_6H_4)_6Ge_2$, 2, 130
(m-$MeC_6H_4)_6Ge_2$, 2, 130
(p-$MeC_6H_4)_6Ge_2$, 2, 130
$Ge_2C_{42}H_{42}O$ (m-$MeC_6H_4)_6Ge_2O$, 2, 101
(p-$MeC_6H_4)_6Ge_2O$, 2, 101
$Ge_2C_{42}H_{42}S$ $(PhCH_2)_6Ge_2S$, 2, 120
(p-$MeC_6H_4)_6Ge_2S$, 2, 120
$Ge_2C_{48}H_{42}$ $(PhCH{=}CH)_6Ge_2$, 2, 130
$Ge_2C_{50}H_{40}$ $(Ph_3GeCPh{=})_2$, 2, 48
$Ge_2C_{72}H_{54}S$ $(PhC_6H_4)_6Ge_2S$, 2, 120
$Ge_3C_5H_{12}Cl_6$ $Me_3GeCH_2CH(GeCl_3)_2$, 2, 46
$Ge_3C_6H_{18}Cl_3N$ $(Me_2GeCl)_3N$, 2, 127
$Ge_3C_6H_{18}S_3$ $(Me_2GeS)_3$, 2, 120
$Ge_3C_8H_{24}$ $Me(GeMe_2)_3Me$, 2, 131
$Ge_3C_9H_{27}N$ $(Me_3Ge)_3N$, 2, 126
$Ge_3C_9H_{27}N_3$ $(Me_2GeNMe)_3$, 2, 126
$Ge_3C_9H_{27}O_4P$ $(Me_3Ge)_3PO_4$, 2, 97
$Ge_3C_{11}H_{30}$ $Me_3GeCH_2CH(GeMe_3)_2$, 2, 46
$Ge_3C_{18}H_{42}O_3$ $(Pr_2GeO)_3$, 2, 103
(i-$Pr_2GeO)_3$, 2, 103*
$Ge_3C_{18}H_{42}S_3$ (i-$Pr_2GeS)_3$, 2, 120
$Ge_3C_{20}H_{32}$ (p-$Me_3GeC_6H_4)_2GeMe_2$, 2, 46
$Ge_3C_{24}H_{54}S_3$ $(Bu_2GeS)_3$, 2, 120
$Ge_3C_{24}H_{58}N_3$ $(Bu_2GeNH)_3$, 2, 127
$Ge_3C_{36}H_{30}O_3$ $(Ph_2GeO)_3$, 2, 103
$Ge_3C_{36}H_{30}S_3$ $(Ph_2GeS)_3$, 2, 120
$Ge_3C_{41}H_{42}$ $(Ph_3Ge)_2GeMeBu$, 2, 131
$Ge_3C_{42}H_{42}O_3$ (m-$Tolyl_2GeO)_3$, 2, 103
$Ge_3C_{48}H_{40}S_2$ $(Ph_3GeS)_2GePh_2$, 2, 120
$Ge_3C_{54}H_{45}N$ $(Ph_3Ge)_3N$, 2, 127
$Ge_3C_{54}H_{46}$ $(Ph_3Ge)_3GeH$, 2, 133
$Ge_4C_8H_{24}$ $Me_2GeGeMe_2GeMe_2GeMe_2$, 2, 131
$Ge_4C_{10}H_{30}$ $Me(GeMe_2)_4Me$, 2, 131
$Ge_4C_{24}H_{56}O_4$ $(Pr_2GeO)_4$, 2, 103
$Ge_4C_{28}H_{42}$ p-(p-$Me_3GeC_6H_4GeMe_2)_2C_6H_4$, 2, 46
$Ge_4C_{48}H_{40}$ $Ph_2GeGePh_2GePh_2GePh_2$, 2, 131
$Ge_4C_{48}H_{40}I_2$ $Ph_8Ge_4I_2$, 2, 133
$Ge_4C_{48}H_{40}O$ Ph_8Ge_4O, 2, 133
$Ge_4C_{48}H_{40}O_4$ $(Ph_2GeO)_4$, 2, 103
$Ge_4C_{55}H_{48}$ $(Ph_3Ge)_3GeMe$, 2, 131
$Ge_4C_{60}H_{50}$ $Ph_{10}Ge_4$, 2, 131
$Ge_5C_{60}H_{50}$ Cyclic-$(Ph_2Ge)_5$, 2, 131
$Ge_6C_{12}H_{36}$ Cyclic-$(Me_2Ge)_6$, 2, 131
$Ge_6C_{36}H_{30}$ $(PhGe)_6$, 2, 132
$Ge_6C_{72}H_{60}$ $(Ph_2Ge)_6$, 2, 131
$Ge_7C_{16}H_{48}$ $Me(GeMe_2)_7Me$, 2, 131
$GeAgC_{72}H_{60}P_3$ $Ph_3GeAg{\cdot}3PPh_3$, 2, 139
$(GeAs_2C_{42}H_{34}N_2)_n$ (1,3-$Ph_2GeN{=}AsPh_2$-$C_6H_4AsPh_2{=}N)_n$, 2, 128*, 3, 369
$GeAuC_{21}H_{24}P$ $Ph_3GeAuPMe_3$, 2, 139
$GeAuC_{36}H_{30}P$ $Ph_3GeAuPPh_3$, 2, 139
$GeBC_4H_9F_3N$ $Me_3GeCN{\cdot}BF_3$, 2, 94
$GeBC_4H_{12}F_3O$ $Me_3GeOMe{\cdot}BF_3$, 2, 107
$GeBC_{40}H_{42}N$ $Me_4N[Ph_3GeBPh_3]$, 2, 139
$GeBC_{55}H_{48}P$ $MePh_3P[Ph_3GeBPh_3]$, 2, 139
$GeBC_{61}H_{52}P$ $[Ph_3PCH_2GePh_3]BPh_4$, 2, 45
$GeCrC_{12}H_{14}O_3$ $(Me_3GePh)Cr(CO)_3$, 1, 93
$GeCoC_{27}H_{26}O$ $(PhCH_2)_3Ge[C_5H_5Co(CO)]$, 2, 142
$GeCrC_2H_6O_4$ Me_2GeCrO_4, 2, 97
$GeCrC_{12}H_{14}O_3$ $Me_3GePhCr(CO)_3$, 2, 140
$GeCrC_{24}H_{34}O_3$ i-Bu_3Ge[indenyl-$Cr(CO)_3$], 2, 140
$GeCr_2C_{44}H_{64}O_6$ $Cyclohexyl_2Ge[C_8H_{17}C_5H_4$-$Cr(CO)_3]_2$, 1, 93, 2, 140
$GeCuC_{72}H_{60}P_3$ $Ph_3GeCu{\cdot}3PPh_3$, 2, 139
$GeFeC_8H_9NO_4$ $[Me_3GeNCFe(CO)_4]$, 1, 524, 2, 141
$GeFeC_9H_{11}FO_2$ $Me_2GeF[C_5H_5Fe(CO)_2]$, 1, 533, 2, 141*
$GeFeC_{25}H_{20}O_2$ $Ph_3Ge[C_5H_5Fe(CO)_2]$, 1, 533, 2, 141*
$GeFeC_{28}H_{24}$ $Ph_3GeC_5H_4FeC_5H_5$, 1, 537, 2, 141
$GeFeC_{30}H_{25}BrO_2$ (1-$C_{10}H_7)_2GeBr[MeEt$-$C_5H_3Fe(CO)_2]$, 1, 534, 2, 141
$GeFeC_{33}H_{43}ClO_2$ $(PhCH_2)_2GeCl[C_{12}H_{25}$-$C_5H_4Fe(CO)_2]$, 1, 534, 2, 141
$GeFe_3C_{49}H_{36}O_6$ BuGe[9-fluorenyl-Fe-$(CO)_2]_3$, 2, 141
$GeHf_3C_{41}H_{44}O_{12}$ n-$C_8H_{17}Ge[Me_2C_5H_3Hf$-$(CO)_4]_3$, 1, 42, 2, 139
$GeHgC_4H_{11}Br$ Me_3GeCH_2HgBr, 2, 52
$GeHgC_4H_{11}Cl$ Me_3GeCH_2HgCl, 2, 52
$GeKC_{18}H_{15}$ Ph_3GeK, 2, 135
$GeLiC_4H_{12}N$ $Me_3GeNMeLi$, 2, 128
$GeLiC_7H_{17}$ Et(i-$C_5H_{11})GeHLi$, 2, 137
$GeLiC_{10}H_{15}$ p-$Me_3GeCH_2C_6H_4Li$, 2, 52
$GeLiC_{17}H_{15}$ MePh(1-$C_{10}H_7)GeLi$, 2, 137

$GeLiC_{18}H_{15}$ Ph_3GeLi, 2, 134
$GeLiC_{20}H_{19}$ $Ph_2(PhCH_2CH_2)GeLi$, 2, 137
$GeLiC_{24}H_{19}$ p-$Ph_3GeC_6H_4Li$, 2, 52
$GeLiC_{30}H_{47}$ Ph_2(n-$C_{18}H_{37}$)GeLi, 2, 137*
$GeMgC_4H_{11}Cl$ Me_3GeCH_2MgCl, 2, 51
$GeMgC_8H_{15}Br$ $Et_3GeC{\equiv}CMgBr$, 2, 51*
$GeMgC_9H_{13}Br$ p-$Me_3GeC_6H_4MgBr$, 2, 51
$GeMgC_9H_{17}Br$ $Et_3GeCH_2C{\equiv}CMgBr$, 2, 51*
$GeMgC_9H_{21}Cl$ $Et_3Ge(CH_2)_3MgCl$, 2, 51
$GeMgC_{10}H_{23}Br$ $Et_3Ge(CH_2)_4MgBr$, 2, 51
$GeMgC_{12}H_{19}Br$ p-$Et_3GeC_6H_4MgBr$, 2, 51
$GeMgC_{14}H_{27}Br$ $Bu_3GeC{\equiv}CMgBr$, 2, 52*
$GeMgC_{15}H_{33}Cl$ $Bu_3Ge(CH_2)_3MgCl$, 2, 52
$GeMgC_{18}H_{15}Cl$ $Ph_3GeMgCl$, 2, 137
$GeMgC_{19}H_{17}Cl$ Ph_3GeCH_2MgCl, 2, 52
$GeMgC_{21}H_{21}Cl$ (o-$MeC_6H_4)_3GeMgCl$, 2, 137
(m-$MeC_6H_4)_3GeMgCl$, 2, 137
(p-$MeC_6H_4)_3GeMgCl$, 2, 137
$GeMnC_{15}H_{23}O_3$ $Et_3GeMn(CO)_3$-($CH{=}CHCH{=}CHCH_2CH_2$), 1, 184
$GeMnC_{20}H_{18}BrO_3$ $Ph_2GeBrMn(CO)_3(CH_2{=}C$-$MeCH{=}CH_2$), 2, 141
$GeMnC_{22}H_{23}O_3$ $Ph_2EtGeMn(CO)_3CH_2{=}CMe$-$CH{=}CH_2$), 1, 184
$GeMnC_{33}H_{35}O_3$ $(PhCH_2CH_2)_3GeMn(CO)_3$-($CH{=}CHCH{=}CHCH_2CH_2$), 1, 184
$GeMoC_{11}H_{15}O_5$ $Et_3GeMo(CO)_5$, 2, 140
$GeMoC_{14}H_{21}O_5$ $Pr_3GeMo(CO)_5$, 2, 140
$GeMoC_{15}H_{23}O_3$ $Et_3GeMo(CO)_3\cdot$1,3-C_6H_8, 2, 140
$GeMoC_{16}H_{27}O_3$ $Pr_3GeMo(CO)_3\cdot$1,3-C_4H_6, 2, 140
$GeMoC_{17}H_{10}BrO_5$ $Ph_2GeBrMo(CO)_5$, 2, 140
$GeMoC_{23}H_{15}O_5$ $Ph_3GeMo(CO)_5$, 2, 140
$GeMoC_{26}H_{21}O_5$ $(PhCH_2)_3GeMo(CO)_5$, 2, 140
$GeMoC_{29}H_{27}O_5$ $(PhCH_2CH_2)_3GeMo(CO)_5$, 2, 140
$GeMoC_{30}H_{29}O_3$ $(PhCH_2)_3GeMo(CO)_3\cdot$1,3-$C_6H_8$, 2, 140
$GeMoC_{32}H_{21}O_5$ 2-Indenyl$_3GeMo(CO)_5$, 2, 140
$GeMoC_{33}H_{35}O_3$ $(PhCH_2CH_2)_3GeMo(CO)_3\cdot$1,3-$C_6H_8$, 2, 140
$GeMoC_{35}H_{25}O_4$ 2-Indenyl$_3GeMo(CO)_3\cdot$1,3-C_5H_4O, 2, 141
$GeMoC_{35}H_{63}O_5$ $(C_{10}H_{21})_3GeMo(CO)_5$, 2, 140
$GeMoC_{38}H_{71}O_3$ $(C_{10}H_{21})_3GeMo(CO)_3\cdot$1,3-$C_5H_8$, 2, 140
$GeMoC_{53}H_{99}O_5$ $(C_{16}H_{33})_3GeMo(CO)_5$, 2, 140
$GeMoC_{60}H_{111}O_4$ $(C_{16}H_{33})_3GeMo(CO)_3\cdot$1,3-$C_5Me_4O$, 2, 141
$GeNaC_{18}H_{15}$ Ph_3GeNa, 2, 135
$GeNa_2C_{12}H_{10}$ Ph_2GeNa_2, 2, 137
$GeNiC_7H_{10}O_3$ $[Et_2GeNi(CO)_3]$, 2, 142
$GeOsC_{13}H_{14}O_2$ $(CH_2{=}CH)_3Ge[C_5H_5Os(CO)_2]$, 1, 563, 2, 141
$GeOsC_{13}H_{20}O_2$ $Me_2PrGe[MeC_5H_4Os(CO)_2]$, 2, 141
$GeOsC_{35}H_{52}O_5$ $(AcC_6H_{10})_3Ge[BuC_5H_4Os$-$(CO)_2]$, 1, 563, 2, 141
$GePbC_7H_{21}N$ $Me_3PbNMeGeMe_3$, 2, 128*, 587
$GePb_4C_{72}H_{60}$ $(Ph_3Pb)_4Ge$, 2, 136, 593
$GePdC_{20}H_{16}Br_2O$ $[MeC_5H_4GeBr_2$(9-fluorenyl)PdCO], 1, 687, 2, 142*
$GePdC_{26}H_{23}BrO$ $[(MeC_5H_4)_2GeBr$(9-fluorenyl)PdCO], 1, 687, 2, 142*
$GeReC_3H_9O_4$ Me_3GeReO_4, 2, 97
$GeRuC_{19}H_{15}ClO_2$ $Ph_2GeCl[C_5H_5Ru(CO)_2]$, 2, 141
$GeRu_3C_{38}H_{50}O_6$ (n-C_8H_{17})$Ge[MeEtC_5H_3$-$Ru(CO)_2]_3$, 1, 559, 2, 141
$GeRu_3C_{83}H_{132}O_6$ 2,4-$Me_2C_6H_3Ge[C_{18}H_{37}$-$C_5H_4Ru(CO)_2]_3$, 1, 559, 2, 141*
$GeSiC_2H_3Cl_7$ $Cl_3GeCHClCH_2SiCl_3$, 2, 92
$Cl_3GeCH_2CHClSiCl_3$, 2, 92
$GeSiC_2H_4Cl_6$ $Cl_3GeCH_2CH_2SiCl_3$, 2, 92
$GeSiC_3H_7Cl_5$ $Cl_3GeCH_2CH_2SiMeCl_2$, 2, 92
$(GeSiC_4H_{11}O)_n$ $(Me_3GeCH_2SiO)_n$, 2, 49
$GeSiC_6H_{18}O$ $Me_3GeOSiMe_3$, 2, 106
$GeSiC_6H_{18}O_4S$ $Me_3GeSO_4SiMe_3$, 2, 97
$GeSiC_6H_{18}S$ $Me_3GeSSiMe_3$, 2, 124
$GeSiC_6H_{18}Se$ $Me_3GeSeSiMe_3$, 2, 124
$GeSiC_7H_{20}$ $Me_3GeCH_2SiMe_3$, 2, 49
$GeSiC_7H_{20}O_2$ $Me_3GeCH_2SiMe(OMe)_2$, 2, 49
$GeSiC_7H_{21}N$ $Me_3GeNMeSiMe_3$, 2, 128
$GeSiC_8H_{20}$ $Me_3GeCH{=}CHSiMe_2$, 2, 49
$Me_3GeC({=}CH_2)SiMe_3$, 2, 49
$GeSiC_8H_{22}$ $Me_3GeCH_2CH_2SiMe_3$, 2, 49
$GeSiC_9H_{24}O$ $Me_3GeOSiEt_3$, 2, 110

$Et_3GeOSiMe_3$, 2, 111

$GeSiC_{12}H_{30}$ $Et_3GeSiEt_3$, 2, 136

$GeSiC_{13}H_{24}$ p-$Me_3GeCH_2C_6H_4SiMe_3$, 2, 49

$GeSiC_{15}H_{28}O$ p-$Et_3GeC_6H_4OSiMe_3$, 2, 49

$GeSiC_{16}H_{30}$ p-$Et_3GeC_6H_4CH_2SiMe_3$, 2, 49

$GeSiC_{24}H_{30}$ $Ph_3GeSiEt_3$, 2, 136

$GeSiC_{26}H_{30}F_2$ $Ph_3GeCF{=}CFSiEt_3$, 2, 49

$GeSiC_{36}H_{30}$ $Ph_3GeSiPh_3$, 2, 136

$GeSi_2C_8H_{24}O_2$ $Me_2Ge(OSiMe_3)_2$, 2, 111

$GeSi_2C_9H_{27}N$ $Me_3GeN(SiMe_3)_2$, 2, 128

$GeSi_3C_{10}H_{30}O_3$ $MeGe(OSiMe_3)_3$, 2, 111

$GeSi_4C_{14}H_{42}N_2$ $Me_2Ge[N(SiMe_3)_2]_2$, 2, 128

$GeSi_4C_{27}H_{25}$ (p-$Me_3SiC_6H_4)_4Ge$, 2, 6

$GeSnC_6H_{18}O$ $Me_3GeOSnMe_3$, 2, 110, 394

$GeSnC_6H_{18}S$ $Me_3GeSSnMe_3$, 2, 124, 473

$GeSnC_6H_{18}Se$ $Me_3GeSeSnMe_3$, 2, 124, 473

$GeSnC_7H_{21}N$ $Me_3GeNMeSnMe_3$, 2, 128, 476

$GeSnC_{13}H_{24}$ p-$Me_3GeCH_2C_6H_4SnMe_3$, 2, 50

$GeSnC_{28}H_{28}$ $Ph_2GeCH_2CH_2SnPh_2CH_2CH_2$, 2, 50

$GeSnC_{36}H_{30}$ $Ph_3GeSnPh_3$, 2, 136, 495

$GeSnC_{36}H_{30}S$ $Ph_3GeSSnPh_3$, 2, 124, 473

$GeSnC_{36}H_{30}Se$ $Ph_3GeSeSnPh_3$, 2, 124, 473

$GeSnC_{38}H_{34}$ $Ph_3GeCH_2CH_2SnPh_3$, 2, 50

$GeSnC_{44}H_{38}$ p-$Ph_3GeC_6H_4CH_2CH_2SnPh_3$, 2, 50

$GeSnC_{52}H_{48}$ $(Ph_3SnCH_2CH_2)_2GePh_2$, 2, 50

$GeSn_4C_{72}H_{60}$ $(Ph_3Sn)_4Ge$, 2, 136, 495

$Ge_2SnC_{64}H_{56}$ (p-$Ph_3GeC_6H_4CH_2CH_2)_2SnPh_2$, 2, 50

$Ge_2Sn_2C_{62}H_{62}$ p-(p-$Ph_3GeC_6H_4CH_2CH_2SnMe_2)_2C_6H_4$, 2, 50

$Ge_2Sn_2C_{70}H_{62}$ p-$(Ph_3SnCH_2CH_2GePh_2)_2C_6H_4$, 2, 50

$GeTiC_{37}H_{64}O_4$ (n-$C_8H_{17})_3Ge$[i-$BuC_5H_4Ti(CO)_4]$, 2, 139

$GeTi_2C_{32}H_{40}O_8$ $Bu_2Ge[MeEtC_5H_3Ti(CO)_4]_2$, 1, 35, 2, 139

$GeW_2C_{32}H_{28}O_6$ $(PhCH_2)_2Ge[MeC_5H_4W(CO)_3]_2$, 2, 140

$GeW_3C_{54}H_{36}O_9$ Cyclohexyl-Ge[fluorenyl-$W(CO)_3]_3$, 2, 140

$GeZr_3C_{50}H_{54}O_{12}$ $PhCH_2CH_2Ge[MeBuC_5H_3Zr(CO)_4]_3$, 1, 41, 2, 139*

$Ge_2As_2C_{47}H_{88}N_2$ 4,4'-$(Bu_3GeN{=}AsEt_2C_6H_4)_2CMe_2$, 2, 128, 3, 369

$Ge_2As_4C_{90}H_{74}N_4$ p-$[(Ph_3As{=}N)_2GePh]_2C_6H_4$, 2, 128, 3, 369

$Ge_2AuC_{44}H_{50}N$ $Et_4N[(Ph_3Ge)_2Au]$, 2, 139

$Ge_2CdC_{12}H_{30}$ $(Et_3Ge)_2Cd$, 2, 134

$Ge_2FeC_{46}H_{38}$ $(Ph_3GeC_5H_4)_2Fe$, 1, 540, 2, 141

$Ge_2HgC_{12}H_{30}$ $(Et_3Ge)_2Hg$, 2, 134

$Ge_2Nb_2C_{40}H_{54}O_6$ $[(CH_2{=}CH)_2Ge(C_8H_{17}C_5H_4)Nb(CO)_3]_2$, 2, 139

$Ge_2Nb_2C_{52}H_{54}O_6$ $[Ph_2Ge(C_6H_{13}C_5H_4)Nb(CO)_3]_2$, 2, 139

$Ge_2Sb_2C_{66}H_{54}N_2$ m-$(Ph_3GeN{=}SbPh_2)_2C_6H_4$, 2, 128, 3, 752

$Ge_2Ta_2C_{78}H_{110}O_6$ $[(MeC_5H_4)_2Ge(C_{18}H_{37}C_5H_4)Ta(CO)_3]_2$, 1, 54, 2, 139

$Ge_2TiC_{46}H_{38}Cl_2$ $(Ph_3GeC_5H_4)_2TiCl_2$, 1, 35, 2, 139

$Ge_3AsC_9H_{27}O_4$ $(Me_3Ge)_3AsO_4$, 2, 97

$Ge_3BC_9H_{27}O_3$ $(Me_3Ge)_3BO_3$, 2, 97

$Ge_3LiC_{54}H_{45}$ $(Ph_3Ge)_3GeLi$, 2, 133

$Ge_3SiC_{54}H_{45}Br$ $(Ph_3Ge)_3SiBr$, 2, 136

$Ge_3SiC_{54}H_{45}Cl$ $(Ph_3Ge)_3SiCl$, 2, 136

$Ge_3SiC_{54}H_{46}$ $(Ph_3Ge)_3SiH$, 2, 136

$Ge_3SiC_{54}H_{46}O$ $(Ph_3Ge)_3SiOH$, 2, 136

$Ge_3SiC_{54}H_{47}N$ $(Ph_3Ge)_3SiNH_2$, 2, 136

$Ge_3SiC_{56}H_{50}$ $(Ph_3Ge)_3SiEt$, 2, 136

$Ge_4Ag_4C_{72}H_{100}I_4P_4$ $(Et_3GePPh_2{\cdot}AgI)_4$, 2, 127

$Ge_4PbC_{72}H_{60}$ $(Ph_3Ge)_4Pb$, 2, 136, 593

$Ge_4Si_2C_{72}H_{60}$ $(Ph_3Ge)_4Si_2$ (?), 2, 136

$Ge_6Si_2C_{110}H_{94}Br_2Cl_2$ $[(Ph_3Ge)_3SiBr]_2{\cdot}(CH_2Cl)_2$, 2, 136

$GeAsMnC_{12}H_{14}I_2O_4$ $\{I_2EtGeMn[As(CH{=}CH_2)_3](CO)_4\}$, 3, 6

$GeAs_2Mn_2C_{58}H_{44}O_8$ $\{(PhCH_2)_2Ge[Mn(AsPh_3)(CO)_4]_2\}$, 3, 23

$GeAs_3Mn_3C_{122}H_{228}O_{12}$ $CH_2{=}CHGe\{Mn[As(C_{12}H_{25})_3](CO)_4\}_3$, 3, 29

$GeBLiC_{36}H_{30}$ $Li[Ph_3GeBPh_3]$, 2, 139

$GeCrSiC_6H_{18}O_4$ $Me_3GeCrO_4SiMe_3$, 2, 97

$GeHgSiC_8H_{22}$ $Me_3GeCH_2HgCH_2SiMe_3$, 2, 52

$GeHgSiC_{12}H_{30}$ $Et_3GeHgSiEt_3$, 2, 137

$GeHgSi_2C_{16}H_{40}$ $Et_3GeHgSi_2Et_5$, 2, 137

$GeMnSbC_{21}H_{32}I_2O_4$ $I_2(C_8H_{17})GeMn(CO)_4[Sb(CH_2CH{=}CH_2)_3]$, 3, 658

$Ge_2AuLiC_{36}H_{30}$ $Li[(Ph_3Ge)_2Au]$, 2, 139

$Ge_3LiSiC_{54}H_{45}$ $(Ph_3Ge)_3SiLi$, 2, 136

$HfC_7H_9I_3$ $(EtC_5H_4)HfI_3$, 1, 42

$HfC_9H_7F_3$ Indenyl-HfF_3, 1, 42

$HfC_{10}H_{10}Cl_2$ $(C_5H_5)_2HfCl_2$, 1, 42

$HfC_{14}H_{18}F_2$ $(EtC_5H_4)_2HfF_2$, 1, 42

$HfC_{15}H_{12}F_2O$ Indenyl(phenoxy)-HfF_2, 1, 42

$Hf_2C_{14}H_{14}I_4O$ $[(CH_2{=}CHC_5H_4)HfI_2]_2O$, 1, 42

$HfPbC_{27}H_{20}O_4$ $Ph_3PbC_5H_5Hf(CO)_4$, 2, 594

$HfPbC_{43}H_{34}O_4$ $(1\text{-}C_{10}H_7)_3PbHf(CO)_4\text{-}(C_5H_3Et_2)$, 1, 42, 2, 594

$HfSnC_{34}H_{28}O_4$ Indenyl-$Hf(CO)_4Sn(C_6H_4\text{-}Me\text{-}o)_3$, 1, 42, 2, 500

$Hf_2SnC_{54}H_{84}O_8$ $(C_{16}H_{33})_2Sn[Me_2C_5H_3Hf(CO)_4]_2$, 2, 500

$Hf_3GeC_{41}H_{44}O_{12}$ $[(Me_2C_5H_3)Hf(CO)_4]_3\text{-}GeC_8H_{17}$, 1, 42, 2, 139

$HgAsC_6H_3BrClO$ 4,2-$Br(ClHg)C_6H_3AsO$, 3, 235

$HgAsC_6H_3Cl_2O$ 4,2-$Cl(ClHg)C_6H_3AsO$, 3, 235

$HgAsC_6H_4ClO$ 4-$ClHgC_6H_4AsO$, 3, 235

$HgAsC_7H_6ClO_4$ 2,4-$HO_2C(ClHg)C_6H_3As(OH)_2$, 3, 222

$HgAsC_8H_5Cl_2O_3$ 2,4,6-$Cl_2(AcOHg)C_6H_2AsO$, 3, 238

$HgAsC_8H_6BrO_3$ 4,2-$Br(AcOHg)C_6H_3AsO$, 3, 238

$HgAsC_8H_6ClO_3$ $Cl(AcOHg)C_6H_3AsO$ isomers, 3, 238

$HgAsC_8H_6NO_4S$ $O_2N(AcOHg)C_6H_3AsS$ isomers, 3, 265

$HgAsC_8H_7O_2S$ 4-$AcOHgC_6H_4AsS$, 3, 265

$HgAsC_8H_7O_3$ 4-$AcOHgC_6H_4AsO$, 3, 239

$HgAsC_8H_8NO_6$ $O_2N(AcOHg)C_6H_3As(OH)_2$ isomers, 3, 223

$HgAsC_9H_9O_6$ $HO_2C(AcOHg)C_6H_3As(OH)_2$ isomers, 3, 223

$HgAsC_9H_{10}NO_6$ 2,4,5-$Me(AcOHg)(O_2N)\text{-}C_6H_2As(OH)_2$, 3, 223

$HgAsC_{10}H_{11}O_3$ 2,4,6-$Me_2(AcOHg)C_6H_2AsO$, 3, 243

$HgCo_2C_{20}H_{12}O_4$ $\{Co(CO)_2[PhC{=}C(Ph)CH{=}CH]Co(CO)_2Hg\}_n$, 1, 600

$HgCo_2C_{30}H_{40}O_6$ $\{[Me_3CC{=}CHCH{=}C(CMe_3)CO]\text{-}Co(CO)_2\}_2Hg$, 1, 601

$HgCo_2C_{62}H_{40}O_6$ $\{[PhC{=}C(Ph)C(Ph){=}C(Ph)CO]\text{-}Co(CO)_2\}_2Hg$, 1, 601

$HgCrC_{18}H_{16}I_3$ $[(C_6H_6)(Ph_2)Cr]I\cdot HgI_2$, 1, 87

$HgCr_3C_{72}H_{60}Cl_5$ $\{[(Ph_2)_2Cr]Cl\}_3\cdot HgCl_2$, 1, 83

$HgFeC_7H_5ClO_2$ $C_5H_5Fe(CO)_2HgCl$, 1, 207

$(HgFeC_{10}H_8)_n$ $\{Fcd\text{-}Hg\}_n$, 1, 531

$HgFeC_{10}H_8Cl_2$ Cl-Fcd-HgCl, 1, 529

$HgFeC_{10}H_9Cl$ FcHgCl, 1, 529

$HgFeC_{12}H_{12}O_2$ FcHgOAc, 1, 529

$HgFe_2C_{14}H_{10}O_4$ $Hg[C_5H_5Fe(CO)_2]_2$, 1, 207

$HgFe_2C_{20}H_{16}Br_2$ $(BrFcd)_2Hg$, 1, 531

$HgFe_2C_{20}H_{16}Cl_2$ $(ClFcd)_2Hg$, 1, 531

$HgFe_2C_{20}H_{18}$ Fc_2Hg, 1, 530

$HgGeC_4H_{11}Br$ Me_3GeCH_2HgBr, 2, 52

$HgGeC_4H_{11}Cl$ Me_3GeCH_2HgCl, 2, 52

$HgGe_2C_{12}H_{30}$ $(Et_3Ge)_2Hg$, 2, 134

$HgMnC_8H_4ClO_3$ $(ClHgC_5H_4)Mn(CO)_3$, 1, 183

$HgMnC_9H_6ClO_3$ $[Me(ClHg)C_5H_3]Mn(CO)_3$, 1, 183

$HgMn_2C_{16}H_8O_6$ $Hg[C_5H_4Mn(CO)_3]_2$, 1, 183

$HgReC_8H_4ClO_3$ $(ClHgC_5H_4)Re(CO)_3$, 1, 192

$HgSbC_8H_6BrO_3$ 4,2-$Br(AcOHg)C_6H_3SbO$, 3, 702

$HgSbC_8H_7O_3$ 4-$(AcOHg)C_6H_4SbO$, 3, 702

$HgSbC_9H_9O_3$ 2,4-$Me(AcOHg)C_6H_3SbO$, 3, 703

$HgSnC_{21}H_{31}IN_2O_2$ $Pr_3SnO_2CCH_2I\cdot(3\text{-}C_5H_4N)_2Hg$, 2, 425

$HgSnC_{24}H_{24}O_4$ $Ph_3SnO_2CCH_2CH_2CO_2HgEt$, 2, 425

$HgSnC_{24}H_{37}IN_2O_2$ $Bu_3SnO_2CCH_2I\cdot(3\text{-}C_5H_4N)_2Hg$, 2, 425

$HgSnC_{25}H_{38}INO_2$ $Bu_3SnO_2CCH_2I\cdot 3\text{-}C_5H_4NHgPh$, 2, 425

$HgSnC_{25}H_{39}IN_2O_2$ $Bu_3SnO_2CCH_2CH_2I\cdot(3\text{-}C_5H_4N)_2Hg$, 2, 425

$HgSnC_{27}H_{43}BrN_2O_2$ $Pr_3SnO_2CCH_2Br\cdot(p\text{-}Me_2NC_6H_4)_2Hg$, 2, 425

$HgSnC_{27}H_{43}IN_2O_2$ $Pr_3SnO_2CCH_2I\cdot(p\text{-}Me_2NC_6H_4)_2Hg$, 2, 425

$HgSnC_{30}H_{47}Cl_2IN_2O_2$ $Bu_3SnO_2CCH_2I\cdot[2,4\text{-}Cl(Me_2N)C_6H_3]_2Hg$, 2, 425

$HgSnC_{30}H_{49}BrN_2O_2$ $Bu_3SnO_2CCH_2Br\cdot(p-Me_2NC_6H_4)_2Hg$, 2, 425

$HgSnC_{30}H_{49}IN_2O_2$ $Bu_3SnO_2CCH_2I\cdot(p-Me_2NC_6H_4)_2Hg$, 2, 425

$HgSnC_{32}H_{53}IN_2O_2$ $Bu_3SnO_2CCH_2I\cdot[2,4-Me(Me_2N)C_6H_3]_2Hg$, 2, 425

$HgSnC_{35}H_{46}O_6S_2$ $EtHgO_3SC_{10}H_6CH_2C_{10}H_6SO_3SnBu_3$, 2, 352

$HgSn_2C_{26}H_{42}I_2N_2O_4$ $(Me_3SnO_2CCH_2I\cdot p-Me_2NC_6H_4)_2Hg$, 2, 425

$HgSn_2C_{30}H_{59}O_4P$ $(Bu_3SnO)_2P(O)(OHgPh)$, 2, 351

$HgW_2C_{16}H_{10}O_6$ $Hg[C_5H_5W(CO)_3]_2$, 1, 120

$Hg_2AsC_5H_{13}O_3$ $MeAsO(OHgEt)_2$, 3, 547
$PrAsO(OHgMe)_2$, 3, 547

$Hg_2AsC_6H_{15}O_3$ $EtAsO(OHgEt)_2$, 3, 547
$BuAsO(OHgMe)_2$, 3, 548

$Hg_2AsC_{10}H_{14}ClO_3$ $4-ClC_6H_4AsO(OHgEt)_2$, 3, 548

$Hg_2AsC_{13}H_{13}O_3$ $MeAsO(OHgPh)_2$, 3, 547

$Hg_2AsC_{15}H_{17}O_3$ $PrAsO(OHgPh)_2$, 3, 547

$Hg_2AsC_{16}H_{19}O_3$ $EtAsO(OHgC_6H_4Me-p)_2$, 3, 547

$Hg_2AsC_{18}H_{12}ClN_2O_7$ $o-ClC_6H_4AsO(OHgC_6H_4NO_2-p)_2$, 3, 409

$Hg_2AsC_{19}H_{14}ClN_2O_7$ $o-ClC_6H_4CH_2AsO(OHgC_6H_4NO_2-p)_2$, 3, 549

$Hg_2AsC_{19}H_{17}O_3$ $2-MeC_6H_4AsO(OHgPh)_2$, 3, 549

$Hg_2AsC_{20}H_{19}O_3$ $PhAsO(OHgC_6H_4Me-4)_2$, 3, 400, 549

$Hg_2AsC_{30}H_{21}Cl_4$ 5,5-Ph[2-(2-$ClHgC_6H_4$)-C_6H_4]-dibenzarsolium$[HgCl_3]^-$, 3, 568

$Hg_2AsC_{32}H_{26}Cl_4N$ 5,5-($p-Me_2NC_6H_4$)[2-($ClHgC_6H_4$)C_6H_4]-dibenzarsolium $[HgCl_3]^-$, 3, 569

$Hg_2Co_4C_{24}H_{20}O_{12}$ $\{[Me_3CC=C(CMe_3)CH=CH][Co(CO)_2HgCo(CO)_4]_2\}$, 1, 599

$Hg_2Co_4C_{28}H_{12}O_{12}$ $\{[PhC=C(Ph)CH=CH][Co(CO)_2HgCo(CO)_4]_2\}$, 1, 599

$Hg_2FeC_{10}H_8Cl_2$ $Fcd(HgCl)_2$, 1, 530

$Hg_2FeC_{14}H_{14}O_4$ $Fcd(HgOAc)_2$, 1, 530

$Hg_2FeC_{16}H_{11}Cl_2NO_2$ $(ClHg)_2(p-O_2NC_6H_4)$-ferrocene, 1, 530

$Hg_2MnC_8H_3Cl_2O_3$ $[(ClHg)_2C_5H_3]Mn(CO)_3$, 1, 183

$Hg_2MnC_9H_5Cl_2O_3$ $[Me(ClHg)_2C_5H_2]Mn(CO)_3$, 1, 184

$Hg_2ReC_8H_3Cl_2O_3$ $[(ClHg)_2C_5H_3]Re(CO)_3$, 1, 192

$Hg_3Co_6C_{48}H_{24}O_{16}$ $Hg\{Co(CO)_2[PhC=C(Ph)CH=CH]Co(CO)_2HgCo(CO)_4\}_2$, 1, 600

$Hg_4AsC_{14}H_{16}NO_{11}$ Tetrakis(acetoxymercuri)arsanilic acid, 3, 504

$Hg_4As_2C_{32}H_{28}O_9$ $[(4-AcOHgC_6H_4)_2As]_2O$, 3, 218

$HgCoFeC_{11}H_5O_6$ $C_5H_5Fe(CO)_2HgCo(CO)_4$, 1, 529

$HgCo_2Si_4C_{26}H_{40}O_6$ $\{[Me_3SiC=CHCH=C(SiMe_3)CO]Co(CO)_2\}_2Hg$, 1, 601

$HgGeSiC_8H_{22}$ $Me_3GeCH_2HgCH_2SiMe_3$, 2, 52

$HgGeSiC_{12}H_{30}$ $Et_3GeHgSiEt_3$, 2, 137

$HgGeSi_2C_{16}H_{40}$ $Et_3GeHgSi_2Et_5$, 2, 137

$HgNaSnC_{38}H_{65}O_6$ $Bu_3SnO_2CC_{23}H_{33}O_4NaHgEt$, 2, 425

$HoC_{10}H_{10}Cl$ $(C_5H_5)_2HoCl$, 1, 2

$HoC_{17}H_{29}Cl_2O_3$ $C_5H_5HoCl_2\cdot3THF$, 1, 1

$InAsC_6H_{18}$ $Me_3AsInMe_3$, 3, 4

$InSbC_{21}H_{24}$ $Ph_3Sb\cdot InMe_3$, 3, 662

$In_2As_2C_{28}H_{32}$ $(Ph_2AsInMe_2)_2$, 3, 298

$IrC_7H_5O_2$ $C_5H_5Ir(CO)_2$, 1, 629

$IrC_{10}H_{10}Br_3$ $[(C_5H_5)_2Ir]Br_3$, 1, 630

$IrC_{10}H_{11}$ $C_5H_5Ir(C_5H_6)$, 1, 629

$IrC_{10}H_{13}Cl_2O_2$ Duroquinone-$IrHCl_2$, 1, 630

$IrC_{10}H_{13}I_2O_2$ Duroquinone-$IrHI_2$, 1, 630

$(IrC_{10}H_{13}O_3)_n$ (Duroquinone-IrHO)$_n$, 1, 630

$IrC_{13}H_{17}$ 1,5-Cyclooctadienecyclopentadienyl-Ir, 1, 630

$IrC_{13}H_{19}O_2$ Acetylacetonato-1,5-cyclooctadiene-Ir, 1, 631*

$IrC_{14}H_{24}I_2NO_2$ Me_4N^+diiododuroquinoneiridate, 1, 630

$IrC_{15}H_{17}O_2$ Cyclopentadienylduroquinone-Ir, 1, 629

$IrC_{20}H_{22}ClN_2O_2$ [Duroquinonebis(pyridine)-Ir]Cl, 1, 630

$IrC_{22}H_{14}N_7O_{12}$ $[(C_5H_5)_2Ir]\{N[C_6H_2(NO_2)_3]_2\}$, 1, 630

$IrC_{25}H_{21}ClN_3O$ $[(p-MeC_6H_4NC)_3Ir(CO)]Cl$,

1, 631
$IrC_{32}H_{28}ClN_4$ [(p-$MeC_6H_4NC)_4$Ir]Cl, 1, 631
$IrC_{38}H_{33}ClIOP_2$ [MeIr(CO)$(PPh_3)_2$]ClI, 1, 629
$IrC_{40}H_{35}ClIOP_2$ [CH_2=$CHCH_2$Ir(CO)-$(PPh_3)_2$]ClI, 1, 629
$IrC_{40}H_{35}ClIO_3P_2$ [MeO_2CCH_2Ir(CO)-$(PPh_3)_2$]ClI, 1, 629
$IrC_{40}H_{35}Cl_2OP_2$ [CH_2=$CHCH_2$Ir(CO)-$(PPh_3)_2$]Cl_2, 1, 629
$Ir_2C_{16}H_{26}Cl_4$ 1,5-Cyclooctadiene-Ir-HCl_2 dimer, 1, 630
$Ir_2C_{18}H_{30}O_2$ Di-μ-methoxybis(1,5-cyclooctadiene)-Ir_2, 1, 631
$IrAsC_{20}H_{15}I_3O_2$ [Ir($AsPh_3$)$(CO)_2I_3$], 3, 20
$IrAsC_{25}H_{20}I_4O$ [Ph_4As][Ir(CO)I_4], 3, 333
$IrAsC_{25}H_{20}I_5O$ [Ph_4As][Ir(CO)I_5], 3, 333
$IrAsC_{25}H_{23}I_4NO$ [Ph_4As][Ir(CO)(NH_3)I_4], 3, 333
$IrAsC_{26}H_{20}I_3O_2$ [Ph_4As][Ir$(CO)_2I_3$], 3, 333
$IrAsC_{26}H_{20}I_4O_2$ [Ph_4As][Ir$(CO)_2I_4$], 3, 333
$IrAsC_{27}H_{27}I_4NO$ [Ph_4As][Ir(CO)($EtNH_2$)-I_4], 3, 333
$IrAsC_{28}H_{29}I_2NO$ [Ph_4As][Ir(CO)(Me_2CH-NH_2)I_2], 3, 333
$IrAs_2C_{13}H_{30}Cl_3O$ [Ir$(AsEt_3)_2$(CO)Cl_3], 3, 8
$IrAs_2C_{21}H_{30}Cl_3O$ [Ir$(Et_2AsPh)_2$(CO)Cl_3], 3, 47
$IrAs_2C_{37}H_{30}I_3O$ [Ir$(AsPh_3)_2$(CO)I_3], 3, 20
$IrAs_2C_{37}H_{31}Cl_2O$ [IrH$(AsPh_3)_2$(CO)Cl_2], 3, 20
$IrAs_2C_{44}H_{42}BrO_2$ {Ir[As$(C_6H_4Me$-4$)_3]_2$-$(CO)_2$Br}, 3, 28
$IrAs_2C_{50}H_{40}Br_2O_2$ $[Ph_4As]_2$[Ir$(CO)_2$-Br_2], 3, 335
$IrAs_2C_{52}H_{40}Br_3O_4$ $[Ph_4As]_2$[Ir$(CO)_4$-Br_3], 3, 335
$IrCrC_{14}H_{16}N_6S_4$ [$(C_5H_5)_2$Ir][Cr$(NH_3)_2$-$(SCN)_4$], 1, 630
$IrPbC_{38}H_{71}O$ $(C_{16}H_{33})_2PbC_5H_5$IrCO, 2, 595
$IrSb_2C_{19}H_{42}ClO$ [Ir$(SbPr_3)_2$(CO)Cl], 3, 658
$IrSb_2C_{37}H_{24}Cl_7O$ Ir[Sb$(C_6H_4Cl$-p$)_3]_2$-(CO)Cl], 3, 660
$IrSnC_{10}H_{14}ClO$ EtSnClEtC_5H_4IrCO, 2, 503
$IrSnC_{13}H_{17}O$ (MeC_5H_4)Ir(CO)Sn(CH_2-CH=$CH_2)_2$, 1, 631, 2, 503
$IrSnC_{50}H_{95}O$ ($C_8H_{17}C_5H_4$)Ir(CO)Sn-$(C_{18}H_{37})_2$, 1, 631, 2, 503
$KGeC_{18}H_{15}$ Ph_3GeK, 2, 135
$KWC_9H_5N_2O_2$ K[W$(CN)_2(C_5H_5)(CO)_2$], 1, 123
$K_2MnC_8H_4$ K_2[Mn(C≡CH)$_4$], 1, 133
$K_2MnC_{12}H_{12}$ K_2[Mn(C≡CMe)$_4$], 1, 133
$K_2MnC_{32}H_{20}$ K_2[Mn(C≡CPh)$_4$], 1, 133
$K_3CrC_{12}H_6$ K_3[Cr(C≡CH)$_6$], 1, 57
$K_3WC_{12}H_9O_3$ K_3[W$(CO)_3$(C≡CMe)$_3$], 1, 119
$LaC_{15}H_{15}$ $(C_5H_5)_3$La, 1, 4
$LaC_{18}H_{21}$ $(MeC_5H_4)_3$La, 1, 4
$LiCrC_{14}H_{14}O_3$ Li[Cr$(C_6H_4OMe$-o$)_2$O], 1, 58
$LiCrC_{23}H_{20}$ Li[Cr(C_5H_5)Ph_3], 1, 58
$LiCrC_{24}H_{30}N_3$ Li[Cr$(C_6H_4NMe_2$-p$)_3$], 1, 58
$LiCrC_{27}H_{22}$ Li[(indenyl)triphenyl-Cr], 1, 58
$LiCrC_{28}H_{28}O_4$ Li[Cr$(C_6H_4OMe$-o$)_4$], 1, 58
$LiGeC_4H_{12}N$ Me_3GeNMeLi, 2, 128
$LiGeC_7H_{17}$ Et(i-C_5H_{11})GeHLi, 2, 137
$LiGeC_{10}H_{15}$ p-$Me_3GeCH_2C_6H_4$Li, 2, 52
$LiGeC_{17}H_{15}$ MePh(1-$C_{10}H_7$)GeLi, 2, 131
$LiGeC_{18}H_{15}$ Ph_3GeLi, 2, 134
$LiGeC_{20}H_{19}$ Ph_2($PhCH_2CH_2$)GeLi, 2, 137
$LiGeC_{24}H_{19}$ p-$Ph_3GeC_6H_4$Li, 2, 52
$LiGeC_{30}H_{47}$ Ph_2(n-$C_{18}H_{37}$)GeLi, 2, 137*
$LiMnC_3H_9$ Li[$MnMe_3$], 1, 133
$LiMnC_6H_{15}$ Li[$MnEt_3$], 1, 133
$LiMnC_{18}H_{15}$ Li[$MnPh_3$], 1, 133
$LiMnC_{21}H_{21}$ Li[Mn$(C_6H_4Me$-p$)_3$], 1, 133
$LiMnC_{21}H_{21}O_3$ Li[Mn$(C_6H_4OMe$-p$)_3$], 1, 133
$LiPbC_{18}H_{15}$ Ph_3PbLi, 2, 591
$LiSbC_{36}H_{30}$ Li[$SbPh_6$], 3, 711
$LiSnC_3H_9$ Me_3SnLi, 2, 496
$LiSnC_6H_{15}$ Et_3SnLi, 2, 498

$LiSnC_7H_{19}O$ $Me_3SnLi\cdot OEt_2$, 2, 496
$LiSnC_9H_{13}$ p-$Me_3SnC_6H_4Li$, 2, 254
$LiSnC_{10}H_{15}$ p-$Me_3SnCH_2C_6H_4Li$, 2, 254
$LiSnC_{12}H_{27}$ Bu_3SnLi, 2, 496
$LiSnC_{18}H_{15}$ Ph_3SnLi, 2, 497
$LiSnC_{21}H_{21}$ $(MeC_6H_4)_3SnLi$ isomers, 2, 498
$LiSnC_{22}H_{25}O$ $Ph_3SnLi\cdot OEt_2$, 2, 497
$LiSnC_{24}H_{37}$ p-(Cyclohexyl$_3$Sn)C_6H_4Li, 2, 254
$Li_2CrC_{30}H_{25}$ $Li_2[Cr(Ph_5]$, 1, 58
$Li_2Mn_2C_{40}H_{35}N_6O_5$ $Li_2[Mn_2(C_5H_5)(CO)_5$-$Py_6]$, 1, 173
$Li_2SnC_{12}H_{10}$ Ph_2SnLi_2, 2, 498
$Li_2WC_{12}H_{11}$ $Li_2[WHPh_2]$, 1, 119
$Li_2WC_{36}H_{30}$ $Li_2[WPh_6]$ complex with Et_2O, 1, 119
$Li_3CrC_6H_{18}$ $Li_3[CrMe_6]$ complex with dioxan, 1, 58
$Li_3CrC_{12}H_8BrO_2$ $Li[Cr(C_6H_4OLi$-o$)_2Br$, 1, 58
$Li_3CrC_{24}H_{38}BrO_5$ $Li[Cr(C_6H_4OLi$-o$)_2Br]\cdot 3Et_2O$, 1, 58
$Li_3CrC_{30}H_{26}$ $Li_3[CrHPh_5]$, 1, 58
$Li_3CrC_{36}H_{24}Br_6$ $Li_3[Cr(C_6H_4Br$-p$)_6]$, 1, 59
$Li_3CrC_{36}H_{24}Cl_6$ $Li_3[Cr(C_6H_4Cl$-m$)_6]$, 1, 59
$Li_3CrC_{36}H_{27}Cl_3$ $Li_3[Cr(C_6H_4Cl$-m$)_3Ph_3]$, 1, 58
$Li_3CrC_{36}H_{30}$ $Li_3[CrPh_6]$, 1, 58
$Li_3CrC_{42}H_{42}$ $Li_3[Cr(C_6H_4Me$-p$)_6]$, 1, 59
$Li_3CrC_{42}H_{42}O_6$ $Li_3[Cr(C_6H_4OMe$-p$)_6$, 1, 59
$Li_3CrC_{48}H_{60}N_6$ $Li_3[Cr(C_6H_4NMe_2$-p$)_6]$, 1, 59
$Li_3CrC_{72}H_{54}$ $Li_3[Cr(C_6H_4Ph$-p$)_6]$, 1, 59
$Li_3WC_{36}H_{30}$ $Li_3[WPh_6]$ complex with Et_2O, 1, 119
$Li_4CrC_{36}H_{24}BrCl_6$ $Li_4[Cr(C_6H_4Cl$-m$)_6$-$Br]$, 1, 59
$Li_4CrC_{36}H_{24}Br_7$ $Li_4[Cr(C_6H_4Br$-p$)_6Br]$, 1, 59
$LiAuGe_2C_{36}H_{30}$ $Li[(Ph_3Ge)_2Au]$, 2, 139
$LiBGeC_{36}H_{30}$ $Li[Ph_3GeBPh_3]$, 2, 139
$LiGe_3SiC_{54}H_{45}$ $(Ph_3Ge)_3SiLi$, 2, 136
$LiSnTeC_{18}H_{15}$ $Ph_3SnTeLi$, 2, 473
$LuC_{10}H_{10}Cl$ $(C_5H_5)_2LuCl$, 1, 2
$LuC_{17}H_{29}Cl_2O_3$ $C_5H_5LuCl_2\cdot 3THF$, 1, 1
$MgAsC_2H_6Br$ $Me_2AsMgBr$, 3, 297
$MgGeC_4H_{11}Cl$ Me_3GeCH_2MgCl, 2, 51
$MgGeC_8H_{15}Br$ $Et_3GeC\equiv CMgBr$, 2, 51*
$MgGeC_9H_{13}Br$ p-$Me_3GeC_6H_4MgBr$, 2, 51
$MgGeC_9H_{17}Br$ $Et_3GeCH_2C\equiv CMgBr$, 2, 51*
$MgGeC_9H_{21}Cl$ $Et_3Ge(CH_2)_3MgCl$, 2, 51
$MgGeC_{10}H_{23}Br$ $Et_3Ge(CH_2)_4MgBr$, 2, 51
$MgGeC_{12}H_{19}Br$ p-$Et_3GeC_6H_4MgBr$, 2, 51
$MgGeC_{14}H_{27}Br$ $Bu_3GeC\equiv CMgBr$, 2, 52*
$MgGeC_{15}H_{33}Cl$ $Bu_3Ge(CH_2)_3MgCl$, 2, 52
$MgGeC_{18}H_{15}Cl$ $Ph_3GeMgCl$, 2, 137
$MgGeC_{19}H_{17}Cl$ Ph_3GeCH_2MgCl, 2, 52
$MgGeC_{21}H_{21}Cl$ $(MeC_6H_4)_3GeMgCl$ isomers, 2, 137
$MgPbC_{12}H_{19}Br$ p-$Et_3PbC_6H_4MgBr$, 2, 550
$MgPbC_{27}H_{33}Br$ Mesityl$_3$PbMgBr, 2, 593
$MgSnC_4H_{11}Cl$ Me_3SnCH_2MgCl, 2, 252
$MgSnC_9H_{13}Br$ p-$Me_3SnC_6H_4MgBr$, 2, 252
$MgSnC_{10}H_{15}Br$ p-$Me_3SnCH_2C_6H_4MgBr$, 2, 252
$MgSnC_{12}H_{19}Br$ p-$Et_3SnC_6H_4MgBr$, 2, 252
$MgSn_2C_{36}H_{30}$ $(Ph_3Sn)_2Mg$, 2, 495
$Mg_2AsC_6H_5Br_2$ $PhAs(MgBr)_2$, 3, 297
$MnCH_3I$ MeMnI, 1, 131
MnC_2H_6 Me_2Mn, 1, 131
MnC_4H_2 $Mn(C\equiv CH)_2$ complex with NH_3, 1, 133*
$(MnC_5HO_3)_n$ $[HC\equiv CMn(CO)_3]_n$, 1, 182
$MnC_6F_3O_5$ $CF_3Mn(CO)_5$, 1, 134
$MnC_6H_3O_5$ $MeMn(CO)_5$, 1, 134
MnC_6H_5I PhMnI, 1, 131
$MnC_6H_5O_4S$ Methylthiomethyl-$Mn(CO)_4$, 1, 177
$MnC_6H_6NO_5$ $MeCOMn(CO)_4(NH_3)$, 1, 141
$MnC_6H_7Cl_2$ $(MeC_5H_4)MnCl_2$, 1, 173
$MnC_7F_3O_6$ $CF_3COMn(CO)_5$, 1, 138
$MnC_7F_5O_5$ $C_2F_5Mn(CO)_5$, 1, 134
$MnC_7HClF_3O_5$ $HCF_2CFClMn(CO)_5$, 1, 135
$HCClFCF_2Mn(CO)_5$, 1, 135
$MnC_7HCl_2F_2O_5$ $HCCl_2CF_2Mn(CO)_5$, 1, 136
$MnC_7HF_4O_5$ $HCF_2CF_2Mn(CO)_5$, 1, 135
$MnC_7H_3O_6$ $AcMn(CO)_5$, 1, 138
$MnC_7H_4ClO_4$ 3-Chloroallyl-$Mn(CO)_4$, 1, 177
$MnC_7H_4NO_3$ Pyrrolyl-$Mn(CO)_3$, 1, 181

$MnC_7H_5F_6NO_3P$ $[C_5H_5Mn(CO)_2(NO)]PF_6$, 1, 171

$MnC_7H_5O_4$ Allyl-Mn(CO)$_4$, 1, 176

$MnC_7H_5O_4S$ $(C_5H_5)Mn(CO)_2(SO_2)$, 1, 171

$MnC_7H_5O_5$ $EtMn(CO)_5$, 1, 135

$MnC_7H_7O_2S$ $(C_5H_5)Mn(CO)_2(H_2S)$, 1, 171

$MnC_7H_8NO_2$ $(C_5H_5)Mn(CO)_2(NH_3)$, 1, 168

$MnC_8F_5O_5$ $CF_3CF{=}CFMn(CO)_5$, 1, 136

$CF_2{=}CFCF_2Mn(CO)_5$, 1, 136

$MnC_8F_5O_6$ $C_2F_5COMn(CO)_5$, 1, 139

$MnC_8F_7O_5$ $C_3F_7Mn(CO)_5$, 1, 135

$(CF_3)_2CFMn(CO)_5$, 1, 137

$MnC_8HF_4O_6$ $HCF_2CF_2COMn(CO)_5$, 1, 139, 140*

$MnC_8H_3ClF_3O_5$ $MeCF_2CClFMn(CO)_5$, 1, 137

$MnC_8H_3F_4O_5$ $MeCF_2CF_2Mn(CO)_5$, 1, 137

$MnC_8H_3O_6$ $CH_2{=}CHCOMn(CO)_5$, 1, 139

$MnC_8H_4ClO_5$ $ClCH{=}CHCH_2Mn(CO)_5$, 1, 137

$MnC_8H_4ClO_5S$ $(ClSO_2C_5H_4)Mn(CO)_3$, 1, 149

$MnC_8H_4IO_3$ $(IC_5H_4)Mn(CO)_3$, 1, 149

$MnC_8H_5O_3$ $C_5H_5Mn(CO)_3$, 1, 141

$MnC_8H_5O_5$ $CH_2{=}CHCH_2Mn(CO)_5$, 1, 135

$MnC_8H_5O_5S$ $(HOSOC_5H_4)Mn(CO)_3$, 1, 149

$MnC_8H_5O_6$ $EtCOMn(CO)_5$, 1, 140

$MnC_8H_5O_7$ $EtO_2CMn(CO)_5$, 1, 132

$MnC_8H_6NO_3$ $(H_2NC_5H_4)Mn(CO)_3$, 1, 149

(2-Methylpyrrolyl)-Mn(CO)$_3$, 1, 181

$MnC_8H_7O_4$ Methallyl-Mn(CO)$_4$, 1, 177

2-Butenyl-Mn(CO)$_4$, 1, 177

$MnC_8H_7O_4S$ $(MeC_5H_4)Mn(CO)_2(SO_2)$, 1, 171

$MnC_8H_7O_5$ i-PrMn(CO)$_5$, 1, 137

$MnC_8H_7O_5S$ 2-(Methylthio)ethyl-Mn-(CO)$_5$, 1, 136

MnC_8N_6 $Mn[C(CN)_3]_2$, 1, 131

$MnC_9ClF_8O_5$ $Cl(CF_2)_4Mn(CO)_5$, 1, 137

$MnC_9F_7O_6$ $CF_3CF_2CF_2COMn(CO)_5$, 1, 139

$(CF_3)_2CFCOMn(CO)_5$, 1, 140

$MnC_9HF_6O_5$ $CF_3CH{=}C(CF_3)Mn(CO)_5$, 1, 137

$MnC_9HF_8O_5$ $H(CF_2)_4Mn(CO)_5$, 1, 137

$MnC_9H_3O_6$ $\overline{OCH{=}CHCH{=}C}Mn(CO)_5$, 1, 137

$MnC_9H_4ClO_4$ $(ClCOC_5H_4)Mn(CO)_3$, 1, 144

$MnC_9H_4N_3O_4$ $(N_3COC_5H_4)Mn(CO)_3$, 1, 149

$MnC_9H_5O_4$ $(HCOC_5H_4)Mn(CO)_3$, 1, 141

$MnC_9H_5O_5$ $(HO_2CC_5H_4)Mn(CO)_3$, 1, 149

$MnC_9H_5O_8$ $HO_2C(CH_2)_2COMn(CO)_5$, 1, 139

$MnC_9H_6BrO_3$ $[(C_6H_6)Mn(CO)_3]Br$, 1, 179

$MnC_9H_6ClO_3$ $(ClCH_2C_5H_4)Mn(CO)_3$, 1, 145

$MnC_9H_6ClO_5$ $MeCCl{=}CHCH_2Mn(CO)_5$, 1, 137

$MnC_9H_6ClO_7$ $[(C_6H_6)Mn(CO)_3]ClO_4$, 1, 179

$MnC_9H_6DO_3$ (Cyclohexadienyl-d$_1$)-Mn-(CO)$_3$, 1, 177

$MnC_9H_6IO_3$ $[(C_6H_6)Mn(CO)_3]I$, 1, 179

$MnC_9H_6NO_2$ $C_6H_6Mn(CO)_2CN$, 1, 179

$MnC_9H_7N_2O_4$ $(H_2NNHCOC_5H_4)Mn(CO)_3$, 1, 149

$MnC_9H_7O_3$ $(MeC_5H_4)Mn(CO)_3$, 1, 143

Cyclohexadienyl-Mn(CO)$_3$, 1, 177

$MnC_9H_7O_5$ $MeCH{=}CHCH_2Mn(CO)_5$, 1, 136

$CH_2{=}C(Me)CH_2Mn(CO)_5$, 1, 137

$MnC_9H_7O_5S$ $(MeOSOC_5H_4)Mn(CO)_3$, 1, 149

$MnC_9H_7O_6$ $Me_2CHCOMn(CO)_5$, 1, 140

$PrCOMn(CO)_5$, 1, 140

$MnC_9H_8NO_2$ $(C_5H_5)Mn(CO)_2(CNMe)$, 1, 169

$MnC_9H_8NO_3$ (Dimethylpyrrolyl)-Mn(CO)$_3$, 1, 181

$MnC_9H_9O_2$ $(C_5H_5)(C_2H_4)Mn(CO)_2$, 1, 168

$MnC_9H_9O_4$ 3-Methyl-2-butenyl-Mn(CO)$_4$, 1, 177

$MnC_9H_9O_5S$ $(C_5H_5)Mn(CO)_2[(\text{-}CH_2O)_2SO]$, 1, 171

$MnC_9H_{11}O_3S$ $(C_5H_5)Mn(CO)_2(Me_2SO)$, 1, 171

$MnC_9H_{12}NO_2$ $(C_5H_5)Mn(CO)_2(Me_2NH)$, 1, 168

$MnC_9H_{13}N_2O_2$ $[(Me_3C)C_5H_4]Mn(NO)_2$, 1, 172

$MnC_9N_3O_5$ $(NC)_3CMn(CO)_5$, 1, 132

$MnC_{10}ClF_8O_6$ $Cl(CF_2)_4COMn(CO)_5$, 1, 140

$MnC_{10}HF_8O_6$ $H(CF_2)_4COMn(CO)_5$, 1, 140

$MnC_{10}H_3O_7$ $\overline{OCH{=}CHCH{=}C}COMn(CO)_5$, 1, 140

$MnC_{10}H_6ClO_4$ $[1,2\text{-}ClCO(Me)C_5H_3]Mn(CO)_3$, 1, 152

$[1,3\text{-}ClCO(Me)C_5H_3]Mn(CO)_3$, 1, 152

$MnC_{10}H_7Br_2O_3$ $(CH_2BrCHBrC_5H_4)Mn(CO)_3$, 1, 149

$MnC_{10}H_7Cl_2O_3$ $[(ClCH_2)_2C_5H_3]Mn(CO)_3$, 1, 153

$MnC_{10}H_7O_3$ $(CH_2{=}CHC_5H_4)Mn(CO)_3$, 1, 145

$MnC_{10}H_7O_4$ $(AcC_5H_4)Mn(CO)_3$, 1, 145

$[1,2\text{-}HCO(Me)C_5H_3]Mn(CO)_3$, 1, 153

$[1,3\text{-}HCO(Me)C_5H_3]Mn(CO)_3$, 1, 153

$MnC_{10}H_7O_5$ $(MeO_2CC_5H_4)Mn(CO)_3$, 1, 149
[1,2-$HO_2C(Me)C_5H_3$]$Mn(CO)_3$, 1, 153
[1,3-$HO_2C(Me)C_5H_3$]$Mn(CO)_3$, 1, 153
$MnC_{10}H_7O_8$ $MeO_2C(CH_2)_2COMn(CO)_5$, 1, 140
$MnC_{10}H_8ClO_7$ [$(MePh)Mn(CO)_3$]ClO_4, 1, 179
$MnC_{10}H_8IO_3$ [$(MePh)Mn(CO)_3$]I, 1, 179
$MnC_{10}H_8NO_4$ [$MeC(=NOH)C_5H_4$]$Mn(CO)_3$, 1, 145
$MnC_{10}H_8N_3O_3$ [$MeCH(N_3)C_5H_4$]$Mn(CO)_3$, 1, 149
$MnC_{10}H_9N_2O_3$ [$MeC(=NNH_2)C_5H_4$]$Mn(CO)_3$, 1, 146
$MnC_{10}H_9O_3$ $(EtC_5H_4)Mn(CO)_3$, 1, 145
(Methylcyclohexadienyl)-$Mn(CO)_3$, 1, 177, 178
$MnC_{10}H_9O_4$ [$MeCH(OH)C_5H_4$]$Mn(CO)_3$, 1, 146
$(MeOCH_2C_5H_4)Mn(CO)_3$, 1, 146
$MnC_{10}H_9O_5$ $Me_2C{=}CHCH_2Mn(CO)_5$, 1, 137
$MnC_{10}H_9O_6$ $Me_2CHCH_2COMn(CO)_5$, 1, 140
$MnC_{10}H_{10}$ $(C_5H_5)_2Mn$, 1, 174
$MnC_{10}H_{11}O$ $(C_5H_5)(CH_2{=}CHCH{=}CH_2)Mn(CO)$, 1, 168
$MnC_{10}H_{11}O_3$ $(C_5H_5)Mn(CO)_2(Me_2CO)$, 1, 171
$MnC_{10}H_{13}O_3S$ $(MeC_5H_4)Mn(CO)_2(Me_2SO)$, 1, 171
$MnC_{10}H_{14}NO_2$ $(C_5H_5)Mn(CO)_2(PrNH_2)$, 1, 168
$(C_5H_5)Mn(CO)_2(Me_3N)$, 1, 168
$MnC_{10}H_{15}N_2O_2$ $(C_5H_5)Mn(CO)_2$[$H_2N(CH_2)_3$-NH_2], 1, 168
$MnC_{10}H_{17}O_3S_2$ $(C_5H_5)Mn(CO)(Me_2SO)_2$, 1, 171
$MnC_{10}H_{28}N_6$ $(C_5H_5)_2Mn(NH_3)_6$, 1, 174
$MnC_{11}F_5O_5$ $C_6F_5Mn(CO)_5$, 1, 137
$MnC_{11}H_5Cl_2O_3$ Dichlorodihydropentalenyl-$Mn(CO)_3$, 1, 164
$MnC_{11}H_5F_6O_2$ $(C_5H_5)(CF_3C{\equiv}CCF_3)Mn(CO)_2$, 1, 168
$MnC_{11}H_5O_5$ $PhMn(CO)_5$, 1, 136
$MnC_{11}H_7O_3$ Dihydropentalenyl-$Mn(CO)_3$, 1, 163
$MnC_{11}H_7O_4$ Tetrahydro-4-oxopentalenyl-$Mn(CO)_3$, 1, 164
$MnC_{11}H_7O_5$ $(HO_2CCH{=}CHC_5H_4)Mn(CO)_3$, 1, 146
$MnC_{11}H_7O_7$ [1,2-$HO_2C(HO_2CCH_2)C_5H_3$]Mn-$(CO)_3$, 1, 154
$MnC_{11}H_9O_3$ $(CH_2{=}CHCH_2C_5H_4)Mn(CO)_3$, 1, 147
$(MeCH{=}CHC_5H_4)Mn(CO)_3$, 1, 149
Tetrahydropentalenyl-$Mn(CO)_3$, 1, 163
$MnC_{11}H_9O_4$ $(EtCOC_5H_4)Mn(CO)_3$, 1, 147
[$Ac(Me)C_5H_3$]$Mn(CO)_3$, 1, 154
$MnC_{11}H_9O_5$ $(HO_2CCH_2CH_2C_5H_4)Mn(CO)_3$, 1, 149
[1,x-$Me(MeO_2C)C_5H_3$]$Mn(CO)_3$, 1, 158
$MnC_{11}H_9O_8$ $EtO_2C(CH_2)_2COMn(CO)_5$, 1, 140
$MnC_{11}H_{10}NO_2$ $(EtPh)Mn(CO)_2CN$, 1, 179
$(Me_2C_6H_4)Mn(CO)_2CN$, 1, 179
$MnC_{11}H_{11}$ $C_5H_5MnC_6H_6$, 1, 172
$MnC_{11}H_{11}N_2O_3$ [$Me(H_2NN{=}CMe)C_5H_3$]$Mn(CO)_3$, 1, 154
$MnC_{11}H_{11}O_3$ $(PrC_5H_4)Mn(CO)_3$, 1, 149
$(MeEtC_5H_3)Mn(CO)_3$, 1, 158
$MnC_{11}H_{11}O_4$ $(EtOCH_2C_5H_4)Mn(CO)_3$, 1, 149
[$EtCH(OH)C_5H_4$]$Mn(CO)_3$, 1, 149
$MnC_{11}H_{12}$ $C_5H_5Mn(C_5H_4Me)$, 1, 174
$MnC_{11}H_{12}NO_3$ [$Me(MeCHNH_2)C_5H_3$]$Mn(CO)_3$, 1, 154
(2,3,4,5-Tetramethylpyrrolyl)-$Mn(CO)_3$, 1, 181
$MnC_{11}H_{13}O_2S$ $(C_5H_5)Mn(CO)_2$[$(CH_2)_4S$], 1, 171
$MnC_{11}H_{13}O_3$ $(C_5H_5)Mn(CO)_2(THF)$, 1, 171
$MnC_{11}H_{13}O_3S$ $(C_5H_5)Mn(CO)_2$[$(CH_2)_4SO$], 1, 171
$MnC_{11}H_{14}NO_2$ $(C_5H_5)Mn(CO)_2$[$CH_2(CH_2)_3NH$], 1, 168
$MnC_{11}H_{15}N_2O_2$ $(C_5H_5)Mn(CO)_2$[$HN(CH_2CH_2)_2$-NH], 1, 168
$MnC_{11}H_{15}O_2S$ $(C_5H_5)Mn(CO)_2(Et_2S)$, 1, 171
$MnC_{11}H_{16}O_5P$ $(MeC_5H_4)Mn(CO)_2$[$P(OMe)_3$], 1, 169
$MnC_{12}F_5O_6$ Pentafluorobenzoyl-$Mn(CO)_5$, 1, 140
$MnC_{12}H_5O_6$ $BzMn(CO)_5$, 1, 139
$MnC_{12}H_7O_3$ Indenyl-$Mn(CO)_3$, 1, 165
$MnC_{12}H_7O_5$ $PhCH_2Mn(CO)_5$, 1, 136

$p\text{-}MeC_6H_4Mn(CO)_5$, 1, 137
$MnC_{12}H_8NO_5$ $[Me(NCCH_2O_2C)C_5H_3]Mn(CO)_3$, 1, 158
$MnC_{12}H_9N_2O_3$ $[\overline{CH{=}C(Me)NHN{=}C}C_5H_4]Mn(CO)_3$, 1, 150
$MnC_{12}H_9O_3$ (Dihydromethylpentalenyl)-$Mn(CO)_3$, 1, 164
4,5-Dihydroindenyl-$Mn(CO)_3$, 1, 165
$MnC_{12}H_9O_4$ $[CH{\equiv}CC(OH)(Me)C_5H_4]Mn(CO)_3$, 1, 150
$[1,x\text{-}Me(CH_2{=}CHCO)C_5H_3]Mn(CO)_3$, 1, 158
Dihydromethyloxopentalenyl-$Mn(CO)_3$ isomers, 1, 164, 165
$MnC_{12}H_9O_5$ $(AcCH_2COC_5H_4)Mn(CO)_3$, 1, 147
$[HO_2CCH{=}C(Me)C_5H_4]Mn(CO)_3$, 1, 150
$[HO_2CCH{=}CH(Me)C_5H_3]Mn(CO)_3$ isomers, 1, 155
$MnC_{12}H_9O_6$ $(HO_2CCH_2CH_2COC_5H_4)Mn(CO)_3$, 1, 150
$MnC_{12}H_{10}$ Ph_2Mn, 1, 131
$MnC_{12}H_{10}N$ $(C_5H_5)Mn(CNPh)_3$, 1, 168
$MnC_{12}H_{10}NO_2$ $(C_5H_5)Mn(CO)_2(Py)$, 1, 169
$MnC_{12}H_{10}NO_5$ $MeCOMn(CO)_4(PhNH_2)$, 1, 141
$MnC_{12}H_{11}N_2O_4$ $[AcNHN{=}C(Me)C_5H_4]Mn(CO)_3$, 1, 150
$MnC_{12}H_{11}O_3$ $[Me(CH_2{=}CMe)C_5H_3]Mn(CO)_3$, 1, 155
(Methyltetrahydropentalenyl)-$Mn(CO)_3$, 1, 164
(4,5,6,7-Tetrahydroindenyl)-$Mn(CO)_3$, 1, 166
$MnC_{12}H_{11}O_4$ $[CH_2{=}CHC(OH)(Me)C_5H_4]Mn(CO)_3$, 1, 147
$(CH_2{=}CHCH_2OCH_2C_5H_4)Mn(CO)_3$, 1, 150
$[Ac(Et)C_5H_3]Mn(CO)_3$, 1, 155
(Tetrahydro-4-hydroxy-6-methylpentalenyl)-$Mn(CO)_3$, 1, 165
(4,5,6,7-Tetrahydro-4-hydroxyindenyl)-$Mn(CO)_3$, 1, 166
$MnC_{12}H_{11}O_5$ $[HO_2C(CH_2)_3C_5H_4]Mn(CO)_3$, 1, 150
$[HO_2CCH_2CH(Me)C_5H_4]Mn(CO)_3$, 1, 150
$[Me(HO_2CCH_2CH_2)C_5H_3]Mn(CO)_3$ isomers, 1, 155
$MnC_{12}H_{11}O_8$ $PrO_2C(CH_2)_2COMn(CO)_5$, 1, 140
$MnC_{12}H_{12}ClO_7$ $[(1,3,5\text{-}Me_3C_6H_3)Mn(CO)_3]ClO_4$, 1, 179
$MnC_{12}H_{12}IO_3$ $[(1,3,5\text{-}Me_3C_6H_3)Mn(CO)_3]I$, 1, 179
$MnC_{12}H_{12}NO_2$ $(1,3,5\text{-}Me_3C_6H_3)Mn(CO)_2CN$, 1, 179
$MnC_{12}H_{13}$ $(MeC_5H_4)MnC_6H_6$, 1, 172
$MnC_{12}H_{13}O_3$ $(BuC_5H_4)Mn(CO)_3$, 1, 147
$(tert\text{-}BuC_5H_4)Mn(CO)_3$, 1, 147
$(Et_2C_5H_3)Mn(CO)_3$, 1, 156
$[Me(Me_2CH)C_5H_3]Mn(CO)_3$, 1, 158
(1,3,5-Trimethylcyclohexadienyl)-$Mn(CO)_3$, 1, 178
$MnC_{12}H_{13}O_4$ $(PrOCH_2C_5H_4)Mn(CO)_3$, 1, 150
$MnC_{12}H_{14}$ $(MeC_5H_4)_2Mn$, 1, 174
$(MePh)MnC_5H_6$, 1, 180
$MnC_{12}H_{16}NO_2$ $(C_5H_5)Mn(CO)_2(Pip)$, 1, 169
$MnC_{12}H_{16}NO_5$ $MeCOMn(CO)_4(cyclo\text{-}C_6H_{11}\text{-}NH_2)$, 1, 141
$MnC_{12}H_{18}ClN_6O_4$ $[(MeNC)_6Mn]ClO_4$, 1, 182
$MnC_{12}H_{18}IN_6$ $[(MeNC)_6Mn]I$, 1, 182
$MnC_{12}H_{18}I_3N_6$ $[(MeNC)_6Mn]I_3$, 1, 182
$MnC_{12}H_{23}OP_2$ $(C_5H_5)Mn(CO)(PMe_3)_2$, 1, 169
$MnC_{13}H_5F_4O_5$ $PhCF_2CF_2Mn(CO)_5$, 1, 137
$MnC_{13}H_7O_4S$ $(\overline{SCH{=}CHCH{=}C}COC_5H_4)Mn(CO)_3$, 1, 150
$MnC_{13}H_7O_5$ $(\overline{OCH{=}CHCH{=}C}COC_5H_4)Mn(CO)_3$, 1, 150
$MnC_{13}H_7O_6$ $p\text{-}MeC_6H_4COMn(CO)_5$, 1, 140
$PhCH_2COMn(CO)_5$, 1, 140
$MnC_{13}H_8ClO_7$ $[(Naphthalene)Mn(CO)_3]ClO_4$, 1, 179
$MnC_{13}H_8O_3$ Azulene-$Mn(CO)_3$, 1, 166
$MnC_{13}H_9O_3$ Benzocyclohexadienyl-$Mn(CO)_3$, 1, 178
$MnC_{13}H_9O_4$ 3-Phenylallyl-$Mn(CO)_4$, 1, 177
$MnC_{13}H_{11}O_3$ (Tetrahydrodimethylpentalenyl)-$Mn(CO)_3$, 1, 165
(Tetrahydroazulenyl)-$Mn(CO)_3$, 1, 166
$MnC_{13}H_{11}O_4$ $[AcCH{=}C(Me)C_5H_4]Mn(CO)_3$, 1, 147
$[CH{\equiv}CCH_2C(OH)(Me)C_5H_4]Mn(CO)_3$, 1, 148

$[EtCO(CH_2{=}CH)C_5H_3]Mn(CO)_3$, 1, 156
(1,4,5,6,7,8-Hexahydro-4-oxoazulenyl)-$Mn(CO)_3$, 1, 166
$MnC_{13}H_{11}O_6$ $[HO_2C(CH_2)_3COC_5H_4]Mn(CO)_3$, 1, 150
$MnC_{13}H_{12}NO_2$ $(C_5H_5)Mn(CO)_2(PhNH_2)$, 1, 169
$(MeC_5H_4)Mn(CO)_2(Py)$, 1, 169
(Tetrahydronaphthalene)-$Mn(CO)_2CN$, 1, 179
$MnC_{13}H_{13}N_2O_4$ $\{Me[AcNHN{=}C(Me)]C_5H_3\}$-$Mn(CO)_3$, 1, 158
$MnC_{13}H_{13}O_2$ $(C_5H_5)(CH_2CH_2CH{=}CHCH{=}CH)$-$Mn(CO)_2$, 1, 168
$MnC_{13}H_{13}O_3$ $[Me(MeCH{=}CMe)C_5H_3]Mn(CO)_3$, 1, 158
$MnC_{13}H_{13}O_4$ 1,4,5,6,7,8-Hexahydro-4-hydroxyazulenyl)-$Mn(CO)_3$, 1, 166*
$MnC_{13}H_{13}O_5$ $[HO_2C(CH_2)_4C_5H_4]Mn(CO)_3$, 1, 150
$MnC_{13}H_{14}F_6P$ $[C_5H_5Mn(C_7H_6Me)]PF_6$, 1, 173
$MnC_{13}H_{15}O_3$ $[Me(i\text{-}Bu)C_5H_3]Mn(CO)_3$, 1, 158
$(Et_2MeC_5H_2)Mn(CO)_3$, 1, 159
(Pentamethylcyclopentadienyl)-Mn-$(CO)_3$, 1, 160
$MnC_{13}H_{15}O_4$ $(BuOCH_2C_5H_4)Mn(CO)_3$, 1, 150
$[Me(EtMeCOH)C_5H_3]Mn(CO)_3$, 1, 158
$MnC_{13}H_{17}N_4$ $(C_5H_5)Mn(CO)_2$(hexamethylenetetramine), 1, 169
$MnC_{13}H_{18}NO_2$ $(C_5H_5)Mn(CO)_2$-$[MeCH(CH_2)_2CH(Me)NH]$, 1, 169
$(MeC_5H_4)Mn(CO_2)(Pip)$, 1, 169
$MnC_{13}H_{18}NO_5$ $MeCOMn(CO)_4(cyclo\text{-}C_6H_{11}$-$NHMe)$, 1, 141
$MnC_{13}H_{20}NO_2$ $(C_5H_5)Mn(CO)_2(n\text{-}C_6H_{13}NH_2)$, 1, 169
$MnC_{13}H_{25}O_7P_2$ $(MeC_5H_4)Mn(CO)[P(OMe)_3]_2$, 1, 169
$MnC_{14}H_9O_3$ $(PhC_5H_4)Mn(CO)_3$, 1, 148
$MnC_{14}H_9O_4$ (Acetylindenyl)-$Mn(CO)_3$, 1, 166
$MnC_{14}H_{10}NO_2$ $(C_5H_5)Mn(CO)_2(PhNC)$, 1, 169
$MnC_{14}H_{11}O_4$ $[CH_2{=}CHC{\equiv}CC(OH)(Me)C_5H_4]$-$Mn(CO)_3$, 1, 150
$MnC_{14}H_{13}N_2O_5$ $[Ac_2NN{=}C(Me)C_5H_4]Mn(CO)_3$, 1, 150
$MnC_{14}H_{13}O_2$ (Bicyclo[2.2.1]hepta-2,5-diene)cyclopentadienyl-$Mn(CO)_2$, 1, 168
$MnC_{14}H_{13}O_5$ $[EtO_2CCH{=}C(Me)C_5H_4]Mn(CO)_3$, 1, 151
$MnC_{14}H_{14}ClN_2O_4$ $\{Me[ClCH_2CH_2CONHN{=}C$-$(Me)]C_5H_3\}Mn(CO)_3$, 1, 158
$MnC_{14}H_{14}NO_6$ $[Me(EtO_2CCH_2NHCO)C_5H_3]$-$Mn(CO)_3$, 1, 158
$MnC_{14}H_{15}O_3$ $(cyclo\text{-}C_6H_{11}C_5H_4)Mn(CO)_3$, 1, 151
$MnC_{14}H_{15}O_4$ $[Ac(Me_3C)C_5H_3]Mn(CO)_3$, 1, 158
$[Et_2(Ac)C_5H_2]Mn(CO)_3$, 1, 159
$MnC_{14}H_{15}O_6$ $[EtO_2CCH_2C(OH)(Me)C_5H_4]Mn$-$(CO)_3$, 1, 151
$MnC_{14}H_{16}NO_2$ (Cyclohexyl-NC)$(C_5H_5)Mn$-$(CO)_2$, 1, 168
$MnC_{14}H_{17}O_3$ $(Et_3C_5H_2)Mn(CO)_3$, 1, 159
$MnC_{14}H_{18}$ $(EtC_5H_4)_2Mn$, 1, 175
$(EtPh)Mn(C_5H_5Me)$, 1, 180
$MnC_{15}H_9O_4$ $(BzC_5H_4)Mn(CO)_3$, 1, 148
$MnC_{15}H_{10}NO_4$ $(PhNHCOC_5H_4)Mn(CO)_3$, 1, 151
$(BzNHC_5H_4)Mn(CO)_3$, 1, 151
$MnC_{15}H_{10}N_3O_3$ $[PhCH(N_3)C_5H_4]Mn(CO)_3$, 1, 151
$MnC_{15}H_{11}N_2O_3$ $[PhC({=}NNH_2)C_5H_4]Mn(CO)_3$, 1, 151
$MnC_{15}H_{11}O_3$ (Phenylcyclohexadienyl)-$Mn(CO)_3$, 1, 178
$MnC_{15}H_{11}O_4$ $[PhCH(OH)C_5H_4]Mn(CO)_3$, 1, 151
$MnC_{15}H_{12}NO_5S$ $(p\text{-}MeC_6H_4NHSO_2C_5H_4)Mn$-$(CO)_3$, 1, 151
$MnC_{15}H_{12}O_3$ $(PhCH_2C_5H_5)Mn(CO)_3$, 1, 151
$MnC_{15}H_{13}O_4$ $[CH_2{=}CHC{\equiv}CC(OMe)(Me)C_5H_4]$-$Mn(CO)_3$, 1, 151
$MnC_{15}H_{18}NO_2$ $(Et_3C_6H_3)Mn(CO)_2CN$, 1, 179
$MnC_{15}H_{19}O_3$ $(Et_3MeC_5H)Mn(CO)_3$, 1, 160
(Hexamethylcyclohexadienyl)-$Mn(CO)_3$, 1, 178
$MnC_{16}H_9O_3$ Fluorenyl-$Mn(CO)_3$, 1, 166
$MnC_{16}H_{10}ClO_4$ $[Me(p\text{-}ClC_6H_4CO)C_5H_3]Mn$-

$(CO)_3$, 1, 158
$MnC_{16}H_{11}N_4O_7$ {[2,4-$(O_2N)_2C_6H_3NHN$=C-(Me)]C_5H_4}$Mn(CO)_3$, 1, 145
{Me[2,4-$(O_2N)_2C_6H_3NHN$=CH]C_5H_3}Mn-$(CO)_3$ isomers, 1, 153
$MnC_{16}H_{11}O_4$ $(PhCH_2COC_5H_4)Mn(CO)_3$, 1, 151
$(MeC_6H_4COC_5H_4)Mn(CO)_3$ isomers, 1, 151
[Bz(Me)C_5H_3]$Mn(CO)_3$ isomers, 1, 156, 158
$MnC_{16}H_{12}NO_2$ $(C_5H_5)Mn(CO)_2$(quinoline), 1, 169
$MnC_{16}H_{12}NO_4$ {Me[PhC(=NOH)]C_5H_3}Mn-$(CO)_3$ isomers, 1, 156
[Me(PhNHCO)C_5H_3]$Mn(CO)_3$, 1, 156
$MnC_{16}H_{13}O_4$ $(PhCH_2OCH_2C_5H_4)Mn(CO)_3$, 1, 151
$MnC_{16}H_{13}O_7$ [(3,4-Dicarboxycyclo-1-hexenyl)cyclopentadienyl]-$Mn(CO)_3$, 1, 148
$MnC_{16}H_{15}O_3$ (1,3,4,7-Tetramethylindenyl)-$Mn(CO)_3$, 1, 166
$MnC_{16}H_{15}O_4$ [CH_2=CHC≡CC(OEt)(Me)C_5H_4]-$Mn(CO)_3$, 1, 151
$MnC_{16}H_{17}O_3$ Octahydrofluorenyl-$Mn(CO)_3$, 1, 167
$MnC_{16}H_{18}$ $(CH_2=CHCH_2C_5H_4)_2Mn$, 1, 175
$MnC_{16}H_{19}O_4$ [$Et_3(Ac)C_5H$]$Mn(CO)_3$, 1, 160
$MnC_{16}H_{21}O_3$ (2-Octylcyclopentadienyl)-$Mn(CO)_3$, 1, 148
[$(Me_3C)_2C_5H_3$]$Mn(CO)_3$, 1, 158
$(Et_4C_5H)Mn(CO)_3$, 1, 160
$MnC_{16}H_{26}O_2P$ $(C_5H_5)Mn(CO)_2(PPr_3)$, 1, 169
$MnC_{17}H_{11}O_3$ Tetrahydrophenylpentalenyl-$Mn(CO)_3$, 1, 165
$MnC_{17}H_{11}O_4$ $(PhCH=CHCOC_5H_4)Mn(CO)_3$, 1, 151
$MnC_{17}H_{12}NO_5$ $PhCOMn(CO)_4(PhNH_2)$, 1, 141
$MnC_{17}H_{13}N_2O_4$ [PhC(=NNHAc)C_5H_4]Mn-$(CO)_3$, 1, 151
[BzNHN=C(Me)C_5H_4]$Mn(CO)_3$, 1, 151
$MnC_{17}H_{13}O_3$ [Me(CH_2=CPh)C_5H_3]$Mn(CO)_3$, 1, 158
$MnC_{17}H_{13}O_4$ [CH_2=CHC(Ph)(OH)C_5H_4]Mn-$(CO)_3$, 1, 151
$MnC_{17}H_{13}O_5$ [PhCH(OAc)C_5H_4]$Mn(CO)_3$, 1, 152
$MnC_{17}H_{15}$ $C_5H_5Mn(Ph_2)$, 1, 172
$MnC_{17}H_{15}O_4$ {[PhCH(OEt)]C_5H_4}$Mn(CO)_3$, 1, 151
[Me(MePhCOH)C_5H_3]$Mn(CO)_3$, 1, 157
$MnC_{17}H_{17}Cl_2N_2O_5$ {Me[$ClCH_2CH_2CO)_2$NN=C-(Me)]C_5H_3}$Mn(CO)_3$, 1, 158
$MnC_{17}H_{17}O_3$ Hexenyltetrahydropentalenyl-$Mn(CO)_3$, 1, 165
$MnC_{17}H_{17}O_4$ [CH_2=CHC≡CC(OPr)(Me)C_5H_4]-$Mn(CO)_3$, 1, 152
$MnC_{17}H_{18}NO_5$ $PhCOMn(CO)_4$(cyclo-C_6H_{11}-NH_2), 1, 141
$MnC_{17}H_{22}NO_2$ (n-$C_8H_{17}Ph)Mn(CO)_2CN$, 1, 179
$MnC_{18}H_{13}O_4$ [PhC≡CC(OH)(Me)C_5H_4]Mn-$(CO)_3$, 1, 152
$MnC_{18}H_{14}$ Bis(indenyl)-Mn, 1, 175
$MnC_{18}H_{15}N_2O_3$ [Me(PhCH=NN=CMe)C_5H_3]-$Mn(CO)_3$, 1, 159
$MnC_{18}H_{15}N_2O_4$ [Me(BzNHN=CMe)C_5H_3]Mn-$(CO)_3$, 1, 159
$MnC_{18}H_{15}O_4$ [p-$MeC_6H_4C(CH=CH_2)$(OH)-C_5H_4]$Mn(CO)_3$, 1, 152
$MnC_{18}H_{16}F_6P$ [$C_5H_5Mn(C_7H_6Ph)$]PF_6, 1, 173
$MnC_{18}H_{17}N_2O_3$ [Me($PhCH_2NHN$=CMe)C_5H_3]-$Mn(CO)_3$, 1, 159
$MnC_{18}H_{17}O_3$ (3-Cyclohexylindenyl)-Mn-$(CO)_3$, 1, 166
$MnC_{18}H_{19}O_4$ [CH_2=CHC≡CC(OBu)(Me)C_5H_4] $Mn(CO)_3$, 1, 152
$MnC_{18}H_{22}$ (2,4,6-$Me_3C_6H_2)_2$-Mn, 1, 131
Bis(4,5,6,7-tetrahydroindenyl)Mn, 1, 176
$MnC_{18}H_{23}O_4$ Acetyltetraethylcyclopentadienyl-$Mn(CO)_3$, 1, 160
$MnC_{18}H_{25}O_3$ (n-$C_{10}H_{21}C_5H_4)Mn(CO)_3$, 1, 149
(Pentaethylcyclopentadienyl)$Mn(CO)_3$, 1, 160
$MnC_{18}H_{30}IN_6$ [$(EtNC)_6Mn$]I, 1, 182
$MnC_{18}H_{30}I_3N_6$ [$(EtNC)_6Mn$]I_3, 1, 182
$MnC_{19}H_{15}O_2S$ $(C_5H_5)Mn(CO)_2(Ph_2S)$, 1, 171

$MnC_{19}H_{17}O_4$ $[Me_3C(Bz)C_5H_3]Mn(CO)_3$, 1, 159

$MnC_{19}H_{32}O_2P$ $(C_5H_5)Mn(CO)_2(PBu_3)$, 1, 169

$MnC_{20}H_{13}N_2O_4$ [(2-NC-1,2-dihydroquinolin-1-yl-CO)(Me)C_5H_3]$Mn(CO)_3$ isomers, 1, 157

$MnC_{20}H_{15}O_3$ [1,2-Et($C_{10}H_7$)C_5H_3]Mn-$(CO)_3$, 1, 157

$MnC_{20}H_{17}O_3$ (1,8-Diethylfluorenyl)-$Mn(CO)_3$, 1, 167

$MnC_{20}H_{17}O_3S$ $(MeC_5H_4)Mn(CO)_2(Ph_2SO)$, 1, 171

$MnC_{20}H_{19}N_4O_7$ {Et_2[2,4-$(O_2N)_2C_6H_3$NH-N=CMe]C_5H_2}$Mn(CO)_3$, 1, 159

$MnC_{20}H_{25}O_3$ [(cyclo-C_6H_{11})$_2C_5H_3$]Mn-$(CO)_3$, 1, 159

$MnC_{20}H_{29}O_3$ $[(Me_3C)_3C_5H_2]Mn(CO)_3$, 1, 159

$MnC_{20}H_{30}$ Bis(pentamethylcyclopentadienyl)-Mn, 1, 175

$MnC_{21}H_{15}O_2$ $(C_5H_5)(PhC{\equiv}CPh)Mn(CO)_2$, 1, 168

$MnC_{21}H_{15}O_3$ $(Ph_2CHC_5H_4)Mn(CO)_3$, 1, 149

$MnC_{21}H_{15}O_4$ $[Ph_2C(OH)C_5H_4]Mn(CO)_3$, 1, 152

$MnC_{21}H_{19}O_3$ [Tetrahydro(4-phenylbutyl)pentalenyl]-$Mn(CO)_3$, 1, 165

$MnC_{22}H_{15}N_4O_7$ {Me[2,4-$(O_2N)_2C_6H_3$NH-N=C(Ph)]C_5H_3}$Mn(CO)_3$ isomers, 1, 156

$MnC_{22}H_{17}O_2$ $(MeC_5H_4)Mn(CO)_2(PhC{\equiv}CPh)$, 1, 168

$MnC_{22}H_{17}O_3$ $[Ph_2CH(Me)C_5H_3]Mn(CO)_3$, 1, 157

$MnC_{22}H_{17}O_4$ $[Me(Ph_2COH)C_5H_3]Mn(CO)_3$, 1, 159

$MnC_{22}H_{18}$ $(PhC_5H_4)_2Mn$, 1, 175

$MnC_{22}H_{22}$ Bis(3-ethylindenyl)-Mn, 1, 176

$MnC_{23}H_{35}O_6$ n-$C_{17}H_{35}COMn(CO)_5$, 1, 140

$MnC_{24}H_{17}N_2O_5$ [Bz_2NN=C(Me)C_5H_4]Mn-$(CO)_3$, 1, 152

$MnC_{24}H_{18}O_5P$ $MeCOMn(CO)_4(PPh_3)$, 1, 141

$MnC_{24}H_{18}O_8P$ $MeCOMn(CO)_4[P(OPh)_3]$, 1, 141

$MnC_{25}H_{19}N_2O_5$ [Me(Bz_2NN=CMe)C_5H_3]Mn-$(CO)_3$, 1, 159*

$MnC_{25}H_{20}O_2P$ $(C_5H_5)Mn(CO)_2(PPh_3)$, 1, 169

$MnC_{25}H_{20}O_5P$ $(C_5H_5)Mn(CO)_2[P(OPh_3]$, 1, 169

$MnC_{26}H_{15}N_3O_4P$ $(NC)_3CMn(CO)_4PPh_3$, 1, 132

$MnC_{26}H_{15}N_3O_7P$ $(NC)_3CMn(CO)_4P(OPh)_3$, 1, 132

$MnC_{26}H_{18}$ Bis(fluorenyl)-Mn, 1, 176

$MnC_{26}H_{20}N_3$ $(C_5H_5)Mn(NCPh)_3$, 1, 168*

$MnC_{26}H_{22}O_2P$ $(MeC_5H_4)Mn(CO)_2(PPh_3)$, 1, 169

$MnC_{26}H_{22}O_5P$ $(MeC_5H_4)Mn(CO)_2[P(OPh)_3]$, 1, 170

$MnC_{26}H_{30}$ Bis(1,3,4,7-tetramethylindenyl)-Mn, 1, 176

$MnC_{26}H_{33}N_3O_4P$ $(NC)_3CMn(CO)_4P(C_6H_{11})_3$, 1, 132

$MnC_{26}H_{34}$ Bis(octahydrofluorenyl)-Mn, 1, 176

$MnC_{26}H_{42}$ (i-$C_8H_{17}C_5H_4$)$_2$Mn, 1, 175

$MnC_{27}H_{24}N_3O_{10}$ [(p-$C_6H_{13}C_6H_4$Ph)Mn-$(CO)_3$]$OC_6H_2(NO_2)_3$, 1, 179

$MnC_{30}H_{34}$ Bis(3-cyclohexylindenyl)-Mn, 1, 176

$MnC_{30}H_{50}$ (n-$C_{10}H_{21}C_5H_4$)$_2$Mn, 1, 175

$MnC_{32}H_{21}O_4$ (Hydroxytetraphenylcyclopentadienyl)-$Mn(CO)_3$, 1, 160

$MnC_{32}H_{29}OP_2$ $(C_5H_5)Mn(CO)(Ph_2PCH_2CH_2$-$PPh_2)$, 1, 169

$MnC_{32}H_{60}O_4P$ (n-$C_8H_{17}C_5H_4$)Mn(CO)[P(O-$C_6H_{13})_3$], 1, 170

$MnC_{33}H_{23}O_4$ (Methoxytetraphenylcyclopentadienyl)-$Mn(CO)_3$, 1, 160

$MnC_{33}H_{29}O_2P_2$ $(C_5H_5)Mn(CO)_2(Ph_2PCH_2$-$CH_2PPh_2)$, 1, 169

$MnC_{33}H_{31}OP_2$ $(MeC_5H_4)Mn(CO)(Ph_2PCH_2$-$CH_2PPh_2)$, 1, 170

$MnC_{33}H_{36}O_8P$ $(EtPrC_5H_3)Mn(CO)_2[P(OC_6H_4$-$OMe)_3]$, 1, 170

$MnC_{34}H_{30}$ [1,2-Et($C_{10}H_7$)C_5H_3]$_2$Mn, 1, 175

$MnC_{34}H_{31}O_2P_2$ $(MeC_5H_4)Mn(CO)_2(Ph_2PCH_2$-$CH_2PPh_2)$, 1, 170

$MnC_{34}H_{34}$ Bis(1,8-diethylfluorenyl)-Mn, 1, 176

$MnC_{34}H_{46}$ Bis(2-octylindenyl)-Mn, 1, 176

$MnC_{34}H_{58}$ $(C_{12}H_{25}C_5H_4)_2Mn$, 1, 175

$MnC_{35}H_{40}O_2P$ $(MeC_5H_4)Mn(CO)_2[P(C_6H_4\text{-}Pr)_3]$, 1, 170

$MnC_{41}H_{33}O_5P_2$ $MeO_2CMn(CO)_3(PPh_3)_2$, 1, 132

$MnC_{42}H_{24}Cl_6N_6I$ $[(p\text{-}ClC_6H_4NC)_6Mn]I$, 1, 182

$MnC_{42}H_{24}Cl_6N_6I_3$ $[(p\text{-}ClC_6H_4NC)_6Mn]I_3$, 1, 182

$MnC_{42}H_{24}Cl_7N_6O_4$ $[(p\text{-}ClC_6H_4NC)_6Mn]\text{-}ClO_4$, 1, 182

$MnC_{42}H_{30}ClN_6O_4$ $[(PhNC)_6Mn]ClO_4$, 1, 182

$MnC_{42}H_{30}IN_6$ $[(PhNC)_6Mn]I$, 1, 182

$MnC_{42}H_{30}I_3N_6$ $[(PhNC)_6Mn]I_3$, 1, 182

$MnC_{42}H_{35}OP_2$ $(C_5H_5)Mn(CO)(PPh_3)_2$, 1, 169

$MnC_{42}H_{35}O_5P_2$ $EtO_2CMn(CO)_3(PPh_3)_2$, 1, 132

$MnC_{42}H_{35}O_7P_2$ $(C_5H_5)Mn(CO)[P(OPh_3]_2$, 1, 169

$MnC_{42}H_{66}IN_6$ $[(C_6H_{11}NC)_6Mn]I$, 1, 182

$MnC_{42}H_{66}I_3N_6$ $[(C_6H_{11}NC)_6Mn]I_3$, 1, 182

$MnC_{43}H_{37}O_7P_2$ $(MeC_5H_4)Mn(CO)[P(OPh)_3]_2$ 1, 170

$MnC_{45}H_{41}O_5P_2$ $n\text{-}C_5H_{11}O_2CMn(CO)_3(PPh_3)_2$ 1, 132

$MnC_{45}H_{84}O_5P$ $(Me_2C_5H_3)Mn(CO)_2P(O\text{-}C_{12}H_{25}\text{-}n)_3$, 1, 170

$MnC_{47}H_{37}O_5P_2$ $PhCH_2O_2CMn(CO)_3(PPh_3)_2$, 1, 132

$MnC_{48}H_{42}BrN_6O_6$ $[(p\text{-}MeOC_6H_4NC)_6Mn]Br$, 1, 183

$MnC_{48}H_{42}ClN_6O_4$ $[(p\text{-}MeC_6H_4NC)_6Mn]ClO_4$, 1, 182
$[(PhCH_2NC)_6Mn]ClO_4$, 1, 183

$MnC_{48}H_{42}ClN_6O_6$ $[(p\text{-}MeOC_6H_4NC)_6Mn]Cl$, 1, 183

$MnC_{48}H_{42}ClN_6O_9$ $[(p\text{-}MeOC_6H_4NC)_6Mn]\text{-}ClO_3$, 1, 183

$MnC_{48}H_{42}ClN_6O_{10}$ $[(p\text{-}MeOC_6H_4NC)_6Mn]\text{-}ClO_4$, 1, 183

$MnC_{48}H_{42}F_6N_6O_6P$ $[(p\text{-}MeOC_6H_4NC)_6Mn]\text{-}PF_6$, 1, 183

$MnC_{48}H_{42}IN_6$ $[(p\text{-}MeC_6H_4NC)_6Mn]I$, 1, 182
$[(PhCH_2NC)_6Mn]I$, 1, 183

$MnC_{48}H_{42}IN_6O_6$ $[(p\text{-}MeOC_6H_4NC)_6Mn]I$, 1, 183

$MnC_{48}H_{42}I_3N_6$ $[(p\text{-}MeC_6H_4NC)_6Mn]I_3$, 1, 182
$[(PhCH_2NC)_6Mn]I_3$, 1, 183

$MnC_{48}H_{42}I_3N_6O_6$ $[(p\text{-}MeOC_6H_4NC)_6Mn]I_3$, 1, 183

$MnC_{48}H_{43}N_6O_7$ $[(p\text{-}MeOC_6H_4NC)_6Mn]OH$, 1, 183

$MnC_{49}H_{43}N_6O_9$ $[(p\text{-}MeOC_6H_4NC)_6Mn]CO_3H$, 1, 183

$MnC_{56}H_{63}O_7P_2$ $(Me_2C_5H_3)Mn(CO)[P(OC_6H_4\text{-}Et)_3]_2$, 1, 170

$MnC_{61}H_{73}OP_2$ $(MeC_5H_4)Mn(CO)[P(C_6H_2\text{-}Me_3)_3]_2$, 1, 170

$Mn_2C_{12}H_{10}N_2O_4$ $[C_5H_5Mn(CO)(NO)]_2$, 1, 172

$Mn_2C_{13}ClN_3O_{10}$ (6-Chloro-s-triazine-2,4-diyl)$[Mn(CO)_5]_2$, 1, 137*

$Mn_2C_{13}F_6O_{10}$ $(OC)_5Mn(CF_2)_3Mn(CO)_5$, 1, 138

$Mn_2C_{13}H_3F_3O_{10}$ $(OC)_5MnCF(Me)CF_2Mn\text{-}(CO)_5$, 1, 137

$Mn_2C_{13}H_6O_{10}$ $(OC)_5Mn(CH_2)_3Mn(CO)_5$, 1, 138

$Mn_2C_{14}H_8O_{10}$ $(OC)_5MnCH(Me)CH_2CH_2Mn\text{-}(CO)_5$, 1, 138

$Mn_2C_{15}F_6O_{12}$ $(CF_2)_3[COMn(CO)_5]_2$, 1, 139

$Mn_2C_{15}H_9Cl_2O_5P$ $(OC)_3MnC_5H_4PCl_2{=}Mn\text{-}(CO)_2C_5H_5$, 1, 163

$Mn_2C_{15}H_{15}N_3O_3$ $(C_5H_5)_3Mn_2(NO)_3$, 1, 172

$Mn_2C_{16}H_4O_{10}$ $p\text{-}C_6H_4[Mn(CO)_5]_2$, 1, 138

$Mn_2C_{16}H_8O_{12}$ $(CH_2)_4[COMn(CO)_5]_2$, 1, 140

$Mn_2C_{17}H_8O_7$ $CO[C_5H_4Mn(CO)_3]_2$, 1, 161

$Mn_2C_{17}H_{10}O_6$ $CH_2[C_5H_4Mn(CO)_3]_2$, 1, 161

$Mn_2C_{17}H_{20}N_2O_4$ $[(C_5H_5)Mn(CO)_2]_2(H_2N\text{-}CH_2)_2CH_2$, 1, 169

$Mn_2C_{18}H_4O_{12}$ $p\text{-}C_6H_4[COMn(CO)_5]_2$, 1, 140

$Mn_2C_{18}H_{12}O_6$ $[\text{-}CH_2C_5H_4Mn(CO)_3]_2$, 1, 161

$Mn_2C_{18}H_{16}O_4$ $[(MePh)Mn(CO)_2]_2$, 1, 179

$Mn_2C_{18}H_{20}N_2O_4$ $[(C_5H_5)Mn(CO)_2]_2[HN(CH_2CH_2)_2NH]$, 1, 169

$Mn_2C_{20}H_{12}O_6$ [(Dimethyleneethylene)-dicyclopentadienylene]-$[Mn(CO)_3]_2$, 1, 161

$Mn_2C_{20}H_{14}O_6$ [(Dimethylvinylene)dicyclopentadienylene]-$[Mn(CO)_3]$, 1, 161

$Mn_2C_{20}H_{16}O_7$ $O\{[CH(Me)C_5H_4]Mn(CO)_3\}_2$, 1, 161

$Mn_2C_{20}H_{16}O_8$ [(1,2-Dihydroxy-1,2-dimethylethylene)biscyclopentadienylene]-$[Mn(CO)_3]_2$, 1, 161

$Mn_2C_{20}H_{18}O_4$ 1,3-Cyclohexadiene$[C_5H_5Mn(CO)_2]_2$, 1, 168*

$Mn_2C_{20}H_{20}O_4$ $[(o\text{-}Me_2C_6H_4)Mn(CO)_2]_2$, 1, 179

$Mn_2C_{22}H_{16}O_8$ $\{[\equiv CC(OH)(Me)C_5H_4]Mn(CO)_3\}_2$, 1, 161

$Mn_2C_{22}H_{18}N_2O_6$ $\{N{=}C(Me)(Me)C_5H_3Mn(CO)_3\}_2$, 1, 162

$Mn_2C_{22}H_{24}O_4$ $[(1,3,5\text{-}Me_3C_6H_3)Mn(CO)_2]_2$, 1, 179

$Mn_2C_{24}H_{14}O_9$ [(4,5-Dicarboxy-1-cyclohexene-1,2-ylene)biscyclopentadienylene]-$[Mn(CO)_3]_2$ anhydride, 1, 162

$Mn_2C_{24}H_{20}O_8$ Adipoylbis(methylcyclopentadienylene)-$[Mn(CO)_3]_2$, 1, 162

$Mn_2C_{24}H_{20}O_{10}$ $[(EtPh)Mn(CO)_3]_2(COO^-)_2$, 1, 179

$Mn_2C_{24}H_{24}$ Tris(cyclooctatetraene)-Mn_2, 1, 181

$Mn_2C_{25}H_{20}O_5$ (2-Benzyl-2,4,6-cycloheptatrienyl 2,4,6-cycloheptatrienyl ether)$[Mn(CO)_2]_2$, 1, 180

$Mn_2C_{26}H_{16}O_6$ Bis(azulene)-$[Mn(CO)_3]_2$, 1, 180

$Mn_2C_{26}H_{16}O_8$ $p\text{-}C_6H_4\{[CO(Me)C_5H_3]Mn(CO)_3\}_2$, 1, 162

$Mn_2C_{26}H_{24}N_4O_8$ (Acetylmethylcyclopentadienyl)-$Mn(CO)_3$ succinyldihydrazone, 1, 162

$Mn_2C_{30}H_{18}O_7$ [(2-Oxo-1,1-diphenylethylene)biscyclopentadienylene]-$[Mn(CO)_3]_2$, 1, 162

$Mn_2C_{30}H_{20}O_8$ $\{[\text{-}C(Ph)(OH)C_5H_4]Mn(CO)_3\}_2$, 1, 162

$Mn_2C_{32}H_{22}O_7$ [(2-Oxo-1,1-diphenylethylene)bis(2-methylcyclopentadienylene)]-$[Mn(CO)_3]_2$, 1, 162

$Mn_2C_{32}H_{24}O_8$ [(1,2-Dihydroxy-1,2-diphenylethylene)bis(2-methylcyclopentadienylene)]-$[Mn(CO)_3]_2$, 1, 163

$Mn_2C_{32}H_{28}O_6$ Bis(4,6,8-trimethylazulene)$[Mn(CO)_3]_2$, 1, 180

$Mn_6C_{30}H_{30}N_6O_6$ $(C_5H_5)_6Mn_6(NO)_6$, 1, 172

$MnAlC_7H_4Cl_4O_5$ $[(C_2H_4)Mn(CO)_5]AlCl_4$, 1, 180

$MnAl_6C_{12}H_{12}Cl_{18}$ $(C_6H_6)_2Mn(AlCl_3)_6$, 1, 180

$MnAsC_{14}H_{22}O_2$ $[(MeC_5H_4)Mn(AsEt_3)(CO)_2]$, 3, 8

$MnAsC_{15}H_{24}O_2$ $(EtC_5H_4)Mn(CO)_2(AsEt_3)$, 1, 170

$MnAsC_{17}H_{22}O_5$ Indenyl-$Mn(CO)_2[As(OEt)_3]$, 1, 170

$MnAsC_{18}H_{14}O_4$ $(4\text{-}MeC_6H_4)_2AsMn(CO)_4$, 3, 302

$(4\text{-}MeC_6H_4)_2AsMn(CO)_4$, 3, 302

$MnAsC_{22}H_{15}BrO_4$ $[Mn(AsPh_3)(CO)_4]Br$, 3, 22

$MnAsC_{22}H_{15}ClO_4$ $[Mn(AsPh_3)(CO)_4]Cl$, 3, 22

$MnAsC_{22}H_{15}IO_4$ $[Mn(AsPh_3)(CO)_4]I$, 3, 22

$MnAsC_{22}H_{15}O_4$ $[Mn(AsPh_3)(CO)_4]$, 3, 22

$MnAsC_{25}H_{20}O_2$ $(C_5H_5)Mn(CO)_2(AsPh_3)$, 1, 170, 3, 23

$MnAsC_{26}H_{22}O_2$ $(MeC_5H_4)Mn(CO)_2(AsPh_3)$, 1, 170, 3, 23

$MnAsC_{28}H_{26}O_2$ $(C_5H_5)Mn(CO)_2[As(CH_2Ph)_3]$, 1, 170, 3, 27

$MnAsC_{28}H_{50}O_2$ $[Me(n\text{-}C_8H_{17})C_5H_3]Mn(CO)_2(AsBu_3)$, 1, 170

$MnAsC_{38}H_{72}O_2$ $[Me(C_{18}H_{37})C_5H_5]Mn(CO)_2(AsBu_3)$, 1, 15

$MnAsC_{51}H_{48}O_2$ $(n\text{-}C_8H_{17}C_5H_4)Mn(CO)_2[As(C_6H_4Ph\text{-}p)_3]$, 1, 170

$MnAs_2C_9H_{18}BrO_3$ $[Mn(AsMe_3)_2(CO)_3Br]$, 3, 4

$MnAs_2C_{12}H_{16}Br_2O_2$ $[Mn(2\text{-}Me_2AsC_6H_4AsMe_2)(CO)_2Br_2]$, 3, 87

$MnAs_2C_{12}H_{16}O_2$ $[Mn(2\text{-}Me_2AsC_6H_4AsMe_2)$-

$(CO)_2]$, 3, 86
$MnAs_2C_{13}H_{16}BrO_3$ $[Mn(2\text{-}Me_2AsC_6H_4\text{-}AsMe_2)(CO)_3Br]$, 3, 86
$MnAs_2C_{13}H_{16}IO_3$ $[Mn(2\text{-}Me_2AsC_6H_4AsMe_2)\text{-}(CO)_3I]$, 3, 86
$MnAs_2C_{13}H_{16}O_3$ $[Mn(2\text{-}Me_2AsC_6H_4AsMe_2)\text{-}(CO)_3]$, 3, 86
$MnAs_2C_{17}H_{21}O_2$ $(C_5H_5)Mn(CO)_2[o\text{-}C_6H_4\text{-}(AsMe_2)_2]$, 1, 170
$MnAs_2C_{18}H_{23}O_2$ $(MeC_5H_4)Mn(CO)_2[o\text{-}C_6H_4\text{-}(AsMe_2)_2]$, 1, 170
$MnAs_2C_{39}H_{30}BrO_3$ $[Mn(AsPh_3)_2(CO)_3]Br$, 3, 23
$MnAs_2C_{39}H_{30}ClO_3$ $[Mn(AsPh_3)_2(CO)_3]Cl$, 3, 23
$MnAs_2C_{39}H_{30}IO_3$ $[Mn(AsPh_3)_2(CO)_3]I$, 3, 23
$MnAs_2C_{45}H_{42}IO_3$ $\{Mn[As(CH_2Ph)_3]_2(CO)_3\text{-}I\}$, 3, 27
$MnAs_2C_{50}H_{87}O$ Fluorenyl-$Mn(CO)[As\text{-}(C_6H_{13}\text{-}n)_3]_2$, 1, 170
$MnAs_2C_{57}H_{66}IO_3$ $\{Mn[As(C_6H_2Me_3)_3]_2\text{-}(CO)_3I\}$, 3, 29
$MnAs_2C_{86}H_{171}O_7$ $(n\text{-}C_8H_{17}C_5H_4)Mn(CO)\text{-}[As(OC_{12}H_{25})_3]_2$, 1, 170
$MnAs_3C_{13}H_{27}BrO_2$ $\{Mn[MeAs(CH_2CH_2CH_2\text{-}AsMe_2)_2](CO)_2\}Br$, 3, 105
$MnAs_3C_{13}H_{27}IO_2$ $\{Mn[MeAs(CH_2CH_2CH_2\text{-}AsMe_2)_2](CO)_2\}I$, 3, 105
$MnB_7H_4F_4O_5$ $[(C_2H_4)Mn(CO)_5]BF_4$, 1, 180
$MnBC_9H_6F_4O_3$ $[(C_6H_6)Mn(CO)_3]BF_4$, 1, 179
$MnBC_{12}H_8F_2O_5$ $[AcCH{=}C(OBF_2)C_5H_4]Mn\text{-}(CO)_3$, 1, 149
$MnBC_{18}H_{30}F_4N_6$ $[(EtNC)_6Mn]BF_4$, 1, 182
$MnBC_{33}H_{26}O_3$ $[(C_6H_6)Mn(CO)_3]BPh_4$, 1, 179
$MnBC_{36}H_{38}N_6$ $[(MeNC)_6Mn]BPh_4$, 1, 182
$MnBC_{42}H_{66}F_4N_6$ $[(C_6H_{11}NC)_6Mn]BF_4$, 1, 182
$MnBC_{48}H_{42}F_4N_6$ $[(PhCH_2NC)_6Mn]BF_4$, 1, 183
$MnBC_{48}H_{42}F_4N_6O_6$ $[(p\text{-}MeOC_6H_4NC)_6Mn]BF_4$ 1, 183
$MnBC_{48}H_{56}$ [Bis(hexamethylbenzene)-$Mn]BPh_4$, 1, 131, 180
$MnBC_{72}H_{62}N_6O_6$ $[(p\text{-}MeOC_6H_4NC)_6Mn]BPh_4$, 1, 183
$MnBaC_8H_4$ $Ba[Mn(C{\equiv}CH)_4]$, 1, 133
$MnBaC_{32}H_{20}$ $Ba[Mn(C{\equiv}CPh)_4]$, 1, 133
$MnCoC_{15}H_{10}O_5$ $[(C_5H_5)_2Co][Mn(CO)_5]$, 1, 587
$MnCrC_{13}H_{12}N_6O_3S_4$ $[(C_6H_6)Mn(CO)_3]\text{-}[Cr(NH_3)_2(SCN)_4]$, 1, 179
$MnCrC_{17}H_{12}O_5$ $[(C_6H_6)_2Cr][Mn(CO)_5]$, 1, 79, 94*
$MnCrC_{19}H_{24}N_6O_3S_4$ [(Hexamethylbenzene)-$Mn(CO)_3][Cr(NH_3)_2(SCN)_4]$, 1, 179
$MnFeC_{12}H_5DF_6O_7P$ $[C_5H_5Fe(CO)_2Mn(CO)_5]\text{-}D{\cdot}PF_6$, 1, 184
$MnFeC_{12}H_5O_7$ $(C_5H_5)Fe(CO)_2Mn(CO)_5$, 1, 184
$MnFeC_{12}H_6F_6O_7P$ $[C_5H_5Fe(CO)_2Mn(CO)_5]\text{-}H{\cdot}PF_6$, 1, 184
$MnFeC_{15}H_{11}O_4$ $C_6H_6Mn(CO)_2Fe(C_5H_5)(CO)_2$, 1, 184
$MnFeC_{19}H_{13}O_4$ $FcCOC_5H_4Mn(CO)_3$, 1, 184
$MnFeC_{19}H_{15}O_3$ $FcCH_2C_5H_4Mn(CO)_3$, 1, 185
$MnGeC_{15}H_{23}O_3$ $Et_3GeMn(CO)_3\text{-}(CH{=}CHCH{=}CHCH_2CH_2)$, 1, 184
$MnGeC_{20}H_{18}BrO_3$ $Ph_2GeBrMn(CO)_3(CH_2{=}C\text{-}MeCH{=}CH_2)$, 2, 141
$MnGeC_{22}H_{23}O_3$ $Ph_2EtGeMn(CO)_3(CH_2{=}CMe\text{-}CH{=}CH_2)$, 1, 184
$MnGeC_{33}H_{35}O_3$ $(PhCH_2CH_2)_3GeMn(CO)_3\text{-}(CH{=}CHCH{=}CHCH_2CH_2)$, 1, 184
$MnHgC_8H_4ClO_3$ $(ClHgC_5H_4)Mn(CO)_3$, 1, 183
$MnHgC_9H_6ClO_3$ $[Me(ClHg)C_5H_3]Mn(CO)_3$, 1, 183
$MnHg_2C_8H_3Cl_2O_3$ $[(ClHg)_2C_5H_3]Mn(CO)_3$, 1, 183
$MnHg_2C_9H_5Cl_2O_3$ $[Me(ClHg)_2C_5H_2]Mn(CO)_3$, 1, 184
$MnK_2C_8H_4$ $K_2[Mn(C{\equiv}CH)_4]$, 1, 133
$MnK_2C_{12}H_{12}$ $K_2[Mn(C{\equiv}CMe)_4]$, 1, 133
$MnK_2C_{32}H_{20}$ $K_2[Mn(C{\equiv}CPh)_4]$, 1, 133
$MnLiC_3H_9$ $Li[MnMe_3]$, 1, 133
$MnLiC_6H_{15}$ $Li[MnEt_3]$, 1, 133
$MnLiC_{18}H_{15}$ $Li[MnPh_3]$, 1, 133
$MnLiC_{21}H_{21}$ $Li[Mn(C_6H_4Me\text{-}p)_3]$, 1, 133
$MnLiC_{21}H_{21}O_3$ $Li[Mn(C_6H_4OMe\text{-}p)_3]$, 1, 133

$MnMoC_{14}H_7O_7$ [Cycloheptatrienyl-Mo-$(CO)_2$][$Mn(CO)_5$], 1, 118
$MnNa_2C_8H_4$ $Na_2[Mn(C{=}CH)_4]$, 1, 133
$MnNiC_{12}H_7O_6$ $(MeC_5H_4)Ni(CO)Mn(CO)_5$, 1, 185
$MnPbC_8H_9O_5$ $Me_3PbMn(CO)_5$, 2, 594
$MnPbC_{11}H_{15}O_5$ $Et_3PbMn(CO)_5$, 2, 594
$MnPbC_{14}H_{15}O_5$ $(CH_2{=}CHCH_2)_3PbMn(CO)_5$, 2, 594
$MnPbC_{17}H_{23}O_3$ $(CH_2{=}CHCH_2)_3PbMn(CO)_3$-$(CH_2{=}CHCH{=}CHMe)$, 1, 184, 2, 594*
$MnPbC_{23}H_{15}O_5$ $Ph_3PbMn(CO)_5$, 2, 594
$MnPbC_{23}H_{21}O_5$ $(MeC_5H_4)_3PbMn(CO)_5$, 2, 594
$MnPbC_{26}H_{21}O_5$ (m-$MeC_6H_4)_3PbMn(CO)_5$, 2, 594
$MnPbC_{29}H_{25}O_4$ (m-$MeC_6H_4)_3PbMn(CO)_3\cdot$-$C_5H_4O$, 2, 594
$MnPbC_{29}H_{27}O_5$ $(PhCH_2CH_2)_3PbMn(CO)_5$, 2, 594
$MnPbC_{29}H_{51}O_5$ $(C_8H_{17})_3PbMn(CO)_5$, 2, 594
$MnPbC_{32}H_{33}O_5$ Mesityl$_3PbMn(CO)_5$, 2, 594
$MnPbC_{32}H_{59}O_3$ (n-$C_8H_{17})_3PbMn(CO)_3$-$(CH_2{=}CHCH{=}CHMe)$, 1, 184, 2, 594*
$MnPbC_{35}H_{35}O_3$ $(PhCH_2CH_2)_3PbMn(CO)_3\cdot$-$C_8H_8$, 2, 594
$MnPbC_{35}H_{37}O_4$ Mesityl$_3PbMn(CO)_3\cdot C_5H_4O$, 2, 594
$MnPbC_{44}H_{47}O_7$ $(BuO)_3PbMn(CO)_3$-(PhC=CPhCPh=CPhCO), 1, 184
$MnPbC_{50}H_{35}O_4$ $Ph_3PbMn(CO)_3$-(PhC=CPhCPh=CPhCO), 2, 594
$MnPbC_{50}H_{41}O_4$ $(MeC_5H_4)_3PbMn(CO)_3$-(PhC=CPhCPh=CPhCO), 2, 594
$MnPbC_{59}H_{53}O_4$ (2,4,6-$Me_3C_6H_2)_3PbMn$-$(CO)_3$(PhC=CPhCPh=CPhCO), 1, 184
$MnSbC_{18}H_{30}O_5$ (n-$C_8H_{17}C_5H_4)Mn(CO)_2$-[$Sb(OMe)_3$], 1, 170
$MnSbC_{21}H_{33}FO_3$ [$MnSb(C_6H_{11})_3(CO)_3F$], 3, 666
$MnSbC_{22}H_{15}BrO_4$ [$Mn(SbPh_3)(CO)_4$]Br, 3, 663
$MnSbC_{22}H_{15}IO_4$ [$Mn(SbPh_3)(CO)_4I$], 3, 663
$MnSbC_{22}H_{15}O_4$ [$Mn(SbPh_3)(CO)_4$], 3, 663
$MnSbC_{24}H_{21}BrO_3$ [$MnSb(C_6H_4Me)_3(CO)_3$-Br], 3, 667
$MnSbC_{25}H_{20}O_2$ $(C_5H_5)Mn(CO)_2(SbPh_3)$, 1, 170, 3, 664
$MnSbC_{29}H_{48}O_5$ $(BuC_5H_4)Mn(CO)_2$[Sb(O-$C_6H_{11})_3$], 1, 170
$MnSbC_{37}H_{68}O_2$ (n-$C_{12}H_{25}C_5H_4)Mn(CO)_2$-[$Sb(C_6H_{13})_3$], 1, 170, 3, 666
$MnSbC_{42}H_{54}O_2$ $(Bu_2C_5H_3)Mn(CO)_2$[Sb-$(C_6H_2Me_3)_3$], 1, 171, 3, 667
$MnSbC_{45}H_{82}O_3$ [$(AcC_5H_4)MnSb(C_{12}H_{25})_3$-$(CO)_2$], 3, 668
$MnSbC_{46}H_{62}O_5$ [$Me(C_{17}H_{35})C_5H_3$]$Mn(CO)_2$-[$Sb(OC_6H_4Me$-p$)_3$], 1, 171
$MnSbC_{48}H_{42}O_2$ $(MeBuC_5H_3)Mn(CO)_2$[Sb-$(C_6H_4Ph)_3$], 1, 171
$MnSb_2C_9H_{18}IO_3$ [$Mn(SbMe_3)_2(CO)_3I$], 3, 655
$MnSb_2C_{12}H_{16}Br_2O_2$ Mn[o-$C_6H_4(SbMe_2)_2$]-$(CO)_2Br_2$, 3, 678
$MnSb_2C_{12}H_{16}Cl_2O_2$ Mn[o-$C_6H_4(SbMe_2)$]-$(CO)_2Cl_2$, 3, 678
$MnSb_2C_{13}H_{16}IO_3$ Mn[o-$C_6H_4(SbMe_2)_2$]-$(CO)_3I$, 3, 678
$MnSb_2C_{39}H_{30}ClO_3$ [$Mn(SbPh_3)_2(CO)_3Cl$], 3, 663
$MnSb_2C_{39}H_{78}BrO_3$ {Mn[$Sb(C_6H_{13})_3]_2$-$(CO)_3Br$}, 3, 666
$MnSb_2C_{40}H_{30}NO_3S$ [$Mn(SbPh_3)_2(CO)_3SCN$], 3, 663
$MnSb_2C_{43}H_{73}O$ $(MeC_5H_4)Mn(CO)$[Sb-$(C_6H_{11})_3]_2$, 1, 170, 3, 666
$MnSb_2C_{48}H_{47}O$ $(C_5H_5)Mn(CO)[Sb(CH_2Ph)_3]_2$, 1, 170, 3, 666
$MnSb_2C_{75}H_{150}ClO_3$ {Mn[$Sb(C_{12}H_{25})_3]_2$-$(CO)_3Cl$}, 3, 668
$MnSb_2C_{83}H_{69}O$ $(MeBuC_5H_3)Mn[Sb(C_6H_4Ph)_3]_2$-(CO), 3, 668
$MnSiC_9H_9Cl_2O_3$ $MeCl_2SiMn(CO)_3(C_5H_6)$, 1, 184
$MnSiC_{10}H_{15}O_3$ $Me_3SiMn(CO)_3(CH_2{=}CHCH{=}CH_2)$, 1, 184
$MnSiC_{28}H_{27}O_3$ $(PhCH_2)_3SiMn(CO)_3(CH_2{=}CH$-$CH{=}CH_2)$, 1, 184
$MnSiC_{56}H_{107}O_3$ (n-$C_{16}H_{33})_3SiMn(CO)_3$-$(CH_2{=}CMeCH{=}CH_2)$, 1, 184
$MnSnC_8H_8Cl_3O_3$ $Cl_3SnMn(CO)_3(CH_2{=}CH$-

CH=CHMe), 1, 184
$MnSnC_{10}H_{13}O_5$ $MeEt_2SnMn(CO)_5$, 2, 500
$MnSnC_{11}H_5Br_2O_5$ $PhSnBr_2Mn(CO)_5$, 2, 501
$MnSnC_{11}H_5Cl_2O_5$ $PhSnCl_2Mn(CO)_5$, 2, 501
$MnSnC_{11}H_9O_5$ $(CH_2=CH)_3SnMn(CO)_5$, 2, 501
$MnSnC_{13}H_{19}O_7$ $(EtO)_3SnMn(CO)_3$-(CH=CHCH=CHO), 1, 184
$MnSnC_{14}H_{21}O_3$ $MeEt_2SnMn(CO)_3$·1,3-C_6H_8, 2, 500
$MnSnC_{17}H_{10}BrO_5$ $Ph_2SnBrMn(CO)_5$, 2, 501
$MnSnC_{17}H_{10}ClO_5$ $Ph_2SnClMn(CO)_5$, 2, 501
$MnSnC_{17}H_{21}O_3$ $(CH_2=CH)_3SnMn(CO)_3$(1,3-cyclooctadiene), 1, 184, 2, 501
$MnSnC_{17}H_{27}O_5$ $Bu_3SnMn(CO)_5$, 2, 501
$MnSnC_{20}H_{15}O_5$ $(C_5H_5)_3SnMn(CO)_5$, 2, 501*
$MnSnC_{23}H_{15}O_5$ $Ph_3SnMn(CO)_5$, 2, 499
$MnSnC_{23}H_{39}O_3$ $Bu_3SnMn(CO)_3$(1,3-cyclooctadiene), 1, 184, 2, 501
$MnSnC_{25}H_{23}O_4$ $(C_5H_5)_3SnMn(CO)_3$·-CH=CMeCOCMe=CH, 2, 501*
$MnSnC_{26}H_{21}O_2$ $Ph_3SnMn(CO)_2(C_6H_6)$, 1, 184
$MnSnC_{29}H_{27}O_5$ (2,4-$Me_2C_6H_3)_3SnMn(CO)_5$, 2, 501
$MnSnC_{32}H_{31}O_4$ (2,4-$Me_2C_6H_3)_3SnMn(CO)_3$-(CH=CHCH=CHCO), 1, 184, 2, 501
$MnSnC_{32}H_{33}O_5$ $Mesityl_3SnMn(CO)_5$, 2, 501
$MnSnC_{35}H_{41}O_3$ $Mesityl_3SnMn(CO)_3$·1,3-C_5H_8, 2, 501
$MnSnC_{40}H_{30}O_4P$ $Ph_3SnMn(CO)_4PPh_3$, 2, 501
$MnSnC_{44}H_{27}O_5$ (2-Fluorenyl)$_3$-SnMn-$(CO)_5$, 2, 501
$MnSnC_{50}H_{35}O_4$ $Ph_3SnMn(CO)_3$-(PhC=CPhCPh=CPhCO), 1, 184, 2, 501
$MnSnC_{51}H_{39}O_4$ (2-Fluorenyl)$_3$SnMn-$(CO)_3$·C_4Me_4CO, 2, 501*
$Mn_2AsC_{10}F_6IO_8$ $(CF_3)_2As[Mn(CO)_4]_2I$, 3, 301
$Mn_2As_2C_{12}F_{12}O_8$ $[Mn(CO)_4As(CF_3)_2]_2$, 3, 301
$Mn_2As_2C_{12}H_{12}O_8$ $[Mn(CO)_4AsMe_2]_2$, 3, 302
$Mn_2As_2C_{20}H_{30}O_8$ $[Mn(AsEt_3)(CO)_4]_2$, 3, 8
$Mn_2As_2C_{24}H_{26}O_4$ $[C_5H_5Mn(CO)_2]_2$(2-Me_2-$AsC_6H_4AsMe_2$), 3, 87
$Mn_2As_2C_{26}H_{30}O_4$ $[(MeC_5H_4)Mn(CO)_2]_2$-(2-$Me_2AsC_6H_4AsMe_2$), 3, 87
$Mn_2As_2C_{32}H_{20}O_8$ $[Mn(CO)_4AsPh_2]_2$, 3, 302
$Mn_2As_2C_{36}H_{28}O_8$ $[Mn(CO)_4As(C_6H_4Me$-m$)_2]_2$, 3, 302
$Mn_2As_4C_{28}H_{32}O_6$ [Mn(2-$Me_2AsC_6H_4AsMe_2$)-$(CO)_3]_2$, 3, 86
$Mn_2As_6C_{26}H_{54}O_4$ {Mn[MeAs($CH_2CH_2CH_2$-$AsMe_2)_2](CO)_2\}_2$, 3, 105
$Mn_2As_8C_{32}H_{72}O_2$ {Mn[($Me_2AsCH_2CH_2CH_2)_3$-As](CO)}$_2$, 3, 110
$Mn_2Bi_2C_{44}H_{30}O_8$ $[Mn(BiPh_3)(CO)_4]_2$, 3, 820
$Mn_2CdC_{96}H_{84}Br_4N_{12}$ [(p-$MeC_6H_4NC)_6Mn]_2$-$CdBr_4$, 1, 182
$Mn_2CdC_{96}H_{84}Br_4N_{12}O_{12}$ [(p-$MeOC_6H_4NC)_6$-$Mn]_2CdBr_4$, 1, 183
$Mn_2FeC_{28}H_{16}O_8$ Fcd$[COC_5H_4Mn(CO)_3]_2$, 1, 185
$Mn_2FeC_{30}H_{20}O_8$ Fcd$[CO(Me)C_5H_3Mn(CO)_3]_2$, 1, 185
$Mn_2HgC_{16}H_8O_6$ $Hg[C_5H_4Mn(CO)_3]_2$, 1, 183
$Mn_2Li_2C_{40}H_{35}N_6O_5$ $Li_2[Mn_2(C_5H_5)(CO)_5$-$Py_6]$, 1, 173
$Mn_2Na_2C_{25}H_{20}N_3O_5$ $Na_2[Mn_2(C_5H_5)(CO)_5$-$Py_3]$, 1, 173
$Mn_2Na_2C_{40}H_{35}N_6O_5$ $Na_2[Mn_2(C_5H_5)(CO)_5$-$Py_6]$, 1, 173
$Mn_2Na_2C_{46}H_{47}N_6O_5$ $Na_2[Mn_2(C_5H_5)(CO)_5$-(4-Picoline)$_6$], 1, 173
$Mn_2PbC_{12}H_6O_{10}$ $Me_2Pb[Mn(CO)_5]_2$, 2, 594
$Mn_2PbC_{14}H_{10}O_{10}$ $Et_2Pb[Mn(CO)_5]_2$, 2, 594
$Mn_2PtC_{14}H_{10}Cl_6N_2O_6$ $[C_5H_5Mn(CO)_2(NO)]_2$-$PtCl_6$, 1, 171
$Mn_2Sb_2C_{42}H_{30}F_2O_6$ $[Mn(SbPh_3)(CO)_3F]_2$, 3, 663
$Mn_2Sb_4C_{26}H_{32}O_6$ {Mn[o-$C_6H_4(SbMe_2)_2$]-$(CO)_3\}_2$, 3, 678
$Mn_2SnC_{12}H_6O_{10}$ $Me_2Sn[Mn(CO)_5]_2$, 2, 501*
$Mn_2SnC_{22}H_{10}O_{10}$ $Ph_2Sn[Mn(CO)_5]_2$, 2, 501
$MnAsGeC_{12}H_{14}I_2O_4$ {$I_2EtGeMn$[As-$(CH=CH_2)_3](CO)_4\}$, 3, 6
$MnAsPbC_{16}H_{14}Cl_2O_4$ {$Cl_2PhPbMn$[As-

$(CH{=}CH_2)_3](CO)_4\}$, 3, 6

$MnAsPbC_{34}H_{42}O_4$ $\{Pr_3PbMn[As(CH_2Ph)_3]$-$(CO)_4\}$, 3, 27

$MnAsSnC_{10}H_{18}O_4$ $[Me_3SnMn(AsMe_3)(CO)_4]$, 3, 4

$MnAsSnC_{21}H_{21}ClO_4$ $Me_2Sn(Cl)Mn[As$-$(C_5H_5)_3](CO)_4$, 3, 17

$MnAsSnC_{40}H_{30}O_4$ $[Ph_3SnMn(AsPh_3)(CO)_4]$, 2, 501, 3, 23

$MnAs_3CuC_{16}H_{27}O_5$ $\{Cu[MeC(CH_2AsMe_2)_3]$-$Mn(CO)_5\}$, 3, 107

$MnGeSbC_{21}H_{32}I_2O_4$ $\{I_2(C_8H_{17})GeMn(CO)_4$-$[Sb(CH_2CH{=}CH_2)_3]\}$, 3, 658

$MnPbSbC_{28}H_{20}Br_2O_4$ $\{Br_2PhPb[Mn(SbPh_3)$-$(CO)_4]\}$, 3, 663

$MnPbSbC_{28}H_{30}O_4$ $[Ph_3PbMn(CO)_4(SbEt_3)]$, 3, 656

$MnSbSiC_{58}H_{104}Br_2O_4$ $\{Br_2PhSiMn[Sb$-$(C_{16}H_{33})_3](CO)_4\}$, 3, 668

$MnSbSnC_{10}H_{15}Br_3O_4$ $[Br_3SnMn(CO)_4$-$(SbEt_3)]$, 3, 656

$MnSbSnC_{34}H_{21}Br_3O_4$ $\{Br_3SnMn[Sb$-$(C_{10}H_7)_3](CO)_4\}$, 3, 668

$MnSbSnC_{37}H_{72}O_4$ $Pr_3Sn\{Mn(CO)_4[Sb$-$(C_8H_{17})_3]\}$, 3, 667

$MnSbSnC_{41}H_{43}BrO_4$ $Br(C_5H_5)_2SnMn(CO)_4$-$Sb(C_6H_4CHMe_2)_3$, 3, 667

$Mn_2As_2GeC_{58}H_{44}O_8$ $\{(PhCH_2)_2Ge[Mn$-$(AsPh_3)(CO)_4]_2\}$, 3, 23

$Mn_2As_2SiC_{36}H_{52}O_8$ $(Me_3CCH_2)_2Si[Mn$-$(CO)_4As(CH_2CH{=}CH_2)_3]_2$, 3, 11

$Mn_2As_2SiC_{120}H_{216}O_8$ $(2$-$EtC_6H_4)_2Si$-$[Mn(CO)_4As(C_{16}H_{33})_3]_2$, 3, 29

$Mn_2As_2SnC_{30}H_{52}O_8$ $Et_2Sn[Mn(AsPr_3)$-$(CO)_4]_2$, 3, 13

$Mn_2As_2SnC_{78}H_{84}O_8$ $(2,5$-$Me_2C_6H_3)_2Sn$-$[Mn(CO)_4As(C_6H_2Me_3)_3]_2$, 3, 29

$Mn_2Sb_2SiC_{70}H_{68}O_8$ $(PhCH_2)_2Si[Mn(CO)_4$-$Sb(CH_2CH_2Ph)_3]_2$, 3, 667

$Mn_2Sb_2SiC_{80}H_{104}O_8$ $(C_{18}H_{37})_2Si[Mn$-$(SbPh_3)(CO)_4]_2$, 3, 663

$Mn_3As_3Au_3C_{26}H_{27}O_{15}$ $\{[MeC(CH_2AsMe_2)_3]$-$[AuMn(CO)_5]_3\}$, 3, 107

$Mn_3As_3GeC_{122}H_{228}O_{12}$ $CH_2{=}CHGe[Mn(CO)_4$-$As(C_{12}H_{25})_3]_3$, 3, 29

$Mn_3As_3SnC_{81}H_{68}O_{12}$ $PhSn[Mn(CO)_4As$-$(C_6H_4Me$-$o)_3]_3$, 3, 27

$Mn_3Sb_3SiC_{60}H_{50}O_{12}$ $CH_2{=}CHCH_2Si[Mn(CO)_4$-$Sb(C_5H_5)_3]_3$, 3, 659

$Mn_3Sb_3SiC_{106}H_{108}O_{12}$ 2-Fluorenyl-Si-$[Mn(CO)_4Sb(C_6H_4CHMe_2)_3]_3$, 3, 667

$Mn_3Sb_3SnC_{55}H_{88}O_{12}$ $PhCH_2Sn[Mn(CO)_4Sb$-$(CH_2CHMe_2)_3]_3$, 3, 659

$Mn_3Sb_3SnC_{69}H_{50}O_{12}$ $CH_2{=}CHCH_2Sn[Mn$-$(SbPh_3)(CO)_4]_3$, 3, 663

$MoC_5H_5ClO_2$ $C_5H_5MoO_2Cl$, 1, 104

$MoC_5H_5Cl_2O$ $C_5H_5Mo(O)Cl_2$, 1, 104

$C_5H_5Mo(O)Cl_2$, 1, 104

$MoC_7H_2O_5$ $(C_2H_2)Mo(CO)_5$, 1, 116

$MoC_7H_4O_5$ $(C_2H_4)Mo(CO)_5$, 1, 109

$MoC_7H_5NO_3$ $C_5H_5Mo(CO)_2NO$, 1, 105

$MoC_8H_4O_5$ $(MeC{\equiv}CH)Mo(CO)_5$, 1, 116

$MoC_8H_5BrO_3$ $C_5H_5Mo(CO)_3Br$, 1, 745

$MoC_8H_5ClO_3$ $C_5H_5Mo(CO)_3Cl$, 1, 745

$MoC_8H_5DO_3$ $C_5H_5Mo(CO)_3D$, 1, 96

$MoC_8H_5IO_3$ $C_5H_5Mo(CO)_3I$, 1, 745

$MoC_8H_6N_2O_2$ $(CH_2{=}CHCN)_2Mo(CO)_2$, 1, 109

$MoC_8H_6O_3$ $C_5H_5Mo(CO)_3H$, 1, 95

$MoC_8H_6O_4$ Butadiene-$Mo(CO)_4$, 1, 109

$MoC_8H_6O_5$ $(MeCH{=}CH_2)Mo(CO)_5$, 1, 109

$MoC_9D_6O_3$ $C_6D_6Mo(CO)_3$, 1, 107

$MoC_9H_5FO_3$ $(FPh)Mo(CO)_3$, 1, 106

$MoC_9H_5F_3O_3$ $(C_5H_5)Mo(CO)_3CF_3$, 1, 96

$MoC_9H_5F_6O_4P$ $[C_5H_5Mo(CO)_4]PF_6$, 1, 98

$MoC_9H_6O_3$ $C_6H_6Mo(CO)_3$, 1, 106

$MoC_9H_6O_5$ Butadiene-$Mo(CO)_5$, 1, 109

$MoC_9H_7IO_2$ [Heptatrienium-$Mo(CO)_2$]I, 1, 113

$MoC_9H_8IO_3$ $(PyMeI)Mo(CO)_3$, 1, 115

$MoC_9H_8O_3$ $(C_5H_5)Mo(CO)_3Me$, 1, 96

$MoC_9H_9N_3O_3$ $(MeNC)_3Mo(CO)_3$, 1, 116

$MoC_9H_{10}O_2S$ $C_5H_5Mo(CO)_2CH_2SMe$, 1, 103

$MoC_{10}H_5F_3O_4$ $(C_5H_5)Mo(CO)_3COCF_3$, 1, 97

$MoC_{10}H_5F_3O_5$ $C_5H_5Mo(CO)_3O_2CCF_3$, 1, 745

$MoC_{10}H_6F_4O_3$ $(C_5H_5)Mo(CO)_3CF_2CF_2H$, 1, 97

$MoC_{10}H_6O_4$ (Cycloheptatrienone)-$Mo(CO)_3$, 1, 112

$MoC_{10}H_7ClO_3$ (7-Chlorocycloheptatriene)-$Mo(CO)_3$, 1, 112

$MoC_{10}H_7ClO_7$ [Cycloheptatrienium-Mo-$(CO)_3]ClO_4$, 1, 113

$MoC_{10}H_7I_3O_3$ [Cycloheptatrienium-Mo-$(CO)_3]I_3$, 1, 113

$MoC_{10}H_8O_3$ $(PhMe)Mo(CO)_3$, 1, 107
Cycloheptatriene-$Mo(CO)_3$, 1, 111
$MoC_{10}H_8O_4$ $(C_5H_5)Mo(CO)_3Ac$, 1, 97
$MoC_{10}H_9F_6O_3P$ $[C_5H_5Mo(C_2H_4)(CO)_3]PF_6$, 1, 98
$MoC_{10}H_{10}Br_2$ $(C_5H_5)_2MoBr_2$, 1, 105
$MoC_{10}H_{10}Br_4$ $[(C_5H_5)_2MoBr_3]Br$, 1, 106
$MoC_{10}H_{10}Cl_2$ $(C_5H_5)_2MoCl_2$, 1, 105
$MoC_{10}H_{10}Cl_4$ $[(C_5H_5)_2MoCl_3]Cl$, 1, 106
$MoC_{10}H_{10}I_2$ $(C_5H_5)_2MoI_2$, 1, 105
$MoC_{10}H_{10}O_2$ $(C_5H_5)(\pi\text{-}CH_2{=}CHCH_2)Mo(CO)_2$ 1, 102
$MoC_{10}H_{10}O_3$ $(C_5H_5)Mo(CO)_3Et$, 1, 97
$MoC_{10}H_{10}O_3S$ $(C_5H_5)Mo(CO)_3CH_2SMe$, 1, 97
$MoC_{10}H_{11}IN_2O$ $[C_5H_5Mo(CNMe)_2(CO)]I$, 1, 98
$MoC_{10}H_{12}$ $(C_5H_5)_2MoH_2$, 1, 105
$MoC_{10}H_{12}O_2$ (Butadiene)$_2Mo(CO)_2$, 1, 109
$MoC_{10}H_{12}O_4$ $(MeCH{=}CH_2)_2Mo(CO)_4$, 1, 109
$MoC_{10}H_{13}Cl$ $[(C_5H_5)_2MoH_3]Cl$, 1, 105
$MoC_{11}H_5F_7O_3$ $(C_5H_5)Mo(CO)_3CF_2CF_2CF_3$, 1, 97
$MoC_{11}H_8O_4$ Bicyclo[2.2.1]hepta-2,5-diene-$Mo(CO)_4$, 1, 115
$MoC_{11}H_{10}O_3$ $(C_5H_5)Mo(CO)_3CH_2CH{=}CH_2$, 1, 97
$(p\text{-}Me_2C_6H_4)Mo(CO)_3$, 1, 107
1,3,5-Cyclooctatriene-$Mo(CO)_3$, 1, 113
(7-Methylcycloheptatriene)-$Mo(CO)_3$, 1, 113
$MoC_{11}H_{10}O_4$ $(C_5H_5)Mo(CO)_3COEt$, 1, 97
(7-Methoxycycloheptatriene)-$Mo(CO)_3$, 1, 112
$MoC_{11}H_{10}O_5$ $(EtC{\equiv}CEt)Mo(CO)_5$, 1, 116
$MoC_{11}H_{11}$ $C_6H_6MoC_5H_5$, 1, 103
$MoC_{11}H_{11}F_6O_3P$ $[C_5H_5Mo(CH_2{=}CHMe)(CO)_3]PF_6$, 1, 99
$MoC_{11}H_{11}IO_3$ $(Me_2CHC_5H_4)Mo(CO)_3I$, 1, 745
$MoC_{11}H_{11}NO_3$ (6-Me_2N-fulvene)-$Mo(CO)_3$, 1, 103
$MoC_{11}H_{12}O_3$ $(EtC_5H_4)Mo(CO)_3Me$, 1, 97
$(C_5H_5)Mo(CO)_3i\text{-}Pr$, 1, 97
$MoC_{12}H_5F_7O_4$ $(C_5H_5)Mo(CO)_3COCF_2CF_2CF_3$, 1, 97
$MoC_{12}H_5NO_5$ $(PhNC)Mo(CO)_5$, 1, 116
$MoC_{12}H_8O_4$ Cyclooctatetraene-$Mo(CO)_4$, 1, 114
$MoC_{12}H_9N_3O_3$ $(CH_2{=}CHCN)_3Mo(CO)_3$, 1, 109
$MoC_{12}H_{10}O_3$ Cyclononatetraene-$Mo(CO)_3$, 1, 115
5,6-Dimethylenebicyclo[2.2.1]hept-2-ene-$Mo(CO)_3$, 1, 115
$MoC_{12}H_{10}O_4$ Bicyclo[2.2.2]octa-2,5-diene-$Mo(CO)_4$, 1, 114
$MoC_{12}H_{11}ClO$ $[(C_6H_6)Mo(C_5H_5)(CO)]Cl$, 1, 99
$MoC_{12}H_{11}F_6OP$ $[C_5H_5Mo(C_6H_6)(CO)]PF_6$, 1, 104
$MoC_{12}H_{12}$ $(C_6H_6)_2Mo$, 1, 108
$MoC_{12}H_{12}I$ $(C_6H_6)_2MoI$, 1, 108
$MoC_{12}H_{12}O_3$ $(1,3,5\text{-}Me_3C_6H_3)Mo(CO)_3$, 1, 107
(7-Ethylcycloheptatriene)-$Mo(CO)_3$, 1, 112
Cyclononatriene-$Mo(CO)_3$, 1, 114
$MoC_{12}H_{12}O_4$ 1,5-Cyclooctadiene-Mo-$(CO)_4$, 1, 114
$MoC_{12}H_{13}IO_3$ $[(EtMeCH)C_5H_4]Mo(CO)_3I$, 1, 745
$MoC_{12}H_{14}O_3$ $(i\text{-}PrC_5H_4)Mo(CO)_3Me$, 1, 97
$MoC_{12}H_{15}N_3O_3$ $(EtNC)_3Mo(CO)_3$, 1, 116
$MoC_{13}H_7NO_6$ $(p\text{-}MeOC_6H_4NC)Mo(CO)_5$, 1, 116
$MoC_{13}H_8O_3$ Azulene-$Mo(CO)_3$, 1, 112
$MoC_{13}H_{10}INO_3$ (Quinoline·MeI)-$Mo(CO)_3$, 1, 116
$MoC_{13}H_{14}O_3$ (3,7,7-Trimethylcycloheptatriene)-$Mo(CO)_3$, 1, 112
$MoC_{13}H_{15}IO_3$ $[(Et_2CH)C_5H_4]Mo(CO)_3I$, 1, 745
$[(MePrCH)C_5H_4]Mo(CO)_3I$, 1, 745
$MoC_{13}H_{16}N_2O_3$ [6,6-$(Me_2N)_2$-fulvene]-$Mo(CO)_3$, 1, 103
$MoC_{13}H_{16}O_3$ $(i\text{-}PrC_5H_4)Mo(CO)_3Et$, 1, 97
$MoC_{13}H_{23}IN_3O_2P$ $\{C_5H_5Mo(CO)_2[P(NMe_2)_3]\}I$, 1, 98
$MoC_{14}H_{10}O_4$ $(\pi\text{-}BzO)_2Mo$, 1, 108
$MoC_{14}H_{12}O_2$ π-Cycloheptatrienyl-$(C_5H_5)Mo(CO)_2$, 1, 103

$MoC_{14}H_{16}$ $(MePh)_2Mo$, 1, 108

$MoC_{14}H_{16}O_2$ (1,3-Cyclohexadiene)$_2$-Mo-$(CO)_2$, 1, 111

$MoC_{14}H_{16}O_3$ (2,3,7,7-Tetramethylcycloheptatriene)-Mo$(CO)_3$, 1, 112

$MoC_{14}H_{16}O_4$ 3-Ethyl-1,5-cyclooctadiene-Mo$(CO)_4$, 1, 114

$MoC_{14}H_{25}IN_2OP$ Cycloheptatrienyl-Mo-(CO)$[P(NMe_2)_3]I$, 1, 113

$MoC_{15}H_8O_3$ Biphenylene-Mo$(CO)_3$, 1, 107

$MoC_{15}H_{11}IO_4$ $(p\text{-}MeOC_6H_4C_5H_4)Mo(CO)_3I$, 1, 745

$MoC_{15}H_{12}O_3$ (7-Cyclopentadienylcycloheptatriene)-Mo$(CO)_3$, 1, 112

$MoC_{15}H_{18}F_6O_3$ $(EtCF{=}CHF)_3Mo(CO)_3$, 1, 109

$MoC_{15}H_{18}O_3$ Hexamethylbenzene-Mo$(CO)_3$, 1, 107

$MoC_{16}H_{13}IO_3$ $[(MePhCH)C_5H_4]Mo(CO)_3I$, 1, 745

$MoC_{16}H_{14}O_4$ $(\pi\text{-}o\text{-}MeC_6H_4COO)_2Mo$, 1, 109

$MoC_{16}H_{14}O_6$ $(\pi\text{-}MeOC_6H_4COO)_2Mo$, 1, 109

$MoC_{16}H_{16}O_4$ 7,8-Dimethyltricyclo-$[4.2.2.0^{2,5}]$deca-3,9-diene-Mo-$(CO)_4$, 1, 115*

$MoC_{17}H_{12}INO_3$ (Acridine methiodide)-Mo$(CO)_3$, 1, 116

$MoC_{17}H_{14}O_3$ (Bicycloheptatrienyl)-Mo-$(CO)_3$, 1, 112

3-Methyl-7-phenylcycloheptatriene)-Mo$(CO)_3$, 1, 112

$MoC_{17}H_{18}O_7$ $[(EtO_2C)_2CHC_7H_7]Mo(CO)_3$, 1, 112

$MoC_{18}H_{10}N_2O_4$ $(PhNC)_2Mo(CO)_4$, 1, 116

$MoC_{18}H_{14}Cl_3$ Bis(indenyl)-MoCl_3, 1, 106

$MoC_{18}H_{14}F_6O_4$ 2,3,5,6-Tetramethyl-7,8-bis(trifluoromethyl)bicyclo[2.2.2]octatriene-Mo$(CO)_4$, 1, 115

$MoC_{18}H_{15}$ Ph_3Mo, 1, 95

$MoC_{18}H_{20}O_2$ Bis(1,3,6-cyclooctatriene)-Mo$(CO)_2$, 1, 113

$MoC_{20}H_{14}Cl_2O_4$ Bis(chlorocyclooctatetraene)-Mo$(CO)_4$, 1, 114

$MoC_{20}H_{16}O_4$ Bis(cyclooctatetraene)-Mo$(CO)_4$, 1, 114

$MoC_{20}H_{20}$ $(C_5H_5)_4Mo$, 1, 106

$MoC_{21}H_{28}O_4$ 1-Ethyl-2-octylbicyclo-[2.2.1]heptadiene-Mo$(CO)_4$, 1, 115

$MoC_{22}H_{20}O_4$ Bis(methylcyclooctatetraene)-Mo$(CO)_4$, 1, 114

$MoC_{23}H_{34}O_3$ $[2,5,1,6\text{-}Bu_2(Pr)_2C_6H_2]Mo$-$(CO)_3$, 1, 107

$MoC_{24}H_{15}N_3O_3$ $(PhNC)_3Mo(CO)_3$, 1, 116

$MoC_{24}H_{32}O_4$ $[(t\text{-}Bu)_3$-cyclooctatetraene]-Mo$(CO)_4$, 1, 114

$MoC_{26}H_{19}O_3P$ $(Ph_3PC_5H_4)Mo(CO)_3$, 1, 103

$MoC_{27}H_{21}N_3O_3$ $(p\text{-}MeC_6H_4NC)_3Mo(CO)_3$, 1, 116

$MoC_{34}H_{22}N_2O_4$ $(PhCH{=}CPhCN)_2Mo(CO)_4$, 1, 109

$MoC_{35}H_{27}$ $(1,2,3,4\text{-}Ph_4C_6H_2)MoC_5H_5$, 1, 104

$MoC_{42}H_{18}Cl_{12}N_6$ $(2,4\text{-}Cl_2C_6H_3NC)_6Mo$, 1, 117

$MoC_{42}H_{24}Cl_6N_6$ $(m\text{-}ClC_6H_4NC)_6Mo$, 1, 116

$MoC_{42}H_{30}N_6$ $(PhNC)_6Mo$, 1, 116

$MoC_{48}H_{36}Cl_6N_6$ $[Cl(Me)C_6H_3NC]_6Mo$ isomers, 1, 117

$MoC_{48}H_{42}N_6$ $(p\text{-}MeC_6H_4NC)_6Mo$, 1, 116

$MoC_{57}H_{32}Cl_8O$ $[(p\text{-}ClC_6H_4)_4$-cyclobutadiene]$(p\text{-}ClC_6H_4C{\equiv}CC_6H_4Cl\text{-}p)_2MoCO$, 1, 110

$MoC_{57}H_{40}O$ $(PhC{=}CPhCPh{=}CPh)(PhC{\equiv}CPh)_2$-MoCO, 1, 110

$MoC_{58}H_{40}O_2$ $(PhC{=}CPhCPh{=}CPh)_2Mo(CO)_2$, 1, 110

$MoC_{59}H_{40}O_3$ $(PhC{=}CPhCPh{=}CPhC{=}O)$-$(PhC{=}CPhCPh{=}CPh)Mo(CO)_2$, 1, 110

$MoC_{60}H_{40}O_4$ $(PhC{=}CPhCPh{=}CPhC{=}O)_2Mo$-$(CO)_2$, 1, 111

$MoC_{70}H_{50}$ $(Ph_5$-cyclopentadienyl$)_2$-Mo, 1, 105

$MoC_{70}H_{50}Br$ $(Ph_5$-cyclopentadienyl$)_2$-MoBr, 1, 105

$Mo_2C_{10}H_{10}O_4$ $(C_5H_5MoO_2)_2$, 1, 104

$Mo_2C_{10}H_{10}O_5$ $(C_5H_5MoO_2)_2O$, 1, 104

$Mo_2C_{12}H_{20}$ $[(\pi\text{-}CH_2{=}CHCH_2)_2Mo]_2$, 1, 106

$Mo_2C_{14}H_{22}S_4$ $[C_5H_5Mo(SMe)_2]_2$, 1, 102

$Mo_2C_{16}H_7BrO_6$ 1-Bromoazulene-$[Mo(CO)_3]_2$, 1, 112

$Mo_2C_{16}H_{10}O_6$ $[C_5H_5Mo(CO)_3]_2$, 1, 100

$Mo_2C_{16}H_{16}O_4S_2$ $[C_5H_5Mo(CO)_2SMe]_2$, 1, 102

$Mo_2C_{16}H_{17}O_4P$ $\{[C_5H_5Mo(CO)_2H][Me_2PMo(CO)_2C_5H_5]\}$, 1, 101

$Mo_2C_{18}H_8O_6$ Biphenylene-$[Mo(CO)_3]_2$, 1, 108

$Mo_2C_{18}H_{10}F_{12}S_4$ $\{C_5H_5Mo[CF_3C(S)=C(S)-CF_3]\}_2$, 1, 102

$Mo_2C_{18}H_{14}O_6$ $[(MeC_5H_4)Mo(CO)_3]_2$, 1, 100

$Mo_2C_{18}H_{22}O_4P_2$ $[C_5H_5Mo(PMe_2)(CO)_2]_2$, 1, 101

$Mo_2C_{19}H_{10}F_6O_6$ $[(C_5H_5)Mo(CO)_3CF_2-]_2-CF_2$, 1, 97

$Mo_2C_{19}H_{14}O_6$ 4,6,8-Trimethylazulene-$[Mo(CO)_3]_2$, 1, 112

$Mo_2C_{20}H_{14}O_6$ (Bicycloheptatrienyl)-$[Mo(CO)_3]_2$, 1, 112

$Mo_2C_{20}H_{14}O_7$ (Dicycloheptatrienyl ether)-$[Mo(CO)_3]_2$, 1, 112

$Mo_2C_{20}H_{18}O_6$ $[(EtC_5H_4)Mo(CO)_3]_2$, 1, 100

$Mo_2C_{21}H_{10}F_6O_8$ $[(C_5H_5)Mo(CO)_3OCCF_2-]_2-CF_2$, 1, 97

$Mo_2C_{21}H_{18}O_6$ 1,4,7-Me_2(i-Pr)-azulene-$[Mo(CO)_3]_2$, 1, 112

$Mo_2C_{22}H_{16}O_4$ $PhC{\equiv}CH[Me(CO)_2C_5H_5]_2$, 1, 101

$Mo_2C_{22}H_{20}O_6$ $[(CH_2{=}CMeC_5H_4)(Me_2CHC_5H_4)-Mo_2(CO)_6]$, 1, 101

$Mo_2C_{22}H_{22}O_6$ $[(PrC_5H_4)Mo(CO)_3]_2$, 1, 100

$[(i\text{-}PrC_5H_4)Mo(CO)_3]_2$, 1, 100-1

$Mo_2C_{24}H_{14}O_6$ [Indenyl-$Mo(CO)_3]_2$, 1, 101*

$Mo_2C_{24}H_{26}O_6$ $[(sec\text{-}BuC_5H_4)Mo(CO)_3]_2$, 1, 101

$Mo_2C_{26}H_{30}O_6$ $[(Et_2CHC_5H_4)Mo(CO)_3]_2$, 1, 101

$[(MePrCHC_5H_4)Mo(CO)_3]_2$, 1, 101

$Mo_2C_{28}H_{20}O_4$ $PhC{\equiv}CPh[Mo(CO)_2C_5H_5]_2$, 1, 101

$Mo_2C_{28}H_{30}O_6$ $[(cyclo\text{-}C_6H_{11}C_5H_4)Mo-(CO)_3]_2$, 1, 101

$[(cyclo\text{-}C_6H_{11}C_5H_4)Mo(CO)_3]_2$, 1, 101

$Mo_2C_{32}H_{26}O_8$ $[(p\text{-}MeOC_6H_4CH_2C_5H_4)Mo-(CO)_3]_2$, 1, 101

$Mo_2C_{42}H_{30}O_6$ $[(Ph_2CHC_5H_4)Mo(CO)_3]_2$, 1, 101

$Mo_2C_{62}H_{40}Br_2O_6$ $[(PhC{=}CPhCPh{=}CPh)Mo-(CO)_3Br]_2$, 1, 111

$Mo_2C_{62}H_{40}I_2O_6$ $[(PhC{=}CPhCPh{=}CPh)Mo-(CO)_3I]_2$, 1, 111

$Mo_2C_{74}H_{50}O_4$ $[(PhC{=}CPhCPh{=}CPh)Mo-(CO)_2]_2(PhC{\equiv}CPh)$, 1, 110

$Mo_3C_{54}H_{45}O_3P_3$ $[C_5H_5Mo(PPh_2)(CO)]_3$, 1, 102

$MoAsC_{10}H_5F_6O_3$ $C_5H_5Mo(CO)_3As(CF_3)_2$, 3, 300

$MoAsC_{11}H_{15}O_5$ $[Mo(AsEt_3)(CO)_5]$, 3, 9

$MoAsC_{23}H_{15}O_5$ $Mo(AsPh_3)(CO)_5$, 3, 23

$MoAs_2C_{12}H_{16}I_2O_2$ $[Mo(2\text{-}Me_2AsC_6H_4-AsMe_2)(CO)_2I_2]$, 3, 87

$MoAs_2C_{13}H_{16}Br_2O_3$ $[Mo(2\text{-}Me_2AsC_6H_4-AsMe_2)(CO)_3Br_2]$, 3, 88

$MoAs_2C_{13}H_{16}I_2O_3$ $[Mo(2\text{-}Me_2AsC_6H_4AsMe_2)-(CO)_3I_2]$, 3, 88

$MoAs_2C_{14}H_{16}O_4$ $[Mo(2\text{-}Me_2AsC_6H_4AsMe_2)-(CO)_4]$, 3, 87

$MoAs_2C_{16}H_{30}O_4$ $[Mo(AsEt_3)_2(CO)_4]$, 3, 9

$MoAs_2C_{40}H_{30}O_4$ $Mo(AsPh_3)_2(CO)_4$, 3, 23

$MoAs_3C_{14}H_{27}O_3$ $\{Mo[MeAs(CH_2CH_2CH_2-AsMe_2)_2](CO)_3\}$, 3, 105

$MoAs_3C_{21}H_{45}O_3$ $[Mo(AsEt_3)_3(CO)_3]$, 3, 9

$MoAs_3C_{57}H_{45}O_3$ $Mo(AsPh_3)_3(CO)_3$, 3, 23

$MoAs_4C_{22}H_{32}Br_2O_2$ $[Mo(2\text{-}Me_2AsC_6H_4-AsMe_2)_2(CO)_2Br]Br$, 3, 88

$MoAs_4C_{22}H_{32}Br_4O_2$ $[Mo(2\text{-}Me_2AsC_6H_4-AsMe_2)_2(CO)_2Br]Br_3$, 3, 88

$MoAs_4C_{22}H_{32}I_2O_2$ $[Mo(2\text{-}Me_2AsC_6H_4AsMe_2)_2-(CO)_2I]I$, 3, 88

$MoAs_4C_{22}H_{32}O_2$ $[Mo(2\text{-}Me_2AsC_6H_4AsMe_2)_2-(CO)_2]$, 3, 87

$MoBC_{10}H_9F_4O_3$ $[C_5H_5Mo(C_2H_4)(CO)_3]BF_4$, 1, 98

$MoBC_{32}H_{28}NO_3$ $[C_5H_5Mo(CO)_3NH_3]BPh_4$, 1, 99

$MoBC_{32}H_{29}N_2O_3$ $[C_5H_5Mo(CO)_3H_2NNH_2]BPh_4$, 1, 99

$MoBC_{34}H_{27}O_3$ [Cycloheptatrienium-$Mo-(CO)_3]BPh_4$, 1, 113

$MoBC_{36}H_{32}$ $[(C_6H_6)_2Mo]BPh_4$, 1, 108

$MoBiC_{11}H_{15}O_5$ $[Mo(BiEt_3)(CO)_5]$, 3, 816

$MoBiC_{23}H_{15}O_5$ [Mo(BiPh$_3$)(CO)$_5$], 3, 820
$MoBi_2C_{16}H_{30}O_4$ [Mo(BiEt$_3$)$_2$(CO)$_4$], 3, 816
$MoCoC_{18}H_{15}O_3$ [(C$_5$H$_5$)$_2$Co][C$_5$H$_5$Mo(CO)$_3$], 1, 118
$MoCrC_{12}H_{14}N_7O_3S_4$ [C$_5$H$_5$Mo(CO)$_3$NH$_3$]-[Cr(NH$_3$)$_2$(SCN)$_4$], 1, 99
$MoCrC_{12}H_{15}N_8O_3S_4$ [C$_5$H$_5$Mo(CO)$_3$H$_2$NNH$_2$]-[Cr(NH$_3$)$_2$(SCN)$_4$], 1, 99
$MoCrC_{14}H_{13}N_6O_3S_4$ [Cycloheptatrienium-Mo(CO)$_3$][Cr(NH$_3$)$_2$(SCN)$_4$], 1, 113
$MoCrC_{14}H_{16}ClN_6S_4$ (C$_5$H$_5$)$_2$MoCl[Cr(NH$_3$)$_2$-(SCN)$_4$], 1, 118
$MoCrC_{20}H_{14}O_6$ (Bicyclohepta-2,4,6-trienyl)[Cr(CO)$_3$][Mo(CO)$_3$], 1, 94
$MoFeC_{15}H_{10}O_5$ (C$_5$H$_5$)Fe(CO)$_2$Mo(C$_5$H$_5$)-(CO)$_3$, 1, 118
$MoFeC_{16}H_7BrO_6$ 1-Bromoazulene-[Fe-(CO)$_3$][Mo(CO)$_3$], 1, 118
$MoGeC_{11}H_{15}O_5$ Et$_3$GeMo(CO)$_5$, 2, 140
$MoGeC_{14}H_{21}O_5$ Pr$_3$GeMo(CO)$_5$, 2, 140
$MoGeC_{15}H_{23}O_3$ Et$_3$GeMo(CO)$_3$·1,3-C$_6$H$_8$, 2, 140
$MoGeC_{16}H_{27}O_3$ Pr$_3$GeMo(CO)$_3$·1,3-C$_4$H$_6$, 2, 140
$MoGeC_{17}H_{10}BrO_5$ Ph$_2$GeBrMo(CO)$_5$, 2, 140
$MoGeC_{23}H_{15}O_5$ Ph$_3$GeMo(CO)$_5$, 2, 140
$MoGeC_{26}H_{21}O_5$ (PhCH$_2$)$_3$GeMo(CO)$_5$, 2, 140
$MoGeC_{29}H_{27}O_5$ (PhCH$_2$CH$_2$)$_3$GeMo(CO)$_5$, 2, 140
$MoGeC_{30}H_{29}O_3$ (PhCH$_2$)$_3$GeMo(CO)$_3$·1,3-C$_6$H$_8$, 2, 140
$MoGeC_{32}H_{21}O_5$ 2-Indenyl$_3$GeMo(CO)$_5$, 2, 140
$MoGeC_{33}H_{35}O_3$ (PhCH$_2$CH$_2$)$_3$GeMo(CO)$_3$·-1,3-C$_6$H$_8$, 2, 140
$MoGeC_{35}H_{25}O_4$ (2-Indenyl)$_3$GeMo(CO)$_3$·-CH=CHCH=CHCO, 2, 141
$MoGeC_{35}H_{63}O_5$ (C$_{10}$H$_{21}$)$_3$GeMo(CO)$_5$, 2, 140
$MoGeC_{38}H_{71}O_3$ (C$_{10}$H$_{21}$)$_3$GeMo(CO)$_3$·1,3-C$_5$H$_8$, 2, 140
$MoGeC_{53}H_{99}O_5$ (C$_{16}$H$_{33}$)$_3$GeMo(CO)$_5$, 2, 140
$MoGeC_{60}H_{111}O_4$ (C$_{16}$H$_{33}$)$_3$GeMo(CO)$_3$-(CMe=CMeCMe=CMeCO), 1, 141
$MoKC_9H_5N_2O_2$ K[Mo(CN)$_2$(C$_5$H$_5$)(CO)$_2$], 1, 103
$MoK_3C_9H_3O_3$ K$_3$[Mo(CO)$_3$(C≡CH)$_3$], 1, 95
$MoK_3C_{12}H_9O_3$ K$_3$[Mo(CO)$_3$(C≡CMe)$_3$], 1, 95
$MoK_3C_{27}H_{15}O_3$ K$_3$[Mo(CO)$_3$(C≡CPh)$_3$], 1, 95
$MoLiC_8H_5O_3$ Li[C$_5$H$_5$Mo(CO)$_3$], 1, 96
$MoMnC_{14}H_7O_7$ [Cycloheptatrienyl-Mo-(CO)$_2$][Mn(CO)$_5$], 1, 118
$MoPbC_{26}H_{20}O_3$ C$_5$H$_5$Mo(CO)$_3$PbPh$_3$, 1, 118, 2, 594
$MoSbC_9H_{10}ClO_5$ Mo(Et$_2$SbCl)(CO)$_5$, 3, 684
$MoSbC_{11}H_{15}O_5$ [Mo(SbEt$_3$)(CO)$_5$], 3, 656
$MoSbC_{23}H_{15}O_5$ [Mo(SbPh$_3$)(CO)$_5$], 3, 664
$MoSb_2C_{12}H_{20}Cl_2O_4$ Mo(Et$_2$SbCl)$_2$(CO)$_4$, 3, 684
$MoSb_2C_{14}H_{25}ClO_4$ [Mo(SbEt$_3$)(Et$_2$SbCl)-(CO)$_4$], 3, 656, 684
$MoSb_2C_{16}H_{30}O_4$ [Mo(SbEt$_3$)$_2$(CO)$_4$], 3, 656
$MoSb_2C_{40}H_{30}O_4$ [Mo(SbPh$_3$)$_2$(CO)$_4$], 3, 664
$MoSb_3C_{19}H_{40}ClO_3$ [Mo(SbEt$_3$)$_2$(Et$_2$SbCl)-(CO)$_3$], 3, 656, 684
$MoSb_3C_{21}H_{45}O_3$ [Mo(SbEt$_3$)$_3$(CO)$_3$], 3, 656
$MoSb_3C_{57}H_{45}O_3$ [Mo(SbPh$_3$)$_3$(CO)$_3$], 3, 664
$MoSiC_{10}H_{12}O_4$ Me$_2$Si(CH=CH$_2$)$_2$Mo(CO)$_4$, 1, 109
$MoSnC_8H_5Cl_3O_3$ C$_5$H$_5$Mo(CO)$_3$SnCl$_3$, 1, 117
$MoSnC_{14}H_{10}Cl_2O_3$ C$_5$H$_5$Mo(CO)$_3$SnPhCl$_2$, 1, 117, 2, 500
$MoSnC_{26}H_{20}O_3$ C$_5$H$_5$Mo(CO)$_3$SnPh$_3$, 1, 117, 2, 500
$MoWC_{16}H_{10}O_6$ C$_5$H$_5$Mo(CO)$_6$WC$_5$H$_5$, 1, 118
$Mo_2As_2C_{12}H_{12}O_8$ [(OC)$_4$MoAsMe$_2$]$_2$, 3, 300
$Mo_2As_2C_{14}H_{12}O_{10}$ [(OC)$_5$MoAsMe$_2$]$_2$, 3, 300
$Mo_2As_2C_{18}H_{10}F_{12}O_4$ [C$_5$H$_5$Mo(CO)$_2$As-

$(CF_3)_2]_2$, 3, 300
$Mo_2As_2C_{18}H_{22}O_4$ $[C_5H_5Mo(CO)_2(AsMe_2)]_2$, 1, 102, 3, 301
$Mo_2As_2C_{32}H_{20}O_8$ $[Mo(CO)_4AsPh_2]_2$, 3, 301
$Mo_2PtC_{22}H_{22}Cl_6O_6$ $[C_5H_5Mo(CH_2{=}CHMe)$-$(CO)_3]_2PtCl_6$, 1, 99
$Mo_2SnC_{16}H_{10}Cl_2O_6$ $[C_5H_5Mo(CO)_3]_2SnCl_2$, 1, 117
$Mo_2SnC_{18}H_{16}O_6$ $Me_2Sn[C_5H_5Mo(CO)_3]_2$, 1, 117, 2, 500
$Mo_2SnC_{28}H_{20}O_6$ $Ph_2Sn[C_5H_5Mo(CO)_3]_2$, 1, 117, 2, 500
$Mo_3PbC_{28}H_{24}O_9$ $[(MeC_5H_4)Mo(CO)_3]_3PbMe$, 1, 118, 2, 594
$Mo_3SnC_{86}H_{136}O_{10}$ $AcC_6H_{10}Sn[(C_{18}H_{37}$-$C_5H_4)Mo(CO)_3]_3$, 1, 117*, 2, 500
$NaGeC_{18}H_{15}$ Ph_3GeNa, 2, 135
$NaPbC_6H_{15}$ Et_3PbNa, 2, 593
$NaPbC_{18}H_{15}$ Ph_3PbNa, 2, 593
$NaPbC_{18}H_{33}$ $Cyclohexyl_3PbNa$, 2, 593
$NaSnC_2H_7$ Me_2HSnNa, 2, 498
$NaSnC_2H_8N$ $Me_2Sn(NH_2)Na$, 2, 498
$NaSnC_3H_9$ Me_3SnNa, 2, 498
$NaSnC_6H_{15}$ Et_3SnNa, 2, 498
$NaSnC_{18}H_{15}$ Ph_3SnNa, 2, 498
$NaSnC_{35}H_{35}O$ $Ph_2(Ph_3C)SnNa \cdot Et_2O$, 2, 498
$NaSn_2C_{10}H_{25}$ Et_5Sn_2Na, 2, 498
$Na_2GeC_{12}H_{10}$ Ph_2GeNa_2, 2, 137
$Na_2MnC_8H_4$ $Na_2[Mn(C{\equiv}CH)_4]$, 1, 133
$Na_2Mn_2C_{25}H_{20}N_3O_5$ $Na_2[Mn_2(C_5H_5)(CO)_5$-$Py_3]$, 1, 173
$Na_2Mn_2C_{40}H_{35}N_6O_5$ $Na_2[Mn_2(C_5H_5)(CO)_5$-$Py_6]$, 1, 173
$Na_2Mn_2C_{46}H_{47}N_6O_5$ $Na_2[Mn_2(C_5H_5)(CO)_5$-(4-Picoline)$_6]$, 1, 173
$Na_2SnC_2H_6$ Me_2SnNa_2, 2, 496
$Na_2SnC_4H_{10}$ Et_2SnNa_2, 2, 498
$Na_2SnC_{12}H_{10}$ Ph_2SnNa_2, 2, 498
$Na_2Sn_2C_4H_{12}$ $(Me_2SnNa)_2$, 2, 498
$Na_2Sn_2C_8H_{20}$ $(Et_2SnNa)_2$, 2, 498
$NbC_9H_5O_4$ $C_5H_5Nb(CO)_4$, 1, 52
$NbC_{10}H_{10}$ $(C_5H_5)_2Nb$, 1, 52
$NbC_{10}H_{10}Br_3$ $(C_5H_5)_2NbBr_3$, 1, 53
$NbC_{10}H_{10}Cl_2$ $(C_5H_5)_2NbCl_2$, 1, 52
$NbC_{10}H_{10}Cl_3$ $(C_5H_5)_2NbCl_3$, 1, 53
$NbC_{10}H_{11}Br_2O$ $(C_5H_5)_2NbBr_2(OH)$, 1, 53
$NbC_{10}H_{11}Cl_2O$ $(C_5H_5)_2NbCl_2(OH)$, 1, 53
$NbC_{20}H_{20}$ $(C_5H_5)_4Nb$, 1, 53
$NbC_{26}H_{18}Cl_3$ Difluorenyl-$NbCl_3$, 1, 53
$NbBC_{10}H_{14}Cl$ $(C_5H_5)_2NbClBH_4$, 1, 53
$Nb_2Ge_2C_{40}H_{54}O_6$ $[(CH_2{=}CH)_2Ge(C_8H_{17}$-$C_5H_4Nb)(CO)_3]_2$, 2, 139
$Nb_2Ge_2C_{52}H_{54}O_6$ $[Ph_2Ge(C_6H_{13}C_5H_4Nb)$-$(CO)_3]_2$, 2, 139
$Nb_2Pb_2C_{88}H_{158}O_6$ $[(C_{16}H_{33})_2PbNb(CO)_3$-$(C_5H_4Bu)]_2$, 2, 594*
$Nb_2Sn_2C_{72}H_{94}O_6$ $\{[Bu(i\text{-}C_8H_{17})C_5H_3]Nb$-$(CO)_3Sn(C_6H_3Me_2\text{-}2,4)_2\}_2$, 2, 500
$NdC_{15}H_{15}$ $(C_5H_5)_3Nd$, 1, 4
$NdC_{18}H_{21}$ $(MeC_5H_4)_3Nd$, 1, 4
NiC_4H_{10} Et_2Ni, 1, 638
$NiC_4H_{14}N_4$ $[Ni(C{\equiv}CH)_2 \cdot 4NH_3]$, 1, 636
$(NiC_4N_4O)_n$ $[NC)_3CNi(NO)]_n$, 1, 638
$NiC_5H_3NO_3$ $(MeNC)Ni(CO)_3$, 1, 662
$NiC_5H_5Cl_2$ $C_5H_5NiCl_2$, 1, 650
$NiC_5H_5I_2$ $C_5H_5NiI_2$, 1, 650
NiC_5H_5NO C_5H_5NiNO, 1, 644
$NiC_6F_6O_2$ $(CF_3C{\equiv}CCF_3)Ni(CO)_2$, 1, 661
$NiC_6H_4N_2$ Cyclobutadiene-$Ni(CN)_2$, 1, 657
NiC_6H_5IO $C_5H_5Ni(CO)I$, 1, 639
$NiC_6H_5NO_3$ $(EtNC)Ni(CO)_3$, 1, 662
NiC_6H_6ClNO $[(C_6H_6)Ni(NO)]Cl$, 1, 655
$NiC_6H_6N_2$ $(CH_2{=}CHCN)_2Ni$, 1, 655
$NiC_6H_6N_2O_2$ $(MeNC)_2Ni(CO)_2$, 1, 662
NiC_6H_7NO $(MeC_5H_4)NiNO$, 1, 644
$NiC_6H_8O_2$ $(CH_2{=}CHCHO)_2Ni$, 1, 655
NiC_6H_{10} $(\pi\text{-}CH_2{=}CHCH_2)_2Ni$, 1, 654
$NiC_7H_5F_3O$ $C_5H_5Ni(CO)CF_3$, 1, 640
NiC_7H_9NO $(EtC_5H_4)NiNO$, 1, 644
$NiC_7H_9N_3O$ $(MeNC)_3Ni(CO)$, 1, 662
$NiC_8H_4N_4$ $(NCCH{=}CHCN)_2Ni$, 1, 656
$NiC_8H_5F_5O$ $C_5H_5Ni(CO)C_2F_5$, 1, 640
NiC_8H_8 Cyclooctatetraene-Ni, 1, 660
$NiC_8H_9NO_3$ $(BuNC)Ni(CO)_3$, 1, 662
NiC_8H_{10} $\pi\text{-}CH_2{=}CHCH_2NiC_5H_5$, 1, 641
$NiC_8H_{10}N_2O_2$ $(EtNC)_2Ni(CO)_2$, 1, 662
$NiC_8H_{10}N_4$ $(EtNC)_2Ni(CN)_2$, 1, 662
$NiC_8H_{11}NO$ $(PrC_5H_4)NiNO$, 1, 645
NiC_8H_{12} Cyclooctadiene-Ni, 1, 660
$NiC_8H_{12}Cl$ $[C_6H_6NiMe_2]Cl$, 1, 638
$NiC_8H_{12}Cl_2$ Tetramethylcyclobutadiene-

$NiCl_2$, 1, 657
$NiC_8H_{12}N_4$ $(MeNC)_4Ni$, 1, 662
NiC_8H_{14} $[\pi\text{-}CH_2{=}C(Me)CH_2]_2Ni$, 1, 654
NiC_8N_6 $Ni[C(CN)_3]_2$, 1, 638
$NiC_9H_5F_5O$ $C_5H_5Ni(CO)CF{=}CFCF_3$, 1, 640
$NiC_9H_5F_6S_2$ $C_5H_5\overline{NiSC(CF_3){=}C(CF_3)S}$, 1, 650
$NiC_9H_5F_7O$ $C_5H_5Ni(CO)CF_2CF_2CF_3$, 1, 640
$NiC_9H_7F_3O$ $C_5H_5Ni(CO)CH_2CF{=}CF_2$, 1, 640
NiC_9H_7NO Indenyl-NiNO, 1, 645
NiC_9H_{12} $\pi\text{-}CH_2{=}CHCH(Me)NiC_5H_5$, 1, 641
$NiC_9H_{13}NO$ $[1,3\text{-}Me(Me_2CH)C_5H_3]NiNO$, 1, 644
$(Et_2C_5H_3)NiNO$, 1, 645
$NiC_9H_{21}Cl_2NO_3$ $[1,5\text{-}Me_2N(Me)\text{-}2,3\text{-hexa-dien-5-ol}]NiCl_2\cdot 2H_2O$, 1, 661
$NiC_{10}H_6N_2$ $[PhCH{=}C(CN)_2]Ni$, 1, 656
$NiC_{10}H_{10}$ $(C_5H_5)_2Ni$, 1, 651
$NiC_{10}H_{10}Br_2$ Dibromocyclopentenyl-NiC_5H_5, 1, 650
$NiC_{10}H_{10}Br_3$ $[(C_5H_5)_2Ni]Br_3$, 1, 652
$NiC_{10}H_{10}Cl_4$ Tetrachlorocyclopentyl-NiC_5H_5, 1, 650
$NiC_{10}H_{10}I_3$ $[(C_5H_5)_2Ni]I_3$, 1, 652
$NiC_{10}H_{10}I_4$ Tetraiodocyclopentyl-Ni-C_5H_5, 1, 650
$NiC_{10}H_{12}$ $C_5H_5Ni\overline{CHCH{=}CHCH_2CH_2}$, 1, 641
$(C_5H_6)_2Ni$, 1, 641
$NiC_{10}H_{13}F_3O_2$ $(C_6H_{13}CF{=}CF_2)Ni(CO)_2$, 1, 661
$NiC_{10}H_{15}N_3O$ $(EtNC)_3Ni(CO)$, 1, 662
$NiC_{11}H_5N_3$ $[PhC(CN){=}C(CN)_2]Ni$, 1, 656
$NiC_{11}H_{16}$ $\pi\text{-}PrCH{=}CHCH_2NiC_5H_5$, 1, 641
π-2-Pentenyl-$Ni(C_5H_4Me)$, 1, 642
$NiC_{11}H_{17}NO$ $(n\text{-}C_6H_{13}C_5H_4)NiNO$, 1, 645
$NiC_{11}H_{20}IP$ $C_5H_5Ni(PEt_3)I$, 1, 639
$NiC_{12}H_{10}$ Ph_2Ni, 1, 638
$NiC_{12}H_{10}ClF_3$ 6,6,7,7-ClF_3-bicyclo-[3.2.0]heptenyl-NiC_5H_5, 1, 642
$NiC_{12}H_{10}F_4$ 6,6,7,7-F_4-bicyclo-[3.2.0]heptenyl-NiC_5H_5, 1, 642
$NiC_{12}H_{10}F_6$ $(CF_3C_5H_5)_2Ni$, 1, 659
$NiC_{12}H_{14}$ $(MeC_5H_4)_2Ni$, 1, 652
$NiC_{12}H_{16}$ Methylcyclopentenyl-Ni-(C_5H_4Me), 1, 642
$(MeC_5H_5)_2Ni$, 1, 659
$NiC_{12}H_{18}$ Di-π-cyclohexenyl-Ni, 1, 654
1,5,9-Cyclododecatriene-Ni, 1, 660*
$NiC_{12}H_{18}N_2O_2$ $(BuNC)_2Ni(CO)_2$, 1, 662
$NiC_{12}H_{22}Cl_2N_2$ $(Me_2CHCH_2CH_2NC)_2NiCl_2$, 1, 662
$(n\text{-}C_5H_{11}NC)_2NiCl_2$, 1, 662
$NiC_{14}H_{10}Cl_2N_2$ $(PhNC)_2NiCl_2$, 1, 662
$NiC_{14}H_{10}Cl_6$ 2,3-$(CCl_3)_2$-bicyclo[2.2.1]-hepta-2π,5-dien-7-yl-NiC_5H_5, 1, 643
$NiC_{14}H_{10}F_6$ 2,3-$(CF_3)_2$-bicyclo[2.2.1]-hepta-2π,5-dien-7-yl-NiC_5H_5, 1, 643
$NiC_{14}H_{14}N_2O_2$ 2,3-Dicarbamoylbicyclo-[2.2.1]hepta-2π,5-dien-7-yl-NiC_5H_5, 1, 643
$NiC_{14}H_{14}O_2$ 2-(MeO_2C)-bicyclo[2.2.1]-hepta-2π,5-dien-7-yl-NiC_5H_5, 1, 643
$NiC_{14}H_{15}NO$ $(PhMe_2CC_5H_4)NiNO$, 1, 645
$NiC_{14}H_{20}$ $(Me_2C_5H_4)_2Ni$, 1, 659
$NiC_{14}H_{34}P_2$ $(C_2H_4)Ni(PEt_3)_2$, 1, 657
$NiC_{15}H_{16}$ $\pi\text{-}PhCH{=}CHCH_2Ni(C_5H_4Me)$, 1, 642
$NiC_{15}H_{16}O_2$ 2-(EtO_2C)-bicyclo[2.2.1]-hepta-2π,5-dien-7-yl-NiC_5H_5, 1, 643
$NiC_{16}H_8N_2$ [9-Fluorenylidene=$C(CN)_2$]-Ni, 1, 656
$NiC_{16}H_{10}N_2$ $[Ph_2C{=}C(CN)_2]Ni$, 1, 656
$NiC_{16}H_{10}N_4$ $(PhNC)_2Ni(CN)_2$, 1, 662
$NiC_{16}H_{12}N_3O_7$ $[(C_5H_5)_2Ni]OC_6H_2(NO_2)_3$, 1, 652
$NiC_{16}H_{16}O_4$ 2,3-$(MeO_2C)_2$-bicyclo-[2.2.1]hepta-2π,5-dien-7-yl-NiC_5H_5, 1, 643
$NiC_{16}H_{18}O_4$ 2,3-$(MeO_2C)_2$-bicyclo-[2.2.1]hept-2π-en-7-yl-NiC_5H_5, 1, 643
$NiC_{16}H_{20}O_2$ (1,5-Cyclooctadiene)(2,5-xyloquinone)-Ni, 1, 659
(1,5-Cyclooctadiene)(2,6-xyloquinone)-Ni, 1, 659
$NiC_{16}H_{22}$ $(Me_2CHC_5H_4)_2Ni$, 1, 653
$NiC_{16}H_{22}N_4$ $[Ni(C{\equiv}CPh)_2\cdot 4NH_3]$, 1, 636
$NiC_{16}H_{24}$ Dicyclooctadiene-Ni, 1, 660
$NiC_{16}H_{26}$ $\pi\text{-}C_8H_{17}CH{=}CHCH_2NiC_5H_5$, 1, 641

$NiC_{18}H_{26}$ π-$C_8H_{17}CH{=}CHCH_2NiC_5H_5$, 1, 641
Di-π-octadienyl-Ni, 1, 654
$NiC_{18}H_{27}N_3O$ $(BuNC)_3Ni(CO)$, 1, 662
$NiC_{18}H_{32}P_2$ $[(HC{\equiv}C)_2Ni(PEt_3)_2]$, 1, 635
$NiC_{17}H_{14}N_2$ [o-(Phenylazo)phenyl]-NiC_5H_5, 1, 650
$NiC_{17}H_{20}O_2$ (Norbornadiene)(duroquinone)-Ni, 1, 660
$NiC_{17}H_{22}O_2$ (1,5-Cyclooctadiene)(trimethylquinone)-Ni, 1, 659
$NiC_{17}H_{28}$ π-$CH_2CH{=}CH(Me)$-$Ni(C_5H_4C_8H_{17})$, 1, 642
$NiC_{18}H_{14}$ Bis(indenyl)-Ni, 1, 653
$NiC_{18}H_{14}N_2$ 2,3-$(NCCH{=}CH)_2$-bicyclo-[2.2.1]hepta-2π,5-dien-7-yl-NiC_5H_5, 1, 643
$(PhCH{=}CHCN)_2Ni$, 1, 656
$NiC_{18}H_{18}$ (π-$PhCH{=}CHCH_2)_2Ni$, 1, 654
$NiC_{18}H_{20}O_2$ (Cyclooctatetraene)(duroquinone)-Ni, 1, 660
7,1,4-s-Bu$(Me)_2$-azulene-$Ni(CO)_2$, 1, 661
$NiC_{18}H_{20}O_3$ (2,4,6-Cyclooctatrienone)-(duroquinone)-Ni, 1, 660
$NiC_{18}H_{20}O_4$ 2,3-$(EtO_2C)_2$-bicyclo-[2.2.1]hepta-2π,5-dien-7-yl-NiC_5H_5, 1, 643
$NiC_{18}H_{22}$ $(2,4,6\text{-}Me_3C_6H_2)_2Ni$, 1, 638
8,9-Dihydro-4,5,6,7-tetramethylindenyl-NiC_5H_5, 1, 642
$NiC_{18}H_{22}O_2$ (1,3,5-Cyclooctatriene)-(duroquinone)-Ni, 1, 659
$NiC_{18}H_{24}N_4$ $[Ni(C{\equiv}CPh)_2{\cdot}H_2NCH_2CH_2NH_2{\cdot}$-$2NH_3]$, 1, 636
$NiC_{18}H_{24}O_2$ (1,5-Cyclooctadiene)(duroquinone)-Ni, 1, 659
$NiC_{18}H_{26}$ 4,5,6,7-Me_4-4,5,6,7,8,9-hexahydroindenyl-NiC_5H_5, 1, 642
$(Me_3CC_5H_4)_2Ni$, 1, 653
$NiC_{18}H_{30}Cl_5IP_2$ $[C_6Cl_5Ni(PEt_3)_2]I$, 1, 633
$NiC_{18}H_{30}Cl_6P_2$ $[C_6Cl_5Ni(PEt_3)_2]Cl$, 1, 633
$NiC_{18}H_{34}BrP$ $(MeC_5H_4)Ni(PBu_3)Br$, 1, 639
$NiC_{18}H_{36}P_2$ $[(MeC{\equiv}C)_2Ni(PEt_3)_2]$, 1, 635
$NiC_{19}H_{37}BrP_2$ $[o\text{-}MeC_6H_4Ni(PEt_3)_2]Br$, 1, 633
$NiC_{19}H_{37}ClP_2$ $[o\text{-}MeC_6H_4Ni(PEt_3)_2]Cl$, 1, 633
$NiC_{20}H_{20}O_4$ 2,3-$(MeO_2CCH{=}CH)_2$-bicyclo-[2.2.1]hepta-2π,5-dien-7-yl-NiC_5H_5, 1, 643
$NiC_{20}H_{24}O_2$ 2,3-$(EtOCH{=}CH)_2$-bicyclo-[2.2.1]hepta-2π,5-dien-7-yl-NiC_5H_5, 1, 643
(Tricyclo[5.2.1.$0^{1,5}$]deca-3,8-diene)-duroquinone-Ni, 1, 660
$NiC_{20}H_{24}O_4$ 2,3-$(Me_2CHO_2C)_2$-bicyclo-[2.2.1]hepta-2π,5-dien-7-yl-NiC_5H_5, 1, 643
Bis(duroquinone)-Ni, 1, 659
$NiC_{21}H_{20}ClP$ [π-$CH_2{=}CHCH_2Ni(PPh_3)Cl$], 1, 654
$NiC_{21}H_{34}IP$ Indenyl-$Ni(PBu_3)I$, 1, 639
$NiC_{21}H_{41}BrP_2$ $[2,4,6\text{-}Me_3C_6H_2Ni(PEt_3)_2]$-Br, 1, 633
$NiC_{21}H_{41}ClP_2$ $[2,4,6\text{-}Me_3C_6H_2Ni(PEt_3)_2]$-Cl, 1, 633
$NiC_{22}H_{10}Cl_8O_4$ $[(C_5H_5)_2Ni]$chloranilide chloranil complex, 1, 652
$NiC_{22}H_{22}ClP$ [π-$CH_2{=}C(Me)CH_2Ni(PPh_3)$-Cl], 1, 655
$NiC_{22}H_{37}BrP_2$ $[1\text{-}C_{10}H_7Ni(PEt_3)_2]Br$, 1, 633
$NiC_{23}H_{20}ClP$ $C_5H_5Ni(PPh_3)Cl$, 1, 639
$NiC_{23}H_{20}IP$ $C_5H_5Ni(PPh_3)I$, 1, 639
$NiC_{23}H_{37}NP_2S$ $[1\text{-}C_{10}H_7Ni(PEt_3)_2]SCN$, 1, 633
$NiC_{24}H_{21}N_2P$ $(CH_2{=}CHCN)_2Ni(PPh_3)$, 1, 657
$NiC_{24}H_{23}P$ $C_5H_5Ni(PPh_3)Me$, 1, 640
$NiC_{24}H_{40}P_2$ $[Ph_2Ni(PEt_3)_2]$, 1, 635
$NiC_{25}H_{20}F_5P$ $C_5H_5Ni(PPh_3)C_2F_5$, 1, 640
$NiC_{25}H_{21}P$ $C_5H_5Ni(PPh_3)C{\equiv}CH$, 1, 640
$NiC_{25}H_{25}P$ $C_5H_5Ni(PPh_3)Et$, 1, 640
$NiC_{26}H_{10}N_8$ $Ni[C(CN)_2C(Ph){=}C(CN)_2]_2$, 1, 638
$NiC_{26}H_{18}F_6$ 2,3-$(p\text{-}CF_3C_6H_4)_2$-bicyclo-[2.2.1]hepta-2π,5-dien-7-yl-NiC_5H_5, 1, 643
$NiC_{26}H_{24}O_2$ 2,3-$(p\text{-}MeOC_6H_4)_2$-bicyclo-[2.2.1]hepta-2π,5-dien-7-yl-NiC_5H_5,

1, 643

$NiC_{26}H_{30}Cl_6P_2$ $[C_6Cl_5Ni(PEt_2Ph)_2]Cl$, 1, 633

$NiC_{26}H_{34}BrClP_2$ $[o\text{-}ClC_6H_4Ni(PEt_2Ph)_2]Br$, 1, 633

$NiC_{26}H_{34}Br_2P_2$ $[o\text{-}BrC_6H_4Ni(PEt_2Ph)_2]Br$, 1, 633

$NiC_{26}H_{39}BrP_2$ [9-Phenanthryl-Ni-$(PEt_3)_2$]Br, 1, 634

$NiC_{26}H_{44}N_6$ $(Me_2CHCH_2CH_2NC)_4Ni(CN)_2$, 1, 663*

$(n\text{-}C_5H_{11}NC)_4Ni(CN)_2$, 1, 663

$NiC_{26}H_{52}N_6O_6P_2$ $(C_2H_4)Ni\{P[N(CH_2CH_2)_2O]_3\}_2$, 1, 657

$NiC_{27}H_{20}F_5P$ $C_5H_5Ni(PPh_3)CF{=}CFCF{=}CF_2$, 1, 640

$NiC_{27}H_{27}BrP_2$ $MeNi[(\text{-}CH_2PPh_2)_2]Br$, 1, 633

$NiC_{27}H_{28}ClP$ $(MeC_5H_4)Ni[P(C_6H_4Me)_3]Cl$, 1, 639

$NiC_{27}H_{34}ClF_3P_2$ $[o\text{-}CF_3C_6H_4Ni(PEt_2Ph)_2]Cl$, 1, 633

$NiC_{27}H_{37}BrOP_2$ $[o\text{-}MeOC_6H_4Ni(PEt_2Ph)_2]Br$, 1, 633

$NiC_{27}H_{37}BrP_2$ $[o\text{-}MeC_6H_4Ni(PEt_2Ph)_2]Br$, 1, 633

$NiC_{28}H_{16}Cl_4N_4$ $(p\text{-}ClC_6H_4NC)_4Ni$, 1, 663

$NiC_{28}H_{20}Br_2$ Tetraphenylcyclobutadiene-$NiBr_2$, 1, 658

$NiC_{28}H_{20}N_4$ $(PhNC)_4Ni$, 1, 663

$NiC_{28}H_{22}O_2$ Tetraphenylcyclobutadiene-$Ni(OH)_2$, 1, 658

$NiC_{28}H_{24}F_5IP_2$ $[C_2F_5Ni(\text{-}CH_2PPh_2)_2]I$, 1, 633

$NiC_{28}H_{24}O_4$ 2,3-$(p\text{-}MeO_2CC_6H_4)_2$-bicyclo[2.2.1]hepta-2π,5-dien-7-yl-NiC_5H_5, 1, 643

$NiC_{28}H_{30}$ $[(PhMe_2C)C_5H_4]_2Ni$, 1, 653

$NiC_{28}H_{35}ClP_2$ $[PhC\equiv CNi(PEt_2Ph)_2]Cl$, 1, 633

$NiC_{28}H_{40}P_2$ $[(PhC\equiv C)_2Ni(PEt_3)_2]$, 1, 635

$NiC_{29}H_{24}F_7IP_2$ $[C_3F_7Ni(\text{-}CH_2PPh_2)_2]I$, 1, 633

$NiC_{29}H_{25}P$ $C_5H_5Ni(PPh_3)Ph$, 1, 640

$NiC_{29}H_{41}BrP_2$ [2,4,6-$Me_3C_6H_2Ni(PEt_2Ph)_2$]Br, 1, 633

$NiC_{29}H_{41}ClP_2$ [2,4,6-$Me_3C_6H_2Ni(PEt_2Ph)_2$]Cl, 1, 633

$NiC_{30}H_{24}F_8P_2$ $\overline{CF_2CF_2CF_2CF_2Ni}(\text{-}CH_2PPh_2)_2$, 1, 638

$NiC_{30}H_{37}BrCl_2P_2$ $[1\text{-}C_{10}H_7Ni(PEt_2Ph)_2]BrCl_2$, 1, 634

$NiC_{30}H_{37}BrP_2$ $[1\text{-}C_{10}H_7Ni(PEt_2Ph)_2]Br$, 1, 633

$NiC_{30}H_{37}Br_3P_2$ $[1\text{-}C_{10}H_7Ni(PEt_2Ph)_2]Br_3$, 1, 633

$NiC_{30}H_{37}IP_2$ $[1\text{-}C_{10}H_7Ni(PEt_2Ph)_2]I$, 1, 633

$NiC_{30}H_{37}NO_2P_2$ $[1\text{-}C_{10}H_7Ni(PEt_2Ph)_2]NO_2$, 1, 634

$NiC_{30}H_{52}P_2$ $[Ph_2Ni(PPr_3)_2]$, 1, 635

[(2,4,6-$Me_3C_6H_2)_2Ni(PEt_3)_2$], 1, 635

$NiC_{31}H_{25}P$ $C_5H_5Ni(PPh_3)C\equiv CPh$, 1, 640

$NiC_{31}H_{37}NP_2S$ $[1\text{-}C_{10}H_7Ni(PEt_2Ph)_2]SCN$, 1, 634

$NiC_{32}H_{26}O_4$ Tetraphenylcyclobutadiene-$Ni(OAc)_2$, 1, 658

$NiC_{32}H_{28}N_4$ $(p\text{-}MeC_6H_4NC)_4Ni$, 1, 663

$NiC_{32}H_{39}BrP_2$ $[2\text{-}PhC_6H_4Ni(PEt_2Ph)_2]Br$, 1, 634

$NiC_{32}H_{64}N_6P_2$ $(C_2H_4)Ni\{P[\overline{N(CH_2)_4CH_2}]_3\}_2$, 1, 657

$NiC_{34}H_{28}N_6$ $(PhCH_2NC)_4Ni(CN)_2$, 1, 663

$NiC_{34}H_{39}BrP_2$ [9-Phenanthryl-$Ni(PEt_2Ph)_2$]Br, 1, 634

$NiC_{34}H_{44}P_2$ $[(o\text{-}MeC_6H_4)_2Ni(PEt_2Ph)_2]$, 1, 635

$NiC_{34}H_{52}P_2$ $[(PhC\equiv C)_2Ni(PPr_3)_2]$, 1, 635

$NiC_{36}H_{36}N_4O_4$ $(p\text{-}EtOC_6H_4NC)_4Ni$, 1, 663

$NiC_{36}H_{40}P_2$ $[(PhC\equiv C)_2Ni(PEt_2Ph)_2]$, 1, 635

$NiC_{37}H_{62}O_3$ (1,5-Cyclooctadiene)(vitamine E quinone)-Ni, 1, 659

$NiC_{38}H_{34}P_2$ $(C_2H_4)Ni(PPh_3)_2$, 1, 657

$NiC_{38}H_{37}BrP_2$ $[1\text{-}C_{10}H_7Ni(PEtPh_2)_2]Br$, 1, 634

$NiC_{38}H_{52}P_2$ [(2,4,6-$Me_3C_6H_2)_2Ni(PEt_2Ph)_2$], 1, 635

$NiC_{38}H_{70}P_2$ $(C_2H_4)Ni[P(C_6H_{11})_3]_2$, 1, 657

$NiC_{39}H_{70}BrP$ Indenyl-$Ni[P(C_{10}H_{21})_3]Br$, 1, 639

$NiC_{40}H_{36}P_2$ $(MeC\equiv CMe)Ni(PPh_3)_2$, 1, 661

$NiC_{42}H_{36}$ Tris(stilbene)-Ni, 1, 661
$NiC_{42}H_{36}N_2P_2$ $(CH_2{=}CHCN)_2Ni(PPh_3)_2$, 1, 657
$NiC_{42}H_{38}O_2P_2$ $(CH_2{=}CHCHO)_2Ni(PPh_3)_2$, 1, 657
$NiC_{44}H_{28}N_4$ $(\beta\text{-}C_{10}H_7NC)_4Ni$, 1, 663
$NiC_{44}H_{38}P_2$ $(PhCH{=}CH_2)Ni(PPh_3)_2$, 1, 657
$NiC_{44}H_{40}P_2$ $[(PhC{\equiv}C)_2Ni(PEtPh_2)_2]$, 1, 635
$NiC_{44}H_{42}P_2$ 1,5-Cyclooctadiene-Ni-$(PPh_3)_2$, 1, 657
$NiC_{44}H_{46}O_{12}P_2$ $(C_2H_4)Ni[P(OC_6H_4\text{-}OMe\text{-})_3]_2$, 1, 657
$NiC_{44}H_{48}P_2$ $[(2\text{-}PhC_6H_4)_2Ni(PEt_2Ph)_2]$, 1, 635
$NiC_{45}H_{40}P_2$ $[PhC(Me){=}CH_2]Ni(PPh_3)_2$, 1, 657
$NiC_{46}H_{37}BrP_2$ $[1\text{-}C_{10}H_7Ni(PPh_3)_2]Br$, 1, 634
$NiC_{46}H_{55}P_3$ $[(PhC{\equiv}C)_2Ni(PEt_2Ph)_3]$, 1, 635
$NiC_{50}H_{40}P_2$ $(PhC{\equiv}CPh)Ni(PPh_3)_2$, 1, 661
$NiC_{50}H_{42}P_2$ $(PhCH{=}CHPh)Ni(PPh_3)_2$, 1, 657
$NiC_{52}H_{40}P_2$ $[(PhC{\equiv}C)_2Ni(PPh_3)_2]$, 1, 635
$NiC_{58}H_{40}O_2$ $[PhC{=}C(Ph)C(Ph){=}C(Ph)CO]_2$-Ni, 1, 659
$NiC_{74}H_{58}O_6P_2$ $(C_2H_4)Ni[P(OC_6H_4Ph\text{-}o)_3]_2$ 1, 657
$Ni_2C_6H_{10}Br_2$ $(\pi\text{-}CH_2{=}CHCH_2NiBr)_2$, 1, 654
$Ni_2C_6H_{10}Cl_2$ $(\pi\text{-}CH_2{=}CHCH_2NiCl)_2$, 1, 654
$Ni_2C_6H_{10}I_2$ $(\pi\text{-}CH_2{=}CHCH_2NiI)_2$, 1, 654
$Ni_2C_{12}H_{10}O_2$ $[C_5H_5Ni(CO)]_2$, 1, 645
$Ni_2C_{12}H_{12}$ $HC{\equiv}CH(NiC_5H_5)_2$, 1, 647
$Ni_2C_{12}H_{16}S_2$ $(C_5H_5NiSMe)_2$, 1, 649
$Ni_2C_{13}H_{14}$ $MeC{\equiv}CH(NiC_5H_5)_2$, 1, 647
$Ni_2C_{14}H_{10}F_6$ $CF_3C{\equiv}CCF_3(NiC_5H_5)_2$, 1, 647
$Ni_2C_{14}H_{14}O_2$ $[(MeC_5H_4)Ni(CO)]_2$, 1, 646
$MeO_2CC{\equiv}CH(NiC_5H_5)_2$, 1, 647
$Ni_2C_{14}H_{16}$ $MeC{\equiv}CMe(NiC_5H_5)_2$, 1, 647
$Ni_2C_{14}H_{20}S_2$ $(C_5H_5NiSEt)_2$, 1, 649
$Ni_2C_{14}H_{22}P_2$ $(C_5H_5NiPMe_2)_2$, 1, 649
$Ni_2C_{15}H_{16}O_2$ $EtO_2CC{\equiv}CH(NiC_5H_5)_2$, 1, 647
$Ni_2C_{15}H_{18}$ $PrC{\equiv}CH(NiC_5H_5)_2$, 1, 647
$Ni_2C_{16}H_{16}O_4$ $MeO_2CC{\equiv}CCO_2Me(NiC_5H_5)_2$, 1, 647
$Ni_2C_{16}H_{20}$ $EtC{\equiv}CEt(NiC_5H_5)_2$, 1, 647
$BuC{\equiv}CH(NiC_5H_5)_2$, 1, 648
$Ni_2C_{18}H_{16}$ $PhC{\equiv}CH(NiC_5H_5)_2$, 1, 648
$Ni_2C_{20}H_{24}O_4$ $Me_2CHO_2CC{\equiv}CCO_2CHMe_2$-$(NiC_5H_5)_2$, 1, 648
$Ni_2C_{20}H_{28}$ $PrC{\equiv}CPr[Ni(C_5H_4Me)]_2$, 1, 648
$Ni_2C_{22}H_{14}F_6$ $CF_3C{\equiv}CCF_3$(indenyl-Ni)$_2$, 1, 648
$Ni_2C_{22}H_{20}S_2$ $(C_5H_5NiSPh)_2$, 1, 747
$Ni_2C_{24}H_{20}$ $PhC{\equiv}CPh(NiC_5H_5)_2$, 1, 648
$Ni_2C_{24}H_{20}N_2$ $(C_5H_5)_2(PhNC)_2Ni_2$, 1, 650
$Ni_2C_{24}H_{24}$ $EtC{\equiv}CEt$(indenyl-Ni)$_2$, 1, 648
$Ni_2C_{26}H_{20}$ $PhC{\equiv}CC{\equiv}CPh(NiC_5H_5)_2$, 1, 648
$Ni_2C_{32}H_{20}$ $Ni[Ni(C{\equiv}CPh)_4]$, 1, 637
$Ni_2C_{34}H_{30}P_2$ $(C_5H_5NiPPh_2)_2$, 1, 649
$Ni_2C_{36}H_{36}N_4$ $[Ni(H_2NCH_2CH_2NH_2)_2]$-$[Ni(C{\equiv}CPh)_4]$, 1, 637
$Ni_2C_{44}H_{30}Br_2O_2$ $[\pi\text{-}PhCC(Ph)C(Ph)Ni(CO)$-$Br]_2$, 1, 654
$Ni_2C_{44}H_{30}Cl_2O_2$ $[\pi\text{-}PhCC(Ph)C(Ph)Ni(CO)$-$Cl]_2$, 1, 654
$Ni_3C_{17}H_{15}O_2$ $(C_5H_5Ni)_3(CO)_2$, 1, 649
$Ni_4C_{36}H_{30}$ $[(C_5H_5Ni)_2PhC{\equiv}C\text{-}]_2$, 1, 648
$NiAl_5C_{12}H_{12}Cl_{15}$ $(C_6H_6)_2Ni\cdot 5AlCl_3$, 1, 655
$NiAsC_9H_{15}O_3$ $Ni(AsEt_3)(CO)_3$, 3, 9
$NiAsC_{21}H_{15}O_3$ $Ni(AsPh_3)(CO)_3$, 3, 24
$NiAs_2C_{12}H_{16}O_2$ $Ni(2\text{-}Me_2AsC_6H_4AsMe_2)$-$(CO)_2$, 3, 88
$NiAs_2C_{14}H_{30}O_2$ $Ni(AsEt_3)_2(CO)_2$, 3, 9
$NiAs_2C_{28}H_{24}O_2$ $Ni(Ph_2AsCH_2\text{-})_2(CO)_2$, 3, 102
$NiAs_2C_{32}H_{24}O_2$ $Ni(2\text{-}Ph_2AsC_6H_4AsPh_2)$-$(CO)_2$, 3, 103
$NiAs_2C_{38}H_{30}O_2$ $Ni(AsPh_3)_2(CO)_2$, 3, 24
$NiBC_{34}H_{30}$ $[(C_5H_5)_2Ni]BPh_4$, 1, 652
$NiBiC_9H_{15}O_3$ $Ni(BiEt_3)(CO)_3$, 3, 816
$NiCrC_{14}H_{16}N_6S_4$ $[(C_5H_5)_2Ni][Cr(NH_3)_2$-$(SCN)_4]$, 1, 652
$NiCr_2C_{21}H_{14}O_9$ $[(MeC_5H_4)Cr(CO)_3]_2Ni$-$(CO)_3$, 1, 94
$NiFeC_{13}H_{10}O_3$ $C_5H_5Ni(CO)_2Fe(CO)C_5H_5$,

1, 547

$NiFeC_{33}H_{25}Br_4$ $\{C_5H_5Ni[PhC{=}C(Ph)C(Ph){=}C(Ph)]\}FeBr_4$, 1, 658

$NiFe_2C_{17}H_{10}O_7$ $[C_5H_5Fe(CO)_2]_2Ni(CO)_3$, 1, 548

$NiGeC_7H_{10}O_3$ $[Et_2GeNi(CO)_3]$, 2, 142

$NiK_2C_8H_4$ $K_2[Ni(C{\equiv}CH)_4]$, 1, 637

$NiK_2C_8H_{10}N_2$ $K_2[Ni(C{\equiv}CH)_4]\cdot 2NH_3$, 1, 637

$NiK_2C_{12}H_{12}$ $K_2[Ni(C{\equiv}CMe)_4]$, 1, 637

$NiK_2C_{12}H_{18}N_2$ $K_2[Ni(C{\equiv}CMe)_4]\cdot 2NH_3$, 1, 637

$NiK_2C_{32}H_{20}$ $K_2[Ni(C{\equiv}CPh)_4]$, 1, 637

$NiK_4C_8H_4$ $K_4[Ni(C{\equiv}CH)_4]$, 1, 637

$NiMnC_{12}H_7O_6$ $(MeC_5H_4)Ni(CO)Mn(CO)_5$, 1, 185

$NiNa_2C_8H_{10}N_2$ $Na_2[Ni(C{\equiv}CH)_4]\cdot 2NH_3$, 1, 637

$NiNa_2C_{32}H_{20}$ $Na_2[Ni(C{\equiv}CPh)_4]$, 1, 637

$NiNa_2C_{32}H_{26}N_2$ $Na_2[Ni(C{\equiv}CPh)_4]\cdot 2NH_3$, 1, 637

$NiPbC_7H_8Cl_2O$ $MePbCl_2C_5H_5NiCO$, 2, 595

$NiPbC_8H_7I_3O_2$ $(AcC_5H_4)Ni(CO)PbI_3$, 1, 663

$NiPbC_8H_9Br_3O$ $(EtC_5H_4)Ni(CO)PbBr_3$, 1, 663

$NiPbC_{10}H_{15}BrO$ $(EtC_5H_4)Ni(CO)PbMe_2Br$, 1, 663, 2, 595

$NiPbC_{15}H_{14}I_2O_2$ $(AcC_5H_4)Ni(CO)Pb(C_6H_4Me\text{-}o)I_2$, 1, 663, 2, 595

$NiSbC_7H_{10}ClO_3$ $Ni(Et_2SbCl)(CO)_3$, 3, 684

$NiSbC_9H_{15}O_3$ $Ni(SbEt_3)(CO)_3$, 3, 656

$NiSbC_{21}H_{15}O_3$ $Ni(SbPh_3)(CO)_3$, 3, 664

$NiSb_2C_{10}H_{20}Cl_2O_2$ $Ni(Et_2SbCl)_2(CO)_2$, 3, 684

$NiSb_2C_{12}H_{25}ClO_2$ $Ni(SbEt_3)(Et_2SbCl)(CO)_2$, 3, 657, 684

$NiSb_2C_{14}H_{30}O_2$ $Ni(SbEt_3)_2(CO)_2$, 3, 657

$NiSb_2C_{38}H_{30}O_2$ $Ni(SbPh_3)_2(CO)_2$, 3, 664

$NiSb_4C_{72}H_{60}$ $Ni(SbPh_3)_4$, 3, 664

$NiSi_2C_{24}H_{30}$ $\{[Me_2(C_5H_5)Si]C_5H_4\}_2Ni$, 1, 653

$NiSnC_{10}H_{13}F_3O$ $(Et_2C_5H_3)Ni(CO)SnF_3$, 1, 663

$NiSnC_{10}H_{16}O$ $(MeC_5H_4)Ni(CO)SnMe_3$, 2, 503

$NiSnC_{17}H_{14}O$ Indenyl-$Ni(CO)Sn(C_6H_4\text{-}Me\text{-}o)F_2$, 2, 503

$NiSnC_{18}H_{15}O$ $C_5H_5Ni(CO)SnPh_2$, 2, 503

$NiSnC_{38}H_{31}FO$ $(C_{13}H_9)_2SnF(Et_2C_5H_3)NiCO$, 2, 503

$Ni_2As_2C_{14}H_{22}$ $(C_5H_5NiAsMe_2)_2$, 1, 649, 3, 304

$Ni_2Co_2C_{32}H_{20}O_6$ $(C_5H_5Ni)_2(PhC{\equiv}CC{\equiv}CPh)[Co(CO)_3]_2$, 1, 618

$Ni_2FeC_{21}H_{16}O_3$ $[(C_5H_5Ni)_2(PhC{\equiv}CH)]Fe(CO)_3$, 1, 547

$Ni_2FeC_{27}H_{20}O_3$ $[(C_5H_5Ni)_2(PhC{\equiv}CPh)]Fe(CO)_3$, 1, 547

$Ni_2FeC_{29}H_{20}O_3$ $(C_5H_5Ni)_2(PhC{\equiv}CC{\equiv}CPh)Fe(CO)_3$, 1, 548

$Ni_2FeC_{29}H_{22}O_3$ $[C_5H_5Ni(PhC{\equiv}CH)]_2Fe(CO)_3$, 1, 548

$Ni_2Fe_2C_{24}H_{16}O_6$ $[(C_5H_5Ni)_2(PhC{\equiv}CH)][Fe(CO)_3]_2$, 1, 548

$Ni_2Fe_2C_{30}H_{20}O_6$ $[(C_5H_5Ni)_2(PhC{\equiv}CPh)][Fe(CO)_3]_2$, 1, 548

$Ni_2K_6C_{16}H_8$ $K_6[Ni(C{\equiv}CH)_4]_2$, 1, 637

$Ni_2K_6C_{24}H_{24}$ $K_6[Ni(C{\equiv}CMe)_4]_2$, 1, 637

$Ni_2K_6C_{64}H_{40}$ $K_6[Ni(C{\equiv}CPh)_4]_2$, 1, 637

$Ni_2PbC_{21}H_{29}ClO_2$ $MePbCl(C_5H_3Et_2NiCO)_2$, 2, 595

$Ni_2SnC_{20}H_{28}O_2$ $Pr_2Sn(MeC_5H_4NiCO)_2$, 2, 503

$Ni_4CrC_{21}H_{13}O_9$ $[(C_6H_6)_2Cr][Ni_4(CO)_9H]$, 1, 95

$OsC_{10}H_{10}$ $(C_5H_5)_2Os$, 1, 560

$OsC_{10}H_{11}F_6OP$ $[(C_5H_5)_2Os(OH)]PF_6$, 1, 560

$OsC_{11}H_{10}O_3$ (Methylcycloheptatriene)-$Os(CO)_3$, 1, 563

$OsC_{11}H_{12}O_3$ (3-Methyl-2,4,6-heptatriene)-$Os(CO)_3$, 1, 562

$OsC_{12}H_{12}$ $C_5H_5OsC_5H_4CH{=}CH_2$, 1, 561

$OsC_{12}H_{12}O$ $C_5H_5OsC_5H_4Ac$, 1, 561

$OsC_{12}H_{14}$ $C_5H_5OsC_5H_4Et$, 1, 561

$(C_6H_6)Os(CH{=}CHCH{=}CHCH_2CH_2)$, 1, 562

$OsC_{12}H_{14}O$ $C_5H_5Os[C_5H_4CH(OH)Me]$, 1, 561

$OsC_{13}H_{14}O_3$ (Trimethylcycloheptatriene) $Os(CO)_3$, 1, 563

$OsC_{14}H_{10}O_4$ $[C_5H_5Os(CO)_2]_2$, 1, 560

$OsC_{14}H_{14}O_2$ $(AcC_5H_4)_2Os$, 1, 562
$OsC_{14}H_{15}Cl_3$ $(Cl_3CC_5H_5)(m\text{-}Me_2C_6H_4)Os$, 1, 562
$OsC_{14}H_{16}O_2$ $C_5H_5Os[C_5H_4CH(OAc)Me]$, 1, 561
$OsC_{17}H_{14}O$ $C_5H_5OsC_5H_4Bz$, 1, 561
$OsC_{17}H_{15}N_3$ $C_5H_5Os[C_5H_4CH(N_3)Ph]$, 1, 561
$OsC_{17}H_{16}$ $C_5H_5Os(C_5H_4CH_2Ph)$, 1, 562
$OsC_{17}H_{16}O$ $C_5H_5Os[C_5H_4CH(OH)Ph]$, 1, 562
$OsC_{18}H_{14}O_3$ (p-Tolylcyclooctatetraene)-$Os(CO)_3$, 1, 563
$OsC_{31}H_{42}O_4$ $[1,8\text{-}(Me_2CHO_2C)_2C_{10}H_6]$-$(C_8H_{17}C_5H_5)Os$, 1, 562
$OsC_{51}H_{47}ClP_4$ $MeOs[(Ph_2P)_2CH_2]_2Cl$, 1, 560
$OsC_{53}H_{51}ClP_4$ $MeOs[(Ph_2PCH_2\text{-})_2]_2Cl$, 1, 560
$OsC_{54}H_{53}ClP_4$ $EtOs[(Ph_2PCH_2\text{-})_2]_2Cl$, 1, 560
$OsC_{55}H_{52}ClP_4$ $EtOs[(Ph_2P)_2CH_2]_2Cl\cdot 1/2\ C_6H_6$, 1, 560
$Os_2C_{12}H_8O_6$ $[CH{=}C(Me)C(Me){=}CHOs(CO)_3]$-$Os(CO)_3$, 1, 562
$OsAs_3C_{55}H_{46}BrO$ $[OsH(AsPh_3)_3(CO)Br]$, 3, 24
$OsAs_3C_{55}H_{46}ClO$ $[OsH(AsPh_3)_3(CO)Cl]$, 3, 24
$OsB_2C_{66}H_{64}$ $[(1,3,5\text{-}Me_3C_6H_3)_2Os]$-$[BPh_4]_2$, 1, 562
$OsGeC_{13}H_{14}O_2$ $C_5H_5Os(CO)_2Ge(CH{=}CH_2)_3$, 1, 563, 2, 141
$OsGeC_{13}H_{20}O_2$ $(MeC_5H_4)Os(CO)_2GeMe_2Pr$, 2, 141
$OsGeC_{35}H_{52}O_5$ $(BuC_5H_4)Os(CO)_2Ge(C_6H_{10}\text{-}Ac)_3$, 1, 563, 2, 141
$OsSiC_{13}H_{18}$ $C_5H_5OsC_5H_4SiMe_3$, 1, 563
$PbCH_3I_3$ $MePbI_3$, 2, 562
$PbCH_6O_3$ $MePb(OH)_3$, 2, 568
$PbC_2H_5Cl_3$ $EtPbCl_3$, 2, 562
$PbC_2H_5I_3$ $EtPbI_3$, 2, 562
$PbC_2H_6Br_2$ Me_2PbBr_2, 2, 560
$PbC_2H_6Cl_2$ Me_2PbCl_2, 2, 560
PbC_2H_6O Me_2PbO, 2, 568
PbC_2H_8 Me_2PbH_2, 2, 551
$PbC_2H_8O_3$ $EtPb(OH)_3$, 2, 568
$PbC_3H_7I_3$ $PrPbI_3$, 2, 562
$PbC_3H_8O_3$ $CH_2{=}CHCH_2Pb(OH)_3$, 2, 569
PbC_3H_9Br Me_3PbBr, 2, 552
PbC_3H_9Cl Me_3PbCl, 2, 552
PbC_3H_9F Me_3PbF, 2, 555
PbC_3H_9I Me_3PbI, 2, 555
$PbC_3H_9NO_3$ Me_3PbNO_3, 2, 564
PbC_3H_{10} Me_3PbH, 2, 550
$PbC_3H_{10}O$ Me_3PbOH, 2, 568
$PbC_3H_{10}O_3$ $PrPb(OH)_3$, 2, 568
i-$PrPb(OH)_3$, 2, 562
PbC_4F_{12} $(CF_3)_4Pb$ (?), 2, 531
$PbC_4H_4Cl_2O_4S$ $(ClCH{=}CH)_2PbSO_4$, 2, 565
$PbC_4H_4Cl_4$ $(ClCH{=}CH)_2PbCl_2$, 2, 561
$PbC_4H_6Br_2$ $(CH_2{=}CH)_2PbBr_2$, 2, 561
$PbC_4H_6Cl_2$ $(CH_2{=}CH)_2PbCl_2$, 2, 561
$PbC_4H_8Cl_2$ $Et(CH_2{=}CH)PbCl_2$, 2, 561
$PbC_4H_9I_3$ $BuPbI_3$, 2, 562
PbC_4H_{10} Et_2Pb, 2, 592
$PbC_4H_{10}Br_2$ Et_2PbBr_2, 2, 560
$PbC_4H_{10}Cl_2$ Et_2PbCl_2, 2, 559
$PbC_4H_{10}N_2O_6$ $Et_2Pb(NO_3)_2$, 2, 565
$PbC_4H_{10}O$ Et_2PbO, 2, 568
$PbC_4H_{10}O_2$ Me_3PbO_2CH, 2, 574
$PbC_4H_{10}O_3S$ Et_2PbSO_3, 2, 565
$PbC_4H_{10}O_3Se$ Et_2PbSeO_3, 2, 565
$PbC_4H_{10}O_4S$ Et_2PbSO_4, 2, 565
$PbC_4H_{11}Br$ $Me_2EtPbBr$, 2, 555
$PbC_4H_{11}Cl$ $Me_2EtPbCl$, 2, 555
PbC_4H_{12} Me_4Pb, 2, 519
Et_2PbH_2, 2, 551
$PbC_4H_{12}O_2$ $Et_2Pb(OH)_2$, 2, 567
$PbC_4H_{12}O_3$ $BuPb(OH)_3$, 2, 569
$PbC_4H_{12}S$ Me_3PbSMe, 2, 584
$PbC_4H_{12}Se$ $Me_3PbSeMe$, 2, 585
$PbC_5H_9Cl_3O_2$ $Me_3PbO_2CCCl_3$, 2, 577
$PbC_5H_9F_5$ $Me_3PbC_2F_5$, 2, 542
$PbC_5H_{10}O_3$ Et_2PbCO_3, 2, 580
$PbC_5H_{11}ClO_2$ $Me_3PbO_2CCH_2Cl$, 2, 577
PbC_5H_{12} $Me_3PbCH{=}CH_2$, 2, 537
$PbC_5H_{12}O_2$ Me_3PbOAc, 2, 574
$PbC_5H_{13}NOS$ $Me_3PbSCH_2CONH_2$, 2, 584
PbC_5H_{14} Me_3PbEt, 2, 533
$PbC_6H_5Br_3$ $PhPbBr_3$, 2, 562
$PbC_6H_5Cl_3$ $PhPbCl_3$, 2, 562
$PbC_6H_6O_2$ $PhPbO_2H$, 2, 569
$PbC_6H_8Cl_2S$ 2-Thienyl-$EtPbCl_2$, 2, 561
PbC_6H_9Cl $(CH_2{=}CH)_3PbCl$, 2, 557

$PbC_6H_{10}Cl_2$ $(\overline{CH_2CH_2C}H)_2PbCl_2$, 2, 560
$PbC_6H_{10}N_2$ $Et_2Pb(CN)_2$, 2, 563
$PbC_6H_{11}Cl$ $Et(CH_2{=}CH)_2PbCl$, 2, 557
$PbC_6H_{12}OS_2$ $PrPbS_2COEt$, 2, 585
$PbC_6H_{13}Cl$ $Et_2(CH_2{=}CH)PbCl$, 2, 557
$PbC_6H_{14}Br_2$ Pr_2PbBr_2, 2, 560
$PbC_6H_{14}Cl_2$ $Et_2Pb(CH_2Cl)_2$, 2, 542
Pr_2PbCl_2, 2, 560
$PbC_6H_{14}O_2S$ $Me_3PbSCH_2CO_2Me$, 2, 584
$PbC_6H_{14}O_4S$ Pr_2PbSO_4, 2, 565
$PbC_6H_{15}Br$ Et_3PbBr, 2, 553
$PbC_6H_{15}Cl$ Et_3PbCl, 2, 552
$PbC_6H_{15}I$ Et_3PbI, 2, 555
$PbC_6H_{15}NOS$ $Me_2EtPbSCH_2CONH_2$, 2, 584
$PbC_6H_{15}NO_3$ Et_3PbNO_3, 2, 564
PbC_6H_{16} $Me_3PbCHMe_2$, 2, 533
Me_2PbEt_2, 2, 533
Et_3PbH, 2, 551
$PbC_6H_{16}O$ Et_3PbOH, 2, 567
$PbC_6H_{17}O_4P$ $Et_3PbH_2PO_4$, 2, 564
$PbC_7H_7Br_3O$ p-$MeOC_6H_4PbBr_3$, 2, 562
$PbC_7H_8O_2$ p-$MeC_6H_4PbO_2H$, 2, 569
$PbC_7H_8O_3$ p-$MeOC_6H_4PbO_2H$, 2, 569
$PbC_7H_{10}O_3$ $PhCH_2Pb(OH)_3$, 2, 568
$PbC_7H_{15}N$ Et_3PbCN, 2, 563
$PbC_7H_{15}NO$ Et_3PbNCO, 2, 563
$PbC_7H_{15}NS$ Et_3PbNCS, 2, 563
$PbC_7H_{15}NSe$ $Et_3PhNCSe$, 2, 563
$PbC_7H_{17}Br$ $Et_2PrPbBr$, 2, 555
$PbC_7H_{17}Cl$ Et_3PbCH_2Cl, 2, 542
$Et_2PrPbCl$, 2, 555
$PbC_7H_{17}ClOS$ $Me_3PbCl\cdot O\overline{S(CH_2)_3C}H_2$, 2, 552
PbC_7H_{18} Me_3PbBu, 2, 533
$Me_3PbCHMeEt$, 2, 533
Me_3PbCMe_3, 2, 533
$MePbEt_3$, 2, 534
$PbC_7H_{18}ClNO$ $Me_3PbCl\cdot AcNMe_2$, 2, 552
$PbC_7H_{18}O$ Et_3PbOMe, 2, 571
$PbC_7H_{18}O_2$ $Me_3PbO_2CMe_3$, 2, 570
$PbC_7H_{19}NO_2S$ $Et_3PbNHSO_2Me$, 2, 586
PbC_8D_{20} $(C_2D_5)_4Pb$, 2, 528
PbC_8F_{12} $(CF_2{=}CF)_4Pb$, 2, 531
$PbC_8H_8Cl_2$ $Ph(CH_2{=}CH)PbCl_2$, 2, 561
$PbC_8H_9Cl_3O_2$ $(CH_2{=}CH)_3PbO_2CCCl_3$, 2, 577
$PbC_8H_{10}Cl_2$ $EtPhPbCl_2$, 2, 560
$PbC_8H_{10}Cl_2O_4$ $(ClCH{=}CH)_2Pb(OAc)_2$, 2, 580
$PbC_8H_{10}Cl_6O_4$ $Et_2Pb(O_2CCCl_3)_2$, 2, 581
PbC_8H_{12} $(CH_2{=}CH)_4Pb$, 2, 530
$PbC_8H_{12}O_2$ $(CH_2{=}CH)_3PbOAc$, 2, 574
$PbC_8H_{13}ClS$ 2-Thienyl-Et_2PbCl, 2, 557
PbC_8H_{14} $Me_3PbC_5H_5$, 2, 537
$EtPb(CH{=}CH_2)_3$, 2, 538
$PbC_8H_{14}Cl_2O_4$ $Et_2Pb(O_2CCH_2Cl)_2$, 2, 581
$PbC_8H_{15}Cl_3O_2$ $Et_3PbO_2CCCl_3$, 2, 577
PbC_8H_{16} $Et_2Pb(CH{=}CH_2)_2$, 2, 538
$Et_3PbC{\equiv}CH$, 2, 540
$PbC_8H_{16}O_2S_2$ $Et_2Pb(SAc)_2$, 2, 585
$PbC_8H_{16}O_4$ $Et_2Pb(OAc)_2$, 2, 580
$PbC_8H_{17}BrO_2$ $Et_3PbO_2CCH_2Br$, 2, 577
$PbC_8H_{17}ClO_2$ $Et_3PbO_2CCH_2Cl$, 2, 577
$PbC_8H_{17}FO_2$ $Et_3PbO_2CCH_2F$, 2, 577
PbC_8H_{18} $Et_3PbCH{=}CH_2$, 2, 537, 597
$PbC_8H_{18}Cl_2$ Bu_2PbCl_2, 2, 560
$PbC_8H_{18}OS$ Et_3PbSAc, 2, 585
$PbC_8H_{18}O_2$ i-$BuCO_2PbMe_3$, 2, 574
Et_3PbOAc, 2, 574
$PbC_8H_{18}O_3$ Et_3PbO_2Ac (?), 2, 570
PbC_8H_{20} Et_4Pb, 2, 521
i-Pr_2PbMe_2, 2, 533
$PbC_8H_{20}N_3O_5$ $[Et_4Pb(NO)_2]NO_3$, 2, 566
$PbC_8H_{20}O$ Et_3PbOEt, 2, 571
$PbC_8H_{20}O_3$ $Me_3PbO(CH_2CH_2O)_2Me$, 2, 571
$PbC_8H_{20}S$ Et_3PbSEt, 2, 585
PbC_9H_{14} Me_3PbPh, 2, 533
$PbC_9H_{14}Se$ $Me_3PbSePh$, 2, 585
$PbC_9H_{15}Cl$ $(\overline{CH_2CH_2C}H)_3PbCl$, 2, 556
PbC_9H_{16} $Me_3PbC_5H_4Me$, 2, 537
PbC_9H_{18} $Et_3PbC{\equiv}CMe$, 2, 540
$PbC_9H_{18}O_2$ $Et_3PbO_2CCH{=}CH_2$, 2, 576
$PbC_9H_{19}ClO_2$ $Et_3PbO_2CCH_2CH_2Cl$, 2, 577
$Et_2PrPbO_2CCH_2Cl$, 2, 577
PbC_9H_{20} $Et_3PbCH_2CH{=}CH_2$, 2, 537
$PbC_9H_{20}O_2$ Et_3PbO_2CEt, 2, 574
$PbC_9H_{21}Br$ Pr_3PbBr, 2, 556
$PbC_9H_{21}Cl$ Pr_3PbCl, 2, 556
$PbC_9H_{21}NOS$ Et_2(i-Pr)$PbSCH_2CONH_2$, 2, 585
PbC_9H_{22} $MeEt_2PbBu$, 2, 534
$MeEt_2PhCHMeEt$, 2, 534
$PbC_9H_{22}O$ Pr_3PbOH, 2, 568
$PbC_{10}H_8Cl_2S$ 2-Thienyl-$PhPbCl_2$, 2, 561

$PbC_{10}H_8O_2$ 1-$C_{10}H_7PbO_2H$, 2, 569
2-$C_{10}H_7PbO_2H$, 2, 569
$PbC_{10}H_{10}$ $(C_5H_5)_2Pb$, 2, 592
$PbC_{10}H_{10}Cl_6O_4$ $(CH_2CH_2CH)_2Pb(O_2CCCl_3)_2$, 2, 582
$PbC_{10}H_{11}Cl$ $(Ph(CH_2{=}CH)_2PbCl$, 2, 557
$PbC_{10}H_{11}ClS_2$ 2-$Thienyl_2EtPbCl$, 2, 557
$PbC_{10}H_{12}Cl_4O_4$ $(CH_2CH_2CH)_2Pb(O_2CCHCl_2)_2$ 2, 582
$PbC_{10}H_{14}Cl_2O_4$ $(CH_2CH_2CH)_2Pb(O_2CCH_2Cl)_2$ 2, 582
$PbC_{10}H_{15}Cl$ $Et_2PhPbCl$, 2, 555
$PbC_{10}H_{16}O_4$ $(CH_2CH_2CH)_2Pb(OAc)_2$, 2, 580
$PbC_{10}H_{17}Cl_2NO_6$ $(ClCH{=}CH)_2Pb(OAc)_2 \cdot$-$NH_4OAc$, 2, 580
$PbC_{10}H_{18}$ $Et_3PbC{\equiv}CCH{=}CH_2$, 2, 540
$PbC_{10}H_{18}S$ 2-Thienyl-$PbEt_3$, 2, 536
$PbC_{10}H_{19}ClO_2$ $Et_3PbO_2CCCl{=}CHMe$, 2, 576
$PbC_{10}H_{20}$ $Et_3PbC{\equiv}CEt$, 2, 540
$PbC_{10}H_{20}O_2$ $Et_3PbO_2CCMe{=}CH_2$, 2, 576
$PbC_{10}H_{20}O_4$ $Et_3PbO_2CCO_2Et$, 2, 578
$Pr_2Pb(OAc)_2$, 2, 580
$PbC_{10}H_{21}ClO_2$ $Et_2BuPbO_2CCH_2Cl$, 2, 577
$PbC_{10}H_{21}N$ Pr_3PbCN, 2, 563
$PbC_{10}H_{22}$ $Et_3PbCH_2CMe{=}CH_2$, 2, 537
$Et_3PbCH{=}CMe_2$, 2, 537
$PbC_{10}H_{22}O_2$ Et_2PrPbO_2CEt, 2, 574
$PbC_{10}H_{23}ClOS$ $Et_3PbCl \cdot OS(CH_2)_3CH_2$, 2, 553
$PbC_{10}H_{24}$ $Me_2Pb(CH_2CHMe_2)_2$, 2, 533
$Me_2Pb(CHMeEt)_2$, 2, 533
$Et_3PbCHMeEt$, 2, 534
Et_2PbPr_2, 2, 534
$PbC_{10}H_{24}O_2$ $Et_3PbO_2CMe_3$, 2, 570
$PbC_{10}H_{25}N$ $Et_3PbNHCH_2CHMe_2$, 2, 586
$PbC_{10}H_{25}NO_2S$ $Pr_3PbNHSO_2Me$, 2, 587
$PbC_{11}H_{15}Cl_3O_2$ $(CH_2CH_2CH)_3PbO_2CCCl_3$, 2, 577

$PbC_{11}H_{16}$ p-$CH_2{=}CHC_6H_4PbMe_3$, 2, 537, 597

$PbC_{11}H_{18}$ $MeEt_2PbPh$, 2, 534
$PbC_{11}H_{18}O_3$ Et_3Pb 2-furanecarboxylate, 2, 576

$PbC_{11}H_{20}$ $Et_3PbC_5H_5$, 2, 537
$Et_3PbC{\equiv}CCMe{=}CH_2$, 2, 540
$PbC_{11}H_{21}Cl_3O_2$ $Pr_3PbO_2CCCl_3$, 2, 577
$PbC_{11}H_{23}BrO_2$ $Pr_3PbO_2CCH_2Br$, 2, 577
$PbC_{11}H_{23}ClO_2$ $Pr_3PbO_2CCH_2Cl$, 2, 577
$PbC_{11}H_{23}IO_2$ $Pr_3PbO_2CCH_2I$, 2, 577
$PbC_{11}H_{24}O_2$ $Et_3PbO_2CCH_2CHMe_2$, 2, 574
Pr_3PbOAc, 2, 574
$PbC_{11}H_{25}NOS$ $Me_2(n\text{-}C_7H_{15})PbSCH_2CONH_2$, 2, 584
$PbC_{11}H_{26}O_3$ $Et_3PbO(CH_2CH_2O)_2Me$, 2, 571
$PbC_{12}H_8Br_2N_2O_4$ $(m\text{-}O_2NC_6H_4)_2PbBr_2$, 2, 561
$PbC_{12}H_8Cl_2N_2O_4$ $(m\text{-}O_2NC_6H_4)_2PbCl_2$, 2, 561
$PbC_{12}H_8I_2N_2O_4$ $(m\text{-}O_2NC_6H_4)_2PbI_2$, 2, 561
$PbC_{12}H_8N_2O_5$ $(O_2NC_6H_4)_2PbO$, 2, 569
$PbC_{12}H_8N_4O_{10}$ $(m\text{-}O_2NC_6H_4)_2Pb(NO_3)_2$, 2, 566
$PbC_{12}H_{10}$ Ph_2Pb, 2, 592
$PbC_{12}H_{10}Br_2$ Ph_2PbBr_2, 2, 560
$PbC_{12}H_{10}Cl_2$ Ph_2PbCl_2, 2, 559
$PbC_{12}H_{10}Cl_2O_4S_2$ 2-$Thienyl_2Pb(O_2CCH_2$-$Cl)_2$, 2, 582
$PbC_{12}H_{10}F_2$ Ph_2PbF_2, 2, 560
$PbC_{12}H_{10}I_2$ Ph_2PbI_2, 2, 561
$PbC_{12}H_{10}N_2O_6$ $Ph_2Pb(NO_3)_2$, 2, 565
$PbC_{12}H_{10}N_6$ $Ph_2Pb(N_3)_2$, 2, 563
$PbC_{12}H_{10}O$ Ph_2PbO, 2, 569
$PbC_{12}H_{10}S$ Ph_2PbS, 2, 585
$PbC_{12}H_{12}O_2$ $Ph_2Pb(OH)_2$, 2, 569
$PbC_{12}H_{14}$ $PhPb(CH{=}CH_2)_3$, 2, 539
$(C_5H_4Me)_2Pb$, 2, 592
$PbC_{12}H_{14}N_2O_2$ $(H_2NC_6H_4)_2Pb(OH)_2$, 2, 569
$PbC_{12}H_{14}O_6$ $PhPb(OAc)_3$, 2, 583
$PbC_{12}H_{15}I_3N_2$ $EtPbI_3 \cdot 2C_5H_5N$, 2, 562
$PbC_{12}H_{16}$ $Me_2Pb(C_5H_5)_2$, 2, 537
$PbC_{12}H_{16}S_2$ 2-$Thienyl_2PbEt_2$, 2, 536
$PbC_{12}H_{18}$ p-$Me_3PbC_6H_4CMe{=}CH_2$, 2, 537
$PbC_{12}H_{19}Br$ p-$BrC_6H_4PbEt_3$, 2, 542
$PbC_{12}H_{20}$ $(CH_2CH_2CH)_4Pb$, 2, 528
$(CH_2{=}CHCH_2)_4Pb$ (?), 2, 531
$(CH_2{=}CMe)_4Pb$, 2, 531
Et_3PbPh, 2, 534
$PbC_{12}H_{20}S$ Et_3PbSPh, 2, 585

$PbC_{12}H_{21}NO_2S$ $Et_3PbNHSO_2Ph$, 2, 586

$PbC_{12}H_{22}$ $Et_3PbC_5H_4Me$, 2, 537

$PbC_{12}H_{22}N_2O_2S$ p-$H_2NC_6H_4SO_2NHPbEt_3$, 2, 586

$PbC_{12}H_{23}NO_4S$ p-$H_2NC_6H_4SO_3PbEt_3 \cdot H_2O$, 2, 564

$PbC_{12}H_{24}O_2$ $Pr_3PbO_2CCH{=}CH_2$, 2, 576

$PbC_{12}H_{25}ClO_2$ $Pr_3PbO_2CCH_2CH_2Cl$, 2, 577

$PbC_{12}H_{26}O_2$ n-$C_5H_{11}CO_2PbEt_3$, 2, 574

Pr_3PbO_2CEt, 2, 574

$PbC_{12}H_{26}O_2S$ $C_6H_{11}SO_2PbEt_3$, 2, 564

$PbC_{12}H_{27}Br$ Bu_3PbBr, 2, 556

$PbC_{12}H_{27}Cl$ Bu_3PbCl, 2, 556

$(Me_2CHCH_2)_3PbCl$, 2, 556

$PbC_{12}H_{28}$ Pr_4Pb, 2, 525

i-Pr_4Pb, 2, 528

$PbC_{12}H_{28}N_3O_5$ $[Pr_4Pb(NO)_2]NO_3$, 2, 566

$PbC_{12}H_{28}O$ Bu_3PbOH, 2, 569

$PbC_{12}H_{29}N$ $Et_3PbCH_2CH_2NEt_2$, 2, 545

$PbC_{12}H_{29}O_4P$ $Et_3PbO_2P(OC_5H_{11})OMe$, 2, 564

$PbC_{13}H_{16}O_6$ p-$MeC_6H_4Pb(OAc)_3$, 2, 583

$PbC_{13}H_{16}O_7$ p-$MeOC_6H_4Pb(OAc)_3$, 2, 583

$PbC_{13}H_{19}BrO_2$ o-$BrC_6H_4CO_2PbEt_3$, 2, 577

m-$BrC_6H_4CO_2PbEt_3$, 2, 577

p-$BrC_6H_4CO_2PbEt_3$, 2, 577

$PbC_{13}H_{19}ClO_2$ p-$ClC_6H_4CO_2PbEt_3$, 2, 577

$PbC_{13}H_{19}IO_2$ o-$IC_6H_4CO_2PbEt_3$, 2, 577

m-$IC_6H_4CO_2PbEt_3$, 2, 577

p-$IC_6H_4CO_2PbEt_3$, 2, 577

$PbC_{13}H_{19}NO_3S$ $Et_3PbNOCC_6H_4SO_2$, 2, 586

$PbC_{13}H_{19}NO_4$ o-$Et_3PbO_2CC_6H_4NO_2$, 2, 578

m-$Et_3PbO_2CC_6H_4NO_2$, 2, 578

p-$Et_3PbO_2CC_6H_4NO_2$, 2, 578

$PbC_{13}H_{20}O_2$ Et_3PbOBz, 2, 574

$PbC_{13}H_{20}O_2S$ o-$HOC_6H_4COSPbEt_3$, 2, 585

$PbC_{13}H_{20}O_3$ Et_3PbO_2Bz (?), 2, 570

o-$HOC_6H_4CO_2PbEt_3$, 2, 578

$PbC_{13}H_{21}NO_2$ o-$Et_3PbO_2CC_6H_4NH_2$, 2, 578

$PbC_{13}H_{22}$ $PhCH_2PbEt_3$, 2, 534

1-Cyclopentenyl-$C{\equiv}CPbEt_3$, 2, 540

$PbC_{13}H_{22}O$ Et_3PbOCH_2Ph, 2, 571

$PbC_{13}H_{22}O_2S$ p-$MeC_6H_4SO_2PbEt_3$, 2, 564

$PbC_{13}H_{22}O_3S$ o-$MeC_6H_4SO_3PbEt_3$, 2, 564

p-$MeC_6H_4SO_3PbEt_3$, 2, 564

$PbC_{13}H_{23}NO_2S$ o-$MeC_6H_4SO_2NHPbEt_3$, 2, 586

p-$MeC_6H_4SO_2NHPbEt_3$, 2, 586

$PbC_{13}H_{23}NO_3S$ 5,2-$Me(H_2N)C_6H_3SO_3PbEt_3$, 2, 564

p-$MeC_6H_4NHSO_3PhEt_3$, 2, 564

$PbC_{13}H_{26}O_2$ $Pr_3PbO_2CCH{=}CHMe$, 2, 576

$PbC_{13}H_{28}O_2$ Pr_3PbO_2CPr, 2, 574

$PbC_{13}H_{31}N$ $Et_3Pb(CH_2)_3NEt_2$, 2, 545

$Et_3PbCHEtNEt_2$, 2, 545

$PbC_{14}H_8N_6O_{12}$ $[m\text{-}O_2NC_6H_4C(NO_2)_2]_2Pb$ (?), 2, 592

$PbC_{14}H_{11}ClS_2$ 2-$Thienyl_2PhPbCl$, 2, 557

$PbC_{14}H_{12}O_4$ $Ph_2Pb(O_2CH)_2$, 2, 580

$PbC_{14}H_{13}Cl$ $Ph_2(CH_2{=}CH)PbCl$, 2, 557

$PbC_{14}H_{14}Cl_2$ $(m\text{-}MeC_6H_4)_2PhCl_2$, 2, 561

$(p\text{-}MeC_6H_4)_2PbCl_2$, 2, 561

$PbC_{14}H_{14}Cl_2O_2$ $(o\text{-}MeOC_6H_4)_2PbCl_2$, 2, 561

$(p\text{-}MeOC_6H_4)_2PbCl_2$, 2, 561

$PbC_{14}H_{14}I_2O_2$ $(p\text{-}MeOC_6H_4)_2PbI_2$, 2, 561

$PbC_{14}H_{14}N_2O_6$ $(p\text{-}MeC_6H_4)_2Pb(NO_3)_2$, 2, 566

$PbC_{14}H_{14}O$ $(p\text{-}MeC_6H_4)_2PbO$, 2, 569

$PbC_{14}H_{14}S_3$ 2-$Thienyl_3PbEt$, 2, 536

$PbC_{14}H_{15}Cl$ $EtPh_2PbCl$, 2, 555

$PbC_{14}H_{15}Cl_4NO_2$ o-$Et_3PbNOCC_6Cl_4CO$, 2, 586

$PbC_{14}H_{18}O_8$ 2,4-$(MeO)_2C_6H_3Pb(OAc)_3$, 2, 583

2,5-$(MeO)_2C_6H_3Pb(OAc)_3$, 2, 583

$PbC_{14}H_{19}NO_2$ o-$Et_3PbNOCC_6H_4CO$, 2, 586

$PbC_{14}H_{20}$ $Me_2Pb(C_5H_4Me)_2$, 2, 537

$Et_2Pb(C_5H_5)_2$, 2, 538

$Et_3PbC{\equiv}CPh$, 2, 540

$PbC_{14}H_{20}N_2O_2$ o-$Et_3PbNNHOCC_6H_4CO$, 2, 586

$PbC_{14}H_{22}$ p-$Et_3PbC_6H_4CH{=}CH_2$, 2, 537, 597

$PbC_{14}H_{22}O_2$ $Et_3PbO_2CCH_2Ph$, 2, 574

$PbC_{14}H_{22}O_3$ 2-$(C_4H_3O)CH_2CH{=}CHCO_2PbEt_3$, 2, 576

o-$MeOC_6H_4CO_2PbEt_3$, 2, 578

$PbC_{14}H_{22}O_8$ $[(EtO_2C)_2CH]_2Pb$, 2, 592

$PbC_{14}H_{23}NO_2$ o-$Et_3PbO_2CC_6H_4NHMe$, 2, 578

$PbC_{14}H_{23}NO_2S$ $Et_3PbNPhSO_2CH{=}CH_2$, 2, 586

$PbC_{14}H_{24}$ $(MeCH{=}CHCH_2)_3PbCH{=}CH_2$, 2, 539

$PbC_{14}H_{25}N$ p-$Et_3PbC_6H_4NMe_2$, 2, 545

$PbC_{14}H_{25}NO_2S$ p-$MeC_6H_4SO_2NHPbPrEt_2$, 2, 586
$PbC_{14}H_{27}Cl_3O_2$ $Bu_3PbO_2CCCl_3$, 2, 577
$PbC_{14}H_{29}BrO_2$ $Bu_3PbO_2CCH_2Br$, 2, 577
$PbC_{14}H_{29}ClO_2$ $Bu_3PbO_2CCH_2Cl$, 2, 577
$PbC_{14}H_{29}IO_2$ $Bu_3PbO_2CCH_2I$, 2, 577
$PbC_{14}H_{30}O_2$ $Pr_3PbO_2CCH_2CHMe_2$, 2, 574
Bu_3PbOAc, 2, 575
$(Me_2CHCH_2)_3PbOAc$, 2, 575
$PbC_{14}H_{31}NOS$ $Bu_3PbSCH_2CONH_2$, 2, 585
$PbC_{15}H_{26}O$ $Et_3PbOCMe_2Ph$, 2, 571
$PbC_{15}H_{26}O_2$ $Et_3PbO_2CMe_2Ph$, 2, 570
$PbC_{15}H_{27}NO_2S$ $Pr_3PbNHSO_2Ph$, 2, 587
$PbC_{15}H_{28}N_2O_2S$ p-$H_2NC_6H_4SO_2NHPbPr_3$, 2, 587
$PbC_{15}H_{24}$ p-$Et_3PbC_6H_4CMe{=}CH_2$, 2, 537, 597
$PbC_{15}H_{20}O_2$ $Et_3PbO_2CC{\equiv}CPh$, 2, 576
$PbC_{15}H_{22}O_2$ $Et_3PbO_2CCH{=}CHPh$, 2, 576
$PbC_{15}H_{31}ClO_2$ $Bu_3PbO_2CCH_2CH_2Cl$, 2, 577
$PbC_{15}H_{32}O_2$ Bu_3PbO_2CEt, 2, 575
$Me_2CHCH_2PbO_2CEt$, 2, 575
$PbC_{16}H_{12}S_4$ 2-$Thienyl_4Pb$, 2, 531
$PbC_{16}H_{13}ClS$ 2-Thienyl-Ph_2PbCl, 2, 557
$PbC_{16}H_{14}Cl_2O_4$ $Ph_2Pb(O_2CCH_2Cl)_2$, 2, 582
$PbC_{16}H_{16}$ $Ph_2Pb(CH{=}CH_2)_2$, 2, 538
$PbC_{16}H_{16}O_2S_2$ $Ph_2Pb(SAc)_2$, 2, 585
$PbC_{16}H_{16}O_4$ $Ph_2Pb(OAc)_2$, 2, 579
$PbC_{16}H_{16}O_6$ 1-$C_{10}H_7Pb(OAc)_3$, 2, 583
2-$C_{10}H_7Pb(OAc)_3$, 2, 583
$PbC_{16}H_{18}Br_2$ $(Me_2C_6H_3)_2PbBr_2$, 2, 561
$PbC_{16}H_{18}N_2O_4$ (3-$C_5H_4NCO_2)_2PbEt_2$, 2, 581
$PbC_{16}H_{18}N_2O_6$ (p-$O_2NC_6H_4O)_2PbEt_2$, 2, 571
$PbC_{16}H_{20}$ Et_2PbPh_2, 2, 534
$PbC_{16}H_{20}N_2$ (p-$Me_2NC_6H_4)_2Pb$, 2, 592
$PbC_{16}H_{20}O_4S_2$ 2-$Thienyl_2Pb(O_2CCHMe_2)_2$, 2, 581
$PbC_{16}H_{22}O_3S$ 2-$C_{10}H_7SO_3PbEt_3$, 2, 564
$PbC_{16}H_{22}O_7$ p-$MeOC_6H_4Pb(O_2CEt)_3$, 2, 583
$PbC_{16}H_{23}NO_3S$ 1,4-$H_2NC_{10}H_6SO_3PbEt_3$, 2, 564*
$PbC_{16}H_{24}$ $Et_2Pb(C_5H_4Me)_2$, 2, 538
$PbC_{16}H_{25}NO_3S$ o-$Pr_3PbNOCC_6H_4SO_2$, 2, 587
$PbC_{16}H_{27}NO_2$ o-$H_2NC_6H_4CO_2PbPr_3$, 2, 578
$PbC_{16}H_{28}O_3S$ o-$MeC_6H_4SO_3PbPr_3$, 2, 565
p-$MeC_6H_4SO_3PbPr_3$, 2, 565
$PbC_{16}H_{29}NO_2S$ p-$MeC_6H_4SO_2NHPbPr_3$, 2, 587
$PbC_{16}H_{30}O_4$ $Et_3PbO_2CCH{=}CHCO_2C_6H_{13}$, 2, 576
$PbC_{16}H_{32}$ $(C_6H_{13})_2Pb(CH{=}CH_2)_2$, 2, 538
$PbC_{16}H_{32}O_2$ $Bu_3PbO_2CCH{=}CHMe$, 2, 576
$PbC_{16}H_{34}Cl_2$ $(C_8H_{17})_2PbCl_2$, 2, 560
$PbC_{16}H_{34}O$ $(C_8H_{17})_2PbO$, 2, 569
$PbC_{16}H_{34}O_2$ $(Me_2CHCH_2)_3PbO_2CPr$, 2, 575
$PbC_{16}H_{36}$ Bu_4Pb, 2, 526
i-Bu_4Pb, 2, 528
$PbC_{17}H_{14}IN$ Ph_2(2-$C_5H_4N)PbI$, 2, 557
$PbC_{17}H_{20}$ $Ph_2Pb(CH_2)_4CH_2$, 2, 596
$PbC_{17}H_{22}NO_2$ Et_3Pb 4,2-methylquinolinecarboxylate, 2, 576
$PbC_{17}H_{24}O_3$ $Et_3PbO_2CCH{=}CHCH_2Bz$, 2, 576
$PbC_{17}H_{25}NO_2$ o-$Pr_3PbNOCC_6H_4CO$, 2, 587
$PbC_{17}H_{36}O_2$ (n-$C_5H_{11})_3PbOAc$, 2, 575
(i-$C_5H_{11})_3PbOAc$, 2, 575
$PbC_{17}H_{38}$ (i-$C_5H_{11})_3PbEt$, 2, 535
$PbC_{18}H_{12}N_4O_9$ (m-$O_2NC_6H_4)_3PbNO_3$, 2, 566
$PbC_{18}H_{14}Cl_2O_4$ $(ClCH{=}CH)_2Pb(OBz)_2$, 2, 580
$PbC_{18}H_{14}Cl_{10}O_2$ i-$Pr_2Pb(OC_6Cl_5)_2$, 2, 572
$PbC_{18}H_{14}O_2$ o-$Ph_2PbC_6H_4CO_2$, 2, 578
$PbC_{18}H_{15}Br$ Ph_3PbBr, 2, 554
$PbC_{18}H_{15}Cl$ Ph_3PbCl, 2, 554
$PbC_{18}H_{15}I$ Ph_3PbI, 2, 555
$PbC_{18}H_{15}NO_3$ Ph_3PbNO_3, 2, 565
$PbC_{18}H_{15}N_3$ Ph_3PbN_3, 2, 562
$PbC_{18}H_{16}$ Ph_3PbH, 2, 551
$PbC_{18}H_{16}O$ Ph_3PbOH, 2, 568
$PbC_{18}H_{18}Br_2O_4$ (m-$BrC_6H_4CO_2)_2PbEt_2$, 2, 581
$PbC_{18}H_{18}Cl_2O_2$ (p-$EtO_2CC_6H_4)_2PbCl_2$, 2, 561
$PbC_{18}H_{18}Cl_2O_4$ (p-$ClC_6H_4CO_2)_2PbEt_2$, 2, 581
$PbC_{18}H_{18}N_2O_8$ (m-$O_2NC_6H_4CO_2)_2PbEt_2$, 2, 581
$PbC_{18}H_{20}$ $Ph_2Pb(CH_2CH{=}CH_2)_2$, 2, 538
$PbC_{18}H_{20}Br_2$ $Ph_2(CH_2{=}CHCH_2)PbC_3H_5Br$;

(?), 2, 542
$PbC_{18}H_{20}O_4$ $Et_2Pb(OBz)_2$, 2, 580
$Ph_2Pb(O_2CEt)_2$, 2, 580
$(p\text{-}MeC_6H_4)_2Pb(OAc)_2$, 2, 581
$PbC_{18}H_{20}O_6$ $Ph_2Pb(O_2CCHOHMe)_2$, 2, 582
$(o\text{-}MeOC_6H_4)_2Pb(OAc)_2$, 2, 581
$PbC_{18}H_{22}Br_2$ $Mesityl_2PbBr_2$, 2, 561
$PbC_{18}H_{22}O_2$ $Ph_2(CH_2{=}CHCH_2)PbC_3H_5(OH)_2$ (?), 2, 544
$PbC_{18}H_{24}O_4$ $Ph_2Pb(OCH_2CH_2OMe)_2$, 2, 572
$PbC_{18}H_{24}O_6S_2$ $(o\text{-}MeC_6H_4SO_3)_2PbEt_2$, 2, 565
$(p\text{-}MeC_6H_4SO_3)_2PbEt_2$, 2, 565
$PbC_{18}H_{25}O_4P$ $Et_3PbO_2P(OPh)_2$, 2, 564
$PbC_{18}H_{26}O_6$ $PhPb(O_2CCHMe_2)_3$, 2, 583
$PbC_{18}H_{27}Cl$ $Cyclohexyl_2PhPbCl$, 2, 556
$PbC_{18}H_{33}Br$ $Cyclohexyl_3PbBr$, 2, 556
$PbC_{18}H_{33}Cl$ $Cyclohexyl_3PbCl$, 2, 556
$PbC_{18}H_{33}I$ $Cyclohexyl_3PbI$, 2, 553
$PbC_{18}H_{34}O$ $Cyclohexyl_3PbOH$, 2, 569
$PbC_{18}H_{34}O_4$ $Et_3PbO_2CCH{=}CHCO_2CH_2CHEtBu$, 2, 576
$PbC_{18}H_{37}O_4P$ $Et_3PbO_2P(OC_6H_{11})_2$, 2, 564
$PbC_{19}H_{15}ClO_2$ $(o\text{-}HO_2CC_6H_4)Ph_2PbCl$, 2, 558
$PbC_{19}H_{15}N$ Ph_3PbCN, 2, 563
$PbC_{19}H_{15}NS$ Ph_3PbNCS, 2, 563
$PbC_{19}H_{16}O_3$ $Ph_2(o\text{-}HO_2CC_6H_4)PbOH$, 2, 569*
$PbC_{19}H_{18}O$ Ph_3PbOMe, 2, 572
$PbC_{19}H_{18}S$ Ph_3PbSMe, 2, 585
$PbC_{19}H_{22}O_6$ $p\text{-}MeC_6H_4Pb(O_2CCMe{=}CH_2)_3$, 2, 583*
$2\text{-}C_{10}H_7Pb(O_2CEt)_3$, 2, 583
$PbC_{19}H_{25}NO_2$ $o\text{-}Et_3PbO_2CC_6H_4NHPh$, 2, 578
$PbC_{19}H_{26}BrNO_2S$ $p\text{-}MeC_6H_4SO_2(p\text{-}BrC_6H_4)\text{-}NPbEt_3$, 2, 586
$PbC_{19}H_{26}ClN$ $Ph_2(Et_2NCH_2CH_2CH_2)PbCl$, 2, 558
$PbC_{19}H_{26}ClNO_2S$ $p\text{-}MeC_6H_4SO_2(p\text{-}ClC_6H_4)\text{-}NPbEt_3$, 2, 586
$PbC_{19}H_{26}O_2$ $2\text{-}C_{10}H_7CH_2CH_2CO_2PbEt_3$, 2, 574
$PbC_{19}H_{27}Cl_2N$ $Ph_2(Et_2NCH_2CH_2CH_2)PbCl\cdot HCl$, 2, 558
$PbC_{19}H_{27}NO_2S$ $p\text{-}MeC_6H_4SO_2N(Ph)PbEt_3$, 2, 586
$PbC_{19}H_{34}O_3S$ $p\text{-}MeC_6H_4SO_3PbBu_3$, 2, 565
$o\text{-}MeC_6H_4SO_3Pb(CH_2CHMe_2)_3$, 2, 565
$p\text{-}MeC_6H_4SO_3Pb(CH_2CHMe_2)_3$, 2, 565
$PbC_{19}H_{43}O_4P$ $Et_3PbO_2P(OC_5H_{11})OC_8H_{17}$, 2, 564

$PbC_{20}H_{16}S_2$ $2\text{-}Thienyl_2PbPh_2$, 2, 536
$PbC_{20}H_{17}ClO_2$ $Ph_2(o\text{-}MeO_2CC_6H_4)PbCl$, 2, 558
$PbC_{20}H_{18}$ $Ph_3PbCH{=}CH_2$, 2, 538, 597
$PbC_{20}H_{18}OS$ Ph_3PbSAc, 2, 585
$PbC_{20}H_{18}O_2$ Ph_3PbOAc, 2, 575*
$PbC_{20}H_{18}O_3$ Ph_3PbO_2Ac (?), 2, 570
$PbC_{20}H_{20}$ $EtPbPh_3$, 2, 535
$PbC_{20}H_{20}O_4$ $Ph_2Pb(O_2CCMe{=}CH_2)_2$, 2, 582, 597
$PbC_{20}H_{20}S$ Ph_3PbSEt, 2, 585
$PbC_{20}H_{23}NO_2$ $p\text{-}(Phthalimido)C_6H_4PbEt_3$, 2, 545
$PbC_{20}H_{24}O_2$ Et_3Pb 9-fluorenecarboxylate, 2, 576
$PbC_{20}H_{24}O_4$ $(p\text{-}MeC_6H_4CO_2)_2PbEt_2$, 2, 580
$Ph_2Pb(O_2CPr)_2$, 2, 580
$Ph_2Pb(O_2CCHMe_2)_2$, 2, 580
$PbC_{20}H_{24}O_6$ $Ph_2Pb(O_2CCHOHEt)_2$, 2, 582
$(p\text{-}MeOC_6H_4)_2Pb(O_2CEt)_2$, 2, 581
$PbC_{20}H_{26}O_2$ $Et_3PbO_2CCHPh_2$, 2, 574
$PbC_{20}H_{29}O_4P$ $Et_3PbO_2P(OC_6H_4Me)_2$, 2, 564
$PbC_{20}H_{40}$ $(C_8H_{17})_2Pb(CH{=}CH_2)_2$, 2, 538
$PbC_{20}H_{40}O_4$ $(C_8H_{17})_2Pb(OAc)_2$, 2, 580
$PbC_{20}H_{42}O_2$ $(C_6H_{13})_3PbOAc$, 2, 575
$PbC_{20}H_{44}$ $(n\text{-}C_5H_{11})_4Pb$, 2, 528
$(C_5H_{11})_4Pb$, 2, 528
$PbC_{20}H_{45}O_4P$ $Et_3PbO_2P(OC_{12}H_{25})OEt$, 2, 564*
$PbC_{21}H_{18}$ $Ph_3PbC{\equiv}CMe$, 2, 540
$Ph_3PbCH_2C{\equiv}CH$, 2, 540
$PbC_{21}H_{18}O_2$ $Ph_3PbO_2CCH{=}CH_2$, 2, 576, 597
$PbC_{21}H_{20}$ $Ph_3PbCH_2CH{=}CH_2$, 2, 538
$PbC_{21}H_{21}Br$ $(m\text{-}MeC_6H_4)_3PbBr$, 2, 556
$(p\text{-}MeC_6H_4)_3PbBr$, 2, 556
$PbC_{21}H_{21}Cl$ $Ph_3Pb(CH_2)_3Cl$ (?), 2, 542
$(m\text{-}MeC_6H_4)_3PbCl$, 2, 556
$(p\text{-}MeC_6H_4)_3PbCl$, 2, 556

$PbC_{21}H_{21}ClO_3$ $(o\text{-}MeOC_6H_4)_3PbCl$, 2, 558
$(p\text{-}MeOC_6H_4)_3PbCl$, 2, 558
$PbC_{21}H_{21}I$ $(m\text{-}MeC_6H_4)_3PbI$, 2, 556
$(p\text{-}MeC_6H_4)_3PbI$, 2, 556
$PbC_{21}H_{21}IO_3$ $(o\text{-}MeOC_6H_4)_3PbI$, 2, 558
$PbC_{21}H_{22}$ $PrPbPh_3$, 2, 535
$PbC_{21}H_{22}S$ Ph_3PbSPr, 2, 585
$PbC_{21}H_{28}NO_2$ Et_3Pb 9,2-Ethylcarbazolecarboxylate, 2, 576
$PbC_{22}H_{16}$ $Ph_3PbC{\equiv}CC{\equiv}CH$, 2, 540
$PbC_{22}H_{18}O_4$ $Ph_3PbO_2CCH{=}CHCO_2H$, 2, 576
$PbC_{22}H_{18}S$ 2-Thienyl-$PbPh_3$, 2, 536
$PbC_{22}H_{20}$ $Ph_3PbC{\equiv}CEt$, 2, 540
$PbC_{22}H_{20}O_2$ $Ph_3PbO_2CCMe{=}CH_2$, 2, 576, 597
$PbC_{22}H_{20}O_4$ 2,2[$(p\text{-}MeOC_6H_4)_2Pb$]$(C_4H_3O)_2$, 2, 536
$PbC_{22}H_{22}O_2$ $Ph_3PbCH_2CO_2Et$, 2, 543
$Ph_3PbO_2CCHMe_2$, 2, 575*
$PbC_{22}H_{24}N_2O_8$ $Ph_2Pb[O_2CCH_2N(CH_2CO_2H)\text{-}CH_2]_2$, 2, 582
$PbC_{22}H_{24}O_2$ $Ph_3PbO_2CMe_3$, 2, 570
$PbC_{22}H_{24}O_8$ $(p\text{-}EtO_2CC_6H_4)_2Pb(OAc)_2$, 2, 581
$PbC_{22}H_{24}S$ Ph_3PbSBu, 2, 585
$PbC_{22}H_{28}O_4$ $Ph_2Pb(O_2CCH_2CHMe_2)_2$, 2, 580
$(p\text{-}MeC_6H_4)_2Pb(O_2CCHMe_2)_2$, 2, 581
$PbC_{22}H_{28}O_6$ 1-$C_{10}H_7Pb(O_2CCHMe_2)_3$, 2, 583
$PbC_{22}H_{33}NO_2S$ $p\text{-}MeC_6H_4SO_2N(Ph)PbPr_3$, 2, 587
$PbC_{22}H_{34}$ $(C_6H_{11})_3PbC{\equiv}CC{\equiv}CH$, 2, 540
$PbC_{22}H_{34}O_3S$ 2-$C_{10}H_7SO_3PbBu_3$, 2, 565
$PbC_{22}H_{49}O_4P$ $Et_3PbO_2P(OCH_2CHEtBu)_2$, 2, 564
$(i\text{-}C_8H_{17}O)_2PO_2PbEt_3$, 2, 564*
$PbC_{23}H_{18}$ $Ph_3PbC{\equiv}CC{\equiv}CMe$, 2, 540
$PbC_{23}H_{18}N_2O_4$ Ph_3Pb orotate, 2, 576
$PbC_{23}H_{19}N$ 2-Pyridyl-$PbPh_3$, 2, 536
$PbC_{23}H_{22}O_4$ $Ph_3PbO_2CCH_2CO_2Et$, 2, 578
$PbC_{23}H_{48}O_2$ $(C_7H_{15})_3PbOAc$, 2, 575
$PbC_{24}D_{20}$ $(C_6D_5)_4Pb$, 2, 529
$PbC_{24}H_{15}F_5$ $Ph_3PbC_6F_5$, 2, 542
$PbC_{24}H_{18}$ o,o'-$\overline{C_6H_4C_6H_4}PbPh_2$, 2, 596
$PbC_{24}H_{18}Cl_2$ $(p\text{-}ClC_6H_4)_2PbPh_2$, 2, 542
$PbC_{24}H_{18}N_2O_4$ Ph_2Pb di-3-pyridinecarboxylate, 2, 582
$PbC_{24}H_{19}Br$ $p\text{-}BrC_6H_4PbPh_3$, 2, 542
$PbC_{24}H_{19}Cl$ $p\text{-}ClC_6H_4PbPh_3$, 2, 542
$PbC_{24}H_{20}$ Ph_4Pb, 2, 526
$PbC_{24}H_{20}N_4O_8$ Ph_2Pb diuracil acetate, 2, 582
$PbC_{24}H_{20}O$ $o\text{-}HOC_6H_4PbPh_3$, 2, 543
$PbC_{24}H_{20}O_4$ $(2\text{-}C_{10}H_7)_2Pb(OAc)_2$, 2, 581
$PbC_{24}H_{20}S$ Ph_3PbSPh, 2, 585
$PbC_{24}H_{21}N$ $o\text{-}H_2NC_6H_4PbPh_3$, 2, 545
$p\text{-}H_2NC_6H_4PbPh_3$, 2, 546
$PbC_{24}H_{27}Br$ $(PhCH_2CH_2)_3PbBr$, 2, 555
$PbC_{24}H_{27}IO_3$ $(o\text{-}EtOC_6H_4)_3PbI$, 2, 558
$PbC_{24}H_{28}$ $(C_5H_4Me)_4Pb$, 2, 531
$PbC_{24}H_{32}$ Cyclohexyl$_2PbPh_2$, 2, 535
$PbC_{24}H_{32}O_4$ $Ph_2Pb[O_2C(CH_2)_4Me]_2$, 2, 580
$PbC_{24}H_{37}Cl$ $m\text{-}ClC_6H_4Pb(C_6H_{11})_3$, 2, 542
$p\text{-}ClC_6H_4Pb(C_6H_{11})_3$, 2, 542
$PbC_{24}H_{38}$ Cyclohexyl$_3PbPh$, 2, 535
Cyclohexyl$_3PbC{\equiv}CC{\equiv}CEt$, 2, 540
$PbC_{24}H_{42}$ $(C_6H_{13})_4Pb$, 2, 528
$PbC_{24}H_{44}$ Cyclohexyl$_4Pb$, 2, 529
$PbC_{24}H_{51}Cl$ $(C_8H_{17})_3PbCl$, 2, 556
$PbC_{24}H_{52}O$ $(C_8H_{17})_3PbOH$, 2, 569
$PbC_{24}H_{52}O_2$ $Bu_2Pb(OC_8H_{17})_2$, 2, 572
$PbC_{25}H_{19}Cl$ 9-Fluorenyl-Ph_2PbCl, 2, 557
$PbC_{25}H_{20}OS$ Ph_3PbSBz, 2, 585
$PbC_{25}H_{20}O_2$ $p\text{-}Ph_3PbC_6H_4CO_2H$, 2, 543
Ph_3PbOBz, 2, 575*
$PbC_{25}H_{21}Br$ $p\text{-}BrC_6H_4CH_2PbPh_3$, 2, 542
$PbC_{25}H_{22}$ $PhCH_2PbPh_3$, 2, 534
1-Cyclopentenyl-$C{\equiv}CPbPh_3$, 2, 540
$PbC_{25}H_{22}O$ $o\text{-}HOCH_2C_6H_4PbPh_3$, 2, 543
$m\text{-}HOCH_2C_6H_4PbPh_3$, 2, 543
$p\text{-}HOCH_2C_6H_4PbPh_3$, 2, 543
$o\text{-}MeC_6H_4PbPh_3$, 2, 543
$p\text{-}MeC_6H_4PbPh_3$, 2, 543
$PbC_{25}H_{22}S$ Ph_3PbSCH_2Ph, 2, 585
$PbC_{25}H_{23}N$ $p\text{-}Ph_3PbC_6H_4NHMe$, 2, 546
$PbC_{25}H_{24}O_4$ 2-[$(p\text{-}MeOC_6H_4)_3Pb$]C_4H_3O, 2, 536
$PbC_{25}H_{31}N$ $Ph_3Pb(CH_2)_3NEt_2$, 2, 545
$PbC_{25}H_{40}$ Cyclohexyl$_3PbCH_2Ph$, 2, 534
p-(Cyclohexyl$_3$Pb)C_6H_4Me, 2, 535
Cyclohexyl$_3PbC{\equiv}CC{\equiv}CPr$, 2, 540
$PbC_{25}H_{40}O$ Cyclohexyl$_3PbC_6H_4OMe$, 2, 543

$PbC_{26}H_{18}$ 9-Fluorenyl$_2$Pb, 2, 592
$PbC_{26}H_{20}$ $Ph_3PbC{\equiv}CPh$, 2, 540
$PbC_{26}H_{20}O_4$ $Ph_2Pb(OBz)_2$, 2, 580
$PbC_{26}H_{22}$ p-$Ph_3PbC_6H_4CH{=}CH_2$, 2, 538, 597
$PbC_{26}H_{22}O_2$ p-$Ph_3PbC_6H_4CO_2Me$, 2, 543
$PbC_{26}H_{23}N$ o,o′-$\overline{C_6H_4NEtC_6H_4Pb}Ph_2$, 2, 596*, 638*
$PbC_{26}H_{24}$ p-$EtC_6H_4PbPh_3$, 2, 535
(p-$MeC_6H_4)_2PbPh_2$, 2, 535
1-Cyclohexenyl-$C{\equiv}CPbPh_3$, 2, 540*
$PbC_{26}H_{24}O$ m-$HOCH_2CH_2C_6H_4PbPh_3$, 2, 543
p-$HOCH_2CH_2C_6H_4PbPh_3$, 2, 543
p-$HOCHMeC_6H_4PbPh_3$, 2, 543
$PbC_{26}H_{25}N$ o-$Ph_3PbC_6H_4NMe_2$, 2, 545
$PbC_{26}H_{30}O_2$ $Et_3PbO_2CCPh_3$, 2, 574
$PbC_{26}H_{34}IN$ $Ph_3Pb(CH_2)_3NEt_2MeI$, 2, 545
$PbC_{26}H_{38}N_2O_4$ (o-$BuNHC_6H_4CO_2)_2PbEt_2$, 2, 581
$PbC_{26}H_{42}$ Cyclohexyl$_3PbC{\equiv}CC{\equiv}Bu$, 2, 540
$PbC_{26}H_{43}N$ p-(Cyclohexyl$_3$Pb)$C_6H_4NMe_2$, 2, 545
$PbC_{26}H_{54}O_2$ $(C_8H_{17})_3PbOAc$, 2, 575
$PbC_{26}H_{55}O_4P$ $Et_3PbO_2P(OC_{18}H_{25})OEt$, 2, 564
$PbC_{27}H_{20}O_6$ $PhPb(OBz)_3$, 2, 583*
$PbC_{27}H_{21}NO$ 8-Quinolinoxy-$PbPh_3$, 2, 572
$PbC_{27}H_{22}$ 3-Indenyl-$PbPh_3$, 2, 538
$PbC_{27}H_{22}O_2$ p-$Ph_3PbO_2CC_6H_4CH{=}CH_2$, 2, 576, 597
$PbC_{27}H_{24}$ 2,4-$CH_2{=}CH(Me)C_6H_3PbPh_3$, 2, 538, 597
4,2-$CH_2{=}CH(Me)C_6H_3PbPh_3$, 2, 538, 597
$PbC_{27}H_{26}O_2$ $Ph_3PbO_2CMe_2Ph$, 2, 570
$PbC_{27}H_{33}Br$ Mesityl$_3$PbBr, 2, 556
$PbC_{27}H_{33}I$ Mesityl$_3$PbI, 2, 556
$PbC_{27}H_{37}NO_4S$ $Ph_3Pb(CH_2)_3NEt_2 \cdot Me_2SO_4$, 2, 545
$PbC_{27}H_{46}IN$ p-(Cyclohexyl$_3$Pb)$C_6H_4NMe_3I$, 2, 545
$PbC_{28}H_{20}$ $Ph_3PbC{\equiv}CC{\equiv}CPh$, 2, 540
$PbC_{28}H_{20}O_4$ (p-$OHCC_6H_4)_4Pb$, 2, 532
$PbC_{28}H_{22}O_7$ p-$MeOC_6H_4Pb(OBz)_3$, 2, 583
$PbC_{28}H_{24}$ (p-$CH_2{=}CHC_6H_4)_2PbPh_2$, 2, 512, 513, 538, 547
$PbC_{28}H_{24}Br_4$ (o-$BrC_6H_4CH_2)_4Pb$, 2, 531
$PbC_{28}H_{24}Cl_4$ (o-$ClC_6H_4CH_2)_4Pb$, 2, 531
$PbC_{28}H_{24}F_4$ (o-$FC_6H_4CH_2)_4Pb$, 2, 531
$PbC_{28}H_{28}$ $(PhCH_2)_4Pb$, 2, 528
(o-$MeC_6H_4)_4Pb$, 2, 529
(m-$MeC_6H_4)_4Pb$, 2, 529
(p-$MeC_6H_4)_4Pb$, 2, 529
(p-$MeC_6H_4)_3PbCH_2Ph$, 2, 534
$PbC_{28}H_{28}O_3$ (o-$MeOC_6H_4)_3PbCH_2Ph$, 2, 543
$PbC_{28}H_{28}O_4$ (o-$MeOC_6H_4)_4Pb$, 2, 531
(p-$MeOC_6H_4)_4Pb$, 2, 531
$PbC_{28}H_{29}NO_2S$ $Ph_3PbC_6H_4SO_2NEt_2$, 2, 544
$PbC_{28}H_{29}NO_3S$ p-$Me_3CC_6H_4NHSO_3PbPh_3$, 2, 565
$PbC_{28}H_{30}N_2$ (p-$Me_2NC_6H_4)_2PbPh_2$, 2, 546
$PbC_{28}H_{31}NO_4S$ $Ph_3PbC_6H_4NMe_2 \cdot Me_2SO_4$, 2, 545
$PbC_{28}H_{60}$ (n-$C_7H_{15})_4Pb$, 2, 529
$PbC_{28}H_{61}O_4P$ $Et_3PbO_2P(OC_{20}H_{41})OEt$, 2, 564*
$PbC_{29}H_{25}NO_3$ p-$(Ph_3PbO_2C)C_6H_4NHCOCMe{=}CH_2$, 2, 578, 597
$PbC_{29}H_{28}$ p-[(p-$MeC_6H_4)_3Pb]C_6H_4CH{=}CH_2$, 2, 539, 597
$PbC_{29}H_{28}O_2$ $Ph_3PbCH(CO_2Et)CH_2Ph$, 2, 543
$PbC_{29}H_{28}O_3$ p-[(p-$MeOC_6H_4)_3Pb]C_6H_4$-$CH{=}CH_2$, 2, 597
$PbC_{30}H_{21}Cl$ (1-$C_{10}H_7)_3PbCl$, 2, 556
$PbC_{30}H_{22}N_2O_2$ 8-Quinolinoxy$_2PbPh_2$, 2, 572
$PbC_{30}H_{23}BrN_2O$ 6,3-HO(p-$BrC_6H_4N_2)C_6H_3$-$PbPh_3$, 2, 546
$PbC_{30}H_{23}ClN_2O$ 6,3-HO(p-$ClC_6H_4N_2)C_6H_3$-$PbPh_3$, 2, 546
$PbC_{30}H_{23}IN_2O$ 6,3-HO(p-$IC_6H_4N_2)C_6H_3$-$PbPh_3$, 2, 546
$PbC_{30}H_{24}$ 3-Indenyl$_2PbPh_2$, 2, 538
$PbC_{30}H_{25}NO_2S$ $Ph_3PbC_6H_4SO_2NHPh$, 2, 544
$PbC_{30}H_{25}O_2P$ $Ph_3PbO_2PPh_2$, 2, 565
$PbC_{30}H_{25}P$ Ph_3PbPPh_2, 2, 587
$PbC_{30}H_{26}$ 7-$C_{18}H_{11}PbPh_3$, 2, 535
$Me_2\overline{PbCPh{=}CPhCPh{=}C}Ph$, 2, 596
$PbC_{30}H_{26}Br_2$ $Me_2(BrCPh{=}CPhCPh{=}CPh)PbBr$, 2, 557
$PbC_{30}H_{28}O_2$ p,p′-[(p-$CH_2{=}CHC_6H_4)_2Pb$]-$(C_6H_4OMe)_2$, 2, 544
$PbC_{30}H_{28}O_4$ $Ph_3PbO_2CCH(CH_2Ph)CO_2Et$, 2, 578

$PbC_{30}H_{30}$ p-Cyclohexyl-$C_6H_4PbPh_3$, 2, 535

$PbC_{30}H_{38}O_2$ $Ph_3PbO_2CC_{11}H_{23}$, 2, 575*

$PbC_{31}H_{22}O_6$ 1-$C_{10}H_7Pb(OBz)_3$, 2, 583

$PbC_{31}H_{24}$ 9-Fluorenyl-$PbPh_3$, 2, 538

$PbC_{31}H_{24}N_2O_3$ 6,3-HO(p-$HO_2CC_6H_4N_2$)-$C_6H_3PbPh_3$, 2, 546

$PbC_{31}H_{25}N_3O_3$ 4,3-MeO(p-$O_2NC_6H_4N_2$)-$C_6H_3PbPh_3$, 2, 546*

6,3-MeO(p-$O_2NC_6H_4N_2$)$C_6H_3PbPh_3$, 2, 546*

$PbC_{31}H_{26}$ $Ph_2CHPbPh_3$, 2, 534

$PbC_{31}H_{34}O_3$ (p-$EtOC_6H_4$)$_3PbCH_2Ph$, 2, 543

$PbC_{32}H_{24}$ 9-Phenanthryl-$PbPh_3$, 2, 535

$PbC_{32}H_{28}$ (p-CH_2=CHC_6H_4)$_4$Pb, 2, 531

$PbC_{32}H_{28}BrN_3$ 2,5-Me_2N(p-$BrC_6H_4N_2$)-$C_6H_3PbPh_3$, 2, 546

$PbC_{32}H_{28}ClN_3$ 2,5-Me_2N(p-$ClC_6H_4N_2$)-$C_6H_3PbPh_3$, 2, 546

$PbC_{32}H_{28}IN_3$ 2,5-Me_2N(p-$IC_6H_4N_2$)C_6H_3-$PbPh_3$, 2, 546

$PbC_{32}H_{28}N_4O_2$ 2,5-Me_2N(p-$O_2NC_6H_4N_2$)-$C_6H_3PbPh_3$, 2, 546

$PbC_{32}H_{36}$ $(PhCH_2CH_2)_4Pb$, 2, 528

$PbC_{32}H_{36}O_4$ (o-$EtOC_6H_4$)$_4$Pb, 2, 531

(p-$EtOC_6H_4$)$_4$Pb, 2, 532

$PbC_{32}H_{40}N_4$ (p-$Me_2NC_6H_4$)$_4$Pb, 2, 532

$PbC_{32}H_{42}O_2$ $Ph_3PbO_2CC_{13}H_{27}$, 2, 575*

$PbC_{32}H_{66}I_2$ $(C_{16}H_{33})_2PbI_2$, 2, 560

$PbC_{32}H_{68}$ $(C_8H_{17})_4Pb$, 2, 529

$PbC_{33}H_{29}N_3O_2$ 2,5-Me_2N(p-$HO_2CC_6H_4N_2$)-$C_6H_3PbPh_3$, 2, 546

$PbC_{34}H_{26}N_2O$ o-(2,1-$HOC_{10}H_6N_2$)C_6H_4-$PbPh_3$, 2, 545

p-(2,1-$HOC_{10}H_6N_2$)$C_6H_4PbPh_3$, 2, 546

$PbC_{34}H_{46}O_2$ $Ph_3PbO_2CC_{15}H_{31}$, 2, 575*

$PbC_{34}H_{57}O_4P$ (p-$C_8H_{17}C_6H_4O$)$_2PO_2PbEt_3$, 2, 564*

$PbC_{36}H_{26}N_6O_5$ 6,3,5-HO(p-$O_2NC_6H_4N_2$)$_2$-$C_6H_2PbPh_3$, 2, 546*

$PbC_{36}H_{36}$ [2,4-Me(CH_2=CH)C_6H_3]$_4$Pb, 2, 531

$(PbC_{36}H_{36})_n$ (CH_2=$CHCH_2C_6H_4$)$_4$Pb (?), 2, 531

$PbC_{36}H_{36}O_8$ [p-$(CH_2O)_2CHC_6H_4$]$_4$Pb, 2, 532

$PbC_{36}H_{40}$ (p-Cyclohexyl-C_6H_4)$_2PbPh_2$, 2, 535

$PbC_{36}H_{44}$ Mesityl$_4$Pb, 2, 529

$PbC_{36}H_{50}O_2$ $Ph_3PbO_2CC_{17}H_{35}$, 2, 575*

$PbC_{36}H_{52}I_4N_4$ (p-$IMe_3NC_6H_4$)$_4$Pb, 2, 532

$PbC_{36}H_{75}Cl$ (n-$C_{12}H_{25}$)$_3PbCl$, 2, 556

$PbC_{36}H_{75}NO_3$ $(C_{12}H_{25})_3PbNO_3$, 2, 565

$PbC_{37}H_{30}$ Ph_3CPbPh_3 (?), 2, 534

$PbC_{37}H_{30}O_2$ $Ph_3PbO_2CPh_3$, 2, 570

$PbC_{38}H_{28}$ 9-Fluorenyl$_2PbPh_2$, 2, 538

$PbC_{38}H_{78}O_2$ (n-$C_{12}H_{25}$)$_3PbOAc$, 2, 575

$PbC_{39}H_{74}O_{18}$ PhPb[O$(CH_2CH_2O)_5$Me]$_3$, 2, 572

$PbC_{40}H_{28}$ 9-Phenanthryl$_2PbPh_2$, 2, 535

$PbC_{40}H_{36}O_3$ (o-$MeOC_6H_4$)$_3PbCPh_3$, 2, 543

$PbC_{40}H_{76}O_{19}$ p-$MeOC_6H_4$Pb[O$(CH_2CH_2O)_5$-Me]$_3$, 2, 572

$PbC_{40}H_{84}$ $(C_{10}H_{21})_4Pb$, 2, 529

$PbC_{42}H_{87}Cl$ (n-$C_{14}H_{29}$)$_3PbCl$, 2, 556

$PbC_{48}H_{32}$ (7-$C_{18}H_{11}$)$_2PbPh_2$, 2, 535*

$PbC_{48}H_{80}O_4$ $Ph_2Pb(O_2CC_{17}H_{35})_2$, 2, 580

$PbC_{48}H_{99}Cl$ (n-$C_{16}H_{33}$)$_3PbCl$, 2, 556

$PbC_{48}H_{100}$ $(C_{12}H_{25})_4Pb$, 2, 529

$PbC_{52}H_{108}O_{18}$ $(C_8H_{17})_2$Pb[$(OCH_2CH_2)_8$-OEt]$_2$, 2, 572*

$PbC_{54}H_{111}Cl$ (n-$C_{18}H_{37}$)$_3PbCl$, 2, 556

$PbC_{56}H_{116}$ (n-$C_{14}H_{29}$)$_4$Pb, 2, 529

$PbC_{64}H_{132}$ (n-$C_{16}H_{33}$)$_4$Pb, 2, 529

$Pb_2C_5H_{16}$ Me_5Pb_2H, 2, 551

$Pb_2C_6H_{18}$ Me_6Pb_2, 2, 587

$Pb_2C_6H_{18}S$ $(Me_3Pb)_2S$, 2, 584

$Pb_2C_7H_7Br_5$ $(C_7H_7)Pb_2Br_5$, 2, 562

$Pb_2C_7H_{20}$ $(Me_3Pb)_2CH_2$, 2, 548

$Pb_2C_9H_{23}Br$ $Me_3PbCH_2CH_2PbMe_2CH_2CH_2Br$, 2, 548

$Pb_2C_{10}H_{26}$ Et_5Pb_2H, 2, 551*

$Pb_2C_{12}H_{30}$ Et_6Pb_2, 2, 588

$Pb_2C_{12}H_{30}FO_3P$ $(Et_3Pb)_2PO_3F$, 2, 564

$Pb_2C_{12}H_{30}O$ $(Et_3Pb)_2O$, 2, 567

$Pb_2C_{12}H_{30}O_4S$ $(Et_3Pb)_2SO_4$, 2, 564

$Pb_2C_{12}H_{30}S$ $(Et_3Pb)_2S$, 2, 584

$Pb_2C_{13}H_{30}O_3$ $(Et_3Pb)_2CO_3$, 2, 574

$Pb_2C_{14}H_{30}$ $(Et_3PbC\equiv)_2$, 2, 548

$Pb_2C_{16}H_{30}$ $(Et_3PbC\equiv C)_2$, 2, 548

$Pb_2C_{16}H_{32}O_4$ $(Et_3PbO_2CCH=)_2$, 2, 576

$Pb_2C_{16}H_{38}$ $(Et_3PbCHMe)_2$, 2, 548

$Pb_2C_{18}H_{30}$ $(Et_3PbC\equiv CC\equiv)_2$, 2, 548

$Pb_2C_{18}H_{31}N_3O_8$ 1,3,2,4,6$(Et_3PbO)_2C_6H(NO_2)_3$, 2, 571*
$Pb_2C_{18}H_{38}O_4$ $(Et_3PbO_2CCH_2CH_2)_2$, 2, 574
$Pb_2C_{18}H_{42}$ Pr_6Pb_2, 2, 590
$Pb_2C_{22}H_{44}O_4$ $(Et_3Pb)_2$ d-camphorate, 2, 578*
$Pb_2C_{24}H_{22}O_3$ $(Ph_2PbOH)_2O$, 2, 569
$Pb_2C_{24}H_{46}O_4$ p-$(Et_3PbO_2CMe_2)_2C_6H_4$ (?), 2, 570
$Pb_2C_{24}H_{54}$ Bu_6Pb_2, 2, 590
$Pb_2C_{26}H_{44}N_2O_4S_2$ $(Et_3PbNPhSO_2CH_2)_2$, 2, 586
$Pb_2C_{28}H_{26}O_5$ $(Ph_2PbOAc)_2O$, 2, 569, 581
$Pb_2C_{28}H_{56}O_4$ $(Bu_3PbO_2CCH{=})_2$, 2, 576
$Pb_2C_{32}H_{34}O_5$ (i-$PrCO_2PbPh_2)_2O$, 2, 569
$Pb_2C_{36}H_{30}$ Ph_6Pb_2, 2, 589
$Pb_2C_{36}H_{30}O_2$ $(Ph_3PbO)_2$, 2, 570
$Pb_2C_{36}H_{30}S$ $(Ph_3Pb)_2S$, 2, 585
$Pb_2C_{36}H_{66}$ $Cyclohexyl_6Pb_2$, 2, 588
$Pb_2C_{38}H_{30}$ $(Ph_3PbC{\equiv})_2$, 2, 548
$Pb_2C_{38}H_{66}$ $(Cyclohexyl_3PbC{\equiv})_2$, 2, 548
$Pb_2C_{39}H_{36}$ $(Ph_3PbCH_2)_2CH_2$, 2, 548
$Pb_2C_{40}H_{30}$ $(Ph_3PbC{\equiv}C)_2$, 2, 548
$Pb_2C_{40}H_{32}O_3$ 2,3-$(Ph_3Pb)_2$ succinic anhydride, 2, 549
$Pb_2C_{40}H_{32}O_4$ $(Ph_3PbO_2CCH{=})_2$, 2, 576
$Pb_2C_{40}H_{38}$ $(Ph_3PbCH_2CH_2)_2$, 2, 548
$Pb_2C_{40}H_{66}$ $(Cyclohexyl_3PbC{\equiv}C)_2$, 2, 548
$Pb_2C_{42}H_{30}$ $(Ph_3PbC{\equiv}CC{\equiv})_2$, 2, 548
$Pb_2C_{42}H_{34}$ p-$(Ph_3Pb)_2C_6H_4$, 2, 549
$Pb_2C_{42}H_{36}Br_6$ (o-$BrC_6H_4CH_2)_6Pb_2$, 2, 590
$Pb_2C_{42}H_{36}Cl_6$ (o-$ClC_6H_4CH_2)_6Pb_2$, 2, 590
$Pb_2C_{42}H_{42}$ (o-$MeC_6H_4)_6Pb_2$, 2, 590
(m-$MeC_6H_4)_6Pb_2$, 2, 590
(p-$MeC_6H_4)_6Pb_2$, 2, 590
$Pb_2C_{42}H_{42}O_6$ (o-$MeOC_6H_4)_6Pb_2$, 2, 590
(p-$MeOC_6H_4)_6Pb_2$, 2, 590
$Pb_2C_{42}H_{66}$ $(Cyclohexyl_3PbC{\equiv}CC{\equiv})_2$, 2, 548
$Pb_2C_{44}H_{42}$ [(o-$MeC_6H_4)_3PbC{\equiv}]_2$, 2, 549
[(p-$MeC_6H_4)_3PbC{\equiv}]_2$, 2, 549
$Pb_2C_{48}H_{54}O_6$ (o-$EtOC_6H_4)_6Pb_2$, 2, 590
(p-$EtOC_6H_4)_6Pb_2$, 2, 590
$Pb_2C_{50}H_{54}$ $[(PhCH_2CH_2)_3PbC{\equiv}]_2$, 2, 548
$Pb_2C_{54}H_{66}$ $Mesityl_6Pb_2$, 2, 590
$Pb_2C_{60}H_{42}$ (1-$C_{10}H_7)_6Pb_2$, 2, 590, 597
$Pb_2C_{60}H_{46}O_2$ {4-[3,4-$Ph_3Pb(HO)C_6H_3$]-$C_6H_4\}_2$, 2, 549
$Pb_3C_{24}H_{50}O_7$ $(Et_3Pb)_3$ citrate, 2, 578*
$Pb_5C_{72}H_{60}$ $(Ph_3Pb)_4Pb$, 2, 592
$Pb_5C_{84}H_{84}$ [(p-$MeC_6H_4)_3Pb]_4Pb$, 2, 592
$PbAsC_{12}H_{22}NO_3$ p-$H_2NC_6H_4AsO_3H(PbEt_3)$, 2, 564
$PbAsC_{13}H_{14}ClO_3$ $Ph_2Pb(HO_3AsMe)Cl$, 2, 565
$PbAsC_{14}H_{16}ClO_2$ $Ph_2Pb(O_2AsMe_2)Cl$, 2, 566, 3, 374
$PbAsC_{15}H_{18}ClO_3$ $Ph_2Pb(HO_3AsPr)Cl$, 2, 566
$PbAsC_{20}H_{21}O_2$ $Ph_3PbO_2AsMe_2$, 2, 565, 3, 374
$PbAs_2C_{18}H_{22}O_6$ $Ph_2Pb(HO_3AsCH_2CH{=}CH_2)_2$, 2, 566
$PbAs_2C_{24}H_{20}N_2O_{10}$ (m-$O_2NC_6H_4AsO_3H)_2$-$PbPh_2$, 2, 566
$PbAs_2C_{24}H_{22}O_6$ $Ph_2Pb(HO_3AsPh)_2$, 2, 566
$PbAs_2C_{26}H_{26}O_6$ $Ph_2Pb(HO_3AsCH_2Ph)_2$, 2, 566
$PbCoC_{22}H_{15}O_4$ $Ph_3PbCo(CO)_4$, 2, 595
$PbCr_2C_{32}H_{32}O_6$ $(C_5H_5)_2Pb[Me_2CHC_5H_4Cr(CO)_3]_2$, 2, 594
$PbFeC_{24}H_{36}O_8$ $[Bu_2PbFe(CO)_4]_2$, 2, 595
$PbFe_2C_{16}H_{16}O_4$ $Me_2Pb[C_5H_5Fe(CO)_2]_2$, 1, 534, 2, 595
$PbGeC_7H_{21}N$ $Me_3PbNMeGeMe_3$, 2, 128*, 587
$PbGe_4C_{72}H_{60}$ $(Ph_3Ge)_4Pb$, 2, 136, 593
$PbHfC_{27}H_{20}O_4$ $Ph_3Pb(C_5H_5)Hf(CO)_4$, 2, 594
$PbHfC_{43}H_{34}O_4$ (1-$C_{10}H_7)_3Pb(C_5H_3Et_2)Hf(CO)_4$, 1, 42, 2, 594
$PbIrC_{38}H_{71}O$ $(C_{16}H_{33})_2PbC_5H_5IrCO$, 2, 595
$PbLiC_{18}H_{15}$ Ph_3PbLi, 2, 591
$PbMgC_{12}H_{19}Br$ p-$Et_3PbC_6H_4MgBr$, 2, 550
$PbMgC_{27}H_{33}Br$ $Mesityl_3PbMgBr$, 2, 593
$PbMnC_8H_9O_5$ $Me_3PbMn(CO)_5$, 2, 594
$PbMnC_{11}H_{15}O_5$ $Et_3PbMn(CO)_5$, 2, 594
$PbMnC_{14}H_{15}O_5$ $(CH_2{=}CHCH_2)_3PbMn(CO)_5$, 2, 594
$PbMnC_{17}H_{23}O_3$ $(CH_2{=}CHCH_2)_3PbMn(CO)_3\cdot C_5H_8$, 1, 184, 2, 594*

$PbMnC_{23}H_{15}O_5$ $Ph_3PbMn(CO)_5$, 2, 594

$PbMnC_{23}H_{21}O_5$ $(MeC_5H_4)_3PbMn(CO)_5$, 2, 594

$PbMnC_{26}H_{21}O_5$ $(m\text{-}MeC_6H_4)_3PbMn(CO)_5$, 2, 594

$PbMnC_{29}H_{25}O_4$ $(m\text{-}MeC_6H_4)_3PbMn(CO)_3 \cdot C_5H_4O$, 2, 594

$PbMnC_{29}H_{27}O_5$ $(PhCH_2CH_2)_3PbMn(CO)_5$, 2, 594

$PbMnC_{29}H_{51}O_5$ $(C_8H_{17})_3PbMn(CO)_5$, 2, 594

$PbMnC_{32}H_{33}O_5$ $Mesityl_3PbMn(CO)_5$, 2, 594

$PbMnC_{32}H_{59}O_3$ $(C_8H_{17})_3PbMn(CO)_3 \cdot C_5H_8$, 1, 184, 2, 594*

$PbMnC_{35}H_{35}O_3$ $(PhCH_2CH_2)_3PbMn(CO)_3 \cdot C_8H_8$, 2, 594

$PbMnC_{35}H_{37}O_4$ $Mesityl_3PbMn(CO)_3 \cdot C_5H_4O$, 2, 594

$PbMnC_{50}H_{35}O_4$ $Ph_3PbMn(CO)_3 \cdot C_5Ph_4O$, 2, 594

$PbMnC_{50}H_{41}O_4$ $(MeC_5H_4)_3PbMn(CO)_3 \cdot C_5Ph_4O$, 2, 594

$PbMnC_{59}H_{53}O_4$ $Mesityl_3PbMn(CO)_3$-(PhC=CPhCPh=CPhCO), 1, 184

$PbMn_2C_{12}H_6O_{10}$ $Me_2Pb[Mn(CO)_5]_2$, 2, 594

$PbMn_2C_{14}H_{10}O_{10}$ $Et_2Pb[Mn(CO)_5]_2$, 2, 594

$PbMoC_{26}H_{20}O_3$ $Ph_3PbC_5H_5Mo(CO)_3$, 1, 118, 2, 594

$PbMo_3C_{28}H_{24}O_9$ $MePb[C_5H_4MeMo(CO)_3]_3$, 1, 118, 2, 594

$PbNaC_6H_{15}$ Et_3PbNa, 2, 593

$PbNaC_{18}H_{15}$ Ph_3PbNa, 2, 593

$PbNaC_{18}H_{33}$ $Cyclohexyl_3PbNa$, 2, 593

$PbNa_2C_{18}H_{14}O_9S_2$ $Ph_2PbO \cdot (NaO_3S)_2C_6H_2(OH)_2$ (?), 2, 566*

$PbNiC_7H_8Cl_2O$ $MePbCl_2(C_5H_5)NiCO$, 2, 595

$PbNiC_{10}H_{15}BrO$ $Me_2PbBr(EtC_5H_4)NiCO$, 1, 663, 2, 595

$PbNiC_{15}H_{14}I_2O_2$ $o\text{-}MeC_6H_4PbI_2(C_5H_4Ac)$-NiCO, 1, 663, 2, 595

$PbNi_2C_{21}H_{29}ClO_2$ $MePbCl(C_5H_3Et_2NiCO)_2$, 2, 595

$PbPdC_{16}H_{24}Cl_2O$ Cyclohexyl-$PbCl_2(C_5H_3Et_2)PdCO$, 1, 687, 2, 595

$PbPdC_{18}H_{19}ClO$ Et_2PbCl(9-fluorenyl)-PdCO, 2, 595

$PbPdC_{28}H_{46}O$ $Cyclohexyl_3Pb(C_5H_3Et_2)$-PdCO, 1, 687, 2, 595

$PbPtC_{24}H_{34}O$ Et_2Pb(2-indenyl-PtCO$)_2$, 2, 595

$PbPtC_{35}H_{42}O$ $Mesityl_3Pb(C_5H_4Et)PtCO$, 2, 595

$PbPtC_{36}H_{51}FO$ $Ph_2PbF(C_5H_4C_{18}H_{37})PtCO$, 1, 733, 2, 595

$PbPtC_{42}H_{56}O$ $Ph_3Pb(C_5H_4C_{18}H_{37})PtCO$, 2, 595

$PbReC_{23}H_{15}O_5$ $Ph_3PbRe(CO)_5$, 2, 594

$PbRhC_{30}H_{43}O$ $(C_8H_{17})_2Pb$(9-fluorenyl)-RhCO, 2, 595

$PbSiC_6H_{18}O$ $Me_3PbOSiMe_3$, 2, 572

$PbSiC_7H_{20}$ $Me_3PbCH_2SiMe_3$, 2, 550

$PbSiC_9H_{24}O$ $Me_3PbOSiEt_3$, 2, 572
 $Et_3PbOSiMe_3$, 2, 572

$PbSiC_{12}H_{30}O$ $Et_3PbOSiEt_3$, 2, 572

$PbSiC_{36}H_{30}O$ $Ph_3PbOSiPh_3$, 2, 572

$PbSi_2C_9H_{27}N$ $Me_3PbN(SiMe_3)_2$, 2, 587

$PbSi_3C_{12}H_{33}Cl$ $(Me_3SiCH_2)_3PbCl$, 2, 558

$PbSi_4C_{16}H_{44}$ $(Me_3SiCH_2)_4Pb$, 2, 532

$PbSnC_{36}H_{30}S$ $Ph_3PbSSnPh_3$, 2, 473, 585

$PbSnC_{44}H_{38}$ $p\text{-}Ph_3SnCH_2CH_2C_6H_4PbPh_3$, 2, 251

$PbSn_2C_{34}H_{40}N_2O_8$ $[(CH_2{=}CH)_3SnO_2CCH_2]_2NCH_2CH_2N(CH_2CO_2)_2PbPh_2$, 2, 429

$PbSn_2C_{64}H_{56}$ $(p\text{-}Ph_3SnCH_2CH_2C_6H_4)_2PbPh_2$, 2, 251

$PbSn_4C_{72}H_{60}$ $(Ph_3Sn)_4Pb$, 2, 495, 593

$PbTiC_{16}H_{22}O_4$ $Et_3Pb(C_5H_4Me)Ti(CO)_4$, 1, 35, 2, 594

$PbTi_2C_8H_{20}Cl_4$ $Et_4Pb \cdot 2TiCl_4$, 2, 524

$PbTi_2C_{32}H_{24}O_8$ $Ph_2Pb[MeC_5H_4Ti(CO)_4]_2$, 2, 594

$PbWC_{26}H_{20}O_3$ $Ph_3PbC_5H_5W(CO)_3$, 1, 130, 2, 594

$PbWC_{45}H_{58}O_3$ $Ph_3Pb(MeC_5H_3C_{18}H_{37})W(CO)_3$, 1, 130*, 2, 594*

$PbZr_3C_{45}H_{50}$ Cyclohexyl-$Pb[Et_2C_5H_3Zr(CO)_4]_3$, 2, 594

$Pb_2AsC_{25}H_{23}Cl_2O_3$ $(Ph_2PbCl)_2O_3AsMe$, 2, 565

$Pb_2As_2C_{26}H_{26}Cl_2O_5$ $(Ph_2PbClO_2AsMe)_2O$, 2, 566*

$Pb_2FeC_{40}H_{30}O_4$ $(Ph_3Pb)_2Fe(CO)_4$, 2, 595

$Pb_2FeC_{40}H_{66}O_4$ $[(Cyclohexyl)_3Pb]_2Fe(CO)_4$, 2, 595

$Pb_2Fe_2C_{12}H_{12}O_8$ $[Me_2PbFe(CO)_4]_2$, 2, 595

$Pb_2Fe_2C_{16}H_{20}O_8$ $[Et_2PbFe(CO)_4]_2$, 2, 595

$Pb_2Fe_2C_{20}H_{28}O_8$ $[Pr_2PbFe(CO)_4]_2$, 2, 595

$Pb_2Fe_2C_{32}H_{20}O_8$ $[Ph_2PbFe(CO)_4]_2$, 2, 595

$Pb_2Nb_2C_{88}H_{158}O_6$ $[(C_{16}H_{33})_2Pb(C_5H_4Bu)-Nb(CO)_3]_2$, 2, 594*

$Pb_2SnC_{64}H_{56}$ $(p-Ph_3PbC_6H_4CH_2CH_2)_2SnPh_2$, 2, 251

$Pb_2Sn_2C_{62}H_{62}$ $p-(p-Ph_3PbC_6H_4CH_2CH_2-Me_2Sn)_2C_6H_4$, 2, 251

$Pb_2Ta_2C_{44}H_{38}O_6$ $[(MeC_6H_4)_2Pb(C_5H_5)Ta-(CO)_3]_2$, 2, 594

$Pb_2V_2C_{64}H_{54}O_6$ $[(Me_2C_6H_3)_2Pb$(9-fluorenyl)$V(CO)_3]_2$, 1, 52, 2, 594

$Pb_4GeC_{72}H_{60}$ $(Ph_3Pb)_4Ge$, 2, 136, 593

$Pb_4SnC_{72}H_{60}$ $(Ph_3Pb)_4Sn$, 2, 495, 593

$PbAsMnC_{16}H_{14}Cl_2O_4$ $Cl_2PhPbMn[As-(CH{=}CH_2)_3](CO)_4$, 3, 6

$PbAsMnC_{34}H_{42}O_4$ $\{Pr_3PbMn[As(CH_2Ph)_3]-(CO)_4\}$, 3, 27

$PbMnSbC_{28}H_{20}Br_2O_4$ $Br_2PhPb[Mn(SbPh_3)-(CO)_4]$, 3, 663

$PbMnSbC_{28}H_{30}O_4$ $Ph_3Pb[Mn(CO)_4(SbEt_3)]$, 3, 656

$PbSnTeC_{36}H_{30}$ $Ph_3PbTeSnPh_3$, 2, 473, 585

PdC_5H_5Br C_5H_5PdBr, 1, 670

PdC_5H_5Cl C_5H_5PdCl, 1, 670

PdC_5H_5NO C_5H_5PdNO, 1, 670

$PdC_6H_6Cl_2N_2$ $[(CH_2{=}CHCN)_2PdCl_2]$, 1, 685

PdC_6H_{10} $(\pi-CH_2{=}CHCH_2)_2Pd$, 1, 681

$PdC_6H_{10}Cl_2$ $(-CH_2CH{=}CH_2)_2PdCl_2$, 1, 685

$PdC_6H_{16}S_2$ $Me_2Pd(MeSCH_2-)_2$, 1, 665

$PdC_7H_8Br_2$ Bicyclo[2.2.1]heptadiene-$PdBr_2$, 1, 685

$PdC_7H_8Cl_2$ Bicyclo[2.2.1]heptadiene-$PdCl_2$, 1, 685

$PdC_8H_8Cl_2$ Cyclooctatetraene-$PdCl_2$, 1, 685

PdC_8H_{10} $\pi-CH_2{=}CHCH_2PdC_5H_5$, 1, 669

$PdC_8H_{12}Br_2$ 1,5-Cyclooctadiene-$PdBr_2$, 1, 685

$PdC_8H_{12}Cl_2$ 1,5-Cyclooctadiene-$PdCl_2$, 1, 685

$PdC_8H_{12}O_2$ Acetylacetonato(π-allyl)-Pd, 1, 680

PdC_8H_{14} $(\pi-CH_2{=}CMeCH_2)_2Pd$, 1, 681

$PdC_8H_{14}N_2$ $(Me_2CHNC)_2Pd$, 1, 687

$PdC_8H_{20}S_2$ $Me_2Pd(EtSCH_2-)_2$, 1, 665

PdC_9H_{12} $\pi-CH_2{=}CHCH_2Pd(C_5H_4Me)$, 1, 669

$PdC_9H_{14}O_2$ Acetylacetonato(π-2-methallyl)-Pd, 1, 680

$PdC_{10}H_{10}$ $(C_5H_5)_2Pd$, 1, 670

$PdC_{10}H_{10}ClN$ $[\pi-CH_2{=}CHCH_2Pd(NCPh)Cl]$, 1, 681

$PdC_{10}H_{10}Cl_2$ Tricyclo[$4.2.2.0^{2,5}$]deca-3,7,9-triene-$PdCl_2$, 1, 686

$PdC_{10}H_{12}$ π-Cyclopentenyl-PdC_5H_5, 1, 669

$PdC_{10}H_{12}Cl_2$ Tricyclo[$4.2.2.0^{2,5}$]deca-3,9-diene-$PdCl_2$, 1, 686

Tricyclo[$5.2.1.0^{2,6}$]deca-3,8-diene-$PdCl_2$, 1, 686

$PdC_{10}H_{14}$ $\pi-CH_2{=}CHCH_2Pd(C_5H_4Et)$, 1, 669

$PdC_{10}H_{14}ClNO$ $\pi-MeOCH_2CH{=}CHCH_2Pd(Py)Cl$, 1, 681

$PdC_{10}H_{18}$ 1,5-Cyclooctadiene-$PdMe_2$, 1, 668

$PdC_{11}H_{11}FN_2$ MePd(Bipy)F, 1, 664

$PdC_{11}H_{14}$ $\pi-CH_2{=}CHCH_2Pd(C_5H_4CH_2CH{=}CH_2)$, 1, 669

π-Cyclohexenyl-PdC_5H_5, 1, 669

$PdC_{11}H_{15}ClO_2$ Acetylacetonato$\{\pi-CH_2{=}C-[C({=}CH_2)CH_2Cl]CH_2\}$-Pd, 1, 680

$PdC_{11}H_{16}$ $\pi-CH_2{=}CHCH_2Pd(C_5H_4Pr)$, 1, 669

$PdC_{11}H_{16}O$ $\pi-CH_2{=}CMeCH(CH_2OMe)PdC_5H_5$, 1, 669

$PdC_{11}H_{18}O_3$ Acetylacetonato-$[\pi-CH_2{=}CMe-CH(CH_2OMe)]$-Pd, 1, 680

Acetylacetonato-$\{\pi-CH_2{=}CHCH[CH(OMe)-Me]\}$-Pd, 1, 680

$PdC_{12}H_{14}N_2$ $Me_2Pd(Bipy)$, 1, 665

$PdC_{12}H_{16}$ π-2-Cyclohexenyl-$Pd(C_5H_4Me)$,

1, 669

$PdC_{12}H_{16}ClNO$ [π-AcCHC(Me)CH_2Pd(picoline)Cl], 1, 681

$PdC_{12}H_{16}Cl_2$ 7,8-Dimethyltricyclo[4.2.2.0^{2,5}]deca-3,9-diene-$PdCl_2$, 1, 686

$PdC_{12}H_{16}Cl_2O_2$ 7,8-($HOCH_2$)$_2$-tricyclo[4.2.2.0^{2,5}]deca-3,9-diene-$PdCl_2$, 1, 686

$PdC_{12}H_{18}O$ π-CH_2=CHCH[C(OMe)Me_2]PdC_5H_5 1, 669

$PdC_{12}H_{20}O_3$ Acetylacetonato[π-CH_2=C-(Me)C(Me)(CH_2OMe)]-Pd, 1, 680

$PdC_{13}H_{18}$ π-CH_2=C(Me)CH(CH=CMe_2)PdC_5H_5 1, 669

$PdC_{13}H_{20}O$ π-CH_2=C(Me)CH[C(OMe)Me_2]Pd-C_5H_5, 1, 669

$PdC_{13}H_{20}O_3$ Acetylacetonato[π-CH_2=C-(Me)CH(CH=CMe_2)]-Pd, 1, 680

$PdC_{13}H_{22}O_2$ Acetylacetonato{π-CH_2=C-(Me)CH[C(OMe)Me_2]}-Pd, 1, 680

$PdC_{13}H_{33}BrP_2$ MePd(PEt_3)$_2$Br, 1, 664

$PdC_{14}H_8Cl_2I_2N_2$ (p-ClC_6H_4NC)$_2PdI_2$, 1, 687

$PdC_{14}H_{10}I_2N_2$ (PhNC)$_2PdI_2$, 1, 687

$PdC_{14}H_{10}N_2$ (PhNC)$_2$Pd, 1, 687

$PdC_{14}H_{11}F_7N_2$ Me(n-C_3F_7)Pd(Bipy), 1, 665

$PdC_{14}H_{14}Cl_2O_4$ 7,8-(MeO_2C)$_2$-tricyclo[4.2.2.0^{2,5}]deca-3,7,9-triene-$PdCl_2$, 1, 686

$PdC_{14}H_{16}Cl_2O_4$ 7,8-(MeO_2C)$_2$-tricyclo[4.2.2.0^{2,5}]deca-3,9-diene-$PdCl_2$, 1, 686

$PdC_{14}H_{22}ClNO$ [π-Me_2(MeO)CCH=CHC(Me_2)-Pd(Py)Cl], 1, 681

$PdC_{14}H_{22}N_2$ (cyclo-C_6H_{11}NC)$_2$Pd, 1, 687

$PdC_{14}H_{22}O$ π-Me_2C=CHCH[C(OMe)Me_2]Pd-C_5H_5, 1, 669

$PdC_{14}H_{33}BrOP_2$ AcPd(PEt_3)$_2$Br, 1, 664

$PdC_{14}H_{33}ClOP_2$ AcPd(PEt_3)$_2$Cl, 1, 664

$PdC_{14}H_{33}IOP_2$ AcPd(PEt_3)$_2$I, 1, 664

$PdC_{14}H_{33}NP_2$ MePd(PEt_3)$_2$CN, 1, 664

$PdC_{14}H_{33}NP_2S$ MePd(PEt_3)$_2$SCN, 1, 664

$PdC_{14}H_{36}P_2$ Me_2Pd(PEt_3)$_2$, 1, 665

$PdC_{16}H_8F_{14}N_2$ (n-C_3F_7)$_2$Pd(Bipy), 1, 665

$PdC_{16}H_{12}Cl_2$ 1,2;5,6-Dibenzocyclooctatetraene-$PdCl_2$, 1, 686

$PdC_{16}H_{14}Br_2N_2O_2$ (p-$MeOC_6H_4$NC)$_2PdBr_2$, 1, 687

$PdC_{16}H_{14}Cl_2N_2$ (HNC_6H_4CH=C)$_2$Pd·2HCl, 1, 669

$PdC_{16}H_{14}I_2N_2$ (p-MeC_6H_4NC)$_2PdI_2$, 1, 687

$PdC_{16}H_{14}I_2N_2O_2$ (p-$MeOC_6H_4$NC)$_2PdI_2$, 1, 687

$PdC_{16}H_{14}N_2$ (p-MeC_6H_4NC)$_2$Pd, 1, 687

$PdC_{16}H_{14}N_2O_2$ (p-$MeOC_6H_4$NC)$_2$Pd, 1, 687

$PdC_{16}H_{16}Cl_2$ (PhCH=CH_2)$_2PdCl_2$, 1, 685

$PdC_{16}H_{20}ClNO$ [π-Me_2(MeO)CCH=CHCH_2Pd-(quinoline)Cl], 1, 681

$PdC_{16}H_{20}O$ Methoxytricyclo[5.2.1.0^{2,6}]-decenyl-PdC_5H_5, 1, 670

$PdC_{18}H_{24}ClNO$ {[Methoxytricyclo[5.2.1.0^{2,6}]decenyl](p-$MeC_6H_4NH_2$)-PdCl}, 1, 667

$PdC_{18}H_{24}Cl_2$ Bis(spiro[4.4]nona-1,3-diene)-$PdCl_2$, 1, 686

$PdC_{18}H_{35}BrP_2$ PhPd(PEt_3)$_2$Br, 1, 664

$PdC_{19}H_{34}BrF_3P_2$ p-$CF_3C_6H_4$Pd(PEt_3)$_2$Br, 1, 644

$PdC_{21}H_{41}BrP_2$ 2,4,6-$Me_3C_6H_2$Pd(PEt_3)$_2$Br, 1, 664

$PdC_{24}H_{24}ClOP$ π-AcCH=CMeCH_2Pd(PPh_3)Cl, 1, 681

$PdC_{25}H_{39}F_3P_2$ Ph(p-$CF_3C_6H_4$)Pd(PEt_3)$_2$, 1, 665

$PdC_{28}H_{20}Br_2$ Ph_4-cyclobutadiene-$PdBr_2$, 1, 682

$PdC_{28}H_{20}Cl_2$ Ph_4-cyclobutadiene-$PdCl_2$, 1, 681

$PdC_{28}H_{20}I_2$ Ph_4-cyclobutadiene-PdI_2, 1, 682

$PdC_{28}H_{30}P_2$ Me_2Pd(Ph_2PCH_2-)$_2$, 1, 665

$PdC_{28}H_{38}N_2O_4P_2$ (p-$O_2NC_6H_4$C≡C)$_2$Pd-(PEt_3)$_2$, 1, 665

$PdC_{28}H_{40}P_2$ (PhC≡C)$_2$Pd(PEt_3)$_2$, 1, 665

$PdC_{28}H_{50}N_2P_2$ (p-$Me_2NC_6H_4$)$_2$Pd(PEt_3)$_2$, 1, 665

$PdC_{33}H_{25}Br$ {C_5H_5Pd[PhC=C(Ph)C(Ph)=C-(Ph)]}Br, 1, 670

$PdC_{36}H_{31}Cl$ 3,5,7,2,4,6,8-Me_3(Ph)$_4$-cyclooctatetraenyl-CH_2PdCl, 1, 668

$PdC_{36}H_{31}I$ 3,5,7,2,4,6,8-$Me_3(Ph)_4$-cyclooctatetraenyl-CH_2PdI, 1, 668

$PdC_{37}H_{31}NS$ 3,5,7,2,4,6,8-$Me_3(Ph)_4$-cyclooctatetraenyl-CH_2PdSCN, 1, 668

$PdC_{38}H_{36}P_2$ $Me_2Pd(PPh_3)_2$, 1, 665

$PdC_{40}H_{39}Cl$ 3,5,7,2,4,6,8-$Et_3(Ph)_4$-cyclooctatetraenyl-CH_2CH_2PdCl, 1, 668

$PdC_{40}H_{39}I$ 3,5,7,2,4,6,8-$Et_3(Ph)_4$-cyclooctatetraenyl-CH_2CH_2PdI, 1, 668

$PdC_{41}H_{39}NS$ 3,5,7,2,4,6,8-$Et_3(Ph)_4$-cyclooctatetraenyl-CH_2CH_2PdSCN, 1, 668

$PdC_{61}H_{41}Cl_9NO_9P_3$ $(PhNC)Pd[P(OC_6H_4$-Cl-p$)_3]_3$, 1, 687

$PdC_{62}H_{43}Cl_9NOP_3$ $(p\text{-}MeOC_6H_4NC)Pd$-$[P(C_6H_4Cl\text{-}p)_3]_3$, 1, 687

$PdC_{62}H_{43}Cl_9NO_9P_3$ $(p\text{-}MeC_6H_4NC)Pd$-$[P(OC_6H_4Cl\text{-}p)_3]_3$, 1, 687

$PdC_{62}H_{43}Cl_9NO_{10}P_3$ $(p\text{-}MeOC_6H_4NC)Pd$-$[P(OC_6H_4Cl\text{-}p)_3]_3$, 1, 687

$PdC_{62}H_{52}NO_{10}P_3$ $(p\text{-}MeOC_6H_4NC)Pd$-$[P(OPh)_3]_3$, 1, 687

$Pd_2C_4H_8Cl_4$ $[Pd(C_2H_4)Cl_2]_2$, 1, 683

$Pd_2C_5H_5Cl_2O_2$ π-CH_2=$CHCH_2PdCl_2Pd(CO)_2$, 1, 679*

$Pd_2C_6H_7Cl_2O_2$ π-CH_2=$C(Me)CH_2PdCl_2Pd$-$(CO)_2$, 1, 679*

π-MeCH=$CHCH_2PdCl_2Pd(CO)_2$, 1, 679*

$Pd_2C_6H_8Br_2Cl_2$ (π-CHBr=$CHCH_2PdCl)_2$, 1, 674

$Pd_2C_6H_8Cl_4$ (π-CH_2=$CClCH_2PdCl)_2$, 1, 674

(π-CHCl=$CHCH_2PdCl)_2$, 1, 674

$Pd_2C_6H_{10}Br_2$ (π-CH_2=$CHCH_2PdBr)_2$, 1, 672

$Pd_2C_6H_{10}Cl_2$ (π-CH_2=$CHCH_2PdCl)_2$, 1, 672

$Pd_2C_8H_{12}Cl_4$ (CH_2=CHCH=$CH_2PdCl_2)_2$, 1, 673

[π-CH_2=$CHCH(CH_2Cl)PdCl]_2$, 1, 673

$Pd_2C_8H_{14}Cl_2$ [π-CH_2=$C(Me)CH_2PdCl]_2$, 1, 672

(π-MeCH=$CHCH_2PdCl)_2$, 1, 673

$Pd_2C_8H_{16}Cl_4$ $[(Me_2C$=$CH_2)PdCl_2]_2$, 1, 684

$Pd_2C_9H_{12}Cl_4$ π-CH_2=$C[C(CH_2Cl)$=$CH_2]$-$CH_2PdCl_2PdCH_2CCl$=CH_2-π, 1, 679

$Pd_2C_{10}H_{16}Cl_4$ [π-CH_2=$CHCH(CHClMe)PdCl]_2$, 1, 674

[π-CH_2=$CClC(Me_2)PdCl]_2$, 1, 674

$Pd_2C_{10}H_{18}Br_2O_2$ [π-CH_2=$CHCH(CH_2OMe)Pd$-$Br]_2$, 1, 676

$Pd_2C_{10}H_{18}Cl_2$ [π-CH_2=$C(Me)CH(Me)PdCl]_2$, 1, 674

$Pd_2C_{10}H_{18}Cl_2O_2$ [π-CH_2=$CHCH(CH_2OMe)Pd$-$Cl]_2$, 1, 676

$Pd_2C_{12}H_{16}Br_4$ {π-CH_2=$C[C(=CH_2)CH_2Br]$-$CH_2PdBr\}_2$, 1, 674

$Pd_2C_{12}H_{16}Cl_2I_2$ {π-CH_2=$C[C(=CH_2)CH_2Cl]$-$CH_2PdI\}_2$, 1, 674

$Pd_2C_{12}H_{16}Cl_4$ {π-CH_2=$C[C(=CH_2)CH_2Cl]$-$CH_2PdCl\}_2$, 1, 674

$Pd_2C_{12}H_{18}Cl_2$ (π-2-Cyclohexenyl-PdCl$)_2$, 1, 678

$Pd_2C_{12}H_{18}Cl_2O_2$ [π-$CH_2C(Me)CH(Ac)PdCl]_2$, 1, 675

$Pd_2C_{12}H_{20}Cl_4$ [π-CH_2=$CMeC(Me)(CH_2Cl)Pd$-$Cl]_2$, 1, 674

$Pd_2C_{12}H_{22}Br_2O_2$ [π-CH_2=$C(Me)CH(CH_2OMe)$-$PdBr]_2$, 1, 676

$Pd_2C_{12}H_{22}Cl_2$ [π-CH_2=$C(Me)CH(Et)PdCl]_2$, 1, 674

[π-MeCH=$C(Me)CH(Me)PdCl]_2$, 1, 674

[π-CH_2=$C(Me)C(Me_2)PdCl]_2$, 1, 674

[π-CH_2=$CHCH_2Pd(CH_2$=$CHMe)Cl]_2$, 1, 681

$Pd_2C_{12}H_{22}Cl_2O_2$ [π-CH_2=$C(Me)CH(CH_2OMe)$-$PdCl]_2$, 1, 676

$Pd_2C_{12}H_{22}I_2O_2$ [π-CH_2=$C(Me)CH(CH_2OMe)$-$PdI]_2$, 1, 676

$Pd_2C_{14}H_{16}Cl_4$ (Cycloheptatriene-$PdCl_2)_2$, 1, 684

$Pd_2C_{14}H_{20}Cl_4$ (Cyclohepta-1,3-diene-$PdCl_2)_2$, 1, 684

$Pd_2C_{14}H_{22}Cl_2$ (2-Methyl-π-2-cyclohexenyl-PdCl$)_2$, 1, 678

(π-2-Cycloheptenyl-PdCl$)_2$, 1, 678

$Pd_2C_{14}H_{22}N_2O_2S_2$ [π-CH$_2$=C(Me)CH(CH$_2$OMe)Pd(SCN)]$_2$, 1, 676
$Pd_2C_{14}H_{24}Cl_4$ (Methylcyclohexene-PdCl$_2$)$_2$, 1, 684
(Cycloheptene-PdCl$_2$)$_2$, 1, 684
$Pd_2C_{14}H_{26}Br_2O_2$ [π-CH$_2$=C(Me)C(Me)(CH$_2$OMe)PdBr]$_2$, 1, 676
{π-CH$_2$=CHCH[C(OMe)Me$_2$]PdBr}$_2$, 1, 676
$Pd_2C_{14}H_{26}Cl_2$ [π-MeCH=C(Et)CH(Me)PdCl]$_2$, 1, 674
[π-CH$_2$=C(Me)CH(CHMe$_2$)PdCl]$_2$, 1, 674
[π-CH$_2$=C(Me)CH(Pr)PdCl]$_2$, 1, 674
[π-MeCH=C(Et)CH(Me)PdCl]$_2$, 1, 675
$Pd_2C_{14}H_{26}Cl_2O_2$ {π-CH$_2$=CHCH[C(OMe)Me$_2$]PdCl}$_2$, 1, 676
[π-CH$_2$=C(Me)CH(CH$_2$OEt)PdCl]$_2$, 1, 676
[π-CH$_2$=C(Me)C(Me)(CH$_2$OMe)PdCl]$_2$, 1, 676
$Pd_2C_{14}H_{26}I_2O_2$ {π-CH$_2$=CHCH[C(OMe)Me$_2$]PdI}$_2$, 1, 676
$Pd_2C_{16}H_{16}Cl_4$ [(PhCH=CH$_2$)-PdCl$_2$]$_2$, 1, 684
$Pd_2C_{16}H_{22}Cl_2$ (π-Cyclooctadienyl-PdCl)$_2$, 1, 679
$Pd_2C_{16}H_{26}Cl_2$ [π-CH$_2$=C(Me)CH(CH=CMe$_2$)-PdCl]$_2$, 1, 675
$Pd_2C_{16}H_{28}Cl_2$ (π-1-Cycloheptenylmethyl-PdCl)$_2$, 1, 678
$Pd_2C_{16}H_{28}Cl_4$ [π-Me$_2$C=CHCH(CClMe$_2$)PdCl]$_2$, 1, 675
(Cyclooctene-PdCl$_2$)$_2$, 1, 684
$Pd_2C_{16}H_{30}Cl_2$ [π-CH$_2$=C(Me)CH(CH$_2$CHMe$_2$)-PdCl]$_2$, 1, 675
[π-CH$_2$=C(CH$_2$CMe$_3$)CH$_2$PdCl]$_2$, 1, 675
[π-CH$_2$=C(Me)CH(CMe$_3$)PdCl]$_2$, 1, 675
$Pd_2C_{16}H_{30}Cl_2O_2$ {π-CH$_2$=C(Me)CH[C(OMe)Me$_2$]PdCl}$_2$, 1, 676
[π-CH$_2$=C(Me)CH(CH$_2$OCHMe$_2$)PdCl]$_2$, 1, 676
$Pd_2C_{16}H_{32}Cl_4$ (Diisobutene-PdCl$_2$)$_2$, 1, 684
$Pd_2C_{18}H_{18}Cl_2$ [π-CH$_2$=C(Ph)CH$_2$PdCl]$_2$, 1, 675
[π-CH$_2$=CHCH(Ph)PdCl]$_2$, 1, 675
$Pd_2C_{18}H_{20}Cl_4$ {[PhC(Me)=CH$_2$]-PdCl$_2$}$_2$, 1, 684
[(PhCH=CHMe)-PdCl$_2$]$_2$, 1, 684
$Pd_2C_{18}H_{30}Br_2O_2$ Di-μ-bromobis(8-methoxy-4-cyclooctenyl)-Pd$_2$, 1, 667
$Pd_2C_{18}H_{30}Cl_2O_2$ Di-μ-chlorobis(8-methoxy-4-cyclooctenyl)-Pd$_2$, 1, 667
$Pd_2C_{18}H_{32}Cl_2$ (π-1-Cyclooctenylmethyl-PdCl)$_2$, 1, 679
$Pd_2C_{18}H_{34}Br_2O_2$ {π-Me$_2$C=CHCH[C(OMe)Me$_2$]-PdBr}$_2$, 1, 676
$Pd_2C_{18}H_{34}Cl_2O_2$ {π-Me$_2$C=CHCH[C(OMe)Me$_2$]-PdCl}$_2$, 1, 676
{π-Me$_2$C=CHCH[CH(Me)OEt]PdCl}$_2$, 1, 676
$Pd_2C_{18}H_{34}I_2O_2$ {π-Me$_2$C=CHCH[C(OMe)Me$_2$]-PdI}$_2$, 1, 676
$Pd_2C_{20}H_{22}Cl_2$ [π-CH$_2$=C(Ph)CH(Me)PdCl]$_2$, 1, 676
$Pd_2C_{20}H_{24}Cl_4$ {[PhC(Me)=CHMe]-PdCl$_2$}$_2$, 1, 684
$Pd_2C_{22}H_{30}Cl_2O_2$ Di-μ-chlorobis(methoxytricyclo[5.2.1.0^{2,6}]decenyl)-Pd$_2$, 1, 667
$Pd_2C_{24}H_{34}Cl_2O_2$ Di-μ-chlorobis(ethoxytricyclo[5.2.1.0^{2,6}]decenyl)-Pd$_2$, 1, 667
$Pd_2C_{24}H_{46}Cl_2$ (π-Triisobutenyl-PdCl)$_2$, 1, 676
$Pd_2C_{26}H_{38}Cl_2O_2$ Di-μ-chlorobis(propoxytricyclo[5.2.1.0^{2,6}]decenyl)-Pd$_2$, 1, 667
Di-μ-chlorobis(isopropoxytricyclo-[5.2.1.0^{2,6}]decenyl)-Pd$_2$, 1, 667
$Pd_2C_{30}H_{26}Cl_2$ [π-CH$_2$=C(Ph)CH(Ph)PdCl]$_2$, 1, 676
$Pd_2C_{30}H_{70}P_2S_2$ [MePd(PBu$_3$)SEt]$_2$, 1, 664
$Pd_2C_{36}H_{32}Cl_2$ 1,3,5,7,2,4,6,8-Me$_4$(Ph)$_4$-cyclooctatetraene-[PdCl]$_2$, 1, 682*
$Pd_2C_{36}H_{68}Cl_4O_4$ (Oleic acid-PdCl$_2$)$_2$, 1, 684
(Elaidic acid-PdCl$_2$)$_2$, 1, 684
$Pd_2C_{40}H_{38}Cl_2$ [(Diethyltetraphenyl-1,3,5,7-cyclooctatetraenylene)diethylene]-(PdCl)$_2$ (?), 1, 683*
$Pd_2C_{40}H_{76}Cl_4O_4$ (Ethyl oleate-PdCl$_2$)$_2$, 1, 684
$Pd_2C_{56}H_{42}Cl_2O_2$ [HOC(Ph)C(Ph)=C(Ph)C(Ph)PdCl]$_2$, 1, 678
$Pd_2C_{58}H_{42}N_2O_2S_2$ [HOC(Ph)C(Ph)=C(Ph)C(Ph)Pd(SCN)]$_2$, 1, 678

$Pd_2C_{58}H_{46}Cl_2O_2$ [MeOC(Ph)=C(Ph)C-(Ph)=C(Ph)PdCl]$_2$, 1, 668
[MeOC(Ph)C(Ph)=C(Ph)C(Ph)PdCl]$_2$, 1, 677
$Pd_2C_{60}H_{50}Br_2O_2$ [EtOC(Ph)=C(Ph)C-(Ph)=C(Ph)PdBr]$_2$, 1, 668
[EtOC(Ph)C(Ph)=C(Ph)C(Ph)PdBr]$_2$, 1, 677
$Pd_2C_{60}H_{50}Cl_2O_2$ [EtOC(Ph)=C(Ph)C-(Ph)=C(Ph)PdCl]$_2$, 1, 667
[EtOC(Ph)C(Ph)=C(Ph)C(Ph)PdCl]$_2$, 1, 677
$Pd_2C_{60}H_{50}I_2O_2$ [EtOC(Ph)=C(Ph)C(Ph)=C-(Ph)PdI]$_2$, 1, 668
[EtOC(Ph)C(Ph)=C(Ph)C(Ph)PdI]$_2$, 1, 677
$Pd_2C_{62}H_{50}N_2O_2S_2$ [EtOC(Ph)=C(Ph)C-(Ph)=C(Ph)Pd(SCN)]$_2$, 1, 668
[EtOC(Ph)C(Ph)=C(Ph)C(Ph)Pd(SCN)]$_2$, 1, 677
$Pd_2C_{62}H_{54}Cl_2O_2$
[Me_2CHOC(Ph)C(Ph)=C(Ph)C(Ph)PdCl]$_2$, 1, 677
$Pd_3C_6H_{10}Br_4$ Tetra-μ-bromobis(π-allyl)-Pd_3, 1, 679
$Pd_3C_8H_9Cl_5N_2O_2$ [C(=O)CH=CHN=CPd]$_2$-Pd·5HCl, 1, 669
$Pd_3C_8H_{13}Cl_5N_2O_2$ [C(=O)CH=CHNHCHPd]$_2$-Pd·5HCl, 1, 669
$Pd_3C_{42}H_{30}Cl_4$ Tetra-μ-chlorobis(triphenylcyclopropenyl)-Pd_3, 1, 680
$PdAsC_{11}H_{19}BrP$ MePd(o-$Me_2AsC_6H_4PMe_2$)Br, 1, 664
$PdAs_2C_{12}H_{22}$ Me_2Pd(2-$Me_2AsC_6H_4AsMe_2$), 1, 665, 3, 91
$PdAs_2C_{14}H_{36}$ $Me_2Pd(AsEt_3)_2$, 1, 665
$PdBaC_6H_2N_2$·3/4NH_3 Ba[Pd(C≡CH)$_2$-(CN)$_2$]·3/4NH_3, 1, 666
$PdBaC_{18}H_{10}N_2$·1.5NH_3 Ba[Pd(C≡CPh)$_2$-(CN)$_2$]·1.5NH_3, 1, 666
$PdFeC_{13}H_{17}Br_4$ (π-1,5-Cyclooctadiene-PdC_5H_5)$FeBr_4$, 1, 670
$PdFeC_{33}H_{25}Br_4$
{C_5H_5Pd[PhC=C(Ph)C(Ph)=CPh]}$FeBr_4$, 1, 670
$PdGeC_{20}H_{16}Br_2O$ π-Fluorenyl-Pd(CO)Ge-(C_5H_4Me)Br_2, 1, 687, 2, 142*
$PdGeC_{26}H_{23}BrO$ π-Fluorenyl-Pd(CO)Ge-(C_5H_4Me)$_2$Br, 1, 687*, 2, 142*
$PdK_2C_4H_2$ K_2[Pd(C≡CH)$_2$], 1, 666
$PdK_2C_6H_6$ K_2[Pd(C≡CMe)$_2$], 1, 666
$PdK_2C_{16}H_{10}$ K_2[Pd(C≡CPh)$_2$], 1, 666
$PdPbC_{16}H_{24}Cl_2O$ ($Et_2C_5H_3$)Pd(CO)Pb-(C_6H_{11})Cl_2, 1, 687, 2, 595
$PdPbC_{18}H_{19}ClO$ Et_2PbCl(9-fluorenyl)-PdCO, 2, 595
$PdPbC_{28}H_{46}O$ ($Et_2C_5H_3$)Pd(CO)Pb(C_6H_{11})$_3$, 1, 687, 2, 595
$PdSb_2C_6H_{18}Cl_2$ [Pd($SbMe_3$)$_2Cl_2$], 3, 655
$PrC_{15}H_{15}$ (C_5H_5)$_3$Pr, 1, 4
$PtC_2H_5Cl_3$ H[Pt(C_2H_4)Cl_3], 1, 721
PtC_2H_7BrClN [Pt(NH_3)(C_2H_4)BrCl] cis and trans isomers, 1, 707
$PtC_2H_7Br_2N$ [Pt(NH_3)(C_2H_4)Br_2] cis and trans isomers, 1, 707
$PtC_2H_7ClN_2O_3$ cis-[Pt(NH_3)(C_2H_4)Cl-(NO_3)], 1, 707
$PtC_2H_7Cl_2N$ [Pt(NH_3)(C_2H_4)Cl_2] cis and trans isomers, 1, 707
$PtC_2H_8BrCl_2N$ NH_4[Pt(C_2H_4)BrCl_2], 1, 721
$PtC_2H_8Br_2ClN$ NH_4[Pt(C_2H_4)Br_2Cl], 1, 721
$PtC_2H_8Cl_3N$ NH_4[Pt(C_2H_4)Cl_3], 1, 721
$PtC_3H_8Cl_3N$ [($H_3\overset{+}{N}CH_2CH=CH_2$)$\overset{-}{Pt}Cl_3$], 1, 722
PtC_3H_9Br Me_3PtBr, 1, 696
PtC_3H_9Cl Me_3PtCl, 1, 695
$PtC_3H_9Cl_4N$ H[Pt(CH_2=CHCH_2NH_2·HCl)Cl_3], 1, 722
$PtC_3H_9D_9IN_3$ [(CD_3)$_3$Pt(NH_3)$_3$]I, 1, 698
$PtC_3H_9NO_3$ Me_3PtNO_3, 1, 696
$PtC_3H_{10}O$ Me_3PtOH, 1, 696
$PtC_3H_{11}N_3$ CH_2=CHCH_2NH_2Pt(NH_2)$_2$, 1, 725
$PtC_3H_{12}Cl_4N_2$ NH_4[Pt(CH_2=CHCH_2NH_2·HCl)-Cl_3], 1, 723
$PtC_3H_{15}IN_2$ Me_3Pt(NH_3)$_2$I, 1, 697
$PtC_3H_{16}Br_2N_4$ [Pt(NH_3)$_3$(CH_2=CHCH_2NH_2)-Br_2], 1, 725
$PtC_3H_{18}BrN_3$ Me_3Pt(NH_3)$_3$Br, 1, 697
$PtC_3H_{18}ClN_3$ Me_3Pt(NH_3)$_3$Cl, 1, 697
$PtC_3H_{18}IN_3$ Me_3Pt(NH_3)$_3$I, 1, 697
$PtC_4H_8Cl_2$ [Pt(C_2H_4)$_2Cl_2$], 1, 703
$PtC_4H_9Cl_2N$ trans-[Pt(NH_3)(butadiene)-

$Cl_2]$, 1, 709

$PtC_4H_{10}Cl_3N$ $[(H_3\overset{+}{N}CH_2CH_2CH{=}CH_2)\overset{-}{Pt}Cl_3]$, 1, 722

$PtC_4H_{11}Cl_2N$ trans-$[Pt(Me_2NH)(C_2H_4)Cl_2]$, 1, 707

PtC_4H_{12} Me_4Pt, 1, 699

$PtC_4H_{12}O$ Me_3PtOMe, 1, 696

$PtC_4H_{19}ClN_4S$ $[Me_3Pt(NH_3)_2(H_2NCSNH_2)]Cl$, 1, 698

$PtC_5H_{10}Cl_2$ $[Pt(C_2H_4)(CH_2{=}CHMe)Cl_2]$, 1, 704

$PtC_5H_{12}Cl_3N$ $\{H_3\overset{+}{N}(CH_2)_3CH{=}CH_2]\overset{-}{Pt}Cl_3\}$, 1, 722

$[(EtH_2\overset{+}{N}CH_2CH{=}CH_2)\overset{-}{Pt}Cl_3]$, 1, 722

$PtC_5H_{17}IN_2$ $Me_3Pt(H_2NCH_2-)_2I$, 1, 697

$PtC_5H_{20}ClN_5S_2$ $[Me_3Pt(NH_3)(H_2NCSNH_2)_2]Cl$, 1, 698

$PtC_5H_{20}IN_5S_2$ $[Me_3Pt(NH_3)(H_2NCSNH_2)_2]I$, 1, 698

PtC_6H_5IO $C_5H_5Pt(CO)I$, 1, 702

PtC_6H_{10} $(\pi\text{-}CH_2CHCH_2)_2Pt$, 1, 703

$PtC_6H_{10}Cl_2$ $[Pt(CH_2{=}CHCH_2-)_2Cl_2]$, 1, 704

$PtC_6H_{10}Cl_2O$ $\{Pt[(CH_2{=}CHCH_2)_2O]Cl_2\}$, 1, 704

$PtC_6H_{10}I_2$ $[Pt(CH_2{=}CHCH_2-)_2I_2]$, 1, 705

$PtC_6H_{11}Br_2N$ $\{Pt[(CH_2{=}CHCH_2)_2NH]Br_2\}$, 1, 704

$PtC_6H_{11}Cl_2N$ $\{Pt[(CH_2{=}CHCH_2)_2NH]Cl_2\}$, 1, 704

$PtC_6H_{12}Cl_2$ $[Pt(C_2H_4)(CH_2{=}CHEt)Cl_2]$, 1, 704

$PtC_6H_{12}Cl_3N$ $[H_2\overset{+}{N}(CH_2CH{=}CH_2)_2\overset{-}{Pt}Cl_3]$, 1, 722

$PtC_6H_{17}Cl_2N_3$ $Pt(NH_3)_2[(CH_2{=}CHCH_2)_2NH]Cl_2$, 1, 725

$PtC_6H_{21}ClN_6S_3$ $[Me_3Pt(H_2NCSNH_2)_3]Cl$, 1, 698

$PtC_7H_8Br_2$ Cycloheptatriene-$PtBr_2$, 1, 703

Bicyclo[2.2.1]hepta-2,5-diene-$PtBr_2$, 1, 705

$PtC_7H_8Cl_2$ Cycloheptatriene-$PtCl_2$, 1, 704

Bicyclo[2.2.1]hepta-2,5-diene-$PtCl_2$, 1, 705*

$PtC_7H_8Cl_2N_2O_3$ trans$\{Pt(C_2H_4)[\overline{ON{=}CHCH{=}C(NO_2)CH{=}CH}]Cl_2\}$, 1, 710

$PtC_7H_8Cl_3NO$ trans-$[Pt(C_2H_4)(\overline{ON{=}CHCH{=}CClCH{=}CH})Cl_2]$, 1, 710

$PtC_7H_8I_2$ Bicyclo[2.2.1]hepta-2,5-diene-PtI_2, 1, 705

$PtC_7H_9Br_2N$ $[Pt(Py)(C_2H_4)Br_2]$ cis and trans isomers, 1, 707, 708

$PtC_7H_9Cl_2N$ $[Pt(Py)(C_2H_4)Cl_2]$ cis and trans isomers, 1, 708*

$PtC_7H_{10}Br_2ClN$ $Py{\cdot}H[Pt(C_2H_4)Br_2Cl]$, 1, 721

$PtC_7H_{10}Br_3N$ $Py{\cdot}H[Pt(C_2H_4)Br_3]$, 1, 721

$PtC_7H_{10}Cl_3N$ $Py{\cdot}H[Pt(C_2H_4)Cl_3]$, 1, 721

$PtC_7H_{12}ClN_3O_3$ $[Pt(NH_3)(Py)(C_2H_4)(NO_3)Cl]$, 1, 725

$PtC_7H_{15}Cl_2N$ $[Pt(pip)(C_2H_4)Cl_2]$ cis and trans isomers, 1, 708

$PtC_7H_{16}Cl_3N$ $[(Et_2H\overset{+}{N}CH_2CH{=}CH_2)\overset{-}{Pt}Cl_3]$, 1, 722

$PtC_7H_{18}ClD_3P_2$ $CD_3Pt(PMe_3)_2Cl$ cis and trans isomers, 1, 690

$PtC_7H_{21}BrP_2$ trans-$MePt(PMe_3)_2Br$, 1, 689

$PtC_7H_{21}ClP_2$ $MePt(PMe_3)_2Cl$ cis and trans isomers, 1, 689

$PtC_7H_{21}IP_2$ trans-$MePt(PMe_3)_2I$, 1, 689

$PtC_7H_{21}NO_2P_2$ trans-$MePt(PMe_3)_2NO_2$, 1, 689

$PtC_7H_{21}NO_3P_2$ trans-$MePt(PMe_3)_2NO_3$, 1, 689

$PtC_7H_{25}IN_4$ $Me_3Pt(H_2NCH_2CH_2NH_2)_2I$, 1, 698

$PtC_8H_8Cl_2$ Cyclooctatetraene-$PtCl_2$, 1, 705

$PtC_8H_8Cl_3NO_2$ $H[Pt(m\text{-}O_2NC_6H_4CH{=}CH_2)Cl_3]$, 1, 723

$H[Pt(p\text{-}O_2NC_6H_4CH{=}CH_2)Cl_3]$, 1, 723

$PtC_8H_8Cl_4$ $H[Pt(m\text{-}ClC_6H_4CH{=}CH_2)Cl_3]$, 1, 723

$PtC_8H_8I_2$ Cyclooctatetraene-PtI_2, 1, 705

$PtC_8H_9Cl_3$ $H[Pt(PhCH{=}CH_2)Cl_3]$, 1, 723

PtC_8H_{10} $CH_2{=}CHCH_2PtC_5H_5$, 1, 702

$PtC_8H_{10}Cl_3N$ trans-$Pt(p\text{-}H_2NC_6H_4Cl)(C_2H_4)Cl_2$, 1, 708

$PtC_8H_{11}Cl_2N$ $Pt(Py)(C_3H_6)Cl_2$, 1, 708

$Pt(NH_3)(PhCH{=}CH_2)Cl_2$ cis and trans

isomers, 1, 709
$PtC_8H_{11}Cl_2NO$ trans-$Pt(Py)(CH_2{=}CHCH_2OH)Cl_2$, 1, 709
trans-$Pt(C_2H_4)[ON{=}CHCH{=}C(Me)CH{=}CH]Cl_2$, 1, 710
$PtC_8H_{11}Cl_2NO_2$ trans-$Pt(C_2H_4)[ON{=}CHCH{=}C(OMe)CH{=}CH]Cl_2$, 1, 710
$PtC_8H_{11}I$ MePd(bicyclo[2.2.1]hepta-2,5-diene)I, 1, 689
$PtC_8H_{12}Br_2$ 1,5-Cyclooctadiene-$PtBr_2$, 1, 705
$PtC_8H_{12}Cl_2$ 1,5-Cyclooctadiene-$PtCl_2$, 1, 705
$PtC_8H_{12}Cl_3N$ $NH_4[Pt(PhCH{=}CH_2)Cl_3]$, 1, 723
$PtC_8H_{12}I_2$ 1,5-Cyclooctadiene-PtI_2, 1, 705
PtC_8H_{14} $C_5H_5PtMe_3$, 1, 702
$PtC_8H_{16}Cl_2$ $[Pt(CH_2{=}CHEt)_2Cl_2]$, 1, 704
$PtC_8H_{16}Cl_3N$ $[EtH\overset{+}{N}(CH_2CH{=}CH_2)_2PtCl_3]$, 1, 722
$PtC_8H_{17}Cl_2NO$ trans-$Pt(Pip)(CH_2{=}CHCH_2OH)Cl_2$, 1, 709
$PtC_8H_{17}Cl_2NO_2$ $Pt(NH_3)[Me_2C(OH)C{\equiv}]_2Cl_2$, 1, 727
$PtC_8H_{17}IN_2$ $Me_3Pt(NH_3)(Py)I$, 1, 697
$PtC_8H_{18}D_6P_2$ cis-$(CD_3)_2Pt(PMe_3)_2$, 1, 694
$PtC_8H_{19}Br_2P$ cis-$Pt(C_2H_4)(PEt_3)Br_2$, 1, 712
$PtC_8H_{19}Cl_2P$ cis-$Pt(C_2H_4)(PEt_3)Cl_2$, 1, 712
$PtC_8H_{20}Cl_2N_2O_2$ $Pt(NH_3)_2[Me_2C(OH)C{\equiv}]_2Cl_2$, 1, 727*
$PtC_8H_{20}S_2$ $Me_2Pt(-CH_2SEt)_2$, 1, 694
$PtC_8H_{21}NP_2S$ trans-$MePt(PMe_3)_2SCN$, 1, 689
$PtC_8H_{24}P_2$ cis-$Me_2Pt(PMe_3)_2$, 1, 694
$PtC_9H_{11}Cl_2N$ $Pt(Py)(butadiene)Cl_2$ cis and trans isomers, 1, 709
$PtC_9H_{11}Cl_2NO_3$ trans-$Pt(C_2H_4)[ON{=}CHCH{=}C(CO_2Me)CH{=}CH]Cl_2$, 1, 710
$PtC_9H_{11}Cl_3$ $H[Pt(p\text{-}MeC_6H_4CH{=}CH_2)Cl_3]$, 1, 723
$PtC_9H_{11}Cl_3O$ $H[Pt(m\text{-}MeOC_6H_4CH{=}CH_2)Cl_3]$, 1, 723
$H[Pt(p\text{-}MeOC_6H_4CH{=}CH_2)Cl_3]$, 1, 723
$PtC_9H_{13}Cl_2N$ $Pt(p\text{-}H_2NC_6H_4Me)(C_2H_4)Cl_2$ cis and trans isomers, 1, 708
cis-$Pt(Py)(CH_2{=}CHEt)Cl_2$, 1, 709
$PtC_9H_{13}Cl_2NO$ trans-$Pt(Py)(CH_2{=}CHCH_2OMe)Cl_2$, 1, 709
$PtC_9H_{15}Br_2N$ $Pt[(CH_2{=}CHCH_2)_3N]Br_2$, 1, 704
$PtC_9H_{19}Cl_2NO$ trans-$Pt(Pip)(CH_2{=}CHCH_2OMe)Cl_2$, 1, 709
$PtC_9H_{20}Cl_3N$ $(Et_3\overset{+}{N}CH_2CH{=}CH_2)\overset{-}{Pt}Cl_3$, 1, 722
$PtC_9H_{21}Cl_2OP$ cis-$Pt(CH_2{=}CHCH_2OH)(PEt_3)Cl_2$, 1, 712
$PtC_9H_{21}Cl_2P$ cis-$Pt(C_3H_6)(PEt_3)Cl_2$, 1, 712
$PtC_9H_{25}ClP_2$ trans-$PrPt(PMe_3)_2Cl$, 1, 690
$PtC_{10}H_{12}Br_2$ Tricyclo[5.2.1.0^{2,6}]deca-3,8-diene-$PtBr_2$, 1, 705
$PtC_{10}H_{12}Cl_2$ Tricyclo[5.2.1.0^{2,6}]deca-3,8-diene-$PtCl_2$, 1, 705
$PtC_{10}H_{12}I_2$ Tricyclo[5.2.1.0^{2,6}]deca-3,8-diene-PtI_2, 1, 705
$PtC_{10}H_{13}I$ EtPt(cyclooctatetraene)I, 1, 689
$PtC_{10}H_{14}$ Cyclooctatetraene-$PtMe_2$, 1, 693
$PtC_{10}H_{14}ClO_4$ Bis(acetylacetonato)-PtCl, 1, 726
$PtC_{10}H_{15}Cl_2N$ trans-$Pt(H_2NCHMePh)(C_2H_4)Cl_2$, 1, 708
$PtC_{10}H_{15}Cl_2NO$ trans-$Pt(p\text{-}H_2NC_6H_4Me)(CH_2{=}CHCH_2OH)Cl_2$, 1, 709
$PtC_{10}H_{16}Cl_2$ Limonene-$PtCl_2$, 1, 706*
$PtC_{10}H_{18}$ 1,5-Cyclooctadiene-$PtMe_2$, 1, 693
$PtC_{10}H_{23}Cl_2P$ cis-$Pt(CH_2{=}CHEt)(PEt_3)Cl_2$, 1, 712
$PtC_{10}H_{25}NP_2$ trans-$PrPt(PMe_3)_2CN$, 1, 690
$PtC_{10}H_{28}P_2$ cis-$Et_2Pt(PMe_3)_2$, 1, 694
$PtC_{11}H_{13}IN_2$ $MePtPy_2I$, 1, 689
$PtC_{11}H_{15}I$ MePt(tricyclo[5.2.1.0^{2,6}]-decadiene)I, 1, 689
$PtC_{11}H_{17}Cl_2N$ trans-$Pt(H_2NCHMeCH_2Ph)(C_2H_4)Cl_2$, 1, 708
trans-$Pt(H_2NCHMePh)(C_3H_6)Cl_2$, 1, 709

$PtC_{11}H_{17}Cl_2NO$ trans-$Pt(p\text{-}H_2NC_6H_4Me)$-$(CH_2{=}CHCH_2OMe)Cl_2$, 1, 709*

$PtC_{11}H_{24}Cl_3N$ $(n\text{-}C_8H_{17}H_2\overset{+}{N}CH_2CH{=}CH_2)\bar{P}tCl_3$, 1, 722

$PtC_{11}H_{25}Cl_2P$ cis-$Pt(C_2H_4)(PPr_3)Cl_2$, 1, 712

$PtC_{12}H_8Cl_2$ Acenaphthylene-$PtCl_2$, 1, 706

$PtC_{12}H_{12}N_2S_2$ Tricyclo$[5.2.1.0^{2,6}]$-deca-3,8-diene-$Pt(SCN)_2$, 1, 705

$PtC_{12}H_{16}N_2$ Me_2PtPy_2, 1, 693

$PtC_{12}H_{19}Cl_2N$ $Pt(n\text{-}C_5H_{11}$-pyridine)-$(C_2H_4)Cl_2$ cis and trans isomers, 1, 708

trans-$Pt(H_2NCHMePh)$(cis-2-butene)Cl_2, 1, 709

trans-$Pt(H_2NCHMePh)$(trans-2-butene)-Cl_2, 1, 709

$PtC_{12}H_{22}$ 1,5-Cyclooctadiene-$PtEt_2$, 1, 693

$PtC_{12}H_{23}Cl_2N$ $Pt(Pip)(Me_3CC{\equiv}CMe)Cl_2$, 1, 727

$PtC_{12}H_{23}N_9O_7S_3$ $[Me_3Pt(H_2NCSNH_2)_3]O$-$C_6H_2(NO_2)_3$, 1, 698

$PtC_{12}H_{25}Cl_3$ H[Pt(1-dodecene)Cl_3], 1, 724

$PtC_{12}H_{26}Cl_3N$ $(n\text{-}C_8H_{17}H_2\overset{+}{N}CH_2CH_2CH{=}CH_2)$-$\bar{P}tCl_3$, 1, 722

$PtC_{12}H_{30}P_2$ $Me_2Pt(\text{-}CH_2PEt_2)_2$, 1, 694

$PtC_{12}H_{32}P_2$ cis-$Pr_2Pt(PMe_3)_2$, 1, 694

$PtC_{13}H_{10}D_9IN_2$ $(CD_3)_3PtPy_2I$, 1, 698

$PtC_{13}H_{11}Cl_2N_3O_5$ trans-$Pt(O_2NC_6H_4$-$CH{=}CH_2)[ON{=}CHCH{=}C(NO_2)CH{=}CH]Cl_2$ m- and p-isomers, 1, 711

$PtC_{13}H_{11}Cl_3N_2O_3$ trans-$Pt(O_2NC_6H_4$-$CH{=}CH_2)(ON{=}CHCH{=}CClCH{=}CH)Cl_2$ m- and p-isomers, 1, 711

trans-$Pt(ClC_6H_4CH{=}CH_2)$-$[ON{=}CHCH{=}C(NO_2)CH{=}CH]Cl_2$ m- and p-isomers, 1, 711

$PtC_{13}H_{11}Cl_4NO$ trans-$Pt(ClC_6H_4CH{=}CH_2)$-$(ON{=}CHCH{=}CClCH{=}CH)Cl_2$ m- and p-isomers, 1, 711

$PtC_{13}H_{12}Cl_2N_2O_3$ trans-$Pt(O_2NC_6H_4$-$CH{=}CH_2)(ON{=}CHCH{=}CHCH{=}CH)Cl_2$ m- and p-isomers, 1, 710

trans-$Pt(PhCH{=}CH_2)$-$[ON{=}CHCH{=}C(NO_2)CH{=}CH]Cl_2$, 1, 711

$PtC_{13}H_{12}Cl_3NO$ trans-$Pt(ClC_6H_4CH{=}CH_2)$-$(ON{=}CHCH{=}CHCH{=}CH)Cl_2$ m- and p-isomers, 1, 710

trans-$Pt(PhCH{=}CH_2)(ON{=}CHCH{=}CClCH{=}CH)$-$Cl_2$, 1, 710

$PtC_{13}H_{13}Cl_2NO$ trans-$Pt(PhCH{=}CH_2)$-$(ON{=}CHCH{=}CHCH{=}CH)Cl_2$, 1, 710

$PtC_{13}H_{14}Cl_2N_2$ $CH_2CH_2CH_2Pt(Bipy)Cl_2$, 1, 747

$PtC_{13}H_{14}Cl_3N$ $Py{\cdot}H[Pt(PhCH{=}CH_2)Cl_3]$, 1, 724

$PtC_{13}H_{16}Cl_2N_2$ $CH_2CH_2CH_2Pt(Py)_2Cl_2$, 1, 747

$PtC_{13}H_{17}IN_2$ $Me_3Pt(2,2'\text{-}BiPy)I$, 1, 697

$PtC_{13}H_{19}Br_2NO_2$ $Pt(Py)[Me_2C(OH)C{\equiv}]_2$-$Br_2$, 1, 727

$PtC_{13}H_{19}ClN_2$ Me_3PtPy_2Cl, 1, 697

$PtC_{13}H_{19}Cl_2NO_2$ $Pt(Py)[Me_2C(OH)C{\equiv}]_2$-$Cl_2$, 1, 727

$PtC_{13}H_{19}IN_2$ Me_3PtPy_2I, 1, 697

$PtC_{13}H_{25}Br_2NO_2$ $Pt(Pip)[Me_2C(OH)C{\equiv}]_2$-$Br_2$, 1, 727*

$PtC_{13}H_{25}Cl_2N$ $Pt(Pip)(Me_3CC{\equiv}CEt)Cl_2$, 1, 727

$PtC_{13}H_{25}Cl_2NO_2$ $Pt(Pip)[Me_2C(OH)C{\equiv}]_2$-$Cl_2$, 1, 727

$PtC_{13}H_{28}Cl_3N$ $[n\text{-}C_8H_{17}H_2\overset{+}{N}(CH_2)_3CH{=}CH_2]$-$\bar{P}tCl_3$, 1, 722

$PtC_{13}H_{30}ClD_3P_2$ trans-$CD_3Pt(PEt_3)_2Cl$, 1, 690

$PtC_{13}H_{33}BrP_2$ trans-$MePt(PEt_3)_2Br$, 1, 690

$PtC_{13}H_{33}ClP_2$ $MePt(PEt_3)_2Cl$ cis and trans isomers, 1, 690

$PtC_{13}H_{33}IP_2$ trans-$MePt(PEt_3)_2I$, 1, 690

$PtC_{13}H_{33}NO_2P_2$ trans-$MePt(PEt_3)_2NO_2$, 1, 690

$PtC_{13}H_{33}NO_3P_2$ trans-$MePt(PEt_3)_2NO_3$, 1, 690

$PtC_{14}H_{10}Cl_2N_2$ $Pt(CNPh)_2Cl_2$, 1, 733

$PtC_{14}H_{14}Br_6$ (Cycloheptatrienium)$_2$-$PtBr_6$, 1, 702

$PtC_{14}H_{14}Cl_2N_2O_3$ trans-$Pt(MeC_6H_4CH{=}CH_2)$-$[ON{=}CHCH{=}C(NO_2)CH{=}CH]Cl_2$ m- and p-isomers, 1, 711

trans-$Pt(O_2NC_6H_4CH{=}CH_2)$-

($O\overline{N=CHCH=CMeCH=C}H)Cl_2$ m- and p-isomers, 1, 711

$PtC_{14}H_{14}Cl_2N_2O_4$ trans-$Pt(MeOC_6H_4$-$CH=CH_2)[ON=CHCH=C(NO_2)CH=CH]Cl_2$ m- and p-isomers, 1, 711

trans-$Pt(O_2NC_6H_4CH=CH_2)$-$[ON=CHCH=C(OMe)CH=CH]Cl_2$ m- and p-isomers, 1, 711

$PtC_{14}H_{14}Cl_3NO$ trans-$Pt(MeC_6H_4CH=CH_2)$-$(ON=CHCH=CClCH=CH)Cl_2$ m- and p-isomers, 1, 711

trans-$Pt(ClC_6H_4CH=CH_2)$-$(ON=CHCH=CMeCH=CH)Cl_2$ m- and p-isomers, 1, 711

$PtC_{14}H_{14}Cl_3NO_2$ trans-$Pt(MeOC_6H_4$-$CH=CH_2)(ON=CHCH=CClCH=CH)Cl_2$ m- and p-isomers, 1, 711

trans-$Pt(ClC_6H_4CH=CH_2)$-$[ON=CHCH=C(OMe)CH=CH]Cl_2$ m- and p-isomers, 1, 711

$PtC_{14}H_{14}Cl_6$ (Cycloheptatrienium)$_2$-$PtCl_6$, 1, 702

$PtC_{14}H_{15}Cl_2NO$ trans-$Pt(MeC_6H_4CH=CH_2)$-$(ON=CHCH=CHCH=CH)Cl_2$ m- and p-isomers, 1, 710

trans-$Pt(PhCH=CH_2)$-$(ON=CHCH=CMeCH=CH)Cl_2$, 1, 711

$PtC_{14}H_{15}Cl_2NO_2$ trans-$Pt(MeOC_6H_4$-$CH=CH_2)(ON=CHCH=CHCH=CH)Cl_2$ m- and p-isomers, 1, 710

trans-$Pt(C_2H_4)$-$[ON=CHCH=C(OCH_2Ph)CH=CH]Cl_2$, 1, 710

trans-$Pt(PhCH=CH_2)$-$[ON=CHCH=C(OMe)CH=CH]Cl_2$, 1, 711

$PtC_{14}H_{21}Cl_2NO_2$ $Pt(Py)[Me_2C(OMe)C\equiv C$-$C(OH)Me_2]Cl_2$, 1, 728

$PtC_{14}H_{27}Cl_2N$ $Pt(Pip)(Me_3CC\equiv CCHMe_2)Cl_2$, 1, 727

$PtC_{14}H_{27}Cl_2NO$ $Pt(Pip)[Me_3CC\equiv CC(OH)$-$Me_2]Cl_2$, 1, 728

$PtC_{14}H_{27}Cl_2NO_2$ $Pt(Pip)[Me_2C(OMe)C\equiv C$-$C(OH)Me_2]Cl_2$, 1, 728

$PtC_{14}H_{31}Cl_2P$ cis-$Pt(C_2H_4)(PBu_3)Cl_2$, 1, 712

$PtC_{14}H_{33}BrOP_2$ $AcPt(PEt_3)_2Br$, 1, 691

$PtC_{14}H_{33}ClOP_2$ $AcPt(PEt_3)_2Cl$, 1, 691

$PtC_{14}H_{33}IOP_2$ $AcPt(PEt_3)_2I$, 1, 691

$PtC_{14}H_{33}NP_2$ trans-$MePt(PEt_3)_2CN$, 1, 690

$PtC_{14}H_{33}NP_2S$ trans-$MePt(PEt_3)_2SCN$, 1, 690

$PtC_{14}H_{35}ClP_2$ trans-$EtPt(PEt_3)_2Cl$, 1, 69

$PtC_{14}H_{35}IP_2$ trans-$EtPt(PEt_3)_2I$, 1, 690

$PtC_{14}H_{36}Cl_2P_2$ $Me_2Pt(PEt_3)_2Cl_2$, 1, 695

$PtC_{14}H_{36}I_2P_2$ $Me_2Pt(PEt_3)_2I_2$, 1, 695

$PtC_{14}H_{36}P_2$ cis-$Bu_2Pt(PMe_3)_2$, 1, 694

cis-$Me_2Pt(PEt_3)_2$, 1, 694

$PtC_{15}H_{14}Cl_2N_2O_5$ trans-$Pt(O_2NC_6H_4$-$CH=CH_2)[ON=CHCH=C(CO_2Me)CH=CH]Cl_2$ m- and p-isomers, 1, 711, 712

$PtC_{15}H_{14}Cl_3NO_3$ trans-$Pt(ClC_6H_4CH=CH_2)$-$[ON=CHCH=C(CO_2Me)CH=CH]Cl_2$ m- and p-isomers, 1, 711

$PtC_{15}H_{15}Cl_2NO_3$ trans-$Pt(PhCH=CH_2)$-$[ON=CHCH=C(CO_2Me)CH=CH]Cl_2$, 1, 711

$PtC_{15}H_{17}Cl_2NO$ trans-$Pt(MeC_6H_4CH=CH_2)$-$(ON=CHCH=CMeCH=CH)Cl_2$ m- and p-isomers, 1, 711

$PtC_{15}H_{17}Cl_2NO_2$ trans-$Pt(MeOC_6H_4CH=CH_2)$-$(ON=CHCH=CMeCH=CH)Cl_2$ m- and p-isomers, 1, 711

trans-$Pt(MeC_6H_4CH=CH_2)$-$[ON=CHCH=C(OMe)CH=CH]Cl_2$ m- and p-isomers, 1, 711

$PtC_{15}H_{17}Cl_2NO_3$ trans-$Pt(MeOC_6H_4CH=CH_2)$-$[ON=CHCH=C(OMe)CH=CH]Cl_2$ m- and p-isomers, 1, 711

$PtC_{15}H_{19}I$ [o-MeC_6H_4Pt(1,5-cyclooctadiene)I], 1, 689

[p-MeC_6H_4Pt(1,5-cyclooctadiene)I], 1, 689

$PtC_{15}H_{22}Cl_3N$ $Pt(N=CHCH=CClCH=CH)$-$(Me_3CC\equiv)_2Cl_2$, 1, 727

$PtC_{15}H_{23}Br_2NO_2$ $Pt(p$-$MeC_6H_4NH_2)[Me_2C$-$(OH)C\equiv]_2Br_2$, 1, 727

$PtC_{15}H_{23}Cl_2N$ $Pt(Py)(Me_3CC\equiv)_2Cl_2$, 1, 727

$PtC_{15}H_{23}Cl_2NO_2$ $Pt(Py)[EtMeC(OH)C\equiv]_2$-$Cl_2$, 1, 727

$Pt(Py)[Me_2C(OMe)C\equiv]_2Cl_2$, 1, 728

$Pt(p$-$MeC_6H_4NH_2)[Me_2C(OH)C\equiv]_2Cl_2$, 1, 727

$PtC_{15}H_{29}Cl_2N$ $Pt(Pip)(Me_3CC\equiv)_2Cl_2$, 1, 727

$PtC_{15}H_{29}Cl_2NO_2$ $Pt(Pip)[EtMeC(OH)C\equiv]_2Cl_2$, 1, 728

$Pt(Pip)[Me_2C(OMe)C\equiv]_2Cl_2$, 1, 728

$PtC_{16}H_{16}Cl_2$ $Pt(CH_2{=}CHPh)_2Cl_2$, 1, 704

$PtC_{16}H_{17}Cl_2NO_3$ trans-$Pt(MeC_6H_4CH{=}CH_2)[ON{=}CHCH{=}C(CO_2Me)CH{=}CH]Cl_2$ m- and p-isomers, 1, 712

$PtC_{16}H_{17}Cl_2NO_4$ trans-$Pt(MeOC_6H_4CH{=}CH_2)[ON{=}CHCH{=}C(CO_2Me)CH{=}CH]Cl_2$ m- and p-isomers, 1, 712

$PtC_{16}H_{19}Cl_2N$ trans-$Pt(H_2NCHMePh)(PhCH{=}CH_2)Cl_2$, 1, 709

$PtC_{16}H_{21}Cl_2N_3$ $Pt(Py)_2[(CH_2{=}CHCH_2)_2NH]Cl_2$, 1, 725

$PtC_{16}H_{24}ClNO$ (2-Methoxy-5-cyclooctenyl)(p-toluidene)-PtCl, 1, 700

$PtC_{16}H_{25}Cl_2N$ trans-$Pt(H_2NCHMePh)(cyclooctene)Cl_2$, 1, 709

$Pt(N{=}CHCH{=}CMeCH{=}CH)(Me_3CC{\equiv}CCMe_3)Cl_2$, 1, 727

$PtC_{16}H_{25}Cl_2NO$ $Pt(p\text{-}MeC_6H_4NH_2)[Me_3CC{\equiv}CC(OH)Me_2]Cl_2$, 1, 728

$PtC_{16}H_{25}Cl_2NO_2$ $Pt(p\text{-}MeC_6H_4NH_2)[Me_2C(OMe)C{\equiv}CC(OH)Me_2]Cl_2$, 1, 728

$PtC_{16}H_{27}Cl_2N$ trans-$Pt(C_9H_{19}\text{-pyridine})(C_2H_4)Cl_2$, 1, 708

$PtC_{16}H_{32}P_2$ trans-$(CH{\equiv}C)_2Pt(PEt_3)_2$, 1, 694

$PtC_{16}H_{40}P_2$ cis-$Et_2Pt(PEt_3)_2$, 1, 694

$PtC_{17}H_{17}IN_2$ p-$MeC_6H_4PtPy_2I$, 1, 689

$PtC_{17}H_{27}Cl_2N$ $Pt(p\text{-}MeC_6H_4NH_2)(Me_3CC\equiv)_2Cl_2$, 1, 727

trans-$Pt(H_2NCHMeCH_2Ph)(cyclooctene)Cl_2$, 1, 709

$PtC_{17}H_{27}Cl_2NO_2$ $Pt(p\text{-}MeC_6H_4NH_2)[Me_2C(OMe)C\equiv]_2Cl_2$, 1, 728

$Pt(p\text{-}MeC_6H_4NH_2)[EtMeC(OH)C\equiv]_2Cl_2$, 1, 728

$PtC_{17}H_{28}Cl_2N_2O_3$ trans-$Pt(n\text{-}C_{10}H_{21}CH{=}CH_2)[ON{=}CHCH{=}C(NO_2)CH{=}CH]Cl_2$, 1, 712

$PtC_{17}H_{28}Cl_3NO$ trans-$Pt(n\text{-}C_{10}H_{21}CH{=}CH_2)(ON{=}CHCH{=}CClCH{=}CH)Cl_2$, 1, 712

$PtC_{17}H_{29}Cl_2NO$ trans-$Pt(n\text{-}C_{10}H_{21}CH{=}CH_2)(ON{=}CHCH{=}CHCH{=}CH)Cl_2$, 1, 712

$PtC_{18}H_{20}I_2N_2$ $Me(o\text{-}MeC_6H_4)PtPy_2I_2$, 1, 693

$PtC_{18}H_{24}ClNO$ (9-Methoxytricyclo-$[5.2.1.0^{2,6}]$-3-decen-8-yl)(p-toluidine)-PtCl, 1, 700

$PtC_{18}H_{27}Cl_2NO_2$ $Pt(Pip)[MePhC(OH)C{\equiv}CC(OH)Me_2]Cl_2$, 1, 728

$PtC_{18}H_{31}Cl_2NO$ trans-$Pt(n\text{-}C_{10}H_{21}CH{=}CH_2)[ON{=}CHCH{=}C(Me)CH{=}CH]Cl_2$, 1, 712

$PtC_{18}H_{31}Cl_2NO_2$ trans-$Pt(n\text{-}C_{10}H_{21}CH{=}CH_2)[ON{=}CHCH{=}C(OMe)CH{=}CH]Cl_2$, 1, 712

$PtC_{18}H_{35}BrP_2$ trans-$PhPt(PEt_3)_2Br$, 1, 690

$PtC_{18}H_{35}ClP_2$ $PhPt(PEt_3)_2Cl$ cis and trans isomers, 1, 690, 691

$PtC_{18}H_{36}P_2$ trans-$(MeC{\equiv}C)_2Pt(PEt_3)_2$, 1, 694

$PtC_{19}H_{18}$ Bicyclo[2.2.1]hepta-2,5-diene-$PtPh_2$, 1, 693

$PtC_{19}H_{29}Cl_2NO$ $Pt(Pip)[Me_3CC{\equiv}CC(OH)MePh]Cl_2$, 1, 728

$PtC_{19}H_{31}Cl_2NO_3$ trans-$Pt(n\text{-}C_{10}H_{21}CH{=}CH_2)[ON{=}CHCH{=}C(CO_2Me)CH{=}CH]Cl_2$, 1, 712

$PtC_{19}H_{35}BrOP_2$ $BzPt(PEt_3)_2Br$, 1, 691

$PtC_{19}H_{35}ClOP_2$ $BzPt(PEt_3)_2Cl$, 1, 691

$PtC_{19}H_{35}IOP_2$ $BzPt(PEt_3)_2I$, 1, 691

$PtC_{19}H_{37}ClP_2$ trans-$PhCH_2Pt(PEt_3)_2Cl$, 1, 691

cis-o-$MeC_6H_4Pt(PEt_3)_2Cl$, 1, 691

cis-p-$MeC_6H_4Pt(PEt_3)_2Cl$, 1, 691

$PtC_{19}H_{45}IP_2$ trans-$MePt(PPr_3)_2I$, 1, 690

$PtC_{20}H_{19}Cl_2NO_2$ trans-$Pt(PhCH{=}CH_2)[ON{=}CHCH{=}C(OCH_2Ph)CH{=}CH]Cl_2$, 1, 712

$PtC_{20}H_{22}$ 1,5-Cyclooctadiene-$PtPh_2$, 1, 693

$PtC_{20}H_{26}Cl_3N$ $Et_3NH[Pt(PhC{\equiv}CPh)Cl_3]$, 1, 733

$PtC_{20}H_{31}Cl_2N$ $Pt(Pip)(Me_3CC{\equiv}CCMe_2Ph)Cl_2$, 1, 727

$PtC_{20}H_{48}I_2P_2$ $Me_2Pt(PPr_3)_2I_2$, 1, 695

$PtC_{21}H_{22}$ Bicyclo[2.2.1]hepta-2,5-diene-$Pt(C_6H_4Me\text{-}p)_2$, 1, 693

Bicyclo[2.2.1]hepta-2,5-diene-Pt-

$(C_6H_4Me\text{-}o)_2$, 1, 693

$PtC_{21}H_{41}BrP_2$ cis-$(2,4,6\text{-}Me_3C_6H_2)Pt(PEt_3)_2Br$, 1, 691

$PtC_{21}H_{46}Cl_3N$ $[n\text{-}C_{16}H_{33}NMe_3][Pt(C_2H_4)Cl_3]$, 1, 721

$PtC_{22}H_{20}N_2$ Ph_2PtPy_2, 1, 693

$PtC_{22}H_{22}$ Cyclooctatetraene-$Pt(C_6H_4Me\text{-}o)_2$, 1, 693

Cyclooctatetraene-$Pt(C_6H_4Me\text{-}p)_2$, 1, 693

$PtC_{22}H_{26}$ 1,5-Cyclooctadiene-$Pt(C_6H_4Me\text{-}o)_2$, 1, 693

$PtC_{22}H_{34}P_2$ $Ph_2Pt(\text{-}CH_2PEt_2)_2$, 1, 694

$PtC_{23}H_{19}Cl_3$ $[PhC{=}C(Ph)C(Ph)][Pt(C_2H_4)Cl_3]$, 1, 722

$PtC_{23}H_{29}Cl_2NO_2$ $Pt(Pip)[MePhC(OH)C{\equiv}]_2Cl_2$, 1, 728

$PtC_{24}H_{24}N_2$ $(MeC_6H_4)_2PtPy_2$ o- and p-isomers, 1, 693

$PtC_{24}H_{38}Cl_2P_2$ trans-$(p\text{-}ClC_6H_4)_2Pt(PEt_3)_2$, 1, 694

$PtC_{24}H_{40}Cl_2P_2$ $Ph_2Pt(PEt_3)_2Cl_2$, 1, 695

$PtC_{24}H_{40}I_2P_2$ $Ph_2Pt(PEt_3)_2I_2$, 1, 695

$PtC_{24}H_{40}P_2$ $Ph_2Pt(PEt_3)_2$ cis- and trans-isomers, 1, 694

$PtC_{26}H_{44}P_2$ $(o\text{-}MeC_6H_4)_2Pt(PEt_3)_2$ cis- and trans-isomers, 1, 694

cis-$(p\text{-}MeC_6H_4)_2Pt(PEt_3)_2$, 1, 694

cis-$(PhCH_2)_2Pt(PEt_3)_2$, 1, 694

$PtC_{27}H_{22}$ Bicyclo[2.2.1]hepta-2,5-diene-$Pt(C_{10}H_7\text{-}\alpha)_2$, 1, 693

$PtC_{28}H_{26}$ 1,5-Cyclooctadiene-$Pt(C_{10}H_7\text{-}\alpha)_2$, 1, 693

$PtC_{28}H_{40}P_2$ trans-$(PhC{\equiv}C)_2Pt(PEt_3)_2$, 1, 694

$PtC_{34}H_{38}Cl_3N$ $Et_3NH\{Pt[Ph_2C(OH)C{\equiv}]_2Cl_3\}$, 1, 733

$PtC_{37}H_{33}IP_2$ trans-$MePt(PPh_3)_2I$, 1, 690

$PtC_{37}H_{33}I_3P_2$ $MePt(PPh_3)_2I_3$, 1, 690

$PtC_{38}H_{30}Cl_6$ $(Ph_3C)_2PtCl_6$, 1, 701

$PtC_{38}H_{32}P_2$ $[(Ph_3P)_2Pt(HC{\equiv}CH)]$, 1, 729

$PtC_{38}H_{33}NP_2S$ trans-$MePt(PPh_3)_2SCN$, 1, 690

$PtC_{38}H_{35}IP_2$ trans-$EtPt(PPh_3)_2I$, 1, 690*

$PtC_{38}H_{36}P_2$ $Me_2Pt(PPh_3)_2$, 1, 694

$PtC_{39}H_{35}NP_2S$ trans-$EtPt(PPh_3)_2SCN$, 1, 690*

$PtC_{40}H_{30}F_6P_2$ $(Ph_3P)_2Pt(CF_3C{\equiv}CCF_3)$, 1, 729

$PtC_{43}H_{37}IP_2$ $p\text{-}MeC_6H_4Pt(PPh_3)_2I$, 1, 691

$PtC_{44}H_{29}ClF_6P_2$ $[(p\text{-}FC_6H_4)_3P]_2Pt(p\text{-}ClC_6H_4C{\equiv}CH)$, 1, 730

$PtC_{44}H_{30}F_6P_2$ $[(p\text{-}FC_6H_4)_3P]_2Pt(PhC{\equiv}CH)$, 1, 730

$PtC_{44}H_{35}BrP_2$ $(Ph_3P)_2Pt(p\text{-}BrC_6H_4C{\equiv}CH)$, 1, 729

$PtC_{44}H_{35}ClP_2$ $(Ph_3P)_2Pt(p\text{-}ClC_6H_4C{\equiv}CH)$, 1, 729

$(Ph_3P)_2Pt(o\text{-}ClC_6H_4C{\equiv}CH)$, 1, 729

$PtC_{44}H_{35}FP_2$ $(Ph_3P)_2Pt(p\text{-}FC_6H_4C{\equiv}CH)$, 1, 729

$(Ph_3P)_2Pt(o\text{-}FC_6H_4C{\equiv}CH)$, 1, 729

$PtC_{44}H_{35}NO_2P_2$ $(Ph_3P)_2Pt(O_2NC_6H_4C{\equiv}CH)$ m- and p-isomers, 1, 729

$PtC_{44}H_{36}P_2$ $(Ph_3P)_2Pt(PhC{\equiv}CH)$, 1, 729

$PtC_{45}H_{38}OP_2$ $(Ph_3P)_2Pt(CH{\equiv}CC_6H_4OMe)$ o-, m-, and p-isomers, 1, 729

$PtC_{45}H_{38}P_2$ $(Ph_3P)_2Pt(p\text{-}MeC_6H_4C{\equiv}CH)$, 1, 729

$PtC_{46}H_{48}O_2P_2$ $(Ph_3P)_2Pt[MeEtC(OH)C{\equiv}]_2$, 1, 729

$PtC_{48}H_{38}P_2$ (Acenaphthylene)$Pt(PPh_3)_2$, 1, 713

$PtC_{48}H_{40}P_2$ cis-$Ph_2Pt(PPh_3)_2$, 1, 694

$PtC_{50}H_{34}F_6P_2$ $[(p\text{-}FC_6H_4)_3P]_2Pt(PhC{\equiv}CPh)$, 1, 730

$PtC_{50}H_{38}Br_2P_2$ $(Ph_3P)_2Pt(p\text{-}BrC_6H_4C{\equiv})_2$, 1, 729

$PtC_{50}H_{38}Cl_2P_2$ $(Ph_3P)_2Pt(p\text{-}ClC_6H_4C{\equiv})_2$, 1, 730

$PtC_{50}H_{39}ClP_2$ $(Ph_3P)_2Pt(p\text{-}ClC_6H_4C{\equiv}CPh)$, 1, 730

$PtC_{50}H_{40}N_2O_4P_2$ $(p\text{-}O_2NC_6H_4CH{=})_2Pt(PPh_3)_2$, 1, 713

$PtC_{50}H_{40}P_2$ $(Ph_3P)_2Pt(PhC{\equiv}CPh)$, 1, 729

$PtC_{50}H_{42}P_2$ $(PhCH{=}CHPh)Pt(PPh_3)_2$, 1, 713

$PtC_{50}H_{44}P_2$ cis-$(p\text{-}MeC_6H_4)_2Pt(PPh_3)_2$, 1, 694

$PtC_{50}H_{47}ClP_2$ $[(p\text{-}MeC_6H_4)_3P]_2Pt(p\text{-}ClC_6H_4C{\equiv}CH)$, 1, 730

$PtC_{50}H_{47}NO_2P_2$ $[(p\text{-}MeC_6H_4)_3P]_2Pt(m\text{-}O_2NC_6H_4C{\equiv}CH)$, 1, 730

$PtC_{50}H_{48}P_2$ $[(p\text{-}MeC_6H_4)_3P]_2Pt(PhC{\equiv}CH)$, 1, 730

$PtC_{51}H_{50}OP_2$ $[(p\text{-}MeC_6H_4)_3P]_2Pt(o\text{-}MeO\text{-}C_6H_4C{\equiv}CH)$, 1, 730

$PtC_{52}H_{44}O_2P_2$ $(Ph_3P)_2Pt(p\text{-}MeOC_6H_4C{\equiv})_2$, 1, 730

$PtC_{52}H_{44}P_2$ $(Ph_3P)_2Pt(p\text{-}MeC_6H_4C{\equiv})_2$, 1, 730

$(Ph_3P)_2Pt(o\text{-}MeC_6H_4C{\equiv})_2$, 1, 730

$PtC_{56}H_{52}P_2$ $[(p\text{-}MeC_6H_4)_3P]_2Pt(PhC{\equiv}CPh)$, 1, 730

$Pt_2C_4H_4Cl_8$ $[Pt(trans\text{-}ClCH{=}CHCl)Cl_2]_2$, 1, 714

$Pt_2C_4H_8Cl_4$ $[Pt(C_2H_4)Cl_2]_2$, 1, 713

$Pt_2C_4H_{12}Cl_4N_2$ μ-Butadiene-$[cis\text{-}Pt(NH_3)Cl_2]_2$, 1, 717

μ-Butadiene-$[trans\text{-}Pt(NH_3)Cl_2]_2$, 1, 717

μ-H_2NNH_2-$[Pt(C_2H_4)Cl_2]_2$, 1, 717

$Pt_2C_4H_{14}Cl_6N_2$ $(NH_4)_2[Pt_2(butadiene)Cl_6]$, 1, 724

$Pt_2C_6H_{12}Cl_4$ $[Pt(CH_2{=}CHMe)Cl_2]_2$, 1, 714

$Pt_2C_6H_{14}Br_4N_2$ Di-μ-$(CH_2{=}CHCH_2NH_2)$-$[PtBr_2]_2$, 1, 718

$Pt_2C_6H_{14}Cl_4N_2$ μ-Butadiene-μ′-ethylenediamine-$[PtCl_2]_2$, 1, 717

Di-μ-$(CH_2{=}CHCH_2NH_2)$-$[PtCl_2]_2$, 1, 718

$Pt_2C_6H_{16}Cl_4N_2$ μ-H_2NNH_2-$[Pt(C_3H_6)Cl_2]_2$, 1, 717

μ-$H_2NCH_2CH_2NH_2$-$[Pt(C_2H_4)Cl_2]_2$, 1, 717

μ-1,5-Hexadiene-$[Pt(NH_3)Cl_2]_2$, 1, 717

$Pt_2C_6H_{16}Cl_6N_2$ $({-}CH_2NH_3)_2[Pt_2(butadiene)Cl_6]$, 1, 724

$Pt_2C_6H_{17}Cl_4N_3$ $[Pt(NH_3)_2(CH_2{=}CHCH_2)_2NH][PtCl_4]$, 1, 725

$Pt_2C_6H_{18}$ $(Me_3Pt)_2$, 1, 696

$Pt_2C_6H_{18}Cl_6N_2$ $\{{-}CH_2NH_2{\cdot}H[Pt(C_2H_4)Cl_3]\}_2$, 1, 721

$Pt_2C_6H_{22}O_6S$ $(Me_3Pt)_2SO_4{\cdot}2H_2O$, 1, 696

$Pt_2C_8H_{12}Cl_4$ $[Pt(CH_2{=}CHCH{=}CH_2)Cl_2]_2$, 1, 714

$Pt_2C_8H_{16}Cl_4$ $[Pt(CH_2{=}CHEt)Cl_2]_2$, 1, 714

$[Pt(cis\text{-}MeCH{=}CHMe)Cl_2]_2$, 1, 714

$[Pt(trans\text{-}MeCH{=}CHMe)Cl_2]_2$, 1, 714

$[Pt(Me_2C{=}CH_2)Cl_2]_2$, 1, 714

$Pt_2C_8H_{18}Cl_4N_2$ Di-μ-$(MeCH{=}CHCH_2NH_2)$-$[PtCl_2]_2$, 1, 718

$Pt_2C_8H_{20}Cl_4N_2$ μ-$H_2NNH_2[Pt(C_4H_8)Cl_2]_2$, 1, 717

$Pt_2C_{10}H_{20}Cl_4$ $[Pt(cis\text{-}EtCH{=}CHMe)Cl_2]_2$, 1, 715

$Pt_2C_{10}H_{22}Cl_4N_2$ Di-μ-$(CH_2{=}CHCH_2NHEt)$-$[PtCl_2]_2$, 1, 718

$Pt_2C_{10}H_{24}Cl_4N_2$ μ-$H_2NNH_2[Pt(pentene)Cl_2]_2$, 1, 717

$Pt_2C_{12}H_{10}O_2$ $(C_5H_5PtCO)_2$, 1, 702

$Pt_2C_{12}H_{20}$ Cyclooctatetraene-$(PtMe_2)_2$, 1, 693

$Pt_2C_{12}H_{20}Br_4$ $[Pt(cyclohexene)Br_2]_2$, 1, 715

$Pt_2C_{12}H_{20}Cl_4$ $[Pt(cyclohexene)Cl_2]_2$, 1, 715

$Pt_2C_{12}H_{22}Cl_4N_2$ $\{Pt[(CH_2{=}CHCH_2)_2NH]_2\}[PtCl_4]$, 1, 725

$Pt_2C_{12}H_{24}Cl_4$ $[Pt(cis\text{-}Me_2CHCH{=}CHMe)Cl_2]_2$, 1, 715

$[Pt(trans\text{-}Me_2CHCH{=}CHMe)Cl_2]_2$, 1, 715

$[Pt(BuCH{=}CH_2)Cl_2]_2$, 1, 715

$Pt_2C_{12}H_{42}I_2N_6$ $[(Me_3Pt)_2(H_2NCH_2CH_2NH_2)_3I_2]$, 1, 698

$Pt_2C_{14}H_{16}Cl_4N_2$ μ-Butadiene-$[trans\text{-}Pt(Py)Cl_2]_2$, 1, 717

$Pt_2C_{16}H_{14}Cl_4N_2O_4$ $[Pt(O_2NC_6H_4CH{=}CH_2)Cl_2]_2$ m- and p-isomers, 1, 715

$Pt_2C_{16}H_{14}Cl_6$ $[Pt(m\text{-}ClC_6H_4CH{=}CH_2)Cl_2]_2$, 1, 715

$Pt_2C_{16}H_{16}Br_4$ $[Pt(PhCH{=}CH_2)Br_2]_2$, 1, 715

$Pt_2C_{16}H_{16}Cl_4$ $[Pt(PhCH{=}CH_2)Cl_2]_2$, 1, 715

$Pt_2C_{16}H_{20}Cl_4N_2$ μ-1,5-Hexadiene-$[Pt(Py)Cl_2]_2$, 1, 717

$Pt_2C_{16}H_{21}Cl_4N_3$ $[Pt(Py)_2(CH_2{=}CHCH_2)_2NH][PtCl_4]$, 1, 725

$Pt_2C_{16}H_{28}Cl_4$ $[Pt(Me_3CC{\equiv}CEt)Cl_2]_2$, 1, 731

$Pt_2C_{16}H_{28}I_2N_2$ $[Me_3Pt(Py)I]_2$, 1, 698

$Pt_2C_{16}H_{32}Cl_4$ $[Pt(1\text{-}octene)Cl_2]_2$, 1, 715

$Pt_2C_{16}H_{32}O_4$ Acetylacetonato-PtMe$_3$ dimer, 1, 698, 699
$Pt_2C_{16}H_{36}Cl_4P_2$ μ-Butadiene[Pt(PEt$_3$)-Cl$_2$]$_2$, 1, 718
$Pt_2C_{18}H_{16}Cl_4$ [Pt(indene)Cl$_2$]$_2$, 1, 715
$Pt_2C_{18}H_{20}Cl_4$ [Pt(p-MeC$_6$H$_4$CH=CH$_2$)Cl$_2$]$_2$, 1, 715
$Pt_2C_{18}H_{20}Cl_4O_2$ [Pt(MeOC$_6$H$_4$CH=CH$_2$)-Cl$_2$]$_2$ m- and p-isomers, 1, 715
$Pt_2C_{18}H_{30}Cl_2O_2$ Di-μ-chlorobis(2-methoxy-5-cyclooctenyl)-Pt$_2$, 1, 700*
$Pt_2C_{18}H_{30}I_2O_2$ Di-μ-iodobis(2-methoxy-5-cyclooctenyl)-Pt$_2$, 1, 700*
$Pt_2C_{18}H_{32}Cl_4$ [Pt(Me$_3$CC≡CCHMe$_2$)Cl$_2$]$_2$, 1, 731
$Pt_2C_{20}H_{28}Cl_4O_4$ {Pt[Me$_2$C(OH)C≡CC≡CC-(OH)Me$_2$]Cl$_2$}$_2$, 1, 731*
$Pt_2C_{20}H_{32}Cl_4$ [Pt(pinene)Cl$_2$]$_2$, 1, 715
$Pt_2C_{20}H_{36}Br_4$ [Pt(Me$_3$CC≡CCMe$_3$)Br$_2$]$_2$, 1, 731
$Pt_2C_{20}H_{36}Cl_4$ [Pt(Me$_3$CC≡CCMe$_3$)Cl$_2$]$_2$, 1, 731
$Pt_2C_{20}H_{40}O_4$ [Me$_3$Pt(EtCOCHCOEt)]$_2$, 1, 699
$Pt_2C_{22}H_{30}Cl_2O_2$ Di-μ-chlorobis(9-methoxytricyclo[5.2.1.0^{2,6}]-3-decen-8-yl)-Pt$_2$, 1, 700
$Pt_2C_{22}H_{30}I_2O_2$ Di-μ-iodobis(9-methoxy-tricyclo[5.2.1.0^{2,6}]-3-decen-8-yl)-Pt$_2$, 1, 700
$Pt_2C_{22}H_{38}Cl_2O_2$ [Pt(C$_{10}$H$_{16}$OMe)Cl]$_2$, 1, 700
$Pt_2C_{22}H_{46}Cl_4N_2$ Di-μ-(CH$_2$=CHCH$_2$NH-C$_8$H$_{17}$)[PtCl$_2$]$_2$, 1, 718
$Pt_2C_{22}H_{48}Cl_4P_2$ μ-Butadiene[Pt(PPr$_3$)-Cl$_2$]$_2$, 1, 718
$Pt_2C_{24}H_{30}N_2O_2S_2$ Di-μ-thiocyanatobis-(9-methoxytricyclo[5.2.1.0^{2,6}]-3-decen-8-yl)-Pt$_2$, 1, 700
$Pt_2C_{24}H_{34}Cl_2O_2$ Di-μ-chlorobis(9-ethoxytricyclo[5.2.1.0^{2,6}]-3-decen-8-yl)-Pt$_2$, 1, 700
$Pt_2C_{24}H_{48}Cl_4$ [Pt(1-dodecene)Cl$_2$]$_2$, 1, 715
$Pt_2C_{24}H_{48}O_4$ [Me$_3$Pt(PrCOCHCOPr)]$_2$, 1, 699
[Me$_3$Pt(Me$_2$CHCOCHCOCHMe$_2$]$_2$, 1, 699
$Pt_2C_{26}H_{38}Cl_2O_2$ Di-μ-chlorobis(9-propoxytricyclo[5.2.1.0^{2,6}]-3-decen-8-yl)-Pt$_2$, 1, 700
$Pt_2C_{26}H_{38}Cl_4N_2$ μ-1,5-Cyclooctadiene-[Pt(PhCH$_2$CHMeNH$_2$)Cl$_2$]$_2$, 1, 717
$Pt_2C_{26}H_{40}O_2S_2$ Di-μ-ethylthiolatobis-(9-methoxytricyclo[5.2.1.0^{2,6}]-3-decen-8-yl)-Pt$_2$, 1, 700
$Pt_2C_{26}H_{54}Cl_4N_2$ Di-μ-[CH$_2$=CH(CH$_2$)$_3$NH-C$_8$H$_{17}$][PtCl$_2$]$_2$, 1, 718
$Pt_2C_{28}H_{20}Cl_4N_4$ [Pt(CNPh)$_4$][PtCl$_4$], 1, 733
$Pt_2C_{28}H_{24}Cl_4$ [Pt(trans-PhCH=CHPh)Cl$_2$]$_2$, 1, 715
$Pt_2C_{28}H_{28}$ Cyclooctatetraene-[Pt-(C$_5$H$_5$)$_2$]$_2$, 1, 693, 702
$Pt_2C_{28}H_{56}O_4$ [Me$_3$Pt(BuCOCHCOBu)]$_2$, 1, 699
[Me$_3$Pt(Me$_2$CHCH$_2$COCHCOCH$_2$CHMe$_2$)]$_2$, 1, 699
$Pt_2C_{28}H_{60}Cl_4P_2$ μ-Butadiene[Pt(PBu$_3$)-Cl$_2$]$_2$, 1, 718
$Pt_2C_{30}H_{40}Cl_4$ [Pt(Me$_3$CC≡CCMe$_2$Ph)Cl$_2$]$_2$, 1, 731
$Pt_2C_{32}H_{20}Cl_4$ [Pt(PhC≡CC≡CPh)Cl$_2$]$_2$, 1, 731
$Pt_2C_{32}H_{28}$ Cyclooctatetraene-(PtPh$_2$)$_2$, 1, 693
$Pt_2C_{32}H_{64}O_4$ [Me$_3$Pt(n-C$_5$H$_{11}$COCHCO-C$_5$H$_{11}$-n)]$_2$, 1, 699
$Pt_2C_{34}H_{24}Cl_4$ [Pt(PhC≡CCH$_2$C≡CPh)Cl$_2$]$_2$, 1, 731
$Pt_3C_4H_{18}Cl_6N_4$ [Pt(NH$_3$)$_4$][Pt$_2$(butadiene)Cl$_6$], 1, 724
$Pt_3C_4H_{20}Br_6N_4$ [Pt(NH$_3$)$_4$][Pt(C$_2$H$_4$)-Br$_3$]$_2$, 1, 721
$Pt_3C_4H_{20}Cl_4N_6O_4$ [Pt(NH$_3$)$_4$][Pt(C$_2$H$_4$)-(NO$_2$)Cl$_2$]$_2$, 1, 721
$Pt_3C_6H_{22}Cl_6N_4$ [Pt(NH$_3$)$_4$][Pt$_2$(1,5-hexadiene)Cl$_6$], 1, 724
$Pt_3C_{16}H_{28}Cl_6N_4$ [Pt(NH$_3$)$_4$][Pt(PhCH=CH$_2$)-Cl$_3$]$_2$, 1, 724
$Pt_3C_{18}H_{30}Cl_6O_3$ Tris(4-methyl-3-penten-2-one)-[PtCl$_2$]$_3$, 1, 726
$Pt_3C_{24}H_{28}Br_6N_4$ [PtPy$_4$][Pt(C$_2$H$_4$)Br$_3$]$_2$, 1, 721

$Pt_3C_{24}H_{28}Cl_6N_4$ $[PtPy_4][Pt(C_2H_4)Cl_3]_2$, 1, 721
$Pt_4C_{12}H_{36}I_4$ $(Me_3PtI)_4$, 1, 695
$PtAsC_{14}H_{13}Br_2O$ $Pt(Ph_2AsMe)(CO)Br_2$, 3, 55
$PtAsC_{14}H_{13}Cl_2O$ $Pt(Ph_2AsMe)(CO)Cl_2$, 3, 55
$PtAsC_{14}H_{13}I_2O$ $Pt(Ph_2AsMe)(CO)I_2$, 3, 55
$PtAsC_{25}H_{20}Br_3O$ $[Ph_4As][Pt(CO)Br_3]$, 3, 333
$PtAsC_{25}H_{20}Cl_3O$ $[Ph_4As][Pt(CO)Cl_3]$, 3, 333
$PtAsC_{25}H_{20}I_3O$ $[Ph_4As][Pt(CO)I_3]$, 3, 333
$PtAs_2C_{18}H_{35}Br$ trans-$PhPt(AsEt_3)_2Br$, 1, 691
$PtAs_2C_{19}H_{37}Cl$ trans-$PhCH_2Pt(AsEt_3)_2Cl$, 1, 691
$PtAs_2C_{24}H_{40}$ $Ph_2Pt(AsEt_3)_2$ cis- and trans-isomers, 1, 694
$PtAs_2C_{28}H_{40}$ trans-$(PhC{\equiv}C)_2Pt(AsEt_3)_2$, 1, 694*
$PtAs_2C_{30}H_{52}$ trans-$Ph_2Pt(AsPr_3)_2$, 1, 694
$PtAs_2C_{40}H_{30}F_6$ $(Ph_3As)_2Pt(CF_3C{\equiv}CCF_3)$, 1, 730
$PtCoC_{10}H_{26}Cl_5N_4$ $[Co(H_2NCH_2CH_2NH_2)_2Cl_2][Pt(cyclohexene)Cl_3]$, 1, 723
$PtCoC_{13}H_{24}Cl_5N_4$ $[Co(H_2NCH_2CH_2NH_2)_2Cl_2][Pt(indene)Cl_3]$, 1, 724
$PtKC_2H_4Br_2Cl$ $K[Pt(C_2H_4)Br_2Cl]$, 1, 721
$PtKC_2H_4Br_3$ $K[Pt(C_2H_4)Br_3]$, 1, 721
$PtKC_2H_6Cl_3O$ $K[Pt(C_2H_4)Cl_3]{\cdot}H_2O$, 1, 719
$PtKC_3H_6Cl_3O$ $K[Pt(CH_2{=}CHCH_2OH)Cl_3]$, 1, 723
$PtKC_3H_8Cl_3O$ $K[Pt(C_3H_6)Cl_3]{\cdot}H_2O$, 1, 723
$PtKC_4H_8Cl_3$ $K[Pt(cis\text{-}2\text{-}butene)Cl_3]$, 1, 723
$PtKC_4H_8Cl_3O$ $K[Pt(CH_2{=}CHCH_2OMe)Cl_3]$, 1, 723
$PtKC_6H_{10}Cl_3$ $K[Pt(cyclohexene)Cl_3]$, 1, 723
$PtKC_8H_8Cl_3$ $K[Pt(PhCH{=}CH_2)Cl_3]$, 1, 723
$PtKC_8H_{14}Br_3O_2$ $K\{Pt[Me_2C(OH)C{\equiv}]_2Br_3\}$, 1, 732
$PtKC_8H_{14}Cl_2NO_4$ $K\{Pt[Me_2C(OH)C{\equiv}]_2(NO_2)Cl_2\}$, 1, 732
$PtKC_8H_{14}Cl_3O_2$ $K\{Pt[Me_2C(OH)C{\equiv}CC(OH)Me_2]Cl_3\}$, 1, 731
$PtKC_9H_8Cl_3$ $K[Pt(indene)Cl_3]$, 1, 724
$PtKC_9H_{16}Cl_3O$ $K\{Pt[Me_3CC{\equiv}CC(OH)Me_2]Cl_3\}$, 1, 732
$PtKC_9H_{16}Cl_3O_2$ $K\{Pt[Me_2C(OH)C{\equiv}CC(OMe)Me_2]Cl_3\}$, 1, 732
$PtKC_{10}H_{14}ClO_4$ $K[Pt(AcCHAc)_2Cl]$, 1, 701
$PtKC_{10}H_{18}Cl_3$ $K[Pt(Me_3CC{\equiv}CCMe_3)Cl_3]$, 1, 732
$PtKC_{10}H_{18}Cl_3O_2$ $K\{Pt[Me_2C(OMe)C{\equiv}]_2Cl_3\}$, 1, 732
$K\{Pt[EtMeC(OH)C{\equiv}]_2Cl_3\}$, 1, 732
$PtKC_{13}H_{16}Cl_3O_2$ $K\{Pt[MePhC(OH)C{\equiv}CC(OH)Me_2]Cl_3\}$, 1, 732
$PtKC_{14}H_{12}Cl_3$ $K[Pt(Ph_2C{=}CH_2)Cl_3]$, 1, 724
$PtKC_{14}H_{18}Cl_3O$ $K\{Pt[Me_3CC{\equiv}CC(OH)PhMe]Cl_3\}$, 1, 733
$PtKC_{18}H_{18}Cl_3O_2$ $K\{Pt[MePhC(OH)C{\equiv}]_2Cl_3\}$, 1, 733
$PtKC_{18}H_{34}Cl_3O_2$ $K\{Pt[Me(C_6H_{13})C(OH)C{\equiv}]_2Cl_3\}$, 1, 732
$PtK_2C_4H_2$ $K_2[Pt(C{\equiv}CH)_2]$, 1, 701
$PtK_2C_6H_6$ $K_2[Pt(C{\equiv}CMe)_2]$, 1, 701
$PtK_2C_8H_4$ $K_2[Pt(C{\equiv}CH)_4]$, 1, 701
$PtK_2C_{12}H_{12}$ $K_2[Pt(C{\equiv}CMe)_4]$, 1, 701
$PtK_2C_{16}H_{10}$ $K_2[Pt(C{\equiv}CPh)_2]$, 1, 701
$PtK_2C_{32}H_{20}$ $K_2[Pt(C{\equiv}CPh)_4]$, 1, 701
$PtMn_2C_{14}H_{10}Cl_6N_2O_6$ $[C_5H_5Mn(CO)_2(NO)]_2PtCl_6$, 1, 171
$PtMo_2C_{22}H_{22}Cl_6O_6$ $[C_5H_5Mo(CH_2{=}CHMe)(CO)_3]_2PtCl_6$, 1, 99
$PtNaC_2H_4Cl_3$ $Na[Pt(C_2H_4)Cl_3]$, 1, 721
$PtNaC_4H_6Cl_3$ $Na[Pt(butadiene)Cl_3]$, 1, 723
$PtNaC_5H_{10}Cl_3$ $Na[Pt(EtCH{=}CHMe)Cl_3]$, 1, 723
$PtNaC_8H_{14}Cl_3O_2$ $Na\{Pt[Me_2C(OH)C{\equiv}]_2Cl_3\}$, 1, 732
$PtNaC_9H_{16}Cl_3O$ $Na\{Pt[Me_3CC{\equiv}CC(OH)Me_2]Cl_3\}$, 1, 732
$PtNaC_{10}H_{18}Cl_3$ $Na[Pt(Me_3CC{\equiv}CCMe_3)Cl_3]$, 1, 732

$PtNaC_{10}H_{18}Cl_3O_2$ Na{Pt[EtMeC(OH)C≡]$_2$-Cl_3}, 1, 732
Na{Pt[Me_2C(OMe)C≡]$_2$$Cl_3$}, 1, 732
$PtPbC_8H_9Cl_3O$ (EtC_5H_4)Pt(CO)$PbCl_3$, 1, 733*
$PtPbC_{24}H_{24}O_2$ Et_2Pb(2-indenyl-PtCO)$_2$, 2, 595
$PtPbC_{35}H_{42}O$ Mesityl$_3$Pb(C_5H_4Et)PtCO, 2, 595
$PtPbC_{38}H_{51}FO$ ($C_{18}H_{37}C_5H_4$)Pt(CO)PbPh_2F, 1, 733*, 2, 595
$PtPbC_{42}H_{56}O$ Ph_3Pb($C_5H_4C_{18}H_{37}$)PtCO, 2, 595
$PtSb_2C_{12}H_{18}Cl_2$ Pt[Sb(CH=CH_2)$_3$]$_2$$Cl_2$, 3, 655
$PtW_2C_{20}H_{20}Cl_{10}$ [(C_5H_5)$_2$WCl_2]$_2$PtCl_6, 1, 124
$Pt_2AsC_{42}H_{62}BrP_2$ Ph_2As[PhPt(PPr_3)]$_2$Br, 3, 304
$Pt_2K_2C_4H_6Cl_6$ K_2[Pt_2(butadiene)Cl_6], 1, 724
$Pt_2K_2C_6H_{10}Cl_6$ K_2[Pt_2(1,5-hexadiene)-Cl_6], 1, 724
$Pt_2Sb_2C_6H_{18}Cl_4$ [Pt(SbMe_3)Cl_2]$_2$, 3, 655
$ReC_6H_3O_5$ MeRe(CO)$_5$, 1, 187
$ReC_6H_4F_6O_4P$ [C_2H_4Re(CO)$_4$]PF_6, 1, 190
$ReC_7F_5O_5$ C_2F_5Re(CO)$_5$, 1, 187
$ReC_7HCl_2F_2O_5$ HCCl_2CF_2Re(CO)$_5$, 1, 187
$ReC_7H_3O_6$ AcRe(CO)$_5$, 1, 187
$ReC_7H_5O_5$ EtRe(CO)$_5$, 1, 187
$ReC_8F_5O_6$ C_2F_5CORe(CO)$_5$, 1, 187
$ReC_8H_5O_3$ C_5H_5Re(CO)$_3$, 1, 187
$ReC_8H_5O_6$ EtCORe(CO)$_5$, 1, 187
$ReC_8H_5O_6S$ ($HO_3SC_5H_4$)Re(CO)$_3$, 1, 188
$ReC_8H_7Cl_3N$ (p-MeC_6H_4NC)ReCl_3, 1, 191
$ReC_8H_7I_3N$ (p-MeC_6H_4NC)ReI_3, 1, 191
$ReC_8H_7O_5$ C_3H_7Re(CO)$_5$, 1, 187
$ReC_9F_7O_6$ C_3F_7CORe(CO)$_5$, 1, 187
$ReC_{10}H_3F_8O_5$ Me(CF_2)$_4$Re(CO)$_5$, 1, 187
$ReC_{10}H_7O_4$ (AcC_5H_4)Re(CO)$_3$, 1, 188
$ReC_{10}H_{10}Br_3$ [(C_5H_5)$_2$ReBr_2]Br, 1, 189
$ReC_{10}H_{10}Cl_3$ [(C_5H_5)$_2$ReCl_2]Cl, 1, 189
$ReC_{10}H_{10}D$ (C_5H_5)$_2$ReD, 1, 189
$ReC_{10}H_{11}$ (C_5H_5)$_2$ReH, 1, 189
$ReC_{10}H_{12}Cl$ [(C_5H_5)$_2$ReH_2]Cl, 1, 189
$ReC_{10}H_{16}ClO_2$ [Me_2C(OH)C≡CH]$_2$ReCl, 1, 191
$ReC_{11}H_5O_5$ PhRe(CO)$_5$, 1, 187
$ReC_{12}H_5O_6$ BzRe(CO)$_5$, 1, 187
$ReC_{12}H_7O_3$ Indenyl-Re(CO)$_3$, 1, 188
$ReC_{12}H_7O_5$ PhCH_2Re(CO)$_5$, 1, 187
$ReC_{12}H_{11}O_2$ (C_5H_6)Re(C_5H_5)(CO)$_2$, 1, 188*
$ReC_{12}H_{12}I_3$ [(C_6H_6)$_2$Re]I_3, 1, 190
$ReC_{12}H_{13}$ Cyclohexadienyl-Re(C_6H_6), 1, 190
$ReC_{12}H_{13}O_2$ Cyclopentadienylcyclopentene-Re(CO)$_2$, 1, 189
$ReC_{14}H_{15}Cl_3N$ (PhNMe_2)R(C_5H_4CCl_3), 1, 189
$ReC_{14}H_{15}O_3$ (cyclo-$C_6H_{11}C_5H_4$)Re(CO)$_3$, 1, 188
$ReC_{14}H_{16}Cl_3N$ (Me_2NPh)Re(C_5H_5CCl_3), 1, 190
$ReC_{15}H_9O_4$ (BzC_5H_4)Re(CO)$_3$, 1, 188
$ReC_{15}H_{18}ClO_7$ [(Hexamethylbenzene)Re-(CO)$_3$]ClO_4, 1, 190
$ReC_{15}H_{18}I_3O_3$ [(Hexamethylbenzene)Re-(CO)$_3$]I_3, 1, 190*
$ReC_{15}H_{19}O_3$ Hexamethylcyclohexadienyl-Re(CO)$_3$, 1, 190
$ReC_{16}H_{12}Cl$ (PhC≡CH)$_2$ReCl, 1, 191
$ReC_{16}H_{21}O_3$ (n-$C_8H_{17}C_5H_4$)Re(CO)$_3$, 1, 188
$ReC_{17}H_{11}N_2O_5$ MeO_2CRe(CO)$_3$(o-phenanthroline), 1, 187
$ReC_{18}H_{24}F_6P$ [($Me_3C_6H_3$)$_2$Re]PF_6, 1, 190
$ReC_{18}H_{30}I_3N_6$ [(EtNC)$_6$Re]I_3, 1, 192
$ReC_{18}H_{30}I_3N_6$ [(EtNC)$_6$Re]I_3, 1, 192
$ReC_{20}H_{14}IN_2O_4$ [(p-MeC_6H_4NC)$_2$Re(CO)$_4$]I, 1, 192
$ReC_{20}H_{24}Cl_2$ Bis(dicyclopentadiene)-ReCl_2, 1, 191
$ReC_{24}H_{22}$ (Ph$CH_2C_5H_4$)$_2$Re, 1, 189
$ReC_{27}H_{22}BrNOP$ [(p-MeC_6H_4NC)(CO)(Ph_3P)-Re]Br, 1, 191
$ReC_{27}H_{22}ClNOP$ [(p-MeC_6H_4NC)(CO)(Ph_3P)-Re]Cl, 1, 191
$ReC_{28}H_{31}ClO_2P$ [Me_2C(OH)C≡CH]$_2$Re(PPh_3)-Cl, 1, 191
$ReC_{38}H_{39}ClP$ [Bis(dicyclopentadiene)Rh-(PPh_3)]Cl, 1, 191
$ReC_{41}H_{33}O_5P_2$ MeO_2CRe(CO)$_3$(PPh_3)$_2$, 1,

187
$ReC_{44}H_{37}ClINO_4P_2$ $[(p\text{-}MeC_6H_4NC)(Ph_3P)_2ReI]ClO_4$, 1, 191
$ReC_{44}H_{37}I_2NP_2$ $[(p\text{-}MeC_6H_4NC)(Ph_3P)_2ReI]I$, 1, 191
$ReC_{44}H_{37}I_4NP_2$ $(p\text{-}MeC_6H_4NC)(Ph_3P)_2ReI]I_3$, 1, 191
$ReC_{48}H_{42}ClN_6O_4$ $[(p\text{-}MeC_6H_4NC)_6Re]ClO_4$, 1, 192
$ReC_{48}H_{42}IN_6$ $[(p\text{-}MeC_6H_4NC)_6Re]I$, 1, 192
$ReC_{48}H_{42}I_3N_6$ $[(p\text{-}MeC_6H_4NC)_6Re]I_3$, 1, 192
$ReC_{48}H_{42}N_7O_3$ $[(p\text{-}MeC_6H_4NC)_6Re]NO_3$, 1, 192
$ReC_{53}H_{44}ClN_2OP_2$ $(p\text{-}MeC_6H_4NC)_2(CO)(Ph_3P)_2Re]Cl$, 1, 192
$Re_2C_{18}H_{16}O_4$ $[(MePh)Re(CO)_2]_2$, 1, 190
$ReAlC_{12}H_{12}Cl_4$ $[(C_6H_6)_2Re]AlCl_4$, 1, 190
$ReAlC_{18}H_{24}Cl_4$ $[(Me_3C_6H_3)_2Re]AlCl_4$, 1, 190
$ReAs_2C_{39}H_{30}ClO_3$ $[Re(AsPh_3)_2(CO)_3Cl]$, 3, 25
$ReAs_3C_{39}H_{30}IO_3$ $[Re(AsPh_3)_2(CO)_3I]$, 3, 25
$ReBC_{36}H_{32}$ $[(C_6H_6)_2Re]BPh_4$, 1, 190
$ReBC_{42}H_{44}$ $[(Me_3C_6H_3)_2Re]BPh_4$, 1, 190
$ReBC_{68}H_{57}INP_2$ $[(p\text{-}MeC_6H_4NC)(Ph_3P)_2ReI]BPh_4$, 1, 191
$ReBC_{72}H_{62}N_6$ $[p\text{-}MeC_6H_4NC)_6Re]BPh_4$, 1, 192
$ReBiC_{17}H_{10}O_5$ $Ph_2BiRe(CO)_5$, 3, 833*
$ReCrC_{19}H_{24}N_6O_3S_4$ [(Hexamethylbenzene)$Re(CO)_3][Cr(NH_3)_2(SCN)_4]$, 1, 190*
$ReCrC_{22}H_{30}N_6S_4$ $[(Me_3C_6H_3)_2Re][Cr(NH_3)_2(SCN)_4]$, 1, 190
$ReFeC_{12}H_5O_7$ $C_5H_5Fe(CO)_2Re(CO)_5$, 1, 192
$ReHgC_8H_4ClO_3$ $(ClHgC_5H_4)Re(CO)_3$, 1, 192
$ReHg_2C_8H_3Cl_2O_3$ $[(ClHg)_2C_5H_3]Re(CO)_3$, 1, 192
$RePbC_{23}H_{15}O_5$ $Ph_3PbRe(CO)_5$, 2, 594
$ReSnC_{11}H_5Br_2O_5$ $PhSnBr_2Re(CO)_5$, 2, 502
$ReSnC_{23}H_{15}O_5$ $Ph_3SnRe(CO)_5$, 2, 501
$Re_2SnC_{22}H_{10}O_{10}$ $Ph_2Sn[Re(CO)_5]_2$, 2, 502
$Re_3SnC_{21}H_5O_{15}$ $PhSn[Re(CO)_5]_3$, 2, 502
$RhC_5H_5Br_2$ $C_5H_5RhBr_2$, 1, 619
$RhC_5H_5I_2$ $C_5H_5RhI_2$, 1, 620
$RhC_7H_5O_2$ $C_5H_5Rh(CO)_2$, 1, 620
RhC_9H_{13} $C_5H_5Rh(C_2H_4)_2$, 1, 620
$RhC_9H_{15}O_2$ $(AcCHAc)Rh(C_2H_4)_2$, 1, 625
$RhC_{10}H_{10}Br_3$ $[(C_5H_5)_2Rh]Br_3$, 1, 622
$RhC_{10}H_{10}I_3$ $[(C_5H_5)_2Rh]I_3$, 1, 622
$RhC_{10}H_{11}$ $C_5H_5Rh(CH{=}CHCH{=}CHCH_2)$, 1, 620
$RhC_{11}H_{15}$ $C_5H_5Rh(CH_2{=}CHCH_2CH_2CH{=}CH_2)$, 1, 620
$RhC_{12}H_{15}O_2$ Bicyclo[2.2.1]hepta-2,5-diene-Rh(AcCHAc), 1, 626*
$(RhC_{12}H_{17}Cl_2)_n$ π-Cyclododeca-1,5,9-trienyl-$RhCl_2$ polymer, 1, 622
$RhC_{13}H_{13}$ Cyclooctatetraenecyclopentadienyl-Rh, 1, 621
$RhC_{13}H_{17}$ 1,5-Cyclooctadienecyclopentadienyl-Rh, 1, 621
$RhC_{13}H_{19}O_2$ 1,5-Cyclooctadiene-Rh(AcCHAc), 1, 626
$RhC_{13}H_{23}BrN$ [1,5-Cyclooctadiene-Rh-(Pip)]Br, 1, 625
$RhC_{13}H_{23}ClN$ [1,5-Cyclooctadiene-Rh-(Pip)]Cl, 1, 625
$RhC_{13}H_{23}IN$ [1,5-Cyclooctadiene-Rh-(Pip)]I, 1, 625
$RhC_{14}H_5F_{12}O$ $C_5H_5Rh[CF_3\text{-}C{=}C(CF_3)C(CF_3){=}C(CF_3)CO]$, 1, 620
$RhC_{14}H_{17}ClN$ [Bicyclo[2.2.1]hepta-2,5-diene-Rh($p\text{-}MeC_6H_4NH_2$)]Cl, 1, 625*
$RhC_{15}H_{15}$ $C_5H_5Rh(C_5H_5C_5H_5)$, 1, 621
$RhC_{15}H_{17}O_2$ (C_5H_5)-duroquinone-Rh, 1, 621
$RhC_{15}H_{21}ClN$ [1,5-Cyclooctadiene-($p\text{-}MeC_6H_4NH_2$)-Rh]Cl, 1, 625
$RhC_{16}H_{14}F_6P$ $[(C_5H_5)(PhC_5H_4)Rh]PF_6$, 1, 622
$RhC_{16}H_{15}$ $C_5H_5Rh(C_5H_5Ph)$, 1, 621
$RhC_{16}H_{23}O_2$ (3-Allylpentane-2,4-dionato)-1,5-cyclooctadiene-Rh, 1, 626
$RhC_{17}H_5F_{18}$ Cyclopentadienyl[hexakis-(trifluoromethyl)-2,5-cyclohexadiene-1,4-diyl]-Rh, 1, 621
$RhC_{17}H_{14}BrN_2O_3$ $[(p\text{-}MeOC_6H_4NC)_2Rh(CO)]$-

Br, 1, 627
$RhC_{17}H_{14}ClN_2O_3$ [(p-MeOC$_6$H$_4$NC)$_2$Rh-(CO)]Cl, 1, 627
$RhC_{18}H_{24}I_3$ [(1,3,5-Me$_3$C$_6$H$_3$)$_2$Rh]I$_3$, 1, 623
$RhC_{20}H_{20}ClN_2$ [1,5-Cyclooctadienephenanthroline-Rh]Cl, 1, 625
$RhC_{20}H_{21}ClP$ [(Bicyclo[2.2.1]hepta-2,5-diene)(MePh$_2$P)Rh]Cl, 1, 625
$RhC_{20}H_{22}ClN_2O_2$ [Duroquinone-RhPy$_2$]Cl, 1, 625
$RhC_{22}H_{19}$ (PhC$_5$H$_4$)Rh(C$_5$H$_5$Ph), 1, 621
$RhC_{24}H_{36}F_6P$ [(C$_6$Me$_6$)$_2$Rh]PF$_6$, 1, 623
$RhC_{24}H_{36}F_{12}P_2$ [(C$_6$Me$_6$)$_2$Rh](PF$_6$)$_2$, 1, 623
$RhC_{25}H_{23}ClP$ [(Bicyclo[2.2.1]hepta-2,5-diene)(Ph$_3$P)Rh]Cl, 1, 625
$RhC_{26}H_{27}ClP$ [(1,5-Cyclooctadiene)-(Ph$_3$P)Rh]Cl, 1, 625
$RhC_{26}H_{35}Cl_2N_2$ π-Cyclododeca-1,5,9-trienylbis(p-toluidine)-RhCl$_2$, 1, 622
$RhC_{26}H_{43}$ (C$_8$H$_{17}$C$_5$H$_5$)Rh(C$_5$H$_4$C$_8$H$_{17}$), 1, 621
$RhC_{26}H_{57}ClIOP_2$ MeRh(PBu$_3$)$_2$(CO)ClI, 1, 619
$RhC_{27}H_{57}ClIO_2P_2$ AcRh(PBu$_3$)$_2$(CO)ClI, 1, 619
$RhC_{27}H_{57}I_2O_2P_2$ AcRh(PBu$_3$)$_2$(CO)I$_2$, 1, 619
$RhC_{28}H_{20}ClN_4$ [(PhNC)$_4$Rh]Cl, 1, 628
$RhC_{28}H_{59}ClIO_3P_2$ MeO$_2$CCH$_2$Rh(PBu$_3$)$_2$-(CO)ClI, 1, 619
$RhC_{28}H_{59}I_2O_3P_2$ MeO$_2$CCH$_2$Rh(PBu$_3$)$_2$(CO)-I$_2$, 1, 619
$RhC_{32}H_{28}ClN_4O_4$ [(p-MeC$_6$H$_4$NC)$_4$Rh]ClO$_4$, 1, 628
$RhC_{32}H_{28}ClN_4O_8$ [(p-MeOC$_6$H$_4$NC)$_4$Rh]-ClO$_4$, 1, 628
$RhC_{32}H_{28}F_6N_4O_4P$ [(p-MeOC$_6$H$_4$NC)$_4$Rh]PF$_6$, 1, 628
$RhC_{32}H_{61}BrClOP_2$ PhCH$_2$Rh(PBu$_3$)$_2$(CO)-ClBr, 1, 619
$RhC_{38}H_{56}BrP_2$ (1-C$_{10}$H$_7$)$_2$Rh(PPr$_3$)$_2$Br, 1, 619
$RhC_{40}H_{44}BrP_2$ (1-C$_{10}$H$_7$)$_2$Rh(PEt$_2$Ph)$_2$Br, 1, 619
$RhC_{50}H_{40}BrN_2P_2$ [(PhNC)$_2$Rh(PPh$_3$)$_2$]Br, 1, 627
$RhC_{50}H_{40}ClN_2P_2$ [(PhNC)$_2$Rh(PPh$_3$)$_2$]Cl, 1, 627
$RhC_{50}H_{40}IN_2P_2$ [(PhNC)$_2$Rh(PPh$_3$)$_2$]I, 1, 627
$RhC_{52}H_{44}ClN_2O_2P_2$ [(p-MeOC$_6$H$_4$NC)$_2$Rh-(PPh$_3$)$_2$]Cl, 1, 627
$RhC_{52}H_{44}ClN_2O_6P_2$ [(p-MeOC$_6$H$_4$NC)$_2$Rh-(PPh$_3$)$_2$]ClO$_4$, 1, 627
$RhC_{52}H_{44}ClN_2O_{12}P_2$ {(p-MeOC$_6$H$_4$NC)$_2$Rh-[P(OPh)$_3$]$_2$}ClO$_4$, 1, 627
$RhC_{52}H_{44}IN_2O_2P_2$ [(p-MeOC$_6$H$_4$NC)$_2$Rh-(PPh$_3$)$_2$]I, 1, 627
$Rh_2C_8H_{16}Cl_2$ [(C$_2$H$_4$)$_2$RhCl]$_2$, 1, 623*
$Rh_2C_{12}H_{11}ClO_6$ Ac$_2$C(CH$_2$CH=CH$_2$)Rh$_2$-(CO)$_4$Cl, 1, 626
$Rh_2C_{12}H_{20}Cl_2$ μ-Cl$_2$-bis(1,5-hexadiene)-Rh$_2$, 1, 624
$Rh_2C_{12}H_{24}Cl_2$ [(MeCH=CH$_2$)$_2$RhCl]$_2$, 1, 623
$Rh_2C_{14}H_{10}O_4$ [C$_5$H$_5$Rh(CO)$_2$]$_2$, 1, 620
$Rh_2C_{14}H_{16}Cl_2$ μ-Cl$_2$-bis(bicyclo[2.2.1]-hepta-2,5-diene)-Rh$_2$, 1, 624
$Rh_2C_{16}H_{16}Cl_2$ μ-Cl$_2$-bis(cycloocta-tetraene)-Rh$_2$, 1, 624
$Rh_2C_{16}H_{24}Br_2$ μ-Br$_2$-bis(1,5-cycloocta-diene)-Rh$_2$, 1, 624
$Rh_2C_{16}H_{24}Cl_2$ μ-Cl$_2$-bis(1,5-cycloocta-diene)-Rh$_2$, 1, 624
$Rh_2C_{16}H_{24}I_2$ μ-I$_2$-bis(1,5-cycloocta-diene)-Rh$_2$, 1, 624
$Rh_2C_{18}H_{20}$ μ-Cyclooctatetraene-(Rh-C$_5$H$_6$)$_2$, 1, 626
$Rh_2C_{20}H_{24}Cl_2$ μ-Cl$_2$-bis(dicyclopenta-diene)-Rh$_2$, 1, 624
$Rh_2C_{20}H_{24}Cl_2O_4$ μ-Cl$_2$-bis(duroquinone)-Rh$_2$, 1, 624
$Rh_2C_{20}H_{30}O_4$ Di-μ-acetoxybis(1,5-cyclo-octadiene)-Rh$_2$, 1, 624
$Rh_2C_{24}H_{20}Cl_2O_6$ Di-μ-chlorobis(tricy-clo[4.2.2.0^{2,5}]deca-3,7-diene-9,10-dicarboxylic anhydride)-Rh$_2$, 1, 624
$Rh_2C_{26}H_{32}Cl_2N_2$ [1,5-Cyclooctadiene-(BiPy)Rh][1,5-cyclooctadiene-RhCl$_2$], 1, 625
$Rh_2C_{28}H_{32}Cl_2O_8$ Di-μ-chlorobis(9,10-di-

methoxycarbonyltricyclo-[4.2.2.0^{2,5}]deca-3,7-diene)-Rh_2, 1, 624

$Rh_2C_{30}H_{30}$ $(C_5H_5)_6Rh_2$, 1, 622

$RhAsC_{24}H_{22}O_2$ [AcCH=C(Me)ORh($AsPh_3$)-(CO)], 3, 26

$RhAsC_{25}H_{23}Cl$ [(Bicyclo[2.2.1]hepta-2,5-diene)(Ph_3As)Rh]Cl, 1, 625

$RhAs_2C_{37}H_{24}Cl_7O$ Rh[As(C_6H_4Cl-p)$_3$]$_2$-(CO)Cl, 3, 17

$RhAs_2C_{37}H_{24}Cl_9O$ Rh[As(C_6H_4Cl-p)$_3$]$_2$-(CO)Cl_3, 3, 17

$RhAs_2C_{37}H_{30}ClI_2O$ Rh($AsPh_3$)$_2$(CO)ClI_2, 3, 25

$RhAs_2C_{37}H_{30}ClO$ Rh($AsPh_3$)$_2$(CO)Cl, 3, 25

$RhAs_2C_{37}H_{30}Cl_3O$ Rh($AsPh_3$)(CO)Cl_3, 3, 25

$RhAs_2C_{37}H_{30}I_3O$ Rh($AsPh_3$)$_2$(CO)I_3, 3, 26

$RhAs_2C_{43}H_{42}ClO$ Rh[As(C_6H_4Me-p)$_3$]$_2$-(CO)Cl, 3, 28

$RhAs_2C_{43}H_{42}ClO_7$ Rh[As(C_6H_4OMe-p)$_3$]$_2$-(CO)Cl, 3, 28

$RhAs_2C_{43}H_{42}Cl_3O$ Rh[As(C_6H_4Me-p)$_3$]$_2$-(CO)Cl_3, 3, 28

$RhAs_2C_{52}H_{44}ClN_2O_6$ [(p-$MeOC_6H_4NC$)$_2$Rh-($AsPh_3$)$_2$]ClO_4, 1, 627

$RhAs_4C_{55}H_{42}ClO$ {Rh[(2-$Ph_2AsC_6H_4$)$_3$As]-(CO)}Cl, 3, 112

$RhBC_{38}H_{48}N_2$ [1,5-Cyclooctadiene-Rh-(EtNHCH_2-)$_2$]BPh_4, 1, 625

$RhCrC_{14}H_{16}N_6S_4$ [(C_5H_5)$_2$Rh][Cr(NH_3)$_2$-(SCN)$_4$], 1, 622

$RhPbC_{30}H_{43}O$ (C_8H_{17})$_2$Pb(9-fluorenyl)-RhCO, 2, 595

$RhPtC_{24}H_{36}Cl_6$ [(C_6Me_6)$_2$Rh]$PtCl_6$, 1, 623

$RhSbC_{25}H_{23}Cl$ [(Bicyclo[2.2.1]hepta-2,5-diene)(Ph_3Sb)Rh]Cl, 1, 625

$RhSb_2C_{37}H_{24}Cl_7O$ Rh[Sb(C_6H_4Cl-p)$_3$]$_2$-(CO)Cl, 3, 660

$RhSb_2C_{37}H_{30}ClI_2O$ Rh($SbPh_3$)$_2$(CO)ClI_2, 3, 665

$RhSb_2C_{37}H_{30}ClO$ Rh($SbPh_3$)$_2$(CO)Cl, 3, 665

$RhSb_2C_{37}H_{30}Cl_3O$ Rh($SbPh_3$)$_2$(CO)Cl_3, 3, 665

$RhSb_2C_{37}H_{30}I_3O$ Rh($SbPh_3$)$_2$(CO)I_3, 3, 665

$RhSb_2C_{38}H_{30}ClO_2$ Rh($SbPh_3$)$_2$(CO)$_2$Cl, 3, 665

$RhSb_2C_{43}H_{42}ClO$ Rh[Sb(C_6H_4Me)$_3$]$_2$(CO)Cl, 3, 667

$RhSb_2C_{43}H_{42}Cl_3O$ Rh[Sb(C_6H_4Me)$_3$]$_2$(CO)-Cl_3, 3, 667

$RhSb_2C_{52}H_{44}ClN_2O_6$ [(p-$MeOC_6H_4NC$)$_2$Rh-($SbPh_3$)$_2$]ClO_4, 1, 628

$RhSb_3C_{56}H_{45}ClO_2$ Rh($SbPh_3$)$_3$(CO)$_2$Cl, 3, 665

$RhSb_3C_{60}H_{52}O_3$ AcCH=C(Me)ORh($SbPh_3$)$_3$-(CO), 3, 665

$RhSnC_{12}H_{13}O$ $Me_2SnC_9H_7RhCO$, 2, 503

$RhSnC_{15}H_{12}ClO$ Indenyl-Rh(CO)Sn(C_5H_5)-Cl, 1, 629, 2, 503

$RhSnC_{20}H_{34}FO$ ($Me_2C_5H_3$)Rh(CO)Sn($C_{12}H_{25}$)F, 1, 629, 2, 503*

$RuC_7H_5DO_2$ $C_5H_5Ru(CO)_2D$, 1, 550

$RuC_7H_5IO_2$ [$C_5H_5Ru(CO)_2$]I, 1, 550

$RuC_7H_6O_2$ $C_5H_5Ru(CO)_2H$, 1, 550

$RuC_7H_6O_3$ 1,3-Butadiene-Ru(CO)$_3$, 1, 557

$RuC_7H_7ClO_3$ [CH_2=CHCH(Me)]Ru(CO)$_3$Cl, 1, 555

$RuC_7H_8Br_2$ Bicyclo[2.2.1]hepta-2,5-diene-$RuBr_2$, 1, 557

$RuC_7H_8Cl_2$ Bicyclo[2.2.1]hepta-2,5-diene-$RuCl_2$, 1, 557

$RuC_8H_8O_2$ $C_5H_5Ru(CO)_2Me$, 1, 550

$RuC_8H_{12}Br_2$ 1,5-Cyclooctadiene-$RuBr_2$, 1, 557

$RuC_8H_{12}Br_2N_4$ [(MeNC)$_4RuBr_2$], 1, 558

$RuC_8H_{12}Cl_2$ 1,5-Cyclooctadiene-$RuCl_2$, 1, 557

$RuC_8H_{12}Cl_2N_4$ [(MeNC)$_4RuCl_2$], 1, 558

$RuC_8H_{12}I_2$ 1,5-Cyclooctadiene-RuI_2, 1, 557

$RuC_9H_8O_3$ 1,3,5-Hexatriene-Ru(CO)$_3$, 1, 557

$RuC_9H_{10}O_2$ $C_5H_5Ru(CO)_2Et$, 1, 550

$RuC_9H_{13}N_5O$ [(MeNC)$_4$Ru(OH)(CN)], 1, 558

$RuC_{10}H_{10}$ (C_5H_5)$_2$Ru, 1, 551

$RuC_{10}H_{10}ClO_4$ [(C_5H_5)$_2$Ru]ClO_4, 1, 551

$RuC_{10}H_{10}I_3$ [(C_5H_5)$_2$Ru]I_3, 1, 551

$RuC_{10}H_{10}I_5$ [(C_5H_5)$_2$Ru]I_5, 1, 552

$RuC_{10}H_{10}O_3$ $C_5H_5Ru(CO)_2COEt$, 1, 550
$RuC_{10}H_{12}N_6$ $[(MeNC)_4Ru(CN)_2]$, 1, 558
$RuC_{11}H_9ClO_3$ $ClCH_2$-cycloheptatriene-$Ru(CO)_3$, 1, 557
$RuC_{11}H_{10}O$ $C_5H_5RuC_5H_4CHO$, 1, 552
$RuC_{11}H_{10}O_2$ $C_5H_5RuC_5H_4CO_2H$, 1, 552
$RuC_{12}H_{10}O_4$ $Ru(C_5H_4CO_2H)_2$, 1, 554
$RuC_{12}H_{12}$ $C_5H_5RuC_5H_4CH{=}CH_2$, 1, 552
$RuC_{12}H_{12}Cl_2O_8$ $[(C_6H_6)_2Ru](ClO_4)_2$, 1, 556
$RuC_{12}H_{12}F_{12}P_2$ $[(C_6H_6)_2Ru][PF_6]_2$, 1, 556
$RuC_{12}H_{12}O$ $C_5H_5RuC_5H_4Ac$, 1, 552
$RuC_{12}H_{14}$ $C_5H_5RuC_5H_4Et$, 1, 552
Di(π-cyclohexadienyl)-Ru, 1, 555
$(C_6H_6)Ru(C_5H_5Me)$, 1, 555
(C_6H_6)-cyclohexadiene-Ru, 1, 555
$RuC_{12}H_{14}O$ $C_5H_5RuC_5H_4CH_2CH_2OH$, 1, 552
$C_5H_5RuC_5H_4CH(OH)Me$, 1, 553
$RuC_{12}H_{20}Br_2N_4$ $[(EtNC)_4RuBr_2]$, 1, 558
$RuC_{12}H_{20}Cl_2N_4$ $[(EtNC)_4RuCl_2]$, 1, 558
$RuC_{13}H_{16}O_3$ 3,5,7-Decatriene-$Ru(CO)_3$, 1, 557
$RuC_{13}H_{35}ClP_4$ $MeRu[(Me_2PCH_2-)_2]_2Cl$, 1, 549
$RuC_{13}H_{35}IP_4$ $MeRu[(Me_2PCH_2-)_2]_2I$, 1, 549
$RuC_{14}H_{14}O_2$ $Ru(C_5H_4Ac)_2$, 1, 554
$RuC_{14}H_{16}O_2$ $C_5H_5RuC_5H_4CH(OAc)Me$, 1, 553
$RuC_{14}H_{20}N_6$ $[(EtNC)_4Ru(CN)_2]$, 1, 558
$RuC_{16}H_{12}N_3O_7$ $[(C_5H_5)_2Ru]OC_6H_2(NO_2)_3$, 1, 552
$RuC_{16}H_{13}N_2O_5$ $[(C_5H_5)_2Ru]OC_6H_3(NO_2)_2$, 1, 552
$RuC_{16}H_{22}$ (1,5-Cyclooctadiene)(1,3,5-cyclooctatriene)-Ru, 1, 557
$RuC_{17}H_{12}O$ Dihydro-8-oxocyclopent[a]-indenyl-RuC_5H_5, 1, 555
$RuC_{17}H_{14}O$ $C_5H_5RuC_5H_4Bz$, 1, 553
$RuC_{17}H_{15}ClNO_5S$ $[(C_5H_5)_2Ru]O_3SC_6H_2$-(Me)(Cl)(NO_2), 1, 552
$RuC_{17}H_{15}NO$ $C_5H_5RuC_5H_4CONHPh$, 1, 553
$RuC_{17}H_{15}N_3$ $C_5H_5RuC_5H_4CH(N_3)Ph$, 1, 553
$RuC_{17}H_{16}$ $C_5H_5RuC_5H_4CH_2Ph$, 1, 553
$RuC_{17}H_{16}O$ $[PhCH(OH)C_5H_4]RuC_5H_5$, 1, 554
$RuC_{18}H_{14}$ Bis(indenyl)-Ru, 1, 554
$RuC_{18}H_{22}$ Bis(4,5,6,7-tetrahydroindenyl)-Ru, 1, 555
$RuC_{18}H_{24}F_{12}P_2$ $[(1,3,5\text{-}Me_3C_6H_3)_2Ru]$-$[PF_6]_2$, 1, 556
$RuC_{18}H_{37}ClP_4$ $PhRu[(Me_2PCH_2-)_2]_2Cl$, 1 549
$RuC_{18}H_{38}P_4$ $PhRu[(Me_2PCH_2-)_2]_2H$, 1, 549
$RuC_{19}H_{15}INO_4S$ $[(C_5H_5)_2Ru]^+$7-iodo-8-hydroxyquinoline-5-sulfonate, 1, 552
$RuC_{24}H_{16}N_6O_{14}$ $[(C_6H_6)_2Ru][OC_6H_2(NO_2)_3]_2$, 1, 556
$RuC_{24}H_{18}O_2$ $C_5H_5RuC_5H_3Bz_2$, 1, 554
$Ru(C_5H_4Bz)_2$, 1, 554
$RuC_{24}H_{20}Cl_2O_8$ $[(Ph_2)_2Ru](ClO_4)_2$, 1, 556
$RuC_{24}H_{20}I_2$ $[(Ph_2)_2Ru]I_2$, 1, 556
$RuC_{24}H_{22}$ Bis(1-endo-phenyl-π-cyclohexadienyl)-Ru, 1, 555
$RuC_{31}H_{48}O_3$ 1,2,3,5,6-Bu_5-cyclooctatetraene-$Ru(CO)_3$, 1, 557
$RuC_{32}H_{28}Br_2I_9N_4$ $[(p\text{-}MeC_6H_4NC)_4RuBr_2$-$I_9]$, 1, 558
$RuC_{32}H_{28}Br_2N_4$ $[(p\text{-}MeC_6H_4NC)_4RuBr_2]$, 1, 558
$RuC_{32}H_{28}Cl_2N_4$ $[(p\text{-}MeC_6H_4NC)_4RuCl_2]$, 1, 558
$RuC_{32}H_{28}I_2N_4$ $[(p\text{-}MeC_6H_4NC)_4RuI_2]$, 1, 558
$RuC_{32}H_{28}I_2N_4O_4$ $[(p\text{-}MeOC_6H_4NC)_4RuI_2]$, 1, 559
$RuC_{36}H_{24}N_6O_{14}$ $[(Ph_2)_2Ru][OC_6H_2(NO_2)_3]_2$, 1, 556
$RuC_{40}H_{36}I_2N_4O_4$ $[(p\text{-}MeOC_6H_4NC)_4RuI_2]\cdot$-$4/3C_6H_6$, 1, 559
$RuC_{51}H_{47}ClP_4$ $MeRu[(Ph_2P)_2CH_2]_2Cl$, 1, 549
$RuC_{53}H_{51}BrP_4$ $MeRu[(Ph_2PCH_2-)_2]_2Br$, 1, 549
$RuC_{53}H_{51}ClP_4$ $MeRu[(Ph_2PCH_2-)_2]_2Cl$, 1, 549
$RuC_{53}H_{52}P_4$ $MeRu[(Ph_2PCH_2-)_2]_2H$, 1, 549
$RuC_{54}H_{51}NP_4S$ $MeRu[(Ph_2PCH_2-)_2]_2SCN$, 1, 549
$RuC_{54}H_{53}ClP_4$ $EtRu[(Ph_2PCH_2-)_2]_2Cl$, 1, 549

$RuC_{54}H_{53}IP_4$ $EtRu[(Ph_2PCH_2-)_2]_2I$, 1, 549
$RuC_{54}H_{54}P_4$ $EtRu[(Ph_2PCH_2-)_2]_2H$, 1, 549
$RuC_{55}H_{52}ClP_4$ $EtRu[(Ph_2P)_2CH_2]_2Cl\cdot 1/2\ C_6H_6$, 1, 549
$RuC_{55}H_{55}BrP_4$ $PrRu[(Ph_2PCH_2-)_2]_2Br$, 1, 549
$RuC_{55}H_{55}ClP_4$ $PrRu[(Ph_2PCH_2-)_2]_2Cl$, 1, 549
$RuC_{56}H_{54}IP_4$ $MeRu[(Ph_2PCH_2-)_2]_2I\cdot 1/2\ C_6H_6$, 1, 549
$RuC_{61}H_{62}P_4$ $PrRu[(Ph_2PCH_2-)_2]_2H\cdot C_6H_6$, 1, 549
$Ru_2C_{14}H_{10}O_4$ $[C_5H_5Ru(CO)_2]_2$, 1, 550
$Ru_2C_{32}H_{38}O_4S$ [Tetrahydronaphthalene-$(MeC_5H_4)Ru]_2SO_4$, 1, 550
$Ru_2C_{34}H_{30}O$ $[C_5H_5RuC_5H_4CH(Ph)]_2O$, 1, 554
$RuAs_2C_{22}H_{30}Cl_2O_2$ $Ru(Et_2AsPh)_2(CO)_2Cl_2$, 3, 47
$RuAs_2C_{38}H_{30}I_2O_2$ $Ru(AsPh_3)_2(CO)_2I_2$, 3, 26
$RuAs_3C_{55}H_{46}ClO$ $RuH(AsPh_3)_3(CO)Cl$, 3, 26
$RuAs_4C_{21}H_{35}Cl$ $MeRu(2\text{-}Me_2AsC_6H_4\text{-}AsMe_2)_2Cl$, 3, 94
$RuB_2C_{12}H_{12}F_8$ $[(C_6H_6)_2Ru](BF_4)_2$, 1, 556
$RuB_2C_{60}H_{52}$ $[(C_6H_6)_2Ru][BPh_4]_2$, 1, 556
$RuB_2C_{66}H_{64}$ $[(1,3,5\text{-}Me_3C_6H_3)_2Ru][BPh_4]_2$, 1, 556
$RuB_2C_{72}H_{60}$ $[(Ph_2)_2Ru][BPh_4]_2$, 1, 556
$RuFeC_{21}H_{18}O$ $FcCOC_5H_4RuC_5H_5$, 1, 546
$RuGeC_{19}H_{15}ClO_2$ $C_5H_5Ru(CO)_2GePh_2Cl$, 2, 141
$RuSb_2C_{38}H_{30}I_2O_2$ $[Ru(SbPh_3)_2(CO)_2I_2]$, 3, 665
$RuSiC_{13}H_{18}$ $(Me_3SiC_5H_4)RuC_5H_5$, 1, 559
$Ru_3As_3C_{57}H_{45}Cl_6O_3$ $[Ru(AsPh_3)(CO)Cl_2]_3$, 3, 26
$Ru_3GeC_{38}H_{50}O_6$ $[(EtMeC_5H_3)Ru(CO)_2]_3\text{-}GeC_8H_{17}$, 1, 559, 2, 141
$Ru_3GeC_{83}H_{132}O_6$ $[(C_{18}H_{37}C_5H_4)Ru(CO)_2]_3\text{-}Ge(C_6H_3Me_2)$, 1, 559, 2, 141*
$SbCBr_2F_3$ CF_3SbBr_2, 3, 693
$SbCF_3I_2$ CF_3SbI_2, 3, 689
$SbCH_2Cl_3$ $ClCH_2SbCl_2$, 3, 693
$SbCH_3Br_2$ $MeSbBr_2$, 3, 693
$SbCH_3Cl_2$ $MeSbCl_2$, 3, 693
$SbCH_5$ $MeSbH_2$, 3, 681
SbC_2BrF_6 $(CF_3)_2SbBr$, 3, 686
SbC_2ClF_6 $(CF_3)_2SbCl$, 3, 686
$SbC_2Cl_3F_6$ $(CF_3)_2SbCl_3$, 3, 756
SbC_2F_6I $(CF_3)_2SbI$, 3, 683
$SbC_2H_2Cl_3$ $ClCH{=}CHSbCl_2$, 3, 693
$SbC_2H_3Br_2$ $CH_2{=}CHSbBr_2$, 3, 693
$SbC_2H_3Cl_2$ $CH_2{=}CHSbCl_2$, 3, 693
$SbC_2H_5Cl_2$ $EtSbCl_2$, 3, 693
SbC_2H_6Br Me_2SbBr, 3, 684
SbC_2H_6Cl Me_2SbCl, 3, 686
SbC_2H_7 Me_2SbH, 3, 680
$SbC_3Br_2D_9$ $(CD_3)_3SbBr_2$, 3, 729
$SbC_3Br_2F_9$ $(CF_3)_3SbBr_2$, 3, 729
$SbC_3Cl_2D_9$ $(CD_3)_3SbCl_2$, 3, 729
$SbC_3Cl_2F_9$ $(CF_3)_3SbCl_2$, 3, 723
$SbC_3H_2Cl_2NS$ $SCH{=}CHN{=}CSbCl_2$, 3, 693
$SbC_3H_2Cl_4NS$ $SCH{=}CHN{=}CSbCl_4$, 3, 769
SbC_3H_2NOS $SCH{=}CHN{=}CSbO$, 3, 702
$SbC_3H_4F_9O_3$ $H[(CF_3)_3Sb(OH)_3]$, 3, 739
$SbC_3H_4NO_3S$ $SCH{=}CHN{=}CSbO_3H_2$, 3, 795
$SbC_3H_5Br_2$ $CH_2{=}CHCH_2SbBr_2$, 3, 693
$SbC_3H_6Br_2Cl_3$ $(ClCH_2)_3SbBr_2$, 3, 729
$SbC_3H_6Cl_3$ $(ClCH_2)_3Sb$, 3, 669
$SbC_3H_6F_3$ $Me_2(CF_3)Sb$, 3, 673
SbC_3H_9 Me_3Sb, 3, 854
$SbC_3H_9Br_2$ Me_3SbBr_2, 3, 723
$SbC_3H_9Cl_2$ Me_3SbCl_2, 3, 724
$SbC_3H_9F_2$ Me_3SbF_2, 3, 729
$SbC_3H_9I_2$ Me_3SbI_2, 3, 729
$SbC_3H_9N_2O_6$ $Me_3Sb(ONO_2)_2$, 3, 641
SbC_3H_9O Me_3SbO, 3, 743
$(SbC_3H_9O_2)_n$ Poly(Me_3Sb peroxide), 3, 747
$SbC_3H_9O_4S$ Me_3SbSO_4, 3, 741
$SbC_3H_{11}O_2$ $Me_3Sb(OH)_2$, 3, 740
SbC_3N_3 $Sb(CN)_3$, 3, 669
$SbC_4H_4Cl_3$ $(ClCH{=}CH)_2SbCl$, 3, 686
$SbC_4H_4Cl_5$ $(ClCH{=}CH)_2SbCl_3$, 3, 754
$SbC_4H_5N_2O_3$ $N{=}CHCH{=}CHN{=}CSbO_3H_2$, 3, 795
SbC_4H_6Br $(CH_2{=}CH)_2SbBr$, 3, 686
SbC_4H_6Cl $(CH_2{=}CH)_2SbCl$, 3, 686
$SbC_4H_6F_5$ $Me_2(CF_3CF_2)Sb$, 3, 673
SbC_4H_6I $(CH_2{=}CH)_2SbI$, 3, 686

$SbC_4H_8Cl_3$ $(MeCHCl)_2SbCl$, 3, 686
SbC_4H_9 $Me_2(CH_2{=}CH)Sb$, 3, 673
SbC_4H_9BrN $Me_3SbBrCN$, 3, 739
$SbC_4H_9Cl_2O_2$ $(MeCHCl)_2SbO_2H$, 3, 763
$SbC_4H_9Cl_3I$ $Me_3Sb{\cdot}CCl_3I$, 3, 854
$SbC_4H_{10}Br$ Et_2SbBr, 3, 686
$SbC_4H_{10}Cl$ Et_2SbCl, 3, 684
$SbC_4H_{12}Br$ $[Me_4Sb]Br$, 3, 714
$SbC_4H_{12}Br_3$ $[Me_4Sb]Br_3$, 3, 714
$SbC_4H_{12}Cl$ $[Me_4Sb]Cl$, 3, 714
$SbC_4H_{12}I$ $[Me_4Sb]I$, 3, 712
$SbC_4H_{12}I_3$ $[Me_4Sb]I_3$, 3, 714
$SbC_5H_5Cl_5$ $(CH{=}CHCH{=}\overset{+}{N}HCH{=}CSbCl_4)Cl^-$, 3, 769
$SbC_5H_5Cl_5O$ $[CH{=}CHC(OH){=}\overset{+}{N}HCH{=}CSbCl_4]\text{-}Cl^-$, 3, 769
$SbC_5H_7N_2O_3$ 2-H_2N-5-pyridinestibonic acid, 3, 795
$SbC_5H_9Cl_2$ Cyclopentyl-$SbCl_2$, 3, 693
$SbC_5H_{10}F_3$ $Et_2(CF_3)Sb$, 3, 673
$SbC_5H_{11}N_2S_4$ $MeSb[SC(S)NHMe]_2$, 3, 705
SbC_5H_{15} Me_5Sb, 3, 708
SbC_6F_9 $(CF_2{=}CF)_3Sb$, 3, 669
$SbC_6H_4BrCl_2$ 4-$BrC_6H_4SbCl_2$, 3, 694
$SbC_6H_4BrCl_4$ 4-$BrC_6H_4SbCl_4$, 3, 769
SbC_6H_4BrO 4-BrC_6H_4SbO, 3, 702
$SbC_6H_4ClI_2$ $ClC_6H_4SbI_2$ isomers, 3, 694
SbC_6H_4ClO ClC_6H_4SbO isomers, 3, 702
$SbC_6H_4Cl_2F$ 4-$FC_6H_4SbCl_2$, 3, 694
$SbC_6H_4Cl_2I$ 4-$IC_6H_4SbCl_2$, 3, 694
$SbC_6H_4Cl_2NO_2$ 4-$O_2NC_6H_4SbCl_2$, 3, 690
$SbC_6H_4Cl_3$ $ClC_6H_4SbCl_2$ isomers, 3, 690, 694
$SbC_6H_4Cl_4I$ 4-$IC_6H_4SbCl_4$, 3, 769
$SbC_6H_4Cl_4NO_2$ $O_2NC_6H_4SbCl_4$ isomers, 3, 769, 770
$SbC_6H_4Cl_5$ $ClC_6H_4SbCl_4$ isomers, 3, 770
SbC_6H_4FO 4-FC_6H_4SbO, 3, 702
SbC_6H_4IO 4-IC_6H_4SbO, 3, 702
$SbC_6H_4NO_3$ 4-$O_2NC_6H_4SbO$, 3, 702
SbC_6H_5 $(PhSb)_n$, 3, 682
$SbC_6H_5Cl_2$ $PhSbCl_2$, 3, 690
$SbC_6H_5Cl_4$ $PhSbCl_4$, 3, 767
$SbC_6H_5Cl_4O$ $HOC_6H_4SbCl_4$ isomers, 3, 770
$SbC_6H_5I_2$ $PhSbI_2$, 3, 691
SbC_6H_5O $PhSbO$, 3, 700
$SbC_6H_6BrO_3$ $BrC_6H_4SbO_3H_2$ isomers, 3, 780
$SbC_6H_6ClO_3$ $ClC_6H_4SbO_3H_2$ isomers, 3, 780
$SbC_6H_6Cl_2N$ $H_2NC_6H_4SbCl_2$ isomers, 3, 691
$SbC_6H_6Cl_2NO_2S$ 4-$H_2NO_2SC_6H_4SbCl_2$, 3, 691
$SbC_6H_6Cl_3$ $(ClCH{=}CH)_3Sb$, 3, 655
$SbC_6H_6Cl_4NO_2S$ 4-$H_2NSO_2C_6H_4SbCl_4$, 3, 770
$SbC_6H_6Cl_5$ $(ClCH{=}CH)_3SbCl_2$, 3, 724
$SbC_6H_6FO_3$ 4-$FC_6H_4SbO_3H_2$, 3, 780
$SbC_6H_6IO_3$ 4-$IC_6H_4SbO_3H_2$, 3, 780
$SbC_6H_6NO_3S$ 4-$H_2NO_2SC_6H_4SbO$, 3, 702
$SbC_6H_6NO_5$ $O_2NC_6H_4SbO_3H_2$ isomers, 3, 786, 788
SbC_6H_7 $PhAsH_2$, 3, 681
$SbC_6H_7BrNO_3$ 3,4-$H_2N(Br)C_6H_3SbO_3H_2$, 3, 792
$SbC_6H_7ClNO_5S$ 2,4-$Cl(H_2NO_2S)C_6H_3SbO_3H_2$, 3, 792
$SbC_6H_7Cl_5N$ 4-$(HCl{\cdot}H_2N)C_6H_4SbCl_4$, 3, 770
$SbC_6H_7O_3$ $PhSbO_3H_2$, 3, 776
$SbC_6H_7O_4$ 4-$HOC_6H_4SbO_3H_2$, 3, 781
$SbC_6H_8NO_3$ 4-$H_2NC_6H_4SbO_3H_2$, 3, 786
$SbC_6H_8NO_5S$ $H_2NO_2SC_6H_4SbO_3H_2$ isomers, 3, 782*, 783
SbC_6H_9 $(CH_2{=}CH)_3Sb$, 3, 655
$SbC_6H_9Br_2$ $(CH_2{=}CH)_3SbBr_2$, 3, 729
$SbC_6H_9Cl_2$ $(CH_2{=}CH)_3SbCl_2$, 3, 729
$SbC_6H_9Cl_5N$ $NH_4[PhSbCl_5]$, 3, 767
$SbC_6H_9F_2$ $(CH_2{=}CH)_3SbF_2$, 3, 729
$SbC_6H_9I_2$ $(CH_2{=}CH)_3SbI_2$, 3, 729
$SbC_6H_9O_3$ Ac_3Sb, 3, 669
SbC_6H_{11} $Et_2(CH{\equiv}C)Sb$, 3, 673
$(CH_2{=}CH)_2EtSb$, 3, 673
$SbC_6H_{13}Br_2$ $Et_2(CH_2{=}CH)SbBr_2$, 3, 729
$SbC_6H_{14}Br$ $(Me_2CH)_2SbBr$, 3, 686
SbC_6H_{15} Et_3Sb, 3, 655
Me_2BuSb, 3, 673
$SbC_6H_{15}Br_2$ Et_3SbBr_2, 3, 729
$SbC_6H_{15}Cl_2$ Et_3SbCl_2, 3, 724
$SbC_6H_{15}I_2$ Et_3SbI_2, 3, 730
$SbC_6H_{15}O_4S$ Et_3SbSO_4, 3, 741

$SbC_6H_{15}S$ Et_3SbS, 3, 749
$SbC_6H_{15}Se$ Et_3SbSe, 3, 750
$SbC_7H_4Cl_2N$ $4\text{-}NCC_6H_4SbCl_2$, 3, 694
$SbC_7H_4I_2N$ $4\text{-}NCC_6H_4SbI_2$, 3, 694
$SbC_7H_5Cl_4O_3$ $3,4\text{-}HO(HO_2C)C_6H_3SbCl_4$, 3, 770
$SbC_7H_6F_3O_3$ $3\text{-}(CF_3)C_6H_4SbO_3H_2$, 3, 778
$SbC_7H_6NO_3$ $4\text{-}NCC_6H_4SbO_3H_2$, 3, 778
$SbC_7H_6NO_3S$ $4\text{-}SCNC_6H_4SbO_3H_2$, 3, 788
$SbC_7H_6NO_7$ $3,4\text{-}O_2N(HO_2C)C_6H_3SbO_3H_2$, 3, 792
$SbC_7H_7Cl_2$ $MeC_6H_4SbCl_2$ isomers, 3, 691, 694
$SbC_7H_7Cl_2O$ $MeOC_6H_4SbCl_2$ isomers, 3, 694
$SbC_7H_7Cl_4$ $4\text{-}MeC_6H_4SbCl_4$, 3, 768
$SbC_7H_7Cl_4N_2O$ $4\text{-}H_2NNHCOC_6H_4SbCl_4$, 3, 770
$SbC_7H_7Cl_4O$ $4\text{-}MeOC_6H_4SbCl_4$, 3, 771
$SbC_7H_7I_2$ $4\text{-}MeC_6H_4SbI_2$, 3, 694
SbC_7H_7O MeC_6H_4SbO isomers, 3, 700, 702
$SbC_7H_7O_2$ $MeOC_6H_4SbO$ isomers, 3, 702
$SbC_7H_7O_5$ $HO_2CC_6H_4SbO_3H_2$ isomers, 3, 778
$SbC_7H_8ClO_3$ $Cl(Me)C_6H_3SbO_3H_2$ isomers, 3, 792
$SbC_7H_8NO_4$ $4\text{-}H_2NOCC_6H_4SbO_3H_2$, 3, 778
$SbC_7H_8NO_5$ $3,5\text{-}H_2N(HO_2C)C_6H_3SbO_3H_2$, 3, 792
$O_2N(Me)C_6H_3SbO_3H_2$ isomers, 3, 792
$SbC_7H_8NO_5S$ $2\text{-}HO_2CCH_2S\text{-}5\text{-}$pyridinestibonic acid, 3, 795
$SbC_7H_8NO_6$ $2,4\text{-}O_2N(HOCH_2)C_6H_3SbO_3H_2$, 3, 792
$SbC_7H_9N_2O_3$ $4\text{-}[H_2NC(=NH)]C_6H_4SbO_3H_2$, 3, 778
$SbC_7H_9N_2O_4$ $4\text{-}H_2NCONHC_6H_4SbO_3H_2$, 3, 788
$SbC_7H_9O_3$ $MeC_6H_4SbO_3H_2$ isomers, 3, 777, 778
$SbC_7H_9O_4$ $HOCH_2C_6H_4SbO_3H_2$ isomers, 3, 778
$4\text{-}MeOC_6H_4SbO_3H_2$, 3, 781
$SbC_7H_{10}NO_3$ $3,4\text{-}H_2N(Me)C_6H_3SbO_3H_2$, 3, 792
$SbC_7H_{10}NO_4$ $2,4\text{-}H_2N(HOCH_2)C_6H_3SbO_3H_2$, 3, 792
$3,4\text{-}H_2N(MeO)C_6H_3SbO_3H_2$, 3, 792
$SbC_7H_{10}N_3O_5S$ $4\text{-}[H_2NC(=NH)NHO_2S]C_6H_4\text{-}SbO_3H_2$, 3, 783
$SbC_7H_{11}Cl_5N$ $NH_4[4\text{-}MeC_6H_4SbCl_5]$, 3, 768
$SbC_7H_{12}I$ $[(CH_2=CH)_3MeSb]I$, 3, 714
$SbC_7H_{15}Br_2$ $Et_2(CH_2=CMe)SbBr_2$, 3, 730
$Et_2(MeCH=CH)SbBr_2$, 3, 730
$SbC_7H_{15}Cl_2$ $Et_2(MeCH=CH)SbCl_2$, 3, 730
$SbC_7H_{18}I$ $[Me_3BuSb]I$, 3, 714
$SbC_8H_5F_9N$ $(CF_3)_3Sb\cdot Py$, 3, 654
$SbC_8H_7Cl_2O$ $4\text{-}AcC_6H_4SbCl_2$, 3, 694
$SbC_8H_7Cl_2O_2S$ $p\text{-}HO_2CCH_2SC_6H_4SbCl_2$, 3, 692
$SbC_8H_7Cl_4O_2$ $4\text{-}HO_2CCH_2C_6H_4SbCl_4$, 3, 771
$SbC_8H_7Cl_4O_2S$ $HO_2CCH_2SC_6H_4SbCl_4$ isomers, 3, 771
$SbC_8H_7Cl_4O_3$ $4\text{-}HO_2CCH_2OC_6H_4SbCl_4$, 3, 771
$SbC_8H_7Cl_4O_4S$ $4\text{-}HO_2CCH_2SO_2C_6H_4SbCl_4$, 3, 771
$SbC_8H_7INO_4$ $7,8\text{-}I(HO)\text{-}5\text{-}$quinoline-$SbO_3H_2$, 3, 795
$SbC_8H_7I_2O_2S$ $4\text{-}HO_2CCH_2SC_6H_4SbI_2$, 3, 694
$SbC_8H_7O_2$ $4\text{-}AcC_6H_4SbO$, 3, 702
$SbC_8H_7O_3S$ $4\text{-}HO_2CCH_2SC_6H_4SbO$, 3, 703
$SbC_8H_8ClO_5S$ $4,2\text{-}Cl(HO_2CCH_2S)C_6H_3Sb\text{-}O_3H_2$, 3, 792
$SbC_8H_8Cl_4NOS$ $4\text{-}H_2NCOCH_2SC_6H_4SbCl_4$, 3, 771
$SbC_8H_8NO_4S$ 3,4-Dihydro-3-oxo-6-benzothiazinestibonic acid, 3, 795
$SbC_8H_9ClNO_4$ $3,4\text{-}Cl(AcNH)C_6H_3SbO_3H_2$, 3, 792
$SbC_8H_9Cl_2$ $4\text{-}EtC_6H_4SbCl_2$, 3, 694
$SbC_8H_9Cl_4O$ $4\text{-}EtOC_6H_4SbCl_4$, 3, 771
$SbC_8H_9Cl_4OS$ $4\text{-}HOCH_2CH_2SC_6H_4SbCl_4$, 3, 771
SbC_8H_9O $4\text{-}EtC_6H_4SbO$, 3, 703
$SbC_8H_9O_4$ $4\text{-}AcC_6H_4SbO_3H_2$, 3, 778
$SbC_8H_9O_5$ $MeO_2CC_6H_4SbO_3H_2$ isomers, 3, 778, 779
$AcOC_6H_4SbO_3H_2$ isomers, 3, 781
$3,4\text{-}HO(Ac)C_6H_3SbO_3H_2$, 3, 793
$SbC_8H_9O_5S$ $HO_2CCH_2SC_6H_4SbO_3H_2$ isomers, 3, 783
$SbC_8H_9O_6$ $4\text{-}HO_2CCH_2OC_6H_4SbO_3H_2$, 3, 781

$SbC_8H_9O_7S$ 4-$HO_2CCH_2SO_2C_6H_4SbO_3H_2$, 3, 783
$SbC_8H_{10}Cl_2NO_2S$ 4-$Me_2NO_2SC_6H_4SbCl_2$, 3, 694
$SbC_8H_{10}NO_4$ 3,4-$H_2N(Ac)C_6H_3SbO_3H_2$, 3, 793
$AcNHC_6H_4SbO_3H_2$ isomers, 3, 788
$SbC_8H_{10}NO_4S$ 4-$H_2NCOCH_2SC_6H_4SbO_3H_2$, 3, 783
$SbC_8H_{10}NO_5$ 4-$H_2NOCCH_2OC_6H_4SbO_3H_2$, 3, 781
3,4-$HO(AcNH)C_6H_3SbO_3H_2$, 3, 793
SbC_8H_{11} Me_2PhSb, 3, 673
$SbC_8H_{11}O$ Me_2SbOPh, 3, 697
$SbC_8H_{11}O_3$ 2,4-$Me_2C_6H_3SbO_3H_2$, 3, 793
$SbC_8H_{11}O_4$ $EtOC_6H_4SbO_3H_2$ isomers, 3, 781
$SbC_8H_{11}O_4S$ 4-$HOCH_2CH_2SC_6H_4SbO_3H_2$, 3, 783
$SbC_8H_{12}Br$ $[(CH_2{=}CH)_4Sb]Br$, 3, 714
$SbC_8H_{12}BrI_2$ $[(CH_2{=}CH)_4Sb]Br \cdot I_2$, 3, 714
$SbC_8H_{12}I$ $[(CH_2{=}CH)_4Sb]I$, 3, 714
$SbC_8H_{12}I_3$ $[(CH_2{=}CH)_4Sb]I_3$, 3, 714
$SbC_8H_{12}NO_3$ 4-$Me_2NC_6H_4SbO_3H_2$, 3, 788
$SbC_8H_{12}NO_5S$ 4-$Me_2NO_2SC_6H_4SbO_3H_2$, 3, 783
2,5,4-$Me_2(H_2NO_2S)C_6H_2SbO_3H_2$, 3, 793
SbC_8H_{15} $(Me_2CH)_2(CH{\equiv}C)Sb$, 3, 673
$SbC_8H_{15}Br_2$ $Et(MeCH{=}CH)_2SbBr_2$, 3, 730
$SbC_8H_{18}Br$ Bu_2SbBr, 3, 684
$(Me_2CHCH_2)_2SbBr$, 3, 686
$SbC_8H_{18}Cl$ $(Me_3C)_2SbCl$, 3, 686
$SbC_8H_{18}Cl_3$ Bu_2SbCl_3, 3, 756
$SbC_8H_{20}Br$ $[Et_4Sb]Br$, 3, 714
$SbC_8H_{20}I$ $[Et_4Sb]I$, 3, 715
$[Et_4Sb]I$, 3, 715
$SbC_9H_5Cl_2N_4O$ m-$(N{=}CClN{=}CClN{=}CNH)C_6H_4$-$SbO$, 3, 703
$SbC_9H_5Cl_4N_4$ m-$(N{=}CClN{=}CClN{=}CNH)C_6H_4$-$SbCl_2$, 3, 692*
$SbC_9H_7ClN_5O$ m-$[N{=}CClN{=}C(NH_2)N{=}CNH]$-$C_6H_4SbO$, 3, 703
$SbC_9H_7Cl_2N_4O_3$ 4-$(N{=}CClN{=}CClN{=}CNH)$-$C_6H_4SbO_3H_2$, 3, 788
$SbC_9H_7Cl_2N_4O_4$ 4,3-$HO(N{=}CClN{=}CClN{=}CNH)$-$C_6H_3SbO_3H_2$, 3, 793
$SbC_9H_7N_2O_6$ 7,8-$O_2N(HO)$-5-quinoline-SbO_3H_2, 3, 795
$SbC_9H_8NO_4$ 8-HO-5-quinoline-SbO_3H_2, 3, 796
$SbC_9H_9ClN_5O_3$ 4-$[N{=}CClN{=}C(NH_2)N{=}CNH]$-$C_6H_4SbO_3H_2$, 3, 788
$SbC_9H_9Cl_2N_6$ $N{=}C(NH_2)N{=}C(NH_2)N{=}CNH$-$C_6H_4SbCl_2$ isomers, 3, 695
$SbC_9H_9Cl_2N_6O$
4,3-$HO[N{=}C(NH_2)N{=}C(NH_2)N{=}CNH]C_6H_3$-$SbCl_2$, 3, 692
$SbC_9H_9Cl_2O_2$ 4-$EtO_2CC_6H_4SbCl_2$, 3, 692
$SbC_9H_9Cl_4O_2$ 2-$HO_2CCH_2CH_2C_6H_4SbCl_4$, 3, 772
$SbC_9H_9Cl_4O_3$ 3,4-$HO(EtO_2C)C_6H_3SbCl_4$, 3, 772
$SbC_9H_9N_2O_4$ 7,8-$H_2N(HO)$-5-quinoline-SbO_3H_2, 3, 795
$SbC_9H_9N_2O_5S_2$ 4-$(SCH{=}CHN{=}CNHO_2S)C_6H_4$-$SbO_3H_2$, 3, 784
$SbC_9H_9N_6O$ m-$[N{=}C(NH_2)N{=}C(NH_2)N{=}CNH]$-$C_6H_4SbO$, 3, 703
$SbC_9H_9O_7S$ 4,2-$HO_2C(HO_2CCH_2S)C_6H_3Sb$-$O_3H_2$, 3, 793
$SbC_9H_{10}Cl_2N_7$
3,4-$H_2N[N{=}C(NH_2)N{=}C(NH_2)N{=}CNH]C_6H_3$-$SbCl_2$, 3, 695
$SbC_9H_{10}N_5O_3S$ 4-$[N{=}C(NH_2)N{=}C(NH_2)N{=}CS]$ $C_6H_4SbO_3H_2$, 3, 784
$SbC_9H_{10}N_7O_5$
3,4-$O_2N[N{=}C(NH_2)N{=}C(NH_2)N{=}CNH]C_6H_3$-$SbO_3H_2$, 3, 793
$SbC_9H_{11}N_6O_3$ 4-$[N{=}C(NH_2)N{=}C(NH_2)N{=}CNH]$ $C_6H_4SbO_3H_2$, 3, 789
$SbC_9H_{11}O_5$ $EtO_2CC_6H_4SbO_3H_2$ isomers, 3, 779, 781
$SbC_9H_{11}O_5S$ 2-$[MeCH(CO_2H)S]C_6H_4SbO_3H_2$, 3, 784
2-$(HO_2CCH_2CH_2S)C_6H_4SbO_3H_2$, 3, 784
$SbC_9H_{11}O_6$ 4-$MeO_2CCH_2OC_6H_4SbO_3H_2$, 3, 781
$SbC_9H_{12}Cl_4N$ 4-$Me_2NCH_2C_6H_4SbCl_4$, 3, 772
$SbC_9H_{12}NOS_2$ 4-$H_2NC_6H_4SbSCH_2CH(CH_2OH)S$, 3, 809
$SbC_9H_{12}NO_4$ 4-$Me_2NOCC_6H_4SbO_3H_2$, 3, 779
$SbC_9H_{12}NO_5$ 4-$HOCH_2CH_2NHOCC_6H_4SbO_3H_2$,

3, 779
$4\text{-}[HO_2CCH(NH_2)CH_2]C_6H_4SbO_3H_2$, 3, 779
$SbC_9H_{12}NS_2$ $4\text{-}H_2NC_6H_4SbSCH_2CH(Me)S$, 3, 809
SbC_9H_{13} $Me_2(PhCH_2)Sb$, 3, 673
Cyclopentenyl-$(CH_2{=}CH)_2Sb$, 3, 673
$SbC_9H_{13}Cl_5N$ $4\text{-}Me_3\overset{+}{N}C_6H_4SbCl_5^-$, 3, 772
SbC_9H_{15} $(CH_2{=}CHCH_2)_3Sb$, 3, 657
$(MeCH{=}CH)_3Sb$, 3, 658
$[CH_2{=}C(Me)]_3Sb$, 3, 658
$SbC_9H_{15}Br_2$ $(MeCH{=}CH)_3SbBr_2$ isomers, 3, 725
$(CH_2{=}CHCH_2)_3SbBr_2$, 3, 730
$(CH_2{=}CMe)_3SbBr_2$, 3, 730
$SbC_9H_{15}Cl_2$ $(CH_2{=}CHCH_2)_3SbCl_2$, 3, 730
$(CH_2{=}CMe)_3SbCl_2$, 3, 730
$(MeCH{=}CH)_3SbCl_2$ isomers, 3, 730
$SbC_9H_{15}F_2$ $(CH_2{=}CMe)_3SbF_2$, 3, 730
$(MeCH{=}CH)_3SbF_2$ isomers, 3, 731
$SbC_9H_{15}I_2$ $(CH_2{=}CMe)_3SbI_2$, 3, 730
$(MeCH{=}CH)_3SbI_2$ isomers, 3, 731
SbC_9H_{21} Pr_3Sb, 3, 658
$(Me_2CH)_3Sb$, 3, 669
Bu_2MeSb, 3, 673
$SbC_9H_{21}Br_2$ $(Me_2CH)_3SbBr_2$, 3, 731
Pr_3SbBr_2, 3, 731
$SbC_9H_{21}S$ Pr_3SbS, 3, 749
$SbC_9H_{21}Se$ Pr_3SbSe, 3, 750
$SbC_{10}H_7Br_4$ $2\text{-}C_{10}H_7SbBr_4$, 3, 772
$SbC_{10}H_7Cl_2$ $C_{10}H_7SbCl_2$ isomers, 3, 695, 696
$SbC_{10}H_7Cl_4$ $C_{10}H_7SbCl_4$ isomers, 3, 772
$SbC_{10}H_7O$ $2\text{-}C_{10}H_7SbO$, 3, 703
$SbC_{10}H_8NO_5$ $4,1\text{-}O_2NC_{10}H_6SbO_3H_2$, 3, 794
$SbC_{10}H_9O_3$ $C_{10}H_7SbO_3H_2$ isomers, 3, 794
$SbC_{10}H_9O_4S_2$ $4\text{-}HO_2CCH_2SC_6H_4SbSCH_2C(O)O$, 3, 811
$SbC_{10}H_{10}N_3O_5S$ $4\text{-}(N{=}CHCH{=}CHN{=}CNHO_2S)\text{-}C_6H_4SbO_3H_2$, 3, 784
$SbC_{10}H_{11}Cl_4O_2$ $4\text{-}EtO_2CCH_2C_6H_4SbCl_4$, 3, 773
$SbC_{10}H_{11}Cl_4O_3$ $4\text{-}EtO_2CCH_2OC_6H_4SbCl_4$, 3, 773
$SbC_{10}H_{11}O_8$ $PhSb(O)(OH)OCH(CO_2H)CHOH\text{-}CO_2H$, 3, 777
$SbC_{10}H_{12}NO_4S_2$ $4\text{-}H_2NC_6H_4Sb(SCH_2CO_2H)_2$, 3, 704
$SbC_{10}H_{12}NO_6S$ $2,4\text{-}HO_2CCH_2S(AcNH)C_6H_3\text{-}SbO_3H_2$, 3, 793
$SbC_{10}H_{13}N_2S_4$ $PhSb[SC(S)NHMe]_2$, 3, 705
$SbC_{10}H_{13}O_5S$ $4\text{-}EtO_2CCH_2SC_6H_4SbO_3H_2$, 3, 784
$SbC_{10}H_{14}NO_5$ $4\text{-}[HO(CH_2)_3NHOC]C_6H_4SbO_3H_2$, 3, 779
$SbC_{10}H_{15}$ $(CH_2{=}CH)_5Sb$, 3, 708
$SbC_{10}H_{19}$ $(Me_3C)_2(CH{\equiv}C)Sb$, 3, 674
$Et_2(CH_2{=}CH)_3Sb$, 3, 708
$SbC_{10}H_{20}Br$ $[Et_2(MeCH{=}CH)_2Sb]Br$ isomers, 3, 715
$SbC_{11}H_{10}Cl_4N$ $[PyH]^+[3\text{-}ClC_6H_4SbCl_3]^-$, 3, 690
$SbC_{11}H_{11}Cl_3N$ $[PyH]^+[PhSbCl_3]^-$, 3, 690
$SbC_{11}H_{11}Cl_5N$ $[Py{\cdot}H][PhSbCl_5]$, 3, 768
$SbC_{11}H_{11}N_2O_5S$ $4\text{-}(2\text{-}Pyridyl\text{-}NHO_2S)C_6H_4\text{-}SbO_3H_2$, 3, 784
$SbC_{11}H_{11}O_4S_3$ $4\text{-}HO_2CCH_2SC_6H_4\text{-}SbSCH_2CH(CO_2H)S$, 3, 809
$SbC_{11}H_{12}N_3O_3$ $4\text{-}[N{=}C(Me)N{=}C(Me)N{=}C]\text{-}C_6H_4SbO_3H_2$, 3, 791
$SbC_{11}H_{12}N_3O_5S$ $4\text{-}(4\text{-}Me\text{-}2\text{-}pyrimidyl\text{-}NHO_2S)C_6H_4SbO_3H_2$, 3, 784
$SbC_{11}H_{13}ClNO_2S_2$ $2,4\text{-}Cl(AcNH)C_6H_3\text{-}SbSCH_2CH(CH_2OH)S$, 3, 809
$SbC_{11}H_{13}Cl_4O_2$ $2\text{-}EtO_2CCH_2CH_2C_6H_4SbCl_4$, 3, 773
$SbC_{11}H_{13}O_4$ $4\text{-}MeC_6H_4Sb(OAc)_2$, 3, 704
$SbC_{11}H_{14}NO_2S_2$ $4\text{-}AcNHC_6H_4\text{-}SbSCH_2CH(CH_2OH)S$, 3, 809
$SbC_{11}H_{14}NO_6$ $4\text{-}[HO_2CCH(NHAc)CH_2]C_6H_4\text{-}SbO_3H_2$, 3, 779
$SbC_{11}H_{15}N_6O_3$ $4\text{-}[N{=}C(NHMe)N{=}C(NHMe)N{=}C\text{-}NH]C_6H_4SbO_3H_2$, 3, 789
$SbC_{11}H_{16}Cl_4N$ $4\text{-}Et_2NCH_2C_6H_4SbCl_4$, 3, 773
$SbC_{11}H_{17}N_2O_5$ $HOCH_2CH_2NHCH_2CH_2NHOCC_6H_4\text{-}SbO_3H_2$ isomers, 3, 779
$SbC_{11}H_{23}$ $Bu_2(CH_2{=}CHCH_2)Sb$, 3, 674
$SbC_{11}H_{27}O_2$ $Me_3Sb(OCMe_3)_2$, 3, 744
$SbC_{11}H_{27}O_4$ $Me_3Sb(OOCMe_3)_2$, 3, 746
$SbC_{12}H_7Cl_4N_2O_3$ $2,3,4\text{-}Cl_2(2,3\text{-}Cl_2C_6H_3\text{-}N{=}N)C_6H_2SbO_3H_2$, 3, 793
$SbC_{12}H_8Br_2Cl$ $(4\text{-}BrC_6H_4)_2SbCl$, 3, 686

$SbC_{12}H_8Br_2Cl_3$ $(4\text{-}BrC_6H_4)_2SbCl_3$, 3, 756

$SbC_{12}H_8Cl$ 5-Cl-dibenzostibole, 3, 801

$SbC_{12}H_8ClF_2$ $(4\text{-}FC_6H_4)_2SbCl$, 3, 686

$SbC_{12}H_8ClI_2$ $(4\text{-}IC_6H_4)_2SbCl$, 3, 686

$SbC_{12}H_8Cl_3$ $(ClC_6H_4)_2SbCl$ isomers, 3, 686, 687

$SbC_{12}H_8Cl_3F_2$ $(4\text{-}FC_6H_4)_2SbCl_3$, 3, 756

$SbC_{12}H_8Cl_3I_2$ $(4\text{-}IC_6H_4)_2SbCl_3$, 3, 756

$SbC_{12}H_8Cl_3N_2O_4$ $(O_2NC_6H_4)_2SbCl_3$ isomers, 3, 756

$SbC_{12}H_8Cl_4NO_2$ $(4\text{-}ClC_6H_4)(4\text{-}O_2NC_6H_4)\text{-}SbCl_3$, 3, 756

$SbC_{12}H_8Cl_5$ $(4\text{-}ClC_6H_4)_2SbCl_3$, 3, 756

$SbC_{12}H_9BrCl_3$ $Ph(4\text{-}BrC_6H_4)SbCl_3$, 3, 756

$SbC_{12}H_9BrCl_5N_2$ $[4\text{-}BrC_6H_4N_2][PhSbCl_5]$, 3, 768

$SbC_{12}H_9Br_2O_2$ $(4\text{-}BrC_6H_4)_2SbO_2H$, 3, 763

$SbC_{12}H_9ClNO_4$ $(4\text{-}ClC_6H_4)(4\text{-}O_2NC_6H_4)\text{-}SbO_2H$, 3, 763

$SbC_{12}H_9Cl_2I_2O$ $(4\text{-}IC_6H_4)_2Sb(OH)Cl_2$, 3, 760

$SbC_{12}H_9Cl_2O_2$ $(ClC_6H_4)_2SbO_2H$ isomers, 3, 763

$SbC_{12}H_9Cl_2O_3$ $(\overline{OCH{=}CHCH{=}C})_3SbCl_2$, 3, 731

$SbC_{12}H_9Cl_3I$ $Ph(4\text{-}IC_6H_4)SbCl_3$, 3, 756

$SbC_{12}H_9Cl_3NO_2$ $Ph(4\text{-}O_2NC_6H_4)SbCl_3$, 3, 756

$SbC_{12}H_9Cl_3N_3O_2$ $[PhN_2][4\text{-}O_2NC_6H_4SbCl_3]$, 3, 690

$SbC_{12}H_9Cl_4$ $Ph(4\text{-}ClC_6H_4)SbCl_3$, 3, 756

$SbC_{12}H_9Cl_5IN_2$ $[4\text{-}IC_6H_4N_2][PhSbCl_5]$, 3, 768

$SbC_{12}H_9Cl_5N_3O_2$ $[O_2NC_6H_4N_2][PhSbCl_5]$ isomers, 3, 768

$SbC_{12}H_9Cl_6N_2$ $[4\text{-}ClC_6H_4N_2][PhSbCl_5]$, 3, 768

$SbC_{12}H_9F_2O_2$ $(4\text{-}FC_6H_4)_2SbO_2H$, 3, 763

$SbC_{12}H_9I_2O_2$ $(4\text{-}IC_6H_4)_2SbO_2H$, 3, 763

$SbC_{12}H_9N_2O_6$ $(O_2NC_6H_4)_2SbO_2H$ isomers, 3, 763

$SbC_{12}H_9O$ 2-PhC_6H_4SbO, 3, 703

$SbC_{12}H_9O_3$ $(\overline{OCH{=}CHCH{=}C})_3Sb$, 3, 669

$SbC_{12}H_9S_3$ $(\overline{SCH{=}CHCH{=}C})_3Sb$, 3, 669

$SbC_{12}H_{10}BrO_2$ $Ph(4\text{-}BrC_6H_4)SbO_2H$, 3, 764

$SbC_{12}H_{10}Cl$ Ph_2SbCl, 3, 684*

$SbC_{12}H_{10}ClO$ 2-$PhC_6H_4Sb(OH)Cl$, 3, 696

$SbC_{12}H_{10}ClO_2$ $Ph(4\text{-}ClC_6H_4)SbO_2H$, 3, 764

$SbC_{12}H_{10}Cl_3$ Ph_2SbCl_3, 3, 754

$SbC_{12}H_{10}Cl_3N_2$ $[PhN_2][PhSbCl_3]$, 3, 690, 691

$SbC_{12}H_{10}Cl_3O_2$ $(2\text{-}ClC_6H_4)_2Sb(OH)_2Cl$, 3, 760

$SbC_{12}H_{10}Cl_5N_2$ $[PhN_2][PhSbCl_5]$, 3, 768

$SbC_{12}H_{10}I$ Ph_2SbI, 3, 687

$SbC_{12}H_{10}IO_2$ $Ph(4\text{-}IC_6H_4)SbO_2H$, 3, 764

$SbC_{12}H_{10}NO_4$ $Ph(2\text{-}O_2NC_6H_4)SbO_2H$, 3, 764

$SbC_{12}H_{10}NO_5S$ 4-$(4\text{-}O_2NC_6H_4S)C_6H_4SbO_3H_2$, 3, 785

$SbC_{12}H_{10}NO_7S$ 4-$(4\text{-}O_2NC_6H_4SO_2)C_6H_4\text{-}SbO_3H_2$, 3, 785

$SbC_{12}H_{10}NS_2$ 4-$H_2NC_6H_4\overline{SbSC_6H_4S}$, 3, 809
3-(p-$H_2NC_6H_4$)-2,4,3-dithiastibobicyclo[3.3.1]nona-5,7,9-triene, 3, 813

$SbC_{12}H_{11}$ Ph_2AsH, 3, 680

$SbC_{12}H_{11}Cl_3N_3O_2S$ $[PhN_2][4\text{-}H_2NO_2SC_6H_4\text{-}SbCl_3]$, 3, 691

$SbC_{12}H_{11}N_2O_4$ 4-$(4\text{-}HOC_6H_4N{=}N)C_6H_4Sb\text{-}O_3H_2$, 3, 789

$SbC_{12}H_{11}N_2O_5$ 4-$[2,4\text{-}(HO)_2C_6H_3N{=}N]C_6H_4\text{-}SbO_3H_2$, 3, 789

$SbC_{12}H_{11}O_2$ Ph_2SbO_2H, 3, 761

$SbC_{12}H_{11}O_3$ 2-$PhC_6H_4SbO_3H_2$, 3, 794'

$SbC_{12}H_{11}O_5S$ 2,1-$HO_2CCH_2SC_{10}H_6SbO_3H_2$, 3, 794

$SbC_{12}H_{12}NO_3S$ 4-$(4\text{-}H_2NC_6H_4S)C_6H_4SbO_3H_2$, 3, 785

$SbC_{12}H_{12}NO_5S$ 4-$(4\text{-}H_2NC_6H_4SO_2)C_6H_4\text{-}SbO_3H_2$, 3, 785

$SbC_{12}H_{13}Cl_5N$ $[Py{\cdot}H][4\text{-}MeC_6H_4SbCl_5]$, 3, 768

$SbC_{12}H_{13}N_2O_5S$ 4-$(4\text{-}H_2NC_6H_4SO_2NH)C_6H_4\text{-}SbO_3H_2$, 3, 789

$SbC_{12}H_{13}N_6O_2S_2$
2-[p-$(\overline{N{=}C(NH_2)N{=}C(NH_2)N{=}C}NH]C_6H_4$-1,3,2-dithiastibolane-4-CO_2H, 3, 809

$SbC_{12}H_{13}O_6S_3$ 4-$HO_2CCH_2SC_6H_4Sb(SCH_2\text{-}CO_2H)_2$, 3, 704

$SbC_{12}H_{15}N_6OS_2$
5,2-HO[p-($N=C(NH_2)N=C(NH_2)N=CNH$]-C_6H_4-1,3,2-dithiantimonane, 3, 813
$SbC_{12}H_{15}N_6OS_2$
4,2-$HOCH_2$[p-($N=C(NH_2)N=C(NH_2)N=CNH$)-C_6H_4]-1,3,2-dithiastibolane, 3, 810
$SbC_{12}H_{15}N_6S_2$
4,2-Me[p-($N=C(NH_2)N=C(NH_2)N=CNH$)-C_6H_4]-1,3,2-dithiastibolane, 3, 810
$SbC_{12}H_{16}N_3O_2S_4$ 4-$O_2NC_6H_4Sb$[SC(S)-NMe_2]$_2$, 3, 705
$SbC_{12}H_{18}Br$ Ph(n-C_6H_{13})SbBr, 3, 687
$SbC_{12}H_{18}Cl_4N$ 4-$Et_2NCH_2CH_2C_6H_4SbCl_4$, 3, 773
$SbC_{12}H_{18}Cl_4NO$ 4-$Et_2NCH_2CH_2OC_6H_4SbCl_4$, 3, 773
$SbC_{12}H_{18}Cl_4NS$ 4-$Et_2NCH_2CH_2SC_6H_4SbCl_4$, 3, 773
$SbC_{12}H_{20}Br$ [$(MeCH=CH)_4Sb$]Br isomers, 3, 815
[$(CH_2=CMe)_4Sb$]Br, 3, 815
$SbC_{12}H_{20}BrI_2$ [$(MeCH=CH)_4Sb$]Br·I_2 isomers, 3, 715
[$(CH_2=CMe)_4Sb$]Br·I_2, 3, 715
$SbC_{12}H_{20}Cl$ [$(CH_2=CHCH_2)_4Sb$]Cl, 3, 715
$SbC_{12}H_{20}I$ [$(MeCH=CH)_4Sb$]I isomers, 3, 715
$SbC_{13}H_{20}I_3$ [$(MeCH=CH)_4Sb$]I_3 isomers, 3, 715
[$(CH_2=CMe)_4Sb$]I_3, 3, 716
$SbC_{12}H_{21}$ $(EtCH=CH)_3Sb$, 3, 669
$SbC_{12}H_{22}Cl$ (cyclo-C_6H_{11})$_2$SbCl, 3, 687
$SbC_{12}H_{27}$ Bu_3Sb, 3, 659
$(Me_2CHCH_2)_3Sb$, 3, 659
$(Me_3C)_3Sb$, 3, 669
$SbC_{12}H_{27}Br_2$ Bu_3SbBr_2, 3, 731
$(Me_2CHCH_2)_3SbBr_2$, 3, 731
$SbC_{12}H_{27}O$ Bu_3SbO, 3, 743
$SbC_{12}H_{27}S$ Bu_3SbS, 3, 749
$SbC_{12}H_{27}Se$ Bu_3SbSe, 3, 750
$SbC_{12}H_{28}Br$ [Pr_4Sb]Br, 3, 716
$SbC_{13}H_{10}NO_3S$ 2-Ph-benzothiazole-6-SbO_3H_2, 3, 796
$SbC_{13}H_{10}NO_4$ 2-Ph-benzoxazole-6-SbO_3H_2, 3, 796
$SbC_{13}H_{11}Cl_3NO_2$ (4-$O_2NC_6H_4$)(4-MeC_6H_4)-$SbCl_3$, 3, 757
$SbC_{13}H_{11}Cl_3N_3O_2$ [4-$MeC_6H_4N_2$][4-O_2N-$C_6H_4SbCl_3$], 3, 690
$SbC_{13}H_{11}Cl_3N_3O_3$ [4-$MeOC_6H_4N_2$][4-O_2N-$C_6H_4SbCl_3$], 3, 690
$SbC_{13}H_{11}Cl_4$ (4-ClC_6H_4)(4-MeC_6H_4)$SbCl_3$, 3, 757
$SbC_{13}H_{11}Cl_4O$ (4-ClC_6H_4)(4-$MeOC_6H_4$)-$SbCl_3$, 3, 757
$SbC_{13}H_{11}N_2O_6$ 4-[4,3-HO(HO_2C)$C_6H_3N=N$]-$C_6H_4SbO_3H_2$, 3, 789
$SbC_{13}H_{12}ClO_2$ (4-ClC_6H_4)(4-MeC_6H_4)-SbO_2H, 3, 764
$SbC_{13}H_{12}ClO_3$ (4-ClC_6H_4)(4-$MeOC_6H_4$)-SbO_2H, 3, 764
$SbC_{13}H_{12}Cl_3$ Ph(4-MeC_6H_4)$SbCl_3$, 3, 757
$SbC_{13}H_{12}Cl_3N_2$ [$MeC_6H_4N_2$][$PhSbCl_3$], 3, 691
$SbC_{13}H_{12}Cl_3N_2O$ [$MeOC_6H_4N_2$][$PhSbCl_3$], 3, 691
$SbC_{13}H_{12}Cl_5N_2$ [4-$MeC_6H_4N_2$][$PhSbCl_5$], 3, 768
[PhN_2][4-$MeC_6H_4SbCl_5$], 3, 768
$SbC_{13}H_{12}NO_4$ (4-$O_2NC_6H_4$)(4-MeC_6H_4)-SbO_2H, 3, 764
4-$BzNHC_6H_4SbO_3H_2$, 3, 789
4-$BzNHC_6H_4SbO_3H_2$, 3, 789

$SbC_{13}H_{13}$ Ph_2MeSb, 3, 674
$SbC_{13}H_{13}N_2O_4S$ 4-(4-$HOC_6H_4NHCSNH$)C_6H_4-SbO_3H_2, 3, 789
$SbC_{13}H_{13}N_2O_6S_2$ 4-($HO_3SC_6H_4NHCSNH$)C_6H_4-SbO_3H_2 isomers, 3, 789, 790
$SbC_{13}H_{13}N_6O_4S_2$
2-[p-($N=C(NH_2)N=C(NH_2)N=CNH$)C_6H_4]-1,3,2-dithiastibolane-4,5-$(CO_2H)_2$, 3, 810
$SbC_{13}H_{13}O_2$ Ph(4-MeC_6H_4)SbO_2H, 3, 764
$SbC_{13}H_{14}N_3O_5S_2$ 4-(4-$H_2NO_2SC_6H_4NHCSNH$)-$C_6H_4SbO_3H_2$, 3, 790
$SbC_{13}H_{15}N_6O_4S_2$ [$N=C(NH_2)N=C(NH_2)N=CNH$]-$C_6H_4Sb(SCH_2CO_2H)_2$ isomers, 3, 704*
$SbC_{13}H_{18}NO_3S_2$ 4-$AcNHC_6H_4$-$SbSCH_2C(CH_2OH)_2CH_2S$, 3, 813
$SbC_{13}H_{25}$ $Et_2(MeCH=CH)_3Sb$ isomers, 3, 708
$Et_2(CH_2=CMe)_3Sb$, 3, 709
$SbC_{14}H_9N_2S_2$ 2-Ph-1,3,2-dithiastibolo-

[4,5-b]quinoxaline, 3, 811
$SbC_{14}H_9O_5$ 1-Anthraquinonestibonic acid, 3, 794
$SbC_{14}H_{11}Br_2O_2$ $(4-BrC_6H_4)_2SbOAc$, 3, 698
$SbC_{14}H_{11}Cl_2O_2$ $(4-ClC_6H_4)_2SbOAc$, 3, 698
$SbC_{14}H_{11}Cl_2O_3$ $(2-ClC_6H_4)_2Sb(O)OAc$, 3, 766
$SbC_{14}H_{11}Cl_3N_3O_3$ $[4-AcC_6H_4N_2][4-O_2N-C_6H_4SbCl_3]$, 3, 690
$SbC_{14}H_{11}F_2$ $(3-FC_6H_4)_2(CH_2{=}CH)Sb$, 3, 674
$SbC_{14}H_{11}I_2O_2$ $(4-IC_6H_4)_2SbOAc$, 3, 698
$SbC_{14}H_{12}Cl_3$ $[4,2-Cl(Me)C_6H_3]_2SbCl$, 3, 687
$SbC_{14}H_{12}Cl_3O_2$ $[5,2-Cl(MeO)C_6H_3]_2SbCl$, 3, 687
$SbC_{14}H_{12}Cl_5$ $[2,4-Me(Cl)C_6H_3]_2SbCl_3$, 3, 757
$SbC_{14}H_{12}Cl_5N_2O_2$ $[2-MeO_2CC_6H_4N_2][PhSbCl_5]$, 3, 768
$SbC_{14}H_{13}$ $Ph_2(CH_2{=}CH)Sb$, 3, 674
$SbC_{14}H_{13}Cl_3NO_3$ $(4-O_2NC_6H_4)(4-EtOC_6H_4)SbCl_3$, 3, 757
$SbC_{14}H_{13}N_2O_5S$ $4-(HO_2CC_6H_4NHCSNH)C_6H_4-SbO_3H_2$ isomers, 3, 790
$SbC_{14}H_{13}O_2$ Ph_2SbOAc, 3, 698
$SbC_{14}H_{13}O_3$ $Ph_2Sb(O)OAc$, 3, 766
$SbC_{14}H_{14}Br_3$ $(PhCH_2)_2SbBr_3$, 3, 757
$SbC_{14}H_{14}Cl$ $(4-MeC_6H_4)_2SbCl$, 3, 687
$SbC_{14}H_{14}ClO_2$ $(MeOC_6H_4)_2SbCl$ isomers, 3, 687
$SbC_{14}H_{14}Cl_3$ $(4-MeC_6H_4)_2SbCl_3$, 3, 755
$(PhCH_2)_2SbCl_3$, 3, 757
$SbC_{14}H_{14}Cl_3N_2$ $[4-MeC_6H_4N_2][4-MeC_6H_4-SbCl_3]$, 3, 691, 692
$SbC_{14}H_{14}Cl_3N_2O$ $[EtOC_6H_4N_2][PhSbCl_3]$, 3, 691
$[4-MeOC_6H_4N_2][4-MeC_6H_4SbCl_3]$, 3, 692
$SbC_{14}H_{14}Cl_3N_4O_2$ $[4-Me_2NC_6H_4N_2][4-O_2N-C_6H_4SbCl_3]$, 3, 690
$SbC_{14}H_{14}Cl_3O$ $Ph(4-EtOC_6H_4)SbCl_3$, 3, 757
$(4-MeOC_6H_4)(4-MeC_6H_4)SbCl_3$, 3, 757
$SbC_{14}H_{14}Cl_5N_2$ $[2,4-Me_2C_6H_3N_2][PhSb-Cl_5]$, 3, 768
$SbC_{14}H_{14}Cl_5N_2O$ $[4-EtOC_6H_4N_2][PhSbCl_5]$, 3, 768
$SbC_{14}H_{14}NO_5$ $(4-EtOC_6H_4)(4-O_2NC_6H_4)-SbO_2H$, 3, 764
$SbC_{14}H_{15}Br_2$ $EtPh_2SbBr_2$, 3, 731
$SbC_{14}H_{15}Cl_2O$ $(4-MeC_6H_4)_2Sb(OH)Cl_2$, 3, 760
$SbC_{14}H_{15}Cl_3N_3$ $[4-Me_2NC_6H_4N_2][PhSbCl_3]$, 3, 691
$SbC_{14}H_{15}Cl_3N_3O_3S$ $[4-EtOC_6H_4N_2][4-H_2NO_2-SC_6H_4SbCl_3]$, 3, 691
$SbC_{14}H_{15}Cl_5N_3$ $[4-Me_2NC_6H_4N_2][PhSbCl_5]$, 3, 768
$SbC_{14}H_{15}O_2$ $(4-MeC_6H_4)_2SbO_2H$, 3, 762
$SbC_{14}H_{15}O_3$ $Ph(4-EtOC_6H_4)SbO_2H$, 3, 764
$(4-MeOC_6H_4)(4-MeC_6H_4)SbO_2H$, 3, 764
$SbC_{14}H_{16}Cl_3N_4O_2S$ $[4-Me_2NC_6H_4N_2][4-H_2-NO_2SC_6H_4SbCl_3]$, 3, 691
$SbC_{15}H_{12}Cl_4N$ Quinolinium$[3-ClC_6H_4-SbCl_3]^-$, 3, 690
$SbC_{15}H_{13}Cl_2O_2$ $2-(4-EtO_2CC_6H_4)C_6H_4-SbCl_2$, 3, 695
$SbC_{15}H_{13}N_6S_2$ 2-[p-($\overline{N{=}C(NH_2)N{=}C(NH_2)N{=}C}NH)C_6H_4$]-1,3,2-benzodithiastibole, 3, 810
$SbC_{15}H_{14}Cl_5N_2O_2$ $[4-EtO_2CC_6H_4N_2][Ph-SbCl_5]$, 3, 768
$SbC_{15}H_{15}$ $(C_5H_5)_3Sb$, 3, 659
$SbC_{15}H_{15}Br_2$ $Ph_2(MeCH{=}CH)SbBr_2$ isomers, 3, 731
$Ph_2(CH_2{=}CMe)SbBr_2$, 3, 731
$SbC_{15}H_{17}$ Ph_2PrSb, 3, 674
$Ph_2(Me_2CH)Sb$, 3, 674
$SbC_{15}H_{17}Cl_3N_3$ $[4-Me_2NC_6H_4N_2][4-MeC_6H_4-SbCl_3]$, 3, 692
$SbC_{15}H_{22}N_3O_6$ $4-[EtO_2CN(CH_2CH_2)_2NCH_2CO-NH]C_6H_4SbO_3H_2$, 3, 790
$SbC_{15}H_{24}Br$ $Ph(n-C_9H_{19})SbBr$, 3, 687
$SbC_{15}H_{24}I$ $Ph(n-C_9H_{19})SbI$, 3, 687
$SbC_{15}H_{25}$ $(MeCH{=}CH)_5Sb$ isomers, 3, 709
$(CH_2{=}CMe)_5Sb$, 3, 709
$SbC_{15}H_{25}N_2O_9$ $4-[HOCH_2(CHOH)_4CH_2N(Me)-COCH_2NH]C_6H_4SbO_3H_2$, 3, 790
$SbC_{16}H_{12}Cl_3$ $[2,4-Cl(CH_2{=}CH)C_6H_3]_2SbCl$, 3, 687*
$Ph(C_{10}H_7)SbCl_3$ isomers, 3, 757
$SbC_{16}H_{15}Cl_2O_2$ $2-(EtO_2CCH_2C_6H_4)C_6H_4SbCl_2$,

$SbC_{16}H_{15}Cl_2O_2$
3, 695
$SbC_{16}H_{15}N_2O_5$ 2,4-Ac(4-$AcC_6H_4N{=}N)C_6H_3$-SbO_3H_2, 3, 793
$(SbC_{16}H_{15}O_2)_n$ $[Ph_2SbO_2CC(Me)CH_2]_n$, 3, 698
$SbC_{16}H_{16}ClO_2$ 4-MeC_6H_4(4-$EtO_2CC_6H_4$)-SbCl, 3, 687
$SbC_{16}H_{16}Cl_3O_2$ (4-$EtO_2CC_6H_4$)(4-MeC_6H_4)$SbCl_3$, 3, 758
$SbC_{16}H_{17}Br_2O_2$ $Ph_2(EtO_2CCH_2)SbBr_2$, 3, 731
$SbC_{16}H_{17}Cl_2O_2$ $Ph_2(EtO_2CCH_2)SbCl_2$, 3, 731
$SbC_{16}H_{17}O_2$ $Ph_2(EtO_2CCH_2)Sb$, 3, 674
(4-$MeC_6H_4)_2SbOAc$, 3, 698
$SbC_{16}H_{17}O_4$ (4-$MeOC_6H_4)_2SbOAc$, 3, 698
$SbC_{16}H_{18}ClO_2$ (2-$EtOC_6H_4)_2SbCl$, 3, 687
$SbC_{16}H_{18}Cl_3$ $(Me_2C_6H_3)_2SbCl_3$ isomers, 3, 758
$SbC_{16}H_{18}Cl_3O_2$ (2-$EtOC_6H_4)_2SbCl_3$, 3, 755
$SbC_{16}H_{19}$ Ph_2BuSb, 3, 674
$Ph_2(Me_2CHCH_2)Sb$, 3, 674
$Ph_2(Me_3C)Sb$, 3, 674
$SbC_{16}H_{19}Cl_2O$ (2,4-$Me_2C_6H_3)_2Sb(OH)Cl_2$, 3, 760
$SbC_{16}H_{19}O_2$ (2,4-$Me_2C_6H_3)_2SbO_2H$, 3, 764
$SbC_{16}H_{19}O_4$ $Ph_2(EtO_2CCH_2)Sb(OH)_2$, 3, 740
(2-$EtOC_6H_4)_2SbO_2H$, 3, 764
$SbC_{16}H_{20}Br$ $[Me_2(PhCH_2)_2Sb]Br$, 3, 712
$SbC_{16}H_{20}I$ $[Me_2(PhCH_2)_2Sb]I$, 3, 716
$SbC_{16}H_{21}N_2O_5S_2$ [2,5,4-$Me_2(H_2NO_2S)$-$C_6H_2]_2SbOH$, 3, 696
$SbC_{16}H_{36}Cl$ $[Bu_4Sb]Cl$, 3, 716
$SbC_{17}H_{16}Cl_4N$ Py·H$[Ph_2SbCl_4]$, 3, 754
$SbC_{17}H_{17}N_6S_2$
3-[p-($N{=}C(NH_2)N{=}C(NH_2)N{=}CNH)C_6H_4$]-1,5-dihydrobenzo-2,4,3-dithiastibepin, 3, 814
$SbC_{17}H_{21}$ $Ph_2(n\text{-}C_5H_{11})Sb$, 3, 674
$Ph_2(Me_2CHCH_2CH_2)Sb$, 3, 674
$SbC_{17}H_{22}I$ $\{Me_3[PhCH_2CH(Ph)]Sb\}I$, 3, 713
$SbC_{18}H_9Cl_8$ (2,4-$Cl_2C_6H_3)_3SbCl_2$, 3, 732
$SbC_{18}H_{12}Br_2Cl_3$ (4-$ClC_6H_4)_3SbBr_2$, 3, 732
(4-$BrC_6H_4)_2$(4-$ClC_6H_4)SbCl_2$, 3, 732
$SbC_{18}H_{12}Br_3$ (4-$BrC_6H_4)_3Sb$, 3, 659
$SbC_{18}H_{12}Br_3Cl_2$ (4-$BrC_6H_4)_3SbCl_2$, 3, 725
$SbC_{18}H_{12}Br_3F_2$ (4-$BrC_6H_4)_3SbF_2$, 3, 732
$SbC_{18}H_{12}Br_3I_2$ (4-$BrC_6H_4)_3SbI_2$, 3, 732
$SbC_{18}H_{12}Br_5$ (4-$BrC_6H_4)_3SbBr_2$, 3, 732
$SbC_{18}H_{12}Cl_2I_3$ (4-$IC_6H_4)_3SbCl_2$, 3, 732
$SbC_{18}H_{12}Cl_3$ (4-$ClC_6H_4)_3Sb$, 3, 660
(2-$ClC_6H_4)_3Sb$, 3, 669
$SbC_{18}H_{12}Cl_3F_2$ (4-$ClC_6H_4)_3SbF_2$, 3, 732
$SbC_{18}H_{12}Cl_3I_2$ (4-$ClC_6H_4)_3SbI_2$, 3, 732
$SbC_{18}H_{12}Cl_4NO_2$ (4-$ClC_6H_4)_2$(4-$O_2NC_6H_4$)-$SbCl_2$, 3, 732
$SbC_{18}H_{12}Cl_5$ (4-$ClC_6H_4)_3SbCl_2$, 3, 726
(2-$ClC_6H_4)_3SbCl_2$, 3, 732
$SbC_{18}H_{12}N_3O_6$ (4-$O_2NC_6H_4)_3Sb$, 3, 670
$SbC_{18}H_{13}$ 5-Ph-dibenzostibole, 3, 801
$SbC_{18}H_{14}Br$ Ph_2(4-$BrC_6H_4)Sb$, 3, 674
$SbC_{18}H_{14}BrCl_4N_2$ [4-$BrC_6H_4N_2][Ph_2SbCl_4]$, 3, 755
$SbC_{18}H_{14}Br_3$ (4-$BrC_6H_4)Ph_2SbBr_2$, 3, 732
$SbC_{18}H_{14}Cl$ Ph_2(4-$ClC_6H_4)Sb$, 3, 674
$SbC_{18}H_{14}Cl_4IN_2$ $[IC_6H_4N_2][Ph_2SbCl_4]$, 3, 755
$SbC_{18}H_{14}Cl_4N_3O_2$ $[O_2NC_6H_4N_2][Ph_2SbCl_4]$ isomers, 3, 755
$SbC_{18}H_{14}Cl_5N_2$ $[ClC_6H_4N_2][Ph_2SbCl_4]$ isomers, 3, 754
$SbC_{18}H_{15}$ Ph_3Sb, 3, 660
$SbC_{18}H_{15}Br_2$ Ph_3SbBr_2, 3, 726
$SbC_{18}H_{15}ClF$ Ph_3SbClF, 3, 732
$SbC_{18}H_{15}Cl_2$ Ph_3SbCl_2, 3, 726
$SbC_{18}H_{15}Cl_2N_2$ $[PhN_2][Ph_2SbCl_2]$, 3, 685
$SbC_{18}H_{15}Cl_4N_2$ $[PhN_2][Ph_2SbCl_4]$, 3, 754
$SbC_{18}H_{15}F_2$ Ph_3SbF_2, 3, 732
$SbC_{18}H_{15}I_2$ Ph_3SbI_2, 3, 732
$SbC_{18}H_{15}O$ Ph_3SbO, 3, 743
$(SbC_{18}H_{15}O_2)_n$ Poly(Ph_3Sb peroxide), 3, 747
$SbC_{18}H_{15}O_3$ (4-$HOC_6H_4)_3Sb$, 3, 670
$SbC_{18}H_{15}O_4S$ Ph_3SbOSO_3, 3, 665
$SbC_{18}H_{15}S$ Ph_3SbS, 3, 749
$SbC_{18}H_{15}Se$ Ph_3SbSe, 3, 750
$SbC_{18}H_{16}BrO$ $Ph_3Sb(OH)Br$, 3, 738

$SbC_{18}H_{16}BrO_2$ $Ph_2(4\text{-}BrC_6H_4)Sb(OH)_2$, 3, 740

$SbC_{18}H_{17}O_2$ $Ph_3Sb(OH)_2$, 3, 740

$SbC_{18}H_{17}O_4$ $Ph_3Sb(OOH)_2$, 3, 746

$SbC_{18}H_{18}Cl_3O_4$ $(4\text{-}EtO_2CC_6H_4)_2SbCl_3$, 3, 758

$SbC_{18}H_{19}O_6$ $Ph_2Sb(OAc)_3$, 3, 766

$SbC_{18}H_{21}$ $Ph_2(cyclo\text{-}C_6H_{11})Sb$, 3, 675

$SbC_{18}H_{22}BrO_2$ 2,4,4,5,5,2,2-Br(Me)$_4$-(Ph)$_2$-2,2-dihydro-1,3,2-dioxastibolane, 3, 807

$SbC_{18}H_{23}$ $Ph_2(n\text{-}C_6H_{13})Sb$, 3, 675

$SbC_{18}H_{33}$ $(cyclo\text{-}C_6H_{11})_3Sb$, 3, 666

$SbC_{18}H_{33}Cl_2$ $(cyclo\text{-}C_6H_{11})_3SbCl_2$, 3, 732

$SbC_{18}H_{33}S$ $(cyclo\text{-}C_6H_{11})_3SbS$, 3, 749

$SbC_{18}H_{34}ClO$ $(cyclo\text{-}C_6H_{11})_3Sb(OH)Cl$, 3, 738

$SbC_{18}H_{39}$ $[Me_2CH(CH_2)_3]_3Sb$, 3, 670
$(C_6H_{13})_3Sb$, 3, 666, 670

$SbC_{19}H_{12}BrO_2$ 4-(2-Br-dibenzostibol-5-yl)$C_6H_4CO_2H$, 3, 801

$SbC_{19}H_{12}NO_5$ 4-(3-O_2N-dibenzostibol-5-yl)-$C_6H_4CO_2H$, 3, 801

$SbC_{19}H_{13}BrClO_2$ $(4\text{-}HO_2CC_6H_4)$[4,2-Br-(Ph)C_6H_3]SbCl, 3, 688

$SbC_{19}H_{13}BrCl_3O_2$ $(4\text{-}HO_2CC_6H_4)$[4,2-Br-(Ph)C_6H_3]$SbCl_3$, 3, 758

$SbC_{19}H_{13}O_2$ p-(Dibenzostibol-5-yl)-$C_6H_4CO_2H$, 3, 801

$SbC_{19}H_{14}Br$ 2,5-Br(4-MeC_6H_4)-dibenzostibole, 3, 802

$SbC_{19}H_{14}BrCl_2O$ 10,10,10,2-Cl_2(p-Br-C_6H_4)(Me)-phenoxantimonin, 3, 812

$SbC_{19}H_{14}BrO$ 2,10-Me(p-BrC_6H_4)-phenoxantimonin, 3, 812

$SbC_{19}H_{14}Br_3$ 2,5-Br(p-MeC_6H_4)-dibenzostibole 5,5-dibromide, 3, 801

$SbC_{19}H_{14}F_3$ $Ph_2(3\text{-}F_3CC_6H_4)Sb$, 3, 675

$SbC_{19}H_{14}NO_2$ 3,5-O_2N(p-MeC_6H_4)-dibenzostibole, 3, 802

$SbC_{19}H_{15}$ 5-p-Tolyldibenzostibole, 3, 802

$SbC_{19}H_{15}BrCl$ $(4\text{-}MeC_6H_4)$[4,3-Br(Ph)-C_6H_3]SbCl, 3, 688

$SbC_{19}H_{15}BrCl_3$ $(4\text{-}MeC_6H_4)$[4,2-Br(Ph)-C_6H_3]$SbCl_3$, 3, 758

$SbC_{19}H_{15}Br_2Cl_2$ $(4\text{-}BrC_6H_4)_2(4\text{-}MeC_6H_4)$-$SbCl_2$, 3, 733

$SbC_{19}H_{15}Cl_2O$ 10,10,10,2-Cl_2(Ph)(Me)-phenoxantimonin, 3, 812

$SbC_{19}H_{15}Cl_3NO_2$ $(4\text{-}MeC_6H_4)$[2-(4-O_2N-C_6H_4)C_6H_4]$SbCl_3$, 3, 758

$SbC_{19}H_{15}Cl_4$ $(4\text{-}ClC_6H_4)_2(4\text{-}MeC_6H_4)$-$SbCl_2$, 3, 733

$SbC_{19}H_{15}O$ 2,10-Me(Ph)-phenoxantimonin, 3, 812

$SbC_{19}H_{15}O_2$ Ph_2SbOBz, 3, 698

$SbC_{19}H_{16}Cl$ $(4\text{-}MeC_6H_4)(2\text{-}PhC_6H_4)SbCl$, 3, 685

$SbC_{19}H_{16}ClO_2$ $(EtO_2CC_6H_4)(C_{10}H_7)SbCl$ isomers, 3, 688

$SbC_{19}H_{16}Cl_3$ $(4\text{-}MeC_6H_4)(2\text{-}PhC_6H_4)SbCl_3$, 3, 758

$SbC_{19}H_{16}Cl_3O_2$ $(4\text{-}EtO_2CC_6H_4)(C_{10}H_7)$-$SbCl_3$ isomers, 3, 758

$SbC_{19}H_{16}N$ 3,5-H_2N(4-MeC_6H_4)-dibenzostibole, 3, 802

$SbC_{19}H_{17}$ $Ph_2(2\text{-}MeC_6H_4)Sb$, 3, 675
$Ph_2(PhCH_2)Sb$, 3, 675

$SbC_{19}H_{17}Cl_4N_2$ $[MeC_6H_4N_2][Ph_2SbCl_4]$ isomers, 3, 755

$SbC_{19}H_{17}Cl_4N_2O$ $[MeOC_6H_4N_2][Ph_2SbCl_4]$, 3, 755

$SbC_{19}H_{17}O$ $Ph_2(MeOC_6H_4)Sb$ isomers, 3, 675

$SbC_{19}H_{17}O_4$ $Ph_2(4\text{-}HO_2CC_6H_4)Sb(OH)_2$, 3, 740

$SbC_{19}H_{18}Cl$ $[MePh_3Sb]Cl$, 3, 716

$SbC_{19}H_{18}ClO$ $Ph_3Sb(OMe)Cl$, 3, 738

$SbC_{19}H_{18}I$ $[MePh_3Sb]I$, 3, 716

$SbC_{19}H_{22}N_2O_3S_2$ N,N′-Bis{p-[4-(hydroxymethyl)-1,3,2-dithiastibolan-2-yl]-phenyl}urea, 3, 810

$SbC_{20}H_{14}Cl_2NO$ p-[10,10,2-Cl_2(Me)-10-phenoxantimoninyl]benzonitrile, 3, 812

$SbC_{20}H_{14}Cl_3$ $(C_{10}H_7)_2SbCl_3$ isomers, 3, 758

$SbC_{20}H_{14}NO$ p-(2-Me-10-phenoxantimoninyl)benzonitrile, 3, 812

$SbC_{20}H_{15}Br_2O_2$ 5,5,5-Br_2(4-MeC_6H_4)-dibenzostibole-5-CO_2H, 3, 802

$SbC_{20}H_{15}O_2$ p-(3-Methyl-5-dibenzosti-

bolyl)-$C_6H_4CO_2H$, 3, 802
5-p-Tolyl-3-dibenzostibole-CO_2H, 3, 803
$SbC_{20}H_{15}O_3$ p-(3-MeO-dibenzostibol-5-yl)-$C_6H_4CO_2H$, 3, 803
p-(2-Me-10-phenoxantimoninyl)-C_6H_4-CO_2H, 3, 812
$SbC_{20}H_{16}NO_4S_2$ 4-$H_2NC_6H_4Sb(SC_6H_4CO_2$-H-o$)_2$, 3, 705
$SbC_{20}H_{17}$ Ph_2(4-CH_2=CHC_6H_4)Sb, 3, 675
Poly[p-(diphenylstibino)styrene], 3, 680
$SbC_{20}H_{17}BrNO$ Ph_3Sb=$NCOCH_2Br$, 3, 751
$SbC_{20}H_{17}ClNO$ Ph_3Sb=$NCOCH_2Cl$, 3, 751
$SbC_{20}H_{17}INO$ Ph_3Sb=$NCOCH_2I$, 3, 751
$SbC_{20}H_{17}O$ 3,5-MeO(p-MeC_6H_4)-dibenzostibole, 3, 803
$SbC_{20}H_{17}O_5$ p-[5,5,3-$(HO)_2$(MeO)-dibenzostibol-5-yl]$C_6H_4CO_2H$, 3, 803
$SbC_{20}H_{18}BrCl_2$ (4-$MeC_6H_4)_2$(4-BrC_6H_4)-$SbCl_2$, 3, 733
$SbC_{20}H_{18}Cl$ (4-$MeC_6H_4)_2$(4-ClC_6H_4)Sb, 3, 675
$SbC_{20}H_{18}Cl_3O$ (4-MeC_6H_4)[2-(4-$MeOC_6H_4$)-C_6H_4]$SbCl_3$, 3, 758
$SbC_{20}H_{18}NO$ Ph_3Sb=NAc, 3, 751
$SbC_{20}H_{19}$ Ph_2(3,5-$Me_2C_6H_3$)Sb, 3, 675
$SbC_{20}H_{19}Br_2O$ Ph_2(4-$EtOC_6H_4$)$SbBr_2$, 3, 733
$SbC_{20}H_{19}Cl_4N_2O$ [$EtOC_6H_4N_2$][Ph_2SbCl_4] isomers, 3, 755
$SbC_{20}H_{19}O$ Ph_2(4-$EtOC_6H_4$)Sb, 3, 675
Ph_2(4-$EtOC_6H_4$)Sb, 3, 675
$SbC_{20}H_{19}O_3$ Ph_3Sb(OAc)OH, 3, 741
$SbC_{20}H_{20}Cl_4N_3$ [4-$Me_2NC_6H_4N_2$][Ph_2Sb-Cl_4], 3, 755
$SbC_{20}H_{21}O_3$ Ph_2(4-$EtOC_6H_4$)$Sb(OH)_2$, 3, 740
$SbC_{20}H_{23}$ (2-$MeC_6H_4)_2$(cyclo-C_6H_9)Sb, 3, 675
$SbC_{20}H_{25}O_2$ (4-$MeOC_6H_4)_2$(Me_2CHCH_2-CH=CH)Sb, 3, 675
$SbC_{20}H_{27}$ Ph_2(n-C_8H_{17})Sb, 3, 675
$SbC_{21}H_{12}F_9$ (3-$CF_3C_6H_4)_3$Sb, 3, 670
$SbC_{21}H_{16}BrO_2$ p-(2-Br-dibenzostibol-5-yl)-$C_6H_4CO_2Et$, 3, 803
$SbC_{21}H_{16}Br_3O_2$ p-(2,5,5-Br_3-dibenzostibol-5-yl)-$C_6H_4CO_2Et$, 3, 803
$SbC_{21}H_{16}NO_4$ p-(3-O_2N-dibenzostibol-5-yl)-$C_6H_4CO_2Et$, 3, 803
$SbC_{21}H_{17}BrClO_2$ (4-$EtO_2CC_6H_4$)[4,2-Br-(Ph)C_6H_3]SbCl, 3, 688
$SbC_{21}H_{17}BrCl_3O_2$ ($EtO_2CC_6H_4$)[4,2-Br-(Ph)C_6H_3]$SbCl_3$, 3, 758
$SbC_{21}H_{17}ClNO_4$ (4-$EtO_2CC_6H_4$)[2-(4-O_2N-C_6H_4)C_6H_4]SbCl, 3, 688
$SbC_{21}H_{17}Cl_3NO_4$ (4-$EtO_2CC_6H_4$)[2-(4-O_2N-C_6H_4)C_6H_4]$SbCl_3$, 3, 758
$SbC_{21}H_{17}O_2$ $Ph_2SbO_2CC_6H_4CH$=CH_2-p, 3, 698
p-(Dibenzostibol-5-yl)-$C_6H_4CO_2Et$, 3, 804
$SbC_{21}H_{17}O_3$ 5-p-Tolyldibenzostibol-3-yloxyacetic acid, 3, 804
p-(5-Oxodibenzostibol-5-yl)-C_6H_4-CO_2Et, 3, 804
$SbC_{21}H_{18}ClO_2$ (4-$EtO_2CC_6H_4$)(2-PhC_6H_4)-SbCl, 3, 688
$SbC_{21}H_{18}Cl_3$ [Cl(Me)$C_6H_3]_3$Sb isomers, 3, 670
$SbC_{21}H_{18}Cl_3O_2$ (4-$EtO_2CC_6H_4$)(PhC_6H_4)-$SbCl_3$ isomers, 3, 758
$SbC_{21}H_{18}Cl_3O_3$ [5,2-Cl(MeO)$C_6H_3]_3$Sb, 3, 670
$SbC_{21}H_{18}Cl_5$ [Cl(Me)$C_6H_3]_3SbCl_2$ isomers, 3, 733
$SbC_{21}H_{18}Cl_5O_3$ [5,2-Cl(MeO)$C_6H_3]_3SbCl_2$, 3, 733
$SbC_{21}H_{18}NO_6$ (4-$EtO_2CC_6H_4$)[2-(4-O_2N-C_6H_4)C_6H_4]SbO_2H, 3, 765
$SbC_{21}H_{19}$ Ph_2(4-CH_2=$CHCH_2C_6H_4$)Sb, 3, 676
$SbC_{21}H_{19}Br_4$ Ph_2[4-($BrCH_2CHBrCH_2$)C_6H_4]-$SbBr_2$, 3, 733
$SbC_{21}H_{19}N_6O_6S_4$ 4-[N=C(NH_2)N=C(NH_2)N=C-NH]$C_6H_4Sb(SC_6H_4SO_3H$-p$)_2$, 3, 705
$SbC_{21}H_{19}O_4$ (2-PhC_6H_4)(4-$EtO_2CC_6H_4$)-SbO_2H, 3, 765
$SbC_{21}H_{19}O_5$ 5,5,5-$(HO)_2$(p-MeC_6H_4)-dibenzostibol-3-yloxyacetic acid, 3, 804
$SbC_{21}H_{21}$ $(PhCH_2)_3$Sb, 3, 666
$(MeC_6H_4)_3$Sb isomers, 3, 666, 670

$Ph_2(2,4,6\text{-}Me_3C_6H_2)Sb$, 3, 676
$SbC_{21}H_{21}Br_2$ $(MeC_6H_4)_3SbBr_2$ isomers, 3, 733
$(PhCH_2)_3SbBr_2$, 3, 733
$SbC_{21}H_{21}Br_2O_3$ $(MeOC_6H_4)_3SbBr_2$ isomers, 3, 733
$SbC_{21}H_{21}Cl_2$ $(MeC_6H_4)_3SbCl_2$ isomers, 3, 727, 734
$(PhCH_2)_3SbCl_2$, 3, 734
$SbC_{21}H_{21}Cl_2O_3$ $(MeOC_6H_4)_3SbCl_2$ isomers, 3, 733, 734
$SbC_{21}H_{21}Cl_4N_2$ $[4\text{-}MeC_6H_4N_2][(4\text{-}MeC_6H_4)_2SbCl_4]$, 3, 755
$SbC_{21}H_{21}Cl_4N_4$ $[4\text{-}MeC_6H_4N_2]_2[4\text{-}MeC_6H_4SbCl_4]$, 3, 692
$SbC_{21}H_{21}F_2$ $(MeC_6H_4)_3SbF_2$ isomers, 3, 734
$SbC_{21}H_{21}I_2$ $(4\text{-}MeC_6H_4)_3SbI_2$, 3, 734
$SbC_{21}H_{21}I_2O_3$ $(MeOC_6H_4)_3SbI_2$ isomers, 3, 734
$SbC_{21}H_{21}O$ $(4\text{-}MeC_6H_4)_3SbO$, 3, 743
$(PhCH_2)_3SbO$, 3, 743
$SbC_{21}H_{21}O_3$ $(MeOC_6H_4)_3Sb$ isomers, 3, 670
$SbC_{21}H_{21}S$ $(4\text{-}MeC_6H_4)_3SbS$, 3, 749
$SbC_{21}H_{21}Se$ $(4\text{-}MeC_6H_4)_3SbSe$, 3, 750
$SbC_{21}H_{24}ClO_2$ $(4\text{-}EtO_2CC_6H_4)[4\text{-}(C_6H_{11})\text{-}C_6H_4]SbCl$, 3, 688
$SbC_{21}H_{24}Cl_3O_2$ $(4\text{-}EtO_2CC_6H_4)(4\text{-}C_6H_{11}\text{-}C_6H_4)SbCl_3$, 3, 758
$SbC_{21}H_{24}N_2O_8$ $(CH_2{=}CHCH_2)_2C(CONHC_6H_4\text{-}SbO_3H_2)_2$, 3, 799
$SbC_{21}H_{25}$ $Ph_2(MeCH{=}CH)_3Sb$ isomers, 3, 710
$Ph_2(CH_2{=}CMe)_3Sb$, 3, 710
$SbC_{21}H_{27}O_6$ Me_3Sb bis(isochromanyl peroxide), 3, 746
$SbC_{21}H_{29}$ $Ph_2(n\text{-}C_9H_{19})Sb$, 3, 676
$SbC_{21}H_{31}O_4$ $Me_3Sb(OOCMe_2Ph)_2$, 3, 746
$SbC_{21}H_{45}$ $(n\text{-}C_7H_{15})_3Sb$, 3, 670
$SbC_{21}H_{45}I_2$ $(n\text{-}C_7H_{15})_3SbI_2$, 3, 734
$SbC_{21}H_{45}O$ $(n\text{-}C_7H_{15})_3SbO$, 3, 743
$SbC_{22}H_{17}$ $Ph_2(1\text{-}C_{10}H_7)Sb$, 3, 676
$SbC_{22}H_{19}O_2$ 5-p-Tolyldibenzostibol-3-yl-CO_2Et, 3, 804
p-(3-Me-dibenzostibol-5-yl)-C_6H_4-CO_2Et, 3, 804
$SbC_{22}H_{19}O_3$ p-(3-MeO-dibenzostibol-5-yl)-$C_6H_4CO_2Et$, 3, 804
p-(2-Me-10-phenoxantimoninyl)-C_6H_4-CO_2Et, 3, 813
$SbC_{22}H_{20}ClO_2$ $(4\text{-}EtO_2CC_6H_4)[5,2\text{-}Me(Ph)\text{-}C_6H_3]SbCl$, 3, 688
$(4\text{-}MeC_6H_4)[2\text{-}(4\text{-}EtO_2CC_6H_4)C_6H_4]SbCl$, 3, 688
$SbC_{22}H_{20}ClO_3$ $(4\text{-}EtO_2CC_6H_4)[2\text{-}(4\text{-}MeO\text{-}C_6H_4)C_6H_4]SbCl$, 3, 688
$(4\text{-}EtO_2CC_6H_4)[3,4\text{-}Ph(MeO)C_6H_3]SbCl$, 3, 688
$SbC_{22}H_{20}Cl_3O_2$ $(4\text{-}MeC_6H_4)[2\text{-}(4\text{-}EtO_2C\text{-}C_6H_4)C_6H_4]SbCl_3$, 3, 759
$(4\text{-}EtO_2CC_6H_4)[2,5\text{-}Ph(Me)C_6H_3]SbCl_3$, 3, 759
$SbC_{22}H_{20}Cl_3O_3$ $(4\text{-}EtO_2CC_6H_4)[2\text{-}(4\text{-}MeO\text{-}C_6H_4)C_6H_4]SbCl_3$, 3, 759
$(4\text{-}EtO_2CC_6H_4)[Ph(MeO)C_6H_3]SbCl_3$ isomers, 3, 759
$SbC_{22}H_{21}O_4$ $Ph_3Sb(OAc)_2$, 3, 741
$SbC_{22}H_{21}O_5$ $(4\text{-}EtO_2CC_6H_4)[2\text{-}(4\text{-}MeO\text{-}C_6H_4)C_6H_4]SbO_2H$, 3, 765
p-[5,5,3-$(HO)_2(MeO)$-dibenzostibol-5-yl]$C_6H_4CO_2Et$, 3, 805
$SbC_{22}H_{22}Cl_2NO_4$ $(2\text{-}EtOC_6H_4)_2(4\text{-}O_2NC_6H_4)\text{-}SbCl_2$, 3, 734
$SbC_{22}H_{22}Cl_3O_2$ $(2\text{-}EtOC_6H_4)_2(2\text{-}ClC_6H_4)\text{-}SbCl_2$, 3, 734
$SbC_{22}H_{23}Cl_2$ $(2,4\text{-}Me_2C_6H_3)_2PhSbCl_2$, 3, 734
$SbC_{22}H_{24}BrO_2$ $Ph_3Sb(Br)OOCMe_3$, 3, 747
$SbC_{22}H_{24}ClO$ $Ph(2,4\text{-}Me_2C_6H_3)_2Sb(OH)Cl$, 3, 738
$SbC_{22}H_{25}O_3$ $Ph_3Sb(OH)OOCMe_3$, 3, 747
$SbC_{22}H_{26}BrIP$ $[Me_3SbCH_2PPh_3]BrI$, 3, 716
$SbC_{22}H_{26}Br_2P$ $[Me_3SbCH_2PPh_3]Br_2$, 3, 716
$SbC_{23}H_{17}$ $(PhC{\equiv}C)_2(4\text{-}MeC_6H_4)Sb$, 3, 676
$SbC_{23}H_{17}O_2$ $Ph(4\text{-}HO_2CC_6H_4)(C_{10}H_7)Sb$ isomers, 3, 677
$SbC_{23}H_{19}$ $(C_{10}H_7)_2(MeCH{=}CH)Sb$, 3, 676
$SbC_{23}H_{19}N_6O_4S_2$ 4-[$N{=}C(NH_2)N{=}C(NH_2)N{=}C$-$NH]C_6H_4Sb(SC_6H_4CO_2H\text{-}p)_2$, 3, 705
$SbC_{23}H_{21}O_3$ 5-p-Tolyldibenzostibol-3-yl-OCH_2CO_2Et, 3, 805
$SbC_{23}H_{22}Cl_3O_3$ $(4\text{-}MeC_6H_4)[2\text{-}(4\text{-}EtO_2C\text{-}CH_2OC_6H_4)C_6H_4]SbCl_3$, 3, 759
$SbC_{23}H_{23}ClNO$ $(MeC_6H_4)_3Sb{=}NCOCH_2Cl$ iso-

mers, 3, 751
$SbC_{23}H_{23}Cl_2O_2$ 3-$EtOCOC_6H_4$(4-MeC_6H_4)$_2$-$SbCl_2$, 3, 728
$SbC_{23}H_{23}O_5$ [2-(4-$EtO_2CCH_2OC_6H_4$)C_6H_4]-(4-MeC_6H_4)SbO_2H, 3, 765
5,5,5-$(HO)_2$(p-MeC_6H_4)-dibenzostibol-3-yl-OCH_2CO_2Et, 3, 805
$SbC_{23}H_{24}NO$ $(MeC_6H_4)_3Sb$=NAc isomers, 3, 751
$SbC_{23}H_{25}O_3$ (4-MeC_6H_4)$_3$Sb(OAc)OH, 3, 741
$SbC_{24}H_{15}$ $(PhC{\equiv}C)_3Sb$, 3, 670
$SbC_{24}H_{17}O$ Ph_2(2-dibenzofuryl)Sb, 3, 676*
$SbC_{24}H_{18}Cl$ (4-PhC_6H_4)$_2$SbCl, 3, 688
$(SbC_{24}H_{19})_n$ Poly(p-phenylenetriphenylantimony), 3, 711
$SbC_{24}H_{19}Cl_4N_4$ [4-$PhN_2C_6H_4N_2$][Ph_2SbCl_4], 3, 755
$SbC_{24}H_{19}O$ Ph_2(4-$PhOC_6H_4$)Sb, 3, 676
$SbC_{24}H_{20}Br$ [Ph_4Sb]Br, 3, 713, 717
$SbC_{24}H_{20}Br_3$ [Ph_4Sb]Br_3, 3, 717
$SbC_{24}H_{20}Cl$ [Ph_4Sb]Cl, 3, 717
$SbC_{24}H_{20}ClO_4$ [Ph_4Sb]ClO_4, 3, 718
$SbC_{24}H_{20}F$ [Ph_4Sb]F, 3, 717
$SbC_{24}H_{20}I$ [Ph_4Sb]I, 3, 717, 718
$SbC_{24}H_{20}I_3$ [Ph_4Sb]I_3, 3, 718
$SbC_{24}H_{20}NO_2S$ $Ph_3Sb{=}NO_2SPh$, 3, 751
$SbC_{24}H_{20}NO_3$ [Ph_4Sb]NO_3, 3, 718
$SbC_{24}H_{21}O$ [Ph_4Sb]OH, 3, 717
$SbC_{24}H_{22}N$ Ph_2[4-(MeC=CHCH=CMeN)C_6H_4]-Sb, 3, 676
$SbC_{24}H_{24}Cl_2N_3O_3$ (4-$AcNHC_6H_4$)$_3SbCl_2$, 3, 735
$SbC_{24}H_{25}$ Ph_2(4-$C_6H_{11}C_6H_4$)Sb, 3, 676
$SbC_{24}H_{25}O_4$ $Ph_3Sb(O_2CEt)_2$, 3, 741
$SbC_{24}H_{27}$ $(PhCH_2CH_2)_3Sb$, 3, 667
$SbC_{24}H_{27}Br_2O_3$ (2-$EtOC_6H_4$)$_3SbBr_2$, 3, 735
$SbC_{24}H_{27}Cl_2$ $(Me_2C_6H_3)_3SbCl_2$ isomers, 3, 735
$SbC_{24}H_{27}Cl_2O_3$ (2-$EtOC_6H_4$)$_3SbCl_2$, 3, 735
$SbC_{24}H_{27}I_2O_3$ (2-$EtOC_6H_4$)$_3SbI_2$, 3, 735
$SbC_{24}H_{27}O_2$ 4,4,5,5,2,2,2-$Me_4(Ph)_3$-2,2-dihydro-1,3,2-dioxastibolane, 3, 807
$SbC_{24}H_{27}O_3$ $(EtOC_6H_4)_3Sb$ isomers, 3, 670, 671
$SbC_{24}H_{29}O_2$ $EtPh_3SbO$-$OCMe_3$, 3, 722
$SbC_{24}H_{29}O_3$ $Ph_3Sb(OEt)OOCMe_3$, 3, 747
$SbC_{24}H_{30}N_3$ (4-$Me_2NC_6H_4$)$_3$Sb, 3, 671
$SbC_{24}H_{50}Cl$ (n-$C_{12}H_{25}$)$_2$SbCl, 3, 688
$SbC_{24}H_{51}$ $(C_8H_{17})_3Sb$, 3, 667
(n-C_5H_{11}CHEt)$_3$Sb, 3, 671
$SbC_{25}H_{19}O_2$ Ph(2-PhC_6H_4)(4-$HO_2CC_6H_4$)Sb, 3, 677
$SbC_{25}H_{19}O_3$ Ph(2-PhC_6H_4)(4-$HO_2CC_6H_4$)-SbO, 3, 744
$SbC_{25}H_{21}$ Ph(2-PhC_6H_4)(4-MeC_6H_4)Sb, 3, 677
$SbC_{25}H_{21}Br_2$ Ph(4-MeC_6H_4)(2-PhC_6H_4)-$SbBr_2$, 3, 735
$SbC_{25}H_{21}Cl_2$ Ph(4-MeC_6H_4)(2-PhC_6H_4)-$SbCl_2$, 3, 735
$SbC_{25}H_{21}O_2$ [Ph_4Sb]O_2CH, 3, 717
$SbC_{25}H_{22}NO_2S$ $Ph_3Sb{=}NO_2SC_6H_4Me$-p, 3, 751, 752
$SbC_{25}H_{25}O_2$ Ph(4-$HO_2CC_6H_4$)(4-$C_6H_{11}C_6H_4$)-Sb, 3, 677
$SbC_{26}H_{17}$ $(PhC{\equiv}C)_2$(1-$C_{10}H_7$)Sb, 3, 676
$SbC_{26}H_{17}Cl_4N_2O_2$ [2-Anthraquinone-N_2]-[Ph_2SbCl_4], 3, 755
$SbC_{26}H_{18}ClO_4$ [2-(4-$HO_2CC_6H_4$)C_6H_4]$_2$-SbCl, 3, 688
$SbC_{26}H_{19}O_4S_3$ (SCH=CHCH=C)$_3$Sb(OBz)$_2$, 3, 742
$SbC_{26}H_{21}O_3$ Ph(2-$HO_2CC_6H_4$)[3,4-Ph(MeO)-C_6H_3]Sb, 3, 678
$SbC_{26}H_{21}O_4$ [Ph_4Sb]O_2CCO_2H, 3, 718
$SbC_{26}H_{23}O_2$ [Ph_4Sb]OAc, 3, 716
$SbC_{26}H_{29}O_4$ $Ph_3Sb(O_2CCHMe_2)_2$, 3, 742
$SbC_{26}H_{33}O_3$ $Ph_3Sb(OBu)OOCMe_3$, 3, 747
$SbC_{26}H_{33}O_4$ $Ph_3Sb(OOCMe_3)_2$, 3, 746
$SbC_{26}H_{51}O_4$ (cyclo-C_6H_{11})$_3$Sb$(OOCMe_3)_2$, 3, 747
$SbC_{26}H_{53}$ (n-$C_{12}H_{25}$)$_2$(CH_2=CH)Sb, 3, 676
$SbC_{27}H_{23}Br_2O_2$ Ph(4-$EtO_2CC_6H_4$)(2-Ph-C_6H_4)$SbBr_2$, 3, 735
$SbC_{27}H_{23}Cl_2O_2$ Ph(4-$EtO_2CC_6H_4$)(2-Ph-C_6H_4)$SbCl_2$, 3, 735
$SbC_{27}H_{23}O_2$ Ph(2-PhC_6H_4)(4-$EtO_2CC_6H_4$)-Sb, 3, 678
$SbC_{27}H_{23}O_3$ Ph_4Sb pyruvate, 3, 718

$SbC_{27}H_{25}O_2$ Ph_4Sb propionate, 3, 718

$SbC_{27}H_{27}Cl_2O_6$ $(4\text{-}EtO_2CC_6H_4)_3SbCl_2$, 3, 735

$SbC_{27}H_{29}Cl_2O_2$ $Ph(4\text{-}EtO_2CC_6H_4)(4\text{-}C_6H_{11}\text{-}C_6H_4)SbCl_2$, 3, 736

$SbC_{27}H_{33}$ $(2,4,6\text{-}Me_3C_6H_2)_3Sb$, 3, 667

$[3\text{-}(Me_2CH)C_6H_4]_3Sb$, 3, 667

$SbC_{27}H_{51}$ $(3,3,5\text{-}Me_3\text{-cyclohexyl})_3Sb$, 3, 671

$SbC_{28}H_{27}O_2$ $[Ph_4Sb]O_2CPr$, 3, 717

$[Ph_4Sb]O_2CCHMe_2$, 3, 718

$SbC_{28}H_{28}Br$ $[(4\text{-}MeC_6H_4)_4Sb]Br$, 3, 719

$SbC_{28}H_{28}Cl$ $[(4\text{-}MeC_6H_4)_4Sb]Cl$, 3, 719

$SbC_{28}H_{28}ClO_4$ $[(4\text{-}MeC_6H_4)_4Sb]ClO_4$, 3, 721

$SbC_{28}H_{28}F$ $[(4\text{-}MeC_6H_4)_4Sb]F$, 3, 720

$SbC_{28}H_{28}NO_3$ $[(4\text{-}MeC_6H_4)_4Sb]NO_3$, 3, 721

$SbC_{28}H_{29}O$ $[(4\text{-}MeC_6H_4)_4Sb]OH$, 3, 720

$SbC_{29}H_{29}O_2$ Ph_4Sb isovalerate, 3, 718

Ph_4Sb valerate, 3, 718

$[(4\text{-}MeC_6H_4)_4Sb]O_2CH$, 3, 720

$SbC_{29}H_{39}O_4$ $(PhCH_2)_3Sb(OOCMe_3)_2$, 3, 747

$SbC_{30}H_{21}$ $(C_{10}H_7)_3Sb$ isomers, 3, 668, 671

5-Ph-5,5′-spirobi(dibenzostibole), 3, 806

$SbC_{30}H_{21}Br_2$ $(1\text{-}C_{10}H_7)_3SbBr_2$, 3, 736

$SbC_{30}H_{21}Cl_2$ $(2\text{-}C_{10}H_7)_3SbCl_2$, 3, 736

$SbC_{30}H_{21}Cl_4N_4$ $[2\text{-}C_{10}H_7N_2^{+}]_2[2\text{-}C_{10}H_7\text{-}SbCl_4]^{--}$, 3, 696

$SbC_{30}H_{21}F_2$ $(1\text{-}C_{10}H_7)_3SbF_2$, 3, 736

$SbC_{30}H_{21}O$ $(1\text{-}C_{10}H_7)_3SbO$, 3, 744

$SbC_{30}H_{23}$ $(2\text{-}PhC_6H_4)_2PhSb$, 3, 676

$SbC_{30}H_{23}Cl_2$ $Ph(2\text{-}PhC_6H_4)_2SbCl_2$, 3, 728

$SbC_{30}H_{24}NO_3$ $[Ph_4Sb]OC_6H_4NO_2\text{-}p$, 3, 718

$SbC_{30}H_{25}$ Ph_5Sb, 3, 810

$SbC_{30}H_{25}O_3S$ $[Ph_4Sb]O_3SPh$, 3, 716

$SbC_{30}H_{27}O_7$ Ph_4Sb citrate, 3, 717

$SbC_{30}H_{28}Cl_3O_2$ $[(4\text{-}MeC_6H_4)_4Sb]O_2CCCl_3$, 3, 721

$SbC_{30}H_{29}Cl_2O_2$ $[(4\text{-}MeC_6H_4)_4Sb]O_2CCHCl_2$, 3, 720

$SbC_{30}H_{29}O_4$ $[(4\text{-}MeC_6H_4)_4Sb]O_2CCO_2H$, 3, 721

$SbC_{30}H_{30}ClO_2$ $[(4\text{-}MeC_6H_4)_4Sb]O_2CCH_2Cl$, 3, 719

$SbC_{30}H_{31}O_2$ $[(4\text{-}MeC_6H_4)_4Sb]OAc$, 3, 719

$SbC_{30}H_{33}O_4$ $Ph_3Sb(OOCHCH{=}CH(CH_2)_2CH_2)_2$, 3, 746

$SbC_{30}H_{42}N_3$ $(4\text{-}Et_2NC_6H_4)_3Sb$, 3, 671

$SbC_{30}H_{63}$ $(C_{10}H_{21})_3Sb$, 3, 671

$SbC_{31}H_{23}Br_2Cl_2$ $4\text{-}MeC_6H_4[5,2\text{-}Br(Ph)\text{-}C_6H_3]_2SbCl_2$, 3, 736

$SbC_{31}H_{25}Cl_2$ $4\text{-}MeC_6H_4(2\text{-}PhC_6H_4)_2SbCl_2$, 3, 736

$SbC_{31}H_{25}O_2$ $[Ph_4Sb]OBz$, 3, 717

$SbC_{31}H_{25}O_3$ Ph_4Sb salicylate, 3, 718

$SbC_{31}H_{32}BrO_2$ $[(4\text{-}MeC_6H_4)_4Sb]O_2CCH_2\text{-}CH_2Br$, 3, 719

$[(4\text{-}MeC_6H_4)_4Sb]O_2CCHBrMe$, 3, 719

$SbC_{31}H_{32}ClO_2$ $[(4\text{-}MeC_6H_4)_4Sb]O_2CCH_2\text{-}CH_2Cl$, 3, 719

$[(4\text{-}MeC_6H_4)_4Sb]O_2CCHClMe$, 3, 719

$SbC_{31}H_{33}O_2$ $(4\text{-}MeC_6H_4)_4Sb$ propionate, 3, 721

$SbC_{31}H_{33}O_3$ $(4\text{-}MeC_6H_4)_4Sb$ lactate, 3, 720

$SbC_{32}H_{25}O_4$ Ph_4Sb phthalate, 3, 718

$Ph_3Sb(OBz)_2$, 3, 742

$SbC_{32}H_{27}O$ 5,5,5-EtO(Ph)(2-PhC_6H_4)-dibenzostibole, 3, 805

$SbC_{32}H_{27}O_2$ Ph_4Sb toluate isomers, 3, 718

2,2,2,4,5-Ph_5-2,2-dihydro-1,3,2-dioxastibolane, 3, 807

$SbC_{32}H_{31}O_4$ $(4\text{-}MeC_6H_4)_4Sb$ fumarate, 3, 720

$(4\text{-}MeC_6H_4)_4Sb$ maleate, 3, 720

$SbC_{32}H_{35}O_2$ $[(4\text{-}MeC_6H_4)_4Sb]O_2CPr$, 3, 719

$[(4\text{-}MeC_6H_4)_4Sb]O_2CHMe_2$, 3, 720

$SbC_{33}H_{25}Br_2Cl_2O_2$ $(4\text{-}EtO_2CC_6H_4)[2,4\text{-}Br(Ph)C_6H_3]_2SbCl_2$, 3, 736

$SbC_{33}H_{31}O_3$ $[(4\text{-}MeC_6H_4)_4Sb]$ furoate, 3, 720

$SbC_{33}H_{33}O_4$ $(4\text{-}MeC_6H_4)_4Sb$ mesaconate, 3, 720

$SbC_{33}H_{37}O_2$ $(4\text{-}MeC_6H_4)_4Sb$ isovalerate, 3, 720

$(4\text{-}MeC_6H_4)_4Sb$ valerate, 3, 720

$SbC_{34}H_{25}$ Pentaphenylstibole, 3, 800

$SbC_{34}H_{25}O$ Pentaphenylstibole oxide, 3, 800

$SbC_{34}H_{30}N_3O_7$ $(4\text{-}MeC_6H_4)_4Sb$ picrate, 3, 721

$SbC_{34}H_{31}Cl_2O_3S$ $[(4\text{-}MeC_6H_4)_4Sb]O_3S\text{-}C_6H_3Cl_2\text{-}2,4$, 3, 720

$SbC_{34}H_{32}NO_5S$ $[(4\text{-}MeC_6H_4)_4Sb]O_3SC_6H_4\text{-}NO_2\text{-}m$, 3, 721

$SbC_{34}H_{33}O_3S$ $[(4\text{-}MeC_6H_4)_4Sb]O_3SPh$, 3, 719

$SbC_{35}H_{27}O_2$ Ph_4Sb α-naphthoate, 3, 718

$SbC_{35}H_{31}Cl_2O_2$ $4\text{-}EtO_2CC_6H_4[4,2\text{-}Me(Ph)\text{-}C_6H_3]_2SbCl_2$, 3, 736

$SbC_{35}H_{31}Cl_2O_4$ $4\text{-}EtO_2CC_6H_4[2\text{-}(4\text{-}MeO\text{-}C_6H_4)C_6H_4]_2SbCl_2$, 3, 736

$SbC_{35}H_{32}BrO_2$ $[(4\text{-}MeC_6H_4)_4Sb]O_2CC_6H_4\text{-}Br$ isomers, 3, 719

$SbC_{35}H_{32}ClO_2$ $[(4\text{-}MeC_6H_4)_4Sb]O_2CC_6H_4\text{-}Cl$ isomers, 3, 719

$SbC_{35}H_{32}IO_2$ $[(4\text{-}MeC_6H_4)_4Sb]O_2CC_6H_4I$ isomers, 3, 720

$SbC_{35}H_{32}NO_4$ $[(4\text{-}MeC_6H_4)_4Sb]O_2CC_6H_4\text{-}NO_2\text{-}m$, 3, 721

$SbC_{35}H_{33}ClNO_5S$ $[(4\text{-}MeC_6H_4)_4Sb]O_3S\text{-}C_6H_2(Cl)(Me)NO_2$, 3, 719

$SbC_{35}H_{33}O_2$ $[(4\text{-}MeC_6H_4)_4Sb]OBz$, 3, 719

$SbC_{35}H_{33}O_3$ $[(4\text{-}MeC_6H_4)_4Sb]O_2CC_6H_4OH$ isomers, 3, 720

$SbC_{35}H_{33}O_6S$ $[(4\text{-}MeC_6H_4)_4Sb]O_2CC_6H_3\text{-}(SO_3H)OH$, 3, 721

$SbC_{35}H_{35}O_3S$ $[(4\text{-}MeC_6H_4)_4Sb]O_3SC_6H_4\text{-}Me\text{-}p$, 3, 721

$SbC_{36}H_{21}O_3$ Tris(dibenzofuryl)stibine, 3, 671

$SbC_{36}H_{27}$ $(PhC_6H_4)_3Sb$ isomers, 3, 668, 671

$SbC_{36}H_{27}Br_2$ $(2\text{-}PhC_6H_4)_3SbBr_2$, 3, 736

$SbC_{36}H_{27}Br_2O_3$ $(4\text{-}PhOC_6H_4)_3SbBr_2$, 3, 736

$SbC_{36}H_{27}Cl_2$ $(2\text{-}PhC_6H_4)_3SbCl_2$, 3, 737

$SbC_{36}H_{27}Cl_2O_3$ $(4\text{-}PhOC_6H_4)_3SbCl_2$, 3, 737

$SbC_{36}H_{27}I_2O_3$ $(4\text{-}PhOC_6H_4)_3SbI_2$, 3, 737

$SbC_{36}H_{27}O_3$ $(4\text{-}PhOC_6H_4)_3Sb$, 3, 671

$SbC_{36}H_{29}O_2$ $(2\text{-}PhC_6H_4)_3Sb(OH)_2$, 3, 740

$SbC_{36}H_{33}O_4$ $(4\text{-}MeC_6H_4)_4Sb$ phthalate, 3, 721

$SbC_{36}H_{33}O_6$ Ph_3Sb bis(isochromanyl peroxide), 3, 746

$SbC_{36}H_{35}O_2$ $(4\text{-}MeC_6H_4)_4Sb^-$ toluate isomers, 3, 721

$SbC_{36}H_{35}O_3$ $[(4\text{-}MeC_6H_4)_4Sb]O_2CC_6H_4\text{-}OMe\text{-}p$, 3, 720

$SbC_{36}H_{36}N_3$ $\{4\text{-}[M\overline{eC=CHCH=C(Me)N}]\text{-}C_6H_4\}_3Sb$, 3, 668, 671

$SbC_{36}H_{37}O_4$ $Ph_3Sb(OOCMe_2Ph)_2$, 3, 746

$SbC_{36}H_{75}$ $(C_{12}H_{25})_3Sb$, 3, 668

$SbC_{37}H_{33}Cl_2O_4$ $4\text{-}MeC_6H_4[2\text{-}(4\text{-}EtO_2C\text{-}C_6H_4)C_6H_4]_2SbCl_2$, 3, 737

$SbC_{37}H_{35}O_4$ $[(4\text{-}MeC_6H_4)_4Sb]O_2CC_6H_4\text{-}OAc\text{-}o$, 3, 719

$SbC_{38}H_{37}O_3$ $[(4\text{-}MeC_6H_4)_4Sb]O_2CCH=CH\text{-}C_6H_4OMe$ isomers, 3, 720, 721

$SbC_{38}H_{37}O_4$ Ph_3Sb bis(tetrahydronaphthyl peroxide), 3, 747

$SbC_{38}H_{39}O_2$ $[(4\text{-}MeC_6H_4)_4Sb]O_2CC_6H_2Me_3$, 3, 721

$SbC_{39}H_{37}Cl_2O_6$ $4\text{-}MeC_6H_4[2\text{-}(4\text{-}EtO_2CCH_2O\text{-}C_6H_4)C_6H_4]_2SbCl_2$, 3, 737

$SbC_{41}H_{43}O_4$ $(PhCH_2)_3Sb$ bis(tetrahydronaphthyl peroxide), 3, 747

$SbC_{42}H_{87}$ $(C_{14}H_{29})_3Sb$, 3, 671

$SbC_{48}H_{28}N_8$ $[Ph_4Sb][(NC)_2CC_6H_4C(CN)_2]_2$, 3, 718

$SbC_{48}H_{99}$ $(C_{16}H_{33})_3Sb$, 3, 668

$SbC_{59}H_{51}O_4$ $(PhCH_2)_3Sb(OOCPh_3)_2$, 3, 747

$SbC_{60}H_{123}$ $(C_{20}H_{41})_3Sb$, 3, 671

$Sb_2C_4F_{12}$ $(CF_3)_2SbSb(CF_3)_2$, 3, 681

$Sb_2C_4H_{12}$ $Me_2SbSbMe_2$, 3, 681

$Sb_2C_6H_8O_6$ $p\text{-}C_6H_4(SbO_3H_2)_2$, 3, 797

$Sb_2C_6H_{12}$ $Me_2SbC\equiv CSbMe_2$, 3, 678

$Sb_2C_6H_{18}Cl_2O$ $(Me_3SbCl)_2O$, 3, 748*

$Sb_2C_6H_{18}Cl_2O_9$ $(Me_3SbClO_4)_2O$, 3, 748*

$Sb_2C_6H_{18}Se_4$ 3,3,3,6,6,6-Me_6-3,3,6,6-tetrahydro-1,2,4,5,3,6-tetraseladiantimonane, 3, 814

$Sb_2C_7H_{15}I$ $[Me_2SbC\equiv CSbMe_3]I$, 3, 714

$Sb_2C_8H_{20}Br_4$ $[EtSbBr]^+[Et_3SbBr_3]^-$, 3, 707

$Sb_2C_{10}H_{12}O_{10}S_2$ $2,4\text{-}(HO_2CCH_2S)_2C_6H_2\text{-}1,5\text{-}(SbO_3H_2)_2$, 3, 797

$Sb_2C_{10}H_{16}$ $C_6H_4(SbMe_2)_2$ isomers, 3, 678

$Sb_2C_{10}H_{20}$ $Et_2SbC\equiv CSbEt_2$, 3, 678

$Sb_2C_{12}H_8O_4S$ 4,4'-$SO_2(C_6H_4SbO)_2$, 3, 703

$Sb_2C_{12}H_{12}O_6$ p-$(C_6H_4SbO_3H_2)_2$, 3, 797

$Sb_2C_{12}H_{12}O_6S$ 4,4'-$S(C_6H_4SbO_3H_2)_2$, 3, 797

$Sb_2C_{12}H_{12}O_6S_2$ p-$(-SC_6H_4SbO_3H_2)_2$, 3, 797

$Sb_2C_{12}H_{12}O_8S$ 4,4'-$SO_2(C_6H_4SbO_3H_2)_2$, 3, 797

$Sb_2C_{12}H_{26}I_2$ $[MeEt_2SbC{\equiv}CSbEt_2Me]I_2$, 3, 716

$Sb_2C_{13}H_{11}NO_6$ Acridine-3,6-distibonic acid, 3, 797

$Sb_2C_{13}H_{14}N_2O_6S$ 4,4'-$CS(NHC_6H_4SbO_3H_2)_2$, 3, 785, 797

$Sb_2C_{13}H_{14}O_6$ 4,4'-$CH_2(C_6H_4SbO_3H_2)_2$, 3, 797

$Sb_2C_{14}H_{16}O_8$ 3,3'-Dimethoxy-p-biphenyldistibonic acid, 3, 798

$Sb_2C_{14}H_{28}$ $(Me_2CH)_2SbC{\equiv}CSb(CHMe_2)_2$, 3, 679

$Sb_2C_{16}H_{20}N_2O_6$ N,N'-Bis(p-stibonophenyl)piperazine, 3, 798

$Sb_2C_{16}H_{34}I_2$ $[Me(Me_2CH)_2SbC{\equiv}]_2I_2$, 3, 716

$Sb_2C_{17}H_{20}N_2O_8$ $EtCH(CONHC_6H_4SbO_3H_2)_2$, 3, 798

$Sb_2C_{18}H_{36}$ $(Me_3C)_2SbC{\equiv}CSb(CMe_3)_2$, 3, 679

$Sb_2C_{19}H_{24}N_2O_8$ $Et_2C(CONHC_6H_4SbO_3H_2)_2$, 3, 798

$Sb_2C_{20}H_{20}$ $(C_5H_5)_2SbSb(C_5H_5)_2$, 3, 682

$Sb_2C_{20}H_{26}N_4O_8$ N,N'-Bis(p-stibonophenylcarbamoylmethyl)piperazine, 3, 798

$Sb_2C_{21}H_{20}N_4O_8S_2$ 3,5-$(H_2O_3SbC_6H_4NHCS\text{-}NH)_2C_6H_3CO_2H$ isomers, 3, 799

$Sb_2C_{24}H_{16}Cl_4O$ $[(4\text{-}ClC_6H_4)_2Sb]_2O$, 3, 699

$Sb_2C_{24}H_{16}F_4O$ $[(4\text{-}FC_6H_4)_2Sb]_2O$, 3, 699

$Sb_2C_{24}H_{16}N_4O_9$ $[(4\text{-}O_2NC_6H_4)_2Sb]_2O$, 3, 699*

$Sb_2C_{24}H_{20}$ $Ph_2SbSbPh_2$, 3, 682

$Sb_2C_{24}H_{20}Cl_4O$ $(Ph_2SbCl_2)_2O$, 3, 766

$Sb_2C_{24}H_{20}O$ $(Ph_2Sb)_2O$, 3, 699

$Sb_2C_{24}H_{20}S$ $(Ph_2Sb)_2S$, 3, 704

$Sb_2C_{26}H_{16}Cl_4$ $(4\text{-}ClC_6H_4)_2SbC{\equiv}CSb\text{-}(C_6H_4Cl\text{-}4)_2$, 3, 679

$Sb_2C_{26}H_{20}$ $Ph_2SbC{\equiv}CSbPh_2$, 3, 679

$Sb_2C_{28}H_{20}$ $(Ph_2SbC{\equiv}C\text{-})_2$, 3, 679

$Sb_2C_{28}H_{28}O$ $[(4\text{-}MeC_6H_4)_2Sb]_2O$, 3, 699

$Sb_2C_{28}H_{28}O_3$ $[(PhCH_2)_2SbO]_2O$, 3, 766

$Sb_2C_{30}H_{24}$ m-$C_6H_4(SbPh_2)_2$, 3, 679

$Sb_2C_{30}H_{28}$ $(4\text{-}MeC_6H_4)_2SbC{\equiv}CSb(C_6H_4\text{-}Me\text{-}4)_2$, 3, 679

$Sb_2C_{32}H_{36}O_5$ 2,2'-Oxybis(2,2-dihydro-4,5-dimethyl-2,2-diphenyl-1,3,2-dioxastibolane), 3, 807

$Sb_2C_{36}H_{30}Br_2O$ $(Ph_3SbBr)_2O$, 3, 748

$Sb_2C_{36}H_{30}Cl_2O$ $(Ph_3SbCl)_2O$, 3, 748

$Sb_2C_{36}H_{30}N_2O_7$ $(Ph_3SbNO_3)_2O$, 3, 748*

$Sb_2C_{36}H_{44}O_5$ 2,2'-Oxybis(2,2-dihydro-4,4,5,5-tetramethyl-2,2-diphenyl-1,3,2-dioxastibolane), 3, 807

$Sb_2C_{36}H_{66}Cl_2O$ $[(cyclo\text{-}C_6H_{11})_3SbCl]_2O$, 3, 748

$Sb_2C_{39}H_{30}Cl_6O_6S_3$ (2,4,6,6-Cl_4-s-trithiane-2,4-diyl)bis[triphenylstibonium chloride]S,S,S',S',S'',S''-hexaoxide, 3, 721

$Sb_2C_{42}H_{42}Cl_2O$ $[(PhCH_2)_3SbCl]_2O$, 3, 748

$Sb_2C_{44}H_{48}O_5$ $(Ph_3SbOOCMe_3)_2O$, 3, 747

$Sb_2C_{48}H_{40}NO_7S_2$ $[Ph_4Sb]_2ON(SO_3)_2$, 3, 719

$Sb_2C_{48}H_{40}O_4S$ $[Ph_4Sb]_2SO_4$, 3, 719

$Sb_2C_{56}H_{56}O_4S$ $[(4\text{-}MeC_6H_4)_4Sb]_2SO_4$, 3, 721

$Sb_3C_6H_{18}N$ $(Me_2Sb)_3N$, 3, 706

$Sb_3C_{11}H_{27}$ $(Me_2SbCH_2CH_2CH_2)_2SbMe$, 3, 680

$Sb_3C_{21}H_{45}O_4$ $(n\text{-}C_7H_{15})_3SbO \cdot Sb_2O_3$, 3, 744

$Sb_3C_{27}H_{27}N_6O_9S_3$ 1,3,5-$C_6H_3(NHCSNHC_6H_4\text{-}SbO_3H_2)_3$, 3, 799

$SbAlC_6H_{15}Cl_3$ $Et_3Sb \cdot AlCl_3$, 3, 656

$SbAlC_8H_{20}Cl_4$ $[Et_4Sb]AlCl_4$, 3, 714

$SbAlC_9H_{21}I_3$ $Pr_3Sb \cdot AlI_3$, 3, 658

$SbAlC_9H_{24}$ $Me_3Sb \cdot AlEt_3$, 3, 654

$SbAlC_{12}H_{28}I_4$ $[Pr_4Sb]AlI_4$, 3, 716

$SbAuC_{14}H_{20}$ $PhC{\equiv}CAu \cdot SbEt_3$, 3, 656

$SbBC_2H_8$ Me_2SbBH_2, 3, 706

$SbBC_3H_{12}$ $Me_3Sb \cdot BH_3$, 3, 854

$SbBC_{19}H_{18}F_4$ $[MePh_3Sb]BF_4$, 3, 716

$SbBC_{23}H_{30}$ $[Me_4Sb][BMePh_3]$, 3, 714

$SbBC_{24}H_{20}F_4$ $[Ph_4Sb]BF_4$, 3, 718
$SbBC_{48}H_{40}$ $[Ph_4Sb][BPh_4]$, 3, 718
$SbCoC_{18}H_{15}BrN_2O_2$ $Co(SbPh_3)(NO)_2Br$, 3, 662
$SbCoC_{20}H_{12}Cl_3NO_3$ $Co[Sb(C_6H_4Cl\text{-}p)_3]\text{-}(NO)(CO)_2$, 3, 660
$SbCoC_{20}H_{15}NO_3$ $Co(SbPh_3)(CO)_2(NO)$, 3, 662
$SbCoC_{23}H_{15}F_5O_3$ $C_2F_5Co(CO)_3(SbPh_3)$, 1, 566, 3, 662
$SbCoC_{23}H_{21}NO_3$ $Co[Sb(C_6H_4Me)_3](CO)_2\text{-}(NO)$, 3, 667
$SbCrC_{23}H_{15}O_5$ $Cr(SbPh_3)(CO)_5$, 3, 662
$SbCuC_{20}H_{17}BrCl_2NO$ $Ph_3Sb{=}NCOCH_2Br\cdot Cu\text{-}Cl_2$, 3, 751
$SbCuC_{20}H_{17}Cl_2INO$ $Ph_3Sb{=}NCOCH_2I\cdot CuCl_2$, 3, 751
$SbCuC_{20}H_{17}Cl_3NO$ $Ph_3Sb{=}NCOCH_2Cl\cdot CuCl_2$, 3, 751
$SbCuC_{20}H_{18}Cl_2NO$ $Ph_3Sb{=}NAc\cdot CuCl_2$, 3, 751
$SbCuC_{23}H_{23}Cl_3NO$ $(MeC_6H_4)_3Sb{=}NCOCH_2\text{-}Cl\cdot CuCl_2$ isomers, 3, 751
$SbCuC_{23}H_{24}Cl_2NO$ $(MeC_6H_4)_3Sb{=}NAc\cdot Cu\text{-}Cl_2$ isomers, 3, 751
$SbCuC_{24}H_{20}Cl_2NO_2S$ $Ph_3Sb{=}NO_2SPh\cdot CuCl_2$, 3, 751
$SbCuC_{25}H_{22}Cl_2NO_2S$ $Ph_3Sb{=}NSO_2C_6H_4\text{-}Me\text{-}p\cdot CuCl_2$, 3, 751
$SbFeC_{18}H_{15}BrN_2O_2$ $Fe(SbPh_3)(NO)_2Br$, 3, 662
$SbFeC_{18}H_{15}IN_2O_2$ $Fe(SbPh_3)(NO)_2I$, 3, 663
$SbFeC_{19}H_{15}N_2O_3$ $Fe(SbPh_3)(CO)(NO)_2$, 3, 662
$SbFeC_{21}H_{15}I_2O_3$ $Fe(SbPh_3)(CO)_3I_2$, 3, 662
$SbFeC_{22}H_{15}O_4$ $Fe(SbPh_3)(CO)_4$, 3, 662
$SbFeC_{22}H_{21}N_2O_3$ $Fe[Sb(C_6H_4Me)_3](CO)\text{-}(NO)_2$, 3, 667
$SbFeC_{28}H_{23}O_2$ $[\overline{CH{=}CH(CH{=}CH)_2CH{=}CH}]Fe\text{-}(CO)_2(SbPh_3)$, 1, 501, 3, 662
$SbFeC_{39}H_{30}O_3P$ $Fe(SbPh_3)(PPh_3)(CO)_3$, 3, 662
$SbFe_2C_{35}H_{25}O_5S_2$ $Fe_2(Ph_3Sb)(CO)_5(SPh)_2$, 3, 663
$SbGaC_6H_{18}$ $Me_3Sb\cdot GaMe_3$, 3, 655
$SbHgC_4H_{12}I_3$ $[Me_4Sb]HgI_3$, 3, 714
$SbHgC_8H_6BrO_3$ $4,2\text{-}Br(AcOHg)C_6H_3SbO$, 3, 702
$SbHgC_8H_7O_3$ $4\text{-}(AcOHg)C_6H_4SbO$, 3, 702
$SbHgC_9H_9O_3$ $2,4\text{-}Me(AcOHg)C_6H_3SbO$, 3, 703
$SbHgC_{20}H_{17}BrCl_2NO$ $Ph_3Sb{=}NCOCH_2Br\cdot Hg\text{-}Cl_2$, 3, 751
$SbHgC_{20}H_{17}Cl_2INO$ $Ph_3Sb{=}NCOCH_2I\cdot HgCl_2$, 3, 751
$SbHgC_{20}H_{17}Cl_3NO$ $Ph_3Sb{=}NCOCH_2Cl\cdot HgCl_2$, 3, 751
$SbHgC_{20}H_{18}Cl_2NO$ $Ph_3Sb{=}NAc\cdot HgCl_2$, 3, 751
$SbHgC_{23}H_{23}Cl_3NO$ $(MeC_6H_4)_3Sb{=}NCOCH_2\text{-}Cl\cdot HgCl_2$ isomers, 3, 751
$SbHgC_{23}H_{24}Cl_2NO$ $(MeC_6H_4)_3Sb{=}NAc\cdot HgCl_2$ isomers, 3, 751
$SbHgC_{24}H_{20}Cl_2NO_2S$ $Ph_3Sb{=}NO_2SPh\cdot HgCl_2$, 3, 751
$SbHgC_{24}H_{20}Cl_3$ $[Ph_4Sb]HgCl_3$, 3, 718
$SbHgC_{25}H_{22}Cl_2NO_2S$ $Ph_3Sb{=}NO_2SC_6H_4\text{-}Me\text{-}p\cdot HgCl_2$, 3, 751
$SbInC_{21}H_{24}$ $Ph_3Sb\cdot InMe_3$, 3, 662
$SbLiC_{36}H_{30}$ $Li[SbPh_6]$, 3, 711
$SbMnC_{18}H_{15}N_3O_3$ $Mn(SbPh_3)(NO)_3$, 3, 663
$SbMnC_{18}H_{30}O_5$ $(n\text{-}C_8H_{17}C_5H_4)Mn(CO)_2\text{-}[Sb(OMe)_3]$, 1, 170
$SbMnC_{21}H_{33}FO_3$ $Mn[Sb(C_6H_{11})_3](CO)_3F$, 3, 666
$SbMnC_{22}H_{15}BrO_4$ $[Mn(SbPh_3)(CO)_4]Br$, 3, 663
$SbMnC_{22}H_{15}IO_4$ $[Mn(SbPh_3)(CO)_4I]$, 3, 663
$SbMnC_{22}H_{15}O_4$ $Mn(SbPh_3)(CO)_4$, 3, 663
$SbMnC_{24}H_{21}BrO_3$ $Mn[Sb(C_6H_4Me)_3](CO)_3Br$, 3, 667
$SbMnC_{25}H_{20}O_2$ $(C_5H_5)Mn(CO)_2(SbPh_3)$, 1, 170, 3, 664
$SbMnC_{29}H_{48}O_5$ $(BuC_5H_4)Mn(CO)_2[Sb\text{-}(OC_6H_{11})_3]$, 1, 170
$SbMnC_{37}H_{68}O_2$ $(n\text{-}C_{12}H_{25}C_5H_4)Mn(CO)_2\text{-}[Sb(C_6H_{13})_3]$, 1, 170, 3, 666
$SbMnC_{42}H_{54}O_2$ $(Bu_2C_5H_3)Mn(CO)_2[Sb\text{-}(C_6H_2Me_3)_3]$, 1, 171, 3, 667
$SbMnC_{45}H_{82}O_3$ $(AcC_5H_4)Mn(CO)_2[Sb\text{-}(C_{12}H_{25}\text{-}n)_3]$, 1, 170, 3, 668

$SbMnC_{46}H_{62}O_5$ $[Me(C_{17}H_{35})C_5H_3]Mn(CO)_2$-$[Sb(OC_6H_4Me$-p$)_3]$, 1, 171

$SbMnC_{48}H_{42}O_2$ $(MeBuC_5H_3)Mn(CO)_2[Sb$-$(C_6H_4Ph)_3]$, 1, 171

$SbMoC_9H_{10}ClO_5$ $Mo(Et_2SbCl)(CO)_5$, 3, 684

$SbMoC_{11}H_{15}O_5$ $Mo(SbEt_3)(CO)_5$, 3, 656

$SbMoC_{23}H_{15}O_5$ $Mo(SbPh_3)(CO)_5$, 3, 664

$SbNaC_{13}H_{10}N_2O_9S_2$ 4,5-Dicarboxy-2,2-dihydro-2-oxo-2-[4-(thiazol-2-ylsulfamoyl)phenyl]-1,3,2-dioxastibolane Na salt, 3, 806

$SbNa_2C_6H_6NO_5S$ 4-$H_2NO_2SC_6H_4SbO(ONa)_2$, 3, 782

$SbNiC_7H_{10}ClO_3$ $Ni(Et_2SbCl)(CO)_3$, 3, 684

$SbNiC_9H_{15}O_3$ $Ni(SbEt_3)(CO)_3$, 3, 656

$SbNiC_{21}H_{15}O_3$ $Ni(SbPh_3)(CO)_3$, 3, 664

$SbPtC_{11}H_{20}Cl_2N$ $Pt(SbEt_3)(Py)Cl_2$, 3, 657

$SbPtC_{16}H_{30}Cl_2N$ $Pt(SbPr_3)(MeC_6H_4NH_2)$-Cl_2, 3, 658

$SbRhC_{25}H_{23}Cl$ [(Bicyclo[2.2.1]hepta-2,5-diene)$(Ph_3Sb)Rh]Cl$, 1, 625

$SbSiC_5H_{15}O$ $Me_2SbOSiMe_3$, 3, 697

$SbSiC_6H_{18}ClO$ $Me_3Sb(OSiMe_3)Cl$, 3, 738

$SbSiC_7H_{21}O$ $Me_4SbOSiMe_3$, 3, 712

$SbSiC_{11}H_{27}$ $Bu_2SbSiMe_3$, 3, 706

$SbSiC_{15}H_{19}O$ $Ph_2SbOSiMe_3$, 3, 697

$SbSiC_{21}H_{24}ClO$ $Ph_3Sb(OSiMe_3)Cl$, 3, 738

$SbSiC_{27}H_{29}O$ $[Ph_4Sb]OSiMe_3$, 3, 718

$SbSi_2C_9H_{27}O_2$ $Me_3Sb(OSiMe_3)_2$, 3, 744

$SbSi_2C_{24}H_{33}O_2$ $Ph_3Sb(OSiMe_3)_2$, 3, 744

$SbSi_3C_{12}H_{33}$ $(Me_3SiCH_2)_3Sb$, 3, 669

$SbSi_3C_{12}H_{33}Br_2$ $(Me_3SiCH_2)_3SbBr_2$, 3, 731

$SbSi_3C_{30}H_{45}$ $(4$-$MeC_6H_4SiMe_2CH_2)_3Sb$, 3, 671

$SbSnC_{18}H_{25}$ $Et_3SnSbPh_2$, 3, 706

$SbSnC_{21}H_{31}$ $Pr_3SnSbPh_2$, 3, 706

$SbSnC_{24}H_{37}$ $Bu_3SnSbPh_2$, 3, 706

$SbTlC_{12}H_{20}Br_4$ $[(MeCH{=}CH)_4Sb]TlBr_4$ isomers, 3, 715

$[(CH_2{=}CMe)_4Sb]TlBr_4$, 3, 715

$SbZn_2C_6H_{15}Cl_4$ $Et_3Sb(ZnCl_2)_2$, 3, 657

$Sb_2CoC_{37}H_{30}NO_2$ $Co(SbPh_3)_2(CO)(NO)$, 3, 662

$Sb_2Co_2C_{42}H_{30}O_6$ $[Co(SbPh_3)(CO)_3]_2$, 3, 662

$Sb_2Co_2C_{43}H_{30}O_7$ $[Co(SbPh_3)_2(CO)_3]$-$[Co(CO)_4]$, 3, 662

$Sb_2FeC_{39}H_{30}O_3$ $Fe(SbPh_3)_2(CO)_3$, 3, 662

$Sb_2Ge_2C_{66}H_{54}N_2$ m-$C_6H_4(Ph_2Sb{=}NGePh_3)_2$, 2, 128, 3, 752

$Sb_2IrC_{19}H_{42}ClO$ $Ir(SbPr_3)_2(CO)Cl$, 3, 658

$Sb_2IrC_{36}H_{31}Cl_2$ $IrH(SbPh_3)_2Cl_2$, 3, 663

$Sb_2IrC_{37}H_{24}Cl_7O$ $Ir[Sb(C_6H_4Cl$-p$)_3]_2$-$(CO)Cl$, 3, 660

$Sb_2MnC_9H_{18}IO_3$ $Mn(SbMe_3)_2(CO)_3I$, 3, 655

$Sb_2MnC_{12}H_{16}Br_2O_2$ $Mn[o$-$C_6H_4(SbMe_2)_2]$-$(CO)_2Br_2$, 3, 678

$Sb_2MnC_{12}H_{16}Cl_2O_2$ $Mn[o$-$C_6H_4(SbMe_2)_2]$-$(CO)_2Cl_2$, 3, 678

$Sb_2MnC_{13}H_{16}IO_3$ $Mn[o$-$C_6H_4(SbMe_2)_2](CO)_3I$, 3, 678

$Sb_2MnC_{36}H_{30}IN_2O_2$ $Mn(SbPh_3)_2(NO)_2I$, 3, 663

$Sb_2MnC_{39}H_{30}ClO_3$ $Mn(SbPh_3)_2(CO)_3Cl$, 3, 663

$Sb_2MnC_{39}H_{78}BrO_3$ $Mn[Sb(C_6H_{13})_3]_2(CO)_3$-$Br$, 3, 666

$Sb_2MnC_{40}H_{30}NO_3S$ $Mn(SbPh_3)_2(CO)_3SCN$, 3, 663

$Sb_2MnC_{43}H_{73}O$ $(MeC_5H_4)Mn(CO)[Sb$-$(C_6H_{11})_3]_2$, 1, 170, 3, 666

$Sb_2MnC_{48}H_{47}O$ $(C_5H_5)Mn(CO)[Sb(CH_2$-$Ph)_3]_2$, 1, 170, 3, 666

$Sb_2MnC_{75}H_{150}ClO_3$ $Mn[Sb(C_{12}H_{25})_3]_2$-$(CO)_3Cl$, 3, 668

$Sb_2MnC_{83}H_{69}O$ $(MeBuC_5H_3)Mn[Sb(C_6H_4$-$Ph)_3]_2(CO)$, 3, 668

$Sb_2Mn_2C_{42}H_{30}F_2O_6$ $[Mn(SbPh_3)(CO)_3F]_2$, 3, 663

$Sb_2MoC_{12}H_{20}Cl_2O_4$ $Mo(Et_2SbCl)_2(CO)_4$, 3, 684

$Sb_2MoC_{14}H_{25}ClO_4$ $Mo(SbEt_3)(Et_2SbCl)$-$(CO)_4$, 3, 656, 684

$Sb_2MoC_{16}H_{30}O_4$ $Mo(SbEt_3)_2(CO)_4$, 3, 656

$Sb_2MoC_{40}H_{30}O_4$ $Mo(SbPh_3)_2(CO)_4$, 3, 664

$Sb_2NiC_{10}H_{20}O_2$ $Ni(Et_2SbCl)_2(CO)_2$, 3, 684

$Sb_2NiC_{12}H_{25}ClO_2$ $Ni(SbEt_3)(Et_2SbCl)$-

$\mathrm{(CO)_2}$, 3, 657, 684
$\mathrm{Sb_2NiC_{14}H_{30}O_2}$ $\mathrm{Ni(SbEt_3)_2(CO)_2}$, 3, 657
$\mathrm{Sb_2NiC_{36}H_{30}BrNO}$ $\mathrm{Ni(SbPh_3)_2(NO)Br}$, 3, 664
$\mathrm{Sb_2NiC_{36}H_{30}INO}$ $\mathrm{Ni(SbPh_3)_2(NO)I}$, 3, 664
$\mathrm{Sb_2NiC_{38}H_{30}O_2}$ $\mathrm{Ni(SbPh_3)_2(CO)_2}$, 3, 664
$\mathrm{Sb_2PdC_6H_{18}Cl_2}$ $\mathrm{Pd(SbMe_3)_2Cl_2}$, 3, 655
$\mathrm{Sb_2PdC_{12}H_{30}Cl_2}$ $\mathrm{Pd(SbEt_3)_2Cl_2}$, 3, 657
$\mathrm{Sb_2PdC_{18}H_{42}Cl_2}$ $\mathrm{Pd(SbPr_3)_2Cl_2}$, 3, 658
$\mathrm{Sb_2PdC_{36}H_{30}Cl_2}$ $\mathrm{Pd(SbPh_3)_2Cl_2}$, 3, 664
$\mathrm{Sb_2PdC_{36}H_{30}I_2}$ $\mathrm{Pd(SbPh_3)_2I_2}$, 3, 664
$\mathrm{Sb_2PdC_{38}H_{30}N_2S_2}$ $\mathrm{Pd(SbPh_3)_2(SCN)_2}$, 3, 664
$\mathrm{Sb_2Pd_2C_{12}H_{30}Cl_4}$ $\mathrm{[Pd(SbEt_3)Cl_2]_2}$, 3, 657
$\mathrm{Sb_2PtC_{12}H_{18}Cl_2}$ $\mathrm{Pt[Sb(CH{=}CH_2)_3]_2Cl_2}$, 3, 655
$\mathrm{Sb_2PtC_{12}H_{30}Br_2}$ $\mathrm{Pt(SbEt_3)_2Br_2}$, 3, 657
$\mathrm{Sb_2PtC_{12}H_{30}Cl_2}$ $\mathrm{Pt(SbEt_3)_2Cl_2}$, 3, 657
$\mathrm{Sb_2PtC_{12}H_{30}I_2}$ $\mathrm{Pt(SbEt_3)_2I_2}$, 3, 657
$\mathrm{Sb_2PtC_{12}H_{30}N_2O_4}$ $\mathrm{Pt(SbEt_3)_2(NO_2)_2}$, 3, 657
$\mathrm{Sb_2PtC_{18}H_{42}Br_2}$ $\mathrm{Pt(SbPr_3)_2Br_2}$, 3, 658
$\mathrm{Sb_2PtC_{18}H_{42}Cl_2}$ $\mathrm{Pt(SbPr_3)_2Cl_2}$, 3, 658
$\mathrm{Sb_2PtC_{24}H_{54}Cl_2}$ $\mathrm{Pt(SbBu_3)_2Cl_2}$, 3, 659
$\mathrm{Sb_2PtC_{36}H_{30}Cl_2}$ $\mathrm{Pt(SbPh_3)_2Cl_2}$, 3, 664
$\mathrm{Sb_2PtC_{38}H_{30}N_2S_2}$ $\mathrm{Pt(SbPh_3)_2(SCN)_2}$, 3, 664
$\mathrm{Sb_2Pt_2C_6H_{18}Cl_4}$ $\mathrm{[Pt(SbMe_3)Cl_2]_2}$, 3, 655
$\mathrm{Sb_2Pt_2C_{12}H_{30}Cl_4}$ $\mathrm{[Pt(SbEt_3)Cl_2]_2}$, 3, 657
$\mathrm{Sb_2Pt_2C_{18}H_{42}Cl_4}$ $\mathrm{[Pt(SbPr_3)Cl_2]_2}$, 3, 659
$\mathrm{Sb_2ReC_{36}H_{30}Br_3O}$ $\mathrm{ReO(SbPh_3)_2Br_3}$, 3, 664
$\mathrm{Sb_2ReC_{36}H_{30}Cl_3O}$ $\mathrm{ReO(SbPh_3)_2Cl_3}$, 3, 664
$\mathrm{Sb_2RhC_{37}H_{24}Cl_7O}$ $\mathrm{Rh[Sb(C_6H_4Cl\text{-}p)_3]_2\text{-}(CO)Cl}$, 3, 660
$\mathrm{Sb_2RhC_{37}H_{30}ClI_2O}$ $\mathrm{Rh(SbPh_3)_2(CO)ClI_2}$, 3, 665
$\mathrm{Sb_2RhC_{37}H_{30}ClO}$ $\mathrm{Rh(SbPh_3)_2(CO)Cl}$, 3, 665
$\mathrm{Sb_2RhC_{37}H_{30}Cl_3O}$ $\mathrm{Rh(SbPh_3)_2(CO)Cl_3}$, 3, 665
$\mathrm{Sb_2RhC_{37}H_{30}I_3O}$ $\mathrm{Rh(SbPh_3)_2(CO)I_3}$, 3, 665
$\mathrm{Sb_2RhC_{38}H_{30}ClO_2}$ $\mathrm{Rh(SbPh_3)_2(CO)_2Cl}$, 3, 665
$\mathrm{Sb_2RhC_{43}H_{42}ClO}$ $\mathrm{Rh[Sb(C_6H_4Me)_3]_2(CO)Cl}$, 3, 667
$\mathrm{Sb_2RhC_{43}H_{42}Cl_3O}$ $\mathrm{Rh[Sb(C_6H_4Me)_3]_2(CO)\text{-}Cl_3}$, 3, 667
$\mathrm{Sb_2RhC_{52}H_{44}ClN_2O_6}$ $\mathrm{[(p\text{-}MeOC_6H_4NC)_2Rh\text{-}(SbPh_3)_2]ClO_4}$, 1, 628
$\mathrm{Sb_2RuC_{12}H_{30}Cl_3NO}$ $\mathrm{Ru(SbEt_3)_2(NO)Cl_3}$, 3, 657
$\mathrm{Sb_2RuC_{36}H_{30}Cl_3NO}$ $\mathrm{Ru(SbPh_3)_2(NO)Cl_3}$, 3, 665
$\mathrm{Sb_2RuC_{38}H_{30}I_2O_2}$ $\mathrm{Ru(SbPh_3)_2(CO)_2I_2}$, 3, 665
$\mathrm{(Sb_2SiC_{22}H_{26}N_2)_n}$ $\mathrm{\{N{=}SbMe_2\text{-}m\text{-}C_6H_4\text{-}Sb\text{-}Me_2{=}NSiPh_2\}_n}$, 3, 752
$\mathrm{Sb_2Si_2C_{32}H_{42}N_2O}$ $\mathrm{\{[Me_2PhSb{=}NSiMe_2\text{-}]\text{-}C_6H_4\}_2O}$, 3, 752
$\mathrm{Sb_3MoC_{19}H_{40}ClO_3}$ $\mathrm{Mo(SbEt_3)_2(Et_2SbCl)\text{-}(CO)_3}$, 3, 656, 684
$\mathrm{Sb_3MoC_{21}H_{45}O_3}$ $\mathrm{Mo(SbEt_3)_3(CO)_3}$, 3, 656
$\mathrm{Sb_3MoC_{57}H_{45}O_3}$ $\mathrm{Mo(SbPh_3)_3(CO)_3}$, 3, 664
$\mathrm{Sb_3OsC_{54}H_{45}Br_3}$ $\mathrm{Os(SbPh_3)_3Br_3}$, 3, 664
$\mathrm{Sb_3RhC_{54}H_{45}NO}$ $\mathrm{Rh(SbPh_3)_3(NO)}$, 3, 665
$\mathrm{Sb_3RhC_{56}H_{45}ClO_2}$ $\mathrm{Rh(SbPh_3)_3(CO)_2Cl}$, 3, 665
$\mathrm{Sb_3RhC_{60}H_{52}O_3}$ $\mathrm{AcCH{=}C(Me)ORh(SbPh_3)_3\text{-}(CO)}$, 3, 665
$\mathrm{Sb_3SnC_{60}H_{50}N_3}$ $\mathrm{(Ph_3Sb{=}N)_3SnPh}$, 2, 481, 3, 752
$\mathrm{Sb_4AgC_{72}H_{60}ClO_4}$ $\mathrm{[Ag(SbPh_3)_4]ClO_4}$, 3, 665
$\mathrm{Sb_4Mn_2C_{26}H_{32}O_6}$ $\mathrm{\{Mn[o\text{-}C_6H_4(SbMe_2)_2]\text{-}(CO)_3\}_2}$, 3, 678
$\mathrm{Sb_4NiC_{72}H_{60}}$ $\mathrm{Ni(SbPh_3)_4}$, 3, 664
$\mathrm{Sb_4OsC_{72}H_{60}Br_2}$ $\mathrm{Os(SbPh_3)_4Br_2}$, 3, 664
$\mathrm{Sb_6Mn_2C_{26}H_{54}O_4}$ $\mathrm{\{Mn[(Me_2SbCH_2CH_2CH_2)_2\text{-}SbMe](CO)_2\}_2}$, 3, 680
$\mathrm{SbAlSi_4C_{16}H_{48}O_4}$ $\mathrm{[Me_4Sb][Al(OSiMe_3)_4]}$, 3, 712
$\mathrm{SbFeSi_4C_{16}H_{48}O_4}$ $\mathrm{[Me_4Sb][Fe(OSiMe_3)_4]}$, 3, 713
$\mathrm{SbGaSi_4C_{16}H_{48}O_4}$ $\mathrm{[Me_4Sb][Ga(OSiMe_3)_4]}$, 3, 712

$SbGeMnC_{21}H_{32}I_2O_4$ $I_2(C_8H_{17})GeMn(CO)_4$-[$Sb(CH_2CH{=}CH_2)_3$], 3, 658
$SbMnPbC_{28}H_{20}Br_2O_4$ $Br_2PhPb[Mn(SbPh_3)$-$(CO)_4]$, 3, 663
$SbMnPbC_{28}H_{30}O_4$ $Ph_3PbMn(CO)_4(SbEt_3)$, 3, 656
$SbMnSiC_{58}H_{104}Br_2O_4$ $Br_2PhSiMn[Sb$-$(C_{16}H_{33})_3](CO)_4$, 3, 668
$SbMnSnC_{10}H_{15}Br_3O_4$ $Br_3SnMn(CO)_4(Sb$-$Et_3)$, 3, 656
$SbMnSnC_{34}H_{21}Br_3O_4$ $Br_3SnMn[Sb(C_{10}H_7)_3]$ $(CO)_4$, 3, 668
$SbMnSnC_{37}H_{72}O_4$ $Pr_3SnMn(CO)_4[Sb$-$(C_8H_{17})_3]$, 3, 667
$SbMnSnC_{41}H_{43}BrO_4$ $Br(C_5H_5)_2SnMn(CO)_4$-$[Sb(C_6H_4CHMe_2)_3]$, 3, 667
$Sb_2Co_2HgC_{42}H_{30}O_6$ $Hg[Co(SbPh_3)(CO)_3]_2$, 3, 662
$Sb_2Fe_2HgC_{40}H_{30}N_2O_6$ $Hg[Fe(SbPh_3)(CO)_2$-$(NO)]_2$, 3, 663
$Sb_2Fe_2PtC_{50}H_{40}Cl_6O_4$ $[C_5H_5Fe(CO)_2(Sb$-$Ph_3)]_2PtCl_6$, 1, 209
$Sb_2Mn_2SiC_{70}H_{68}O_8$ $(PhCH_2)_2Si\{Mn[Sb$-$(CH_2CH_2Ph)_3](CO)_4\}_2$, 3, 667
$Sb_2Mn_2SiC_{80}H_{104}O_8$ $(C_{18}H_{37})_2Si[Mn(Sb$-$Ph_3)(CO)_4]_2$, 3, 663
$Sb_3Mn_3SiC_{60}H_{50}O_{12}$ $CH_2{=}CHCH_2Si\{Mn$-$(CO)_4[Sb(C_5H_5)_3]\}_3$, 3, 659
$Sb_3Mn_3SiC_{106}H_{108}O_{12}$ 2-Fluorenyl-Si-$\{Mn(CO)_4[Sb(C_6H_4CHMe_2)_3]\}_3$, 3, 667
$Sb_3Mn_3SnC_{55}H_{88}O_{12}$ $PhCH_2Sn\{Mn(CO)_4$-$[Sb(CH_2CHMe_2)_3]\}_3$, 3, 659
$Sb_3Mn_3SnC_{69}H_{50}O_{12}$ $CH_2{=}CHCH_2Sn[Mn(Sb$-$Ph_3)(CO)_4]_3\}$, 3, 663
$Sb_4AgBC_{72}H_{60}F_4$ $[Ag(SbPh_3)_4]BF_4$, 3, 665
$ScC_4H_{10}Br$ Et_2ScBr, 1, 1
$ScC_4H_{11}O$ Et_2ScOH, 1, 1
$ScC_{10}H_{25}O$ $Et_3Sc{\cdot}Et_2O$, 1, 1
$ScC_{15}H_{15}$ $(C_5H_5)_3Sc$, 1, 3
$SiAsC_5H_{15}O$ $Me_2AsOSiMe_3$, 3, 209
$SiAsC_6H_{18}N$ $Me_2AsN(Me)SiMe_3$, 3, 286
$SiAsC_{20}H_{21}$ $Me_2AsSiPh_3$, 3, 298
$SiAsC_{21}H_{23}O_2$ 4-$[Me(MeO)_2Si]C_6H_4AsPh_2$, 3, 71
$SiAsC_{22}H_{25}$ $Et_2AsSiPh_3$, 3, 299
$SiAsC_{22}H_{26}Br$ $[Ph_3(Me_3SiCH_2)As]Br$, 3, 329
$SiAsC_{28}H_{28}N_3O_7$ $[Ph_3(Me_3SiCH_2)As]OC_6H_2$-$(NO_2)_3$, 3, 329
$SiAs_2C_{38}H_{68}N_2O_2$ $[Bu_3As{=}N]_2Si(C_6H_4$-OMe-$p)_2$, 3, 369
$SiCo_2C_{11}H_{10}O_6$ $(HC{\equiv}CSiMe_3)Co_2(CO)_6$, 1, 605
$SiCo_2C_{17}H_{14}O_6$ $(PhC{\equiv}CSiMe_3)Co_2(CO)_6$, 1, 606
$SiCo_4C_{15}H_{10}O_{10}$ $(HC{\equiv}CSiMe_3)Co_4(CO)_{10}$, 1, 608*
$SiCo_4C_{21}H_{14}O_{10}$ $(PhC{\equiv}CSiMe_3)Co_4(CO)_{10}$, 1, 608
$SiCrC_{12}H_{14}O_3$ $(Me_3SiPh)Cr(CO)_3$, 1, 93, 738
$SiFeC_8H_9NO_4$ $(Me_3SiNC)Fe(CO)_4$, 1, 524
$SiFeC_{10}H_{14}O_2$ $C_5H_5Fe(CO)_2SiMe_3$, 1, 533
$SiFeC_{13}H_{18}$ $FcSiMe_3$, 1, 535
  FcSiHEtMe, 1, 535
$(SiFeC_{14}H_{16}O_2)_n$ $[FcCOCH_2CH_2\dot{S}i(Me)O]_n$, 1, 546
$SiFeC_{14}H_{20}$ $FcSiHEt_2$, 1, 535
  $FcCH_2SiMe_3$, 1, 536
$SiFeC_{14}H_{20}O$ $FcSi(OEt)Me_2$, 1, 535
$SiFeC_{15}H_{20}$ $FcSi(CH_2CH{=}CH_2)Me_2$, 1, 535
$SiFeC_{15}H_{20}O$ $FcCOCH_2SiMe_3$, 1, 536
$SiFeC_{15}H_{22}$ $FcCH(Me)SiMe_3$, 1, 536
$SiFeC_{16}H_{22}O$ $FcCOCH_2CH_2SiMe_3$, 1, 536
$SiFeC_{16}H_{24}$ $FcSiEt_3$, 1, 535
  $Fc(CH_2)_3SiMe_3$, 1, 536
$SiFeC_{16}H_{24}O$ $Fc(CH_2)_4Si(OH)Me_2$, 1, 536
$SiFeC_{17}H_{22}O_3$ MeO_2C-Fcd-$COCH_2SiMe_3$, 1, 541
$SiFeC_{17}H_{24}O$ $FcCO(CH_2)_3SiMe_3$, 1, 536
$SiFeC_{17}H_{26}$ $Fc(CH_2)_4SiMe_3$, 1, 536
  Me-Fcd-$SiEt_3$, 1, 541
$SiFeC_{17}H_{26}O$ $FcCH(OH)(CH_2)_3SiMe_3$, 1, 536
$SiFeC_{18}H_{20}$ $FcSiPhMe_2$, 1, 535
$SiFeC_{18}H_{28}O$ $FcC(OH)(Me)(CH_2)_3SiMe_3$, 1, 536
$SiFeC_{19}H_{19}F_3$ $FcSi(C_6H_4CF_3$-$m)Me_2$, 1, 535
$SiFeC_{19}H_{21}F_9$ $FcSi(CH_2CH_2CF_3)_3$, 1, 535
$SiFeC_{19}H_{30}$ $FcSi(C_6H_{13})EtMe$, 1, 533
  Me_2CH-Fcd-$SiEt_3$, 1, 541
$SiFeC_{20}H_{32}$ Me_3C-Fcd-$SiEt_3$, 1, 541

$SiFeC_{23}H_{30}O$ $FcC(OH)(Ph)(CH_2)_3SiMe_3$, 1, 536
$SiFeC_{28}H_{24}$ $FcSiPh_3$, 1, 535
$SiFeC_{28}H_{48}$ $FcSi(C_6H_{13}\text{-}n)_3$, 1, 535
$SiGeC_2H_3Cl_7$ $Cl_3GeCHClCH_2SiCl_3$, 2, 92
$Cl_3GeCH_2CHClSiCl_3$, 2, 92
$SiGeC_2H_4Cl_6$ $Cl_3GeCH_2CH_2SiCl_3$, 2, 92
$SiGeC_3H_7Cl_5$ $Cl_3GeCH_2CH_2SiMeCl_2$, 2, 92
$(SiGeC_4H_{11}O)_n$ $(Me_3GeCH_2SiO)_n$, 2, 49
$SiGeC_6H_{18}O$ $Me_3GeOSiMe_3$, 2, 106
$SiGeC_6H_{18}O_4S$ $Me_3GeSO_4SiMe_3$, 2, 97
$SiGeC_6H_{18}S$ $Me_3GeSSiMe_3$, 2, 124
$SiGeC_6H_{18}Se$ $Me_3GeSeSiMe_3$, 2, 124
$SiGeC_7H_{20}$ $Me_3GeCH_2SiMe_3$, 2, 49
$SiGeC_7H_{20}O_2$ $Me_3GeCH_2SiMe(OMe)_2$, 2, 49
$SiGeC_7H_{21}N$ $Me_3GeNMe(SiMe_3)$, 2, 128
$SiGeC_8H_{20}$ $Me_3GeCH{=}CHSiMe_3$, 2, 49
$Me_3GeC({=}CH_2)SiMe_3$, 2, 49
$SiGeC_8H_{22}$ $Me_3GeCH_2CH_2SiMe_3$, 2, 49
$SiGeC_9H_{24}O$ $Me_3GeOSiEt_3$, 2, 110
$Et_3GeOSiMe_3$, 2, 111
$SiGeC_{12}H_{30}$ $Et_3GeSiEt_3$, 2, 136
$SiGeC_{13}H_{24}$ $p\text{-}Me_3SiC_6H_4CH_2GeMe_3$, 2, 49
$SiGeC_{15}H_{28}O$ $p\text{-}Et_3GeC_6H_4OSiMe_3$, 2, 49
$SiGeC_{16}H_{30}$ $p\text{-}Et_3GeC_6H_4CH_2SiMe_3$, 2, 49
$SiGeC_{24}H_{30}$ $Ph_3GeSiEt_3$, 2, 136
$SiGeC_{26}H_{30}F_2$ $Ph_3GeCF{=}CFSiEt_3$, 2, 49
$SiGeC_{36}H_{30}$ $Ph_3GeSiPh_3$, 2, 136
$SiGe_3C_{54}H_{45}Br$ $(Ph_3Ge)_3SiBr$, 2, 136
$SiGe_3C_{54}H_{45}Cl$ $(Ph_3Ge)_3SiCl$, 2, 136
$SiGe_3C_{54}H_{46}$ $(Ph_3Ge)_3SiH$, 2, 136
$SiGe_3C_{54}H_{46}O$ $(Ph_3Ge)_3SiOH$, 2, 136
$SiGe_3C_{54}H_{47}N$ $(Ph_3Ge)_3SiNH_2$, 2, 136
$SiGe_3C_{56}H_{50}$ $(Ph_3Ge)_3SiEt$, 2, 136
$SiMnC_9H_9Cl_2O_3$ $MeCl_2SiMn(CO)_3(C_5H_6)$, 1, 184
$SiMnC_{10}H_{15}O_3$ $Me_3SiMn(CO)_3(CH_2{=}CH\text{-}CH{=}CH_2)$, 1, 184
$SiMnC_{28}H_{27}O_3$ $(PhCH_2)_3SiMn(CO)_3\text{-}(CH_2{=}CHCH{=}CH_2)$, 1, 184
$SiMnC_{56}H_{107}O_3$ $(n\text{-}C_{16}H_{33})_3SiMn(CO)_3\text{-}(CH_2{=}CMeCH{=}CH_2)$, 1, 184
$SiMoC_{10}H_{12}O_4$ $Me_2Si(CH{=}CH_2)_2Mo(CO)_4$, 1, 109
$SiOsC_{13}H_{18}$ $C_5H_5OsC_5H_4SiMe_3$, 1, 563
$SiPbC_6H_{18}O$ $Me_3PbOSiMe_3$, 2, 572
$SiPbC_7H_{20}$ $Me_3PbCH_2SiMe_3$, 2, 550
$SiPbC_9H_{24}O$ $Me_3PbOSiEt_3$, 2, 572
$Et_3PbOSiMe_3$, 2, 572
$SiPbC_{12}H_{30}O$ $Et_3PbOSnEt_3$, 2, 572
$SiPbC_{36}H_{30}O$ $Ph_3PbOSiPh_3$, 2, 572
$SiRuC_{13}H_{18}$ $(Me_3SiC_5H_4)RuC_5H_5$, 1, 559
$SiSbC_5H_{15}O$ $Me_2SbOSiMe_3$, 3, 697
$SiSbC_6H_{18}ClO$ $Me_3Sb(OSiMe_3)Cl$, 3, 738
$SiSbC_7H_{21}O$ $Me_4SbOSiMe_3$, 3, 712
$SiSbC_{11}H_{27}$ $Bu_2SbSiMe_3$, 3, 706
$SiSbC_{15}H_{19}O$ $Ph_2SbOSiMe_3$, 3, 697
$SiSbC_{21}H_{24}ClO$ $Ph_3Sb(OSiMe_3)Cl$, 3, 738
$SiSbC_{27}H_{29}O$ $[Ph_4Sb]OSiMe_3$, 3, 718
$(SiSb_2C_{22}H_{26}N_2)_n$ $\{N{=}SbMe_2\text{-}m\text{-}C_6H_4\text{-}Sb\text{-}Me_2{=}NSiPh_2\}_n$, 3, 752
$SiSnC_5H_{14}Cl_2$ $Me_3SnCH_2SiMeCl_2$, 2, 249
$SiSnC_5H_{14}O$ $Me_3SnCH_2Si(O)Me$, 2, 249
$SiSnC_6H_{17}Br$ $Me_3SiCH_2SnMe_2Br$, 2, 301
$SiSnC_6H_{18}O$ $Me_3SnOSiMe_3$, 2, 394
$SiSnC_6H_{18}S$ $Me_3SnSSiMe_3$, 2, 473
$SiSnC_7H_{20}$ $Me_3SiCH_2SnMe_3$, 2, 249
$SiSnC_7H_{20}O_2$ $Me_3SnCH_2SiMe(OMe)_2$, 2, 249
$SiSnC_8H_{19}Cl_3$ $Et_3SnCH_2CH_2SiCl_3$, 2, 250
$SiSnC_9H_{24}O$ $Me_3SnOSiEt_3$, 2, 394
$Et_3SnOSiMe_3$, 2, 394
$SiSnC_{10}H_{26}O$ $Me_3SnO(CH_2)_3SiMe_2Et$, 2, 393
$SiSnC_{11}H_{28}O_3$ $Et_3SnCH_2CH_2Si(OMe)_3$, 2, 250
$SiSnC_{12}H_{22}$ $p\text{-}Me_3SiC_6H_4SnMe_3$, 2, 249
$SiSnC_{12}H_{29}Br$ $Me_3SiCH_2SnBu_2Br$, 2, 301
$SiSnC_{12}H_{30}O$ $Et_3SnOSiEt_3$, 2, 394
$Pr_3SnOSiMe_3$, 2, 394
$SiSnC_{13}H_{24}$ $p\text{-}Me_3SiC_6H_4CH_2SnMe_3$, 2, 249
$SiSnC_{14}H_{29}F_3O_2$ $Bu_2(Me_3SiCH_2)SnO_2CCF_3$, 2, 432
$SiSnC_{15}H_{36}$ $Bu_3SnCH_2SiHMe_2$, 2, 250
$Bu_3SnSiMe_3$, 2, 495
$SiSnC_{15}H_{36}O$ $Pr_3SnOSiEt_3$, 2, 394
$Bu_3SnOSiMe_3$, 2, 394
$SiSnC_{17}H_{40}O$ $Bu_3SnCH_2SiMe_2OEt$, 2, 250
$SiSnC_{18}H_{42}$ $Bu_3SnCH_2CHMeCH_2SiHMe_2$, 2, 250
$SiSnC_{22}H_{26}$ $Ph_3SnCH_2SiMe_3$, 2, 250
$SiSnC_{23}H_{44}O$ $i\text{-}Bu_3SnOSiMe(CH_2CHMe_2)Ph$, 2, 394

$\mathrm{SiSnC_{27}H_{28}}$ $\mathrm{p\text{-}Me_3SiC_6H_4SnPh_3}$, 2, 251

$\mathrm{SiSnC_{28}H_{28}}$ $\mathrm{Ph_2SnCH_2CH_2SiPh_2CH_2CH_2}$, 2, 251

$\mathrm{SiSnC_{36}H_{30}}$ $\mathrm{Ph_3SiSnPh_3}$, 2, 495

$\mathrm{SiSnC_{36}H_{30}O}$ $\mathrm{Ph_3SnOSiPh_3}$, 2, 394

$\mathrm{SiSnC_{38}H_{34}}$ $\mathrm{Ph_3SnCH_2CH_2SiPh_3}$, 2, 250

$\mathrm{SiSn_2C_8H_{24}O_2}$ $\mathrm{(Me_3SnO)_2SiMe_2}$, 2, 394

$\mathrm{SiSn_2C_{28}H_{64}O_2}$ $\mathrm{(i\text{-}Bu_3SnO)_2SiEt_2}$, 2, 394*

$\mathrm{SiSn_2C_{48}H_{40}O_2}$ $\mathrm{(Ph_3SnO)_2SiPh_2}$, 2, 394

$\mathrm{SiSn_2C_{52}H_{48}}$ $\mathrm{(Ph_3SnCH_2CH_2)_2SiPh_2}$, 2, 251

$\mathrm{SiTiC_{13}H_{19}ClO}$ $\mathrm{(C_5H_5)_2Ti(OSiMe_3)Cl}$, 1, 34

$\mathrm{SiTiC_{23}H_{23}ClO}$ $\mathrm{(C_5H_5)_2Ti(OSiMePh_2)Cl}$, 1, 35

$\mathrm{SiTiC_{28}H_{25}ClO}$ $\mathrm{(C_5H_5)_2Ti(OSiPh_3)Cl}$, 1, 35

$\mathrm{SiTi_2C_{80}H_{120}O_8}$ $\mathrm{\{[(n\text{-}C_{12}H_{25})_2C_5H_3]Ti(CO)_4\}_2Si(CH_2Ph)_2}$, 1, 35

$\mathrm{SiWC_{10}H_{12}O_4}$ $\mathrm{Me_2Si(CH{=}CH_2)_2W(CO)_4}$, 1, 126

$\mathrm{SiWC_{16}H_{24}O_3}$ $\mathrm{(EtC_5H_4)W(CO)_3SiEt_3}$, 1, 129

$\mathrm{SiZr_2C_{32}H_{40}O_8}$ $\mathrm{[(MeEtC_5H_3)Zr(CO)_4]_2SiBu_2}$, 1, 41

$\mathrm{Si_2AsC_3H_9Cl_4O_3}$ $\mathrm{MeAsO(OSiMeCl_2)_2}$, 3, 544

$\mathrm{Si_2AsC_5H_{15}Cl_2O_3}$ $\mathrm{MeAsO(OSiMe_2Cl)_2}$, 3, 544

$\mathrm{Si_2AsC_8H_{24}N}$ $\mathrm{Me_2AsN(SiMe_3)_2}$, 3, 287

$\mathrm{Si_2AsC_9H_{15}Cl_4O_3}$ $\mathrm{CH_2{=}CHCH_2AsO[OSiCl_2(CH_2CH{=}CH_2)]_2}$, 3, 545

$\mathrm{Si_2AsC_{11}H_{17}Cl_2N_2O_3}$ $\mathrm{MeAsO[OSiMe(\text{-}NCH{=}CHCH{=}CH)Cl]_2}$, 3, 544

$\mathrm{Si_2AsC_{13}H_{25}Cl_4O_3}$ $\mathrm{MeAsO[OSiCl_2(C_6H_{11})]_2}$, 3, 544

$\mathrm{Si_2AsC_{17}H_{21}Cl_4O_3}$ $\mathrm{MeAsO[OSiCl_2(CH_2CH_2Ph)]_2}$, 3, 545

$\mathrm{Si_2AsC_{18}H_{15}Cl_4O_3}$ $\mathrm{PhAsO(OSiCl_2Ph)_2}$, 3, 546

$\mathrm{Si_2AsC_{23}H_{25}Cl_4O_3}$ $\mathrm{PhCH_2AsO[OSiCl_2(CH_2CH_2Ph)]_2}$, 3, 546

$\mathrm{Si_2AsC_{30}H_{25}Cl_2O_3}$ $\mathrm{PhAsO(OSiClPh_2)_2}$, 3, 546

$\mathrm{Si_2AsC_{37}H_{33}}$ $\mathrm{MeAs(SiPh_3)_2}$, 3, 299

$\mathrm{Si_2AsC_{42}H_{35}O_2}$ $\mathrm{PhAs(OSiPh_3)_2}$, 3, 255

$\mathrm{Si_2AsC_{42}H_{35}O_3}$ $\mathrm{PhAsO(OSiPh_3)_2}$, 3, 546

$\mathrm{Si_2As_2C_{36}H_{30}O_4}$ Hexaphenyl-1,3,5,7-tetraoxa-2,6-diarsa-4,8-disilacyclooctane, 3, 648

$\mathrm{Si_2CoC_{46}H_{38}}$ $\mathrm{(Ph_3SiC_5H_4)_2Co}$, 1, 618

$\mathrm{Si_2Co_2C_{14}H_{18}O_6}$ $\mathrm{(Me_3SiC{\equiv}CSiMe_3)Co_2(CO)_6}$, 1, 606

$\mathrm{Si_2Co_2C_{26}H_{40}O_6}$ $\mathrm{\{[Me_3SiC{=}CHCH{=}C(SiMe_3)CO]Co(CO)_2\}_2}$, 1, 601

$\mathrm{Si_2FeC_{12}H_{16}O_3}$ $\mathrm{MeSi(OH)\text{-}Fcd\text{-}SiMe(OH)O}$, 1, 544

$\mathrm{Si_2FeC_{12}H_{18}O_4}$ $\mathrm{Fcd[SiMe(OH)_2]_2}$, 1, 538

$\mathrm{Si_2FeC_{14}H_{20}Cl_2}$ $\mathrm{Fcd(SiMe_2Cl)_2}$, 1, 538

$\mathrm{Si_2FeC_{14}H_{20}O}$ $\mathrm{Me_2Si\text{-}Fcd\text{-}SiMe_2O}$, 1, 544

$\mathrm{Si_2FeC_{14}H_{20}O_4}$ Bis(trimethylsilyl)cyclopentadienone-$\mathrm{Fe(CO)_3}$, 1, 484, 546

$\mathrm{Si_2FeC_{16}H_{24}Cl_2}$ $\mathrm{FcSiMe_2(CH_2)_3SiCl_2Me}$, 1, 535

$\mathrm{Fcd(SiMe_2CH_2Cl)_2}$, 1, 538

$\mathrm{Si_2FeC_{16}H_{24}O_3}$ $\mathrm{Me(EtO)Si\text{-}Fcd\text{-}Si(OEt)MeO}$, 1, 544

$\mathrm{Si_2FeC_{16}H_{26}}$ $\mathrm{Fcd(SiMe_3)_2}$, 1, 538

$\mathrm{Fcd(SiHMeEt)_2}$, 1, 538

$\mathrm{Si_2FeC_{16}H_{28}N_2}$ $\mathrm{Fcd[Si(CH_2NH_2)Me_2]_2}$, 1, 538

$\mathrm{Si_2FeC_{18}H_{29}Cl}$ $\mathrm{FcSiMe_2(CH_2)_3SiClEtMe}$, 1, 536

$\mathrm{Si_2FeC_{18}H_{30}}$ $\mathrm{Fcd(SiMe_2Et)_2}$, 1, 538

$\mathrm{Fcd(SiHEt_2)_2}$, 1, 538

$\mathrm{Si_2FeC_{18}H_{30}O_2}$ $\mathrm{Fcd[Si(OEt)Me_2]_2}$, 1, 538

$\mathrm{Si_2FeC_{19}H_{32}}$ $\mathrm{Me_3Si\text{-}Fcd\text{-}SiEt_3}$, 1, 541

$\mathrm{Si_2FeC_{20}H_{26}O}$ $\mathrm{FcSiMe_2OSiPhMe_2}$, 1, 535

$(\mathrm{Si_2FeC_{20}H_{28}O_3})_n$ $\mathrm{\{SiMe_2CH_2CH_2CO\text{-}Fcd\text{-}COCH_2CH_2SiMe_2O\}_n}$, 1, 545

$(\mathrm{Si_2FeC_{20}H_{32}O})_n$ $\mathrm{\{SiMe_2(CH_2)_3\text{-}Fcd\text{-}(CH_2)_3SiMe_2O\}_n}$, 1, 545

$\mathrm{Si_2FeC_{20}H_{34}}$ $\mathrm{FcSiMe_2(CH_2)_3SiEt_2Me}$, 1, 536

$\mathrm{Si_2FeC_{20}H_{34}O_4}$ $\mathrm{Fcd[SiMe(OEt)_2]_2}$, 1, 538

$\mathrm{Si_2FeC_{21}H_{36}}$ $\mathrm{FcSiMe_2(CH_2)_3SiEt_3}$, 1, 536

$(\mathrm{Si_2FeC_{22}H_{32}O_3})_n$ $\mathrm{[SiMe_2(CH_2)_3CO\text{-}Fcd\text{-}CO(CH_2)_3SiMe_2O]_n}$, 1, 545

$\mathrm{Si_2FeC_{22}H_{34}O_2}$ $\mathrm{Fcd(COCH_2CH_2SiMe_3)_2}$, 1, 538

$(\mathrm{Si_2FeC_{22}H_{36}O})_n$ $\mathrm{\{SiMe_2(CH_2)_4\text{-}Fcd\text{-}}$

$(CH_2)_4SiMe_2O\}_n$, 1, 546
$Si_2FeC_{22}H_{38}$ $Fcd(SiEt_3)_2$, 1, 538
$Fcd[(CH_2)_3SiMe_3]_2$, 1, 538
$Si_2FeC_{23}H_{40}O$ $FcC(OH)[(CH_2)_3SiMe_3]_2$, 1, 536
$Si_2FeC_{24}H_{30}$ $Fcd(SiMe_2C_5H_5)_2$, 1, 539
$Si_2FeC_{24}H_{38}O_2$ $Fcd[CO(CH_2)_3SiMe_3]_2$, 1, 538
$Si_2FeC_{24}H_{40}N_2$ $Fcd[SiMe_2\overline{N(CH_2)_4C}H_2]_2$, 1, 539
$Si_2FeC_{24}H_{42}$ $FcSiMe_2(CH_2)_3SiBu_2Me$, 1, 536
$Fcd[(CH_2)_4SiMe_3]_2$, 1, 539
$Si_2FeC_{26}H_{28}O_4$ Diphenylbis(trimethylsilyl)cyclopentadienone-$Fe(CO)_3$, 1, 485, 546
$Si_2FeC_{28}H_{28}F_6$ $Fcd[Si(C_6H_4CF_3\text{-}m)Me_2]_2$, 1, 539
$Si_2FeC_{28}H_{32}F_{18}$ $Fcd[Si(CH_2CH_2CF_3)_3]_2$, 1, 746
$Si_2FeC_{28}H_{50}$ $Fcd(SiPr_3)_2$, 1, 539
$Si_2FeC_{32}H_{32}N_2O_4$ $Fcd[SiMe_2CH_2N(CO)_2\text{-}C_6H_4]_2$, 1, 539
$Si_2FeC_{46}H_{38}$ $Fcd(SiPh_3)_2$, 1, 540
$Si_2FeC_{46}H_{86}$ $Fcd[Si(C_6H_{13})_3]_2$, 1, 540
$Si_2FeC_{82}H_{158}$ $Fcd[Si(C_{12}H_{25})_3]_2$, 1, 540
$Si_2Fe_2C_{16}H_{20}O_6$ $(Me_3SiC{\equiv}CH)_2Fe_2(CO)_6$, 1, 518
$Si_2Fe_2C_{28}H_{28}O_6$ $(PhC{\equiv}CSiMe_3)_2Fe_2(CO)_6$, 1, 518
$Si_2Fe_2C_{29}H_{28}O_7$ $(PhC{\equiv}CSiMe_3)_2Fe_2(CO)_7$, 1, 518
$Si_2Fe_2C_{30}H_{38}O_3$ $(FcCOCH_2CH_2SiMe_2)_2O$, 1, 543
$Si_2Fe_2C_{30}H_{42}O$ $[Fc(CH_2)_3SiMe_2]_2O$, 1, 543
$Si_2Fe_2C_{32}H_{42}O_3$ $[FcCO(CH_2)_3SiMe_2]_2O$, 1, 543
$Si_2Fe_2C_{32}H_{46}O$ $[Fc(CH_2)_4SiMe_2]_2O$, 1, 543
$Si_2Fe_2C_{40}H_{46}O_4$ $p\text{-}(Me_3SiCH_2CH_2CO\text{-}Fcd\text{-}CO)_2C_6H_4$, 1, 543
$Si_2Fe_3C_{30}H_{28}O_8$ $(PhC{\equiv}CSiMe_3)_2Fe_3(CO)_8$, 1, 518
$Si_2GeC_8H_{24}O_2$ $Me_2Ge(OSiMe_3)_2$, 2, 111
$Si_2GeC_9H_{27}N$ $Me_3GeN(SiMe_3)_2$, 2, 128
$Si_2Ge_4C_{72}H_{60}$ $(Ph_3Ge)_4Si_2$ (?), 2, 130
$Si_2Ge_6C_{110}H_{94}Br_2Cl_2$ $[(Ph_3Ge)_3SiBr]_2\cdot(CH_2Cl)_2$, 2, 136
$Si_2NiC_{24}H_{30}$ $\{[Me_2(C_5H_5)Si]C_5H_4\}_2Ni$, 1, 653
$Si_2PbC_9H_{27}N$ $Me_3PbN(SiMe_3)_2$, 2, 587
$Si_2SbC_9H_{27}O_2$ $Me_3Sb(OSiMe_3)_2$, 3, 744
$Si_2SbC_{24}H_{33}O_2$ $Ph_3Sb(OSiMe_3)_2$, 3, 744
$Si_2Sb_2C_{32}H_{42}N_2O$ $\{[Me_2PhSb{=}NSiMe_2]\text{-}C_6H_4\}_2O$, 3, 752
$Si_2SnC_8H_{22}Br_2$ $(Me_3SiCH_2)_2SnBr_2$, 2, 327
$Si_2SnC_8H_{22}I_2$ $(Me_3SiCH_2)_2SnI_2$, 2, 327
$Si_2SnC_8H_{22}O$ $Me_2Sn(CH_2SiMe_2)_2O$, 2, 249
$(Me_3SiCH_2)_2SnO$, 2, 374
$(Si_2SnC_8H_{22}O)_n$ $[Me_2Sn(CH_2SiMe_2)_2O]_n$, 2, 249
$(Si_2SnC_8H_{22}S)_n$ $[(Me_3SiCH_2)_2SnS]_n$, 2, 454*
$Si_2SnC_8H_{24}$ $Me_2Sn(CH_2SiHMe_2)_2$, 2, 249
$Si_2SnC_8H_{24}O_2$ $Me_2Sn(OSiMe_3)_2$, 2, 406
$Si_2SnC_9H_{25}Br$ $Me(Me_3SiCH_2)_2SnBr$, 2, 301
$Si_2SnC_9H_{25}I$ $Me(Me_3SiCH_2)_2SnI$, 2, 301
$Si_2SnC_9H_{27}N$ $Me_3SnN(SiMe_3)_2$, 2, 476
$Si_2SnC_{10}H_{28}$ $Me_2Sn(CH_2SiMe_3)_2$, 2, 249
$Si_2SnC_{12}H_{28}$ $(CH_2{=}CH)_2Sn(CH_2SiMe_3)_2$, 2, 250
$Si_2SnC_{12}H_{30}O$ $Me_2Sn(CH_2CH_2CH_2Me_2Si)_2O$, 2, 250
$Si_2SnC_{12}H_{31}I$ $Bu(Me_3SiCH_2)_2SnI$, 2, 301
$Si_2SnC_{12}H_{32}$ $Me_2Sn(CH_2CH_2CH_2SiHMe_2)_2$ 2, 250
$Si_2SnC_{12}H_{32}O_2$ $Me_2Sn(CH_2SiMe_2OEt)_2$, 2, 249
$Si_2SnC_{14}H_{34}O$ $Bu_2Sn(CH_2SiMe_2)_2O$, 2, 250
$Si_2SnC_{14}H_{36}$ $Me_2Sn(CH_2CHMeCH_2SiHMe_2)_2$, 2, 250
$Bu_2Sn(CH_2SiHMe_2)_2$, 2, 250
$Si_2SnC_{14}H_{36}O_2$ $Bu_2Sn(OSiMe_3)_2$, 2, 406
$Si_2SnC_{16}H_{40}$ $Bu_2Sn(CH_2SiMe_3)_2$, 2, 250
$Si_2SnC_{18}H_{28}$ $Me_2Sn(CH_2SiHMePh)_2$, 2, 249
$Si_2SnC_{18}H_{28}O_2$ $Ph_2Sn(OSiMe_3)_2$, 2, 406
$Si_2SnC_{18}H_{44}O_2$ $Bu_2Sn(CH_2Me_2SiOEt)_2$, 2, 250

$Si_2SnC_{18}H_{45}N$ $Bu_3SnN(SiMe_3)_2$, 2, 476
$Si_2SnC_{19}H_{45}Cl$ $Bu_3SnCH_2SiMe_2CH_2CH_2Si$-$Me_2Cl$, 2, 250
$Si_2SnC_{20}H_{32}$ $Ph_2Sn(CH_2SiMe_3)_2$, 2, 251
$Si_2SnC_{26}H_{26}Cl_2O_2$ $Me_2Sn(OSiPh_2Cl)_2$, 2, 406
$Si_2SnC_{30}H_{38}$ $(p\text{-}Me_3SiC_6H_4)_2SnPh_2$, 2, 251
$Si_2SnC_{30}H_{52}O_2$ $Bu_2Sn(OSiMePhCH_2CHMe_2)_2$, 2, 406
$Si_2SnC_{38}H_{36}O_2$ $Me_2Sn(OSiPh_3)_2$, 2, 406
$Si_2SnC_{44}H_{48}O_2$ $Bu_2Sn(OSiPh_3)_2$, 2, 406
$Si_2SnC_{48}H_{40}O_2$ $Ph_2Sn(OSiPh_3)_2$, 2, 406
$Si_2SnC_{50}H_{44}O_2$ $(PhCH_2)_2Sn(OSiPh_3)_2$, 2, 406
$Si_2Sn_2C_{10}H_{30}O_3$ $(Me_2SnOSiMe_3)_2O$, 2, 406
$Si_2Sn_2C_{14}H_{38}O_3$ $(Et_2SnOSiMe_3)_2O$, 2, 406
$Si_2Sn_2C_{18}H_{46}O_3$ $(Pr_2SnOSiMe_3)_2O$, 2, 406
$Si_2Sn_2C_{22}H_{54}O_3$ $(Bu_2SnOSiMe_3)_2O$, 2, 406
$Si_2Sn_2C_{30}H_{70}O$ $(Bu_3SnCH_2Me_2Si)_2O$, 2, 250
$Si_2Ta_2C_{104}H_{190}O_6$ $[(Bu_2C_5H_3)Ta(CO)_3$-$Si(C_{18}H_{37})_2]_2$, 1, 54
$Si_2TiC_{46}H_{40}O_2$ $(C_5H_5)_2Ti(OSiPh_3)_2$, 1, 35
$Si_2V_2C_{36}H_{46}O_6$ $[(BuC_5H_4)V(CO)_3Si(CH_2$-$CH{=}CH_2)_2]_2$, 1, 52
$Si_3AsC_{12}H_{33}$ $(Me_3SiCH_2)_3As$, 3, 16
$Si_3AsC_{12}H_{33}Br_2$ $(Me_3SiCH_2)_3AsBr_2$, 3, 344
$Si_3AsC_{12}H_{33}I_2$ $(Me_3SiCH_2)_3AsI_2$, 3, 344
$Si_3AsC_{13}H_{36}I$ $[Me(Me_3SiCH_2)_3As]I$, 3, 318
$Si_3AsC_{14}H_{38}I$ $[Et(Me_3SiCH_2)_3As]I$, 3, 319
$Si_3AsC_{21}H_{51}$ $(BuMe_2SiCH_2)_3As$, 3, 21
$Si_3AsC_{27}H_{57}$ $(cyclo\text{-}C_6H_{11}Me_2SiCH_2)_3As$, 3, 32
$Si_3BiC_{12}H_{33}$ $(Me_3SiCH_2)_3Bi$, 3, 822
$Si_3Co_2C_{19}H_{30}O_4$ $(HC{\equiv}CSiMe_3)_3Co_2(CO)_4$, 1, 609
$Si_3FeC_{16}H_{26}O_2$ $Me_2\overline{Si\text{-}Fcd\text{-}SiMe_2OSiMe_2O}$, 1, 545
$Si_3FeC_{23}H_{34}O$ Me_3Si-Fcd-$SiMe_2OSiPhMe_2$, 1, 541
$Si_3FeC_{27}H_{41}NO$ $Me_2[\overline{CH_2(CH_2)_4N}]Si$-Fcd-$SiMe_2OSiMe_2Ph$, 1, 541
$Si_3FeC_{28}H_{36}O$ Me_2PhSi-Fcd-$SiMe_2OSiPh$-Me_2, 1, 541
$Si_3FeC_{28}H_{52}$ Tris(triethylsilyl)ferrocene, 1, 542
$Si_3GeC_{10}H_{30}O_3$ $MeGe(OSiMe_3)_3$, 2, 111
$Si_3PbC_{12}H_{33}Cl$ $(Me_3SiCH_2)_3PbCl$, 2, 558
$Si_3SbC_{12}H_{33}$ $(Me_3SiCH_2)_3Sb$, 3, 669
$Si_3SbC_{12}H_{33}Br_2$ $(Me_3SiCH_2)_3SbBr_2$, 3, 731
$Si_3SbC_{30}H_{45}$ $(4\text{-}MeC_6H_4SiMe_2CH_2)_3Sb$, 3, 671
$Si_3SnC_{10}H_{30}O_3$ $MeSn(OSiMe_3)_3$, 2, 408
$Si_3SnC_{12}H_{33}I$ $(Me_3SiCH_2)_3SnI$, 2, 301
$Si_3TiC_{14}H_{32}O_3$ $C_5H_5Ti(OSiMe_3)_3$, 1, 34
$Si_4AsC_9H_{27}O_6$ $\{As(Me)(O)O\}\{Si(Et)(H)$-$O\}_4$, 3, 545
$Si_4AsC_{16}H_{44}I$ $[(Me_3SiCH_2)_4As]I$, 3, 321
$Si_4FeC_{20}H_{38}O_2$ $Fcd(SiMe_2OSiMe_3)_2$, 1, 538
$Si_4FeC_{30}H_{38}Cl_4O_2$ $Fcd(SiMe_2OSiMe_2C_6H_3$-$Cl_2)_2$, 1, 539
$Si_4FeC_{30}H_{42}O_2$ $Fcd(SiMe_2OSiMe_2Ph)_2$, 1, 539
$Si_4FeC_{32}H_{46}O_2$ $Fe[C_5H_3(Me)SiMe_2OSiMe_2$-$Ph]_2$, 1, 539
$Si_4FeC_{36}H_{70}O_2$ $Fcd\{C(OH)[(CH_2)_3Si$-$Me_3]_2\}_2$, 1, 539
$Si_4FeC_{50}H_{50}O_2$ $Fcd(SiMe_2OSiPh_3)_2$, 1, 540
$Si_4Fe_2C_{24}H_{28}O_4$ $\{\overline{OSi(Me)\text{-}Fcd\text{-}Si(Me)(O)}\}_2$, 1, 543
$Si_4Fe_2C_{40}H_{50}O$ $(Me_2PhSi$-Fcd-$SiMe_2)_2O$, 1, 543
$Si_4GeC_{14}H_{42}N_2$ $Me_2Ge[N(SiMe_3)_2]_2$, 2, 128
$Si_4GeC_{36}H_{52}$ $(p\text{-}Me_3SiC_6H_4)_4Ge$, 2, 6
$Si_4PbC_{16}H_{44}$ $(Me_3SiCH_2)_4Pb$, 2, 532
$Si_4SnC_{16}H_{44}$ $(Me_3SiCH_2)_4Sn$, 2, 175
$Si_4SnC_{36}H_{52}$ $(p\text{-}Me_3SiC_6H_4)_4Sn$, 2, 175
$Si_4Sn_2C_{38}H_{90}O$ $(Bu_3SnCH_2SiMe_2CH_2CH_2$-$Me_2Si)_2O$, 2, 250
$Si_5Fe_2C_{32}H_{52}O_2$ $(Me_3Si$-Fcd-$SiMe_2O)_2$-$SiMe_2$, 1, 544
$Si_5Fe_2C_{42}H_{56}O_2$ $(Me_2PhSi$-Fcd-$SiMe_2O)_2$-

$Si_5Fe_2C_{42}H_{56}O_2$
$SiMe_2$, 1, 544
$Si_6FeC_{26}H_{54}O_6$ Fcd($SiMe_2OSiMe_2OSiMe_2$-OEt)$_2$, 1, 539
$Si_6FeC_{34}H_{54}O_4$ Fcd($SiMe_2OSiMe_2OSiMe_2$-Ph)$_2$, 1, 539*
$Si_6Fe_2C_{44}H_{62}O_3$ ($Me_2PhSiOSiMe_2$-Fcd-$SiMe_2$)$_2$O, 1, 543
$Si_6Fe_3C_{54}H_{70}O_2$ (Me_2PhSi-Fcd-$SiMe_2O$-$SiMe_2$)$_2$Fcd, 1, 544
$Si_7Fe_3C_{61}H_{78}O_3$ (Me_2PhSi-Fcd-$SiMe_2O$)$_3$-SiMe, 1, 544
$Si_8FeC_{38}H_{66}O_6$ Fcd[$SiMe_2(OSiMe_2)_2OSi$-Me_2Ph]$_2$, 1, 540
$Si_9Fe_4C_{80}H_{100}O_4$ (Me_2PhSi-Fcd-$SiMe_2O$)$_4$-Si, 1, 544
$Si_{12}AsC_{13}H_{39}O_{20}$ {As(Me)(O)O}{Si(Me)-$O_{3/2}$}$_{12}$, 3, 545
$Si_{12}AsC_{25}H_{75}O_{14}$ {AsMe(O)O}{Si(Me)$_2$-O}$_{12}$, 3, 545
$Si_{12}AsC_{33}H_{78}NO_{15}$ 8-Hydroxy-5-quinoline-As(O)O{Si(Et)(H)O}$_{12}$, 3, 547
($Si_{13}FeC_{42}H_{98}O_{12}$)$_n$ {{$SiMe_2(CH_2)_3$-Fcd-$(CH_2)_3SiMe_2O$}{$SiMe_2O$}$_{11}$}$_n$, 1, 546
$SiAs_2Mn_2C_{36}H_{52}O_8$ {$(Me_3CCH_2)_2Si${Mn-[As($CH_2CH=CH_2$)$_3$](CO)$_4$}$_2$}, 3, 11
$SiAs_2Mn_2C_{120}H_{216}O_8$ (2-EtC_6H_4)$_2$Si-{Mn(CO)$_4$[As($C_{16}H_{33}$)$_3$]}$_2$, 3, 29
$SiGeHgC_8H_{22}$ $Me_3GeCH_2HgCH_2SiMe_3$, 2, 52
$SiGeHgC_{12}H_{30}$ $Et_3GeHgSiEt_3$, 2, 137
$SiGe_3LiC_{54}H_{45}$ (Ph_3Ge)$_3$SiLi, 2, 136
$SiMnSbC_{58}H_{104}Br_2O_4$ {$Br_2PhSiMn$[Sb-($C_{16}H_{33}$)$_3$](CO)$_4$}, 3, 668
$SiMn_2Sb_2C_{70}H_{68}O_8$ ($PhCH_2$)$_2$Si{Mn[Sb-(CH_2CH_2Ph)$_3$](CO)$_4$}$_2$, 3, 667
$SiMn_2Sb_2C_{80}H_{104}O_8$ ($C_{18}H_{37}$)$_2$Si[Mn(Sb-Ph_3)(CO)$_4$]$_2$, 3, 663
$SiMn_3Sb_3C_{60}H_{50}O_{12}$ $CH_2=CHCH_2Si${Mn-(CO)$_4$[Sb(C_5H_5)$_3$]}$_3$, 3, 659
$SiMn_3Sb_3C_{106}H_{108}O_{12}$ 2-Fluorenyl-Si-{Mn(CO)$_4$[Sb($C_6H_4CHMe_2$)$_3$]}$_3$, 3, 667
$Si_2Co_2HgC_{26}H_{40}O_8$ [$Me_3SiC=CHCH=C$(Si-Me_3)$CoCo(CO)_2$]$_2$Hg, 1, 601
$Si_2GeHgC_{16}H_{40}$ $Et_3GeHgSi_2Et_5$, 2, 137
$Si_3AsHgC_{12}H_{33}Cl_2$ (Me_3SiCH_2)$_3$As·$HgCl_2$, 3, 16
$Si_4AlSbC_{16}H_{48}O_4$ [Me_4Sb][Al($OSiMe_3$)$_4$], 3, 712
$Si_4FeSbC_{16}H_{48}O_4$ [Me_4Sb][Fe($OSiMe_3$)$_4$], 3, 713
$Si_4GaSbC_{16}H_{48}O_4$ [Me_4Sb][Ga($OSiMe_3$)$_4$], 3, 712
$SmC_{10}H_{10}Cl$ (C_5H_5)$_2$SmCl, 1, 2
$SmC_{15}H_{15}$ (C_5H_5)$_3$Sm, 1, 4
$SmC_{17}H_{29}Cl_2O_3$ $C_5H_5SmCl_2$·3THF, 1, 1
$SnCH_2Br_4$ $BrCH_2SnBr_3$, 2, 335
$SnCH_2Cl_4$ $ClCH_2SnCl_3$, 2, 335
$SnCH_3Br_3$ $MeSnBr_3$, 2, 331
$SnCH_3Cl_3$ $MeSnCl_3$, 2, 328
$SnCH_3I_3$ $MeSnI_3$, 2, 328
$SnCH_4O_2$ $MeSnO_2H$, 2, 376
$SnCH_6$ $MeSnH_3$, 2, 263
$SnC_2H_2Cl_3F$ $CH_2=CFSnCl_3$, 2, 335
$SnC_2H_2Cl_4$ $ClCH=CHSnCl_3$, 2, 335
$SnC_2H_3Br_3$ $CH_2=CHSnBr_3$, 2, 334
$SnC_2H_3Cl_3$ $CH_2=CHSnCl_3$, 2, 334
$SnC_2H_4Br_2Cl_2$ ($BrCH_2$)$_2SnCl_2$, 2, 324
$SnC_2H_4Cl_4$ $MeCHClSnCl_3$, 2, 335
$SnC_2H_4O_2$ $CH_2=CHSnO_2H$, 2, 376
$SnC_2H_5Br_3$ $EtSnBr_3$, 2, 331
$SnC_2H_5Cl_3$ $EtSnCl_3$, 2, 328
$SnC_2H_5I_3$ $EtSnI_3$, 2, 329
SnC_2H_6 $CH_2=CHSnH_3$, 2, 263
(SnC_2H_6)$_n$ (Me_2Sn)$_n$, 2, 491
$SnC_2H_6Br_2$ Me_2SnBr_2, 2, 304
$SnC_2H_6Cl_2$ Me_2SnCl_2, 2, 303, 513
$SnC_2H_6Cl_2O_8$ $Me_2Sn(ClO_4)_2$, 2, 349
$SnC_2H_6D_2$ Me_2SnD_2, 2, 265
$SnC_2H_6I_2$ Me_2SnI_2, 2, 305
$SnC_2H_6I_2O_6$ $Me_2Sn(IO_3)_2$, 2, 349
$SnC_2H_6N_2O_6$ $Me_2Sn(NO_3)_2$, 2, 348
SnC_2H_6O Me_2SnO, 2, 367, 513, 516
$SnC_2H_6O_2$ $EtSnO_2H$, 2, 376
$SnC_2H_6O_3S$ Me_2SnSO_3, 2, 349
$SnC_2H_6O_4S$ Me_2SnSO_4, 2, 349
SnC_2H_7Br EtH_2SnBr, 2, 267
$SnC_2H_7O_4P$ Me_2SnPO_4H, 2, 348
SnC_2H_8 Me_2SnH_2, 2, 256, 512
$EtSnH_3$, 2, 258
$SnC_2H_8O_2$ $Me_2Sn(OH)_2$, 2, 372
$SnC_2H_{10}N_2$ $Me_2Sn(NH_2)_2$, 2, 475, 513
$SnC_2H_{14}O_6P_2$ $Me_2Sn(H_2PO_2·H_2O)_2$, 2, 348
$SnC_3H_5Br_3$ Cyclopropyl-$SnBr_3$, 2, 331
$CH_2=CHCH_2SnBr_3$, 2, 334

$SnC_3H_5Cl_3$ Cyclopropyl-$SnCl_3$, 2, 331
$MeCH{=}CHSnCl_3$, 2, 334
$CH_2{=}CMeSnCl_3$, 2, 334
$SnC_3H_6Br_4$ $(BrCH_2)_3SnBr$, 2, 295
$SnC_3H_6ClF_3$ $Me_2(CF_3)SnCl$, 2, 297
$SnC_3H_6Cl_4$ $(ClCH_2)_3SnCl$, 2, 295
$SnC_3H_6O_2$ $CH_2{=}CHCH_2SnO_2H$, 2, 376
$SnC_3H_7Br_5$ $CH_2{=}CHCH_2SnBr_3{\cdot}2HBr$, 2, 334
$SnC_3H_7ClO_2$ $Me_2Sn(Cl)O_2CH$, 2, 343
$SnC_3H_7Cl_3$ i-$PrSnCl_3$, 2, 331
$SnC_3H_7I_3$ $PrSnI_3$, 2, 331
SnC_3H_8 $Me_2Sn{=}CH_2$, 2, 254
$SnC_3H_8Br_2$ $Me_2(BrCH_2)SnBr$, 2, 297
$SnC_3H_8Cl_2$ $Me_2(ClCH_2)SnCl$, 2, 297
$MeEtSnCl_2$, 2, 316
SnC_3H_8FI $Me_2(ICH_2)SnF$, 2, 297
$SnC_3H_8I_2$ $Me_2(ICH_2)SnI$, 2, 297
SnC_3H_9Br Me_3SnBr, 2, 270
SnC_3H_9Cl Me_3SnCl, 2, 269, 514
$SnC_3H_9ClO_3Se$ $Me_3SnCl{\cdot}SeO_3$, 2, 270
$SnC_3H_9ClO_4$ Me_3SnClO_4, 2, 348
SnC_3H_9D Me_3SnD, 2, 265
SnC_3H_9F Me_3SnF, 2, 269*
SnC_3H_9I Me_3SnI, 2, 271
$SnC_3H_9NO_3$ Me_3SnNO_3, 2, 348
$SnC_3H_9N_3$ Me_3SnN_3, 2, 345
SnC_3H_{10} Me_3SnH, 2, 255
$PrSnH_3$, 2, 263
$SnC_3H_{10}O$ Me_3SnOH, 2, 358
$SnC_3H_{15}ClN_2$ $Me_3SnCl{\cdot}2NH_3$, 2, 270
$SnC_4H_4Cl_4$ (cis-$ClCH{=}CH)_2SnCl_2$, 2, 325
(trans-$ClCH{=}CH)_2SnCl_2$, 2, 325
$SnC_4H_6Br_2$ $(CH_2{=}CH)_2SnBr_2$, 2, 323
$SnC_4H_6Cl_2$ $(CH_2{=}CH)_2SnCl_2$, 2, 321
$SnC_4H_6F_2$ $(CH_2{=}CH)_2SnF_2$, 2, 323
$SnC_4H_6I_2$ $(CH_2{=}CH)_2SnI_2$, 2, 323
SnC_4H_6O $(CH_2{=}CH)_2SnO$, 2, 372
$SnC_4H_8Br_4$ $(BrCH_2)_4Sn$, 2, 174
$SnC_4H_8Cl_2I_2$ $(ClCH_2CH_2)_2SnI_2$, 2, 325*
$SnC_4H_8Cl_2O$ $(MeCHCl)_2SnO$, 2, 374
$SnC_4H_8Cl_4$ $(ClCH_2)_4Sn$, 2, 174
$(MeCHCl)_2SnCl_2$, 2, 325
$SnC_4H_8F_4$ $Me_2(F_2CHCF_2)SnH$, 2, 266
$SnC_4H_8I_4$ $(ICH_2)_4Sn$, 2, 174
$SnC_4H_8O_4$ $Me_2Sn(O_2CH)_2$, 2, 435
$SnC_4H_8O_5$ $Me_2Sn(O_2C)_2{\cdot}H_2O$, 2, 445
$(SnC_4H_9)_n$ $(BuSn)_n$, 2, 491
SnC_4H_9Br $Me_2(CH_2{=}CH)SnBr$, 2, 293
$SnC_4H_9Br_3$ $BuSnBr_3$, 2, 331
SnC_4H_9Cl $Me_2(CH_2{=}CH)SnCl$, 2, 293
$SnC_4H_9ClO_2$ $Me_2Sn(Cl)OAc$, 2, 343
$SnC_4H_9Cl_3$ $BuSnCl_3$, 2, 329
i-$BuSnCl_3$, 2, 331
$SnC_4H_9F_3$ Me_3SnCF_3, 2, 217*
SnC_4H_9I $Me_2(CH_2{=}CH)SnI$, 2, 293
$SnC_4H_9I_3$ $BuSnI_3$, 2, 331
SnC_4H_9N Me_3SnNC, 2, 345
SnC_4H_9NS Me_3SnNCS, 2, 345
$(SnC_4H_{10})_n$ $(Et_2Sn)_n$, 2, 489
$SnC_4H_{10}Br_2$ Et_2SnBr_2, 2, 307
$SnC_4H_{10}Cl_2$ Et_2SnCl_2, 2, 306, 513
$SnC_4H_{10}Cl_2O_2$ $(CH_2{=}CH)_2SnCl_2{\cdot}2H_2O$, 2, 321
$SnC_4H_{10}F_2$ Et_2SnF_2, 2, 316
$SnC_4H_{10}I_2$ Et_2SnI_2, 2, 308
$SnC_4H_{10}N_2O_6$ $Et_2Sn(NO_3)_2$, 2, 350
$SnC_4H_{10}O$ Et_2SnO, 2, 368, 514
$SnC_4H_{10}O_2$ $BuSnO_2H$, 2, 376
Me_3SnO_2CH, 2, 411
$SnC_4H_{10}O_4S$ Et_2SnSO_4, 2, 351
$SnC_4H_{10}S_2$ $Me_2\overline{SnSCH_2CH_2S}$, 2, 463
$SnC_4H_{11}Br$ Me_3SnCH_2Br, 2, 219
$SnC_4H_{11}Cl$ Me_3SnCH_2Cl, 2, 219
$Me_2EtSnCl$, 2, 287
$SnC_4H_{11}I$ Me_3SnCH_2I, 2, 219
Me_2EtSnI, 2, 287
SnC_4H_{12} Me_4Sn, 2, 158
Me_2EtSnH, 2, 263
Et_2SnH_2, 2, 257, 512
$BuSnH_3$, 2, 263
i-$BuSnH_3$, 2, 263
$SnC_4H_{12}O$ Me_3SnCH_2OH, 2, 226
Me_3SnOMe, 2, 384
$SnC_4H_{12}O_2$ $Me_2Sn(OMe)_2$, 2, 396
$SnC_4H_{12}O_6S_2$ $Me_2Sn(O_3SMe)_2$, 2, 349
$SnC_5H_5Cl_3$ $C_5H_5SnCl_3$, 2, 334
$SnC_5H_9Br_3$ Cyclopentyl-$SnBr_3$, 2, 331
$SnC_5H_9Cl_3O_2$ $Me_3SnO_2CCCl_3$, 2, 421
$SnC_5H_9F_5$ $Me_3SnC_2F_5$, 2, 219
SnC_5H_{10} $Me_3SnC{\equiv}CH$, 2, 212
$SnC_5H_{10}Br_2$ $\overline{CH_2(CH_2)_4Sn}Br_2$, 2, 508
$SnC_5H_{10}Cl_2$ $\overline{CH_2(CH_2)_4Sn}Cl_2$, 2, 508
$SnC_5H_{10}Cl_2O_2$ $Me_3SnO_2CCHCl_2$, 2, 421
$SnC_5H_{10}Cl_4$ $ClCH_2(MeCHCl)_2SnCl$, 2, 297

$SnC_5H_{11}ClO_2$ $Me_3SnO_2CCH_2Cl$, 2, 421
$Et_2Sn(Cl)O_2CH$, 2, 343
$SnC_5H_{11}Cl_3$ i-$C_5H_{11}SnCl_3$, 2, 331
$SnC_5H_{11}NS$ Me_3SnCH_2SCN, 2, 224
$SnC_5H_{11}N_3$ N-(1,2,4-Triazolyl)-$SnMe_3$, 2, 478
SnC_5H_{12} $Me_3SnCH{=}CH_2$, 2, 194, 510
$SnC_5H_{12}O_2$ Me_3SnOAc, 2, 411
$SnC_5H_{12}S_2$ $Me_2\overline{SnS(CH_2)_3S}$, 2, 463
$SnC_5H_{13}Br$ $MeEt_2SnBr$, 2, 287
$SnC_5H_{13}I$ i-$PrMe_2SnI$, 2, 287
SnC_5H_{14} Me_3SnEt, 2, 179
$SnC_5H_{14}O_2$ Me_3SnO_2Et, 2, 379
$SnC_5H_{15}N$ Me_3SnNMe_2, 2, 474
$Me_3SnNHEt$, 2, 475
$SnC_5H_{15}O_3P$ $MeP(OMe)O_2SnMe_3$, 2, 348
$SnC_6H_4Br_3I$ p-$IC_6H_4SnBr_3$, 2, 335
$SnC_6H_4Br_4$ p-$BrC_6H_4SnBr_3$, 2, 335
$SnC_6H_4Cl_4$ p-$ClC_6H_4SnCl_3$, 2, 335
$SnC_6H_5BrO_2$ p-$BrC_6H_4SnO_2H$, 2, 376
$SnC_6H_5ClO_2$ p-$ClC_6H_4SnO_2H$, 2, 376
$SnC_6H_5Cl_3$ $PhSnCl_3$, 2, 330
$SnC_6H_5I_3$ $PhSnI_3$, 2, 332
$SnC_6H_5N_9$ $PhSn(N_3)_3$, 2, 346
$(SnC_6H_5O_4P)_n$ $(PhSnPO_4)_n$, 2, 357
$(SnC_6H_5P)_n$ $(PhSnP)_n$, 2, 484
$SnC_6H_6Br_2$ $PhHSnBr_2$, 2, 267
$SnC_6H_6Cl_2S$ 2-Thienyl$(CH_2{=}CH)SnCl_2$, 2, 321
$SnC_6H_6Cl_4$ (cis-$ClCH{=}CH)_3SnCl$, 2, 296
(trans-$ClCH{=}CH)_3SnCl$, 2, 296
$SnC_6H_6Cl_6O_4$ $Me_2Sn(O_2CCCl_3)_2$, 2, 443
$SnC_6H_6F_6$ $Me_2Sn(CF{=}CF_2)_2$, 2, 207
$SnC_6H_6F_6O_4$ $Me_2Sn(O_2CCF_3)_2$, 2, 443
$SnC_6H_6F_{10}$ $Me_2Sn(C_2F_5)_2$, 2, 220
$SnC_6H_6N_2S_2$ $(CH_2{=}CH)_2Sn(NCS)_2$, 2, 345
$SnC_6H_6O_2$ $PhSnO_2H$, 2, 376
SnC_6H_8 $PhSnH_3$, 2, 264
$SnC_6H_8Br_2F_6$ $Me_2Sn(C_2HBrF_3)_2$, 2, 220
$SnC_6H_8Cl_2N_2$ $(CNCH_2CH_2)_2SnCl_2$, 2, 322
$SnC_6H_8Cl_4O_4$ $Me_2Sn(O_2CCHCl_2)_2$, 2, 443
$SnC_6H_8F_8$ $Me_2Sn(CF_2CHF_2)_2$, 2, 220
SnC_6H_9Br $(CH_2{=}CH)_3SnBr$, 2, 292
SnC_6H_9Cl $(CH_2{=}CH)_3SnCl$, 2, 292
$SnC_6H_9Cl_3$ 1-Cyclohexenyl-$SnCl_3$, 2, 334*
SnC_6H_9F $(CH_2{=}CH)_3SnF$, 2, 292
SnC_6H_9I $(CH_2{=}CH)_3SnI$, 2, 292
$SnC_6H_{10}Br_2$ $(\overline{CH_2CH_2CH})_2SnBr_2$, 2, 316
$(CH_2{=}CHCH_2)_2SnBr_2$, 2, 321
$(MeCH{=}CH)_2SnBr_2$, 2, 323*
(cis-$MeCH{=}CH)_2SnBr_2$, 2, 323*
(trans-$MeCH{=}CH)_2SnBr_2$, 2, 323*
$(CH_2{=}CMe)_2SnBr_2$, 2, 323
$SnC_6H_{10}Br_2O_4$ $Me_2Sn(O_2CCH_2Br)_2$, 2, 443
$SnC_6H_{10}Cl_2$ Cyclopropyl$_2SnCl_2$, 2, 316
$(MeCH{=}CH)_2SnCl_2$, 2, 323
(cis-$MeCH{=}CH)_2SnCl_2$, 2, 323
(trans-$MeCH{=}CH)_2SnCl_2$, 2, 323
$(CH_2{=}CMe)_2SnCl_2$, 2, 323
$SnC_6H_{10}Cl_2O_4$ $Me_2Sn(O_2CCH_2Cl)_2$, 2, 443
$SnC_6H_{10}F_2$ $(CH_2{=}CHCH_2)_2SnF_2$, 2, 323
$SnC_6H_{10}I_2$ $(\overline{CH_2CH_2CH})_2SnI_2$, 2, 316
$SnC_6H_{10}O$ $(CH_2{=}CH)_3SnOH$, 2, 364
$(CH_2{=}CHCH_2)_2SnO$, 2, 372
$SnC_6H_{11}Br_3$ Cyclohexyl-$SnBr_3$, 2, 331
$SnC_6H_{11}ClO_2$ $Me_2(CH_2{=}CH)SnO_2CCH_2Cl$, 2, 431
$SnC_6H_{11}Cl_3$ $Me_2CHCH_2C({=}CH_2)SnCl_3$, 2, 334*
SnC_6H_{12} $Me_3SnC{\equiv}CMe$, 2, 212
$Me_2Sn(CH{=}CH_2)_2$, 2, 200
$SnC_6H_{12}Cl_2$ $Bu(CH_2{=}CH)SnCl_2$, 2, 323
$SnC_6H_{12}Cl_2N_2O_2$ $(H_2NOCCH_2CH_2)_2SnCl_2$, 2, 326
$SnC_6H_{12}Cl_4$ $(ClCH_2)_2Sn(CHClMe)_2$, 2, 221
$(MeCHCl)_3SnCl$, 2, 296
$(C_2H_4Cl)_3SnCl$, 2, 296
$SnC_6H_{12}N_2$ N-Imidazolyl-$SnMe_3$, 2, 478
$SnC_6H_{12}O_2$ $Me_3SnO_2CCH{=}CH_2$, 2, 418, 510
$SnC_6H_{12}O_3$ $BuSn(O)OAc$, 2, 449
$SnC_6H_{12}O_4$ $Et_2Sn(O_2CH)_2$, 2, 435
$SnC_6H_{13}Cl$ $Me_2(MeCH{=}CHCH_2)SnCl$, 2, 293
$Et_2(CH_2{=}CH)SnCl$, 2, 293
$SnC_6H_{13}ClO_2$ $Et_2Sn(Cl)OAc$, 2, 343
$SnC_6H_{13}Cl_3$ $Bu(ClCH_2)_2SnCl$, 2, 297
$C_6H_{13}SnCl_3$, 2, 331
SnC_6H_{14} $Me_3Sn\overline{CHCH_2CH_2}$, 2, 179
$Me_3SnCH{=}CHMe$, 2, 195
$Me_3SnCH_2CH{=}CH_2$, 2, 194, 510
$Me_3SnCMe{=}CH_2$, 2, 199
$Me_2EtSnCH{=}CH_2$, 2, 200
$Me_2(MeCH{=}CHCH_2)SnH$, 2, 263

$SnC_6H_{14}Br_2$ Pr_2SnBr_2, 2, 309, 610*, 611*
i-Pr_2SnBr_2, 2, 316
$SnC_6H_{14}Cl_2$ Pr_2SnCl_2, 2, 308
i-Pr_2SnCl_2, 2, 316
$SnC_6H_{14}I_2$ $(ICH_2)_2SnEt_2$, 2, 221
Pr_2SnI_2, 2, 309
i-Pr_2SnI_2, 2, 316
$SnC_6H_{14}O$ Pr_2SnO, 2, 368, 610*
i-Pr_2SnO, 2, 372
$SnC_6H_{14}O_2$ Me_3SnO_2CEt, 2, 411
$SnC_6H_{15}Br$ Et_3SnBr, 2, 274
$SnC_6H_{15}Cl$ Et_3SnCl, 2, 272, 514
$SnC_6H_{15}D$ Et_3SnD, 2, 265
$SnC_6H_{15}F$ Et_3SnF, 2, 272
$SnC_6H_{15}FO$ $Et_2Sn(OEt)F$, 2, 341
$SnC_6H_{15}I$ Me_2BuSnI, 2, 287
i-$BuMe_2SnI$, 2, 287
s-$BuMe_2SnI$, 2, 287
Et_3SnI, 2, 275
$SnC_6H_{15}NO_2$ $Me_3SnO_2CNMe_2$, 2, 424
$Me_3SnO_2CCHMeNH_2$, 2, 424
$SnC_6H_{15}NO_3$ Et_3SnNO_3, 2, 349
$SnC_6H_{15}NS_2$ $Me_3SnS_2CNMe_2$, 2, 471
SnC_6H_{16} Me_2SnEt_2, 2, 177
Et_3SnH, 2, 256
Pr_2SnH_2, 2, 258, 512
$SnC_6H_{16}ClNO$ $Me_3SnCl \cdot HCONMe_2$, 2, 270
$SnC_6H_{16}O$ Et_3SnOH, 2, 359
$SnC_6H_{16}O_2$ $Me_2Sn(OEt)_2$, 2, 396
$Et_2Sn(OMe)_2$, 2, 396
$SnC_6H_{16}O_6S_2$ $Et_2Sn(O_3SMe)_2$, 2, 351
$SnC_6H_{18}Cl_2O_6P_2$ $Me_2Sn[OP(OMe)_2Cl]_2$, 2, 348
$SnC_6H_{18}N_2$ $Me_2Sn(NMe_2)_2$, 2, 475
$SnC_6H_{18}O_6P_2$ $[MeP(OMe)O_2]_2SnMe_2$, 2, 348
$SnC_7H_7Cl_3$ $(ClCH_2)PhSnCl_2$, 2, 324
o-$MeC_6H_4SnCl_3$, 2, 330
p-$MeC_6H_4SnCl_3$, 2, 332
$SnC_7H_7Cl_3O$ p-$MeOC_6H_4SnCl_3$, 2, 335
$SnC_7H_8Br_4$ 3-(5-Bromotricyclo-[2.2.1.0^{2,6}]heptyl)tin tribromide, 2, 332
$SnC_7H_8Cl_2$ $(CH_2{=}CH)C_5H_5SnCl_2$, 2, 323
$SnC_7H_8Cl_2S$ 2,5-Methylthienyl$(CH_2{=}CH)$-$SnCl_2$, 2, 321
$SnC_7H_8Cl_4$ 3-(5-Chlorotricyclo-[2.2.1.0^{2,6}]heptyl)tin trichloride, 2, 332
$SnC_7H_8O_2$ o-$MeC_6H_4SnO_2H$, 2, 376
p-$MeC_6H_4SnO_2H$, 2, 376
SnC_7H_{12} $Me_3SnC{\equiv}CCH{=}CH_2$, 2, 210
$SnC_7H_{13}BrO_2$ $Me_2(CH_2{=}CH)SnO_2CCHMeBr$, 2, 431
$SnC_7H_{13}Cl$ $\overline{CH_2(CH_2)_4Sn}(CH{=}CH_2)Cl$, 2, 508
$SnC_7H_{13}N$ N-$(Me_3Sn)C_4H_4N$, 2, 478
SnC_7H_{14} $Me_2(CH_2{=}CH)Sn\overline{CHCH_2CH_2}$, 2, 200
$SnC_7H_{14}BrN$ $Et_2(NCCH_2CH_2)SnBr$, 2, 298
$SnC_7H_{14}IN$ $Et_2(NCCH_2CH_2)SnI$, 2, 298
$SnC_7H_{14}O_2$ $Me_3SnO_2CCMe{=}CH_2$, 2, 418, 510
$SnC_7H_{14}O_3$ $BuSn(O)O_2CEt$, 2, 449
$SnC_7H_{15}ClO_2$ $Pr_2Sn(Cl)O_2CH$, 2, 343
$SnC_7H_{15}F_3$ Et_3SnCF_3, 2, 220
$SnC_7H_{15}I$ Cyclopentyl$(Me_2)SnI$, 2, 287
$SnC_7H_{15}N$ $Et_2(NCCH_2CH)_2SnH$, 2, 266
Et_3SnCN, 2, 345
$SnC_7H_{15}NO$ Et_3SnNCO, 2, 345
$SnC_7H_{15}NS$ Et_3SnNCS, 2, 345
SnC_7H_{16} $Me_3SnCH_2CH{=}CHMe$, 2, 199
$Me_3SnCH_2CMe{=}CH_2$, 2, 199
$SnC_7H_{16}Br_2$ $Et(Me_3CCH_2)SnBr_2$, 2, 316
$SnC_7H_{16}Cl_2$ $Et(Me_3CCH_2)SnCl_2$, 2, 316
$PrBuSnCl_2$, 2, 316
$SnC_7H_{16}O_2$ Et_3SnO_2CH, 2, 412
$SnC_7H_{17}Cl$ Et_3SnCH_2Cl, 2, 220
$SnC_7H_{17}ClOS$ $Me_3SnCl \cdot O\overline{S(CH_2)_3CH_2}$, 2, 270
$SnC_7H_{17}I$ n-$C_5H_{11}Me_2SnI$, 2, 287
i-$C_5H_{11}Me_2SnI$, 2, 287
i-Pr_2MeSnI, 2, 288
Et_2PrSnI, 2, 288
SnC_7H_{18} Me_3SnBu, 2, 179
$Me_3SnCH_2CHMe_2$, 2, 179
$MeSnEt_3$, 2, 181
$SnC_7H_{18}ClNO$ $Me_3SnCl \cdot MeCONMe_2$, 2, 270
$SnC_7H_{18}O$ Me_3SnOBu, 2, 384
Et_3SnOMe, 2, 382
$SnC_7H_{18}O_2$ $Me_3SnO_2CMe_3$, 2, 379
$SnC_7H_{18}O_3S_3$ $MeSn(SCH_2CH_2OH)_3$, 2, 468*
$SnC_7H_{18}S$ Et_3SnSMe, 2, 458
$SnC_7H_{19}N$ $Et_2[NH_2(CH_2)_3]SnH$, 2, 266
Me_3SnNEt_2, 2, 475

$SnC_7H_{19}NO_2S$ $MeSO_2NHSnEt_3$, 2, 480
$SnC_7H_{21}N_3$ $MeSn(NMe_2)_3$, 2, 475
$SnC_7H_{21}O_9P_3$ $(MePO_2OMe)_3SnMe$, 2, 349
SnC_8F_{12} $(CF_2{=}CF)_4Sn$, 2, 174
SnC_8H_4 $(HC{\equiv}C)_4Sn$, 2, 173
$SnC_8H_4Cl_4S_2$ 2,5-Chlorothienyl$_2SnCl_2$, 2, 321
$SnC_8H_6ClF_5$ $Me_2(C_6F_5)SnCl$, 2, 297
$SnC_8H_6Cl_2S_2$ 2-Thienyl$_2SnCl_2$, 2, 321
$SnC_8H_6F_6$ $(CH_2{=}CH)_2Sn(CF{=}CF_2)_2$, 2, 208
$SnC_8H_6S_2$ $(2\text{-}C_4H_3S)_2Sn$, 2, 491
$SnC_8H_7Cl_3O_2$ o-$Cl_3SnC_6H_4CO_2Me$, 2, 335
$SnC_8H_8Cl_2$ $Ph(CH_2{=}CH)SnCl_2$, 2, 323
$SnC_8H_8O_3$ $PhSn(O)OAc$, 2, 450
$SnC_8H_8O_4$ o-$MeO_2CC_6H_4SnO_2H$, 2, 376
SnC_8H_9ClS 2-Thienyl$(CH_2{=}CH)_2SnCl$, 2, 291
$SnC_8H_9Cl_3$ $Ph(ClCH_2)_2SnCl$, 2, 298
$PhCH_2CH_2SnCl_3$, 2, 331
$SnC_8H_{10}Cl_3N$ p-$Me_2NC_6H_4SnCl_3$, 2, 335
$SnC_8H_{10}F_6$ $Et_2Sn(CF{=}CF_2)_2$, 2, 207
$SnC_8H_{10}F_6O_4$ $Et_2Sn(O_2CCF_3)_2$, 2, 443
$SnC_8H_{10}N_2S_2$ $(CH_2CH_2CH)_2Sn(NCS)_2$, 2, 346
$SnC_8H_{10}O_2$ o-$Me_2SnOC_6H_4O$, 2, 396
$SnC_8H_{11}O_3P$ Me_2SnO_3PPh, 2, 348
SnC_8H_{12} $(CH_2{=}CH)_4Sn$, 2, 171
$Et_2Sn(C{\equiv}CH)_2$, 2, 213
$SnC_8H_{12}N_2O_4$ Me_3Sn-orotate, 2, 416
$SnC_8H_{12}O_5$ Et_2Sn epoxysuccinate, 2, 445
$SnC_8H_{13}Cl_3O_2$ $PhSnCl_3\cdot 2MeOH$, 2, 330
$SnC_8H_{13}I_3O$ Acetylcyclohexyl-SnI_3, 2, 335
$SnC_8H_{13}N$ 2-Pyridyl-$SnMe_3$, 2, 192
SnC_8H_{14} $Me_3SnC_5H_5$, 2, 199
$Me(CH_2CH_2CH)Sn(CH{=}CH_2)_2$, 2, 200
$Me_3SnC{\equiv}CCH{=}CHMe$, 2, 212
$Me_3SnC{\equiv}CCMe{=}CH_2$, 2, 212
$CH{=}CHCH_2CH_2CH{=}CHSnMe_2$, 2, 507*
$SnC_8H_{14}BrN$ $Me_3SnBr\cdot Py$, 2, 271
$SnC_8H_{14}Br_2$ $(MeCH{=}CHCH_2)_2SnBr_2$, 2, 324
$(CH_2{=}CHCHMe)_2SnBr_2$, 2, 324
$SnC_8H_{14}Br_2O_4$ $(EtO_2CCH_2)_2SnBr_2$, 2, 326
$(MeO_2CCH_2CH_2)_2SnBr_2$, 2, 326
$SnC_8H_{14}ClN$ $Me_3SnCl\cdot Py$, 2, 270
$SnC_8H_{14}IN$ $Me_3SnI\cdot Py$, 2, 271
$SnC_8H_{14}N_6$ N-(1,2,4-triazolyl)$_2SnEt_2$, 2, 478
$SnC_8H_{15}Br$ $Bu(CH_2{=}CH)_2SnBr$, 2, 294
$SnC_8H_{15}Cl$ $Bu(CH_2{=}CH)_2SnCl$, 2, 294
$SnC_8H_{15}F_3$ $Et_3SnCF{=}CF_2$, 2, 207
$SnC_8H_{15}F_3O_2$ $Me_2BuSnO_2CCF_3$, 2, 431
$SnC_8H_{15}F_5$ $Et_3SnC_2F_5$, 2, 221
SnC_8H_{16} $Et_2Sn(CH{=}CH_2)_2$, 2, 201
$Me_3SnC{\equiv}CPr$, 2, 212
$Me_3SnC{\equiv}CCHMe_2$, 2, 212
$Et_3SnC{\equiv}CH$, 2, 212
$SnC_8H_{16}Cl_2O_2$ $Et_3SnO_2CCHCl_2$, 2, 421
$SnC_8H_{16}Cl_4$ $(MeCHCl)_4Sn$, 2, 174
$(PrCHCl)_2SnCl_2$, 2, 325
$SnC_8H_{16}N_2S_3$ 2,5-$Et_3SnS(1,3,4\text{-}C_2N_2S)$-SH, 2, 460
$SnC_8H_{16}O_3$ $BuSn(O)O_2CPr$, 2, 449
$SnC_8H_{16}O_4$ $Et_2Sn(OAc)_2$, 2, 435, 513
$SnC_8H_{17}BrO_2$ $Et_3SnO_2CCH_2Br$, 2, 421
$SnC_8H_{17}ClO_2$ $Et_3SnO_2CCH_2Cl$, 2, 421
$Pr_2Sn(Cl)OAc$, 2, 343
$SnC_8H_{17}Cl_3$ n-$C_8H_{17}SnCl_3$, 2, 332
$SnC_8H_{17}FO_2$ $Et_3SnO_2CCH_2F$, 2, 421
$SnC_8H_{17}I$ Cyclohexyl$(Me_2)SnI$, 2, 287
$SnC_8H_{17}IO_2$ $Et_3SnO_2CCH_2I$, 2, 421
SnC_8H_{18} $CH_2{=}CH(CH_2)_3SnMe_3$, 2, 199, 510*
$MeCH{=}CHCH_2CH_2SnMe_3$, 2, 199
$MeCH{=}CHCHMeSnMe_3$, 2, 199
$Et_3SnCH{=}CH_2$, 2, 195
$(SnC_8H_{18})_n$ $(Bu_2Sn)_n$, 2, 491, 517
$(t\text{-}Bu_2Sn)_n$, 2, 491
$SnC_8H_{18}BrCl$ $Bu_2SnClBr$, 2, 317
$SnC_8H_{18}Br_2$ Bu_2SnBr_2, 2, 312
i-Bu_2SnBr_2, 2, 317
s-Bu_2SnBr_2, 2, 317
$SnC_8H_{18}Cl_2$ $(MeCHCl)_2SnEt_2$, 2, 221
Bu_2SnCl_2, 2, 309, 513, 517
i-Bu_2SnCl_2, 2, 312
s-Bu_2SnCl_2, 2, 317
t-Bu_2SnCl_2, 2, 317
$SnC_8H_{18}F_2$ Bu_2SnF_2, 2, 317
$SnC_8H_{18}I_2$ Bu_2SnI_2, 2, 312
i-Bu_2SnI_2, 2, 317
s-Bu_2SnI_2, 2, 317
t-Bu_2SnI_2, 2, 317
$SnC_8H_{18}N_2S_4$ $Me_2Sn(2\text{-}SC_3H_6NS)_2$, 2, 464

$SnC_8H_{18}N_4$ 3,1-Amino(1,2,4-triazolyl)-$SnMe_3$, 2, 478
$SnC_8H_{18}O$ Bu_2SnO, 2, 309, 513, 516, 517
s-Bu_2SnO, 2, 372
i-Bu_2SnO, 2, 372
$(Me_3C)_2SnO$, 2, 372
$SnC_8H_{18}O_2$ Et_3SnOAc, 2, 409
$Me_2BuSnOAc$, 2, 430
$SnC_8H_{18}O_2S$ $Et_3SnSCH_2CO_2H$, 2, 461, 516
$SnC_8H_{18}O_3$ $Et_3SnO_2CCH_2OH$, 2, 423
$SnC_8H_{18}O_3S$ Bu_2SnSO_3, 2, 353
$SnC_8H_{18}O_4S$ Bu_2SnSO_4, 2, 353
$SnC_8H_{19}Cl$ $(MeCHCl)SnEt_3$, 2, 221
Bu_2HSnCl, 2, 267
$SnC_8H_{19}ClO$ $Bu_2Sn(OH)Cl$, 2, 340
$SnC_8H_{19}FO$ Bu_2SnFOH (?), 2, 317
$SnC_8H_{19}I$ Et_2BuSnI, 2, 288
$SnC_8H_{19}O_3P$ Bu_2SnHPO_3, 2, 353
SnC_8H_{20} Et_4Sn, 2, 160
n-$C_5H_{11}SnMe_3$, 2, 179
i-$C_5H_{11}SnMe_3$, 2, 179
$Me_2Sn(CHMe_2)_2$, 2, 180
Bu_2SnH_2, 2, 259, 512
i-Bu_2SnH_2, 2, 260
$SnC_8H_{20}O$ Et_3SnOEt, 2, 382
$SnC_8H_{20}O_2$ $Bu_2Sn(OH)_2$ (?), 2, 372
$Et_3SnOOEt$, 2, 379
$Me_2Sn(OPr)_2$, 2, 396
$Et_2Sn(OEt)_2$, 2, 396
$Pr_2Sn(OMe)_2$, 2, 397
$SnC_8H_{20}O_3S$ Et_3SnO_3SEt, 2, 350
$SnC_8H_{20}O_6S_2$ $Et_2Sn(O_3SEt)_2$, 2, 351
$Pr_2Sn(O_3SMe)_2$, 2, 351
i-$Pr_2Sn(O_3SMe)_2$, 2, 351
$SnC_8H_{20}S$ Et_3SnSEt, 2, 458
$SnC_8H_{21}N$ Et_3SnNMe_2, 2, 475
$SnC_8H_{22}ClN$ $Et_3SnCl\cdot NHMe_2$, 2, 273
$SnC_8H_{22}O_6P_2$ $[MeP(OMe)O_2]_2SnEt_2$, 2, 350
$SnC_8H_{24}Cl_2N_2$ $Bu_2SnCl_2\cdot 2NH_3$, 2, 311
$SnC_9H_7Cl_3$ Indenyl-$SnCl_3$, 2, 334
$SnC_9H_9F_5$ $Me_3SnC_6F_5$, 2, 219
$SnC_9H_{11}Cl_3$ $PhSn(CH_2Cl)_3$, 2, 223
$SnC_9H_{11}I_3$ $PhSn(CH_2I)_3$, 2, 223
$SnC_9H_{12}BrN_3$ $(NCCH_2CH_2)_3SnBr$, 2, 298
$SnC_9H_{12}ClN_3$ $(NCCH_2CH_2)_3SnCl$, 2, 298
$SnC_9H_{12}FN_3$ $(NCCH_2CH_2)_3SnF$, 2, 298
$SnC_9H_{12}IN_3$ $(NCCH_2CH_2)_3SnI$, 2, 298
$SnC_9H_{13}Cl$ m-$ClC_6H_4SnMe_3$, 2, 219
p-$ClC_6H_4SnMe_3$, 2, 219
$SnC_9H_{13}Br$ p-$BrC_6H_4SnMe_3$, 2, 218
$SnC_9H_{13}F$ o-$FC_6H_4SnMe_3$, 2, 219
$SnC_9H_{13}NO_2$ Me_3Sn 3-pyridinecarboxylate, 2, 416
$SnC_9H_{13}N_3$ N-(1,2,3-Benzotriazolyl)-$SnMe_3$, 2, 478
SnC_9H_{14} Me_3SnPh, 2, 177
$SnC_9H_{14}N_2O_4$ Me_3Sn uracil acetate, 2, 416
$SnC_9H_{15}Br$ $(\overline{CH_2CH_2C}H)_3SnBr$, 2, 282
$SnC_9H_{15}Cl$ $(\overline{CH_2CH_2C}H)_3SnCl$, 2, 282
$(MeCH{=}CH)_3SnCl$, 2, 292
$(CH_2{=}CHCH_2)_3SnCl$, 2, 292
$(CH_2{=}CMe)_3SnCl_2$, 2, 292
$SnC_9H_{15}F$ $(\overline{CH_2CH_2C}H)_3SnF$, 2, 282
$SnC_9H_{15}I$ $(\overline{CH_2CH_2C}H)_3SnI$, 2, 282
$SnC_9H_{15}F_5O_2$ $Et_3SnO_2CC_2F_5$, 2, 421
$Me_2BuSnO_2CC_2F_5$, 2, 431
$SnC_9H_{15}N$ $Me_3SnNHPh$, 2, 475
SnC_9H_{16} $Me(\overline{CH_2CH_2C}H)_2SnCH{=}CH_2$, 2, 200
$SnC_9H_{17}Cl_3O_3$ $(C_2H_4Cl)_3SnO_2CCHOHMe$, 2, 432
SnC_9H_{18} $Me_3SnC{\equiv}CBu$, 2, 212
$Et_3SnC{\equiv}CMe$, 2, 212
$SnC_9H_{18}BrCl_3$ $Et_3SnCHBrCH_2CCl_3$, 2, 221
$SnC_9H_{18}Br_2O_2$ $Et_3SnO_2CCHBrCH_2Br$, 2, 421
$SnC_9H_{18}Cl_4$ $Et_3SnCHClCH_2CCl_3$, 2, 221
$SnC_9H_{18}O$ $Et_3SnOCH_2C{\equiv}CH$, 2, 388
$SnC_9H_{18}O_2$ $Et_3SnO_2CCH{=}CH_2$, 2, 418, 510
$Pr_2\overline{SnCH_2CH_2CO_2}$, 2, 432
$SnC_9H_{18}O_3$ $BuSn(O)O_2CBu$, 2, 449
$SnC_9H_{19}BrO_2$ $Et_3SnO_2CCH_2CH_2Br$, 2, 421
$Et_3SnO_2CCHBrMe$, 2, 421
$SnC_9H_{19}ClO_2$ $Bu_2Sn(Cl)O_2CH$, 2, 343
$Et_3SnO_2CCH_2CH_2Cl$, 2, 421
$Et_3SnO_2CCHClMe$, 2, 421
$SnC_9H_{19}Cl_3$ $Et_3SnCH_2CH_2CCl_3$, 2, 221
$SnC_9H_{19}I$ $Pr_2(CH_2{=}CHCH_2)SnI$, 2, 293
$SnC_9H_{19}N$ $Et_3SnCH_2CH_2CN$, 2, 224
SnC_9H_{20} $Et_3SnCH_2CH{=}CH_2$, 2, 195
$SnC_9H_{20}Cl_2$ $Bu_2(ClCH_2)SnCl$, 2, 297
$SnC_9H_{20}O$ Et_3SnCH_2Ac, 2, 230
$SnC_9H_{20}O_4$ $Et_3SnCH_2CO_2Me$, 2, 234

$Et_2(MeO_2CCHMeCH_2)SnH$, 2, 266
Et_3SnO_2CEt, 2, 412
$SnC_9H_{21}Br$ Pr_3SnBr, 2, 282*, 611*
i-Pr_3SnBr, 2, 282
$MeBu_2SnBr$, 2, 288
$Et_2(Me_3CCH_2)SnBr$, 2, 288
$SnC_9H_{21}Cl$ Pr_3SnCl, 2, 275
i-Pr_3SnCl, 2, 275
$SnC_9H_{21}F$ Pr_3SnF, 2, 282
$SnC_9H_{21}I$ Pr_3SnI, 2, 282*, 611*
i-Pr_3SnI, 2, 282
$MeBu_2SnI$, 2, 288
s-Bu_2MeSnI, 2, 288
$SnC_9H_{21}NO$ $Et_3SnCH_2CH_2CONH_2$, 2, 234
$SnC_9H_{21}NO_2$ $Et_3SnO_2CCH_2CH_2NH_2$, 2, 424
SnC_9H_{22} Et_3SnPr, 2, 182
i-$PrSnEt_3$, 2, 182
Pr_3SnH, 2, 258
i-Pr_3SnH, 2, 263
$SnC_9H_{22}O$ $Et_3Sn(CH_2)_3OH$, 2, 226
Pr_3SnOH, 2, 360
i-Pr_3SnOH, 2, 364
Et_3SnOPr, 2, 384
$SnC_9H_{22}S$ $Et_3SnSCHMe_2$, 2, 458
$SnC_9H_{23}N$ $Me_3SnN(CHMe_2)_2$, 2, 475
$SnC_9H_{24}IOP$ $Me_3SnI \cdot OPEt_3$, 2, 271
$SnC_{10}H_7Cl_3$ 1-$C_{10}H_7SnCl_3$, 2, 332
$SnC_{10}H_8O_2$ 1-$C_{10}H_7SnO_2H$, 2, 376
$SnC_{10}H_9ClS_2$ 2-Thienyl$_2(CH_2=CH)SnCl$, 2, 291
$SnC_{10}H_9F_9$ $BuSn(CF=CF_2)_3$, 2, 208
$SnC_{10}H_{10}$ $(C_5H_5)_2Sn$, 2, 490
$SnC_{10}H_{10}O_4$ Me_2Sn phthalate, 2, 445
$SnC_{10}H_{11}Br$ $Ph(CH_2=CH)_2SnBr$, 2, 295
$SnC_{10}H_{11}Br_2N$ $Ph(CNCH_2CH_2CH_2)SnBr_2$, 2, 322
$SnC_{10}H_{11}Cl$ $Ph(CH_2=CH)_2SnCl$, 2, 295
$SnC_{10}H_{11}F$ $Ph(CH_2=CH)_2SnF$, 2, 295
$SnC_{10}H_{11}O_3P$ $(CH_2=CH)_2SnO_3PPh$, 2, 351
$SnC_{10}H_{12}Br_2O_2$ $Ph(MeO_2CCH_2CH_2)SnBr_2$, 2, 326
$SnC_{10}H_{12}O_3$ $PhSn(O)O_2CCHMe_2$, 2, 450
$SnC_{10}H_{12}S$ 2-Thienyl-$Sn(CH=CH_2)_3$, 2, 192
$SnC_{10}H_{13}NO_4$ p-$O_2NC_6H_4CO_2SnMe_3$, 2, 424
$SnC_{10}H_{14}BrCl$ p-$ClC_6H_4SnEt_2Br$, 2, 297
$SnC_{10}H_{14}Br_2$ p-$BrC_6H_4SnEt_2Br$, 2, 297
$SnC_{10}H_{14}Cl_2$ $BuPhSnCl_2$, 2, 317
$SnC_{10}H_{14}N_2$ N-(Benzimidazolyl)$SnMe_3$, 2, 478
$SnC_{10}H_{14}O_2$ p-$Me_3SnC_6H_4CO_2H$, 2, 234
Me_3SnOBz, 2, 411
$SnC_{10}H_{15}Br$ p-$BrC_6H_4CH_2SnMe_3$, 2, 219
$SnC_{10}H_{15}Cl$ p-$ClC_6H_4CH_2SnMe_3$, 2, 219
$Et_2PhSnCl$, 2, 288
$SnC_{10}H_{15}F_7O_2$ $Et_3SnO_2CC_3F_7$, 2, 421
$Me_2BuSnO_2CC_3F_7$, 2, 431
$SnC_{10}H_{15}NO_2$ p-$H_2NC_6H_4CO_2SnMe_3$, 2, 424
$SnC_{10}H_{16}$ Me_3SnCH_2Ph, 2, 179
m-$MeC_6H_4SnMe_3$, 2, 179
p-$MeC_6H_4SnMe_3$, 2, 180
1-Cyclopentenyl-$C \equiv CSnMe_3$, 2, 212
$SnC_{10}H_{16}Cl_4O_4$ i-$Pr_2Sn(O_2CCHCl_2)_2$, 2, 443
$SnC_{10}H_{16}O$ p-$Me_3SnC_6H_4OMe$, 2, 226
$Et_2PhSnOH$, 2, 364
$SnC_{10}H_{16}O_4$ $Me_2Sn(O_2CCMe=CH_2)_2$, 2, 439, 510
$SnC_{10}H_{17}Cl_3O_2$ $PhSnCl_3 \cdot 2EtOH$, 2, 330
$SnC_{10}H_{17}F_5O_2$ $Me_2(n\text{-}C_5H_{11})SnO_2CC_2F_5$, 2, 431
$SnC_{10}H_{17}N$ $Me_3SnNMePh$, 2, 475
$SnC_{10}H_{17}O_2P$ Me_3SnO_2PMePh, 2, 348
$SnC_{10}H_{18}$ $MeSn(CHCH_2CH_2)_3$, 2, 181
$BuSn(CH=CH_2)_3$, 2, 203
$Et_3SnC \equiv CCH=CH_2$, 2, 211
$CH=CHCH_2CH_2CH=CHSnEt_2$, 2, 507*
$SnC_{10}H_{18}Br_2$ Cyclopentyl$_2SnBr_2$, 2, 318
$SnC_{10}H_{18}Br_2O_4$ $(EtO_2CCHMe)_2SnBr_2$, 2, 326
$SnC_{10}H_{18}Cl_2$ $(CH_2=CHCHMeCH_2)_2SnCl_2$, 2, 324
$SnC_{10}H_{18}Cl_2O_4$ i-$Pr_2Sn(O_2CCH_2Cl)_2$, 2, 443
$(C_3H_6Cl)_2Sn(OAc)_2$, 2, 443
$SnC_{10}H_{18}N_2$ $Et_2Sn(CH_2CH_2CN)_2$, 2, 224
$SnC_{10}H_{18}N_2S$ 2,5-Dimercapto(1,3,4-thiadiazolyl)$SnBu_2$, 2, 464
$SnC_{10}H_{18}N_2S_2$ $Bu_2Sn(NCS)_2$, 2, 346
$SnC_{10}H_{18}O_4$ Et_2Sn adipate, 2, 445
$SnC_{10}H_{18}O_6$ $BuSn(OAc)_3$, 2, 449
$SnC_{10}H_{18}O_6S_3$ $BuSn(SCH_2CO_2H)_3$, 2, 469
$SnC_{10}H_{19}Br$ $Bu(CH_2=CHCH_2)_2SnBr$, 2, 294
n-$C_6H_{13}(CH_2=CH)_2SnBr$, 2, 294

$SnC_{10}H_{19}NO_2$ N-(Succinimido)$SnEt_3$, 2, 480

$SnC_{10}H_{20}$ $Me_2Sn(CH_2CH{=}CHMe)_2$, 2, 200
$Et_2Sn(CH_2CH{=}CH_2)_2$, 2, 201, 510
$Et_3SnCH{=}CHCH{=}CH_2$, 2, 200

$SnC_{10}H_{20}O$ $Et_3SnOCH_2CH_2C{\equiv}CH$, 2, 388
$Et_3SnOCHMeC{\equiv}CH$, 2, 388

$SnC_{10}H_{20}O_2$ $Et_3SnO_2CCMe{=}CH_2$, 2, 418, 510, 511

$SnC_{10}H_{20}O_2S$ $Bu_2\overline{SnO_2CCH_2S}$, 2, 447

$SnC_{10}H_{20}O_4$ $Et_2Sn(CH_2CO_2Me)_2$, 2, 235

$SnC_{10}H_{20}O_5$ $Bu_2Sn(O_2C)_2{\cdot}H_2O$, 2, 445

$SnC_{10}H_{21}Br$ $Bu_2(CH_2{=}CH)SnBr$, 2, 293

$SnC_{10}H_{21}BrO_2$ $Et_2(EtO_2CCH_2CHMe)SnBr$, 2, 300
$Bu_2Sn(Br)OAc$, 2, 343

$SnC_{10}H_{21}Cl$ 5-Hexenyl-Et_2SnCl, 2, 293*
$Bu_2(CH_2{=}CH)SnCl$, 2, 293
i-$Bu_2(CH_2{=}CH)SnCl$, 2, 294

$SnC_{10}H_{21}ClO_2$ $Bu_2Sn(Cl)OAc$, 2, 343

$SnC_{10}H_{21}I$ $Bu_2(CH_2{=}CH)SnI$, 2, 293

$SnC_{10}H_{22}$ $Et_3SnCH_2CMe{=}CH_2$, 2, 200
$\overline{CH{=}CHCH_2CH_2CH{=}CHS}nPr_2$, 2, 507*

$SnC_{10}H_{22}Br_2$ $(i\text{-}C_5H_{11})_2SnBr_2$, 2, 318
$(Me_3CCH_2)_2SnBr_2$, 2, 318

$SnC_{10}H_{22}Cl_2$ $(C_5H_{11})_2SnCl_2$, 2, 317
$(i\text{-}C_5H_{11})_2SnCl_2$, 2, 317
$(Me_3CCH_2)_2SnCl_2$, 2, 318

$SnC_{10}H_{22}I_2$ $(C_5H_{11})_2SnI_2$, 2, 317
$(i\text{-}C_5H_{11})_2SnI_2$, 2, 318

$SnC_{10}H_{22}O$ $(n\text{-}C_5H_{11})_2SnO$, 2, 372
$(Me_2EtC)_2SnO$, 2, 372

$SnC_{10}H_{22}OS$ $Bu_2\overline{SnOCH_2CH_2S}$ (?), 2, 517

$SnC_{10}H_{22}O_2$ $Et_3SnCH_2CH_2OAc$, 2, 226
$Et_3SnCH_2CO_2Et$, 2, 234
$Et_3SnCH_2CH_2CO_2Me$, 2, 234
$Bu_2HSnOAc$, 2, 267
$Bu_2\overline{SnOCH_2CH_2O}$, 2, 405
Et_3SnO_2CPr, 2, 412
$Et_2BuSnOAc$, 2, 430
Pr_3SnO_2CH, 2, 412

$SnC_{10}H_{22}O_2S$ $Et_3SnSCH_2CO_2Et$, 2, 461

$SnC_{10}H_{22}S_2$ $Bu_2\overline{SnSCH_2CH_2S}$, 2, 463

$SnC_{10}H_{23}Br$ $n\text{-}C_8H_{17}Me_2SnBr$, 2, 287

$SnC_{10}H_{23}Cl$ $n\text{-}C_6H_{13}Et_2SnCl$, 2, 288

$SnC_{10}H_{23}ClOS$ $Et_3SnCl{\cdot}O\overline{S(CH_2)_3C}H_2$, 2, 274

$SnC_{10}H_{23}I$ $n\text{-}C_8H_{17}Me_2SnI$, 2, 287
$n\text{-}C_6H_{13}Et_2SnI$, 2, 287
Pr_2BuSnI, 2, 289

$SnC_{10}H_{23}NO$ $Et_3SnCH_2CHMeCONH_2$, 2, 234

$SnC_{10}H_{24}$ Me_2SnBu_2, 2, 180
$Me_2Sn(CHEtMe)_2$, 2, 180
$Me_2Sn(CMe_3)_2$, 2, 180
$MeSnPr_3$, 2, 181
Et_3SnBu, 2, 182
i-Pr_2SnEt_2, 2, 183

$SnC_{10}H_{24}O$ $Et_3SnCH_2CH_2OEt$, 2, 226
$Et_3SnOCH_2CHMe_2$, 2, 384
$Et_3SnOCHMeEt$, 2, 384
$Et_3SnOCMe_3$, 2, 384
Pr_3SnOMe, 2, 384
i-Pr_3SnOMe, 2, 385

$SnC_{10}H_{24}O_2$ $Et_3SnOOCMe_3$, 2, 379
$Me_2Sn(OBu)_2$, 2, 396
$Et_2Sn(OPr)_2$, 2, 396
$Pr_2Sn(OEt)_2$, 2, 397
$Bu_2Sn(OMe)_2$, 2, 397

$SnC_{10}H_{24}O_2$ $Bu_2Sn(OMe)_2$, 2, 514, 517

$SnC_{10}H_{24}O_3S$ Pr_3SnO_3SMe, 2, 351
i-Pr_3SnO_3SMe, 2, 351

$SnC_{10}H_{24}O_4$ $Me_2Sn(O_2CMe_3)_2$, 2, 379

$SnC_{10}H_{24}O_4S_4$ $Bu_2Sn(SO_2SMe)_2$, 2, 354

$SnC_{10}H_{24}O_6S_2$ $Pr_2Sn(O_3SEt)_2$, 2, 351
$Bu_2Sn(O_3SMe)_2$, 2, 354

$SnC_{10}H_{24}S$ Et_3SnSBu, 2, 458
$Et_3SnSCH_2CHMe_2$, 2, 458
$Et_3SnSCMe_3$, 2, 458

$SnC_{10}H_{24}S_2$ $Me_2Sn(SBu)_2$, 2, 462

$SnC_{10}H_{25}N$ Et_3SnNEt_2, 2, 475

$SnC_{10}H_{25}O_3P$ $EtP(OEt)O_2SnEt_3$, 2, 349

$SnC_{10}H_{25}P$ Et_3SnPEt_2, 2, 482

$SnC_{10}H_{26}ClN$ $Et_3SnCl{\cdot}NHEt_2$, 2, 273

$SnC_{10}H_{26}N_2$ $Me_2Sn(NEt_2)_2$, 2, 475

$SnC_{10}H_{26}O_6P_2$ $(EtPO_2OEt)_2SnMe_2$, 2, 348

$SnC_{10}H_{27}N_3$ $BuSn(NMe_2)_3$, 2, 476

$SnC_{10}H_{30}Br_2Cl_2N_2$ $Me_2SnBr_2{\cdot}2Me_4NCl$, 2, 305

$SnC_{10}H_{30}Cl_4N_2$ $Me_2SnCl_2{\cdot}2Me_4NCl$, 2, 304

$SnC_{11}H_9Cl_3$ 1,4-$MeC_{10}H_6SnCl_3$, 2, 332

$SnC_{11}H_{11}Cl_3N_2$ $MeSnCl_3{\cdot}$2,2′-bipyridyl, 2, 328

$SnC_{11}H_{13}Br_3N_2$ $MeSnBr_3{\cdot}2Py$, 2, 331

$SnC_{11}H_{13}Cl$ $PhCH_2(CH_2{=}CH)_2SnCl$, 2, 293

$SnC_{11}H_{13}Cl_3N_2$ $MeSnCl_3 \cdot 2Py$, 2, 328
$SnC_{11}H_{14}$ $Me_3SnC \equiv CPh$, 2, 212
$SnC_{11}H_{15}Br_2N$ $(CH_2=CHCH_2)_2SnBr_2 \cdot Py$, 2, 322
$SnC_{11}H_{15}I$ $CH_2(CH_2)_4SnPhI$, 2, 508
$SnC_{11}H_{16}$ $Me_3SnCH=CHPh$, 2, 194
p-$Me_3SnC_6H_4CH=CH_2$, 2, 199, 510
$SnC_{11}H_{18}O$ $Et_2(PhCH_2)SnOH$, 2, 364
$SnC_{11}H_{18}O_2$ $(CH_2CH_2CH)_3SnOAc$, 2, 413
$SnC_{11}H_{19}N$ p-$Me_3SnC_6H_4NMe_2$, 2, 241
$SnC_{11}H_{20}$ $Et_3SnC_5H_5$, 2, 200
$Et_3SnC \equiv CCHCH_2CH_2$, 2, 213
$Et_3SnC \equiv CCH=CHMe$, 2, 213
$Et_3SnC \equiv CCMe=CH_2$, 2, 213
$SnC_{11}H_{20}N_2$ 2-Pyrazoline-$C \equiv CSnEt_3$, 2, 191
$SnC_{11}H_{20}O_2$ $Et_3SnOCH_2C_4H_3O$, 2, 387
$SnC_{11}H_{21}BrO_4$ $Pr(MeO_2CCH_2CH_2)_2SnBr$, 2, 300
$SnC_{11}H_{21}Cl_3O_2$ $Pr_3SnO_2CCCl_3$, 2, 422
$(i\text{-}C_3H_6Cl)_3SnOAc$, 2, 432
$SnC_{11}H_{21}F_3O_2$ $Pr_3SnO_2CCF_3$, 2, 422
$SnC_{11}H_{21}I$ (Cyclopentyl)$_2$MeSnI, 2, 288
$SnC_{11}H_{22}$ $Et_3SnC \equiv CCHMe_2$, 2, 212
Cyclopentyl-Pr_2SnH, 2, 263
$SnC_{11}H_{22}BrN$ $Bu_2(NCCH_2CH_2)SnBr$, 2, 298
$SnC_{11}H_{22}ClN$ $i\text{-}Bu_2(NCCH_2CH_2)SnCl$, 2, 298
$SnC_{11}H_{22}Cl_2O_2$ $Pr_3SnO_2CCHCl_2$, 2, 422
$SnC_{11}H_{22}O$ $Et_3SnOCMe_2C \equiv CH$, 2, 388
$SnC_{11}H_{22}O_2$ $Et_3SnC(CO_2Et)=CHMe$, 2, 207
$Bu_2SnCH_2CH_2CO_2$, 2, 432
$SnC_{11}H_{22}O_2S$ $Bu_2SnO_2CCH_2CH_2S$, 2, 447
$Bu_2{}^{113}SnO_2CCH_2CH_2S$, 2, 447
$(Pr^{14}CH_2)_2SnO_2CCH_2CH_2S$, 2, 447
$Bu_2SnO_2CCH_2{}^{14}CH_2S$, 2, 447
$SnC_{11}H_{23}Br$ $Bu_2(CH_2=CHCH_2)SnBr$, 2, 294
$SnC_{11}H_{23}BrO_2$ $Pr_3SnO_2CCH_2Br$, 2, 422
$Et_2[Br(CH_2)_5]SnOAc$, 2, 432
$SnC_{11}H_{23}ClO_2$ $Pr_3SnO_2CCH_2Cl$, 2, 422
$Bu_2(ClCH_2)SnOAc$, 2, 432
$SnC_{11}H_{23}IO_2$ $Pr_3SnO_2CCH_2I$, 2, 422
$SnC_{11}H_{23}NO_2S$ $Bu_2SnO_2CCHNH_2CH_2S$, 2, 447
$SnC_{11}H_{24}$ $Pr_3SnCH=CH_2$, 2, 201
$SnC_{11}H_{24}O_2$ $Et_3Sn(CH_2)_3OAc$, 2, 226
$Et_3SnCH_2CO_2Pr$, 2, 234
$Et_3SnCH_2CH_2CO_2Et$, 2, 234
$Et_3SnCH_2CHMeCO_2Me$, 2, 234
Bu_2SnOCH_2CHMeO, 2, 405
Pr_3SnOAc, 2, 412
$SnC_{11}H_{24}O_3$ $Bu_2Sn(OMe)OAc$, 2, 407
$SnC_{11}H_{24}S_2$ $Bu_2SnS(CH_2)_3S$, 2, 463
$SnC_{11}H_{25}ClO$ $i\text{-}Bu_2(HOCH_2CH_2CH_2)SnCl$, 2, 300
$SnC_{11}H_{25}I$ $(n\text{-}C_5H_{11})_2MeSnI$, 2, 288
$SnC_{11}H_{26}$ $n\text{-}C_8H_{17}SnMe_3$, 2, 179
$Et_3SnCH_2CMe_3$, 2, 182
$EtSnPr_3$, 2, 184
$SnC_{11}H_{26}O$ Pr_3SnOEt, 2, 384
$SnC_{11}H_{26}O_2$ $Et_3Sn(CH_2)_3OCH_2CH_2OH$, 2, 226
$SnC_{11}H_{26}O_3$ $i\text{-}Pr_3SnOAc \cdot H_2O$, 2, 413
$SnC_{11}H_{26}S$ $i\text{-}C_5H_{11}SSnEt_3$, 2, 458
$SnC_{11}H_{27}N$ Me_3SnNBu_2, 2, 475
$SnC_{11}H_{29}Br_2N$ $Me_3SnBr \cdot Et_4NBr$, 2, 271
$SnC_{11}H_{33}Cl_3N_2$ $Me_3SnCl \cdot 2Me_4NCl$, 2, 270
$SnC_{12}Cl_2F_{10}$ $(C_6F_5)_2SnCl_2$, 2, 325
$SnC_{12}H_5F_9$ $PhSn(CF=CF_2)_3$, 2, 209
$SnC_{12}H_6Cl_4S_3$ 2,5-Chlorothienyl$_3$SnCl, 2, 291
$SnC_{12}H_6Cl_6$ Trichlorobiphenyl-$SnCl_3$, 2, 335
$SnC_{12}H_8Br_2Cl_2$ $(p\text{-}BrC_6H_4)_2SnCl_2$, 2, 325
$SnC_{12}H_8Br_2O$ $(p\text{-}BrC_6H_4)_2SnO$, 2, 374
$SnC_{12}H_8Cl_2I_2$ $(p\text{-}IC_6H_4)_2SnCl_2$, 2, 325
$SnC_{12}H_8Cl_2O$ $(p\text{-}ClC_6H_4)_2SnO$, 2, 374
$SnC_{12}H_8Cl_4$ $(p\text{-}ClC_6H_4)_2SnCl_2$, 2, 325
$SnC_{12}H_8F_2O$ $(p\text{-}FC_6H_4)_2SnO$, 2, 374
$SnC_{12}H_8I_2O$ $(p\text{-}IC_6H_4)_2SnO$, 2, 374
$SnC_{12}H_8NO_5$ $(p\text{-}O_2NC_6H_4)_2SnO$, 2, 374
$SnC_{12}H_9ClS_3$ 2-Thienyl$_3$SnCl, 2, 291
$SnC_{12}H_9Cl_3O$ $p\text{-}PhOC_6H_4SnCl_3$, 2, 335
$(SnC_{12}H_{10})_n$ $(Ph_2Sn)_n$, 2, 490
$SnC_{12}H_{10}Br_2$ Ph_2SnBr_2, 2, 319
$SnC_{12}H_{10}Cl_2$ Ph_2SnCl_2, 2, 313, 514
$SnC_{12}H_{10}F_2$ Ph_2SnF_2, 2, 319
$SnC_{12}H_{10}I_2$ Ph_2SnI_2, 2, 319
$SnC_{12}H_{10}O$ Ph_2SnO, 2, 370, 514, 518
$SnC_{12}H_{10}O_4$ $(CH_2=CH)_2Sn$ phthalate, 2, 445
$SnC_{12}H_{11}ClO$ $Ph_2Sn(OH)Cl$, 2, 340
$SnC_{12}H_{12}$ $(MeC \equiv C)_4Sn$, 2, 173

Ph_2SnH_2, 2, 262, 512
$SnC_{12}H_{12}S_2$ (2-Thienyl)$_2$Sn(CH=CH$_2$)$_2$, 2, 192
$SnC_{12}H_{14}$ (CH$_2$=CH)$_3$SnPh, 2, 206
(C$_5$H$_4$Me)$_2$Sn, 2, 491
$SnC_{12}H_{14}Cl_2N_2$ Me$_2$SnCl$_2$·2,2'-bipyridyl, 2, 304
$SnC_{12}H_{14}N_2S_2$ 2,3-(Et$_2$SnS$_2$)quinoxaline, 2, 464
$SnC_{12}H_{14}O_6$ PhSn(OAc)$_3$, 2, 450
$SnC_{12}H_{16}$ Me$_2$Sn(C$_5$H$_5$)$_2$, 2, 200
$SnC_{12}H_{16}Br_2N_2$ Me$_2$SnBr$_2$·2Py, 2, 305
$SnC_{12}H_{16}Cl_2N_2$ Me$_2$SnCl$_2$·2Py, 2, 304
$SnC_{12}H_{16}I_2N_2$ Me$_2$SnI$_2$·2Py, 2, 305
$SnC_{12}H_{16}N_4$ (NCCH$_2$CH$_2$)$_4$Sn, 2, 175
$SnC_{12}H_{17}BrO_2$ Et$_2$(p-BrC$_6$H$_4$)SnOAc, 2, 432
$SnC_{12}H_{17}ClO_2$ Et$_2$(p-ClC$_6$H$_4$)SnOAc, 2, 432
$SnC_{12}H_{17}Cl_3N_2$ Me$_2$SnCl$_2$·2Py·HCl, 2, 304
$SnC_{12}H_{18}$ p-Me$_3$SnC$_6$H$_4$CMe=CH$_2$, 2, 199
$SnC_{12}H_{18}Cl_2$ 1-Cyclohexenyl$_2$SnCl$_2$, 2, 324
$SnC_{12}H_{18}Cl_6O_4$ Bu$_2$Sn(O$_2$CCCl$_3$)$_2$, 2, 443
$SnC_{12}H_{18}F_6$ Bu$_2$Sn(CF=CF$_2$)$_2$, 2, 208
$SnC_{12}H_{18}F_{10}$ Bu$_2$Sn(C$_2$F$_5$)$_2$, 2, 222
$SnC_{12}H_{18}O_2$ Et$_2$PhSnOAc, 2, 430
$SnC_{12}H_{19}Br$ p-BrC$_6$H$_4$SnEt$_3$, 2, 221
$SnC_{12}H_{19}BrO$ p-BrC$_6$H$_4$OSnEt$_3$, 2, 389
$SnC_{12}H_{19}ClO$ p-ClC$_6$H$_4$OSnEt$_3$, 2, 389
$SnC_{12}H_{19}I$ p-IC$_6$H$_4$SnEt$_3$, 2, 221
$SnC_{12}H_{19}NO_2$ Et$_3$Sn 3-pyridinecarboxylate, 2, 416
$SnC_{12}H_{19}NO_3$ p-Et$_3$SnOC$_6$H$_4$NO$_2$, 2, 391
$SnC_{12}H_{20}$ (CH$_2$CH$_2$CH)$_4$Sn, 2, 163
(CH$_2$=CHCH$_2$)$_4$Sn, 2, 172
(MeCH=CH)$_4$Sn, 2, 173
(CH$_2$=CMe)$_4$Sn, 2, 173
Et$_3$SnPh, 2, 178
Bu$_2$Sn(C≡CH)$_2$, 2, 214
$SnC_{12}H_{20}Cl_4O_4$ Bu$_2$Sn(O$_2$CCHCl$_2$)$_2$, 2, 443
$SnC_{12}H_{20}F_8$ Bu$_2$Sn(CF$_2$CF$_2$H)$_2$, 2, 222
$SnC_{12}H_{20}N_2$ Me$_3$SnN=CPhNMe$_2$, 2, 480
$SnC_{12}H_{20}N_2O$ (2-C$_5$H$_4$N)CH=NOSnEt$_3$, 2, 387
(3-C$_5$H$_4$N)CH=NOSnEt$_3$, 2, 387
(4-C$_5$H$_4$N)CH=NOSnEt$_3$, 2, 387
Me$_3$SnNPhCONMe$_2$, 2, 480
$SnC_{12}H_{20}N_2S$ Me$_3$SnNPhCSNMe$_2$, 2, 480
$SnC_{12}H_{20}O$ o-HOC$_6$H$_4$SnEt$_3$, 2, 226
Et$_3$SnOPh, 2, 384
$SnC_{12}H_{20}O_2$ p-HOC$_6$H$_4$OSnEt$_3$, 2, 390
$SnC_{12}H_{20}O_2S$ Et$_3$SnO$_2$SPh, 2, 350
$SnC_{12}H_{20}O_4$ Me$_2$Sn(CHAc$_2$)$_2$, 2, 230
Et$_2$Sn(O$_2$CCMe=CH$_2$)$_2$, 2, 439, 511
Bu$_2$Sn maleate, 2, 442
Bu$_2$Sn fumarate, 2, 445
$SnC_{12}H_{20}O_5$ Bu$_2$Sn epoxysuccinate, 2, 445
$SnC_{12}H_{20}S$ Et$_3$SnSPh, 2, 458
$SnC_{12}H_{21}BrO_6$ (MeO$_2$CCH$_2$CH$_2$)$_3$SnBr, 2, 299
$SnC_{12}H_{21}N$ 2-Pyridyl-CH$_2$SnEt$_3$, 2, 192
$SnC_{12}H_{22}$ n-C$_6$H$_{13}$Sn(CH=CH$_2$)$_3$, 2, 204
Et$_3$SnC≡CCMeCH$_2$CH$_2$, 2, 213
$SnC_{12}H_{22}Br_2$ Cyclohexyl$_2$SnBr$_2$, 2, 318
$SnC_{12}H_{22}Cl_2$ Cyclohexyl$_2$SnCl$_2$, 2, 318
[Me$_2$CHCH$_2$C(=CH$_2$)]$_2$SnCl$_2$, 2, 324
$SnC_{12}H_{22}Cl_2O_4$ Bu$_2$Sn(O$_2$CCH$_2$Cl)$_2$, 2, 443
$SnC_{12}H_{22}I_2$ Cyclohexyl$_2$SnI$_2$, 2, 318
$SnC_{12}H_{22}N_2$ Et$_3$SnC=CHNHN=CCMe=CH$_2$ (?), 2, 191
Pr$_2$Sn(CH$_2$CH$_2$CN)$_2$, 2, 224
$SnC_{12}H_{22}N_6$ N-(1,2,4-Triazolyl)$_2$SnBu$_2$, 2, 479
$SnC_{12}H_{22}O_4$ Et$_3$SnC(CO$_2$Me)=CHCO$_2$Me, 2, 208
Bu$_2$Sn succinate, 2, 445
$SnC_{12}H_{23}Br$ n-C$_8$H$_{17}$(CH$_2$=CH)$_2$SnBr, 2, 294
$SnC_{12}H_{23}NO_2$ Pr$_3$SnO$_2$CCH$_2$CN, 2, 424
Pr$_2$(NCCH$_2$CH$_2$CH$_2$)SnOAc, 2, 432
$SnC_{12}H_{24}$ Cyclopentyl$_2$SnMe$_2$, 2, 180
Pr$_2$Sn(CH$_2$CH=CH$_2$)$_2$, 2, 202
i-Pr$_2$Sn(CH$_2$CH=CH$_2$)$_2$, 2, 202
Bu$_2$Sn(CH=CH$_2$)$_2$, 2, 196
$SnC_{12}H_{24}N_2$ N-(Imidazolyl)SnPr$_3$, 2, 478
$SnC_{12}H_{24}N_2S_4$ Bu$_2$Sn(S$_2$CNHCH$_2$)$_2$, 2, 472
$SnC_{12}H_{24}O$ 2-Ketocyclohexyl-SnEt$_3$,

$SnC_{12}H_{24}O$

2, 230
$SnC_{12}H_{24}O_2$ $Pr_3SnO_2CCH{=}CH_2$, 2, 413, 511
$SnC_{12}H_{24}O_2S_2$ $Bu_2Sn(SAc)_2$, 2, 471
$SnC_{12}H_{24}O_4$ $Et_2Sn(CH_2CO_2Et)_2$, 2, 235
$Et_2Sn(CH_2CH_2CO_2Me)_2$, 2, 235
$Pr_2Sn(CH_2CO_2Me)_2$, 2, 236
$Et_2Sn(OCH_2CH_2OEt)O_2CCH{=}CHMe$, 2, 407
$Bu_2Sn(OAc)_2$, 2, 433, 514
i-$Bu_2Sn(OAc)_2$, 2, 438
$SnC_{12}H_{24}O_4S_2$ $Bu_2Sn(SCH_2CO_2H)_2$, 2, 465
$SnC_{12}H_{24}O_6$ $Bu_2Sn(O_2CCH_2OH)_2$, 2, 444
$SnC_{12}H_{25}BrO_2$ $Pr_2(EtO_2CCH_2CHMe)SnBr$, 2, 300*
$Bu_2(MeO_2CCH_2CH_2)SnBr$, 2, 300
$Pr_3SnO_2CCHMeBr$, 2, 422
$SnC_{12}H_{25}ClO_2$ i-$Bu_2(MeO_2CCH_2CH_2)SnCl$, 2, 300
$SnC_{12}H_{25}F_3$ $C_{12}H_{25}SnF_3$, 2, 332
$SnC_{12}H_{25}N$ $Pr_3SnCH_2CH_2CN$, 2, 224
$SnC_{12}H_{26}$ $Et_3Sn(CH_2)_4CH{=}CH_2$, 2, 201
$Pr_3SnCH_2CH{=}CH_2$, 2, 202
i-$Pr_3SnCH_2CH{=}CH_2$, 2, 202
$SnC_{12}H_{26}Br_2$ $(n\text{-}C_6H_{13})_2SnBr_2$, 2, 318
$SnC_{12}H_{26}Cl_2$ $(n\text{-}C_6H_{13})_2SnCl_2$, 2, 318
$SnC_{12}H_{26}I_2$ $(n\text{-}C_6H_{13})_2SnI_2$, 2, 318
$SnC_{12}H_{26}O$ $Pr_3SnCH{=}CHCH_2OH$, 2, 208
Pr_3SnCH_2Ac, 2, 230
$(n\text{-}C_6H_{13})_2SnO$, 2, 372
$Et_3SnOC_6H_{11}$, 2, 384
$SnC_{12}H_{26}O_2$ $Et_3Sn(CH_2)_3OCH_2CHCH_2O$, 2, 226
$Et_3Sn(CH_2)_3CO_2Et$, 2, 234
$Et_3SnCHMeCH_2CO_2Et$, 2, 234
$Pr_3SnCH_2CO_2Me$, 2, 235
$C_{12}H_{25}SnO_2H$, 2, 376
$Bu_2SnOCH_2CH_2CHMeO$, 2, 405*
Pr_3SnO_2CEt, 2, 412
$Me_2(n\text{-}C_8H_{17})SnOAc$, 2, 430
$Et_2(n\text{-}C_6H_{13})SnOAc$, 2, 430
$SnC_{12}H_{26}O_3$ $Bu_2SnOCH_2CH_2OCH_2CH_2O$, 2, 405
$SnC_{12}H_{27}Br$ Bu_3SnBr, 2, 277
$Me_2(n\text{-}C_{10}H_{21})SnBr$, 2, 287
$Et(Me_3CCH_2)_2SnBr$, 2, 289
$SnC_{12}H_{27}Cl$ Bu_3SnCl, 2, 276, 514, 516
i-Bu_3SnCl, 2, 283
t-Bu_3SnCl, 2, 283
$SnC_{12}H_{27}ClS$ $Bu_2Sn(SBu)Cl$, 2, 342
$SnC_{12}H_{27}D$ i-Bu_3SnD, 2, 265
$SnC_{12}H_{27}F$ Bu_3SnF, 2, 282
i-Bu_3SnF, 2, 283
$SnC_{12}H_{27}I$ Bu_3SnI, 2, 283
$Et_2(n\text{-}C_8H_{17})SnI$, 2, 288
$Et(n\text{-}C_5H_{11})_2SnI$, 2, 289
$SnC_{12}H_{27}NO$ $Pr_3SnCH_2CH_2CONH_2$, 2, 235
$SnC_{12}H_{27}NO_3$ Bu_3SnNO_3, 2, 351
$SnC_{12}H_{27}N_3$ Bu_3SnN_3, 2, 346
$SnC_{12}H_{28}$ Pr_4Sn, 2, 162
i-Pr_4Sn, 2, 163
$(n\text{-}C_5H_{11})_2SnMe_2$, 2, 180
$Me_2Sn(CMe_2Et)_2$, 2, 180
n-$C_6H_{13}SnEt_3$, 2, 182
Et_2SnBu_2, 2, 183
$Et_2Sn(CH_2CHMe_2)_2$, 2, 183
$Pr_2Sn(CHMe_2)_2$, 2, 185
Bu_3SnH, 2, 258
i-Bu_3SnH, 2, 260
$SnC_{12}H_{28}O$ $Et_3SnCH_2CH_2OBu$, 2, 226
i-$BuOCH_2CH_2SnEt_3$, 2, 226
Bu_3SnOH, 2, 361, 516
Pr_3SnOPr, 2, 385
$SnC_{12}H_{28}O_2$ $Et_2Sn(OBu)_2$, 2, 396
$Pr_2Sn(OPr)_2$, 2, 397
$Bu_2Sn(OEt)_2$, 2, 397
$SnC_{12}H_{28}O_2S_2$ $Bu_2Sn(SCH_2CH_2OH)_2$, 2, 465
$SnC_{12}H_{28}S$ $Et_3SnSC_6H_{13}$, 2, 458
$SnC_{12}H_{28}S_2$ $Bu_2Sn(SEt)_2$, 2, 462
$SnC_{12}H_{29}N$ $Pr_3Sn(CH_2)_3NH_2$, 2, 241
$SnC_{12}H_{29}NO_2$ $Et_3Sn(CH_2)_3OCH_2CHOHCH_2NH_2$, 2, 226
$SnC_{12}H_{30}ClOP$ $Et_3SnCl{\cdot}OPEt_3$, 2, 274
$SnC_{12}H_{30}IOP$ $Et_3SnI{\cdot}OPEt_3$, 2, 275
$SnC_{12}H_{30}N_2$ $Et_2Sn(NEt_2)_2$, 2, 476
$Bu_2Sn(NMe_2)_2$, 2, 476
$SnC_{12}H_{30}O_6P$ $[EtP(OEt)O_2]_2SnEt_2$, 2, 350
$SnC_{12}H_{33}Cl_5N_2$ $BuSnCl_3{\cdot}2Me_4NCl$, 2, 329
$SnC_{12}H_{34}Cl_4N_2$ $Et_2SnCl_2{\cdot}2Me_4NCl$, 2, 307
$SnC_{12}H_{34}Cl_4N_2O_2$ $Me_2SnCl_2{\cdot}2Me_3(HOCH_2\text{-}CH_2)NCl$, 2, 304
$SnC_{13}H_{11}Br_2Cl_5N_4$ $MeSnCl_3{\cdot}2$ p-$BrC_6H_4\text{-}N_2Cl$, 2, 328
$SnC_{13}H_{11}Cl_5N_6O_4$ $MeSnCl_3{\cdot}2$ o-$O_2NC_6H_4$-

N_2Cl, 2, 328
$SnC_{13}H_{12}BrF$ $Ph_2(BrCH_2)SnF$, 2, 298
$SnC_{13}H_{12}Br_2$ $Ph_2(BrCH_2)SnBr$, 2, 298
$SnC_{13}H_{12}Cl_2$ $Ph_2(ClCH_2)SnCl$, 2, 298
$SnC_{13}H_{13}Br$ $MePh_2SnBr$, 2, 288
$SnC_{13}H_{13}Cl_3N_2$ $MeSnCl_3 \cdot 9,10'$-phenanthroline, 2, 328
$SnC_{13}H_{14}Cl_2$ i-Pr(2-$C_{10}H_7$)$SnCl_2$, 2, 316, 513
$SnC_{13}H_{16}$ 1-$C_{10}H_7SnMe_3$, 2, 180
2-$C_{10}H_7SnMe_3$, 2, 180
$SnC_{13}H_{17}Cl_3N_2$ $BuSnCl_3 \cdot 2,2'$-bipyridyl, 2, 329
$SnC_{13}H_{20}O_2$ p-$HO_2CC_6H_4SnEt_3$, 2, 234
Et_3SnOBz, 2, 409
$SnC_{13}H_{20}O_3$ $Et_3SnOOBz$ (?), 2, 379
$Et_2Sn(OEt)OBz$, 2, 407
$SnC_{13}H_{21}Br$ p-$BrC_6H_4CH_2SnEt_3$, 2, 221
$SnC_{13}H_{21}Cl$ m-$ClC_6H_4CH_2SnEt_3$, 2, 220
p-$ClC_6H_4CH_2SnEt_3$, 2, 220
$SnC_{13}H_{21}F$ p-$FC_6H_4CH_2SnEt_3$, 2, 220
$SnC_{13}H_{21}NO$ $Et_3SnN(Ph)CHO$, 2, 480
$SnC_{13}H_{21}NS$ $Et_3SnSCH{=}NPh$, 2, 471
$SnC_{13}H_{22}$ p-$MeC_6H_4SnEt_3$, 2, 183
$SnC_{13}H_{22}O$ p-$MeOC_6H_4SnEt_3$, 2, 226
Et_3SnOCH_2Ph, 2, 384
p-$MeC_6H_4OSnEt_3$, 2, 384
$SnC_{13}H_{22}O_2S$ m-$MeC_6H_4SO_2SnEt_3$, 2, 350
$SnC_{13}H_{22}O_4$ Bu_2Sn itaconate, 2, 445
$SnC_{13}H_{22}O_4S_3$ $Bu_2Sn(O_2CCH_2S)_2C{=}S$, 2, 445
$SnC_{13}H_{22}S$ Et_3SnSCH_2Ph, 2, 458
o-$MeC_6H_4SSnEt_3$, 2, 458
m-$MeC_6H_4SSnEt_3$, 2, 458
p-$MeC_6H_4SSnEt_3$, 2, 458
$SnC_{13}H_{23}NO_2S$ p-$MeC_6H_4SO_2NHSnEt_3$, 2, 480
$SnC_{13}H_{24}$ $BuSn(\overline{CHCH_2CH_2})_3$, 2, 185
$BuSn(CH_2CH{=}CH_2)_3$, 2, 203
$SnC_{13}H_{24}O_4$ Et_2Sn azelate, 2, 445
$SnC_{13}H_{24}O_5$ $Et_3SnC(CO_2Me){=}COMeCO_2Me$, 2, 208
$Bu_2Sn(OH)O_2CCHCHCO_2Me$, 2, 441
$SnC_{13}H_{24}O_6$ $BuSn(O_2CEt)_3$, 2, 449
$SnC_{13}H_{25}F_3O_2$ Me(n-C_5H_{11})$_2SnO_2CCF_3$, 2, 431
$SnC_{13}H_{25}I$ $Cyclohexyl_2MeSnI$, 2, 288
$SnC_{13}H_{26}O_2$ $Pr_3SnO_2CCMe{=}CH_2$, 2, 418, 511
$SnC_{13}H_{27}I$ Cyclopentyl Bu_2SnI, 2, 289
$SnC_{13}H_{27}N$ $Pr_3Sn(CH_2)_3CN$, 2, 224
Bu_3SnCN, 2, 346
Bu_3SnNC, 2, 346
$SnC_{13}H_{27}NO$ $Et_3SnCONHC_6H_{11}$, 2, 234
$SnC_{13}H_{27}NO_3$ $Bu_2(H_2NOCCH_2CH_2)SnOAc$, 2, 432
$SnC_{13}H_{27}NS$ Bu_3SnNCS, 2, 346
$SnC_{13}H_{28}O$ $Pr_3SnCH_2CH_2Ac$, 2, 230
i-$Pr_3SnOCMe{=}CHMe$, 2, 230
$SnC_{13}H_{28}O_2$ $Pr_3SnCH_2CO_2Et$, 2, 235
$Pr_3SnCH_2CH_2CO_2Me$, 2, 235
Pr_3SnO_2CPr, 2, 412
Bu_3SnO_2CH, 2, 413
$SnC_{13}H_{29}Cl$ Bu_3SnCH_2Cl, 2, 221
$SnC_{13}H_{29}ClO_2$ $Bu_2[HOCH_2CH_2O(CH_2)_3]$-SnCl, 2, 300
$SnC_{13}H_{29}I$ Bu_2(i-C_5H_{11})SnI, 2, 289
$SnC_{13}H_{29}NO_2$ $Et_3SnNBuCO_2Et$, 2, 480
$SnC_{13}H_{30}$ n-$C_{10}H_{21}SnMe_3$, 2, 179
$MeSnBu_3$, 2, 181
Pr_3SnBu, 2, 184
$SnC_{13}H_{30}O$ Bu_3SnOMe, 2, 382
$SnC_{13}H_{30}O_2S_2$ Bu_3SnSO_2SMe, 2, 352
$SnC_{13}H_{30}O_3S$ i-Bu_3SnO_3SMe, 2, 354
$Bu_3SnO_2S(OMe)$ (?), 2, 352
$SnC_{13}H_{30}O_6$ $MeEtSn(OCH_2CH_2OCH_2CH_2OMe)_2$, 2, 396
$SnC_{13}H_{30}S$ $Et_3SnSC_7H_{15}$, 2, 458
$SnC_{13}H_{30}S_3$ $MeSn(SBu)_3$, 2, 468*
$SnC_{13}H_{31}N$ Pr_3SnNEt_2, 2, 476
$SnC_{13}H_{33}I_3O_6P_2$ $MeSnI[PI(OEt)_3]_2$, 2, 328
$SnC_{13}H_{33}N_3$ $MeSn(NEt_2)_3$, 2, 475
$SnC_{14}H_6F_{10}$ $Me_2Sn(C_6F_5)_2$, 2, 220
$SnC_{14}H_{10}F_4$ $Ph_2SnC_2F_4$ (?), 2, 223
$SnC_{14}H_{11}ClS_2$ 2-$Thienyl_2PhSnCl$, 2, 291
$SnC_{14}H_{11}Cl_3$ $PhCH{=}\overline{CPhSnCl_3}$, 2, 334
$SnC_{14}H_{12}$ o,o'-$\overline{C_6H_4CH_2CH_2C_6H_4Sn}$, 2, 508
$SnC_{14}H_{12}Br_2Cl_2$ (p-$BrC_6H_4CH_2)_2SnCl_2$, 2, 324
$SnC_{14}H_{12}Br_4$ (p-$BrC_6H_4CH_2)_2SnBr_2$, 2, 325
$SnC_{14}H_{12}Cl_2$ o,o'-$\overline{C_6H_4CH_2CH_2C_6H_4Sn}Cl_2$,

2, 508

$SnC_{14}H_{12}Cl_2F_2$ (p-$FC_6H_4CH_2)_2SnCl_2$, 2, 324

$SnC_{14}H_{12}Cl_4$ (m-$ClC_6H_4CH_2)_2SnCl_2$, 2, 324

(p-$ClC_6H_4CH_2)_2SnCl_2$, 2, 324

$SnC_{14}H_{12}IN$ $Ph_2(NCCH_2)SnI$, 2, 299

$SnC_{14}H_{12}O$ o,o'-$C_6H_4CH_2CH_2C_6H_4SnO$, 2, 508

$SnC_{14}H_{12}O_2$ (PhCH=CPh)SnO_2H, 2, 376

$SnC_{14}H_{12}S$ o,o'-$C_6H_4CH_2CH_2C_6H_4SnS$, 2, 508

$SnC_{14}H_{12}S_3$ (2-Thienyl)$_3SnCH{=}CH_2$, 2, 193

$SnC_{14}H_{13}Br$ $Ph_2(CH_2{=}CH)SnBr$, 2, 294

$SnC_{14}H_{13}Cl$ $Ph_2(CH_2{=}CH)SnCl$, 2, 294

$SnC_{14}H_{13}F$ $Ph_2(CH_2{=}CH)SnF$, 2, 294

$SnC_{14}H_{13}I$ $Ph_2(CH_2{=}CH)SnI$, 2, 294

$SnC_{14}H_{14}$ o,o'-$C_6H_4CH_2CH_2C_6H_4SnH_2$, 2, 508

$SnC_{14}H_{14}Br_2$ $(PhCH_2)_2SnBr_2$, 2, 316

(p-$MeC_6H_4)_2SnBr_2$, 2, 320

$SnC_{14}H_{14}Br_2O_2$ (p-$MeOC_6H_4)_2SnBr_2$, 2, 326

$SnC_{14}H_{14}Br_6$ $(C_7H_7)_2SnBr_6$, 2, 322

$SnC_{14}H_{14}Cl_2$ $(PhCH_2)_2SnCl_2$, 2, 305, 513

(o-$MeC_6H_4)_2SnCl_2$, 2, 319

(p-$MeC_6H_4)_2SnCl_2$, 2, 320

$SnC_{14}H_{14}Cl_2O_2$ (o-$MeOC_6H_4)_2SnCl_2$, 2, 326

(p-$MeOC_6H_4)_2SnCl_2$, 2, 326

$SnC_{14}H_{14}Cl_4N_2$ $(ClCH{=}CH)_2SnCl_2$·2 Py, 2, 325

$SnC_{14}H_{14}Cl_6$ $(C_7H_7)_2SnCl_6$, 2, 322

$SnC_{14}H_{14}I_2$ $(PhCH_2)_2SnI_2$, 2, 316

$SnC_{14}H_{14}O$ $(PhCH_2)_2SnO$, 2, 368, 513, 516

(o-$MeC_6H_4)_2SnO$, 2, 372

(m-$MeC_6H_4)_2SnO$, 2, 373

(p-$MeC_6H_4)_2SnO$, 2, 374

$SnC_{14}H_{14}O_2$ o,o'-$Me_2SnO(C_6H_4)_2O$, 2, 396

$SnC_{14}H_{14}O_3$ (p-$MeOC_6H_4)_2SnO$, 2, 374

$SnC_{14}H_{14}S_2$ $Ph_2SnSCH_2CH_2S$, 2, 463

$SnC_{14}H_{15}Cl$ $EtPh_2SnCl$, 2, 289

$SnC_{14}H_{16}Cl_2N_2$ Me_2SnCl_2·1,10-phenanthroline, 2, 304

$SnC_{14}H_{16}O$ $EtPh_2SnOH$, 2, 364

$SnC_{14}H_{16}O_2$ $Ph_2Sn(OMe)_2$, 2, 399, 514

$SnC_{14}H_{16}O_6$ (o-$HOC_6H_4CO_2)_2SnMe_2$, 2, 444

$SnC_{14}H_{17}N_2$ t-$BuSnBr_3$·2,2'-bipyridyl, 2, 331*

$SnC_{14}H_{18}Cl_2N_2$ Et_2SnCl_2·2,2'-bipyridyl, 2, 307

$SnC_{14}H_{18}O_4P_2$ $Me_2Sn(O_2PPhH)_2$, 2, 348*

$SnC_{14}H_{19}Br_5O_2$ $Bu_2Sn(OH)OC_6Br_5$, 2, 402*

$SnC_{14}H_{19}ClF_2$ $Et_3SnC_6H_4CF{=}CFCl$, 2, 207

$SnC_{14}H_{19}Cl_3N_2$ $BuSnCl_3$·2Py, 2, 329

$SnC_{14}H_{19}Cl_5O_2$ $Bu_2Sn(OH)OC_6Cl_5$, 2, 402

$SnC_{14}H_{19}NO_2$ N-(Phthalimido)$SnEt_3$, 2, 480

$SnC_{14}H_{20}$ $Et_2Sn(C_5H_5)_2$, 2, 201

$Et_3SnC{\equiv}CPh$, 2, 211

$SnC_{14}H_{20}Br_2N_2$ Et_2SnBr_2·2Py, 2, 308

$SnC_{14}H_{20}Cl_2O_2$ (i-$C_3H_6Cl)_2PhSnOAc$, 2, 432

$SnC_{14}H_{20}Cl_4$ $ClC_6H_4(C_4H_8Cl)_2SnCl$, 2, 298

$SnC_{14}H_{20}I_2N_2$ Me_2SnI_2·2 H_2NPh, 2, 305

Me_2SnI_2·2 MeC_5H_4N, 2, 305

Et_2SnI_2·2 Py, 2, 308

$SnC_{14}H_{20}N_2O_8$ $(CH_2{=}CH)_2Sn[H(O_2CCH_2)_2NCH_2]_2$, 2, 446

$SnC_{14}H_{21}Br_3O_2$ Bu_2(2,4,6-$Br_3C_6H_2O)SnOH$, 2, 402

$SnC_{14}H_{21}Cl_3O_2$ Bu_2(2,4,6-$Cl_3C_6H_2O)SnOH$, 2, 402

$SnC_{14}H_{21}Cl_3O_4$ $PhSnCl_3$·2 $C_4H_8O_2$, 2, 330

$SnC_{14}H_{22}$ $Et_3SnCH{=}CHPh$, 2, 200

p-$Et_3SnC_6H_4CH{=}CH_2$, 2, 201, 510

$SnC_{14}H_{22}O_2$ o-$Bu_2SnOC_6H_4O$, 2, 405

$SnC_{14}H_{23}Br$ $Bu_2PhSnBr$, 2, 289

$SnC_{14}H_{23}Cl$ $Bu_2PhSnCl$, 2, 289

$SnC_{14}H_{24}$ m-$MeC_6H_4CH_2SnEt_3$, 2, 182

p-$MeC_6H_4CH_2SnEt_3$, 2, 182

$Et_3SnCH_2CH_2Ph$, 2, 182

$(CH_2{=}CH)_2Sn(CH_2CHMeCH{=}CH_2)_2$, 2, 206

$SnC_{14}H_{24}O$ p-$Et_3SnC_6H_4CHOHMe$, 2, 226

$SnC_{14}H_{24}O_2$ p-$MeOC_6H_4CH_2OSnEt_3$, 2, 390

$SnC_{14}H_{24}O_4$ $Pr_2Sn(O_2CCMe{=}CH_2)_2$, 2, 439, 511

$SnC_{14}H_{24}O_6$ $Bu_2Sn(O_2CCHCH_2O)_2$, 2, 444*

$SnC_{14}H_{25}F_5O_2$ $Me(n\text{-}C_5H_{11})_2SnO_2CC_2F_5$, 2, 431

$SnC_{14}H_{25}N$ $p\text{-}Et_3SnC_6H_4NMe_2$, 2, 241

$SnC_{14}H_{25}O_2P$ Et_3SnO_2PEtPh, 2, 350

$SnC_{14}H_{25}P$ $Et_3SnPEtPh$, 2, 482

$SnC_{14}H_{26}$ $n\text{-}C_8H_{17}Sn(CH{=}CH_2)_3$, 2, 204

$CH{=}CHCH_2CH_2CH{=}CHSnBu_2$, 2, 507*

$SnC_{14}H_{26}N_2$ $i\text{-}Bu_2Sn(CH_2CH_2CN)_2$, 2, 224

$SnC_{14}H_{26}N_2S_4$ $(2\text{-}C_3H_4NS)_2S_2SnBu_2$, 2, 464

$SnC_{14}H_{26}O_4$ $Et_3SnC(CO_2Et){=}CHCO_2Et$, 2, 208

Et_2Sn sebacate, 2, 445

Bu_2Sn adipate, 2, 445

$(SnC_{14}H_{26}O_4)_n$ $(Bu_2Sn$ adipate$)_n$, 2, 445

$SnC_{14}H_{26}O_4S$ $Bu_2Sn(O_2CCH_2CH_2)_2S$, 2, 445

$SnC_{14}H_{26}O_5$ $Et_3SnC(CO_2Me){=}COEtCO_2Me$, 2, 208

$Bu_2Sn(OMe)O_2CCH{=}CHCO_2Me$, 2, 407

$SnC_{14}H_{27}Br$ $n\text{-}C_{10}H_{21}(CH_2{=}CH)_2SnBr$, 2, 294

$SnC_{14}H_{27}Cl_3O_2$ $Bu_3SnO_2CCCl_3$, 2, 422

$SnC_{14}H_{27}F_3$ $Bu_3SnCF{=}CF_2$, 2, 208

$SnC_{14}H_{27}F_5$ $Bu_3SnC_2F_5$, 2, 221

$SnC_{14}H_{28}$ $Cyclohexyl_2SnMe_2$, 2, 180

$Bu_2Sn(CHCH_2CH_2)_2$, 2, 185

2-Cyclohexenyl-$CH_2CH_2SnEt_3$, 2, 200

$Bu_2Sn(CH_2CH{=}CH_2)_2$, 2, 203

$Bu_3SnC{\equiv}CH$, 2, 213

$Sn(CH_2CH_2CMe_2CH_2CH_2)_2$, 2, 506, 514

$SnC_{14}H_{28}Br_2$ $CH_2CH_2CMe_2CH_2CH_2Sn(C_7H_{14}\text{-}Br)Br$, 2, 508

$SnC_{14}H_{28}Br_4$ $(BrCH_2CH_2CMe_2CH_2CH_2)_2\text{-}SnBr_2$, 2, 325

$SnC_{14}H_{28}Cl_2$ $CH_2CH_2CMe_2CH_2CH_2Sn\text{-}(C_7H_{14}Cl)Cl$, 2, 508

$SnC_{14}H_{28}Cl_2O_2$ $Bu_3SnO_2CCHCl_2$, 2, 422

$SnC_{14}H_{28}I_2$ $CH_2CH_2CMe_2CH_2CH_2Sn(C_7H_{14}\text{-}I)I$, 2, 508

$SnC_{14}H_{28}O_2$ $Pr_3SnC(CO_2Me){=}CHMe$, 2, 208

$Bu_2Sn(OCH_2CH{=}CH_2)_2$, 2, 398

$SnC_{14}H_{28}O_4$ $Et_2Sn(CH_2CO_2Pr)_2$, 2, 235

$Et_2Sn(CH_2CHMeCO_2Me)_2$, 2, 235

$Pr_2Sn(CH_2CO_2Et)_2$, 2, 236

$Pr_2Sn(CH_2CH_2CO_2Me)_2$, 2, 236

$SnC_{14}H_{28}O_5$ $Bu_2Sn(OMe)O_2CCH_2CH_2CO_2Me$, 2, 407

$SnC_{14}H_{28}O_6$ $Bu_2Sn(O_2CCHOHMe)_2$, 2, 444

$SnC_{14}H_{28}S_3$ $2,5\text{-}Bu_3SnS(1,3,4\text{-}C_2N_2S)SH$, 2, 460

$SnC_{14}H_{29}BrO_2$ $Bu_3SnO_2CCH_2Br$, 2, 422

$SnC_{14}H_{29}Br_5N_2$ $PhSnBr_3\cdot 2Me_4NBr$, 2, 332

$SnC_{14}H_{29}ClO_2$ $Bu_3SnO_2CCH_2Cl$, 2, 422

$SnC_{14}H_{29}Cl_2N_2O_4P$ $PhSnPO_4\cdot 2Me_4NCl$, 2, 357

$SnC_{14}H_{29}Cl_5N_2$ $PhSnCl_3\cdot 2Me_4NCl$, 2, 330

$SnC_{14}H_{29}I$ Cyclohexyl-Bu_2SnI, 2, 289

$SnC_{14}H_{29}IO_2$ $Bu_3SnO_2CCH_2I$, 2, 422

$SnC_{14}H_{29}N$ Bu_3SnCH_2CN, 2, 224

$SnC_{14}H_{29}N_3$ N-(1,2,4-Triazolyl)$SnBu_3$, 2, 478

$SnC_{14}H_{30}$ $Et_3Sn(CH_2)_6CH{=}CH_2$, 2, 201

$Bu_3SnCH{=}CH_2$, 2, 195

$i\text{-}Bu_3SnCH{=}CH_2$, 2, 203

$SnC_{14}H_{30}Br_2$ $(C_7H_{15})_2SnBr_2$, 2, 318

$SnC_{14}H_{30}N_2S_4$ $(C_6H_{13})_2Sn(S_2CNH_2)_2$, 2, 472

$SnC_{14}H_{30}OS$ Bu_3SnSAc, 2, 471

$SnC_{14}H_{30}OS_2$ $i\text{-}Pr_3SnS_2COBu$, 2, 471

Bu_3SnS_2COMe, 2, 471

$SnC_{14}H_{30}O_2$ $Pr_3SnCH_2CO_2Pr$, 2, 235

Bu_3SnOAc, 2, 410, 513

$i\text{-}Bu_3SnOAc$, 2, 414

$Me_2(n\text{-}C_{10}H_{21})SnOAc$, 2, 430

$Et_2(n\text{-}C_8H_{17})SnOAc$, 2, 430

$Et_2(C_8H_{17})SnOAc$, 2, 430

$Et(n\text{-}C_5H_{11})_2SnOAc$, 2, 430

$SnC_{14}H_{30}O_2S$ $Bu_3SnO_2CCH_2SH$, 2, 423

$Bu_3SnSCH_2CO_2H$, 2, 461, 516

$SnC_{14}H_{30}O_3$ Bu_3SnOCO_2Me, 2, 413

$Bu_3SnO_2CCH_2OH$, 2, 423

$SnC_{14}H_{31}Br$ $Me_2(C_{12}H_{25})SnBr$, 2, 287

$Bu(Me_3CCH_2)_2SnBr$, 2, 289

$SnC_{14}H_{31}I$ $Me_2(C_{12}H_{25})SnI$, 2, 287

$Bu_2(n\text{-}C_6H_{13})SnI$, 2, 289

$SnC_{14}H_{31}NO_2$ $Bu_2Sn(OCH_2CH_2)_2{=}NEt$, 2, 404

$SnC_{14}H_{31}NO_4$ $Bu_2Sn(OAc)_2\cdot NH_2Et$, 2, 434

$SnC_{14}H_{32}$ $n\text{-}C_8H_{17}SnEt_3$, 2, 182

$(n\text{-}C_5H_{11})_2SnEt_2$, 2, 183

$Et_2Sn(CH_2CMe_3)_2$, 2, 183

$EtSnBu_3$, 2, 184
$i\text{-}Pr_2SnBu_2$, 2, 185
$SnC_{14}H_{32}Cl_2$ $(Et_2SnClCH_2CH_2CH_2)_2$, 2, 301
$SnC_{14}H_{32}O$ Bu_3SnOEt, 2, 385
$SnC_{14}H_{32}O_2$ $Pr_2Sn(OBu)_2$, 2, 397
$Bu_2Sn(OPr)_2$, 2, 397
$Bu_2Sn(OCHMe_2)_2$, 2, 398
$SnC_{14}H_{32}O_4$ $Et_2Sn[(CH_2)_3OCH_2CH_2OH]_2$, 2, 226
$SnC_{14}H_{32}O_6S_2$ $(i\text{-}C_6H_{13})_2Sn(O_3SMe)_2$, 2, 354
$SnC_{14}H_{32}S$ $Bu_3SnCH_2CH_2SH$, 2, 239
$SnC_{14}H_{33}N$ Bu_3SnNMe_2, 2, 476
$SnC_{14}H_{34}Cl_4N_2$ $(CH_2{=}CHCH_2)_2SnCl_2\cdot 2Me_4NCl$, 2, 323
$SnC_{14}H_{34}N_2$ $Me_2Sn(NPr_2)_2$, 2, 475
$SnC_{14}H_{39}Cl_3N_2$ $Et_3SnCl\cdot 2Me_4NCl$, 2, 274
$SnC_{15}H_{14}IN$ $Ph_2(NCCH_2CH_2)SnI$, 2, 299
$SnC_{15}H_{14}O_2$ $Ph_2SnCH_2CH_2CO_2$, 2, 432
$SnC_{15}H_{15}Br$ $Ph_2(CH_2{=}CHCH_2)SnBr$, 2, 294
$SnC_{15}H_{15}Cl$ $(C_5H_5)_3SnCl$, 2, 292
$SnC_{15}H_{16}O_2$ $Ph(p\text{-}CH_2{=}CHCH_2C_6H_4)Sn(OH)_2$, 2, 372
$SnC_{15}H_{16}S_2$ $Ph_2SnS(CH_2)_3S$, 2, 463
$SnC_{15}H_{17}BrO$ $Ph_2(HOCH_2CH_2CH_2)SnBr$, 2, 300
$SnC_{15}H_{17}Cl_5N_4$ $MeSnCl_3\cdot 2p\text{-}MeC_6H_4N_2Cl$, 2, 328
$SnC_{15}H_{18}$ $p\text{-}PhC_6H_4SnMe_3$, 2, 180
$SnC_{15}H_{19}N$ Me_3SnNPh_2, 2, 475
$SnC_{15}H_{19}O_2P$ $Me_3SnO_2PPh_2$, 2, 348
$SnC_{15}H_{19}P$ Me_3SnPPh_2, 2, 482
$SnC_{15}H_{20}$ $(CH_2CH_2CH)_3SnPh$, 2, 185
$SnC_{15}H_{20}ClNO$ 8-(Et_3SnO)-5-chloroquinoline, 2, 387
$SnC_{15}H_{20}O_6$ $PhSn(O_2CEt)_3$, 2, 450
$SnC_{15}H_{21}Cl_5O$ $Pr_3SnOC_6Cl_5$, 2, 389
$SnC_{15}H_{21}NO$ 8-(Et_3SnO)quinoline, 2, 387
$SnC_{15}H_{22}$ 1-Indenyl-$SnEt_3$, 2, 200
$SnC_{15}H_{22}O_2$ $Et_3SnO_2CCH{=}CHPh$, 2, 418, 511
$SnC_{15}H_{24}$ $CH_2{=}CHCH_2C_6H_4SnEt_3$, 2, 201
$p\text{-}CH_2{=}CMeC_6H_4SnEt_3$, 2, 201, 510
$SnC_{15}H_{25}Br$ $p\text{-}BrC_6H_4SnPr_3$, 2, 221
$SnC_{15}H_{25}NO_2$ Pr_3Sn 3-pyridinecarboxylate, 2, 416
$Pr_2(4,2\text{-}EtC_5H_3N)SnOAc$, 2, 432
$SnC_{15}H_{26}$ $Et_3SnCH_2CHMePh$, 2, 182
Pr_3SnPh, 2, 184
$i\text{-}Pr_3SnPh$, 2, 185
$SnC_{15}H_{26}O$ Pr_3SnOPh, 2, 385
$SnC_{15}H_{26}O_2$ $Et_3SnOOCMe_2Ph$, 2, 379
$p\text{-}MeOC_6H_4CHMeOSnEt_3$, 2, 390
$SnC_{15}H_{26}S$ Pr_3SnSPh, 2, 459
$i\text{-}Pr_3SnSPh$, 2, 459
$SnC_{15}H_{27}Br$ $Cyclopentyl_3SnBr$, 2, 284
$SnC_{15}H_{27}F_7$ $Bu_3SnC_3F_7$, 2, 221
$SnC_{15}H_{28}O_2$ $Bu_3SnO_2CC{\equiv}CH$, 2, 418
$SnC_{15}H_{28}O_3$ $Bu_2Sn(OMe)$ sorbate, 2, 407
$SnC_{15}H_{28}O_4S_2$ $Bu_2Sn(O_2CCH_2S)_2CMe_2$ (?), 2, 445
$SnC_{15}H_{28}O_5$ $Et_3SnC(CO_2Et){=}COMeCO_2Et$, 2, 208
$Et_3SnC(CO_2Me){=}COPrCO_2Me$, 2, 208
$SnC_{15}H_{28}O_6$ $PrSn(CH_2CH_2CO_2Me)_3$, 2, 236
$SnC_{15}H_{28}O_7$ $Me_3SnOC(CH_2CO_2Et)_2CO_2Et$, 2, 390
$SnC_{15}H_{29}NO_2$ $Bu_3SnO_2CCH_2CN$, 2, 424
$SnC_{15}H_{30}Cl_3NO$ $Bu_3SnN{=}C(OMe)CCl_3$, 2, 480
$SnC_{15}H_{30}N_2$ N-(Pyrazolyl)$SnBu_3$, 2, 478
N-(Imidazolyl)$SnBu_3$, 2, 478
$SnC_{15}H_{30}O_2$ $Et_3SnC{\equiv}CCH_2OCHMeOBu$, 2, 215
Cyclohexenylperoxy-$SnPr_3$, 2, 379
$Bu_3SnO_2CCH{=}CH_2$, 2, 417, 511
$SnC_{15}H_{30}O_3$ Bu_3SnO_2CAc, 2, 423
$SnC_{15}H_{31}BrO_2$ $Bu_3SnO_2CCH_2CH_2Br$, 2, 422
$Bu_3SnO_2CCHBrMe$, 2, 422
$SnC_{15}H_{31}ClO_2$ $Bu_3SnO_2CCHClMe$, 2, 422
$SnC_{15}H_{31}IO_2$ $Bu_3SnO_2CCH_2CH_2I$, 2, 422
$SnC_{15}H_{31}N$ $Bu_3SnCH_2CH_2CN$, 2, 224
$SnC_{15}H_{32}$ $Bu_3SnCHCH_2CH_2$, 2, 186
$Pr_3SnCH{=}CHBu$, 2, 202*
$Bu_3SnCH_2CH{=}CH_2$, 2, 202
$Bu_2SnCH_2CH_2CMe_2CH_2CH_2$, 2, 507
$SnC_{15}H_{32}O$ Bu_3SnCH_2Ac, 2, 230
$SnC_{15}H_{32}O_2$ $Pr_3Sn(CH_2)_3CO_2Et$, 2, 235
$Pr_3SnCHMeCH_2CO_2Et$, 2, 235
$Bu_3SnCH_2CH_2CO_2H$, 2, 236
$Me_3SnO_2CC_{11}H_{23}$, 2, 411
Bu_3SnO_2CEt, 2, 413
$SnC_{15}H_{32}O_2S$ $i\text{-}C_7H_{15}O_2CCH_2SSnEt_3$, 2,

461
$SnC_{15}H_{32}O_3$ $Bu_3SnO_2CCHMeOH$, 2, 423
$SnC_{15}H_{33}Br$ $(Me_3CCH_2)_3SnBr$, 2, 284
$(i\text{-}C_5H_{11})_3SnBr$, 2, 283
$SnC_{15}H_{33}Cl$ $(C_5H_{11})_3SnCl$, 2, 283
$(i\text{-}C_5H_{11})_3SnCl$, 2, 283
$(Me_3CCH_2)_3SnCl$, 2, 284
$SnC_{15}H_{33}F$ $(i\text{-}C_5H_{11})_3SnF$, 2, 283
$SnC_{15}H_{33}I$ $(C_5H_{11})_3SnI$, 2, 283
$(i\text{-}C_5H_{11})_3SnI$, 2, 283
$(Me_3CCH_2)_3SnI$, 2, 284
$SnC_{15}H_{33}NO_2$ $Bu_2Sn(OCHMeCH_2)_2{=}NMe$, 2, 404
$SnC_{15}H_{34}$ $n\text{-}C_{12}H_{25}SnMe_3$, 2, 179
$n\text{-}C_9H_{19}SnEt_3$, 2, 183
$Et(i\text{-}C_5H_{11})SnPr$, 2, 184
$SnC_{15}H_{34}O$ $Bu_3Sn(CH_2)_3OH$, 2, 226
$SnC_{15}H_{34}S$ $Bu_3Sn(CH_2)_3SH$, 2, 239
$Me_3SnSC_{12}H_{25}$, 2, 458
$SnC_{15}H_{36}IOP$ $Pr_3SnI \cdot OPEt_3$, 2, 282
$SnC_{16}H_6F_{10}$ $(CH_2{=}CH)_2Sn(C_6F_5)_2$, 2, 221
$SnC_{16}H_8Cl_4S_4$ 2,5-Chlorothienyl$_4$Sn, 2, 171
$SnC_{16}H_{10}Cl_6O_4$ $Ph_2Sn(O_2CCCl_3)_2$, 2, 443
$SnC_{16}H_{10}F_6$ $Ph_2Sn(CF{=}CF_2)_2$, 2, 209
$SnC_{16}H_{10}F_{10}$ $Et_2Sn(C_6F_5)_2$, 2, 221
$SnC_{16}H_{12}N_4S_6$ $[2,5\text{-}HS(1,3,4\text{-}C_2N_2S)S]_2\text{-}SnPh_2$, 2, 464*
$SnC_{16}H_{12}S_4$ 2-Thienyl$_4$Sn, 2, 171
$SnC_{16}H_{13}ClS$ 2-Thienyl-Ph_2SnCl, 2, 291
$SnC_{16}H_{14}Br_4O_2$ $Et_2SnOC_6H_2Br_2C_6H_2Br_2O$, 2, 402
$SnC_{16}H_{14}Cl_2O_4$ $Ph_2Sn(O_2CCH_2Cl)_2$, 2, 443
$SnC_{16}H_{14}N_2S_4$ $Me_2Sn(2\text{-}SC_7H_4NS)_2$, 2, 464
$SnC_{16}H_{14}O_5$ $(o\text{-}MeO_2CC_6H_4)_2SnO$, 2, 374
$SnC_{16}H_{15}Cl_3N_2$ $PhSnCl_3 \cdot 2$ Py, 2, 330
$SnC_{16}H_{16}$ $Ph_2Sn(CH{=}CH_2)_2$, 2, 197
$SnC_{16}H_{16}FN$ $Ph_2(NCCH_2CH_2CH_2)SnF$, 2, 299
$SnC_{16}H_{16}IN$ $Ph_2(NCCH_2CH_2CH_2)SnI$, 2, 299
$SnC_{16}H_{16}O_4$ $(p\text{-}HO_2CC_6H_4)_2SnMe_2$, 2, 234
$Me_2Sn(OBz)_2$, 2, 435
$Ph_2Sn(OAc)_2$, 2, 438
$SnC_{16}H_{17}IO_2$ $Ph_2(MeO_2CCH_2CH_2)SnI$, 2, 300
$SnC_{16}H_{18}$ $MeSn(C_5H_5)_3$, 2, 200
o,o'-$Et_2SnC_6H_4C_6H_4$, 2, 506
o,o'-$Me_2SnC_6H_4CH_2CH_2C_6H_4$, 2, 507*
$SnC_{16}H_{18}Br_2Cl_4N_4$ $Et_2SnCl_2 \cdot 2$ p-$BrC_6H_4\text{-}N_2Cl$, 2, 307
$SnC_{16}H_{18}Cl_2$ $(o\text{-}MeC_6H_4CH_2)_2SnCl_2$, 2, 316
$(m\text{-}MeC_6H_4CH_2)_2SnCl_2$, 2, 316
$(p\text{-}MeC_6H_4CH_2)_2SnCl_2$, 2, 316
$SnC_{16}H_{18}Cl_2O_2$ $(p\text{-}MeOC_6H_4CH_2)_2SnCl_2$, 2, 326
$(p\text{-}EtOC_6H_4)_2SnCl_2$, 2, 326
$SnC_{16}H_{18}Cl_6N_4$ $Et_2SnCl_2 \cdot 2$ p-$ClC_6H_4N_2Cl$, 2, 307
$SnC_{16}H_{18}N_2O_4$ Et_2Sn di-3-pyridinecarboxylate, 2, 439
$SnC_{16}H_{18}O_3$ $(p\text{-}EtOC_6H_4)_2SnO$, 2, 374
$SnC_{16}H_{19}Br$ $BuPh_2SnBr$, 2, 289
$SnC_{16}H_{19}Cl$ $BuPh_2SnCl$, 2, 289
$SnC_{16}H_{20}$ $(p\text{-}MeC_6H_4)_2SnMe_2$, 2, 180
Et_2SnPh_2, 2, 183
$SnC_{16}H_{20}Br_2N_2$ $(CH_2{=}CHCH_2)_2SnBr_2 \cdot 2$ Py, 2, 322
$SnC_{16}H_{20}Cl_2N_2$ $Et_2SnCl_2 \cdot 9,10$-phenanthroline, 2, 307
$(p\text{-}Me_2NC_6H_4)_2SnCl_2$, 2, 327
$SnC_{16}H_{20}Cl_4N_4$ $Et_2SnCl_2 \cdot 2$ PhN_2Cl, 2, 307
$SnC_{16}H_{20}O_6S_2$ $Ph_2Sn(O_3SEt)_2$, 2, 357
$SnC_{16}H_{20}S_2$ $(p\text{-}MeC_6H_4S)_2SnMe_2$, 2, 462
$SnC_{16}H_{22}Cl_2N_2$ $Pr_2SnCl_2 \cdot 2,2'$-bipyridyl, 2, 309
$SnC_{16}H_{22}N_2$ $Ph_2Sn(NMe_2)_2$, 2, 476
$SnC_{16}H_{22}N_2S_2$ 2,3-(Bu_2SnS_2)quinoxaline, 2, 464
$SnC_{16}H_{22}O$ $2\text{-}C_{10}H_7OSnEt_3$, 2, 384
$Et_3SnOC_{10}H_7$, 2, 514
$SnC_{16}H_{22}O_4$ p-$Bu_2Sn(O_2C)_2C_6H_4$ (?), 2, 445
$SnC_{16}H_{22}O_4P_2$ $Me_2Sn(O_2PMePh)_2$, 2, 348
$SnC_{16}H_{22}O_4S$ $BuPhSn(O_2CCH_2CH_2)_2S$, 2, 446
$SnC_{16}H_{22}S$ $2\text{-}C_{10}H_7SSnEt_3$, 2, 458
$SnC_{16}H_{24}Br_2N_2$ $Pr_2SnBr_2 \cdot 2$ Py, 2, 309
$SnC_{16}H_{24}I_2N_2$ $Pr_2SnI_2 \cdot 2$ Py, 2, 309
$SnC_{16}H_{24}N_4$ $(NCCHMeCH_2)_4Sn$, 2, 175

SnC$_{16}$H$_{26}$ p-(Et$_3$SnCH$_2$CH$_2$)C$_6$H$_4$CH=CH$_2$, 2, 200
SnC$_{16}$H$_{26}$N$_2$O$_4$ Bu$_2$Sn(CHCNCO$_2$Me)$_2$, 2, 224
(N-Succinimido)$_2$SnBu$_2$, 2, 481
SnC$_{16}$H$_{26}$O$_2$ Pr$_3$SnOBz, 2, 412
i-Pr$_3$SnOBz, 2, 413
Pr$_2$(PhCH$_2$CH$_2$)SnOAc, 2, 430
SnC$_{16}$H$_{26}$O$_6$S 3,5-HO$_3$S(HO)C$_6$H$_3$CO$_2$Sn-Pr$_3$ (?), 2, 429
SnC$_{16}$H$_{27}$N 4-Pyridyl-CH$_2$CH$_2$SnPr$_3$, 2, 192
SnC$_{16}$H$_{28}$O$_4$ Bu$_2$Sn(O$_2$CCMe=CH$_2$)$_2$, 2, 439, 511
i-Bu$_2$Sn(O$_2$CCMe=CH$_2$)$_2$, 2, 439, 511*
(n-C$_6$H$_{13}$)$_2$Sn(O$_2$CCH=)$_2$, 2, 446
SnC$_{16}$H$_{28}$O$_5$ Bu$_2$Sn(OMe)O$_2$CCH=CHCO$_2$CH$_2$-CH=CH$_2$, 2, 407
SnC$_{16}$H$_{28}$O$_8$S$_2$ Bu$_2$Sn[SCH(CO$_2$H)CH$_2$CO$_2$H]$_2$, 2, 466
SnC$_{16}$H$_{28}$S Pr$_3$SnSCH$_2$Ph, 2, 458
o-MeC$_6$H$_4$SSnPr$_3$, 2, 459
p-MeC$_6$H$_4$SSnPr$_3$, 2, 459
i-Pr$_3$SnSCH$_2$Ph, 2, 459
p-MeC$_6$H$_4$SSnCHMe$_2$, 2, 459
SnC$_{16}$H$_{29}$N 2,4-Bu(Et$_3$SnCH$_2$)pyridine, 2, 192
SnC$_{16}$H$_{30}$ n-C$_{10}$H$_{21}$Sn(CH=CH$_2$)$_3$, 2, 204
Bu$_3$SnC≡CCH=CH$_2$, 2, 213
SnC$_{16}$H$_{30}$Cl$_2$O$_4$S$_2$ Bu$_2$Sn(SCH$_2$CO$_2$CH$_2$CH$_2$-Cl)$_2$, 2, 465
SnC$_{16}$H$_{30}$N 3-Pyrryl-SnBu$_3$, 2, 189
SnC$_{16}$H$_{30}$O$_2$ Et$_3$SnO$_2$CCHCH(CH=CMe$_2$)CMe$_2$, 2, 418
SnC$_{16}$H$_{30}$O$_4$S (Me$_2$EtC)$_2$Sn(O$_2$CCH$_2$CH$_2$)$_2$S, 2, 446
SnC$_{16}$H$_{30}$O$_5$ Bu$_2$Sn(OH)O$_2$CCH=CHCO$_2$Bu, 2, 441
SnC$_{16}$H$_{30}$O$_6$ BuSn(CH$_2$CH$_2$CO$_2$Me)$_3$, 2, 236
i-BuSn(CH$_2$CH$_2$CO$_2$Me)$_3$, 2, 237
BuSn(O$_2$CPr)$_3$, 2, 449
BuSn(O$_2$CCHMe$_2$)$_3$, 2, 449
SnC$_{16}$H$_{31}$I Cyclohexyl$_2$BuSnI, 2, 289
SnC$_{16}$H$_{31}$N N-(Bu$_3$Sn)C$_4$H$_4$N, 2, 478
SnC$_{16}$H$_{31}$NO$_2$ Succinimido-SnBu$_3$, 2, 480
SnC$_{16}$H$_{32}$ Bu$_2$Sn(CH$_2$CMe=CH$_2$)$_2$, 2, 203
SnC$_{16}$H$_{32}$O$_2$ Et$_3$SnC≡CCH$_2$CH$_2$OCHMeOBu, 2, 215
Et$_3$SnC≡CCHMeOCHMeOBu, 2, 215
Bu$_3$SnO$_2$CCH=CHMe, 2, 419
Bu$_3$SnO$_2$CCMe=CH$_2$, 2, 417, 511
i-Bu$_3$SnO$_2$CCMe=CH$_2$, 2, 419, 511
SnC$_{16}$H$_{32}$O$_2$S$_4$ Bu$_2$Sn(S$_2$COCHMe$_2$)$_2$, 2, 471
SnC$_{16}$H$_{32}$O$_4$ i-Bu$_2$Sn(CH$_2$CH$_2$CO$_2$Me)$_2$, 2, 237*
Et$_2$Sn(O$_2$CC$_5$H$_{11}$)$_2$, 2, 436
SnC$_{16}$H$_{33}$NO$_3$ Bu$_3$SnO$_2$CCH$_2$NHAc, 2, 424
SnC$_{16}$H$_{34}$ Bu$_3$SnCH$_2$CH$_2$CH=CH$_2$, 2, 202
Bu$_3$SnCH$_2$CMe=CH$_2$, 2, 202
Bu$_2$(n-C$_6$H$_{13}$)SnCH=CH$_2$, 2, 203
SnC$_{16}$H$_{34}$Br$_2$ (C$_8$H$_{17}$)$_2$SnBr$_2$, 2, 318
SnC$_{16}$H$_{34}$Cl$_2$ (n-C$_8$H$_{17}$)$_2$SnCl$_2$, 2, 313, 517
(BuEtCHCH$_2$)$_2$SnCl$_2$, 2, 318, 517
SnC$_{16}$H$_{34}$I$_2$ (C$_8$H$_{17}$)$_2$SnI$_2$, 2, 318
(BuEtCHCH$_2$)$_2$SnI$_2$, 2, 319
SnC$_{16}$H$_{34}$N$_2$O$_2$ Bu$_2$Sn(ON=CHPr)$_2$, 2, 404*
SnC$_{16}$H$_{34}$O (n-C$_8$H$_{17}$)$_2$SnO, 2, 370
(BuEtCHCH$_2$)$_2$SnO, 2, 372
(C$_8$H$_{17}$)$_2$SnO, 2, 514, 517
SnC$_{16}$H$_{34}$OS Bu$_3$SnCH$_2$CH$_2$SAc, 2, 239
SnC$_{16}$H$_{34}$OS$_2$ Bu$_3$SnS$_2$COCHMe$_2$, 2, 471
SnC$_{16}$H$_{34}$O$_2$ Bu$_3$SnCH$_2$CO$_2$Et, 2, 236
Bu$_3$SnCH$_2$CH$_2$CO$_2$Me, 2, 236
Bu$_3$SnO$_2$CPr, 2, 413
Bu$_3$SnO$_2$CCHMe$_2$, 2, 413
Me$_2$(n-C$_{12}$H$_{25}$)SnOAc, 2, 430
SnC$_{16}$H$_{35}$Cl i-Bu$_2$(n-C$_8$H$_{17}$)SnCl, 2, 289
SnC$_{16}$H$_{35}$I Et$_2$(n-C$_{12}$H$_{25}$)SnI, 2, 288
SnC$_{16}$H$_{36}$ Bu$_4$Sn, 2, 163
i-Bu$_4$Sn, 2, 165
s-Bu$_4$Sn, 2, 170
(i-C$_5$H$_{11}$)$_3$SnMe, 2, 181
n-C$_{10}$H$_{21}$SnEt$_3$, 2, 183
i-BuSnBu$_3$, 2, 186
t-Bu$_2$SnBu$_2$, 2, 186
(C$_8$H$_{17}$)$_2$SnH$_2$, 2, 264
SnC$_{16}$H$_{36}$O Bu$_3$SnOBu, 2, 385
Bu$_3$SnOCMe$_3$, 2, 385
(C$_5$H$_{11}$)$_3$SnOMe, 2, 386
SnC$_{16}$H$_{36}$O$_2$ Bu$_3$SnO$_2$CMe$_3$, 2, 378
Bu$_2$Sn(OBu)$_2$, 2, 398
SnC$_{16}$H$_{36}$O$_4$ Bu$_2$Sn(O$_2$CMe$_3$)$_2$, 2, 380

$SnC_{16}H_{36}O_6S_2$ $Bu_2Sn(O_3SBu)_2$, 2, 354

$SnC_{16}H_{36}S$ $Pr_3SnSC_7H_{15}$, 2, 458

i-$Pr_3SnSC_7H_{15}$, 2, 459

$SnC_{16}H_{36}S_2$ $Bu_2Sn(SCH_2CHMe_2)_2$, 2, 462

$SnC_{16}H_{36}S_3$ $MeSn(SCHMePr)_3$, 2, 468*

$BuSn(SBu)_3$, 2, 469

$SnC_{16}H_{37}N$ Bu_3SnNEt_2, 2, 476

$SnC_{16}H_{37}O_2PS_2$ $Bu_3SnS_2P(OEt)_2$, 2, 351

$SnC_{16}H_{37}O_4P$ $Bu_3SnO_2P(OEt)_2$, 2, 351

$SnC_{16}H_{38}O_6P_2$ $[PrP(OPr)O_2]_2SnEt_2$, 2, 350

$SnC_{16}H_{40}N_4$ $Sn(NEt_2)_4$, 2, 476

$SnC_{17}H_{17}Cl_3N_2$ o-$MeC_6H_4SnCl_3$·2 Py, 2, 330

p-$MeC_6H_4SnCl_3$·2 Py, 2, 332

$SnC_{17}H_{18}$ 9-Phenanthryl $SnMe_3$, 2, 180

4,1-$MeC_{10}H_6Sn(CH{=}CH_2)_3$, 2, 206

$CH_2{=}CHSn(C_5H_5)_3$, 2, 206

Cyclopentyl-Ph_2SnH, 2, 264

$SnC_{17}H_{18}O_2$ $Ph_2(CH_2{=}CHCH_2)SnOAc$, 2, 430

$SnC_{17}H_{20}$ $\overline{CH_2(CH_2)_4Sn}Ph_2$, 2, 507

$SnC_{17}H_{21}Cl_5N_4O_2$ $MeSnCl_3$·2 p-$EtOC_6H_4$-N_2Cl, 2, 328

$SnC_{17}H_{22}O_5S$ Pr_2Sn (S-benzoyl)thiomalate, 2, 445

$SnC_{17}H_{27}I$ Bu_2(p-$CH_2{=}CHCH_2C_6H_4)SnI$, 2, 294

$SnC_{17}H_{28}$ $Pr_3SnCH{=}CHPh$, 2, 201

$SnC_{17}H_{28}O$ Pr_3SnCH_2Bz, 2, 230

$SnC_{17}H_{30}$ $PhCH_2CH_2SnPr_3$, 2, 184

$SnC_{17}H_{30}N_2O$ p-$Et_3SnON{=}CHC_6H_4NEt_2$, 2, 391

$SnC_{17}H_{30}O_3$ Bu_3Sn 2-furanecarboxylate, 2, 416

$SnC_{17}H_{31}P$ $Pr_3SnPEtPh$, 2, 483

$SnC_{17}H_{32}$ $Bu_3SnC_5H_5$, 2, 202

$SnC_{17}H_{32}O_4$ $Pr_3SnC(CO_2Et){=}CHCO_2Et$, 2, 208

$SnC_{17}H_{32}O_4S_2$ $Bu_2Sn(O_2CCH_2CH_2S)_2CMe_2$, 2, 446

$SnC_{17}H_{32}O_5$ $Bu_2Sn(OMe)O_2CCH{=}CHCO_2Bu$, 2, 407

$SnC_{17}H_{36}O_2$ Bu_3SnO_2CBu, 2, 413

i-$BuCO_2SnBu_3$, 2, 413

$SnC_{17}H_{34}O_2$ $Et_3SnC{\equiv}CCMe_2OCHMeOBu$, 2, 216

$Bu_3SnCHAc_2$, 2, 230

$SnC_{17}H_{36}OS$ $Bu_3Sn(CH_2)_3SAc$, 2, 239

$SnC_{17}H_{36}O_2$ i-$Bu_3Sn(CH_2)_3OAc$, 2, 227

i-$Bu_3SnCH_2CHMeCO_2Me$, 2, 237

(n-$C_5H_{11})_3SnOAc$, 2, 414

(i-$C_5H_{11})_3SnOAc$, 2, 414

$SnC_{17}H_{37}NS_2$ $Bu_3SnS_2CNEt_2$, 2, 471

$SnC_{17}H_{38}$ $EtSn(CH_2CMe_3)_3$, 2, 184

i-$C_5H_{11}SnBu_3$, 2, 186

$SnC_{17}H_{38}O$ $Bu_3SnCH_2CH_2CMe_2OH$, 2, 226

$Bu_3SnOCHEt_2$, 2, 385

$SnC_{18}ClF_{15}$ $(C_6F_5)_3SnCl$, 2, 296

$SnC_{18}H_{11}Cl_5O_2$ $Ph_2Sn(OH)OC_6Cl_5$, 2, 402

$SnC_{18}H_{12}BrCl_3$ (p-$ClC_6H_4)_3SnBr$, 2, 296

$SnC_{18}H_{12}BrF_3$ (p-$FC_6H_4)_3SnBr$, 2, 296

$SnC_{18}H_{12}ClF_3$ (p-$FC_6H_4)_3SnCl$, 2, 296

$SnC_{18}H_{12}Cl_2N_2$ 2-$Quinolyl_2SnCl_2$, 2, 321

6-$Quinolyl_2SnCl_2$, 2, 321

$SnC_{18}H_{12}Cl_3I$ (p-$ClC_6H_4)_3SnI$, 2, 296

$SnC_{18}H_{12}Cl_4$ (p-$ClC_6H_4)_3SnCl$, 2, 296

$SnC_{18}H_{13}Br_3O$ (p-$BrC_6H_4)_3SnOH$, 2, 366

$SnC_{18}H_{13}Cl_3O$ (p-$ClC_6H_4)_3SnOH$, 2, 366

$SnC_{18}H_{13}F_3$ (p-$FC_6H_4)_3SnH$, 2, 266

$SnC_{18}H_{14}S_3$ (2-Thienyl)$_3SnPh$, 2, 193

$SnC_{18}H_{14}S_6$ $PhSn(SC_4H_3S)_3$, 2, 469

$SnC_{18}H_{15}Br$ Ph_3SnBr, 2, 280

$SnC_{18}H_{15}Cl$ Ph_3SnCl, 2, 278, 514

$SnC_{18}H_{15}D$ Ph_3SnD, 2, 265

$SnC_{18}H_{15}F$ Ph_3SnF, 2, 284

$SnC_{18}H_{15}I$ Ph_3SnI, 2, 280

$SnC_{18}H_{15}NO_2$ Ph_3SnNO_2, 2, 355

$SnC_{18}H_{15}NO_3$ Ph_3SnNO_3, 2, 355

$SnC_{18}H_{15}N_3$ Ph_3SnN_3, 2, 346

$SnC_{18}H_{15}O_3P$ Ph_3SnPO_3, 2, 355

Ph_2SnO_3PPh, 2, 356

$SnC_{18}H_{15}O_5P_2$ $PhSn(O_2PPh)(O_3PPh)$, 2, 357

$SnC_{18}H_{16}$ $Me_2Sn(C{\equiv}CPh)_2$, 2, 212

Ph_3SnH, 2, 260

$SnC_{18}H_{16}O$ Ph_3SnOH, 2, 363

$SnC_{18}H_{16}O_4$ $(PhCH_2)_2Sn(O_2CCH{=})_2$, 2, 445

$SnC_{18}H_{17}O_4P$ $Ph_3SnH_2PO_4$, 2, 355

$SnC_{18}H_{18}$ $\overline{CH{=}CHCH_2CH_2CH{=}CHSn}Ph_2$, 2, 507*

$SnC_{18}H_{18}Br_2O_4$ (p-$EtO_2CC_6H_4)_2SnBr_2$,

2, 326
$SnC_{18}H_{18}Cl_2O_4$ $(m\text{-}EtO_2CC_6H_4)_2SnCl_2$, 2, 326
$(p\text{-}EtO_2CC_6H_4)_2SnCl_2$, 2, 326
$SnC_{18}H_{18}I_2O_4$ $(p\text{-}EtO_2CC_6H_4)_2SnI_2$, 2, 326
$SnC_{18}H_{18}O_5$ $(m\text{-}EtO_2CC_6H_4)_2SnO$, 2, 374
$(p\text{-}EtO_2CC_6H_4)_2SnO$, 2, 374
$SnC_{18}H_{20}$ $Ph_2Sn(CH_2CH{=}CH_2)_2$, 2, 205
$SnC_{18}H_{20}INO$ N-(2-oxopyrolidino)CH_2-CH_2SnPh_2I, 2, 291
$SnC_{18}H_{20}O_2$ $(p\text{-}CH_2{=}CHCH_2C_6H_4)_2SnO\cdot H_2O$, 2, 373
$SnC_{18}H_{20}O_4$ $(PhCH_2)_2Sn(OAc)_2$, 2, 435
$Et_2Sn(OBz)_2$, 2, 436
$(MeC_6H_4)_2Sn(OAc)_2$, 2, 438
$SnC_{18}H_{20}O_4S_2$ $(p\text{-}MeO_2CC_6H_4S)_2SnMe_2$, 2, 465
$SnC_{18}H_{22}$ $BuPh_2SnCH{=}CH_2$, 2, 203
$SnC_{18}H_{22}Cl_2$ $Mesityl_2SnCl_2$, 2, 320
$SnC_{18}H_{22}N_2O_4$ Pr_2Sn di-3-pyridinecarboxylate, 2, 439
$SnC_{18}H_{22}N_2O_6$ $(p\text{-}O_2NC_6H_4O)_2SnPr_2$, 2, 404
$SnC_{18}H_{22}O$ $Mesityl_2SnO$, 2, 373
$SnC_{18}H_{23}Br$ $(n\text{-}C_6H_{13})Ph_2SnBr$, 2, 289
$SnC_{18}H_{24}$ $(PhCH_2)_2SnEt_2$, 2, 181
Pr_2SnPh_2, 2, 185
$SnC_{18}H_{24}Cl_4N_4$ $Et_2SnCl_2\cdot 2\ p\text{-}MeC_6H_4N_2Cl$, 2, 307
$SnC_{18}H_{24}I_2N_2$ $Me_2SnI_2\cdot 2\ o\text{-}MeC_6H_4NH_2$, 2, 305
$SnC_{18}H_{24}O_2$ $Et_2Sn(OCH_2Ph)_2$, 2, 396
$SnC_{18}H_{24}S_2$ $Pr_2Sn(SPh)_2$, 2, 462
$SnC_{18}H_{25}O_2P$ $Et_3SnO_2PPh_2$, 2, 350
$SnC_{18}H_{25}P$ Et_3SnPPh_2, 2, 482
$SnC_{18}H_{26}Br_2N_2$ $Bu_2SnBr_2\cdot$2,2'-bipyridyl, 2, 312
$SnC_{18}H_{26}Cl_2N_2$ $Bu_2SnCl_2\cdot$2,2'-bipyridyl, 2, 311
$SnC_{18}H_{26}I_2N_2$ $Bu_2SnI_2\cdot$2,2'-bipyridyl, 2, 312
$SnC_{18}H_{26}N_2$ $Et_3SnN(Ph)NHPh$, 2, 475
$SnC_{18}H_{26}O_6$ $PhSn(O_2CPr)_3$, 2, 450
$PhSn(O_2CCHMe_2)_3$, 2, 450
$SnC_{18}H_{27}Cl$ 1-Cyclohexenyl$_3$SnCl, 2, 292
$SnC_{18}H_{27}Cl_5O$ $Bu_3SnOC_6Cl_5$, 2, 389
$SnC_{18}H_{27}F_5$ $Bu_3SnC_6F_5$, 2, 221
$SnC_{18}H_{28}$ $Bu_2Sn(C_5H_5)_2$, 2, 203
$SnC_{18}H_{28}I_2N_2$ $Pr_2SnI_2\cdot 2\ PhNH_2$, 2, 309
$SnC_{18}H_{28}O$ Indenyl-Bu_2SnOMe, 2, 386
$SnC_{18}H_{28}O_4$ (2-Furfuryloxy)$_2SnBu_2$, 2, 401
$SnC_{18}H_{28}O_5$ $(o\text{-}MeO_2CC_6H_4CO_2)Bu_2SnOMe$, 2, 407
$SnC_{18}H_{28}O_8$ $Bu_2{}^{113}Sn(O_2CCH{=}CHCO_2Me)_2$, 2, 440
$Bu_2Sn(O_2CCH{=}CHCO_2{}^{14}CH_3)_2$, 2, 440
$(Pr^{14}CH_2)_2Sn(O_2CCH{=}CHCO_2Me)_2$, 2, 440
$SnC_{18}H_{30}$ $p\text{-}Pr_3SnC_6H_4CH_2CH{=}CH_2$, 2, 202
$SnC_{18}H_{30}Br_2O_2$ 2,6,4-$Br_2(Me_3C)C_6H_2O$-$SnBu_2OH$, 2, 402*
$SnC_{18}H_{30}O$ $Pr_3SnCH_2CH_2Bz$, 2, 230
$SnC_{18}H_{30}O_2$ 1,4,2-$Bu_2\overline{SnO(Me_3C)C_6H_3O}$, 2, 405
$SnC_{18}H_{31}Cl$ $o\text{-}ClC_6H_4SnBu_3$, 2, 222
$p\text{-}ClC_6H_4SnBu_3$, 2, 222
$SnC_{18}H_{31}ClO$ $p\text{-}ClC_6H_4OSnBu_3$, 2, 389
$SnC_{18}H_{31}ClO_2$ 2,4-$Cl(Me_3C)C_6H_3O(Bu_2)$-SnOH, 2, 402
$SnC_{18}H_{31}Cl_3O_2$ $(n\text{-}C_4H_8Cl)_3Sn$ sorbate, 2, 432
$SnC_{18}H_{31}NO_2$ Bu_3Sn 3-pyridinecarboxylate, 2, 416
$SnC_{18}H_{31}N_3$ N-(1,2,3-Benzotriazolyl)-$SnBu_3$, 2, 478
$SnC_{18}H_{32}$ Bu_3SnPh, 2, 178
$SnC_{18}H_{32}O$ Bu_3SnOPh, 2, 385
$SnC_{18}H_{32}O_4$ $Bu_2Sn(CHAc_2)_2$, 2, 231
$(i\text{-}C_5H_{11})_2Sn(O_2CCMe{=}CH_2)_2$, 2, 439, 511*
$SnC_{18}H_{32}O_6$ (2-Furfuryl-$O_2CCH{=}CHCO_2$)-Bu_2SnOMe, 2, 407
$SnC_{18}H_{32}S$ Bu_3SnSPh, 2, 459
$SnC_{18}H_{32}S_3$ $PhSn(SBu)_3$, 2, 469
$SnC_{18}H_{33}Br$ Cyclohexyl$_3$SnBr, 2, 277
$SnC_{18}H_{33}Cl$ Cyclohexyl$_3$SnCl, 2, 284
$[Me_2CHCH_2C({=}CH_2)]_3SnCl$, 2, 292
$SnC_{18}H_{33}F$ Cyclohexyl$_3$SnF, 2, 284
$SnC_{18}H_{33}F_5O_2$ $(i\text{-}C_5H_{11})_3SnO_2CC_2F_5$, 2, 422
$(SnC_{18}H_{34}O_4)_n$ $(Bu_2Sn\ \text{sebacate})_n$, 2, 445

$SnC_{18}H_{34}O_5$ $Pr_3SnC(CO_2Et)=COMeCO_2Et$, 2, 208
$SnC_{18}H_{35}N_3$ $PhSn(NEt_2)_3$, 2, 476
$SnC_{18}H_{36}$ Cyclopentyl$_2$SnBu$_2$, 2, 186
$[CH_2=CH(CH_2)_3]_2SnBu_2$, 2, 203
$SnC_{18}H_{36}O$ 2-Ketocyclohexyl-SnBu$_3$, 2, 230
$SnC_{18}H_{36}O_2$ $Et_2(CH_2=CH)SnO_2CC_{11}H_{23}$, 2, 430
$SnC_{18}H_{36}O_2S_4$ $Bu_2Sn(S_2COBu)_2$, 2, 471
$SnC_{18}H_{36}O_4$ $Bu_2Sn[(CH_2)_3OAc]_2$, 2, 226
i-$Bu_2Sn(CH_2CHMeCO_2Me)_2$, 2, 237
(n-$C_8H_{17})_2Sn(O_2CH)_2$, 2, 438
$SnC_{18}H_{36}O_5$ $Et_2Sn(OEt)O_2C(CH_2)_8CO_2Et$, 2, 407
$SnC_{18}H_{36}O_6$ $Et_2Sn(OCHMeCO_2Bu)_2$, 2, 403
$Bu_2Sn(OCH_2CHCH_2OCOMeH$, 2, 403
$Bu_2Sn(OCHMeCO_2Et)_2$, 2, 403
$SnC_{18}H_{38}$ $Bu_3Sn(CH_2)_4CH=CH_2$, 2, 202
$(C_5H_{11})_3SnCH_2CH=CH_2$, 2, 203
(i-$C_5H_{11})_3SnCH_2CH=CH_2$, 2, 203
$SnC_{18}H_{38}Cl_2$ $(C_9H_{19})_2SnCl_2$, 2, 319
(i-$C_9H_{19})_2SnCl_2$, 2, 319
$SnC_{18}H_{38}N_2S_4$ $Bu_2Sn(S_2CNEt_2)_2$, 2, 471
$SnC_{18}H_{38}O$ $Bu_3Sn(CH_2)_4Ac$, 2, 230
(i-$C_9H_{19})_2SnO$, 2, 372
Cyclohexyl-OSnBu$_3$, 2, 385
$SnC_{18}H_{38}O_2$ $Et_3Sn(CH_2)_{10}CO_2Me$, 2, 234
$Bu_3SnOCMe_2CH_2Ac$, 2, 390
$Bu_3SnO_2CC_5H_{11}$, 2, 413
i-$C_5H_{11}CO_2SnBu_3$, 2, 413
Et_2(n-$C_{12}H_{25})SnOAc$, 2, 430
$Et_2(C_{12}H_{25})SnOAc$, 2, 430
$SnC_{18}H_{38}O_2S$ $Et_3SnSCH_2CO_2C_{10}H_{21}$, 2, 461, 516
$SnC_{18}H_{38}O_4P_2S_4$
$Bu_2Sn(S_2POCH_2CMe_2CH_2O)_2$, 2, 353
$SnC_{18}H_{38}S$ $Et_2(CH_2=CH)SnSC_{12}H_{25}$, 2, 458
$SnC_{18}H_{39}Cl$ $(C_6H_{13})_3SnCl$, 2, 284
$SnC_{18}H_{40}$ (n-$C_8H_{17})_2SnMe_2$, 2, 180
$(BuEtCHCH_2)_2SnMe_2$, 2, 180
n-$C_{12}H_{25}SnEt_3$, 2, 183
n-$C_6H_{13}SnBu_3$, 2, 186
$Bu_2Sn(CH_2CMe_3)_2$, 2, 186
(n-$C_6H_{13})_3SnH$, 2, 263
$SnC_{18}H_{40}N_2O$ i-$Pr_3SnON=CMe(CH_2)_3NEt_2$, 2, 391
$SnC_{18}H_{40}N_2O_2$ Bu_3SnOAc·piperazine, 2, 410
$SnC_{18}H_{40}O$ $Bu_3SnCH_2CH_2OCH_2CHMe_2$, 2, 226
i-$Bu_3SnCH_2CH_2OBu$, 2, 226
$SnC_{18}H_{40}O_2$ $(C_8H_{17})_2Sn(OMe)_2$, 2, 399
$SnC_{18}H_{40}S_2$ $Bu_2Sn(SC_5H_{11})_2$, 2, 462
$SnC_{18}H_{42}IOP$ $Bu_3SnI\cdot OPEt_3$, 2, 283*
$SnC_{18}H_{42}O_6P_2$ $[PrP(OPr)O_2]_2SnPr_2$, 2, 351
$SnC_{18}H_{42}O_8P_2$ $(C_7H_{15})_2Sn(PO_4Me_2)_2$, 2, 354
$SnC_{18}H_{46}Br_4N_2$ $Me_2SnBr_2\cdot 2\,Et_4NBr$, 2, 305
$SnC_{18}H_{52}$ (n-$C_6H_{17})_3SnH$, 2, 263
$SnC_{19}H_3F_{15}$ $MeSn(C_6F_5)_3$, 2, 220
$SnC_{19}H_{14}O_2$ o-$Ph_2SnC_6H_4CO_2$, 2, 432
$SnC_{19}H_{15}ClO_2$ (o-$HO_2CC_6H_4)Ph_2SnCl$, 2, 300
$SnC_{19}H_{15}N$ Ph_3SnCN, 2, 346
$SnC_{19}H_{15}NO$ Ph_3SnNCO, 2, 346
$SnC_{19}H_{15}NS$ Ph_3SnNCS, 2, 346
$SnC_{19}H_{16}$ $Ph_2Sn=CHPh$ (?), 2, 254
$SnC_{19}H_{16}O_2$ Ph_3SnO_2CH, 2, 415
$SnC_{19}H_{16}O_3$ Ph_3SnHCO_3, 2, 415
$SnC_{19}H_{18}$ Ph_3SnMe, 2, 181
$SnC_{19}H_{18}O$ Ph_3SnOMe, 2, 386
$SnC_{19}H_{18}O_2S_2$ S-Ph_3Sn thionomethylsulfonate, 2, 356*
$SnC_{19}H_{18}O_3S$ Ph_3SnO_3SMe, 2, 356
$SnC_{19}H_{18}S$ Ph_3SnSMe, 2, 459
$SnC_{19}H_{24}$ 9-Fluorenyl-SnEt$_3$, 2, 200
$BuSn(C_5H_5)_3$, 2, 203
$SnC_{19}H_{24}O_2$ $C_5H_{11}Ph_2SnOAc$, 2, 430
$SnC_{19}H_{26}O_5S$ Bu_2Sn (S-benzoyl)thiomalate, 2, 445
$SnC_{19}H_{27}N$ $Et_3SnN(CH_2Ph)Ph$, 2, 475
$SnC_{19}H_{27}NO_2$ Et_3SnN(1-$C_{10}H_7)CO_2Et$, 2, 480
$SnC_{19}H_{27}N_4S$ Et_3Sn dithizonate, 2, 486
$SnC_{19}H_{28}$ 1-$C_{10}H_7Sn(CHMe_2)_3$, 2, 185
2-$C_{10}H_7Sn(CHMe_2)_3$, 2, 185
$SnC_{19}H_{28}O_4S_2$ $Bu_2Sn(O_2CCH_2S)_2CHPh$, 2, 445
$SnC_{19}H_{31}I$ p-$CH_2=CHCH_2C_6H_4$(i-$C_5H_{11})_2SnI$, 2, 294

$SnC_{19}H_{31}NS_2$ $Bu_3SnSC_7H_4NS$, 2, 460
$SnC_{19}H_{32}N_2$ N-(Benzopyrazolyl)$SnBu_3$, 2, 478
N-(Benzimidazolyl)$SnBu_3$, 2, 478
$SnC_{19}H_{32}O_2$ Tetralineperoxy-Pr_3Sn, 2, 379
Bu_3SnOBz, 2, 414
$SnC_{19}H_{32}O_3$ o-$HOC_6H_4CO_2SnBu_3$, 2, 423
$SnC_{19}H_{32}O_6S$ 3,6-$HO_3S(HO)C_6H_3CO_2SnBu_3$ (?), 2, 429
$SnC_{19}H_{33}NO_2$ o-$H_2NC_6H_4CO_2SnBu_3$, 2, 424
p-$H_2NC_6H_4CO_2SnBu_3$, 2, 424
$SnC_{19}H_{34}O$ Bu_3SnOCH_2Ph, 2, 385
$SnC_{19}H_{35}NO_3S$ $PhSO_2NBu(Bu_2)SnOMe$, 2, 481
$SnC_{19}H_{36}O_4S_2$ $(Me_3CCH_2)_2Sn(O_2CCH_2$-$CH_2S)_2CMe_2$, 2, 517
$SnC_{19}H_{36}O_6$ $BuSn(CH_2CHMeCO_2Me)_3$, 2, 236
i-$BuSn(CH_2CHMeCO_2Me)_3$, 2, 237
$BuSn(O_2CBu)_3$, 2, 449
$SnC_{19}H_{38}O_2$ (i-$C_5H_{11})_3SnO_2CCMe{=}CH_2$, 2, 419, 511
$SnC_{19}H_{38}O_2S$ $(C_8H_{17})_2SnO_2CCH_2{}^{14}CH_2S$, 2, 447*
$(C_8H_{17})_2{}^{113}SnO_2CCH_2CH_2S$, 2, 447
$SnC_{19}H_{40}O$ $Bu_3SnOCHMeCH_2CH_2CH{=}CHMe$, 2, 388
$SnC_{19}H_{40}O_2$ $Et_3Sn(CH_2)_{10}CO_2Et$, 2, 234
n-$C_6H_{13}CO_2SnBu_3$, 2, 413
$SnC_{19}H_{42}$ n-$C_{10}H_{21}SnPr_3$, 2, 184
n-$C_7H_{15}SnBu_3$, 2, 186
$SnC_{19}H_{42}N_2O$ $Bu_3SnON{=}CMeCHMeCH_2NMe_2$, 2, 391
$SnC_{19}H_{42}S$ $Pr_3SnSC_{10}H_{21}$, 2, 459
i-$Pr_3SnSC_{10}H_{21}$, 2, 459
$SnC_{20}H_{12}F_6O_2$ (p-$FC_6H_4)_3SnO_2CCF_3$, 2, 432
$SnC_{20}H_{14}BrCl$ (1-$C_{10}H_7)_2SnClBr$, 2, 320
$SnC_{20}H_{14}Br_2$ (1-$C_{10}H_7)_2SnBr_2$, 2, 320
$SnC_{20}H_{14}Cl_2$ (1-$C_{10}H_7)_2SnCl_2$, 2, 315, 320
$SnC_{20}H_{14}I_2$ (1-$C_{10}H_7)_2SnI_2$, 2, 320
$SnC_{20}H_{14}O$ (1-$C_{10}H_7)_2SnO$, 2, 373
(2-$C_{10}H_7)_2SnO$, 2, 373
$SnC_{20}H_{15}Cl$ $Ph_3SnC{\equiv}CCl$, 2, 216
$SnC_{20}H_{15}Cl_3O_2$ $Ph_3SnO_2CCCl_3$, 2, 422
(p-$ClC_6H_4)_3SnOAc$, 2, 432
$SnC_{20}H_{15}F_3$ $Ph_3SnCF{=}CF_2$, 2, 209
$SnC_{20}H_{16}$ $Ph_3SnC{\equiv}CH$, 2, 214
$SnC_{20}H_{16}S_2$ (2-Thienyl)$_2SnPh_2$, 2, 193
$SnC_{20}H_{17}ClO_2$ Ph_2(o-$MeO_2CC_6H_4)SnCl$, 2, 300
$Ph_3SnO_2CCH_2Cl$, 2, 422
$SnC_{20}H_{17}N$ Ph_3SnCH_2CN, 2, 224
$SnC_{20}H_{17}N_3$ N-(1,2,3-Triazolyl)$SnPh_3$, 2, 479
N-(1,2,4-Triazolyl)$SnPh_3$, 2, 479
$SnC_{20}H_{18}$ $Ph_3SnCH{=}CH_2$, 2, 196, 510, 511
$SnC_{20}H_{18}Cl_{10}O_2$ $Bu_2Sn(OC_6Cl_5)_2$, 2, 402
$SnC_{20}H_{18}F_{10}$ $Bu_2Sn(C_6F_5)_2$, 2, 222
$SnC_{20}H_{18}N_2O_2$ (8-Quinolyloxy)$_2SnMe_2$, 2, 401
$SnC_{20}H_{18}O_2$ Ph_3SnOAc, 2, 410
$SnC_{20}H_{18}O_3$ Ph_3SnO_2Ac or $Ph_2(PhO)SnOAc$, 2, 380
$SnC_{20}H_{19}I$ $(PhCH_2CH_2)Ph_2SnI$, 2, 289
$SnC_{20}H_{20}$ $(C_5H_5)_4Sn$, 2, 173
$EtSnPh_3$, 2, 184
$SnC_{20}H_{20}O$ $Ph_3SnCH_2CH_2OH$, 2, 227
$SnC_{20}H_{20}O_4$ (p-$CH_2{=}CHC_6H_4CO_2)_2SnMe_2$, 2, 439
$Ph_2Sn(O_2CCMe{=}CH_2)_2$, 2, 440, 511
$SnC_{20}H_{20}O_8$ $Me_2Sn[C_6H_4CH(CO_2H)_2]_2$, mixture with tricarboxylic acid, 2, 238
$SnC_{20}H_{21}N$ Ph_3SnNMe_2, 2, 476
$SnC_{20}H_{22}Br_6O_2$ $Bu_2Sn(OC_6H_2Br_3)_2$, 2, 402
$SnC_{20}H_{22}N_2S_4$ $Pr_2Sn(2$-$SC_7H_5NS)_2$, 2, 464
$SnC_{20}H_{22}O_4S$ $Ph_2Sn(O_2CCH_2CH_2CH_2)_2S$, 2, 446
$SnC_{20}H_{22}O_8$ $Me_2SnC_6H_4CH(CO_2H)_2C_6H_4CH$-$MeCO_2H$, mixture with tetracarboxylic acid, 2, 238
$SnC_{20}H_{24}$ o,o'-$C_6H_4C_6H_4SnBu_2$, 2, 506
$SnC_{20}H_{24}Cl_2O_2$ (p-$ClC_6H_4)_2CHCO_2SnEt_3$, 2, 421
$SnC_{20}H_{24}O_4$ $Ph_2Sn(CH_2CH_2CO_2Me)_2$, 2, 238
$(PhCH_2CH_2)_2Sn(OAc)_2$, 2, 436
$Pr_2Sn(OBz)_2$, 2, 436
i-$Pr_2Sn(OBz)_2$, 2, 436

$SnC_{20}H_{24}O_4S_2$ $(p\text{-}MeC_6H_4O_2CCH_2S)_2SnMe_2$, 2, 465
$SnC_{20}H_{26}Cl_2$ $(p\text{-}Me_2CHC_6H_4CH_2)_2SnCl_2$, 2, 316
$SnC_{20}H_{26}N_4S_2$ $Bu_2Sn(NCS)_2\cdot 2,2'$-bipyridyl, 2, 346
$SnC_{20}H_{26}O$ $(p\text{-}Me_2CHC_6H_4CH_2)_2SnO$, 2, 372
$SnC_{20}H_{26}S_2$ $Bu_2SnSC_6H_4C_6H_4S$ (?), 2, 463
$SnC_{20}H_{27}N$ N-Carbazolyl-$CH_2CH_2SnEt_3$, 2, 191*
$SnC_{20}H_{28}$ $Et_2Sn(CH_2CH_2Ph)_2$, 2, 183
Bu_2SnPh_2, 2, 178
$SnC_{20}H_{28}Br_2N_2$ $Bu_2SnBr_2\cdot$phenanthroline, 2, 312
$SnC_{20}H_{28}Cl_2N_2$ $Bu_2SnCl_2\cdot$phenanthroline, 2, 311
$SnC_{20}H_{28}Cl_4N_4O_2$ $Et_2SnCl_2\cdot 2\ p\text{-}EtOC_6H_4\text{-}N_2Cl$, 2, 307
$SnC_{20}H_{28}I_2N_2$ $Bu_2SnI_2\cdot$phenanthroline, 2, 312
$SnC_{20}H_{28}O_2$ $Bu_2Sn(OPh)_2$, 2, 398
$SnC_{20}H_{28}S_2$ $Bu_2Sn(SPh)_2$, 2, 462
$SnC_{20}H_{30}Br_2$ $(EtPrC_5H_3)_2SnBr_2$, 2, 324
$SnC_{20}H_{30}N_2$ $Ph_2Sn(NEt_2)_2$, 2, 476
$SnC_{20}H_{30}N_2O_4S_2$ $(PhSO_2NH)_2SnBu_2$, 2, 481
$SnC_{20}H_{30}N_2S_2$ $(o\text{-}H_2NC_6H_4S)_2SnBu_2$, 2, 467
$SnC_{20}H_{30}O_4P_2$ $Et_2Sn(O_2PEtPh)_2$, 2, 350
$SnC_{20}H_{31}Cl_3O_3$ $2,4,6\text{-}Cl_3C_6H_2OCH_2CO_2\text{-}SnBu_3$, 2, 429
$SnC_{20}H_{31}NO_2$ N-(Phthalimido)$SnBu_3$, 2, 480
$SnC_{20}H_{32}$ $Bu_3SnC{\equiv}CPh$, 2, 213
$SnC_{20}H_{32}O_4$ $Bu_2Sn(O_2CCH{=}CHCO_2Et)_2$, 2, 440
$SnC_{20}H_{34}$ $p\text{-}Bu_3SnC_6H_4CH{=}CH_2$, 2, 202, 510
$SnC_{20}H_{34}Cl_4N_2$ $Ph_2SnCl_2\cdot 2\ Me_4NCl$, 2, 315
$SnC_{20}H_{34}O$ Bu_3SnCH_2Bz, 2, 230
$SnC_{20}H_{34}O_2$ $Bu_3SnO_2CCH_2Ph$, 2, 413
$SnC_{20}H_{35}NOS$ $Bu_3SnNPhCSOMe$, 2, 480
$SnC_{20}H_{35}NO_2$ $Bu_3SnNPhCO_2Me$, 2, 480
$SnC_{20}H_{36}$ Cyclopentyl$_4$Sn, 2, 170
(3-Cyclohexenyl-$CH_2CH_2)_2SnEt_2$, 2, 201
$SnC_{20}H_{36}O$ $Bu_3SnOCHMePh$, 2, 385
$SnC_{20}H_{36}O_2$ Cyclohexyl$_3$SnOAc, 2, 414
$SnC_{20}H_{36}O_4$ $(n\text{-}C_6H_{13})_2Sn(O_2CCMe{=}CH_2)_2$, 2, 439, 511*
$(C_8H_{17})_2Sn(O_2CCH{=})_2$, 2, 446
$(BuEtCHCH_2)_2Sn(O_2CCN{=})_2$, 2, 446
$SnC_{20}H_{36}O_4S_2$ $Bu_2Sn(SCH_2CO_2CH_2CMe{=}CH_2)_2$, 2, 465
$SnC_{20}H_{36}O_5$ $(BuEtCHCH_2)_2Sn(O_2CCH_2)_2O$, 2, 517
$SnC_{20}H_{36}O_6$ $Bu_2Sn(CHAcCO_2Et)_2$, 2, 231
$SnC_{20}H_{37}Cl_3O_4S$ $Bu_3SnSC(CCl_3)(CO_2Et)_2$, 2, 461
$SnC_{20}H_{37}N_3O_4$ N-(4,5-Dicarboxyethyl-triazolyl)$SnBu_3$, 2, 478
$SnC_{20}H_{37}O_2P$ Bu_3SnO_2PEtPh, 2, 352
$SnC_{20}H_{38}N_2O_4S_2$ $Bu_2Sn[SCH_2CO(NOC_4H_8)]_2$, 2, 466
$SnC_{20}H_{38}O_4S_2$ $Bu_2Sn[SCH_2CH_2O_2C(CH_2)_3]_2$, 2, 465
$SnC_{20}H_{40}$ Cyclohexyl$_2$$SnBu_2$, 2, 186
$SnC_{20}H_{40}O_4$ $Bu_2Sn(O_2CC_5H_{11})_2$, 2, 437
$(n\text{-}C_8H_{17})_2Sn(OAc)_2$, 2, 438
$SnC_{20}H_{40}O_6$ $Bu_2Sn(OCMe_2CO_2Et)_2$, 2, 403
$SnC_{20}H_{41}NO_2$ $Bu_2Sn(OCHMeCH_2)_2{=}NC_6H_{11}$, 2, 404
$SnC_{20}H_{42}$ $CH_2{=}CH(CH_2)_6SnBu_3$, 2, 202
$(n\text{-}C_6H_{13})_3SnCH{=}CH_2$, 2, 204
$SnC_{20}H_{42}Cl_2$ $(C_{10}H_{21})_2SnCl_2$, 2, 319
$SnC_{20}H_{42}I_2$ $(C_{10}H_{21})_2SnI_2$, 2, 319
$SnC_{20}H_{42}O$ $Bu_3SnOCHMeCH_2CH_2CH{=}CMe_2$, 2, 388
$SnC_{20}H_{42}O_2$ $Bu_3SnO_2CC_7H_{15}$, 2, 413
$(n\text{-}C_6H_{13})_3SnOAc$, 2, 414
$SnC_{20}H_{42}O_3$ $Bu_2Sn(OH)O_2CC_{11}H_{23}$, 2, 437*
$SnC_{20}H_{42}O_4S$ $(C_{10}H_{21})_2SnSO_4$, 2, 354
$SnC_{20}H_{42}O_8P_2S_4$ $Bu_2Sn[O_2CCH_2S_2P(OEt)_2]_2$, 2, 444
$SnC_{20}H_{43}NOS$ $n\text{-}C_5H_{11}NHOCCH_2CH_2SSnBu_3$, 2, 461
$SnC_{20}H_{44}$ $(n\text{-}C_5H_{11})_4Sn$, 2, 165
$(i\text{-}C_5H_{11})_4Sn$, 2, 165
$(MePrCH)_4Sn$, 2, 170
$(Et_2CH)_4Sn$, 2, 170
$(Me_2EtC)_4Sn$, 2, 170

$(Me_3CCH_2)_4Sn$, 2, 170
$(C_6H_{13})_2SnBu_2$, 2, 186
$n\text{-}C_8H_{17}Sn(CH_2CHMe_2)_3$, 2, 187
$SnC_{20}H_{44}O_2$ $(C_8H_{17})_2Sn(OEt)_2$, 2, 399
$SnC_{20}H_{45}Cl_2N$ $Bu_2SnCl_2 \cdot C_{12}H_{25}NH_2$, 2, 311
$SnC_{20}H_{46}N_2O_4S_2$ $Bu_2Sn(NHO_2SC_6H_{13})_2$, 2, 481
$SnC_{21}H_{12}BrF_{12}$ $(m\text{-}CF_3C_6H_4)_3SnBr$, 2, 296
$SnC_{21}H_{12}F_8O_2$ $(p\text{-}FC_6H_4)_3SnO_2CC_2F_5$, 2, 432
$SnC_{21}H_{16}O_2$ $Ph_3SnO_2CC{\equiv}CH$, 2, 419
$SnC_{21}H_{17}Cl_3$ $Ph_3SnCH{=}CHCCl_3$, 2, 209
$SnC_{21}H_{17}NO_2$ $Ph_3SnO_2CCH_2CN$, 2, 424
$SnC_{21}H_{18}$ $Ph_3SnCH_2C{\equiv}CH$, 2, 214
$Ph_3SnC{\equiv}CMe$, 2, 214
$Ph_3SnCH{=}C{=}CH_2$, 2, 214
$SnC_{21}H_{18}BrCl_3$ $Ph_3SnCHBrCH_2CCl_3$, 2, 222
$SnC_{21}H_{18}Br_3Cl$ $(p\text{-}BrC_6H_4CH_2)_3SnCl$, 2, 295
$SnC_{21}H_{18}ClF_3$ $(o\text{-}FC_6H_4CH_2)_3SnCl$, 2, 295
$(p\text{-}FC_6H_4CH_2)_3SnCl$, 2, 295
$SnC_{21}H_{18}Cl_3F$ $(o\text{-}ClC_6H_4CH_2)_3SnF$, 2, 295
$SnC_{21}H_{18}Cl_4$ $(o\text{-}ClC_6H_4CH_2)_3SnCl$, 2, 295
$(p\text{-}ClC_6H_4CH_2)_3SnCl$, 2, 295
$SnC_{21}H_{18}N_2$ N-(Imidazolyl)$SnPh_3$, 2, 479
$SnC_{21}H_{18}O_2$ $Ph_3SnO_2CCH{=}CH_2$, 2, 419, 511
$SnC_{21}H_{18}O_4$ $Ph_3SnO_2CCH_2CO_2H$, 2, 423
$SnC_{21}H_{19}N$ $Ph_3SnCH_2CH_2CN$, 2, 224
$SnC_{21}H_{20}$ $Ph_3SnCHCH_2CH_2$, 2, 186
$Ph_3SnCH_2CH{=}CH_2$, 2, 196
$PhSn(C_5H_5)_3$, 2, 206
$SnC_{21}H_{20}O$ $Ph_3SnCH{=}CHCH_2OH$, 2, 209
$Ph_2(p\text{-}CH_2{=}CHCH_2C_6H_4)SnOH$ (?), 2, 365
$SnC_{21}H_{20}O_2$ Ph_3SnO_2CEt, 2, 415
$SnC_{21}H_{20}O_3$ Ph_3SnO_3CEt or $Ph_2Sn(OPh)\text{-}O_2CEt$, 2, 380
$Ph_3SnO_2CCHOHMe$, 2, 423
$SnC_{21}H_{21}Br$ $(o\text{-}MeC_6H_4)_3SnBr$, 2, 284
$(m\text{-}MeC_6H_4)_3SnBr$, 2, 285
$(p\text{-}MeC_6H_4)_3SnBr$, 2, 285
$SnC_{21}H_{21}Cl$ $(PhCH_2)_3SnCl$, 2, 271
$(o\text{-}MeC_6H_4)_3SnCl$, 2, 280
$(m\text{-}MeC_6H_4)_3SnCl$, 2, 285
$(p\text{-}MeC_6H_4)_3SnCl$, 2, 281
$Ph_2(PhCHMeCH_2)SnCl$, 2, 289
$SnC_{21}H_{21}ClO$ $Ph_3SnCH_2CHOHCH_2Cl$, 2, 223
$SnC_{21}H_{21}ClO_3$ $(p\text{-}MeOC_6H_4)_3SnCl$, 2, 299
$SnC_{21}H_{21}F$ $(PhCH_2)_3SnF$, 2, 282
$SnC_{21}H_{21}I$ $(o\text{-}MeC_6H_4)_3SnI$, 2, 285
$SnC_{21}H_{21}NO$ $Ph_3SnCH_2CH_2CONH_2$, 2, 237
$SnC_{21}H_{22}$ $PrSnPh_3$, 2, 185
$(o\text{-}MeC_6H_4)_3SnH$, 2, 264
$(m\text{-}MeC_6H_4)_3SnH$, 2, 264
$(p\text{-}MeC_6H_4)_3SnH$, 2, 264
$SnC_{21}H_{22}O$ $Ph_3Sn(CH_2)_3OH$, 2, 227
$(PhCH_2)_3SnOH$, 2, 364
$(o\text{-}MeC_6H_4)_3SnOH$, 2, 366
$(m\text{-}MeC_6H_4)_3SnOH$, 2, 366
$(p\text{-}MeC_6H_4)_3SnOH$, 2, 366
$SnC_{21}H_{23}Cl_2P$ $Me_2SnCl_2 \cdot Ph_3P{=}CH_2$, 2, 304
$SnC_{21}H_{31}O_2P$ $Pr_3SnO_2PPh_2$, 2, 351
$SnC_{21}H_{31}P$ Pr_3SnPPh_2, 2, 483
$SnC_{21}H_{32}O_2$ $Bu_3SnO_2CC{\equiv}CPh$, 2, 419
$SnC_{21}H_{32}O_4S_2$ $Bu_2Sn(O_2CCH_2CH_2S)_2CHPh$, 2, 446
$SnC_{21}H_{32}O_5S_2$ $m\text{-}HOC_6H_4CH(SCH_2CH_2CO_2)_2\text{-}SnBu_2$, 2, 446
$SnC_{21}H_{34}$ 1-Indenyl-$SnBu_3$, 2, 202
$SnC_{21}H_{34}O_2$ $Bu_3SnO_2CCH{=}CHPh$, 2, 418*
$p\text{-}Bu_3SnO_2CC_6H_4CH{=}CH_2$, 2, 418*, 511*
$SnC_{21}H_{34}O_4$ $o\text{-}Bu_3SnO_2CC_6H_4CO_2Me$, 2, 423
$SnC_{21}H_{35}NO_3$ $p\text{-}Bu_3SnO_2CC_6H_4NHAc$, 2, 424
$SnC_{21}H_{36}$ $p\text{-}Bu_3SnC_6H_4CH_2CH{=}CH_2$, 2, 202
$SnC_{21}H_{36}O_2$ $Bu_3SnO_2CCH_2CH_2Ph$, 2, 413
$SnC_{21}H_{38}O_2$ $Bu_3SnO_2CMe_2Ph$, 2, 380
$SnC_{21}H_{38}O_6$ $Bu_2Sn(O_2CCH{=}CHCO_2Me)O_2C\text{-}C_7H_{15}$ (?), 2, 441
$SnC_{21}H_{39}NO$ $o\text{-}Bu_3SnOC_6H_4CH_2NMe_2$, 2, 391
$SnC_{21}H_{39}NO_2$ $(C_6H_{11})_3SnO_2CCHMeNH_2$, 2, 424
$SnC_{21}H_{39}NO_2S$ $p\text{-}MeC_6H_4SO_2NEtSnBu_3$, 2, 480
$SnC_{21}H_{41}N_3O$ $2,4,6\text{-}(Me_2NCH_2)_3C_6H_2OSn\text{-}$

Et_3, 2, 391
$SnC_{21}H_{42}$ $Bu_3SnCEt(CO_2Et)_2$, 2, 236
$SnC_{21}H_{44}O_2$ $C_8H_{17}CO_2SnBu_3$, 2, 413
$SnC_{21}H_{44}O_3$ $Bu_2Sn(OMe)O_2CC_{11}H_{23}$, 2, 407
$SnC_{21}H_{45}I$ $(n\text{-}C_7H_{15})_3SnI$, 2, 284
$SnC_{22}H_9F_{15}$ $BuSn(C_6F_5)_3$, 2, 222
$SnC_{22}H_{16}$ $Ph_3SnC{\equiv}CC{\equiv}CH$, 2, 214
$SnC_{22}H_{18}$ $Ph_3SnC{\equiv}CCH{=}CH_2$, 2, 214
$SnC_{22}H_{18}Cl_2$ $(1\text{-}C_{10}H_7CH_2)_2SnCl_2$, 2, 316
$SnC_{22}H_{18}Cl_2N_2$ $Ph_2SnCl_2 \cdot 2,2'$-bipyridyl, 2, 315
$SnC_{22}H_{18}O$ 2-Furyl-$SnPh_3$, 2, 191
$SnC_{22}H_{18}S$ 2-Thienyl-$SnPh_3$, 2, 193
$SnC_{22}H_{19}N$ N-$Ph_3SnC_4H_4N$, 2, 479
$SnC_{22}H_{20}$ $Ph_2Sn(C_5H_5)_2$, 2, 205
$(CH_2{=}CHCH_2)_2Sn(C{\equiv}CPh)_2$, 2, 213
$Ph_3SnCHMeC{\equiv}CH$, 2, 214
$Ph_3SnCMe{=}C{=}CH_2$, 2, 214
$SnC_{22}H_{20}O_2$ $Ph_3SnO_2CCMe{=}CH_2$, 2, 417, 511
$SnC_{22}H_{20}O_2S_2$ $(2\text{-Thienyl})_2Sn(C_6H_4\text{-}OMe\text{-}p)_2$, 2, 193
$SnC_{22}H_{20}O_4$ $(2\text{-Furfuryloxy})_2SnPh_2$, 2, 401
$(p\text{-}CH_2{=}CHC_6H_4CO_2)_2Sn(CH{=}CH_2)_2$, 2, 439
$SnC_{22}H_{21}Cl_3S_3$ $(p\text{-}ClC_6H_4CH_2S)_3SnMe$, 2, 468*
$SnC_{22}H_{21}N$ $Ph_3Sn(CH_2)_3CN$, 2, 224
$Ph_3SnCHMeCH_2CN$, 2, 224
$SnC_{22}H_{21}NO_3$ $Ph_3SnO_2CCH_2NHAc$, 2, 424
$SnC_{22}H_{22}$ o-$CH_2CH_2C_6H_4CH_2CH_2SnPh_2$, 2, 507*
$Ph_3SnCH_2CH{=}CHMe$, 2, 205
$Ph_3SnCH_2CMe{=}CH_2$, 2, 205
$SnC_{22}H_{22}N_2O_2$ $(8\text{-Quinolyloxy})_2SnEt_2$, 2, 401
$SnC_{22}H_{22}O_2$ $Ph_3SnCH_2CH_2OAc$, 2, 227
$Ph_3SnCH_2CH_2CO_2Me$, 2, 237
$Ph_3Sn(CH_2)_3CO_2H$, 2, 237
$SnC_{22}H_{22}O_3$ o-$HOC_6H_4CO_2SnEtPhC_6H_4Me$, 2, 431
$SnC_{22}H_{24}$ $BuSnPh_3$, 2, 186
$SnC_{22}H_{24}N_2O_8$ $Ph_2Sn[H(O_2CCH_2)_2NCH_2]_2$, 2, 446
$SnC_{22}H_{24}O$ $Ph_3Sn(CH_2)_4OH$, 2, 227
$Ph_3SnCH_2CHMeCH_2OH$, 2, 227
$SnC_{22}H_{24}O_2$ $Ph_3SnO_2CMe_3$, 2, 380
$SnC_{22}H_{24}O_3S_3$ $(p\text{-}HOC_6H_4CH_2S)_3SnMe$, 2, 468*
$SnC_{22}H_{25}N$ Ph_3SnNEt_2, 2, 476
$SnC_{22}H_{26}Cl_2O_4$ $(o\text{-}ClC_6H_4CO_2)_2SnBu_2$, 2, 443
$SnC_{22}H_{26}N_2S_2$ $Bu_2Sn(2\text{-}SC_7H_4NS)_2$, 2, 464
$SnC_{22}H_{26}O_5$ $Ph_2Sn(OMe)O_2CCH_2C({=}CH_2)\text{-}CO_2Bu$, 2, 407
$SnC_{22}H_{28}$ $Cyclopentyl_2SnPh_2$, 2, 187
$Et_2Sn(C_6H_4CH_2CH{=}CH_2)_2$, 2, 201
$Ph_2Sn[(CH_2)_3CH{=}CH_2]_2$, 2, 205
$SnC_{22}H_{28}O_2S_2$ $Bu_2Sn(SBz)_2$, 2, 471
$SnC_{22}H_{28}O_4$ $Ph_2Sn(CH_2CHMeCO_2Me)_2$, 2, 238
$Bu_2Sn(OBz)_2$, 2, 437
$SnC_{22}H_{28}O_6$ $(o\text{-}HOC_6H_4CO_2)_2SnBu_2$, 2, 444
$SnC_{22}H_{30}O_5$ $(Et_2SnOBz)_2O$, 2, 436
$SnC_{22}H_{30}N_2O_4$ $(o\text{-}H_2NC_6H_4CO_2)_2SnBu_2$, 2, 444
$(p\text{-}H_2NC_6H_4CO_2)_2Sn(CH_2CHMe_2)_2$, 2, 444
$SnC_{22}H_{32}$ $(PhCH_2)_2SnBu_2$, 2, 181
$(PhCH_2CH_2)_2SnPr_2$, 2, 184
$SnC_{22}H_{32}N_2O_8$ $[(CH_2{=}CH)_3SnO_2CCH_2(HO_2C\text{-}CH_2)NCH_2]_2$, 2, 429
$SnC_{22}H_{32}O_2$ $Bu_2Sn(OCH_2Ph)_2$, 2, 397
$SnC_{22}H_{32}O_6S_2$ $(p\text{-}MeC_6H_4SO_3)_2SnBu_2$, 2, 354
$SnC_{22}H_{32}S_2$ $Bu_2Sn(SCH_2Ph)_2$, 2, 462
$(p\text{-}MeC_6H_4S)_2SnBu_2$, 2, 463
$SnC_{22}H_{34}$ $1\text{-}C_{10}H_7SnBu_3$, 2, 186
$2\text{-}C_{10}H_7SnBu_3$, 2, 186
$SnC_{22}H_{34}N_2O_4S_2$ $(p\text{-}MeC_6H_4SO_2NH)_2SnBu_2$, 2, 481
$SnC_{22}H_{34}O_3S$ $2\text{-}C_{10}H_7SO_3SnBu_3$, 2, 352
$SnC_{22}H_{34}O_3S_6$ $PhCH_2Sn(S_2COBu)_3$, 2, 471
$SnC_{22}H_{34}O_6P_2$ $[PhP(OPr)O_2]_2SnEt_2$, 2, 350
$SnC_{22}H_{34}O_6S_2$ $2,3\text{-}MeO(HO)C_6H_3CH(SCH_2\text{-}CH_2CO_2)_2SnBu_2$, 2, 446
$SnC_{22}H_{36}I_2N_2$ $Me_2SnI_2 \cdot 2\ PhNEt_2$, 2, 305
$SnC_{22}H_{36}O_4$ $Bu_2Sn(O_2CCH{=}CHCO_2CHMe_2)_2$, 2, 440

$SnC_{22}H_{36}O_8S_2$ $Bu_2Sn(O_2CCH_2SCH_2CO_2CH_2\text{-}CH{=}CH_2)_2$, 2, 444
$SnC_{22}H_{36}O_{10}$ $Bu_2Sn(O_2CCH{=}CHCO_2CH_2CHOH\text{-}Me)_2$, 2, 440
$SnC_{22}H_{38}N_2O_5$ 6,2,4-(s-Bu)$(O_2N)_2C_6H_2\text{-}OSnBu_3$, 2, 391
$SnC_{22}H_{38}O_4$ $(C_8H_{17})_2Sn(O_2CCH{=}CH)_2$, 2, 446
$SnC_{22}H_{40}O_4$ $(n\text{-}C_7H_{15})_2Sn(O_2CCMe{=}CH_2)_2$, 2, 440, 511*
$SnC_{22}H_{40}O_6S_2$ $(C_4H_7OCH_2O_2CCH_2S)_2SnBu_2$, 2, 464
$SnC_{22}H_{40}O_8$ $Bu_2Sn[CH(CO_2Et)_2]_2$, 2, 236
$SnC_{22}H_{42}O_2$ $(C_6H_{11})_3SnO_2CMe_3$, 2, 380
$SnC_{22}H_{42}O_4S$ $(C_8H_{17})_2Sn(O_2CCH_2CH_2)_2S$, 2, 446
$SnC_{22}H_{44}O_2$ 9-Decalinperoxy-$SnBu_3$, 2, 380
$(n\text{-}C_6H_{13})_3SnO_2CMe{=}CH_2$, 2, 419, 511
$Bu_2(CH_2{=}CH)SnO_2CC_{11}H_{23}$, 2, 430
$SnC_{22}H_{44}O_3$ $Bu_3SnCHAcCO_2C_6H_{13}$, 2, 231
$SnC_{22}H_{44}O_4S_2$ $Bu_2Sn(SCH_2CH_2CO_2Bu)_2$, 2, 466
$SnC_{22}H_{46}N_2O_2S_2$ $Bu_2Sn(SCH_2CONHC_5H_{11})_2$, 2, 466
$SnC_{22}H_{46}O_2$ $Bu_3SnO_2CC_9H_{19}$, 2, 414
$SnC_{22}H_{46}O_2S$ $i\text{-}C_8H_{17}O_2CCH_2SSnBu_3$, 2, 461
$SnC_{22}H_{48}$ $(n\text{-}C_9H_{19})_2SnEt_2$, 2, 183
$(i\text{-}C_5H_{11})_3Sn(CH_2)_6Me$, 2, 187
$SnC_{22}H_{48}N_4O_2$ $Et_2Sn[ON{=}CMe(CH_2)_3\text{-}NEt_2]_2$, 2, 404
$Bu_2Sn[ON{=}CMe(CH_2)_3NMe_2]_2$, 2, 404
$SnC_{22}H_{48}O_2$ $(C_8H_{17})_2Sn(OPr)_2$, 2, 399
$SnC_{22}H_{48}S_2$ $Bu_2Sn(SC_7H_{15})_2$, 2, 462
$SnC_{23}H_{18}$ $Ph_3SnC{\equiv}CC{\equiv}CMe$, 2, 215
$SnC_{23}H_{18}O_2S$ Ph_3Sn 2-thiophenecarboxylate, 2, 416
$SnC_{23}H_{18}O_3$ Ph_3Sn 2-furanecarboxylate, 2, 416
$SnC_{23}H_{19}N$ 2-Pyridyl-$SnPh_3$, 2, 192
3-Pyridyl-$SnPh_3$, 2, 192
$SnC_{23}H_{20}$ $Ph_3SnC_5H_5$, 2, 197
$Ph_3SnC{\equiv}CCMe{=}CH_2$, 2, 215
$SnC_{23}H_{20}O_2$ endo-2,5-$O_2C_5H_5SnPh_3$, 2, 229
$SnC_{23}H_{20}O_6$ $o\text{-}AcOC_6H_4CO_2(AcO)SnPh_2$, 2, 444
$EtSn(OBz)_3$, 2, 449
$SnC_{23}H_{21}Cl_3O_2$ $(PhCH_2)_3SnO_2CCCl_3$, 2, 421
$SnC_{23}H_{22}O_3$ $Ph_3SnO_2CCH_2COEt$, 2, 423
$SnC_{23}H_{24}$ Cyclopentyl-$SnPh_3$, 2, 187
$Ph_3SnCH_2CH{=}CMe_2$, 2, 205
$SnC_{23}H_{24}O_2$ $Ph_3Sn(CH_2)_3OAc$, 2, 227
$Ph_3SnCH{=}CHC(O_2H)Me_2$ + $Ph_3SnCH_2CH\text{-}(O_2H)CMe{=}CH_2$, 2, 229, 511
$Ph_3SnCH_2CH_2CO_2Et$, 2, 237
$Ph_3Sn(CH_2)_3CO_2Me$, 2, 237
$Ph_3SnCH_2CHMeCO_2Me$, 2, 237
$Ph_3Sn(CH_2)_4CO_2H$, 2, 237
$(PhCH_2)_3SnOAc$, 2, 411
$(m\text{-}MeC_6H_4)_3SnOAc$, 2, 415
$(p\text{-}MeC_6H_4)_3SnOAc$, 2, 415
$SnC_{23}H_{25}NO$ $Ph_3Sn(CH_2)_3NHAc$, 2, 241
$SnC_{23}H_{26}$ $i\text{-}C_5H_{11}SnPh_3$, 2, 187
$SnC_{23}H_{26}O$ $Ph_3Sn(CH_2)_5OH$, 2, 227
$SnC_{23}H_{28}O_5$ $EtPh_2Sn[CH(CO_2Et)COCH_2CO_2\text{-}Et]$ (?), 2, 230
$SnC_{23}H_{34}O_2$ 1-$C_{10}H_7CO_2SnBu_3$, 2, 414
$SnC_{23}H_{37}NO_3$ $p\text{-}Bu_3SnO_2CC_6H_4NHCOCMe{=}CH_2$, 2, 424
$SnC_{23}H_{40}O_2$ 2,4,5-$Me_2(2\text{-}MeC_6H_{10})C_6H_2O\text{-}SnBu_2OH$, 2, 398
$SnC_{23}H_{40}O_3$ $o\text{-}Bu_3SnO_2CC_6H_4OBu$, 2, 423
$SnC_{23}H_{40}O_8$ $Bu_2Sn[C(CO_2Et)_2]_2CH_2$, 2, 236*
$SnC_{23}H_{41}N_2O$ $p\text{-}Bu_3SnON{=}CC_6H_4NEt_2$, 2, 391
$SnC_{23}H_{46}O_4$ $Bu_3SnCH(CO_2Bu)_2$, 2, 236
$SnC_{23}H_{48}O_2$ $Bu_3SnO_2CC_{10}H_{21}$, 2, 414
$(C_7H_{15})_3SnOAc$, 2, 414
$SnC_{23}H_{48}O_2S$ $Bu_3SnO_2CCH_2CH_2SCHEtC_5H_{11}$, 2, 423
$SnC_{23}H_{49}I$ $(n\text{-}C_{10}H_{21})_2PrSnI$, 2, 289
$SnC_{24}F_{20}$ $(C_6F_5)_4Sn$, 2, 174
$SnC_{24}H_5F_{15}$ $PhSn(C_6F_5)_3$, 2, 223
$SnC_{24}H_{10}F_{10}$ $Ph_2Sn(C_6F_5)_2$, 2, 223
$SnC_{24}H_{12}Cl_8$ Trichlorobiphenyl$_2SnCl_2$, 2, 325
$SnC_{24}H_{15}Br_5O$ $Ph_3SnOC_6Br_5$, 2, 389
$SnC_{24}H_{15}Cl_5O$ $Ph_3SnOC_6Cl_5$, 2, 389
$SnC_{24}H_{15}F_5$ $Ph_3SnC_6F_5$, 2, 223
$SnC_{24}H_{16}$ 5,5'-Spirobi[dibenzostannole],

2, 506
$SnC_{24}H_{16}Cl_4$ $(p\text{-}ClC_6H_4)_4Sn$, 2, 174
$SnC_{24}H_{16}F_4$ $(p\text{-}FC_6H_4)_4Sn$, 2, 174
$SnC_{24}H_{17}Br_3O$ 2,4,6-$Br_3C_6H_2OSnPh_3$, 2, 389
$SnC_{24}H_{17}Cl$ o,o'-$C_6H_4C_6H_4Sn(C_6H_4Ph)Cl$, 2, 508
$SnC_{24}H_{17}Cl_3O$ 2,4,6-$Cl_3C_6H_2OSnPh_3$, 2, 389
$SnC_{24}H_{18}$ o,o'-$C_6H_4C_6H_4SnPh_2$, 2, 506
$SnC_{24}H_{18}Br_2$ $(o\text{-}PhC_6H_4)_2SnBr_2$, 2, 320
$(p\text{-}PhC_6H_4)_2SnBr_2$, 2, 320
$SnC_{24}H_{18}Cl_2$ $(o\text{-}PhC_6H_4)_2SnCl_2$, 2, 320
$(p\text{-}PhC_6H_4)_2SnCl_2$, 2, 320
$SnC_{24}H_{18}Cl_2O_2$ $(o\text{-}PhOC_6H_4)_2SnCl_2$, 2, 326
$SnC_{24}H_{18}I_2O_2$ $(o\text{-}PhOC_6H_4)_2SnI_2$, 2, 326
$SnC_{24}H_{18}O$ $(p\text{-}PhC_6H_4)_2SnO$, 2, 373
$SnC_{24}H_{18}O_2$ $(p\text{-}PhOC_6H_4)_2Sn$, 2, 491
$SnC_{24}H_{19}Br$ $p\text{-}BrC_6H_4SnPh_3$, 2, 222
$SnC_{24}H_{19}BrO$ $o\text{-}BrC_6H_4OSnPh_3$, 2, 389
$p\text{-}BrC_6H_4OSnPh_3$, 2, 389
$SnC_{24}H_{19}Cl$ $p\text{-}ClC_6H_4SnPh_3$, 2, 222
$SnC_{24}H_{19}ClO$ $o\text{-}ClC_6H_4OSnPh_3$, 2, 389
$p\text{-}ClC_6H_4OSnPh_3$, 2, 389
$SnC_{24}H_{19}F$ $p\text{-}FC_6H_4SnPh_3$, 2, 222
$SnC_{24}H_{19}I$ $p\text{-}IC_6H_4SnPh_3$, 2, 222
$SnC_{24}H_{19}NO_3$ $o\text{-}NO_2C_6H_4OSnPh_3$, 2, 391
$m\text{-}NO_2C_6H_4OSnPh_3$, 2, 391
$p\text{-}NO_2C_6H_{11}OSnPh_3$, 2, 391
$SnC_{24}H_{19}N_3$ N-(1,2,3-Benzotriazolyl)-$SnPh_3$, 2, 479
$SnC_{24}H_{20}$ Ph_4Sn, 2, 166
$SnC_{24}H_{20}Cl_2N_2$ Ph_2SnCl_2·phenanthroline, 2, 315
$SnC_{24}H_{20}O$ $o\text{-}HOC_6H_4SnPh_3$, 2, 227
$m\text{-}HOC_6H_4SnPh_3$, 2, 227
$p\text{-}HOC_6H_4SnPh_3$, 2, 228
$SnC_{24}H_{20}O_2$ $(o\text{-}HOC_6H_4)_2SnPh_2$, 2, 228
$(m\text{-}HOC_6H_4)_2SnPh_2$, 2, 228
$(p\text{-}HOC_6H_4)_2SnPh_2$, 2, 228
$SnC_{24}H_{20}O_2S$ Ph_3SnO_2SPh, 2, 356*
$SnC_{24}H_{20}O_3$ $(m\text{-}HOC_6H_4)_3SnPh$, 2, 228
$SnC_{24}H_{20}O_3S$ Ph_3SnO_3SPh, 2, 356
$SnC_{24}H_{20}O_4$ $(m\text{-}HOC_6H_4)_4Sn$, 2, 175
$(o\text{-}PhOC_6H_4)_2Sn(OH)_2$, 2, 374
$(SnC_{24}H_{20}P_2)_n$ $[Ph_2PSnPhPPh]_n$, 2, 484
$SnC_{24}H_{20}S$ Ph_3SnSPh, 2, 459
$SnC_{24}H_{20}S_2$ $Ph_2Sn(SPh)_2$, 2, 463
$SnC_{24}H_{20}Se$ $Ph_3SnSePh$, 2, 459
$SnC_{24}H_{21}N$ $p\text{-}Ph_3SnC_6H_4NH_2$, 2, 241
$SnC_{24}H_{21}NO_2S$ Ph_3SnNHO_2SPh, 2, 481
$SnC_{24}H_{21}NO_3S$ Ph_3SnO_3SNHPh, 2, 356
$SnC_{24}H_{22}O_7$ $(o\text{-}MeO_2CC_6H_4)_3SnOH$, 2, 366
$SnC_{24}H_{24}O$ 2-Ketocyclohexyl $SnPh_3$, 2, 231
$Ph(p\text{-}CH_2{=}CHCH_2C_6H_4)_2SnOH$ (?), 2, 365
$SnC_{24}H_{24}O_4$ $(p\text{-}CH_2{=}CHC_6H_4CO_2)_2Sn(CH_2\text{-}CH{=}CH_2)_2$, 2, 438
$SnC_{24}H_{25}NO$ $Ph_3SnCH_2CH_2N(CH_2)_3CO$, 2, 191
$Ph_3Sn(CH_2)_3OCH_2CH_2CN$, 2, 227
$SnC_{24}H_{26}$ Cyclohexyl-$SnPh_3$, 2, 188
$Ph_3SnCH_2CMe{=}CMe_2$, 2, 205
$(p\text{-}MeC_6H_4)_3SnCH_2CH{=}CH_2$, 2, 206
$SnC_{24}H_{26}N_2O_2$ (8-Quinolyloxy)$_2SnPr_2$, 2, 401
$SnC_{24}H_{26}O$ $Ph_3Sn(CH_2)_4Ac$, 2, 231
$SnC_{24}H_{26}O_2$ $Ph_3SnCH{=}CMeC(O_2H)Me_2$ + $Ph_3SnCH_2CMe(O_2H)CMe{=}CH_2$, 2, 229
$Ph_3Sn(CH_2)_4CO_2Me$, 2, 237
$(MeC_6H_4)_3SnO_2CEt$, 2, 415
$SnC_{24}H_{27}Cl$ $(o\text{-}MeC_6H_4CH_2)_3SnCl$, 2, 282
$(m\text{-}MeC_6H_4CH_2)_3SnCl$, 2, 282
$(p\text{-}MeC_6H_4CH_2)_3SnCl$, 2, 282
$SnC_{24}H_{27}ClO_3$ $(p\text{-}MeOC_6H_4CH_2)_3SnCl$, 2, 299
$SnC_{24}H_{27}NO$ $Ph_3SnCH_2CH_2N(CH_2)_3CHOH$, 2, 191
$SnC_{24}H_{28}$ $n\text{-}C_6H_{13}SnPh_3$, 2, 187
$Bu_2Sn(C{\equiv}CPh)_2$, 2, 214
$SnC_{24}H_{30}$ o,o'-$C_6H_4C_6H_4Sn$-cyclohexyl$_2$, 2, 506
$SnC_{24}H_{32}$ Cyclohexyl$_2SnPh_2$, 2, 188
$(p\text{-}CH_2{=}CHCH_2C_6H_4)_2SnPr_2$, 2, 202
$SnC_{24}H_{32}O_4$ $Ph_2Sn[(CH_2)_3OCH_2CHCH_2O]_2$, 2, 228
$(p\text{-}MeC_6H_4CO_2)_2SnBu_2$, 2, 437
$SnC_{24}H_{32}O_4S_2$ $(o\text{-}MeO_2CC_6H_4S)_2SnBu_2$, 2, 467
$SnC_{24}H_{32}O_6$ $(o\text{-}MeO_2CC_6H_4O)_2SnBu_2$, 2, 403
$SnC_{24}H_{33}N_4P$ $Ph_3SnN{=}P(NMe_2)_3$, 2, 481
$SnC_{24}H_{36}$ 1-Cyclohexenyl$_4Sn$, 2, 173

i-$Bu_2Sn(CH_2CH_2Ph)_2$, 2, 184
$SnC_{24}H_{36}O_2$ o-$Bu_3SnOC_6H_4OPh$, 2, 390
$SnC_{24}H_{36}O_7P_2$ $Ph_2SnP_2O_5(OC_6H_{13})_2$, 2, 356
$SnC_{24}H_{36}S_2$ $Et_2Sn(SC_6H_4CMe_3)_2$, 2, 516
$SnC_{24}H_{37}Br$ p-BrC_6H_4Sn-cyclohexyl$_3$, 2, 222
$SnC_{24}H_{37}Cl$ m-ClC_6H_4Sn-cyclohexyl$_3$, 2, 222
p-ClC_6H_4Sn-cyclohexyl$_3$, 2, 222
$SnC_{24}H_{37}F$ p-FC_6H_4Sn-cyclohexyl$_3$, 2, 222
$SnC_{24}H_{37}NO_2$ $Bu_3SnN(1\text{-}C_{10}H_7)CO_2Me$, 2, 480
$SnC_{24}H_{37}NO_3S$ p-$Bu_3SnO_3SC_6H_4NHPh$, 2, 352
$SnC_{24}H_{37}OP$ $Bu_3SnP(O)Ph_2$, 2, 483
$SnC_{24}H_{37}O_2P$ $Bu_3SnO_2PPh_2$, 2, 352
$SnC_{24}H_{37}P$ Bu_3SnPPh_2, 2, 483
$SnC_{24}H_{38}$ Cyclohexyl$_3$SnPh, 2, 188
$SnC_{24}H_{38}N_2O_4S_2$ $Bu_2Sn(NEtO_2SPh)_2$, 2, 481
$SnC_{24}H_{40}N_2O_8S$ $Ph_2SnSO_4\cdot 2\ NH_4O_2CC_5H_{11}$, 2, 357
$SnC_{24}H_{40}O_4$ o-$Bu_3SnO_2CC_6H_4CO_2Bu$, 2, 423
$SnC_{24}H_{40}O_8$ $Bu_2Sn(CAc_2CO_2Et)_2$, 2, 232
$Bu_2Sn(O_2CCH{=}CHCO_2Bu)_2$, 2, 440
$SnC_{24}H_{40}O_8S_2$ $Et_2Sn(O_3SC_{10}H_{15}O)_2$, 2, 351
$SnC_{24}H_{42}$ p-(i-$C_5H_{11})_3SnC_6H_4CH_2CH{=}CH_2$, 2, 203
$SnC_{24}H_{42}O_4$ $(C_8H_{17})_2Sn(O_2CCMe{=}CH)_2$, 2, 446
$SnC_{24}H_{44}$ Cyclohexyl$_4$Sn, 2, 166
$[Me_2CHCH_2C({=}CH_2)]_4Sn$, 2, 173
$SnC_{24}H_{44}N_2O$ p-$(C_5H_{11})_3SnON{=}CHC_6H_4$-$NMe_2$, 2, 391
$SnC_{24}H_{44}O_4$ (n-$C_8H_{17})_2Sn(O_2CCMe{=}CH_2)_2$, 2, 511*
$SnC_{24}H_{44}O_4S_2$ $Bu_2Sn(SCH_2CO_2C_6H_{11})_2$, 2, 405
$SnC_{24}H_{44}O_6$ $Bu_2Sn(CHAcCO_2Bu)_2$, 2, 231
$SnC_{24}H_{44}O_8$ $[Bu_2Sn(OAc)O_2CCH{=}]_2$ (?), 2, 441
$SnC_{24}H_{46}$ Cyclohexyl$_3SnC_6H_{13}$, 2, 187
$SnC_{24}H_{47}N_3O$ 2,4,6-$(Me_2NCH_2)_3C_6H_2$-$OSn(CHMe_2)_3$, 2, 391
$SnC_{24}H_{48}O_2$ $Bu_3SnO_2CC_{11}H_{21}$, 2, 419
$SnC_{24}H_{48}O_2S_2$ S-$Bu_2Sn(SOCC_7H_{15})_2$, 2, 471
$SnC_{24}H_{48}O_4$ $Bu_2Sn(O_2CC_7H_{15})_2$, 2, 437
$Bu_2{}^{113}Sn(O_2CC_7H_{15})_2$, 2, 437
$(Pr^{14}CH_2)_2Sn(O_2CCHBuEt)_2$, 2, 437
(n-$C_8H_{17})_2Sn(O_2CPr)_2$, 2, 438
$SnC_{24}H_{50}Br_2$ $(C_{12}H_{25})_2SnBr_2$, 2, 319
$SnC_{24}H_{50}Cl_2$ $(C_{12}H_{25})_2SnCl_2$, 2, 319
$SnC_{24}H_{50}I_2$ (n-$C_{12}H_{25})_2SnI_2$, 2, 319
$SnC_{24}H_{50}O$ $(C_{12}H_{25})_2SnO$, 2, 370
$SnC_{24}H_{50}O_2$ $Bu_3SnO_2CC_{11}H_{23}$, 2, 414
$SnC_{24}H_{50}O_2S$ n-$C_{10}H_{21}O_2CCH_2SSnBu_3$, 2, 461, 516
$SnC_{24}H_{50}O_4P_2S_4$ $Bu_2Sn(S_2\overline{POCHPrCHEtCH_2O})_2$, 2, 353
$(C_8H_{17})_2Sn(S_2\overline{POCHMeCH_2CH_2O})_2$, 2, 354
$SnC_{24}H_{51}Cl$ (n-$C_8H_{17})_3SnCl$, 2, 284
$(C_8H_{17})_3SnCl$, 2, 517
$(BuEtCHCH_2)_3SnCl$, 2, 284
$SnC_{24}H_{51}I$ (n-$C_8H_{17})_3SnI$, 2, 284
$SnC_{24}H_{51}NOS$ $Bu_3SnS(CH_2)_3CONBu_2$, 2, 461
$SnC_{24}H_{51}PS_3$ $(C_{12}H_{25})_2Sn(HPS_3)$ (?), 2, 354
$SnC_{24}H_{52}$ (n-$C_6H_{13})_4Sn$, 2, 165
(i-$C_6H_{13})_4Sn$, 2, 170
(n-$C_{10}H_{21})_2SnEt_2$, 2, 183
$Bu_2Sn(CH_2CHBuEt)_2$, 2, 186
$SnC_{24}H_{52}O$ $Bu_3SnOC_{12}H_{25}$, 2, 385
$SnC_{24}H_{52}OS$ $Bu_2Sn(OBu)SC_{12}H_{25}$, 2, 517
$SnC_{24}H_{52}O_2$ $(C_8H_{17})_2Sn(OBu)_2$, 2, 399
$SnC_{24}H_{52}O_3S$ $Bu_3SnO_3SC_{12}H_{25}$, 2, 352
$SnC_{24}H_{52}S$ $Bu_3SnSC_{12}H_{25}$, 2, 459
$SnC_{24}H_{52}S_2$ $Bu_2Sn(SC_8H_{17})_2$, 2, 462
$SnC_{24}H_{54}O_4P_2S_4$ $Bu_2Sn[S_2P(OBu)_2]_2$, 2, 352
$SnC_{25}H_7F_{15}$ p-$MeC_6H_4Sn(C_6F_5)_3$, 2, 223
$SnC_{25}H_{18}Cl_2O_3$ o-$HOC_6H_4CO_2Sn(C_6H_4Cl)_2$-Ph, 2, 432
$SnC_{25}H_{19}Br$ 9-Fluorenyl-Ph_2SnBr, 2, 295
$SnC_{25}H_{19}Cl$ 9-Fluorenyl-Ph_2SnCl, 2, 295
$SnC_{25}H_{19}F_3$ m-$Ph_3SnC_6H_4CF_3$, 2, 223
$SnC_{25}H_{19}NS_2$ $Ph_3SnSC_7H_4NS$, 2, 460
$SnC_{25}H_{20}Cl_4$ $PhSnCl_3\cdot Ph_3CCl$, 2, 330
$SnC_{25}H_{20}N_2$ N-(Benzimidazolyl)$SnPh_3$,

2, 479
$SnC_{25}H_{20}OS$ Ph_3SnSBz, 2, 472
$SnC_{25}H_{20}O_2$ p-$HO_2CC_6H_4SnPh_3$, 2, 238
Ph_3SnOBz, 2, 415
$SnC_{25}H_{20}O_3$ o-$HOC_6H_4CO_2SnPh_3$, 2, 423
$SnC_{25}H_{20}O_6S$ 3,5-$HO_3S(HO)C_6H_3CO_2SnPh_3$, 2, 429
$SnC_{25}H_{21}N_3S_6$ $BuSn(SC_7H_4NS)_3$, 2, 469
$SnC_{25}H_{22}$ $PhCH_2SnPh_3$, 2, 182
o-$MeC_6H_4SnPh_3$, 2, 189
p-$MeC_6H_4SnPh_3$, 2, 189
$SnC_{25}H_{22}O$ o-$Ph_3SnC_6H_4OMe$, 2, 227
p-$Ph_3SnC_6H_4OMe$, 2, 228
o-$Ph_3SnC_6H_4CH_2OH$, 2, 228
p-$Ph_3SnC_6H_4CH_2OH$, 2, 228
$SnC_{25}H_{22}O_3S$ p-$MeC_6H_4SO_3SnPh_3$, 2, 356
$SnC_{25}H_{23}N$ 2-Pyridyl-$CH_2CH_2SnPh_3$, 2, 192
4-Pyridyl-$CH_2CH_2SnPh_3$, 2, 192*
$SnC_{25}H_{23}NO_2S$ p-$MeC_6H_4SO_2NHSnPh_3$, 2, 481
$SnC_{25}H_{24}O_6S_3$ $(o\text{-}MeO_2CC_6H_4S)_3SnMe$, 2, 468*
$SnC_{25}H_{26}O_2$ $Ph_3SnC{\equiv}CCH(OEt)_2$, 2, 216
$SnC_{25}H_{28}$ $(p\text{-}MeC_6H_4)_3SnCH_2CMe{=}CH_2$, 2, 206
$SnC_{25}H_{28}O_2$ $Ph_3Sn(CH_2)_4CO_2Et$, 2, 237
$SnC_{25}H_{30}$ $C_7H_{15}SnPh_3$, 2, 188
$SnC_{25}H_{30}O_2$ $Ph_3SnCH_2CH_2CH(OEt)_2$, 2, 231
$SnC_{25}H_{30}O_3S_3$ $(p\text{-}HOC_6H_4CH_2S)_3SnBr$, 2, 469
$SnC_{25}H_{31}N$ $Ph_3Sn(CH_2)_3NEt_2$, 2, 241
$SnC_{25}H_{38}O_2$ p-Cyclohexyl$_3SnC_6H_4CO_2H$, 2, 237
$SnC_{25}H_{40}$ m-MeC_6H_4Sn-cyclohexyl$_3$, 2, 188
p-MeC_6H_4Sn-cyclohexyl$_3$, 2, 188
$SnC_{25}H_{40}O$ m-$MeOC_6H_4Sn$-cyclohexyl$_3$, 2, 227
p-$MeOC_6H_4Sn$-cyclohexyl$_3$, 2, 227
$SnC_{25}H_{44}O_3$ $Bu_2(BuEtCHCH_2O_2CCHPh)Sn$-OMe, 2, 386
$SnC_{25}H_{46}O_6$ $Bu_2Sn(O_2CCH{=}CHCO_2Me)O_2C$-$C_{11}H_{23}$ (?), 2, 441
$SnC_{25}H_{48}$ Cyclohexyl$_3SnC_7H_{15}$, 2, 188
$SnC_{25}H_{50}O_2$ $(n\text{-}C_7H_{15})_3SnO_2CCMe{=}CH_2$, 2, 419, 511
$SnC_{25}H_{52}O_2$ $Bu_3SnO_2CC_{12}H_{25}$, 2, 414
$SnC_{25}H_{54}O$ $(n\text{-}C_8H_{17})_3SnOMe$, 2, 386
$SnC_{26}H_{14}F_{10}$ $(p\text{-}MeC_6H_4)_2Sn(C_6F_5)_2$, 2, 223
$SnC_{26}H_{17}Cl_3$ $(p\text{-}ClC_6H_4)_3SnC{\equiv}CPh$, 2, 216
$SnC_{26}H_{20}$ $Ph_3SnC{\equiv}CPh$, 2, 214
$SnC_{26}H_{20}O_2S_2$ $Ph_2Sn(SBz)_2$, 2, 472
$SnC_{26}H_{22}$ $Ph_3SnCH{=}CHPh$, 2, 204
p-$Ph_3SnC_6H_4CH{=}CH_2$, 2, 197, 510
o,o'-$C_6H_4C_6H_4Sn(C_6H_4Me)_2$, 2, 506
o,o'-$C_6H_4CH_2CH_2C_6H_4SnPh_2$, 2, 507
$SnC_{26}H_{22}Cl_2$ $(Ph_2CH)_2SnCl_2$, 2, 316
$SnC_{26}H_{22}Cl_4$ o-$MeC_6H_4SnCl_3\cdot Ph_3CCl$, 2, 330
p-$MeC_6H_4SnCl_3\cdot Ph_3CCl_3$, 2, 332
$SnC_{26}H_{22}N_2O_4$ $(o\text{-}H_2NC_6H_4CO_2)_2SnPh_2$, 2, 444
$SnC_{26}H_{23}N$ o,o'-$C_6H_4NEtC_6H_4SnPh_2$, 2, 507*, 638*
$SnC_{26}H_{23}NS_2$ $Ph_3SnS_2CNHCH_2Ph$, 2, 472
$SnC_{26}H_{24}$ $(p\text{-}PhC_6H_4)_2SnMe_2$, 2, 180
$(PhCH_2)_2SnPh_2$, 2, 181
$PhCH_2CH_2SnPh_3$, 2, 184
p-$EtC_6H_4SnPh_3$, 2, 189
2,4-$Me_2C_6H_3SnPh_3$, 2, 189
2,5-$Me_2C_6H_3SnPh_3$, 2, 189
2,6-$Me_2C_6H_3SnPh_3$, 2, 189
$(p\text{-}MeC_6H_4)_2SnPh_2$, 2, 189
1-Cyclohexenyl-$C{\equiv}CSnPh_3$, 2, 214
$SnC_{26}H_{24}I_2N_2$ $Me_2SnI_2\cdot 2\ MeC_9H_6N$, 2, 305
$SnC_{26}H_{24}O$ $Ph_3SnCH_2CH_2OPh$, 2, 227
o-$Ph_3SnC_6H_4CH_2OMe$, 2, 228
$SnC_{26}H_{24}O_2$ $(p\text{-}MeOC_6H_4)_2SnPh_2$, 2, 228
$SnC_{26}H_{24}S_2$ $Ph_2Sn(SCH_2Ph)_2$, 2, 463
$SnC_{26}H_{25}Cl_3O_4S$ $Ph_3SnSC(CCl_3)(CO_2Et)_2$, 2, 461
$SnC_{26}H_{25}N$ o-$Ph_3SnC_6H_4NMe_2$, 2, 241
m-$Ph_3SnC_6H_4NMe_2$, 2, 241
p-$Ph_3SnC_6H_4NMe_2$, 2, 241
$SnC_{26}H_{26}$ Cyclohexyl-$C{\equiv}CSnPh_3$, 2, 214
$SnC_{26}H_{26}O_4P_2$ $Me_2Sn(O_2PPh_2)_2$, 2, 348
$SnC_{26}H_{26}P_2$ $Me_2Sn(PPh_2)_2$, 2, 482
$SnC_{26}H_{30}N_2O_2$ (8-Quinolyloxy)$_2SnBu_2$, 2, 401

$SnC_{26}H_{30}N_2O_2$
(8-Quinolyloxy)$_2$SnCH$_2$CHMe$_2$, 2, 401
$SnC_{26}H_{30}N_2S_4$ (4,2-PhC$_3$HNS)$_2$S$_2$SnBu$_2$, 2, 464
$SnC_{26}H_{30}O_2$ (PhCH$_2$CH$_2$)$_3$SnOAc, 2, 412
$SnC_{26}H_{32}$ n-C$_8$H$_{17}$SnPh$_3$, 2, 188
1-Indenyl$_2$SnBu$_2$, 2, 203
$SnC_{26}H_{32}I_2N_2$ Pr$_2$SnI$_2$·2 MeC$_9$H$_6$N, 2, 309
$SnC_{26}H_{32}O_4$ (p-CH$_2$=CHC$_6$H$_4$CO$_2$)$_2$SnBu$_2$, 2, 439, 511*
$SnC_{26}H_{34}IN$ Ph$_3$Sn(CH$_2$)$_3$NEt$_2$·MeI, 2, 241
$SnC_{26}H_{34}O_4S_2$ S-Me$_2$Sn[SC$_6$H$_4$CH$_2$O$_2$C-(CH$_2$)$_4$]$_2$, 2, 465
$SnC_{26}H_{36}O_4S_2$ Me$_2$Sn(SCHPhCH$_2$O$_2$CPr)$_2$, 2, 465
Bu$_2$Sn(SCH$_2$CH$_2$OBz)$_2$, 2, 465
$SnC_{26}H_{36}O_6S_2$ Bu$_2$Sn(CHAcO$_2$SPh)$_2$, 2, 232
$SnC_{26}H_{38}N_2S_4$ Me$_2$Sn[SC(SCH$_2$Ph)=NBu]$_2$, 2, 516
$SnC_{26}H_{38}O_3$ Bu$_3$SnO$_2$CCPh$_2$OH, 2, 423
$SnC_{26}H_{40}$ p-(C$_6$H$_{11}$)$_3$SnC$_6$H$_4$CH=CH$_2$, 2, 510
(C$_7$H$_{15}$)$_2$SnPh$_2$, 2, 188
$SnC_{26}H_{42}$ p-EtC$_6$H$_4$Sn-cyclohexyl$_3$, 2, 188
$SnC_{26}H_{42}Cl_2N_2$ (C$_8$H$_{17}$)SnCl$_2$·2,2'-bipyridyl, 2, 313
$SnC_{26}H_{43}N$ p-(C$_6$H$_{11}$)$_3$SnC$_6$H$_4$NMe$_2$, 2, 241
$SnC_{26}H_{44}I_2N_2$ Pr$_2$SnI$_2$·2 PhNEt$_2$, 2, 309
$SnC_{26}H_{44}O_8$ Bu$_2$Sn(O$_2$CCH=CHCO$_2$C$_5$H$_{11}$)$_2$, 2, 440
(C$_8$H$_{17}$)$_2$^{113}Sn(O$_2$CCH=CHCO$_2$Me)$_2$, 2, 441
(C$_8$H$_{17}$)$_2$SnO$_2$CCH=CHCO$_2$^{14}CH$_3$)$_2$, 2, 441
$SnC_{26}H_{48}O_6$ Bu$_2$Sn(CPrAcCO$_2$Et)$_2$, 2, 232
Bu$_2$Sn(O$_2$CCH=CHCO$_2$Et)O$_2$CC$_{11}$H$_{23}$, 2, 441
$SnC_{26}H_{50}$ Cyclohexyl$_3$SnC$_8$H$_{17}$, 2, 188*
$SnC_{26}H_{52}O_4$ Me$_2$Sn(O$_2$CC$_{11}$H$_{23}$)$_2$, 2, 435
$SnC_{26}H_{54}O_2$ Bu$_3$SnO$_2$CC$_{13}$H$_{27}$, 2, 414
(n-C$_8$H$_{17}$)$_3$SnOAc, 2, 414
$SnC_{26}H_{54}O_2S$ Bu$_3$SnSCH$_2$CH$_2$CO$_2$C$_{11}$H$_{23}$, 2, 461
$SnC_{26}H_{54}O_4P_2S_4$ (C$_8$H$_{17}$)$_2$Sn-(S$_2$POCH$_2$CMe$_2$CH$_2$O)$_2$, 2, 354
$SnC_{26}H_{56}$ (n-C$_{10}$H$_{21}$)$_2$SnPr$_2$, 2, 185
$SnC_{27}H_{17}F_9$ (m-CF$_3$C$_6$H$_4$)$_3$SnPh, 2, 223
$SnC_{27}H_{17}N_3S_6$ PhSn(SC$_7$H$_4$NS)$_3$, 2, 469
$SnC_{27}H_{18}ClN_3$ 2-Quinolyl$_3$SnCl, 2, 291
6-Quinolyl$_3$SnCl, 2, 291
$SnC_{27}H_{19}Br_2NO$ 8,5,7-(Ph$_3$SnO)dibromoquinoline, 2, 387
$SnC_{27}H_{20}O_2$ Ph$_3$SnO$_2$CC≡CPh, 2, 419
$SnC_{27}H_{21}F_5$ (p-MeC$_6$H$_4$)$_3$SnC$_6$F$_5$, 2, 223
$SnC_{27}H_{21}NO$ 8-Quinolyloxy-SnPh$_3$, 2, 387
$SnC_{27}H_{22}$ 1-Indenyl-SnPh$_3$, 2, 205
$SnC_{27}H_{22}O_2$ p-Ph$_3$SnO$_2$CC$_6$H$_4$CH=CH$_2$, 2, 419, 511
$SnC_{27}H_{22}O_3$ Ph$_3$SnC$_7$H$_7$(CO)$_2$O, 2, 237
$SnC_{27}H_{23}NO_3$ p-Ph$_3$SnO$_2$CC$_6$H$_4$NHAc, 2, 424
$SnC_{27}H_{24}$ p-Ph$_3$SnC$_6$H$_4$CH$_2$CH=CH$_2$, 2, 205
$SnC_{27}H_{24}O$ Ph$_3$SnCH$_2$CH$_2$Bz, 2, 231
$SnC_{27}H_{24}O_2$ p-Ph$_3$SnOC$_6$H$_4$COEt, 2, 390
$SnC_{27}H_{24}O_2S$ Ph$_3$SnSCH$_2$CO$_2$CH$_2$Ph, 2, 461
$SnC_{27}H_{24}O_3$ Isochromameperoxy-SnPh$_3$, 2, 380
$SnC_{27}H_{26}$ (PhCH$_2$)$_3$SnPh, 2, 181
Mesityl-SnPh$_3$, 2, 189
$SnC_{27}H_{26}O_2$ Ph$_3$SnO$_2$CMe$_2$Ph, 2, 380
$SnC_{27}H_{26}O_3S_3$ (p-HOC$_6$H$_4$CH$_2$S)$_3$SnPh, 2, 469
$SnC_{27}H_{27}ClO_6$ (p-EtO$_2$CC$_6$H$_4$)$_3$SnCl, 2, 299
$SnC_{27}H_{28}IN$ p-Ph$_3$SnC$_6$H$_4$NMe$_3$I, 2, 241
$SnC_{27}H_{28}O$ (p-CH$_2$=CHCH$_2$C$_6$H$_4$)$_3$SnOH, 2, 366
$SnC_{27}H_{33}Br$ (2,4,6-Me$_3$C$_6$H$_2$)$_3$SnBr, 2, 285
$SnC_{27}H_{33}I$ (2,4,6-Me$_3$C$_6$H$_2$)$_3$SnI, 2, 285
$SnC_{27}H_{34}$ n-C$_9$H$_{19}$SnPh$_3$, 2, 189
Mesityl$_3$SnH, 2, 264
$SnC_{27}H_{34}O_5$ Ph$_3$Sn(OCH$_2$CH$_2$)$_4$OMe, 2, 390
$SnC_{27}H_{37}NO_4S$ Ph$_3$Sn(CH$_2$)$_3$NEt$_2$·Me$_2$SO$_4$, 2, 241
$SnC_{27}H_{38}O_2$ p,p'-Bu$_3$SnO$_2$C(C$_6$H$_4$)$_2$CH=CH$_2$, 2, 419, 511

$SnC_{27}H_{42}$ p-Cyclohexyl$_3$SnC$_6$H$_4$CH$_2$-CH=CH$_2$, 2, 204
$SnC_{27}H_{44}$ p-Cyclohexyl$_3$SnC$_6$H$_4$CHMe$_2$, 2, 188
$SnC_{27}H_{44}O_2$ Cyclohexyl$_3$SnO$_2$CMe$_2$Ph, 2, 380
$SnC_{27}H_{46}IN$ p-Cyclohexyl$_3$SnC$_6$H$_4$NMe$_3$I, 2, 241
$SnC_{27}H_{52}$ Cyclohexyl$_3$SnC$_9$H$_{19}$, 2, 188
$SnC_{27}H_{53}N_3O$ 2,4,6-(Me$_2$NCH$_2$)$_3$C$_6$H$_2$O-SnBu$_3$, 2, 391
$SnC_{27}H_{54}O_2$ (C$_8$H$_{17}$)$_3$SnO$_2$CCH=CH$_2$, 2, 419
$SnC_{27}H_{54}O_3$ Bu$_2$Sn(OMe) oleate, 2, 407
$SnC_{27}H_{55}N_3O_3S_3$ i-PrSn[S(CH$_2$)$_3$CONH-Bu]$_3$, 2, 469
$SnC_{27}H_{56}O_2$ Bu$_3$SnO$_2$CC$_{14}$H$_{29}$, 2, 414
$SnC_{28}H_{16}F_{12}$ (m-CF$_3$C$_6$H$_4$)$_4$Sn, 2, 174
(p-CF$_3$C$_6$H$_4$)$_4$Sn, 2, 174
$SnC_{28}H_{18}$ Di(9-phenanthryl)tin, 2, 491
$SnC_{28}H_{18}Cl_2$ 9-Phenanthryl$_2$SnCl$_2$, 2, 320
$SnC_{28}H_{20}$ Ph$_3$SnC≡CC≡CPh, 2, 214
Ph$_2$Sn(C≡CPh)$_2$, 2, 215
$SnC_{28}H_{20}O_4$ (p-CHOC$_6$H$_4$)$_4$Sn, 2, 175
$SnC_{28}H_{20}S_2$ (2-Thienyl)$_2$Sn(1-C$_{10}$H$_7$)$_2$, 2, 193
$SnC_{28}H_{22}$ 1-C$_{10}$H$_7$SnPh$_3$, 2, 189
$SnC_{28}H_{22}N_2S_4$ (PhCH$_2$)$_2$Sn(2-SC$_7$H$_4$NS)$_2$, 2, 464
$SnC_{28}H_{22}N_4S_2$ (PhCH$_2$)$_2$Sn(2-SC$_7$H$_5$N$_2$)$_2$, 2, 464*
$SnC_{28}H_{22}O$ (PhCH=CPh)$_2$SnO, 2, 372
$SnC_{28}H_{22}O_3S$ 1-C$_{10}$H$_7$SO$_3$SnPh$_3$, 2, 356
$SnC_{28}H_{22}S$ 2-C$_{10}$H$_7$SSnPh$_3$, 2, 459
$SnC_{28}H_{23}NO_3$ o-Ph$_3$SnO$_2$CC$_6$H$_4$NHCOCH=CH$_2$, 2, 424, 511
o-Ph$_3$SnO$_2$CC$_6$H$_4$CONHCH=CH$_2$, 2, 429, 511
$SnC_{28}H_{24}$ (p-CH$_2$=CHC$_6$H$_4$)$_2$SnPh$_2$, 2, 198
BuSn(C≡CPh)$_3$, 2, 214
$SnC_{28}H_{24}O_2S_4$ (MeC$_6$H$_4$)$_2$Sn(S$_2$COPh)$_2$, 2, 472
$SnC_{28}H_{26}N_2$ (o,o'-C$_6$H$_4$NEtC$_6$H$_4$)$_2$Sn, 2, 506*, 638*
$SnC_{28}H_{26}O_2$ Tetralinperoxy-SnPh$_3$, 2, 380
(PhCH$_2$)$_3$SnOBz, 2, 412
$SnC_{28}H_{26}O_3$ m-Ph$_3$SnC$_6$H$_4$OCH$_2$CO$_2$Et, 2, 228
$SnC_{28}H_{27}NO$ p-Ph$_3$SnCH$_2$CH$_2$C$_6$H$_4$NHAc, 2, 241
$SnC_{28}H_{27}N_3O_3$ Ph$_3$SnO$_2$CCH$_2$CH$_2$CON(CH$_2$-CH$_2$CN)$_2$, 2, 429
$SnC_{28}H_{28}$ (PhCH$_2$)$_4$Sn, 2, 160
(o-MeC$_6$H$_4$)$_4$Sn, 2, 169
(m-MeC$_6$H$_4$)$_4$Sn, 2, 169
(p-MeC$_6$H$_4$)$_4$Sn, 2, 169
(o-MeC$_6$H$_4$)$_3$SnCH$_2$Ph, 2, 182
p-Ph$_3$SnC$_6$H$_4$CMe$_3$, 2, 189
$SnC_{28}H_{28}O_4$ (o-MeOC$_6$H$_4$)$_4$Sn, 2, 175
(p-MeOC$_6$H$_4$)$_4$Sn, 2, 175
$SnC_{28}H_{30}N_2$ (p-Me$_2$NC$_6$H$_4$)$_2$SnPh$_2$, 2, 242
$SnC_{28}H_{30}O_4P_2$ Et$_2$Sn(O$_2$PPh$_2$)$_2$, 2, 350
$SnC_{28}H_{30}O_6S_3$ (p-MeO$_2$CC$_6$H$_4$S)$_3$SnBu, 2, 469
$SnC_{28}H_{31}NO_4S$ p-Ph$_3$SnC$_6$H$_4$NMe$_2$·Me$_2$SO$_4$, 2, 242
$SnC_{28}H_{32}O_2$ 9-Decalinperoxy-SnPh$_3$, 2, 380
$SnC_{28}H_{32}S_2$ (1-C$_{10}$H$_7$S)$_2$SnBu$_2$, 2, 463
$SnC_{28}H_{36}$ Mesityl$_3$SnMe, 2, 181
9-(Cyclopentyl$_3$Sn)fluorene, 2, 204
$SnC_{28}H_{40}O_4S_2$ Bu$_2$Sn(SCHPhCO$_2$Et)$_2$, 2, 466
$SnC_{28}H_{40}O_6S_2$ Bu$_2$Sn(SCH$_2$CO$_2$CH$_2$CH$_2$OPh)$_2$, 2, 465
$SnC_{28}H_{44}$ (n-C$_8$H$_{17}$)$_2$SnPh$_2$, 2, 188
$SnC_{28}H_{44}Cl_2N_2$ (C$_8$H$_{17}$)$_2$SnCl$_2$·phenanthroline, 2, 313
$SnC_{28}H_{44}S_2$ (p-Me$_3$CC$_6$H$_4$S)$_2$SnBu$_2$, 2, 463
(C$_8$H$_{17}$)$_2$Sn(SPh)$_2$, 2, 463
$SnC_{28}H_{46}$ Cyclohexyl$_3$SnC$_6$H$_4$CMe$_3$, 2, 188
$SnC_{28}H_{46}N_2O_4S_2$ (PhSO$_2$NBu)$_2$SnBu$_2$, 2, 481
$SnC_{28}H_{48}Cl_2N_2$ i-Bu$_2$SnCl$_2$·2 PhNEt$_2$, 2, 317
$SnC_{28}H_{48}O_4S_2$ EtPhSn(SCH$_2$CO$_2$C$_8$H$_{17}$)$_2$, 2, 465
$SnC_{28}H_{52}O_4$ (CH$_2$=CH)$_2$Sn(O$_2$CC$_{11}$H$_{23}$)$_2$, 2, 436*
$SnC_{28}H_{52}O_5$ (C$_{12}$H$_{25}$)$_2$Sn epoxysuccinate,

2, 446

$SnC_{28}H_{52}O_6$ $(i\text{-}C_6H_{13}O_2CCHAc)_2SnBu_2$, 2, 232

$Bu_2Sn(CBuAcCO_2Et)_2$, 2, 232

$Bu_2Sn(O_2CCH{=}CHCO_2Bu)O_2CC_{11}H_{23}$, 2, 441

$SnC_{28}H_{54}$ $Cyclohexyl_3SnC_{10}H_{21}$, 2, 188

$SnC_{28}H_{54}O_6$ $BuSn(O_2CCHEtBu)_3$, 2, 449

$SnC_{28}H_{56}O_4$ $Bu_2Sn(O_2CC_9H_{19})_2$, 2, 437

$SnC_{28}H_{56}O_4S_2$ $Bu_2Sn(SCH_2CO_2C_8H_{17})_2$, 2, 466

$Bu_2Sn(SCH_2CO_2CH_2CHEtBu)_2$, 2, 466

$Bu_2{}^{113}Sn(SCH_2CO_2CH_2CHEtBu)_2$, 2, 466

$(Pr^{14}CH_2)_2Sn(SCH_2CO_2C_8H_{17})_2$, 2, 466

$Bu_2Sn(SCH_2{}^{14}CO_2C_8H_{17})_2$, 2, 466*

$Bu_2Sn(SCHBuCO_2Bu)_2$, 2, 466

$Bu_2Sn(SCEtBuCO_2Et)_2$, 2, 467

$(i\text{-}C_8H_{17}O_2CCH_2S)_2SnCH_2CHMe_2$, 2, 467

$(C_8H_{17})_2Sn(SCH_2CO_2Bu)_2$, 2, 467

$SnC_{28}H_{56}S_2$ $(CH_2{=}CH)_2Sn(SC_{12}H_{25})_2$, 2, 462

$SnC_{28}H_{57}NO_2$ $(C_8H_{17})_2Sn{=}(OCHMeCH_2)_2N\text{-}C_6H_{11}$, 2, 404

$SnC_{28}H_{58}O_2$ $Bu_3SnO_2CC_{15}H_{31}$, 2, 414

$SnC_{28}H_{60}$ $(n\text{-}C_7H_{15})_4Sn$, 2, 166

$(n\text{-}C_8H_{17})_3SnCH_2CHMe_2$, 2, 187

$SnC_{28}H_{60}S_2$ $Me_2Sn(SC_{12}H_{25})SC_{14}H_{29}$, 2, 462

$SnC_{28}H_{61}O_4P$ $Bu_3SnPO_2(OC_8H_{17})_2$, 2, 351

$SnC_{28}H_{70}O_8$ $Bu_2Sn[O(CH_2CH_2O)_3Bu]_2$, 2, 403

$SnC_{29}H_{25}NO_3$ $o\text{-}Ph_3SnO_2CC_6H_4NHCO\text{-}CMe{=}CH_2$, 2, 424, 511

$p\text{-}Ph_3SnO_2CC_6H_4NHCOCMe{=}CH_2$, 2, 424, 511

$SnC_{29}H_{26}$ $(p\text{-}MeC_6H_4)_3SnC{\equiv}CPh$, 2, 215

$SnC_{29}H_{28}$ $p\text{-}(PhCH_2)_3SnC_6H_4CH{=}CH_2$, 2, 510

$SnC_{29}H_{36}O_2$ $(PhCH_2)_3SnO_2CC_7H_{15}$, 2, 411

$SnC_{29}H_{38}$ $Mesityl_3SnEt$, 2, 184

$SnC_{29}H_{56}O_4$ $Bu_2SnOCH_2CHOCH_2O_2C(CH_2)_7\text{-}CH{=}CHC_8H_{17}$, 2, 405

$SnC_{29}H_{60}O_2$ $Bu_3SnO_2CC_{16}H_{33}$, 2, 414

$SnC_{29}H_{62}O_2$ $(C_8H_{17})_3Sn(CH_2)_3OCH_2CH_2OH$, 2, 227

$SnC_{30}H_{20}$ $PhSn(C{\equiv}CPh)_3$, 2, 215

$SnC_{30}H_{21}Br$ $(1\text{-}C_{10}H_7)_3SnBr$, 2, 285

$SnC_{30}H_{21}Cl$ $(1\text{-}C_{10}H_7)_3SnCl$, 2, 285

$SnC_{30}H_{22}Cl_2O_4$ $(p\text{-}ClC_6H_4)_2Sn(O_2CC_6H_4\text{-}CH{=}CH_2\text{-}p)_2$, 2, 440

$SnC_{30}H_{22}N_2O_2$ $(8\text{-}Quinolyloxy)_2SnPh_2$, 2, 401

$SnC_{30}H_{22}O$ $(1\text{-}C_{10}H_7)_3SnOH$, 2, 366

$SnC_{30}H_{24}$ $1\text{-}Indenyl_2SnPh_2$, 2, 206

$SnC_{30}H_{25}O_2P$ $Ph_3SnO_2PPh_2$, 2, 355

$SnC_{30}H_{24}O_4$ $(p\text{-}CH_2{=}CHC_6H_4CO_2)_2SnPh_2$, 2, 440, 511*

$SnC_{30}H_{25}P$ Ph_3SnPPh_2, 2, 483

$Ph_2Sn\cdot PPh_3$, 2, 483

$SnC_{30}H_{26}$ $CPh{=}CPhCPh{=}CPhSnMe_2$, 2, 506*

$SnC_{30}H_{26}Br_2$ $Me_2(BrCPh{=}CPhCPh{=}CPh)SnBr$, 2, 297

$SnC_{30}H_{26}N_2$ $Ph_3SnNPhNHPh$, 2, 476

$SnC_{30}H_{28}$ $(p\text{-}CH_2{=}CHCH_2C_6H_4)_2SnPh_2$, 2, 206, 512

$SnC_{30}H_{30}$ $p\text{-}Ph_3SnC_6H_4C_6H_{11}$, 2, 189

$SnC_{30}H_{36}I_2N_2$ $(p\text{-}IMe_3NC_6H_4)_2SnPh_2$, 2, 242

$SnC_{30}H_{36}O_4S$ $Ph_3SnSCH(CO_2Bu)CH_2CO_2Bu$, 2, 461

$SnC_{30}H_{36}O_8$ $(PhCH_2)_2Sn(O_2CCH{=}CHCO_2Bu)_2$, 2, 440

$SnC_{30}H_{38}O_2S$ $(PhCH_2)_3SnS(CH_2)_4O_2CBu$, 2, 461

$n\text{-}C_{10}H_{21}O_2CCH_2SSnPh_3$, 2, 461

$SnC_{30}H_{39}Cl$ $(p\text{-}Me_2CHC_6H_4CH_2)_3SnCl$, 2, 282

$SnC_{30}H_{40}$ $Mesityl_3SnPr$, 2, 185

$Mesityl_3SnCHMe_2$, 2, 185

$SnC_{30}H_{40}O$ $(p\text{-}Me_2CHC_6H_4CH_2)_3SnOH$, 2, 364

$SnC_{30}H_{40}O_4$ $(p\text{-}CH_2{=}CHC_6H_4CO_2)_2Sn\text{-}(C_6H_{13})_2$, 2, 440

$SnC_{30}H_{40}O_6$ $Bu_2Sn(CHBzCO_2Et)_2$, 2, 232

$SnC_{30}H_{40}S$ $Ph_3SnSC_{12}H_{25}$, 2, 459

$SnC_{30}H_{42}$ $o\text{-}PhC_6H_4Sn\text{-}cyclohexyl_3$, 2, 188

$p\text{-}PhC_6H_4Sn\text{-}cyclohexyl_3$, 2, 188

$SnC_{30}H_{44}O_4$ $(PhCH_2)_2Sn(O_2CCHBuEt)_2$, 2, 435*

$SnC_{30}H_{48}$ $(n\text{-}C_9H_{19})_2SnPh_2$, 2, 189

$SnC_{30}H_{48}S_2$ $(C_8H_{17})_2Sn(SCH_2Ph)_2$, 2, 463

$SnC_{30}H_{56}O_8$ $Bu_2Sn[CH(CO_2Bu)_2]_2$, 2, 236

$SnC_{30}H_{58}O_2$ Bu_3Sn linoleate, 2, 419

$SnC_{30}H_{60}O_2$ Bu_3Sn oleate, 2, 419

$SnC_{30}H_{60}O_3$ Bu_3Sn ricinolate, 2, 429

$SnC_{30}H_{60}O_4S_2$ $Me_2Sn(SCH_2CH_2O_2CC_{11}H_{23})_2$ 2, 465

$(Bu_2Sn(SCH_2CH_2O_2CC_8H_{17})_2$, 2, 465*

$Bu_2Sn(SCH_2CO_2C_9H_{19})_2$, 2, 466

$Bu_2Sn(SCH_2CH_2CO_2CH_2CHEtBu)_2$, 2, 466

$Bu_2Sn[SCH(C_5H_{11})CO_2Bu]_2$, 2, 467

$SnC_{30}H_{62}N_2S_4$ $[(n-C_5H_{11})_2NCS_2]_2SnBu_2$, 2, 471

$SnC_{30}H_{62}O_2$ $Bu_3SnO_2CC_{17}H_{35}$, 2, 414

$SnC_{31}H_{24}$ 9-Fluorenyl-$SnPh_3$, 2, 205

$SnC_{31}H_{24}S_3$ $(1-C_{10}H_7S)_3SnMe$, 2, 468*

$SnC_{31}H_{25}Cl$ $Ph_2(Ph_3C)SnCl$, 2, 288

$SnC_{31}H_{26}$ $Ph_2CHSnPh_3$, 2, 182

$SnC_{31}H_{27}N_3O_3$ (8-Quinolyloxy)$_3$SnBu, 2, 408

$SnC_{31}H_{27}OP$ $Ph_3SnCH_2P(O)Ph_2$, 2, 243

$SnC_{31}H_{28}IP$ $[Ph_3SnPMePh_2]I$, 2, 483

$SnC_{31}H_{30}O_4$ $Ph_3SnC_7H_5(CO_2Et)_2$, 2, 238

$SnC_{31}H_{32}O_4$ $Ph_3SnC_7H_7(CO_2Et)_2$, 2, 237

$SnC_{31}H_{40}O_2$ $(PhCH_2)_3SnO_2CC_9H_{19}$, 2, 411

$SnC_{31}H_{42}$ Mesityl$_3$SnBu, 2, 187

Mesityl$_3SnCH_2CHMe_2$, 2, 187

9-Fluorenyltricyclohexyltin, 2, 204

$SnC_{31}H_{54}O_2S$ o-$SC_6H_4CO_2Sn(C_{12}H_{25})_2$, 2, 447

$SnC_{31}H_{54}S_3$ Tri(α-pinenemercapto)SnMe, 2, 468*

$SnC_{31}H_{56}O_6$ $Bu_2Sn(O_2CCH=CHCO_2Me)O_2C-C_{17}H_{33}$ (?), 2, 441

$SnC_{31}H_{56}O_6S$ 3,6-$HO_3S(HO)C_6H_3CO_2Sn-(C_8H_{17})_3$ (?), 2, 429

$SnC_{31}H_{58}O_6$ $Bu_2Sn(O_2CCH=CHCO_2Me)O_2C-C_{17}H_{35}$ (?), 2, 441

$SnC_{31}H_{60}O_6S_3$ $BuSn(SCH_2CH_2CO_2C_6H_{13})_3$, 2, 469

$SnC_{31}H_{66}O_3S$ $(i-C_{10}H_{21})_3SnO_3SMe$, 2, 354

$SnC_{32}H_{20}$ $(PhC\equiv C)_4Sn$, 2, 173

$SnC_{32}H_{23}Cl$ 9-Fluorenyl$_2$PhSnCl, 2, 295

$SnC_{32}H_{24}$ $(1-C_{10}H_7)_2SnPh_2$, 2, 190

$SnC_{32}H_{26}$ p,p'-$Ph_3Sn(C_6H_4)_2CH=CH_2$, 2, 205, 510

$CPh=CPhCPh=CPhSn(CH=CH_2)_2$, 2, 506

$SnC_{32}H_{26}N_2O_2$ (8-Quinolyloxy)$_2Sn(CH_2-Ph)_2$, 2, 401

(8-Quinolyloxy)$_2Sn(o-MeC_6H_4)_2$, 2, 401

$SnC_{32}H_{27}N$ N-Carbazolyl-$CH_2CH_2SnPh_3$, 2, 191*

$SnC_{32}H_{28}$ $(p-CH_2=CHC_6H_4)_4Sn$, 2, 173

$SnC_{32}H_{28}BrN_3$ 4,3-$(p-BrC_6H_4N=N)Me_2N-C_6H_3SnPh_3$, 2, 242

6,3-$(p-BrC_6H_4N=N)Me_2NC_6H_3SnPh_3$, 2, 242

$SnC_{32}H_{28}ClN_3$ 4,3-$(p-ClC_6H_4N=N)Me_2N-C_6H_4SnPh_3$, 2, 242

$SnC_{32}H_{28}N_4O_2$ 4,3-$(p-O_2NC_6H_4N=N)Me_2N-C_6H_4SnPh_3$, 2, 242

6,3-$(p-O_2NC_6H_4N=N)Me_2NC_6H_4SnPh_3$, 2, 242

$SnC_{32}H_{28}O_4$ $(p-CH_2=CHC_6H_4CO_2)_2Sn-(CH_2Ph)_2$, 2, 439

$SnC_{32}H_{29}Br$ 9-Fluorenyl$_2$(cyclohexyl)-SnBr, 2, 294

$SnC_{32}H_{30}Cl_2NP$ $Ph_2SnCl_2\cdot Ph_3P=NEt$, 2, 315

$SnC_{32}H_{30}Cl_2N_4$ $Ph_2SnCl_2\cdot 4$ Py, 2, 315

$SnC_{32}H_{34}N_6O_6$ $Ph_2Sn[O_2CCH_2CH_2CON(CH_2-CH_2CN)_2]_2$, 2, 444

$SnC_{32}H_{36}$ $(Me_2C_6H_3)_4Sn$, 2. 170

$(p-Me_3CC_6H_4)_2SnPh_2$, 2, 189

$SnC_{32}H_{36}O_4$ $(p-EtOC_6H_4)_4Sn$, 2, 175

$SnC_{32}H_{36}O_4P_2$ $Bu_2Sn(O_2PPh_2)_2$, 2, 353*

$SnC_{32}H_{38}P_2$ $Bu_2Sn(PPh_2)_2$, 2, 483

$SnC_{32}H_{40}N_4$ $(p-Me_2NC_6H_4)_4Sn$, 2, 175

$SnC_{32}H_{40}O_6$ $Bu_2Sn(CBzAc_2)_2$, 2, 231

$SnC_{32}H_{42}N_2O_8S_2$ $(p-Me_2SO_4\cdot Me_2NC_6H_4)_2-SnPh_2$, 2, 242*

$SnC_{32}H_{42}O_2S$ $Ph_3SnSCH_2CO_2C_{12}H_{25}$, 2, 518

$SnC_{32}H_{44}$ Mesityl$_3SnC_5H_{11}$, 2, 187

$SnC_{32}H_{44}O_4$ $Bu_2Sn(CAc_2CH_2Ph)_2$, 2, 231

$SnC_{32}H_{52}$ $(n-C_{10}H_{21})_2SnPh_2$, 2, 189

$SnC_{32}H_{56}O_8$ $Bu_2Sn[O_2CCH=CHCO_2(CH_2)_5-CHMe_2]_2$, 2, 441

$Bu_2Sn(O_2CCH=CHCO_2CH_2CHEtBu)_2$, 2, 441

$(C_8H_{17})_2Sn(O_2CCH=CHCO_2Bu)_2$, 2, 441

$SnC_{32}H_{50}O_{14}$ $Bu_2Sn[OC(CH_2CO_2Et)_2-CO_2Et]_2$, 2, 403

$SnC_{32}H_{57}O_2$ Bu_3Sn abietate, 2, 419

$SnC_{32}H_{58}N_2S_2$ 2,3-$(Bu_3SnS)_2$ quinoxa-

line, 2, 460

$SnC_{32}H_{60}O_8S_2$ $Bu_2Sn[SCH(CO_2Bu)CH_2CO_2Bu]_2$, 2, 466

$SnC_{32}H_{60}O_{12}$ $Bu_2Sn[OCH(CO_2Bu)CHOHCO_2Bu]_2$, 2, 403

$SnC_{32}H_{64}O_2S_2$ S-$Bu_2Sn(SOCC_{11}H_{23})_2$, 2, 471

$SnC_{32}H_{64}O_2S_4$ $(n\text{-}C_{12}H_{25})_2Sn(S_2COCHMe_2)_2$, 2, 472

$SnC_{32}H_{64}O_4$ $Bu_2Sn(O_2CC_{11}H_{23})_2$, 2, 434

i-$Bu_2Sn(O_2CC_{11}H_{23})_2$, 2, 438

$(n\text{-}C_8H_{17})_2Sn(O_2CCHEtBu)_2$, 2, 438

$(n\text{-}C_8H_{17})_2{}^{113}Sn(O_2CCHEtBu)_2$, 2, 438

$SnC_{32}H_{64}O_4S_2$ $(i\text{-}C_9H_{19}O_2CCH_2CH_2S)_2SnBu_2$, 2, 466

$(i\text{-}C_9H_{19}O_2CCHMeS)_2SnBu_2$, 2, 466

$Bu_2Sn[S(CH_2)_7CO_2Bu]_2$, 2, 467

$Bu_2Sn[SCH(C_6H_{13})CO_2Bu]_2$, 2, 467

$Bu_2Sn[SCEt(C_6H_{13})CO_2Et]_2$, 2, 467

$SnC_{32}H_{64}O_7S$ $Bu_3SnO_3SCH(CO_2C_8H_{17})CH_2CO_2C_8H_{17}$, 2, 352

$SnC_{32}H_{66}O$ $(C_{16}H_{33})_2SnO$, 2, 372

$SnC_{32}H_{66}O_2S$ $Bu_2Sn(O_2CC_{11}H_{23})SC_{12}H_{25}$, 2, 447

$SnC_{32}H_{66}O_4P_2S_4$ $(C_8H_{17})_2Sn(S_2\overline{POCHPrCHEtCH_2O})_2$, 2, 354

$SnC_{32}H_{68}$ $(n\text{-}C_8H_{17})_4Sn$, 2, 166

$(BuEtCHCH_2)_4Sn$, 2, 170

$SnC_{32}H_{68}S_2$ $Bu_2Sn(SC_{12}H_{25})_2$, 2, 462

$(t\text{-}C_{12}H_{25}S)_2SnBu_2$, 2, 462

$SnC_{32}H_{70}O_4P_2S_4$ $Bu_2Sn[S_2P(OC_6H_{13})_2]_2$, 2, 352

$SnC_{32}H_{72}Cl_2N_2$ $Bu_2SnCl_2\cdot 2\ (C_{12}H_{25}NH_2)$, 2, 311

$SnC_{33}H_{27}BrO_3$ 1-$(2\text{-}MeOC_{10}H_6)_3SnBr$, 2, 299

$SnC_{33}H_{29}I$ o-$IC_6Ph_4SnMe_3$, 2, 220

$SnC_{33}H_{29}N$ 2,9-$Me_2N(Ph_3Sn)$-fluorene, 2, 242

$SnC_{33}H_{29}N_3O_2$ 6,3-$(p\text{-}HO_2CC_6H_4N{=}N)Me_2NC_6H_3SnPh_3$, 2, 242

$SnC_{33}H_{38}$ $Mesityl_3SnPh$, 2, 190

$SnC_{33}H_{44}O_2$ $(PhCH_2)_3SnO_2CC_{11}H_{23}$, 2, 411

$SnC_{33}H_{46}$ n-$(Mesityl_3Sn)C_6H_{13}$, 2, 187

$SnC_{33}H_{62}O_6$ $(C_8H_{17})_2Sn(O_2CC_{11}H_{23})O_2CCH{=}CHCO_2Me$, 2, 441

$SnC_{33}H_{70}$ $(C_{10}H_{21})_3SnPr$, 2, 185

$SnC_{34}H_{28}O_2$ $(1\text{-}C_{10}H_7)_2Sn(C_6H_4OMe\text{-}p)_2$, 2, 228

$SnC_{34}H_{32}$ o-$MeC_6Ph_4SnMe_3$, 2, 180

$SnC_{34}H_{34}N_2S_4$ $(p\text{-}Me_2CHC_6H_4CH_2)_2Sn(2\text{-}SC_7H_4NS)_2$, 2, 464

$SnC_{34}H_{40}S_2$ $(PhCH_2)_2Sn(SC_6H_4CMe_3)_2$, 2, 516

$SnC_{34}H_{42}O_6S_3$ $BuSn(SCH_2CH_2CO_2CH_2Ph)_3$, 2, 469

$SnC_{34}H_{44}N_2O$ $Bu_3SnN(1\text{-}C_{10}H_7)C(OMe){=}N\text{-}(1\text{-}C_{10}H_7)$, 2, 480

$SnC_{34}H_{46}N_2O_2$ $(8\text{-Quinolyloxy})_2Sn(C_8H_{17})_2$, 2, 401

$SnC_{34}H_{48}N_2O_8S_2$ Et_2Sn dicamphorsulfonate$\cdot$2,2′-bipyridyl, 2, 351

$SnC_{34}H_{48}O_4$ $(p\text{-}CH_2{=}CHC_6H_4CO_2)_2Sn(C_8H_{17})_2$, 2, 440

$SnC_{34}H_{52}O_3$ $Ph_2Sn(OBu)$ oleate, 2, 407

$SnC_{34}H_{56}N_4O_8$ $Bu_2Sn(NPhCO_2CH_2CH_2NEt_2)_2$, 2, 481

$SnC_{34}H_{64}O_4$ $Bu_2Sn(CAc_2CH_2CHEtBu)_2$, 2, 231

$SnC_{34}H_{64}O_4S_2$ $(i\text{-}C_9H_{19}O_2CCH_2S)_2Sn(C_8H_{11})_2$, 2, 467

$SnC_{34}H_{68}O_4S_2$ $(i\text{-}C_9H_{19}O_2CCHMeCH_2S)_2SnBu_2$, 2, 466

$(i\text{-}C_9H_{19}O_2CCHEtS)_2SnBu_2$, 2, 466

$[i\text{-}C_8H_{17}CO_2(CH_2)_4S]_2SnBu_2$, 2, 466

$Bu_2Sn[SCH(C_7H_{15})CO_2Bu]_2$, 2, 467

$SnC_{34}H_{72}S_2$ $Bu_2Sn(SC_{13}H_{27})_2$, 2, 462

$SnC_{35}H_{26}$ 9-Methylanthryl-$C{\equiv}CSnPh_3$, 2, 214*

$SnC_{35}H_{38}O_6$ $(p\text{-}CH_2{=}CHC_6H_4CO_2)_3SnC_8H_{17}$, 2, 450

$SnC_{35}H_{42}O_6S$ S-$(PhCH_2)_3Sn$-difuryl thiomalate, 2, 461

$SnC_{35}H_{50}$ n-$(Mesityl_3Sn)C_8H_{17}$, 2, 188

$SnC_{35}H_{68}O_6S_3$ $EtSn(SCH_2CO_2C_9H_{19}\text{-}i)_3$, 2, 468

$SnC_{36}H_{18}Cl_{10}$ $Trichlorobiphenylyl_3SnCl$, 2, 296

$SnC_{36}H_{24}N_4$ 2-$Quinolyl_4Sn$, 2, 171

6-$Quinolyl_4Sn$, 2, 171

$SnC_{36}H_{26}$ o,o′-$\overline{C_6H_4C_6H_4Sn}(C_6H_4Ph)_2$, 2, 506

$SnC_{36}H_{27}Br$ $(2\text{-}PhC_6H_4)_3SnBr$, 2, 285

$SnC_{36}H_{27}Cl$ $(2\text{-}PhC_6H_4)_3SnCl$, 2, 285

$SnC_{36}H_{28}$ $Indenyl_4Sn$, 2, 173

p,p'-$Ph(C_6H_4)_2SnPh_3$, 2, 189

(o-$PhC_6H_4)_2SnPh_2$, 2, 190

(o-$PhC_6H_4)_3SnH$, 2, 264

$SnC_{36}H_{28}O$ (p-$PhC_6H_4)_3SnOH$, 2, 366

$SnC_{36}H_{28}O_2$ (p-$PhOC_6H_4)_2SnPh_2$, 2, 228

$SnC_{36}H_{30}O_4P_2$ $Ph_2Sn(O_2PPh_2)_2$, 2, 356

$SnC_{36}H_{30}P_2$ $Ph_2Sn(PPh_2)_2$, 2, 484

$SnC_{36}H_{33}BrO_3$ 1-(2-$EtOC_{10}H_6)_3SnBr$, 2, 299

$SnC_{36}H_{33}NO$ 9-(2-Morpholinomethylfluorenyl)$SnPh_3$, 2, 192

$SnC_{36}H_{36}$ (p-CH_2=$CHCH_2C_6H_4)_4Sn$, 2, 173

Di(9-fluorenyl)dicyclopentyltin, 2, 204

$SnC_{36}H_{36}O_8$ ($\overline{OCH_2CH_2OC}HC_6H_4)_4Sn$, 2, 175

$SnC_{36}H_{38}O_4S_2$ $Ph_2Sn[SC_6H_4CH_2O_2C$-$(CH_2)_4]_2$, 2, 468

$SnC_{36}H_{40}$ (p-$C_6H_{11}C_6H_4)_2SnPh_2$, 2, 190

$SnC_{36}H_{40}O_4S_2$ $Ph_2Sn(SCHPhCH_2O_2CPr)_2$, 2, 468

$SnC_{36}H_{44}$ $Mesityl_4Sn$, 2, 170

(p-$Me_3CC_6H_4)_3SnPh$, 2, 190

$SnC_{36}H_{46}O_4P_2S_4$ $Bu_2Sn[S_2P(OC_6H_4Me)_2]_2$, 2, 353

$SnC_{36}H_{46}O_6S_3$ i-$PrSn(O_2CCHMeCH_2SCH_2$-$Ph)_3$, 2, 449

$SnC_{36}H_{48}O_8$ (p-$Me_2CHC_6H_4CH_2)_2Sn(O_2C$-$CH$=$CHCO_2Bu)_2$, 2, 440

$SnC_{36}H_{52}I_4N_4$ (p-$IMe_3NC_6H_4)_4Sn$, 2, 175

$SnC_{36}H_{52}O_8$ $Bu_2Sn[C(CH_2Ph)(CO_2Et)_2]_2$, 2, 236

$SnC_{36}H_{56}O_4$ $Ph_2Sn(O_2CC_{11}H_{23})_2$, 2, 438

$SnC_{36}H_{56}O_4S_2$ $(PhCH_2)_2Sn(SCH_2CH_2CO_2$-$C_8H_{17})_2$, 2, 465

$SnC_{36}H_{60}S_2$ (n-$C_{12}H_{25})_2Sn(SPh)_2$, 2, 463

$SnC_{36}H_{68}O_6$ $Bu_2Sn(O_2CCH$=$CHCO_2C_{12}H_{25})$-$O_2CC_{11}H_{23}$ (?), 2, 441

$SnC_{36}H_{70}Cl_2$ $(C_{18}H_{35})_2SnCl_2$, 2, 324

$SnC_{36}H_{70}O_4P_2S_4$ $Bu_2Sn[S_2P(OC_6H_{10}$-$Me)_2]_2$, 2, 353

$SnC_{36}H_{70}O_6S_3$ $PrSn(SCH_2CH_2O_2CC_8H_{17})_3$, 2, 469

i-$PrSn(SCH_2CH_2O_2CC_8H_{17})_3$, 2, 469

$SnC_{36}H_{72}O_4$ (n-$C_6H_{13})_2Sn(O_2CC_{11}H_{23})_2$, 2, 438

$SnC_{36}H_{72}O_4S_2$ $Bu_2Sn(SCH_2CH_2O_2CC_{11}H_{23})_2$, 2, 465

(i-$C_{12}H_{25}O_2CCH_2S)_2SnBu_2$, 2, 466

[i-$C_9H_{19}O_2C(CH_2)_4S]_2SnBu_2$, 2, 466

[i-$C_9H_{19}O_2CCHPrS]_2SnBu_2$, 2, 466*

$Bu_2Sn[SCH(C_8H_{17})CO_2Bu]_2$, 2, 467

$(C_8H_{17})_2Sn(SCH_2CO_2C_8H_{17})_2$, 2, 467

$(C_8H_{17})_2{}^{113}Sn(SCH_2CO_2C_8H_{17})_2$, 2, 467

$(C_8H_{17})_2Sn(SCH_2{}^{14}CO_2C_8H_{17})_2$, 2, 467

$(C_{12}H_{25})_2Sn(SCH_2CO_2Bu)_2$, 2, 467, 518

$SnC_{36}H_{75}Cl$ (n-$C_{12}H_{25})_3SnCl$, 2, 284

$SnC_{36}H_{76}$ $(Me_3CCH_2CHMeCH_2CH_2)_4Sn$, 2, 170

$SnC_{36}H_{78}O_4P_2S_4$ (n-$C_{12}H_{25})_2Sn[S_2P$-$(OCHMe_2)_2]_2$, 2, 354

$SnC_{37}H_{30}$ Ph_3SnCPh_3, 2, 182

$SnC_{37}H_{30}O_2$ $Ph_3SnO_2CPh_3$, 2, 380

$SnC_{37}H_{43}NS_2$ (p-$Me_2CHC_6H_4CH_2)_3SnSC_7H_4NS$, 2, 460

$SnC_{37}H_{40}O_4P_2S_4$ $(C_6H_{13})MeC_6H_4Sn[S_2P$-$(OPh)_2]_2$, 2, 354

$SnC_{37}H_{78}S_3$ (n-$C_{12}H_{25}S)_3SnMe$, 2, 468*

$SnC_{38}H_{28}$ 9-$Fluorenyl_2SnPh_2$, 2, 206

$SnC_{38}H_{30}N_2O_6$ $(8\text{-}Quinolyloxy)_2Sn(C_6H_4$-$CO_2Et$-p$)_2$, 2, 401

$SnC_{38}H_{32}$ (o-$PhC_6H_4)_2Sn(CH_2Ph)_2$, 2, 181

$(Ph_2CH)_2SnPh_2$, 2, 182

$SnC_{38}H_{40}$ Di(9-fluorenyl)dicyclohexyltin, 2, 204

$SnC_{38}H_{50}O_4$ (p-$Me_2CHC_6H_4CH_2)_3SnO_2C$-$CH$=$CHCO_2Bu$, 2, 429

$SnC_{38}H_{56}O_8S_2$ S-(o-$MeC_6H_4)_2Sn$ di(dibutyl thiomalate), 2, 468

$SnC_{38}H_{60}O_4$ $(PhCH_2)_2Sn(O_2CC_{11}H_{23})_2$, 2, 435

$(C_{12}H_{25})_2Sn(OBz)_2$, 2, 438

$SnC_{38}H_{70}N_6O_2$ [2,4,6-$(Me_2NCH_2)_3C_6H_2O]_2$-$SnBu_2$, 2, 404

$SnC_{38}H_{72}O_6$ CH_2=$CHSn(O_2CC_{11}H_{23})_3$, 2, 449*

$SnC_{38}H_{74}$ Tris[8-(2,6-dimethylocten-2-yl)]SnC_8H_{17}, 2, 204

$SnC_{38}H_{76}O_4S_2$ (i-$C_{12}H_{25}O_2CCH_2CH_2S)_2$-$SnBu_2$, 2, 466

[i-$C_9H_{19}O_2C(CH_2)_5S]_2SnBu_2$, 2, 466
(i-$C_9H_{19}O_2CCHBu)_2SnBu_2$, 2, 466
(i-$C_9H_{19}O_2CCEt_2)_2SnBu_2$, 2, 466
$(C_{12}H_{25})_2Sn(SCH_2CH_2CO_2Bu)_2$, 2, 467
$SnC_{38}H_{79}NO_4$ $Me_2Sn(O_2CC_{11}H_{23})_2 \cdot NH_2$-$C_{12}H_{25}$, 2, 435
$SnC_{38}H_{80}S_3$ $EtSn(SC_{12}H_{25})_3$, 2, 469
$SnC_{39}H_{34}$ $Me_3SnC_6Ph_5$, 2, 180
$SnC_{39}H_{54}O_3$ $(PhCH_2)_3Sn$ ricinolate, 2, 429
$SnC_{39}H_{56}O_2$ $(PhCH_2)_3SnO_2CC_{17}H_{35}$, 2, 412
$SnC_{39}H_{69}NO_5$ $Bu_2Sn(O_2CC_{11}H_{23})_2 \cdot PhNCO$, 2, 434
$SnC_{39}H_{78}O_4S$ $Bu_3SnSCH_2CH(O_2CC_{11}H_{23})$-$CH_2O_2CC_{11}H_{23}$, 2, 461
$SnC_{40}H_{28}$ (1-$C_{10}H_7)_4Sn$, 2, 171
$SnC_{40}H_{30}$ Hexaphenylstannole, 2, 506*
$SnC_{40}H_{36}$ 2-(m-$MeC_6H_4)C_6Ph_4SnMe_3$, 2, 180
2-(p-$MeC_6H_4)C_6Ph_4SnMe_3$, 2, 180
$SnC_{40}H_{38}NO_5$ 9-(2-Morpholinomethylfluorenyl)-$SnPh_3$ malate, 2, 192
$SnC_{40}H_{48}O_4S_2$ $Bu_2Sn(SCPh_2CO_2Et)_2$, 2, 466
$SnC_{40}H_{52}$ (p-$Me_3CC_6H_4)_4Sn$, 2, 170
$SnC_{40}H_{60}O_8$ (o-$BuEtCHCH_2O_2CC_6H_4CO_2)_2$-$SnBu_2$, 2, 444
$SnC_{40}H_{64}N_4O_{16}S_4$ [(p-$Me_2NC_6H_4) \cdot Me_2$-$SO_4]_4Sn$, 2, 175
$SnC_{40}H_{64}O_4S_2$ $Ph_2Sn(SCH_2CH_2O_2CC_{11}H_{23})_2$ 2, 468
$Ph_2Sn(SCH_2CO_2C_{12}H_{25})_2$, 2, 468
$SnC_{40}H_{68}O_2S_2$ (o-$C_8H_{17}C_6H_4OCH_2CH_2S)_2$-$SnBu_2$, 2, 465
$SnC_{40}H_{72}O_8$ $(C_8H_{17})_2Sn(O_2CCH{=}CHCO_2CH_2$-$CHEtBu)_2$, 2, 441
$SnC_{40}H_{76}O_6$ Et_2Sn diricinolate, 2, 440
$SnC_{40}H_{76}O_8S_2$ $Bu_2Sn(SCH_2CO_2CH_2CH_2O_2C$-$C_{11}H_{23})_2$, 2, 465
$SnC_{40}H_{78}O_6S_3$ $BuSn(SCH_2CO_2C_{10}H_{21})_3$, 2, 469
$SnC_{40}H_{80}O_4$ $Et_2Sn(O_2CC_{17}H_{35})_2$, 2, 436
(n-$C_8H_{17})_2Sn(O_2CC_{11}H_{23})_2$, 2, 438
$(BuEtCHCH_2)_2Sn(O_2CC_{11}H_{23})_2$, 2, 438
$SnC_{40}H_{80}O_4S_2$ (n-$C_{12}H_{25}SCH_2CHMeCO_2)_2$-$SnBu_2$, 2, 444
[i-$C_9H_{19}O_2(CH_2)_6S]_2SnBu_2$, 2, 467
$Bu_2Sn[S(CH_2)_{11}CO_2Bu]_2$, 2, 467
$Bu_2S[SCH_2(C_{10}H_{21})CO_2Bu]_2$, 2, 467
$(C_{12}H_{25})_2Sn[S(CH_2)_3CO_2Bu]_2$, 2, 467
$SnC_{40}H_{84}$ (i-$C_{10}H_{21})_4Sn$, 2, 170
$SnC_{40}H_{84}N_4O_4$ $Bu_2Sn(O_2CC_{11}H_{23})_2 \cdot 2$ piperazine, 2, 434
$SnC_{40}H_{84}S_2$ $Bu_2Sn(SC_{16}H_{33})_2$, 2, 462
(t-$C_{16}H_{33}S)_2SnBu_2$, 2, 462
$(C_8H_{17})_2Sn(SC_{12}H_{25})_2$, 2, 463
$SnC_{40}H_{84}S_3$ $BuSn(SC_{12}H_{25})_3$, 2, 469
$SnC_{40}H_{86}O_4P_2S_4$ $Bu_2Sn[S_2P(OC_5H_8Me_3)_2]_2$, 2, 353
$(C_8H_{17})_2Sn[S_2P(OC_6H_{13})_2]_2$, 2, 354
$SnC_{42}H_{32}$ (o-$PhC_6H_4)_3SnPh$, 2, 190
$SnC_{42}H_{35}O_6P_3$ $PhSn(O_2PPh_2)_3$, 2, 357
$SnC_{42}H_{35}P_3$ $PhSn(PPh_2)_3$, 2, 484
$SnC_{42}H_{39}NO_6$ $Ph_2Sn(OAc)_2 \cdot Ph_4NOAc$, 2, 438
$SnC_{42}H_{50}$ (p-$C_6H_{11}C_6H_4)_3SnPh$, 2, 190
$SnC_{42}H_{62}O_2$ (p-$Me_2CHC_6H_4CH_2)_3SnO_2C$-$C_{11}H_{23}$, 2, 412
$SnC_{42}H_{80}O_4S$ $Bu_2Sn(O_2CCH{=}CHCO_2C_{18}H_{35})$-$SC_{12}H_{25}$, 2, 447
$SnC_{42}H_{84}O_4$ (i-$C_9H_{19})_2Sn(O_2CC_{11}H_{23})_2$, 2, 438
$SnC_{42}H_{86}N_2S_4$ (n-$C_{12}H_{25})_2Sn(S_2CNBu_2)_2$, 2, 472
$SnC_{42}H_{87}Cl$ (n-$C_{14}H_{29})_3SnCl$, 2, 284
$SnC_{43}H_{30}$ 9-(1-$C_{10}H_7)_3Sn$-fluorene, 2, 206
$SnC_{43}H_{34}$ p-(p-$PhC_6H_4)_3SnC_6H_4Me$, 2, 190
$SnC_{43}H_{62}O_9$ $Ph_3SnO(CH_2O)_7OCC_{17}H_{33}$, 2, 390
$SnC_{44}H_{36}$ Tri(9-fluorenyl)cyclopentyltin, 2, 204
$SnC_{44}H_{72}O_4$ (p-$Me_2CHC_6H_4CH_2)_2Sn(O_2C$-$C_{11}H_{23})_2$, 2, 435
$SnC_{44}H_{74}N_2O_2$ (p-$H_{29}C_{14}HNOCC_6H_4)_2SnMe_2$, 2, 234*
$SnC_{44}H_{76}S_2$ $(C_{12}H_{25})_2Sn(SC_6H_4CMe_3)_2$, 2, 518
$SnC_{44}H_{80}O_{14}$ $Bu_2Sn[OC(CH_2CO_2Bu)_2CO_2$-$Bu]_2$, 2, 403
$SnC_{44}H_{84}O_4$ $Bu_2Sn(O_2CC_{17}H_{33})_2$, 2, 439
$SnC_{44}H_{84}O_6$ Bu_2Sn diricinolate, 2, 441

Bu_2Sn di(9,10-epoxystearate), 2, 444
$SnC_{44}H_{84}O_{8}$ Bu_2Sn di(9,10,12-epoxy-hydroxystearate), 2, 444
$SnC_{44}H_{88}O_{4}$ $Bu_2Sn(O_2CC_{17}H_{35})_2$, 2, 437
$SnC_{44}H_{88}O_{4}S_{2}$ $Me_2Sn(SCH_2CH_2CO_2C_{18}H_{37})_2$ 2, 465
$SnC_{44}H_{90}N_{2}O_{2}S_{2}$ $Bu_2Sn[SCH_2CON(CH_2CH\text{-}EtBu)_2]_2$, 2, 466
$SnC_{44}H_{92}S_{2}$ $Bu_2Sn(SC_{18}H_{37})_2$, 2, 462
$SnC_{45}H_{32}$ 9-Fluorenyl$_3$SnPh, 2, 206
$SnC_{45}H_{38}$ Tri(9-fluorenyl)cyclohexyl-tin, 2, 204
$SnC_{45}H_{88}O_{6}S_{3}$ i-PrSn$(SCH_2CH_2O_2C\text{-}C_{11}H_{23})_3$, 2, 469
$SnC_{46}H_{32}$ Di(9-fluoronyl)di(1-naphthyl)tin, 2, 206
$SnC_{46}H_{68}O_{9}$ $(PhCH_2)_3Sn(OCH_2)_7O_2C_{17}H_{33}$, 2, 390
$SnC_{46}H_{86}O_{6}$ $Bu_2Sn[OCH(C_6H_{13})CH_2CH{=}CH\text{-}(CH_2)_7CO_2CH_2]_2$, 2, 405*
$SnC_{46}H_{88}O_{6}$ $Bu_2Sn[OCH(C_6H_{13})CH_2CH{=}CH\text{-}(CH_2)_7CO_2Me]_2$, 2, 403*
$SnC_{46}H_{88}O_{8}$ $[(i\text{-}C_8H_{17}O_2C)_2CH]_2SnBu_2$, 2, 236
$SnC_{48}H_{24}Cl_{12}$ Trichlorobiphenyl$_4$Sn, 2, 174
$SnC_{48}H_{36}$ $(o\text{-}PhC_6H_4)_4Sn$, 2, 170
$(m\text{-}PhC_6H_4)_4Sn$, 2, 170
$(p\text{-}PhC_6H_4)_4Sn$, 2, 170
$SnC_{48}H_{36}O_{4}$ $(o\text{-}PhOC_6H_4)_4Sn$, 2, 175
$(p\text{-}PhOC_6H_4)_4Sn$, 2, 175
$SnC_{48}H_{48}$ $Et_3SnC{\equiv}CCH(CPh_3)CH_2CPh_3$, 2, 213
$SnC_{48}H_{60}$ p-Cyclohexylphenyl$_4$Sn, 2, 170
$SnC_{48}H_{78}O_{4}S_{2}$ $(C_{12}H_{25})_2Sn[SC_6H_4CH_2O_2\text{-}C(CH_2)_4]_2$, 2, 467
$SnC_{48}H_{80}O_{4}$ $Ph_2Sn(O_2CC_{17}H_{35})_2$, 2, 438
$SnC_{48}H_{80}O_{4}S_{2}$ $(C_{12}H_{25})_2Sn(SCHPhCH_2O_2\text{-}CPr)_2$, 2, 467
$SnC_{48}H_{92}O_{8}S_{2}$ S-Bu_2Sn di(diethylhexyl thiomalate), 2, 466
$SnC_{48}H_{96}O_{4}S_{2}$ $Bu_2Sn(SCH_2CO_2C_{18}H_{37})_2$, 2, 466
$Bu_2Sn[SCH(C_{14}H_{29})CO_2Bu]_2$, 2, 467
$SnC_{48}H_{100}$ $(n\text{-}C_{12}H_{25})_4Sn$, 2, 170
$SnC_{48}H_{100}O$ $(C_{16}H_{33})_3SnOH$, 2, 365
$SnC_{48}H_{102}O_{4}P_{2}S_{4}$ $(C_8H_{17})_2Sn[S_2P(OC_5H_8\text{-}Me_3)_2]_2$, 2, 354
$SnC_{49}H_{34}$ Tri(9-fluorenyl)-1-naphthyl-tin, 2, 206
$SnC_{49}H_{50}$ $Et_3SnC(CPh_3){=}C{=}CMeCH_2CPh_3$, 2, 213
$SnC_{50}H_{40}O_{2}$ $(p\text{-}Ph_2COHC_6H_4)_2SnPh_2$, 2, 228
$SnC_{50}H_{80}O_{4}$ $(PhCH_2)_2Sn(O_2CC_{17}H_{33})_2$, 2, 439
$SnC_{50}H_{80}O_{6}$ $(PhCH_2)_2Sn$ diricinolate, 2, 440
$SnC_{50}H_{84}O_{4}$ $(PhCH_2)_2Sn(O_2CC_{17}H_{35})_2$, 2, 435
$SnC_{50}H_{96}O_{8}S_{2}$ $[i\text{-}C_8H_{17}CO_2CH_2CH(O_2C\text{-}C_8H_{17}\text{-}i)CH_2S]_2SnBu_2$, 2, 466
$[C_8H_{17}O_2CCHMeCH(CO_2C_8H_{17})S]_2SnBu_2$, 2, 466
$SnC_{50}H_{102}O_{2}$ $(C_{16}H_{33})_3SnOAc$, 2, 414
$SnC_{52}H_{36}$ 9-Fluorenyl$_4$Sn, 2, 173
$SnC_{52}H_{84}O_{4}S_{2}$ $Bu_2Sn(SCH_2CH_2O_2CC_{19}H_{29})_2$, 2, 465
$SnC_{52}H_{88}O_{4}S_{2}$ $Bu_2Sn(SCH_2CO_2CH_2C_{19}H_{31})_2$, 2, 466
$SnC_{52}H_{92}O_{4}S_{2}$ $Bu_2Sn(SCH_2CO_2CH_2C_{19}H_{33})_2$, 2, 466
$SnC_{52}H_{96}O_{10}S_{2}$ $Bu_2Sn(SCH_2CO_2CH_2CH_2O_2C\text{-}C_{17}H_{32}OH)_2$, 2, 465
$SnC_{52}H_{98}N_{2}S_{2}$ $Ph_2Sn(SC_{12}H_{25}\cdot NHBu_2)_2$, 2, 463
$SnC_{52}H_{104}O_{4}$ $(n\text{-}C_8H_{17})_2Sn(O_2CC_{17}H_{35})_2$, 2, 438
$SnC_{52}H_{104}O_{4}S_{2}$ $Bu_2Sn[SCH(C_{16}H_{33})CO_2\text{-}Bu]_2$, 2, 467
$(C_{12}H_{25})_2Sn(SCH_2CH_2O_2CC_{11}H_{23})_2$, 2, 467
$SnC_{54}H_{111}Cl$ $(n\text{-}C_{18}H_{37})_3SnCl$, 2, 284
$SnC_{55}H_{114}S_{3}$ $MeSn(SC_{18}H_{37})_3$, 2, 468*
$SnC_{56}H_{36}$ 9-Phenanthryl$_4$Sn, 2, 171
$SnC_{56}H_{40}$ 5,5'-Spirobi[tetraphenylstannole], 2, 506*
$SnC_{56}H_{82}N_{2}O_{4}S_{2}$ $Bu_2Sn(O_2CC_{11}H_{23})_2\cdot 2$(10-phenothiazine), 2, 434
$SnC_{56}H_{108}O_{8}S_{2}$ S-$(C_8H_{17})_2Sn$ di(diisooctyl thiomalate), 2, 467
$SnC_{56}H_{112}O_{4}S_{2}$ $(C_8H_{17})_2Sn(SCH_2CO_2\text{-}C_{18}H_{37})_2$, 2, 467

$SnC_{56}H_{116}$ $(n\text{-}C_{14}H_{29})_4Sn$, 2, 170
$SnC_{57}H_{58}O_{12}S_3$ S-(i-PrSn) tri(dibenzyl thiomalate), 2, 469*
$SnC_{57}H_{112}O_6$ $i\text{-}PrSn(O_2CC_{17}H_{35})_3$, 2, 449
$SnC_{58}H_{112}O_8S$ S-$(i\text{-}C_5H_{11})_2Sn$ di(didecyl thiomalate), 2, 467*
$SnC_{63}H_{49}I$ 2,2'-$Me_3Sn(C_6Ph_4)_2I$, 2, 220
$SnC_{64}H_{102}O_4P_2S_4$ $Bu_2Sn[S_2P(OC_6H_4\text{-}C_8H_{17})_2]_2$, 2, 353
$SnC_{64}H_{132}$ $(n\text{-}C_{16}H_{33})_4Sn$, 2, 170
$SnC_{72}H_{120}O_{10}$ $Bu_2Sn[OCH_2CH(O_2CC_{11}H_{23})\text{-}CH_2O_2CC_{11}H_{23}]_2$, 2, 403
$SnC_{72}H_{140}$ $Oleyl_4Sn$, 2, 173
$SnC_{72}H_{148}$ $(n\text{-}C_{18}H_{37})_4Sn$, 2, 170
$SnC_{86}N_{62}N_6$ $Me_2Sn(CN)_4 \cdot 2N(C_{10}H_7)_4$, 2, 345
$Sn_2C_2H_4Cl_2O_2$ $(CH_2SnClO)_2$, 2, 340
$Sn_2C_2H_4Cl_4$ $(ClCH_2)_2SnCl_2$, 2, 324
$Sn_2C_2H_6O_3$ $Me_2Sn_2O_3$, 2, 376
$Sn_2C_2H_6O_4$ $(CH_2SnO_2H)_2$, 2, 377
$Sn_2C_2H_6S_3$ $Me_2Sn_2S_3$, 2, 457
$(Sn_2C_2H_6Se_3)_n$ $(Me_2Sn_2Se_3)_n$, 2, 457
$Sn_2C_3H_8O_4$ $CH_2(CH_2SnO_2H)_2$, 2, 377
$Sn_2C_4H_6N_2S_2$ $Me_2Sn(NCS)_2$, 2, 345
$Sn_2C_4H_{10}Br_2$ $(EtSnBr)_2$, 2, 492
$Sn_2C_4H_{10}O_4$ $(CH_2CH_2SnO_2H)_2$, 2, 377
$(Sn_2C_4H_{10}Se_3)_n$ $(Et_2Sn_2Se_3)_n$, 2, 457
$Sn_2C_4H_{12}Cl_2$ $(Me_2SnCl)_2$, 2, 492
$Sn_2C_4H_{12}Cl_2O$ $(Me_2SnCl)_2O$, 2, 338
$Sn_2C_4H_{12}I_2O$ $Me_2SnI_2 \cdot Me_2SnO$, 2, 338
$Sn_2C_4H_{13}BrO_2$ $(Me_2SnBr)O(Me_2SnOH)$, 2, 338
$Sn_2C_4H_{13}ClO_2$ $(Me_2SnCl)O(Me_2SnOH)$, 2, 338
$Sn_2C_4H_{13}IO_2$ $(Me_2SnI)O(Me_2SnOH)$, 2, 338
$Sn_2C_5H_{15}BrO_2$ $(Me_2SnBr)O(Me_2SnOMe)$, 2, 341
$Sn_2C_5H_{15}ClO_2$ $(Me_2SnCl)O(Me_2SnOMe)$, 2, 341
$Sn_2C_5H_{15}IO_2$ $(Me_2SnI)O(Me_2SnOMe)$, 2, 341
$Sn_2C_6H_4Cl_6$ $m\text{-}(Cl_3Sn)_2C_6H_4$, 2, 335
$Sn_2C_6H_{16}I_2O$ $Me_2SnI_2 \cdot Et_2SnO$, 2, 338, 610*
  $Me_2SnO \cdot Et_2SnI_2$, 2, 338, 610*
$Sn_2C_6H_{17}IO_2$ $(Me_2SnI)O(Me_2SnOEt)$, 2, 341
$Sn_2C_6H_{18}$ Me_6Sn_2, 2, 485
$Sn_2C_6H_{18}O$ $(Me_3Sn)_2O$, 2, 358
$Sn_2C_6H_{18}O_4S$ $(Me_3Sn)_2SO_4$, 2, 348
$Sn_2C_6H_{18}S$ $(Me_3Sn)_2S$, 2, 452
$Sn_2C_6H_{18}Se$ $(Me_3Sn)_2Se$, 2, 452
$Sn_2C_6H_{19}BrO$ $Me_3SnBr \cdot Me_3SnOH$, 2, 338
$Sn_2C_6H_{19}O_3P$ $(Me_3Sn)_2HPO_3$ (?), 2, 348
$Sn_2C_6H_{21}ClO_2$ $Me_3SnCl \cdot Me_3SnOH \cdot H_2O$, 2, 338
$Sn_2C_7H_{18}O_3$ $(Me_3Sn)_2CO_3$, 2, 411
$Sn_2C_7H_{20}$ $(Me_3Sn)_2CH_2$, 2, 245
$Sn_2C_7H_{21}N$ $(Me_3Sn)_2NMe$, 2, 475
$Sn_2C_8H_{18}O_5$ $Me_2Sn(O_2CCH_2)_2 \cdot Me_2SnO$, 2, 445
$Sn_2C_8H_{18}$ $(Me_3SnC\equiv)_2$, 2, 245
$Sn_2C_8H_{18}F_4$ $(Me_3SnCF_2)_2$, 2, 245
$(Sn_2C_8H_{18}O_3)_n$ $(Bu_2Sn_2O_3)_n$, 2, 376
$Sn_2C_8H_{18}O_4S$ $(Me_2SnOAc)_2S$, 2, 446
$Sn_2C_8H_{18}O_5$ $(Me_2SnOAc)_2O$, 2, 435
$Sn_2C_8H_{20}Br_2O$ $Me_2SnO \cdot Pr_2SnBr_2$, 2, 339, 610*
  $(Et_2SnBr)_2O$, 2, 339
$Sn_2C_8H_{20}Cl_2$ $(Et_2SnCl)_2$ (?), 2, 492
$Sn_2C_8H_{20}Cl_2O$ $(Et_2SnCl)_2O$, 2, 339
$Sn_2C_8H_{21}BrO_2$ $(Et_2SnBr)O(Et_2SnOH)$, 2, 339
$Sn_2C_8H_{21}ClO_2$ $(Et_2SnCl)O(Et_2SnOH)$, 2, 339
$Sn_2C_8H_{21}IO_2$ $(Et_2SnI)O(Et_2SnOH)$, 2, 339
$Sn_2C_8H_{23}N$ $(Me_3Sn)_2NEt$, 2, 475
$Sn_2C_9H_{23}ClO_2$ $(Et_2SnCl)O(Et_2SnOMe)$, 2, 341
$Sn_2C_9H_{23}IO_2$ $(Et_2SnI)O(Et_2SnOMe)$, 2, 341
$Sn_2C_9H_{24}$ $Me_3SnSnEt_3$, 2, 488
$Sn_2C_{10}H_{16}Cl_2$ $p\text{-}(Me_2SnCl)_2C_6H_4$, 2, 301
$Sn_2C_{10}H_{18}$ $(Me_3SnC\equiv C)_2$, 2, 245
  $p\text{-}(Me_2SnH)_2C_6H_4$, 2, 266, 512
$Sn_2C_{10}H_{18}O_4$ $(Me_3SnO_2CC\equiv)_2$, 2, 427
$Sn_2C_{10}H_{24}Br_2O$ $Et_2SnBr_2 \cdot Pr_2SnO$, 2, 339, 610*
  $Et_2SnO \cdot Pr_2SnBr_2$, 2, 339, 610*
$Sn_2C_{10}H_{25}ClO_2$ $(Et_2SnCl)O(Et_2SnOEt)$, 2, 341

$Sn_2C_{10}H_{25}IO_2$ $(Et_2SnI)O(Et_2SnOEt)$, 2, 341
$Sn_2C_{10}H_{26}$ $(Me_3SnCH_2CH_2)_2$, 2, 245
Et_5Sn_2H, 2, 267
$Sn_2C_{10}H_{28}BrN$ $Me_3SnBr\cdot Me_3SnNEt_2$, 2, 271
$(Sn_2C_{12}H_{10}S_3)_n$ $(Ph_2Sn_2S_3)_n$, 2, 457
$Sn_2C_{12}H_{22}$ p-$(Me_3Sn)_2C_6H_4$, 2, 245
$Sn_2C_{12}H_{26}O_3$ $(n\text{-}C_6H_{13})_2Sn_2O_3$, 2, 376
$Sn_2C_{12}H_{26}O_5$ $(Et_2SnOAc)_2O$, 2, 436
$Sn_2C_{12}H_{28}Br_2O$ $(Pr_2SnBr)_2O$, 2, 340
$Sn_2C_{12}H_{28}Cl_2O$ $(Pr_2SnCl)_2O$, 2, 339
$Sn_2C_{12}H_{29}BrO_2$ $(Pr_2SnBr)O(Pr_2SnOH)$, 2, 340
$Sn_2C_{12}H_{29}ClO_2$ $(Pr_2SnCl)O(Pr_2SnOH)$, 2, 340
$Sn_2C_{12}H_{29}ClO_3$ $(Et_2SnCl)O(Et_2SnO_2\text{-}CMe_3)$, 2, 379
$Sn_2C_{12}H_{29}IO_2$ $(Pr_2SnI)O(Pr_2SnOH)$, 2, 340
$(i\text{-}Pr_2SnI)O(i\text{-}Pr_2SnOH)$, 2, 340
$Sn_2C_{12}H_{30}$ Et_6Sn_2, 2, 486
$Sn_2C_{12}H_{30}O$ $(Et_3Sn)_2O$, 2, 359, 514
$Sn_2C_{12}H_{30}O_2$ $(Et_3SnO)_2$, 2, 378
$Sn_2C_{12}H_{30}O_4S$ $(Et_3Sn)_2SO_4$, 2, 350
$Sn_2C_{12}H_{30}S$ $(Et_3Sn)_2S$, 2, 452
$Sn_2C_{12}H_{30}S_{1.3}$ $(Et_3Sn)_2S_{1.3}$, 2, 452
$Sn_2C_{12}H_{30}S_{1.4}$ $(Et_3Sn)_2S_{1.4}$, 2, 452
$Sn_2C_{12}H_{34}O_3$ $(Et_3Sn)_2O\cdot 2H_2O$, 2, 359
$Sn_2C_{13}H_{24}$ p-$(Me_3Sn)_2C_6H_4CH_2$, 2, 245
$Sn_2C_{13}H_{30}Cl_2$ $[Et_2Sn(Cl)CH_2CH_2]_2CH_2$, 2, 301
$Sn_2C_{13}H_{30}O_3$ $(Et_3Sn)_2CO_3$, 2, 412
$Sn_2C_{13}H_{32}$ $(Et_2HSnCH_2CH_2)_2CH_2$, 2, 266
$Sn_2C_{14}H_{14}N_{12}O$ $[p\text{-}MeSn(N_3)_2C_6H_4]_2O$, 2, 345
$(Sn_2C_{14}H_{14}S_3)_n$ $[(o\text{-}MeC_6H_4)_2Sn_2S_3]_n$, 2, 457
$Sn_2C_{14}H_{22}N_2S_2$ 2,3-$(Me_3SnS)_2$ quinoxaline, 2, 460
$Sn_2C_{14}H_{30}$ $(Et_3SnC{\equiv})_2$, 2, 245
$Sn_2C_{14}H_{34}$ $(Et_2HSnCH_2CH_2CH_2)_2$, 2, 266
$Sn_2C_{14}H_{34}O$ $(n\text{-}C_5H_{11}SnMe_2)_2O$, 2, 364
$Sn_2C_{14}H_{37}Cl_2N$ 2 $Et_3SnCl\cdot NHMe_2$, 2, 273
$Sn_2C_{15}H_{32}O$ $Et_3SnC{\equiv}CCH_2OSnEt_3$, 2, 215
$Sn_2C_{15}H_{36}$ $(Et_3SnCH_2)_2CH_2$, 2, 246
$Sn_2C_{15}H_{39}ClN$ 2 $Et_3SnCl\cdot NMe_3$, 2, 273
$Sn_2C_{15}H_{40}N_2P_2$ $(Me_3SnN{=}PMe_2CH_2CH_2)_2CH_2$, 2, 481
$Sn_2C_{16}H_{22}O_3$ $(Me_2SnOPh)_2O$, 2, 396
$Sn_2C_{16}H_{30}O_4$ $(Et_3SnO_2CC{\equiv})_2$, 2, 427
$Sn_2C_{16}H_{34}O$ $Et_3SnC{\equiv}CCH_2CH_2OSnEt_3$, 2, 215
$Et_3SnC{\equiv}CCHMeOSnEt_3$, 2, 216
$Sn_2C_{16}H_{36}Br_2O$ $(Bu_2SnBr)_2O$, 2, 340
$Sn_2C_{16}H_{36}Cl_2$ $(Bu_2SnCl)_2$, 2, 492
$Sn_2C_{16}H_{36}Cl_2O$ $(Bu_2SnCl)_2O$, 2, 337
$Sn_2C_{16}H_{36}Cl_2S$ $(Bu_2SnCl)_2S$, 2, 342
$Sn_2C_{16}H_{36}F_2O$ $(Bu_2SnF)_2O$, 2, 340
$Sn_2C_{16}H_{36}N_2O_8$ $Bu_2Sn(NO_3)_2\cdot Bu_2SnO$ (?), 2, 352
$Sn_2C_{16}H_{37}BrO_2$ $(Bu_2SnBr)O(Bu_2SnOH)$, 2, 340
$Sn_2C_{16}H_{37}ClO_2$ $(Bu_2SnCl)O(Bu_2SnOH)$, 2, 340
$Sn_2C_{16}H_{37}IO_2$ $(Bu_2SnI)O(Bu_2SnOH)$, 2, 340
$(i\text{-}Bu_2SnI)O(i\text{-}Bu_2SnOH)$, 2, 340
$Sn_2C_{16}H_{38}$ $(Bu_2SnH)_2$, 2, 267
$Sn_2C_{16}H_{38}O_2$ $(Et_3SnOCH_2CH_2)_2$, 2, 393
$Sn_2C_{16}H_{38}O_3$ $(Bu_2SnOH)_2O$, 2, 372
$Sn_2C_{16}H_{38}O_5$ $(Et_2SnO_2CMe_3)_2O$, 2, 379
$Sn_2C_{16}H_{41}Cl_2N$ 2 $Et_3SnCl\cdot NHEt_2$, 2, 273
$Sn_2C_{17}H_{36}O$ $Et_3SnC{\equiv}CCMe_2OSnEt_3$, 2, 216
$Sn_2C_{17}H_{39}ClO_2$ $(Bu_2SnCl)O(Bu_2SnOMe)$, 2, 341
$Sn_2C_{17}H_{40}$ $(Et_3SnCH_2CH_2)_2CH_2$, 2, 246
$Sn_2C_{18}H_{22}O_5$ $(Me_2SnOBz)_2O$, 2, 435
$Sn_2C_{18}H_{30}Cl_4O_2$ p-$(Et_3SnO)_2C_6Cl_4$, 2, 393
$Sn_2C_{18}H_{34}O_2$ m-$(Et_3SnO)_2C_6H_4$, 2, 393, 514
$Sn_2C_{18}H_{35}P$ $(Et_3Sn)_2PPh$, 2, 482
$Sn_2C_{18}H_{36}N_2OS_2$ $(Bu_2SnSCN)_2O$, 2, 346
$Sn_2C_{18}H_{38}O_4$ $(Bu_2SnO_2CH)_2$ (?), 2, 493
$Sn_2C_{18}H_{40}N_2O_4$ $(Et_3SnNCO_2Et)_2$, 2, 475
$Sn_2C_{18}H_{40}O_4$ $(Bu_2SnOAc)O(Bu_2SnOH)$, 2, 436
$Sn_2C_{18}H_{42}$ $(Et_3SnCH_2CH_2CH_2)_2$, 2, 246
Pr_6Sn_2, 2, 488
$Sn_2C_{18}H_{42}O$ $(Pr_3Sn)_2O$, 2, 360, 611*
$(i\text{-}Pr_3Sn)_2O$, 2, 364

$Sn_2C_{18}H_{42}O$ $(MeSnBu_2)_2O$, 2, 364
$Sn_2C_{18}H_{42}O_2$ $(Pr_3SnO)_2$, 2, 379
$Sn_2C_{18}H_{42}O_3$ $(Bu_2SnOMe)_2O$, 2, 397
$Sn_2C_{18}H_{42}S$ $(Pr_3Sn)_2S$, 2, 452
$Sn_2C_{18}H_{42}S_6$ $(Et_3SnCH_2CHMeS_3)_2$, 2, 239*
$Sn_2C_{19}H_{33}NO_4$ $(Et_3Sn)_2$ 2,5-pyridinedicarboxylate, 2, 416
$Sn_2C_{19}H_{42}O_3$ $(Pr_3Sn)_2CO_3$, 2, 412
$Sn_2C_{20}H_{34}N_2S_2$ 2,3-$(Et_3SnS)_2$ quinoxaline, 2, 460
$Sn_2C_{20}H_{36}Cl_6O_4$ $(Bu_2SnO_2CCCl_3)_2$, 2, 493
$Sn_2C_{20}H_{38}Cl_4O_4$ $(Bu_2SnO_2CCHCl_2)_2$, 2, 493
$Sn_2C_{20}H_{40}Cl_2O_4$ $(Bu_2SnO_2CCH_2Cl)_2$, 2, 493
$Sn_2C_{20}H_{42}$ $(Pr_3SnC\equiv)_2$, 2, 246
$Sn_2C_{20}H_{42}N_2S_3$ 2,5-(i-$Pr_3SnS)_2$-1,3,4-thiadiazole, 2, 460
$Sn_2C_{20}H_{42}O$ $(CH_2{=}CHSnBu_2)_2O$, 2, 364
$Sn_2C_{20}H_{42}O_4$ $(Bu_2SnO_2CMe)_2$, 2, 493
$Sn_2C_{20}H_{42}O_5$ $(Bu_2SnOAc)_2O$, 2, 436
$Sn_2C_{20}H_{44}Cl_2O$ $[(n\text{-}C_5H_{11})_2SnCl]_2O$ (?), 2, 340
$Sn_2C_{20}H_{44}O_4$ cyclo-$(Bu_2SnOCH_2CH_2O)_2$, 2, 405
$Sn_2C_{20}H_{45}ClO_3$ $(Bu_2SnCl)O(Bu_2SnO_2CMe_3)$, 2, 380
$Sn_2C_{20}H_{45}IO_2$ $[(i\text{-}C_5H_{11})_2SnI]O[(i\text{-}C_5H_{11})_2SnOH]$, 2, 340
$Sn_2C_{20}H_{46}$ $(Et_3Sn)_2(CH_2)_8$, 2, 246
$Sn_2C_{20}H_{46}O_5$ $(Bu_2SnOCH_2CH_2OH)_2O$, 2, 403
$Sn_2C_{21}H_{24}$ $Me_3SnSnPh_3$, 2, 488
$Sn_2C_{21}H_{36}N_2OS_2$ 2,3,7-$(Et_3SnS)_2$ methoxyquinoxaline, 2, 460
$Sn_2C_{21}H_{36}N_2S_2$ 2,3,7-$(Et_3SnS)_2$ methylquinoxaline, 2, 460
$Sn_2C_{21}H_{44}O_4$ $[Bu_2Sn(OCH_2)_2]_2C$, 2, 405
$Sn_2C_{22}H_{26}N_2O_8$ Di(diethyltin 2,5-pyridinedicarboxylate), 2, 439
$Sn_2C_{22}H_{30}O_7$ $(o\text{-}HOC_6H_4CO_2)_2SnEt_2\cdot Et_2SnO$, 2, 444
$Sn_2C_{22}H_{32}N_2O_7$ $[2,4\text{-}HO(H_2N)C_6H_3CO_2]_2SnEt_2\cdot Et_2SnO$, 2, 444*
$Sn_2C_{22}H_{40}O_3$ $[p\text{-}(Bu_2SnO)_2C_6H_4]O$, 2, 405
$Sn_2C_{22}H_{42}O_4$ $(Pr_3SnO_2CC\equiv)_2$, 2, 427
$Sn_2C_{22}H_{48}O_2$ $(Pr_3Sn)_2CH_2CHCO_2Me$, 2, 246
$Sn_2C_{22}H_{50}O$ $Me_2(n\text{-}C_5H_{11})_4Sn_2O$, 2, 364
$Sn_2C_{22}H_{52}N_2$ $(i\text{-}Bu)_4(H_2NCH_2CH_2CH_2)_2Sn_2$, 2, 488
$Sn_2C_{24}H_{20}Cl_2O$ $(Ph_2SnCl)_2O$ (?), 2, 340
$Sn_2C_{24}H_{30}O_7P_2$ $(Ph_2Sn)_2P_2O_7$, 2, 356
$Sn_2C_{24}H_{30}$ $(p\text{-}MeC_6H_4)_3SnSnMe_3$, 2, 488
  $Et_3SnSnPh_3$, 2, 488
$Sn_2C_{24}H_{54}$ $(Pr_3SnCH_2CH_2CH_2)_2$, 2, 246
  Bu_6Sn_2, 2, 488
  $Et_4(n\text{-}C_8H_{17})_2Sn_2$, 2, 488
$Sn_2C_{24}H_{54}O$ $(Bu_3Sn)_2O$, 2, 361, 513
  $(i\text{-}Bu_3Sn)_2O$, 2, 362
  $(Me_3C)_6Sn_2O$, 2, 365
  $Et_4(n\text{-}C_8H_{17})_2Sn_2O$, 2, 364
$Sn_2C_{24}H_{54}O_5$ $(Bu_2SnO_2CMe_3)_2O$, 2, 380
$Sn_2C_{24}H_{54}S$ $(Bu_3Sn)_2S$, 2, 452
$Sn_2C_{24}H_{54}S_{1.3}$ $(Bu_3Sn)_2S_{1.3}$, 2, 452
$Sn_2C_{24}H_{54}S_{2.3}$ $(Bu_3Sn)_2S_{2.3}$, 2, 452
$Sn_2C_{24}H_{58}N_6O_{18}P_6$ $m\text{-}C_6H_4\{Sn[N{=}P(OMe)_3]_3\}_2$, 2, 481
$Sn_2C_{25}H_{35}Br_2P$ $[Me_3SnCH_2PPh_3][Me_3SnBr_2]$, 2, 243
$Sn_2C_{25}H_{45}NO_4$ $(Pr_3Sn)_2$ 2,5-pyridinedicarboxylate, 2, 416
$Sn_2C_{25}H_{47}NO_2$ $Pr_3SnNPhCO_2SnPr_3$, 2, 480
$Sn_2C_{25}H_{54}O_3$ $(Bu_3Sn)_2CO_3$, 2, 413
$Sn_2C_{26}H_{22}O_4$ $(Ph_2SnO_2CH)_2$ (?), 2, 494
$Sn_2C_{26}H_{34}N_2O_8$ Di(dipropyltin 2,5-pyridinedicarboxylate), 2, 439
$Sn_2C_{26}H_{38}O_5$ $(Pr_2SnOBz)_2O$, 2, 436
$Sn_2C_{26}H_{46}NS_2$ 2,3-$(Pr_3SnS)_2$ quinoxaline, 2, 460
  2,3-(i-$Pr_3SnS)_2$ quinoxaline, 2, 460
$Sn_2C_{26}H_{48}O_9$ $(Bu_2SnO_2CCH{=}CHCO_2Me)_2O$, 2, 441*
$Sn_2C_{26}H_{50}O_4$ 1,2,4-$(HOBu_2SnO)_2C_6H_3CMe_3$, 2, 405*
$Sn_2C_{26}H_{54}$ $(Bu_3SnC\equiv)_2$, 2, 246
$Sn_2C_{26}H_{54}O_4$ $(Bu_3SnO_2C)_2$, 2, 427
$Sn_2C_{26}H_{54}S_3$ 2,5-$(Bu_3SnS)_2$-1,3,4-thiadiazole, 2, 460
$Sn_2C_{30}H_{56}O_8$ $[Bu_2Sn(O_2CCH_2)_2CHMeCH_2]_2$, 2, 445

$Sn_2C_{27}H_{56}O_4$ $(Bu_3SnO_2C)_2CH_2$, 2, 427
$Sn_2C_{28}H_{20}Cl_6O_4$ $(Ph_2SnO_2CCCl_3)_2$, 2, 494
$Sn_2C_{28}H_{20}F_6O_4$ $(Ph_2SnO_2CCF_3)_2$, 2, 494
$Sn_2C_{28}H_{22}Cl_4O_4$ $(Ph_2SnO_2CCHCl_2)_2$, 2, 494
$Sn_2C_{28}H_{24}Cl_2O_4$ $(Ph_2SnO_2CCH_2Cl)_2$, 2, 494
$Sn_2C_{28}H_{26}O_4$ $(Ph_2SnOAc)_2$, 2, 494
$Sn_2C_{28}H_{40}Cl_6O_3$ (2,4,5-$Cl_3C_6H_2OSn$-$Bu_2)_2O$, 2, 402*
$Sn_2C_{28}H_{42}Br_2Cl_2O_3$ (2,5-$ClBrC_6H_3OSn$-$Bu_2)_2O$, 2, 402
$Sn_2C_{28}H_{42}Br_4O_3$ (2,4-$Br_2C_6H_3OSnBu_2)_2O$, 2, 402
$Sn_2C_{28}H_{42}Cl_4O_3$ (2,4-$Cl_2C_6H_3OSnBu_2)_2O$, 2, 402
(2,6-$Cl_2C_6H_3OSnBu_2)_2O$, 2, 402
$Sn_2C_{28}H_{42}N_6O_{16}$ [2,4,6-$(O_2N)_3C_6H_2OSn$-$Bu_2]_2O \cdot H_2O$, 2, 404
$Sn_2C_{28}H_{44}Br_2O_3$ (p-$BrC_6H_4OSnBu_2)_2O$, 2, 402
$Sn_2C_{28}H_{44}N_2O_7$ (p-$O_2NC_6H_4OSnBu_2)_2O$, 2, 404
$Sn_2C_{28}H_{46}Cl_2O_4$ $[(C_6H_{10}Cl)_3Sn]_2$ sebacate (?), 2, 432
$Sn_2C_{28}H_{46}O_3$ $(Bu_2SnOPh)_2O$, 2, 398
$Sn_2C_{28}H_{50}O_9S_2$ $Et_2SnO \cdot Et_2Sn(O_3SC_{10}H_{15}$-$O)_2$, 2, 351
$Sn_2C_{28}H_{54}O_4$ $(Bu_3SnO_2CC\equiv)_2$, 2, 427
$Sn_2C_{28}H_{56}O_4$ cis-$(Bu_3SnO_2CCH=)_2$, 2, 427
trans-$(Bu_3SnO_2CCH=)_2$, 2, 427
$Sn_2C_{28}H_{58}O_4$ $(Bu_3SnO_2CCH_2)_2$, 2, 427
$Sn_2C_{28}H_{58}O_4S$ $(Bu_3Sn)_2$ thiomalate, 2, 427
$Sn_2C_{28}H_{58}O_5$ $(Bu_3Sn)_2$ malate, 2, 427
$Sn_2C_{28}H_{58}O_6$ $(Bu_3SnO_2CCHOH)_2$, 2, 427
$Sn_2C_{29}H_{50}O_5S$ $(Pr_3Sn)_2$(S-Benzoyl)thiomalate, 2, 427
$Sn_2C_{29}H_{52}O_8S_4$ S-$(Bu_2Sn)_2(SCH_2CO_2$-$CH_2)_4C$, 2, 465
$Sn_2C_{30}H_{25}O_4P$ $Ph_3SnPO_4SnPh_2$, 2, 355
$Sn_2C_{30}H_{25}O_9P_3$ $(PhSn)_2(O_3PPh)_3$, 2, 357
$(Sn_2C_{30}H_{25}P_3)_n$ $[(PhPSnPh)_2PPh]_n$, 2, 484
$Sn_2C_{30}H_{44}Cl_2O_4$ (o-$ClC_6H_4CO_2SnBu_2)_2$, 2, 493
(p-$ClC_6H_4CO_2SnBu_2)_2$, 2, 493
$Sn_2C_{30}H_{46}O_4$ $(Bu_2SnOBz)_2$, 2, 493
$Sn_2C_{30}H_{50}O_5$ (o-$MeOC_6H_4OSnBu_2)_2O$, 2, 403
$Sn_2C_{30}H_{54}Cl_4O_2$ p-$(Bu_3SnO)_2C_6Cl_4$, 2, 393
$Sn_2C_{30}H_{58}NO_6P$ $(Bu_3Sn)_2PO_4C_6H_4NO_2$, 2, 351
$Sn_2C_{30}H_{62}O_4$ (Bu_3Sn) adipate, 2, 427
$Sn_2C_{30}H_{66}$ $(Bu_3SnCH_2CH_2CH_2)_2$, 2, 246
$(Me_3CCH_2)_6Sn_2$, 2, 488
$Sn_2C_{30}H_{66}O$ $(i-C_5H_{11})_6Sn_2O$, 2, 365
$(Me_3CCH_2)_6Sn_2O$, 2, 365
$Sn_2C_{30}H_{66}S_{6.6}$ $(Bu_3SnCH_2CHMe)_2S_{6.6}$, 2, 239
$Sn_2C_{31}H_{48}Br_4O_4$ [4,2,6-$(HOBu_2SnO)Br_2$-$C_6H_2]_2CMe_2$, 2, 405*
$Sn_2C_{31}H_{48}Cl_4O_4$ [4,2,6-$(HOBu_2SnO)Cl_2$-$C_6H_2]_2CMe_2$, 2, 405*
$Sn_2C_{31}H_{50}O_3$ [(p-$Bu_2SnOC_6H_4)_2CMe_2]O$, 2, 405
$Sn_2C_{31}H_{58}O_6S$ 3,4-$Bu_3SnO_2C(HO)C_6H_3SO_3$-$SnBu_3$, 2, 352
$Sn_2C_{31}H_{62}O_5S$ $(Bu_3SnO_2C)_2CH_2CHSOCEt$, 2, 427
$Sn_2C_{31}H_{64}O_4$ $(Bu_3Sn)_2$ pimelate, 2, 428
$Sn_2C_{32}H_{50}O_4$ (p-$MeC_6H_4CO_2SnBu_2)_2$, 2, 493
$Sn_2C_{32}H_{58}O_4$ o-$(Bu_3SnO_2C)_2C_6H_4$, 2, 428
p-$(Bu_3SnO_2C)_2C_6H_4$, 2, 428
$Sn_2C_{32}H_{58}O_8S$ $(Bu_2SnO_2CCH=CHCO_2Bu)_2S$, 2, 447
$Sn_2C_{32}H_{66}O_5$ $(Bu_2SnO_2CC_7H_{15})_2O$, 2, 437
$Sn_2C_{32}H_{70}$ $(Bu_3Sn)_2(CH_2)_8$, 2, 246
$Sn_2C_{33}H_{46}N_2O_6S_2$ $(Et_3SnO_3SC_{10}H_5NH_2)_2$-$CH_2$, 2, 350
$Sn_2C_{33}H_{68}O_4$ $(Bu_3Sn)_2$ azelate, 2, 428
$Sn_2C_{33}H_{68}O_4S_2$ $(Bu_3SnO_2CCH_2CH_2S)_2CMe_2$, 2, 428
$Sn_2C_{34}H_{26}O_4S_2$ (2-$Ph_2SnO_2CC_4H_3S)_2$, 2, 494
$Sn_2C_{34}H_{26}O_6$ (2-$Ph_2SnO_2CC_4H_3O)_2$, 2, 494
$Sn_2C_{34}H_{70}O_4$ $(Bu_3Sn)_2$ sebacate, 2, 428
$Sn_2C_{35}H_{62}O_4S$ $(Bu_3Sn)_2$ S-benzoylthiomalate, 2, 427

$Sn_2C_{36}H_{20}Cl_{10}O_3$ $(Ph_2SnOC_6Cl_5)_2O$, 2, 402
$Sn_2C_{36}H_{24}Cl_6$ $(p\text{-}ClC_6H_4)_6Sn_2$, 2, 488
$Sn_2C_{36}H_{26}Br_2Cl_2O_3$ $(2,5\text{-}ClBrC_6H_3OSn\text{-}Ph_2)_2O$, 2, 402
$Sn_2C_{36}H_{28}Br_2O_3$ $(p\text{-}BrC_6H_4OSnPh_2)_2O$, 2, 402
$Sn_2C_{36}H_{30}$ Ph_6Sn_2, 2, 487
$Sn_2C_{36}H_{30}O$ $(Ph_3Sn)_2O$, 2, 362
$Sn_2C_{36}H_{30}O_2$ $(Ph_3SnO)_2$, 2, 380
$Sn_2C_{36}H_{30}O_3Se$ $(Ph_3Sn)_2SeO_3$, 2, 356
$Sn_2C_{36}H_{30}O_4S$ $(Ph_3Sn)_2SO_4$, 2, 355
$Sn_2C_{36}H_{30}S$ $(Ph_3Sn)_2S$, 2, 451
$Sn_2C_{36}H_{30}S_2$ $(Ph_3Sn)_2S_2$, 2, 452
$Sn_2C_{36}H_{30}S_3$ $(Ph_3Sn)_2S_3$, 2, 452
$Sn_2C_{36}H_{30}S_4$ $(Ph_3Sn)_2S_4$, 2, 452
$Sn_2C_{36}H_{30}Se$ $(Ph_3Sn)_2Se$, 2, 452
$Sn_2C_{36}H_{31}O_4P$ $(Ph_3Sn)_2HPO_4$, 2, 355
$Sn_2C_{36}H_{38}$ $o\text{-}(Me_3Sn)_2C_6Ph_4$, 2, 245
$Sn_2C_{36}H_{42}O_4$ $(n\text{-}C_5H_{11}CO_2SnPh_2)_2$, 2, 494
$Sn_2C_{36}H_{50}O_3$ $(2\text{-}Bu_2SnOC_{10}H_7)_2O$, 2, 399
$Sn_2C_{36}H_{62}O_3$ $(p\text{-}Bu_2SnOC_6H_4CMe_3)_2O$, 2, 398
$Sn_2C_{36}H_{64}O_5S$ $(Bu_3SnO_2C)_2CH_2CH(SOCCH_2\text{-}Ph)$, 2, 427
$(Bu_3Sn)_2$ S-toloylthiomalate, 2, 427
$Sn_2C_{36}H_{66}$ $Cyclohexyl_6Sn_2$, 2, 488
$Sn_2C_{36}H_{66}O$ $Cyclohexyl_6Sn_2O$, 2, 365
$Sn_2C_{36}H_{70}O_7$ $(Bu_2SnO_2CCH{=}CHCO_2Bu)O\text{-}(Bu_2SnO_2CC_{11}H_{23})$, 2, 441
$Sn_2C_{36}H_{78}O$ $(C_6H_{13})_6Sn_2O$, 2, 365
$Sn_2C_{37}H_{30}N_2$ $(Ph_3SnN{=})_2C$, 2, 481
$Sn_2C_{37}H_{30}O_3$ $(Ph_3Sn)_2CO_3$, 2, 415
$Sn_2C_{37}H_{58}Cl_6O_2$ $2,2'\text{-}(Bu_3SnOC_6HCl_3)_2\text{-}CH_2$, 2, 393
$Sn_2C_{37}H_{62}Cl_2O_2$ $2,2'\text{-}(5\text{-}ClC_6H_3OSn\text{-}Bu_3)_2CH_2$, 2, 393
$Sn_2C_{37}H_{78}O_4S_2$ $(Bu_3SnO_2CCH_2CH_2S)_2CHPh$, 2, 428
$Sn_2C_{38}H_{24}Cl_6$ $[(p\text{-}ClC_6H_4)_3SnC{\equiv}]_2$, 2, 247
$Sn_2C_{38}H_{28}Cl_2O_4$ $(o\text{-}ClC_6H_4CO_2SnPh_2)_2$, 2, 494
$Sn_2C_{38}H_{30}$ $(Ph_3SnC{\equiv})_2$, 2, 246
$Sn_2C_{38}H_{30}O_4$ $(Ph_3SnO_2C)_2$, 2, 428
$(Ph_2SnOBz)_2$, 2, 494
$Sn_2C_{38}H_{30}O_6$ $(o\text{-}HOC_6H_4CO_2SnPh_2)_2$, 2, 494
$Sn_2C_{38}H_{34}$ $(Ph_3SnCH_2)_2$, 2, 246
$Sn_2C_{38}H_{70}N_2S_2$ $2,3\text{-}[(C_5H_{11})_3SnS]_2$ quinoxaline, 2, 460
$Sn_2C_{39}H_{32}O_4$ $(Ph_3SnO_2C)_2CH_2$, 2, 428
$Sn_2C_{39}H_{36}$ $(Ph_3SnCH_2)_2CH_2$, 2, 246
$Sn_2C_{40}H_{30}$ $(Ph_3SnC{\equiv}C)_2$, 2, 247
$Sn_2C_{40}H_{30}O_4$ $(Ph_3SnO_2CC{\equiv})_2$, 2, 428
$Sn_2C_{40}H_{34}$ $(Ph_3SnCH{=}CH)_2$, 2, 247
$Sn_2C_{40}H_{34}O_4$ $(Ph_3SnO_2CCH_2)_2$, 2, 428
$Sn_2C_{40}H_{36}$ $CH_2CHPhSnPh_2CHPhCH_2SnPh_2$, 2, 507
$Sn_2C_{40}H_{36}N_2S_4$ $(Ph_3SnS_2CNHCH_2)_2$, 2, 472
$Sn_2C_{40}H_{36}O_2$ $(Ph_3Sn)_2CH_2CHCO_2Me$, 2, 247
$Sn_2C_{40}H_{38}$ $(Ph_3SnCH_2CH_2)_2$, 2, 247
$Sn_2C_{40}H_{50}O_4$ $(n\text{-}C_7H_{15}CO_2SnPh_2)_2$, 2, 494
$Sn_2C_{40}H_{62}N_4O_{11}$ $[2,4,5\text{-}(O_2N)_2(C_6H_{11})\text{-}C_6H_2OSnBu_2]_2O$, 2, 404
$Sn_2C_{40}H_{64}Cl_2O_3$ $(2,4\text{-}Cyclohexyl\text{-}Cl\text{-}C_6H_3OSnBu_2)_2O$, 2, 402
$Sn_2C_{40}H_{82}O_4$ $(Bu_2SnO_2CC_{11}H_{23})_2$, 2, 493
$Sn_2C_{41}H_{72}O_9S_2$ $(Bu_2SnSCH_2CO_2CH_2)_2(O_2C\text{-}C_6H_4OH)O_2CC_{11}H_{23}$, 2, 447
$Sn_2C_{42}H_{24}F_{18}$ $(m\text{-}CF_3C_6H_4)_6Sn_2$, 2, 488
$Sn_2C_{42}H_{30}$ $(Ph_3SnC{\equiv}CC{\equiv})_2$, 2, 247
$Sn_2C_{42}H_{30}Cl_4O_2$ $p\text{-}(Ph_3SnO)_2C_6Cl_4$, 2, 393
$Sn_2C_{42}H_{34}$ $p\text{-}(Ph_3Sn)_2C_6H_4$, 2, 247
$Sn_2C_{42}H_{34}O_2$ $p\text{-}(Ph_3SnO)_2C_6H_4$, 2, 393
$Sn_2C_{42}H_{35}O_3P$ $(Ph_3Sn)_2O_3PPh$, 2, 355
$Sn_2C_{42}H_{35}P$ $(Ph_3Sn)_2PPh$, 2, 483
$Sn_2C_{42}H_{36}Br_3O$ $(o\text{-}BrC_6H_4CH_2)_6Sn_2O$, 2, 364
$Sn_2C_{42}H_{36}Cl_3O$ $(o\text{-}ClC_6H_4CH_2)_6Sn_2O$, 2, 364
$Sn_2C_{42}H_{36}F_3O$ $(o\text{-}FC_6H_4CH_2)_6Sn_2O$, 2, 364
$Sn_2C_{42}H_{38}O_3$ $(Ph_3SnCH_2CH_2CO)_2O$, 2, 247
$Sn_2C_{42}H_{42}$ $(o\text{-}MeC_6H_4)_6Sn_2$, 2, 488
$(p\text{-}MeC_6H_4)_6Sn_2$, 2, 488
$Sn_2C_{42}H_{42}O$ $(Ph_3SnCH_2CH_2CH_2)_2O$, 2, 247
$(PhCH_2)_6Sn_2O$, 2, 364

$Sn_2C_{42}H_{42}S$ $[(p\text{-}MeC_6H_4)_3Sn]_2S$, 2, 452

$Sn_2C_{42}H_{42}S_{6.5}$ $(Ph_3SnCH_2CHMe)_2S_{6.5}$, 2, 239

$Sn_2C_{42}H_{46}Br_4P_2$ $[Me_2Sn(CH_2PPh_3)_2]$-$[Me_2SnBr_4]$, 2, 243

$Sn_2C_{42}H_{46}Cl_4P_2$ $[Me_2Sn(CH_2PPh_3)_2]$-$[Me_2SnCl_4]$, 2, 243

$Sn_2C_{42}H_{50}O_{12}S_6$ $(i\text{-}PrSn)_2[(SCH_2CH_2O_2\text{-}C)_2C_6H_4]_3$, 2, 469

$Sn_2C_{42}H_{70}$ p-$(Cyclohexyl_3Sn)_2C_6H_4$, 2, 246

$Sn_2C_{42}H_{90}O$ $(C_7H_{15})_6Sn_2O$, 2, 365

$Sn_2C_{43}H_{36}$ $(Ph_3SnC{\equiv}CCH_2)_2CH_2$, 2, 247

$Sn_2C_{44}H_{34}N_2S_2$ 2,3-$(Ph_3SnS)_2$quinoxaline, 2, 460

$Sn_2C_{44}H_{34}O_4$ $(Ph_3Sn)_2$ phthalate, 2, 428

$Sn_2C_{44}H_{38}$ $(Ph_3Sn)_2CH_2CHPh$, 2, 246
 $(Ph_3SnC{\equiv}CCH_2CH_2)_2$, 2, 247
 p-$(Ph_3Sn)_2C_6H_4CH_2CH_2$, 2, 247

$Sn_2C_{44}H_{42}$ $[(PhCH_2)_3SnC{\equiv}]_2$, 2, 245
 $[(o\text{-}MeC_6H_4)_3SnC{\equiv}]_2$, 2, 247
 $[(p\text{-}MeC_6H_4)_3SnC{\equiv}]_2$, 2, 247

$Sn_2C_{44}H_{42}O_4$ $(Ph_3SnCH_2CH_2CO_2CH_2)_2$, 2, 247

$Sn_2C_{44}H_{46}$ $(Ph_3SnCH_2CHMeCH_2)_2$, 2, 247

$Sn_2C_{44}H_{54}O_{12}S_6$ $(i\text{-}BuSn)_2[(SCH_2CH_2\text{-}O_2C)_2C_6H_4]_3$, 2, 469

$Sn_2C_{44}H_{76}O_{12}$ $[Bu_2Sn(O_2CCH{=}CHCO_2\text{-}C_8H_{17}\text{-}i)O_2CCH{=}]_2$ (?), 2, 441

$Sn_2C_{44}H_{84}O_8$ $[Bu_2Sn(O_2CC_{11}H_{23})O_2C\text{-}CH{=}]_2$ (?), 2, 441

$Sn_2C_{45}H_{68}N_2O$ $Bu_3SnN(1\text{-}C_{10}H_7)C(OSn\text{-}Bu_3){=}N(1\text{-}C_{10}H_7)$, 2, 480

$Sn_2C_{45}H_{68}O_6S_2$ $(Bu_3SnO_3SC_{10}H_6)_2CH_2$, 2, 352

$Sn_2C_{46}H_{34}$ p-$(Ph_3SnC{\equiv}C)_2C_6H_4$, 2, 247

$Sn_2C_{46}H_{38}$ p-$(Ph_3SnCH{=}CH)_2C_6H_4$, 2, 247

$Sn_2C_{46}H_{42}$ p-$(Ph_3SnCH_2CH_2)_2C_6H_4$, 2, 247

$Sn_2C_{46}H_{124}O_8$ $(Bu_2SnO_2CCH{=})_2$ dibehenate (?), 2, 441

$Sn_2C_{47}H_{60}O_{12}S_6$ $(BuSn)_2[(O_2CCH_2CH_2S)_2\text{-}CHPh]_3$, 2, 449

$Sn_2C_{47}H_{86}O_5S$ $[(C_6H_{13})_3SnO_2C]_2CH_2CH\text{-}SBz$, 2, 428

$Sn_2C_{48}H_{38}$ $(p\text{-}Ph_3SnC_6H_4)_2$, 2, 247

$Sn_2C_{48}H_{38}O_7$ $[(o\text{-}PhOC_6H_4)_2SnOH]_2O$, 2, 374

$Sn_2C_{48}H_{54}S_{6.9}$ $[(p\text{-}MeC_6H_4)_3SnCH_2CH\text{-}Me]_2S_{6.9}$, 2, 239

$Sn_2C_{48}H_{102}O$ $(C_8H_{17})_6Sn_2O$, 2, 365

$Sn_2C_{52}H_{58}S_2O_4$ $[(PhCH_2)_3SnSCH_2CH_2CO_2\text{-}CH_2CH_2]_2$, 2, 461

$Sn_2C_{54}H_{42}$ 9,10-$(Ph_3SnC{\equiv}C)_2$anthracene, 2, 247

$Sn_2C_{54}H_{54}S$ $(p\text{-}EtO_2CC_6H_4)_6Sn_2S$, 2, 452

$Sn_2C_{60}H_{42}$ $(C_{10}H_7)_6Sn_2$, 2, 488

$Sn_2C_{60}H_{42}O$ $(1\text{-}C_{10}H_7)_6Sn_2O$, 2, 366

$Sn_2C_{60}H_{78}S$ $(p\text{-}Me_2CHC_6H_4CH_2)_6Sn_2S$, 2, 452

$Sn_2C_{60}H_{126}O$ $(C_{10}H_{21})_6Sn_2O$, 2, 365

$Sn_2C_{72}H_{54}$ $(o\text{-}PhC_6H_4)_6Sn_2$, 2, 488

$Sn_3C_6H_{14}Cl_6O_2$ $(Cl_2SnCH_2)_2SnCl_2O{\cdot}Et_2O$, 2, 340

$Sn_3C_6H_{18}Cl_2O_2$ $Me_2SnCl_2{\cdot}2\ Me_2SnO$, 2, 338

$Sn_3C_6H_{18}S_3$ $(Me_2SnS)_3$, 2, 453

$Sn_3C_6H_{18}Se_3$ $(Me_2SnSe)_3$, 2, 455

$Sn_3C_9H_{27}BrO$ $(Me_3Sn)_3OBr$, 2, 338

$Sn_3C_9H_{27}IO$ $(Me_3Sn)_3OI$, 2, 338

$Sn_3C_9H_{27}N$ $(Me_3Sn)_3N$, 2, 475

$Sn_3C_9H_{27}P$ $(Me_3Sn)_3P$, 2, 482

$Sn_3C_9H_{29}BrO_2$ $Me_3SnBr{\cdot}2\ Me_3SnOH$, 2, 338

$Sn_3C_9H_{29}ClO_2$ $Me_3SnCl{\cdot}2\ Me_3SnOH$, 2, 338

$Sn_3C_9H_{29}IO_2$ $Me_3SnI{\cdot}2\ Me_3SnOH$, 2, 338

$Sn_3C_{12}H_{30}S_3$ $(Et_2SnS)_3$, 2, 454

$Sn_3C_{12}H_{30}S_{3.6}$ $(Et_2SnS_{1.2})_3$, 2, 454

$Sn_3C_{16}H_{40}$ Et_8Sn_3, 2, 491

$[Sn_3C_{18}H_{15}ClS_4]_n$ $[Ph_3Sn_3S_4Cl]_n$, 2, 342

$Sn_3C_{18}H_{42}Cl_2$ $(Pr_2Sn)_3Cl_2$, 2, 492

$Sn_3C_{18}H_{45}N$ $(Et_3Sn)_3N$, 2, 475

$Sn_3C_{18}H_{45}O_4P$ $(Et_3Sn)_3PO_4$ (?), 2, 349

$Sn_3C_{18}H_{49}BrO_3$ $Et_3SnBr{\cdot}2\ Et_3SnOH{\cdot}H_2O$, 2, 338

$Sn_3C_{18}H_{49}ClO_3$ $Et_3SnCl{\cdot}2\ Et_3SnOH{\cdot}H_2O$, 2, 338

$Sn_3C_{18}H_{49}IO_3$ $Et_3SnI{\cdot}2\ Et_3SnOH{\cdot}H_2O$, 2, 338

$Sn_3C_{22}H_{30}N_2O_8$ $[(CH_2{=}CH)_3SnO_2CCH_2]_2N$-

$CH_2CH_2N(CH_2CO_2)_2Sn$, 2, 429
$(Sn_3C_{24}H_{20}S_4)_n$ $(Ph_4Sn_3S_4)_n$, 2, 457
$Sn_3C_{22}H_{48}O_6$ $Pr_2Sn(OAc)_2 \cdot 2\ Pr_2SnO$, 2, 436
$Sn_3C_{24}H_{54}S_3$ $(Bu_2SnS)_3$, 2, 453
$Sn_3C_{24}H_{54}S_{4.2}$ $(Bu_2SnS_{1.4})_3$, 2, 454
$Sn_3C_{24}H_{54}S_{4.8}$ $(Bu_2SnS_{1.6})_3$, 2, 454
$Sn_3C_{26}H_{36}N_2O_8$ $[(CH_2{=}CH)_3SnO_2CCH_2]_2N$-$CH_2CH_2N(CH_2CO_2)_2Sn(CH{=}CH_2)_2$, 2, 429
$Sn_3C_{27}H_{63}N$ $(Pr_3Sn)_3N$, 2, 476
$Sn_3C_{27}H_{63}S_3$ $(Pr_3SnS)_3$, 2, 454
$(i\text{-}Pr_3SnS)_3$, 2, 454
$Sn_3C_{28}H_{60}O_6$ $Bu_2Sn(OAc)_2 \cdot 2\ Bu_2SnO$, 2, 436
$Sn_3C_{30}H_{25}ClS_3$ $Ph_5Sn_3S_3Cl$ (?), 2, 342
$Sn_3C_{30}H_{64}O_6$ $Bu_2Sn(O_2CEt)_2 \cdot 2\ Bu_2SnO$, 2, 437
$Sn_3C_{30}H_{66}S_3$ $[(n\text{-}C_5H_{11})_2SnS]_3$, 2, 454
$(n\text{-}C_6H_{13}BuSnS)_3$, 2, 454
$Sn_3C_{34}H_{40}N_2O_8$ $[(CH_2{=}CH)_3SnO_2CCH_2]_2N$-$CH_2CH_2N(CH_2CO_2)_2SnPh_2$, 2, 429
$Sn_3C_{36}H_{30}O_{13}P_4$ $(PhSnO_3PPh)_3PO_4$, 2, 357
$(Sn_3C_{36}H_{30}P_4)_n$ $[(PhPSnPh)_3P]_n$, 2, 484
$Sn_3C_{36}H_{30}S_3$ $(Ph_2SnS)_3$, 2, 453
$Sn_3C_{36}H_{30}Se_3$ $(Ph_2SnSe)_3$, 2, 455
$Sn_3C_{36}H_{78}S_3$ $[(n\text{-}C_6H_{13})_2SnS]_3$, 2, 454
$Sn_3C_{36}H_{81}O_4P$ $(Bu_3Sn)_3PO_4$, 2, 351
$Sn_3C_{42}H_{42}S_3$ $[(p\text{-}MeC_6H_4)_2SnS]_3$, 2, 454
$Sn_3C_{44}H_{46}O_5$ $3(PhCH_2)_2SnO \cdot AcOH$, 2, 435
$Sn_3C_{48}H_{42}O_{12}S_3$ $[(p\text{-}MeO_2CC_6H_4)_2SnS]_3$, 2, 454
$Sn_3C_{48}H_{54}O_6S_3$ $[(p\text{-}EtOC_6H_4)_2SnS]_3$, 2, 454
$Sn_3C_{48}H_{102}S_3$ $[(C_8H_{17})_2SnS]_3$, 2, 454
$Sn_3C_{46}H_{76}N_2O_8$ $[CH_2N(CH_2CO_2SnBu_3)CH_2$-$CO_2]_2SnPh_2$, 2, 429
$Sn_3C_{52}H_{115}O_{10}P_3$ $(Bu_3Sn)_3P_3O_8(OC_8H_{17})_2$, 2, 352
$Sn_3C_{54}H_{45}O_4P$ $(Ph_3Sn)_3PO_4$, 2, 355
$Sn_3C_{54}H_{45}P$ $(Ph_3Sn)_3P$, 2, 483
$Sn_3C_{54}H_{45}P_3$ $(Ph_2SnPPh)_3$, 2, 484
$Sn_3C_{54}H_{47}O_7P_2$ $(Ph_3Sn)_2H_2P_2O_7$, 2, 355
$Sn_3C_{54}H_{54}O_{12}S_3$ $[(p\text{-}EtO_2CC_6H_4)_2SnS]_3$, 2, 454
$Sn_3C_{24}H_{54}Se_3$ $(Bu_2SnSe)_3$, 2, 455
$Sn_3C_{58}H_{52}N_2O_8$ $[CH_2N(CH_2CO_2SnPh_3)CH_2$-$CO_2]_2SnPh_2$, 2, 429
$Sn_3C_{60}H_{42}S_3$ $[(1\text{-}C_{10}H_7)_2SnS]_3$, 2, 454
$Sn_3C_{60}H_{60}O_6$ $(Ph_3Sn)_3$ citrate, 2, 423
$Sn_3C_{60}H_{78}S_3$ $[(Me_2CHC_6H_4CH_2)_2SnS]_3$, 2, 454
$Sn_3C_{64}H_{56}$ $(p\text{-}Ph_3SnCH_2CH_2C_6H_4)_2SnPh_2$, 2, 247
$(p\text{-}Ph_3SnC_6H_4CH_2CH_2)_2SnPh_2$, 2, 247
$Sn_3C_{66}H_{66}$ $(o\text{-}CH_2CH_2C_6H_4CH_2CH_2SnPh_2)_3$, 2, 507*
$Sn_3C_{72}H_{54}O_6S_3$ $[(o\text{-}PhOC_6H_4)_2SnS]_3$, 2, 454
$Sn_3C_{72}H_{54}S_3$ $[(p\text{-}PhC_6H_4)_2SnS]_3$, 2, 454
$Sn_3C_{72}H_{66}S_3$ $[(C_{10}H_5Me_2)_2SnS]_3$, 2, 454
$Sn_3C_{72}H_{150}S_3$ $[(n\text{-}C_{12}H_{25})_2SnS]_3$, 2, 454
$Sn_4C_8H_{24}Cl_4O_2$ $[(Me_2SnCl)_2O]_2$, 2, 338
$Sn_4C_{12}H_{34}Br_2O_4$ $[(Me_2SnBr)O(Me_2SnOEt)]_2$, 2, 341
$Sn_4C_{12}H_{34}I_2O_4$ $[(Me_2SnI)O(Me_2SnOEt)]_2$, 2, 341
$Sn_4C_{14}H_{38}I_2O_4$ $Me_2SnI_2 \cdot 3\ Et_2SnO \cdot H_2O$, 2, 338, 610*
$EtO(Me_2SnO)_3Et \cdot Et_2SnI_2$, 2, 341*, 610*
$Sn_4C_{16}H_{40}Cl_2$ $(Et_2Sn)_4Cl_2$, 2, 492
$Sn_4C_{16}H_{40}ClO_3$ $Et_2SnCl \cdot 3\ Et_2SnO$, 2, 339
$Sn_4C_{16}H_{42}Br_2O_4$ $[(Et_2SnBr)O(Et_2SnOH)]_2$, 2, 339
$Sn_4C_{16}H_{42}I_2O_4$ $[(Et_2SnI)O(Et_2SnOH)]_2$, 2, 339
$Sn_4C_{16}H_{42}Cl_2O_4$ $[(Et_2SnCl)O(Et_2SnOH)]_2$, 2, 339
$Sn_4C_{16}H_{42}O_5$ $(Et_2SnO)_4 \cdot H_2O$, 2, 372
$Sn_4C_{17}H_{37}Cl_7$ $(BuSnCl)_4 \cdot HCCl_3$, 2, 492
$Sn_4C_{18}H_{46}I_2O_4$ $EtO(Et_2SnO)_3Et \cdot Me_2SnI_2$, 2, 341*
$Sn_4C_{20}H_{50}$ $Et_{10}Sn_4$, 2, 491
$Sn_4C_{24}H_{58}I_2O_4$ $[(Pr_2SnI)O(Pr_2SnOH)]_2$, 2, 340*, 610*
$Sn_4C_{28}H_{66}I_2O_4$ $[(Pr_2SnI)O(Pr_2SnOEt)]_2$, 2, 341, 610*
$Sn_4C_{30}H_{50}O_9$ $(o\text{-}HOC_6H_4CO_2)_2SnEt_2 \cdot 3\ Et_2$-$SnO$, 2, 444
$Sn_4C_{30}H_{70}I_2O_4$ $[(Pr_3SnI)O(Pr_2SnOPr)]_2$,

2, 341, 610*
$Sn_4C_{32}H_{50}$ p-(p-$Me_3SnC_6H_4CH_2CH_2Sn$-$Me_2)_2C_6H_4$, 2, 245
$Sn_4C_{32}H_{54}$ $(Bu_2Sn)_4$, 2, 491
$Sn_4C_{34}H_{54}O_{11}$ (o-$AcOC_6H_4CO_2)_2SnEt_2$·3 Et_2SnO, 2, 444
$Sn_4C_{36}H_{78}O_7$ $Bu_2Sn(OAc)_2$·3 Bu_2SnO, 2, 437
$Sn_4C_{38}H_{82}O_7$ $Bu_2Sn(O_2CEt)_2$·3 Bu_2SnO, 2, 437
$Sn_4C_{40}H_{80}O_8$ $[(Bu_2SnO_2CCH_2)_2]_2$, 2, 493
$Sn_4C_{44}H_{99}O_{10}P_3$ $Bu_2Sn(Bu_3Sn)_3P_3O_{10}$, 2, 352
$Sn_4C_{44}H_{100}N_2O_5$ $Et_2NCH_2CH_2O(Bu_2SnO)_4$-$CH_2CH_2NEt_2$, 2, 404*
$Sn_4C_{48}H_{102}O_7$ $Bu_2Sn(O_2CC_7H_{15})_2$·3 Bu_2-SnO, 2, 437
$Sn_4C_{62}H_{62}$ (p-$Ph_3SnC_6H_4CH_2CH_2SnMe_2)_2$-$C_6H_4$, 2, 247
$Sn_4C_{72}H_{60}P_2$ $(Ph_3Sn)_4P_2$, 2, 483
$Sn_5C_{12}H_{36}$ $(Me_3Sn)_4Sn$, 2, 491
$Sn_5C_{20}H_{52}O_6$ $(Et_2SnO)_5\cdot H_2O$, 2, 372
$Sn_5C_{40}H_{90}Cl_2$ $(Bu_2Sn)_5Cl_2$, 2, 492
$Sn_5C_{72}H_{60}$ $(Ph_3Sn)_4Sn$, 2, 491
$Sn_5C_{118}H_{190}O_{40}$ 2 $Bu_2Sn(O_2CCH{=}CHCO_2$-$C_{12}H_{25})(O_2CCH{=}CHCO_2Bu)$·3 Bu_2Sn-$(O_2CCH{=}CHCO_2CH_2)_2$, 2, 441*
$Sn_6C_{12}H_{36}$ $(Me_2Sn)_6$, 2, 491
$Sn_6C_{90}H_{75}P_3$ $(Ph_3SnPSnPh_2)_3$, 2, 484
$Sn_6C_{56}H_{126}Cl_2O_6$ $BuO(Bu_2SnO)_5Bu\cdot Bu_2$-$SnCl_2$, 2, 341*
$Sn_7C_{60}H_{132}O_{10}$ $Bu_2Sn(OAc)_2$·6 Bu_2SnO, 2, 437
$Sn_8C_{28}H_{76}I_4O_8$ $[EtO(Me_2SnO)_3\cdot Et_2SnI_2]_2$, 2, 341*, 610*
$Sn_8C_{36}H_{92}I_4O_8$ $[EtO(Et_2SnO)_3Et\cdot Me_2Sn$-$I_2]_2$, 2, 341*
$Sn_8C_{68}H_{150}O_{11}$ $Bu_2Sn(OAc)_2$·7 Bu_2SnO, 2, 437
$Sn_{16}C_{132}H_{294}O_{19}$ $Bu_2Sn(OAc)_2$·15 Bu_2-SnO, 2, 437
$SnAlC_2H_5Cl_6$ $EtSnCl_3\cdot AlCl_3$, 2, 329
$SnAlC_2H_6Br_5$ $Me_2SnBr_2\cdot AlBr_3$, 2, 305
$SnAlC_3H_9Br_4$ $Me_3SnBr\cdot AlBr_3$, 2, 271
$SnAlC_3H_9Cl_4$ $Me_3SnCl\cdot AlCl_3$, 2, 270
$SnAlC_4H_{10}Br_5$ $Et_2SnBr_2\cdot AlBr_3$, 2, 308
$SnAlC_4H_{10}Cl_5$ $Et_2SnCl_2\cdot AlCl_3$, 2, 307
$SnAlC_6H_{15}Br_4$ $Et_3SnBr\cdot AlBr_3$, 2, 274
$SnAlC_6H_{15}Cl_4$ $Et_3SnCl\cdot AlCl_3$, 2, 274
$SnAlC_8H_{18}Cl_5$ $Bu_2SnCl_2\cdot AlCl_3$, 2, 311
$SnAlC_8H_{20}Cl_3$ $Et_4Sn\cdot AlCl_3$, 2, 162
$SnAlC_{12}H_{27}Cl_4$ i-$Bu_3SnCl\cdot AlCl_3$, 2, 283
$SnAlC_{24}H_{31}O_3$ $Ph_3Sn[HAl(OEt)_3]$, 2, 253*
$SnAlC_{24}H_{49}O_3$ $Cyclohexyl_3Sn[HAl(OEt)_3]$, 2, 253*
$SnAl_2C_{16}H_{40}$ Et_2Sn·2 $AlEt_3$, 2, 253
$SnAl_2C_{24}H_{42}O_6$ $Ph_2Sn[HAl(OEt)_3]_2$, 2, 253*
$SnAl_3C_{24}H_{53}O_9$ $PhSn[HAl(OEt)_3]_3$, 2, 253*
$SnAsC_2H_7O_4$ Me_2SnAsO_4H, 2, 349
$SnAsC_3H_9F_6$ Me_3SnAsF_6, 2, 348
$(SnAsC_3H_9O_3)_n$ $(Me_2SnO_3AsMe)_n$, 2, 349, 3, 547
$(SnAsC_8H_{11}O_3)_n$ $(Me_2SnO_3AsPh)_n$, 2, 349, 3, 549
$SnAsC_{14}H_{18}O_6$ $Me_2Sn(HO_3AsPh)_2$, 2, 349
$SnAsC_{14}H_{22}NO_5$ p-$O_2NC_6H_4AsO_3SnBu_2$, 2, 353, 3, 548
$SnAsC_{14}H_{22}NO_6$ 3,4-$O_2N(HO)C_6H_3AsO_3SnBu_2$, 2, 353, 3, 475, 548
$SnAsC_{14}H_{23}N_2O_5$ 3,4-$O_2N(H_2N)C_6H_3AsO_3$-$SnBu_2$, 2, 353, 3, 475, 549
$SnAsC_{14}H_{24}NO_3$ p-$H_2NC_6H_4AsO_3SnBu_2$, 2, 353, 3, 428, 549
$SnAsC_{15}H_{19}$ $Me_3SnAsPh_2$, 2, 484, 3, 299
$SnAsC_{15}H_{19}O_2$ $Ph_2AsO_2SnMe_3$, 3, 376
$SnAsC_{16}H_{19}O_3$ Ph_2SnO_3AsBu, 2, 356, 3, 548
$SnAsC_{18}H_{15}O_2$ Ph_3SnAsO_2, 2, 355
$SnAsC_{18}H_{15}O_3$ Ph_3SnAsO_3, 2, 355
$SnAsC_{18}H_{25}$ $Ph_2AsSnEt_3$, 3, 299
$SnAsC_{18}H_{25}O_2$ $Ph_2AsO_2SnEt_3$, 3, 376
$SnAsC_{21}H_{31}$ $Ph_2AsSnPr_3$, 3, 299
$SnAsC_{24}H_{20}NO_5$ p-$O_2NC_6H_4AsO_3HSnPh_3$, 2, 355, 3, 548
$SnAsC_{24}H_{37}$ $Ph_2AsSnBu_3$, 3, 299
$SnAsC_{24}H_{37}O_2$ $Ph_2AsO_2SnBu_3$, 3, 377
$SnAsC_{30}H_{25}O_2$ $Ph_2AsO_2SnPh_3$, 3, 377
$SnAs_2C_6H_{18}O_4$ $Me_2Sn(O_2AsMe_2)_2$, 2, 349, 3, 374
$SnAs_2C_{14}H_{18}O_6$ $[PhAs(O)(OH)O]_2SnMe_2$, 3, 549

$SnAs_2C_{20}H_{28}N_2O_{10}$ $[p\text{-}O_2NC_6H_4AsO_3H]_2$-$SnBu_2$, 2, 353, 3, 548

$SnAs_2C_{20}H_{28}N_2O_{12}$ $[3,4\text{-}O_2N(HO)C_6H_3$-$AsO_3H]_2SnBu_2$, 2, 353, 3, 475, 548

$SnAs_2C_{20}H_{30}N_4O_{10}$ $[3,4\text{-}O_2N(H_2N)C_6H_3$-$AsO_3H]_2SnBu_2$, 2, 353, 3, 475, 549*

$SnAs_2C_{20}H_{30}O_8$ $Ph_2Sn(HO_3AsBu)_2$, 2, 356, 3, 548

$SnBC_3H_9F_4$ Me_3SnBF_4, 2, 269

$SnBC_4H_9F_6$ $Me_3SnCF_3\cdot BF_3$, 2, 218

$SnBC_8H_{21}ClN$ $Et_3SnB(NMe_2)Cl$, 2, 495

$SnBC_{10}H_{27}N_2$ $Et_3SnB(NMe_2)_2$, 2, 495

$SnBC_{18}H_{15}O_2$ Ph_3SnBO_2, 2, 355

$SnBC_{46}H_{73}O_3$ $(p\text{-}Me_2CHC_6H_4CH_2)_3SnOB$-$(OCH_2CHEtBu)_2$, 2, 349*

$SnB_2C_{40}H_{86}O_6$ $Bu_2Sn[OB(OCH_2CHEtBu)_2]_2$, 2, 352

$SnB_2C_{68}H_{126}O_6$ $(p\text{-}Me_2CHC_6H_4CH_2)_2Sn$-$[OB(OC_{12}H_{25})_2]_2$, 2, 349*

$SnB_2C_{88}H_{80}P_2$ $[Me_2Sn(CH_2PPh_3)_2][BPh_4]_2$ 2, 243

$SnCoC_9H_{12}ClO$ $EtSnCl(MeC_5H_4)CoCO$, 1, 618, 2, 502

$SnCoC_{16}H_{27}O_4$ $Bu_3SnCo(CO)_4$, 2, 502

$SnCoC_{17}H_{16}IO$ Indenyl-$SnI(Me_2C_5H_3)$-$CoCO$, 1, 618, 2, 502

$SnCoC_{30}H_{46}ClO_2$ $EtOC_6H_4SnCl(C_{16}H_{33}$-$C_5H_4)CoCO$, 1, 618*, 2, 502

$SnCoC_{33}H_{42}O_3P$ $Bu_3SnCo(CO)_3PPh_3$, 2, 502

$SnCo_2C_{10}H_6O_8$ $Me_2Sn[Co(CO)_4]_2$, 2, 502

$SnCo_2C_{16}H_{18}O_8$ $Bu_2Sn[Co(CO)_4]_2$, 2, 502

$SnCo_2C_{44}H_{36}O_6P_2$ $Me_2Sn[Co(CO)_3PPh_3]_2$, 2, 502

$SnCo_2C_{50}H_{48}O_6P_2$ $Bu_2Sn[Co(CO)_3PPh_3]_2$, 2, 502

$SnCrC_{12}H_{14}O_3$ $Me_3SnPhCr(CO)_3$, 1, 94, 2, 500

$SnCrC_{26}H_{20}O_3$ $Ph_3SnC_5H_5Cr(CO)_3$, 1, 94, 2, 500

$SnCrC_{26}H_{32}N_6PS_4$ $[Me_3SnCH_2PPh_3][Cr$-$(NH_3)_2(SCN)_4]$, 2, 243

$SnCr_2C_{36}H_{40}O_6$ $(EtPrC_5H_3)_2Sn[C_5H_5Cr$-$(CO)_3]_2$, 1, 94, 2, 500

$SnCr_2C_{48}H_{52}N_{12}P_2S_8$ $[Me_2Sn(CH_2PPh_3)_2]$-$[Cr(NH_3)_2(SCN)_4]_2$, 2, 243

$SnFeC_{12}H_{18}O_4$ $Bu_2SnFe(CO)_4$, 2, 502

$SnFeC_{25}H_{20}O_2$ $Ph_3SnC_5H_5Fe(CO)_2$, 1, 533, 2, 502

$SnFe_2C_{16}H_{16}O_4$ $Me_2Sn[C_5H_5Fe(CO)_2]_2$, 1, 533, 2, 502

$SnFeC_{29}H_{33}O_3P$ $Bu_2SnFe(CO)_3PPh_3$ (?), 2, 502

$SnFe_2C_{18}H_{20}O_4$ $Et_2Sn[C_5H_5Fe(CO)_2]_2$, 1, 533, 2, 502

$SnFe_2C_{26}H_{20}O_4$ $[C_5H_5Fe(CO)_2]_2SnPh_2$, 1, 534

$SnGeC_6H_{18}O$ $Me_3GeOSnMe_3$, 2, 110, 394

$SnGeC_6H_{18}S$ $Me_3SnSGeMe_3$, 2, 124, 473

$SnGeC_6H_{18}Se$ $Me_3SnSeGeMe_3$, 2, 124, 473

$SnGeC_7H_{21}N$ $Me_3SnN(Me)GeMe_3$, 2, 128, 476

$SnGeC_{13}H_{24}$ $p\text{-}Me_3GeCH_2C_6H_4SnMe_3$, 2, 50

$SnGeC_{28}H_{28}$ $Ph_2\overline{SnCH_2CH_2GePh_2CH_2C}H_2$, 2, 50

$SnGeC_{36}H_{30}$ $Ph_3SnGePh_3$, 2, 136, 495

$SnGeC_{36}H_{30}S$ $Ph_3SnSGePh_3$, 2, 124, 473

$SnGeC_{36}H_{30}Se$ $Ph_3SnSeGePh_3$, 2, 124, 473

$SnGeC_{38}H_{34}$ $Ph_3GeCH_2CH_2SnPh_3$, 2, 50

$SnGeC_{44}H_{38}$ $p\text{-}Ph_3GeC_6H_4CH_2CH_2SnPh_3$, 2, 50

$SnGe_2C_{64}H_{56}$ $(p\text{-}Ph_3GeC_6H_4CH_2CH_2)_2SnPh_2$, 2, 50

$SnHfC_{34}H_{28}O_4$ $(o\text{-}MeC_6H_4)_3Sn[C_9H_7Hf$-$(CO)_4]$, 1, 42, 2, 500

$SnHf_2C_{54}H_{84}O_8$ $(C_{16}H_{33})_2Sn[Me_2C_5H_3Hf$-$(CO)_4]_2$, 2, 500

$SnHgC_{21}H_{31}IN_2O_2$ $Pr_3SnO_2CCH_2I\cdot(3\text{-}C_5H_4$-$N)_2Hg$, 2, 425

$SnHgC_{22}H_{26}Br_3P$ $[Me_3SnCH_2PPh_3][HgBr_3]$, 2, 243

$SnHgC_{24}H_{24}O_4$ $Ph_3SnO_2CCH_2CH_2CO_2HgEt$, 2, 425

$SnHgC_{24}H_{37}IN_2O_2$ $Bu_3SnO_2CCH_2I\cdot(3\text{-}C_5H_4$-$N)_2Hg$, 2, 425

$SnHgC_{25}H_{38}INO_2$ $Bu_3SnO_2CCH_2I\cdot 3\text{-}C_5H_4N$-$HgPh$, 2, 425

$SnHgC_{25}H_{39}IN_2O_2$ $Bu_3SnO_2CCH_2CH_2I\cdot(3$-$C_5H_4N)_2Hg$, 2, 425

$SnHgC_{27}H_{43}BrN_2O_2$ $Pr_3SnO_2CCH_2Br\cdot(p\text{-}Me_2$-$NC_6H_4)_2Hg$, 2, 425

$SnHgC_{27}H_{43}IN_2O_2$ $Pr_3SnO_2CCH_2I\cdot(p\text{-}Me_2N$-$C_6H_4)_2Hg$, 2, 425

$SnHgC_{30}H_{47}Cl_2IN_2O_2$ $Bu_3SnO_2CCH_2I\cdot(2,4\text{-}ClMe_2NC_6H_3)_2Hg$, 2, 425
$SnHgC_{30}H_{49}BrN_2O_2$ $Bu_3SnO_2CCH_2Br\cdot(p\text{-}Me_2NC_6H_4)_2Hg$, 2, 425
$SnHgC_{30}H_{49}IN_2O_2$ $Bu_3SnO_2CCH_2I\cdot(p\text{-}Me_2NC_6H_4)_2Hg$, 2, 425
$SnHgC_{32}H_{53}IN_2O_2$ $Bu_3SnO_2CCH_2I\cdot(2,4\text{-}MeC_6H_3NMe_2)_2Hg$, 2, 425
$SnHgC_{35}H_{46}O_6S_2$ $Bu_3SnO_3SC_{10}H_6CH_2C_{10}H_6\text{-}SO_3HgEt$, 2, 352
$SnHgC_{40}H_{40}Br_2Cl_2P_2$ $[Me_2Sn(CH_2PPh_3)_2]\text{-}[HgBr_2Cl_2]$, 2, 243
$SnIrC_{10}H_{14}ClO$ $EtSnCl(EtC_5H_4)IrCO$, 2, 503
$SnIrC_{13}H_{17}O$ $(CH_2{=}CHCH_2)_2Sn(MeC_5H_4)\text{-}IrCO$, 1, 631, 2, 503
$SnIrC_{50}H_{95}O$ $(C_{18}H_{37})_2Sn(C_8H_{17}C_5H_4)\text{-}IrCO$, 1, 631, 2, 503
$SnLiC_3H_9$ Me_3SnLi, 2, 496
$SnLiC_3H_9O$ Me_3SnOLi, 2, 394
$SnLiC_6H_{15}$ Et_3SnLi, 2, 498
$SnLiC_7H_{19}O$ $Me_3SnLi\cdot OEt_2$, 2, 496
$SnLiC_9H_{13}$ $p\text{-}Me_3SnC_6H_4Li$, 2, 254
$SnLiC_{10}H_{15}$ $p\text{-}Me_3SnCH_2C_6H_4Li$, 2, 254
$SnLiC_{12}H_{27}$ Bu_3SnLi, 2, 496
$SnLiC_{18}H_{15}$ Ph_3SnLi, 2, 497
$SnLiC_{18}H_{15}S$ Ph_3SnSLi, 2, 473
$SnLiC_{18}H_{15}Se$ $Ph_3SnSeLi$, 2, 473
$SnLiC_{21}H_{21}$ $(o\text{-}MeC_6H_4)_3SnLi$, 2, 498
$(p\text{-}MeC_6H_4)_3SnLi$, 2, 498
$SnLiC_{22}H_{25}O$ $Ph_3SnLi\cdot OEt_2$, 2, 497
$SnLiC_{24}H_{37}$ $p\text{-}Cyclohexyl_3SnC_6H_4Li$, 2, 254
$SnLi_2C_{12}H_{10}$ Ph_2SnLi_2, 2, 498
$SnLi_2C_{12}H_{10}S_2$ $Ph_2Sn(SLi)_2$, 2, 473
$SnMgC_4H_{11}Cl$ Me_3SnCH_2MgCl, 2, 252
$SnMgC_9H_{13}Br$ $p\text{-}Me_3SnC_6H_4MgBr$, 2, 252
$SnMgC_{10}H_{15}Br$ $p\text{-}Me_3SnCH_2C_6H_4MgBr$, 2, 252
$SnMgC_{12}H_{19}Br$ $p\text{-}Et_3SnC_6H_4MgBr$, 2, 252
$SnMnC_{10}H_{13}O_5$ $MeEt_2SnMn(CO)_5$, 2, 500
$SnMnC_{11}H_5Br_2O_5$ $PhSnBr_2Mn(CO)_5$, 2, 501
$SnMnC_{11}H_5Cl_2O_5$ $PhSnCl_2Mn(CO)_5$, 2, 501
$SnMnC_{11}H_9O_5$ $(CH_2{=}CH)_3SnMn(CO)_5$, 2, 501
$SnMnC_{14}H_{21}O_3$ $MeEt_2SnMn(CO)_3\cdot 1,3\text{-}C_6H_8$, 2, 500
$SnMnC_{17}H_{10}BrO_5$ $Ph_2SnBrMn(CO)_5$, 2, 501
$SnMnC_{17}H_{10}ClO_5$ $Ph_2SnClMn(CO)_5$, 2, 501
$SnMnC_{17}H_{21}O_3$ $(CH_2{=}CH)_3SnMn(CO)_3\cdot 1,3\text{-}C_8H_{12}$, 1, 184, 2, 501
$SnMnC_{17}H_{27}O_5$ $Bu_3SnMn(CO)_5$, 2, 501
$SnMnC_{20}H_{15}O_5$ $(C_5H_5)_3SnMn(CO)_5$, 2, 501*
$SnMnC_{23}H_{15}O_5$ $Ph_3SnMn(CO)_5$, 2, 499
$SnMnC_{23}H_{39}O_3$ $Bu_3SnMn(CO)_3\cdot 1,3\text{-}C_8H_{12}$, 1, 184, 2, 501
$SnMnC_{25}H_{23}O_4$ $(C_5H_5)_3SnMn(CO)_3\cdot\text{-}CH{=}CMeCOCMe{=}CH$, 2, 501*
$SnMnC_{26}H_{21}O_2$ $Ph_3SnMn(CO)_2(C_6H_6)$, 1, 184
$SnMnC_{29}H_{27}O_5$ $(2,4\text{-}Me_2C_6H_3)_3SnMn(CO)_5$, 2, 501
$SnMnC_{32}H_{31}O_4$ $(2,4\text{-}Me_2C_6H_3)_3SnMn(CO)_3\cdot\text{-}C_5H_4O$, 1, 184, 2, 501
$SnMnC_{32}H_{33}O_5$ $Mesityl_3SnMn(CO)_5$, 2, 501
$SnMnC_{35}H_{41}O_3$ $Mesityl_3SnMn(CO)_3\cdot 1,3\text{-}C_5H_8$, 2, 501
$SnMnC_{40}H_{30}O_4P$ $Ph_3SnMn(CO)_4PPh_3$, 2, 501
$SnMnC_{44}H_{27}O_5$ $2\text{-}Fluorenyl_3SnMn(CO)_5$, 2, 501
$SnMnC_{50}H_{35}O_4$ $Ph_3SnMn(CO)_3\cdot C_5Ph_4O$, 1, 184, 2, 501
$SnMnC_{51}H_{39}O_4$ $2\text{-}Fluorenyl_3SnMn(CO)_3\cdot\text{-}C_4Me_4O$, 2, 501*
$SnMn_2C_{12}H_6O_{10}$ $Me_2Sn[Mn(CO)_5]_2$, 2, 501*
$SnMn_2C_{22}H_{10}O_{10}$ $Ph_2Sn[Mn(CO)_5]_2$, 2, 501
$SnMoC_2H_6O_4$ Me_2SnMoO_4, 2, 349*
$SnMoC_{14}H_{10}Cl_2O_3$ $PhSnCl_2(C_5H_5)Mo(CO)_3$, 1, 117, 2, 500
$SnMoC_{26}H_{20}O_3$ $Ph_3Sn(C_5H_5)Mo(CO)_3$, 1, 117, 2, 500
$SnMo_2C_{18}H_{16}O_6$ $Me_2Sn[C_5H_5Mo(CO)_3]_2$, 1, 117, 2, 500
$SnMo_2C_{28}H_{20}O_6$ $Ph_2Sn[C_5H_5Mo(CO)_3]_2$, 1, 117, 2, 500
$SnMo_3C_{86}H_{136}O_{10}$ $AcC_6H_{10}Sn[C_{18}H_{37}C_5H_4\text{-}Mo(CO)_3]_3$, 1, 117*, 2, 500
$SnNaC_2H_7$ Me_2HSnNa, 2, 498
$SnNaC_2H_8N$ $Me_2Sn(NH_2)Na$, 2, 498
$SnNaC_3H_9$ Me_3SnNa, 2, 498
$SnNaC_6H_{15}$ Et_3SnNa, 2, 498

$SnNaC_{12}H_{25}O_2$ $Pr_3SnCH_2CH_2CO_2Na$, 2, 235
$SnNaC_{18}H_{15}$ Ph_3SnNa, 2, 498
$SnNaC_{21}H_{19}O_2$ $Ph_3SnCH_2CH_2CO_2Na$, 2, 237
$SnNaC_{24}H_{19}O$ m-$NaOC_6H_4SnPh_3$, 2, 228
$SnNaC_{35}H_{35}O$ $Ph_2(Ph_3C)SnNa \cdot Et_2O$, 2, 498
$SnNa_2C_2H_6$ Me_2SnNa_2, 2, 496
$SnNa_2C_4H_{10}$ Et_2SnNa_2, 2, 498
$SnNa_2C_{12}H_{10}$ Ph_2SnNa_2, 2, 498
$SnNa_2C_{12}H_{22}O_4$ $Pr_2Sn(CH_2CH_2CO_2Na)_2$, 2, 236
$SnNa_3C_9H_{13}O_7$ $(NaO_2CCH_2CH_2)_3SnOH$, 2, 365
$SnNiC_{10}H_{16}O$ $Me_3Sn(MeC_5H_4)NiCO$, 2, 503
$SnNiC_{17}H_{14}O$ o-$MeC_6H_4SnF_2(C_9H_7)NiCO$, 2, 503
$SnNiC_{18}H_{15}O$ $Ph_2Sn(C_5H_5)NiCO$, 2, 503
$SnNiC_{36}H_{31}FO$ $(C_{13}H_9)_2SnF(Et_2C_5H_3)$-$NiCO$, 2, 503
$SnNi_2C_{20}H_{28}O_2$ $Pr_2Sn(MeC_5H_4NiCO)_2$, 2, 503
$SnPbC_{36}H_{30}S$ $Ph_3SnSPbPh_3$, 2, 473, 585
$SnPbC_{44}H_{38}$ p-$Ph_3SnCH_2CH_2C_6H_4PbPh_3$, 2, 251
$SnPb_2C_{64}H_{56}$ (p-$Ph_3PbC_6H_4CH_2CH_2)_2Sn$-$Ph_2$, 2, 251
$SnPb_4C_{72}H_{60}$ $(Ph_3Pb)_4Sn$, 2, 495, 593
$SnReC_{11}H_5Br_2O_5$ $PhSnBr_2Re(CO)_5$, 2, 502
$SnReC_{23}H_{15}O_5$ $Ph_3SnRe(CO)_5$, 2, 501
$SnRe_2C_{22}H_{10}O_{10}$ $Ph_2Sn[Re(CO)_5]_2$, 2, 502
$SnRe_3C_{21}H_5O_{15}$ $PhSn[Re(CO)_5]_3$, 2, 502
$SnRhC_{12}H_{13}O$ $Me_2SnC_9H_7RhCO$, 2, 503
$SnRhC_{15}H_{12}O$ $C_5H_5SnClC_9H_7RhCO$, 1, 629, 2, 503
$SnRhC_{20}H_{34}FO$ $C_{12}H_{25}SnF(Me_2C_5H_3)RhCO$, 1, 629, 2, 503*
$SnSbC_3H_9F_6$ Me_3SnSbF_6, 2, 348
$SnSbC_{18}H_{25}$ $Et_3SnSbPh_2$, 3, 706
$SnSbC_{21}H_{31}$ $Pr_3SnSbPh_2$, 3, 706
$SnSbC_{24}H_{37}$ $Bu_3SnSbPh_2$, 3, 706
$SnSb_3C_{60}H_{50}N_3$ $PhSn(N{=}SbPh_3)_3$, 2, 481, 3, 752
$SnSiC_5H_{14}Cl_2$ $Me_3SnCH_2SiMeCl_2$, 2, 249
$SnSiC_5H_{14}O$ $Me_3SnCH_2Si({=}O)Me$, 2, 249
$SnSiC_6H_{17}Br$ $Me_2SnBrCH_2SiMe_3$, 2, 301
$SnSiC_6H_{18}O$ $Me_3SnOSiMe_3$, 2, 394
$SnSiC_6H_{18}S$ $Me_3SnSSiMe_3$, 2, 473
$SnSiC_7H_{20}$ $Me_3SnCH_2SiMe_3$, 2, 249
$SnSiC_7H_{20}O_2$ $Me_3SnCH_2SiMe(OMe)_2$, 2, 249
$SnSiC_8H_{19}Cl_3$ $Et_3SnCH_2CH_2SiCl_3$, 2, 250
$SnSiC_9H_{24}O$ $Me_3SnOSiEt_3$, 2, 394
$Et_3SnOSiMe_3$, 2, 394
$SnSiC_{10}H_{26}O$ $Me_3SnO(CH_2)_3SiMe_2Et$, 2, 393
$SnSiC_{11}H_{28}O_3$ $Et_3SnCH_2CH_2Si(OMe)_3$, 2, 250
$SnSiC_{12}H_{22}$ p-$Me_3SnC_6H_4SiMe_3$, 2, 249
$SnSiC_{12}H_{29}Br$ $Bu_2SnBrCH_2SiMe_3$, 2, 301
$SnSiC_{12}H_{30}O$ $Et_3SnOSiEt_3$, 2, 394
$Pr_3SnOSiMe_3$, 2, 394
$SnSiC_{13}H_{24}$ p-$Me_3SnCH_2C_6H_4SiMe_3$, 2, 249
$SnSiC_{14}H_{29}F_3O_2$ $Bu_2(Me_3SiCH_2)SnO_2CCF_3$, 2, 432
$SnSiC_{15}H_{36}$ $Bu_3SnCH_2SiHMe_2$, 2, 250
$Bu_3SnSiMe_3$, 2, 495
$SnSiC_{15}H_{36}O$ $Pr_3SnOSiEt_3$, 2, 394
$Bu_3SnOSiMe_3$, 2, 394
$SnSiC_{17}H_{40}O$ $Bu_3SnCH_2SiMe_2OEt$, 2, 250
$SnSiC_{18}H_{42}$ $Bu_3SnCH_2CHMeCH_2SiHMe_2$, 2, 250
$SnSiC_{22}H_{26}$ $Ph_3SnCH_2SiMe_3$, 2, 250
$SnSiC_{23}H_{44}O$ i-$Bu_3SnOSiMe(CH_2CHMe_2)Ph$, 2, 394
$SnSiC_{27}H_{28}$ p-$Ph_3SnC_6H_4SiMe_3$, 2, 251
$SnSiC_{28}H_{28}$ $Ph_2SnCH_2CH_2SiPh_2CH_2CH_2$, 2, 251
$SnSiC_{36}H_{30}$ $Ph_3SnSiPh_3$, 2, 495
$SnSiC_{36}H_{30}O$ $Ph_3SnOSiPh_3$, 2, 394
$SnSiC_{38}H_{34}$ $Ph_3SnCH_2CH_2SiPh_3$, 2, 250
$SnSi_2C_8H_{22}Br_2$ $(Me_3SiCH_2)_2SnBr_2$, 2, 327
$SnSi_2C_8H_{22}I_2$ $(Me_3SiCH_2)_2SnI_2$, 2, 327
$SnSi_2C_8H_{22}O$ $Me_2Sn(CH_2SiMe_2)_2O$, 2, 249
$(Me_3SiCH_2)_2SnO$, 2, 374
$[SnSi_2C_8H_{22}O]_n$ $[Me_2Sn(CH_2SiMe_2)_2O]_n$, 2, 249
$(SnSi_2C_8H_{22}S)_n$ $[(Me_3SiCH_2)_2SnS]_n$, 2, 454*

$SnSi_2C_8H_{24}O_2$ $Me_2Sn(OSiMe_3)_2$, 2, 406
$SnSi_2C_9H_{24}$ $Me_2Sn(CH_2SiHMe_2)_2$, 2, 249
$SnSi_2C_9H_{25}Br$ $Me(Me_3SiCH_2)_2SnBr$, 2, 301
$SnSi_2C_9H_{25}I$ $Me(Me_3SiCH_2)_2SnI$, 2, 301
$SnSi_2C_9H_{27}N$ $Me_3SnN(SiMe_3)_2$, 2, 476
$SnSi_2C_{10}H_{28}$ $Me_2Sn(CH_2SiMe_3)_2$, 2, 249
$SnSi_2C_{12}H_{28}$ $(CH_2{=}CH)_2Sn(CH_2SiMe_3)_2$, 2, 250
$SnSi_2C_{12}H_{30}O$ $Me_2Sn(CH_2CH_2CH_2Me_2Si)_2O$, 2, 250
$SnSi_2C_{12}H_{31}I$ $Bu(Me_3SiCH_2)_2SnI$, 2, 301
$SnSi_2C_{12}H_{32}$ $Me_2Sn(CH_2CH_2CH_2SiHMe_2)_2$, 2, 250
$SnSi_2C_{12}H_{32}O_2$ $Me_2Sn(CH_2SiMe_2OEt)_2$, 2, 249
$SnSi_2C_{14}H_{34}O$ $Bu_2Sn(CH_2SiMe_2)_2O$, 2, 250
$SnSi_2C_{14}H_{36}$ $Me_2Sn(CH_2CHMeCH_2SiHMe_2)_2$, 2, 250
$Bu_2Sn(CH_2SiHMe_2)_2$, 2, 250
$SnSi_2C_{14}H_{36}O_2$ $Bu_2Sn(OSiMe_3)_2$, 2, 406
$SnSi_2C_{16}H_{40}$ $Bu_2Sn(CH_2SiMe_3)_2$, 2, 250
$SnSi_2C_{18}H_{28}$ $Me_2Sn(CH_2SiHMePh)_2$, 2, 249
$SnSi_2C_{18}H_{28}O_2$ $Ph_2Sn(OSiMe_3)_2$, 2, 406
$SnSi_2C_{18}H_{44}O_2$ $Bu_2Sn(CH_2Me_2SiOEt)_2$, 2, 250
$SnSi_2C_{18}H_{45}N$ $Bu_3SnN(SiMe_3)_2$, 2, 476
$SnSi_2C_{19}H_{45}Cl$ $Bu_3SnCH_2SiMe_2CH_2CH_2SiMe_2Cl$, 2, 250
$SnSi_2C_{20}H_{32}$ $Ph_2Sn(CH_2SiMe_3)_2$, 2, 251
$SnSi_2C_{26}H_{26}Cl_2O_2$ $Me_2Sn(OSiPh_2Cl)_2$, 2, 406
$SnSi_2C_{30}H_{36}$ $(p\text{-}Me_3SiC_6H_4)_2SnPh_2$, 2, 251
$SnSi_2C_{30}H_{52}O_2$ $Bu_2Sn(OSiMePhCH_2CHMe_2)_2$, 2, 406
$SnSi_2C_{38}H_{36}O_2$ $Me_2Sn(OSiPh_3)_2$, 2, 406
$SnSi_2C_{44}H_{48}O_2$ $Bu_2Sn(OSiPh_3)_2$, 2, 406
$SnSi_2C_{48}H_{40}O_2$ $Ph_2Sn(OSiPh_3)_2$, 2, 406
$SnSi_2C_{50}H_{44}O_2$ $(PhCH_2)_2Sn(OSiPh_3)_2$, 2, 406
$SnSi_3C_{10}H_{30}O_3$ $MeSn(OSiMe_3)_3$, 2, 408
$SnSi_3C_{12}H_{33}I$ $(Me_3SiCH_2)_3SnI$, 2, 301
$SnSi_4C_{16}H_{44}$ $(Me_3SiCH_2)_4Sn$, 2, 175
$SnSi_4C_{36}H_{52}$ $(p\text{-}Me_3SiC_6H_4)_4Sn$, 2, 175
$SnTiC_{14}H_{18}O_4$ $Me_3Sn(EtC_5H_{11})Ti(CO)_4$, 2, 500
$SnTi_2C_2H_5Cl_{11}$ $EtSnCl_3\cdot 2\ TiCl_4$, 2, 329
$SnTi_2C_4H_{10}Cl_{10}$ $Et_2SnCl_2\cdot 2\ TiCl_4$, 2, 307
$SnTi_2C_6H_{15}Cl_9$ $Et_3SnCl\cdot 2\ TiCl_4$, 2, 274
$SnTi_2C_8H_{20}Cl_4$ $Et_4Sn\cdot 2\ TiCl_4$, 2, 162
$SnTi_2C_{32}H_{24}O_8$ $(PhCH_2)_2Sn[C_5H_5Ti(CO)_4]_2$, 2, 500
$SnVC_{18}H_{15}O_3$ Ph_3SnVO_3, 2, 355
$SnV_2C_2H_6O_6$ $Me_2Sn(VO_3)_2$, 2, 349
$SnWC_2H_6O_4$ Me_2SnWO_4, 2, 349
$SnWC_8H_5Cl_3O_3$ $C_5H_5W(CO)_3SnCl_3$, 1, 129
$SnWC_{14}H_{10}Cl_2O_3$ $PhSnCl_2(C_5H_5)W(CO)_3$, 1, 129, 2, 500
$SnWC_{26}H_{20}O_3$ $Ph_3Sn(C_5H_5)W(CO)_3$, 1, 129, 2, 500
$SnW_2C_{28}H_{20}O_6$ $Ph_2Sn[C_5H_5W(CO)_3]_2$, 1, 130, 2, 500
$SnZr_3C_{41}H_{44}O_{12}$ $n\text{-}C_8H_{17}Sn[Me_2C_5H_3Zr(CO)_4]_3$, 1, 41, 2, 500
$Sn_2AsC_{42}H_{35}O_3$ $(Ph_3Sn)_2O_3AsPh$, 2, 355, 3, 549
$Sn_2BC_{14}H_{36}N$ $(Et_3Sn)_2BNMe_2$, 2, 495
$Sn_2CrC_4H_{12}O_5$ $Me_2SnCrO_4\cdot Me_2SnO$, 2, 349
$Sn_2CrC_{36}H_{30}O_4$ $(Ph_3Sn)_2CrO_4$, 2, 356
$Sn_2Cr_2C_{36}H_{30}O_7$ $(Ph_3Sn)_2Cr_2O_7$, 2, 356
$Sn_2FeC_{28}H_{54}O_4$ $(Bu_3Sn)_2Fe(CO)_4$, 2, 502
$Sn_2FeC_{40}H_{30}O_4$ $(Ph_3Sn)_2Fe(CO)_4$, 2, 502
$Sn_2Fe_2C_{12}H_{12}O_8$ $[Me_2SnFe(CO)_4]_2$, 2, 502
$Sn_2Fe_2C_{16}H_{20}O_8$ $[Et_2SnFe(CO)_4]_2$, 2, 502
$Sn_2Fe_2C_{24}H_{36}O_8$ $[Bu_2SnFe(CO)_4]_2$, 2, 502
$Sn_2Fe_2C_{36}H_{28}O_8$ $[(PhCH_2)_2SnFe(CO)_4]_2$, 2, 502
$Sn_2Fe_2C_{38}H_{30}O_4$ $[Ph_2Sn(C_5H_5)Fe(CO)_2]_2$, 2, 502
$Sn_2Fe_2C_{46}H_{34}O_4$ $[Ph_2Sn(C_9H_7)Fe(CO)_2]_2$, 2, 502
$Sn_2GeC_{52}H_{48}$ $(Ph_3SnCH_2CH_2)_2GePh_2$, 2, 50
$Sn_2Ge_2C_{62}H_{62}$ $(p\text{-}Ph_3GeC_6H_4CH_2CH_2SnMe_2)_2C_6H_4$, 2, 50
$Sn_2Ge_2C_{70}H_{62}$ $p\text{-}(Ph_3SnCH_2CH_2GePh_2)_2C_6H_4$, 2, 50
$Sn_2HgC_{26}H_{42}I_2N_2O_4$ $(Me_3SnO_2CCH_2I\cdot Me_2N\text{-}C_6H_4\text{-}p)_2Hg$, 2, 425
$Sn_2HgC_{30}H_{59}O_4P$ $(Bu_3Sn)_2PhHgPO_4$, 2, 351

$Sn_2MgC_{36}H_{30}$ $(Ph_3Sn)_2Mg$, 2, 495
$Sn_2MoC_{36}H_{30}O_4$ $(Ph_3Sn)_2MoO_4$, 2, 356
$Sn_2NaC_{10}H_{25}$ Et_5Sn_2Na, 2, 498
$Sn_2Na_2C_4H_{12}$ $(Me_2SnNa)_2$, 2, 498
$Sn_2Na_2C_8H_{20}$ $(Et_2SnNa)_2$, 2, 498
$Sn_2Nb_2C_{72}H_{94}O_6$ $[(2,4\text{-}Me_2C_6H_3)_2Sn(Bu\text{-}i\text{-}C_8H_{17}C_5H_3)Nb(CO)_3]_2$, 2, 500
$Sn_2PbC_{34}H_{40}N_2O_8$ $[(CH_2{=}CH)_3SnO_2CCH_2]_2\text{-}NCH_2CH_2N(CH_2CO_2)_2PbPh_2$, 2, 429
$Sn_2PbC_{64}H_{56}$ $(p\text{-}Ph_3SnCH_2CH_2C_6H_4)_2Pb\text{-}Ph_2$, 2, 251
$Sn_2Pb_2C_{62}H_{62}$ $p\text{-}(p\text{-}Ph_3PbC_6H_4CH_2CH_2\text{-}Me_2Sn)_2C_6H_4$, 2, 251
$Sn_2SiC_8H_{24}O_2$ $(Me_3SnO)_2SiMe_2$, 2, 394
$Sn_2SiC_{28}H_{64}O_2$ $(i\text{-}Bu_3SnO)_2SiEt_2$, 2, 394*
$Sn_2SiC_{48}H_{40}O_2$ $(Ph_3SnO)_2SiPh_2$, 2, 394
$Sn_2SiC_{52}H_{48}$ $(Ph_3SnCH_2CH_2)_2SiPh_2$, 2, 251
$Sn_2Si_2C_{10}H_{30}O_3$ $(Me_2SnOSiMe_3)_2O$, 2, 406
$Sn_2Si_2C_{14}H_{38}O_3$ $(Et_2SnOSiMe_3)_2O$, 2, 406
$Sn_2Si_2C_{18}H_{46}O_3$ $(Pr_2SnOSiMe_3)_2O$, 2, 406
$Sn_2Si_2C_{22}H_{54}O_3$ $(Bu_2SnOSiMe_3)_2O$, 2, 406
$Sn_2Si_2C_{30}H_{70}O$ $(Bu_3SnCH_2Me_2Si)_2O$, 2, 250
$Sn_2Si_4C_{38}H_{90}O$ $(Bu_3SnCH_2SiMe_2CH_2CH_2\text{-}Me_2Si)_2O$, 2, 250
$Sn_2Ta_2C_{84}H_{118}O_6$ $[Mesityl_2Sn(C_{16}H_{33}\text{-}C_5H_4)Ta(CO)_3]_2$, 1, 54*, 2, 500
$Sn_2TeC_{36}H_{30}$ $(Ph_3Sn)_2Te$, 2, 452
$Sn_2TeC_{36}H_{30}O_3$ $(Ph_3Sn)_2TeO_3$, 2, 356
$Sn_2V_2C_{38}H_{34}O_6$ $[(CH_2{=}CHCH_2)_2SnC_9H_7V\text{-}(CO)_3]_2$, 2, 500
$Sn_2V_2C_{50}H_{82}O_6$ $[(C_8H_{17})_2Sn(MeC_5H_4)V\text{-}(CO)_3]_2$, 2, 500
$Sn_2V_2C_{72}H_{126}O_6$ $[(n\text{-}C_{12}H_{25})_2Sn(Et_2\text{-}C_5H_3)V(CO)_3]_2$, 1, 52, 2, 500
$Sn_2WC_{36}H_{30}O_4$ $(Ph_3Sn)_2WO_4$, 2, 356
$Sn_3AsC_6H_{18}O_4$ $(Me_2Sn)_3AsO_4$, 2, 348*
$Sn_3AsC_{54}H_{45}$ $(Ph_3Sn)_3As$, 2, 484
$Sn_3AsC_{54}H_{45}O_4$ $(Ph_3Sn)_3AsO_4$, 2, 355
$Sn_3BC_{12}H_{33}$ $(Me_3SnCH_2)_3B$, 2, 253
$Sn_3B_3C_{42}H_{72}N_3$ cyclic$(Et_3SnCH_2CH_2NB\text{-}Ph)_3$, 2, 253
$Sn_3B_3C_{78}H_{72}N_3$ cyclic$(Ph_3SnCH_2CH_2NB\text{-}Ph)_3$, 2, 253
$Sn_3BiC_{54}H_{45}$ $(Ph_3Sn)_3Bi$, 2, 484
$Sn_3FeC_{12}H_{18}N_6O$ $(Me_2Sn)_2Fe(CN)_6\cdot Me_2\text{-}SnO$, 2, 345
$Sn_3FeC_{60}H_{45}N_6$ $(Ph_3Sn)_3Fe(CN)_6$, 2, 346
$Sn_3Fe_2C_{18}H_6N_{12}$ $(Me_2Sn)_3[Fe(CN)_6]_2$, 2, 345*
$Sn_3SbC_{54}H_{45}$ $(Ph_3Sn)_3Sb$, 2, 484
$Sn_3Te_3C_6H_{18}$ $(Me_2SnTe)_3$, 2, 455
$Sn_3VC_{54}H_{45}O_4$ $(Ph_3Sn)_3VO_4$, 2, 355
$Sn_4FeC_{78}H_{60}N_6$ $(Ph_3Sn)_4Fe(CN)_6$, 2, 346
$Sn_4GeC_{72}H_{60}$ $(Ph_3Sn)_4Ge$, 2, 136, 495
$Sn_4PbC_{72}H_{60}$ $(Ph_3Sn)_4Pb$, 2, 495, 593
$Sn_4TiC_{72}H_{60}O_4$ $(Ph_3SnO)_4Ti$, 2, 394
$SnAsMnC_{10}H_{18}O_4$ $Me_3SnMn(AsMe_3)(CO)_4$, 3, 4
$SnAsMnC_{21}H_{21}ClO_4$ $Me_2Sn(Cl)Mn[As\text{-}(C_5H_5)_3](CO)_4$, 3, 17
$SnAsMnC_{40}H_{30}O_4$ $Ph_3SnMn(CO)_4AsPh_3$, 2, 501, 3, 23
$SnAs_2Mn_2C_{30}H_{52}O_8$ $Et_2Sn[Mn(AsPr_3)\text{-}(CO)_4]_2$, 3, 15
$SnAs_2Mn_2C_{78}H_{84}O_8$ $(2,5\text{-}Me_2C_6H_3)_2Sn\text{-}\{Mn[As(C_6H_2Me_3)_3](CO)_4\}_2$, 3, 29
$SnAs_3Mn_3C_{81}H_{68}O_{12}$ $PhSn\{Mn[As(C_6H_4\text{-}Me\text{-}2)_3](CO)_4\}_3$, 3, 27
$SnHgNaC_{38}H_{65}O_6$ $Bu_3SnO_2CC_{23}H_{33}O_4NaHgEt$, 2, 425
$SnLiTeC_{18}H_{15}$ $Ph_3SnTeLi$, 2, 473
$SnMnSbC_{10}H_{15}Br_3O_4$ $Br_3SnMn(CO)_4(SbEt_3)$, 3, 656
$SnMnSbC_{37}H_{72}O_4$ $Pr_3SnMn(CO)_4[Sb\text{-}(C_8H_{17})_3]$, 3, 667
$SnMnSbC_{41}H_{43}BrO_4$ $Br(C_5H_5)_2SnMn(CO)_4\text{-}[Sb(C_6H_4CHMe_2)_3]$, 3, 667
$SnMn_3Sb_3C_{55}H_{88}O_{12}$ $PhCH_2Sn\{Mn(CO)_4\text{-}[Sb(CH_2CHMe_2)_3]\}_3$, 3, 659
$SnMn_3Sb_3C_{69}H_{50}O_{12}$ $CH_2{=}CHCH_2Sn[Mn(Sb\text{-}Ph_3)(CO)_4]_3$, 3, 663
$SnPbTeC_{36}H_{30}$ $Ph_3PbTeSnPh_3$, 2, 473, 585
$TaC_9H_5O_4$ $C_5H_5Ta(CO)_4$, 1, 53
$TaC_{10}H_{10}$ $(C_5H_5)_2Ta$, 1, 54
$TaC_{10}H_{10}Br_3$ $(C_5H_5)_2TaBr_3$, 1, 54
$TaC_{10}H_{10}Cl_3$ $(C_5H_5)_2TaCl_3$, 1, 54

$TaC_{10}H_{13}$ $(C_5H_5)_2TaH_3$, 1, 54
$TaC_{14}H_{10}Cl_5$ Anthracene-$TaCl_5$, 1, 54
$TaC_{14}H_{14}F_3$ $(CH_2{=}CHC_5H_4)_2TaF_3$, 1, 54
$TaC_{20}H_{20}$ $(C_5H_5)_4Ta$, 1, 54
$Ta_2Ge_2C_{76}H_{110}O_6$ $[(n\text{-}C_{18}H_{37}C_5H_4)Ta(CO)_3Ge(C_5H_4Me)_2]_2$, 1, 54, 2, 139
$Ta_2Pb_2C_{44}H_{38}O_6$ $[(MeC_6H_4)_2Pb(C_5H_5)Ta(CO)_3]_2$, 2, 594
$Ta_2Si_2C_{104}H_{190}O_6$ $[(Bu_2C_5H_3)Ta(CO)_3Si(C_{18}H_{37})_2]_2$, 1, 54
$Ta_2Sn_2C_{84}H_{118}O_6$ $[(C_{16}H_{33}C_5H_4)Ta(CO)_3Sn(C_6H_2Me_3)_2]_2$, 1, 54*
$TcC_8H_5O_3$ $C_5H_5Tc(CO)_3$, 1, 185
$TcC_{12}H_{12}F_6P$ $[(C_6H_6)_2Tc]PF_6$, 1, 186
$TcC_{15}H_9O_4$ $(BzC_5H_4)Tc(CO)_3$, 1, 185
$TcC_{15}H_{17}Cl_3$ $(Me_2CHC_6H_5)Tc(C_5H_5CCl_3)$, 1, 186
$Tc_2C_{20}H_{20}$ $[(C_5H_5)_2Tc]_2$, 1, 185
$TcBC_{36}H_{32}$ $[(C_6H_6)_2Tc]BPh_4$, 1, 186
$TeSn_2C_{36}H_{30}$ $(Ph_3Sn)_2Te$, 2, 452
$Te_3Sn_3C_6H_{18}$ $(Me_2SnTe)_3$, 2, 455
$TeLiSnC_{18}H_{15}$ $Ph_3SnTeLi$, 2, 473
$TePbSnC_{36}H_{30}$ $Ph_3PbTeSnPh_3$, 2, 473, 585
ThC_8H_{20} Et_4Th, 1, 1
$ThC_8H_{20}O$ Et_3ThOEt, 1, 1
$ThC_8H_{20}O_2$ $Et_2Th(OEt)_2$, 1, 1
$ThC_8H_{20}O_3$ $EtTh(OEt)_3$, 1, 1
$ThC_{20}H_{20}$ $(C_5H_5)_4Th$, 1, 4
$TiCH_3Br_3$ $MeTiBr_3$, 1, 7
$TiCH_3Cl$ MeTiCl, 1, 5
$TiCH_3Cl_3$ $MeTiCl_3$, 1, 6
TiC_2H_5Cl EtTiCl, 1, 5
$TiC_2H_5Cl_2$ $EtTiCl_2$, 1, 5
$TiC_2H_5Cl_3$ $EtTiCl_3$, 1, 7
TiC_2H_6Cl Me_2TiCl, 1, 6
$TiC_2H_6Cl_2$ Me_2TiCl_2, 1, 8
$TiC_3H_7Cl_2$ i-Pr$TiCl_2$, 1, 5
TiC_3H_9 Me_3Ti, 1, 6
TiC_3H_9Cl Me_3TiCl, 1, 9
$TiC_4H_9Cl_3$ $BuTiCl_3$, 1, 7
$TiC_4H_{10}Cl_2$ Et_2TiCl_2, 1, 8
TiC_4H_{12} Me_4Ti, 1, 10
$TiC_5H_4Br_2$ Cyclopentadienylene-$TiBr_2$, 1, 11
$TiC_5H_4Cl_2$ Cyclopentadienylene-$TiCl_2$, 1, 11
$TiC_5H_4F_2$ Cyclopentadienylene-TiF_2, 1, 11
$TiC_5H_5BrCl_2$ $C_5H_5TiBrCl_2$, 1, 13
$TiC_5H_5Br_3$ $C_5H_5TiBr_3$, 1, 12
$(TiC_5H_5ClO)_n$ Chlorocyclopentadienyldihydroxytitanium oligomers, 1, 15
$TiC_5H_5Cl_2$ $C_5H_5TiCl_2$, 1, 11
$TiC_5H_5Cl_3$ $C_5H_5TiCl_3$, 1, 12
$TiC_5H_5I_3$ $C_5H_5TiI_3$, 1, 13
$(TiC_5H_5O_2)_n$ $[C_5H_5Ti(O)O]_n$, 1, 28
TiC_6H_5Cl PhTiCl, 1, 5
$TiC_6H_5Cl_3$ $PhTiCl_3$, 1, 7
$TiC_6H_7Cl_3$ $(MeC_5H_4)TiCl_3$, 1, 13
$TiC_6H_8Cl_2O$ $C_5H_5(MeO)TiCl_2$, 1, 14
$TiC_7H_7Br_3O$ $(AcC_5H_4)TiBr_3$, 1, 13
$TiC_7H_7Cl_5$ Cycloheptatrienium$[TiCl_5]^-$, 1, 31
$TiC_7H_{10}Cl_2O$ $C_5H_5(EtO)TiCl_2$, 1, 14
$TiC_7H_{11}Cl_3O_4$ $(EtO_2C)_2CHTiCl_3$, 1, 7
$TiC_8H_{12}Br_2O_2$ $C_5H_5(MeOCH_2CH_2O)TiBr_2$, 1, 14
$TiC_8H_{12}Cl_2O$ $C_5H_5(PrO)TiCl_2$, 1, 14
TiC_8H_{14} $C_5H_5TiMe_3$, 1, 11
$TiC_8H_{17}Cl_2$ $C_8H_{17}TiCl_2$, 1, 6
TiC_8H_{18} Bu_2Ti, 1, 5
$TiC_8H_{18}Cl_2$ Bu_2TiCl_2, 1, 8
TiC_8H_{20} Et_4Ti, 1, 10
$TiC_8H_{20}O_2$ $Me_2(i\text{-}PrO)_2Ti$, 1, 9
$TiC_9H_6BrNO_4$ $(C_6H_6)Ti(CO)_3(NO)Br$, 1, 31
$TiC_9H_6O_3$ $(C_6H_6)Ti(CO)_3$, 1, 31
$TiC_9H_{14}Cl_2O$ $C_5H_5(Me_3CO)TiCl_2$, 1, 14
$TiC_{10}H_{10}$ $(C_5H_5)_2Ti$, 1, 17
$TiC_{10}H_{10}Br$ $(C_5H_5)_2TiBr$, 1, 19
$TiC_{10}H_{10}BrCl$ $(C_5H_5)_2TiBrCl$, 1, 26
$TiC_{10}H_{10}Br_2$ $(C_5H_5)_2TiBr_2$. 1, 25
$TiC_{10}H_{10}Br_2Cl_2$ $(C_5H_5)(C_5H_5Br_2)TiCl_2$, 1, 28
$TiC_{10}H_{10}Br_2Cl_4$ $(C_5H_5)(C_5H_5Cl_4)TiBr_2$, 1, 28
$TiC_{10}H_{10}Br_2I_4$ $(C_5H_5)(C_5H_5I_4)TiBr_2$, 1, 28
$TiC_{10}H_{10}Br_4Cl_2$ $(C_5H_5)(C_5H_5Br_4)TiCl_2$, 1, 28
$TiC_{10}H_{10}Br_4F_2$ $(C_5H_5)(C_5H_5Br_4)TiF_2$, 1, 28
$TiC_{10}H_{10}Br_6$ $(C_5H_5)(C_5H_5Br_4)TiBr_2$, 1, 28

$TiC_{10}H_{10}Cl$ $(C_5H_5)_2TiCl$, 1, 18
$TiC_{10}H_{10}Cl_2$ $(C_5H_5)_2TiCl_2$, 1, 22
$TiC_{10}H_{10}Cl_2I_4$ $(C_5H_5)(C_5H_5I_4)TiCl_2$, 1, 28
$TiC_{10}H_{10}Cl_4$ $(C_5H_5)(C_5H_5Cl_2)TiCl_2$, 1, 28
$TiC_{10}H_{10}Cl_6$ $(C_5H_5)(C_5H_5Cl_4)TiCl_2$, 1, 28
$TiC_{10}H_{10}F_2$ $(C_5H_5)_2TiF_2$, 1, 22
$TiC_{10}H_{10}I$ $(C_5H_5)_2TiI$, 1, 19
$TiC_{10}H_{10}I_2$ $(C_5H_5)_2TiI_2$, 1, 25
$TiC_{10}H_{10}I_6$ $(C_5H_5)(C_5H_5I_4)TiI_2$, 1, 28
$TiC_{10}H_{11}BrO$ $(C_5H_5)_2Ti(OH)Br$, 1, 27
$TiC_{10}H_{15}Cl_3$ $(Me_5C_5)TiCl_3$, 1, 13
$TiC_{10}H_{16}O_2$ $[(Me_5C_5)Ti(OH)O]_n$, 1, 16
$TiC_{11}H_{10}$ C_5H_5TiPh, 1, 10
$TiC_{11}H_{11}NO_4$ $(PrC_5H_4)Ti(CO)_3NO$, 1, 17
$TiC_{11}H_{12}Cl_2$ $(C_5H_5)(MeC_5H_4)TiCl_2$, 1, 26
$TiC_{11}H_{13}Cl$ $(C_5H_5)_2MeTiCl$, 1, 21
$TiC_{11}H_{19}ClO_2$ $C_5H_5(PrO)_2TiCl$, 1, 14
$TiC_{11}H_{20}O_3$ $C_5H_5Ti(OEt)_3$, 1, 15
$TiC_{12}H_{10}$ Ph_2Ti, 1, 5
$TiC_{12}H_{10}Br_2$ Ph_2TiBr_2, 1, 9
$TiC_{12}H_{10}Cl_2$ Ph_2TiCl_2, 1, 9
$TiC_{12}H_{10}O_2$ $(C_5H_5)_2Ti(CO)_2$, 1, 30
$TiC_{12}H_{11}Br_3$ $(PhCH_2C_5H_4)TiBr_3$, 1, 13
$TiC_{12}H_{12}$ $(C_6H_6)_2Ti$, 1, 31
$TiC_{12}H_{14}Cl_2$ $(MeC_5H_4)_2TiCl_2$, 1, 26
$(C_5H_5)(EtC_5H_4)TiCl_2$, 1, 27
$TiC_{12}H_{15}$ $(C_5H_5)_2TiEt$, 1, 19
$TiC_{12}H_{15}Cl$ $(C_5H_5)_2EtTiCl$, 1, 21
$TiC_{12}H_{15}ClO$ $(C_5H_5)_2Ti(OEt)Cl$, 1, 27
$TiC_{12}H_{16}$ $(C_5H_5)_2TiMe_2$, 1, 20
$TiC_{12}H_{16}S_2$ $(C_5H_5)_2Ti(SMe)_2$, 1, 30
$TiC_{12}H_{27}Cl$ Bu_3TiCl, 1, 9
$TiC_{12}H_{28}$ Et_2Bu_2Ti, 1, 10
$TiC_{12}H_{28}O_2$ $Bu_2Ti(OEt)_2$, 1, 9
$TiC_{13}H_{15}$ $(C_5H_5)_2TiCH_2CH{=}CH_2$, 1, 19
$TiC_{13}H_{15}ClO$ $(C_5H_5)_2Ti(OCH_2CH{=}CH_2)Cl$, 1, 27
$TiC_{13}H_{17}Cl$ $(C_5H_5)_2PrTiCl$, 1, 21
$TiC_{13}H_{23}BrO_2$ $C_5H_5(BuO)_2TiBr$, 1, 15
$TiC_{13}H_{23}ClO_2$ $C_5H_5(BuO)_2TiCl$, 1, 15
$TiC_{14}H_{10}Cl_6O_4$ $(C_5H_5)_2Ti(O_2CCCl_3)_2$, 1, 29
$TiC_{14}H_{10}F_6O_4$ $(C_5H_5)_2Ti(O_2CCF_3)_2$, 1, 29
$TiC_{14}H_{12}Cl_4O_4$ $(C_5H_5)_2Ti(O_2CCHCl_2)_2$, 1, 29
$TiC_{14}H_{14}Cl_2O_4$ $(C_5H_5)_2Ti(O_2CCH_2Cl)_2$, 1, 29
$TiC_{14}H_{14}F_6O_2$ $(C_5H_5)_2Ti(OCH_2CF_3)_2$, 1, 29
$TiC_{14}H_{16}O_4$ $(C_5H_5)_2Ti(OAc)_2$, 1, 29
$TiC_{14}H_{16}O_4S_2$ $(C_5H_5)_2Ti(O_2CCH_2SH)_2$, 1, 29
$TiC_{14}H_{16}O_6$ $(C_5H_5)_2Ti(O_2CCH_2OH)_2$, 1, 29
$TiC_{14}H_{20}$ $(MeC_5H_4)_2TiMe_2$, 1, 20
$TiC_{14}H_{26}O_3$ $C_5H_5Ti(OPr)_3$, 1, 16
$TiC_{14}H_{32}$ $EtBu_3Ti$, 1, 10
$TiC_{15}H_{15}$ $(C_5H_5)_3Ti$, 1, 30
$TiC_{15}H_{16}Br_2$ $C_5H_5(C_5H_7C_5H_4)TiBr_2$, 1, 27
$TiC_{15}H_{16}Cl_2$ $C_5H_5(C_5H_7C_5H_4)TiCl_2$, 1, 27
$TiC_{15}H_{16}F_2$ $C_5H_5(C_5H_7C_5H_4)TiF_2$, 1, 27
$TiC_{15}H_{16}I_2$ $C_5H_5(C_5H_7C_5H_4)TiI_2$, 1, 27
$TiC_{15}H_{23}Cl_3O_4$ $C_5H_5(BuO)_2TiO_2CCCl_3$, 1, 16
$TiC_{15}H_{26}O_3$ $Ph(i\text{-}PrO)_3Ti$, 1, 8
$TiC_{15}H_{34}O_3$ $i\text{-}PrTi(OBu)_3$, 1, 7
$TiC_{16}H_{10}ClF_5$ $(C_5H_5)_2(C_6F_5)TiCl$, 1, 21
$TiC_{16}H_{11}F_5O$ $(C_5H_5)_2(C_6F_5)TiOH$, 1, 22
$TiC_{16}H_{12}N_3O_7$ $(C_5H_5)_2Ti$ picrate, 1, 19
$TiC_{16}H_{15}ClO$ $(C_5H_5)_2Ti(OPh)Cl$, 1, 27
$TiC_{16}H_{18}Cl_2$ $(CH_2{=}CHCH_2C_5H_4)_2TiCl_2$, 1, 26
$TiC_{16}H_{20}O_3$ $(C_5H_5)_2(EtO)TiO_2CCMe{=}CH_2$, 1, 29
$TiC_{16}H_{36}$ Bu_4Ti, 1, 10
$TiC_{17}H_{25}BrO_2$ $(BuO)_2$(indenyl)-TiBr, 1, 15
$TiC_{17}H_{32}O_3$ $C_5H_5Ti(OBu)_3$, 1, 16
$TiC_{18}H_{10}F_{14}O_4$ $(C_5H_5)_2Ti(O_2CCF_2CF_2CF_3)_2$, 1, 29
$TiC_{18}H_{14}Br$ Diindenyl-TiBr, 1, 19
$TiC_{18}H_{14}Br_2$ Diindenyl-$TiBr_2$, 1, 26
$TiC_{18}H_{14}Cl$ Diindenyl-TiCl, 1, 19
$TiC_{18}H_{14}Cl_2$ Diindenyl-$TiCl_2$, 1, 26
$TiC_{18}H_{14}I$ Diindenyl-TiI, 1, 19
$TiC_{18}H_{14}I_2$ Diindenyl-TiI_2, 1, 26
$TiC_{18}H_{15}Br_2ClO$ $(PhCH_2C_5H_4)Ti(OC_6H_4Cl)$-$Br_2$, 1, 14
$TiC_{18}H_{15}F_5O$ $(C_5H_5)_2(C_6F_5)TiOEt$, 1, 22

$TiC_{18}H_{20}O_4$ $(C_5H_5)_2Ti(O_2CCMe{=}CH_2)_2$, 1, 29
$TiC_{19}H_{34}O_3$ p-$MeC_6H_4Ti(OBu)_3$, 1, 8
$TiC_{19}H_{34}O_4$ p-$MeOC_6H_4Ti(OBu)_3$, 1, 8
$TiC_{20}H_{20}$ $(C_5H_5)_4Ti$, 1, 31
$TiC_{20}H_{28}O_2$ $Ph_2Ti(OBu)_2$, 1, 9
$TiH_{21}H_{34}O_3$ Indenyl-$Ti(OBu)_3$, 1, 16
$TiC_{22}H_{14}Cl_6O_4$ Diindenyl-$Ti(O_2CCCl_3)_2$, 1, 30
$TiC_{22}H_{18}Cl_2N_4$ $(PhN{=}NC_5H_4)_2TiCl_2$, 1, 26
$TiC_{22}H_{20}$ $(C_5H_5)_2TiPh_2$, 1, 20
$TiC_{22}H_{20}S_2$ $(C_5H_5)_2Ti(SPh)_2$, 1, 30
$TiC_{22}H_{34}O_3$ α-$C_{10}H_7Ti(OBu)_3$, 1, 8
$TiC_{23}H_{17}ClN_2O_2$ C_5H_5(8-quinolinato)$_2$-TiCl, 1, 16
$TiC_{23}H_{20}$ $C_5H_5TiPh_3$, 1, 11
$TiC_{24}H_{20}$ Ph_4Ti, 1, 10
$TiC_{24}H_{20}O_4$ $(C_5H_5)_2Ti(OBz)_2$, 1, 30
$TiC_{24}H_{24}$ $(C_5H_5)_2$(m-$MeC_6H_4)_2Ti$, 1, 21
$(C_5H_5)_2$(p-$MeC_6H_4)_2Ti$, 1, 21
$TiC_{25}H_{22}N_2O_3$ C_5H_5(EtO)bis(quinolinato)-Ti, 1, 16
$TiC_{26}H_{30}N_2$ $(C_5H_5)_2$(p-$Me_2NC_6H_4)_2Ti$, 1, 21
$TiC_{30}H_{32}O_2$ Bis(1-isopropyl-5-methylazulene)-$Ti(CO)_2$, 1, 30
$TiC_{38}H_{30}$ $(C_5H_5)_2Ti$ compd. with $PhC{\equiv}CPh$, 1, 18
$TiC_{46}H_{76}O_4$ $(C_5H_5)_2Ti(O_2CC_7H_{14}CH{=}CH$-$C_8H_{17})_2$, 1, 30*
$TiC_{46}H_{80}O_4$ $(C_5H_5)_2Ti(O_2CC_{17}H_{35})_2$, 1, 29
$Ti_2C_{10}H_{10}Cl_4O$ $(C_5H_5TiCl_2)_2O$, 1, 14
$Ti_2C_{12}H_{14}Br_4O$ $[(MeC_5H_4)TiBr_2]_2O$, 1, 14
$Ti_2C_{14}H_{18}I_4O_3$ $(C_5H_5TiI_2OCH_2CH_2)_2O$, 1, 14*
$Ti_2C_{15}H_{20}Cl_4O_2$ $(C_5H_5TiCl_2OCH_2)_2CMe_2$, 1, 14*
$Ti_2C_{18}H_{22}O_9$ $[C_5H_5Ti(OAc)_2]_2O$, 1, 16
$Ti_2H_{20}H_{18}Br_4O_2$ $[(CH_2{=}CHC_5H_4)TiBr_2O]_2$-$C_6H_4$, 1, 14
$Ti_2C_{20}H_{20}Br_2O$ $[(C_5H_5)_2TiBr]_2O$, 1, 28
$Ti_2C_{20}H_{20}Cl_2O$ $[(C_5H_5)_2TiCl]_2O$, 1, 28
$Ti_2C_{24}H_{24}$ (Cyclooctatetraene)$_3$-Ti_2, 1, 31
$Ti_2C_{32}H_{32}$ (Cyclooctatetraene)$_4$-Ti_2, 1, 31
$Ti_3C_{26}H_{30}Br_6O_6$ $[(AcC_5H_4)TiBr_2OCH_2CH_2]_2$-$CHOTiBr_2(C_5H_4Ac)$, 1, 14
$TiAlC_{10}H_{10}Cl_4$ $(C_5H_5)_2TiCl_2AlCl_2$, 1, 32
$TiAlC_{12}H_{15}Cl_3$ $(C_5H_5)_2TiCl_2AlEtCl$, 1, 33
$TiAlC_{12}H_{16}Cl$ $(C_5H_5)_2TiMe_2AlCl$, 1, 33
$TiAlC_{12}H_{16}Cl_2$ $(C_5H_5)_2TiCl_2AlMe_2$, 1, 33
$TiAlC_{12}H_{27}Cl_3$ i-Bu(Cl)$TiCl_2Al$(i-Bu)$_2$, 1, 33
$TiAlC_{14}H_{20}Cl$ $(C_5H_5)_2Ti{\cdot}Et_2AlCl$, 1, 33
$TiAlC_{14}H_{20}Cl_2$ $(C_5H_5)_2TiCl_2AlEt_2$, 1, 33
$TiAlC_{18}H_{28}Cl_2$ $(C_5H_5)_2TiCl_2Al$(i-Bu)$_2$, 1, 34
$TiAlC_{22}H_{24}Br_2$ Diindenyl-$TiBr_2AlEt_2$, 1, 34
$TiAlC_{22}H_{24}Cl_2$ Diindenyl-$TiCl_2AlEt_2$, 1, 34
$TiAlC_{22}H_{24}I_2$ Diindenyl-TiI_2AlEt_2, 1, 34
$TiAl_2C_6H_6Cl_8$ $(C_6H_6)Al_2Ti^{II}Cl_8$, 1, 32
$TiAl_2C_7H_8Cl_8$ $(PhMe)Al_2{}^{II}Cl_8$, 1, 32
$TiAl_2C_9H_{12}Cl_8$ (1,3,5-$Me_3C_6H_3)Al_2$-$Ti^{II}Cl_8$, 1, 32
$TiAl_2C_{26}H_{61}O_4$ $Et_2Al(OBu)_2TiEt(OBu)_2$-$AlEt_2$, 1, 34
$TiBC_{10}H_{10}F_4$ $(C_5H_5)_2TiBF_4$, 1, 19
$TiBC_{10}H_{14}$ $(C_5H_5)_2TiBH_4$, 1, 19
$TiGeC_{37}H_{64}O_4$ (n-$C_8H_{17})_3Ge$[(i-BuC_5H_4)-$Ti(CO)_4$], 2, 139
$TiGe_2C_{46}H_{38}Cl_2$ $(Ph_3GeC_5H_4)_2TiCl_2$, 1, 35, 2, 139
$TiPbC_{16}H_{22}O_4$ $Et_3Pb(MeC_5H_4)Ti(CO)_4$, 1, 35, 2, 594
$TiSiC_{13}H_{19}ClO$ $(C_5H_5)_2Ti(OSiMe_3)Cl$, 1, 34
$TiSiC_{23}H_{23}ClO$ $(C_5H_5)_2Ti(OSiMePh_2)Cl$, 1, 35
$TiSiC_{28}H_{25}ClO$ $(C_5H_5)_2Ti(OSiPh_3)Cl$, 1, 35
$TiSi_2C_{46}H_{40}O_2$ $(C_5H_5)_2Ti(OSiPh_3)_2$, 1, 35
$TiSi_3C_{14}H_{32}O_3$ $C_5H_5Ti(OSiMe_3)_3$, 1, 34
$TiSnC_{14}H_{18}O_4$ $Me_3Sn(EtC_5H_4)Ti(CO)_4$, 2, 500
$Ti_2Al_2C_{28}H_{40}$ $[(C_5H_5)_2TiAlEt_2]_2$, 1, 34

$Ti_2Al_9C_{18}H_{18}Cl_{27}$ $(C_6H_6)_3Ti_2(AlCl_3)_9$, 1, 32
$Ti_2GeC_{32}H_{40}O_8$ $[(EtMeC_5H_3)Ti(CO)_4]_2$-$GeBu_2$, 1, 35, 2, 139
$Ti_2PbC_{32}H_{24}O_8$ $Ph_2Pb[MeC_5H_4Ti(CO)_4]_2$, 2, 594
$Ti_2SiC_{80}H_{120}O_8$ $\{[(n\text{-}C_{12}H_{25})_2C_5H_3]Ti$-$(CO)_4\}_2Si(CH_2Ph)_2$, 1, 35
$Ti_2SnC_{32}H_{24}O_8$ $(PhCH_2)_2Sn[C_5H_5Ti$-$(CO)_4]_2$, 2, 500
$TlAsC_6H_{18}$ $Me_3AsTlMe_3$, 3, 6
$TlBiC_{24}H_{30}$ $Et_3Bi\cdot TlPh_3$, 3, 816
$UC_{15}H_{15}Cl$ $(C_5H_5)_3UCl$, 1, 3
$UC_{20}H_{20}$ $(C_5H_5)_4Cl$, 1, 4
$VC_5H_5Br_2O$ $C_5H_5V(O)Br_2$, 1, 44
$VC_5H_5Br_3$ $C_5H_5VBr_3$, 1, 44
$VC_5H_5Cl_2O$ $C_5H_5V(O)Cl_2$, 1, 44
$VC_5H_5Cl_3$ $C_5H_5VCl_3$, 1, 44
VC_6H_5Cl $PhVCl$, 1, 43
$VC_6H_6Cl_3N_2$ $(CH_2=CHCN)_2VCl_3$, 1, 52
$VC_7H_7Br_3$ $(CH_2=CHC_5H_4)VBr_3$, 1, 44
$VC_7H_9Cl_3$ $(EtC_5H_4)VCl_3$, 1, 44
$VC_7H_{14}Cl_3$ $(C_5H_{11}CH=CH_2)VCl_3$, 1, 52
$VC_9H_5O_4$ $C_5H_5V(CO)_4$, 1, 44
$VC_{10}H_6ClO_4$ $(C_6H_6)V(CO)_4Cl$, 1, 50
$VC_{10}H_6F_6O_4P$ $[(C_6H_6)V(CO)_4]PF_6$, 1, 50
$VC_{10}H_7O_3$ Cycloheptatrienyl-$V(CO)_3$, 1, 49
$VC_{10}H_{10}$ $(C_5H_5)_2V$, 1, 46
$VC_{10}H_{10}Br$ $(C_5H_5)_2VBr$, 1, 48
$VC_{10}H_{10}BrCl$ $(C_5H_5)_2VBrCl$, 1, 49
$VC_{10}H_{10}Br_2$ $(C_5H_5)_2VBr_2$, 1, 49
$VC_{10}H_{10}Cl$ $(C_5H_5)_2VCl$, 1, 47
$VC_{10}H_{10}Cl_2$ $(C_5H_5)_2VCl_2$, 1, 48
$VC_{10}H_{10}I$ $(C_5H_5)_2VI$, 1, 48
$VC_{10}H_{20}Cl_3$ $(C_8H_{17}CH=CH_2)VCl_3$, 1, 52
$VC_{11}H_7O_5$ $(AcC_5H_4)V(CO)_4$, 1, 45
$VC_{11}H_8F_6O_4P$ $[(MePh)V(CO)_4]PF_6$, 1, 50
$VC_{11}H_{13}$ $(C_5H_5)_2VMe$, 1, 48
$VC_{11}H_{15}N_2O_3$ $(EtPrC_5H_3)V(NO)_2CO$, 1, 44
$VC_{12}H_9O_5$ $[Ac(Me)C_5H_3]V(CO)_4$, 1, 45
$(EtCOC_5H_4)V(CO)_4$, 1, 45
$VC_{12}H_{10}Cl$ Ph_2VCl, 1, 43
$VC_{12}H_{10}F_6O_4P$ $[(p\text{-}Me_2C_6H_4)V(CO)_4]PF_6$, 1, 50
$VC_{12}H_{10}I_3O_2$ $[(C_5H_5)_2V(CO)_2]I_3$, 1, 49
$VC_{12}H_{12}$ Cycloheptatrienyl$(C_5H_5)V$, 1, 43
$(C_6H_6)_2V$, 1, 50
$VC_{12}H_{12}F_6O_3P$ $[(1,3,5\text{-}Me_3C_6H_3)V(CO)_3]$-$PF_6$, 1, 50
$VC_{12}H_{15}$ $(C_5H_5)_2VEt$, 1, 48
$VC_{13}H_{11}O_5$ $[Me(EtCO)C_5H_3]V(CO)_4$, 1, 45
$VC_{13}H_{13}O_4$ $(BuC_5H_4)V(CO)_4$, 1, 45
$VC_{13}H_{15}$ $(C_5H_5)_2VCH_2CH=CH_2$, 1, 48
$VC_{13}H_{20}Cl_2O$ $(EtC_5H_4)(cyclo\text{-}C_6H_{11}O)VCl_2$, 1, 44
$VC_{14}H_{16}$ $(MePh)_2V$, 1, 51
$VC_{14}H_{23}N_3O_3P$ $C_5H_5V(CO)_3[P(NMe_2)_3]$, 1, 45
$VC_{14}H_{28}Cl_3$ $(C_5H_{11}CH=CH_2)_2VCl_3$, 1, 52
$VC_{16}H_{15}$ $(C_5H_5)_2VPh$, 1, 46
$VC_{16}H_{16}O_2$ $(C_5H_5)_2V(CO)_2(CH_2=CHCH=CH_2)$, 1, 46
$VC_{17}H_9O_4$ Fluorenyl-$V(CO)_4$, 1, 45
$VC_{17}H_{17}$ $(C_5H_5)_2VCH_2Ph$, 1, 48
$(C_5H_5)_2V(C_6H_4Me)$, 1, 48
$VC_{18}H_{18}O_2$ $(C_5H_5)_2V(CO)_2$-$(\overline{CH=CHCH=CHCH_2CH_2})$, 1, 46
$VC_{18}H_{20}O_2$ $(C_5H_5)_2V(CO)_2(CH_2=CMe$-$CMe=CH_2)$, 1, 46
$VC_{18}H_{24}$ $(1,3,5\text{-}Me_3C_6H_3)_2V$, 1, 51
$VC_{18}H_{24}I$ $(1,3,5\text{-}Me_3C_6H_3)_2VI$, 1, 51
$VC_{18}H_{28}$ $(C_5H_5)_2VBu_2$, 1, 49
$VC_{20}H_{40}Cl_3$ $(C_8H_{17}CH=CH_2)_2VCl_3$, 1, 52
$VC_{34}H_{26}O_2$ $(C_5H_5)_2V(CO)_2$(4,8-diphenylazulene), 1, 46
$V_2C_{14}H_{22}S_4$ $[C_5H_5V(SMe)_2]_2$, 1, 46
$V_2C_{15}F_6O_{12}$ $CF_3CF[V(CO)_6]CF_2V(CO)_6$, 1, 43
$V_2C_{16}H_6O_{10}$ $[(C_6H_6)V(CO)_4][V(CO)_6]$, 1, 50
$V_2C_{17}H_8O_{10}$ $[(MePh)V(CO)_4][V(CO)_6]$, 1, 50
$V_2C_{18}H_{10}F_{12}S_4$ $\{C_5H_5V[CF_3C(S)=C(S)$-$CF_3]\}_2$, 1, 46
$V_2C_{18}H_{10}O_8$ $[(C_5H_5)_2V(CO)_2]V(CO)_6$, 1, 49
$V_2C_{18}H_{10}O_{10}$ $[(p\text{-}Me_2C_6H_4)V(CO)_4][V(CO)_6]$, 1, 50
$V_2C_{19}H_{12}O_{10}$ $[(1,3,5\text{-}Me_3C_6H_3)V(CO)_4]$-$[V(CO)_6]$, 1, 50
$V_2C_{20}H_{15}O_6$ [Cycloheptatriene(cycloheptatrienyl)-V]$[V(CO)_6]$, 1, 50

$V_2C_{20}H_{16}O_6$ $[(MePh)_2V][V(CO)_6]$, 1, 51
$V_2C_{20}H_{18}Br_4O_2$ $m\text{-}C_6H_4[OV(C_5H_4CH{=}CH_2)\text{-}Br_2]_2$, 1, 44
$V_2C_{24}H_{24}O_6$ $[(i\text{-}PrPh)_2V][V(CO)_6]$, 1, 51
$[(1,3,5\text{-}Me_3C_6H_3)_2V][V(CO)_6]$, 1, 51
$V_2C_{30}H_{36}O_6$ $[(Hexamethylbenzene)_2\text{-}V]\text{-}[V(CO)_6]$, 1, 51
$VAlC_{18}H_{24}Cl_4$ $[(1,3,5\text{-}Me_3C_6H_3)_2V]AlCl_4$, 1, 51
$VBC_{36}H_{30}O_2$ $[(C_5H_5)_2V(CO)_2]BPh_4$, 1, 49
$VBC_{36}H_{30}O_4$ $[(p\text{-}Me_2C_6H_4)V(CO)_4]BPh_4$, 1, 50
$VBC_{37}H_{32}O_4$ $[(1,3,5\text{-}Me_3C_6H_3)V(CO)_4]\text{-}BPh_4$, 1, 50
$VCs_2C_8H_5O_3$ $Cs_2[C_5H_5V(CO)_3]$, 1, 46
$VK_2C_8H_5O_3$ $K_2[C_5H_5V(CO)_3]$, 1, 46
$VLi_4C_{36}H_{30}$ Li_4VPh_6 complex with Et_2O, 1, 43
$VNa_2C_8H_5O_3$ $Na_2[C_5H_5V(CO)_3]$,
$VRb_2C_8H_5O_3$ $Rb_2[C_5H_5V(CO)_3]$, 1, 46
$V_2Pb_2C_{64}H_{54}O_6$ [Fluorenyl-$V(CO)_3Pb$-$(C_6H_3Me_2)_2]_2$, 1, 52*, 2, 594
$V_2Si_2C_{36}H_{46}O_6$ $[(BuC_5H_4)V(CO)_3Si\text{-}(CH_2CH{=}CH_2)_2]_2$, 1, 52
$V_2Sn_2C_{36}H_{34}O_6$ [Indenyl-$V(CO)_3Sn$-$(CH_2CH{=}CH_2)_2]_2$, 2, 500
$V_2Sn_2C_{50}H_{82}O_6$ $[(MeC_5H_4)V(CO)_3Sn\text{-}(C_8H_{17})_2]_2$, 2, 500
$V_2Sn_2C_{72}H_{126}O_6$ $[(Et_2C_5H_3)V(CO)_3Sn\text{-}(C_{12}H_{25})_2]_2$, 1, 52, 2, 500
$WC_7H_2O_5$ $(C_2H_2)W(CO)_5$, 1, 128
$WC_7H_4O_5$ $(C_2H_4)W(CO)_5$, 1, 126
$WC_7H_5NO_3$ $C_5H_5W(CO)_2NO$, 1, 123
$WC_8H_3NO_5$ $(CH_2{=}CHCN)W(CO)_5$, 1, 126
$WC_8H_4O_4$ $(C_2H_2)_2W(CO)_4$, 1, 128
$WC_8H_4O_5$ $(MeC{\equiv}CH)W(CO)_5$, 1, 128
$WC_8H_5ClO_3$ $C_5H_5W(CO)_3Cl$, 1, 120
$WC_8H_5DO_3$ $C_5H_5W(CO)_3D$, 1, 120
$WC_8H_6O_3$ $C_5H_5W(CO)_3H$, 1, 119
$WC_8H_6O_4$ $(CH_2{=}CHCH{=}CH_2)W(CO)_4$, 1, 126
$WC_8H_6O_5$ $(MeCH{=}CH_2)W(CO)_5$, 1, 126
$WC_8H_6O_6$ (Methoxymethylcarbenium)W-$(CO)_5$, 1, 124
$WC_8H_8O_4$ $(C_2H_4)_2W(CO)_4$, 1, 126
$WC_9H_5F_6O_4P$ $[C_5H_5W(CO)_4]PF_6$, 1, 121
$WC_9H_5NO_3$ $C_5H_5W(CO)_3CN$, 1, 120
$WC_9H_6O_3$ $(C_6H_6)W(CO)_3$, 1, 124
$WC_9H_6O_5$ $(CH_2{=}CHCH{=}CH_2)W(CO)_5$, 1, 126
$WC_9H_8O_3$ $C_5H_5W(CO)_3Me$, 1, 120
$WC_9H_8O_5$ $(MeCH{=}CHMe)W(CO)_5$, 1, 126
$WC_{10}H_5F_3O_4$ $C_5H_5W(CO)_3OCCF_3$, 1, 121
$WC_{10}H_5F_3O_5$ $C_5H_5W(CO)_3OCOCF_3$, 1, 120
$WC_{10}H_6F_4O_3$ $C_5H_5W(CO)_3CF_2CF_2H$, 1, 121
$WC_{10}H_7BrO_3$ 1-Bromocycloheptatriene-$W(CO)_3$, 1, 127
$WC_{10}H_7ClO_7$ [Cycloheptatrienium-$W(CO)_3$]-ClO_4, 1, 127
$WC_{10}H_8INO_3$ $[C_5H_5W(CO)_3(CNMe)]I$, 1, 122
$WC_{10}H_8O_3$ Cycloheptatriene-$W(CO)_3$, 1, 127
$(MePh)W(CO)_3$, 1, 124
$WC_{10}H_9F_6O_3P$ $[C_5H_5W(CO)_3(C_2H_4)]PF_6$, 1, 121
$WC_{10}H_{10}Br_4$ $[(C_5H_5)_2WBr_3]Br$, 1, 124
$WC_{10}H_{10}Cl_2$ $(C_5H_5)_2WCl_2$, 1, 124
$WC_{10}H_{10}Cl_4$ $[(C_5H_5)_2WCl_3]Cl$, 1, 124
$WC_{10}H_{10}I_2$ $(C_5H_5)_2WI_2$, 1, 124
$WC_{10}H_{10}O_2$ $CH_2{=}CHCH_2W(CO)_2C_5H_5$, 1, 123
$WC_{10}H_{10}O_3$ $(EtC_5H_4)W(CO)_3H$, 1, 120
$C_5H_5W(CO)_3Et$, 1, 121
$WC_{10}H_{10}O_4$ 1,5-Hexadiene-$W(CO)_4$, 1, 126
$WC_{10}H_{12}$ $(C_5H_5)_2WH_2$, 1, 123
$WC_{10}H_{12}O_4$ $(MeCH{=}CH_2)_2W(CO)_4$, 1, 126
$WC_{10}H_{13}Cl$ $[(C_5H_5)_2WH_3]Cl$, 1, 123
$WC_{11}H_8O_4$ $C_5H_5W(CO)_3OCCH{=}CH_2$, 1, 121
Bicyclo[2.2.1]hepta-2,5-diene-$W(CO)_4$, 1, 128
$WC_{11}H_{10}O_3$ $C_5H_5W(CO)_3CH_2CH{=}CH_2$, 1, 121
$(Me_2C_6H_4)W(CO)_3$ isomers, 1, 125
$WC_{11}H_{10}O_4$ $C_5H_5W(CO)_3OCEt$, 1, 121
1-$(HOCH_2)$-cycloheptatriene-$W(CO)_3$, 1, 127
$WC_{11}H_{10}O_5$ $(EtC{\equiv}CEt)W(CO)_5$, 1, 128
$WC_{11}H_{11}F_6O_3P$ $[C_5H_5W(CO)_3(CH_2{=}CHMe)]PF_6$, 1, 121
$WC_{11}H_{12}O_3$ $C_5H_5W(CO)_3CHMe_2$, 1, 121
$WC_{11}H_{15}NO_6$ $Me_4N[W(CO)_5Ac]$, 1, 119
$WC_{12}H_5F_7O_4$ $C_5H_5W(CO)_3OCC_3F_7$, 1, 121
$WC_{12}H_9N_3O_3$ $(CH_2{=}CHCN)_3W(CO)_3$, 1, 126
$WC_{12}H_{10}O_3$ 3a,7a-Dihydroindene-$W(CO)_3$, 1, 128
(5,6-Dimethylenebicyclo[2.2.1]hept-

2-ene)-$W(CO)_3$, 1, 128
$WC_{12}H_{11}F_6OP$ $[C_5H_5W(CO)(C_6H_6)]PF_6$, 1, 122
$WC_{12}H_{12}$ $(C_6H_6)_2W$, 1, 125
$WC_{12}H_{12}I$ $[(C_6H_6)_2W]I$, 1, 125
$WC_{12}H_{12}O_3$ (1,3,5-$Me_3C_6H_3$)$W(CO)_3$, 1, 125
$WC_{12}H_{12}O_4$ 1,5-Cyclooctadiene-$W(CO)_4$, 1, 128
$WC_{12}H_{16}O_4$ $(MeCH{=}CHMe)_2W(CO)_4$, 1, 126
$WC_{12}H_{20}$ $(CH_2{=}CHCH_2)_4W$, 1, 124
$WC_{13}H_8O_6$ Bz(Me)$W(CO)_5$, 1, 119
$WC_{13}H_9NO_5$ 5-Cyanobicyclo[2.2.1]hept-2-ene-$W(CO)_5$, 1, 128
$WC_{13}H_{14}O_3$ p-Cymene-$W(CO)_3$, 1, 125
(1,2-$Et_2C_6H_4$)$W(CO)_3$, 1, 125
$WC_{15}H_{17}F_6OP$ [Mesitylene(C_5H_5)W(CO)]-PF_6, 1, 122
$WC_{15}H_{18}O_3$ Hexamethylbenzene-$W(CO)_3$, 1, 125
$WC_{16}H_{12}O_3$ (1-Phenylcycloheptatriene)-$W(CO)_3$, 1, 127
$WC_{16}H_{17}NO_6$ $Me_4N[W(CO)_5Bz]$, 1, 119
$WC_{16}H_{20}I$ Bis-m-xylene-WI, 1, 126
$WC_{18}H_{20}O_2$ Bis-1,3,6-cyclooctatriene-$W(CO)_2$, 1, 127
$WC_{19}H_{12}O_5$ $(PhCH{=}CHPh)W(CO)_5$, 1, 127
$WC_{19}H_{18}O_3$ 2-Ethylbicyclohepta-2,4,6-trienyl-$W(CO)_3$, 1, 127
$WC_{19}H_{30}O$ $(EtC{\equiv}CEt)_3W(CO)$, 1, 129
$WC_{24}H_{33}F_9O_3$ $(C_5H_{11}CF{=}CF_2)_3W(CO)_3$, 1, 126
$WC_{26}H_{19}O_3P$ $(Ph_3PC_5H_4)W(CO)_3$, 1, 123
$WC_{28}H_{24}O$ $(MeC{\equiv}CPh)_3W(CO)$, 1, 129
$WC_{28}H_{32}O_4$ Bis(butylcyclooctatetraene)-$W(CO)_4$, 1, 128
$WC_{42}H_{24}Cl_6N_6$ $(p\text{-}ClC_6H_4NC)_6W$, 1, 129
$WC_{42}H_{30}N_6$ $(PhNC)_6W$, 1, 129
$WC_{43}H_{30}O$ $(PhC{\equiv}CPh)_3W(CO)$, 1, 129
$WC_{48}H_{36}Cl_6N_6$ [3,2-Cl(Me)$C_6H_3NC]_6W$, 1, 129
$W_2C_{12}H_{14}O_7$ $(PhW)_2O_7H_4$, 1, 118
$W_2C_{16}H_{10}O_6$ $[C_5H_5W(CO)_3]_2$, 1, 122
$W_2C_{16}H_{16}O_4S_2$ $[C_5H_5W(SMe)(CO)_2]_2$, 1, 123
$W_2C_{18}H_{22}O_4P_2$ $[C_5H_5W(PMe_2)(CO)_2]_2$, 1, 123
$W_2C_{18}H_{22}O_8$ $Ph_3W_2O_8H_7$, 1, 118
$W_2C_{22}H_{22}O_6$ [(i-PrC_5H_4)$W(CO)_3]_2$, 1, 122
$W_2C_{26}H_{30}O_6$ $[(Et_2CHC_5H_4)W(CO)_3]_2$, 1, 122
$WAs_2C_{13}H_{16}Br_2O_3$ W(2-$Me_2AsC_6H_4AsMe_2$)-$(CO)_3Br_2$, 3, 97
$WAs_2C_{13}H_{16}Br_3O_3$ [W(2-$Me_2AsC_6H_4AsMe_2$)-$(CO)_3Br_2]Br$, 3, 97
$WAs_2C_{14}H_{16}I_2O_4$ [W(2-$Me_2AsC_6H_4AsMe_2$)-$(CO)_4I]I$, 3, 97
$WAs_2C_{14}H_{16}I_4O_4$ [W(2-$Me_2AsC_6H_4AsMe_2$)-$(CO)_4I]I_3$, 3, 97
$WAs_2C_{14}H_{16}O_4$ [W(2-$Me_2AsC_6H_4AsMe_2$)-$(CO)_4$], 3, 97
$WAs_3C_{14}H_{27}O_3$ W[MeAs$(CH_2CH_2CH_2AsMe_2)_2$]-$(CO)_3$, 3, 107
$WAs_4C_{22}H_{32}O_2$ W(2-$Me_2AsC_6H_4AsMe_2)_2$-$(CO)_2$, 3, 97
$WAs_4C_{22}H_{32}Br_2O_2$ [W(2-$Me_2AsC_6H_4AsMe_2)_2$-$(CO)_2Br]Br$, 3, 97
$WAs_4C_{22}H_{32}Br_4O_2$ [W(2-$Me_2AsC_6H_4AsMe_2)_2$-$(CO)_2Br]Br_3$, 3, 97
$WAs_4C_{22}H_{32}I_2O_2$ [W(2-$Me_2AsC_6H_4AsMe_2)_2$-$(CO)_2I]I$, 3, 97
$WAs_4C_{22}H_{32}I_4O_2$ [W(2-$Me_2AsC_6H_4AsMe_2)_2$-$(CO)_2I]I_3$, 3, 97
$WBC_{10}H_7F_4O_3$ [Cycloheptatrienium-W-$(CO)_3]BF_4$, 1, 127
$WBC_{18}H_{24}F_4$ [(1,3,5-$Me_3C_6H_3)_2W]BF_4$,
$WBC_{32}H_{28}NO_3$ $[C_5H_5W(CO)_3NH_3]BPh_4$, 1, 122
$WBC_{32}H_{29}N_2O_3$ $[C_5H_5W(CO)_3H_2NNH_2]BPh_4$, 1, 122
$WCoC_{16}H_{15}O$ $C_5H_5W(CO)Co(C_5H_5)_2$, 1, 130
$WCoC_{18}H_{15}O_3$ $[(C_5H_5)_2Co][C_5H_5W(CO)_3]$, 1, 588
$WCrC_{12}H_{14}N_7O_3S_4$ $[C_5H_5W(CO)_3NH_3]$[Cr-$(NH_3)_2(SCN)_4$], 1, 122
$WCrC_{12}H_{15}N_8O_3S_4$ $[C_5H_5W(CO)_3H_2NNH_2]$-$[Cr(NH_3)_2(SCN)_4]$, 1, 122
$WKC_9H_5N_2O_2$ $K[W(CN)_2(C_5H_5)(CO)_2]$, 1, 123
$WK_3C_9H_3O_3$ $K_3[W(CO)_3(C{\equiv}CH)_3]$, 1, 119
$WK_3C_{12}H_9O_3$ $K_3[W(CO)_3(C{\equiv}CMe)_3]$, 1, 119
$WLi_2C_{12}H_{11}$ $Li_2[WHPh_2]$, 1, 119
$WLi_2C_{36}H_{30}$ $Li_2[WPh_6]$ complex with

Et_2O, 1, 119
$WLi_3C_{36}H_{30}$ $Li_3[WPh_6]$ complex with Et_2O, 1, 119
$WMoC_{16}H_{10}O_6$ $C_5H_5Mo(CO)_6WC_5H_5$, 1, 118
$WPbC_{26}H_{20}O_3$ $C_5H_5W(CO)_3PbPh_3$, 1, 130, 2, 594
$WPbC_{44}H_{56}O_3$ $(C_{18}H_{37}C_5H_4)W(CO)_3PbPh_3$, 2, 594
$WPbC_{45}H_{58}O_3$ $[Me(C_{18}H_{37})C_5H_3]W(CO)_3$-$PbPh_3$, 1, 130*
$WSiC_{10}H_{12}O_4$ $Me_2Si(CH{=}CH_2)_2W(CO)_4$, 1, 126
$WSiC_{16}H_{24}O_3$ $(EtC_5H_4)W(CO)_3SiEt_3$, 1, 129
$WSnC_8H_5Cl_3O_3$ $C_5H_5W(CO)_3SnCl_3$, 1, 129
$WSnC_{14}H_{10}Cl_2O_3$ $C_5H_5W(CO)_3SnPhCl_2$, 1, 129, 2, 500
$WSnC_{26}H_{20}O_3$ $C_5H_5W(CO)_3SnPh_3$, 1, 129, 2, 500
$W_2As_2C_{12}H_{12}O_8$ $[W(CO)_4AsMe_2]_2$, 3, 301
$W_2As_2C_{14}H_{12}O_{10}$ $(OC)_5W{\leftarrow}As(Me)_2As$-$(Me_2){\rightarrow}W(CO)_5$, 3, 301
$W_2GeC_{32}H_{28}O_6$ $(PhCH_2)_2Ge[MeC_5H_4W$-$(CO)_3]_2$, 2, 140
$W_2HgC_{16}H_{10}O_6$ $Hg[C_5H_5W(CO)_3]_2$, 1, 120
$W_2PtC_{20}H_{20}Cl_{10}$ $[(C_5H_5)_2WCl_2]_2PtCl_6$, 1, 124
$W_2SnC_{16}H_{10}Cl_2O_6$ $[C_5H_5W(CO)_3]_2SnCl_2$, 1, 130
$W_2SnC_{28}H_{20}O_6$ $[C_5H_5W(CO)_3]_2SnPh_2$, 1, 130, 2, 500
$W_3GeC_{54}H_{38}O_9$ Cyclohexyl-Ge[fluorenyl $W(CO)_3]_3$, 2, 140
$YC_4H_{11}O$ Et_2YOH, 1, 1
$YC_{10}H_{25}O$ $Et_3Y{\cdot}Et_2O$, 1, 1
$YC_{15}H_{15}$ $(C_5H_5)_3Y$, 1, 4
$YbC_{10}H_{10}Cl$ $(C_5H_5)_2YbCl$, 1, 1, 2
$YbC_{11}H_{13}O$ $(C_5H_5)_2YbOMe$, 1, 3
$YbC_{12}H_{13}O_2$ $(C_5H_5)_2YbOAc$, 1, 3
$YbC_{12}H_{14}Cl$ $(MeC_5H_4)_2YbCl$, 1, 2
$YbC_{15}H_{15}$ $(C_5H_5)_3Yb$, 1, 4
$YbC_{16}H_{15}O$ $(C_5H_5)_2YbOPh$, 1, 3
$YbC_{17}H_{15}O_2$ $(C_5H_5)_2YbOBz$, 1, 3
$YbC_{17}H_{29}Cl_2O_3$ $C_5H_5YbCl_2{\cdot}3THF$, 1, 1
$ZnAs_2C_{18}H_{34}$ $Bu_2Zn(2\text{-}Me_2AsC_6H_4AsMe_2)$, 3, 97
$ZnAs_2C_{22}H_{16}F_{10}$ $(C_6F_5)_2Zn(2\text{-}Me_2As$-$C_6H_4AsMe_2)$, 3, 98
$ZnAs_2C_{22}H_{26}$ $Ph_2Zn(2\text{-}Me_2AsC_6H_4AsMe_2)$, 3, 98
$ZrC_4H_{10}Cl_2$ Et_2ZrCl_2, 1, 36
$ZrC_5H_4Cl_2$ Cyclopentadienylene-$ZrCl_2$, 1, 38
$ZrC_5H_4F_2$ Cyclopentadienylene-ZrF_2, 1, 38
$ZrC_5H_5Br_3$ $C_5H_5ZrBr_3$, 1, 36
$ZrC_5H_5Cl_3$ $C_5H_5ZrCl_3$, 1, 36
$ZrC_5H_5I_3$ $C_5H_5ZrI_3$, 1, 36
$ZrC_{10}H_{10}$ $(C_5H_5)_2Zr$, 1, 38
$ZrC_{10}H_{10}Br_2$ $(C_5H_5)_2ZrBr_2$, 1, 39
$ZrC_{10}H_{10}Br_4F_2$ $(C_5H_5)(C_5H_5Br_4)ZrF_2$, 1, 40
$ZrC_{10}H_{10}Cl_2$ $(C_5H_5)_2ZrCl_2$, 1, 38
$ZrC_{10}H_{10}Cl_6$ $(C_5H_5)(C_5H_5Cl_4)ZrCl_2$, 1, 40
$ZrC_{10}H_{10}I_2$ $(C_5H_5)_2ZrI_2$, 1, 39
$ZrC_{10}H_{22}$ $(C_5H_{11})_2Zr$, 1, 36
$ZrC_{11}H_9Cl_3$ $(PhC_5H_4)ZrCl_3$, 1, 37
$ZrC_{11}H_{14}O_6$ $C_5H_5Zr(OAc)_3$, 1, 36
$ZrC_{12}H_{12}$ $(C_6H_6)_2Zr$, 1, 41
$ZrC_{12}H_{14}Cl_2$ $(MeC_5H_4)_2ZrCl_2$, 1, 40
$ZrC_{12}H_{15}ClO$ $(C_5H_5)_2Zr(OEt)Cl$, 1, 39
$ZrC_{13}H_9Br_3$ Fluorenyl-$ZrBr_3$, 1, 38
$ZrC_{13}H_{17}ClO$ $(C_5H_5)_2Zr(OPr)Cl$, 1, 39
$ZrC_{14}H_{10}F_6O_4$ $(C_5H_5)_2Zr(O_2CCF_3)_2$, 1, 39
$ZrC_{14}H_{18}O_{14}S_3$ $(C_5H_5)(HO_3SC_5H_4)Zr$-$(OSO_3H)_2{\cdot}Ac_2O$, 1, 40
$ZrC_{15}H_{19}ClO_4$ $C_5H_5Zr[OC(Me){=}CHAc]_2Cl$, 1, 37
$ZrC_{17}H_{13}Cl_2NO_3$ $(PhC_5H_4)(O_2NC_6H_4O)Zr$-$Cl_2$, 1, 37
$ZrC_{17}H_{22}O_6$ $C_5H_5Zr[OC(Me){=}CHAc]_2OAc$, 1, 37
$ZrC_{20}H_{28}O_{13}S_2$ $\{(HO_3SC_5H_4)Zr[OC(Me){=}CH$-$Ac]_2OSO_3H\}{\cdot}Ac_2CH_2$, 1, 37
$ZrC_{20}H_{32}O_6$ $C_5H_5Zr(O_2CBu)_3$, 1, 37
$ZrC_{21}H_{38}Br_2O$ $C_5H_5(n\text{-}C_{16}H_{33}O)ZrBr_2$, 1, 37
$ZrC_{25}H_{23}ClO_4$ $C_5H_5Zr[OC(Me){=}CHBz]_2Cl$, 1, 37
$ZrC_{26}H_{20}O_6$ $C_5H_5Zr(OBz)_3$, 1, 37
$ZrC_{26}H_{44}O_6$ $C_5H_5Zr(O_2CC_6H_{13})_3$, 1, 37
$ZrC_{38}H_{30}$ $(C_5H_5)_2ZrCPh{=}CPhCPh{=}CPh$, 1, 40

$Zr_2C_{20}H_{20}Br_2O$ $[(C_5H_5)_2ZrBr]_2O$, 1, 40
$Zr_2C_{20}H_{20}Cl_2O$ $[(C_5H_5)_2ZrCl]_2O$, 1, 40
$Zr_2C_{22}H_{25}BrO_2$ $(C_5H_5)_2Zr(Br)OZr$-$(C_5H_5)_2OEt$, 1, 41
$Zr_2C_{22}H_{25}ClO_2$ $(C_5H_5)_2Zr(Cl)OZr$-$(C_5H_5)_2OEt$, 1, 41
$Zr_2C_{28}H_{22}Br_4O_2$ (Fluorenyl-$ZrBr_2O$-CH_2-$)_2$, 1, 38
$Zr_2C_{30}H_{38}O_9$ $\{C_5H_5[AcCH{=}C(Me)O]_2Zr\}_2O$, 1, 38
$Zr_2C_{32}H_{32}O_2$ $[(C_5H_5)_2PhZr]_2O \cdot H_2O$, 1, 41
$Zr_2C_{34}H_{34}O$ $[(C_5H_5)_2(MeC_6H_4)Zr]_2O$, 1, 41
$Zr_4C_{52}H_{84}F_4O_4$ $[(n\text{-}C_8H_{17}C_5H_4)Zr(O)F]_4$, 1, 38
$Zr_2SiC_{32}H_{40}O_8$ $[(MeEtC_5H_3)Zr(CO)_4]_2$-$SiBu_2$, 1, 41
$Zr_3GeC_{50}H_{54}O_{12}$ $[(MeBuC_5H_3)Zr(CO)_4]_3$-$GeCH_2CH_2Ph$, 1, 41, 2, 139
$Zr_3PbC_{45}H_{50}O_{12}$ Cyclohexyl-Pb$[Et_2C_5H_3$-$Zr(CO)_4]_3$, 2, 594
$Zr_3SnC_{41}H_{44}O_{12}$ $[(Me_2C_5H_3)Zr(CO)_4]_3$-SnC_8H_{17}, 1, 41, 2, 500

$AsC_7H_{15}O$ $MeCH(CH_2)_4As(O)Me$, 3, 589
$AsC_9H_8NS_2$ $(SCH{=}CHCH{=}C)_2AsCN$, 3, 203
$AsC_9H_{12}N_9O$
{4,3-HO[$N{=}C(NHNH_2)N{=}C(NHNH_2)N{=}CNH$-$NH]C_6H_3As\}_n$, 3, 138
AsC_9H_{13} PrPhAsH, 3, 117
$AsC_9H_{13}NO_3P$ $\{HO\{Me_2C[PH(O)OH]NH\}$-$C_6H_3As\}_n$, 3, 139
$AsC_{10}H_9N_2O_2S$ 4-($SCH{=}CHNHCHNHCO)C_6H_4$-AsO, 3, 243
$AsC_{10}H_{10}N_3O_3$ 4-$N{=}C(NH_2)N{=}CHCH{=}CC_6H_4$-$AsO_3H_2$, 3, 412
$AsC_{10}H_{12}NO_7$ 3,4-$O_2N[HO_2C(CH_2)_3]C_6H_3$-$AsO_3H_2$, 3, 487
$AsC_{11}H_{11}O_3$ $C_{10}H_7CH_2AsO_3H_2$ isomers, 3, 396
$AsC_{11}H_{17}$ MeBuPhAs, 3, 75
$AsC_{12}H_{10}IO_2$ $Ph(IC_6H_4)AsO_2H$ isomers, 3, 381
$AsC_{12}H_{10}NO_3$ Carbazole-3-arsonic acid, 3, 523
$AsC_{12}H_{11}N_2O_3S$ [4,3-HO(4-$H_2NC_6H_4SO_2$-$NH)C_6H_3]_n$, 3, 133
$AsC_{12}H_{13}N_2O_5S$ (4-$H_2NC_6H_4SO_2NH)C_6H_4$-AsO_3H_2, 3, 444
$AsC_{13}H_{10}NO_3S$ 4-(4-$NCC_6H_4S)C_6H_4$-AsO_3H_2, 3, 418
$AsC_{13}H_{18}N_3O_7$ $[MeCH(CH_2)_4AsMe \cdot H]OC_6H_2$-$(NO_2)_3$, 3, 589
$AsC_{14}H_{13}N_2S_4$
1-$C_{10}H_7AsSC(S)NHCH_2CH_2NHC(S)S$, 3, 650
$AsC_{15}H_{14}N_5O_3$
Ph{4-[$N{=}C(NH_2)N{=}C(OH)N{=}CNH]C_6H_4\}$-$AsO_2H$, 3, 385
$AsC_{16}H_{13}N_2O_8S$ 3,4-H_2N(3,4-dihydro-3,4-dioxo-5-sulfo-1-naphthyl)arsanilic acid, 3, 497
$AsC_{16}H_{15}N_2O_{11}S_2$ 4,3-HO[4,1,5-HO-$(HO_3S)_2$-3-naphthylhydrazo]-C_6H_3-AsO_3H_2, 3, 498
$AsC_{16}H_{15}O_4$ $PhAs(OCH_2C{=}CHCH{=}CHO)_2$, 3, 255*
$AsC_{16}H_{19}NO$ $PhN(CH_2CH_2)_2As(Ph)OH$, 3, 619
$AsC_{17}H_{19}ClNO_3S$ 10-Phenoxarsinyl-SCH_2-$CH(NH_2 \cdot HCl)CO_2Et$, 3, 614
$AsC_{18}H_{21}ClNO_4$ 1,1-HO(o-$ClCH_2CH_2CH_2$-C_6H_4)-1,2,3,4-tetrahydroarsinolinium NO_3^-, 3, 592
$AsC_{18}H_{22}Cl$ $Ph_2As(CH_2)_6Cl$, 3, 70
$As_2C_6H_{10}Cl_2S$ $[Me(ClCH{=}CH)As]_2S$, 3, 263
$As_2C_{11}H_{23}I$ $[Et_2AsC{\equiv}CAsMeEt_2]I$, 3, 317
$As_2C_{28}H_{26}N_6O_{16}$ $[MePhAs(OH)CH_2$-$]_2$-$[OC_6H_2(NO_2)_3]_2$, 3, 349
$As_2C_{28}H_{28}Br_4$ $(-CH_2CH_2AsPh_2Br_2)_2$, 3, 345
$As_2C_{42}H_{44}O_3$ $[Ph(4\text{-}EtC_6H_4)(4\text{-}MeC_6H_4)$-$AsOH]_2O$, 3, 363
$As_4C_4H_6O_4$ $(-AsAsCH_2CO_2H)_2$, 3, 305
$AsIrC_{20}H_{15}I_3O_2$ $Ir(AsPh_3)(CO)_2I_3$, 3, 20
$As_2CoC_{48}H_{40}N_4O_{12}$ $[Ph_4As]_2[Co(NO_3)_4]$, 3, 334
$As_2Co_2C_{12}H_{30}Br_4$ $[Co(AsEt_3)Br_2]_2$, 3, 8
$As_2PdC_{12}H_{22}$ $Me_2Pd(Me_2AsC_6H_4AsMe_2)$, 3, 91
$As_2Pt_2C_{24}H_{54}Cl_4$ $[Pt(AsBu_3)Cl_2]_2$, 3, 16
$As_4Pd_2C_{20}H_{32}Cl_4$ [Pd(2-$Me_2AsC_6H_4$-$AsMe_2)_2]PdCl_4$, 3, 91
$As_3Mn_3SnC_{81}H_{68}O_{12}$ $PhSn\{Mn[As(C_6H_4$-Me-o$)_3](CO)_4\}_3$, 3, 27
$As_4Mo_{12}SiC_{96}H_{80}O_{40}$ $[Ph_4As]_4[Si(Mo_3$-$O_{10})_4]$, 3, 336
$BCrNa_2C_{38}H_{30}O_4$ $[(BzONa)_2Cr]BPh_4$, 1, 79
$Co_2C_{14}H_8O_9$ $OCOCH{=}CPrCCo_2(CO)_7$, 1, 747
$Co_2C_{25}H_{24}$ $C_5H_5(C_5H_5CoC_5H_5)_2$, 1, 581
$CrSnC_{26}H_{20}O_3$ $C_5H_5Cr(CO)_3SnPh_3$, 1, 94, 2, 500
$CrBNa_2C_{38}H_{30}O_4$ $[(BzONa)_2Cr]BPh_4$, 1, 79
$(FeC_{12}H_{12})_n$ $(FcCHCH_2-)_n$, 1, 465
$(-FcdCH_2CH_2)_n$, 1, 465
$FeC_{14}H_8Cl_2N_4O_2$ $(p\text{-}ClC_6H_4NC)_2Fe(NO)_2$, 1, 525
$FeC_{20}H_{22}N_4O_7$ $FcCH_2NMe_3$ picrate, 1, 324
$Fe_2C_{10}H_4O_8$ $[HOC{=}CHCH{=}C(OH)Fe(CO)_3]$-$Fe(CO)_3$, 1, 504
$(GeC_8H_{10})_n$ $(p\text{-}Me_2GeC_6H_4-)_n$, 2, 46
$GeC_{10}H_{21}NO$ Pr_3GeNCO, 2, 95
i-Pr_3GeNCO, 2, 95
$GeC_{12}H_{27}N_3$ Bu_3GeN_3, 2, 95
$MnAs_2C_{20}H_{39}O_7$ $(Bu_2C_5H_3)Mn(CO)$-$[As(OMe)_3]_2$, 1, 170

$MoC_8H_8O_4$ $(C_2H_4)_2Mo(CO)_4$, 1, 109
$Mo_2C_{32}H_{26}O_6$ $\{[(MePhCH)C_5H_4]Mo(CO)_3\}_2$, 1, 101
NiC_8H_{14} $(\pi\text{-}MeCH{=}CHCH_2)_2Ni$, 1, 654
$PbC_{12}H_{15}Cl_5O$ $Et_3PbOC_6Cl_5$, 2, 571
$PbC_{12}H_{17}N_3O_7$ $Et_3PbOC_6H_2(NO_2)_2$, 2, 571
$PbC_{12}H_{20}O_2$ $Me_3PbO_2CMe_2Ph$, 2, 570
$PbC_{18}H_{14}S_3$ 2-Thienyl$_3$PbPh, 2, 536
$Pd_2C_{12}H_{18}Cl_4$ (1-Cyclohexenyl-$PdCl_2)_2$, 1, 684
SbC_3F_9 $(CF_3)_3Sb$, 3, 653
$SbC_{20}H_{17}O_2$ Ph(4-MeC_6H_4)(4-$HO_2CC_6H_4$)-Sb, 3, 677
$SnC_2H_4Cl_4$ $(ClCH_2)_2SnCl_2$, 2, 324
$SnC_4H_6N_2S_2$ $Me_2Sn(NCS)_2$, 2, 345
$SnC_8H_{15}F_3O_2$ $Et_3SnO_2CCF_3$, 2, 421
$SnC_{18}H_{29}N$ $Pr_3SnCH(CN)CH_2Ph$, 2, 224
p-$MeC_6H_4(PhCH_2)NSnEt_3$, 2, 475
$SnC_{18}H_{34}O_4S_2$ S-$Bu_2Sn(SCH_2CH_2O_2CCH_2$-$CH_2)_2$, 2, 465
$(Sn_3As_2C_6H_{18}O_4)_n$ $[(Me_2Sn)_3(AsO_4)_2]_n$, 2, 348*

APPENDIX

CORRECTIONS AND ADDITIONS TO VOLUME I

Page	Line*	
14	11	$O(CH_2CH_2O-)_2I$ should read $(C_5H_5TiI_2OCH_2CH_2)_2O$.
	12	$Me_2C(CH_2O-)_2Cl$ should read $(C_5H_5TiCl_2OCH_2)_2CMe_2$.
30	4	$n-C_8H_{17}CH{=}CHC_7H_{15}-$ should read $n-C_8H_{17}CH{=}CHC_7H_{14}-$.
52	17	$[C_{12}H_9V(CO)_3Pb(C_8H_{11})_2]_2$ should read $[C_{13}H_9V(CO)_3Pb(C_8H_9)_2]_2$.
54	2 fb	$[(n-C_{18}H_{33}C_5H_4)Ta(CO)_3Sn(C_6H_2Me_3)_2]$ should read $[n-C_{18}H_{33}C_5H_4)-Ta(CO)_3Sn(C_6H_2Me_3)_2]$.
55	10 fb	$(2,4,6-Me_3C_5H_2)_2Cr$ should read $(2,4,6-Me_3C_6H_2)_2Cr$.
58	4	$CrCl_3 + (o-LiC_6H_4-)_2$ should read $CrCl_3 + o-LiC_6H_4OLi$.
79	3	$[(C_6H_6)_2Cr[C_5H_5Cr(CO)_3$ should read $[(C_6H_6)_2Cr][C_5H_5Cr(CO)_3]$.
89	22	$(C_4H_5S)Cr(CO)_3$ should read $(C_4H_4S)Cr(CO)_3$.
94	18	$[(C_6H_6)_2Cr][Mn(CO)_5]_2$ should read $[(C_6H_6)_2Cr][Mn(CO)_5]$.
95	1	Enneacarbonylhydrotetranickolate should read Nonacarbonylhydro-tetranickelate.
103	27-30	Delete these four lines.
115	10 fb	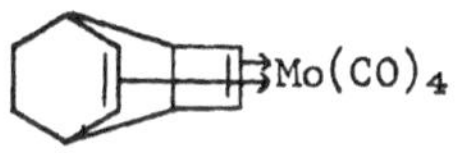 should read 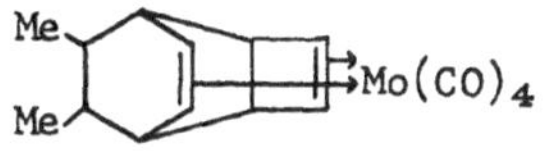 .
117	5 fb	4-Acetylcyclohexyl....(n-octyl... should read 4-Acetylcyclo-hexyl....(octadecyl.... .
	4 fb	$4-AcC_6H_4Sn[(n-C_8H_{17}C_5H_4)Mo(CO)_3]_3$ should read $4-AcC_6H_{10}Sn-[(C_{18}H_{37}C_5H_4)Mo(CO)_3]_3$.
128	8	$Cr(CO)_6$ should read $W(CO)_6$.
130	14	Tricarbonyl(n-octylcyclopentadienyl)... should read Tricar-bonyl(methyloctadecylcyclopentadienyl)... .
	15	$(n-C_8H_{17}C_5H_4)W(CO)_3PbPh_3$ should read $[Me(C_{18}H_{37})C_5H_3]W(CO)_3-PbPh_3$.
	16	$n-C_8H_{17}C_5H_4K$ should read $Me(C_{18}H_{37})C_5H_3K$.
133	7	$Mn(C{\equiv}CH_2)_2\cdot 4NH_3$ should read $Mn(C{\equiv}CH)_2\cdot 4NH_3$.
137	5 fb	1st column - -C=NCCl=NC=N- should read -C=NC(Cl)=NC(=N)- .
140	12	$HCFCF_2$ should read HCF_2CF_2.
159	16	x-BzNN=CMe should read $Bz_2NN{=}CMe$.
166	8 fb	Tricarbonyl(1,4,5,6,7,8-hexahydro-4-hydroxyazulene)... should read Tricarbonyl(1,4,5,6,7,8-hexahydro-4-hydroxyazulenyl)...
168	17	PhHNC should read PhNC.
	18	$[C_5H_5Mn(CO)_2]C_6H_8$ should read $[C_5H_5Mn(CO)_2]_2C_6H_8$.
190	9 fb	I_4^- and $[Cr(NH_3)_2(SCN)_4]^{--}$ should read I_3^- and $[Cr(NH_3)_2(SCN)]^-$.
204	4 fb	6-Chloro-s-triazin-2,4-ylene... should read 6-Chloro-2-tri-azine-2,4-diyl... .
206	3 fb	$K[C_5H_5Fe(CO)CN)_2]$ should read $K[C_5H_5Fe(CO)(CN)_2]$.
215	4	$[C_5H_5Fe(CO)(PPh_2)]_2$ should read $[C_5H_5Fe(CO)(PMe_2)]_2$.

* The "fb" letters following a line number signify that the lines were counted from the bottom of the page.

Page	Line*	
240		2nd formula fb 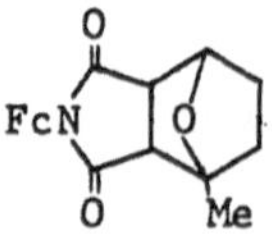 should read 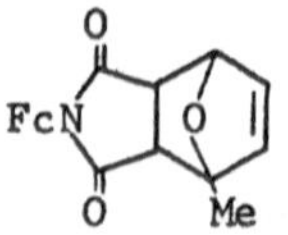 .
246	6	$FcCHCH_2C(Ph)=CCOCH_2$ should read $FcCHCH_2C(Ph)=CHCOCH_2$.
247	10	FcCOCH=CHCH=CHS should read FcCOC=CHCH=CHS.
267		Enter the page number.
273	2	FcCH(OH)C≡CC(OH)Ph should read FcCH(OH)C≡CCH(OH)Ph.
279	11 fb	2nd column - $C_{10}H_2OH$ should read $C_{10}H_{21}OH$.
289	20	ESR (1598) should read ESR of the ketyl (1598).
	5 fb	Insert: Rxn. with K in $MeOCH_2CH_2OMe \rightarrow K^+[FcBz]^{\dot{-}}$ (1598).
293	25	Insert: p-$MeOC_6H_4$ in the 1st column.
294	8	FcCH=CHCOC=CHCH=CHN should read FcCH=CHCOC=CHCH=CHNH.
300	5	$FeCON_3$ should read $FcCON_3$.
355	12	$EtMe_2CH$- should read $EtMe_2C$-.
358	15 fb	-5-bornene-2,3- should read -5-norbornene-2,3- .
	10 fb	-5-bornene-2,3- should read -5-norbornene-2,3- .
440	5 fb	4-[3-(Cyclopentadienyl)propyldihydro... should read 4-[3-(Cyclopentadienyl)propyl]dihydro... .
448	2	4-(2-Carboxyethyl)-1,1′;3,3′-bis(trimethylene)ferrocene should read 4-(2-Carboxyethyl)-3,3′-(γ-oxotrimethylene)-1,1′-trimethyleneferrocene.
449	5 fb	1,1;3,3′;4,5-tris... should read 1,1′;3,3′;4,5-tris... .
450	15 fb	-1,1′;3,3′;4′,5- should read -1,1′;3,3′;4′,5′- .
451	2	3,3′-(α-Oxotrimethylene)-2,2′;3,3′- should read 3,3′-(α-Oxotrimethylene)-1,1′;2,2′- .
479	29-33	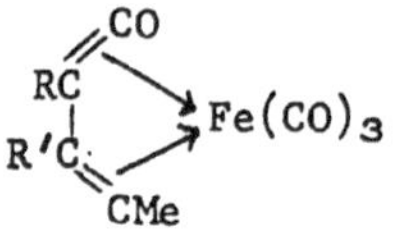 should read 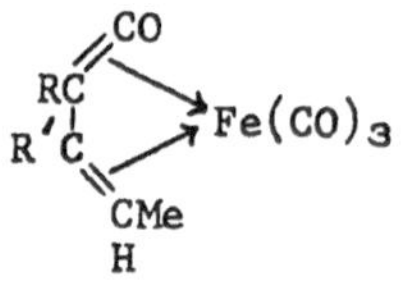 .
482	13	PrCH=C(Bu)C=CHC(Bu)- should read PrCH=C(Bu)CH=CHC(Bu)= .
487	15 fb	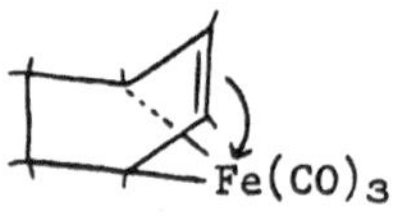 should read 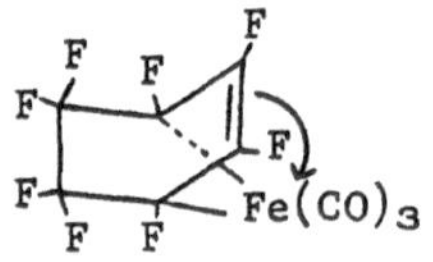 .
489	13 fb	Tricarbonyl(2,4-cycloheptadien-1-one)iron should read Tricarbonyl(2,4-cycloheptadienone)iron.
490	23	Tricarbonyl(2-p-tolyl-1,1′-bicyclo-2,4,6-trienyl)iron should read Tricarbonyl(2-p-tolyl-1,1′-bicyclohepta-2,4,6-trienyl) iron.
494	1	(Bicyclo[2.2.1]hepta-2,5-dien-2-oic acid)tricarbonyliron should read (Bicyclo[2.2.1]hepta-2,5-diene-2-carboxylic acid)tricarbonyliron.
498	3	Tricarbonyl(1,5-dimethylpentadienylium... should read Tricar-

Page	Line*	
		bonyl(1-methylhexadienylium... .
	4	$[(C_7H_{13})Fe(CO)_3]BF_4$ should read $[(C_7H_{11})Fe(CO)_3]BF_4$.
500	3	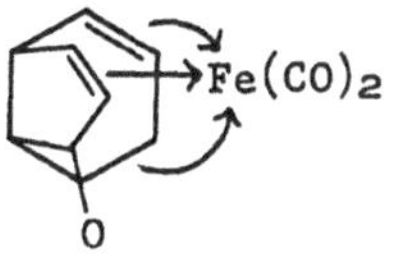 should read 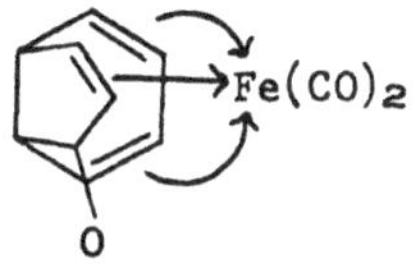
505	1 fb	Insert I below the first formula.
509	10	Hexacarbonyl(2,4,6-cyclooctatrien-1-one)iron should read Hexacarbonyl(2,4,6-cyclooctatrienone)iron.
	8 fb	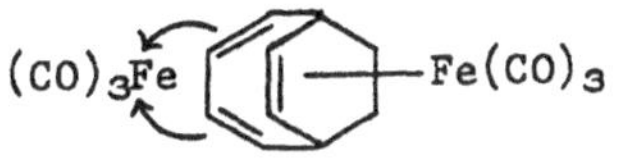 should read 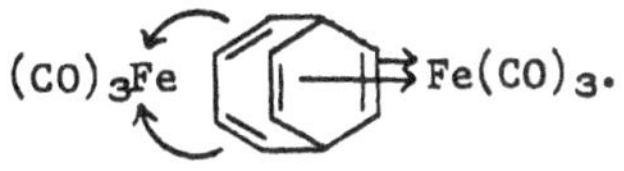
513	15	Diazulenedodecacarbonylpentairon should read Bis(azulene)dodecacarbonylpentairon.
514	16	1,5-Cyclooctadiene-1,3,5-cyclooctatrieneiron should read (1,5-Cyclooctadiene)(1,3,5-cyclooctatriene)iron.
518	21	$(PhC{\equiv}CSiMe_3)Fe_2(CO)_6$ should read $(PhC{\equiv}CSiMe_3)_2Fe_2(CO)_6$.
522	4 fb	$[(PhCH_2NC)_6]FeBr_2$ should read $[(PhCH_2NC)_6Fe]Br_2$.
523	17	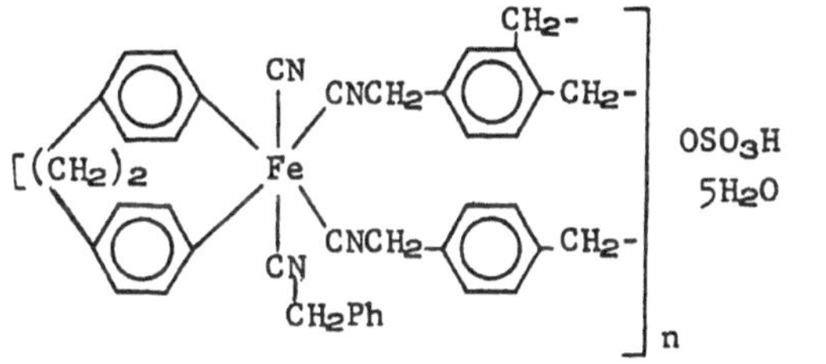 should read 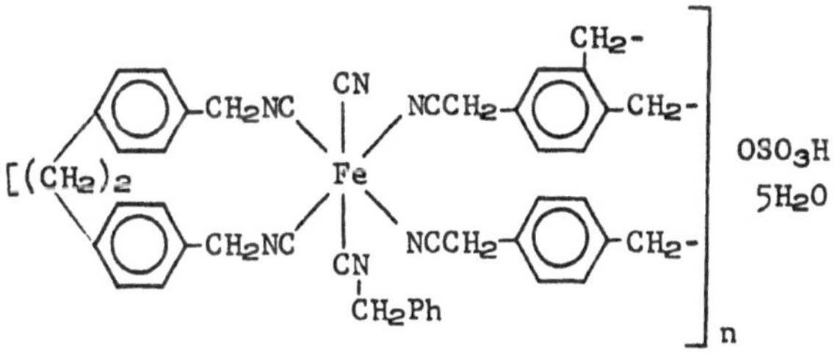
523	9 fb	$[(HO_2CC_6H_4CH_2N)_2(\text{-}COC_6H_3CH_2NC)_3Fe(CN)OSO_3H\cdot 13H_2O]_n$ should read $[(HO_2CC_6H_4CH_2NC)_2(\text{-}COC_6H_3CH_2NC)_3Fe(CN)OSO_3H\cdot 13H_2O]_n$.
539	15	$Fcd(SiMe_2OSiMe_2OMe_2Ph)_2$ should read $Fcd(SiMe_2OSiMe_2OSiMe_2Ph)_2$.
571	10	(h) should read (k).
	19	(h) should read (k).
	25	$PhC{=}C(Ph)C(Ph)COCo(CO)_3$ should read $Ph\overline{C{=}C(Ph)C}(Ph)COCo(CO)_3$.
575	5 fb	$C_5H_5\overline{CoCF_4CF_4CF_4C}F_4(CO)$ should read $C_5H_5\overline{CoCF_2CF_2CF_2C}F_2(CO)$.
588	12	$2,4\text{-}(O_2N)_2C_{10}H_5SO_3^-$ should read $2,4\text{-}(O_2N)_2C_6H_3SO_3^-$.
594	6	$R'R^2R^4R^5$ should read $RR'R^2R^3R^4R^5$.

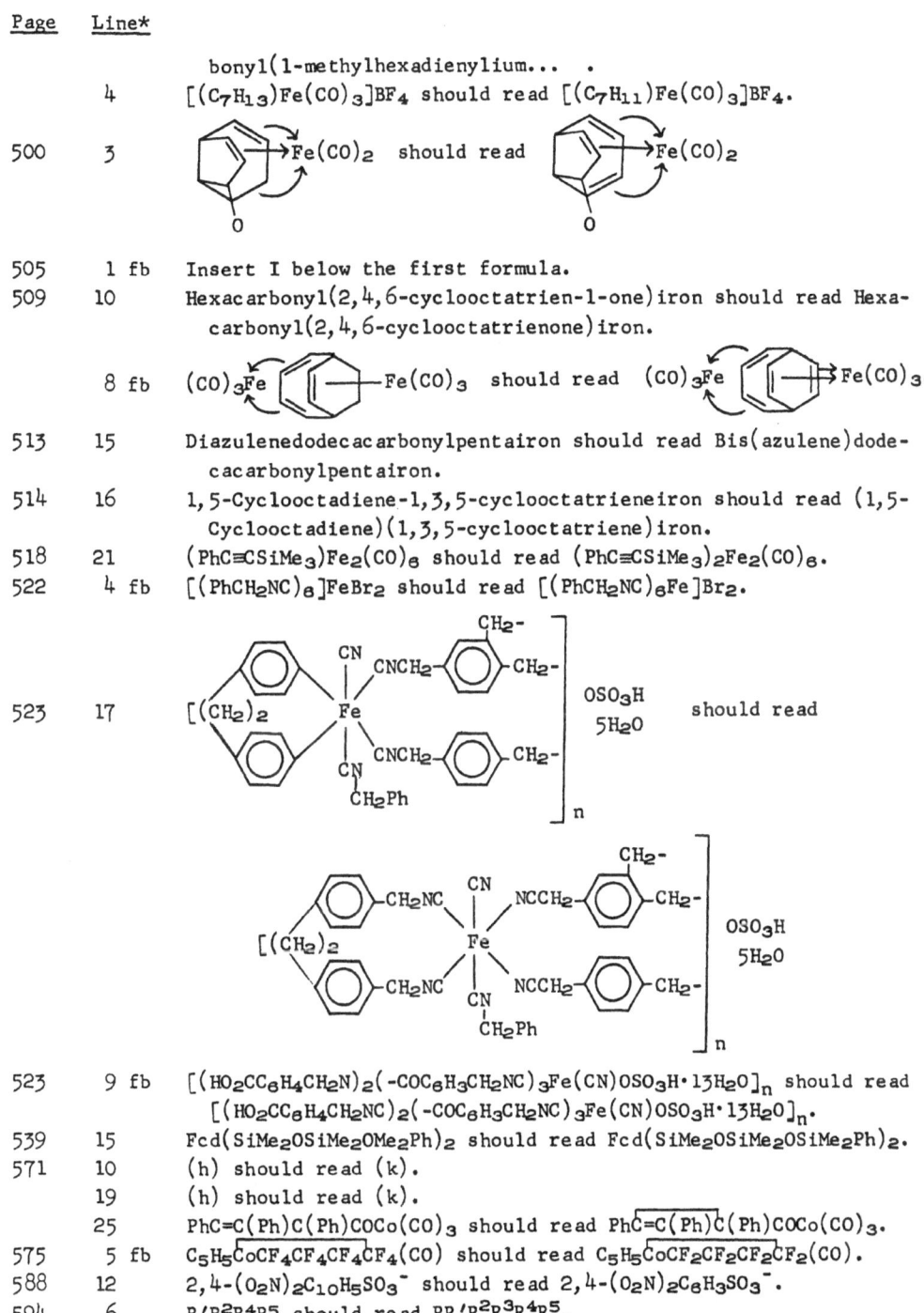

Page	Line*	
594		2nd formula fb - 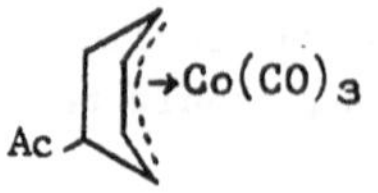 should read 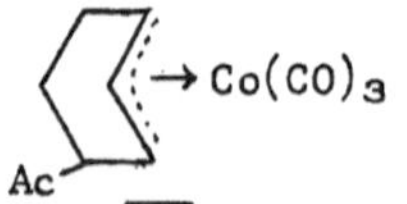 .
		1st formula fb - 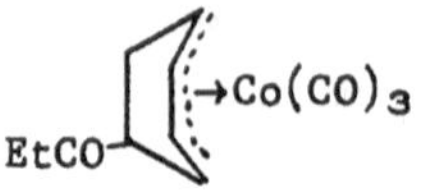 should read 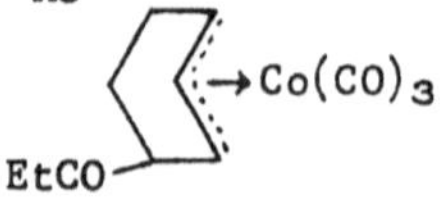 .
602	20	Bicyclo[2.2.1]hepta-2,5-trienehexacarbonyldicobalt should read Bicyclo[2.2.1]hepta-2,5-dienehexacarbonyldicobalt.
605	4	$HC{\equiv}CC(Me){\equiv}CH_2$ should read $HC{\equiv}CC(Me){=}CH_2$.
608	4	$HC{\equiv}CSiMe$ should read $HC{\equiv}CSiMe_3$.
614	10	$[Co_2(O_2C_6H_4NC)_4Co]I_2$ should read $[Co_2(O_2CC_6H_4NC)_4Co]I_2$.
617	11	$[(RNC)_5Co][Co(CO)_3]$ should read $[(PhNC)_5Co][Co(CO)_3]$.
618	13 fb	$(C_{16}H_{33}C_5H_4)Co(CO)SnPhCl$ should read $(C_{16}H_{33}C_5H_4)Co(CO)Sn(C_6H_4OEt)Cl$.
	13 fb	+ $PhSnCl_3$ should read + $EtOC_6H_4SnCl_3$.
623	8 fb	$[(C_2H_2)_2RhCl]_2$ should read $[(C_2H_4)_2RhCl]_2$.
625	17 fb	(b, g) - delete b .
626	1	Biscyclo[2.2.1]hepta-2,5-diene... should read Bicyclo[2.2.1]-hepta-2,5-diene... .
631	6	Acetylacetonato-1,5-cyclooctadienyliridium should read Acetyl-acetonato-1,5-cyclooctadieneiridium.
660	6 fb	$(C_{12}H_{16})Ni$ should read $(C_{12}H_{18})Ni$.
663	14	$(Me_2CHCH_2CH_2)_4Ni(CN)_2$ should read $(Me_2CHCH_2CH_2NC)_4Ni(CN)_2$.
668	15	Halo[2-(3,5,7-triethyl-2,4,6,8-tetraphenyl-1,3,5,7-cyclotetraenyl)ethyl]palladium should read Halo[2-(3,5,7-triethyl-2,4,6,8-tetraphenylcycloocta-1,3,5,7-tetraenyl)ethyl]palladium.
679	16	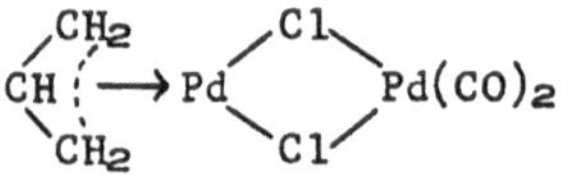 should read 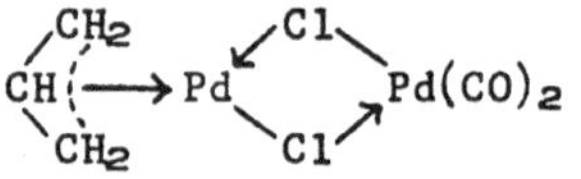 .
	21	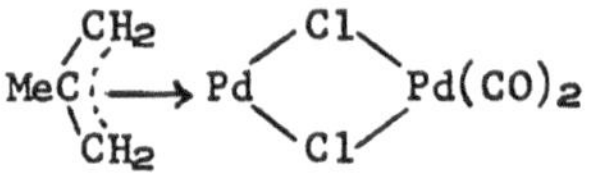 should read 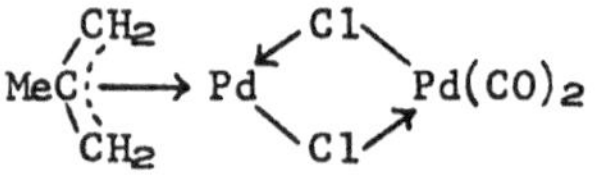 .
	26	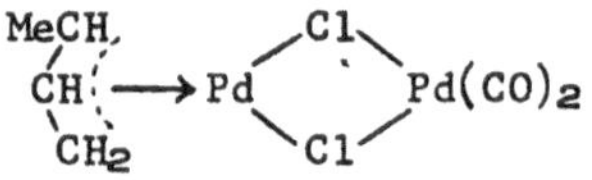 should read 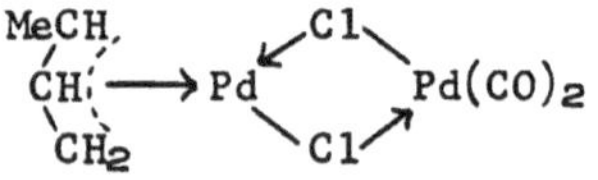 .
680	14	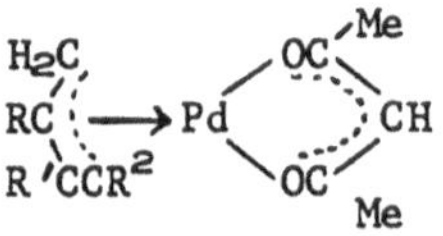 should read 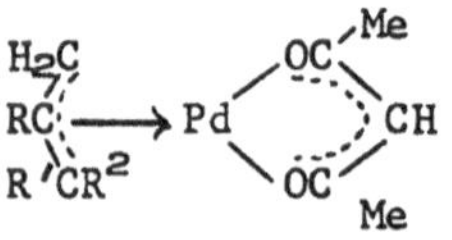 .

Page	Line*	
682	4fb-1fb	Delete this entry.
683	1	(1,3,5,7-Tetraethyl-2,4,6,8-tetraphenylcyclooctatetraene)bis-(palladium chloride) should read [(Diethyltetraphenyl-1,3,5,7-cyclooctatetraenylene)diethylene]bis(palladium chloride).
687	1 fb	$Pd(CO)Ge(C_5H_4Me)Br$ should read $Pd(CO)Ge(C_5H_4Me)_2Br$.
690	10 fb	trans-$EtPt(PPh_3)I$ should read trans-$EtPt(PPh_3)_2I$.
	9 fb	trans-$EtPt(PPh_3)SCN$ should read trans-$EtPt(PPh_3)_2SCN$.
694	8 fb	trans-$(PhC{\equiv}C)_2Pt(AsEt_2)_2$ should read trans-$(PhC{\equiv}C)_2Pt(AsEt_3)_2$.
700	19	$[Pd(C_8H_{12}OMe)Cl]_2$ should read $[Pt(C_8H_{12}OMe)Cl]_2$.
	21	$[Pd(C_8H_{12}OMe)I]_2$ should read $[Pt(C_8H_{12}OMe)I]_2$.
705	8	(i, j) should read (i, j, k).
706	10	$[Pt(C_{10}H_{12})I_2]$ should read $[Pt(C_{10}H_{16})I_2]$.
	12	Acenaphthalene should read acenaphthylene.
708	4	cis-$[Py(Py)(C_2H_4)Cl_2]$ should read cis-$[Pt(Py)(C_2H_4)Cl_2]$.
709	9	trans-$[Pt(p\text{-}H_2C_6H_4Me)(CH_2{=}CHCH_2OMe)Cl_2]$ should read trans-$[Pt(p\text{-}H_2NC_6H_4Me)(CH_2{=}CHCH_2OMe)Cl_2]$.
710	20	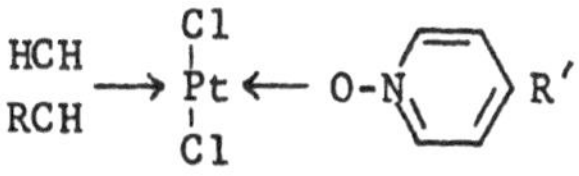 should read 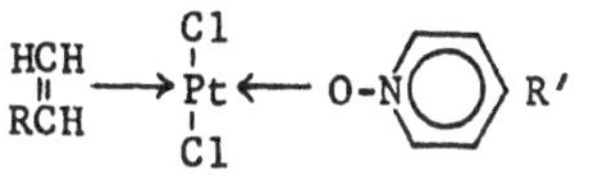 .
727	13	$\{Pt(NH_3)[Me_2C(OH)C{\equiv}]Cl_2\}$ should read $\{Pt(NH_3)[Me_2C(OH)C{\equiv}]_2Cl_2\}$.
	17	$\{Py(Pip)[Me_2C(OH)C{\equiv}]_2Cl_2\}$ should read $\{Pt(Pip)[Me_2C(OH)C{\equiv}]_2Cl_2\}$.
731	20	$\{Pt[Me_2C(OH)C{\equiv}CC{\equiv}C(OH)Me_2]_2Cl_2\}_2$ should read $\{Pt[Me_2C(OH)C{\equiv}C\text{-}CC{\equiv}CC(OH)Me_2]Cl_2\}_2$.
733	20	(Chlorodimesitylplumbyl)... should read Trichloroplumbyl... .
	21	$(EtC_5H_4)Pt(CO)Pb(C_6H_2Me_3)_2Cl$ should read $(EtC_5H_4)Pt(CO)PbCl_3$.
	22	$(C_{18}H_{37}C_5H_4)Pt(CO)PbPh_2Cl$ should read $(C_{18}H_{37}C_5H_4)Pt(CO)PbPh_2F$.
738	21	$\overline{ON{=}C(Me)C{=}C}Ph$ should read $\overline{ON{=}C(Me)CH{=}C}Ph$.
740	7 fb	Decahydrotriphenylene should read Dodecahydrotriphenylene.

CORRECTIONS AND ADDITIONS TO VOLUME II

Page	Line*	
12	9 fb	Correct formula: n-$C_{18}H_{37}$.
47	13 fb	Correct formula: PhCHOH .
	12 fb	Correct formula: PrCHOH .
	11 fb	Move formula: m-C_6H_4 five spaces to left.
	10 fb	Move formula: p-C_6H_4 five spaces to left.
51	18 fb	Correct formula: $Et_3GeC{:}CMgBr$.
	13 fb	Correct formula: $Et_3GeCH_2C{:}CMgBr$.
52	1	Correct formula: $Bu_3GeC{:}CMgBr$.
60	6	Correct formula: $(PhCH_2)_2C_7H_{15}GeH$.
72	11 fb	Correct formula: $(i\text{-}C_5H_{11})_3GeI$.
76	11	Correct formula: $CHOCHMeCH_2$.
103	6	Correct formula: $(i\text{-}Pr_2GeO)_n$.
109	10 fb	Correct formula: $BuEtCHCH_2$.
110	3 fb	In the structural formula an H should be added at N.
116	3 fb	Correct formula: $Pr_2Ge(OPr)O_2CC_7H_{15}$.
119		Table 39, line 4 fb - correct formula: $Bu_3GeO_2CCF_3$.
127	9 fb	Correct formula: $C_3H_3N_2$ = imidazolyl; .
128		Table 44, line 6 fb - correct symbol: Pb instead of Ph .
		Table 45, line 4 - correct formula: $Bu_2Ge[N{:}P(OMe)_3]_2$.
		Table 45, line 4 fb - correct formula: $(1,3\text{-}Ph_2GeN{:}AsPh_2C_6H_4\text{-}AsPh_2{:}N)_n$.
137	13	Correct formula: $Ph_2(n\text{-}C_{18}H_{37})GeLi$.
139	10 fb	Correct formula: $(PhCH_2CH_2)$.
141	10	Correct symbols: R_n = Me_2, F .
	13	After IR spectrum add: (R' = C_5H_5) .
	6 fb	Correct formula: 2,4-$Me_2C_6H_3$.
142		Table 50, line 5 - add: (R' = 9-fluorenyl) .
144		Before Table 51 add: A cyclic tetraorganogermane incorporating tin and germanium in the same ring, 1,1,4,4,1,4-tetraphenylgermastannacyclohexane, is listed in Table 16 on page 50.
		Table 51, after line 5 - insert: (L_2L_2)-tetrabenzo IB 29 m. 244.5-246°, vaporizes at 470° without dec. 315A .
	1 fb	Add: yield 18% IR spectrum 1260A .
147	5	Correct reference 2055 to 2056 .
148		Table 51, line 4 - correct reference 1820 to 1826 .
		Table 51, last line - add reference 1260A .
188	5	Correct formula: C_8H_{17} .
191	8 fb	Correct formula: $C_{12}H_8NCH_2CH_2SnEt_3$.
	7 fb	Correct formula: $C_{12}H_8NCH_2CH_2SnPh_3$.
192	11	Correct formula: 4-$CH_2CH_2SnPh_3$.
202	5	Correct formula: $BuCH{:}CH$.

---

* The "fb" letters following a line number signify that the lines were counted from the bottom of the page.

Page	Line*	
214	12 fb	Correct formula: $9\text{-}MeC_{14}H_8C{:}C$.
217	3 fb	Instead of CF_3SnF should read Me_3SnF .
234	5	Correct formula: $Me_2Sn(C_6H_4CO_2NHR)_2$; $R = C_{14}H_{29}$.
236	3 fb	Correct formula: $R'_2 = [(EtO_2C)_2C]_2CH_2$.
237	2	Correct formula: $i\text{-}Bu_2Sn(CH_2CH_2CO_2Me)_2$.
239	3	Correct formula: $(Et_3SnCH_2CHMeS_3)_2$.
242	9	Correct formula: $Ph_2Sn(C_6H_4NMe\cdot Me_2SO_4\text{-}p)_2$.
253	8	Correct formula: $R_nSn[HAl(OEt)_3]_{4-n}$.
269	10	Add: By thermal dec. of Me_3SnCF_3 (1202).
282	11	Add reference 380, cf. p. 611*.
	14	Add reference 380, cf. p. 611*.
283	3	Correct formula: $Bu_3SnI\cdot OPEt_3$.
293	11	Correct formula: $5\text{-}C_6H_{11}$.
300	4	Correct formula: $Pr_2(EtO_2CCH_2CHMe)SnBr$.
315	11	See also page 320, line 14.
323	13, 11, 9 fb	Correct formulae: $(MeCH{:}CH)_2SnBr_2$.
325	2	Correct formula: $(ClCH_2CH_2)_2SnI_2$.
331	14	Correct formula: $t\text{-}BuSnBr_3\cdot bipy$.
334	7 fb	Correct formula: $Me_2CHCH_2C({:}CH_2)SnCl_3$.
	6 fb	Correct formula: 1-Cyclohexenyl $SnCl_3$.
338	4 fb	Change reference 386 to 382.
340	4	Add formula: $(Pr_2SnI)O(Pr_2SnOH)$; dimeric; change ref. 386 to 382.
341	4, 5, 10, 11 fb	Change reference 386 to 382.
	6 fb	Correct formula: $Bu(Bu_2SnO)_5OBu\cdot Bu_2SnCl_2$.
	5 fb	Correct formula: $Et(Me_2SnO)_3OEt\cdot Et_2SnI_2$.
	4 fb	Correct formula: $Et(Et_2SnO)_3OEt\cdot Me_2SnI_2$.
345	8	Correct formula: $(Me_2Sn)_3[Fe(CN)_6]_2$.
348	3 fb	Correct formula: $2HPhPO_2$.
	1 fb	Correct formula: $(Me_2Sn)_3(AsO_4)_2$.
349	12	Correct formula: MoO_4 .
	13 fb	Correct formula: $(p\text{-}Me_2CHC_6H_4CH_2)_3SnOB(OCH_2CHEtBu)_2$.
	12 fb	Correct formula: $(p\text{-}Me_2CHC_6H_4CH_2)_2Sn[OB(OC_{12}H_{25})_2]_2$.
353	11	Correct formula: $2Ph_2PO_2$.
356	10	Correct formula to $MeSO_2S^-$ and move four spaces to the left.
	11	Correct formula to $PhSO_2^-$ and move four spaces to the left.
394	13 fb	Correct formula: $(i\text{-}Bu_3SnO)_2SiEt_2$.
402	11	Correct formula: $2,4,5\text{-}Cl_3C_6H_2$.
	12 fb	Correct formula: $2,6,5\text{-}Br_2(Me_3C)C_6H_2$.
	10 fb	Correct formula: C_6Br_5 .
403	9 fb	Correct formula: $CH(C_6H_{13})CH_2CH{:}CH(CH_2)_7CO_2Me$.
404	6	Correct formula: $RO(Bu_2SnO)_4R$.
	11	Correct formula: $2PrCH{:}NO$.
405	10	Correct formula: $Bu_2\overline{SnOCH_2CH_2CO}HMe$.

Page	Line*	
405	11	Correct formula: $Bu_2Sn[OCH(C_6H_{13})CH_2CH{:}CH(CH_2)_7CO_2CH_2]_2$.
	10 fb	Correct formula: $1,2,4\text{-}(HOBu_2SnO)_2C_6H_3CMe_3$.
	8 fb	Correct formula: $[4,2,6\text{-}(HOBu_2SnO)Cl_2C_6H_2]_2CMe_2$.
	7 fb	Correct formula: $[4,2,6\text{-}(HOBu_2SnO)Br_2C_6H_2]_2CMe_2$.
418	2 fb	Correct formula: $PhCH{:}CH$.
		Last line - missing formula: $p\text{-}CH_2{:}CHC_6H_4$, under properties add - polymerization rxn.
435	13	Correct formula: $(PhCH_2)_2Sn(O_2CCHEtBu)_2$.
436	15	Correct formula: $(CH_2{:}CH)_2Sn(O_2CC_{11}H_{23})_2$.
437		Following line 14 insert: $Bu_2Sn(OH)O_2CC_{11}H_{23}$ IIIC (f, zz) 1310.
		Last line add: (zz) cf footnote (h), Table 145, page 441.
441		Delete line 15 fb - "$Bu_2Sn(OH)O_2CC_{11}H_{25}$" - insert instead: $2Bu_2SnO_2CR(O_2CR')\cdot 3Bu_2Sn(O_2CCH{:}CHCO_2CH_2)_2$ IIIC n_D^{24} 1.503, d_{20} 1.304 (f, j) 1748 .
	12 fb	Instead of R = Me should read R = $MeO_2CCH{:}CH$.
		Last line - instead of R = R' = Me should read R = R' = $MeO_2C\text{-}CH{:}CH$; after (2712) add: (j) R = $C_{12}H_{25}O_2CCH{:}CH$, R' = $BuO_2C\text{-}CH{:}CH$.
444	9	Correct formula: $2,4\text{-}HO(H_2N)C_6H_3$.
	15	Correct formula: $\overline{OCH_2CH}$.
	14, 15 fb	Delete entry.
	2 fb	Delete footnote (j).
447	8 fb	Correct formula: $(C_8H_{17})_2\overline{SnS^{14}CH_2CH_2CO_2}$.
449	4	Correct formula: $CH_2{:}CHSn(O_2CC_{11}H_{23})_3$.
454	3 fb	Correct formula: $[(Me_3SiCH_2)SnS]_n$.
464	7	Correct formula: $(PhCH_2)_2Sn(SC_7H_5N_2\text{-}2)_2$.
	10 fb	Correct formula: $Ph_2Sn[S(C_2N_2S)SH\text{-}2,5]_2$.
	6 fb	Correct to (a) Mercaptobenzothiazolyl.
465	16	Correct formula: $C_8H_{17}CO_2CH_2CH_2$.
466	10	Correct formula: $Bu_2Sn(SCH_2{}^{14}CO_2C_8H_{17})_2$.
	8 fb	Correct formula: $i\text{-}C_9H_{19}O_2CCHPr$.
467	21	Correct formula: $(i\text{-}C_5H_{11})_2Sn[SCH(CO_2R)CH_2CO_2R]_2$.
468	14 fb	Correct formula: $\underline{MeSn(SR')_3}$.
469	9	Correct formula: $RO_2CCH_2CH(CO_2R)$.
492	11	Delete entry: (II B)... .
		Table 169, line 3 - delete entry: $(MeSnF)_x$... .
501	1	Correct formula: $Me_2Sn[Mn(CO)_5]_2$.
	6	Correct formula: $(C_5H_5)_3SnMn(CO)_5$.
	7	Correct formula: $(C_5H_5)_3SnMn(CO)_3\cdot$diene .
	9	Correct formula: $(2\text{-}C_{13}H_9)_3SnMn(CO)_3\cdot$diene .
	3 fb	Correct first reference to (2342).
503	3	Correct formula: $C_{12}H_{25}SnF[R'RhCO]_3$.
504		Following line 5 fb insert (III E) $SS + SnCl_4 \rightarrow SCl_2$.

Page	Line*	
504		Add new paragraph: Organogermastannacyclohexanes and organosilastannacyclohexanes may be found in Tables 16 and 77 on pages 50 and 251, respectively.
506	3	Add: yield 62% reference 550 .
	7	Add: reference 1260A .
	9	Add: yield 67% before (d) insert - m. 193-195° (1629) .
	11	Correct Ph_2 to Ph, replace reference 527 with 550 .
507	4	Add: reference 1260A .
	5	Delete entry Ph_2P... .
	7-11	Correct reference to 1741 .
	12	Under properties add: cyclic trimer m. 277-280° .
	12 fb	Add: yield 73% (ii) .
		Last line - add: (ii) Rxn. with Br → o,o'-$(BrC_6H_4CH_2)_2$ (538).
510	8	Correct formula: $Me_3Sn(CH_2)_3CH{:}CH_2$.
511		Following line 9 insert: p-$Bu_3SnO_2CC_6H_4CH{:}CH_2$ IIIB copolymer (a), p-$ClC_6H_4CH{:}CH_2$, fungicide 2403 .
		Following line 12 insert: $Bu_2Sn(O_2CC_6H_4CH{:}CH_2\text{-}p)_2$ + (b) IIIB fungicide 2403 .
	15	Correct formula: $R_2Sn(O_2CCMe{:}CH_2)_2$.
		Following line 6 fb insert: $Ph_2Sn(O_2CC_6H_4CH{:}CH_2\text{-}p)_2$ + (a) IIIB fungicide 2403 .
535	8 fb	Correct formula: 7-$C_{18}H_{11}$.
540	12 fb	Correct formula: 1-$C_6H_9C{:}C$.
546	12 fb	Correct formula: $\underline{3\text{-}(p\text{-}O_2NC_6H_4N_2)C_6H_3X\text{-}_n}$.
	9 fb	Correct formula: 3,5-$(p\text{-}O_2NC_6H_4N_2)_2C_6H_2OH$-6 .
551	8	Correct formula: Et_5Pb_2H .
564	12 fb	Correct formula: 1,4-$C_{10}H_6NH_2$.
	5 fb	Instead of (Et + $C_{14}H_{25}$) should read (Et + $C_{12}H_{25}$) .
	4 fb	Instead of (Et + $C_{20}H_{31}$) should read (Et + $C_{20}H_{41}$) .
	3 fb	Instead of $Me_2C_6H_4$ and p-$(i\text{-}C_8H_{17}O)C_6H_4$ should read i-C_8H_{17} and p-i-$C_8H_{17}C_6H_4$, resp.
565	3	Correct formula: Et_2PbY .
566	4	Correct formula: Y = 1/2$(MeAsO_2,Cl)_2O$.
	14	Correct formula: $(HO)_2C_6H_2(SO_3)$.
569	14	Correct formula: $Ph_2(o\text{-}HO_2CC_6H_4)PbOH$.
571		Following line 7 fb insert: R = 1/2$C_6H(NO_2)_3$ --- --- (dd) .
		Last line add: (dd) Use as component in electrical blasting caps (130).
572	5	Correct formula: $(C_8H_{17})_2Pb[(OCH_2CH_2)_6OEt]_2$.
575	10 fb	Correct formula: Ph_3PbO_2CR .
578	5	Correct formula: $(Et_3Pb)_3$ citrate .
	6	Correct formula: $(Et_3Pb)_2$ camphorate .
	16	Delete entry: Et_3Pb-styphnate... .
583	10	Correct formula: $PhPb(OBz)_3$.
594	10	Correct formula: $[(C_{16}H_{33})_2PbRNb(CO)_3]_2$.

Page	Line*	
594	17	Correct formula: $C_{18}H_{37}C_5H_3Me$.
	19, 14 fb	Correct formula: C_5H_8 = 1,3-pentadiene .
596	13, 14, 15	Add reference 1260A.
601		Reference 80 - correct 1460 to 1960.
609		Following reference 315 insert: 315A. --- and Gorsich, R. D., J. Am. Chem. Soc. 80, 1883-6 (1958); CA 52, 16331.
		Ref. 316 - change ibid. to J. Org. Chem.
610		Change reference 379 to 383; 380 to 379.
611		Change reference 383 to 380.
638		Following reference 1260 insert: 1260A. --- and Zuech, E. A., J. Am. Chem. Soc. 82, 2522-4 (1960); CA 54, 21121.
659		Reference 1826- correct to Zh. Obshch. Khim. 32, 1137-46, 1455-60 (1962); CA 58, 1332, 9111.

CORRECTIONS AND ADDITIONS TO VOLUME III

Page	Line*	
1	15	... and primary aromatic amines react at 100° to form triarylarsines should read ... and tertiary aromatic amines react at 100° to form tris(aminoaryl)arsines.
39	5	$[Pt(Ars)Cl_2]$ should read $[Pt(Ars)Br_2]$.
48	1	$C_{10}H_{16}As$ should read $C_{10}H_{18}AsN$.
63	1	$C_{21}H_{19}As$ should read $C_{21}H_{19}AsO$.
75	15	Insert C_{11} in first column.
76	4	C_{10} (Cont.) should read C_{11} (Cont.).
79	13 fb	$C_5H_{12}AsS$ should read $C_5H_{12}As_2S$.
120	5 fb	$3,4\text{-}H_2N(HO)C_6H_4AsH_2$ should read $3,4\text{-}H_2N(HO)C_6H_3AsH_2$.
132	17	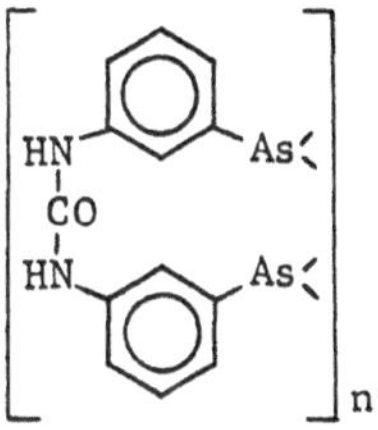 should read 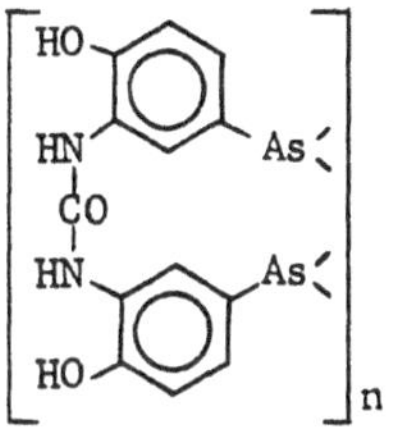
133	6-5 fb	Phosphate, should read Phosphate), $(C_{12}H_{16}AsN_4O_4P)_n$.
134	2	$(C_{13}H_6AsNO)_n$ should read $(C_{13}H_8AsNO)_n$.
142	15	$\{4\text{-}[4\text{-}HO_2CCH(NHBz)C_6H_4As\}$ should be listed with C_{16} as $\{4\text{-}[HO_2CCH(NHBz)CH_2]C_6H_4As\}_n$.
142	2 fb	$[4,3\text{-}HOCCH_2O(4\text{-}Me...$ should read $[4,3\text{-}HO_2CCH_2O(4\text{-}Me...$.
145	23	$(C_{17}H_{18}As_2N_3O_5)_n$ should read $(C_{18}H_{17}As_2N_3O_5)_n$.
	4 fb	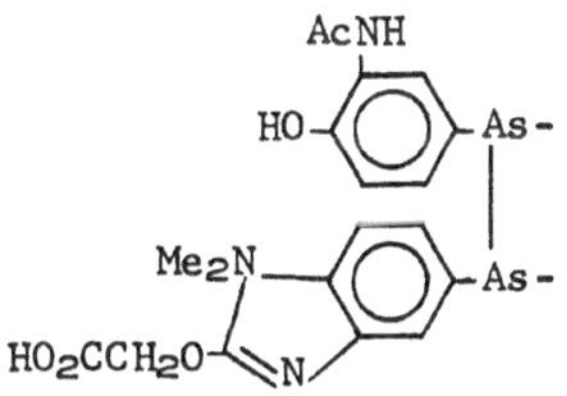 should read 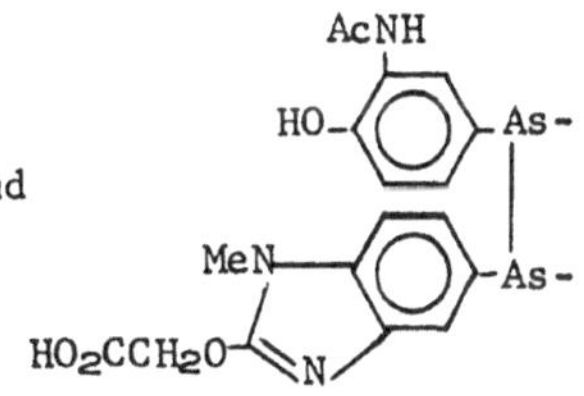
146	2-7	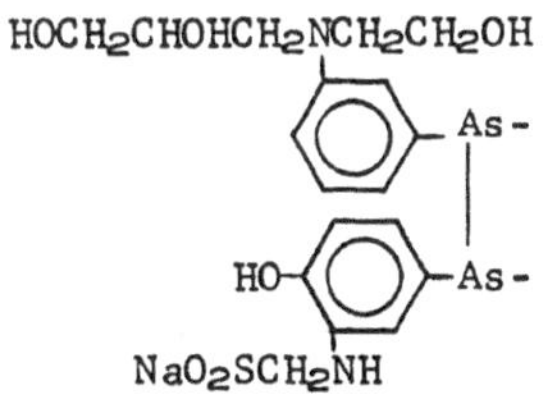 should read 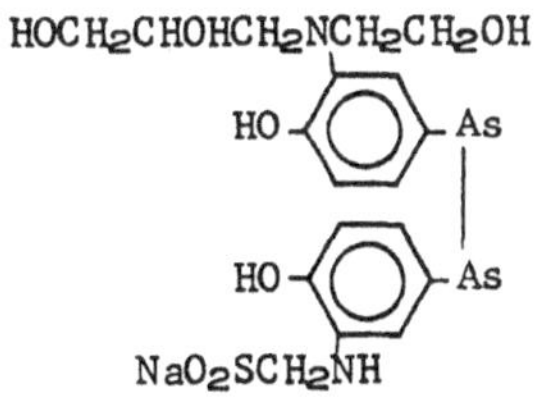

* The "fb" letters following a line number signify that the lines were counted from the bottom of the page.

Page	Line*	
146	16-22	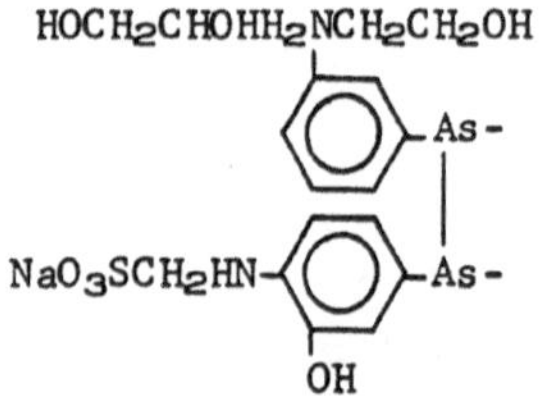 should read 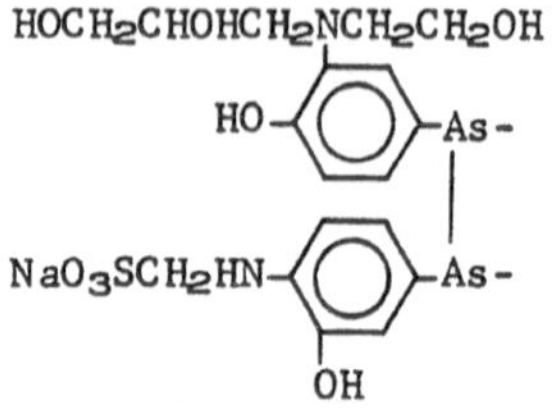
147	12	$(C_{19}H_{20}As_2N_4O_8S)_n$ should read $(C_{19}H_{19}As_2N_4NaO_8S)_n$.
148	1	3-Amino-4,4′-(3-hydroxypropoxy)- should read 3-Amino-4,4′-bis-(3-hydroxypropoxy)-.
	2	$(C_{19}H_{25}As_2N_2NaO_5S)_n$ should read $(C_{19}H_{25}As_2N_2NaO_7S)_n$.
	12	$(C_{10}H_{25}As_2N_2NaO_7S)_n$ should read $(C_{19}H_{25}As_2N_2NaO_7S)_n$.
152	2	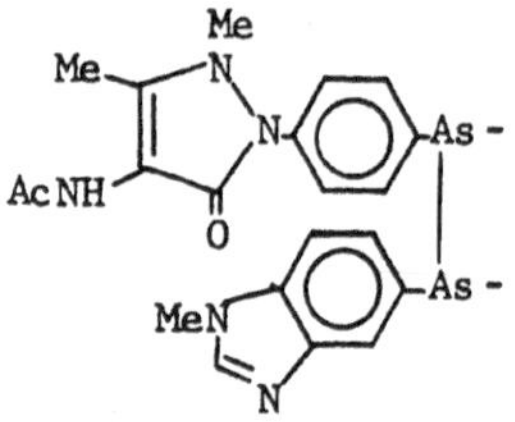 should read 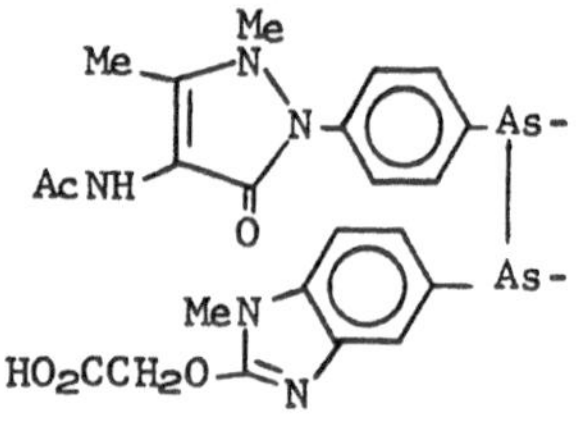
154	7 fb	{3,4-HO[($HOCH_2CH_2$)N]... should read {3,4-HO[($HOCH_2CH_2)_2$N]... .
155	5	{3,4-HO[$HOCH_2CH_2$($HOCH_2COHCH_2$)N]... should read {3,4-HO[$HOCH_2$-CH_2($HOCH_2CHOHCH_2$)N]... .
	13	C_{20} should appear between lines 14 and 15.
169	15	Me[SC(Me)=CHCH=CH)AsCl should read Me[SC(Me)=CHCH=C]AsCl.
206	8	C_1 should read C_3.
	11	C_3 should read C_5.
	13	C_4 should read C_6.
	18	C_5 should read C_7.
	20	C_6 should read C_8.
	21	n-$C_6H_{11}As(CN)_2$ should read n-$C_6H_{13}As(CN)_2$.
224	18	4,3-HO($NaO_3SC_6H_{11}O_5$)$C_6H_3As(OH)_2$ should read 4,3-HO($NaO_3SC_6H_{12}$-O_5)$C_6H_3As(OH)_2$.
	18	$NaO_3SC_6H_{11}O_5$ = Na glucosebisulfite should read $NaO_3SC_6H_{12}O_5$ = Na bisulfitoglucosyl.
225	4 fb	Insert C_{20} between lines 5 and 4 fb.
233	15	$C_{16}H_{13}AsN_2O_{12}S$ should read $C_{16}H_{13}AsN_2O_{12}S_3$.
243	3	8-Methylquinolinyl-5-AsO should read 8-Methylquinolyl-5-AsO.
	5	4-(SCH=CHNHCNHCO)C_6H_4AsO should read 4-(SCH=CHNHCHNHCO)C_6H_4AsO.
	15	4-[N=C(NH_2)N=C(NH_2)CH=CNH]C_6H_3AsO should read 4-[N=C(NH_2)N=C(NH_2)CH=CNH]C_6H_4AsO.
248	1 fb	4-(4-$AcNHC_6H_4SO_2NHCH_2$)C_6H_4AsO should be listed with C_{15} compounds.
249	3	4-[$EtO_2CN(CH_2CH_2)_2NCH_2CONH$]$C_6H_4$AsO should be listed with C_{15} compounds.

Page	Line*	
255	8	$PhAs(OCH_2C{=}CHCH{=}CHO)_2$ should be listed with C_{15} compounds.
260	5 fb	C_{22} should read C_{20}
	4 fb	$Ph_2SC_6H_4CO_2Me$ should read $Ph_2AsSC_6H_4CO_2Me$.
265	15	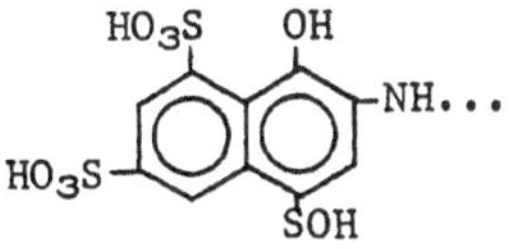 should read 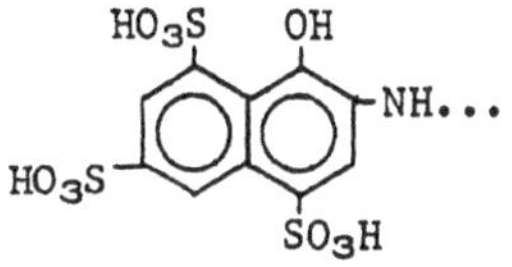
271	6	$4\text{-}[N{=}C(NH_2)N{=}C(NMe_2)N{=}CNH]C_6H_4As{=}(SCH_2CO_2H)_2$ should be listed with C_{15} compounds.
272	10	$MeAs(SC_{19}H_{19})_2$ should read $MeAs(SC_9H_{19})_2$.
273	2	$2\text{-}O_2NC_6H_4As(S\text{-}C_8H_6NS)_2$ should be listed with C_{22} compounds.
	8 fb	$2,5\text{-}HO(AcNH)C_6H_3As(SHMeCO_2Bu)_2$ should read $2,5\text{-}HO(AcNH)C_6H_3As\text{-}[SCH(Me)CO_2Bu]_2$.
274	1 fb	$4\text{-}[N{=}C(NH_2)N{=}C(NH_2)N{=}CNH]C_6H_4As(SC_6H_4CO_2H)_2$ should be listed with C_{23} compounds.
275	8 fb	$3,4\text{-}H_2N(HOCH_2CH_2O)C_6H_3As(SCH_2CO_2CHEtBu)_2$ should be listed with C_{26} compounds.
276	5	$4,3\text{-}HO[N{=}C(NH_2)N{=}C(NH_2)N{=}CNH]C_6H_3As(SC_{10}H_{16}N_3)_2$ should read $4,3\text{-}HO[N{=}C(NH_2)N{=}C(NH_2)N{=}CNH]C_6H_3As(SC_{10}H_{16}N_3O_6)_2$.
	10	$4\text{-}[N{=}C(NH_2)N{=}C(NH_2)N{=}CNH]As(S\text{-}C_{10}H_{16}N_3)_2$ should read $4\text{-}[N{=}C(NH_2)N{=}C(NH_2)N{=}CNH]C_6H_4As(S\text{-}C_{10}H_{16}N_3O_6)_2$.
283	16	$MeAs[SC(S)NHN{=}CHC{=}CHCH{=}CHS]_2$ should read $MeAs[SC(S)NHN{=}CH\text{-}C{=}CHCH{=}CHO]_2$.
284	5	$PhCH_2As[SC(S)NHN{=}CC{=}CHCH{=}CHS]_2$ should read $PhCH_2As[SC(S)NHN{=}CH\text{-}C{=}CHCH{=}CHO]_2$.
	8	$PhAs[SC(S)NMePh]_2$ should be listed with C_{22} compounds.
288	19	$[(CF_3)_2As]NH$ should read $[(CF_3)_2As]_2NH$.
301	11 fb	$C_{10}AsF_6IMnO_8$ should read $C_{10}AsF_6IMn_2O_8$.
304	9	$C_{16}H_{12}As_2Fe_2O_2$ should read $C_{16}H_{22}As_2Fe_2O_2$.
309	4	C_{20} should read C_{19}.
316	25	Insert in column 1-2: $Me[H_2NC(=NH)](HO_2CCH_2CH_2)PhAsSCH_2Ph$.
317	19	Insert in column 1-2: $Me[H_2NC(=NH)](HO_2CCH_2CH_2)(3\text{-}MeOC_6H_4)AsS\text{-}CH_2Ph$.
318	14	$MeEtPh(PhCH_2)As$ should read $MeEtPr(PhCH_2)As$.
319	21	$Me_3(2\text{-}PhC_6H_4CH_2)As$ should be listed with C_{16} compounds.
332	17	$[Ph_4As][Cr(\langle{}^{SCCF_3}_{\overset{\Vert}{SCCF_3}})_3]$ should read $[Ph_4As]_2[Cr(\langle{}^{SCCF_3}_{\overset{\Vert}{SCCF_3}})_3]$.
339	6	Insert C_{27} between lines 6 and 7.
	15	$(MePh_2AsCH_2\text{-})_2$ should be listed with C_{28} compounds.
341	2	$Ph(PhCH_2)[PhCH_2O(CH_2)_3]As$ should read $Ph(PhCH_2)[PhCH_2O(CH_2)_3]_2As$.
	3	$\{Ph(PhCH_2)[PhCH_2O(CH_2)_3]As\}_2$ should read $\{Ph(PhCH_2)[PhCH_2O\text{-}(CH_2)_3]_2As\}_2$.
	4	Insert in 3rd column: H_2PtCl_6.

Page	Line*	
341	10	$4,1,2\text{-}ClC_6H_3(CH_2AsMe_2Ph)_2$ should be listed with C_{24} compounds.
	12	$o\text{-}C_6H_4(CH_2AsMe_2Ph)_2$ should be listed with C_{24} compounds.
	7 fb	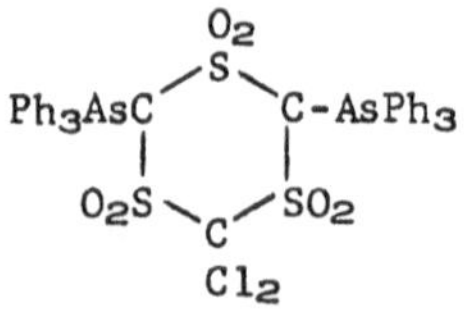 should read 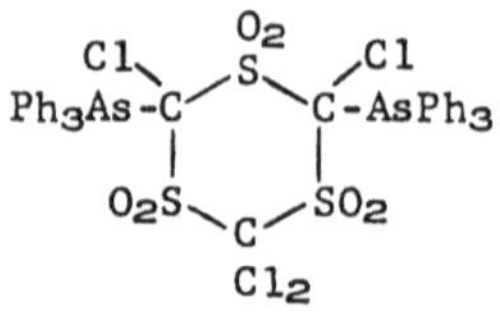
351	14	$Ph_2(1\text{-}C_{10}H_7)As(OH)_2$ should be listed with C_{22} compounds.
357	12	$Ph[HO_2C(CH_2)_2]_2AsO$ should read $Ph[HO_2C(CH_2)_3]_2AsO$.
371	2 fb	$C_{12}H_{12}AsCl$ should read $C_{12}H_{12}AsClO_2$.
373	5 fb	$C_7H_7AsO_2$ should read $C_2H_7AsO_2$.
376	11	$C_{12}H_{11}AsO$ should read $C_{12}H_{11}AsO_2$.
389	4 fb	$C_6H_5AsCl_2N_2O$ should read $C_6H_5AsCl_2N_2O_2$.
411	2	$ROC_6H_4AsO_3H$ should read $ROC_6H_4AsO_3H_2$.
412	2	$ROC_6H_4AsO_3H$ should read $ROC_6H_4AsO_3H_2$.
414	2	$ROC_6H_4AsO_3H$ should read $ROC_6H_4AsO_3H_2$.
422	16	C_{12} should read C_{13}.
428	12	The Ru and Sc... should read The Rb and Cs... .
433	16	$C_{15}H_{16}AsN_4O_5$ should read $C_{15}H_{16}AsN_7O_5$.
	12 fb	$C_{15}H_{25}AsN_{2O9}$ should read $C_{15}H_{25}AsN_2O_9$.
437	10	$MeCH{=}CHCO_2CHCO\text{-}$ should read $MeCH{=}CHCO_2CH_2CO\text{-}$.
	11	$EtO_2CC{=}CMe\text{-}$ should read $EtO_2CCH{=}CMe\text{-}$.
442	3	C_{10} should read C_9.
	22	$2,4\text{-}(O_2N)C_6H_3\text{-}$ should read $2,4\text{-}(O_2N)_2C_6H_3\text{-}$.
443		3rd Formula: 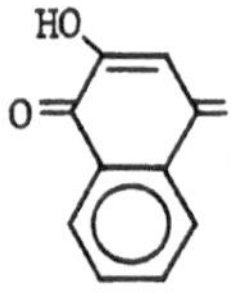 should read 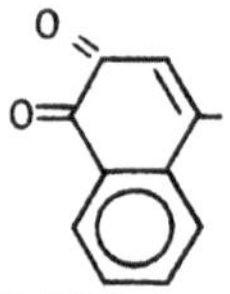
446	15	CH=CH=CH=CHN=C- should read CH=CHCH=CHN=C-.
449	6	$N{=}C(NH_2)N{=}C(NH{=}CHCH{=}CHPh)N{=}C$ should read $N{=}C(NH_2)N{=}C(NHN{=}CHCH{=}CHPh)N{=}C$-.
454	2	$C_{19}H_{21}N_3O_{10}S_2$ should read $C_{19}H_{22}N_3O_9S_2$.
459	6	4-[HNC(Me)=C(CO_2Et)C(Me)=C- should should read 4-[HNC(Me)=C(CO_2Et)C(Me)=C- and should be listed with C_{15} compounds.
	19	4-(8-Hydroxy-8-quinolyl- should read 4-(8-Hydroxy-5-quinolyl-.
463	13	4-{4-[$Et_2N(CH_2)_3$CHN(Me)]C_6H_4- should read 4-{4-[$Et_2N(CH_2)_3$N-(Me)NH]C_6H_4- and should be listed with C_{20} compounds.
465		3rd Formula: Insert H atoms at the $-SO_3$ groups of the naphthalene moiety.
467	4 fb	4-[4-($C_{27}H_{45}$N)C_6H_4- should read 4-[4-($C_{27}H_{45}$NH)C_6H_4-.
473		1st Formula: Insert a Me group at C-4 of the pyrrol ring.

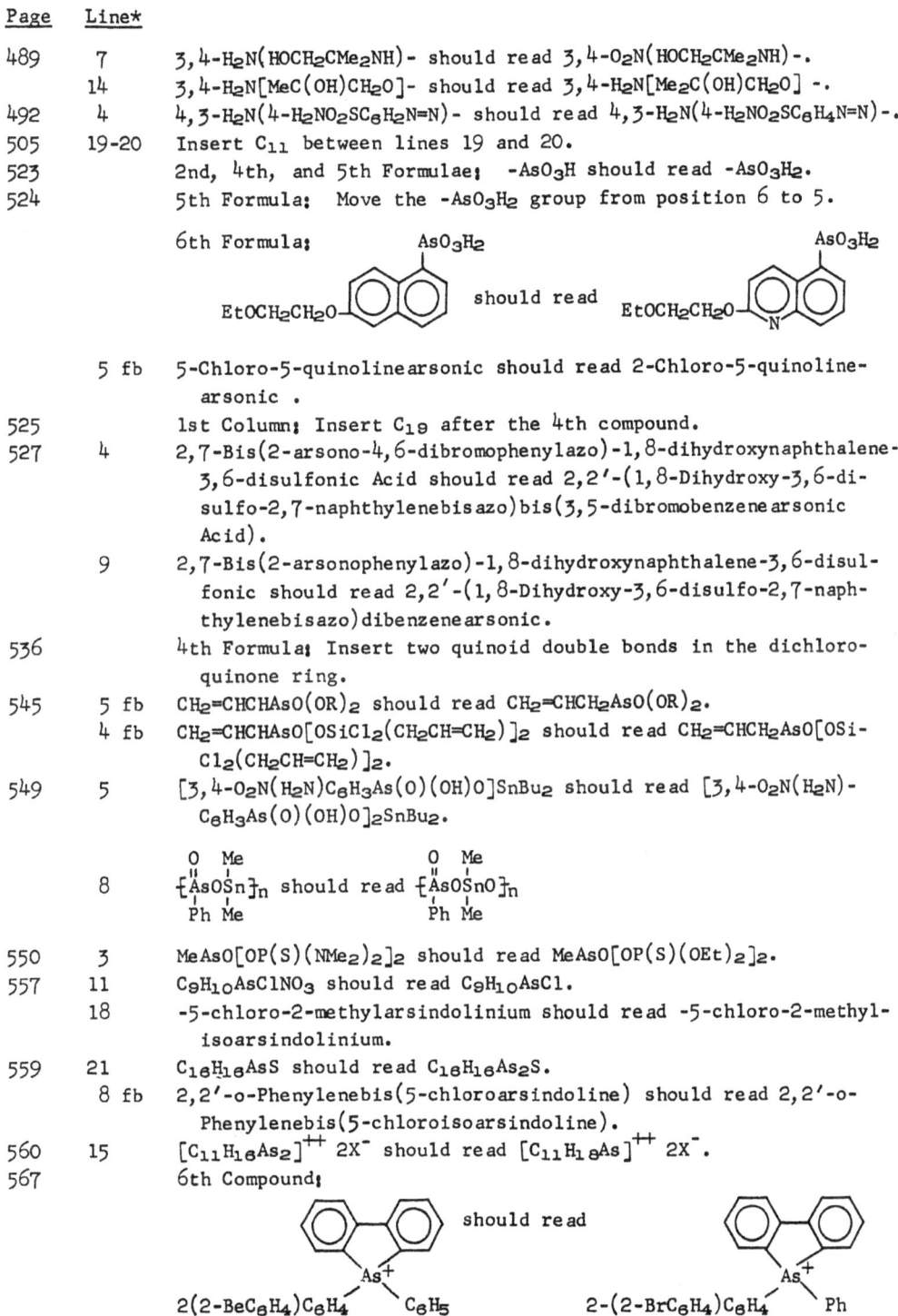

Page	Line*	
489	7	3,4-$H_2N(HOCH_2CMe_2NH)$- should read 3,4-$O_2N(HOCH_2CMe_2NH)$-.
	14	3,4-$H_2N[MeC(OH)CH_2O]$- should read 3,4-$H_2N[Me_2C(OH)CH_2O]$ -.
492	4	4,3-$H_2N(4$-$H_2NO_2SC_6H_2N{=}N)$- should read 4,3-$H_2N(4$-$H_2NO_2SC_6H_4N{=}N)$-.
505	19-20	Insert C_{11} between lines 19 and 20.
523		2nd, 4th, and 5th Formulae; -AsO_3H should read -AsO_3H_2.
524		5th Formula: Move the -AsO_3H_2 group from position 6 to 5.
		6th Formula; 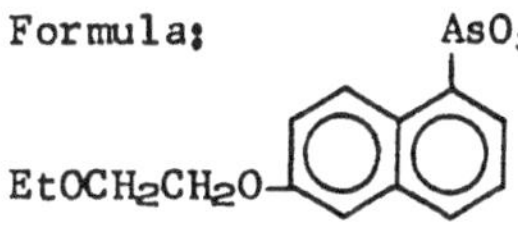 should read 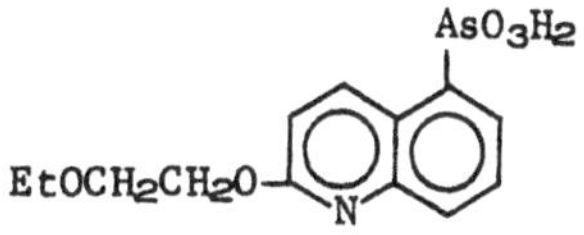
	5 fb	5-Chloro-5-quinolinearsonic should read 2-Chloro-5-quinoline-arsonic .
525		1st Column; Insert C_{19} after the 4th compound.
527	4	2,7-Bis(2-arsono-4,6-dibromophenylazo)-1,8-dihydroxynaphthalene-3,6-disulfonic Acid should read 2,2′-(1,8-Dihydroxy-3,6-di-sulfo-2,7-naphthylenebisazo)bis(3,5-dibromobenzenearsonic Acid).
	9	2,7-Bis(2-arsonophenylazo)-1,8-dihydroxynaphthalene-3,6-disul-fonic should read 2,2′-(1,8-Dihydroxy-3,6-disulfo-2,7-naph-thylenebisazo)dibenzenearsonic.
536		4th Formula; Insert two quinoid double bonds in the dichloro-quinone ring.
545	5 fb	$CH_2{=}CHCHAsO(OR)_2$ should read $CH_2{=}CHCH_2AsO(OR)_2$.
	4 fb	$CH_2{=}CHCHAsO[OSiCl_2(CH_2CH{=}CH_2)]_2$ should read $CH_2{=}CHCH_2AsO[OSi$-$Cl_2(CH_2CH{=}CH_2)]_2$.
549	5	$[3,4$-$O_2N(H_2N)C_6H_3As(O)(OH)O]SnBu_2$ should read $[3,4$-$O_2N(H_2N)$-$C_6H_3As(O)(OH)O]_2SnBu_2$.
	8	$\{\overset{O}{\underset{Ph}{As}}O\overset{Me}{\underset{Me}{Sn}}\}_n$ should read $\{\overset{O}{\underset{Ph}{As}}O\overset{Me}{\underset{Me}{Sn}}O\}_n$
550	3	$MeAsO[OP(S)(NMe_2)_2]_2$ should read $MeAsO[OP(S)(OEt)_2]_2$.
557	11	$C_9H_{10}AsClNO_3$ should read $C_9H_{10}AsCl$.
	18	-5-chloro-2-methylarsindolinium should read -5-chloro-2-methyl-isoarsindolinium.
559	21	$C_{16}H_{16}AsS$ should read $C_{16}H_{16}As_2S$.
	8 fb	2,2′-o-Phenylenebis(5-chloroarsindoline) should read 2,2′-o-Phenylenebis(5-chloroisoarsindoline).
560	15	$[C_{11}H_{16}As_2]^{++}\ 2X^-$ should read $[C_{11}H_{16}As]^{++}\ 2X^-$.
567		6th Compound; 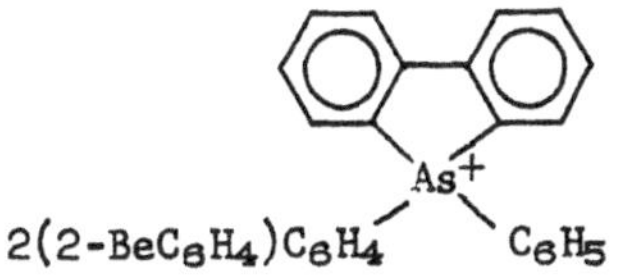 should read 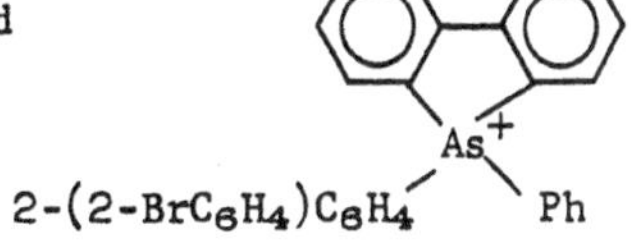

Page	Line*	
575	11 fb	-2,1,3-benzoxadiarsolinium should read -2,1,3-benzoxadiarsolium.
576	9	-2,1,3-benzoxadiarsenolylene) should read -2,1,3-benzoxadiarsolylene).
581	5	-1,3,2-benzodioxarsenole should read -1,3,2-benzodioxarsole.
585	12 fb	2,4-HO(AcNH)$C_6H_4AsSCH_2CH(CH_2OH)S$ should read 2,4-HO(AcNH)C_6H_3-$AsSCH_2CH(CH_2OH)S$.
587		2nd Compound; 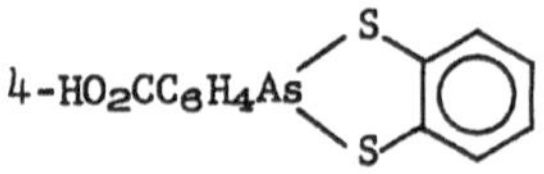 should read 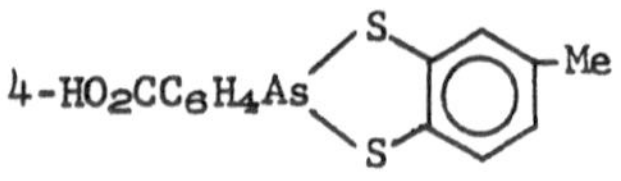
		6th Compound; 4-$H_2NC_6H_4AsSCH_2CH[CH_2O(C_6H_{11}O_5)]S$ should read 4-$H_2NC_6H_4AsSCH_2CH[CH_2O(C_6H_{11}O_5)]S$.
588		1st Compound; 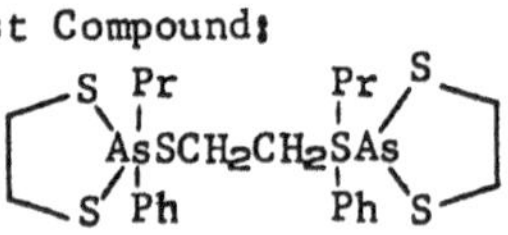 should read 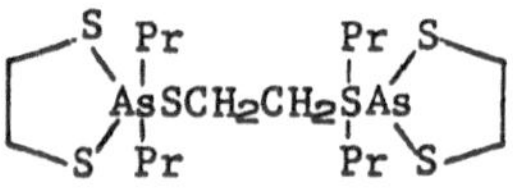
589	22	$C_5H_{11}O_2$ should read $C_5H_{11}AsO_2$.
594	22	-2-methylarsinoline should read -2-methylisoarsinoline.
595	21	1,2,3,4-Tetrachloro-2-phenyl... should read 1,2,3,4-Tetrahydro-2-phenyl... .
596	18	$C_{13}H_{11}AsO_3$ should read $C_{13}H_{11}AsO_2$.
606	16 fb	-5,10-arsanthrene) should read -5,10-arsanthrylene).
608	13 fb	$C_{46}H_{98}As_4X_4$ should read $C_{48}H_{48}As_4X_4$.
611	8	$C_{28}H_{16}AsO_4S_2$ should read $C_{28}H_{16}As_2O_4S_2$. Complete the structural formula by inserting S atoms between As atoms and the COCO bridge.
612	8	10-EtO should appear among C_{14} compounds.
616	2 fb	o-[$Bu_2NC(S)S$]- should appear among C_{21} compounds.
620	20,24	5,10-dihydrophenazarsine should read 5,10-dihydrophenarsazine.
634	10	7-Chloro-8-isopropyl-11-methyl- should read 7-Chloro-8-isopropyl-9-methoxy-11-methyl- .
635	10,15	-7,13-dihydrobenzopentene... should read -7,13-dihydrobenzocyclopenteno... .
642	19	-6-phenyl-6H-benz[c,e]... should read -6-phenyl-6H-dibenz[c,e]..
643	13 fb	$[C_{16}H_{18}As_2]^+X^-$ should read $[C_{16}H_{18}As_2Br]^+X^-$.
649	17	$C_{34}H_{32}AsIN$ should read $C_{33}H_{29}AsIN$.
676	14	$Ph_2(o\text{-}OC_6H_4C_6H_4)Sb$ should read $Ph_2(o\text{-}OC_6H_4C_6H_3)Sb$.
684	4 fb	$C_{12}H_{10}ClSb$ should read $C_{12}H_{10}ClSb$, Ph_2SbCl.
687	19	[2,4-Cl(CH_2=CH)C_6H_3]SbCl should read [2,4-Cl(CH_2=CH)C_6H_3]$_2$SbCl.
692	20	$C_9H_5N_4Sb$ should read $C_9H_5Cl_4N_4Sb$.
693	7	$MeSbBr_3$ should read $MeSbBr_2$.
	4 fb	3rd Column; $C_5H_5(CH_2{=}CH)_2Sb$ should read $C_5H_9(CH_2{=}CH)_2Sb$.
699	16	$C_{24}H_{16}N_4O_9Sb$ should read $C_{24}H_{16}N_4O_9Sb_2$.

Page	Line*	
704		Last Compound: 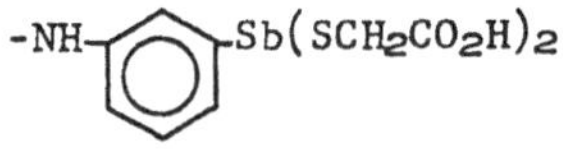 should read 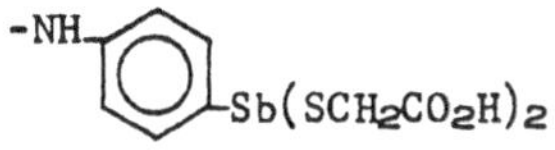
748	6	$C_6H_{18}Cl_2OSb$ should read $C_6H_{18}Cl_2OSb_2$.
	11	$C_6H_{18}Cl_2O_9Sb$ should read $C_6H_{18}Cl_2O_9Sb_2$.
	21	$C_{36}H_{30}N_2O_7Sb$ should read $C_6H_{30}N_2O_7Sb_2$.
770	2-3 fb	Insert C_7 between lines 2 and 3 fb.
782	2	$C_8H_4NO_5SSb$ should read $C_8H_8NO_5SSb$.
784		6th Compound: should be listed with C_{11} compounds.
802		3rd Compound: 5-(4-MeC_6H_4)[$C_{12}H_7Sb$] should read 5-(4-MeC_6H_4)-[$C_{12}H_8Sb$].
		6th Compound: 4-(3-Me[$C_{17}H_7Sb$]-5-)$C_6H_4CO_2H$ should read 4-(3-Me-[$C_{12}H_7Sb$]-5-)$C_6H_4CO_2H$.
813	6 fb	3-(p-Aminophenyl)-2,4,3-dithiastibobicyclo[3.3.1]-5,7,9-triene should read 3-(p-Aminophenyl)-2,4,3-dithiastibobicyclo[3.3.1]-nona-5,7,9-triene.
833	3 fb	$C_{17}H_{10}BiO_{10}Re$ should read $C_{17}H_{10}BiO_5Re$.